Evolution

Evolution

DOUGLAS J. FUTUYMA

State University of New York at Stony Brook

Chapter 19, "Evolution of Genes and Genomes"
by Scott V. Edwards, Harvard University

Chapter 20, "Evolution and Development"
by John R. True, State University of New York at Stony Brook

SINAUER ASSOCIATES, INC. • Publishers
Sunderland, Massachusetts U.S.A.

Front cover
The luxurious plumes of a male bird of paradise (*Paradisaea raggiana*) are the result of sexual selection (see Chapters 11 and 17). Photograph © Art Wolfe/Art Wolfe, Inc.

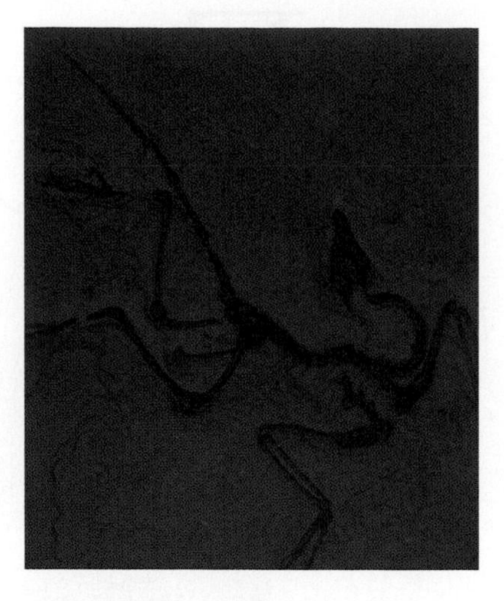

Back cover
Modern birds are almost certainly descended from feathered dinosaurs, such as the famous fossil *Archaeopteryx lithographica*. Long flight feathers were borne by clawed hands and a long tail, characteristic of theropod dinosaurs but not of modern birds (see Chapter 4). Photograph © Tom Stack/Painet, Inc.

Evolution

Sinauer Associates, Inc., 23 Plumtree Road, Sunderland, MA 01375 U.S.A.

FAX: 413-549-1118
www.sinauer.com

Sources of the scientists' photographs appearing in Chapter 1 are gratefully acknowledged:
C. Darwin and A. R. Wallace courtesy of The American Philosophical Library
R. A. Fisher courtesy of Joan Fisher Box
J. B. S. Haldane courtesy of Dr. K. Patau
S. Wright courtesy of Doris Marie Provine
E. Mayr courtesy of Harvard News Service and E. Mayr
G. L. Stebbins, G. G. Simpson, and Th. Dobzhansky courtesy of G. L. Stebbins
M. Kimura courtesy of William Provine

Library of Congress Cataloging-in-Publication Data
Futuyma, Douglas J. 1942-
Evolution / Douglas J. Futuyma
p. cm.
Includes bibliographical references (p.) and index.
ISBN 0-87893-187-2 (hardcover)
1. Evolution (Biology) I. Title.
QH366.2F87 2005
576.8--dc22 2004029808

Brief Contents

Contents

1 Evolutionary Biology 1

2 The Tree of Life: Classification and Phylogeny 17

3 Patterns of Evolution 43

4 Evolution in the Fossil Record 67

13 Evolution of Phenotypic Traits 297

14 Conflict and Cooperation 325

15 Species 353

16 Speciation 379

17 How to Be Fit: Reproductive Success 405

18 Coevolution: Evolving Interactions among Species 429

19 Evolution of Genes and Genomes 449

20 Evolution and Development 473

21 Macroevolution: Evolution above the Species Level 501

22 Evolutionary Science, Creationism, and Society 523

Preface

Since its inception during a sabbatical leave in Australia four years ago, this book has traveled with me to Stony Brook, then to Ann Arbor, and again to Stony Brook, suffering long interruptions along the way. Perhaps that is just as well, for the transformation of evolutionary biology has been even faster in this interval than before, and has resulted in a very different book than might have been—different enough to merit its own title. I had intended to prepare a digest of *Evolutionary Biology* (Third Edition, 1998), rendered of its excesses, and while much of the structure and some of the text of this book descend directly from that tome, it became clear that a new book was in the making. Some topics had to be deleted and many others shortened, while the rapid pace of change in the field required that new topics such as evolutionary genomics be introduced and that almost all topics be updated.

Most importantly, this book is specifically directed toward contemporary undergraduates. That effort will be most immediately evident in the illustrations, but will also be found in the text, where, in the interest of accessibility, I have attempted to make points more explicitly, have (with some reluctance) reduced the quantitative aspects of our science, and have eschewed the Proustian sentences and Elizabethan constructions with which I fain would play. I hope students will enjoy at least some of the results.

I have structured this book, like its recent ancestor, to begin with phylogeny as a framework for inferring history, and history as the natural perspective for evolutionary biology—a perspective that has (quite recently) become almost *de rigeur* in evolutionary studies and beyond. I continue with macroevolutionary patterns (which I believe most intrigues beginning students), emphasizing the evidence for evolution en route. In addition to their intrinsic interest, the historical patterns should excite in the student questions about evolutionary processes, the subject of the next nine chapters. These chapters provide the basis for understanding the evolution of life histories, genetic systems, ecological interactions, genes and genomes, and development. I then return to macroevolution, approached as a synthesis of evolutionary process and pattern.

This book lacks an explicit chapter on human evolution because most of the topics it would contain are distributed throughout. Instead, the final chapter treats what I think are increasingly important, indeed indispensable, topics in an undergraduate course on evolution: the evidence for evolution, the nature of science, and the failings of creationism. These themes recur throughout the book, implicitly and occasionally explicitly, but I believe it will be useful to treat them as a coherent whole. The final chapter ends on a positive note with a brief survey of some of the social applications of evolutionary biology.

The ever-quickening pace of research and the variety of novel techniques, especially in molecular, genomic, and developmental evolutionary biology, make it increasingly difficult for any one person to keep abreast of and be capable of evaluating research across the entire field of evolutionary studies. So I am very grateful to Scott Edwards (Harvard University) and John True (State University of New York at Stony Brook) for joining me in this venture, and contributing chapters on evolution of genes and genomes (Chapter 19) and on evolutionary developmental biology (Chapter 20), respectively. They have brought to these subjects knowledge and critical understanding well beyond any effort I might have made.

I am also grateful to the many people who have made direct or indirect contributions to the content and development of this book. It has profited from my learning of errors in its predecessor that Werner G. Heim, Eric B. Knox, Uzi Ritte, and Robert H. Tamarin generously brought to my attention. Many people have provided information, references, and advice, including Michael Bell, Prosanta Chakraborty, Jerry Coyne, Daniel Dykhuizen, Walter Eanes, Brian Farrell, Daniel Fisher, John Fleagle, Daniel Funk, Douglas Gill, Philip Gingerich, David Houle, David Jablonski, Charles Janson, Lacey Knowles, Jeffrey Levinton, David Mindell, Daniel Stoebel, Randall Susman, John Thompson, Mark Uhen, Brian Verrelli, and Jianzhi Zhang. Surely this list is very incomplete, and I apologize to those whose names I have omitted. Adam Ehmer helped with preparation of some figures, and Massimo Pigliucci read and offered very helpful comments on a draft of Chapter 22. I appreciate the contributions of Elizabeth Frieder, Monica Geber, Matt Gitzendanner, Kenneth Gobalet, Mark Kirkpatrick, Sergei Nuzhdin, Ruth Shaw, and William A. Woods, Jr., who reviewed early outlines and chapter drafts.

I am very grateful to those who provided hospitality and support in Australia, including Mark Burgman and Pauline Ladiges (University of Melbourne), Ary Hoffmann (LaTrobe University), and Ross and Ching Crozier (James Cook University); to John and Gabrielle Barkla, Jeremy Burdon, Brad Congdon, Stuart Dashper, Chris Lester, Michael Mathieson, Susan Myers, Richard Nowotny, Jan Powning, Peter Thrall, Jo Wieneke, and others who helped to make me feel welcome; and to the Fulbright Foundation for fellowship support of my sojourn in Australia. I feel a special debt of gratitude to the friends who sustained me in Ann Arbor, especially Tom Gazi, Deborah Goldberg, Lacey Knowles, Josepha Kurdziel, Don Pelz, Josh Rest, Mark Uhen and Gerry Duprey, and John Vandermeer and Ivette Perfecto, and I am grateful to the faculty, students, and staff of the Department of Ecology and Evolutionary Biology at the University of Michigan for the pleasure of having been one of their number. I must express heartfelt gratitude to the faculty, students, and staff of the Department of Ecology and Evolution at Stony Brook for their support and friendship throughout my odyssey.

Finally, this book is immeasurably better than it might have been, thanks to the wonderfully capable Sinauer team, especially Norma Roche, David McIntyre, Elizabeth Morales, Jefferson Johnson, and the amazing Carol Wigg. Special thanks to Andy Sinauer, who sets the gold standard of quality in publishing, for his continuing faith and support.

DOUGLAS J. FUTUYMA
STONY BROOK, NEW YORK
DECEMBER 2004

To the Student

The great geneticist François Jacob, who won the Nobel Prize in Biology and Medicine for discovering mechanisms by which gene activity is regulated, wrote in 1973 that "there are many generalizations in biology, but precious few theories. Among these, the theory of evolution is by far the most important, because it draws together from the most varied sources a mass of observations which would otherwise remain isolated; it unites all the disciplines concerned with living beings; it establishes order among the extraordinary variety of organisms and closely binds them to the rest of the earth; in short, it provides a causal explanation of the living world and its heterogeneity."

Jacob did not himself do research on evolution, but like most thoughtful biologists, he recognized its pivotal importance in the biological sciences. Today molecular biologists, developmental biologists, and genome biologists, as well as ecologists, behaviorists, anthropologists, and many psychologists, share Jacob's view. Evolution provides an indispensable framework for understanding phenomena ranging from the structure and size of genomes to many features of human behavior. Moreover, evolutionary biology is increasingly recognized for its usefulness: in fields as disparate as public health, agriculture, and computer science, the concepts, methods, and data of evolutionary biology make indispensable contributions to both basic and applied research. Any educated person should know something about evolution, and should understand why it matters that evolution be taught in our schools. For anyone who envisions a career based in the life sciences—whether as physician or as biological researcher—an understanding of evolution is indispensable. As James Watson, co-discoverer of the structure of DNA, wrote, "today, the theory of evolution is an accepted fact for everyone but a fundamentalist minority."

The core of evolutionary biology consists of describing and analyzing the history of evolution and of analyzing its causes and mechanisms. The scope of evolutionary biology is far greater than any other field of biological science, because all organisms, and all their characteristics, are products of a history of evolutionary change. Because of this enormous breadth, courses in evolution generally do not emphasize the details of the evolution of particular groups of organisms—the amount of information would be simply overwhelming. Rather, evolution courses emphasize the general principles of evolution, the hypotheses about the causes of evolutionary change that apply to most or all organisms, and the major patterns of change that have characterized many different groups. In this book, concepts are illustrated with examples drawn from research on a great variety of organisms, but it is less important to know the details of these examples than to understand how the data obtained in those studies bear on a hypothesis.

Determining the causes and patterns of evolution can be difficult, in part because we often are attempting to understand how and why something happened in the past. In this way, evolutionary biology differs from most other biological subjects, which deal with current characteristics of organisms. However, evolutionary biology and other biological disciplines share the fact that we often must make inferences about invisible processes or objects. We cannot see past evolutionary changes in action; but neither can we actually see DNA replicate, nor can we see the hormones that we know regulate growth and reproduction. Rather, we make inferences about these things by (1) posing informed hypotheses about what they are and how they work, then (2) generating predictions (making deductions) from these hypotheses about data that we can actually obtain, and finally (3) judging the validity of each hypothesis by the match between our observations and what we expect to see if the hypothesis were true. This is the "hypothetico-deductive method," of which Darwin was one of the first successful exponents, and which is widely and powerfully used throughout science.

In science, a group of interrelated hypotheses that have been well supported by such tests, and which together explain a wide range of phenomena or observations, is called a *theory*. "Theory" in this sense—the sense in which Jacob and Watson used the word—does not mean mere speculation. "Theory" is a term of honor, reserved for principles such as quantum theory, atomic theory, or cell theory, that are well supported and provide a broad framework of explanation. The emphasis in your course in evolutionary biology, then, will probably be on, first, learning the theory (the principles of evolutionary change that together explain a vast variety of observations about organisms); and second, on learning how to test evolutionary hypotheses with data (the hypothetico-deductive method applied to questions about what has happened, and how it has happened—whether it be the history of corn from its wild state to mass cultivation, or the development of complex societies in some insect species, or the spread of human immunodeficiency virus (HIV).

Many students may find that the emphasis in studying evolution differs from what they have experienced in other biology courses. I suggest that you pay special attention to chapters or passages where the fundamental principles and methods are introduced, be sure you understand them before moving on, and reread these passages after you have gone through one or more later chapters. (You may want to revisit Chapters 2, 9, 10, and 12–13 in particular.) Be sure to emphasize understanding, not memorization, and test your understanding using some of the questions at the end of each chapter or questions that your instructor assigns. I must emphasize that the material in this book builds cumulatively; almost every concept, principle, or major technical term introduced in any chapter is used again in later chapters. You will need to understand the early chapters just as thoroughly for your final exam as for a midterm exam. Evolutionary biology is a unified whole: just as carbohydrate metabolism and amino acid synthesis cannot be divorced in biochemistry, so it is for topics as seemingly different as the phylogeny of species and the theory of genetic drift.

I cannot emphasize too strongly that in every field of science, the unknown greatly exceeds the known. Thousands of research papers on evolutionary topics are published each year, and many of them raise new questions even as they attempt to answer old ones. No one, least of all a scientist, should be afraid to say "I don't know" or "I'm not sure," and that refrain will sound fairly often in this book. To recognize where our knowledge and understanding are uncertain or lacking is to see where research may be warranted and where new research trails might be blazed. I hope that some readers will find evolution so rich a subject, so intellectually challenging, so fertile in insights, and so deep in its implications that they will adopt evolutionary biology as a career. *Felix qui potuit rerum cognoscere causas*, wrote Virgil: "Happy is the person who can learn the nature of things."

1

Evolutionary Biology

Biologists, looking broadly at living things, are stirred to ask thousands of questions. Why do peacocks have such extravagant feathers? Why do some parasites attack only a single species of host, while others can infect many different species? Why do whales have lungs, and why do snakes lack legs? Why does one species of ant have a single chromosome, while some butterflies have more than 200? Why do salamanders have more than 10 times as much DNA as humans, and why does a lily plant have almost twice as much DNA as a salamander? What accounts for the astonishing variety of organisms?

In one of the most breathtaking ideas in the history of science, Charles Darwin proposed that *"all the organic beings which have ever lived on this earth have descended from some one primordial form."* From this idea, it follows that every characteristic of every species—the feathers of a peacock, the number and sequence of its genes, the catalytic abilities of its enzymes, the structure of its cells and organs, its physiological tolerances and nutritional requirements, its life span and reproductive system, its capacity for behavior—is the outcome of an evolutionary history. The evolutionary perspective illuminates every subject in biology, from molecular biology to ecology. Indeed, *evolution is the unifying theory of biology*. "Nothing in biology makes sense," said the geneticist Theodosius Dobzhansky, "except in the light of evolution."

The peacock, *Pavo cristatus*. The extravagant back feathers, which impair this bird's ability to fly, are among the thousands of biological curiosities that evolutionary theory explains. (Photo © Brian Lightfoot/ Nature Picture Library.)

What Is Evolution?

The word *evolution* comes from the Latin *evolvere*, "to unfold or unroll"—to reveal or manifest hidden potentialities. Today "evolution" has come to mean, simply, "change." It is sometimes used to describe changes in individual objects such as stars. **Biological** (or **organic**) **evolution**, however, is *change in the properties of groups of organisms over the course of generations*. The development, or ONTOGENY, of an individual organism is not considered evolution: individual organisms do not evolve. Groups of organisms, which we may call **populations**, undergo *descent with modification*. Populations may become subdivided, so that several populations are derived from a *common ancestral population*. If different changes transpire in the several populations, the populations *diverge*.

The changes in populations that are considered evolutionary are those that are passed via the genetic material from one generation to the next. Biological evolution may be slight or substantial: it embraces everything from slight changes in the proportions of different forms of a gene within a population to the alterations that led from the earliest organism to dinosaurs, bees, oaks, and humans.

No more dramatic example of evolution by natural selection can be imagined than that of today's crisis in antibiotic resistance. Before the 1940s, most people in hospital wards did not have cancer or heart disease. They suffered from tuberculosis, pneumonia, meningitis, typhoid fever, syphilis, and many other kinds of bacterial infection—and they had little hope of being cured (Figure 1.1). Infectious bacterial diseases condemned millions of people in the developed countries to early death. Populations in developing countries bore not only these burdens, but also diseases such as malaria and cholera, even more heavily than they do today.

By the 1960s, the medical situation had changed dramatically. The discovery of antibiotic drugs and subsequent advances in their synthesis led to the conquest of most bacterial diseases, at least in developed countries. Most people today think of tuberculosis as the stuff of operas and dense German novels. The sexual revolution in the 1970s was encouraged by the confidence that sexually transmitted diseases such as gonorrhea and syphilis were merely a temporary inconvenience that penicillin could cure. In 1969, the Surgeon General of the United States proclaimed that it was time to "close the book on infectious diseases."

Or so it seemed. Today we confront not only new infectious diseases such as AIDS, but also a resurgence of old diseases with frightening new faces. The same bacteria are back,

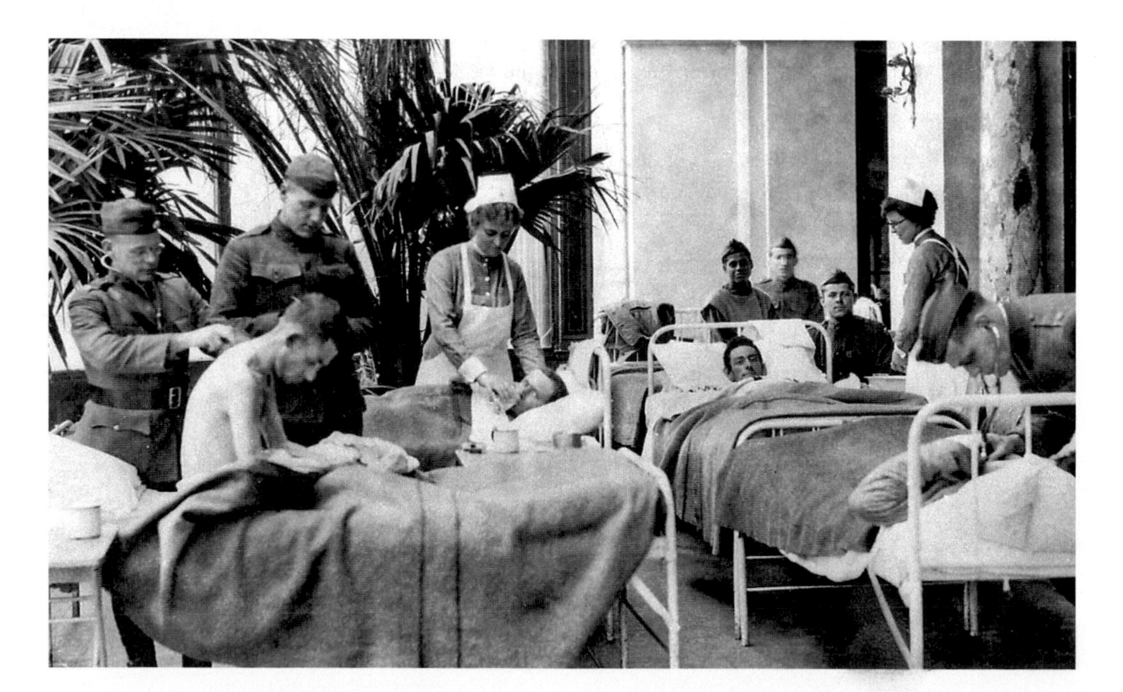

Figure 1.1 A tuberculosis ward at a U.S. Army base hospital in France during World War I. Until recently, it was thought that antibiotics, which came into widespread use after World War II, had conquered this devastating bacterial disease. (Photo courtesy of the National Library of Medicine.)

but now they are resistant to penicillin, ampicillin, erythromycin, vancomycin, fluoroquinolones—all the weapons that were supposed to have vanquished them. Almost every hospital in the world treats casualties in this battle against changing opponents—and unintentionally may make those opponents stronger. They are witnessing, and even instigating, an explosion of evolutionary change (Palumbi 2001).

Staphylococcus aureus, for example, a bacterium that causes many infections in surgical patients, is now almost universally resistant to penicillin, ampicillin, and related drugs. Methicillin was developed as an alternative and worked for a few years, but many *S. aureus* populations became resistant to methicillin, and then to cephalosporins, carbapenems, erythromycin, tetracycline, streptomycin, sulfonamides, and fluoroquinolones. Yet another new drug, vancomycin, seemed to have solved the problem, but it too is becoming less effective.

Drug-resistant strains of *Neisseria gonorrheae,* the bacterium that causes gonorrhea, have steadily increased in abundance; by 1995, more than 40 percent of the gonorrhea cases treated in New York City were resistant to penicillin, tetracycline, or both. Many strains of the pneumonia bacterium are highly resistant to penicillin, and some strains of the cholera bacterium are resistant to a wide variety of antibiotics. Increasingly abundant strains of the tuberculosis bacterium and of the malarial organism are resistant to all available drugs. A person infected with HIV, the human immunodeficiency virus that causes AIDS, often shows indications of drug-resistant virus within 6 to 12 months after drug treatment begins.

As the use of antibiotics increases, so does the incidence of bacteria that are resistant to those antibiotics, so the gains made are almost as quickly lost (Figure 1.2). Why is this happening? Do the drugs cause drug-resistant mutations in the bacteria's genes? Do the mutations occur even without exposure to drugs—are they present in unexposed bacterial populations? How many mutations cause resistance to a drug? How often do they occur? Do the mutations spread from one bacterium to another? Are they spread only among bacteria or viruses of the same species, or can they pass between different species? How is the growth of the organism's population affected by such mutations? Can the evolution of resistance be prevented by using lower doses of drugs? Higher doses? Combinations of different drugs? Can an individual avoid infection by drug-resistant organisms by faithfully following a physician's prescription, or will this work only if everyone else is just as conscientious?

The principles and methods of evolutionary biology have provided some answers to these questions, and to many others that affect society. Evolutionary biologists and other scientists trained in evolutionary principles have traced the transfer of HIV to humans from chimpanzees and mangabey monkeys (Gao et al. 1999; Korber et al. 2000). They have studied the evolution of insecticide resistance in disease-carrying and crop-destroying insects. They have helped to devise methods of nonchemical pest control and have laid the foundations for transferring genetic resistance to diseases and insects from wild plants to crop

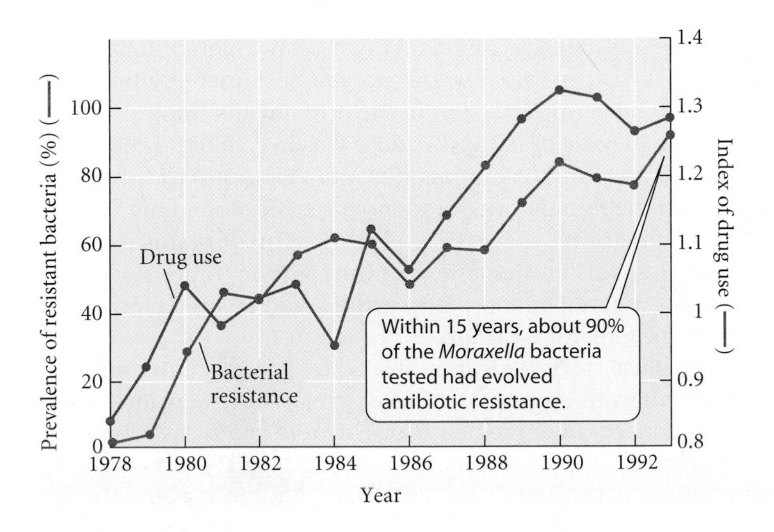

Figure 1.2 Development of drug resistance. The increase in the consumption of penicillin-like antibiotics in a community in Finland between 1978 and 1993 was matched by a dramatic increase in the percentage of resistant isolates of the bacterium *Moraxella catarrhalisis* from middle-ear infections in young children. (After Levin and Anderson 1999.)

plants. Evolutionary principles and knowledge are being used in biotechnology to design new drugs and other useful products. In computer science and artificial intelligence, "evolutionary computation" uses principles taken directly from evolutionary theory to solve mathematically intractable practical problems, such as constructing complex timetables or processing radar data (Meagher and Futuyma 2001; Bull and Wichman 2001).

The importance of evolutionary biology goes far beyond its practical uses, however. The way we think about ourselves can be profoundly shaped by an evolutionary framework. How do we account for human variation—the fact that almost everyone is genetically and phenotypically unique? Are there human races, and if so, how do they differ, and how and when did they develop? Why does sugar taste good? What accounts for behavioral differences between men and women? How did exquisitely complex, useful features such as our hands and our eyes come to exist? What about apparently useless or even potentially harmful characteristics such our wisdom teeth and appendix? Why does noncoding, apparently useless DNA account for more than 98 percent of the human genome? Why do we age, undergo senescence, and eventually die? Why are medical researchers able to use monkeys, mice, and even fruit flies and yeasts as models for processes in the human body? Such questions and their answers lie in the realm of evolutionary biology to which Charles Darwin flung open the door some 150 years ago.

Before Darwin

Darwin's theory of biological evolution is one of the most revolutionary ideas in Western thought, perhaps rivaled only by Newton's theory of physics. It profoundly challenged the prevailing world view, which had originated largely with Plato and Aristotle. Foremost in Plato's philosophy was his concept of the *eidos*, the "form" or "idea," a transcendent ideal form imperfectly imitated by its earthly representations. For example, the reality—the "essence"—of the true equilateral triangle is only imperfectly captured by the triangles we draw or construct. Likewise, horses (or any other species) have an immutable essence, but each individual horse has imperfections. In this philosophy of **essentialism**, variation is accidental imperfection.

Plato's philosophy of essentialism became incorporated into Western philosophy largely through Aristotle, who developed Plato's concept of immutable essences into the notion that species have fixed properties. Later, Christians interpreted the biblical account of Genesis literally and concluded that each species had been created individually by God in the same form it has today. (This belief is known as "special creation.") Christian thought elaborated on Platonic and Aristotelian philosophy, arguing that since existence is good and God's benevolence is complete, He must have bestowed existence on every creature, each with a distinct essence, of which he could conceive. Because order is superior to disorder, God's creation must follow a plan: specifically, a gradation from inanimate objects and barely animate forms of life, through plants and invertebrates, up through ever "higher" forms of life. Humankind, which is both physical and spiritual in nature, formed the link between animals and angels. This "Great Chain of Being," or *scala naturae* (the scale, or ladder, of nature), must be permanent and unchanging, since change would imply that there had been imperfection in the original creation.

As late as the eighteenth century, the role of natural science was to catalogue and make manifest the plan of creation so that we might appreciate God's wisdom. Carolus Linnaeus (1707–1778), who established the framework of modern classification in his *Systema Naturae* (1735), won worldwide fame for his exhaustive classification of plants and animals, undertaken in the hope of discovering the pattern of the creation. Linnaeus classified "related" species into genera, "related" genera into orders, and so on. To him, "relatedness" meant propinquity in the Creator's design.

Belief in the literal truth of the biblical story of creation started to give way as a more materialist view developed in the seventeenth century, starting with Newton's explanations of physical phenomena. The foundations for evolutionary thought were laid by astronomers, who developed theories of the origin of stars and planets, and by geologists, who amassed evidence that the Earth had undergone profound changes, that it had been populated by many creatures now extinct, and that it was very old. The geologists James Hutton and Charles Lyell expounded the principle of **uniformitarianism**, holding that the same processes operated in the past as in the present, and that the observations of geology should therefore be explained by causes that we can now observe. Darwin was greatly influenced by Lyell's teachings, and he adopted uniformitarianism in his thinking about evolution.

In the eighteenth century, several French philosophers and naturalists suggested that species had arisen by natural causes. The most significant pre-Darwinian evolutionary hypothesis, representing the culmination of eighteenth-century evolutionary thought, was proposed by the Chevalier de Lamarck in his *Philosophie Zoologique* (1809). Lamarck proposed that each species originated individually by spontaneous generation from nonliving matter, starting at the bottom of the chain of being. A "nervous fluid" acts within each species, he said, causing it to progress up the chain. Species originated at different times, so we now see a hierarchy of species because they differ in age (Figure 1.3A).

Lamarck argued that species differ from one another because they have different needs, and so use certain of their organs and appendages more than others. The more strongly exercised organs attract more of the "nervous fluid," which enlarges them, just as muscles become strengthened by work. Lamarck, like most people at the time, believed that such alterations, acquired during an individual's lifetime, are inherited—a principle called **inheritance of acquired characteristics**. In the most famous example of Lamarck's theory, giraffes originally had short necks, but stretched their necks to reach foliage above them. Hence their necks were lengthened; longer necks were inherited, and over the course of generations, their necks got longer and longer. This could happen to any and all giraffes, so the entire species could have acquired longer necks because it was composed of individual organisms that changed during their lifetimes (see Figure 1.4A).

JEAN-BAPTISTE PIERRE ANTOINE DE MONET, CHEVALIER DE LAMARCK

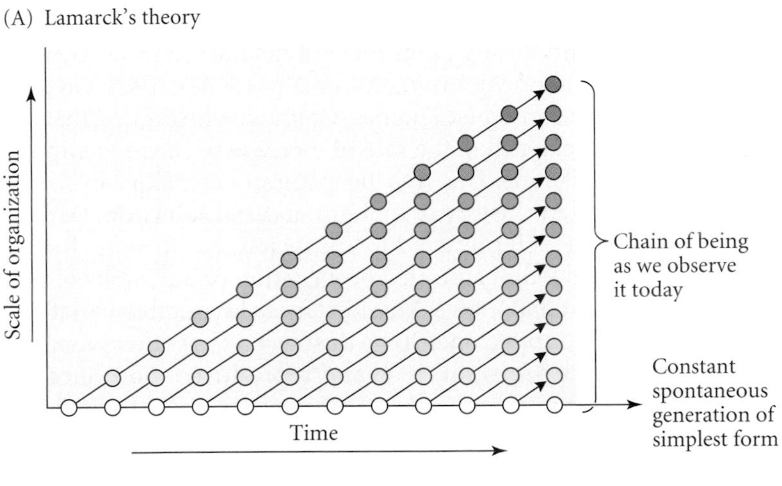

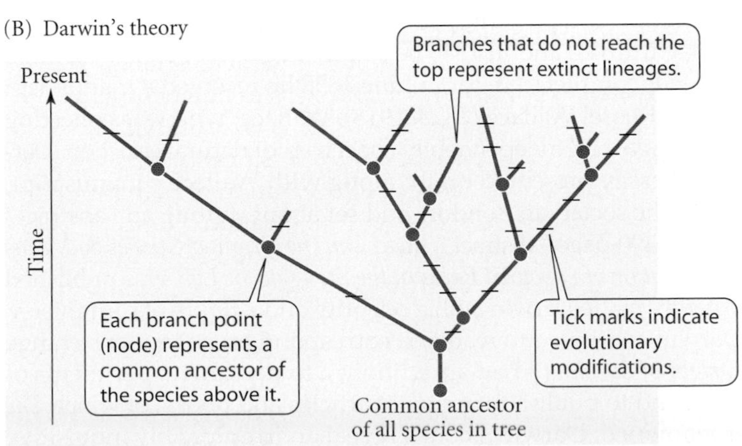

Figure 1.3 (A) Lamarck's theory of organic progression. Over time, species originate by spontaneous generation, and each evolves up the scale of organization, establishing a *scala naturae*, or chain of being, that ranges from newly originated simple forms of life to older, more complex forms. In Lamarck's scheme, species have not originated from common ancestors. (B) Darwin's theory of descent with modification, represented by a phylogenetic tree. Lineages (species) descend from common ancestors, undergoing various modifications in the course of time. Some (such as the leftmost lineages) may undergo less modification from the ancestral condition than others (the rightmost lineages). (A after Bowler 1989.)

CHARLES ROBERT DARWIN

Lamarck's ideas had little impact during his lifetime, partly because they were criticized by respected zoologists, and partly because after the French Revolution, ideas issuing from France were considered suspect in most other countries. Lamarck's ideas of how evolution works were wrong, but he deserves credit for being the first to advance a coherent theory of evolution.

Charles Darwin

Charles Robert Darwin (February 12, 1809–April 19, 1882) was the son of an English physician. He briefly studied medicine at Edinburgh, then turned to studying for a career in the clergy at Cambridge University. He apparently believed in the literal truth of the Bible as a young man. He was passionately interested in natural history and became a companion of the natural scientists on the faculty. His life was forever changed in 1831, when he was invited to serve as a naturalist on the H.M.S. *Beagle*, a ship the British navy was sending to chart the waters of South America.

The *Beagle*'s voyage lasted from December 27, 1831 to October 2, 1836. The ship spent several years traveling along the coast of South America, where Darwin observed the natural history of the Brazilian rain forest and the Argentine pampas, and stopped in the Galápagos Islands (on the Equator off the coast of Ecuador). In the course of the voyage, Darwin became an accomplished naturalist, collected specimens, made innumerable geological and biological observations, and conceived a new (and correct) theory of the formation of coral atolls. Soon after his return, the ornithologist John Gould pointed out that Darwin's specimens of mockingbirds from the Galápagos Islands were so different from one island to another that they represented different species. Darwin then recalled that the giant tortoises, too, differed from one island to the next. These facts, and the similarities between fossil and living mammals that he had found in South America, triggered his conviction that different species had evolved from common ancestors.

Darwin's comfortable finances enabled him to devote the rest of his life exclusively to his biological work (although he was chronically ill for most of his life after the voyage). He set about amassing evidence of evolution and trying to conceive of its causes. On September 28, 1838, he read an essay by the economist Thomas Malthus, who argued that the rate of human population growth is greater than the rate of increase in the food supply, so that unchecked growth must lead to famine. This was the inspiration for Darwin's great idea, one of the most important ideas in the history of thought: **natural selection**. Darwin wrote in his autobiography that "being well prepared to appreciate the struggle for existence which everywhere goes on from long-continued observation of the habits of animals and plants, it at once struck me that under these circumstances favourable variations would tend to be preserved and unfavourable ones to be destroyed." In other words, if individuals of a species with superior features survived and reproduced more successfully than individuals with inferior features, and if these differences were inherited, the average character of the species would be altered.

Mindful of how controversial the subject would be, Darwin then spent twenty years amassing evidence about evolution and pursuing other researches before publishing his ideas. He wrote a private essay in 1844, and in 1856 he finally began a book he intended to call *Natural Selection*. He never completed it, for in June 1858 he received a manuscript from a young naturalist, Alfred Russel Wallace (1823–1913). Wallace, who was collecting specimens in the Malay Archipelago, had independently conceived of natural selection. Darwin had extracts from his 1844 essay presented orally, along with Wallace's manuscript, at a meeting of the major scientific society in London, and set about writing an "abstract" of the book he had intended. The 490-page "abstract," titled *On The Origin of Species by Means of Natural Selection, or The Preservation of Favoured Races in the Struggle for Life*, was published on November 24, 1859; it instantly made Darwin both a celebrity and a figure of controversy.

ALFRED RUSSEL WALLACE

For the rest of his life, Darwin continued to read and correspond on an immense range of subjects, to revise *The Origin of Species* (it had six editions), to perform experiments of all sorts (especially on plants), and to publish many more articles and books, of which *The Descent of Man* is the most renowned. Darwin's books reveal an irrepressibly inquisitive man, fascinated with all of biology, creative in devising hypotheses and in bringing evidence to bear upon them, and profoundly aware that every fact of biology, no matter how seemingly trivial, must fit into a coherent, unified understanding of the world.

Darwin's Evolutionary Theory

The Origin of Species has two major theses. The first is Darwin's theory of **descent with modification**. It holds that all species, living and extinct, have descended, without interruption, from one or a few original forms of life (Figure 1.3B). Species that diverged from a common ancestor were at first very similar, but accumulated differences over great spans of time, so that some are now radically different from one another. Darwin's conception of the course of evolution is profoundly different from Lamarck's, in which the concept of common ancestry plays almost no role.

The second theme of *The Origin of Species* is Darwin's theory of the causal agents of evolutionary change. This was his theory of natural selection: "if variations useful to any organic being ever occur, assuredly individuals thus characterised will have the best chance of being preserved in the struggle for life; and from the strong principle of inheritance, these will tend to produce offspring similarly characterised. This principle of preservation, or the survival of the fittest, I have called natural selection." This theory is a VARIATIONAL THEORY of change, differing profoundly from Lamarck's TRANSFORMATIONAL THEORY, in which individual organisms change (Figure 1.4).

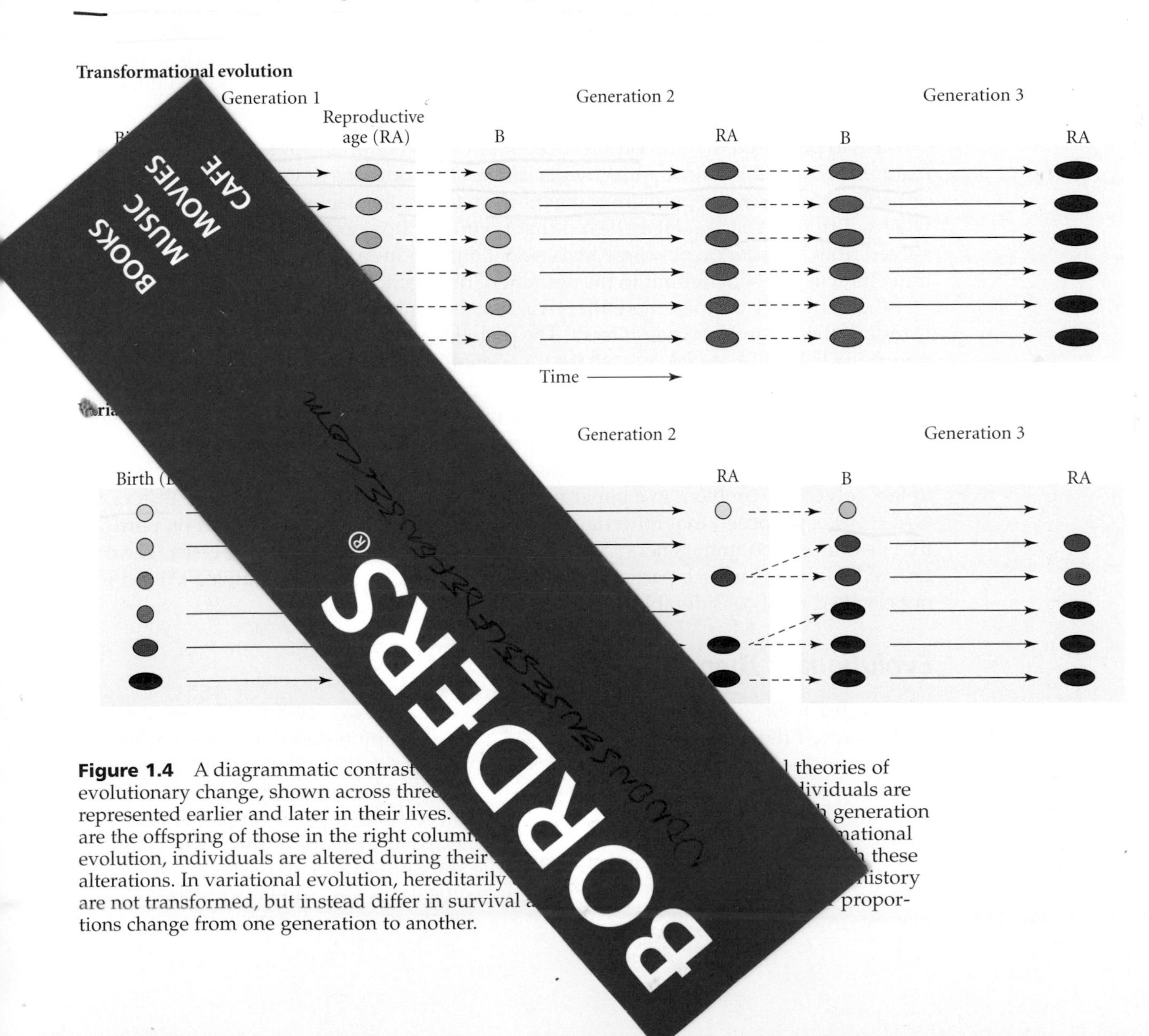

Figure 1.4 A diagrammatic contrast ... theories of evolutionary change, shown across thre... ...ividuals are represented earlier and later in their lives. ... generation are the offspring of those in the right colum... ...national evolution, individuals are altered during their ... these alterations. In variational evolution, hereditarily ... history are not transformed, but instead differ in survival ... proportions change from one generation to another.

What is often called "Darwin's theory of evolution" actually includes five theories (Mayr 1982a):

1. *Evolution as such* is the simple proposition that the characteristics of lineages of organisms change over time. This idea was not original with Darwin, but it was Darwin who so convincingly marshaled the evidence for evolution that most biologists soon accepted that it has indeed occurred.
2. *Common descent* is a radically different view of evolution than the scheme Lamarck had proposed (see Figure 1.3). Darwin was the first to argue that species had diverged from common ancestors and that all of life could be portrayed as one great family tree.
3. *Gradualism* is Darwin's proposition that the differences between even radically different organisms have evolved incrementally, by small steps through intermediate forms. The alternative hypothesis is that large differences evolve by leaps, or SALTATIONS, without intermediates.
4. *Populational change* is Darwin's thesis that evolution occurs by changes in the *proportions* of individuals within a population that have different inherited characteristics (see Figure 1.4B). This concept was a completely original idea that contrasts both with the sudden origin of new species by saltation and with Lamarck's account of evolutionary change by transformation of individuals.
5. *Natural selection* was Darwin's brilliant hypothesis, independently conceived by Wallace, that changes in the proportions of different types of individuals are caused by differences in their ability to survive and reproduce—and that such changes result in the evolution of **adaptations**, features that appear "designed" to fit organisms to their environment. The concept of natural selection revolutionized not only biology, but Western thought as a whole.

Darwin proposed that the various descendants of a common ancestor evolve different features because they are adaptive under different "conditions of life"—different habitats or habits. Moreover, the pressure of competition favors the use of different foods or habitats by different species. He believed that no matter how extensively a species has diverged from its ancestor, new hereditary variations continue to arise, so that given enough time, there is no evident limit to the amount of divergence that can occur.

Where, though, do these hereditary variations come from? This was the great gap in Darwin's theory, and he never filled it. The problem was serious, because according to the prevailing belief in BLENDING INHERITANCE, variation should decrease, not increase. Because offspring are often intermediate between their parents in features such as color or size, it was widely believed that characteristics are inherited like fluids, such as different colors of paint. Blending white and black paints produces gray, but mixing two gray paints doesn't yield black or white: variation decreases. Darwin never knew that Gregor Mendel had in fact solved the problem in a paper published in 1865. Mendel's theory of PARTICULATE INHERITANCE proposed that inheritance is based not on blending fluids, but on particles that pass unaltered from generation to generation—so that variation can persist. The concept of "mutation" in such particles (later called genes) developed only after 1900 and was not clarified until considerably later.

Evolutionary Theories after Darwin

Although *The Origin of Species* raised enormous controversy, by the 1870s most scientists accepted the historical reality of evolution by common descent. There ensued, in the late nineteenth and early twentieth centuries, a "golden age" of paleontology, comparative morphology, and comparative embryology, during which a great deal of information on evolution in the fossil record and on relationships among organisms was amassed. But this consensus did not extend to Darwin's theory of the cause of evolution, natural selection. For about 60 years after the publication of *The Origin of Species*, all but a few faithful Darwinians rejected natural selection, and numerous theories were proposed in its stead. These theories included neo-Lamarckian, orthogenetic, and mutationist theories (Bowler 1989).

NEO-LAMARCKISM includes several theories based on the old idea of inheritance of modifications acquired during an organism's lifetime. Such modifications might have been due, for example, to the direct effect of the environment on development (as in plants that develop thicker leaves if grown in a hot, dry environment). In a famous experiment, August Weismann cut off the tails of mice for many generations and showed that this had no effect on the tail length of their descendants. Extensive subsequent research has provided no evidence that specific hereditary changes can be induced by environmental conditions under which they would be advantageous.

RONALD A. FISHER

Theories of ORTHOGENESIS, or "straight-line evolution," held that the variation that arises is directed toward fixed goals, so that a species evolves in a predetermined direction without the aid of natural selection. Some paleontologists held that such trends need not be adaptive and could even drive species toward extinction. None of the proponents of orthogenesis ever proposed a mechanism for it.

MUTATIONIST theories were advanced by some geneticists who observed that discretely different new phenotypes can arise by a process of mutation. They supposed that such mutant forms constituted new species, and thus believed that natural selection was not necessary to account for the origin of species. Mutationist ideas were advanced by Hugo de Vries, one of the biologists who "discovered" Mendel's neglected paper in 1900, and by Thomas Hunt Morgan, the founder of *Drosophila* genetics. The last influential mutationist was Richard Goldschmidt (1940), an accomplished geneticist who nevertheless erroneously argued that evolutionary change within species is entirely different in kind from the origin of new species and higher taxa. These, he said, originate by sudden, drastic changes that reorganize the whole genome. Although most such reorganizations would be deleterious, a few "hopeful monsters" would be the progenitors of new groups.

J. B. S. HALDANE

The Evolutionary Synthesis

These anti-Darwinian ideas were refuted in the 1930s and 1940s by the **evolutionary synthesis** or **modern synthesis**, forged from the contributions of geneticists, systematists, and paleontologists who reconciled Darwin's theory with the facts of genetics (Mayr and Provine 1980; Smocovitis 1996). Ronald A. Fisher and John B. S. Haldane in England and Sewall Wright in the United States developed a mathematical theory of population genetics, which showed that mutation and natural selection *together* cause adaptive evolution: mutation is not an alternative to natural selection, but is rather its raw material. The study of genetic variation and change in natural populations was pioneered in Russia by Sergei Chetverikov and continued by Theodosius Dobzhansky, who moved from Russia to the United States. In his influential book *Genetics and the Origin of Species* (1937), Dobzhansky conveyed the ideas of the population geneticists to other biologists, thus influencing their appreciation of the genetic basis of evolution.

Other major contributors to the synthesis included the zoologists Ernst Mayr, in *Systematics and the Origin of Species* (1942), and Bernhard Rensch, in *Evolution Above the Species Level* (1959); the botanist G. Ledyard Stebbins, in *Variation and Evolution in Plants* (1950); and the paleontologist George Gaylord Simpson, in *Tempo and Mode in Evolution* (1944) and its successor, *The Major Features of Evolution* (1953). These authors argued persuasively that mutation, recombination, natural selection, and other processes operating *within species* (which Dobzhansky termed **microevolution**) account for the *origin of new species and for the major, long-term features of evolution* (termed **macroevolution**).

SEWALL WRIGHT

Fundamental principles of evolution

The principal claims of the evolutionary synthesis are the foundations of modern evolutionary biology. Although some of these principles have been extended, clarified, or modified since the 1940s, most evolutionary biologists today accept them as fundamen-

ERNST MAYR

G. LEDYARD STEBBINS, GEORGE GAYLORD SIMPSON, AND THEODOSIUS DOBZHANSKY

tally valid. These, then, are the fundamental principles of evolution, to be discussed at length throughout this book.

1. *The phenotype* (observed characteristic) *is different from the genotype* (the set of genes in an individual's DNA); phenotypic differences among individual organisms may be due partly to genetic differences and partly to direct effects of the environment.
2. Environmental effects on an individual's phenotype do not affect the genes passed on to its offspring. In other words, *acquired characteristics are not inherited.*
3. Hereditary variations are based on particles—*genes*—that *retain their identity as they pass through the generations; they do not blend* with other genes. This is true of both discretely varying traits (e.g., brown vs. blue eyes) and continuously varying traits (e.g., body size, intensity of pigmentation). Genetic variation in continuously varying traits is based on several or many discrete, particulate genes, each of which affects the trait slightly ("polygenic inheritance").
4. Genes mutate, usually at a fairly low rate, to equally stable alternative forms, known as *alleles*. The phenotypic effect of such mutations can range from undetectable to very great. The variation that arises by mutation is amplified by recombination among alleles at different loci.
5. *Evolutionary change is a populational process*: it entails, in its most basic form, a change in the relative abundances (proportions or *frequencies*) of individual organisms with different genotypes (hence, often, with different phenotypes) within a population. One genotype may gradually replace other genotypes over the course of generations. Replacement may occur within only certain populations, or in all the populations that make up a species.
6. The rate of mutation is too low for mutation by itself to shift a population from one genotype to another. Instead, the change in genotype proportions within a population can occur by either of two principal processes: random fluctuations in proportions (genetic drift), or nonrandom changes due to the superior survival and/or reproduction of some genotypes compared with others (i.e., natural selection). Natural selection and random genetic drift can operate simultaneously.
7. Even a slight intensity of natural selection can (under certain circumstances) bring about substantial evolutionary change in a realistic amount of time. *Natural selection can account for both slight and great differences among species*, as well as for the earliest stages of evolution of new traits. Adaptations are traits that have been shaped by natural selection.
8. Natural selection can alter populations beyond the original range of variation by increasing the frequency of alleles that, by recombination with other genes that affect the same trait, give rise to new phenotypes.

9. Natural populations are genetically variable, and so can often evolve rapidly when environmental conditions change.
10. Populations of a species in different geographic regions differ in characteristics that have a genetic basis.
11. The differences between different species, and between different populations of the same species, are often based on differences at several or many genes, many of which have a small phenotypic effect. This pattern supports the hypothesis that the differences between species evolve by rather small steps.
12. Differences among geographic populations of a species are often adaptive, and thus are the consequence of natural selection.
13. Phenotypically different genotypes are often found in a single interbreeding population. Species are not defined simply by phenotypic differences. Rather, different species represent distinct "gene pools"; that is, species are groups of interbreeding or potentially interbreeding individuals that do not exchange genes with other such groups.
14. Speciation is the origin of two or more species from a single common ancestor. Speciation usually occurs by the genetic differentiation of geographically segregated populations. Because of the geographic segregation, interbreeding does not prevent incipient genetic differences from developing.
15. Among living organisms, there are many gradations in phenotypic characteristics among species assigned to the same genus, to different genera, and to different families or other higher taxa. Such observations provide evidence that higher taxa arise by the prolonged, sequential accumulation of small differences, rather than by the sudden mutational origin of drastically new "types."
16. The fossil record includes many gaps among quite different kinds of organisms. Such gaps may be explained by the incompleteness of the fossil record. But the fossil record also includes examples of gradations from apparently ancestral organisms to quite different descendants. These data support the hypothesis that the evolution of large differences proceeds incrementally. Hence the principles that explain the evolution of populations and species may be extrapolated to the evolution of higher taxa.

Evolutionary Biology since the Synthesis

Since the evolutionary synthesis, a great deal of research has elaborated and tested its basic principles. Beginning in the 1950s and accelerating since, advances in genetics and molecular biology have virtually revolutionized the study of evolution and have opened entirely new research areas, such as molecular evolution. Molecular biology has provided tools for studying a vast number of evolutionary topics, such as mutation, genetic variation, species differences, development, and the phylogenetic history of life.

MOTOO KIMURA

Since the mid-1960s, evolutionary theory has expanded into areas such as ecology, animal behavior, and reproductive biology, and detailed theories have been developed to explain the evolution of particular kinds of characteristics such as life span, ecological distribution, and social behavior. The study of macroevolution has been renewed by provocative interpretations of the fossil record and by new methods for studying phylogenetic relationships. As molecular methods have become more sophisticated and available, virtually new fields of evolutionary study have developed. Among these fields is MOLECULAR EVOLUTION (analyses of the processes and history of change in genes), in which the NEUTRAL THEORY OF MOLECULAR EVOLUTION has been particularly important. This hypothesis, developed especially by Motoo Kimura (1924–1994), holds that most of the evolution of DNA sequences occurs by genetic drift rather than by natural selection. EVOLUTIONARY DEVELOPMENTAL BIOLOGY is an exciting field devoted to understanding how developmental processes both evolve and constrain evolution. It is closely tied to developmental biology, one of the most rapidly moving fields of biology today. EVOLUTIONARY GENOMICS, concerned with variation and evolution in multiple genes or even entire genomes, is being born. The advances in these fields, though, are complemented by vigorous research, new discoveries, and new ideas about long-standing topics in evolutionary biology, such as the evolution of adaptations and of new species. This is an exciting time to learn about evolution—or to be an evolutionary biologist.

Philosophical Issues

Thousands of pages have been written about the philosophical and social implications of evolution. Darwin argued that every characteristic of a species can vary, and can be altered radically, given enough time. Thus he rejected the essentialism that Western philosophy had inherited from Plato and Aristotle and put variation in its place. Darwin also helped to replace a static conception of the world—one virtually identical to the Creator's perfect creation—with a world of ceaseless change. It was Darwin who extended to living things, including the human species, the principle that change, not stasis, is the natural order.

Above all, Darwin's theory of random, purposeless variation acted on by blind, purposeless natural selection provided a revolutionary new kind of answer to almost all questions that begin with "Why?" Before Darwin, both philosophers and people in general answered "Why?" questions by citing purpose. Since only an intelligent mind, with the capacity for forethought, can have purpose, questions such as "Why do plants have flowers?" or "Why are there apple trees?"—or diseases, or earthquakes—were answered by imagining the possible purpose that God could have had in creating them. This kind of explanation was made completely superfluous by Darwin's theory of natural selection. The adaptations of organisms—long cited as the most conspicuous evidence of intelligent design in the universe—could be explained by purely mechanistic causes. For evolutionary biologists, the flower of a magnolia has a *function*, but not a *purpose*. It was not designed in order to propagate the species, much less to delight us with its beauty, but instead came into existence because magnolias with brightly colored flowers reproduced more prolifically than magnolias with less brightly colored flowers. The unsettling implication of this purely material explanation is that, except in the case of human behavior, we need not invoke, nor can we find any evidence for, any design, goal, or purpose anywhere in the natural world.

It must be emphasized that all of science has come to adopt the way of thought that Darwin applied to biology. Astronomers do not seek the purpose of comets or supernovas, nor chemists the purpose of hydrogen bonds. The concept of purpose plays no part in scientific explanation.

Ethics, Religion, and Evolution

In the world of science, the reality of evolution has not been in doubt for more than a hundred years, but evolution remains an exceedingly controversial subject in the United States and some other countries. The **creationist movement** opposes the teaching of evolution in public schools, or at least demands "equal time" for creationist beliefs. Such opposition arises from the fear that evolutionary science denies the existence of God, and consequently, that it denies any basis for rules of moral or ethical conduct.

Our knowledge of the history and mechanisms of evolution is certainly incompatible with a *literal* reading of the creation stories in the Bible's Book of Genesis, as it is incompatible with hundreds of other creation myths that people have devised. A literal reading of some passages in the Bible is also incompatible with physics, geology, and other natural sciences. But does evolutionary biology deny the existence of a supernatural being or a human soul? No, because science, including evolutionary biology, is silent on such questions. By its very nature, science can entertain and investigate only hypotheses about material causes that operate with at least probabilistic regularity. It cannot test hypotheses about supernatural beings or their intervention in natural events.

Evolutionary biology has provided natural, material causes for the diversification and adaptation of species, just as the physical sciences did when they explained earthquakes and eclipses. The steady expansion of the sciences, to be sure, has left less and less to be explained by the existence of a supernatural creator, but science can neither deny nor affirm such a being. Indeed, some evolutionary biologists are devoutly religious, and many nonscientists, including many priests, ministers, and rabbis, hold both religious beliefs and belief in evolution.

Wherever ethical and moral principles are to be found, it is probably not in science, and surely not in evolutionary biology. Opponents of evolution have charged that evolution by natural selection justifies the principle that "might makes right," and certainly more than one dictator or imperialist has invoked the "law" of natural selection to justify atrocities. But evolutionary theory cannot provide any such precept for behavior. Like any other science, it describes how the world *is*, not how it *should be*. The supposition that what is "natural" is "good" is called by philosophers the NATURALISTIC FALLACY.

Various animals have evolved behaviors that we give names such as cooperation, monogamy, competition, infanticide, and the like. Whether or not these behaviors *ought* to be, and whether or not they are moral, is not a scientific question. The natural world is amoral—it lacks morality altogether. Despite this, the concepts of natural selection and evolutionary progress were taken as a "law of nature" by which Marx justified class struggle, by which the Social Darwinists of the late eighteenth and early nineteenth centuries justified economic competition and imperialism, and by which the biologist Julian Huxley justified humanitarianism (Hofstadter 1955; Paradis and Williams 1989). All these ideas are philosophically indefensible instances of the naturalistic fallacy. Infanticide by lions and langur monkeys does not justify it in humans, and evolution provides no basis for human ethics.

Evolution as Fact and Theory

Is evolution a fact, a theory, or a hypothesis? Biologists often speak of the "theory of evolution," but they usually mean by that something quite different from what nonscientists understand by that phrase.

In science, a **hypothesis** is an informed conjecture or statement of what might be true. A hypothesis may be poorly supported, especially at first, but it can gain support, to the point at which it is effectively a fact. For Copernicus, the revolution of the Earth around the Sun was a hypothesis with modest support; for us, it is a hypothesis with such strong support that we consider it a fact. Most philosophers (and scientists) hold that we do not know anything with absolute certainty. What we call facts are hypotheses that have acquired so much supporting evidence that we act as if they were true.

In everyday use, a "theory" is an unsupported speculation. Like many words, however, this term has a different meaning in science. A **scientific theory** is a mature, coherent body of interconnected statements, based on reasoning and evidence, that explain a variety of observations. Or, to quote the *Oxford English Dictionary*, a theory is "a scheme or system of ideas and statements held as an explanation or account of a group of facts or phenomena; a hypothesis that has been confirmed or established by observation or experiment, and is propounded or accepted as accounting for the known facts; a statement of what are known to be the general laws, principles, or causes of something known or observed." Thus atomic theory, quantum theory, and the theory of plate tectonics are elaborate schemes of interconnected ideas, strongly supported by evidence, that account for a great variety of phenomena.

Given these definitions, evolution is a fact. But *the fact of evolution is explained by evolutionary theory*.

In *The Origin of Species*, Darwin propounded two major hypotheses: that organisms have descended, with modification, from common ancestors; and that the chief cause of modification is natural selection acting on hereditary variation. Darwin provided abundant evidence for descent with modification, and hundreds of thousands of observations from paleontology, geographic distributions of species, comparative anatomy, embryology, genetics, biochemistry, and molecular biology have confirmed this hypothesis since Darwin's time. Thus the hypothesis of descent with modification from common ancestors has long had the status of a scientific fact.

The explanation of *how* modification occurs and *how* ancestors give rise to diverse descendants constitutes the theory of evolution. We now know that Darwin's hypothesis of natural selection on hereditary variation was correct, but we also know that there are more causes of evolution than Darwin realized, and that natural selection and hereditary variation themselves are more complex than he imagined. A body of ideas about the causes of evolution, including mutation, recombination, gene flow, isolation, random genetic drift, the many forms of natural selection, and other factors, constitute our current theory of evolution, or "evolutionary theory." Like all theories in science, it is incomplete, for we do not yet know the causes of all of evolution, and some details may turn out to be wrong. But the main tenets of the theory are well supported, and most biologists accept them with confidence.

Summary

1. Evolution is the unifying theory of the biological sciences. It aims to discover the history of life and the causes of the diversity and characteristics of organisms.
2. Darwin's evolutionary theory, published in *The Origin of Species* in 1859, consisted of two major hypotheses: first, that all organisms have descended, with modification, from common ancestral forms of life, and second, that a chief agent of modification is natural selection.
3. Darwin's hypothesis that all species have descended with modification from common ancestors is supported by so much evidence that it has become as well established a fact as any in biology. His theory of natural selection as the chief cause of evolution was not broadly supported until the "evolutionary synthesis" that occurred in the 1930s and 1940s.
4. The evolutionary theory developed during and since the evolutionary synthesis consists of a body of principles that explain evolutionary change. Among these principles are (a) that genetic variation in phenotypic characters arises by random mutation and recombination; (b) that changes in the proportions of alleles and genotypes within a population may result in replacement of genotypes over generations; (c) that such changes in the proportions of genotypes may occur either by random fluctuations (genetic drift) or by nonrandom, consistent differences among genotypes in survival or reproduction rates (natural selection); and (d) that due to different histories of genetic drift and natural selection, populations of a species may diverge and become reproductively isolated species.
5. Evolutionary biology makes important contributions to other biological disciplines and to social concerns in areas such as medicine, agriculture, computer science, and our understanding of ourselves.
6. The implications of Darwin's theory, which revolutionized Western thought, include the ideas that change rather than stasis is the natural order; that biological phenomena, including those seemingly designed, can be explained by purely material causes rather than by divine creation; and that no evidence for purpose or goals can be found in the living world, other than in human actions.
7. Like other sciences, evolutionary biology cannot be used to justify beliefs about ethics or morality. Nor can it prove or disprove theological issues such as the existence of a deity. Many people hold that, although evolution is incompatible with a literal interpretation of some passages in the Bible, it is compatible with religious belief.

Terms and Concepts

adaptation
creationist movement
descent with modification
essentialism
evolution (biological evolution; organic evolution)
evolutionary synthesis (= modern synthesis)
hypothesis
inheritance of acquired characteristics
macroevolution
microevolution
natural selection
population
theory
uniformitarianism

Suggestions for Further Reading

The readings at the end of each chapter include major works that provide a comprehensive treatment and an entry into the professional literature. The references cited within the text also serve this important function.

No one should fail to read at least part of Darwin's *The origin of species by means of natural selection, or the preservation of favoured races in the struggle for life;* perhaps the Sixth Edition (1872) is best to read. After some adjustment to Victorian prose, you should be enthralled by the craft, detail, completeness, and insight in Darwin's arguments. It is an astonishing book.

Among the best biographies of Darwin are Janet Browne's superb two-volume work, *Charles Darwin: Voyaging* and *Charles Darwin: The power of place* (Knopf, New York, 1995 and 2002, respectively); and *Darwin*, by A. Desmond and J. Moore (Warner Books, New York, 1991), which emphasizes the role played by the religious, philosophical, and intellectual climate of nineteeth-century England on the development of his scientific theories. See also P. J. Bowler, *Charles Darwin: The man and his influence* (Blackwell Scientific, Cambridge, UK, 1990), which emphasizes scientific issues, and J. Bowlby, *Charles Darwin: A biography* (W.W. Norton, New York, 1991), which emphasizes Darwin's personal life.

Important works on the history of evolutionary biology include P. J. Bowler, *Evolution: The history of an idea* (University of California Press, Berkeley, 1989); E. Mayr, *The growth of biological thought: Diversity, evolution, and inheritance* (Harvard University Press, Cambridge, MA, 1982, a detailed, comprehensive history of systematics, evolutionary biology, and genetics that bears the personal stamp of one of the major figures in the evolutionary synthesis); and E. Mayr and W. B. Provine (eds.), *The evolutionary synthesis: Perspectives on the unification of biolog*y (Harvard University Press, Cambridge, MA, 1980), which contains essays by historians and biologists, including some of the major contributors to the synthesis.

Recent books that expose the fallacies of creationism and explain the nature of science and of evolutionary biology include R. T. Pennock, *Tower of Babel: The evidence against the New Creationism* (M.I.T. Press, Cambridge, MA, 1999); B. J. Alters and S. M. Alters, *Defending evolution: A guide to the creation/evolution controversy* (Jones and Bartlett, Sudbury, MA, 2001); and M. Pigliucci, *Denying evolution: Creationism, scientism, and the nature of science* (Sinauer Associates, Sunderland, MA, 2002).

Problems and Discussion Topics

1. How does evolution unify the biological sciences? What other principles might do so?
2. Discuss how a creationist versus an evolutionary biologist might explain some human characteristics, and the implications of their differences. Sample characteristics: eyes; wisdom teeth; individually unique friction ridges (fingerprints); five digits rather than some other number; susceptibility to infections; fever when infected; variation in sexual orientation; limited life span.
3. Analyze and evaluate Ralph Waldo Emerson's couplet,

   Striving to be man, the worm
   Mounts through all the spires of form.

   What pre-Darwinian concepts does it express? What fault in it will a Darwinian find?
4. In February 2001, it was announced that two research groups had effectively completed sequencing the entire human genome. If humans, along with all other forms of life, have evolved from a common ancestor, what evidence of this would be expected in the human genome? In what ways might the history and processes of evolution help us interpret and make sense of genomic sequence data?
5. How might the evolution of antibiotic resistance in pathogenic bacteria be slowed down or prevented? What might you need to know in order to achieve this aim?
6. Should both evolution and creationism be taught as alternative theories in science classes?
7. Based on sources available in a good library, discuss how the "Darwinian revolution" affected one of the following fields: philosophy, literature, psychology, anthropology.

The Tree of Life: Classification and Phylogeny

2

About 2000 million years ago, a bacterium, not unlike the *Escherichia coli* bacteria in our intestines, took up residence within the cell of another bacteria-like organism. This partnership flourished, since each partner evidently provided biochemical services to the other. The "host" in this partnership developed a modern nucleus, chromosomes, and mitotic spindle, while the "guest" bacterium within it evolved into the mitochondrion. This ancestral eukaryote gave rise to diverse one-celled descendants. Some of those descendants later became multicellular when the cells they produced by mitosis remained together and evolved mechanisms of regulating gene expression that enabled groups of cells to form different tissues and organs. One such lineage became the progenitor of green plants, another of the fungi and animals.

How do we classify organisms? Although it lacks legs, this animal is not a snake but a lizard—the eastern glass lizard, *Ophisaurus ventralis.* It is one of about 80 species of glass lizard (family Anguidae), many of which do have legs. The groove along the side of the body is a feature of this family. Snakes have very different scales, skulls, and other internal features. (Photo © John Cancalosi/AGE Fotostock.)

Between about 1000 million and 600 million years ago, a single animal species gave rise to two species that became the progenitors of two quite astonishingly different groups of animals: one evolved into the starfishes, sea urchins, and other echinoderms, and the other into the chordates, including the vertebrates. Most of the species derived from the earliest vertebrate are fishes, but one would prove to be the ancestor of the tetrapods (four-legged vertebrates). About 150 million years after the first land-dwelling tetrapod evolved, some of its descendants stood on the brink of mammalhood. About another 125 million years later, the mammals had diversified into many groups, including the first primates, adapted to life in trees. Some primates became small, some evolved prehensile tails, and one became the ancestor of large, tailless apes. About 14 million years ago, one such ape gave rise to the Asian orangutan on the one hand and an African descendant on the other. The African descendant split into the gorilla and another species. About 6 to 8 million years ago, that species, in turn, divided into one lineage that became today's chimpanzees and into another lineage that underwent rapid evolution of its posture, feet, hands, and brain: our own quite recent ancestor.

This is our best current understanding of some of the high points in our history, in which, metaphorically speaking, the human species developed as one twig in a gigantic tree, the great Tree of Life. (We will look at this history in more detail in Chapters 5 and 7.) With the passage of time, a species is likely to "branch"—to give rise to two species that evolve different modifications of some of their features. Those species branch in turn, and their descendants may be altered further still. By this process of branching and modification, repeated innumerable times over the course of many millions of years, many millions of kinds of organisms have evolved from a single ancestral organism at the very base, or root, of the tree (Figure 2.1).

Figure 2.1 The Tree of Life. This estimate of the relationships among some of the major branches is based mostly on DNA sequences, especially those of genes encoding ribosomal RNA. Of the three empires of life, Archaea and Eucarya appear to have the most recent common ancestor. The majority of taxa in the tree are unicellular. (After Baldauf et al. 2004.)

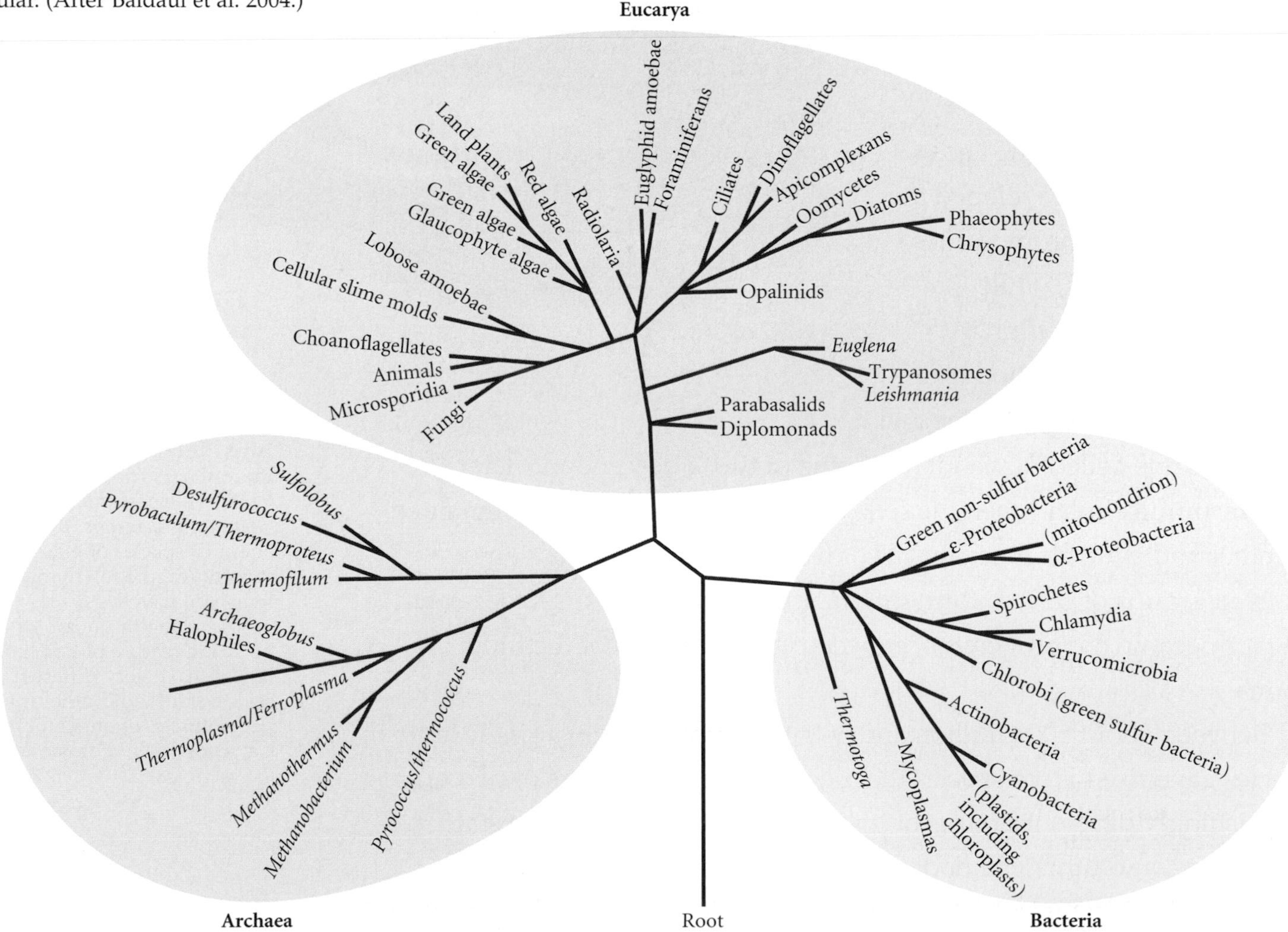

Evolutionary biologists have developed methods of "reconstructing" or "assembling" the tree of life—of estimating the **phylogenetic**, or genealogical, relationships among organisms (i.e., which species share a recent **common ancestor**, which share more distant ancestors, and which share even more distant ancestors). The resulting portrayal of relationships not only is fascinating in itself (have you ever thought of yourself as related to a starfish, a butterfly, a mushroom?) but is also an important foundation for understanding many aspects of evolutionary history, such as the pathways by which various characteristics have evolved.

We cannot directly observe evolutionary history, so we must infer it using deductive logic, like Sherlock Holmes reconstructing the history of a crime. A few points in our brief sketch of human origins, such as the timing of certain events, have been learned from the fossil record. However, most of this history has been determined from studying not fossils, but living organisms. In this chapter, we will become acquainted with some of the methods by which we can infer phylogenetic relationships, and we will see how our understanding of those relationships is reflected in the classification of organisms. In the following chapter, we will examine some common evolutionary patterns that these approaches have helped to elucidate.

Classification

Phylogenetic analysis—the study of relationships among species—has historically been closely associated with the classification and naming of organisms (known as TAXONOMY). Both are among the tasks of the field of SYSTEMATICS.

In the early 1700s, European naturalists believed that God must have created species according to some ordered scheme, as we saw in Chapter 1. It was therefore a work of devotion to discover "the plan of creation" by cataloging the works of the Creator and discovering a "natural," true, classification. The scheme of classification that was adopted then, and is still used today, was developed by the Swedish botanist Carolus Linnaeus (1707–1778). Linnaeus introduced BINOMIAL NOMENCLATURE, a system of two-part names consisting of a genus name and a specific epithet (such as *Homo sapiens*). He proposed a system of grouping species in a HIERARCHICAL CLASSIFICATION of groups nested within larger groups (such as genera nested within families) (Box A). The levels of classification, such as kingdom, phylum, class, order, family, genus, and species, are referred to as **taxonomic categories**, whereas a particular group of organisms assigned to a categorical rank is a **taxon** (plural: **taxa**). Thus the rhesus monkey is placed in the genus *Macaca*, in the family Cercopithecidae, in the order Primates; *Macaca*, Cercopithecidae, and Primates are taxa that exemplify the taxonomic categories genus, family, and order, respectively. Several "intermediate" taxonomic categories, such as SUPERFAMILY and SUBSPECIES, are sometimes used in addition to the more familiar and universal ones. In assigning species to **higher taxa** (those above the species level), Linnaeus used features that he imagined represented propinquity in God's creative scheme. For example, he defined the order Primates by the features "four parallel upper front [incisor] teeth; two pectoral nipples." But without an evolutionary framework, naturalists had no objective basis for classifying mammals by their teeth rather than by, say, their color or size.

Classification took on an entirely different significance after *The Origin of Species* was published in 1859. In the Galápagos Islands, as we saw in Chapter 1, Darwin had observed that different islands harbored similar, but nevertheless distinguishable, mockingbirds. He came to suspect that the different forms of mockingbirds had descended from a single ancestor, acquiring slight differences in the course of descent. But this thought, logically extended, suggested that that ancestor itself had been modified from an ancestor further back in time, which could have given rise to yet other descendants—various South American mockingbirds, for instance. By this logic, some remote ancestor might have been the progenitor of all species of birds, and a still more remote ancestor the progenitor of all vertebrates.

Darwin proposed, then, that on occasion, an ancestral species may split into two descendant species, which at first are very similar to each other, but which **diverge** (become

BOX 2A Taxonomic Practice and Nomenclature

Standardized names for organisms are essential for communication among scientists. To ensure that names are standardized, taxonomy has developed rules of procedure.

Most species are named by taxonomists who are experts on that particular group of organisms. A new species may be one that has never been seen before (e.g., an organism dredged from the deep sea), but many unnamed species are sitting in museum collections, awaiting description. Moreover, a single species often proves, on closer study, to be two or more very similar species. A taxonomist who undertakes a REVISION—a comprehensive analysis—of a group frequently names new species. A species name has legal standing if it is published in a journal, or even in a privately produced publication that is publicly available.

The name of a species consists of its genus and its specific epithet; both are Latin or latinized words. These words are always *italicized* (or underlined), and the genus is always capitalized. In entomology and certain other fields, it is customary to include the name of the AUTHOR (the person who conferred the specific epithet); for example, the corn rootworm beetle *Diabrotica virgifera* LeConte.

Numerous rules govern the construction of species names (e.g., genus and specific epithet must agree in gender: *Rattus norvegicus*, not *Rattus norvegica*, for the brown rat). It is recommended that the name have meaning [e.g., *Vermivora* ("worm eater") *chrysoptera* ("golden-winged") for the golden-winged warbler; *Rana warschewitschii*, "Warschewitsch's frog"], but it need not. Often a taxonomist will honor another person by naming a species after him or her.

The first rule of nomenclature is that no two species of animals, or of plants, can bear the same name. (It is permissible, however, for the same name to be applied to both a plant and an animal genus; for example, *Alsophila* is the name of both a fern genus and a moth genus.) The second is the rule of PRIORITY: the valid name of a taxon is the oldest available name that has been applied to it. Thus it sometimes happens that two authors independently describe the same species under different names; in this case, the valid name is the older one, and the younger name is a junior SYNONYM. Conversely, it may turn out that two or more species have masqueraded under one name; in this case, the name is applied to the species that the author used in his or her description. To prevent the obvious ambiguity that could arise in this way, it has become standard practice for the author to designate a single specimen (the TYPE SPECIMEN, or HOLOTYPE) as the "name-bearer" so that later workers can determine which of several similar species rightfully bears the name. The holotype, usually accompanied by other specimens (PARATYPES) that exemplify the range of variation, is deposited and carefully preserved in a museum or herbarium.

In revising a genus, a taxonomist usually introduces changes in the taxonomy. Some examples of such changes are:

- Species that were placed by previous authors in different genera may be brought together in the same genus because they are shown to be closely related. These species retain their specific epithets, but if they are shifted to a different genus, the author's name is written in parentheses.
- A species may be removed from a genus and placed in a different genus because it is determined not to be closely related to the other members.
- Forms originally described as different species may prove to be the same species, and be synonymized.
- New species may be described.

The rules for naming higher taxa are not all as strict as those for species and genera. In zoology (and increasingly in botany), names of subfamilies, families, and sometimes orders are formed from the stem of the type genus (the first genus described). Most family names of plants end in -aceae. In zoology, subfamily names end in -inae and family names in -idae. Thus *Mus* (from the Latin *mus, muris*, "mouse"), the genus of the house mouse, is the type genus of the family Muridae and the subfamily Murinae; *Rosa* (rose) is the type genus of the family Rosaceae. Endings for categories above family are standardized in some, but not all, groups (e.g., -formes for orders of birds, such as Passeriformes, the perching birds that include *Passer*, the house sparrows). Names of taxa above the genus level are not italicized, but are always capitalized. Adjectives or colloquial nouns formed from these names are not capitalized: thus we may refer to murids or to murid rodents, without capitalization.

more different) in the course of time. Each of those species, in turn, may divide and diverge to yield two descendant species, and the process may be repeated again and again throughout the immensely long history of life. Darwin therefore gave meaning to the notion of "closely related" species: they are descended from a relatively recent common ancestor, whereas more distantly related species are descended from a more remote (i.e., further back in time) common ancestor. The features that species hold in common, such as the vertebrae that all vertebrates possess, were not independently bestowed on each of those species by a Creator, but were inherited from the ancestral species in which the features first evolved. With breathtaking daring, Darwin ventured that all species of organisms, in all their amazing diversity, had descended by endless repetition of such events, through long ages, from perhaps only one common ancestor—the universal ancestor of all life.

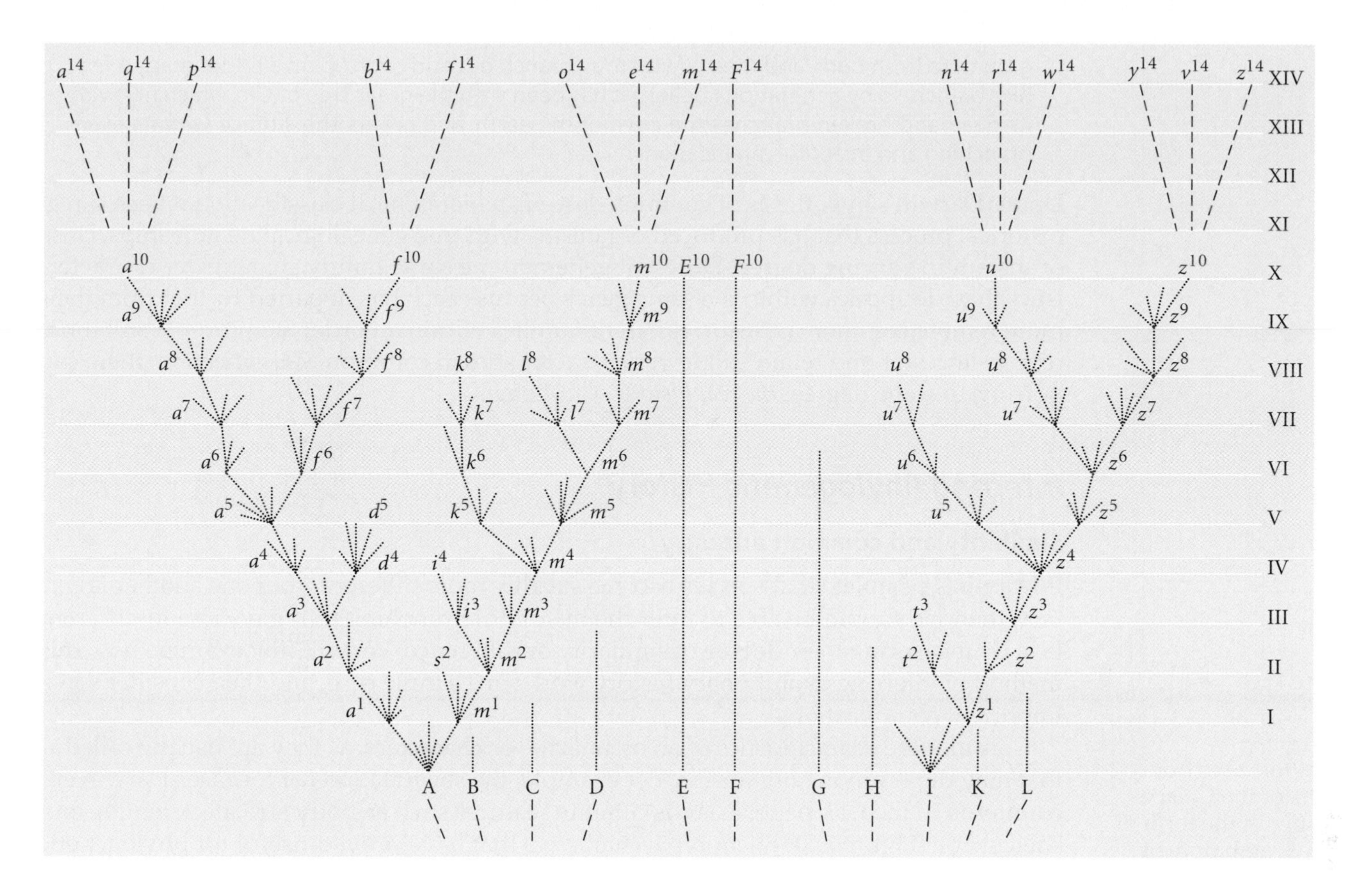

Figure 2.2 Darwin's representation of hypothetical phylogenetic relationships, showing how lineages diverge from common ancestors and give rise to both extinct and extant species. Time intervals (between Roman numerals) represent thousands of generations. Darwin omitted the details of branching for intervals X through XIV. Extant species (at time XIV) can be traced to ancestors A, F, and I; all other original lineages have become extinct. Distance along the horizontal axis represents degree of divergence (as, for example, in body form). Darwin recognized that rates of evolution vary greatly, showing this by different angles in the diagram; for instance, the lineage from ancestor F has survived essentially unchanged. (From Darwin 1859.)

In Darwin's words, all species, extant and extinct, form a great "Tree of Life," or **phylogenetic tree**, in which closely adjacent twigs represent living species derived only recently from their common ancestors, whereas twigs on different branches represent species derived from more ancient common ancestors (Figure 2.2). He expressed this metaphor in some of his most poetic language:

> The affinities of all the beings of the same class have sometimes been represented by a great tree. I believe this simile largely speaks the truth. The green and budding twigs may represent existing species; and those produced during former years may represent the long succession of extinct species. At each period of growth all the growing twigs have tried to branch out on all sides, and to overtop and kill the surrounding twigs and branches, in the same manner as species and groups of species have at all times overmastered other species in the great battle for life. The limbs divided into great branches, and these into lesser and lesser branches, were themselves once, when the tree was young, budding twigs; and this connection of the former and present buds by ramifying branches may well represent the classification of all extinct and living species in groups subordinate to groups. Of the many twigs which flourished when the tree was a mere bush, only two or three, now grown into great branches, yet survive and bear the other branches; so with the species which lived during long-past geological periods, very few have left living and modified descendants. From the very first growth of the tree, many a limb and branch has decayed and dropped off; and these fallen branches of various sizes may represent those whole orders, families, and genera which have now no living representatives, and which are known to us only in a fossil state. As we here and there see a thin, straggling branch springing from a fork low down in a tree, and which by some chance has been favoured and is still alive on its summit, so we occasionally see an animal like the Ornithorhynchus or Lepidosiren,* which in some small degree connects by its affinities two large branches of life, and which has apparently been saved from fatal competition by having inhabited a protected station. As buds give rise by

---

**Ornithorhynchus*, the duck-billed platypus, is a primitive, egg-laying mammal. *Lepidosiren* is a genus of living lungfishes, a group that is closely related to the ancestor of the tetrapod (four-legged) vertebrates, and which is known from ancient fossils.

> growth to fresh buds, and these, if vigorous, branch out and overtop on all sides many a feebler branch, so by generation I believe it has been with the great Tree of Life, which fills with its dead and broken branches the crust of the earth, and covers the surface with its ever-branching and beautiful ramifications.

Under Darwin's hypothesis of common descent, a hierarchical classification reflects a real historical process that has produced organisms with true genealogical relationships, close or distant in varying degree. Different genera in the same family share fewer characteristics than do species within the same genus because each has departed further from their more remote common ancestor; different families within an order stem from a still more remote ancestor and retain still fewer characteristics in common. Classification, then, can portray, to some degree, *the real history of evolution*.

Inferring Phylogenetic History

Similarity and common ancestry

If Darwin's postulate that species become steadily more different from one another is correct, then we should be able to infer the history of branching that gave rise to different taxa by measuring their degree of similarity or difference. We will first examine how this method of inferring evolutionary history works in a simple case, and then consider some important complications.

Consider the characteristics of an organism—or **characters**, as they are usually called—that may differ among organisms. For example, the several kinds of tortoises Darwin encountered in the Galápagos Islands differ in features such as body size, neck length, and shell shape (Figure 2.3). Phenotypic characters that have proved useful for phylogenetic analyses of various organisms have included not only external and internal morphological features, but also differences in behavior, cell structure, biochemistry, and chromosome structure. Today, armed with the knowledge and techniques of molecular biology, biologists often use DNA sequences, in which the identity of the nucleotide base (A, T, C, or G) at a particular site in the sequence may be considered a character. Each character can have different possible **character states**: the nucleotide A versus C, short versus long neck, rounded versus saddle-shaped shell.

As a first step, let us look at a group of four species (1–4) with 10 variable characters of interest (a–j). Our task is to determine which of the species are derived from recent, and which from more ancient, common ancestors. That is, we wish to arrange them into a phylogenetic tree such as that in Figure 2.4A, in which each branch point (NODE) represents the common ancestor of the two lineages. For simplicity, let us suppose that each character can have two states, labeled 0 and 1, and that 0 is the **ancestral** state, found in the common ancestor (Anc1) of the group. State 1 is a **derived** state—that is, a state that has evolved from the ancestral state. (For example, the ancestral state A might be replaced by the derived state C , or the ancestral state red eyes by the derived state yellow eyes, during the evolution of one or more descendant taxa.) The Greek-derived adjectives PLESIOMORPHIC and APOMORPHIC are often used for "ancestral" and "derived," respectively. We use such data on character states to infer the phylogenetic relationships among the species.

Figure 2.4A portrays a hypothetical phylogeny in which four species descend from the ancestor 1. Each evolutionary change, such as evolution from character state a_0 to character state a_1, is indicated by a tick mark along the branch in which it occurs. We refer to the set of species derived from any one common ancestor as a **monophyletic group**. Thus Figure 2.4A shows three

(A)

(B)

Figure 2.3 Giant tortoises from the different islands of the Galápagos archipelago differ in the form of their shells and the length of their necks, but they are members of the same species (*Geochelone elephantophus*). (A) Saddleback tortoise (*G. e. hoodensis*) from Española (formerly Hood) Island. (B) Dome-shelled tortoise (*G. e. vandenburghi*) from Isabela Island. (Photos © François Gohier/Photo Researchers, Inc.)

monophyletic groups: species 2 + 3, species 1 + 2 + 3, and species 1 + 2 + 3 + 4. Suppose that Figure 2.4A does indeed represent the true phylogeny of the four species; can we infer, or estimate, this phylogeny from the data?

We may calculate the similarity of each pair of descendant species as the number of character states they share. Species 1 and 2, for example, both have character states a_0, b_0, c_1, and j_0, as tallied in the matrix of shared character states in Figure 2.4A. In this example, species 2 and 3 are most similar, and evolved from the most recent common ancestor (ancestor 3). Species 1 is more similar to species 2 and species 3, with which it shares the common ancestor (ancestor 2), than it is to species 4, with which the entire group (species 1–3) shares the most remote common ancestor (ancestor 1). In this example, the degree of similarity is a reliable index of recency of common ancestry, and it enables us to infer the monophyletic groups—that is, the phylogeny of these species.

In this hypothetical case, we suppose that we know which character states are ancestral and which are derived. When we measured similarity, we counted both the shared characters that did not evolve during the ancestry of any two species (e.g., the ancestral state a_0 shared by species 1 and 2) and the shared characters that did evolve (c_1 in this instance). If we count only the shared *derived* character states—those that did evolve—we obtain another matrix, in which, again, species 2 and 3 are most similar, and species 1 is more similar to them than to species 4. Shared derived character states are sometimes called **synapomorphies**.

Complications in inferring phylogeny

In Figure 2.4A, the number of character state changes is roughly the same from the ancestor of the group (ancestor 1) to each descendant species. That is, the rate of evolution is about equal among the lineages. This need not be the case, however. In Figure 2.4B, we assume that the rate of evolution between ancestor 3 and species 2 is greater than elsewhere in the phylogeny. Perhaps this difference represents more base pair substitutions in a DNA sequence in species 2 than in the others.* The matrix of shared character states now indicates that species 1 and 3 are most similar. We might therefore be misled into thinking that they are the most closely related species, although they are not. (Remember, degree of relationship means relative recency of common ancestry, not similarity.) In this case, similarity is not an adequate indicator of relationship. But the number of shared *derived* character states accurately indicates that the closest relatives are species 2 and 3: it accurately indicates phylogeny. Why the difference?

Species 1 and 3 share not only one derived character state (c_1), but also six ancestral character states (a_0, b_0, g_0, h_0, i_0, j_0). Because species 2 underwent four evolutionary changes (to g_1, h_1, i_1, and j_1) that species 3 does not share, it is less similar to species 3 than species 1 is, even though it is more closely related. Taxa may be similar because they share ancestral character states or derived character states, but *only the derived character states that are shared among taxa* (i.e., synapomorphies) *indicate monophyletic groups* and enable us to infer the phylogeny successfully. By the same token, derived character states that are restricted to a single lineage, sometimes called **autapomorphies** (such as state j_1 in species 3), do not provide any evidence about its relationship to other lineages.

In the previous examples, each character changed only once across the whole phylogeny. Hence all the taxa sharing a character state inherited it without change from their common ancestor. Such a character state is said to be **homologous** in all the taxa that share it. (Note that we may speak of both homologous character states and homologous characters.) Again, this need not be the case. A character state is **homoplasious** if it has independently evolved two or more times, and so does not have a unique origin. The taxa that share such a character have not all inherited it from their common ancestor. Figure 2.4C shows three homoplasious characters. State g_1 has independently evolved from g_0 in species 1 and 3, and h_1 has independently evolved in species 1 and 2. These two homoplasies are examples

---

*A base pair SUBSTITUTION is the replacement of one nucleotide base pair (e.g., A–T) by the another (e.g., G–C) in an entire population or species. Such substitutions sometimes, but not always, change the amino acid specified by the genetic code, as explained in detail in Chapter 8.

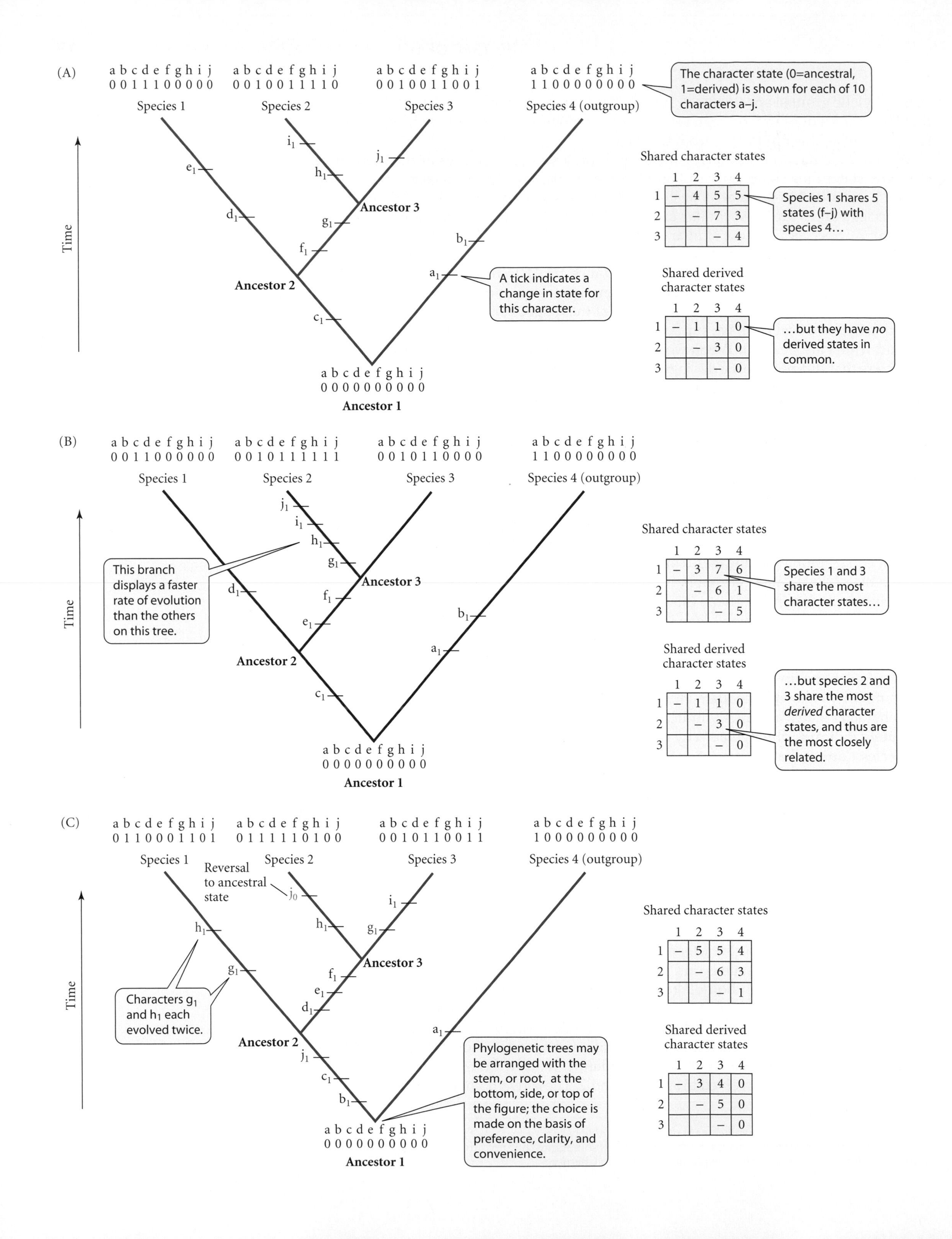
(A)
a b c d e f g h i j
0 0 1 1 1 0 0 0 0 0
Species 1
a b c d e f g h i j
0 0 1 0 0 1 1 1 1 0
Species 2
a b c d e f g h i j
0 0 1 0 0 1 1 0 0 1
Species 3
a b c d e f g h i j
1 1 0 0 0 0 0 0 0 0
Species 4 (outgroup)
The character state (0=ancestral, 1=derived) is shown for each of 10 characters a–j.
Time
e1 d1 i1 h1 j1 g1 f1 c1 b1 a1
Ancestor 3
Ancestor 2
A tick indicates a change in state for this character.
a b c d e f g h i j
0 0 0 0 0 0 0 0 0 0
Ancestor 1
Shared character states
1 2 3 4
1 – 4 5 5
2 – 7 3
3 – 4
Species 1 shares 5 states (f–j) with species 4…
Shared derived character states
1 2 3 4
1 – 1 1 0
2 – 3 0
3 – 0
…but they have no derived states in common.
(B)
a b c d e f g h i j
0 0 1 1 0 0 0 0 0 0
Species 1
a b c d e f g h i j
0 0 1 0 1 1 1 1 1 1
Species 2
a b c d e f g h i j
0 0 1 0 1 1 0 0 0 0
Species 3
a b c d e f g h i j
1 1 0 0 0 0 0 0 0 0
Species 4 (outgroup)
Time
j1 i1 h1 g1 d1 f1 e1 c1 b1 a1
This branch displays a faster rate of evolution than the others on this tree.
Ancestor 3
Ancestor 2
a b c d e f g h i j
0 0 0 0 0 0 0 0 0 0
Ancestor 1
Shared character states
1 2 3 4
1 – 3 7 6
2 – 6 1
3 – 5
Species 1 and 3 share the most character states…
Shared derived character states
1 2 3 4
1 – 1 1 0
2 – 3 0
3 – 0
…but species 2 and 3 share the most derived character states, and thus are the most closely related.
(C)
a b c d e f g h i j
0 1 1 0 0 0 1 1 0 1
Species 1
a b c d e f g h i j
0 1 1 1 1 1 0 1 0 0
Species 2
a b c d e f g h i j
0 0 1 0 1 1 0 0 1 1
Species 3
a b c d e f g h i j
1 0 0 0 0 0 0 0 0 0
Species 4 (outgroup)
Reversal to ancestral state
Time
j0 i1 h1 g1 h1 g1 f1 e1 d1 j1 c1 b1 a1
Characters g1 and h1 each evolved twice.
Ancestor 3
Ancestor 2
Phylogenetic trees may be arranged with the stem, or root, at the bottom, side, or top of the figure; the choice is made on the basis of preference, clarity, and convenience.
a b c d e f g h i j
0 0 0 0 0 0 0 0 0 0
Ancestor 1
Shared character states
1 2 3 4
1 – 5 5 4
2 – 6 3
3 – 1
Shared derived character states
1 2 3 4
1 – 3 4 0
2 – 5 0
3 – 0

◀ **Figure 2.4** An example of phylogenetic analysis applied to three data sets. In each tree, species 1–4 are shown at the top of the tree and their common ancestor at the bottom. The character state (0 or 1) for each of ten characters (a–j) is shown for Ancestor 1 and the extant species. The labeled tick marks along the branches show evolutionary changes in character states. To the right of each tree, the upper matrix shows the total number of character states shared between each pair of species, and the lower matrix shows the number of shared derived character states, as implied by state changes along the branches. (A) A tree with relatively constant evolutionary rates and no homoplasy. (B) The rate of evolution is faster in one branch (Ancestor 3 to species 2) than in other branches. (C) Three characters are homoplasious: character states g_1 and h_1 both independently evolved twice, and character j underwent evolutionary reversal, from j_1 back to the ancestral state j_0. In all these cases, we assume that species 4 is an "outgroup"—i.e., it is more distantly related to the other species than they are to one another.

of **convergent evolution**, the independent origin of a derived character state in two or more taxa. Conversely, character j has evolved from j_0 to j_1 in the evolution of the ancestor 2, and then has undergone **evolutionary reversal** to j_0 in species 2.

Homoplasious characters—those that undergo convergent evolution or evolutionary reversal—provide misleading evidence about phylogeny. In Figure 2.4C, characters g and j would both lead us to mistake species 1 and 3 for the closest relatives; character h erroneously suggests that species 1 and 2 are a monophyletic group. Thus shared derived character states are valid evidence of monophyletic groups only if they are *uniquely* derived.

Many systematists trace the modern practice of inferring phylogenetic relationships to the German entomologist Willi Hennig (1966). Hennig pointed out that taxa may be similar because they share (1) uniquely derived character states, (2) ancestral character states, or (3) homoplasious character states, and that only the similarity due to uniquely derived character states provides evidence for the nested monophyletic groups that make up a phylogenetic tree. Thus, for example, we believe that the tetrapod limb, the amnion, and the feather each evolved only once. We conclude that all tetrapods (vertebrates with four limbs rather than fins) form a single monophyletic group (Figure 2.5). Within the tetrapods, the amniotes form a monophyletic group. Among the amniotes,* all feather-bearing animals (birds) are, again, a single monophyletic group. However, this does not mean that all vertebrates without feathers form a single branch, and indeed they do not. The lack of feathers is an ancestral character state that does not provide evidence that featherless animals are all more closely related to one another than to birds. (Fishes, lizards, and frogs all lack feathers, but so, after all, do all invertebrates.)

Figure 2.5 A phylogeny of some groups of vertebrates, showing monophyletic groups (tetrapods, amniotes, birds) whose members share derived character states that evolved only once.

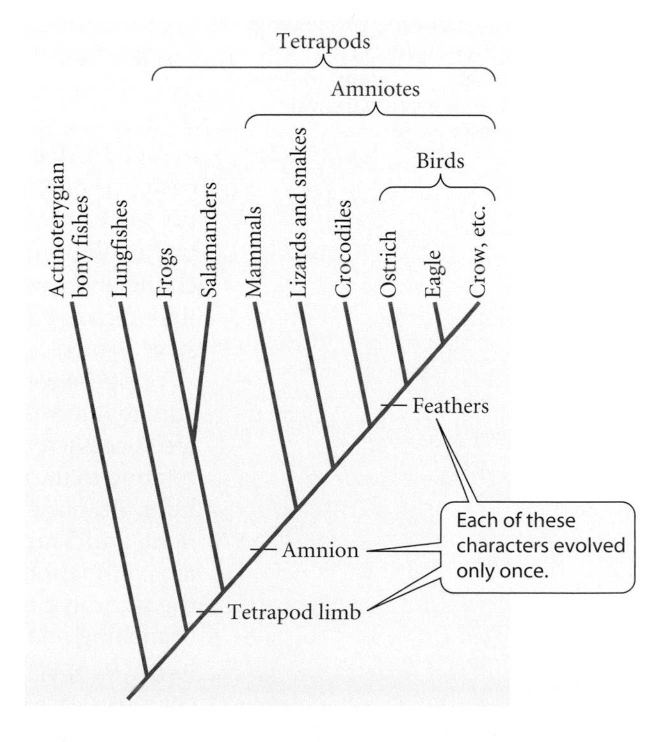

The method of maximum parsimony

Hennig's principle, that uniquely derived character states define monophyletic groups, poses two difficulties: First, how can we tell which state of a character is derived? Second, how can we tell whether it is uniquely derived or homoplasious? For our hypothetical phylogenies in Figure 2.4, we were free to dictate that state 0 was ancestral and that g_1 evolved twice. In real life, we are not given such information—we have to determine it somehow. It might be supposed that the fossil record would tell us what the ancestor's characteristics were, but as we will see, the relationships among fossils and living species have to be interpreted, just like those among living species alone. Besides, the great majority of organisms have a very incomplete fossil record, as we will see in Chapter 4.

*Amniotes are those vertebrates—reptiles, birds, and mammals—characterized by a major adaptation for life on land: the amniotic egg, with its tough shell, protective embryonic membranes (chorion and amnion), and a membranous sac (allantois) for storing embryonic waste products.

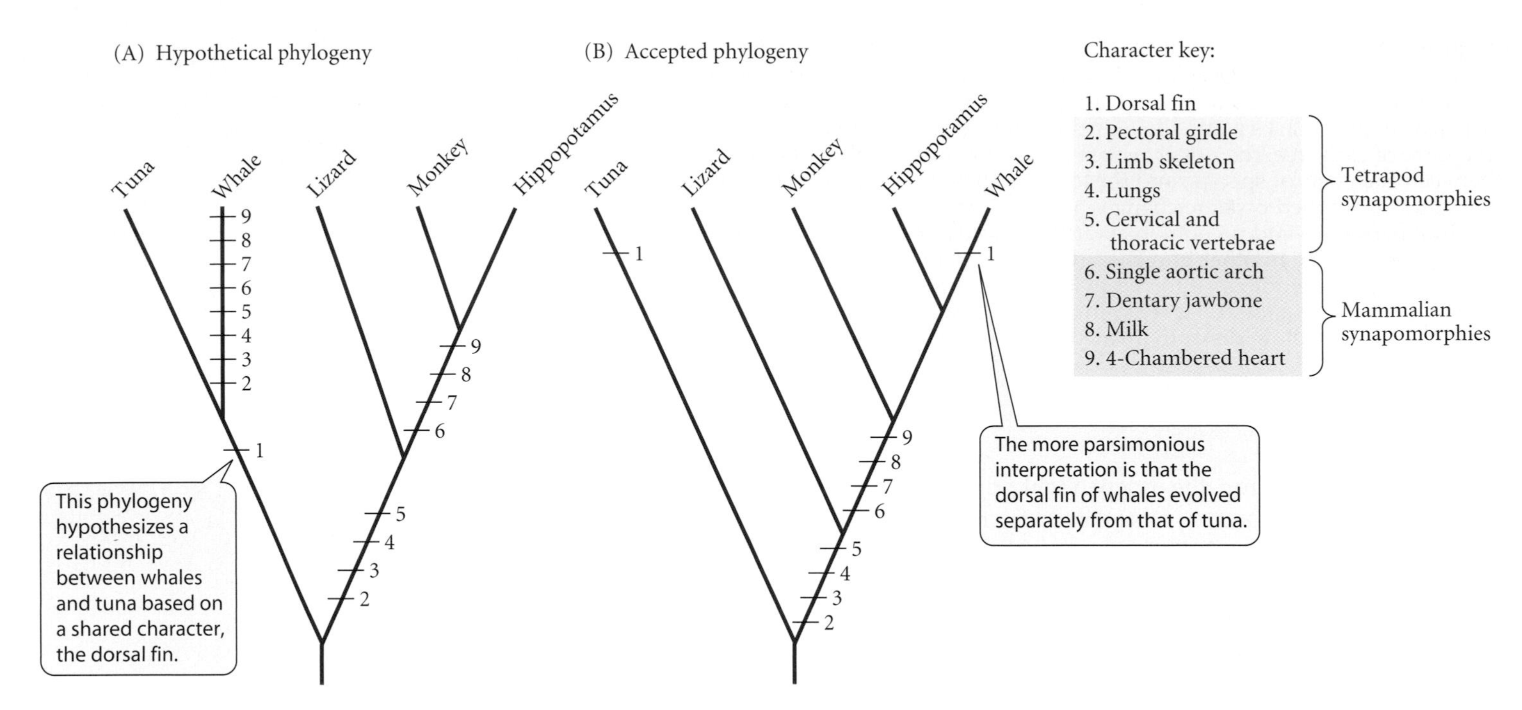

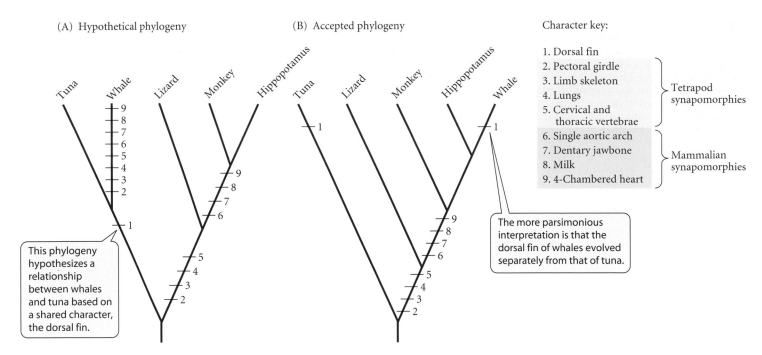

Figure 2.6 Two possible hypotheses for the phylogenetic relationships of whales. (A) A hypothetical phylogeny postulating a close relationship between whales and fishes such as tuna, based on the shared dorsal fin. (B) The accepted phylogeny, in which whales are most closely related to other mammals. Tick marks show the changes in several characters that are implied by each phylogeny. Characters 2–5 are considered uniquely derived synapomorphies of tetrapods, and 6–9 the same for mammals. The accepted phylogeny requires fewer evolutionary changes than the phylogeny in (A), and is therefore a more parsimonious hypothesis.

Some of the methods devised to deal with these problems depend on the concept of **parsimony**. Parsimony is the principle, dating at least from the fourteenth century, that the simplest explanation, requiring the fewest undocumented assumptions, should be preferred over more complicated hypotheses that require more assumptions for which evidence is lacking. The method based on parsimony is among the simplest methods of phylogenetic analysis, and is one of the most widely used.

In phylogenetic analysis, the principle of parsimony suggests that among the various phylogenetic trees that can be imagined for a group of taxa, *the best estimate of the true phylogeny is the one that requires us to postulate the fewest evolutionary changes*. Suppose, for example, we postulate that whales and fishes such as tuna form a monophyletic group because they have a dorsal fin, and that all the other creatures we call mammals form another monophyletic group (Figure 2.6A). This phylogeny would require us to postulate (on the basis of no evidence) that many features shared by whales and other mammals (e.g., four-chambered heart, milk, a single aortic arch), as well as many features shared by whales and other tetrapods such as lizards (e.g., tetrapod limb structure, lungs), each evolved twice. In contrast, if we postulate that whales are descended from the same ancestor as other mammals, then each of these features evolved once, and only the dorsal fin (and a few other features such as body shape) evolved twice (Figure 2.6B). The "extra" evolutionary changes we must postulate in a proposed phylogeny are homoplasious changes: those that we must suppose occurred more than once. Thus parsimony holds that *the best phylogenetic hypothesis is the one that requires us to postulate the fewest homoplasious changes*.

Suppose we wish to determine the relationships among species 1, 2, and 3. We are quite sure that they form a monophyletic group relative to taxa 4 and 5, which we are confident are successively more distantly related. These distantly related taxa are called **outgroups**, relative to the **ingroup**, the monophyletic set of species whose relationships we wish to infer. (In our example, species 1, 2, and 3 might be species of primates, 4 might be a rodent, and 5 might be a marsupial such as a kangaroo. From extensive prior evidence, we are confident that rodents and marsupials are more distantly related to the primates than the various primates are to one another.) As Figure 2.7 shows, there are three possible branching relationships (trees 1–3) among the three ingroup species: the closest relatives might be species 1 and 2, 1 and 3, or 2 and 3. The data matrix in the figure shows the nucleotide bases at seven sites in a homologous DNA sequence from each species.

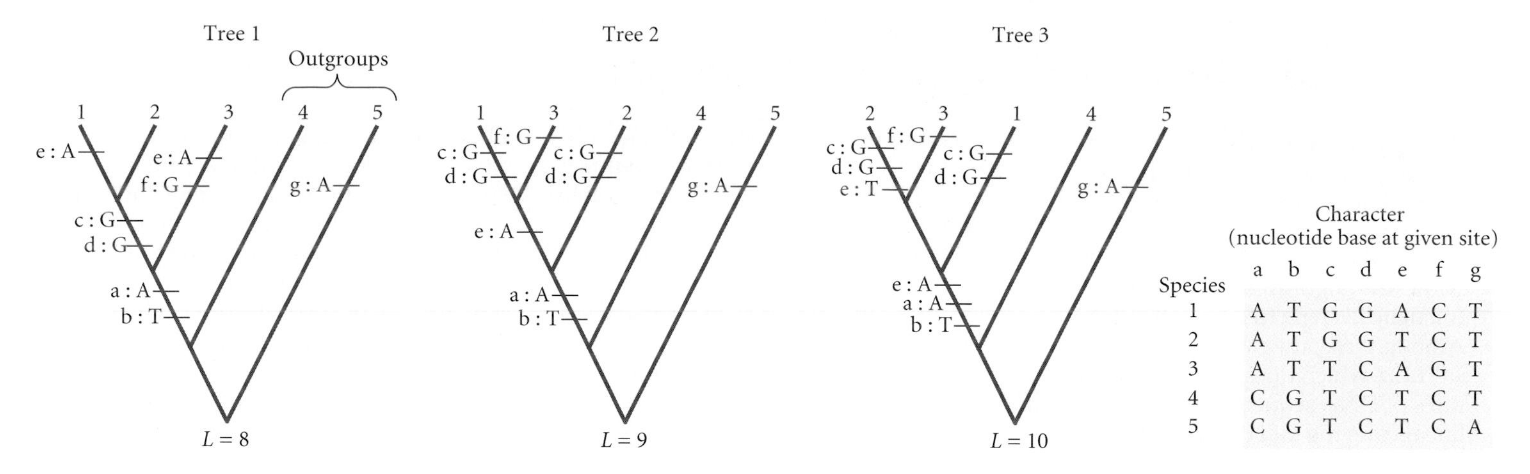

Character (nucleotide base at given site)

Species	a	b	c	d	e	f	g
1	A	T	G	G	A	C	T
2	A	T	G	G	T	C	T
3	A	T	T	C	A	G	T
4	C	G	T	C	T	C	T
5	C	G	T	C	T	C	A

Figure 2.7 Inferring a phylogeny by the method of maximum parsimony. The matrix gives the character states (nucleotide bases) for seven sites (a–g) in a DNA sequence. On each of the three trees representing possible relationships among species 1, 2, and 3, the locations of changes to derived character states are shown. When the lengths (L) of the three trees are compared, we see that tree 1 is the shortest ($L = 8$); that is, it requires us to postulate the fewest character changes.

Our task is to determine which tree structure implies the smallest number of evolutionary changes in the characters. We go about this by plotting, on each tree, the position at which each character must have changed, minimizing the number of state changes. Consider tree 1, and examine first the variation in character *a* among species. The simplest explanation for the fact that species 1, 2, and 3 share state A and species 4 and 5 share state C is that C was replaced by A in the common ancestor of the ingroup; that single change divides the tree into species with A and species with C. We infer that the change was from C to A, rather than vice versa, because if A had been the ancestral character state, two independent changes from A to C in the evolution of species 4 and 5 would have to be postulated. This pattern would not be parsimonious.

Following the same procedure for each character individually, we find that tree 1 requires us to plot evolutionary change in characters c and d in the ancestor of species 1 and 2, providing evidence that they are **sister groups** (groups derived from a common ancestor that is not shared with any other groups). Character e must be convergent, evolving twice from T to A; if this tree is the true branching history, there is no way in which it could have changed only once and yet have the same state in species 1 and 3. Characters f and g are autapomorphies, changing only in the individual species 3 and 5, respectively. They have exactly the same positions in the other possible phylogenetic trees and carry no information about branching sequence.

The same procedure is then followed for every other possible phylogeny. (This very tedious procedure is carried out by computer programs in real phylogenetic analyses, which typically involve more species and many more characters.) The topology of tree 2 implies that character e is not homoplasious, but evolved from T to A in the common ancestor of species 1 and 3, which are sister taxa in this tree. However, both characters c and d must have evolved convergently in this case. Tree 3 likewise implies that characters c and d evolved convergently, and that character e underwent evolutionary reversal from T to A and back to T.

If, having completed this procedure for all three possible branching trees, we count the number of character state changes (the "length" of each tree), we find that tree 1 is shortest, requiring the fewest character state changes. Moreover, more characters support the monophyly of species 1 and 2 (tree 1) than of 1 and 3 or of 2 and 3. According to the parsimony criterion, tree 1 is our *best estimate* of the history of branching (the "true tree"). In any real case, of course, we would want the difference in length between the most parsimonious tree and other possible trees to be much greater before we would have confidence in our estimate.

The method of **maximum parsimony** just described is not the only method for inferring phylogenetic relationships, and although it is the easiest method to describe and one of the simplest to use, it is generally not the most reliable. Several other commonly used methods are mentioned in Box B.

BOX 2B More Phylogenetic Methods

Many methods have been proposed for inferring phylogenies, and their strengths and drawbacks have been extensively argued and analyzed (Felsenstein 2004). Some are "algorithmic" methods that calculate a single tree from the data. Among these, the NEIGHBOR-JOINING METHOD, which does not assume equal rates of DNA sequence evolution among lineages, is the most frequently used. Most practitioners, however, prefer "tree-searching" methods that compare a large sample of trees from among the huge number that are possible for even a few taxa. These methods use various computerized "search" routines to maximize the chance of finding the shortest trees that are compatible with the data.

Some tree-searching programs, such as those based on maximum parsimony, save many of the trees that are examined, so that the shortest tree found can be compared with those that are nearly as short. We are then interested in knowing if the shortest tree (or particular groupings within it) is reliable, or if it differs from other short trees only by chance. This statistical problem is often addressed by a procedure called BOOTSTRAPPING, in which many random subsamples of the characters (e.g., nucleotide sites) are used for repeated phylogenetic analyses. Our confidence in the reliability of a particular grouping is greater if this grouping is consistently found by using these different data sets (bootstrap samples). A group of three or more taxa whose relationships cannot be confidently resolved is often represented by a node with three or more branches—a POLYTOMY. A CONSENSUS TREE is one that portrays both the relationships in which we can be confident and the polytomies that represent "unresolved" relationships.

Some powerful, increasingly popular tree-searching methods are the MAXIMUM LIKELIHOOD (ML) and BAYESIAN methods. They are too complicated to explain here in detail. Both use a model of the evolution of the data (usually DNA sequences). For example, the model might assume that all nucleotide substitutions are equally likely and that a constant substitution rate can be estimated from the data (the "one-parameter model"), or it might assume that different kinds of substitutions occur at different characteristic rates ("Kimura's two-parameter model"). Given the model and a possible tree, the maximum likelihood method calculates the likelihood of observing the data. The best estimate of the phylogeny is the tree that maximizes this likelihood. The more recently developed, and increasingly popular, Bayesian method differs from the maximum likelihood method by maximizing the probability of observing a particular tree, given the model and the data. (Note the seemingly subtle, but nevertheless very important, difference.) Unlike the ML method, the Bayesian method provides and calculates the probabilities of a set of trees so that they can be compared (Huelsenbeck et al. 2001).

An example of phylogenetic analysis

In traditional classifications, the primate superfamily Hominoidea consists of three families: the gibbons (Hylobatidae), the human family (Hominidae), and the great apes (Pongidae). The great apes are the orangutan (*Pongo pygmaeus*) in southeastern Asia; the gorilla (*Gorilla gorilla*) in Africa; and two species of chimpanzees (*Pan*), also in Africa. From anatomical evidence, it has long been accepted that the Hominoidea is a monophyletic group, and that the Hylobatidae are more distantly related to the other species than those species are to one another.

Physically, *Pongo*, *Pan*, and *Gorilla* appear more like one another than like *Homo* (Figure 2.8). Hence the traditional view was that Pongidae is monophyletic and that *Homo* branched off first. However, molecular data have definitively shown that humans and chimpanzees are each other's closest relatives (Ruvolo 1997). Among the many DNA analyses leading to this conclusion is a study by Morris Goodman and his coworkers (Goodman et al. 1989; Bailey et al. 1991), who sequenced more than 10,000 base pairs of a segment of DNA that includes a hemoglobin PSEUDOGENE—a nonfunctional DNA sequence derived early in primate evolution by duplication of a hemoglobin gene (see Chapter 8). The outgroups used were the New World spider monkey, *Ateles*, a distant relative of the Hominoidea, and the Old World rhesus monkey, *Macaca*, which belongs to a more closely related family (Cercopithecidae).

Goodman and colleagues found that the percentage of sequence identity in the ψη-globin pseudogene between pairs of hominoids is

TABLE 2.1 ***Divergence between nucleotide sequences of the ψη-globin pseudogene among orangutan (Pongo), gorilla (Gorilla), chimpanzee (Pan), and human (Homo)[a]***

	Gorilla	Pan	Homo
Pongo	3.39	3.42	3.30
Gorilla		1.82	1.69
Pan			1.56
Homo			0.38

Source: Data from Bailey et al. 1991.

[a]The percentage of divergence between two human sequences is given in the lower right cell. Divergences between *Homo* and other species are calculated using the average of these two sequences. Values are not corrected for multiple substitutions.

(A) Gibbon

(B) Orangutan

(C) Gorilla

(D) Chimpanzee

Figure 2.8 Members of the primate superfamily Hominoidea. (A) The white-handed gibbon, *Hylobates lar*, family Hylobatidae. (B) The orangutan, *Pongo pygmaeus*. (C) The lowland gorilla, *Gorilla gorilla*. (D) The common chimpanzee, *Pan troglodytes*. Two other living hominoid species, *Pan paniscus* (bonobo) and *Homo sapiens* (humans), are not shown. (Photos: A © Steve Bloom/Alamy Images; B © Shaun Cunningham/Alamy Images; C, D © Gerry Ellis/DigitalVision.)

very high, especially between *Homo* and *Pan*, in which less than 2 percent of the base pairs are different (Table 2.1). However, the two species differ in some base pair substitutions and in short insertions and deletions of bases (Figure 2.9). Some sites, such as number 8230 in Figure 2.9A, distinguish the orangutan from the chimpanzee, gorilla, and human. (In this and other examples, the sequence of only one DNA strand is given; the other strand is, of course, complementary.) Fourteen synapomorphies unite chimpanzee and human as sister groups, including deletions of short sequences such as sites 8468–8474. In contrast, only three sites (none of which is shown in this figure) support the hypothesis that chimpanzee and gorilla are the closest relatives.

The phylogeny in Figure 2.9B, with chimpanzee and human as closest relatives and gorilla as sister group to this pair, is eight steps (evolutionary changes) shorter than a tree that separates chimpanzee from human. Many other molecular data sets support this conclusion. Such phylogenetic studies have led taxonomists to alter the classification of apes and humans (Figure 2.9B), as is explained in Chapter 3.

Evaluating phylogenetic hypotheses

How can we judge the validity of our phylogenetic inferences? The estimate of a phylogenetic tree that we obtain from a particular set of data is a phylogenetic *hypothesis* that is *provisionally* accepted (as is any scientific statement). Additional data may lead us to modify or abandon the hypothesis, or may lend further support to it. *The chief way of confirming a phylogenetic hypothesis is to see if it agrees with independent data.* Morphological features and DNA sequences, for example, evolve largely independently of each other and thus provide independent phylogenetic information (see below and Chapter 19). With some exceptions, these two kinds of data usually yield similar estimates of phylogeny (Pat-

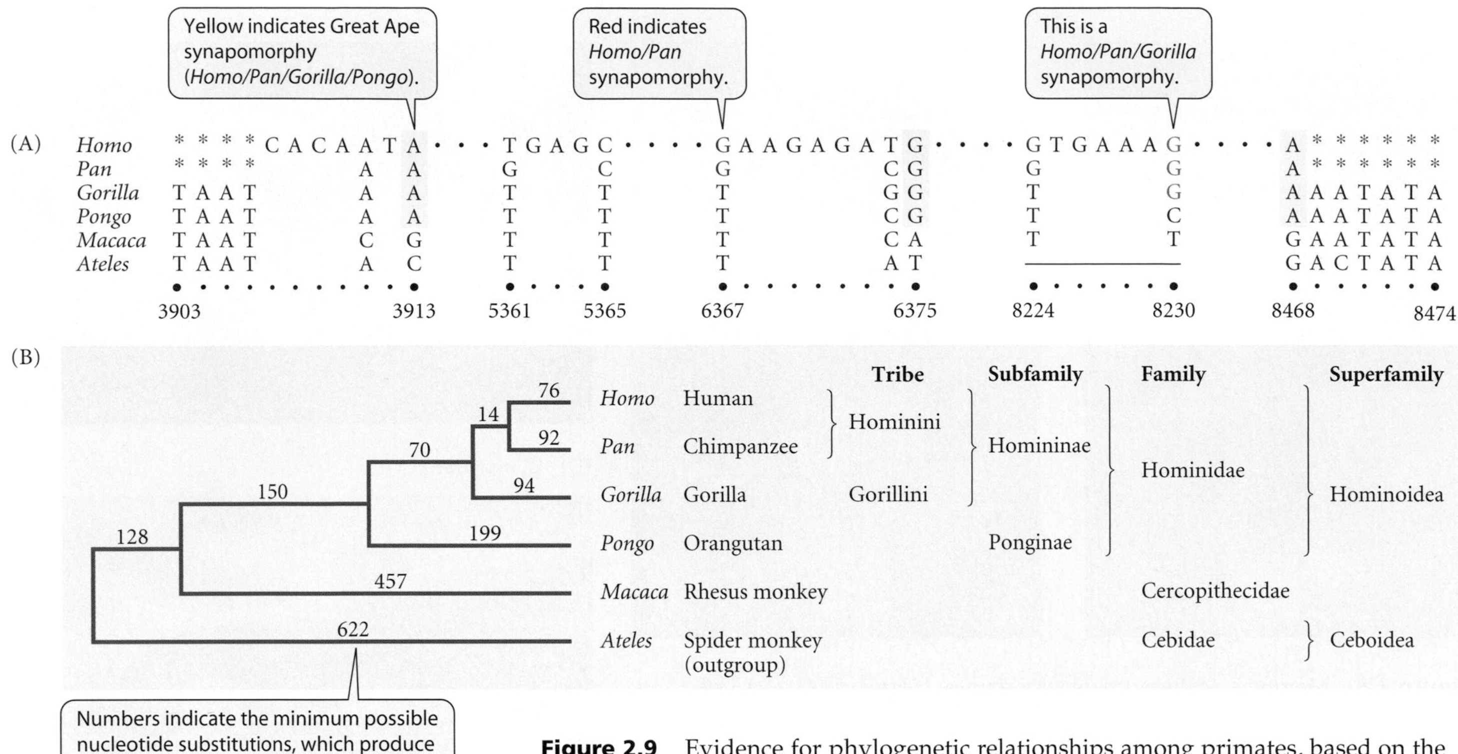

Figure 2.9 Evidence for phylogenetic relationships among primates, based on the ψη-globin pseudogene. (A) Portions of the sequence in six primates. *Macaca*, an Old World monkey, and *Ateles*, a New World monkey, are successively more distantly related outgroups with reference to the Hominoidea. Sequences are identical except as indicated. Using *Ateles* and *Macaca* as outgroups, positions 3913, 6375, and 8468 exemplify synapomorphies of the other four genera, and position 8230 provides a synapomorphy of *Gorilla*, *Pan*, and *Homo*. Synapomorphies of *Pan* and *Homo* include base pair substitutions at positions 5365, 6367, and 8224, and deletions at 3903–3906 and 8469–8474 (red asterisks). Autapomorphies (unshared derived states) include 3911 and 3913 (*Macaca*), 8230 (*Pongo*), 6374 (*Gorilla*), 5361 (*Pan*), and 6374 (*Homo*). (B) The most parsimonious phylogeny based on the ψη-globin sequence, using *Ateles* as an outgroup. The minimal number of changes is indicated along each branch. A tree that split up the *Homo-Pan-Gorilla* group would be 65 steps longer, and one that split *Homo* and *Pan* would be 8 steps longer. The figure includes one of several proposed classifications for humans and apes (Delson et al. 2000). (A after Goodman et al. 1989; B after Shoshani et al. 1996.)

terson et al. 1993). For instance, the phylogenetic relationships among higher taxa of vertebrates based on DNA sequences are the same, with few exceptions, as those inferred from morphological features (Figure 2.10).

We can also test the validity of phylogenetic methods by applying them to phylogenies that are absolutely known. Many investigators have simulated evolution on a computer, allowing computer-generated lineages to branch and their characters to change according to various models of the evolutionary process. (For example, characters might change at random at different average rates, or one of two species derived from a common ancestor might evolve faster than the other.) The investigators then see whether or not the various phylogenetic methods can use the final characteristics of the simulated lineages to yield an accurate history of their branching. Perhaps more convincing to the skeptic are a few studies in which phylogenetic methods have been applied to experimental populations of real organisms that have been split into separate lineages by investigators (creating artificial branching events, or **nodes**) and allowed to evolve. In one such study, David Hillis and coworkers (1992; Cunningham et al. 1998) successively subdivided lineages of T7 bacteriophage and exposed them to a mutation-causing chemical, which caused them to accumulate DNA sequence differences rapidly over the course of

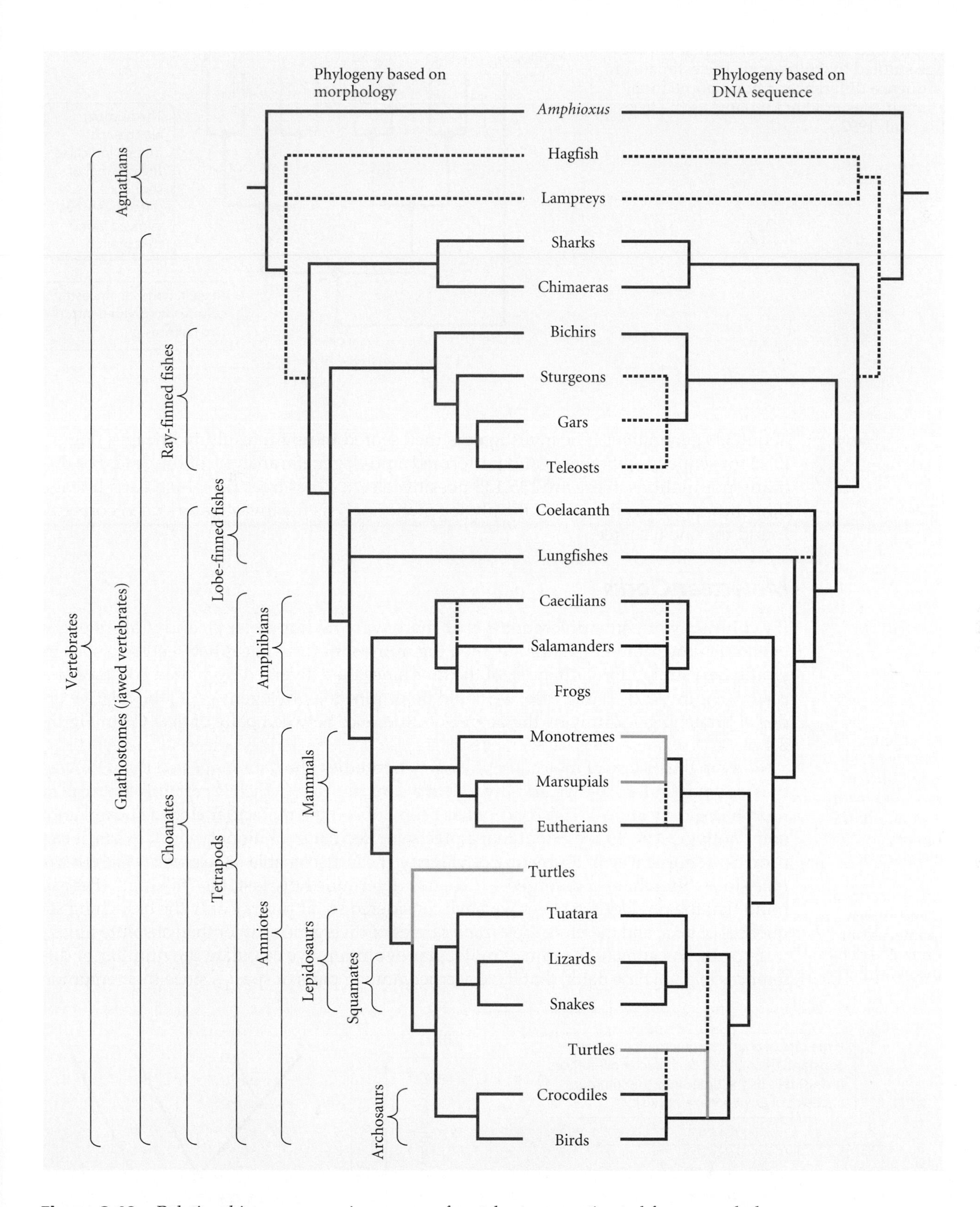

Figure 2.10 Relationships among major groups of vertebrates, as estimated from morphological characters (left) and from DNA sequences (right). On the whole, these two sources of information provide similar estimates of the phylogeny. The molecular data call for only one definite change (turtles) and one probable change (the monotreme/marsupial relationship) in phylogenetic positions relative to the morphological data (branches shown in gray). Uncertain relationships that differ between the two trees, and alternative possible relationships suggested by DNA sequence data, are indicated by dotted lines. (After Meyer and Zardoya 2003.)

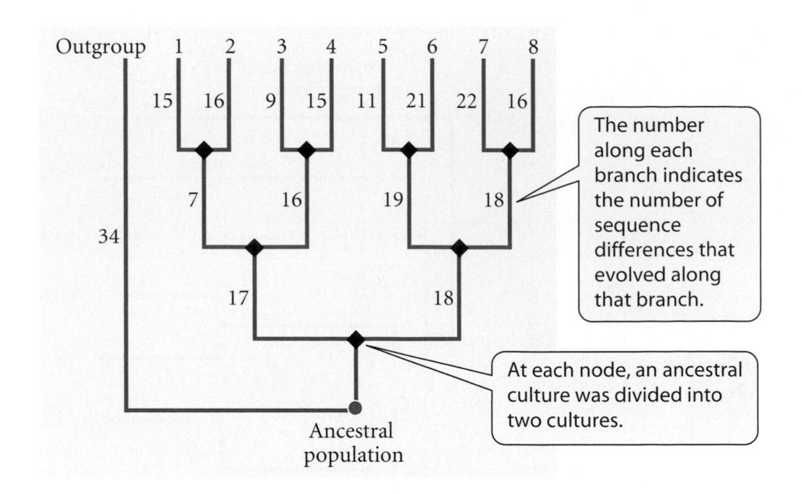

Figure 2.11 The true phylogeny of the experimental population of T7 bacteriophage studied by Hillis et al. The estimate of phylogeny, based on sequence differences among populations at the end of the experiment, was exactly the same as the true phylogeny. (After Hillis et al. 1992.)

about 300 generations. The investigators then scored the eight resulting lineages (Figure 2.11) for sequence differences and performed a phylogenetic analysis of the data. For this many populations, there are 135,135 possible BIFURCATING trees (in which each lineage branches into two others), but the phylogenetic analysis the investigators used correctly found the one true tree.

Molecular Clocks

If evolution were only divergent (i.e., if there were no homoplasy), and if all lineages evolved at the same, constant rate, then the number of differences between two species would be a straightforward index of the time since they diverged from their common ancestor (Figure 2.12). In that case, we could determine the phylogeny—i.e., the relative order of branching—simply by the degree of difference between pairs of taxa (as in Figure 2.4A).

Early in the history of molecular phylogenetic studies, the data suggested that DNA sequences may indeed evolve and diverge at a constant rate (which is certainly not true of morphological features). This concept has been dubbed the **molecular clock** (Zuckerkandl and Pauling 1965). To the extent that a precise molecular evolutionary clock exists, it can provide a simple way of estimating phylogeny. It can also enable us to *estimate the absolute time* since different taxa diverged—if we can determine how fast it is "ticking." (Bear in mind that the phylogenetic trees we have considered so far portray only the branching sequences of taxa, and therefore their *relative* times of divergence, rather than absolute times.)

In order to estimate the rate of molecular evolution, we calculate the number of differences (e.g., in base pairs) that have accrued among pairs of species since their common

Figure 2.12 (A) If divergence occurred at a nearly constant rate, the relative times of divergence of lineages could be determined from the overall differences (or similarities) between taxa, and the phylogeny of the taxa could then be estimated. (B) A hypothetical phylogeny in which evolution occurs at a nearly constant rate, and which therefore could be estimated from the differences between taxa. Each tick mark represents the evolution of a new character state.

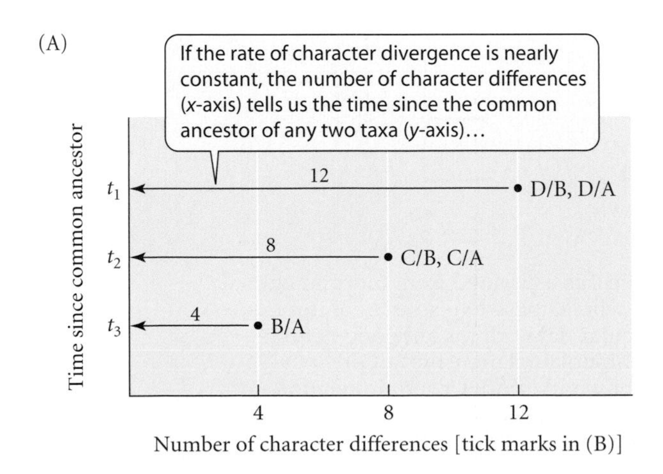

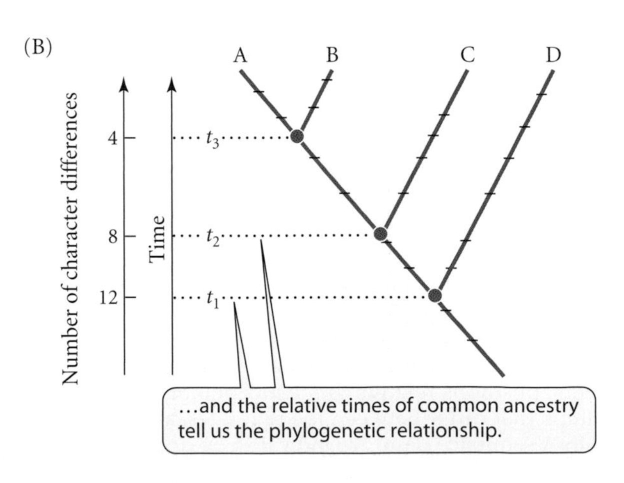

ancestor by parsimoniously plotting on our estimated phylogeny where each change took place (as in Figures 2.4 and 2.7). For example, Figure 2.9B shows 76 changes between *Homo* and its common ancestor with *Pan* (e.g., at site 6374), 14 changes between that common ancestor and the *Gorilla* branch (e.g., at site 5365), and 70 changes between that common ancestor and the *Pongo* branch (e.g., at site 8230). Since these hominoids diverged from the lineage leading to Old World monkeys (Cercopithecidae, represented by the rhesus monkey, *Macaca*), there have been 76 + 14 + 70 + 150 = 310 base pair changes in the lineage leading to *Homo*.

The *average rate of base pair substitution* in any lineage can be estimated if we have an estimate of the absolute time of divergence. For example, the oldest fossils of cercopithecoid monkeys are dated at 25 million years (My) ago, providing a minimal estimate of time since divergence between the rhesus monkey and the hominoids. The number of substitutions per base pair per million years for the rhesus monkey lineage is 457/10,000 base pairs sequenced/25 My = 1.83×10^{-3} per My, or 1.83×10^{-9} per year. From the common ancestor to *Homo*, the average rate has been 310/10,000/25 = 1.24×10^{-3} per My. The average rate at which substitutions have occurred in each lineage is therefore $r = (457 + 310)/2$ = 383.5/10,000/25 My, or 1.534×10^{-3} per My.

In this way, information from the fossil record on the absolute time of divergence of certain taxa may be used to calibrate a molecular clock—to determine its rate—and to estimate the divergence times of other taxa that have not left a good fossil record. For example, suppose the proportion of base pairs that differ between the ψη-globin pseudogene sequences of two primate species is 0.0256. Assuming a molecular clock,

$$D = 2rt$$

where D is the proportion of base pairs that differ between the two sequences, r is the rate of divergence per base pair per My, t is the time (in My) since the species' common ancestor, and the factor 2 represents the two diverging lineages. If $D = 0.0256$ and $r = 0.001534$, as estimated from the data in Figure 2.9, then $t = D/2r = 8.3$, and 8.3 My is our best estimate of when the two species diverged from their common ancestor.

Charles Langley and Walter Fitch (1974) were among the first to use data from fossils to test the molecular clock hypothesis. From the amino acid sequences of seven proteins, they estimated the number of nucleotide differences between pairs of species of mammals (we will see how this can be done in Chapter 8). Langley and Fitch found a strong but inexact correlation between the number of molecular differences and time since divergence (Figure 2.13). Their "clock" could be used for coarse, but not fine, estimates of phylogeny.

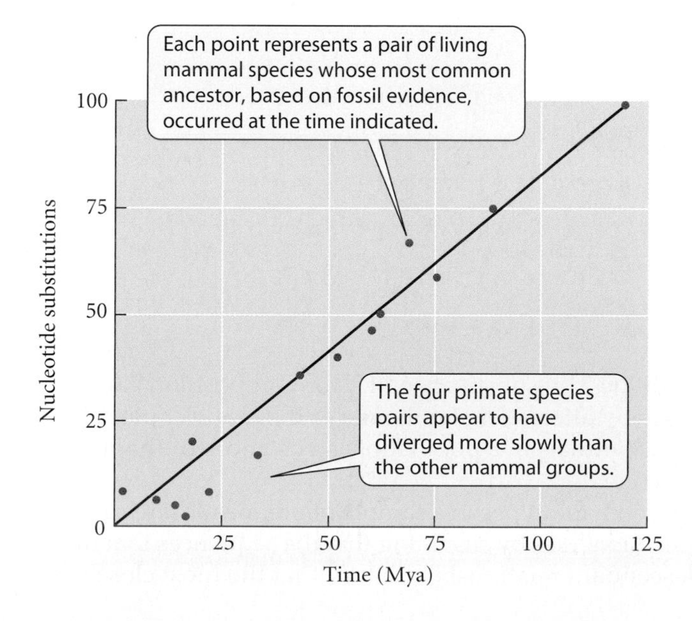

Figure 2.13 Base pair substitutions versus time since divergence, illustrating the approximate constancy of the rate of molecular evolution. Each point represents a pair of living mammal species whose most recent common ancestor, based on fossil evidence, occurred at the time indicated on the x-axis. The y-axis shows the number of base pair substitutions inferred from the difference between the two species in the amino acid sequences of seven proteins. The four green circles represent pairs of primate species. (After Langley and Fitch 1974.)

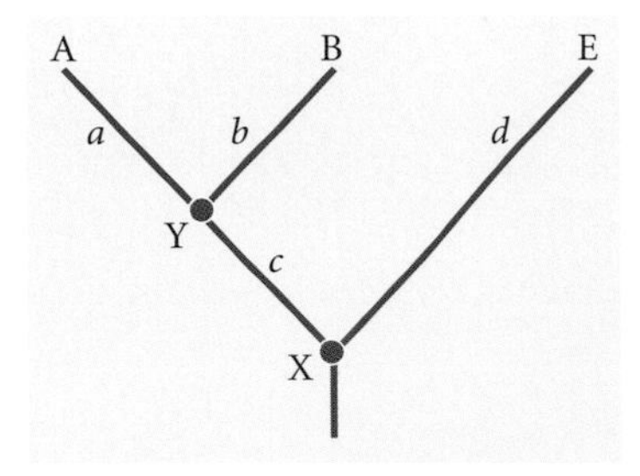

Figure 2.14 The relative rate test for constancy of the rate of molecular divergence. Sequences are obtained for living species A and B and for outgroup species E. Y and X represent ancestral species. Lowercase italic letters represent the number of character differences (e.g., nucleotide changes) along each branch. The genetic distance between A and E is $D_{AE} = a + c + d$. That between B and E is $D_{BE} = b + c + d$. If the rate of nucleotide substitution is constant, then $a = b$, so $D_{AE} = D_{BE}$. If rate constancy holds throughout the tree, the distance between any pair of species that have species X as a common ancestor will equal that between any other such pair of species.

The rate of sequence divergence is not always as nearly constant as in this example, however (Mindell and Thacker 1996; Smith and Peterson 2002). There are several ways of determining whether sequence evolution conforms to a molecular clock, even without information on divergence time from the fossil record. One such method is the RELATIVE RATE TEST (Wilson et al. 1977). We know that the time that has elapsed from any common ancestor (i.e., any branch point on a phylogenetic tree) to each of the living species derived from that ancestor is *exactly the same*. Therefore, if lineages have diverged at a constant rate, the number of changes (sometimes called the GENETIC DISTANCE) along all paths of the phylogenetic tree from one descendant species to another through their common ancestor should be about the same (Figure 2.14). In the hominoid example (see Figure 2.9B), the number of differences between the rhesus monkey and the various hominoids ranges from 806 (to orangutan) to 767 (human). These numbers are so close that they indicate a fairly constant rate of divergence, although the human lineage appears to have slowed down somewhat.

The relative rate test, applied to DNA sequence data from various organisms, has shown that rates of sequence evolution are often quite similar among taxa that are fairly closely related. However, distantly related taxa often have rather different evolutionary rates (Li 1997). For example, the rate of sequence evolution in rodents is two to three times greater than in primates.

It is not necessary to assume a molecular clock in order to estimate phylogenetic relationships among taxa, so molecular clocks are seldom used in phylogenetic analyses. However, they are often useful for estimating approximate dates of divergence, and we will encounter many examples of their use later in this book.

Gene Trees

So far, we have been concerned with inferring phylogenetic trees of species. Using the same principles, we can infer the historical relationships among variant DNA sequences of a gene (**haplotypes**). A phylogeny of genes, often called a **gene tree** or a **gene genealogy**, may include different haplotypes from a single species or more than one species.

One DNA sequence (haplotype) gives rise to another by mutation (as we will see in Chapter 8). The simplest kind of mutation is the replacement of one nucleotide base pair by another at a single site. Consider a hypothetical series of haplotypes that evolve in this way. For illustrative purposes, we'll hypothesize seven haplotypes, arising by a series of base-pair mutations:

	Sites				
Haplotype	1 2 3	4 5 6	7 8 9	10 11 12	13 14 15
3	A T A	C T A	**T** A T	G T T	G C C
2	A T A	C T A	**C** A **T**	G T T	G C C
1	A T A	C T A	C A **C**	G T T	**G** C C
4	A T A	C T A	C A C	G T T	**A** C **C**
5	A T A	C T **A**	C A C	G T T	A C **T**
6	A T **A**	C T **G**	C A C	G T T	A C T
7	A T **G**	C T G	C A C	G T T	A C T

Thus haplotype 3 arises from haplotype 2 by substitution of T for C at position 7 (or vice versa). The new haplotype (3) arises by alteration of only one of the many copies of the gene that are carried by different individuals in a population, so its ancestor (haplotype 2) continues to exist, at least for a while.

Now imagine that we find all seven haplotypes in a sample of organisms. We can correctly arrange them in an UNROOTED GENE TREE by assuming that the sequences that differ least have the closest ancestor-descendant relationship; that is, that the most closely related haplotypes are connected to each other by the smallest possible number of mutations. Haplotype 1 is placed between haplotypes 2 and 4 because a greater number of mutations would be required in order to connect 1 to any other haplotype. In this way, we find the most parsimonious unrooted tree (which doesn't look very treelike):

7——6——5——4——1——2——3

At this point, the tree is "unrooted," meaning that we do not know what the direction of evolution was (i.e., which sequence is closest to the common ancestor of them all). Suppose, moreover, that we find only haplotypes 2 through 7 in our sample. We would infer that haplotype 1 exists, or existed in the past, because it is the necessary step between haplotypes 2 and 4, which differ by two mutations (at positions 9 and 13). It is very unlikely that haplotype 4 arose directly from haplotype 2 (or vice versa), because of the improbability that two mutations would occur in the same individual gene. Therefore, we would postulate haplotype 1 as a hypothetical intermediate, or ancestor.

How can we infer the direction of evolution? Suppose haplotypes 4, 5, 6, and 7 are found in species A, and haplotypes 2 and 3 are found in closely related species B. We already know from the unrooted tree that haplotypes 2 and 3 are close relatives, as are haplotypes 4–7. The most parsimonious inference is that the common ancestor of the two species had a haplotype—such as haplotype 1—that gave rise to these two groups of haplotypes in the two species. This ancestral sequence allows us to ROOT the tree:

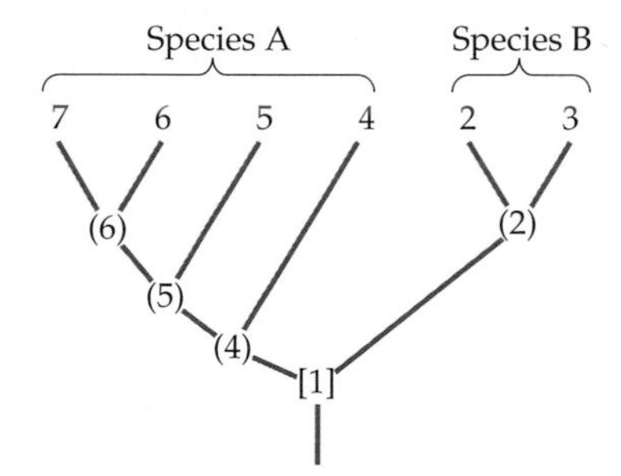

Since haplotype 1 is the postulated ancestral haplotype, we can conclude that the evolutionary sequence of haplotypes was 4 → 5 → 6 → 7 in species 1 and 2 → 3 in species 2. This conclusion remains the same if we should find haplotype 1 in either (or both) of the species.

Difficulties in Phylogenetic Analysis

In practice, it can be quite difficult to resolve relationships among taxa. We will mention some of the difficulties here, although the methods developed to solve them are outside the scope of this book. Nevertheless, it is very important to understand the evolutionary processes that can make phylogenetic analysis difficult; we will encounter examples of these situations quite often.

1. *Scoring characters is difficult.* Several problems can arise in obtaining basic data for phylogenetic analysis. For example, anatomical characters are important in systematics, and they are the only data typically available for extinct organisms. Deciding whether or not organisms have the same character state often requires extensive knowledge of anatomical details and is not an easy or trivial task. Another problem lies in deciding how many independent characters there are. If some mammals have, on each side of each jaw, two incisors, one canine, three premolars, and four molars, and if other mammals (e.g., anteaters) have no teeth at all, does this represent a single character difference (loss of teeth), four differences (loss of four kinds of teeth), or ten differences? A similar problem can arise in DNA sequence data if changes at several sites are likely to evolve in concert, to preserve function. For instance, the structure of the ribosomal RNA molecule includes short sequences whose bases must pair to form "stems," so changes in those sequences are not independent.

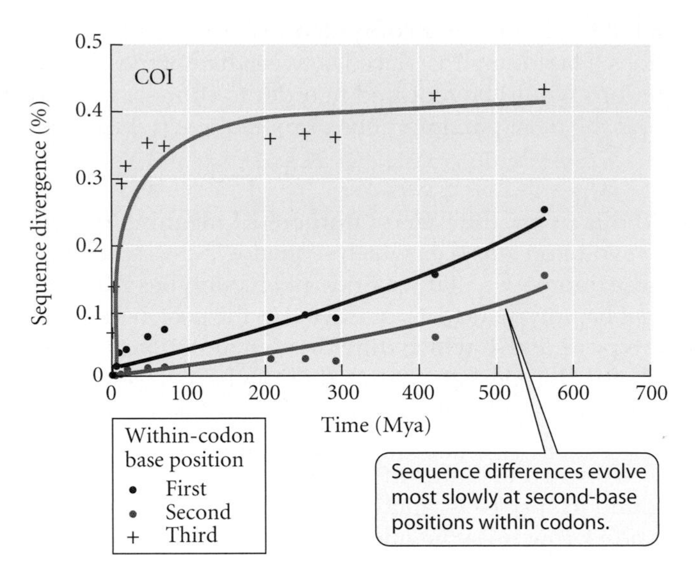

Figure 2.15 The proportion of base pairs in the DNA sequences of the mitochondrial gene *COI* that differ between pairs of vertebrate species, plotted against the time since their most recent common ancestor, based on the fossil record. Sequence differences evolve most rapidly at third-base positions and most slowly at second-base positions within codons (as we will see in Chapter 8). Divergence at third-base positions increases rapidly with time since the common ancestor at first, then levels off due to multiple substitutions at the same sites. Thus these positions provide no phylogenetic information for taxa that diverged more than about 75 million years ago. The species used for this analysis include two cypriniform fish species, one frog, one bird, one marsupial, two rodents, two seals, two baleen whales, and *Homo*. (After Mindell and Thacker 1996)

2. *Homoplasy is very common.* For this reason, a data set may yield several different phylogenetic estimates that are equally good, or almost so. It is unwise to rely on a particular phylogenetic estimate if other phylogenetic hypotheses imply just a few extra evolutionary changes. Instead, one should try to get more data (i.e., on other characters).

3. *The process of evolution often erases the traces of prior evolutionary history.* If the taxa under study diverged long ago or have evolved very rapidly, many of their characteristics will have diverged so greatly that homologous characters may be difficult to discern. For instance, teeth provide important features for determining relationships among many mammals, but they cannot be used to assess the relationships among toothless anteaters. In DNA sequences, multiple substitutions may occur at a site over the course of time, erasing earlier synapomorphies. Mutation at a site from A to C provides a shared derived character only as long as no further changes occur at that site; if in a descendant taxon, C is replaced by G, or by reversal to A, the evidence of common ancestry is lost. Furthermore, as time passes, it becomes more likely that the same substitution will occur in parallel in different lineages. Therefore, convergent evolution and successive substitutions ("multiple hits") cause the amount of sequence divergence between taxa eventually to level off, at a plateau that is attained sooner in rapidly evolving than in slowly evolving sequences (Figure 2.15). For this reason, rapidly evolving DNA sequences are useful for phylogenetic analyses of taxa that have diverged recently (Figure 2.16), whereas slowly evolving sequences are required to assess relationships among taxa that diverged in the distant past.

4. *Some lineages diverge so rapidly that there is little opportunity for the ancestors of each monophyletic group to evolve distinctive synapomorphies* (Figure 2.17). Such "bursts" of divergence of many lineages during a short period, termed EVOLUTIONARY RADIATIONS, are often called ADAPTIVE RADIATIONS because the lineages have often acquired different adaptations. Many families of songbirds, for example, appear to have diversified within a short time, and so their relationships are not well understood.

5. *An accurately estimated gene tree may imply the wrong species phylogeny.* This is a challenging concept, but it has some important implications (see Chapters 11 and 16). Suppose that two successive branching events occur, giving rise first to species A and then to species B and C. Now suppose that the ancestral species that gave rise to A, B, and C carries two different haplotypes of the gene we are studying (is POLYMORPHIC). If, just by chance, haplotype 1 becomes FIXED in species A (that is, the other haplotype is lost from that species by genetic drift; we will see how this can happen in Chapter 10) and haplotype 2 is fixed in the common ancestor of B and C, the gene tree will reflect the

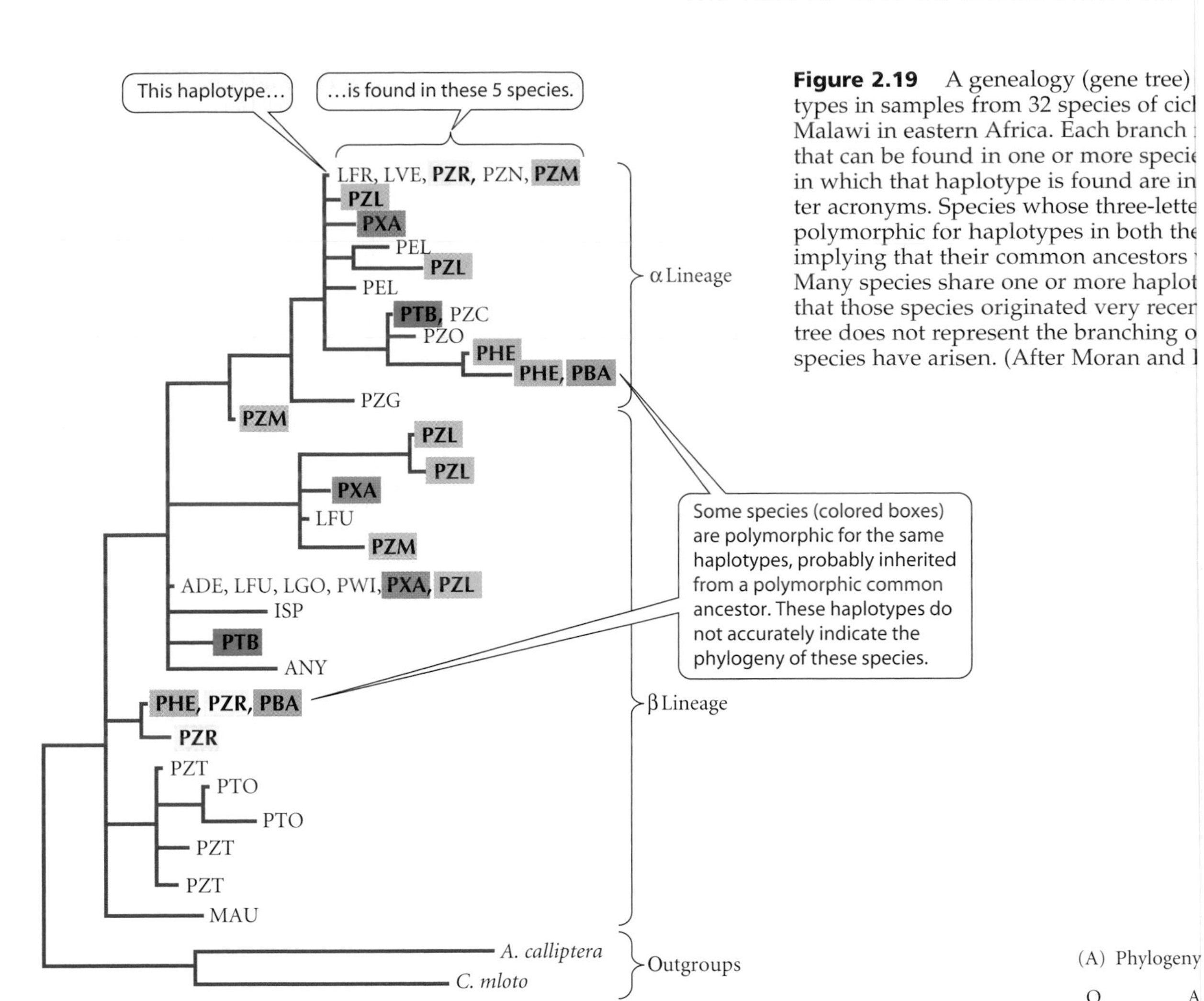
This haplotype...
...is found in these 5 species.
LFR, LVE, PZR, PZN, PZM
α Lineage
β Lineage
Outgroups
A. calliptera
C. mloto
Some species (colored boxes) are polymorphic for the same haplotypes, probably inherited from a polymorphic common ancestor. These haplotypes do not accurately indicate the phylogeny of these species.

Figure 2.19 A genealogy (gene tree) types in samples from 32 species of cicl Malawi in eastern Africa. Each branch that can be found in one or more speci in which that haplotype is found are in ter acronyms. Species whose three-lette polymorphic for haplotypes in both the implying that their common ancestors Many species share one or more haplot that those species originated very recer tree does not represent the branching o species have arisen. (After Moran and

Hybridization and Horizontal Gene Transfer

Many species of plants, and a few species of animals, have arisen from HYBRIDIZATION (interbreeding) between two ancestral species (see Chapter 16). In such cases, part of the phylogeny will be RETICULATED (netlike), rather than strictly branching, and some genes in the hybrid population will be most closely related to genes in each of two other species lineages (Figure 2.20). That is, the species phylogenies based on different genes will differ (be INCONGRUENT). Such incongruence may therefore provide tentative evidence of evolution by hybridization—known as **reticulate evolution**—although incongruence can arise for other reasons.

In contrast to hybridization, which may involve much of the genome, **horizontal** (or **lateral**) **gene transfer** usually incorporates just a few genes of one species into the genome of another species. (These terms contrast with the normal "vertical" transfer of genes from parent to offspring.) For example, a certain "virogene" has been found only in one group of closely related species of cats and in Old World monkeys. This gene would imply that these cats are more closely related to monkeys than they are to other cats, so it is clearly incongruent with the phylogeny indicated by other genes (Figure 2.21). The gene may

Figure 2.20 Hybridization and reticulate evolution. (A, B) Phylogenies inferred for species A, B, C, and an outgroup (O), based on the sequences of two different genes. (C) The incongruence suggests that species B has arisen by hybridization between species A and C, and that the history of this taxon can thus best be portrayed as reticulate (netlike) evolution. (After Sang and Zhong 2000.)

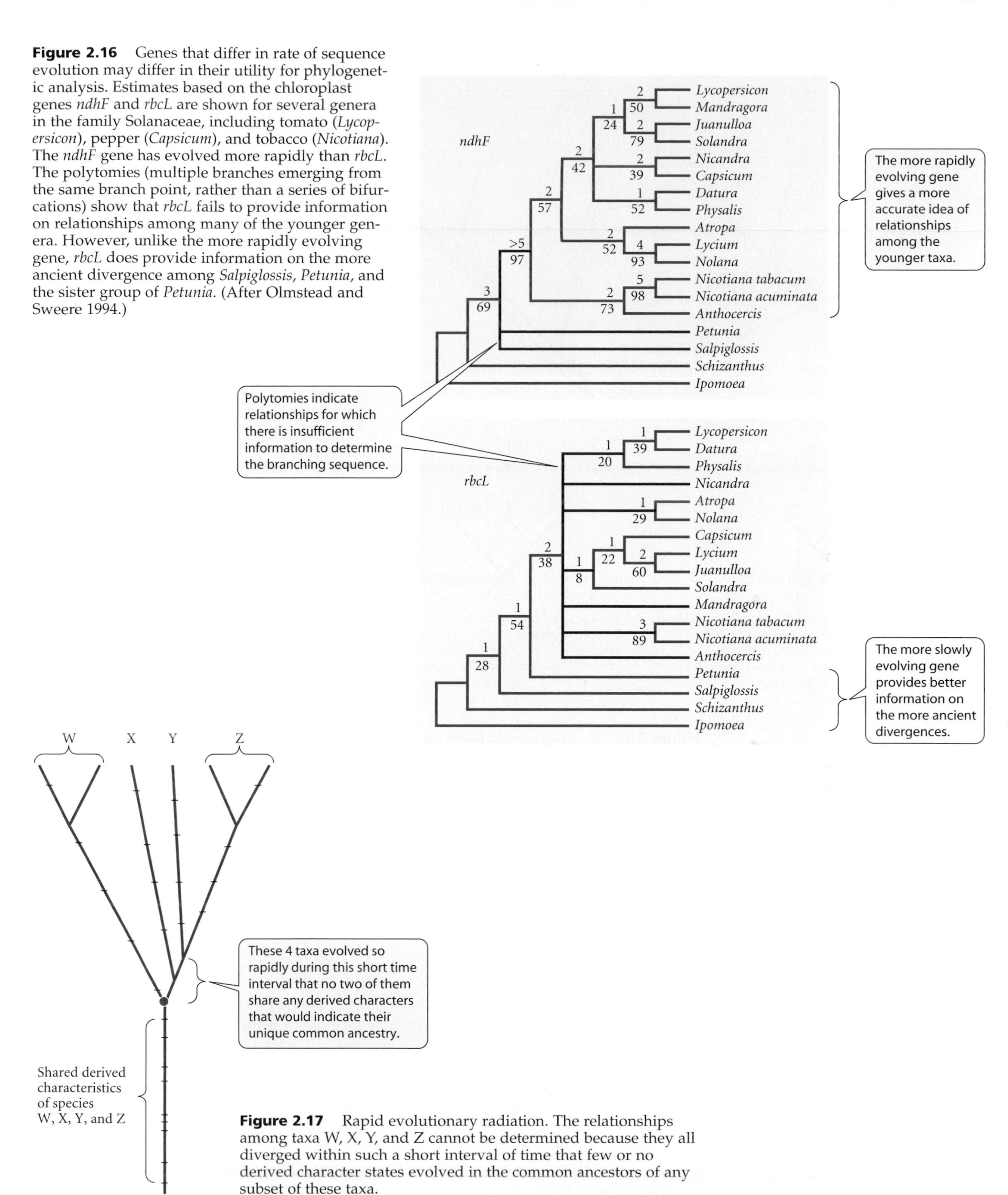
ndhF
rbcL
Lycopersicon
Mandragora
Juanulloa
Solandra
Nicandra
Capsicum
Datura
Physalis
Atropa
Lycium
Nolana
Nicotiana tabacum
Nicotiana acuminata
Anthocercis
Petunia
Salpiglossis
Schizanthus
Ipomoea
The more rapidly evolving gene gives a more accurate idea of relationships among the younger taxa.
Polytomies indicate relationships for which there is insufficient information to determine the branching sequence.
The more slowly evolving gene provides better information on the more ancient divergences.
W
X
Y
Z
These 4 taxa evolved so rapidly during this short time interval that no two of them share any derived characters that would indicate their unique common ancestry.
Shared derived characteristics of species W, X, Y, and Z

Figure 2.16 Genes that differ in rate of sequence evolution may differ in their utility for phylogenetic analysis. Estimates based on the chloroplast genes *ndhF* and *rbcL* are shown for several genera in the family Solanaceae, including tomato (*Lycopersicon*), pepper (*Capsicum*), and tobacco (*Nicotiana*). The *ndhF* gene has evolved more rapidly than *rbcL*. The polytomies (multiple branches emerging from the same branch point, rather than a series of bifurcations) show that *rbcL* fails to provide information on relationships among many of the younger genera. However, unlike the more rapidly evolving gene, *rbcL* does provide information on the more ancient divergence among *Salpiglossis*, *Petunia*, and the sister group of *Petunia*. (After Olmstead and Sweere 1994.)

Figure 2.17 Rapid evolutionary radiation. The relationships among taxa W, X, Y, and Z cannot be determined because they all diverged within such a short interval of time that few or no derived character states evolved in the common ancestors of any subset of these taxa.

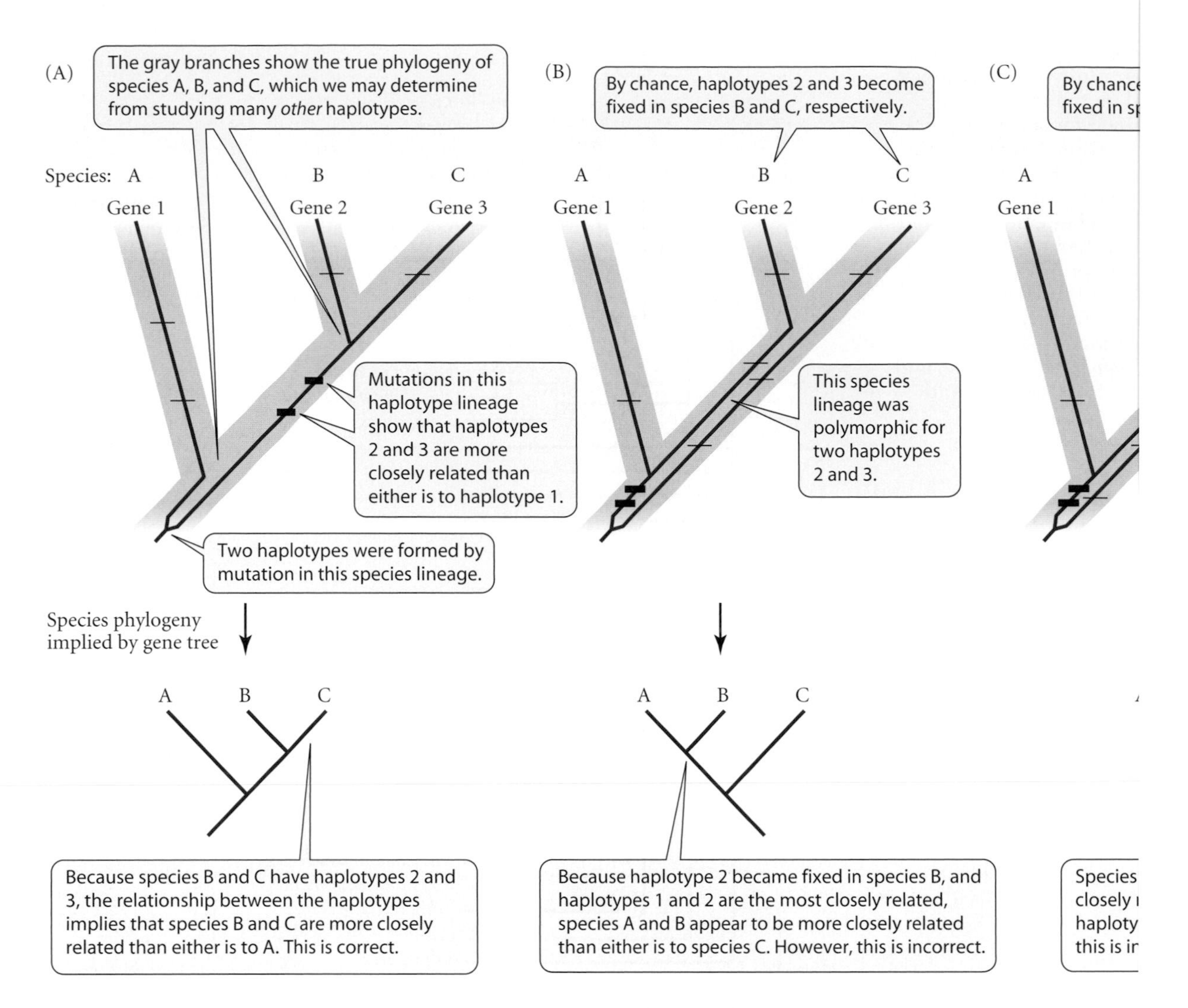

Figure 2.18 A gene tree (red lines) may or may not reflect the phylo
which the genes are sampled (outer envelopes). Species A, B, and C ha
sive branching events. (A) The true phylogeny, which we might infer f
types 2 and 3 form a monophyletic group, as indicated by two shared
(thick tick marks), and the species phylogeny implied by the gene tree
logeny of the species. Thin tick marks show other mutations that distir
types. (B) Haplotypes 1 and 2 form a monophyletic group, and falsely
B are sister species. This is because the common ancestor of species B a
for two gene lineages (2 and 3). The gene lineages that became fixed in
are not sister lineages. (C) The common ancestor of species B and C is
gene lineages 2 and 3, but the fixation of these genes is the reverse of c
falsely implies that species A and C are most closely related. (After Ma

species phylogeny (Figure 2.18A). But if such sorting of haplo
species does not occur, and the common ancestor of B and C
if there is INCOMPLETE SORTING of gene lineages), then the haplo
by chance, in the three species in such a way that the most c
not inherit the same haplotype (Figure 2.18B,C). A phyloger
types will therefore imply incorrect relationships among the
vides a real example, in which each of several closely relatec
appears at several locations on a gene tree of mitochondrial
ing that each species has inherited several gene lineages fro
with other species.

Figu
cats (
branc
horiz
ances
tion.

Estimate the phylogeny of these taxa by plotting the changes on each of the three possible phylogenies for species 1, 2, and 3 and determining which tree requires the fewest evolutionary changes.

2. There is evidence that many of the differences in DNA sequence among species are not adaptive. Other differences among species, both in DNA and in morphology, are adaptive (as we will see in Chapters 12 and 19). Do adaptive and nonadaptive variations differ in their utility for phylogenetic inference? Can you think of ways in which knowledge of a character's adaptive function would influence your judgment of whether or not it provides evidence for relationships among taxa?
3. It is possible for two different genes to imply different phylogenetic relationships among a group of species. What are the possible reasons for this? If there is only one true history of formation of these species, what might we do in order to determine which (if either) gene accurately portrays it? Is it possible for both phylogenetic trees to be accurate even if there has been only one history of species divergence?
4. Explain why rapidly evolving DNA sequences are useful for determining relationships only among taxa that evolved from quite recent common ancestors, and why slowly evolving sequences are useful only for resolving relationships among taxa that diverged much longer ago.
5. Given the principle stated in question 4, suppose you determine the nucleotide sequence of a particular gene from each of several species. (Perhaps, for the sake of argument, these include horse, sheep, giraffe, kangaroo, and human.) How could you tell if this gene has evolved at a rate (not too fast or slow) that would make it useful for estimating the relationships among these animals?
6. Suppose the nucleotide sequence in the gene you used in question 5 has evolved much faster in some lineages (say, horse and human) than in others. Could that affect the estimate of the phylogeny? How? Is there any way you could tell if there was indeed a big difference in the rate of sequence evolution among lineages?
7. What should a biologist do if she or he finds that different methods of analyzing the same data (say, the maximum parsimony method and the maximum likelihood method) provide different estimates of the relationships among certain taxa? What should she do if the different analytical methods give the same estimate, but the estimate differs depending on which of two different genes has been sequenced? (Your answers do not depend on knowing how maximum likelihood works.)
8. Do the quandaries described in the previous question ever occur? Choose a group of organisms that interests you, find recent phylogenetic studies of this group, and see whether such problems have been encountered. (You may use keywords, such as "phylogeny" and "[taxon name, e.g., deer]," in any of several literature-search engines that your instructor can point you to.)

Patterns of Evolution

3

In order to classify organisms, systematists compare their characteristics. These comparisons provide a foundation for a great deal of what we know about evolution. Especially when coupled with the methods of phylogenetic analysis introduced in the previous chapter, they are an indispensable basis for inferring the history of evolution of various groups of organisms, and such histories are often interesting in their own right. The phylogenetic relationships among various birds, plants, or fungi are fascinating to a person who has developed an interest in and knowledge of these organisms—and tracing our own origins back through the last two billion years (see Figure 3.1) cannot fail to stir our imagination.

Diverse adaptations to a dry environment. The adaptations of plants in a California desert include fleshy, leafless stems that store water; dense spines that reduce heat load by reflecting light; small leaves; widely spreading roots that catch rare rainwater; and brief seasonal flowering. Each of these adaptations has evolved independently in many distantly related plants. (Photo © J. A. Kraulis/Masterfile.)

Phylogenies provide more than just the branching relationships among taxa, however. They enable us to infer, with considerable confidence, the history of changes in the characteristics of organisms, even in the absence of a fossil record. In fact, they are the only way of inferring the history of evolution of features that leave little or no fossil trace, such as DNA sequences, biochemical pathways, and behaviors. Phylogenetic and systematic studies enable us to describe past changes in genes, genomes, biochemical and physiological features, development and morphology, and life histories and behavior, as well as associated changes in geographic distribution, habitat associations, and ecological interactions between different species. Figure 3.1 shows the sequence by which our own ancestors acquired important features, such as a skeleton, amnion, middle ear bones, binocular vision, and bipedal lo-

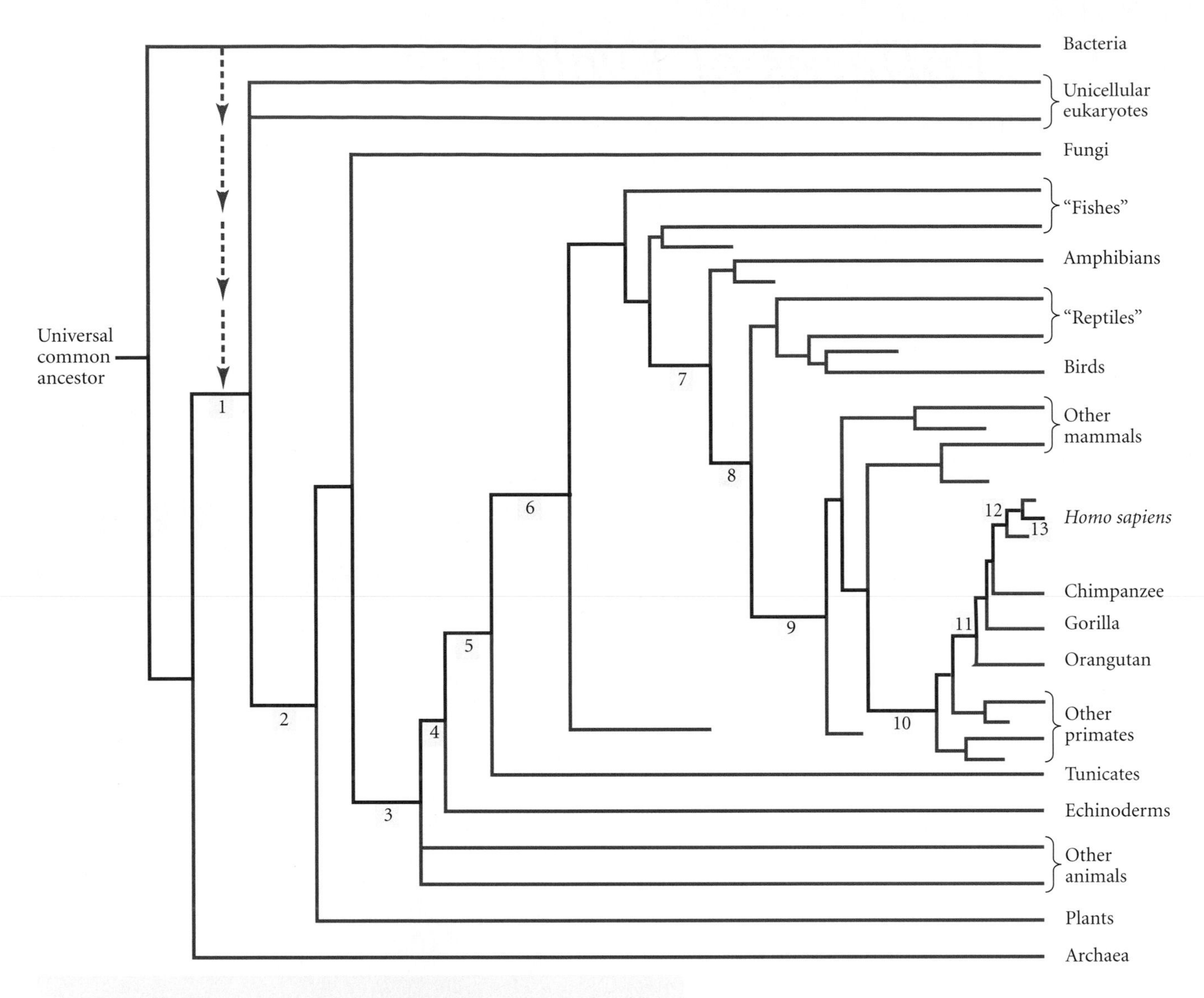

1. Origin of eukaryotes: a symbiotic bacterium becomes the mitochondrion.
2. Multicellularity evolves; cell and tissue differentiation
3. Animals: internal digestive cavity; muscles
4. Deuterostomes: embryonic blastopore develops into anus
5. Chordates: notochord; dorsal nerve cord
6. Vertebrates: bony skeleton
7. Tetrapods: legs
8. Amniotes: amniotic egg; other water-conserving features
9. Mammals: unique jaw joint; middle earbones; milk
10. Primates: binocular vision; arboreality
11. Anthropoid apes: loss of tail
12. Hominins evolve bipedalism
13. *Homo sapiens* spreads from Africa

Figure 3.1 Tracing the path of evolution to *Homo sapiens* from the universal ancestor of all life, in the context of the tree of life. Some of the major events are shown here as character changes mapped onto the phylogenetic tree. Such evolutionary histories are often intrinsically fascinating.

comotion. From our inferences of such changes in many kinds of organisms, it has been possible to find common themes, or **patterns of evolution**.

Because organisms are so diverse, there are few universal "laws" of biology of the kind that are known in physics (Mayr 2004). However, we can make generalizations about what kinds of evolutionary changes have been prevalent—and developing such general statements is one of the chief tasks of science. Furthermore, general patterns of change are the most important phenomena for evolutionary biology to explain. For instance, we know that the size of genomes—the amount of DNA—in various organisms varies greatly. The search for an explanation of genome size might become more interesting (and perhaps easier) if we should find that in many clades, genomes generally become larger over evolutionary time rather than knowing only that the genome of one species is larger than that of another.

Phylogenetic and comparative studies thus provide insights into almost every aspect of evolution and often help us to understand evolutionary processes (Futuyma 2004). This chapter describes a few of the most important patterns of evolution that have emerged from a long history of systematic and phylogenetic analyses, concentrating on morphological and other features at the organismal level. In later chapters, we will encounter phylogenetic insights into other topics, such as the evolution of genomes and behavior.

Evolutionary History and Classification

Darwin's hypothesis of descent with modification from common ancestors provided a scientific foundation for classifying organisms because it explained similarities among species as the result of a real evolutionary history. Many systematists therefore adopted the position that classification should reflect evolution. In almost any discussion of long-term patterns of evolution and biological diversity, it becomes necessary to use classification, since we must refer to named taxa such as Chordata, Vertebrata, Mammalia, or *Homo*. What, then, does it mean for classification to reflect evolution? And does it?

Evolution has two major features: the *branching* of a lineage into two or more descendant lines, called **cladogenesis** (from the Greek *clados*, "branch"), and evolutionary *change* of various characteristics in each of the descendants, called **anagenesis** (from the Greek *ana*, "up," referring to directional change). Some evolutionary changes, such as the evolution of the bird wing from a dinosaur forelimb, are particularly striking and adaptively important. Many traditional classifications have been constructed so as to convey both cladogenesis and anagenesis. For example, birds were placed in a different class (Aves) from other amniote vertebrates because of their wings and other adaptations for flight, and humans were placed in a different family (Hominidae) from other primates because of our erect posture, large brain, and egocentricity. However, it can be difficult to convey both cladogenesis and anagenesis in a classification. Considering the hominoid phylogeny in Figure 2.9B, for example, it is obvious that placing humans in a separate family from the great apes reflects the great divergence in humans' brain size and some other features. But it obscures the fact that humans are more closely related to chimpanzees than gorillas are.

On the basis of phylogenetic relationships, a taxon (a named group of organisms) may be monophyletic, polyphyletic, or paraphyletic (Figure 3.2) A **monophyletic** taxon, as noted previously, is the set of all known descendants from a single common ancestor. For example, the birds (Aves), the beetles (Coleoptera), and the flowering plants (Angiospermae) are believed to be monophyletic groups. A **polyphyletic** taxon includes unrelated lineages that are more closely related to species that are placed in other taxa; modern taxonomists consider polyphyletic taxa to be inappropriate classifications. For instance, a taxon that included whales with fishes would be polyphyletic, since whales stem from ancestors (e.g., the earliest tetrapod, the earliest mammal) that fishes do not share (see Figure 2.6). A **paraphyletic** taxon is a group that is monophyletic except that some descendants of the common ancestor have been placed in other taxa.

Paraphyletic taxa usually lack species that have been placed in another taxon in order to emphasize their distinctive adaptations. For example, the traditional family Pongidae consists of orangutans, gorillas, and chimpanzees; the family is paraphyletic, because humans—the closest relatives of the chimpanzees—have been placed not in the Pongidae,

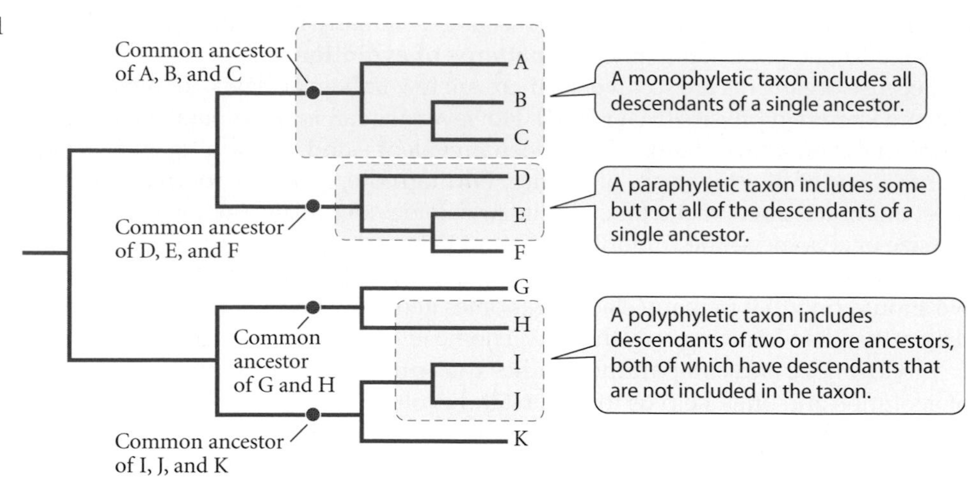

Figure 3.2 Monophyletic, paraphyletic, and polyphyletic groups. Monophyletic groups are preferred by most systematists, who consider paraphyletic and polyphyletic groupings inappropriate in modern taxonomy.

but in the Hominidae. And if birds, with their distinctive adaptations for flight, are placed in the class Aves while dinosaurs and crocodiles are placed in the class Reptilia, then "Reptilia" is paraphyletic. It is sometimes useful to refer to traditional groups, such as "Reptiles," but to enclose their names in quotation marks in order to indicate that they are not used in modern, formal classification.

Increasingly, systematists have adopted a philosophy of classification urged by Willi Hennig, who argued that all taxa should be monophyletic and thus reflect common ancestry. Hennig and his followers argued for the abolition of paraphyletic taxa such as "Pongidae" and "Reptilia." Hennig greatly influenced systematics, first, by proposing a *method* for discovering the real branching pattern of evolutionary history (see Figure 2.4), and second, by articulating an *opinion* on criteria for classification. These two very different proposals have together come to be known as **cladistics**. Branching diagrams constructed by cladistic methods are sometimes referred to as **cladograms**, and monophyletic groups are called **clades**.

Even if we have an accurate (true) estimate of a phylogeny, certain aspects of classification are still difficult. For example, we often find that an extinct group of species early in evolutionary time (which may be termed a **stem group**, such as the mammal-like reptiles, Therapsida) gives rise to a later group with distinctive derived characters (referred to as a **crown group**, such as Mammalia). Any name for the stem group that excludes the crown group will designate a paraphyletic taxon—an unsatisfactory element in classification. Another example is the dinosaur order Theropoda, a stem group from which a crown group, the birds (usually placed in class Aves), evolved (see Figure 4.9).

Furthermore, decisions as to whether the members of a monophyletic group should be SPLIT into several taxa or LUMPED into one taxon may be quite arbitrary. For instance, given the phylogeny shown in Figure 2.9, we might form monophyletic families either by placing the orangutan in one family (Pongidae) and the gorilla, chimpanzee, and human in another family (Hominidae), or by placing all of them in a single family (Hominidae). The latter classification (shown in Figure 2.9B) has been widely adopted; within the single family of humans and great apes, the subfamily Homininae includes the African apes and humans, and the tribe Hominini designates a clade consisting of *Homo sapiens* and various extinct human relatives (which are collectively called "hominins"). Some authors include the chimpanzee and the bonobo (genus *Pan*) in the Hominini.

Inferring the History of Character Evolution

One of the most important uses of phylogenetic information is to reconstruct the history of evolutionary change in interesting characteristics by "mapping" character states on the phylogeny and inferring the state in each common ancestor, using the principle of parsimony (see Chapter 2). That is, we assign to ancestors those character states that require

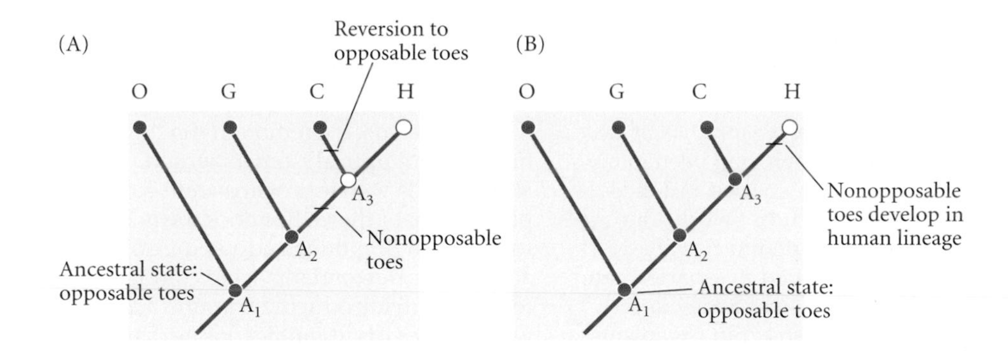

Figure 3.3 Two possible histories of change of a character (opposable versus nonopposable toes) in the Hominoidea (O, orangutan; G, gorilla; C, chimpanzee; H, human). (A) If nonopposable toes (open circles) are hypothesized for A_3, the common ancestor of chimpanzee and human, two state changes (tick marks) must be postulated. (B) If opposable toes are hypothesized for A_3, only one change need be postulated. It is therefore most parsimonious to conclude that humans evolved from an ancestor with opposable toes.

us to postulate the fewest homoplasious evolutionary changes for which we lack independent evidence. This method enables us to infer when (i.e., on which branch or segment of a phylogeny) changes in characters occurred, and thus to trace their history.

Humans, for example, have nonopposable first (great) toes, while the orangutan, gorilla, and chimpanzee have opposable first toes (like our thumbs). In Figure 3.3, we consider two possible evolutionary histories. The common ancestors are labeled A_1, A_2, and A_3 in order of increasing recency. If we assumed (Figure 3.3A) that A_1 and A_2 had opposable toes and that A_3, the immediate common ancestor of chimpanzees and humans, had nonopposable toes, we would have to postulate two changes, with the chimpanzee reverting to the ancestral state. If, however, we assume that A_3, like A_1 and A_2, had opposable toes, we need to infer only one evolutionary change; namely, the shift to nonopposable toes in the human lineage. This is the more parsimonious hypothesis, so our best estimate is that the common ancestor of humans and chimpanzees had opposable toes.

This is an exceedingly simple, even obvious, example, presented only to illustrate the logic by which character evolution can be inferred. More interesting and challenging cases involve convergent evolution. For example, snakes and caecilians (amphibians that superficially resemble earthworms) both lack legs. The phylogeny of vertebrates in Figure 2.10 implies that they independently evolved the legless condition from four-legged (tetrapod) ancestors.

In this book, we will encounter many examples of evolutionary inferences from phylogenetic trees. They are fundamentally important in the study of molecular evolution (as we will see in Chapter 19), and they have even been used to infer the amino acid sequences and functions of ancestral proteins, which have then been experimentally synthesized. Molecular biologist Nils Adey and colleagues (1994) performed such an experiment with a retrotransposon called *L1*. Mammalian genomes carry many copies of this virus-like genetic element, which can copy itself and insert the copies elsewhere in the genome (retrotransposons and other transposable elements will be described in more detail in Chapter 8). The *L1* retrotransposon encodes an enzyme called reverse transcriptase, which allows it to transcribe RNA transcripts of itself into DNA copies, which are then inserted into the genome. Transcription is initiated in an upstream promoter within *L1*, denoted *A*, but in mice, some copies of *L1* have an inactive promoter, denoted *F*. The approximately 200-base-pair sequence of the *F* promoter varies among copies, and it is thought to have been inactivated by various mutations over the course of about 6 million years.

Adey et al. set out to "resurrect" the functional ancestral promoter from which the inactive *F* sequences had evolved. Using a phylogenetic analysis of 30 different *F* sequences, they used the principle of parsimony to construct a best estimate of the ancestral sequence. They then synthesized this sequence and attached it to a "reporter gene," one whose protein product would be detected if the promoter enabled the gene to be transcribed. These recombinant genes were then inserted into mouse cells in tissue culture. The experiment was wildly successful: the amount of protein produced by genes with the synthetic promoter equaled that produced by control genes to which the naturally active *A* promoter was attached. This experiment supports the hypothesis that the inactive *L1* copies evolved from a functional sequence, and it demonstrates the power of phylogenetic analysis to reconstruct the past.

BOX 3A Evidence for Evolution

Systematists have always classified organisms by comparing characteristics among them. The early systematists had amassed an immense body of comparative information even before *The Origin of Species* was published. Their data suddenly made sense in light of Darwin's theory of descent from common ancestors; indeed, Darwin drew on much of this information as evidence for his contention that evolution has occurred. Since Darwin's time, the amount of comparative information has increased greatly, and today includes data not only from the traditional realms of morphology and embryology, but also from cell biology, biochemistry, and molecular biology.

All of this information is consistent with Darwin's hypothesis that living organisms have descended from common ancestors. Indeed, innumerable biological observations are hard to reconcile with the alternative hypothesis, that species have been individually created by a supernatural being, unless that being is credited with arbitrariness, whimsy, or a devious intent to make organisms *look* as if they have evolved. From the comparative data amassed by systematists, we can identify several patterns that confirm the historical reality of evolution, and which make sense only if evolution has occurred.

1. *The hierarchical organization of life.* Before Linnaeus, there had been many attempts to classify species, but those early systems just didn't work. One author, for instance, tried to classify species in complicated five-parted categories—but organisms simply don't come in groups of five. But organisms do fall "naturally" into the hierarchical system of groups within groups that Linnaeus described. A historical process of branching and divergence will yield objects that can be hierarchically ordered, but few other processes will do so. Thus languages can be classified in a hierarchical manner, but elements and minerals cannot.

2. *Homology.* Similarity of structure despite differences in function follows from the hypothesis that the characteristics of organisms have been modified from the characteristics of their ancestors, but it is hard to reconcile with the hypothesis of intelligent design. Design does not require that the same bony elements form the frame of the hands of primates, the digging forelimbs of moles, the wings of bats, birds, and pterosaurs, and the flippers of whales and penguins (see Figure 3.4). Modification of pre-existing structures, not design, explains why the stings of wasps and bees are modified ovipositors, and why only females possess them. All proteins are composed of "left-handed" (L) amino acids, even though the "right-handed" (D) optical isomers would work just as well if proteins were composed only of those. But once the ancestors of all living things adopted L amino acids, their descendants were committed to them; introducing D amino acids would be as disadvantageous as driving on the right in Great Britain or on the left in the United States. Likewise, the nearly universal, arbitrary genetic code (see Chapter 8) makes sense only as a consequence of common ancestry.

3. *Embryological similarities.* Homologous characters include some features that appear during development, but would be unnecessary if the development of an organism were not a modification of its ancestors' ontogeny. For example, tooth primordia appear and then are lost in the jaws of fetal anteaters, and some terrestrial frogs and salamanders pass through a larval stage within the egg, with the features of typically aquatic larvae, but hatch ready for life on land. Early in development, human embryos briefly display branchial pouches similar to the gill slits of fish embryos.

4. *Vestigial characters.* The adaptations of organisms have long been, and still are, cited by creationists as evidence of the Creator's wise beneficence, but no such claim can be made for the features, displayed by almost every

Some Patterns of Evolutionary Change Inferred from Systematics

Many important patterns and principles of evolution have been based on systematic studies over the course of the last century or more. Indeed, systematic studies of living organisms have provided a vast amount of evidence for the reality of evolution (Box A). In this section, we will illustrate some patterns of evolution with examples drawn largely from the tradition of comparative morphology in which they were developed. Many of these principles can also be illustrated at the biochemical or molecular level, as we will see in future chapters.

Most features of organisms have been modified from pre-existing features

One of the most important principles of evolution is that *the features of organisms almost always evolve from pre-existing features of their ancestors;* they do not arise de novo, from nothing. The wings of birds, bats, and pterodactyls are modified forelimbs (Figure 3.4); they do not arise from the shoulders (as in angels), presumably because the ancestors of these animals had no shoulder structures that could be modified for flight. In Chapter 4, we will

BOX 3A ***(continued)***

species, that served a function in the species' ancestors, but do so no longer. Cave-dwelling fishes and other animals display eyes in every stage of degeneration. Flightless beetles retain rudimentary wings, concealed in some species beneath fused wing covers that would not permit the wings to be spread even if there were reason to do so. In *The Descent of Man*, Darwin listed a dozen vestigial features in the human body, some of which occur only as uncommon variations. They included the appendix, the coccyx (four fused tail vertebrae), rudimentary muscles that enable some people to move their ears or scalp, and the posterior molars, or wisdom teeth, that fail to erupt, or do so aberrantly, in many people. At the molecular level, every eukaryote's genome contains numerous nonfunctional DNA sequences, including pseudogenes: silent, nontranscribed sequences that retain some similarity to the functional genes from which they have been derived (see Chapter 19).

5. *Convergence.* There are many examples, such as the eyes of vertebrates and cephalopod molluscs, in which functionally similar features actually differ profoundly in structure (see Figure 3.5). Such differences are expected if structures are modified from features that differ in different ancestors, but are inconsistent with the notion that an omnipotent Creator, who should be able to adhere to an optimal design, provided them. Likewise, evolutionary history is a logical explanation (and creation is not) for cases in which different organisms use very different structures for the same function, such as the various modified structures that enable vines to climb (see Figure 3.10).

6. *Suboptimal design.* The "accidents" of evolutionary history explain many features that no intelligent engineer would be expected to design. For example, the paths followed by food and air cross in the pharynx of terrestrial vertebrates, including humans, so that we risk choking on food. The human eye has a "blind spot," which you can find at about 45° to the right or left of your line of sight. It is caused by the functionally nonsensical arrangement of the axons of the retinal cells, which run forward into the eye and then converge into the optic nerve, which interrupts the retina by extending back through it toward the brain (see Figure 3.5).

7. *Geographic distributions.* The study of systematics includes the geographic distributions of species and higher taxa. This subject, known as biogeography, is treated in Chapter 6. Suffice it to say that the distributions of many taxa make no sense unless they have arisen from common ancestors. For example, many taxa, such as marsupials, are distributed across the southern continents, which is easily understood if they arose from common ancestors that were distributed across the single southern land mass that began to fragment in the Mesozoic.

8. *Intermediate forms.* The hypothesis of evolution by successive small changes predicts the innumerable cases in which characteristics vary by degrees among species and higher taxa. Among living species of birds, we see gradations in beaks; among snakes, some retain a vestige of a pelvic girdle and others have lost it altogether. At the molecular level, the difference in DNA sequences ranges from almost none among very closely related species through increasing degrees of difference as we compare more remotely related taxa.

For each of these lines of evidence, hundreds or thousands of examples could be cited from studies of living species. Even if there were no fossil record, the evidence from living species would be more than sufficient to demonstrate the historical reality of evolution: all organisms have descended, with modification, from common ancestors. We can be even more confident than Darwin, and assert that all organisms that we know of are descended from a single original form of life.

see that the middle ear bones of mammals evolved from jaw bones of reptiles. In Chapter 19, we will encounter examples of proteins that have been modified from ancestral proteins and have new functions. In other words, related organisms have **homologous characters**, which have been inherited (and sometimes modified) from an equivalent organ in the common ancestor. Homologous characters generally have similar genetic and developmental underpinnings, although these foundations sometimes have undergone substantial divergence among species (see Chapter 20).

A *character* may be homologous among species (e.g., "toes"), but a given *character state* may not be (e.g., a certain number of toes). The five-toed state is homologous in humans and crocodiles (as far as we know, both have an unbroken history of pentadactyly as far back as their common ancestor), but the three-toed state in guinea pigs and rhinoceroses is not homologous, for these animals have evolved this condition independently from five-toed ancestors.

A character (or character state) is *defined* as homologous in two species if it has been derived from their common ancestor, but *diagnosing* homology—that is, determining whether or not characters of two species are homologous—can be difficult. The most com-

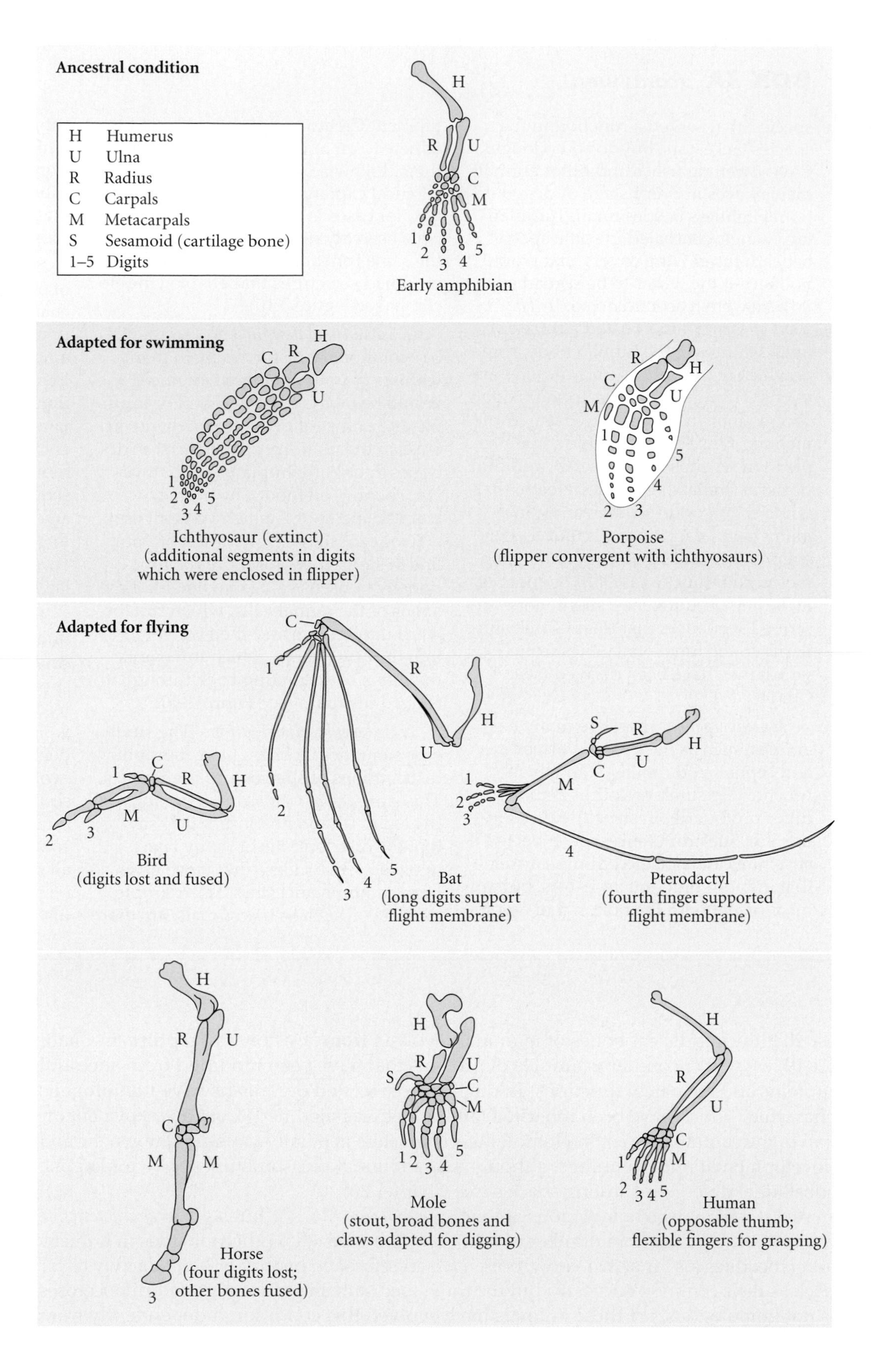

Figure 3.4 The forelimb skeletons of some tetrapod vertebrates. Compared with the "ground plan," as seen in the early amphibian, bones have been lost or fused (e.g., horse, bird), or modified in relative size and shape. Modifications for swimming evolved in the porpoise and ichthyosaur, and for flight in the bird, bat, and pterosaur. All the bones shown are homologous among these organisms, except for the sesamoid bones in the mole and pterosaur; these bones have a different developmental origin from the rest of the limb skeleton. (After Futuyma 1995.)

mon criteria for *hypothesizing* homology of anatomical characters are correspondence of *position* relative to other parts of the body and correspondence of *structure* (the parts of which a complex feature is composed). Correspondence of shape or of function are not useful criteria for homology (consider the forelimbs of a mole and an eagle). Embryological studies often are important for hypothesizing homology. For example, the structural correspondence between the hindlimbs of birds and crocodiles is more evident in the embryo than in the adult because many of the bird's bones become fused as development proceeds. However, the most important criterion for judging whether a character is homologous between species is to see if its distribution on a phylogenetic tree (based on other characters) indicates continuity of inheritance from their common ancestor.

Homoplasy is common

We noted in Chapter 2 that **homoplasy**—the independent evolution of a character or character state in different taxa—includes convergent evolution and evolutionary reversal. Many authors distinguish convergent evolution from parallel evolution. In **convergent evolution** (**convergence**), superficially similar features are formed by different developmental pathways (Lauder 1981). The eyes of vertebrates and cephalopod molluscs (such as squids and octopuses) are an example of convergent evolution (Figure 3.5). Both have a lens and a retina, but their many profound differences indicate that they evolved independently: for example, the axons of the retinal cells arise from the cell bases in cephalopods, but from the cell apices in vertebrates. Darwin (1854), who was the all-time authority on the systematics and anatomy of barnacles, described how the six-plate shell in two barnacle genera has been derived in different ways from the eight-plate ancestral state: in *Chthamalus*, two of the eight plates fail to develop, whereas in *Balanus*, two of the plates fuse with a third (Figure 3.6; see also Figure 4.17).

Parallel evolution (**parallelism**), on the other hand, is thought to involve similar developmental modifications that evolved independently (often in closely related organisms, because they are likely to have similar developmental mechanisms to begin with). A stunning example is the evolution of crustacean feeding structures called maxillipeds. In several diverse crustacean lineages, from one to as many as three pairs of these feeding structures develop not on the head segments that bear the animals' mouthparts, but on their anteriormost thoracic segments—segments that ancestrally bear legs. Michael Averoff and Nipam Patel (1997) studied differences in the expression of two "master" regulatory genes (Hox genes, which will be discussed in great detail in Chapter 21). The two genes they studied, called *Ultrabithorax* (*Ubx*) and *abdominal A* (*abdA*), determine how each

Figure 3.5 The eyes of (A) a vertebrate and (B) a cephalopod mollusc are an extraordinary example of convergent evolution. Despite their many similarities, note the several differences, including interruption of the retina by the optic nerve in the vertebrate, but not in the cephalopod. In vertebrates, the axons (nerve fibers) of the retinal cells run over the surface of the retina and converge into the optic nerve, forming a "blind spot." In cephalopods, the axons run directly from the base of the retinal cells into the optic ganglion. (From Brusca and Brusca 1990).

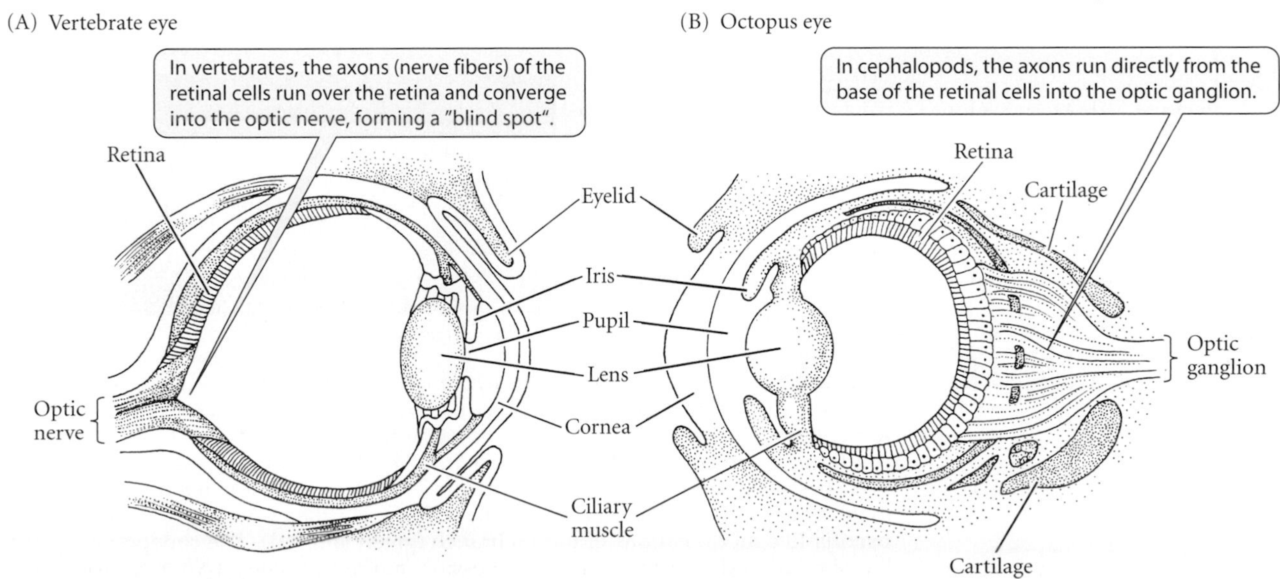
(A) Vertebrate eye
(B) Octopus eye
In vertebrates, the axons (nerve fibers) of the retinal cells run over the retina and converge into the optic nerve, forming a "blind spot".
In cephalopods, the axons run directly from the base of the retinal cells into the optic ganglion.
Retina
Optic nerve
Eyelid
Iris
Pupil
Lens
Cornea
Ciliary muscle
Retina
Cartilage
Optic ganglion
Cartilage

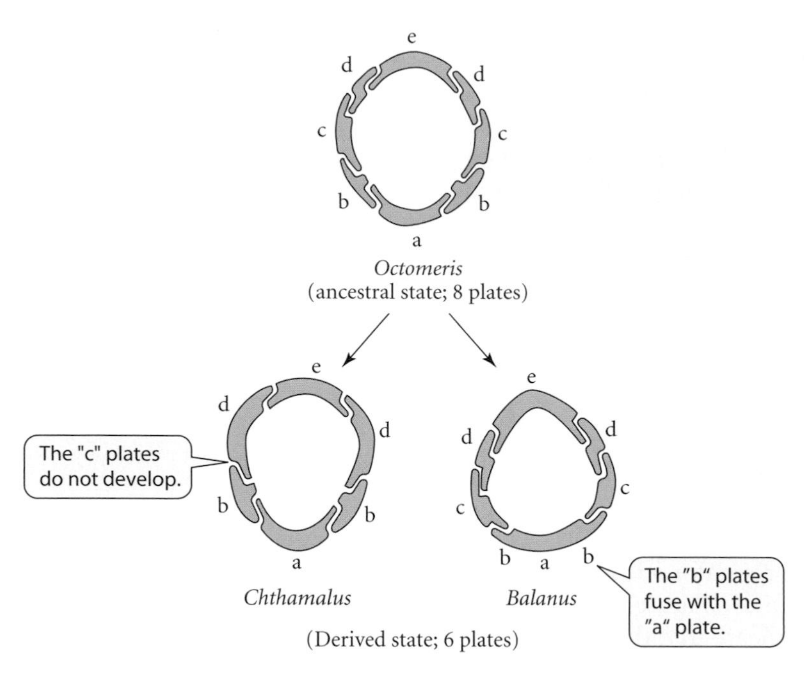

Figure 3.6 Darwin's example of two different ways in which the number of plates that compose the shell of a barnacle has been reduced to six from the ancestral condition of eight in two descendant genera, *Chthamalus* and *Balanus*. These diagrams show the shell in cross section. (After Darwin 1854.)

segment develops in all arthropods, including crustaceans and insects. These two genes are not expressed in head segments (which bear mouthparts), but are expressed in thoracic segments (which bear legs).

Averoff and Patel found that every evolutionary transformation of crustacean legs into maxillipeds corresponds exactly with a loss of expression of *Ubx* and *abdA* in the thoracic segment(s) involved. For example, in both the copepod *Mesocyclops* and the peracarid *Mysidium*, which belong to distantly related orders, the two genes are not expressed in the first thoracic segment, which then develops maxillipeds (Figure 3.7). The genetic and developmental basis of this evolutionary transformation is the same in both groups and occurs in some other crustacean lineages as well.

Evolutionary reversals constitute a return from an "advanced," or derived, character state to a more "primitive," or ancestral, state. For example, almost all frogs lack teeth in the lower jaw, but frogs are descended from ancestors that did have teeth. One genus of frogs, *Amphignathodon*, has "re-evolved" teeth in the lower jaw (Noble 1931). Because the immediate ancestors of *Amphignathodon* lacked teeth, their presence in this genus is a homoplasious reversal to a more remote ancestral condition.

Homoplasious features are often (but not always) adaptations by different lineages to similar environmental conditions. In fact, a correlation between a par-

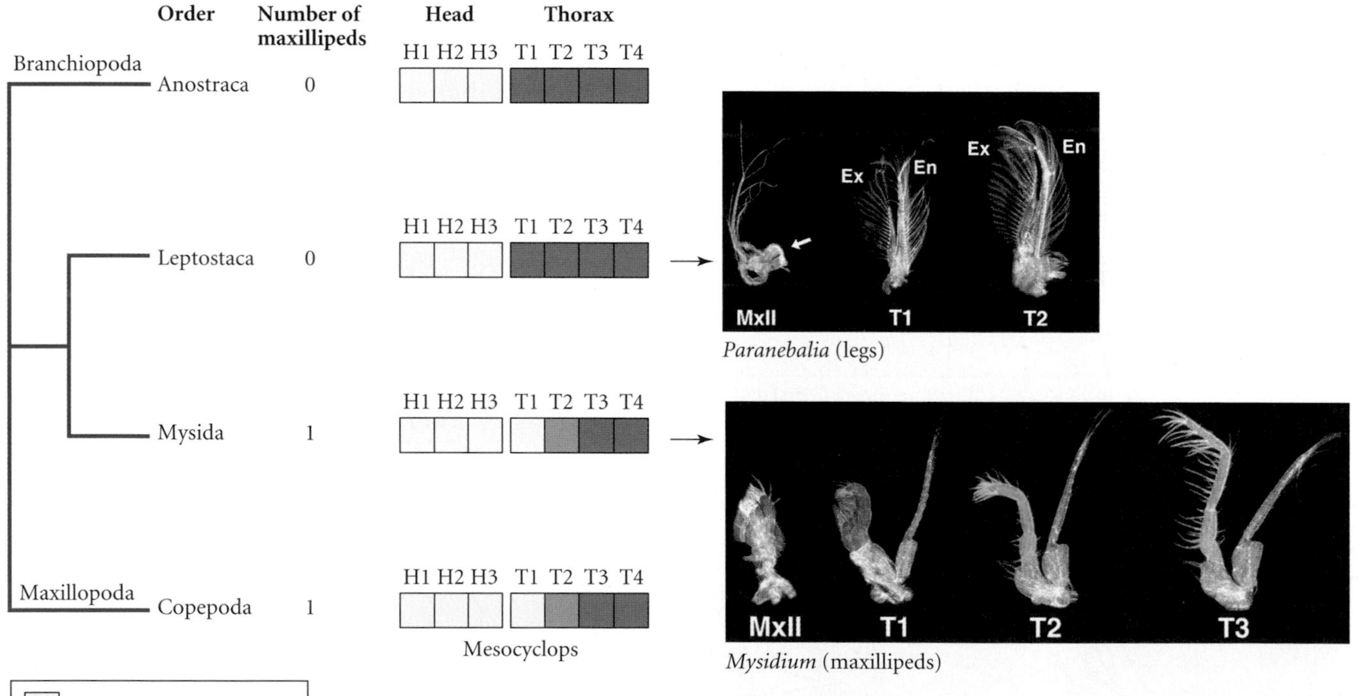

Figure 3.7 Parallel evolution. The evolution of feeding structures (maxillipeds) from thoracic legs in crustaceans is marked by parallel reduction or loss of expression of the genes *Ubx* and *abdA* in the same thoracic segments. The upper photograph (*Paranebalia*, order Leptostraca) shows the ancestral condition: there is a mouthpart (the maxilla, MxII) on head segment H3, and legs (each with two branches, En and Ex) on thoracic segments T1 and T2. In *Mysidium* (order Mysida), thoracic segment T3 has a normal leg (En branch in yellow), but the appendages on T2 and (especially) T1 show maxilla-like modifications of the En branch (green and red). The copepod *Mesocyclops* (not pictured) has similar modifications of gene expression and morphology. (After Averoff and Patel 1997.)

ticular homoplasious character in different groups and a feature of those organisms' environment or niche is often the best initial evidence of the feature's adaptive significance. For example, a long, thin beak has evolved independently in at least six different lineages of nectar-feeding birds. Such a beak enables these birds to reach nectar in the bottoms of the long, tubular flowers in which they often feed (Figure 3.8).

MIMICRY, a condition in which features of one species have specifically evolved to resemble those of another species, provides especially interesting examples of convergent evolution. Two of the most common kinds of mimicry are BATESIAN MIMICRY and MÜLLERIAN MIMICRY, named after the nineteenth-century naturalists who first described the phenomena. Both of these are defenses against predators. A Batesian mimic is a palatable or innocuous animal that has evolved resemblance to an unpalatable or dangerous animal (the *model*); for instance, many harmless flies have bright yellow and black color patterns like those of wasps, and some palatable butterfly species closely resemble other species that are toxic (see Figure 18.23). Predators that learn, from unpleasant experience, to avoid the model also will tend to avoid attacking the mimic.

In Müllerian mimicry, two or more distasteful or dangerous species have evolved similar characteristics, and a predator that associates the features of one with an unpleasant experience avoids attacking both species. There are many examples of mimicry "rings" that involve several species of distantly related distasteful butterflies (Müllerian mimics; see Figure 18.24) as well as palatable butterflies and moths (Batesian mimics) that share the same color pattern. There are many complications in the evolution of mimicry, including a spectrum of intermediate conditions between the Batesian and Müllerian extremes (Mallet and Joron 1999). Studies of mimicry have contributed greatly to our understanding of evolutionary processes, and will be discussed in several chapters of this book.

Figure 3.8 Four bird groups in which similar bill shape has evolved independently as an adaptation for feeding on nectar. (A) A South American honeycreeper, family Thraupidae (*Cyanerpes caeruleus*). (B) *Iiwi vestiaria coccinea*, one of the many Hawaiian honeycreepers, family Fringillidae. The Hawaiian honeycreepers are a well-studied example of adaptive radiation. (C) Hummingbirds, family Trochilidae. This violet sabrewing is from Costa Rica. (*Campylopterus hemileucurus*). (D) Sunbirds, family Nectariniidae. The red-chested sunbird (*Nectarinia pulchella*) is native to the Great Lakes region of Africa. (A, photo © fotolincs/Alamy Images; B, Photo Resource Hawaii/Alamy Images; C, Anthony Mercieca/Photo Researchers, Inc.; D, Photo Resource Hawaii/Alamy Images.)

Rates of character evolution differ

Different characters evolve at different rates, as is evident from the simple observation that any two species differ in some features, but not in others. Some characters, often called **conservative characters**, are retained with little or no change over long periods among the many descendants of an ancestor. For example, humans retain the pentadactyl limb that first evolved in early amphibians (see Figure 3.4); all amphibians and reptiles have two aortic arches, and all mammals have only the left arch. Body size, in contrast, evolves rapidly; within orders of mammals, it often varies at least a hundredfold. As we saw in the previous chapter, the rate of DNA sequence evolution varies among genes, among segments within genes, and among the three positions within amino acid-coding triplets of bases (codons).

Evolution of different characters at different rates within a lineage is called **mosaic evolution**. It is one of the most important principles of evolution, for it says that a species evolves not as a whole, but piecemeal: many of its features evolve quasi-independently. (There are important exceptions; for example, features that function together may evolve in concert.) This principle largely justifies the theory of evolutionary mechanisms, in which we analyze evolution not in terms of whole organisms, but in terms of changes in individual features, or even individual genes underlying such features.

Because of mosaic evolution, it is inaccurate or even wrong to consider one living species more "advanced" than another. The amphibian lineage leading to frogs split from the lineage leading to mammals before the mammalian orders diversified, so in terms of *order of branching*, frogs are an older branch than cows and humans. In that sense, frogs might be termed more primitive. But relative to early Paleozoic amphibians, frogs have both "primitive" features (e.g., five toes on the hind foot, multiple bones in the lower jaw) and advanced features (such as lack of teeth in the lower jaw). Moreover, numerous differences among frog species have evolved in the recent past. For example, one genus has direct development without a tadpole stage, and another gives birth to live young. Humans likewise have both "primitive" characters (five digits on hands and feet, teeth

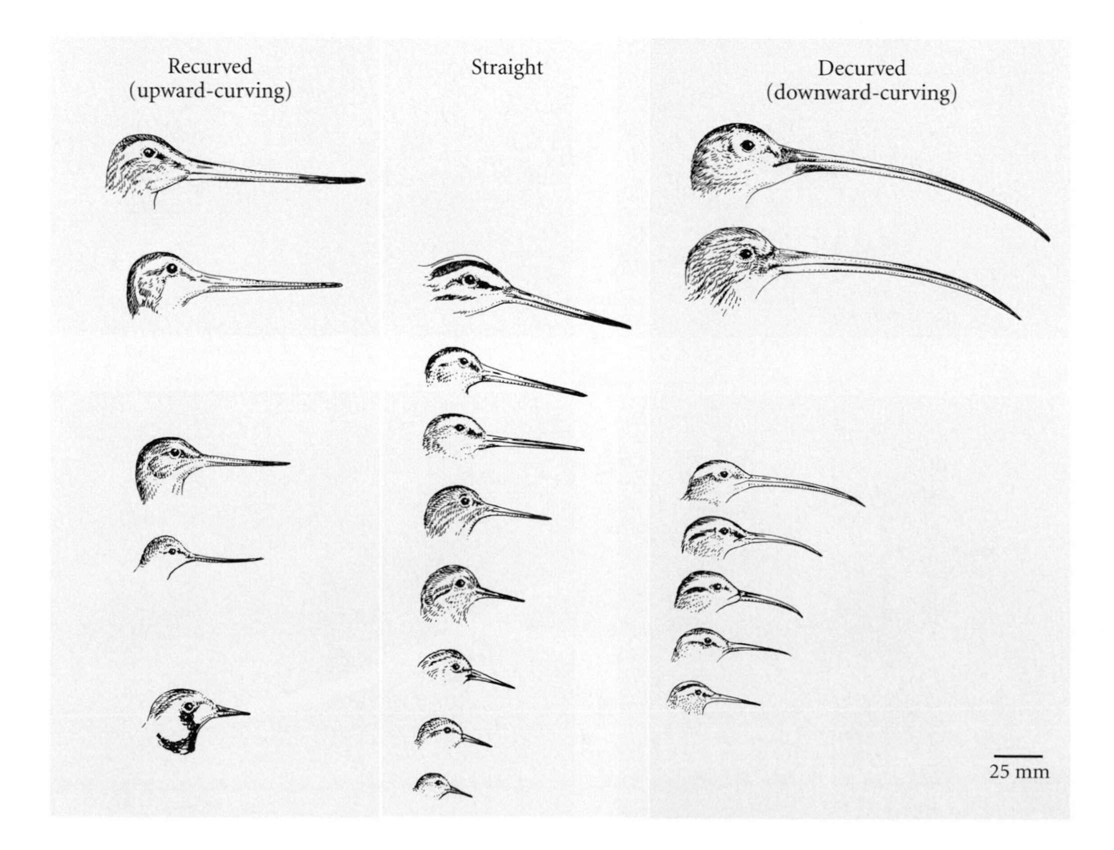

Figure 3.9 Variation in the shape and length of the bill among sandpipers (Scolopacidae). The three vertical series are drawn to scale and are spaced to match the differences in bill length, which range from 18 mm (bottom center) to 166 mm (upper right). Note the gradations in both curvature and length. The phylogenetic relationships among these species are not well resolved, but the variation shows how very different bills could have evolved through small changes. (After Hayman et al. 1986.)

in the lower jaw) and some that are "advanced" compared with frogs (e.g., a single lower jaw bone).

Evolution is often gradual

Darwin argued that evolution proceeds by small successive changes (GRADUALISM) rather than by large "leaps" (SALTATIONS). Whether or not evolution is always gradual is unknown, but the issue is much debated (see Chapter 21). Many higher taxa that diverged in the distant past (e.g., the animal phyla; many orders of insects and of mammals) are very different and are not bridged by intermediate forms, either among living species or in the fossil record. However, the fossil record does document intermediates in the evolution of some higher taxa, as we will see in Chapter 4.

Gradations among living species are commonplace and provide support for gradual evolution. For example, the length and shape of the bill differs greatly among species of sandpipers, but the most extreme forms are bridged by species with intermediate bills (Figure 3.9; another example, in drosophilid flies, is shown in Figure 3.21).

Change in form is often correlated with change in function

One of the reasons a homologous character may differ so greatly among taxa is that its form may have evolved as its function has changed. The sting of a wasp or bee, for example, is a modification of the ovipositor that other members of the Hymenoptera use to insert eggs into plant or arthropod hosts (that is why only female wasps and bees sting). In the many groups of plants that have independently evolved a vinelike climbing habit, the structures that have been modified into climbing organs include roots, leaves, leaflets, stipules, and inflorescences (Figure 3.10).

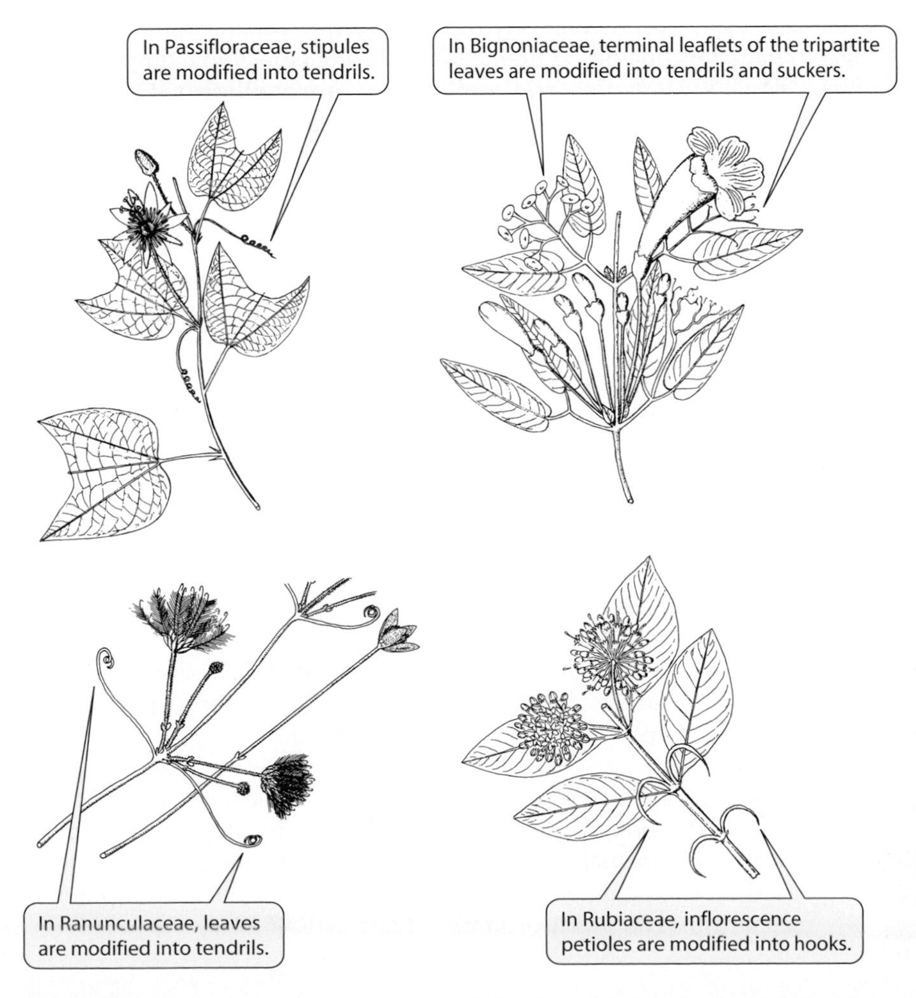

In Passifloraceae, stipules are modified into tendrils.
In Bignoniaceae, terminal leaflets of the tripartite leaves are modified into tendrils and suckers.
In Ranunculaceae, leaves are modified into tendrils.
In Rubiaceae, inflorescence petioles are modified into hooks.

Figure 3.10 Structures modified for climbing in vines from different plant groups show that structures can become modified for new functions, and that different evolutionary paths to the same functional end may be followed in different groups. (After Hutchinson 1969.)

Similarity between species changes throughout ontogeny

Species are often more similar as embryos than as adults. Karl Ernst von Baer noted in 1828 that the features common to a more inclusive taxon (such as the subphylum Vertebrata) often appear in ontogeny (development) before the specific characters of lower-level taxa (such as orders or families). This generalization is now known as VON BAER'S LAW. Probably the most widely known example is the similarity of many tetrapod vertebrate embryos, all of which display gill slits, a notochord, segmentation, and paddle-like limb buds before the features typical of their class or order become apparent (Figure 3.11).

One of Darwin's most enthusiastic supporters, the German biologist Ernst Haeckel, reinterpreted such patterns to mean that "ontogeny recapitulates phylogeny"; that is, that the development of the individual organism repeats the evolutionary history of the adult forms of its ancestors. Haeckel thus supposed that by studying embryology, one could read a species' phylogenetic history, and therefore infer directly phylogenetic relationships among organisms. By the end of the nineteenth century, it was already clear that Haeckel's dictum (known as the "biogenetic law") rather seldom holds (Gould 1977). The gill slits and branchial arches of embryonic mammals and "reptiles" never acquire the form typical of adult fishes. Moreover, various features develop at different rates, relative to one another, in descendants than in their ancestors, and embryos and juvenile stages have stage-specific adaptations of their own (such as the amnion in the embryo). Thus the biogenetic law is certainly not an infallible guide to phylogenetic history. However, embryological similarities provided Darwin with some of his most important evidence of evolution, and they continue to shed important light on how characteristics have been transformed during evolution.

Development underlies some common patterns of morphological evolution

Until a few decades ago, classification and phylogenetic studies relied chiefly on analyses of morphological characters, including their change during embryonic development. In the course of their work, systematists and comparative morphologists documented many common patterns of evolution. Today, one of the most active research areas concerns the genetic and developmental basis of such evolutionary changes (see Chapter 20). Some of these patterns are individualization, heterochrony, allometry, heterotopy, and changes in complexity (Rensch 1959; Müller 1990; Raff 1996; Wagner 1996).

Figure 3.11 Micrographs show the similarities—and differences—among several vertebrate embryos at different stages of development. Each begins with a similar basic structure, although they acquire this structure at different stages and sizes. As the embryos develop, they become less and less alike. (Adapted from Richardson et al. 1998; photo courtesy of M. Richardson.)

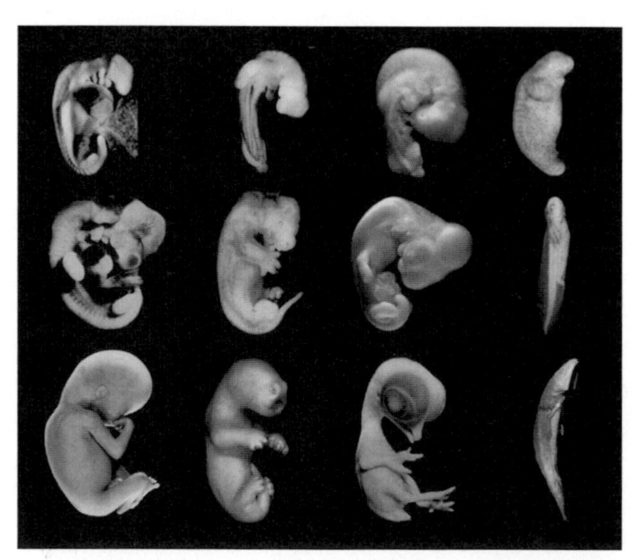

INDIVIDUALIZATION. The bodies of many organisms consist of MODULES—distinct units that have distinct genetic specifications, developmental patterns, locations, and interactions with other modules (Raff 1996). Some modules (e.g., leaves of many plants, teeth of many fishes) lack distinct individual identities and may be considered aspects of a single character. Such structures are termed SERIALLY HOMOLOGOUS if they are arrayed along the body axis (as vertebrae are) and HOMONYMOUS if they are not. An important evolutionary phenomenon is the acquisition of distinct identities by such units, called **individualization** (Wagner 1996; Müller and Wagner 1996), which in turn is an important basis for mosaic evolution. For instance, the teeth of most reptiles are uniform, but they became individualized (differentiated into incisors, canines, premolars, and molars) during the evolution of mammals. Distinct tooth identity was later lost during the evolution of the toothed whales (Figure 3.12).

HETEROCHRONY. **Heterochrony** (Gould 1977; McKinney and McNamara 1991) is broadly defined as *an evolutionary change in the timing or rate of developmental events*. Many phenotypic changes appear to be based on such changes in timing, but several other developmental mechanisms can produce similar changes (Raff 1996).

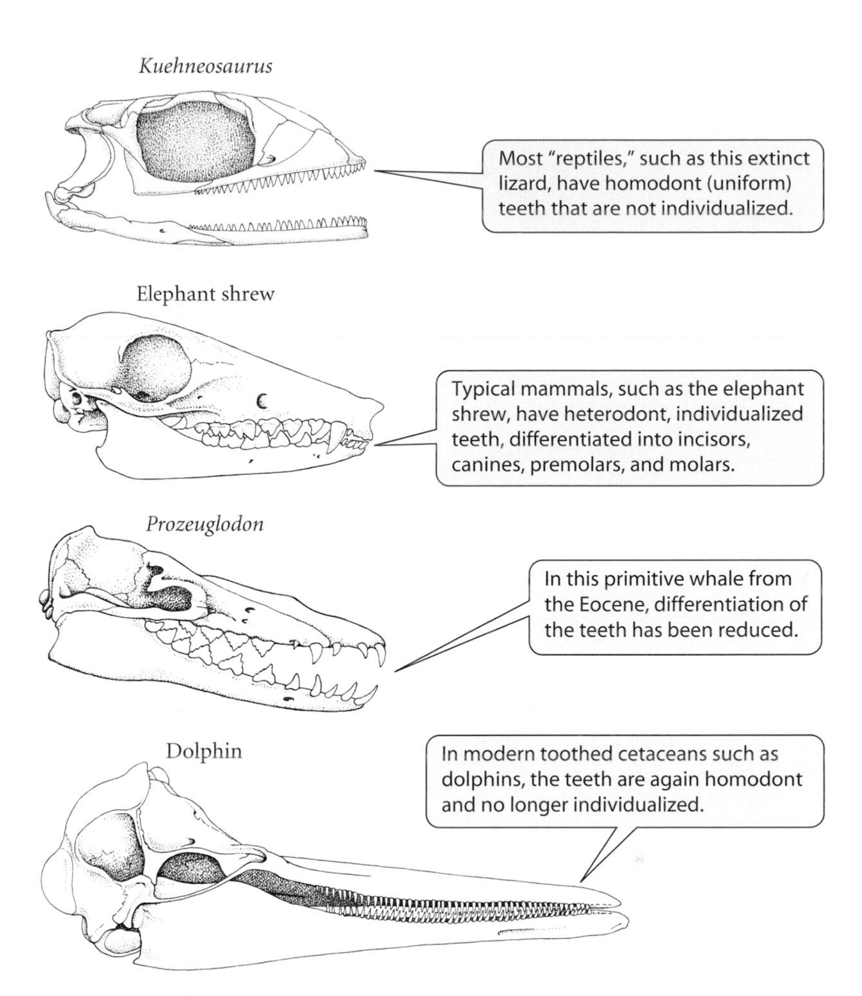

Kuehneosaurus
Most "reptiles," such as this extinct lizard, have homodont (uniform) teeth that are not individualized.
Elephant shrew
Typical mammals, such as the elephant shrew, have heterodont, individualized teeth, differentiated into incisors, canines, premolars, and molars.
Prozeuglodon
In this primitive whale from the Eocene, differentiation of the teeth has been reduced.
Dolphin
In modern toothed cetaceans such as dolphins, the teeth are again homodont and no longer individualized.

Figure 3.12 The teeth of reptiles and mammals provide an example of the acquisition and loss of individualization. The teeth of most reptiles are uniform; teeth became individualized during the evolution of mammals. Distinct tooth identity was later lost in the evolution of the toothed whales. (A after Romer 1966; B–D after Vaughan 1986.)

Relatively *global* heterochronic changes, affecting many characters simultaneously, are illustrated by cases in which the time of development of most SOMATIC features (those other than the gonads and related reproductive structures) is altered relative to the time of maturation of the gonads (i.e., initiation of reproduction). The axolotl, for example, is a salamander that does not undergo metamorphosis, as most salamanders do, but instead reproduces while retaining most of its larval (juvenile) characteristics (Figure 3.13). Such evolution of a more "juvenilized" morphology of the reproductive adult is called **paedomorphosis** (from the Greek *paedos*, "child," and *morphos*, "form"). In contrast, evolution of delayed maturity may result in reproduction at a larger size, associated with the extended development of "hyper-adult" features. Such an evolutionary change is called **peramorphosis**. The large size of the human brain, for example, has been ascribed to humans' extended period of growth (McNamara 1997).

ALLOMETRY. Allometric growth, or **allometry**, refers to the *differential rate of growth of different parts or dimensions of an organism* during its ontogeny. For example, during human growth, the head grows at a lower rate than the body as a whole, and the legs grow at a higher rate. Changes in rates of allometric growth of individual characters— that is, "local" heterochrony—appear to have played an extremely important role in evolution. For example, many evolutionary changes can be described as if local heterochronies had altered the *shape* of one or more characters: an increased rate of elongation of the digits "accounts for" the shape of a bat's wing compared with the forelimbs of other mammals (see Figure 3.4); a change in the growth rate of the lower jaw can "explain" the form of a needlefish or halfbeak (Figure 3.14).

Allometric growth is often described by the equation

$$y = bx^a$$

(A)

(B)

(C)

Figure 3.13 Paedomorphosis in salamanders. (A–B) The tiger salamander (*Ambystoma tigrinum*), like most salamanders, undergoes metamorphosis from (A) an aquatic larva (note the presence of gills) to (B) a terrestrial adult. (C) The adult axolotl (*Ambystoma mexicanum*), with gills and tail fin, resembles the larva of its terrestrial relative. The axolotl remains aquatic throughout its life span. (A, photo by Twan Leenders; B, photo © Painet, Inc.; C, photo by Henk Wallays.)

(A) Flying fish (typical jaws)

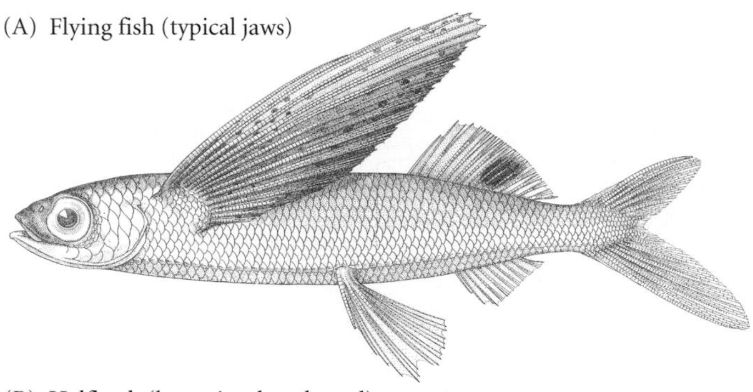

(B) Halfbeak (lower jaw lengthened)

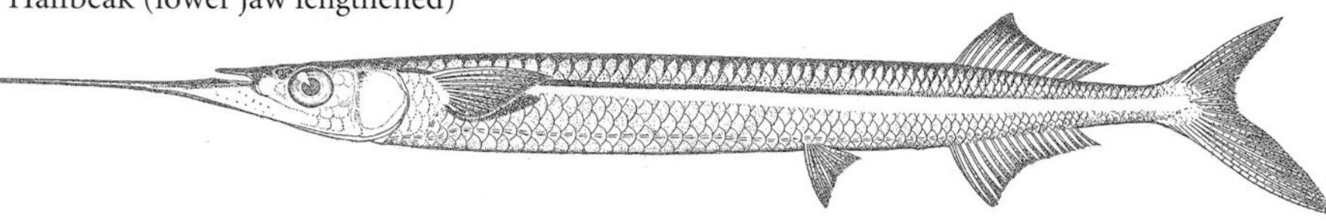

(C) Needlefish (both jaws lengthened)

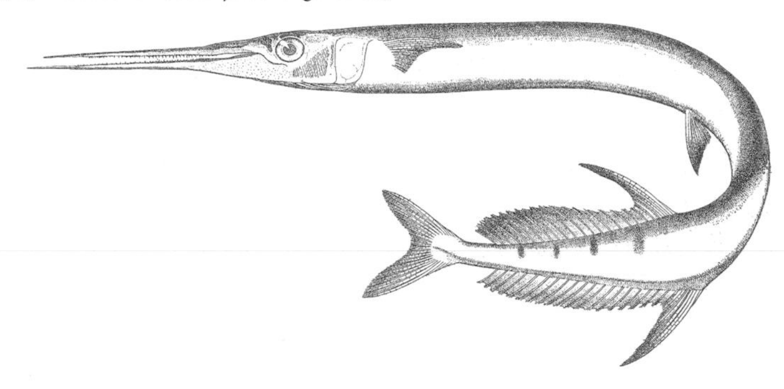

Figure 3.14 Allometric differences in the length of the upper and lower jaws among three closely related families of fishes: (A) flying fish, (B) halfbeak, and (C) needlefish. The differences in form can be accounted for by changes in the rate of jaw growth relative to body growth. (From Jordan and Evermann 1973.)

where y and x are two measurements, such as the height and width of a tooth or the size of the head and the body. (In many studies, x is a measure of body size, such as weight, because many structures change disproportionately with overall size.) The ALLOMETRIC COEFFICIENT, a, describes their relative growth rates. If $a = 1$, growth is ISOMETRIC, meaning that the two structures or dimensions increase at the same rate, and shape does not change. If y increases faster than x, as for human leg length relative to body size or weight, $a > 1$ (positive allometry); if it increases relatively slowly, as for human head size, $a < 1$ (negative allometry) (Figure 3.15). These curvilinear relationships between y and x often appear more linear if transformed to logarithms, yielding the equation for a straight line, $\log y = \log b + a \log x$.

Suppose that in an ancestral form, x represents body size, and y represents the size of some feature that begins to develop (onset) at age α and stops growing (offset) at age β (Alberch et al. 1979). Both paedomorphosis and peramorphosis can result from evolutionary changes in either the *rate* of development or the *duration* of development due to a change in α or β (Figure 3.16). Peramorphosis can result if the duration of development is extended (a change from β to $\beta + \Delta\beta$; Figure 3.16B). For example, the gigantic antlers of the extinct Irish elk (*Megaceros giganteus*), which are larger in relation to body mass than those of any other deer, are a peramorphic feature associated with the animal's extended development to a larger body size (Figure 3.17). Paedomorphosis can be caused by cessation of growth at an earlier age ($\beta - \Delta\beta$) (a form of paedomorphosis called **progenesis**) or by reducing the growth rate of character y (an evolutionary process called **neoteny**). The axolotl (see Figure 3.13C) is a neotenic salamander species that reaches the same size as its metamorphosing relatives, whereas tiny salamanders in the genus *Tho-*

Figure 3.15 Hypothetical curves showing various allometric growth relationships between two body measurements, y and x, according to the equation $y = bx^a$. (A) Arithmetic plots. Curves 1 and 2 show isometric growth ($a = 1$), in which y is a constant multiple (b) of x. Curves 3 and 4 show positive ($a > 1$) and negative ($a < 1$) allometry, respectively. (B) Logarithmic plots of the same curves have a linear form. The slope differences depend on a. Curves 1 and 2 have slopes equal to 1.

(A) Arithmetic plots

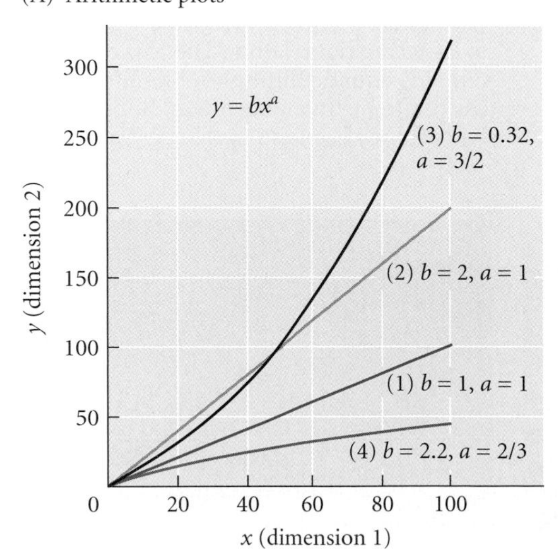

(B) Logarithmic plots

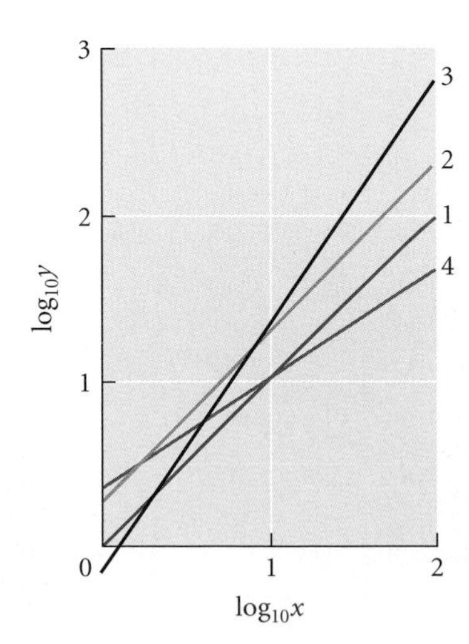

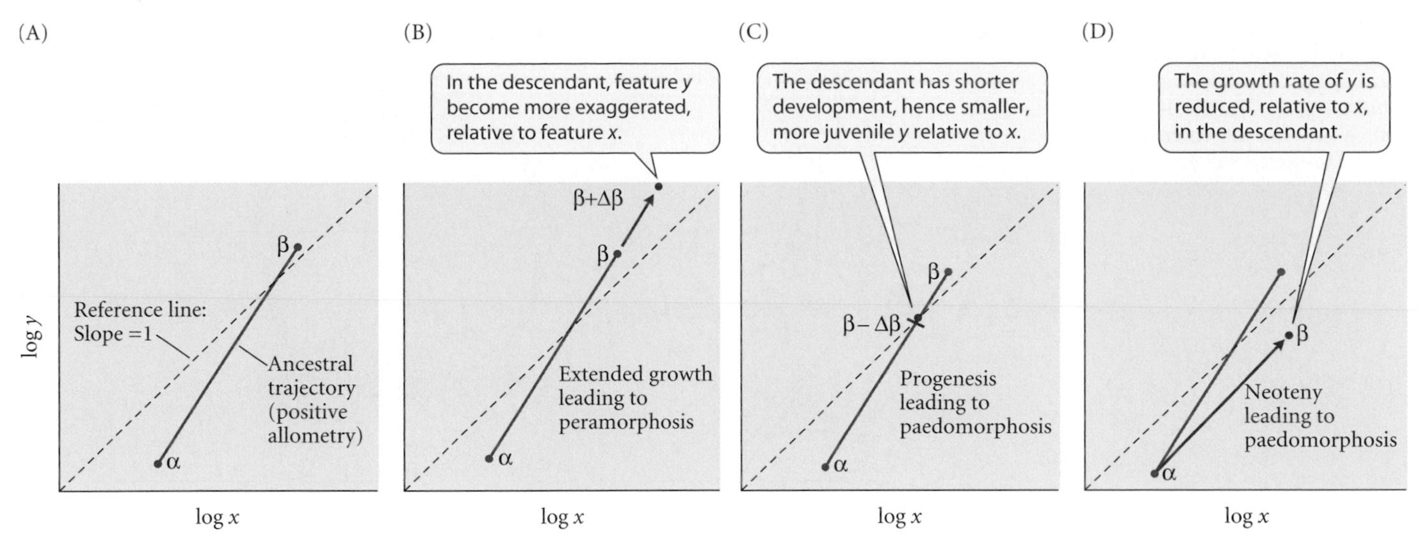
(A)
(B)
(C)
(D)
In the descendant, feature y become more exaggerated, relative to feature x.
The descendant has shorter development, hence smaller, more juvenile y relative to x.
The growth rate of y is reduced, relative to x, in the descendant.
log y
Reference line: Slope =1
Ancestral trajectory (positive allometry)
β
α
β+Δβ
Extended growth leading to peramorphosis
β−Δβ
Progenesis leading to paedomorphosis
Neoteny leading to paedomorphosis
log x
log x
log x
log x

Figure 3.16 A diagrammatic representation of some forms of heterochrony, expressed as logarithmically plotted allometric growth of structures or dimensions y and x. The x-axis may represent body size, the y-axis a character such as leg length. The dashed line, with slope 1, is provided for reference. (A) The blue line shows ontogenetic change in the ancestor from age α to age β. Growth is positively allometric (slope > 1). (B) A longer growth period (extension of growth to age $\beta + \Delta\beta$; purple arrow) results in peramorphosis: an exaggerated structure y in the descendant. (C) Cessation of growth at an earlier age ($\beta - \Delta\beta$) results in a form of paedomorphosis known as progenesis. (D) Reduction in the growth rate of y relative to x, without changing the duration of growth, results in a form of paedomorphosis called neoteny.

rius are progenetic: they have many features characteristic of the juveniles of larger species of salamanders, as if their development had been abbreviated (Figure 3.18).

HETEROTOPY. **Heterotopy** is an evolutionary change in the position within an organism at which a phenotypic character is expressed. Studies of the distribution of gene products have revealed many heterotopic differences among species in sites of gene expression. For instance, certain species of deep-sea squids have organs on the body that house light-emitting bacteria. The light is diffracted through small lenses made up of the same two proteins that compose the lenses of the squids' eyes (Raff 1996).

Figure 3.17 Perhaps the most famous example of allometry and peramorphosis is the largest of deer, the extinct "Irish elk" (*Megaceros giganteus*). Its antlers were larger, relative to body mass, than those of any other deer. (From Millais 1897.)

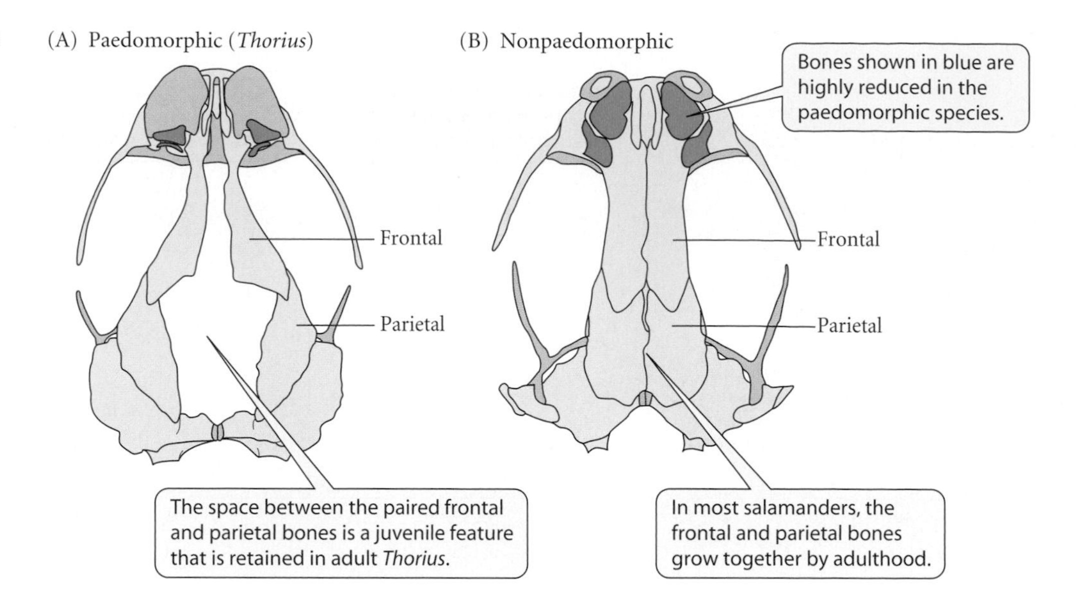
(A) Paedomorphic (Thorius)
(B) Nonpaedomorphic
Bones shown in blue are highly reduced in the paedomorphic species.
Frontal
Parietal
Frontal
Parietal
The space between the paired frontal and parietal bones is a juvenile feature that is retained in adult Thorius.
In most salamanders, the frontal and parietal bones grow together by adulthood.

Figure 3.18 Comparison of the skulls of the progenetic dwarf salamander *Thorius* and a typical nonprogenetic relative, *Pseudoeurycea*. The skull of adult *Thorius* has a number of juvenile features. (After Hanken 1984.)

Heterotopic differences among species are very common in plants. For example, the major photosynthetic organs of most plants are the leaves, but photosynthesis is carried out in the superficial cells of the stem in cacti and many other plants that occupy dry environments. In many unrelated species of lianas (woody vines), roots grow along the aerial stem (Figure 3.19). In some lianas these roots serve as holdfasts, while in others they grow down to the soil from the canopy high above.

The bones of vertebrates provide many examples of heterotopy. For example, many phylogenetically new bones have arisen as SESAMOIDS—bones that develop in tendons or other connective tissues subject to stress (Müller 1990). Many dinosaurs had ossified tendons in the tail, and the giant panda (*Ailuropoda melanoleuca*) is famous for having a "thumb" that is not a true jointed digit, but a single sesamoid (see Figure 21.8).

INCREASES AND DECREASES IN COMPLEXITY. The earliest organisms must have had very few genes and must have been very simple in form. Both phylogenetic and paleontological studies show that there have been great increases in complexity during the history of life, exemplified by the origin of eukaryotes, then of multicellular organisms, and later of the elaborate tissue organization of plants and animals (Chapter 21). Given our impression of overall increases in complexity during evolution, it can be surprising to learn that simplification of morphology—reduction and loss of structures—is one of the most common trends within clades. The primitive "ground plan" of flowers, for example, includes numerous sepals, petals, stamens, and carpels, but the number of one or more of these elements has been reduced in many lineages of flowering plants, and many clades have lost petals and/or sepals altogether. The number of digits in tetrapod vertebrates has been reduced many times (consider the single toe of horses), but has increased only once (in extinct ichthyosaurs). Early lobe-finned fishes had far more skull bones than their amniote descendants (Figure 3.20). Many such changes can be ascribed to increased functional efficiency.

Stems
Roots

Figure 3.19 Plants of the genus *Philodendron*, such as this Jamaican climber, are lianas. Many lianas, which include several genera besides *Philodendron*, have evolved exposed roots that grow from an aerial stem. (Photo © photolibrary.com)

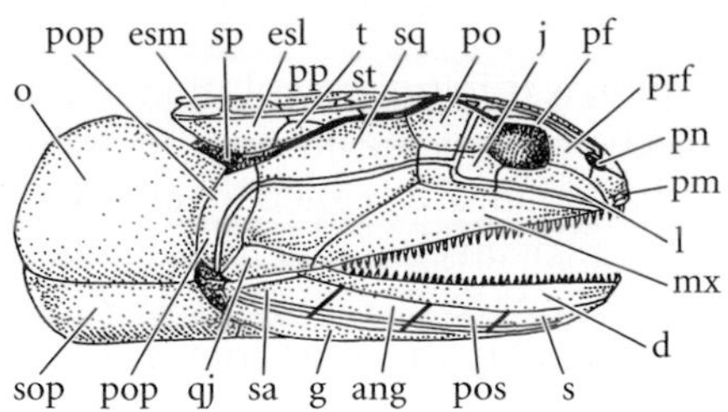

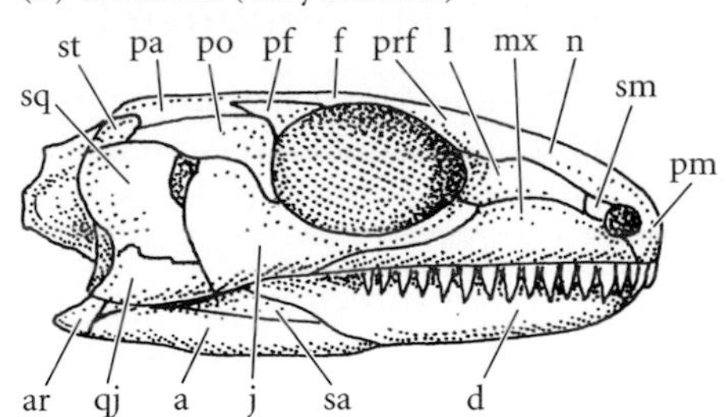

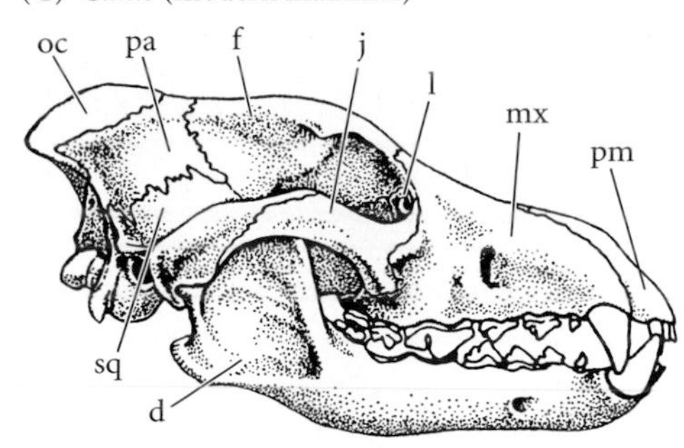

Figure 3.20 An example of reduction and loss of structures during evolution. The number of bones in the skull is higher in early fishes (such as the Devonian *Eusthenopteron*), from which amniotes were derived, than in early amniotes (such as the Permian *Milleretta*). Among the later amniotes are placental mammals, such as the domestic dog (*Canis*), in which the skull has fewer elements still. The reduction in the number of bones in the lower jaw is particularly notable. (After Romer 1966.)

Phylogenetic Analysis Documents Evolutionary Trends

The term **evolutionary trend** can refer to a succession of changes of a character in the same direction, either within a single lineage or, often, in many lineages independently. For example, a phylogenetic analysis indicated that within the fly genus *Zygothrica*, there has been directional evolution toward wider heads in male flies (Figure 3.21)—an evolutionary change that has occurred in at least three other clades of flies. Among flowering plants, many groups independently display trends from low to high chromosome number and DNA content, from many to few flower parts (e.g., stamens or carpels), from separate to fused flower parts, from radial to bilateral symmetry of the flower, from animal to wind pollination, and from woody to herbaceous structure. We will analyze the kinds and causes of evolutionary trends in Chapter 21.

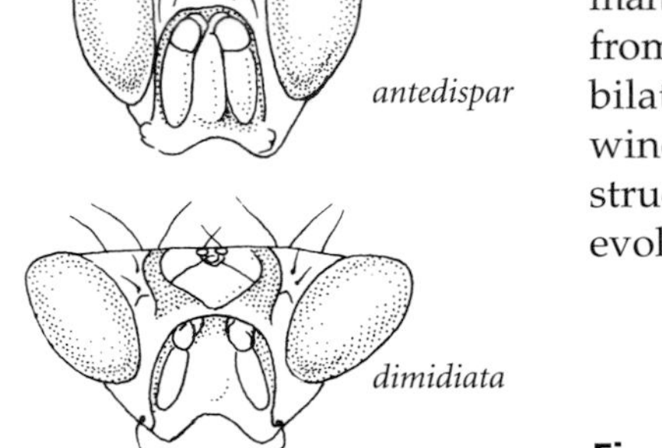

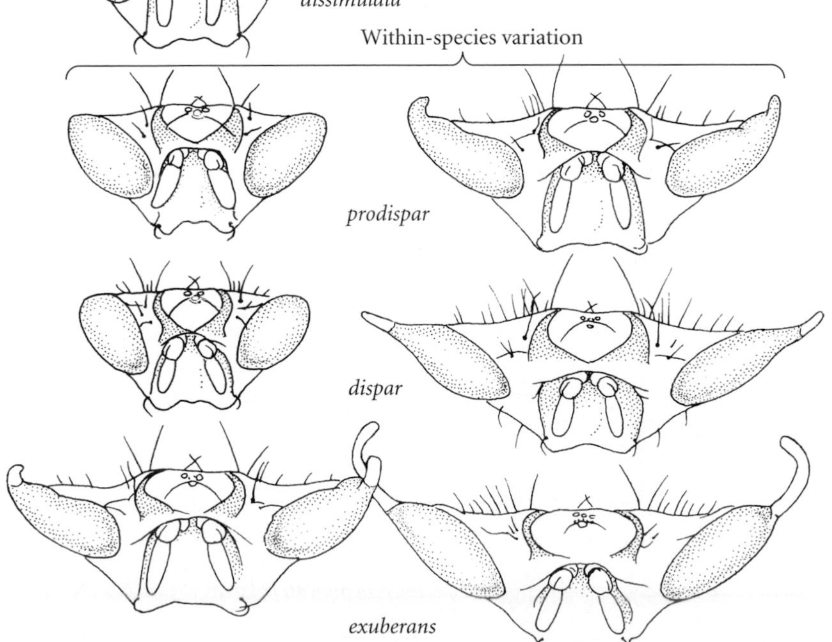

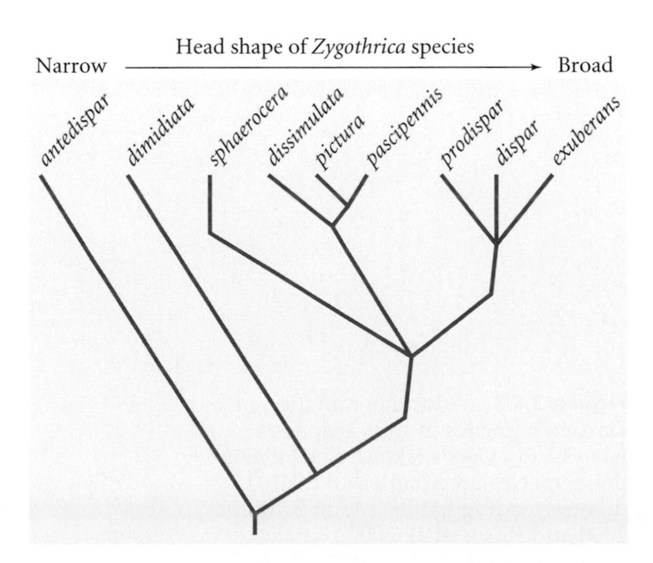

Figure 3.21 Frontal view of the heads of male flies in the *dispar* subgroup of *Zygothrica* (Drosophilidae). Note the gradation in the shape of the head and eyes among species and within *prodispar*, *dispar*, and *exuberans*. These gradations form a phylogenetic series from narrow to broad, as indicated by the phylogenetic analysis, which is based on a number of morphological features. (After Grimaldi 1987.)

Many Clades Display Adaptive Radiation

Evolutionary radiation, as we saw in the previous chapter, is divergent evolution of numerous related lineages within a relatively short time. In most cases, the lineages are modified for different ways of life, and the evolutionary radiation may be called an **adaptive radiation** (Schluter 2000). The characteristics of the members of an evolutionary radiation usually do not show a directional trend in any one direction. Evolutionary radiation, rather than sustained, directional evolutionary trends, is probably the most common pattern of long-term evolution.

Several adaptive radiations have been extensively studied and are cited in many evolutionary contexts. The most famous example is the adaptive radiation of Darwin's finches in the Galápagos archipelago. These finches, which are all descended from a single ancestor that colonized the archipelago from South America, differ in the morphology of the

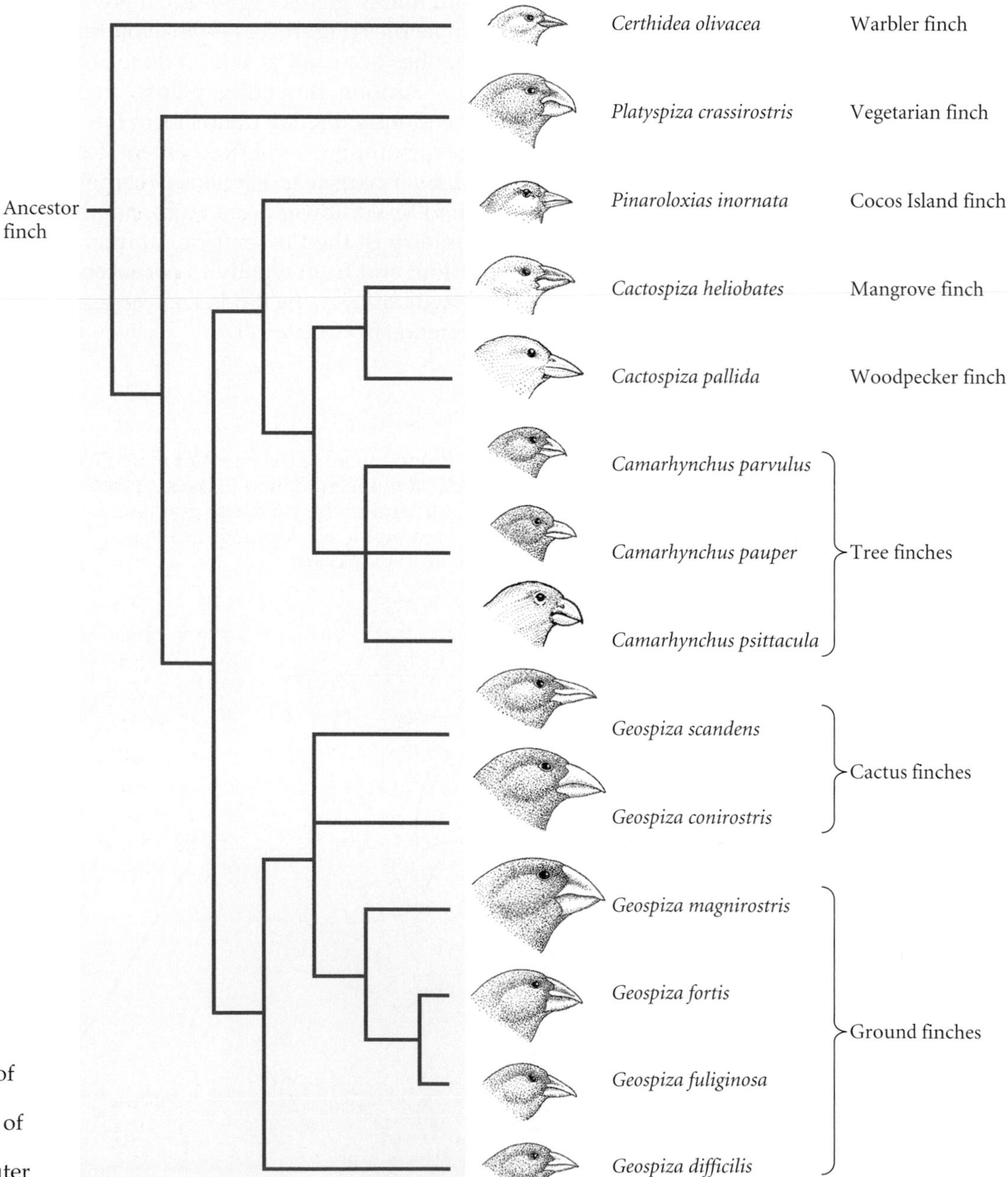

Figure 3.22 Adaptive radiation of Darwin's finches in the Galápagos Islands and Cocos Island. The bills of these species are adapted to their diverse feeding habits. (After Schluter 2000 and Burns et al. 2002.)

(A) *Argyroxiphium sandwicense*

(B) *Wilkesia hobdyi*

(C) *Dubautia menziesii*

Figure 3.23 Some members of the Hawaiian silversword alliance with different growth forms. (A) *Argyroxiphium sandwicense*, a rosette plant that lacks a stem except when flowering (as it is here). (B) *Wilkesia hobdyi*, a stemmed rosette plant. (C) *Dubautia menziesii*, a small shrub. (A, photo by Elizabeth N. Orians; B, photo by Gerald D. Carr; C, photo by Noble Procter/Photo Researchers, Inc.)

bill, which provides adaptation to different diets. Different species of *Geospiza* feed on seeds that vary in size and hardness. Other genera include *Camarhynchus*, which excavates insects from wood, *Certhidea*, which feeds on nectar and insects, and *Cactospiza*, in which one species has the unique habit of using cactus spines as tools to extricate insects from crevices (Figure 3.22).

Relatively few kinds of animals and plants have colonized the Hawaiian Islands, but many of those that have done so have given rise to adaptive radiations. For example, more than 800 species of drosophilid flies (fruit flies) occur in the Hawaiian Islands. In morphology and sexual behavior, they are more diverse than all the drosophilids in the rest of the world combined (Carson and Kaneshiro 1976). Many of the species have bizarre modifications of the mouthparts, legs, and antennae, associated with unusual forms of mating behavior. Among plants, the Hawaiian silverswords and their relatives, three closely related genera of plants in the sunflower family, occupy habitats ranging from exposed lava rock to wet forest, and their growth forms include shrubs, vines, trees, and creeping mats (Figure 3.23). Despite these great differences, most of them can produce fertile hybrids when crossed (Carlquist et al. 2003).

The cichlid fishes in the Great Lakes of eastern Africa have undergone some of the most spectacular adaptive radiations (Kornfield and Smith 2000). Lake Victoria has more than 200 species, Lake Tanganyika at least 140, and Lake Malawi more than 500, and perhaps as many as 1000. The species within each lake vary greatly in coloration, body form, and the form of the teeth and jaws (Figure 3.24), and they are correspondingly diverse in feeding habits. There are specialized feeders on insects, detritus, rock-encrusting algae, phytoplankton, zooplankton, molluscs, baby fishes, and large fishes. Some species feed on the scales of other fishes, and one has the gruesome habit of plucking out other fishes' eyes. The teeth of certain closely related species differ more than among some families of fishes. The species in each lake are a monophyletic group and have diversified very rapidly.

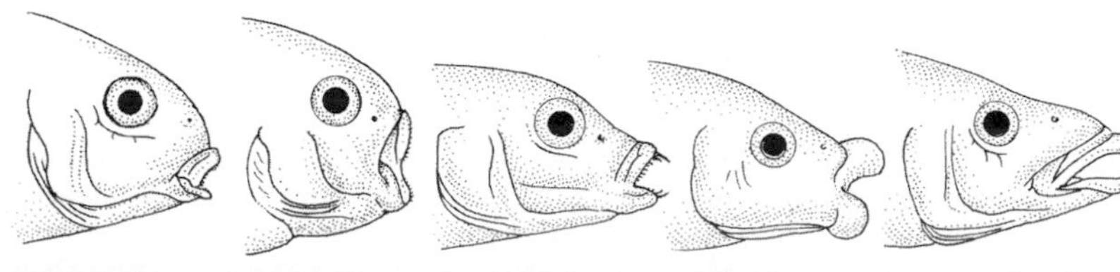

Figure 3.24 A sample of the diverse head shapes among the Cichlidae of the African Great Lakes. The differences in morphology are associated with differences in diet and mode of feeding. (After Fryer and Iles 1972.)

Summary

1. Most traditional classifications of organisms have been devised to reflect both branching (phylogeny) and phenotypic divergence. Many contemporary systematists adopt the "cladistic position" that classification should explicitly reflect phylogenetic relationships, and that all higher taxa should be monophyletic.
2. Phylogenetic analyses have many uses in addition to describing the branching history of life. An important one is inferring the history of evolution of interesting characteristics by parsimoniously "mapping" changes in character states onto a phylogeny that has been derived from other data. Such systematic studies have yielded information on common patterns and principles of character evolution.
3. New features almost always evolve from pre-existing characters.
4. Homoplasy, which is common in evolution, is often a result of similar adaptations in different lineages. It includes convergent evolution, parallel evolution, and reversal.
5. For the most part, different characters evolve piecemeal, at different rates. Conservative characters are those that are retained with little or no change over long time periods; other characters may evolve rapidly and vary widely across a single lineage. This phenomenon is called mosaic evolution.
6. Differences among related species illustrate that large differences often evolve gradually, by small steps. Although this pattern is common, it may not be universal.
7. Changes in structure are often associated with change in a character's function. Different structures may be modified to serve similar functions in different lineages.
8. The evolution of morphological features involves changes in their development. Such evolutionary changes include individualization of repeated structures, alterations in the timing (heterochrony) or site (heterotopy) of developmental events, and increases and decreases in structural complexity. Heterochrony can result in changes in the shape of features, often because of allometric growth.
9. Trends in the evolution of a character may be documented by phylogenetic analysis. Evolutionary trends may occur within a single lineage or repeatedly among different lineages.
10. In an adaptive radiation, numerous related lineages arise in a relatively short time and evolve in many different directions as they adapt to different habitats or ways of life.

Terms and Concepts

adaptive radiation
allometry
anagenesis
clade
cladistics
cladogenesis
cladogram
conservative characters
convergence (convergent evolution)
crown group
evolutionary reversal (reversal)
evolutionary trend
heterochrony
heterotopy
homology (homologous characters)
homoplasy
individualization
monophyletic
mosaic evolution
neoteny
paedomorphosis
parallelism (parallel evolution)
paraphyletic
patterns of evolution
peramorphosis
polyphyletic
progenesis
stem group

Suggestions for Further Reading

Most books and other sources cover only some of the many subjects discussed in this chapter. An incisive and historically important treatment of many patterns of evolution, although not interpreted in an explicitly phylogenetic framework, is Bernhard Rensch's *Evolution above the species level* (Columbia University Press, New York, 1959).

A broad perspective on the uses of phylogenies in studying evolution and ecology is provided by D. R. Brooks and D. A. McLennan in *The nature of diversity: An evolutionary voyage of discovery* (University of Chicago Press, 2002) and, briefly, by D. J. Futuyma, "The fruit of the tree of life: Insights into evolution and ecology" (pp. 25–39 in *Assembling the tree of life*, J. Cracraft and M. J. Donoghue, eds., Oxford University Press, New York, 2004).

The early history of interpretation of morphological evolution in a developmental context was reviewed by Stephen Jay Gould in *Ontogeny and phylogeny* (Harvard University Press, Cambridge, MA, 1977). More contemporary perspectives are provided by M. L. McKinney ad K. J. McNamara in *Heterochrony: The evolution of ontogeny* (Plenum, New York, 1991) and by R. A. Raff in *The shape of life: Genes, development, and the evolution of animal form* (University of Chicago Press, 1996). G. P. Wagner addressed the nature of homology in "The biological homology concept" (*Annual Review of Ecology and Systematics* 20: 51–69, 1989).

Problems and Discussion Topics

1. In Problem 1 in Chapter 2, you were asked to estimate the phylogeny of species 1, 2, and 3, based on the following DNA sequences:

   (1) GCTGATGAGT; (2) ATCAATGAGT; (3) GTTGCAACGT; (4) GTCAATGACA

   Species 4 is an outgroup. Suppose the species are primates that differ in mating system: species 1 and 2 are monogamous, and species 3 and 4 are polygamous. We also happen to know that another polygamous species, 5, is more distantly related to species 1–4 than those species are to one another. Given your best estimate of the phylogenetic history, what has been the probable history of evolution of the mating system?
2. In the absence of a fossil record, how might phylogenetic analysis tell us whether the rate of diversification (increase in number of species) has differed among evolutionary lineages?
3. How would you determine, from phylogenetic analysis, which form of the jaws is ancestral, and which derived, in the three related families of fishes pictured in Figure 3.14? What information would you need?
4. The antlers of the extinct Irish elk (Figure 3.17) were so large that some paleontologists proposed that the species became extinct because the antlers simply became too unwieldy and prevented the elk from escaping predators. Other biologists wondered why such big antlers evolved; some suggested that males with larger antlers could win fights with other males and gain greater access to mates. Stephen Jay Gould (1974) calculated that, given the allometric relationship between antler size and body size among species of deer, the Irish elk's antlers were exactly as large as we would expect in such a large deer. If antler size is a result of a growth correlation with body size, is it necessary to postulate an adaptive advantage for the Irish elk's large antlers in order to account for their great size? And would such a correlation imply that natural selection could not prevent the antlers from becoming so large that they caused the species' extinction?
5. The previous question illustrates questions about evolutionary processes that arise when patterns of evolution are described. Specify some questions about evolutionary processes that might be raised by each of the following patterns: (*a*) modification of different structures to achieve similar adaptive functions (e.g., Figure 3.10); (*b*) evolutionary trends (e.g., Figure 3.21); (*c*) evolutionary reversal (e.g., teeth in the lower jaw of the frog *Amphignathodon*).
6. The first analyses of the human genome offered many insights based on an evolutionary, phylogenetic perspective. After reading these analyses (International Human Genome Sequencing Consortium 2001; Venter et al. 2001) or associated commentaries, describe five questions about or insights into the human genome that could be addressed with the help of phylogenetic analyses.
7. Early in this chapter, the claim was made that phylogenetic information is the basis for describing patterns of evolution, yet some examples of such patterns were presented without phylogenetic trees. Consider the following examples and discuss what phylogenetic evidence or inference was left unstated: (*a*) Teeth in the lower jaw were lost early in frog evolution and regained in one genus, *Amphignathodon*. (*b*) Pentadactyly (five digits) is homologous in humans and crocodiles. (*c*) The sting of a wasp is a derived from an ovipositor, but modified in both structure and function. (*d*) Teeth became individualized in the evolution of mammals, but this pattern was reversed in the whales. (*e*) Paedomorphosis is a derived (not ancestral) condition in salamanders.

Evolution in the Fossil Record

4

Although some of the history of evolution can be inferred from living organisms, only paleontologists—scientists who study the fossil record—can find direct evidence of that history. We can observe, measure, and record the differences and similarities between the anatomies of various living primates, for example, and from these observations we may infer what changes occurred in the human lineage, but it has taken research on fossils to document the details of the evolutionary history in which the human, or hominin, lineage diverged from that of the great apes.

The fossil record tells us of the existence of innumerable creatures that have left no living descendants; of great episodes of extinction and diversification; and of the movements of continents and organisms that explain their present distributions. Only from this record can we obtain an absolute time scale for evolutionary events, as well as evidence of the environmental conditions in which they transpired.

The fossil record provides evidence on two particularly important themes: phenotypic transformations in particular lineages and changes in biological diversity over time. The first of these themes is the chief subject of this chapter.

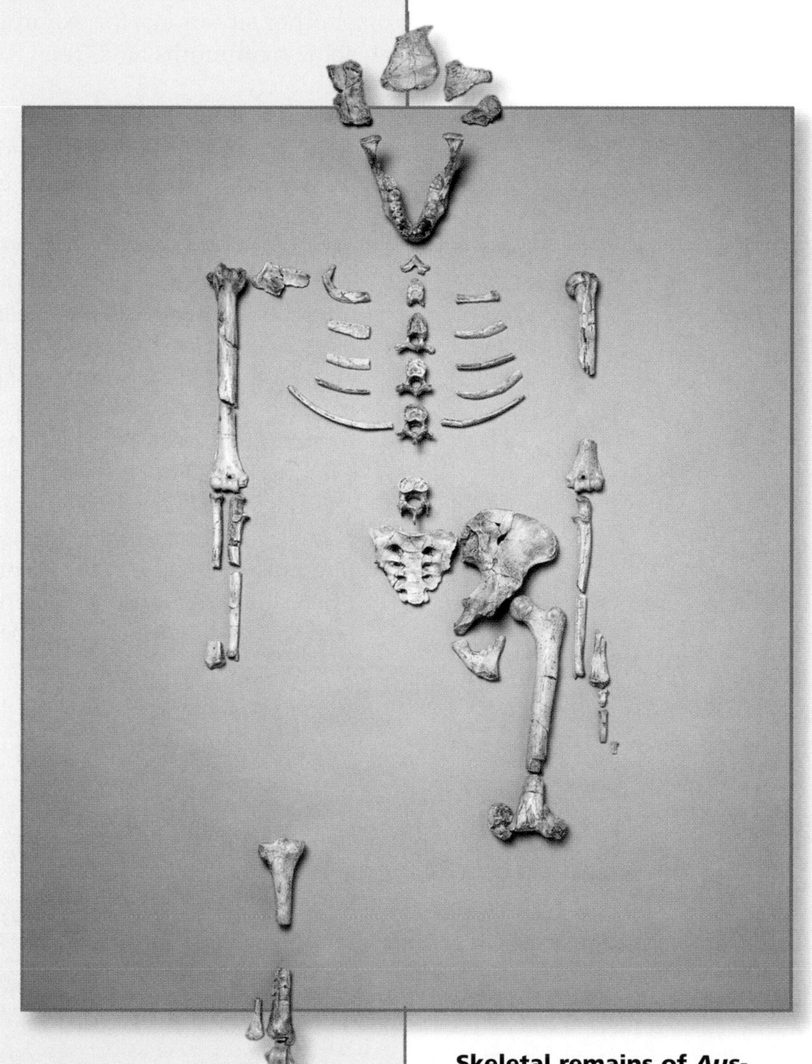

Skeletal remains of *Australopithecus afarensis*. This famous specimen, nicknamed "Lucy," is unusually complete. Studying fossils such as Lucy has enabled scientists to document the evolutionary history of the human lineage. (Photo © The Natural History Museum, London.)

Some Geological Fundamentals

Rock formation

Rocks at the Earth's surface originate as molten material (magma) that is extruded from deep within the Earth. Some of this extrusion occurs via volcanoes, but much rock originates as new crust forms at mid-oceanic ridges (see Figure 4.1). Rock so formed is called IGNEOUS (Latin, "from fire") rock. SEDIMENTARY (Latin, "settle," "sink") rock is formed by the deposition and solidification of sediments, which are usually formed either by the breakdown of older rocks or by precipitation of minerals from water. Under high temperature and pressure, both igneous and sedimentary rocks are altered, forming METAMORPHIC (Greek, "changed form") rocks.

Most fossils are found in sedimentary rocks; they are never found in igneous rocks, and they are usually altered beyond recognition in metamorphic rocks. A few fossils are found in other situations; for example, insects are found in amber (fossilized plant resin), and some mammoths have been found frozen in glacial ice caps.

Plate tectonics

Alfred Wegener first broached the idea of continental drift in 1915, but it was not until the 1960s that both definitive evidence and a theoretical mechanism for continental drift convinced most geologists of its reality. The theory of **plate tectonics** has revolutionized geology.

The lithosphere, the solid outer layer of the Earth bearing both the continents and the crust below the oceans, consists of eight major and a number of minor PLATES that move over the denser, more plastic asthenosphere below. The heat of the Earth's core sets up convection cells within the asthenosphere. At certain regions, such as the mid-oceanic ridge that runs longitudinally down the floor of the Atlantic Ocean, magma from the asthenosphere rises to the surface, cools, and spreads out to form new crust, pushing the existing plates to either side (Figure 4.1). The plates move at velocities of 5–10 centimeters per year. Where two plates come together, the leading edge of one may be forced to plunge under the other (a phenomenon known as SUBDUCTION), rejoining the asthenosphere. The pressure of these collisions is a major cause of mountain building. When a plate moves over a "hot spot" where magma is rising from the asthenosphere, volcanoes may be born, or a continent may be rifted apart. The Great Lakes of eastern Africa lie in such

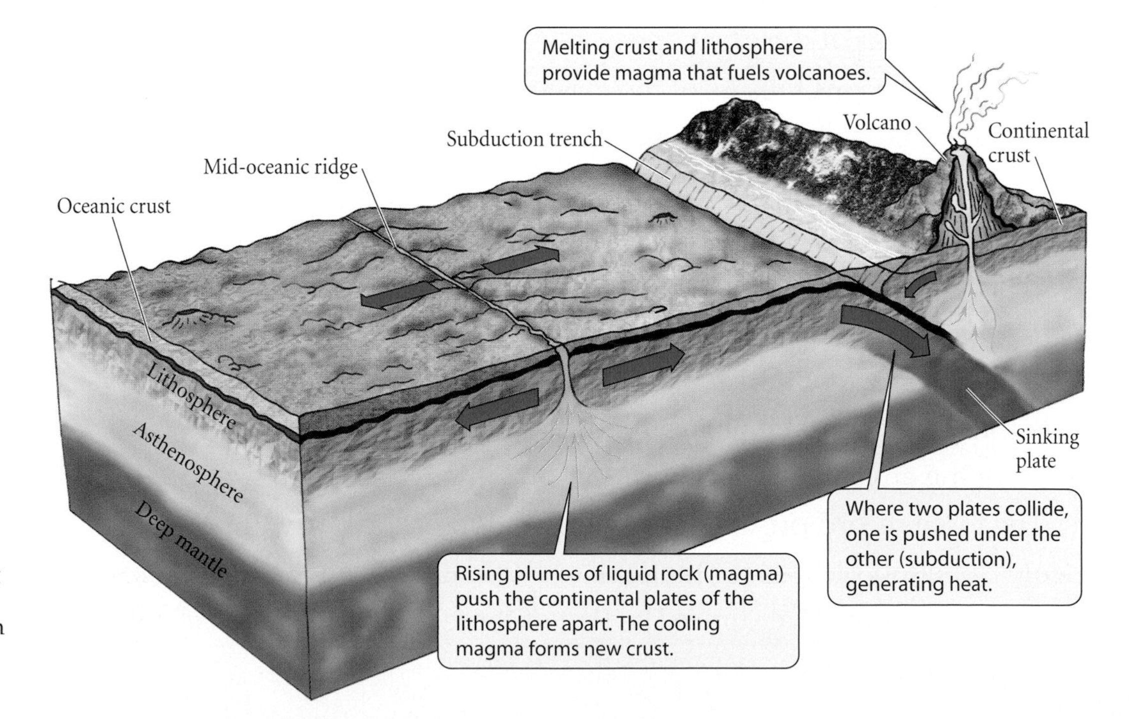

Melting crust and lithosphere provide magma that fuels volcanoes.
Volcano
Continental crust
Subduction trench
Mid-oceanic ridge
Oceanic crust
Lithosphere
Asthenosphere
Deep mantle
Sinking plate
Where two plates collide, one is pushed under the other (subduction), generating heat.
Rising plumes of liquid rock (magma) push the continental plates of the lithosphere apart. The cooling magma forms new crust.

Figure 4.1 Plate tectonic processes. At a mid-oceanic ridge, rising magma creates new lithosphere and pushes the existing plates to either side. When moving lithospheric plates meet, one plunges under the other at a subduction trench, frequently causing earthquakes and mountain-building. Heat generated by subduction melts the lithosphere, causing volcanic activity.

a rift valley; the Hawaiian Islands are a chain of volcanoes that have been formed by the movement of the Pacific plate over a hot spot (see Chapter 6).

Geological time

Astronomers have amassed evidence that the universe originated in a "big bang" about 14 billion years ago (1 billion = 1,000 million) and has been expanding from a central point since then. The Earth and the rest of the solar system are about 4.6 billion years old, but the oldest known rocks on Earth are about 3.8 billion years old. Living things existed by about 3.5 billion years ago; the first evidence of animal life is about 800 million (0.8 billion) years old (see Chapter 5).

Such time spans are hard for us to comprehend. As an analogy, if Earth's age is represented by one year, the first life appears in late March, the first marine animals make their debut in late October, the dinosaurs become extinct and the mammals begin to diversify on December 26, the human and chimpanzee lineages diverge at about 13 hours before midnight on December 31, and Christ is born at about 13 seconds before midnight.

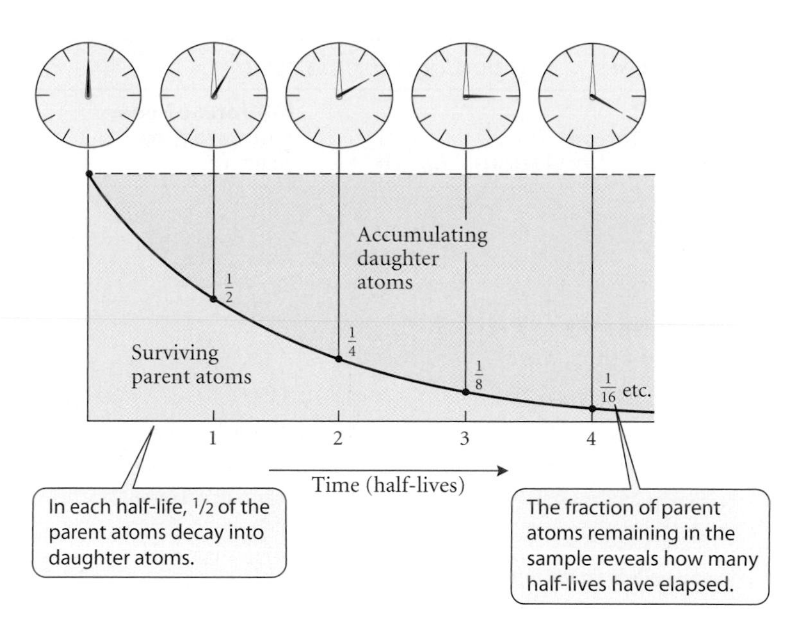
Accumulating daughter atoms
Surviving parent atoms
1/2
1/4
1/8
1/16 etc.
1
2
3
4
Time (half-lives)
In each half-life, 1/2 of the parent atoms decay into daughter atoms.
The fraction of parent atoms remaining in the sample reveals how many half-lives have elapsed.

Figure 4.2 Radiometric dating. The loss of parent atoms and the accumulation of daughter atoms occurs at a constant rate. In each half-life—a unit of time that is specific to each element—half the remaining parent atoms decay into daughter atoms. The relative numbers of the two elements indicate how many half-lives have elapsed. (After Eicher 1976.)

"Absolute" ages of geological events can often be determined by **radiometric dating**, which measures the decay of certain radioactive elements in minerals that form in igneous rock. The decay of radioactive parent atoms into stable daughter atoms (e.g., uranium-235 to lead-207) occurs at a constant rate, such that each element has a specific half-life. The half-life of U-235, for example, is about 0.7 billion years, meaning that in each 0.7 billion year period, half the U-235 atoms present at the beginning of the period will decay into Pb-207. The ratio of parent to daughter atoms in a rock sample thus provides an estimate of the rock's age (Figure 4.2). Only igneous rocks can be dated radiometrically, so a fossil-bearing sedimentary rock must be dated by bracketing it between younger and older igneous formations.

Long before radioactivity was discovered—indeed, before Darwin's time—geologists had established the relative ages (i.e., earlier vs. later) of sedimentary rock formations, based on the principle that younger sediments are deposited on top of older ones. Layers of sediment deposited at different times are called **strata**. Different strata have different characteristics, and they often contain distinctive fossils of species that persisted for a short time and are thus the signatures of the age in which they lived. Using such evidence, geologists can match contemporaneous strata in different localities. In many locations, sediment deposition has not been continuous, and sedimentary rocks have eroded; thus any one area usually has a very intermittent geological record. In general, the older the geological age, the less well it is represented in the fossil record, because erosion and metamorphism have had more opportunity to take their toll.

The geological time scale

Most of the eras and periods of the **geological time scale** (Table 4.1) were named and ordered before Darwin's time, by geologists who did not entertain the idea of evolution. These geological eras and periods were distinguished, and are still most readily recognized in practice, by distinctive fossil taxa. Great changes in faunal composition—the result of mass extinction events—mark many of the boundaries between them. The absolute time of these boundaries is only approximate, and is subject to slight revision as more information accumulates. Phanerozoic time (marked by the first appearance of diverse animals) is divided into three **eras**, each of which is divided into **periods**. We will frequently refer to these divisions, and to the **epochs** into which the Cenozoic periods are divided. Every student of evolution should memorize the sequence of the eras and periods, as well as a few key dates, such as the beginning of the Paleozoic era (and the Cambrian period, 542 million years ago, or 542 Mya), the Mesozoic era (and Triassic period, 251 Mya), the Cenozoic era (and Tertiary period, 65.5 Mya), and the Pleistocene epoch (1.8 Mya).

TABLE 4.1 *The geological time scale*

Era	Period (abbreviation)	Epoch	Millions of years from start to present	Major events
CENOZOIC	Quaternary (Q)	Recent (Holocene)	0.01	Continents in modern positions; repeated glaciations and lowering of sea level; shifts of geographic distributions; extinctions of large mammals and birds; evolution of *Homo erectus* to *Homo sapiens*; rise of agriculture and civilizations
		Pleistocene	1.8	
	Tertiary (T)	Pliocene	5.3	Continents nearing modern positions; increasingly cool, dry climate; radiation of mammals, birds, snakes, angiosperms, pollinating insects, teleost fishes
		Miocene	23.0	
		Oligocene	33.9	
		Eocene	55.8	
		Paleocene	65.5	
MESOZOIC	Cretaceous (K)		145	Most continents separated; continued radiation of dinosaurs; increasing diversity of angiosperms, mammals, birds; mass extinction at end of period, including last ammonoids and dinosaurs
	Jurassic (J)		200	Continents separating; diverse dinosaurs and other "reptiles"; first birds; archaic mammals; "gymnosperms" dominant; evolution of angiosperms; ammonoid radiation; "Mesozoic marine revolution"
	Triassic (Tr)		251	Continents begin to separate; marine diversity increases; "gymnosperms" become dominant; diversification of "reptiles," including first dinosaurs; first mammals
PALEOZOIC	Permian (P)		299	Continents aggregated into Pangaea; glaciations; low sea level; increasing "advanced" fishes; diverse orders of insects; amphibians decline; "reptiles," including mammal-like forms, diversify; major mass extinctions, especially of marine life, at end of period
	Carboniferous (C)		359	Gondwanaland and small northern continents form; extensive forests of early vascular plants, especially lycopsids, sphenopsids, ferns; early orders of winged insects; diverse amphibians; first reptiles
	Devonian (D)		416	Diversification of bony fishes; trilobites diverse; origin of ammonoids, amphibians, insects, ferns, seed plants; mass extinction late in period
	Silurian (S)		444	Diversification of agnathans; origin of jawed fishes (acanthodians, placoderms, Osteichthyes); earliest terrestrial vascular plants, arthropods, insects
	Ordovician (O)		488	Diversification of echinoderms, other invertebrate phyla, agnathan vertebrates; mass extinction at end of period
	Cambrian (€)		542	Marine animals diversify: first appearance of most animal phyla and many classes within relatively short interval; earliest agnathan vertebrates; diverse algae
PROTEROZOIC			2500	Earliest eukaryotes (ca. 1900–1700 Mya); origin of eukaryotic kingdoms; trace fossils of animals (ca. 1000 Mya); multicellular animals from ca. 640 Mya, including possible Cnidaria, Annelida, Arthropoda
ARCHEAN			Lower limit not defined	Origin of life in remote past; first fossil evidence at ca. 3500 Mya; diversification of prokaryotes (bacteria); photosynthesis generates oxygen, replacing earlier oxygen-poor atmosphere; evolution of aerobic respiration

Source: Dates from the International Stratigraphic Chart (International Commission on Stratigraphy, 2004).

The Fossil Record

Some short parts of the fossil record, in certain localities, provide detailed evolutionary histories, and some groups of organisms, such as abundant planktonic protists with hard shells, have left an exceptionally good record. On the whole, however, the fossil record is extremely incomplete (Jablonski et al. 1986). Consequently, the origins of many species and higher taxa have not been well documented. The odds of finding such origins, however, are so low that we are lucky to have informative records at all.

There are several reasons why such records are rare. First, there are many kinds of organisms that rarely become fossilized because they are delicate, or lack hard parts, or occupy environments such as humid forests where decay is rapid. Second, because sediments generally form in any given locality very episodically, they typically contain only a small fraction of the species that inhabited the region over time. Third, if fossils are to be found, the fossil-bearing sediments must first become solidified into rock; the rock must persist for millions of years without being eroded, metamorphosed, or subducted; and the rock must then be exposed and accessible to paleontologists. Finally, the evolutionary changes of interest may not have occurred at the few localities that have strata from the right time; a species that evolved new characteristics elsewhere may appear in a local record fully formed, after having migrated into the area.

The approximately 250,000 described fossil species represent much less than 1 percent of the species that lived in the past. We *know* that the fossil record is extremely incomplete because (1) many time periods are represented by few sedimentary formations worldwide; (2) many lineages are represented only at very widely separated time intervals, even though they must have been present in the meantime; (3) many extinct species of large, conspicuous organisms are known from only one or a few specimens; and (4) new fossil taxa have been found at a steady rate, indicating that many forms have yet to be discovered.

Two of the most basic questions we may ask of the fossil record are, first, whether it provides evidence for evolutionary change—for descent with modification—and second, whether or not evolution is gradual, as Darwin proposed. The answer to the first question is an unequivocal "yes," as the following examples show. However, although the fossil record provides many examples of gradual change, we cannot rule out the possibility that some characteristics have evolved by large, discontinuous changes.

Evolutionary changes within species

A detailed, continuous sedimentary record can be expected only over short geological time spans. Some such records have preserved histories of change within single evolving lineages—single species. In many such cases, characters tend to change gradually (Levinton 2001). For example, each of eight lineages of Ordovician trilobites displays a gradual increase in the number of ribs on the rear dorsal section of the exoskeleton (Figure 4.3; Sheldon 1987).

In an unusually detailed study, Michael Bell and his colleagues (1985) studied Miocene fossils of a stickleback fish, *Gasterosteus doryssus*, in strata that were laid down annually for 110,000 years. They took samples from layers that were on average about 5000 years apart. Three of the characters they studied changed more or less independently and gradually (Figure 4.4A). Two of these features also changed abruptly at one point in time. This shift was caused by the extinction of the local population and immigration of another population—apparently a different species—in which these features were different (Figure 4.4B). An interval of 5000 years between samples is a long time by normal human standards, but is unusually fine-scaled by the standards of geological time.

Origins of higher taxa

In this section, we present several examples of macroevolutionary change—the origin of higher taxa over long periods of geological time. These examples illustrate several important principles, such as homology, change of structure with function, and mosaic evolution, that were introduced in Chapter 2. Moreover, they give the lie to claims by antievo-

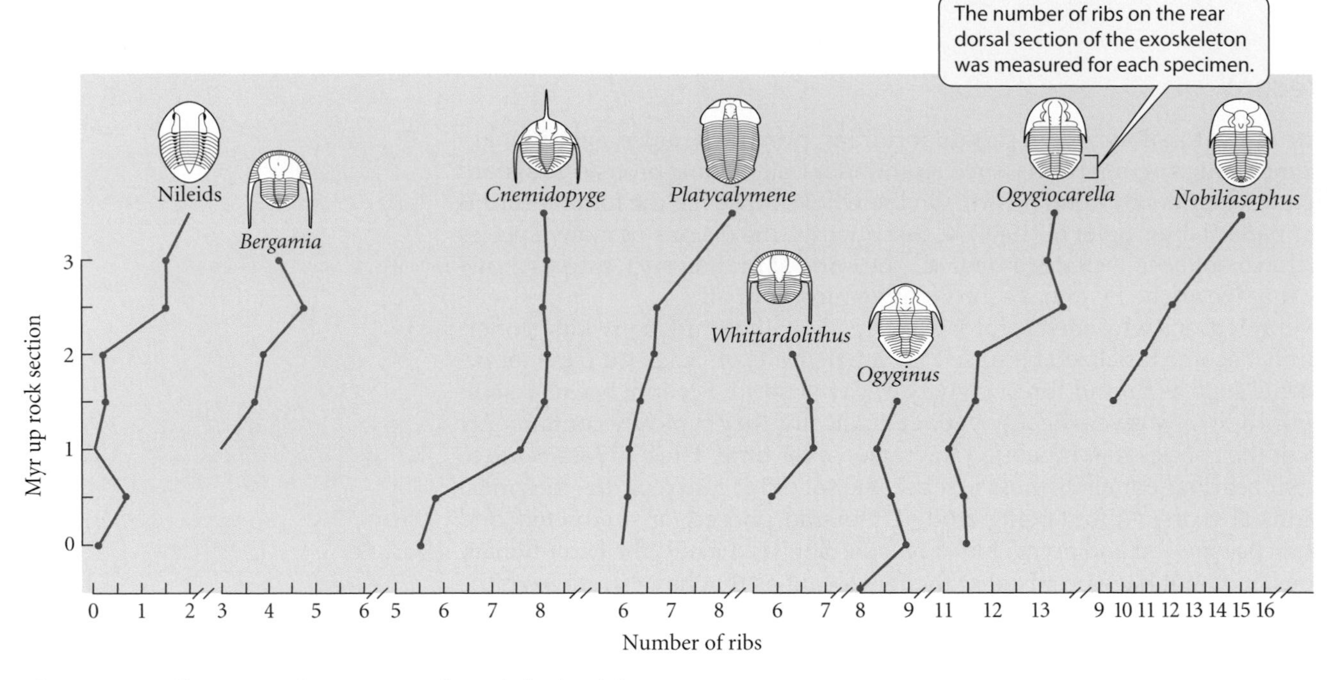

Figure 4.3 Changes in the mean number of ribs in eight lineages of trilobites. Irregular, but mostly gradual, changes are seen in most of the lineages. (After Sheldon 1987.)

lutionists that the fossil record fails to document macroevolution. Anyone educated in biology should be able to counter such charges with examples such as these.

In some cases, a series of taxa through time documents changes in many characters. In other cases, we have only one or a few key intermediate forms. In such instances, we can recognize the intermediate fossil as a transitional stage in the origin of a taxon because even before it is found, some of its critical features are predictable, based on phylogenetic understanding of ancestral and derived character states. For instance, E. O. Wilson, F. M. Carpenter, and W. L. Brown hypothesized what features the ancestor of ants should have had, based on comparisons between primitive species of living ants and related families of wasps. Several years after they published their hypothesis, Cretaceous ants preserved in amber—older than any

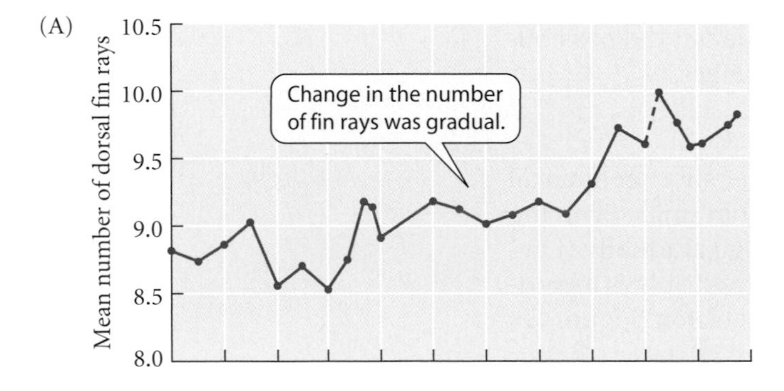

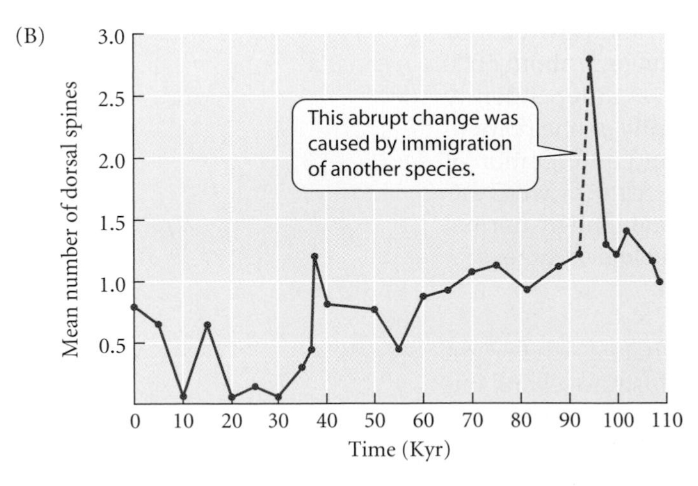

Figure 4.4 Changes in the mean values of characters in fossil sticklebacks, *Gasterosteus doryssus* (photo below). In a study of five characters, three characters, including the number of dorsal fin rays (A), showed independent, gradual change; dorsal spine number (B) was one of two characters displaying abrupt changes in value. Values are given for intervals of 5000 years, over a span of 110,000 years. (After Bell et al. 1985; photo courtesy of M. Bell.)

Gasterosteus doryssus

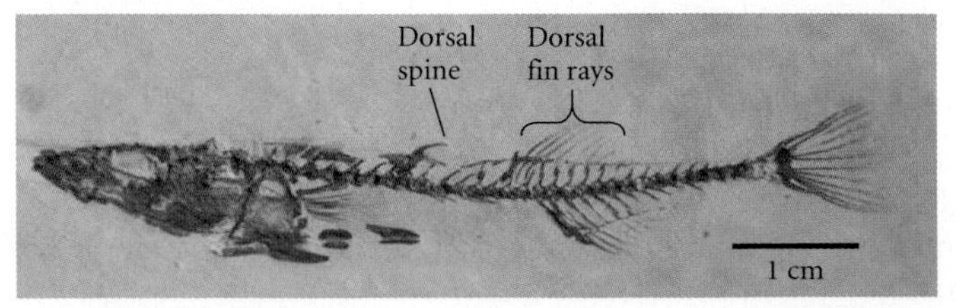

(A) *Sphecomyrma freyi*

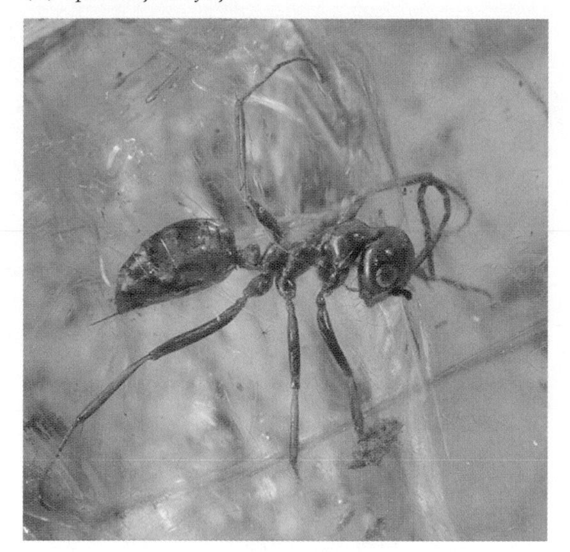

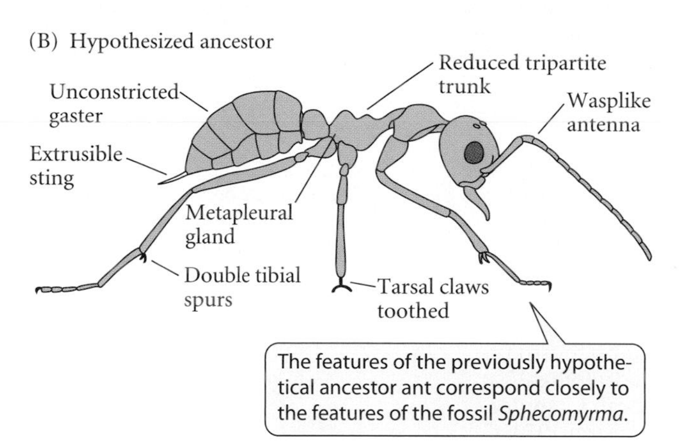

Figure 4.5 A fossil can help confirm an evolutionary hypothesis. (A) Fossilized in amber, this *Sphecomyrma freyi* is a mid-Cretaceous ant that bridges the gap between modern ants and the wasps from which ants are thought to have arisen. (B) Features of the previously hypothesized ancestor of ants that matched those of *Sphecomyrma*. (A courtesy of E. O. Wilson; B after Wilson et al. 1967.)

previously known fossil ants—were found, and they corresponded to the predicted morphology in almost all their features (Figure 4.5).

THE ORIGIN OF AMPHIBIA. The Sarcopterygii, or lobe-finned fishes, appeared in the early Devonian, about 408 Mya. They included lungfishes and coelacanths, a few of which are still alive, as well as extinct groups such as rhipidistians (Figure 4.6A–C). Rhipidistians had a tail fin as well as paired fins that were fleshy, with a skeleton of several large bones (Figure 4.6A). The braincase consisted of two movable units and was surrounded by dermal bones with lateral line canals (Figure 4.6C). The teeth, borne on several skull bones and on several of the bones that made up the lower jaw, had a distinctive internal structure. Rhipidistians had both gills and lungs.

The first definitive amphibians, from the very late Devonian of Greenland, are the ichthyostegids (e.g., *Ichthyostega;* Figure 4.6D–F). In almost every respect, they had the same characters—tail fin, braincase, dermal skull bones and lateral line canals, structure and distribution of teeth—as the rhipidistians. The main difference is that they had larger pectoral and pelvic girdles (compare Figures 4.6A and 4.6D) and fully developed tetrapod limbs. The proximal limb bones are directly homologous, and similar in shape, to those of rhipidistians (compare Figures 4.6B and 4.6E), but they had definitive digits. Ichthyostegids were aquatic, and, unlike almost all later tetrapod vertebrates, which have five or fewer digits, they had a variable, greater number (Clack 2000a). A more terrestrial form that links ichthyostegids and later amphibians, with five-toed feet better adapted for walking on land, has recently been described from deposits that are about 13 million years (Myr) younger (Clack 2002b). The ichthyostegids provide an unusually good example of the origin of a major higher taxon (the Amphibia, and for that matter, the Tetrapoda, or four-legged vertebrates). The transition from rhipidistians to ichthyostegids is marked by mosaic evolution (the limbs evolved faster than the skull or teeth) and by gradual change of individual features (the limbs are intermediate between those of lobe-finned fishes and later amphibians).

THE ORIGIN OF BIRDS. Almost everyone who has studied the origin of birds now agrees that birds are dinosaurs. Not long ago, birds (placed until recently in the class Aves) were defined by their feathers. But because of the many extraordinary fossils that have recently been discovered in China, the distinction between birds and dinosaurs has become arbitrary (Chiappe and Dyke 2002; Xu et al. 2003).

Figure 4.6 (A–C) *Eusthenopteron*, a member of the group of lobe-finned fishes (Rhipidistia) from which tetrapods arose. (A) Skeleton and reconstructed outline. (B) Skeleton of pelvic fin and pelvic girdle. (C) Dorsal view of skull. The dotted lines are canals associated with the lateral line system. (D–F) *Ichthyostega*, a Devonian labyrinthodont amphibian. Although the amphibian's legs and pelvic girdle (E) are well developed, many features, especially those of the skull (F), are very similar to those of rhipidistian fishes. The amphibian's snout is longer, and most of the rhipidistian opercular bones of the (e.g., bone "o" in C) are absent; otherwise the skulls are very similar. Bones labeled for comparison include: j, jugal; l, lacrimal; o, operculum; p, parietal; pm, premaxilla; po, postorbital; pp, postparietal; prf, prefrontal; qj, quadratojugal; sq, squamosal; st, supratemporal; t, tabular. (A, B after Andrews and Westoll 1970; C after Moy-Thomas and Miles 1971; D after Jarvik 1955; E after Coates and Clack 1990; F after Jarvik 1980.)

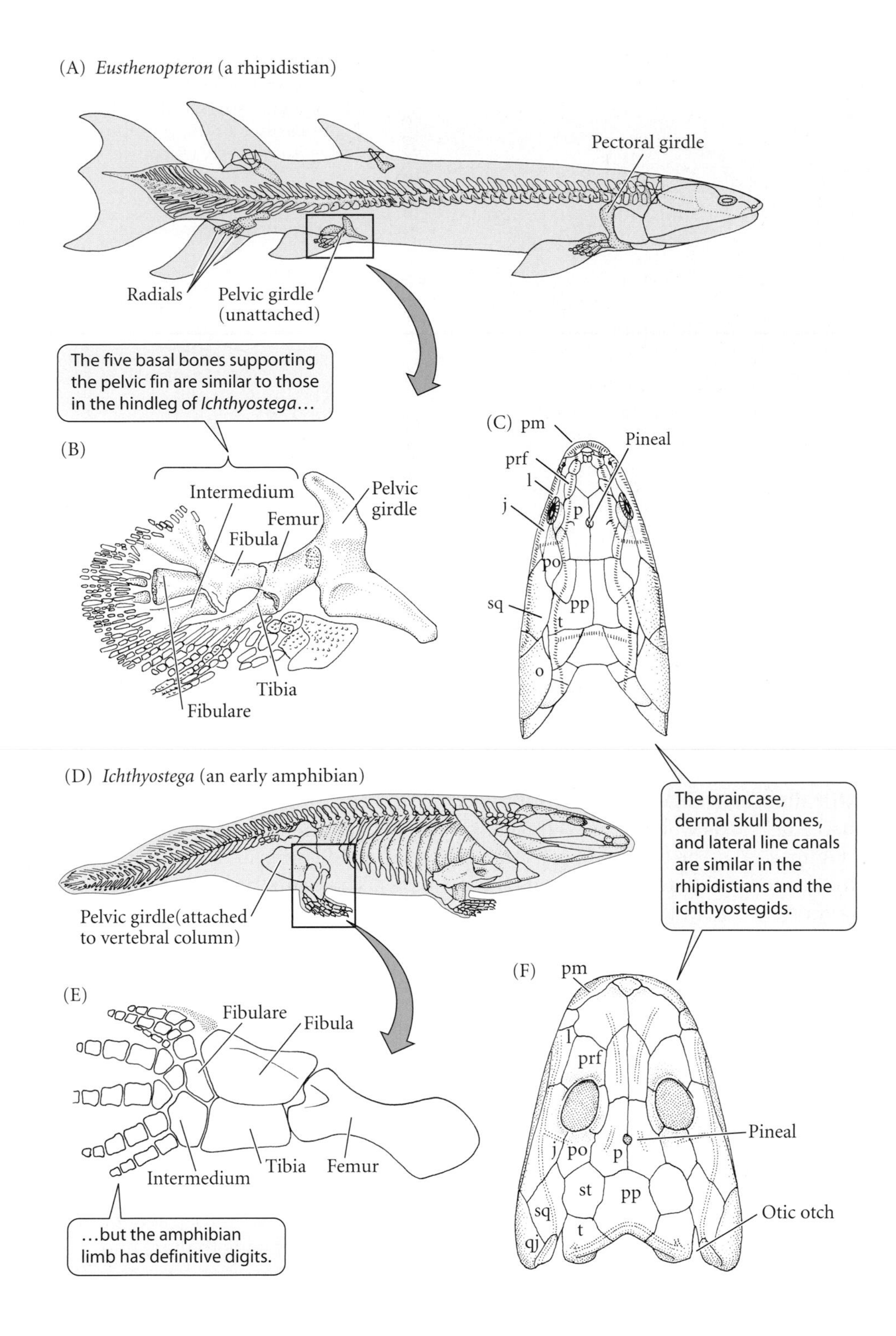

The first intermediate form to be discovered, one of the most famous non-missing links of all time, was *Archaeopteryx*, found in Upper Jurassic strata in Germany (Figure 4.7B). *Archaeopteryx* has only a few of the many modifications of the skeleton of modern birds (Figure 4.8A,B) that accommodate the flying habit. Instead, it closely resembles a small theropod dinosaur (Figure 4.8C) with feathers that enabled it to fly. However, the fossils recently found in China include theropod dinosaurs with a body coat of small feathers (*Sinosauropteryx*), others with long, broad (flight?) feathers on the hand and tail (*Caudipteryx*), and even an extraordinary four-winged dinosaur (*Microraptor gui*) that had flight feathers on all four limbs (Figure 4.7A). The distinctive synapomorphy of *Ar-*

chaeopteryx and other Aves (7 in Figure 4.9) is no longer feathers, but merely their opposable hind toe! Some of the features of modern birds, such as hollow limb bones, evolved in theropods (1 in Figure 4.9) long before *Archaeopteryx*, and other characters, such as fusion of the tail vertebrae (8, in *Confuciusornis*), evolved later. Later still, one subgroup of Aves evolved such features as the keeled breastbone, loss of teeth, and loss of claws on the hand (9) that typify living birds.

We are not sure of the adaptive function of hollow bones in theropods or the body feathers of *Sinosauropteryx*—though perhaps the feathers provided insulation. Certainly, however, these features, in modified form, later became useful in flight. (Hollow bones lighten the load, and some feathers became modified as airfoils.) These features provide an example of evolutionary change of function and of "preadaptation," possession of a fea-

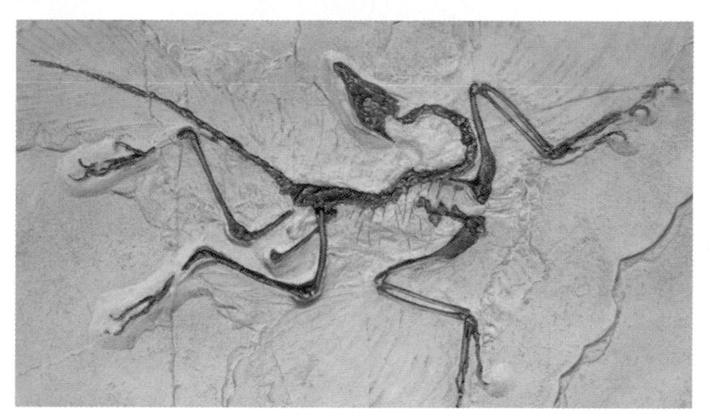

Figure 4.7 Feathered dinosaurs to birds. (A) *Microraptor gui*, a feathered dinosaur from the early Cretaceous (about 140 Mya). This specimen is from a Chinese fossil fauna that in recent years has added greatly to our knowledge of the evolution of birds from dinosaur ancestors. Note the long feathers at the back of both the front and hind limbs. (B) A well-preserved specimen of *Archaeopteryx lithographica*, an early example of a bird, showing the wing feathers and the long tail with feathers on both sides. This fossil is from Germany. (A from X. Xu et al., 2003; B photo © Tom Stack/Painet, Inc.)

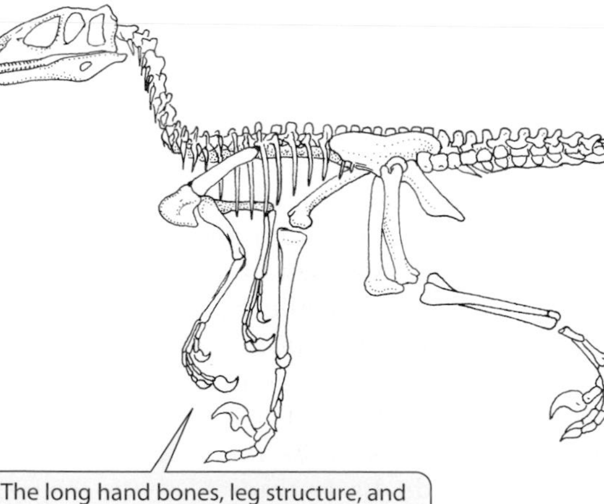

(C) Theropod dinosaur
The long hand bones, leg structure, and shape of the claws are some of the many similarities to *Archaeopteryx*.

Figure 4.8 Skeletal features of (A) *Archaeopteryx*, (B) a modern bird (pigeon), and (C) a theropod dinosaur, *Deinonychus*. Compared with *Archaeopteryx*, the modern bird has (1) an expanded braincase with fused bones, (2) fusion and reduction of the three digits of the hand, (3) fusion of the pelvic bones and several vertebrae into a single structure, (4) fewer tail vertebrae, several of which are fused, (5) a greatly enlarged, keeled sternum (breastbone), and (6) horizontal processes that strengthen the rib cage. Theropod dinosaurs share many features with *Archaeopteryx*, the leg structure being perhaps most evident in this figure. (A, B after Colbert 1980; C after Ostrom 1976.)

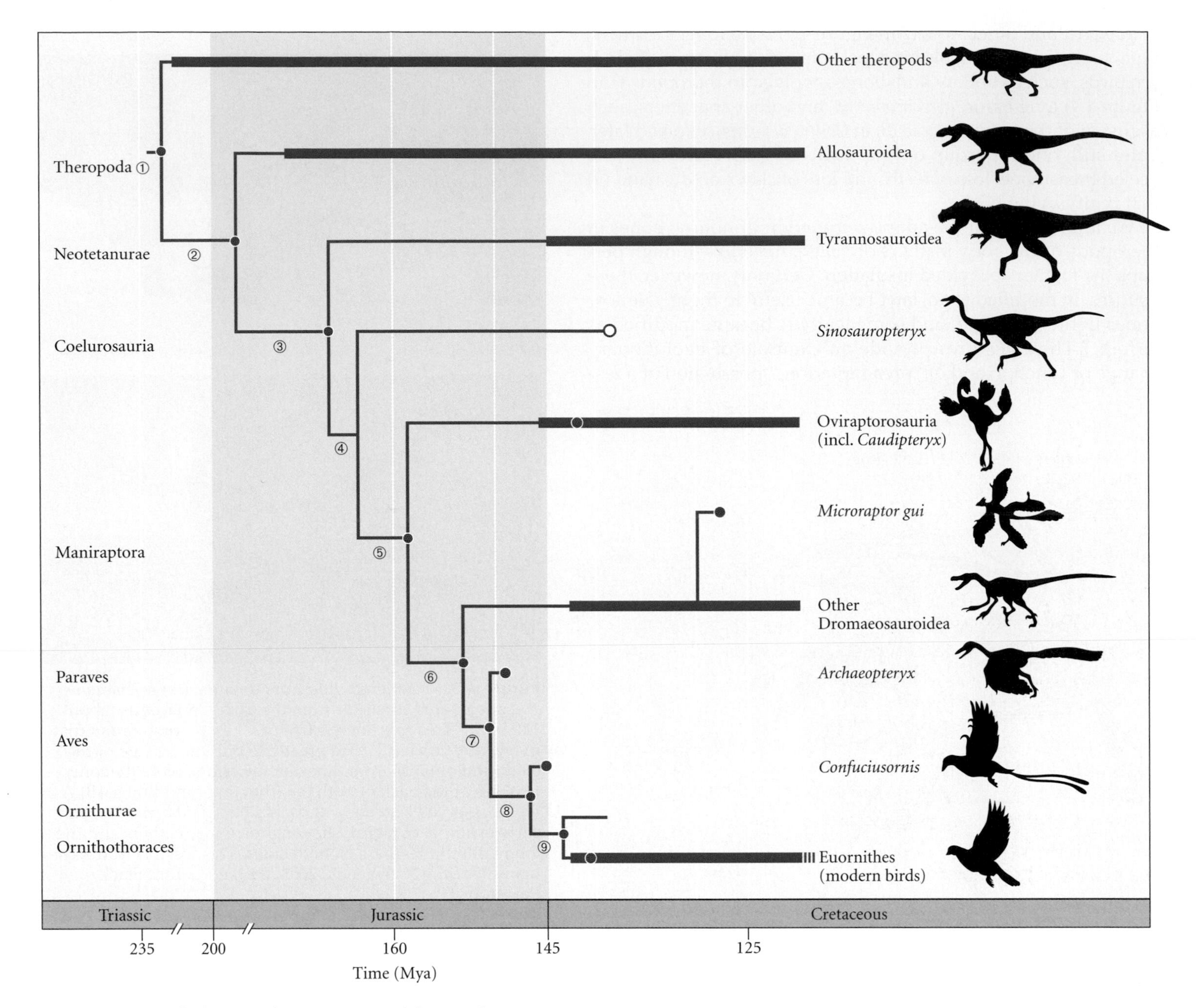

Figure 4.9 A phylogeny of some groups of theropod dinosaurs, showing the hypothesized relationships of *Archaeopteryx* and later birds and the origin of some derived characters (indicated by numbers and names of successively derived clades). (1) Hollow long bones, three-fingered hand, elevated first toe (Theropoda); (2) crescent-shaped wrist bone (Neotetanurae); (3) expanded breastbone (Coelurosauria); (4) small integumentary feathers [open circle] in the coelurosaur *Sinosauropteryx*; (5) vaned feathers in some Maniraptora; (6) sickle-shaped claw on foot in Paraves, which include the four-winged dromaeosaur *M. gui*; (7) opposable hind (first) toe in Aves, including *Archaeopteryx*; (8) short tail with fused vertebrae in more "advanced" birds, Ornithurae; (9) breast keel and eventual loss of teeth and finger claws in "true" birds, Euornithes. Heavier lines indicate the existence of an abundant fossil record. (After Sereno 1999; Chiappe and Dyke 2002; Prum 2003.)

ture that fortuitously plays a different, useful role at a later time (see Chapter 11).

THE ORIGIN OF MAMMALS. The origin of mammals from the earliest amniotes (see Figure 2.4) is one of the most fully documented examples of the evolution of a major taxon (Kemp 1982; Sidor and Hopson 1998). Although some features of living mammals, such as hair and mammary glands, are not fossilized, mammals have diagnostic skeletal features. The lower jaw consists of several bones in reptiles, but only a single bone (the dentary) in mammals. The primary (and in all except the earliest mammals, the exclusive) jaw articulation is between the dentary and the squamosal bones, rather than between the articular and the quadrate bones, as in other tetrapods. Early amniotes have a single bone (the stapes, or stirrup) that transmits sound, whereas mammals have three such bones (hammer, anvil, stirrup) in the middle ear. Mammals' teeth are differentiated into incisors, canines, and multicusped (multi-pointed) premolars and molars, whereas most other tetrapods have uniform, single-cusped teeth. Other features that distinguish most mammals from reptiles include an enlarged braincase, a large space (temporal fenestra) behind the eye socket, and a secondary palate that separates the breathing passage from the mouth cavity.

Soon after the first amniotes originated in the Carboniferous, they gave rise to the Synapsida, a group characterized by an opening (temporal fenestra) behind the eye socket, which probably provided space for enlarged jaw muscles to expand into when contracted (Figure 4.10A). The temporal fenestra became progressively enlarged in later synapsids (Figure 4.10B–D).

Permian synapsids, in the order Therapsida (Figure 4.10B) had large canine teeth, and the center of the palate was recessed, suggesting that the breathing passage was partially separated from the mouth cavity. The hind legs were held rather vertically, more like a mammal than a reptile.

Cynodont therapsids, which lived from the late Permian to the late Triassic, represent several steps in the approach toward mammals. The rear of the skull was compressed, giving it a doglike appearance (Figure 4.10C,D), the dentary was enlarged relative to the other bones of the lower jaw, the cheek teeth had a row of several cusps, and a bony shelf formed a secondary palate that was incomplete in some cynodonts and complete in others. The quadrate was smaller and looser than in previous forms and occupied a socket in the squamosal.

In the advanced cynodonts of the middle and late Triassic (Figure 4.10E), the cheek teeth had not only a linear row of cusps, but also a cusp on the inner side of the tooth. This begins a history of complex cusp patterns in

(A) Synapsid (*Haptodus*)

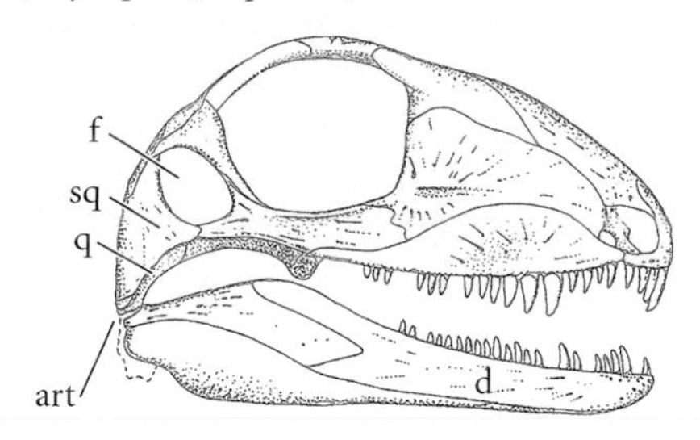
f
sq
q
art
d
Synapsids had large jaw muscles, multiple bones in the lower jaw, and single-cusped (single-point) teeth.

(B) Therapsid (*Biarmosuchus*)

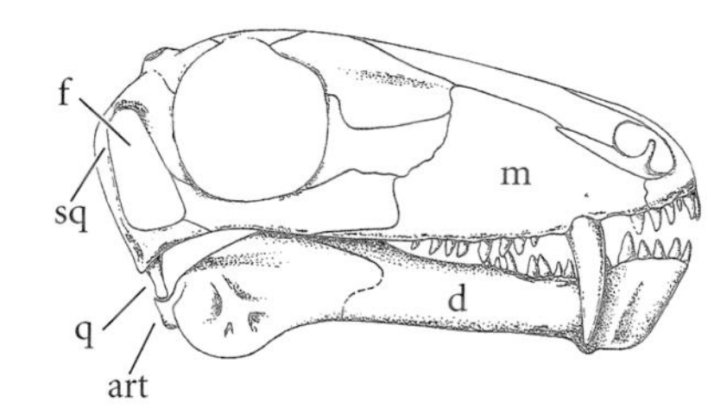
f
sq
m
d
q
art
Synapsids of the order Therapsida had large canine teeth, large maxilla bones, and long faces.

(C) Early cynodont (*Procynosuchus*)

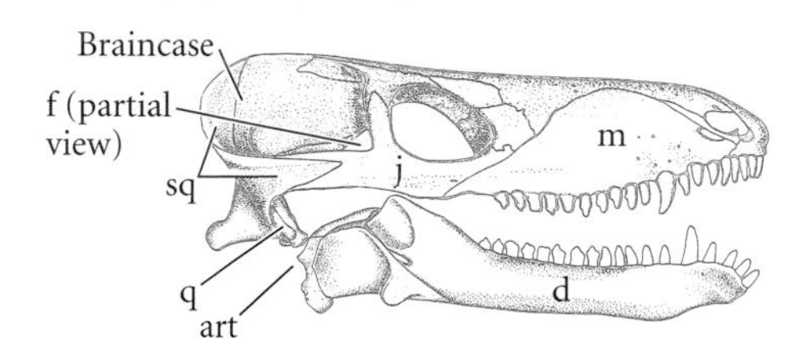
Braincase
f (partial view)
sq
j
m
q
art
d
In cynodont therapsids, the side of the braincase was vertical and the large temporal fenestra was lateral to it.

(D) Cynodont (*Thrinaxodon*)

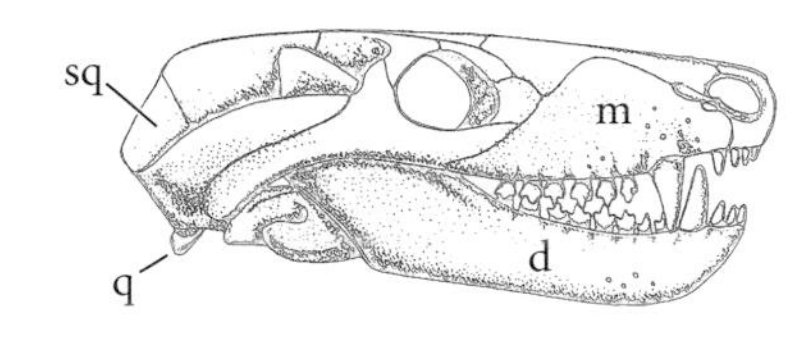
sq
m
d
q
The cynodont dentary bone became enlarged, and the cheek teeth had multiple cusps.

(E) Advanced cynodont (*Probainognathus*)

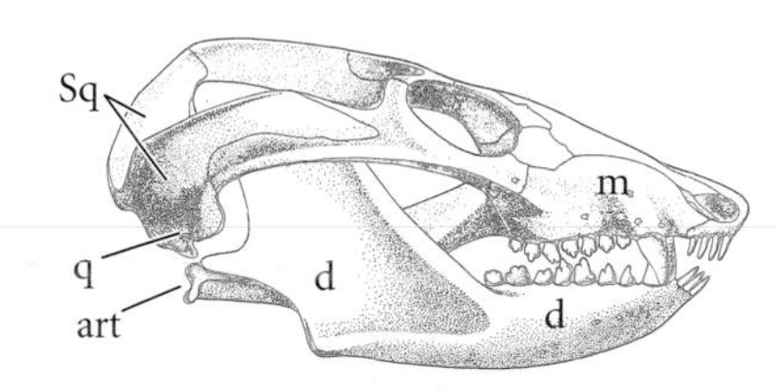
Sq
m
q
art
d
d
In advanced cynodonts, complex cusp patterns enhanced chewing and the dentary (the major jaw bone) formed an articulation with the squamosal.

(F) *Morganucodon*

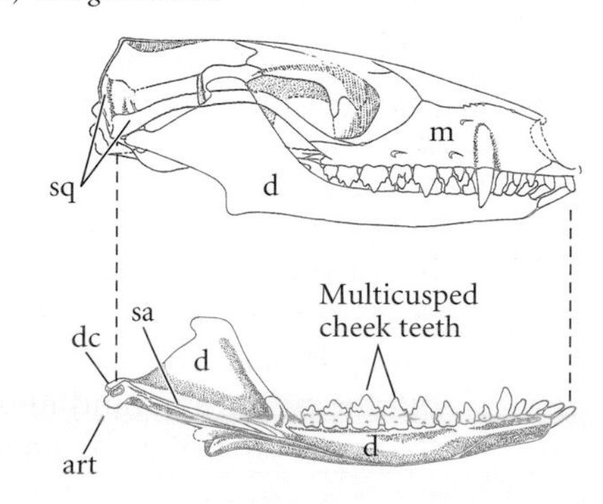
m
sq
d
sa
dc
d
Multicusped cheek teeth
d
art
Morganucodon was almost a mammal, with typical mammalian teeth and a lower jaw composed almost entirely of the dentary. The jaw had a double articulation with the skull.

Figure 4.10 Skulls (medial view) of some stages in evolution from early synapsids to early mammals. (A) A pelycosaur, *Haptodus*. Note temporal fenestra (f), multiple bones in lower jaw, single-cusped teeth, and articular/quadrate (art/q) jaw joint. (B) An early therapsid, *Biarmosuchus*. Note enlarged temporal fenestra. (C) An early cynodont, *Procynosuchus*. The temporal fenestra formed a space between the vertically oriented side of the braincase and a lateral arch formed by the jugal (j) and squamosal (sq). Note the enlarged dentary (d). (D) A later cynodont, *Thrinaxodon*. Note multiple cusps on rear teeth, large upper and lower canine teeth, and greatly enlarged dentary with a vertical extension to which powerful jaw muscles were attached. (E) An advanced cynodont, *Probainognathus*. The cheek teeth had multiple cusps, and the lower jaw articulated with the squamosal by the articular (art), surangular (sa), and posterior process of the dentary (d). (F) *Morganucodon* was almost a mammal. Note multicusped cheek teeth (including inner cusps) and double articulation of lower jaw, including articulation of a dentary condyle (dc) with the squamosal (sq). (After Futuyma 1995; based on Carroll 1988 and various sources.)

the cheek teeth of mammals, which are modified in different lineages for chewing different kinds of food. The lower jaw had not only the old articular/quadrate articulation with the skull, but also an articulation between the dentary and the squamosal. This marks a critical transition between the ancestral condition and the mammalian state.

In *Morganucodon* of the late Triassic and very early Jurassic (Figure 4.10F), the teeth are typical of mammals. *Morganucodon* has both a weak articular/quadrate hinge and a fully developed mammalian articulation between the dentary and the squamosal. The articular and quadrate bones are sunk into the ear region, and together with the stapes, they closely approach the condition in modern mammals, in which these bones transmit sound to the inner ear.

Hadrocodium, a recently described fossil from the early Jurassic, carries the trend from *Morganucodon* to the very brink of mammalhood (Luo et al. 2001). This tiny animal is very similar to *Morganucodon*, but the articular and quadrate bones are fully separated from the jaw joint and fully lodged in the middle ear, and the lower jaw consists entirely of the dentary. Phylogenetically, it is the sister group of the true mammals, first found about 45 Myr later in the late Jurassic.

The fossil record thus shows that most mammalian characters (e.g., posture, tooth differentiation, skull changes associated with jaw musculature, secondary palate, reduction of the elements that became middle ear bones) evolved gradually. Evolution was mosaic, with different characters "advancing" at different rates. No new bones evolved: all the bones of mammals are modified from those of the stem group of "reptiles" (and in turn, from those of amphibians and even lobe-finned fishes). Some major changes in the form of structures are associated with changes in their function. The most striking example is the articular and quadrate bones, which serve for jaw articulation in all other tetrapods, but became the sound-transmitting middle ear bones of mammals. Because the evolution of mammals from synapsids, over the course of more than 130 million years, has been gradual, there is no cutoff point for recognizing mammals: the definition of "Mammalia" is arbitrary.

THE ORIGIN OF CETACEA. Whales and dolphins, traditionally distinguished as the order Cetacea, evolved from terrestrial ancestors. Among living mammals, their closest relatives, based on molecular phylogenetic analyses, appear to be hippopotamuses (Gatesy et al. 1999). Thus cetaceans fall within the order Artiodactyla—even-toed hoofed mammals—along with camels, pigs, and ruminants such as cattle and antelopes.

Compared with basal mammals, living cetaceans are very greatly modified, owing to their adaptations for aquatic life. All share a uniquely shaped tympanic bone that encloses the ear; a nasal opening far back on top of the skull; stiff elbow, wrist, and finger joints, all enclosed in a paddlelike flipper; a rudimentary pelvis (sometimes associated with a hindlimb rudiment) that is disconnected from the vertebral column; and a lack of the fused, differentiated sacral (lower back) vertebrae that land mammals have. Toothed whales have a large cavity (foramen) in the lower jaw (mandible) that contains a sound-transmitting pad of fat.

Philip Gingerich, J. G. M. Thewissen, and their colleagues have recently discovered many Eocene fossils, mostly in Pakistan, that have shed light on the evolutionary history of cetaceans from about 50 to 35 Mya (Figure 4.11) (Gingerich et al. 2001; Thewissen and Bajpai 2001; Thewissen and Williams 2002). The oldest of these, *Pakicetus* (53–48 Mya), was a terrestrial (or perhaps semiaquatic) animal with the distinctive cetacean tympanic bone. The slightly younger *Ambulocetus* (48–47 Mya) was adapted for life in shallow coastal waters. It had short hind legs but large feet, with separate digits that bore small hooves. The mandibular foramen was larger than in pakicetids, starting a steady increase in size. *Ambulocetus* was a predator that had long jaws and teeth with somewhat reduced cusps. In the protocetids (e.g., *Rodhocetus*) (49–39 Mya), the fusion between sacral vertebrae was reduced, tooth form was simpler, and the nasal opening was farther back from the tip of the snout. Protocetids had an artiodactyl ankle bone and small hooves at the tips of the toes, but the pelvis and hindlimbs of some of the aquatic protocetids were too small and weak to bear their weight. Complete adaptation to life in water is shown by the dorudontine

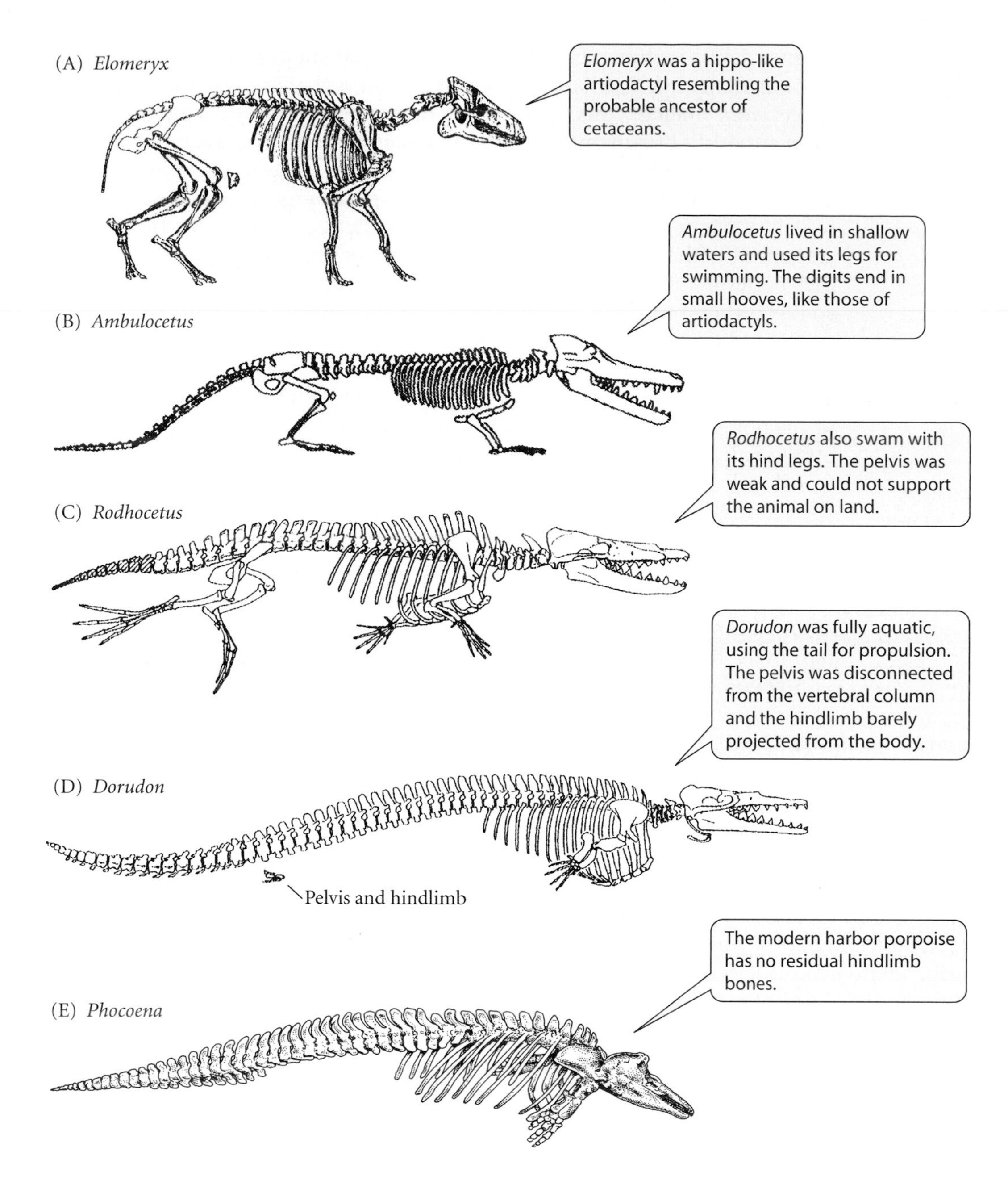

(A) Elomeryx
Elomeryx was a hippo-like artiodactyl resembling the probable ancestor of cetaceans.
Ambulocetus lived in shallow waters and used its legs for swimming. The digits end in small hooves, like those of artiodactyls.
(B) Ambulocetus
Rodhocetus also swam with its hind legs. The pelvis was weak and could not support the animal on land.
(C) Rodhocetus
Dorudon was fully aquatic, using the tail for propulsion. The pelvis was disconnected from the vertebral column and the hindlimb barely projected from the body.
(D) Dorudon
Pelvis and hindlimb
The modern harbor porpoise has no residual hindlimb bones.
(E) Phocoena

Figure 4.11 Reconstruction of stages in the evolution of cetaceans from terrestrial artiodactyl ancestors. (A) The hippopotamus-like Oligocene artiodactyl *Elomeryx* represents the probable ancestral group from which cetaceans evolved. (B) The amphibious *Ambulocetus*. (C) The middle Eocene protocetid *Rodhocetus* had the distinctive ankle bones of artiodactyls but had numerous cetacean characters. (D) *Dorudon*, of the middle to late Eocene, had most of the features of modern cetaceans, although its nonfunctional pelvis and hindlimb were larger. (E) A modern toothed whale, the harbor porpoise, *Phocoena phocoena*. The nostrils, forming a blowhole, are far back on the top of the head, accounting for the peculiar shape of the skull. (A–D after Gingerich 2003 and de Muizon 2001; E art by Nancy Haver.)

basilosaurids, of about 35 Mya, in which the teeth were even simpler, the nostrils were farther back, and the front limbs were flipperlike, with an almost inflexible wrist and elbow. The pelvis and hindlimbs were completely nonfunctional. The small pelvis was disconnected from the vertebral column, and the hind feet and legs barely protruded from the surface of the body. Dorudontines probably had a horizontal tail fin (fluke), and were, indeed, a small step away from modern cetaceans.

The Hominin Fossil Record

DNA sequence differences imply that the chimpanzee and human lineages diverged 5 to 6 million years ago (see Chapter 2). No chimpanzee fossils have been found, but hominin fossils are known from as far back as the late Miocene, between 6 and 7 Mya. (The term "hominin" has been applied to the hominids that are the sister group of the chimpanzees.) The hominin fossil record provides unequivocal evidence of general, more or less unidirectional trends in many characters, such as cranial capacity, a measure of brain size

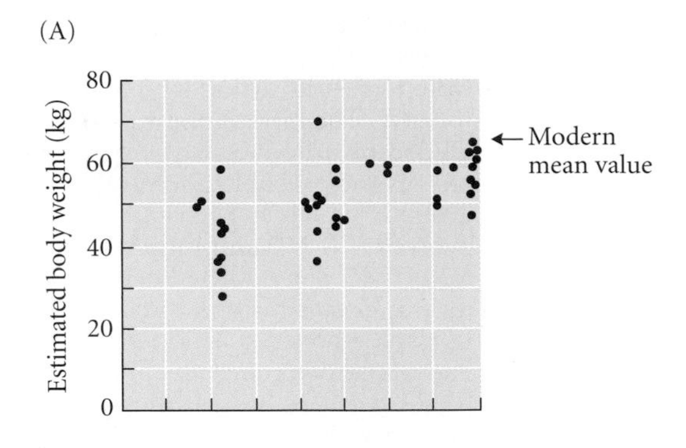

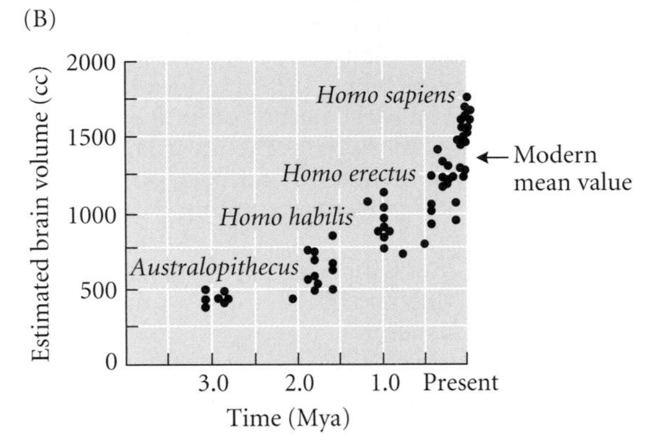

Figure 4.12 Estimated body weights (A) and brain volumes (B) of fossil hominins. There has been a steady, fairly gradual increase in brain volume, even though body size has not increased very much in the last 2 million years. The arrows indicate modern averages. (After Jones et al. 1992.)

(Figure 4.12). It shows beyond any doubt that modern humans evolved through many intermediate steps from ancestors that in most anatomical respects were apelike. Thus, much of the broad sweep of human evolution has been superbly documented. There is disagreement, however, about how many distinct hominin species and genera should be recognized (e.g., Wood and Collard 1999), in part because fossil specimens are too few, and too widely separated in time and space, to characterize variation within species compared with that between species. Moreover, the differences among various hominins are quantitative (differences in degree) and often rather slight, so it is difficult to determine whether specific earlier populations were ancestral to later ones, or were collateral relatives of the actual ancestral lineage. Hence, even if the overall pattern of evolution is clear, the specific phylogenetic relationships among hominin taxa may not be. (Useful summaries of the hominin fossil record can be found in Ciochon and Fleagle 1993 and Strait et al. 1997; see also Begun 2004.)

All early hominins have been found in Africa. The earliest hominin fossil, a skull from Chad in western Africa, was described very recently (Brunet et al. 2002). Named *Sahelanthropus tchadensis*, it is believed to date from the late Miocene, between 6 and 7 Mya. In many respects, such as its small brain, it is the most primitive known hominin, but it also has derived features, such as small canine teeth and a rather flat face, that resemble those of much later hominins. It was almost certainly bipedal (walked on two legs).

Somewhat later hominins include several recently described forms, such as *Orrorin tugenensis* (ca. 6 Mya), *Ardipithecus kadabba* (5.2–5.8 Mya), *Ardipithecus ramidus* (4.4 Mya), and *Australopithecus anamensis* (4.2–3.9 Mya), that are represented by skull fragments, jaws, teeth, and leg and arm bones, as well as a 3.5 Myr old skull named *Kenyanthropus platyops*. The most extensive and informative early fossil material, also about 3.5 Myr old, has been named *Australopithecus afarensis* (see the photograph at the beginning of this chapter).

That these forms are not far removed from the common ancestor of humans and chimpanzees is indicated by their many "primitive" or ancestral features. They had a lower face projecting far beyond the eyes (less so in *platyops*), relatively large canine teeth, long arms relative to the legs, a small brain (about 400 cc), and curved bones in the fingers and toes (which imply that they climbed trees). However, the structure of the pelvis and hindlimb clearly shows that *anamensis* and *afarensis* were bipedal. In fact, "fossilized" foot-

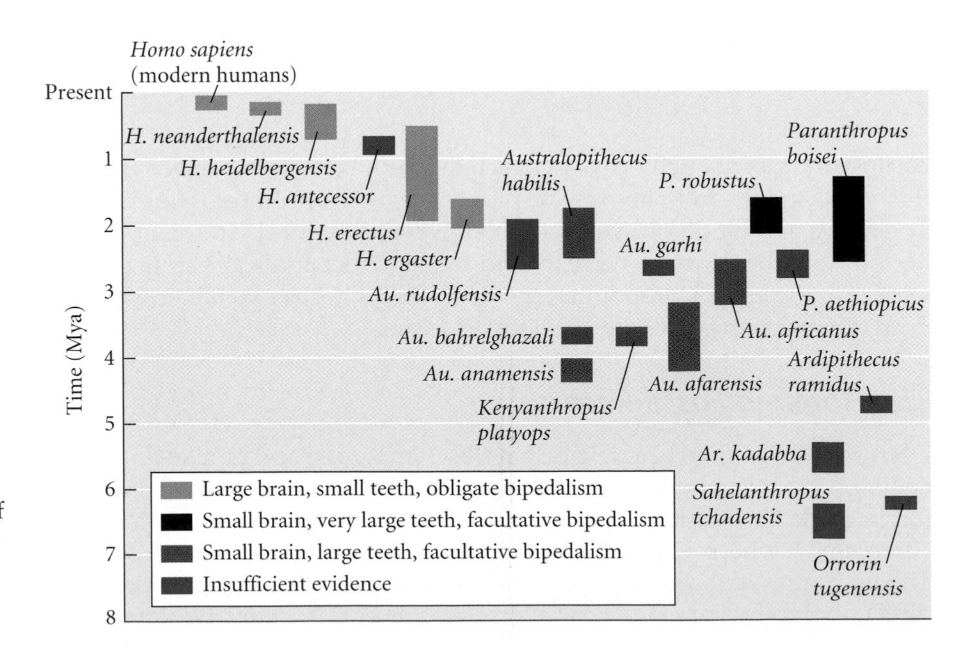

Figure 4.13 The approximate temporal extent of named hominin taxa in the fossil record. The time span indicates either the range of dated fossils or the range of estimated dates for taxa known from few specimens. (After Wood 2002.)

print traces have been found in rock formed from volcanic ash near an *afarensis* site in Tanzania.

Following *afarensis* in the late Pliocene and early Pleistocene, the number of hominin species and the relationships among them have not yet been resolved (Figure 4.13). Most authorities agree that hominin species were quite diverse at this time. A lineage of "robust" australopithecines (*Paranthropus*), of which three species have been named, had large molars and premolars and other features adapted for powerful chewing; they probably fed on tubers and hard plant material. The robust australopithecines may have made stone tools, the oldest of which are 2.6–2.3 Myr old. However, they became extinct without having contributed to the ancestry of modern humans. A more slender form was *Australopithecus africanus*, which is generally thought to have descended from *A. afarensis*, but had a greater cranial capacity (about 450 cc), indicating a larger brain (Figure 4.14B,C). Some of the derived characters of *africanus* resemble those of robust australopithecines, so it may not be in the line of direct ancestry of modern humans.

The earliest fossils that are usually assigned to the genus *Homo* range from about 1.9 to 1.5 Mya (late Pliocene and early Pleistocene). Originally called *Homo habilis*, they are variable enough to be assigned by some authors to two or even three species, *H. habilis*, *H. ergaster*, and *H. rudolfensis*. *Homo habilis* (in the broad sense) is the epitome of a rediscovered missing link (Figure 4.14D), and might better be assigned to *Australopithecus* (Wood and Collard 1999). The oldest specimens are very similar to *Australopithecus africanus*, and the younger ones grade into the later form *Homo erectus*. Compared with *Australopithecus*, *Homo habilis* more nearly resembles modern humans in its greater cranial capacity (610 to nearly 800 cc), flatter face, and shorter tooth row. Although the limbs retain apelike proportions that suggest an ability to climb, the structure of the leg and foot indicates that its bipedal locomotion was more nearly human than that of the australopithecines. *Homo habilis* is associated with stone tools (referred to as Olduwan technology) and with animal bones that bear cut marks and other signs of hominin activity (Potts 1988).

Later hominin fossils, from about 1.6 Mya to about 200,000 years ago (200 Kya), are referred to *Homo erectus*. Most authorities think that *habilis*, *erectus*, and *sapiens* are a single evolutionary lineage. In most respects, *erectus* from the middle Pleistocene onward has fairly modern human features in its skull, in its postcranial anatomy, and in indications of its behavior. The

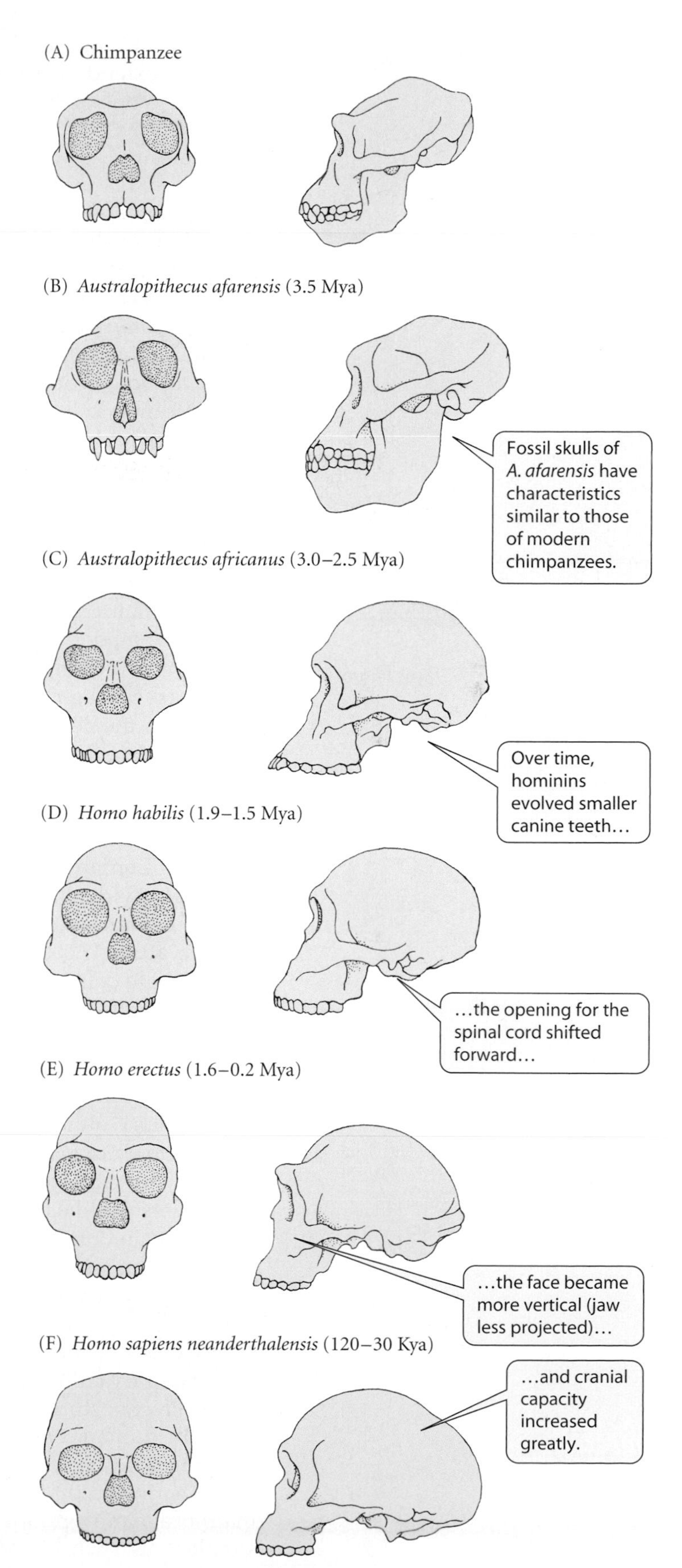
(A) Chimpanzee
(B) *Australopithecus afarensis* (3.5 Mya)
Fossil skulls of *A. afarensis* have characteristics similar to those of modern chimpanzees.
(C) *Australopithecus africanus* (3.0–2.5 Mya)
Over time, hominins evolved smaller canine teeth…
(D) *Homo habilis* (1.9–1.5 Mya)
…the opening for the spinal cord shifted forward…
(E) *Homo erectus* (1.6–0.2 Mya)
…the face became more vertical (jaw less projected)…
(F) *Homo sapiens neanderthalensis* (120–30 Kya)
…and cranial capacity increased greatly.

Figure 4.14 Frontal and lateral reconstructions of the skulls of a chimpanzee and some fossil hominins. (A) *Pan troglodytes*, the chimpanzee. Note large canine teeth, low forehead, prominent face, and brow ridge. (B) *Australopithecus afarensis*. Some of the same features as in the chimpanzee are evident. (C) *Australopithecus africanus* has smaller canines and a higher forehead. (D) *Homo habilis*. The face projects less, and the skull is more rounded, than in earlier forms. (E) *Homo erectus*. Note the still more vertical face and rounded forehead. (F) *Homo (sapiens) neanderthalensis*. The rear of the skull is more rounded than in *H. erectus*, and cranial capacity is greater. (A, B after Jones et al. 1992; C–F after Howell 1978.)

skull is rounded, the face is less projected than in earlier forms, the teeth are smaller, and the cranial capacity is larger, averaging about 1000 cc and evidently increasing over time (Figure 4.14E). At least 1 Mya (perhaps as far back as 1.7 Mya), *erectus* spread from Africa into Asia, extending eastward to China and Java. Throughout its range, *erectus* is associated with stone tools, termed the Acheulian culture, that are more diverse and sophisticated than the Olduwan tools of *H. habilis*. The use of fire was widespread by half a million years ago.

Homo erectus grades into forms that have generally been called "archaic *Homo sapiens*," starting about 400 or 300 Kya. During the history of *sapiens*, mean cranial capacity increased, from about 1175 cc at 200 Kya to its modern mean of 1400 cc. "Archaic *sapiens*" from the middle Pleistocene differs in relatively minor respects from the "anatomically modern *sapiens*" that appeared in the late Pleistocene. The best-known populations of archaic *Homo sapiens* are the Neanderthals of Europe and southwestern Asia, distinguished by some authors as a species, *Homo neanderthalensis*. Neanderthals had dense bones, thick skulls, and projecting brows (Figure 4.14F); contrary to the popular image of a stooping brute, Neanderthals walked fully erect, had brains as large as or even larger than ours (up to 1500 cc), had a fairly elaborate culture that included a variety of stone tools (Mousterian culture), and probably practiced ritualized burial of the dead. Their remains extend from about 120 to 30 Kya.

"Modern *sapiens*," anatomically virtually indistinguishable from today's humans, appeared earlier in Africa (ca. 170 Kya) than elsewhere. Modern humans overlapped with Neanderthals in the Middle East for much of the Neanderthals' history, but abruptly replaced them in Europe about 40,000 years ago. As we shall see in Chapter 6, there is evidence that modern *sapiens* may have replaced archaic *sapiens* (including *neanderthalensis*) without interbreeding—that is, the two lineages may have been distinct biological species. By 12,000 years ago, and possibly earlier, modern humans had spread from northeastern Asia across the Bering Land Bridge to northwestern North America, and thence rapidly throughout the Americas.

"Upper Paleolithic" culture emerged about 40 Kya. The earliest of several successive cultural "styles" in Europe, the Aurignacian, is marked by stone tools more varied and sophisticated than those of the Mousterian culture. Moreover, culture became more than simply utilitarian: art, self-adornment, and possible mythical or religious beliefs are increasingly evident from about 35 Kya onward. Agriculture, which resulted in an enormously increased human population density and began the human transformation of the Earth, is about 11,000 years old. There is, at least at present, no way of knowing which (if any) of these cultural advances were associated with genetic changes in the capacity for reason, imagination, and awareness, but they are not paralleled by any increase in brain size or other anatomical changes.

Throughout hominin evolution, different hominin features evolved at different rates ("mosaic" evolution). On average, brain size (cranial capacity) increased throughout hominin history, although not at a constant rate, and there were progressive changes, from *afarensis* to *africanus* to *erectus* to *sapiens*, in many other features, such as the teeth, face, pelvis, hands, and feet. The very fuzziness of the taxonomic distinctions among the named forms attests to the mosaic and gradual evolution of hominin features. Although many issues remain unresolved, the most important point is fully documented: modern humans evolved from an apelike ancestor.

Why these changes occurred—what advantages they may have provided—is the subject of much speculation but only slight evidence (Lovejoy 1981; Fedigan 1986). What evidence exists is indirect, consisting mostly of inferences from studies of other primates, contemporary cultures, and anatomy and artifacts.

The erect posture and bipedal locomotion are the first major documented changes toward the human condition. A plausible hypothesis is that bipedalism freed the arms for carrying food back to the social unit, especially to an individual's mate and offspring. Food sharing occurs in chimpanzees, which have a complex social structure that includes matrilocal family groups and "friendships." Chimpanzees also make and use a variety of simple tools, such as stone and wooden hammers they use to crack nuts, and twigs fashioned to "fish" termites out of their nests. The advantages gained by using a greater variety of tools may have selected for greater intelligence and brain size. However, many authors, beginning with Darwin, have emphasized that social interactions, such as learning how to provide parental care, forming cooperative liaisons with other group members, detecting cheaters in social exchanges, and competing for resources within and among groups, would place a selective premium on intelligence, learning, and communication—thus selecting for greater intelligence and a larger brain.

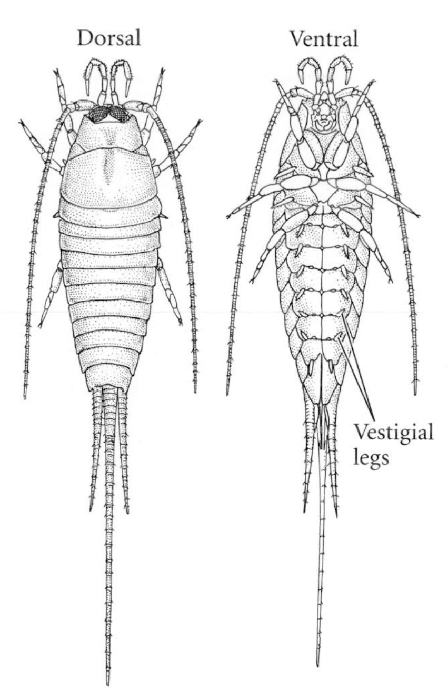

Figure 4.15 A living bristletail (order Archaeognatha). Among extant orders of insects, bristletails have the most primitive features, such as the vestigial legs on the ventral surface of the abdominal segments, derived from the walking legs of ancestral arthropods. (After CSIRO 1991.)

Phylogeny and the Fossil Record

In inferring phylogenetic relationships among living taxa, we conclude that certain taxa share more recent common ancestors than others. If such statements are correct, then there should be some correspondence between the relative times of origin of taxa, as inferred from phylogenetic analysis, and their relative times of appearance in the fossil record. We can expect this correspondence to be imperfect because of the great imperfection of the fossil record; for example, a group that originated in the distant past might be recovered only from recent deposits. Moreover, although a lineage may have branched off early, it may not have acquired its diagnostic characters until much later. The synapsid clade, for example, did not acquire the diagnostic characters of mammals until long after it had diverged from other reptiles.

Nevertheless, there is a strong overall correspondence between phylogenetic branching order and order of appearance in the fossil record. Just by phylogenetic analysis of living species, we infer that the common ancestors of the different orders of mammals, of mammals and "reptiles," of these groups and amphibians, and of all tetrapods and sarcopterygian fishes are sequentially older. The sequence in which these groups appear in the fossil record matches the phylogeny. A striking instance of correspondence is offered by the bristletails (order Archaeognatha), wingless insects that have long been thought to represent the basic "body plan" of ancestral insects (Figure 4.15). Recently, a fossil bristletail was discovered in early Devonian deposits. It is the oldest known fossil insect, and it is as old as would be expected given the assumption that bristletails are phylogenetically more basal than the other insect orders.

Phylogenetic relationships can often be clarified by information from extinct species (Donoghue et al. 1989). Some characters may have been so highly modified that it is difficult to trace their evolutionary transformations, or even to determine their homology. Fossils may provide the crucial missing information. For example, some authors had postulated that mammals and birds are sister taxa, but the fossil record of mammal-like reptiles and dinosaur-like birds showed that this hypothesis was wrong.

Evolutionary Trends

The fossil record presents many instances of evolutionary trends. Among members of the horse family (Equidae), for example, average body size increased steadily over the course of almost 50 Myr (Figure 4.16A). Commonly, some lineages buck the overall trend and undergo reversal, as is the case with body size in some equid lineages (Figure 4.16B). Certain evolutionary changes appear never to have been reversed, however. Ever since *Morganucodon*, for example, mammals have had a single lower jaw bone and have never reverted to the multiple bones of their ancestors.

Although we may have evidence that an evolutionary change has never been reversed, we cannot be sure that reversal is impossible. Some features may never have reverted simply because they are advantageous, or even necessary. For example, even flightless birds such as ostriches and penguins have feathers because feathers provide insulation and are used in sexual and social displays. In most vertebrates, the notochord degenerates after its expression in early embryonic development, but it is retained in the embryo because it induces the development of the central nervous system. Rupert Riedl (1978) suggested that such a character carries a BURDEN, meaning a suite of other features that depend on it for their development or proper function. It is, of course, possible that some changes are

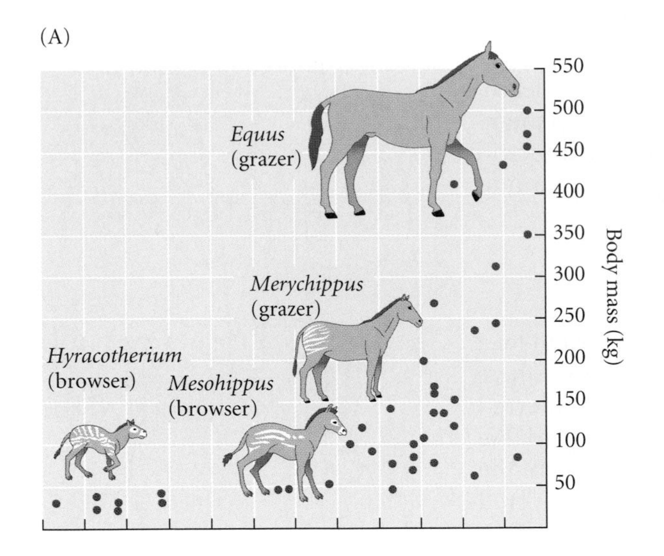

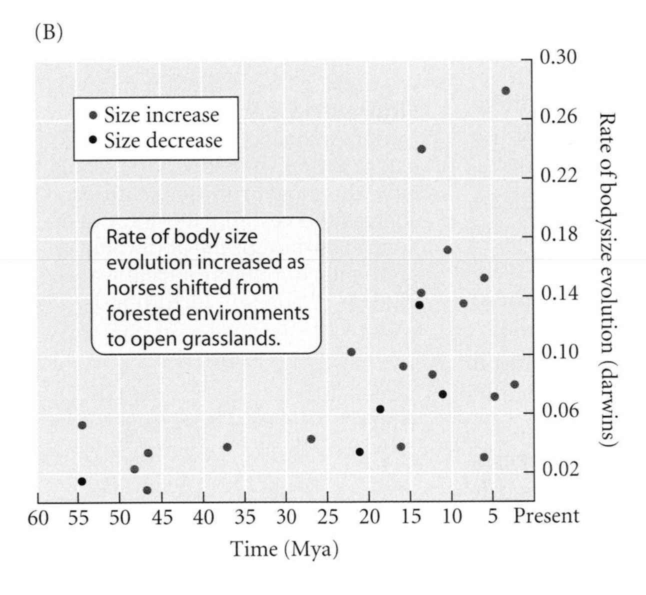

Figure 4.16 Evolution of body mass in the horse family, Equidae. (A) Estimated body masses of 40 species, plotted against geological time. Although some small species occurred throughout the history of the family, the average body size increased over time. (B) Rate and direction of evolution of body mass between various pairs of ancestral and descendant equids, plotted against the midpoint between their geological ages. Most lineages increased in mass (colored circles), but some decreased (black circles). (After MacFadden 1986.)

truly irreversible, perhaps because the developmental foundations of some characters have been lost in evolution. In general, it seems likely that complex characters, once lost, are not regained—a principle called **Dollo's law**.

Many taxa in the fossil record display parallel trends. For example, the horse family is only one of many animal clades in which average body size has increased—a generalization called **Cope's rule**. Multiple lineages often evolve through similar stages, called **grades**. For example, balanomorph barnacles, which first appeared in the Cretaceous, are enclosed by a cone-shaped skeleton made up of a number of slightly overlapping plates. Ancestrally, there were eight plates, but as the balanomorphs diversified in the Cenozoic, the proportion of genera with eight plates steadily declined, and the proportion with fewer plates increased. Several lineages independently evolved through six-plate, four-plate, and even one-plate grades (Figure 4.17). Shells with fewer plates, and hence fewer lines of vulnerability between them, may provide greater protection against predatory snails (Palmer 1982).

Punctuated Equilibria

Although we have described paleontological examples of gradual transitions through intermediate states, these kinds of transitions are by no means universally found in the fossil record. Intermediate stages in the evolution of many higher taxa are not known, and many closely related species in the fossil record are separated by smaller but nonetheless distinct gaps. Most paleontologists have followed Darwin in ascribing these gaps to the great incompleteness of the fossil record. In 1972, Niles Eldredge and Stephen Jay Gould proposed a more complicated, and much more controversial, explanation, which they called **punctuated equilibria**. Their hypothesis applies to the abrupt appearance of closely related species, not to higher taxa.

"Punctuated equilibria" refers to both a *pattern* of change in the fossil record and a *hypothesis* about evolutionary processes. A common pattern, Eldredge and Gould said, is one of long periods in which species exhibit little or no detectable phenotypic change, interrupted by rapid shifts from one such "equilibrium" state to another; that is, **stasis** that is "punctuated" by rapid change (Figure 4.18A). They contrasted this pattern with what they called **phyletic gradualism**, the traditional notion of slow, incremental change (Figure 4.18B).

The fossil record offers examples of both gradual and punctuated patterns. A particularly well-documented instance of phyletic gradualism has been described for changes in the molar teeth of a lineage of grass-feeding voles (*Mimomys occitanus*) in the late Pliocene and Pleistocene (Figure 4.19). Several molar characters of these rodents changed directionally throughout Europe, indicating that gene flow among populations enabled the entire species to respond as a whole to selection for increased tooth height (Chaline and Laurin 1986). In contrast, *Metrarabdotos*, a Miocene genus of ectoprocts (also known as bryozoans, or "moss animals"), clearly shows a pattern of punctuated equilibria (Fig-

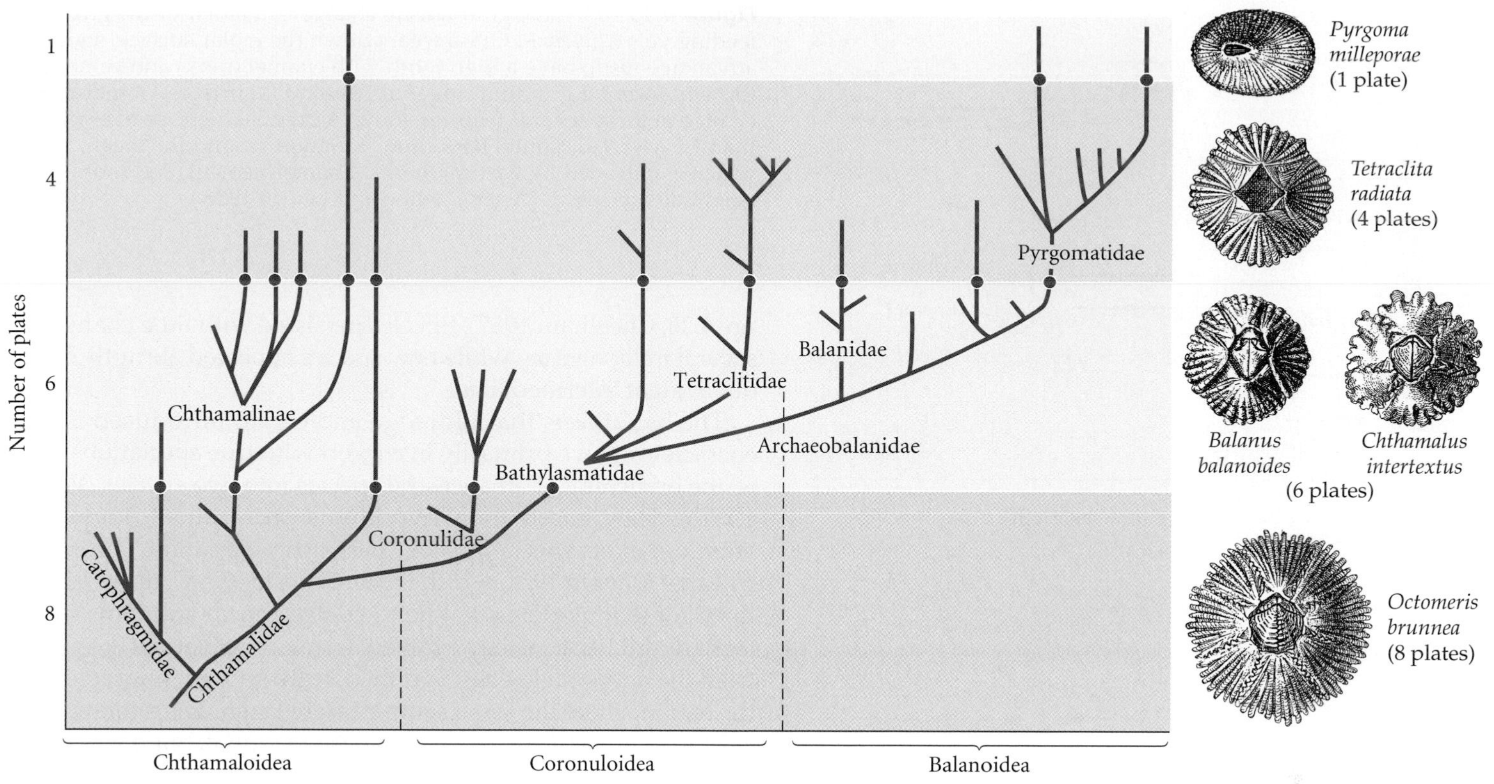

Figure 4.17 A parallel trend. The phylogeny of balanomorph barnacles shows the reduction in the number of shell plates that occurred during the Cenozoic in several independent lineages (see Figure 3.5). *The vertical axis is not time, but grade of organization (plate number).* The drawings are from an extensive monograph on barnacles by Charles Darwin. (After Palmer 1982; drawings by G. Sowerby, from Darwin 1854.)

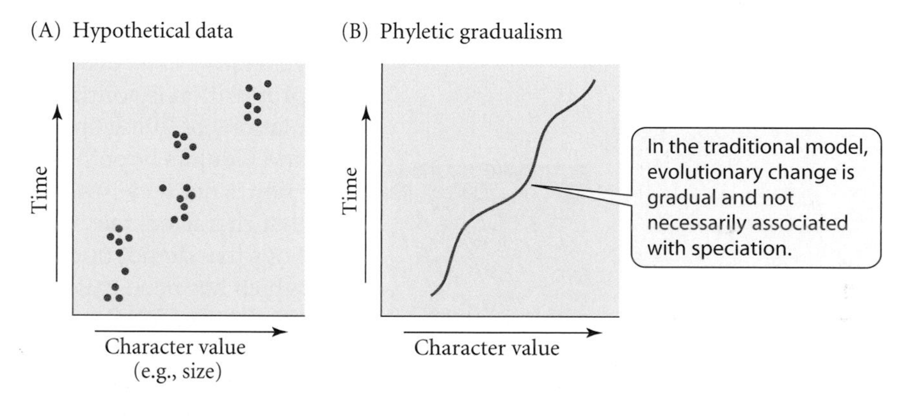

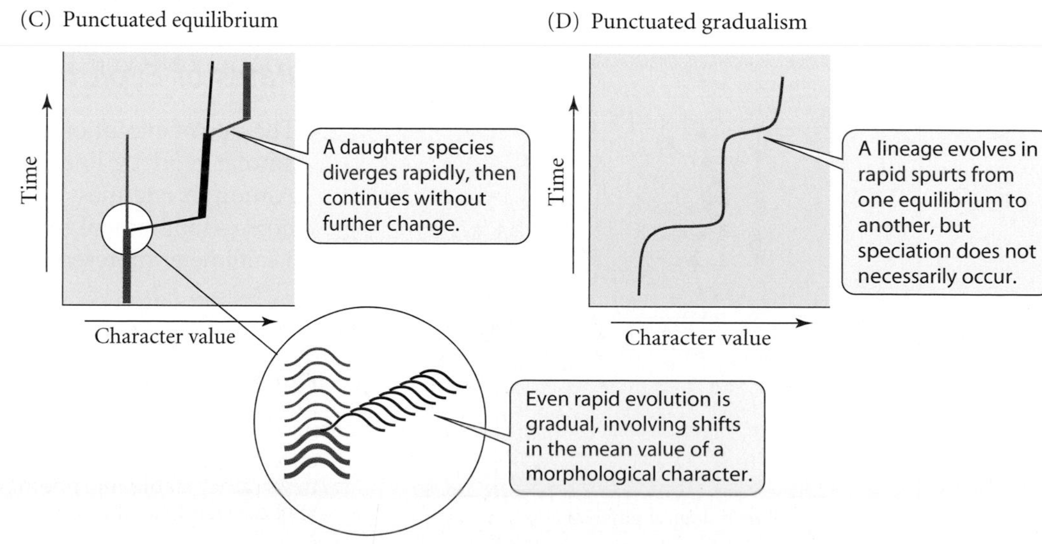

Figure 4.18 Three models of evolution, as applied to a hypothetical set of fossils. (A) Hypothetical values for a character in fossils recorded from different time periods. These data might correspond to any of the models shown in panels B–D. (B) The traditional "phyletic gradualism" model. (C) The "punctuated equilibrium" model of Eldredge and Gould, in which morphological change occurs in new species. Morphological evolution, although rapid, is still gradual, as shown in the inset. (D) The "punctuated gradualism" model of Malmgren et al. (1983).

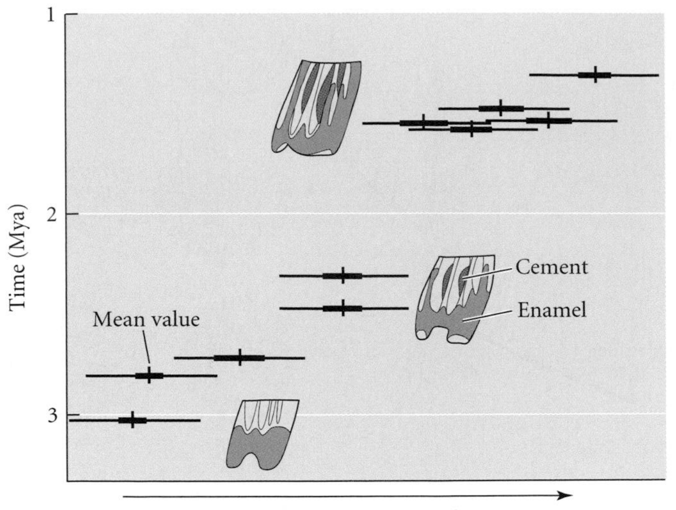

Figure 4.19 Phyletic gradualism: change in a molar of the grass-feeding vole *Mimomys*. Grass wears down the molar surface, so it is advantageous to have a high tooth with enamel (pink) and cement (brown) forming grinding ridges at the tooth's surface. An index of change in these several features shows a gradual increase over more than 1.5 Myr. Horizontal bars show variation around the mean, which is indicated by vertical marks. Enamel, cement, and tooth height all increased. (After Chaline and Laurin 1986.)

ure 4.20; Cheetham 1987). Species persisted with little change for several million years, while new species appeared abruptly, without evident intermediates.

The hypothesis that Eldredge and Gould introduced is that characters evolve primarily in concert with true speciation—that is, the branching of an ancestral species into two species (Figure 4.18C). They based their hypothesis on a model, known as "founder-effect speciation" or "peripatric speciation," proposed by Ernst Mayr in 1954, which we will consider in Chapter 16. The thrust of that model is that new species appear suddenly in the fossil record because they evolved in small populations separated from the ancestral species and then, fully formed, migrated into the region where the fossil samples were taken. The evolutionary change they underwent may have been gradual, but it was rapid and took place "off stage."

This hypothesis would not have been so controversial if Eldredge and Gould had not further proposed that, except in populations that are undergoing speciation, morphological characters generally *cannot* evolve, due to internal genetic "constraints." This proposition is contradicted by considerable evidence from populations of living species (see Chapters 9 and 13), and Eldredge and Gould's hypothesis that evolutionary change requires speciation is not widely accepted. Furthermore, the fossil record shows that characters may evolve between long-stable states in populations that do not undergo speciation (Figure 4.18D). This pattern, which has been called **punctuated gradualism**, is illustrated by the detailed fossil record of abundant shell-bearing protists called foraminiferans (Figure 4.21). It remains to be seen whether punctuation and stasis is the most common pattern in the fossil record (Gould and Eldredge 1993) or not (Levinton 2001).

Rates of Evolution

The rate of evolutionary change varies greatly among characters, among evolving lineages, and within the same lineage over time. Although one may describe the change in, say, the height of a horse's tooth in millimeters (mm) per million years, an increase of 1 millimeter represents far greater change if the original tooth was

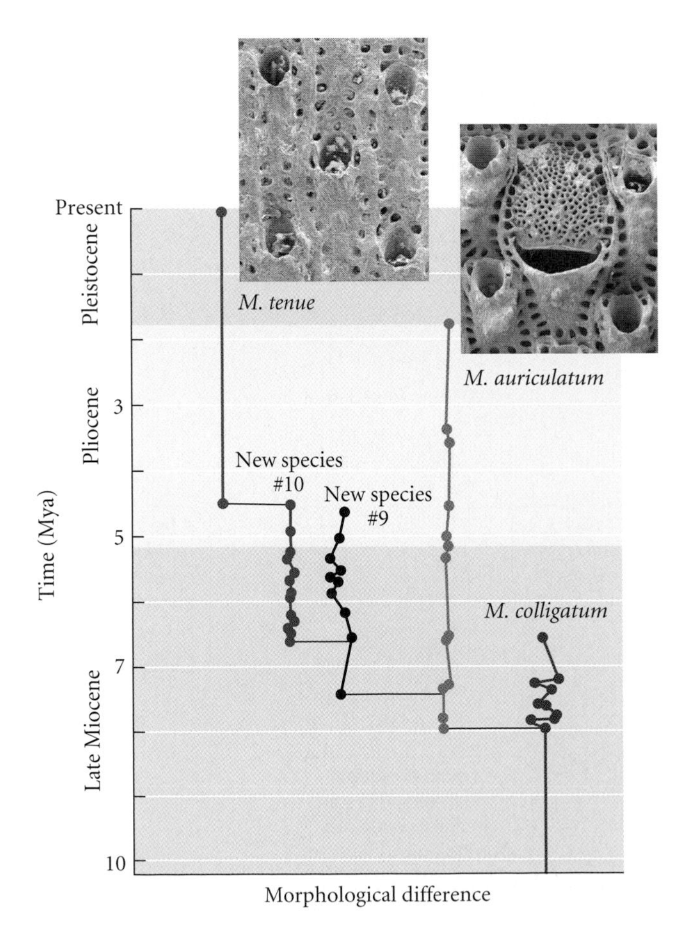

Figure 4.20 Punctuated equilibrium: the phylogeny and temporal distribution of a lineage of ectoprocts (*Metrarabdotos*). The horizontal distance between points represents the amount of morphological difference between samples. The general pattern is one of abrupt shifts to new, rather stable morphologies. (After Cheetham 1987; photos courtesy of Alan Cheetham.)

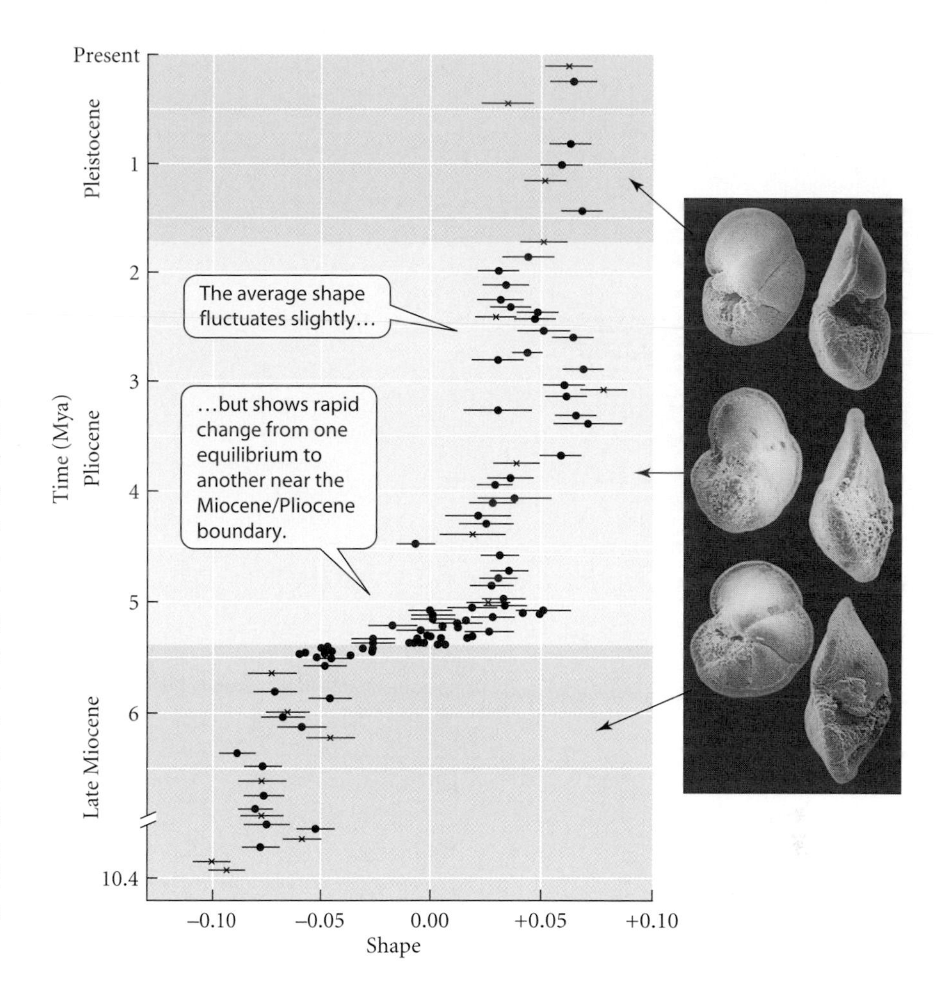

Figure 4.21 "Punctuated gradualism," illustrated by the evolution of shell shape in the foraminiferan lineage *Globorotalia*. Side and edge views of specimens from the late Pleistocene (top), early Pliocene (middle), and late Miocene (bottom) appear at the right. In the graph, a mathematical index of shape is plotted against the ages of fossil samples. There is no evidence that speciation occurred in this lineage. (After Malmgren et al. 1983; photos courtesy of B. Malmgren.)

5 mm high than if it was 50 mm high. Therefore, evolutionary rates are usually described in terms of proportional, rather than absolute, changes by using the logarithm of the measurement rather than the original scale of measurement. J. B. S. Haldane, one of the pioneers of the evolutionary synthesis, proposed a unit of measurement of evolutionary rate that he called the DARWIN, which he defined as a change by a factor of 2.718 (the base of natural logarithms) per million years. The darwin, however, has some drawbacks, and researchers now tend to measure evolutionary rate as the number of standard deviations by which a character mean changes per generation: a unit that paleontologist Philip Gingerich (1993) dubbed the HALDANE. (The standard deviation is a measure of the amount of variation within a population; see Figure 4.22A and Box B in Chapter 9.)

When rates of character evolution have been measured for ancestor-descendant series of dated fossils, the most striking result has been that *average rates of evolution are usually extremely low*. For example, even though the size of the first molar of the early Eocene horse *Hyracotherium grangeri* fluctuated slightly, overall it changed very little over 650,000 years (Figure 4.23A). Its median rate of change was about 0.0000057 haldanes (over a median interval of about 113,000 generations, assuming 2 years per

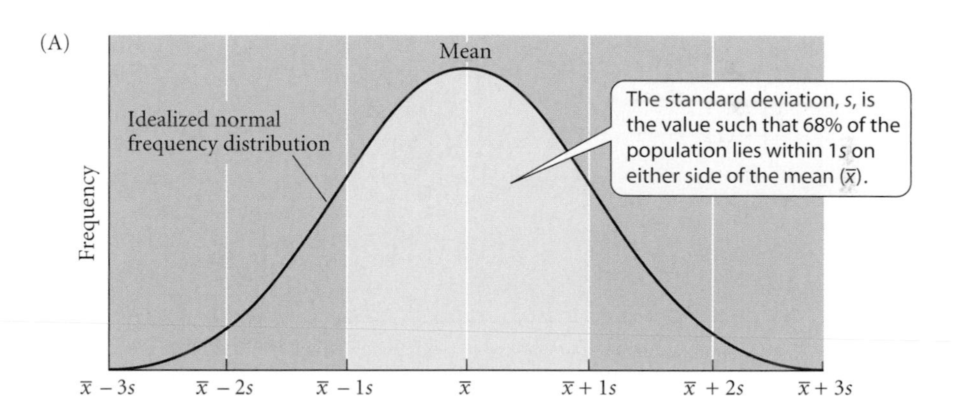

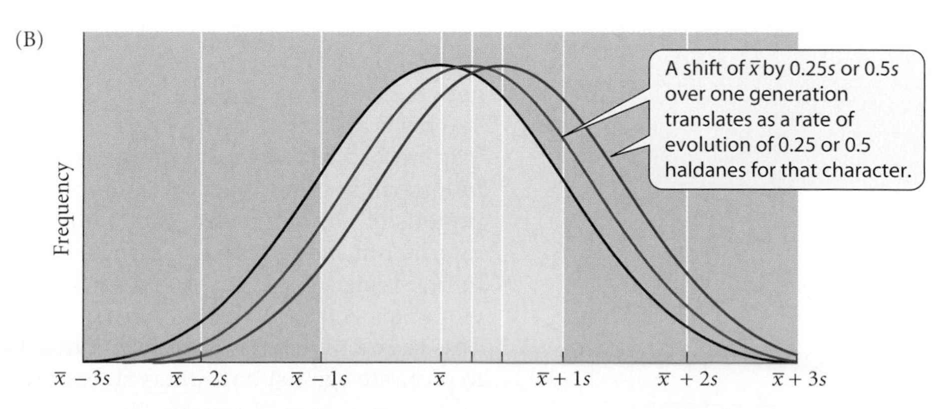

Figure 4.22 (A) In an idealized normal frequency distribution, the standard deviation (s) is the value at which 68% of the population lies within one s above and one s below the mean ($\bar{x}$). 95% of the population lies within 2s, and 99.7% lies within 3s, on either side of the mean. (B) Given the standard deviation of the "original" frequency distribution (black curve), the distribution shown as a red line lies 0.25s, and the distribution shown as a blue line lies 0.5s, away from the original mean. If these shifts in the mean occurred over just one generation, the rate of evolution of the mean character value would be 0.25 haldanes or 0.5 haldanes, respectively.

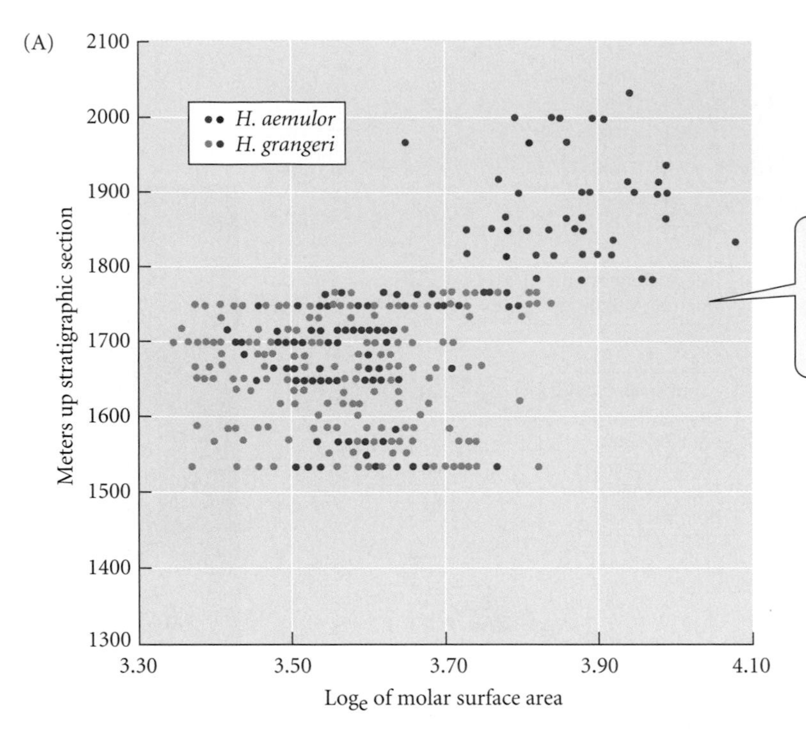

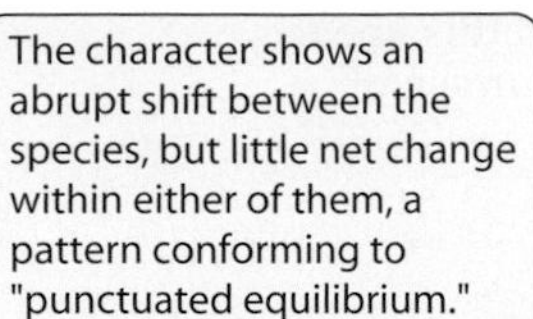

The character shows an abrupt shift between the species, but little net change within either of them, a pattern conforming to "punctuated equilibrium."

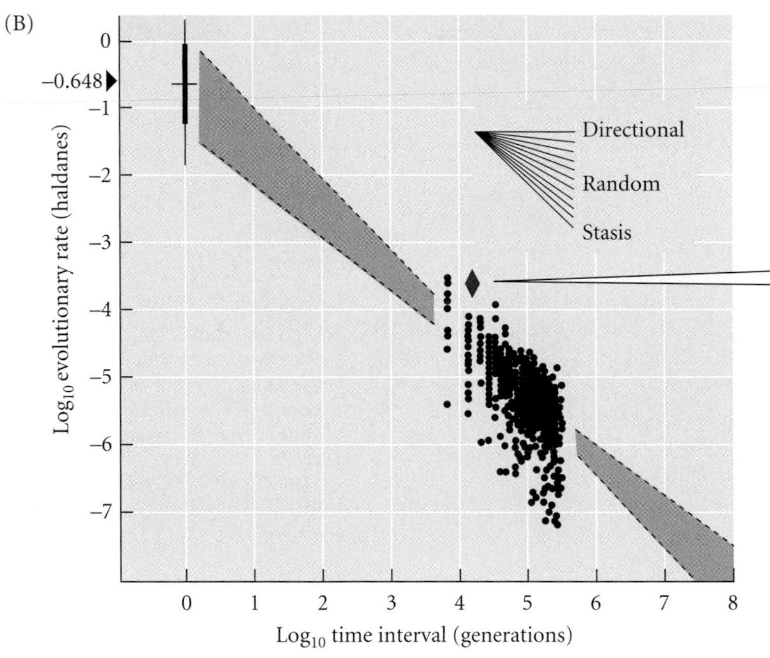

Figure 4.23 Evolution of the size (surface area) of the first molar of Eocene *Hyracotherium*, early members of the horse family. (A) The data points show character values of individual specimens, plotted against depth in a stratigraphic section of rock. (Darker dots indicate multiple species with the same measurements) *H. grangeri* extended for about 650,000 years (from meter 1520 to meter 1760), and evolved abruptly into a form different enough to be designated a different species, *H. aemulor*. (B) Assuming a generation time of 2 years, rates of evolution of molar size were calculated for data points separated by a range of time intervals, and the log of these rates (in haldanes) was plotted against the log interval. Rates calculated over longer intervals are lower, and the slope of this relationship (with confidence interval shown by the dark-shaded area) suggests that the rate for a single generation of evolutionary change would be –0.648 log haldanes, or 0.225 haldanes. (After Gingerich 1993.)

generation). Its maximal rate was 0.00030 haldanes (over a 6745-generation interval) (Gingerich 1993).

This is an extremely low rate of evolution, but it is typical of most such data from the fossil record. However, it is an *average* rate of change over very long intervals of time. This long-term average will conceal rapid evolutionary rates if the rate fluctuates (e.g., long periods of no change alternate with bursts of rapid change) or if the character fluctuates rapidly but shows little net change. Such changes in rate and direction are indeed evident in fine-scaled fossil series, such as the stickleback data portrayed in Figure 4.4. Thus we would expect to find higher rates if measurements are made over shorter intervals of time. This is exactly what is found (Figure 4.23B). So, in order to compare rates, it is necessary to plot rate against time interval. For example, molar size evolved at a rate of 0.00024 haldanes during the 13,500-generation transition from *Hyracotherium grangeri* to a form called *H. aemulon* (Figure 4.23A). This is higher than the maximal rate seen within *H. grangeri* at the same time interval (Figure 4.23B). The history of this lineage exemplifies the pattern of punctuated equilibrium.

Evolutionary changes in the fossil record, which are net changes averaged over long time intervals, display rates that are almost invariably far lower than those observed over the course of a few centuries (or less) in species that have been transported to new regions or otherwise affected by human-induced environmental change. For example, in the course of about 100 generations, North American house sparrows, introduced to the continent from Europe, manifested changes in wing and bill length of up to 0.024 haldanes, and the beak length of soapberry bugs increased at a rate of 0.010 to 0.035 haldanes as they adapted to feeding on introduced plants (Hendry and Kinnison 1999; see Figure 13.3). These rates are two orders of magnitude (a factor of 100) greater than tooth evolution was in the rapid transition between species of *Hyracotherium*. If characteristics evolved in a single direction for thousands or millions of years, even at much lower rates than these, organisms would be vastly more different from one another than they are, and mice would long since have become bigger than the largest dinosaurs. Evolutionary changes can be very rapid, but they are not sustained at high rates for very long.

Summary

1. From both geological and biological evidence, it is clear that the fossil record is extremely incomplete. Nevertheless, a number of evolutionary histories are well known.
2. Unusually detailed records of change within individual species show that characters often evolve gradually, may fluctuate rapidly with little overall change, or may shift rather quickly.
3. The origins of some higher taxa, such as amphibians, birds, mammals, cetaceans, and the genus *Homo*, have been documented in the fossil record. These examples show gradual evolution, entailing both mosaic evolution and gradual change in individual features. The decision of whether to classify intermediate fossils in one taxon or another is often arbitrary.
4. Changes in the form of characters are sometimes associated with major changes in their function.
5. The relative times of origin of taxa, as inferred from phylogenetic analysis, often correspond to their relative times of appearance in the fossil record.
6. Evolutionary trends, which can often be attributed to natural selection, are evident in the fossil record, but such trends may be reversed in related lineages.
7. Species may display very little change for long periods (stasis), and then shift rapidly to new phenotypes. The term "punctuated equilibria" refers both to this pattern and to the hypothesis, not widely accepted, that most changes in morphology occur in association with the evolution of new species (i.e., splitting of lineages).
8. Rates of evolution are very variable. Individual features typically display a low rate of evolution, averaged over long periods of time, but more detailed fossil records show rapid, short-term fluctuations in characters.

Terms and Concepts

Cope's rule
Dollo's law
epoch
era
evolutionary trend
geological time scale (and its component eras, periods, and epochs)
grade
igneous rock
metamorphic rock
period
phyletic gradualism
plate tectonics
punctuated equilibria
punctuated gradualism
radiometric dating
rate of evolution
sedimentary rock
stasis
strata

Suggestions for Further Reading

The second edition of S. M. Stanley's *Earth system history* (W.H. Freeman, New York, 2005) is a thorough introduction to geological processes, Earth history, and major events in the history of life, from a paleontologist's perspective. Other useful works on paleontology include R. L. Carroll's comprehensive and abundantly illustrated *Vertebrate paleontology and evolution* (W.H. Freeman, New York, 1988); and *Paleobiology: A synthesis*, edited by D. E. G. Briggs and P. R. Crowther (Blackwell Publishing, Oxford, 1990), which contains a collection of brief, authoritative essays on numerous topics.

Human evolution: An illustrated introduction (Blackwell Publishing, Oxford, 2005) is a new and very comprehensible introduction to the history of human evolution by R. Lewin.

"Punctuated equilibrium comes of age" (*Nature* 366: 223–227, 1993) by S. J. Gould and N. Eldredge is one of these authors' later essays about their hypothesis. For criticism of the hypothesis of punctuated equilibrium, see the second edition of J. S. Levinton's *Genetics, paleontology, and macroevolution* (Cambridge University Press, Cambridge, UK, 2001).

Problems and Discussion Topics

1. "Time-averaging" refers to the practice of combining fossil samples from different time intervals into a single sample. Refer to Bell's study of sticklebacks (Figure 4.4). What would the data look like if he had collapsed samples spanning 20,000 years into single samples, instead of analyzing separate samples at 5000-year intervals? What conclusions might Bell have drawn about the evolution of dorsal spine number if his only samples had been from 70,000 and 30,000 years ago?
2. The rate of evolution of DNA sequences (and other features) is often calibrated by the age of fossil members of the taxa to which the living species belong (see Chapter 8). How do imperfections in the fossil record affect the estimates of evolutionary rates obtained in this way? Is there any way of setting limits to the range of possible rates?
3. An ideal fossil record would enable researchers to distinguish the patterns of phyletic gradualism, punctuated equilibria, and punctuated gradualism (see Figure 4.18). How would you do so? How do imperfections of the fossil record make it difficult to distinguish these patterns?
4. Creationists deny that the fossil record provides intermediate forms that demonstrate the origins of higher taxa. Of *Archaeopteryx*, some of them say that because it had feathers and flew, it was a bird, not an intermediate form. Evaluate this argument.
5. Consider the hypothesis that Eldredge and Gould advanced to explain the pattern they called punctuated equilibria. What would be the implications for evolution if the hypothesis were true versus false?
6. Changes in a phenotypic character in a population are considered evolutionary changes only if they have a genetic basis. Alterations of organisms' features directly by the environment they experience are not evolution. Because we cannot breed extinct organisms to determine whether differences have a genetic basis, how might we decide which phenotypic changes represent evolution and which do not? Consider (a) the difference in dorsal spine number of sticklebacks at 65 versus 60 Kya (Figure 4.4B), (b) the difference in the same character at 70 versus 25 Kya, and (c) the difference in the shape of the rear teeth in *Morganucodon* (Figure 4.10F) compared with *Procynosuchus* (Figure 4.10C). Can we be more confident that the difference is an evolved one in some cases than in others?
7. What are the possible causes of trends such as those illustrated by barnacles and horses in this chapter? How would you assess the validity of each cause you can think of?

5

A History of Life on Earth

If it were possible to look at the Earth of 3,500,000,000 years ago, to the time of life's beginning, we would see only bacteria-like cells. Among these cells would be our most remote ancestors, unrecognizably different from ourselves. And if we then time-traveled through life's history toward the present, we would see played out before us a drama grander and more splendid than we can imagine: a planetary stage of many scenes, on which emerge and play—and usually die—millions and millions of species with features and roles more astonishing than any writer could conceive.

Scene on the Cambrian sea floor. This artist's rendering imagines the drama of life some 500 million years ago, reconstructed from fossils found in the Burgess Shale (see page 97). Species shown include the predatory clawed arthropod *Anomalocaris* and the burrowing chordate *Pikaia*. The "spiky" *Hallucigenia* and the cuplike *Marrella* are depicted on the seabed. (Art © John Sibbick/NHMPL.)

This chapter describes some high points in the grand history of life, especially the origin, diversification, and extinction of major groups of organisms. Most of this material treats geological and paleontological evidence, but phylogenetic studies of living organisms have also revealed critically important information.

There is much more information in this chapter than you may wish to memorize. You may regard it largely as a source of information, or you may simply enjoy reading a sketch of one of the greatest stories of all time. However, a number of major events or other important points are highlighted in italics; these are points that a well-trained biologist should know. As you read this chapter, moreover, notice examples of the following general patterns:

1. Climates and the distribution of oceans and land masses have changed over time, affecting the geographic distributions of organisms.
2. The taxonomic composition of the biota has changed continually as new forms originated and others became extinct.
3. At several times, extinction rates have been particularly high (so-called **mass extinctions**).
4. Especially after mass extinctions, the diversification of higher taxa has sometimes been relatively rapid.
5. The diversification of higher taxa has included increases both in the number of species and in the variety of their form and ecological habits.
6. Extinct taxa have sometimes been replaced by unrelated but ecologically similar taxa.
7. The ancestral members of related higher taxa are often morphologically more similar to one another, and more ecologically generalized, than their descendants.
8. Of the variety of forms in a higher taxon that were present in the remote past, usually only a few have persisted in the long term.
9. The geographic distributions of many taxa have changed greatly.
10. Over time, the composition of the biota increasingly resembles that of the present.

Before Life Began

Most physicists agree that the current *universe came into existence about 14 billion years ago* (14,000,000,000 years ago) through an explosion (the "big bang") from an infinitely dense point. Elementary particles formed hydrogen shortly after the big bang, and hydrogen ultimately gave rise to the other elements. The collapse of a cloud of "dust" and gas formed our galaxy less than 10 billion years ago. Throughout the history of the universe, material has been expelled into interstellar space, especially during stellar explosions (supernovas), and has condensed into second- and third-generation stars, of which the Sun is one. *Our solar system was formed about 4.6 billion years ago*, according to radiometric dating of meteorites and moon rocks. The Earth is the same age, but because of geological processes such as the subduction of crust (see Figure 4.1), the oldest known rocks on Earth are younger, dating from 3.8 billion years ago. The Earth was probably formed by the collision and aggregation of many smaller bodies, the impact of which contributed enormous heat.

The early Earth formed a solid crust as it cooled, releasing gases that *included water vapor but very little oxygen*. As Earth cooled, oceans of liquid water formed, probably by 4.5 billion years ago, and quickly achieved the salinity of modern oceans. By 4 billion years ago, there were probably many small protocontinents; large land masses may have formed about a billion years later.

The Emergence of Life

The simplest things that might be described as "living" must have developed as complex aggregations of molecules. These aggregations, of course, would have left no fossil record, so it is only through chemical and mathematical theory, laboratory experimentation, and extrapolation from the simplest known living forms that we can hope to develop models of the emergence of life.

Although life is difficult to define, it is generally agreed that an assemblage of molecules is "alive" if it can capture energy from the environment, use that energy to replicate itself, and thus be capable of evolving. In the living things we know, these functions are performed by nucleic acids, which replicate and carry information, and by proteins, which replicate nucleic acids, transduce energy, and generate (and in part constitute) the phenotype. These components are held together in compartments—cells—that are surrounded by a lipid membrane.

Although living or semi-living things might have originated more than once, *we can be quite sure that all organisms we know of stem from a single common ancestor* because they all share certain features that are arbitrary as far as we can tell. For example, organisms synthesize and use only L optical isomers of amino acids as building blocks of proteins; L and D isomers are equally likely to be formed in abiotic synthesis, but a functional protein can be made only of one type or the other. D isomers could have worked just as well. The universality of the presumably rather arbitrary genetic code and of the machinery of protein synthesis are among the other features that imply a monophyletic origin of all organisms.

The most difficult problem in accounting for the origin of life is that in known living systems, only nucleic acids replicate, but their replication requires the action of proteins that are encoded by the nucleic acids. Despite this and other obstacles, progress has been made in understanding some of the likely steps in the origin of life (Orgel 1994; Maynard Smith and Szathmáry 1995; Rasmussen et al. 2004).

First, *simple organic molecules*, the building blocks of complex organic molecules, *can be produced by abiotic chemical reactions*. In a famous experiment, Stanley Miller found that electrical discharges in an atmosphere of methane (CH_4), ammonia (NH_3), hydrogen gas (H_2), and water (H_2O) yield amino acids and compounds such as hydrogen cyanide (HCN) and formaldehyde (H_2CO), which undergo further reactions to yield sugars, amino acids, purines, and pyrimidines (Figure 5.1).

Some such simple molecules must have formed polymers that could replicate. Polymerization (e.g., of amino acids) may have been facilitated by adsorption to clay particles or by concentration due to evaporation. The most likely early replicators were short RNA (or RNA-like) molecules. *RNA has catalytic properties, including self-replication*. Some RNA sequences (RIBOZYMES) can cut, splice, and elongate other oligonucleotides, and short RNA template sequences can self-catalyze the formation of complementary sequences from free nucleotides. Recent experiments have shown that clay particles with RNA adsorbed onto their surface can catalyze the formation of a lipid envelope, which in turn catalyzes the polymerization of amino acids into short proteins. Thus aggregates that have some of the critical components of a proto-cell, including self-replicating RNA, can be formed by chemical processes alone.

Natural selection and evolution can occur in nonliving systems of replicating molecules. When Sol Spiegelman (1970) placed RNAs, RNA polymerase (isolated from a virus, phage Qβ), and nucleotide bases in a cell-free medium, different RNA sequences were replicated by the polymerase at different rates, so that their proportions changed. Because a replicating enzyme from a living system (i.e., the virus) was provided, these experiments do not bear on the very first steps in the origin of life, but they do show that systems of replicating macromolecules can evolve.

Long RNA sequences would not replicate effectively because the mutation rate would be too high for them to maintain any identity. A larger genome might evolve, however, if two or more coupled macromolecules each catalyzed the replication of the other. Replication probably was slow and inexact originally, and only much later acquired the fidelity that modern organisms display. Moreover, many different oligonucleotides undoubtedly could replicate themselves. Before proteins evolved, there were no phenotypes—only genotypes. So *the first "genes" need not have had any particular base pair sequence*. Thus there is no force to the argument, frequently made by skeptics, that the assembly of a particular nucleic acid sequence is extremely improbable (say, 1 chance in 4^{50} for a 50 bp sequence).

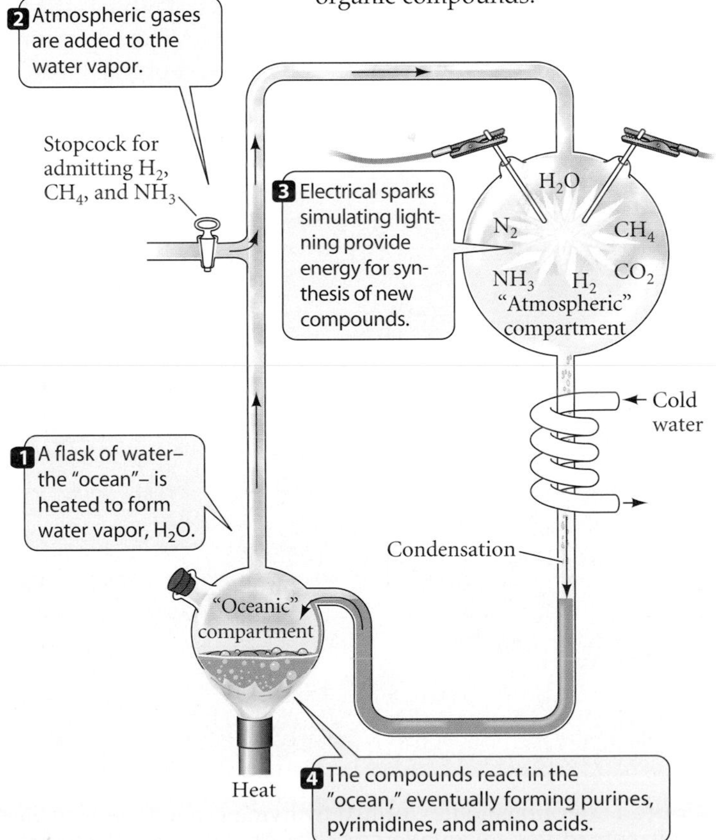
2 Atmospheric gases are added to the water vapor.
Stopcock for admitting H_2, CH_4, and NH_3
3 Electrical sparks simulating lightning provide energy for synthesis of new compounds.
H_2O
N_2
CH_4
NH_3
H_2
CO_2
"Atmospheric" compartment
Cold water
1 A flask of water– the "ocean"– is heated to form water vapor, H_2O.
Condensation
"Oceanic" compartment
Heat
4 The compounds react in the "ocean," eventually forming purines, pyrimidines, and amino acids.

Figure 5.1 The apparatus Miller used to simulate the conditions of early Earth. In his experiments, the continuous cycle of heating and condensation produced prebiotic organic compounds.

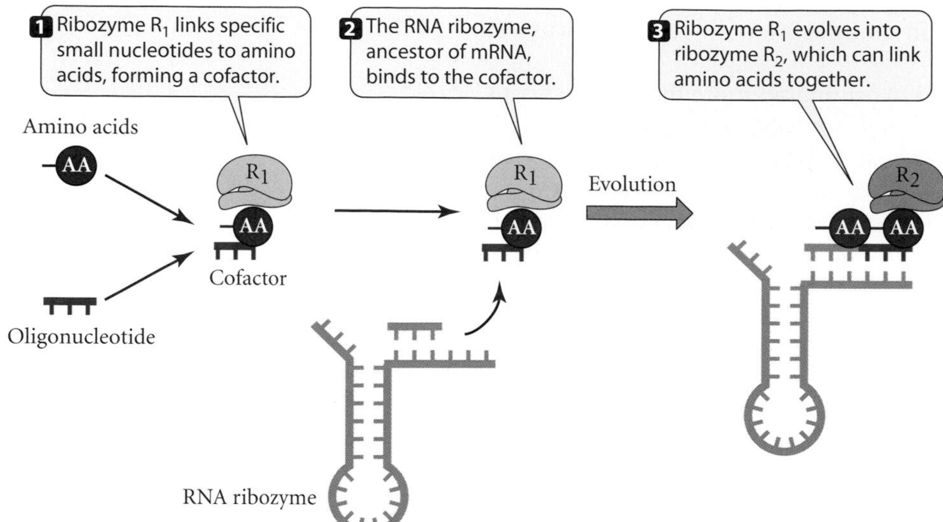

1 Ribozyme R1 links specific small nucleotides to amino acids, forming a cofactor.
2 The RNA ribozyme, ancestor of mRNA, binds to the cofactor.
3 Ribozyme R1 evolves into ribozyme R2, which can link amino acids together.
Amino acids
AA
R1
Cofactor
Oligonucleotide
Evolution
R2
RNA ribozyme

Figure 5.2 A hypothesis for the origin of polypeptide synthesis and the genetic code. An RNA ribozyme (green), an ancestor of mRNA, binds to a cofactor consisting of an amino acid (AA) and a short oligonucleotide, which have been joined by another ribozyme (R_1) that joins specific oligonucleotides to amino acids according to a primitive code. This system evolves to one in which ribozyme R_2, ancestor of the modern ribosome, links amino acids together. (After Maynard Smith and Szathmáry 1995.)

How protein enzymes evolved is perhaps the greatest unsolved problem. Eörs Szathmáry (1993) has suggested that this process began when cofactors, consisting of an amino acid joined to a short oligonucleotide sequence, aided RNA ribozymes in self-replication (Figure 5.2). Many contemporary coenzymes have nucleotide components. The next step may have been the stringing together of several such amino acid–nucleotide cofactors. Ultimately, the ribozyme evolved into the ribosome, the oligonucleotide component of the cofactor into transfer RNA, and the strings of amino acids into catalytic proteins. Such ensembles of macromolecules, packaged within lipid membranes, may have been precursors of the first cells—although many other features evolved between that stage and the only cells we know.

Precambrian Life

Prokaryotes

The Archean, prior to 2.5 billion years ago, and the Proterozoic, from 2.5 billion to 542 million years ago, are together referred to as **Precambrian time**. The oldest known rocks (3.8 billion years) contain carbon deposits that may indicate the existence of life. *The first fairly definite evidence of life dates from 3.5 billion years ago*, in the form of bacteria-like microfossils and layered mounds (stromatolites; Figure 5.3) with the same structure as those formed today along the edges of warm seas by cyanobacteria (blue-green bacteria).

The early atmosphere is generally thought to have had little oxygen, so the earliest organisms were anaerobic. *When photosynthesis evolved* in cyanobacteria and other bacteria, *it introduced oxygen into the atmosphere.* The atmospheric concentration of oxygen increased greatly about 2.2 billion years ago, probably due to geological processes that buried large quantities of organic matter and prevented it from being oxidized (Knoll 2003). As oxygen built up in the atmosphere, many organisms evolved the capacity for aerobic respiration, as well as mechanisms to protect the cell against oxidation.

Figure 5.3 Stromatolites formed by living cyanobacteria in Shark Bay, Australia. Similar structures are found in rocks of Archean age. (Photograph by the author.)

Living things today are classified into three "empires" or "domains": the Eucarya (all eukaryote organisms), and two groups of prokaryotes, the Archaea and Bacteria (see Figure 2.1). *For about 2 billion years—more than half the history of life—the two prokaryotic empires were the only life on Earth.* Today, many Archaea are anaerobic and inhabit extreme environments such as hot springs. [One such species is the source of the DNA polymerase enzyme (Taq polymerase) used for the PCR reaction that is the basis of much of modern molecular biology and biotech-

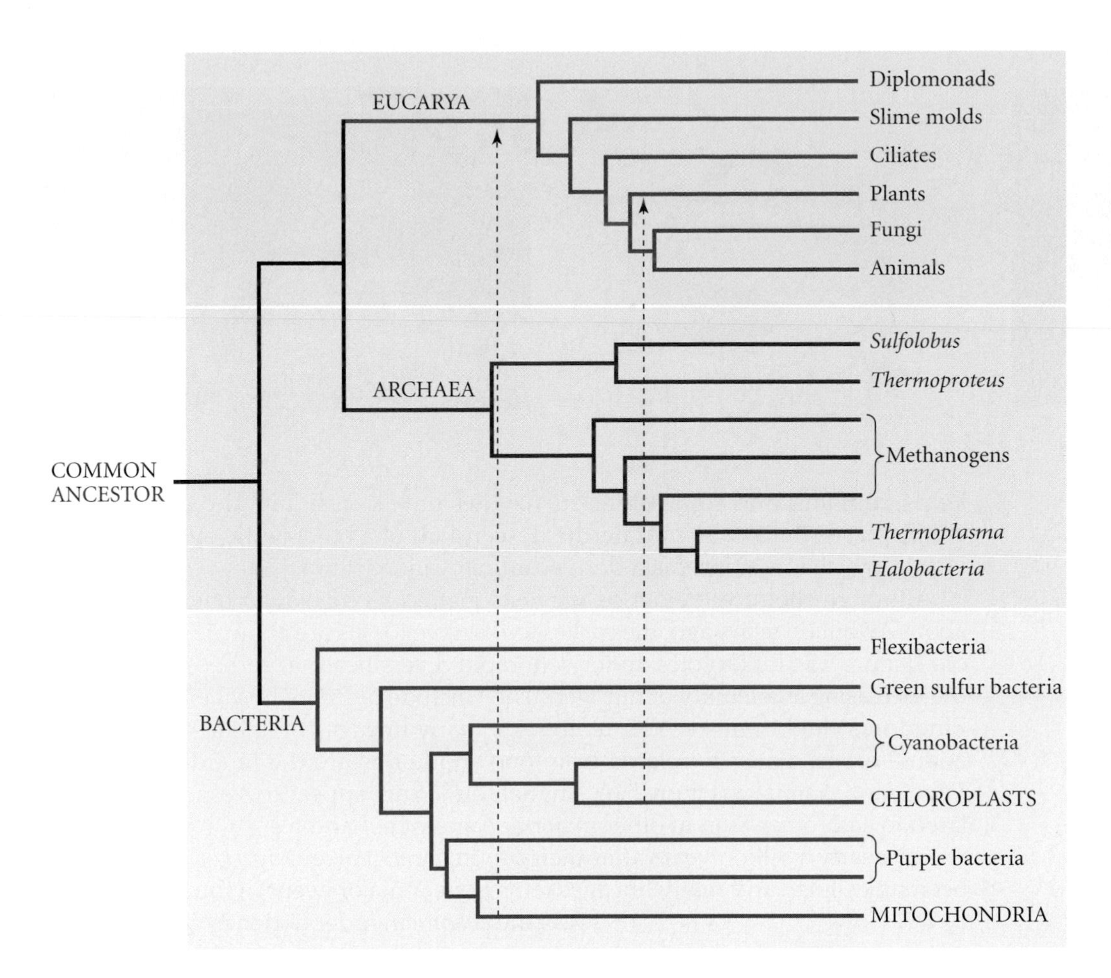

Figure 5.4 The basal branches of the tree of life: a current hypothesis of relationships among the three "empires," and among some lineages within each. Note the origin of mitochondria and chloroplasts from bacterial lineages. The dotted lines indicate their endosymbiotic association with early eukaryotic lineages. (After Knoll 2003 and other sources.)

nology.] The Bacteria are extremely diverse in their metabolic capacities, and many are photosynthetic. Molecular phylogenetic studies of prokaryotes and of the nuclear genes of eukaryotes show that Archaea and Eucarya are more closely related to each other than to Bacteria (Figure 5.4). However, some DNA sequences provide conflicting evidence about relationships both among and within the empires, implying that there was considerable lateral transfer of genes during the early history of life, when well-defined, integrated genomes did not yet exist (Woese 2000). Only later did cells become so complex and integrated that incorporating foreign DNA would reduce their ability to function.

Eukaryotes

A major event in the history of life was the origin of eukaryotes, which are distinguished by such features as a cytoskeleton and a nucleus with distinct chromosomes and a mitotic spindle. Most eukaryotes undergo meiosis, the basis of highly organized recombination and sexual reproduction. Almost all eukaryotes have mitochondria, and many have chloroplasts. Some eukaryotic protists, such as the intestinal parasite *Giardia*, lack mitochondria; however, their nuclear chromosomes harbor genes that are homologous to mitochondrial genes of other eukaryotes, showing that their ancestors must have had mitochondria. (Transfer of genes from mitochondria to the nucleus is a well-documented phenomenon—it has even been recorded in experimental populations of yeast.)

Mitochondria and chloroplasts are descended from bacteria that were ingested, and later became **endosymbionts**, in proto-eukaryotes (Margulis 1993; Maynard Smith and Szathmáry 1995) (see Figure 5.4). These organelles are similar to bacteria both in their ultrastructure and in their DNA sequences. Mitochondria are derived from the purple bacteria (a group that includes *E. coli*). Chloroplasts occur in several separate clades of eukaryotes, indicating that they arose from cyanobacteria at least six to nine times. What is more,

(A)

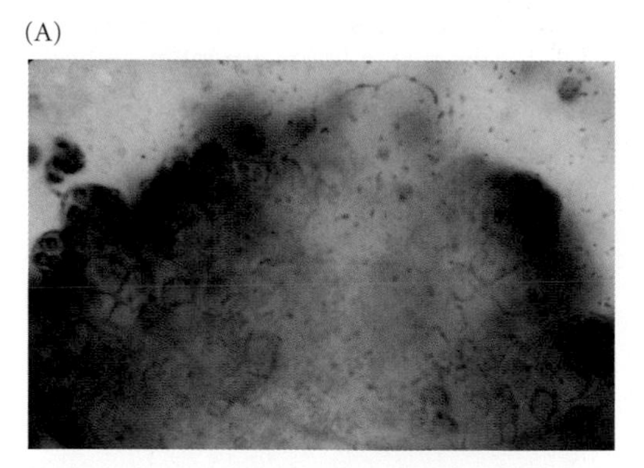

(B)

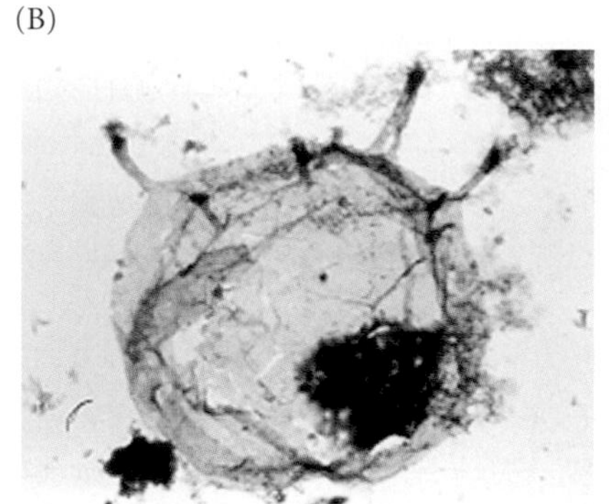

(C)

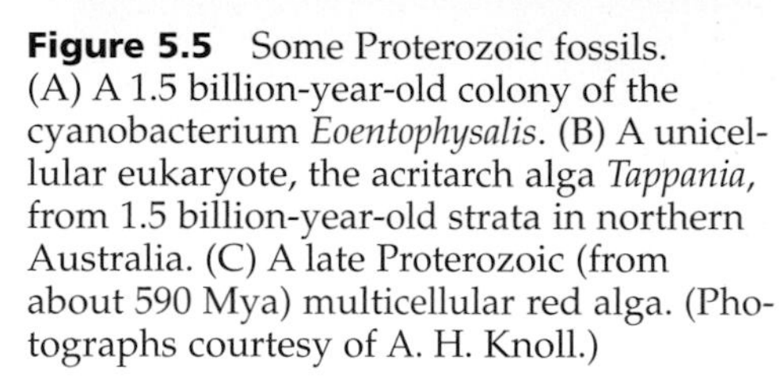

Figure 5.5 Some Proterozoic fossils. (A) A 1.5 billion-year-old colony of the cyanobacterium *Eoentophysalis*. (B) A unicellular eukaryote, the acritarch alga *Tappania*, from 1.5 billion-year-old strata in northern Australia. (C) A late Proterozoic (from about 590 Mya) multicellular red alga. (Photographs courtesy of A. H. Knoll.)

kelps, diatoms, and some other eukaryotes have secondary and even tertiary chloroplasts: their cells contain the descendant of a eukaryotic endosymbiont, within which is a chloroplast derived from cyanobacteria.

Although chemical evidence suggests that eukaryotes may have evolved by about 2.7 billion years ago, the earliest eukaryote fossils are about 1.5 billion years old (Figure 5.5). Eukaryotes underwent rapid diversification, perhaps triggered by the increasing availability of nitrogen. They include far more lineages than the five kingdoms cited in most older textbooks: many lineages of "algae" and "protozoans" are as distantly related from one another as are the kingdoms Plantae, Fungi, and Animalia (Figure 5.6). Animals and fungi appear to be more closely related to each other than to other major lineages of eukaryotes.

For nearly a billion years after their origin, almost all eukaryotes seem to have been unicellular, and many lineages remain so. A major event in the history of life was the *evolution*, independently in several lineages, *of multicellular organisms*. Multicellularity is a prerequisite for large size and for the development of elaborate organ systems. The evolution of tissues and organs required the *evolution of gene regulation*: ways of controlling the expression of different genes in different cells.

Proterozoic life

The oldest fossils of multicellular animals are about 640 Myr old. Among the first animal fossils are the creatures known as the EDIACARAN FAUNA of the late Proterozoic (about 600 Mya) and early Cambrian. Most of these animals were soft-bodied, lacked skeletons, and appear to have been flat creatures that crept or stood on the sea floor (Figure 5.7). They are hard to classify with reference to later animals, but some may have been stem groups of the Cnidaria and the Bilateria.

Figure 5.6 A recent estimate of the phylogeny of some of the major lineages of eukaryotes, inferred from DNA sequences. Only a few of these lineages were recognized as "kingdoms" in traditional classifications. The position of Parabasalia and Diplomonadida, both unicellular parasites that lack mitochondria, is uncertain. (After Baldauf et al. 2000 and McGrath and Katz 2004.)

(A)

(B)

Figure 5.7 Members of the Ediacaran fauna. (A) *Mawsonites spriggi*, may be a cnidarian relative of sea anemones, although this is not certain. (B) The actual relationship of the wormlike *Dickinsonia costata* to later animals is unknown. (A © The Natural History Museum, London; B © Ken Lucas/Visuals Unlimited.)

Paleozoic Life: The Cambrian Explosion

The Paleozoic era began with the Cambrian period, starting about 542 Mya. For the first 10 Myr or so, animal diversity was low; then, during a period of just 10 to 25 million years, *almost all the modern phyla and classes of skeletonized marine animals*, as well as many extinct groups, *appeared in the fossil record*. This interval marks the first appearance of brachiopods (Figure 5.8A), trilobites (Figure 5.8B) and other classes of arthropods, various classes of molluscs, echinoderms, and many others. The remarkably well-preserved Cambrian fauna of the Burgess Shale of British Columbia (Figure 5.9 and the chapter-opening photograph) includes animals rather different from any that succeeded them. The Cambrian diversification included the earliest jawless (agnathan) vertebrates: the recently discovered early Cambrian *Haikouichthys* had eyes, gill pouches, notochord, segmented musculature, and other features resembling those of larval lampreys (Figure 5.10A; Shu et al. 2003), and the late Cambrian conodonts had teeth made of cellular bone (Figure 5.10B). Most of the fundamentally different body plans (often called **Baupläne**, German for "construction plans" or blueprints) known in animals apparently evolved during the Cambrian—perhaps the most dramatic adaptive radiation in the history of life.

The reasons for this diversification, called the **Cambrian explosion**, have been the subject of vigorous debate (Erwin 1991; Lipps and Signor 1992). How and why did so many great changes evolve in such a short time? By applying a molecular clock to the DNA sequence divergence among the living animal phyla, Gregory Wray and colleagues (1996) showed that the phyla actually originated long before their first appearances in the fossil record, perhaps 1000 Mya. Before the "explosion," however, most animals either lacked skeletons or were so small that their Precambrian remains have not been found. The Cambrian explosion thus consists of rapid diversification within clades that had evolved much

(A)

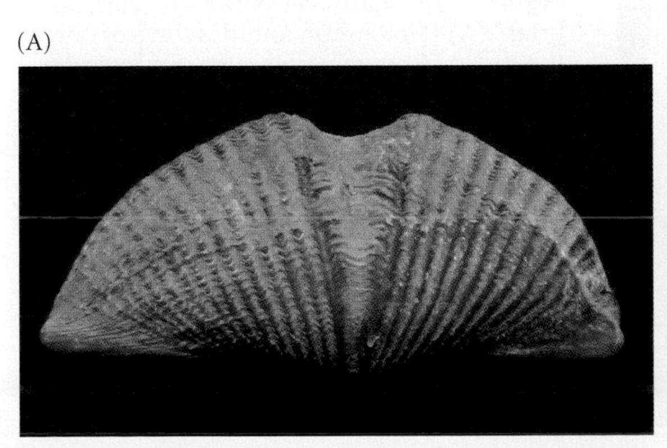

(B)

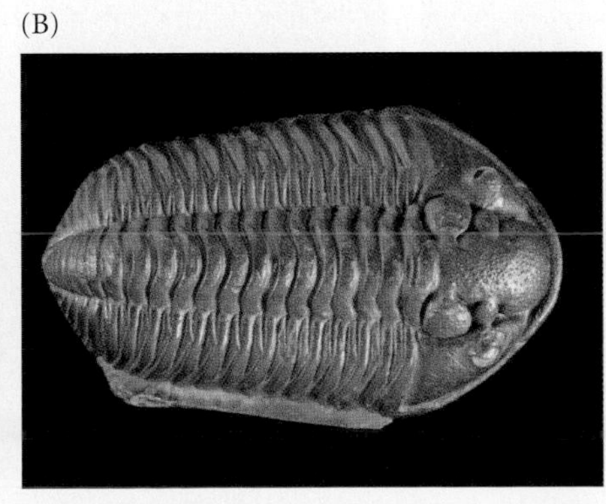

Figure 5.8 Two animal groups that first appeared during the Cambrian explosion. (A) A Devonian brachiopod (*Mucrospirifer*, phylum Brachiopoda). Although brachiopods have clamlike shells, they are not closely related to true bivalves, which are members of the phylum Mollusca. Only a few brachiopod species survive today. (B) A trilobite, *Calymene blumenbachii*, also from the Silurian. Trilobites are an arthropod group, and were very diverse throughout the Paleozoic. They became extinct at the end of the Permian. (A photo by David McIntyre; B © The Natural History Museum, London.)

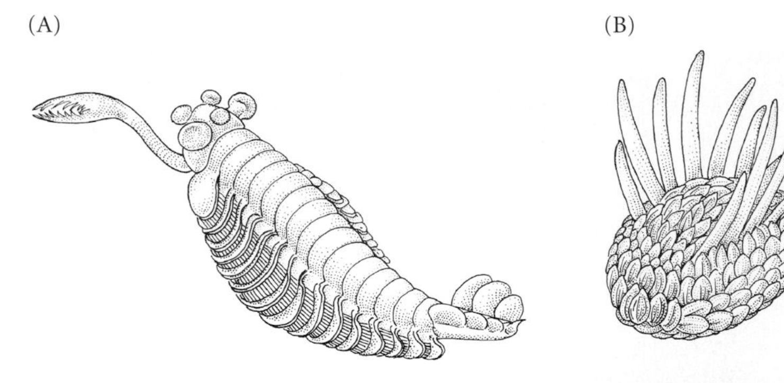

Figure 5.9 Artist's reconstructions of two of the peculiar animals of the Cambrian Burgess Shale. (A) *Opabinia*, probably an arthropod. (B) *Wiwaxia*, perhaps related to annelid worms. Characteristics confirmed in multiple fossils can be illustrated in a single drawing, which therefore often can convey structures more clearly than a photograph of a single fossil specimen. For that reason, taxa are often illustrated by drawings rather than by photographs in this chapter. (Drawings by Marianne Collins, from Gould 1989.)

earlier—diversification that included the evolution of shells and skeletons. A combination of genetic and ecological causes may account for this diversification (Knoll and Carroll 1999; Knoll 2003). Regulatory genes that govern the differentiation of body parts (such as the Hox genes; see Chapter 20) may have undergone major evolutionary changes at this time. Moreover, the extinction of the archaic Ediacaran animals—which may have been brought about by a decrease in oxygen levels—released the survivors from competition, allowing them to diversify just as mammals did after the demise of dinosaurs (see Chapter 7).

Molecular phylogenetic studies show that animals are most closely related to the unicellular choanoflagellates. Sponges (phylum Porifera), which have many choanoflagellate-like cells, are the sister group of the other animals, known collectively as Metazoa (Figure 5.11). The radially symmetrical Cnidaria (jellyfishes, corals) and the Ctenophora (comb jellies) are basal branches relative to the Bilateria—bilaterally symmetrical animals with a head, often equipped with mouth appendages, sensory organs, and a brain. The Bilateria include three major branches: the deuterostomes, in which the blastopore formed during gastrulation becomes the anus, and two groups of protostomes, in which the blastopore becomes the mouth. The largest deuterostome phyla are the Echinodermata (starfishes and relatives) and the Chordata (including the vertebrates, tunicates, and amphioxus). Protostomes form two major clades: the Ecdysozoa (arthropods, nematodes, and some smaller phyla) and the Lophotrochozoa (molluscs, annelid worms, brachiopods, and a variety of other groups).

The end of the Cambrian (500 Mya) was marked by mass extinction. The trilobites, of which there had been more than 90 Cambrian families, were greatly reduced, and several classes of echinoderms became extinct. As Stephen Jay Gould (1989) emphasized, if the early vertebrates had also succumbed, we would not be here today. The same may be said about every point in subsequent time: had our ancestral lineage been among the enormous number of lineages that became extinct, humans would not have evolved, and perhaps no other form of life like us would have, either.

(A)

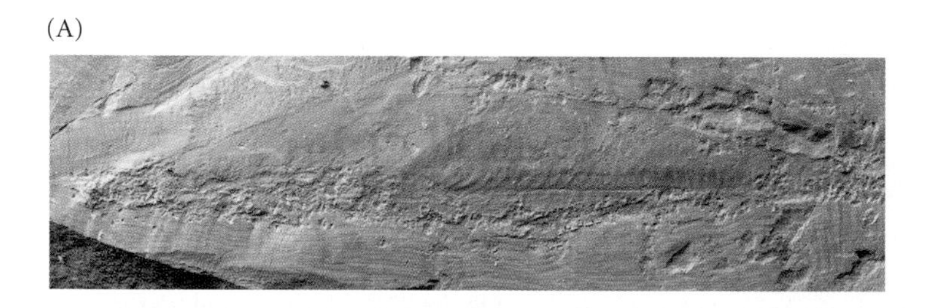

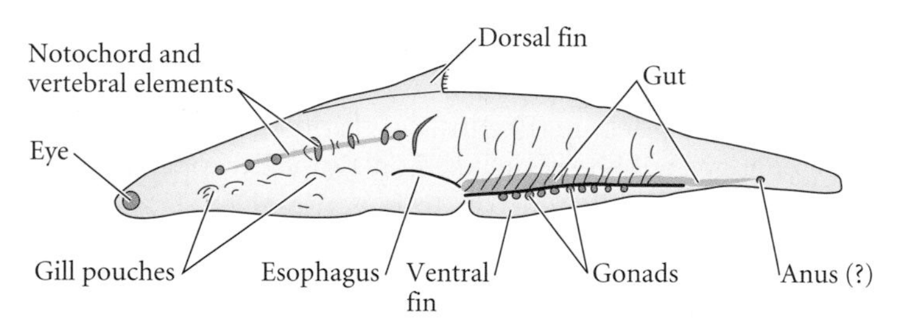

(B)

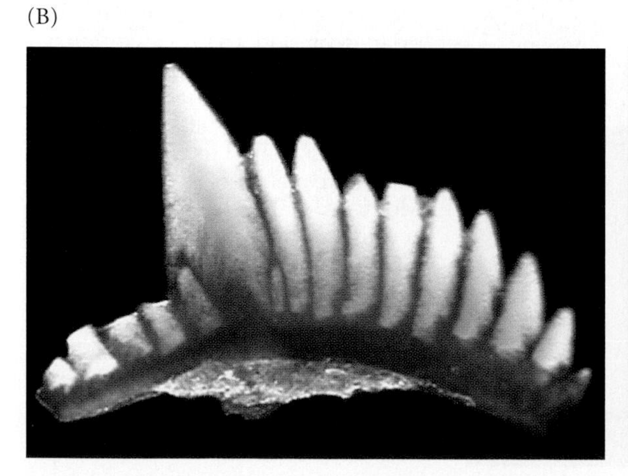

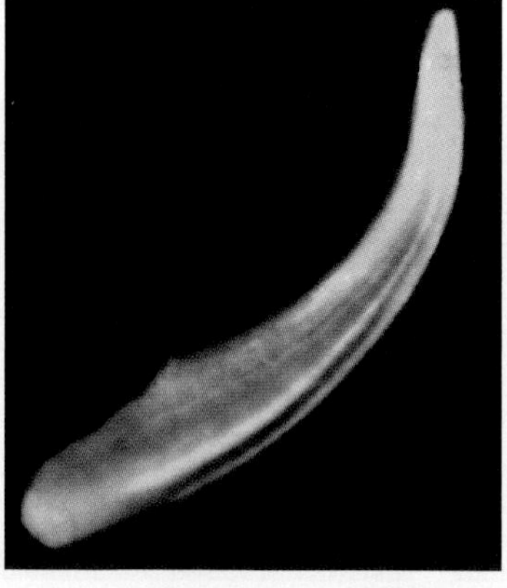

Figure 5.10 Early vertebrates of the Cambrian. (A) Photograph and drawing of one of the earliest known vertebrates, *Haikouichthys*, of the early Cambrian. The drawing calls attention to features interpreted as eye, notochord, vertebral elements, dorsal fin, esophagus, gill pouches, ventral fin, and anus, indicating a postanal tail region characteristic of vertebrates. (B) Bony, toothlike structures of Cambrian conodonts. Conodonts were slender, finless chordates that are thought to be related to agnathans (jawless vertebrates such as lampreys). (A photo courtesy of D.-G. Shu, after Shu et al. 2003; B courtesy of James Davison, M.D.)

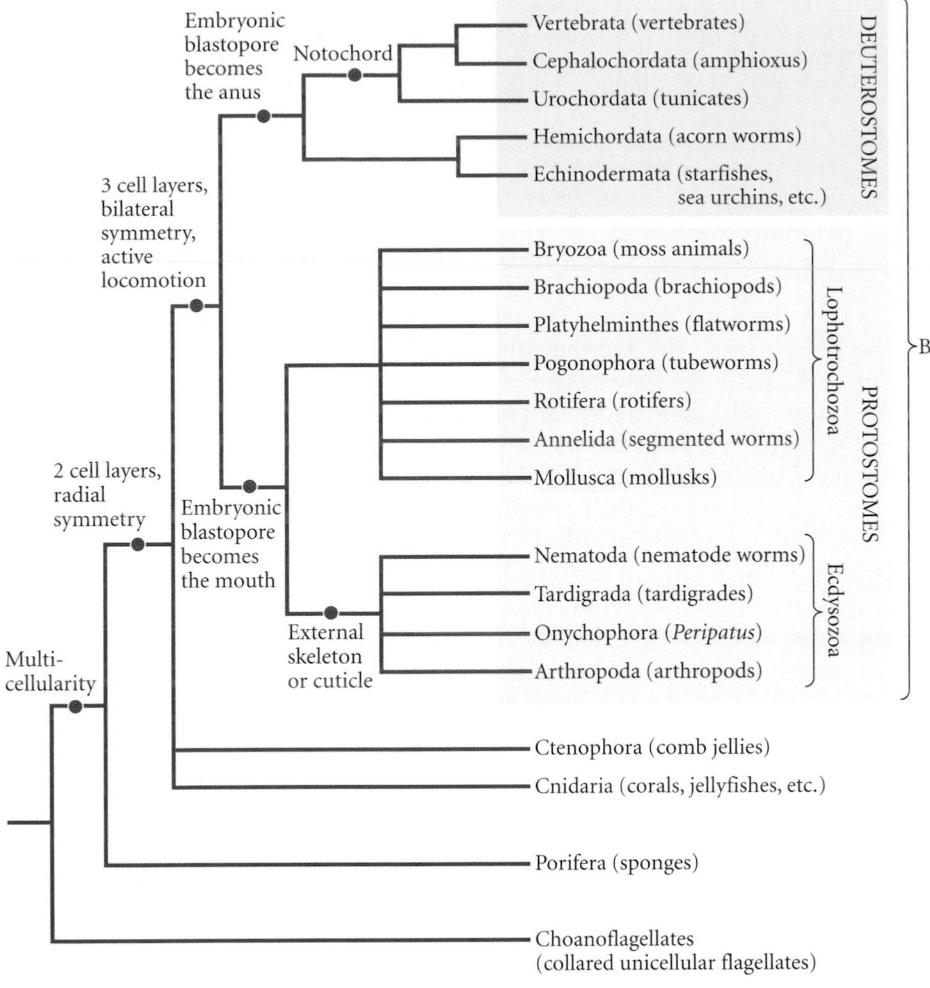

Figure 5.11 A recent estimate of relationships among animal phyla, based on the sequence of genes encoding ribosomal RNA. Some relationships among phyla are uncertain, and are shown as unresolved polytomies (i.e., multiple branches from a single stem, as opposed to a totally dichotomous branching pattern, seen here in the deuterostome lineage). (After Adoutte et al. 2000.)

Paleozoic Life: Ordovician to Devonian

Marine life

Many of the *animal phyla diversified greatly in the Ordovician* (488–444 Mya), giving rise to *many new classes and orders*, including as many as 21 classes of echinoderms (Figure 5.12). These classes differed greatly from one another and from the 5 classes of echinoderms (e.g., starfishes and sea urchins) that still exist. Many of the predominant Cambrian groups did not recover their earlier diversity, so the Ordovician fauna had a very different character from that of the Cambrian. Most Ordovician animals were EPIFAUNAL (i.e., living on the surface of the sea floor), although some bivalves (clams and relatives) evolved an INFAUNAL (burrowing) habit. The major large predators were starfishes and nautiloids (shelled cephalopods; that is, molluscs related to squids). Reefs were built by two groups of corals, with contributions from sponges, bryozoans, and cyanobacteria. The Ordovician ended with a mass extinction that in proportional terms may have been the second largest of all time. It may have been caused by a drop in temperature and a drop in sea level, for there were glaciers at this time in the polar regions of the continents.

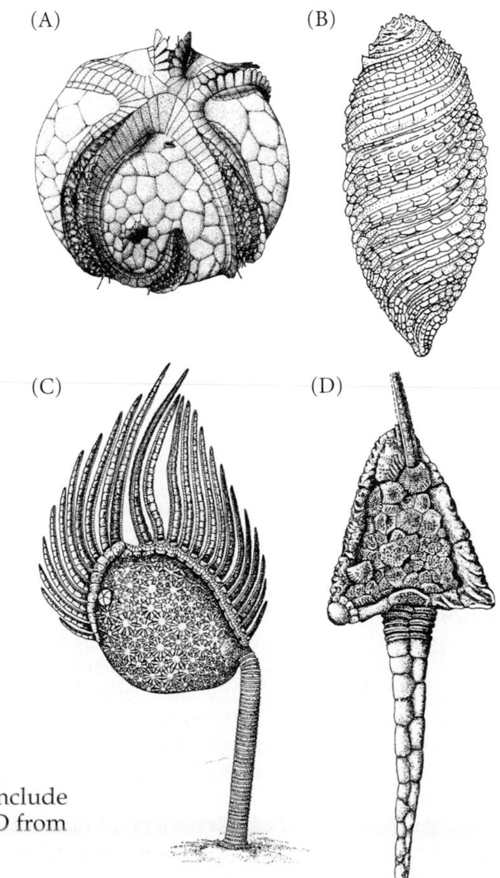

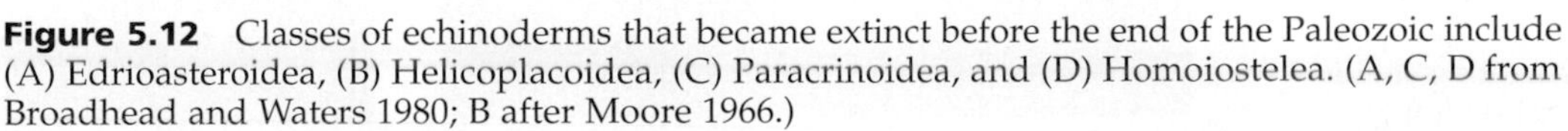
Figure 5.12 Classes of echinoderms that became extinct before the end of the Paleozoic include (A) Edrioasteroidea, (B) Helicoplacoidea, (C) Paracrinoidea, and (D) Homoiostelea. (A, C, D from Broadhead and Waters 1980; B after Moore 1966.)

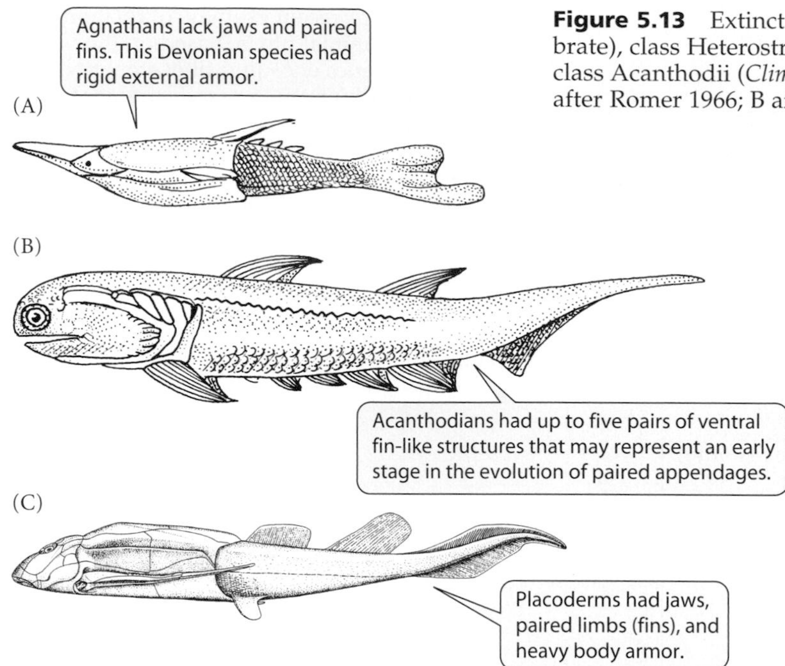

(A)
Agnathans lack jaws and paired fins. This Devonian species had rigid external armor.
(B)
Acanthodians had up to five pairs of ventral fin-like structures that may represent an early stage in the evolution of paired appendages.
(C)
Placoderms had jaws, paired limbs (fins), and heavy body armor.

Figure 5.13 Extinct Paleozoic classes of vertebrates. (A) An agnathan (jawless vertebrate), class Heterostraci (*Pteraspis*, Devonian). (B) A gnathostome (jawed vertebrate), class Acanthodii (*Climatius*, Devonian). (C) A placoderm (*Bothriolepis*, Devonian). (A after Romer 1966; B after Romer and Parsons 1986; C after Carroll 1988.)

One group that originated during the Ordovician and that survived the mass extinction flourished and became diverse in the subsequent environment. The agnathans, armored fishlike vertebrates such as Heterostraci, lacked jaws and paired fins (Figure 5.13A). During the Silurian (439–408 Mya), the jawless agnathans were joined by the *first known gnathostomes*, marine vertebrates with jaws and paired fins (Figure 5.13B,C). These groups continued to flourish as the Silurian was succeeded by a dramatic period in evolutionary history, the Devonian (408–354 Mya). In the same period, the nautiloids gave rise to the ammonoids, shell-bearing cephalopods that are among the most diverse groups of extinct animals (Figure 5.14).

The two subclasses of bony fishes (Osteichthyes) are both recorded first in the late Devonian: the lobe-finned fishes (Sarcopterygii), which included diverse lungfishes and rhipidistians (see Chapter 4), and the ray-

Figure 5.14 Ammonoids and nautiloids. Shells housed the squidlike body of these cephalopod molluscs. (A–C) Three of the diverse forms of ammonoid shells. (A) *Aulacostephanus* (late Jurassic), a microconch. (B) *Nostroceras* (late Cretaceous), demonstrating an unusual coiling pattern. (C) *Maorites* (late Cretaceous); this fossil was found in Antarctica. (D) An orthoconic nautiloid, believed to belong to the same group as the modern *Nautilus*. These animals had noncoiling shells. (Photos © The Natural History Museum, London.)

finned fishes (Actinopterygii), which would later diversify into the largest group of modern fishes, the teleosts.

Terrestrial life

Terrestrial plants, including mosses, liverworts, and tracheophytes (vascular plants), are a monophyletic group that evolved from green algae (Chlorophyta) (Figure 15.15; Kenrick and Crane 1997a,b). Living on land required the evolution of an external surface and spores impermeable to water, structural support, vascular tissue to transport water within the plant body, and internalized sexual organs, protected from desiccation. The *first known*

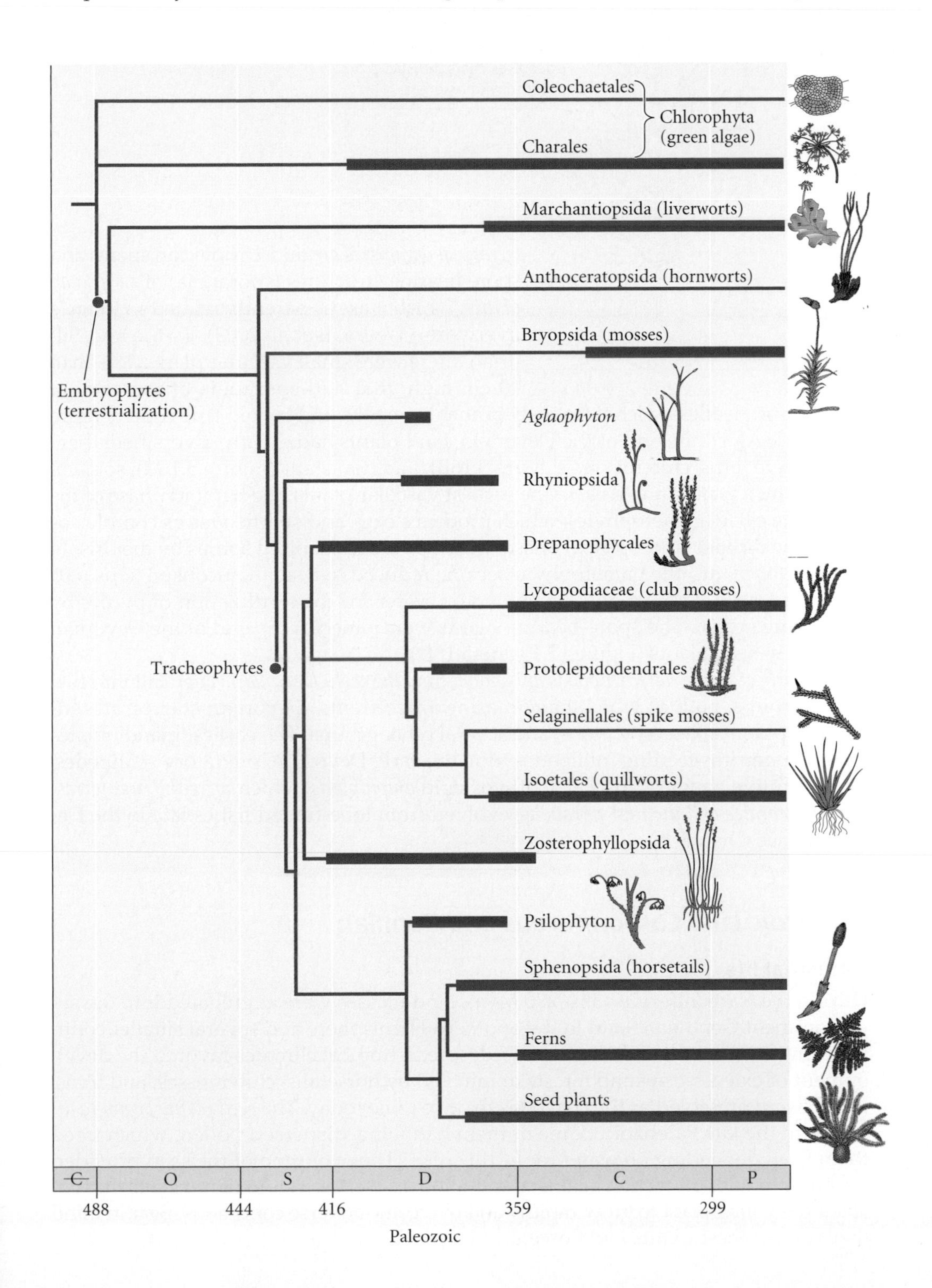

Figure 5.15 The phylogeny and Paleozoic fossil record of major groups of terrestrial plants and their closest relatives among the green algae (Chlorophyta). The broad bars show the known temporal distribution of each group in the fossil record. The Coleochaetales and Charales (two groups of green algae), liverworts, mosses, club mosses, Selaginellales, Isoetales, horsetails, ferns, and seed plants have living representatives. (After Kenrick and Crane 1997a.)

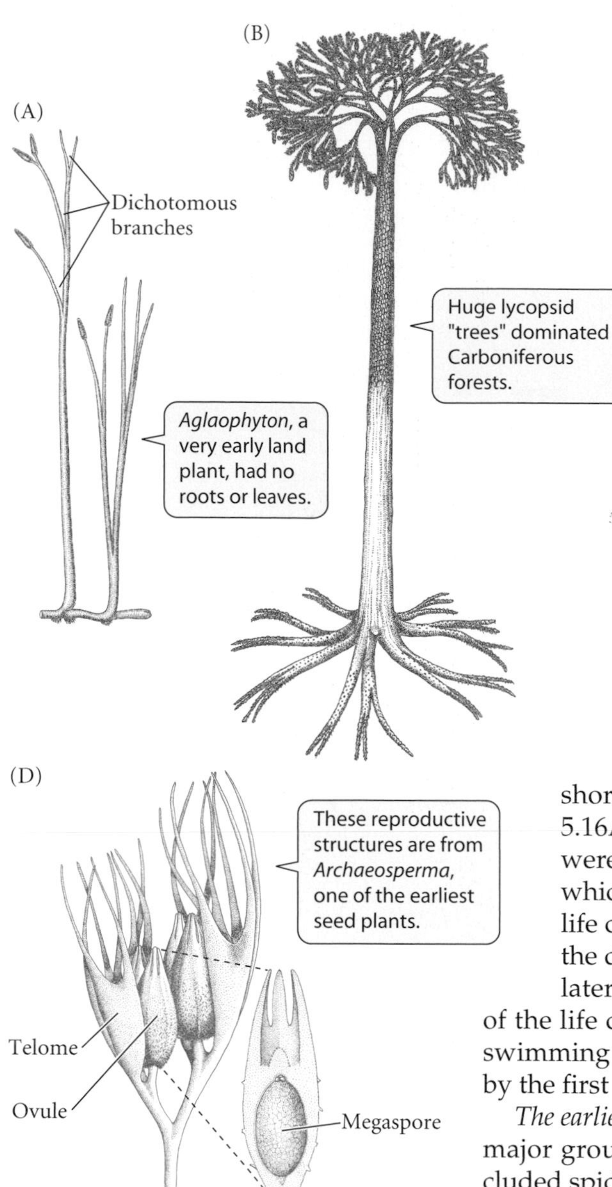
(A)
Dichotomous branches
Aglaophyton, a very early land plant, had no roots or leaves.
(B)
Huge lycopsid "trees" dominated Carboniferous forests.
(C)
The stems of Permian sphenopsids had features characteristic of today's horsetails.
(D)
These reproductive structures are from Archaeosperma, one of the earliest seed plants.
Telome
Ovule
Megaspore
Sagittal section

Figure 5.16 Paleozoic vascular plants, portrayed at different scales. (A) From the Devonian, *Aglaophyton* was less than 10 cm tall. (B) The large Carboniferous club moss *Lepidodendron*. (C) Part of the stem of a Permian sphenopsid tree. (D) Reproductive structure of a Devonian seed plant, *Archaeosperma*. (A from Kidston and Lang 1921; B, D from Stewart 1983; C from Boureau 1964.)

terrestrial organisms are mid-Ordovician spores and spore-bearing structures (sporangia) of *very small plants*, which were apparently related to today's liverworts (Wellman et al. 2003). By the mid-Silurian, there were small vascular plants, less than 10 cm high, that had sporangia at the ends of short, leafless, dichotomously branching stalks and lacked true roots (Figure 5.16A). By the end of the Devonian, land plants had greatly diversified: there were ferns, club mosses (Figure 5.16B), and horsetails (Figure 5.16C), some of which were large trees. In the earliest vascular plants, the haploid phase of the life cycle (gametophyte), which produces eggs and sperm, was as complex as the diploid (sporophyte) phase, which produces haploid spores by meiosis. In later plants, the gametophyte became reduced to a small, inconspicuous part of the life cycle. These plants depended on water for the fertilization of ovules by swimming sperm. The spore-bearing plants were joined at the end of the Devonian by the first seed plants (Figures 5.15 and 5.16D).

The earliest terrestrial arthropods are known from the early Devonian. They fall into two major groups, both of which have marine antecedents. Devonian chelicerates included spiders, mites, scorpions, and several other groups. The earliest mandibulates included detritus-feeding millipedes from the early Devonian, predatory centipedes, and primitive wingless insects. Ichthyostegid *amphibians*, which *were the first terrestrial vertebrates and the first tetrapods*, evolved from lobe-finned fishes late in the Devonian (see Chapter 4).

Paleozoic Life: Carboniferous and Permian

Terrestrial life

During the Carboniferous (354–290 Mya), land masses were aggregated into the supercontinent Gondwanaland in the Southern Hemisphere and several smaller continents in the Northern Hemisphere. Widespread tropical climates favored the development of extensive swamp forests dominated by horsetails, club mosses, and ferns, which were preserved as the coal beds that we mine today. The *seed plants began to diversify* in the late Paleozoic. Some of them had wind-dispersed pollen, which freed them from depending on water for fertilization. The evolution of the seed provided the embryo with protection against desiccation, as well as a store of nutrients that enabled the young plant to grow rapidly and overcome adverse conditions. Bear in mind that none of these plants had flowers.

During the Carboniferous, the *first winged insects evolved*, and they rapidly diversified into many orders, including primitive dragonflies, orthopteroids (roaches, grasshoppers, and relatives) and hemipteroids (leafhoppers and their relatives). In the Permian, the first insect groups with complete metamorphosis (distinct larval and pupal stages) evolved, including beetles, primitive flies (Diptera), and the ancestors of the closely related Trichoptera (caddisflies) and Lepidoptera (moths and butterflies). Some orders of insects became extinct at the end of the Permian.

Amphibians were diverse in the Carboniferous, but most became extinct by the end of the Permian. Several late Carboniferous and early Permian lineages (anthracosaurs) have been variously classified as amphibians or reptiles. They gave rise to the *first known amniotes*, the captorhinomorphs (see Figure 5.19). These primitive amniotes soon gave rise to the *synapsids*, which *included the ancestors of mammals and increasingly acquired mammal-like features* (see Chapter 4). The first amniotes also gave rise to the diapsids, a major reptilian stock whose descendants dominated the Mesozoic landscape.

Aquatic life

During the Permian, the continents approached one another, and by its end they *formed a single world continent*, **Pangaea** (Figure 5.17A). The sea level dropped to its lowest point in history, and climates were greatly altered by the arrangement of land and sea. These changes may have been instrumental in one of the most significant events in the history of life, the *end-Permian mass extinction* (Erwin 1993). It is estimated that in this, *the most massive extinction event in the history of the Earth so far*, at least 52 percent of the families, and perhaps as many as 96 percent of all species, of skeleton-bearing marine invertebrates became extinct over a span of 5 to 8 Myr. Groups such as ammonites, stalked echinoderms, brachiopods, and bryozoans declined greatly, and major taxa such as trilobites and several major groups of corals became extinct. Except for the loss of some orders of insects and of many families of amphibians and mammal-like reptiles, extinctions on land were, curiously, much less pronounced.

Mesozoic Life

The Mesozoic era, divided into the Triassic (251–200 Mya), Jurassic (200–145 Mya), and Cretaceous (145–65.5 Mya) periods, is often called the Age of Reptiles. By its end, the Earth's flora and fauna were acquiring a rather modern cast, but this was preceded by the evolution of some of the most extraordinary creatures of all time. Pangaea began to break up, beginning with the Jurassic formation of the Tethyan Seaway between Asia and Africa, and then the full *separation of northern land masses (Laurasia) from a southern continent (Gondwanaland)* (Figure 5.17B). **Laurasia** began to separate into several fragments during the Jurassic, but northeastern North America, Greenland, and western Europe remained connected until well into the Cretaceous. The southern continent, **Gondwanaland**, consisted of Africa, South America, India, Australia, New Zealand, and Antarctica. *These land masses slowly separated in the late Jurassic and the Cretaceous*, but even then the South Atlantic formed only a narrow seaway between Africa and South America (Figure 5.17C). Throughout the Mesozoic, the sea level rose, and many continental regions were covered by shallow epicontinental seas. Although the polar regions were cool, *most of the Earth enjoyed warm climates*, with global temperatures reaching an all-time high in the mid-Cretaceous, after which substantial cooling occurred.

Marine life

During the Triassic, many of the marine groups that had been decimated during the end-Permian extinction again diversified; ammonoids, for example, had increased from two genera to more than a hundred by the middle Triassic (see Figure 5.14). Planktonic foraminiferans (shelled protists) and modern corals evolved, and bony fishes continued to radiate. Another *mass extinction occurred at the end of the Triassic*, again decimating groups such as ammonoids and bivalves. These groups underwent yet another adaptive radiation during the Jurassic, which marked the beginning of the "*Mesozoic marine revolution*,"

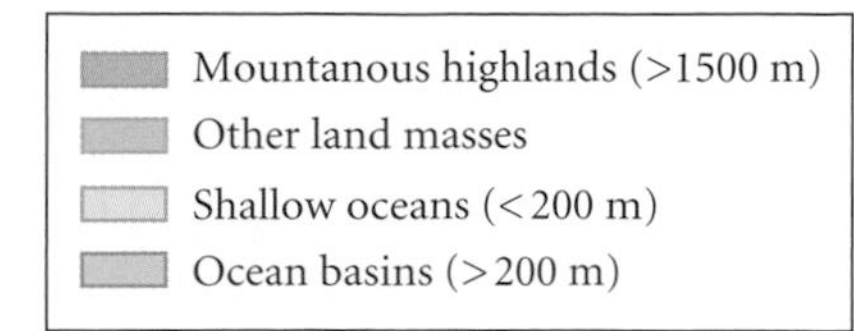

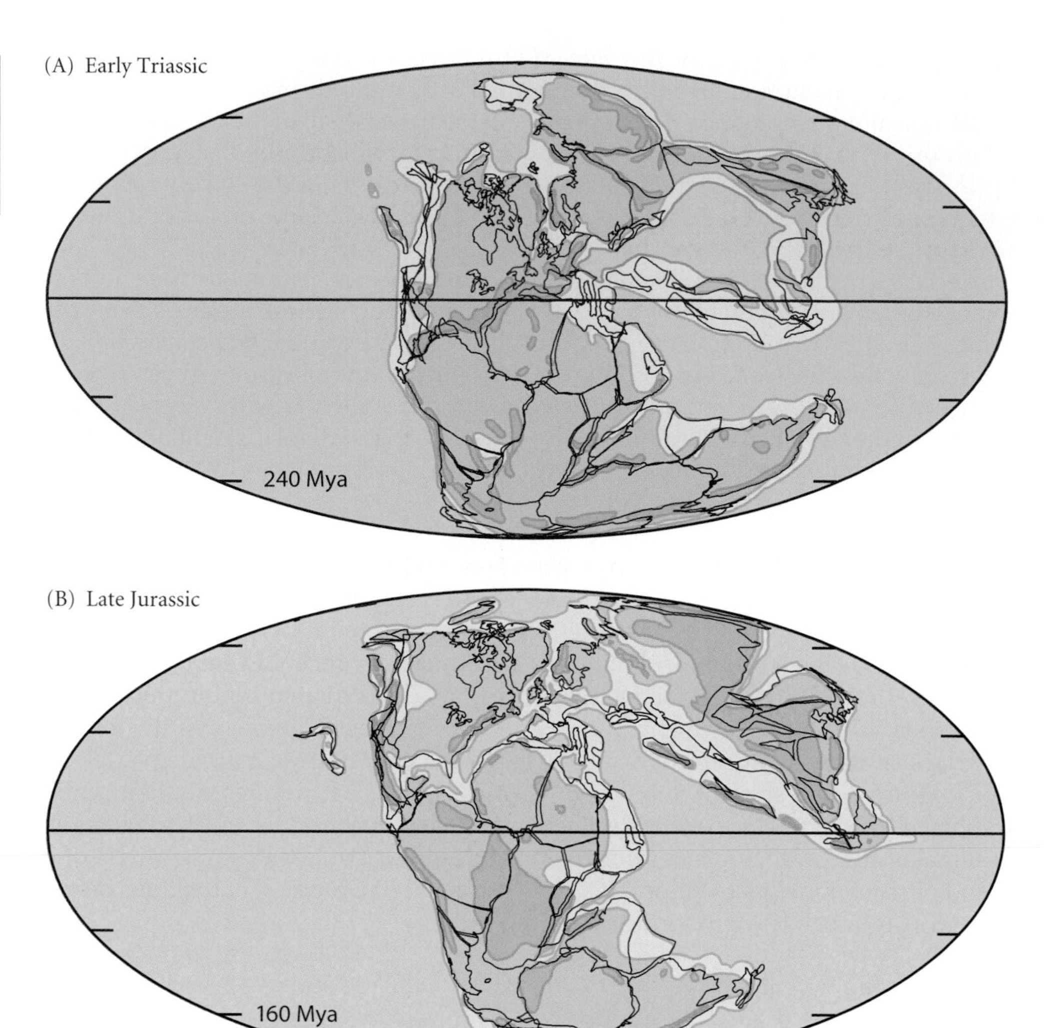

Figure 5.17 The distribution of land masses at several points in geological time. (A) In the earliest Triassic, most land was aggregated into a single mass (Pangaea). (B) Eurasia and North America were fairly separate by the late Jurassic. (C) Gondwanaland had become fragmented into most of the major southern land masses by the late Cretaceous. (D) By the late Oligocene, the land masses were close to their present configurations. The outlines of the modern-day continents are visible in all of the maps; other black lines delineate important tectonic plate boundaries. (Maps © 2004 by C. R. Scotese/ PALEOMAP Project.)

so called because of the evolution of crabs and bony fishes with the ability to crush mollusc shells and of molluscs with the kinds of protective mechanisms, such as thick shells and spines, that characterize many molluscs today (Vermeij 1987). The teleosts, today's dominant group of bony fishes, evolved and began to diversify.

During the Jurassic and Cretaceous, modern groups of gastropods (snails and relatives), bivalves, and bryozoans rose to dominance, gigantic sessile bivalves (rudists) formed reefs, and the seas harbored several groups of marine reptiles. The *end of the Cretaceous is marked by* what is surely the best-known *mass extinction* (often called the *K/T extinction*, using the abbreviations for Cretaceous and Tertiary). Ammonoids, rudists, marine reptiles, and many families of invertebrates and planktonic protists became entirely extinct. The last of the non-avian dinosaurs became extinct at this time. Most paleontologists believe this extinction was caused by the impact of an asteroid or some other extraterrestrial body (see Chapter 7).

(A)

(C) Late Cretaceous

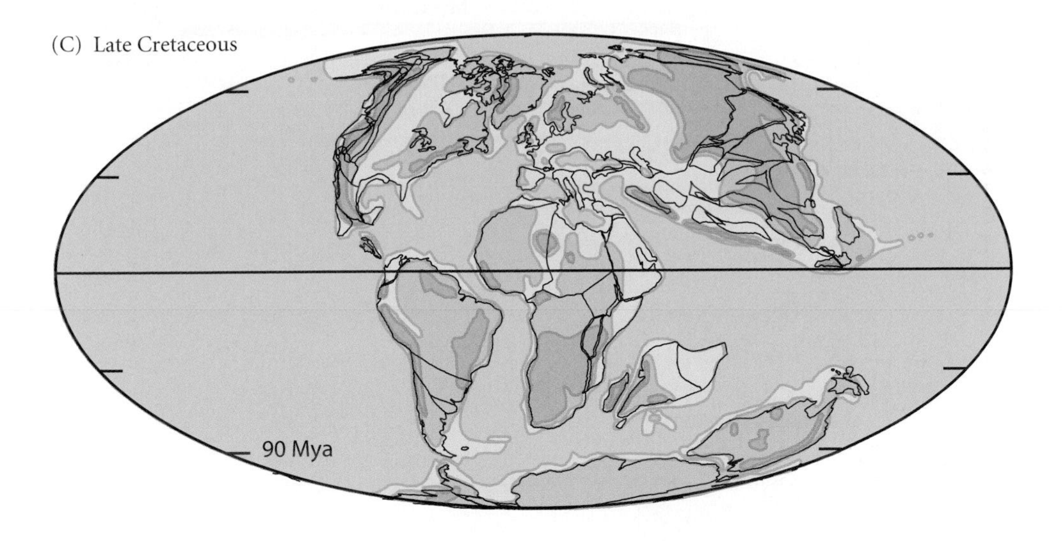

(D) Late Oligocene

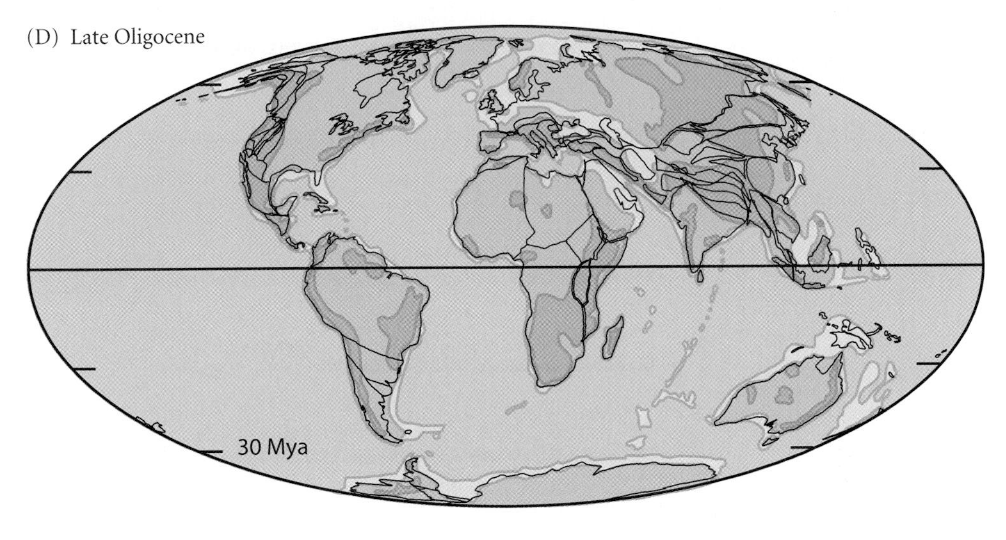

Terrestrial plants and arthropods

For most of the Mesozoic, the flora was dominated by "gymnosperms" (i.e., seed plants that lack flowers). The major groups were the cycads (Figure 5.18A) and the conifers and their relatives, including *Ginkgo*, a Triassic genus that has left one species that survives as a "living fossil" (Figure 5.18B). The *angiosperms, or flower-*

Figure 5.18 Seed plants. (A) A living cycad, a member of a once diverse group. (B) A leaf of the living ginkgo (*Ginkgo biloba*) next to (C) a fossilized *Ginkgo* leaf from the Paleocene. (D) *Protomimosoidea*, a Paleocene/Eocene fossil member of the legume family that includes mimosas and acacias. (A © John Cancalosi/Peter Arnold, Inc./Alamy Images; B photo by David McIntyre; C © The Natural History Museum, London; D courtesy of W. L. Crepet.)

(B)

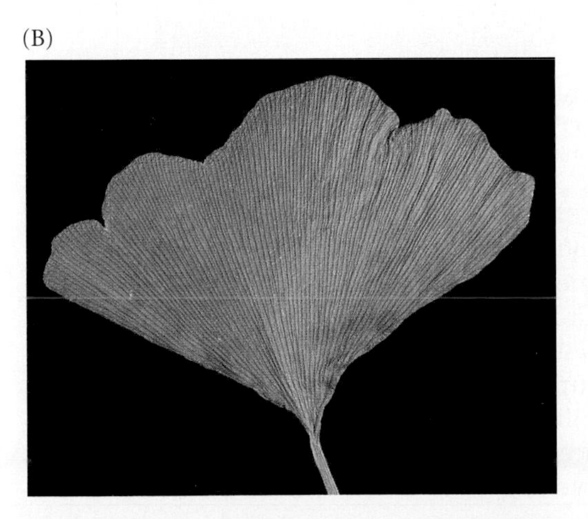

(C)

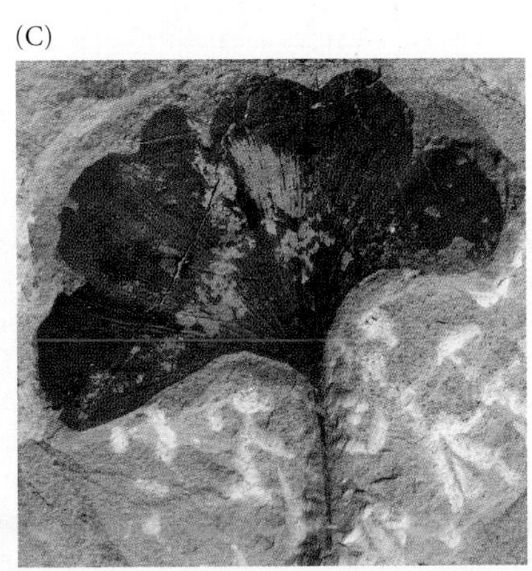

(D)

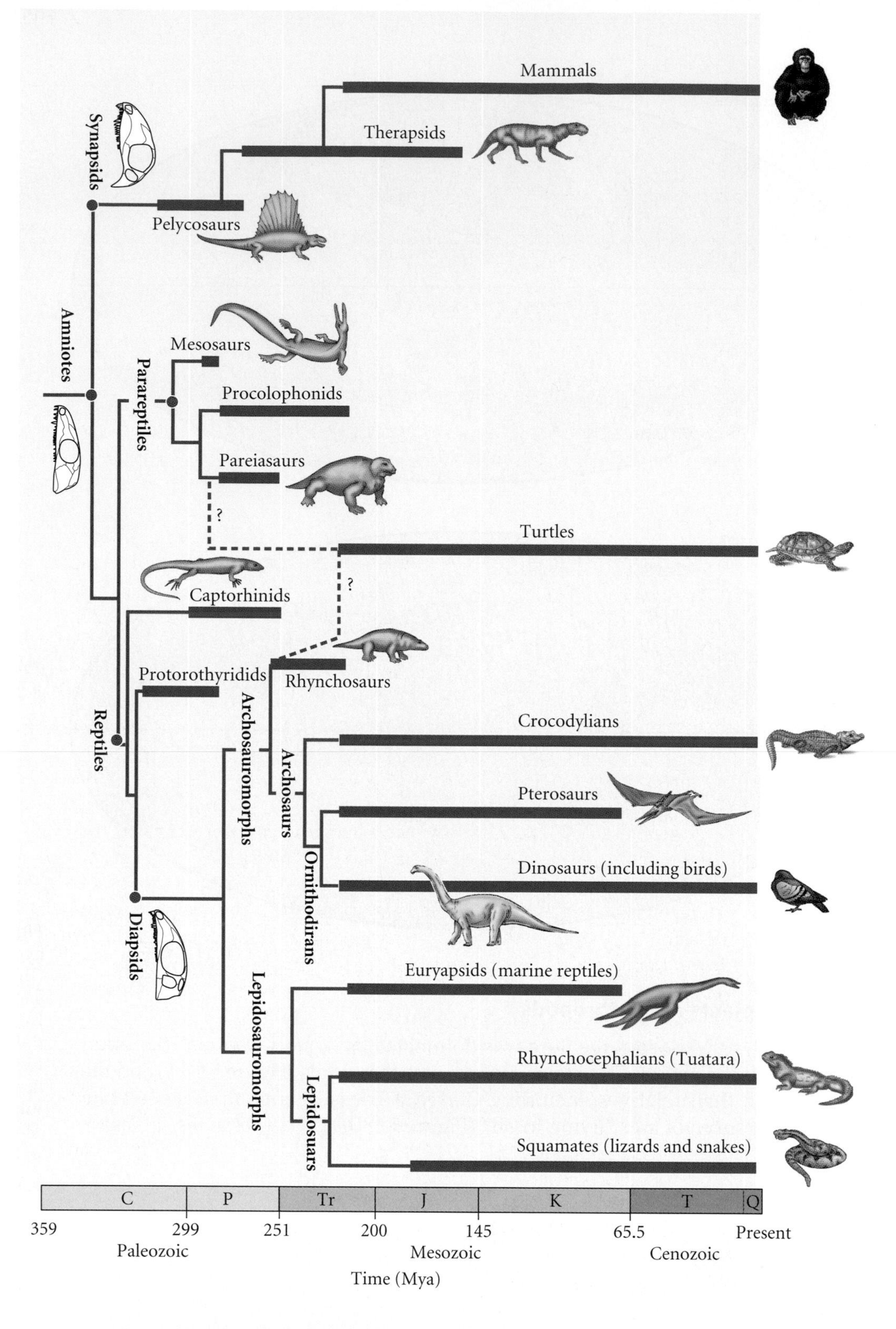

Figure 5.19 Phylogenetic relationships and temporal duration (thick bars) of major groups of amniote vertebrates. Some authors define "reptiles" as one of the two major lineages of amniotes, the other being the synapsids, which includes mammals. (After Lee et al. 2004.)

ing plants (Figure 5.18C), *first appeared in the early Cretaceous.* Many of the anatomical features of angiosperms, including flowerlike structures, had evolved individually in various Jurassic groups of "gymnosperms," some of which were almost certainly pollinated by insects. Beginning about 130 Mya in the Cretaceous, the angiosperms began to increase rapidly in diversity and achieved ecological dominance over the "gymnosperms."

The anatomically most "advanced" groups of insects made their appearance in the Mesozoic. By the late Cretaceous, most families of living insects, including ants and social bees, had evolved. Throughout the Cretaceous and thereafter, insects and angiosperms affected each other's evolution and may have augmented each other's diversity. As dif-

This fossil preserved the outline of the ichthyosaur's skin.

Figure 5.20 An ichthyosaur. The dorsal and tail fins of this extinct marine reptile are superficially similar to those of sharks and porpoises, although the tail fin of porpoises is horizontal. (The Field Museum, negative #GE084968c, Chicago.)

ferent groups of pollinating insects evolved, adaptive modification of flowers to suit different pollinators gave rise to the great floral diversity of modern plants. *It is largely because of the spectacular increase of angiosperms and insects that terrestrial diversity is greater today than ever before.*

Vertebrates

The major groups of amniotes are distinguished by the number of openings in the temporal region of the skull (at least in the stem members of each lineage; Figure 5.19). One such was a group of marine reptiles that flourished from the late Triassic to the end of the Cretaceous and included the dolphinlike ichthyosaurs, which gave birth to live young (Figure 5.20).

The *diapsids*, with two temporal openings, *became the most diverse group of reptiles*. One major diapsid lineage, the lepidosauromorphs, includes the lizards, which became differentiated into modern suborders in the late Jurassic and into modern families in the late Cretaceous. One group of lizards evolved into the snakes. They probably originated in the Jurassic, but their sparse fossil record begins only in the late Cretaceous.

The archosauromorph diapsids were the most spectacular and diverse of the Mesozoic reptiles. Most of the late Permian and Triassic archosaurs were fairly generalized predators a meter or so in length (Figure 5.21). From this generalized body plan, numerous specialized forms evolved. Among the most highly modified archosaurs are the pterosaurs, one of the three vertebrate groups that evolved powered flight. The wing consisted of a membrane extending to the body from the rear edge of a greatly elongated fourth finger (Figure 5.22). One pterosaur was the largest flying vertebrate known; others were as small as sparrows.

Dinosaurs evolved from archosaurs related to the one pictured in Figure 5.21. Dinosaurs are not simply any old large, extinct reptiles, but members of the orders Saurischia and Ornithischia, which are distinguished from each other by the form of the pelvis. Both orders included bipedal forms and quadrupeds that were derived from bipedal ancestors. Both orders arose in the Triassic, but neither order became diverse until the Jurassic.

Dinosaurs became very diverse: more than 39 families are recognized (Figure 5.23). The Saurischia included carnivorous, bipedal theropods and herbivorous, quadrupedal sauropods. Among the noteworthy theropods are *Deinonychus*, with a huge, sharp claw that it probably used to disembowel prey; the renowned *Tyrannosaurus rex* (late Cretaceous), which stood 15 meters high and weighed about 7000 kilograms; and the small theropods from which birds evolved. The sauropods, herbivores with small heads and long necks, include the largest animals that have ever lived on land, such as *Apatosaurus* (= *Brontosaurus*); *Brachiosaurus*, which weighed more than 80,000 kilograms; and *Diplodocus*, which reached about 30 meters in length.

The Ornithischia—herbivores with specialized, sometimes very numerous, teeth—included the well-known stegosaurs, with dorsal plates that probably served for thermoregulation, and the ceratopsians (horned dinosaurs), of which *Triceratops* is the best known.

The extinction of the ceratopsians at the end of the Cretaceous left only one surviving lineage of dinosaurs, which radiated extensively in the late Cretaceous or early Tertiary and today includes about 10,000 surviving species. Aside from

Figure 5.21 *Lagosuchus*, a Triassic thecodont archosaur, showing the generalized body form of the stem group from which dinosaurs evolved. (After Bonaparte 1978.)

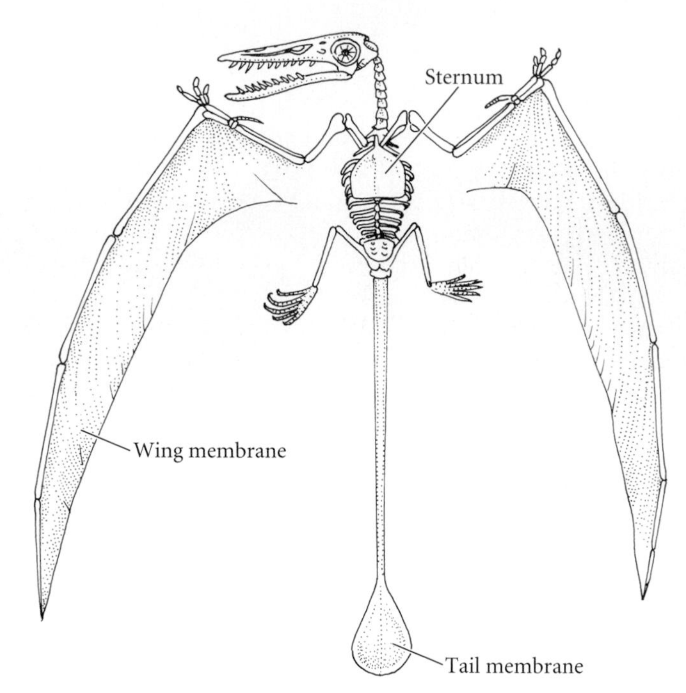

Figure 5.22 A Jurassic pterosaur, *Rhamphorhynchus*, showing the wing membrane supported by the greatly elongated fourth finger, the large breastbone (sternum) to which flight muscles were attached, and the terminal tail membrane found in this genus, possibly used for steering. (After Williston 1925.)

Figure 5.23 The great diversity of dinosaurs. The root of this proposed phylogeny is central, near the top. The two great clades of dinosaurs, Ornithischia (1; left) and Saurischia (2; right), are curved downward to fit the page. The Saurischia included the Sauropoda (3) and Therapoda (4), of which birds (Aves) are the only survivors (5). Ornithischia included stegosaurs (6) and ceratopsians (7). All ornithischian lineages are now extinct. (From Sereno 1999, *Science* 284: 2139. Copyright © AAAS.)

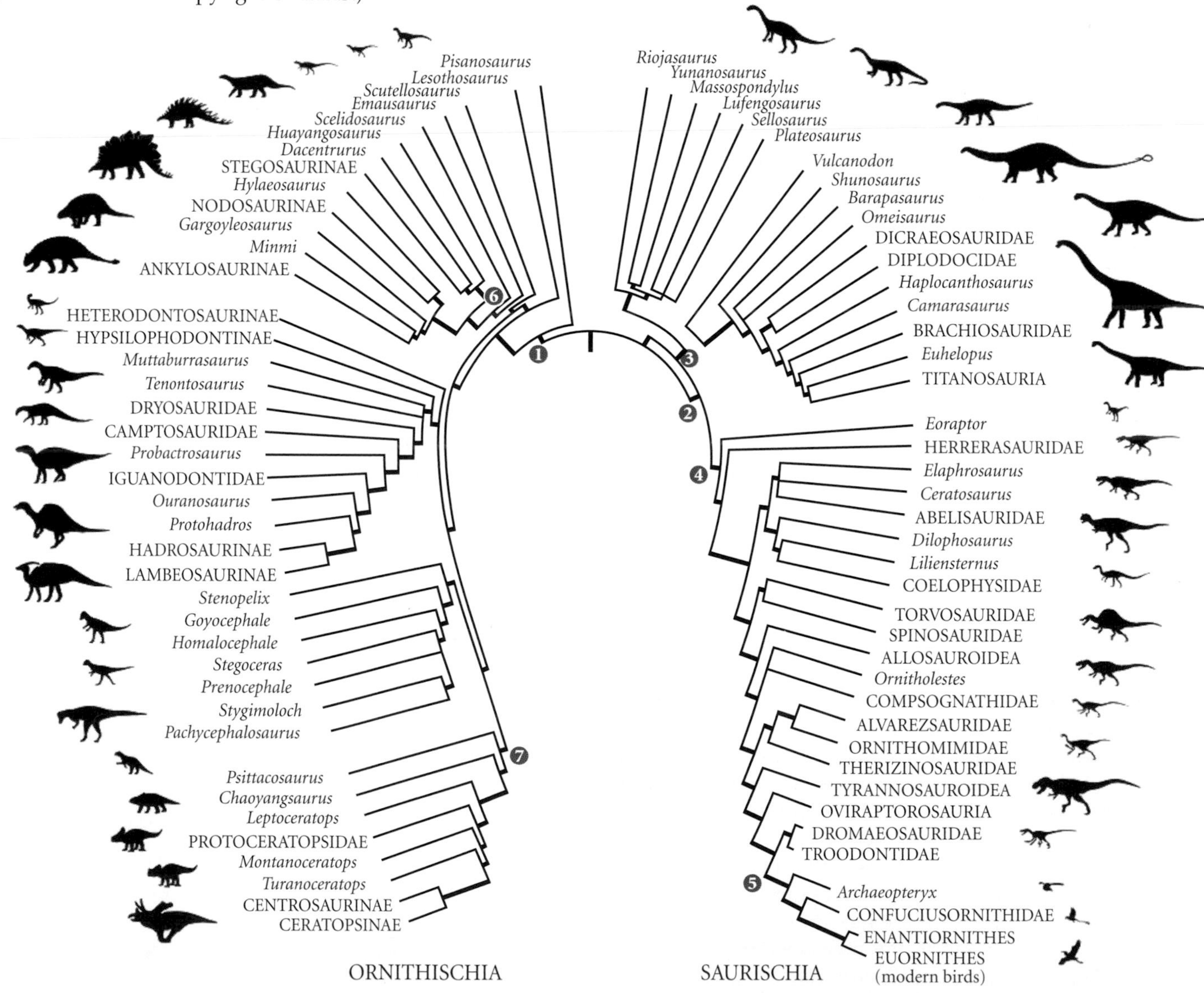

these dinosaurs, more familiarly known as birds, the only living archosaurs are 22 species of crocodilians.

The late Paleozoic *synapsids*, with a single temporal opening, *gave rise to the therapsids*, or "mammal-like reptiles," which flourished until the middle Jurassic. The first therapsid descendants that can be considered borderline mammals were the morganucodonts of the late Triassic and early Jurassic (see Figure 4.10F). Most mammals in later Mesozoic deposits are known only from teeth and jaw fragments, but a number of lineages of small mammals apparently proliferated. Most of them have left no living descendants, but therian mammals, the stem group from which most living mammals are descended, are known from the early Cretaceous. They include one form that appears closely related to the marsupials, but it is not until the *late Cretaceous that the two major subclasses of living mammals, marsupials and eutherians (placental mammals), appear in the fossil record*. Many of the mammals of this time had the primitive "ground plan" of a generalized mammal, resembling living species such as opossums in overall form.

The Cenozoic Era

The Cenozoic era embraces six epochs (Paleocene through Pleistocene). We are really still in the Pleistocene, but the last 10,000 years are often distinguished as a seventh epoch (the Holocene, or Recent). Traditionally, the first five epochs (from 65.5 Mya to 1.8 Mya) are referred to as the Tertiary period, and the Pleistocene and Recent (1.8 Mya–present) as the Quaternary period. Some paleontologists divide the era into Paleogene (65.5–23 Mya) and Neogene (23 Mya–present) periods.

By the beginning of the Cenozoic, North America had moved westward, becoming separated from Europe in the east, but forming the broad *Bering Land Bridge* between Alaska and Siberia, which *remained above sea level* throughout most of the era (see Figure 5.17D). *Gondwanaland broke up* into the separate island continents of South America, Africa, India, and, far to the south, Antarctica plus Australia (which separated in the Eocene). About 18 to 14 Mya, during the Miocene, Africa made contact with southwestern Asia, India collided with Asia (forming the Himalayas), and Australia moved northward, approaching southeastern Asia. During the Pliocene, about 3.5 Mya, *the Isthmus of Panama arose, connecting North and South America for the first time.*

This reconfiguration of continents and oceans contributed to major climatic changes. In the late Eocene and Oligocene, there was *global cooling and drying*; extensive savannahs (sparsely forested grasslands) formed for the first time, and Antarctica acquired glaciers. Sea level fluctuated, dropping drastically in the late Oligocene (about 25 Mya). During the Pliocene, temperatures increased to some extent, but toward its end, *temperatures dropped, and a series of glaciations began* that have persisted throughout the Pleistocene.

Aquatic life

Most of the marine groups that survived the K/T mass extinction proliferated early in the Cenozoic, and some new higher taxa evolved, such as the burrowing sea urchins known as sand dollars. The taxonomic composition of Cenozoic marine communities was quite similar to that of modern ones. Teleost fishes continued to diversify throughout the Cenozoic, becoming by far the most diverse aquatic vertebrates.

During the Pleistocene glaciations, so much water was sequestered in glaciers that the sea level dropped as much as 100 meters below its present level. As many as 70 percent of the mollusc species along the Atlantic coast of North America became extinct at this time; these extinctions were especially pronounced in the tropics.

Terrestrial life

Most of the modern families of angiosperms and insects had become differentiated by the Eocene or earlier, and many fossil insects of the late Eocene and Oligocene belong to genera that still survive. The dominant plants of the savannahs that developed in the Oligocene were grasses (Poaceae), which underwent a major adaptive radiation at this time, and many groups of herbaceous plants, most of which evolved from woody ancestors. Among

Figure 5.24 A phylogeny of living groups of marsupial and placental mammals, based on DNA sequence data. The timing of branch points is based on sequence divergence, calibrated by paleontological data. The data indicate that most orders diverged from one another during the Cretaceous. The dearth of mammalian fossils prior to the K/T boundary, however, suggests that early mammals in all these lineages may have been small and/or uncommon. (After Springer et al. 2003.)

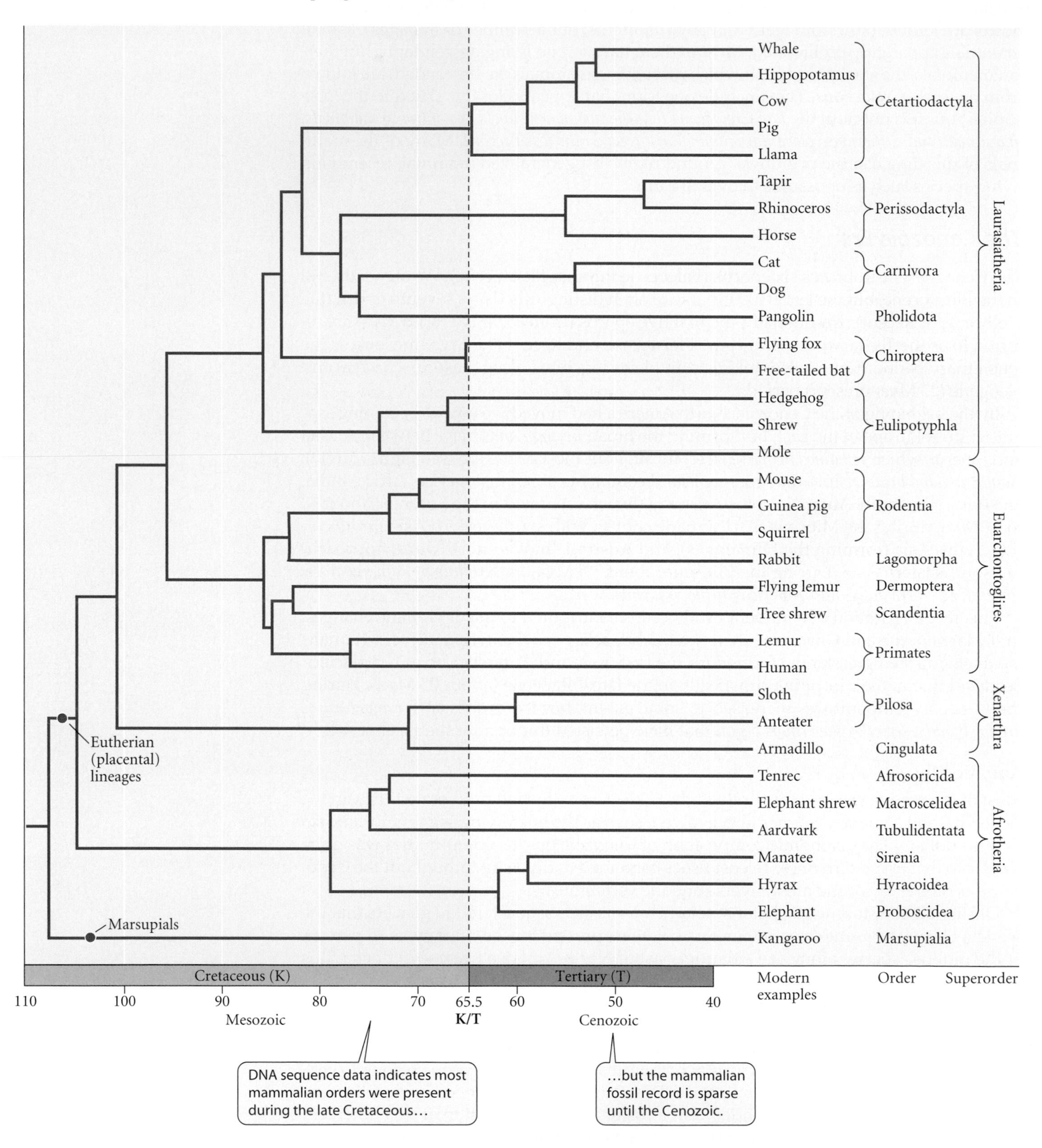

the most important of these is the family Asteraceae, which includes sunflowers, daisies, ragweeds, and many others. It is one of the two largest plant families today.

Many living orders and families of birds are recorded from the Eocene (55.8–33.9 Mya) and Oligocene (33.9–23 Mya). The largest order of birds, the perching birds (Passeriformes) first displayed its great diversity in the Miocene (23–5.3 Mya). Another great adaptive radiation was the snakes, which began an exponential increase in diversity in the Oligocene. Snakes today feed on a great variety of animal prey, from worms and termites to bird eggs and wild pigs, and they include marine, burrowing, and arboreal forms, some of which can glide.

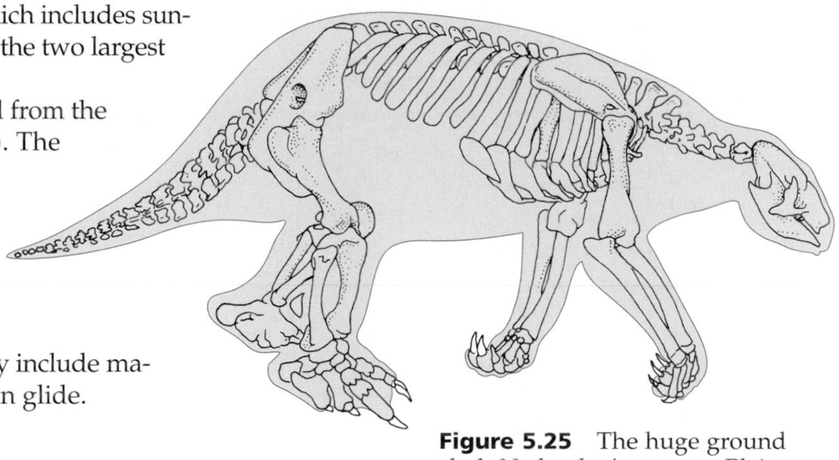

Figure 5.25 The huge ground sloth *Nothrotherium* was a Pleistocene representative of the Xenarthra, one of the earliest branches of the placental mammal phylogeny. (After Stock 1925.)

The adaptive radiation of mammals

Almost all mammalian fossils that can be assigned to modern orders occur after the K/T boundary (65.5 Mya). Nevertheless, DNA sequence differences, analyzed by methods that use multiple fossils to calibrate the rate of sequence evolution, indicate that most of the orders diverged from one another in the Cretaceous, and that the major lineages within each order diverged from about 77 Mya (for the major living lineages of Primates) to 50 Mya (Figure 5.24; Springer et al. 2003). The great difference between DNA-based and fossil-based estimates of the time of diversification suggests that although many lineages of mammals evolved during the Cretaceous, they remained small in size, were relatively uncommon, and perhaps did not evolve their distinctive ecological and morphological features until the Cenozoic. It has often been suggested that the extinction of the large dinosaurs relieved the subjugated mammals from competition and predation and allowed them to undergo adaptive radiation.

Marsupials are known as fossils from all the continents, including Antarctica. Today they are restricted to Australia and South America (with one exception, the North American opossum). The marsupial families that include kangaroos, wombats, and other living Australian marsupials evolved in the mid-Tertiary. In South America, marsupials experienced a great adaptive radiation; some resembled kangaroo rats, others saber-toothed cats. Most South American marsupials became extinct by the end of the Pliocene.

In addition to marsupials, many groups of placental mammals evolved in South America during its long isolation from other continents. These included an ancient placental group, the Xenarthra (or Edentata), which includes the giant ground sloths that survived until the late Pleistocene (Figure 5.25) and the few armadillos, anteaters, and sloths that still persist. At least six orders of hoofed mammals that resembled sheep, rhinoceroses, camels, elephants, horses, and rodents evolved in South America, but declined and became completely extinct after South America became connected to North America in the late Pliocene. The extinction of many South American mammals has been attributed to the ecological impact of North American mammal groups, such as bears, raccoons, weasels, peccaries, and camels, that moved into South America at this time as part of what has been called the "Great American Interchange" of terrestrial organisms.

Among placental orders, one of the earliest and, in many ways, structurally most primitive is the Primates. The earliest fossils assigned to this order are so similar to basal eutherians that it is rather arbitrary whether they be called primates or not (Figure 5.26). The first monkeys are known from the Oligocene, and the first apes (superfamily Hominoidea) from the Miocene, about 22 Mya. The fossil record of hominin evolution, starting about 6 Mya, is described in Chapter 4.

Rodents (Rodentia), related to Primates, are recorded first from the late Paleocene. Perhaps by direct competition, they replaced the Multituberculata, a nonplacental but ecologically similar group that had originated from stem mammals in the late Jurassic (Figure 5.27). The rodents became the most diverse order of mammals, due in part to an extraordinary proliferation of rats and mice (superfamily Muroidea) within the last 10 million years.

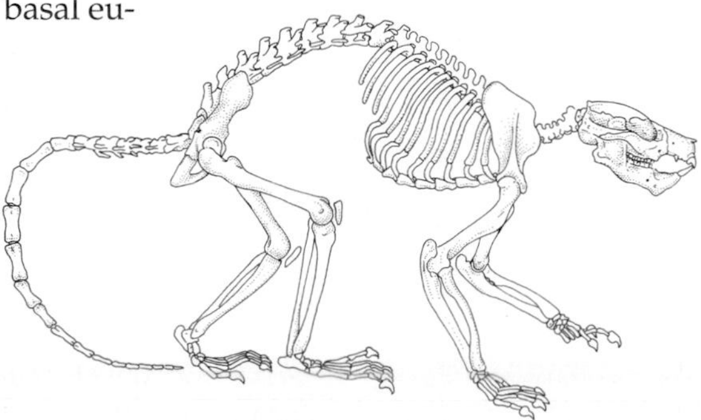

Figure 5.26 An early primate, the Paleocene *Plesiadapis*. (After Simons 1979.)

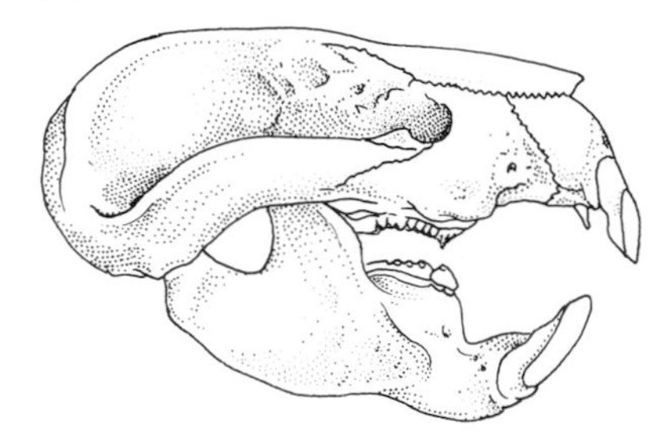

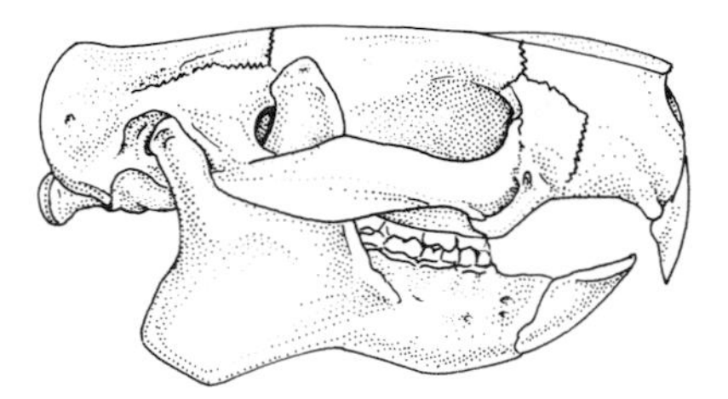

Figure 5.27 Convergent evolution in mammals: (A) A Paleocene multituberculate (*Taeniolabis*) and (B) an Eocene rodent (*Paramys*). Multituberculates, extending from the Cretaceous to the Oligocene, were nonplacental mammals that were ecologically similar to squirrels and other rodents (with similar morphological features such as large, ever-growing incisors). Competition with rodents may have driven them to extinction. (After Romer 1966.)

A monophyletic group of orders ("Afrotheria"), revealed by DNA evidence, includes such very different-looking creatures as the shrewlike tenrecs, the aquatic manatees, and the elephants. This group is first recorded in the Eocene, but represents one of the earliest splits among the placental orders (see Figure 5.24). The elephants (Proboscidea) underwent the greatest diversification, differentiating into at least 40 genera. Woolly mammoths survived through the most recent glaciation, about 10,000 years ago, and two genera (the African and Indian elephants) persist today (Figure 5.28).

A paraphyletic stem group, the condylarths, gave rise in the Eocene to a great radiation of carnivores and of hoofed mammals (ungulates) in the orders Perissodactyla and Artiodactyla, the early members of which differed only slightly. The Perissodactyla, or odd-toed ungulates, were very diverse from the Eocene to the Miocene, then dwindled down to the few extant species of rhinoceroses, horses, and tapirs. The Artiodactyla (now called Cetartiodactyla by some authors) are first known in the Eocene as rabbit-sized animals that had the order's diagnostic ankle bones, but otherwise bore little similarity to the pigs, camels, and ruminants that appeared soon thereafter. In the Miocene, the ruminants began a sustained radiation, mostly in the Old World, that is correlated with the increasing prevalence of grasslands. Among the families that proliferated are the deer, the giraffes and relatives, and the Bovidae, the diverse family of antelopes, sheep, goats, and cattle. Recent research has shown that during the Eocene, one of the artiodactyl lineages became aquatic and evolved into the cetaceans: the dolphins and whales (see Figure 4.11).

Pleistocene events

The last Cenozoic epoch, the *Pleistocene*, embraces a mere 1.8 million years, but because of its recency and drama, it *is critically important for understanding today's organisms*.

By the beginning of the Pleistocene, the continents were situated as they are now. North America was connected in the northwest to eastern Asia by the Bering Land Bridge, in the region where Alaska and Siberia almost meet today (Figure 5.29A). North and South America were connected by the Isthmus of Panama. Except for those species that have become extinct, Pleistocene species are very similar to, or indistinguishable from, the living species that descended from them.

Global temperatures began to drop about 3 million years ago, and then, in the Pleistocene itself, underwent violent fluctuations with a period of about 100,000 years. When temperatures cooled, continental glaciers as thick as 2 kilometers formed at high latitudes,

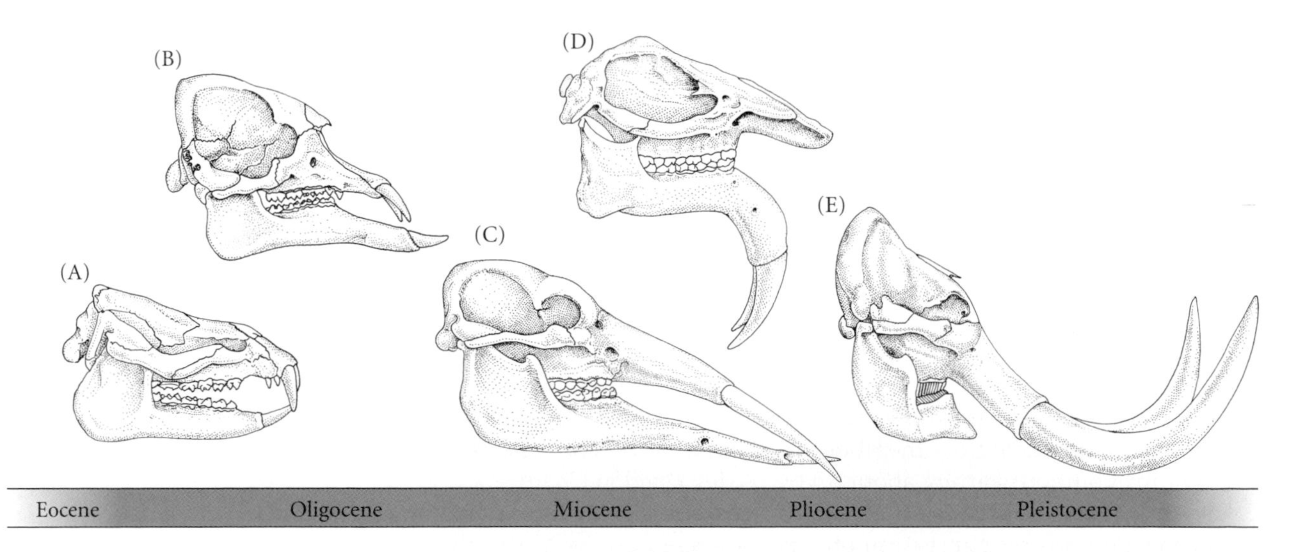

Figure 5.28 Proboscidea, the order of elephants, has only two living genera, but was once very diverse. A few of the extinct forms are (A) an early, generalized proboscidean, *Moeritherium* (late Eocene–early Oligocene); (B) *Phiomia* (early Oligocene); (C) *Gomphotherium* (Miocene); (D) *Deinotherium* (Miocene); and (E) *Mammuthus*, the woolly mammoth (Pleistocene). (After Romer 1966.)

(A)

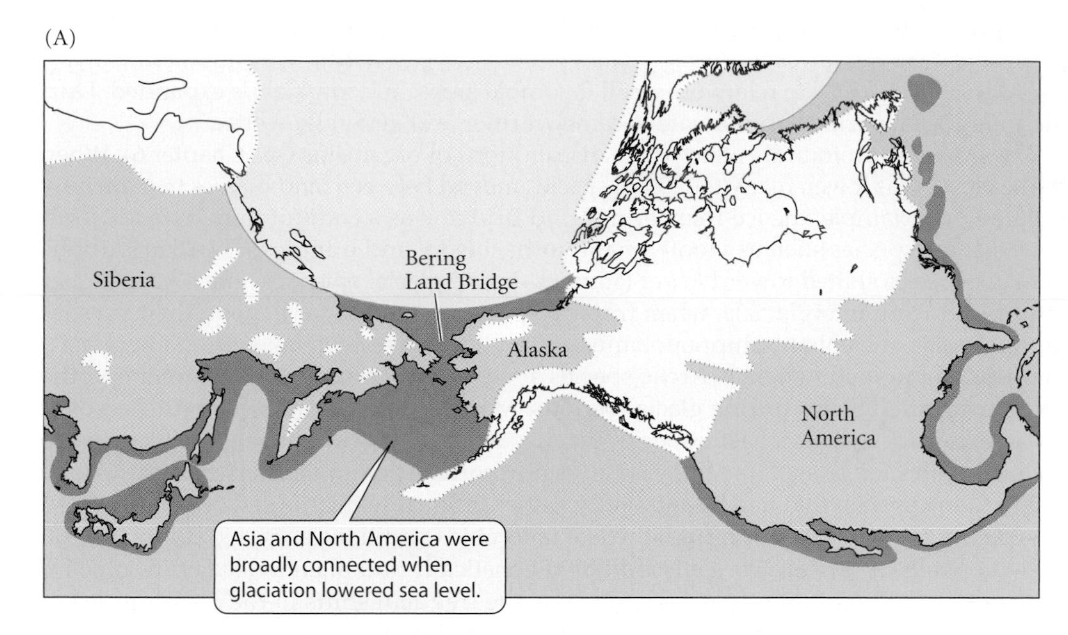

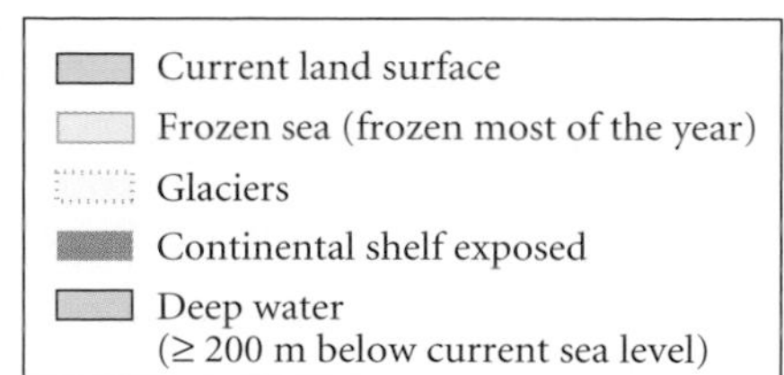

Figure 5.29 Pleistocene glaciers lowered sea levels by at least 100 meters, so that many terrestrial regions that are now separated by oceanic barriers were connected. (A) Eastern Asia and North America were joined by the Bering Land Bridge. Note the extent of the glacier in North America. (B) Indonesia and other islands were connected to either southeastern Asia or Australia. (After Brown and Lomolino 1998.)

(B)

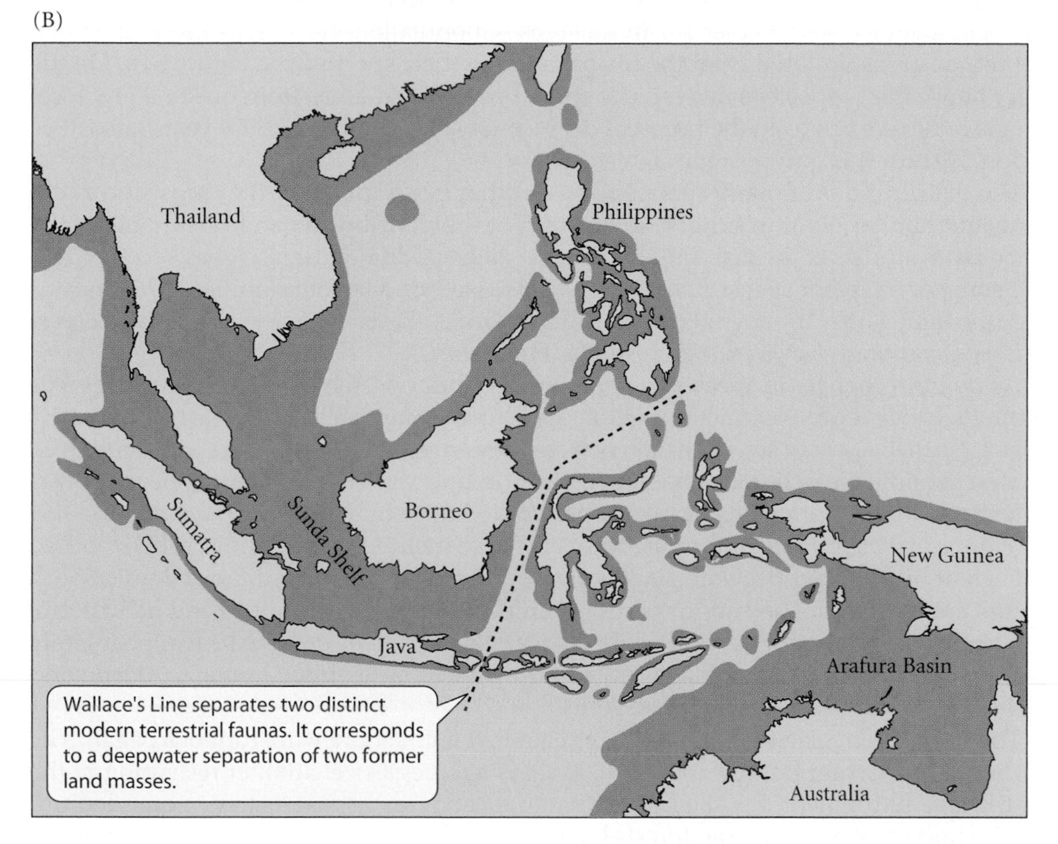

receding during the warmer intervals. *At least four major glacial advances*, and many minor ones, occurred. The most recent glacial stage, termed the Wisconsin in North America and the Riss-Würm in Europe, *reached its maximum about 18,000 years ago* (Figure 5.29A), *and melted back 15,000 to 8000 years ago*. During glacial episodes, *sea level dropped* as much as 100 meters below the present level. This drop exposed parts of the continental shelves, extending many continental margins beyond their present boundaries and *connecting many islands to nearby land masses*. (Japan, for example, was a peninsula of Asia, New Guinea was connected to Australia, and the Malay Archipelago was an extension of Southeast Asia; Figure 5.29B.) Temperatures in equatorial regions were apparently about as high as they are today, so the latitudinal temperature gradient was much steeper than at present. *The global climate during glacial episodes was generally drier*. Thus *mesic and wet forests became restricted* to relatively small favorable areas, and grasslands expanded. During interglacial periods, the climate became warmer and generally wetter.

These events profoundly affected the distributions of organisms (see Chapter 6). When the sea level was lower, many terrestrial species moved between land masses that are now isolated; for example, the ice-free Bering Land Bridge was a conduit from Asia to North America for species such as woolly mammoths, bison, and humans. The distributions of many species shifted toward lower latitudes during glacial episodes and toward higher latitudes during interglacials, when tropical species extended far beyond their present limits. Fossils of elephants, hippopotamuses, and lions have been taken from interglacial deposits in England, whereas Arctic species such as spruce and musk ox inhabited the southern United States during glacial periods. Many species became extinguished over broad areas; for instance, beetle species that occurred in England during the Pleistocene are now restricted to such far-flung areas as northern Africa and eastern Siberia (Coope 1979). Many species that had been broadly, rather uniformly distributed became isolated in separate areas (refuges or **refugia**) where favorable conditions persisted during glacial periods. *Some such isolated populations diverged* genetically and phenotypically, in some instances becoming different species. However, the frequent shifts in the distributions of populations may have prevented many of them from becoming different species by mixing them together (see Chapter 16). In some cases, populations have remained in their glacial refuge areas, isolated from the major range of their species (see Figure 6.8). On the other hand, many species have rapidly spread over broad areas from one or a few local refugia and have achieved their present distributions only in the last 8000 years or so. (See Pielou 1991 for more on postglacial history.)

The distributions of many species shifted rather gradually over the landscape as climate and habitat became suitable at one end of the distribution and unsuitable at the other. From studies of the distribution of fossil pollen of different ages, it has become clear that some of the plant species that are typically associated in communities today moved over the land rather independently, at different rates. Thus *the species composition of ecological communities changed kaleidoscopically* (Figure 5.30).

Aside from changes in species' geographic distributions, the most conspicuous events in the Pleistocene were extinctions. *Many shallow-water marine invertebrate species became extinct, especially tropical species*, which may have been poorly equipped to withstand even modest cooling. No major taxa became entirely extinct, however. On land, the story was different. Although small vertebrates seem not to have suffered, *a very high proportion of large-bodied mammals and birds became extinct*. These included mammoths, saber-toothed cats, giant bison, giant beavers, giant wolves, ground sloths, and all the endemic South American ungulates. They might have succumbed to changes in climate and habitat, but both archaeological evidence and mathematical population models make it appear likely that this "megafaunal extinction" was caused by weapon-wielding humans (Martin and Klein 1984; Alroy 2001; Roberts et al. 2001).

The most recent glaciers had hardly retreated when major new disruptions began. The advent of human agriculture about 11,000 years ago began yet another reshaping of the terrestrial environment. For the last several thousand years, deserts have expanded under the impact of overgrazing, forests have succumbed to fire and cutting, and climates have changed as vegetation has been modified or destroyed. At present, under the impact of an exponentially growing human population and its modern technology, species-rich tropical forests face almost complete annihilation, temperate zone forests and prairies have been eliminated in much of the world, marine communities suffer pollution and appalling overexploitation, and global warming caused by combustion of fossil fuels threatens to change climates and habitats so rapidly that many species are unlikely to adapt (Wilson 1992; Kareiva et al. 1993). An analysis of sample regions covering 20 percent of the Earth's surface projected extinction of 18 to 35 percent of species in those regions, just due to climate change, just within the next 50 years (Thomas et al. 2004). Even if those fig-

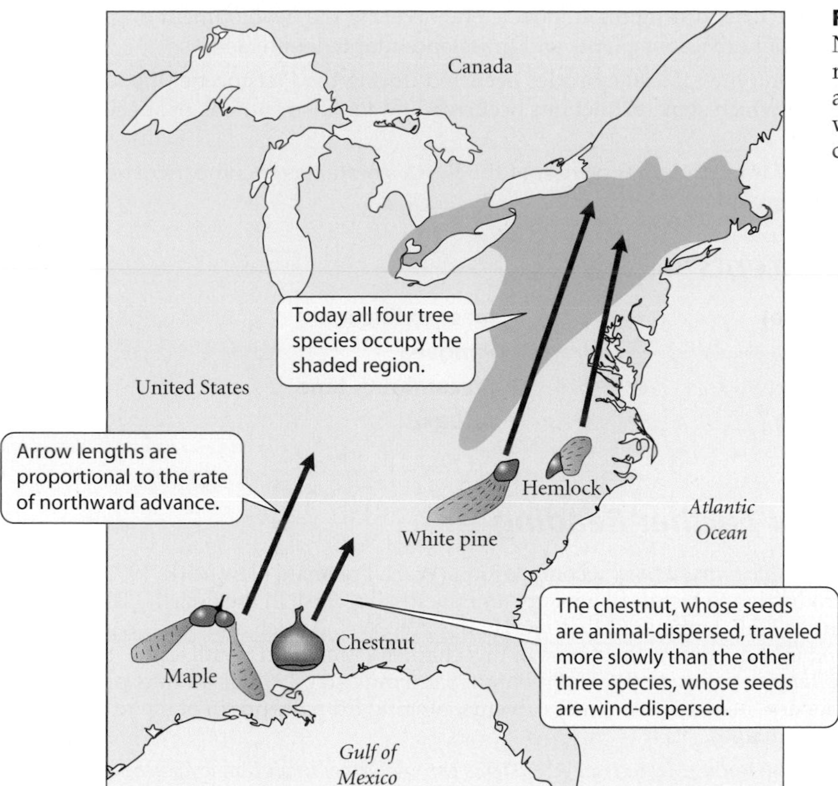

Canada
Today all four tree species occupy the shaded region.
United States
Arrow lengths are proportional to the rate of northward advance.
Hemlock
White pine
Atlantic Ocean
Chestnut
Maple
The chestnut, whose seeds are animal-dispersed, traveled more slowly than the other three species, whose seeds are wind-dispersed.
Gulf of Mexico

Figure 5.30 Different rates of northward spread of four North American tree species from refugia after the most recent glacial episode. After the glacial episode, maple and chestnut moved north from the Gulf region, and white pine and hemlock spread from the mid-Atlantic coastal plain. (After Pielou 1991.)

ures are not totally accurate, it is clear that *one of the greatest mass extinctions of all time will come to pass—unless we act now to prevent it.*

Summary

1. Evidence from living organisms indicates that all living things are descended from a single common ancestor. Some progress has been made in understanding the origin of life, but a great deal remains unknown.
2. The first fossil evidence of life dates from about 3.5 billion years ago, about a billion years after the formation of the Earth. The earliest life forms of which we have evidence were prokaryotes.
3. Eukaryotes evolved at least 1.5 billion years ago. Their mitochondria and chloroplasts evolved from symbiotic bacteria.
4. The fossil record displays an explosive diversification of the animal phyla near the beginning of the Cambrian period, about 542 Mya, but this record might reflect the evolution of hard structures in lineages that had diverged long before then. The causes of this rapid diversification are debated, but may include a combination of genetic and ecological events.
5. Terrestrial plants and arthropods evolved in the late Silurian and early Devonian; amphibians evolved in the late Devonian from lobe-finned fishes.
6. The most devastating mass extinction of all time occurred at the end of the Permian (about 251 million years ago). It profoundly altered the taxonomic composition of the Earth's biota.
7. Seed plants and amniotes ("reptiles") became diverse and ecologically dominant during the Mesozoic era (251–65.5 million years ago). Flowering plants and plant-associated insects diversified greatly from the middle of the Cretaceous onward. A mass extinction (the "K/T extinction") at the end of the Mesozoic included the extinction of the last nonavian dinosaurs.
8. Most of the orders of mammals originated in the late Cretaceous, but underwent adaptive radiation in the early Tertiary (about 65.5–50 million years ago). The extinction of nonavian dinosaurs may have permitted them to diversify.

9. The climate became drier during the Cenozoic era, favoring the development of grasslands and the evolution of herbaceous plants and grassland-adapted animals.
10. A series of glacial and interglacial episodes occurred during the Pleistocene (the last 1.8 million years), during which some extinctions occurred and the distributions of species were greatly altered.
11. With the passage of time, the composition of the Earth's biota has become increasingly similar to its composition today.

Terms and Concepts

Bauplan (pl., Baupläne)
Cambrian explosion
endosymbionts
Gondwanaland
Laurasia
mass extinction
Pangaea
Precambrian time
refugia

Suggestions for Further Reading

S. M. Stanley, *Earth and life through time*, second edition (W. H. Freeman, New York, 1993) is a comprehensive introduction to historical geology and the fossil record. In the Fourth Edition of *Life of the past* (Prentice-Hall, Upper Saddle River, N.J., 1999), W. I. Ausich and N. G. Lane provide a well illustrated introduction to the theme of this chapter. J. Maynard Smith and E. Szathmáry, *The major transitions in evolution* (W. H. Freeman, San Francisco, 1995) is an interpretation by leading evolutionary theoreticians of major events, ranging from the origin of life to the origins of societies and languages.

A. K. Behrensmeyer et al. (eds.), *Terrestrial ecosystems through time: Evolutionary paleoecology of terrestrial plants and animals* (University of Chicago Press, Chicago, 1992) presents detailed summaries of changes in terrestrial environments and communities in the past.

Useful books on the evolution of major taxonomic groups include E. N. K. Clarkson, *Invertebrate paleontology and evolution* (Chapman and Hall, London, 1993); P. Kenrick and P. R. Crane, *The origin and early diversification of land plants* (Smithsonian Institution Press, Washington, 1997); R. L. Carroll, *Vertebrate paleontology and evolution,* (W. H. Freeman, New York, 1988); M. J. Benton (ed.), *The phylogeny and classification of the tetrapods* (Clarendon, Oxford, 1988); D. B. Weishampel, P. Dodson, and H. Osmolska, *The Dinosauria* (University of California Press, Berkeley, 1990).

Problems and Discussion Topics

1. Why, in the evolution of ancestral eukaryotes, might it have been advantageous for separate organisms to become united into a single organism? Can you describe analogous, more recently evolved, examples of intimate symbioses that function as single integrated organisms?
2. Early in the origin of life, as it is presently conceived, there was no distinction between genotype and phenotype. What characterizes this distinction, and at what stage of organization may it be said to have come into being?
3. If we employ the biological species concept (see Chapter 2), when did species first exist? What were organisms before then, if not species? What might the consequences of the emergence of species be for processes of adaptation and diversification?
4. How would you determine whether the morphological diversity of animals has increased, decreased, or remained the same since the Cambrian?
5. Compare terrestrial communities in the Devonian and in the Cretaceous, and discuss what may account for the difference between them in the diversity of plants and animals.
6. What evidence would be necessary to test the hypothesis that the megafaunal extinction in the Pleistocene was caused by humans?
7. Discuss the implications of the spread and change of species distributions after the retreat of the Pleistocene glaciers for evolutionary changes in species and for the species composition of ecological communities (see also Chapters 7 and 16).

The Geography of Evolution

6

Where did humans originate, and by what paths did they spread throughout the world? Why are kangaroos found only in Australia, whereas rats are found worldwide? Why are there so many more species of trees, insects, and birds in tropical than in temperate zone forests?

These questions illustrate the problems that **biogeography**, the study of the geographic distributions of organisms, attempts to solve. ZOOGEOGRAPHY and PHYTOGEOGRAPHY are subdivisions of biogeography, concerning the distributions of animals and plants, respectively. The evolutionary study of organisms' distributions is intimately related to geology, paleontology, systematics, and ecology. For example, geological study of the history of the distributions of land masses and climates often sheds light on the causes of organisms' distributions. Conversely, organisms' distributions have sometimes provided evidence for geological events. In fact, the geographic distributions of organisms were used by some scientists as evidence for continental drift long before geologists agreed that it really happens.

Old World and New World monkeys. African and Asian monkeys such as *Colobus* (left) belong to the taxon Catarrhini. Monkeys from the New World (South and Central America), such as the howler monkey *Alouatta palliata* (below), belong to the entirely distinct taxon Platyrrhini. (*Colobus* © Charles McRae/Visuals Unlimited; *Alouatta* © Roy P. Fontaine/Photo Researchers, Inc.)

In some instances, the geographic distribution of a taxon may best be explained by historical circumstances; in other instances, ecological factors operating at the present time may provide the best explanation. Hence the field of biogeography may be roughly subdivided into **historical biogeography** and **ecological biogeography**. Historical and ecological explanations of geographic distributions are complementary, and both are important (Brown and Lomolino 1998; Myers and Giller 1988; Ricklefs and Schluter 1993).

Biogeographic Evidence for Evolution

Darwin and Wallace were both very interested in biogeography. Wallace devoted much of his later career to the subject and described major patterns of zoogeography that are still valid today. The distributions of organisms provided Darwin with inspiration and with evidence that evolution had occurred. To us, today, the reasons for certain facts of biogeography seem so obvious that they hardly bear mentioning. If someone asks us why there are no elephants in the Hawaiian Islands, we will naturally answer that elephants couldn't get there. This answer assumes that elephants originated somewhere else: namely, on a continent. But in a pre-evolutionary world view, the view of special divine creation that Darwin and Wallace were combating, such an answer would not hold: the Creator could have placed each species anywhere, or in many places at the same time. In fact, it would have been reasonable to expect the Creator to place a species wherever its habitat, such as rain forest, occurred.

Darwin devoted two chapters of *The Origin of Species* to showing that many biogeographic facts that make little sense under the hypothesis of special creation make a great deal of sense if a species (1) has a definite site or region of origin, (2) achieves a broader distribution by dispersal, and (3) becomes modified and gives rise to descendant species in the various regions to which it migrates. (In Darwin's day, there was little inkling that continents might have moved. Today, the movement of land masses also explains certain patterns of distribution.) Darwin emphasized the following points:

First, said Darwin, "*neither the similarity nor the dissimilarity of the inhabitants of various regions can be wholly accounted for by climatal and other physical conditions.*" Similar climates and habitats, such as deserts and rain forests, occur in both the Old and the New World, yet the organisms inhabiting them are unrelated. For example, the cacti (family Cactaceae) are restricted to the New World, but the cactuslike plants in Old World deserts are members of other families (Figure 6.1). All the monkeys in the New World belong to one great group (Platyrrhini), and all Old World monkeys to another (Catarrhini)—even if, like the howler and colobus monkeys shown at the opening of this chapter, they have similar habitats and diets.

Darwin's second point is that "*barriers of any kind, or obstacles to free migration, are related in a close and important manner to the differences between the productions [organisms] of various*

(A)

(B)

(C)

Figure 6.1 Convergent growth form in desert plants. These plants, all leafless succulents with photosynthetic stems, belong to three distantly related families. (A) A North American cactus (family Cactaceae). This species, *Lophocereus schottii*, is native to Baja California. (B) A carrion flower of the genus *Stapelia* (Apocynaceae). These fly-pollinated succulents can be found from southern Africa to east India. (C) A species of *Euphorbia* (Euphorbiaceae) in the Namib Desert of Africa. (A–C © Photo Researchers, Inc. A by Richard Parker; B, by Geoff Bryant; C by Fletcher and Baylis.)

regions." Darwin noted, for instance, that marine species on the eastern and western coasts of South America are very different.

Darwin's "third great fact" is that *inhabitants of the same continent or the same sea are related, although the species themselves differ from place to place*. He cited as an example the aquatic rodents of South America (the coypu and capybara), which are structurally similar to, and related to, South American rodents of the mountains and grasslands, not to the aquatic rodents (beaver, muskrat) of the Northern Hemisphere.

"We see in these facts," said Darwin, "some deep organic bond, throughout space and time, over the same areas of land and water, independently of physical conditions. ... The bond is simply inheritance [i.e., common ancestry], that cause which alone, as far as we positively know, produces organisms quite like each other."

For Darwin, it was important to show that a species had not been created in different places, but had a *single region of origin*. He drew particularly compelling evidence from the inhabitants of islands. First, distant *oceanic islands generally have precisely those kinds of organisms that have a capacity for long-distance dispersal* and lack those that do not. For example, the only native mammals on many islands are bats. Second, *many continental species of plants and animals have flourished on oceanic islands to which humans have transported them*. Thus, said Darwin, "he who admits the doctrine of the creation of each separate species, will have to admit that a sufficient number of the best adapted plants and animals were not created for oceanic islands." Third, most of the species on islands are clearly *related to species on the nearest mainland*, implying that that was their source. This is the case, as Darwin said, for almost all the birds and plants of the Galápagos Islands. Fourth, the *proportion of endemic species on an island is particularly high when the opportunity for dispersal to the island is low*. Fifth, *island species often bear marks of their continental ancestry*. For example, Darwin noted, hooks on seeds are an adaptation for dispersal by mammals, yet on oceanic islands that lack mammals, many endemic plants nevertheless have hooked seeds.

It is a testimony to Darwin's knowledge and insight that all these points hold true today, after nearly a century and a half of research. Our greater knowledge of the fossil record and of geological events such as continental movement and sea level changes has added to our understanding, but has not negated any of Darwin's major points.

Major Patterns of Distribution

The geographic distribution of almost every species is limited to some extent, and many higher taxa are likewise restricted (**endemic**) to a particular geographic region. For example, the salamander genus *Plethodon* is limited to North America, and *Plethodon caddoensis* occupies only the Caddo Mountains of western Arkansas. Some higher taxa, such as the pigeon family (Columbidae), are almost cosmopolitan (found worldwide), whereas others are narrowly endemic (e.g., the kiwi family, Apterygidae, which is restricted to New Zealand; see Figure 6.13).

Wallace and other early biogeographers recognized that many higher taxa have roughly similar distributions, and that the taxonomic composition of the biota is more uniform within certain regions than between them. Based on these observations, Wallace designated several **biogeographic realms** for terrestrial and freshwater organisms that are still widely recognized today (Figure 6.2). These are the *Palearctic* (temperate and tropical Eurasia and northern Africa), the *Nearctic* (North America), the *Neotropical* (South and Central America), the *Ethiopian* (sub-Saharan Africa), the *Oriental* (India and Southeast Asia), and the *Australian* (Australia, New Guinea, New Zealand, and nearby islands). These realms are more the result of Earth's history than of current climate or land mass distribution. For example, WALLACE'S LINE separates islands that, despite their close proximity and similar climate, differ greatly in their fauna. These islands are on two lithospheric plates that approached each other only recently, and they are assigned to two different biogeographic realms: the Oriental and the Australian.

Each biogeographic realm is inhabited by many higher taxa that are much more diverse in that realm than elsewhere, or are even restricted to that realm. For example, the endemic taxa of the Neotropical realm (South America) include the Xenarthra (anteaters

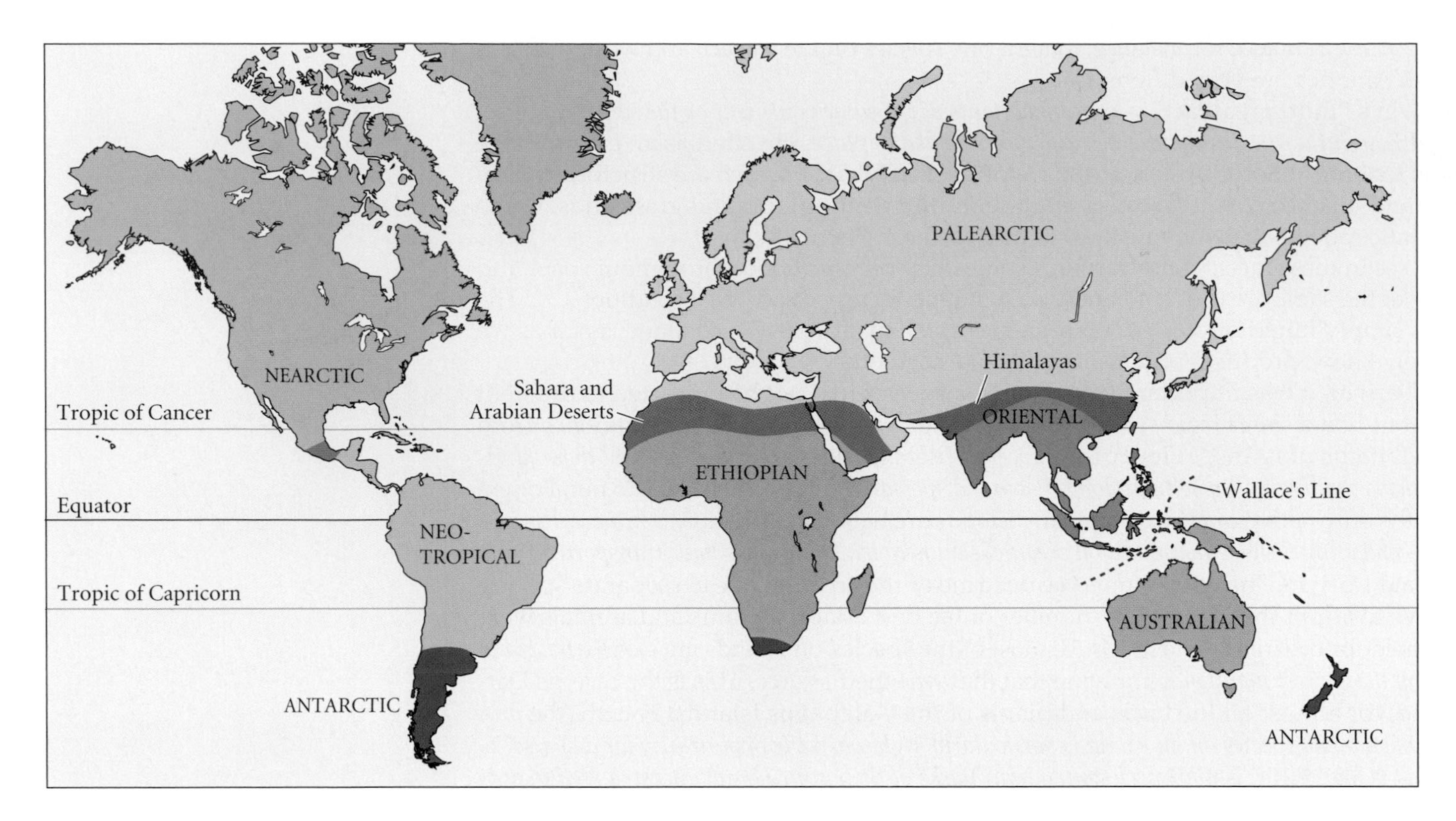

Figure 6.2 Biogeographic realms. The biogeographic realms recognized by A. R. Wallace are the Palearctic, Ethiopian, Oriental, Australian, Nearctic, and Neotropical. Some authors consider parts of southern South America, Africa, and New Zealand to be another realm, the Antarctic.

and allies), platyrrhine primates (such as spider monkeys and marmosets), hummingbirds, a large assemblage of suboscine birds such as flycatchers and antbirds, many families of catfishes, and plant families such as the pineapple family (Bromeliaceae) (Figure 6.3; see also the chapter-opening figure). Within each realm, individual species may have more restricted distributions; regions that differ markedly in habitat, or which are separated by mountain ranges or other barriers, will have rather different sets of species. Thus a biogeographic realm can often be divided into faunal and floral PROVINCES, or regions of endemism (Figure 6.4).

The borders between biogeographic realms (or provinces) cannot be sharply drawn because some taxa infiltrate neighboring realms to varying degrees. In the Nearctic realm

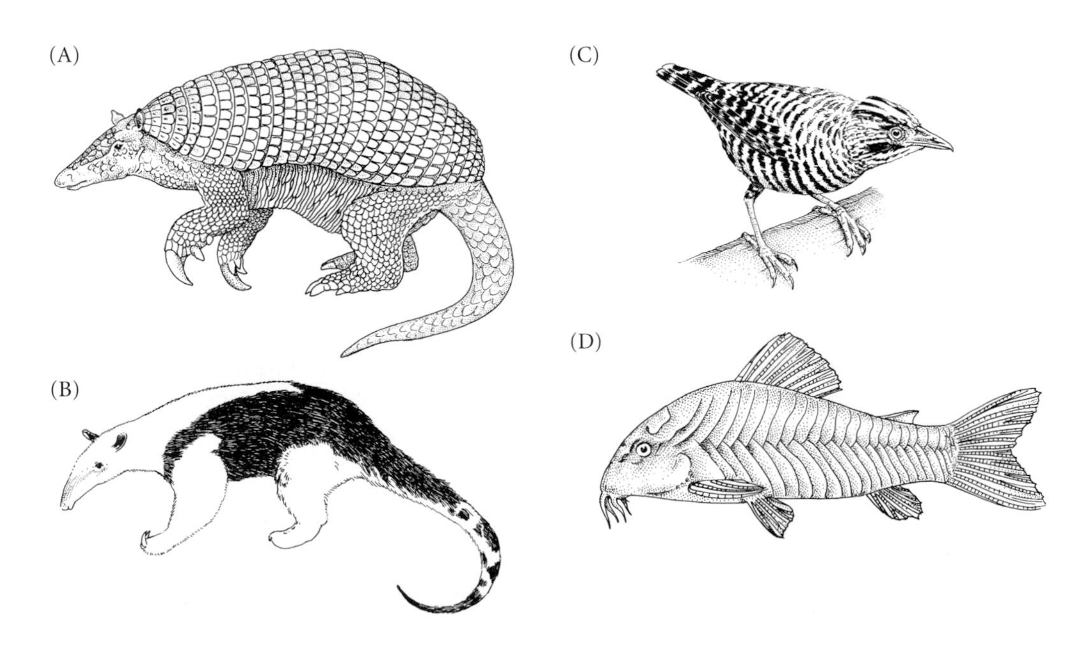

Figure 6.3 Examples of taxa endemic to the Neotropical biogeographic realm. (A) An armadillo (order Xenarthra). (B) An anteater (order Xenarthra). (C) An antshrike (Formicariidae), representing a huge evolutionary radiation of suboscine birds in the Neotropics. (D) An armored catfish (Callichthyidae), one of many families of freshwater catfishes restricted to South America. (A, B after Emmons 1990; C after Haverschmidt 1968; D after Moyle and Cech 1983.)

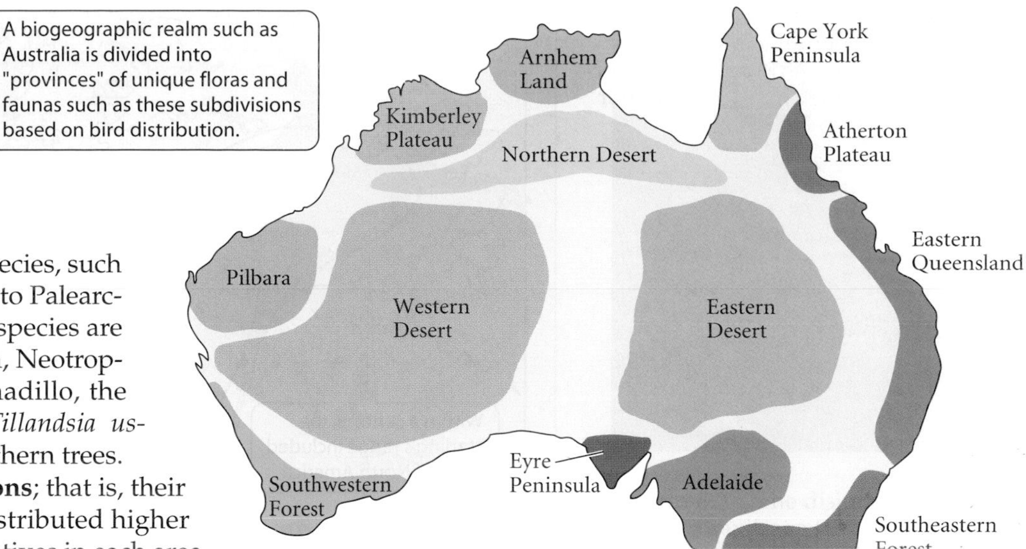

Figure 6.4 Provinces, or regions of endemism, in Australia, based on the pattern of distribution of birds. Distributions of other vertebrates form similar patterns. (After Cracraft 1991.)

(North America), for instance, some species, such as bison, trout, and birches, are related to Palearctic (Eurasian) taxa. But other Nearctic species are related to, and have been derived from, Neotropical stocks: examples include the armadillo, the opossum, and the Spanish moss (*Tillandsia usneoides*), a bromeliad that festoons southern trees.

Some taxa have **disjunct distributions**; that is, their distributions have gaps. Disjunctly distributed higher taxa typically have different representatives in each area they occupy. For example, the mostly flightless birds known as ratites are a monophyletic group that includes the ostrich in Africa, rheas in the Neotropics, the emu and cassowaries of Australia and New Guinea, and the kiwis and the recently extinct moas of New Zealand (see Figure 6.13). Many other taxa are also shared between two or more southern continents, such as lungfishes, marsupials, cichlid fishes (see Figure 6.12), and southern beeches (*Nothofagus*) (Goldblatt 1993). Another common disjunct pattern is illustrated by alligators (*Alligator*), skunk cabbages (*Symplocarpus*), and tulip trees (*Liriodendron*), which are among the many genera that are found both in eastern North America and in temperate eastern Asia, but not in between (Wen 1999). We will investigate the reasons for some of these patterns later in this chapter.

Historical Factors Affecting Geographic Distributions

The geographic distribution of a taxon is affected by both contemporary and historical factors. The limits to the distribution of a species may be set by geological barriers that have not been crossed, or by ecological conditions to which the species is not adapted. In this section, we will focus on the historical processes that have led to the current distribution of a taxon: extinction, dispersal, and vicariance.

The distribution of a species may have been reduced by the extinction of some populations, and that of a higher taxon by the extinction of some constituent species. For example, the horse family, Equidae, originated and became diverse in North America, but it later became extinct there; only the African zebras and the Asian wild asses and horses have survived. (Horses were reintroduced into North America by European colonists.) Likewise, extinction is the cause of the disjunction between related taxa in eastern Asia and eastern North America. During the early Tertiary, many plants and animals spread throughout the northern regions of North America and Eurasia. Their spread was facilitated by a warm, moist climate and by land connections from North America to both Europe and Siberia. Many of these taxa became extinct in western North America in the late Tertiary due to mountain uplift and a cooler, drier climate, and were extinguished in Europe by Pleistocene glaciations (Wen 1999; Sanmartín et al. 2001).

Species expand their ranges by **dispersal** (i.e., movement of individuals). Some authors distinguish two kinds of dispersal: RANGE EXPANSION, or movement across expanses of more or less continuous favorable habitat, and JUMP DISPERSAL, or movement across a barrier (Myers and Giller 1988). Some species of plants and animals can expand their range very rapidly. Within the last 200 years, many species of plants accidentally brought from Europe by humans have expanded across most of North America from New York and New England, and some birds, such as the starling (*Sturnus vulgaris*) and the house sparrow (*Passer domesticus*), have done the same within a century (Figure 6.5). Other species

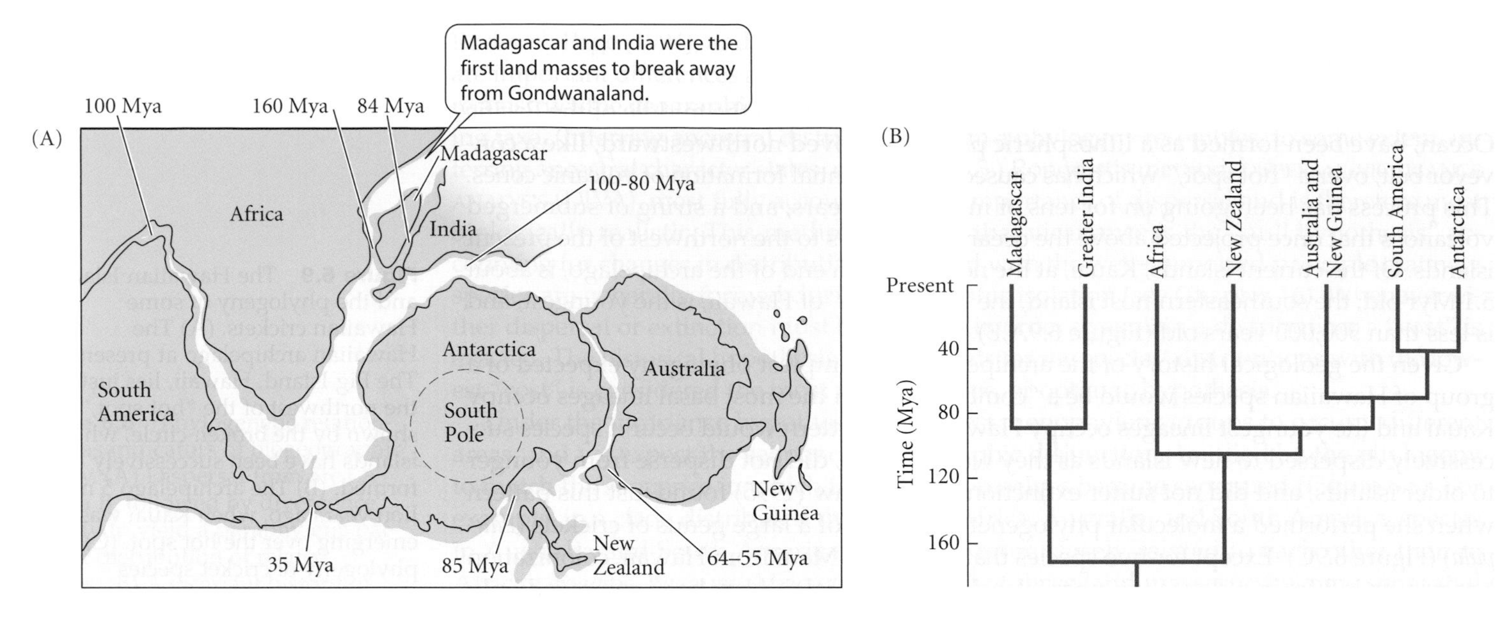

Figure 6.10 (A) A view of Gondwanaland in the early Cretaceous (120 Mya), centered on the present South Pole, indicating the approximate times at which connections among the southern land masses were severed. The current configurations of the continents are shown by the black lines; green areas beyond these lines were also exposed land during the early Cretaceous. (B) A branching diagram, sometimes called an "area cladogram," that attempts to depict the history of the breakup of Gondwanaland. "Greater India" was a large land mass that includes the present subcontinent of India and Sri Lanka. This branching tree does not show how different contiguous areas of some land masses (e.g., South America) separated at different times, as the map does. (A after Cracraft 2001.)

ure 6.10). India became separated from Madagascar 88–63 Mya, and collided with southern Asia about 50 Mya. For many years, biogeographers postulated that many of the endemic Madagascan taxa originated by vicariant separation from their relatives on other southern land masses. However, recent molecular phylogenetic studies indicate that dispersal has played the major role.

Raxworthy et al. (2002) analyzed the phylogeny of chameleons—slow-moving lizards that catch insects with their extraordinary projectile tongues (Figure 6.11A). Chameleons are distributed mostly in Africa, Madagascar, India, and islands in the Indian Ocean. Whereas the vicariance hypothesis would imply that Madagascan and Indian chameleons together should form the sister group of African forms, the phylogeny strongly supports the hypothesis that chameleons originated in Madagascar after the breakup of Gondwanaland and dispersed over water to Africa, India, and the islands (Figure 6.11B). Similar analyses of the lemurs (Primates) and the Madagascan mongoose-like carnivores indicated dispersal in the other direction: the ancestors of both groups colonized Madagascar from Africa, long after these land masses became separated (Yoder et al. 2003).

GONDWANAN DISTRIBUTIONS. Many other intriguing biogeographic problems are posed by taxa that have members on different land masses in the Southern Hemisphere. The simplest hypothesis is, of course, pure vicariance: the breakup of Gondwanaland isolated descendants of a common ancestor. However, phylogenetic analyses show that the story is not that simple, and the histories of some groups are still very controversial. Three examples will make the point.

Cichlids are freshwater fishes found in tropical America, Africa, Madagascar, and India. In molecular phylogenetic analyses by several investigators (e.g., Vences et al. 2001; Sparks 2004), two sister clades have been found, one consisting of Indian and Madagascan species and the other of two monophyletic groups, one in Africa and one in South America. This is exactly the branching pattern predicted by the vicariance hypothesis, since it exactly parallels the separation of these four regions (Figure 6.12). However, in a careful study of rates of DNA sequence evolution, Vences et al. (2001) concluded that the splits between clades of cichlids are much more recent than the splits between land masses; for example, the divergence between the Indian/Madagascan and African/Neotropical clades is no more than 56 million years old, whereas Madagascar and India separated from Africa at least 120 million years ago. Moreover, cichlids are a highly derived group within a huge clade of spiny-finned fishes that are not known before the late Cretaceous, long after the Gondwanan breakup. It seems likely that the cichlids achieved their distribution by dispersal, rather than by being rafted on fragments of Gondwanaland.

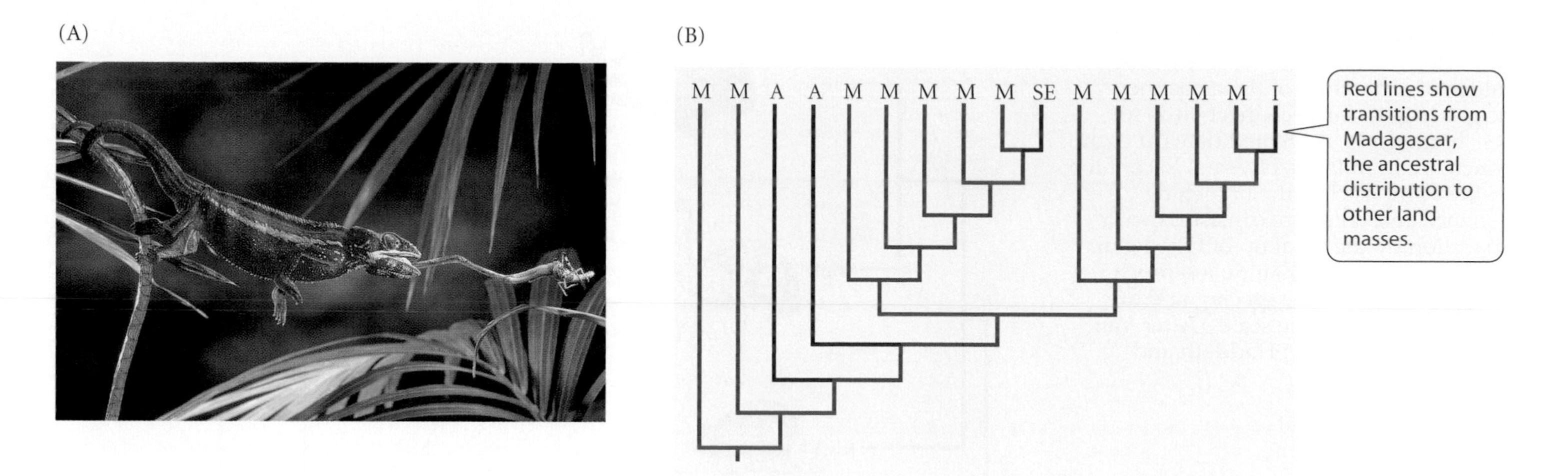
(A)
(B)
M M A A M M M M M SE M M M M M I
Red lines show transitions from Madagascar, the ancestral distribution to other land masses.

Figure 6.11 (A) A Madagascan panther chameleon, *Chamaeleo pardalis*, catches insects with its projectile tongue. (B) A phylogeny of some species of chameleons, showing their distribution in Africa (A), India (I), Madagascar (M), and the Seychelles Islands (SE) in the Indian Ocean. Because the phylogenetic distribution over these areas differs from the sequence by which the areas became separated (see Figure 6.10B), the distribution of chameleons is best explained by dispersal from Madagascar, rather than vicariance caused by the breakup of Gondwanaland. (A © Stephen Dalton/Photo Researchers, Inc.; B after Raxworthy et al. 2002.)

The ratite birds provide some support for Gondwanan vicariance—but only to a point (Haddrath and Baker 2001). These flightless birds, most of which are very large, stem from an ancient ancestor: along with tinamous, they are the sister group of all other living birds. They include not only the extant ostrich, rheas, cassowaries, emu, and kiwis, but also the moas of New Zealand, which were extinguished by indigenous people but have left bones from which DNA can be extracted. Because of their "Gondwanan distribution" and the great age of the clade, the ratites are a prime suspect for vicariance due to the breakup of Gondwanaland. Indeed, a phylogenetic study using complete sequences of the mitochondrial genome provided evidence that the moas diverged first, at about 79 Mya, which is consistent with the early (82 Mya) separation of New Zealand from Gondwanaland (Figure 6.13). The later divergence between the South American and Australian ratites, at about 69 Mya, is consistent with the later separation (at 55–65 Mya) of Australia from South America and Antarctica. But the divergence of the African ostrich (65 Mya) and the New Zealand kiwis (62 Mya) is much later than the separation of their homelands from the rest of Gondwanaland (Africa at 100 Mya and New Zealand at 82 Mya), and they appear to have employed some mode of dispersal.

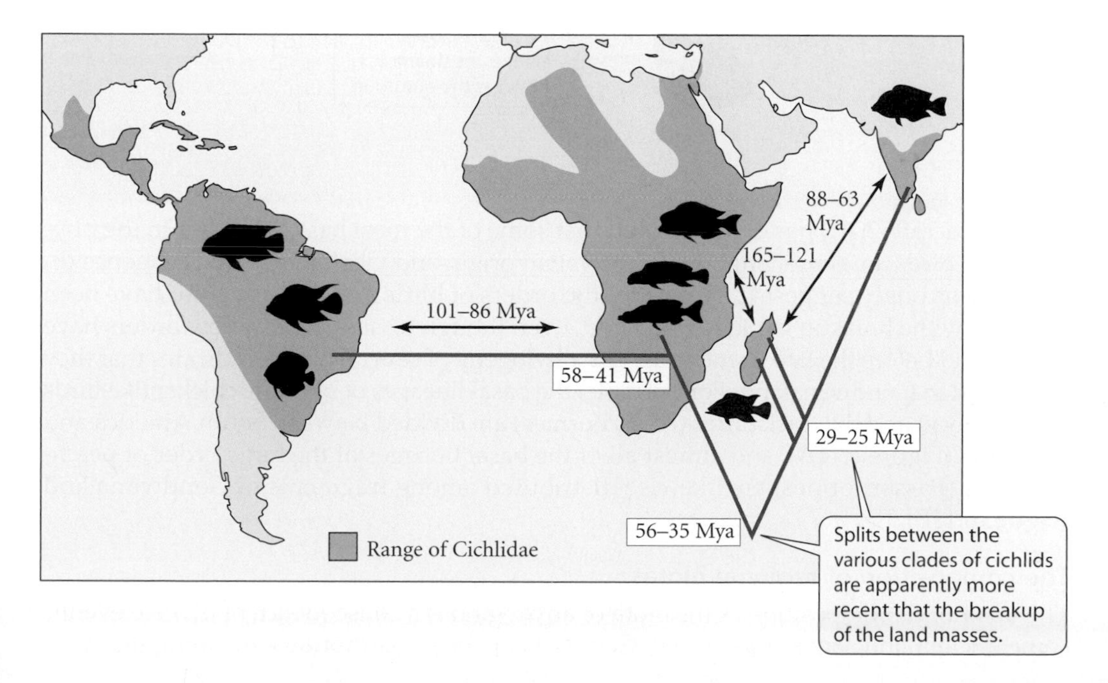
88–63 Mya
165–121 Mya
101–86 Mya
58–41 Mya
29–25 Mya
56–35 Mya
Range of Cichlidae
Splits between the various clades of cichlids are apparently more recent that the breakup of the land masses.

Figure 6.12 A phylogeny (blue tree) of the family Cichlidae, mapped onto its geographic distribution. The boxes indicate the divergence times between clades, estimated from DNA sequence differences; these clade divergence dates are younger than the dates of separation of the land masses (double-headed arrows). (After Vences et al. 2001.)

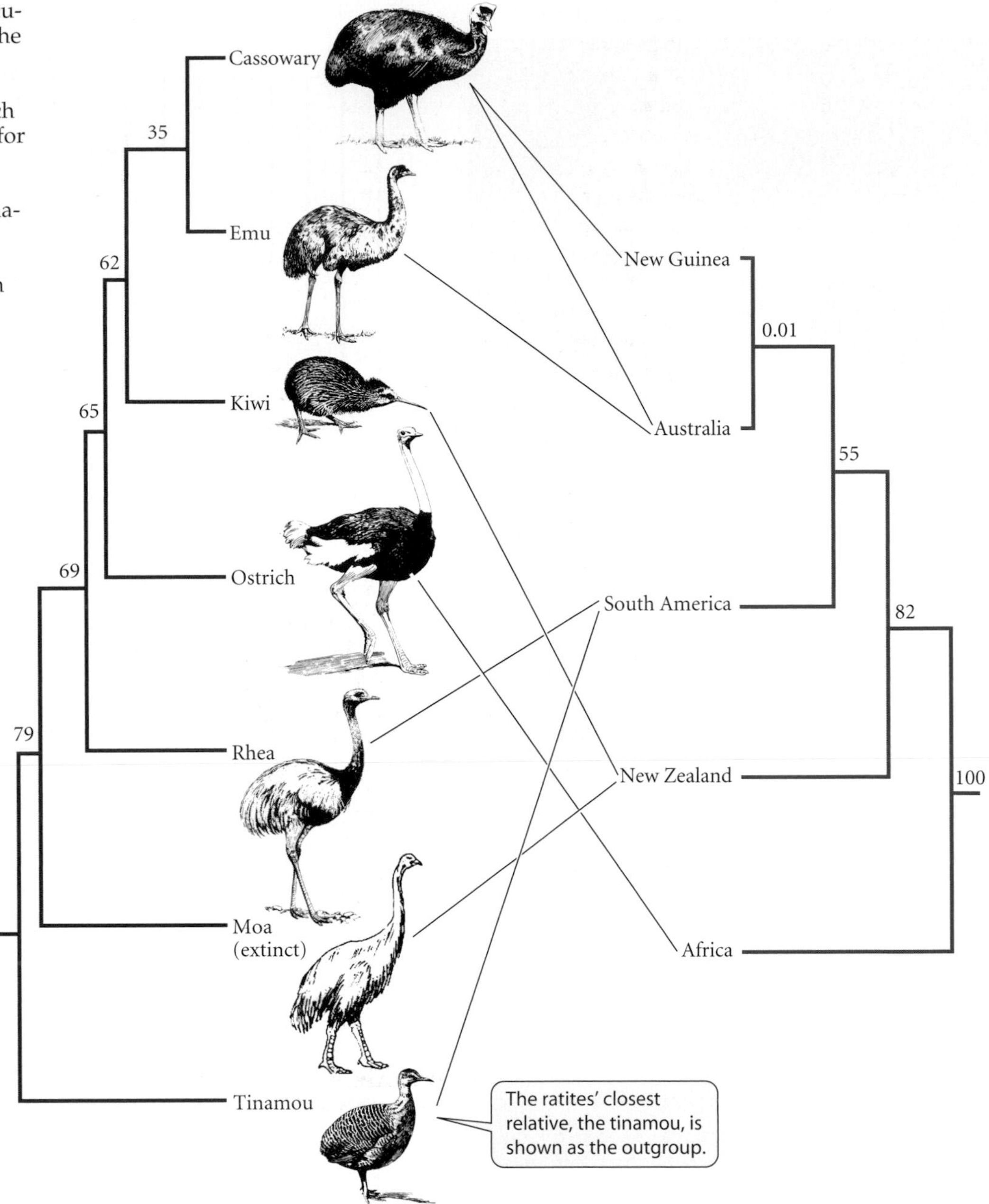

Figure 6.13 A comparison of a molecular phylogeny of the ratites (left) with the history of separation of the areas they occupy (right). Lines connect birds to their homelands; estimated dates at each branch are in millions of years. Except for the kiwi and ostrich, the branching sequence and dates are consistent with separation by the breakup of Gondwanaland. The kiwi and tinamou are much smaller than the other species, and are not drawn to the same scale. (After Van Tyne and Berger 1959; Haddath and Baker 2001.)

Joel Cracraft (2001) has demonstrated that some of the most basal branches in the phylogeny of birds are consistent with Gondwanan origin and vicariance. DNA sequence divergence strongly suggests that most of the orders of birds are old enough to have been affected by the breakup of Gondwanaland, even though fossils of only a few orders have been found before the late Cretaceous. The phylogeny of several orders indicates that they originated in Gondwanaland. For example, the basal lineages of both the chickenlike birds (Galliformes) and the duck order (Anseriformes) are divided between South America and Australia (Figure 6.14A), and almost all of the basal lineages of the huge order of perching birds (Passeriformes) are likewise distributed among fragments of Gondwanaland (Figure 6.14B).

The composition of regional biotas

The taxonomic composition of the biota of any region is a consequence of diverse events, some ancient and some more recent. Certain taxa are **allochthonous**, meaning that they

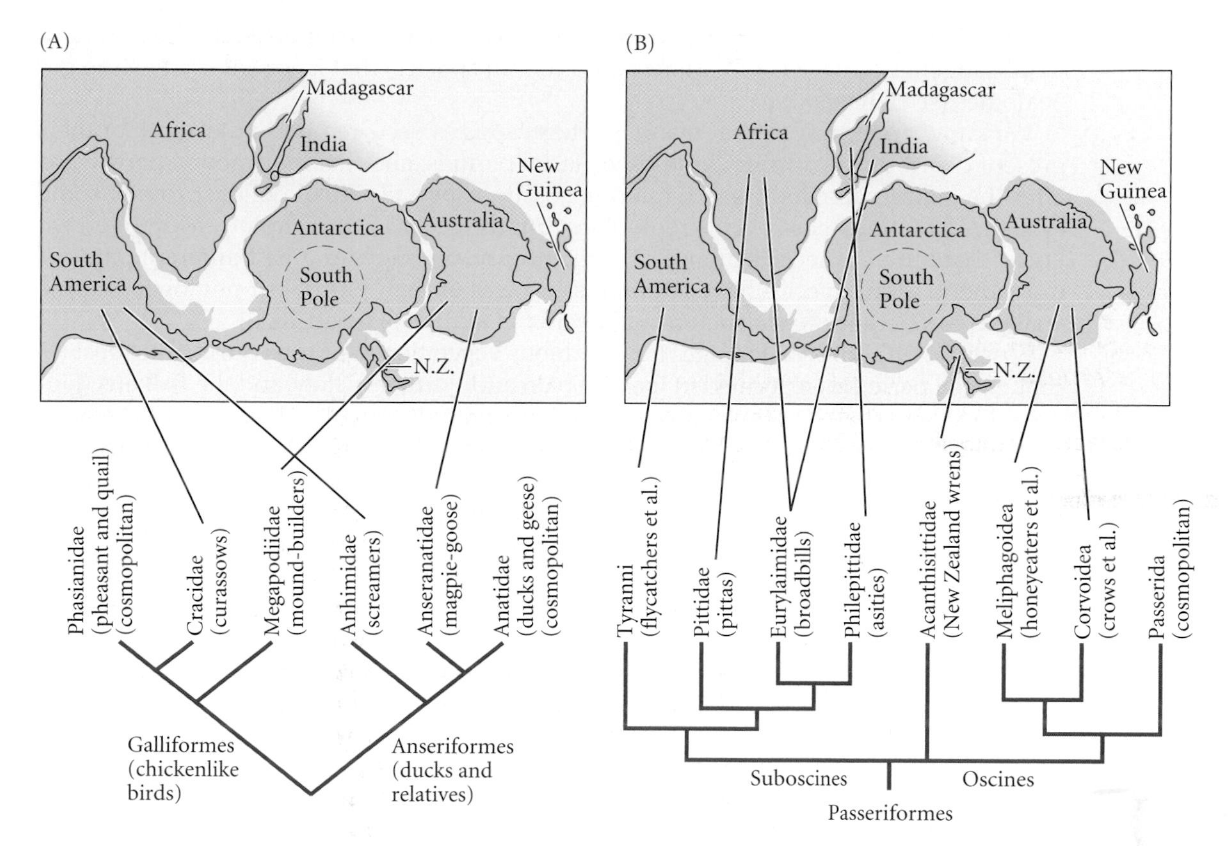

Figure 6.14 Phylogeny of major lineages in three orders of birds, showing their association with land masses, which are pictured as they were situated in the early Cretaceous, in a view centered on the present South Pole. The present continental boundaries are outlined in black; fringing areas shown in green were exposed during the Cretaceous. (A) The orders Galliformes and Anseriformes together form one of the oldest clades of birds. In each order, the basal lineages are divided between South America (curassows, screamers) and Australia (mound-builders, magpie-goose). In each order, a more derived lineage (Phasianidae; Anatidae) has a cosmopolitan (worldwide) distribution. (B) The order Passeriformes (perching or songbirds) has three major clades: suboscines, New Zealand wrens, and oscines. All three of these clades have basal lineages in the southern continents and appear to have originated in Gondwanaland. The relationships among the many families of the cosmopolitan Passerida are too poorly known to determine whether they also originated in a Gondwanan region. (After Cracraft 2001.)

originated elsewhere. Others are **autochthonous**, meaning that they evolved within the region. For example, the biota of South America has (1) some autochthonous taxa that are remnants of the Gondwanan biota and are shared with other southern continents (e.g., lungfishes, rheas); (2) groups that diversified from allochthonous progenitors during the Tertiary, after South America became isolated by continental drift (e.g., New World monkeys, guinea pigs and related rodents); (3) some allochthonous species that entered from North America during the Pleistocene (e.g., the mountain lion, *Panthera concolor*, which also occurs in North America); and (4) a few species that have colonized South America within historical time (e.g., the cattle egret, *Bubulcus ibis*, which apparently arrived from Africa in the 1930s; see Figure 6.6).

Phylogeography

Phylogeography is the description and analysis of the processes that govern the geographic distribution of lineages of genes, especially within species and among closely related species (Avise 2000). These processes include the dispersal of the organisms that carry the genes, so phylogeography provides insight into the past movements of species and the history by which they have attained their present distributions. It relies strongly on phylogenetic analysis of variant genes within species; that is, on inferring gene genealogies (see Chapter 2).

We know, for example, that many northern species occurred far to the south of their present distributions during Pleistocene glacial periods, and that they moved northward after the glaciers receded (see Chapter 5). Fossils, especially fossil pollen, provide some evidence of where these events took place, but the record is incomplete. Moreover, we know that different species occupied different glacial refuges and had different paths of movement. Many species have left no fossil traces of their paths, but phylogeographic analysis can help to reconstruct them (Taberlet et al. 1998; Hewitt 2000).

Fossil pollen shows that refuges for deciduous vegetation in Europe during the most recent glacial period were located in Iberia (Spain and Portugal), Italy, and the Balkans (Figure 6.15A), and that the vegetation expanded most rapidly from the Balkans as the glacier retreated. The grasshopper *Chorthippus parallelus*, sampled from throughout Europe, has unique haplotypes in Iberia and in Italy, whereas the haplotypes found in central and northern Europe are related to those in the Balkans (Figure 6.15B). Thus we can conclude that this herbivorous insect expanded its range chiefly from the Balkans, but did not cross the Pyrenees from Iberia, nor the Alps from Italy (Figure 6.15C). A similar analysis of hedgehogs (*Erinaceus europaeus* and *E. concolor*) indicated, in contrast, that these insectivorous mammals colonized northern Europe from all three refugial areas.

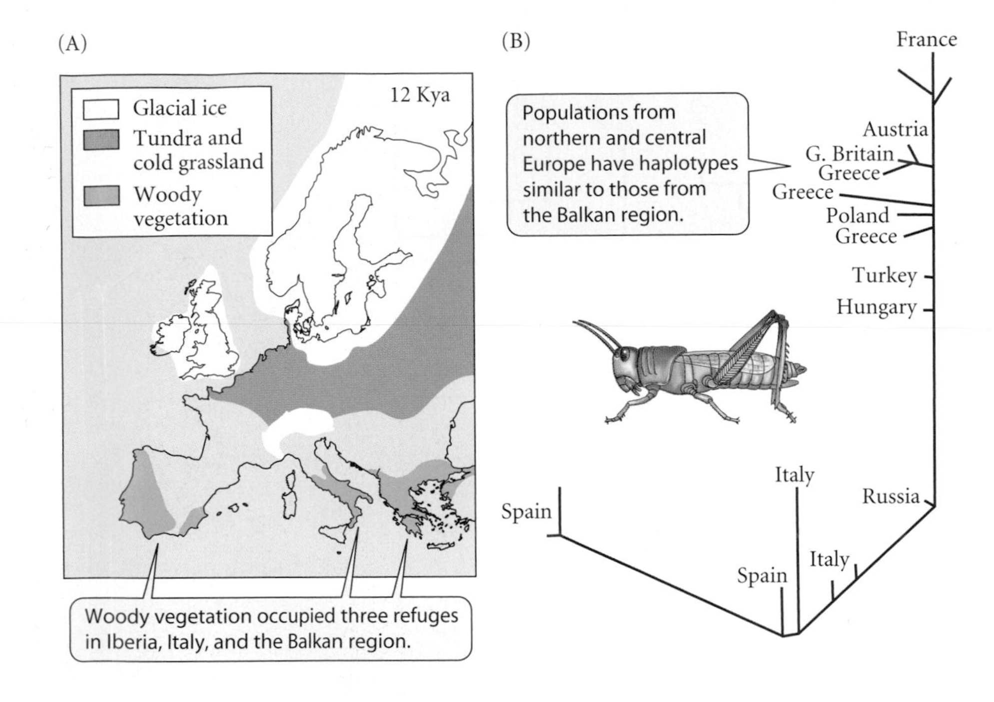
(A)
12 Kya
Glacial ice
Tundra and cold grassland
Woody vegetation
Woody vegetation occupied three refuges in Iberia, Italy, and the Balkan region.
(B)
Populations from northern and central Europe have haplotypes similar to those from the Balkan region.
France
Austria
G. Britain
Greece
Greece
Poland
Greece
Turkey
Hungary
Russia
Italy
Spain
Spain
Italy
(C)
The grasshopper spread through northern and central Europe from the Balkans...
...but did not move across mountain barriers from Iberia and Italy.

Phylogeography has also been applied to our own distribution. We saw in Chapter 4 that *Homo erectus* was broadly distributed throughout Africa and Asia by about a million years ago and had evolved into "archaic *Homo sapiens*" by about 300,000 years ago. How these ancient populations are related to the different human populations of today has been a controversial question (Relethford 2001; Klein and Takahata 2002; Templeton 2002).

Based on the morphology of fossil specimens, advocates of the MULTIREGIONAL HYPOTHESIS hold that archaic *sapiens* populations in Africa, Europe, and Asia all evolved into modern *sapiens*, with gene flow spreading modern traits among the various populations (Figure 6.16A). According to this hypothesis, there should exist genetic differences among

Figure 6.15 The recolonization of Europe from glacial refuges by the grasshopper *Chorthippus parallelus*, inferred from patterns of genetic variation. (A) Europe during the last glacial maximum (about 12,000 years ago). (B) Genetic relationships among contemporary grasshopper populations. The length of a line segment reflects both the number of mutational differences between haplotypes and the difference in proportions of those haplotypes among populations from the areas indicated. (C) Arrows show the inferred spread of *Chorthippus parallelus* after the glacier melted back. (A after Taberlet et al. 1998; B after Cooper et al. 1995; C after Hewitt 2000.)

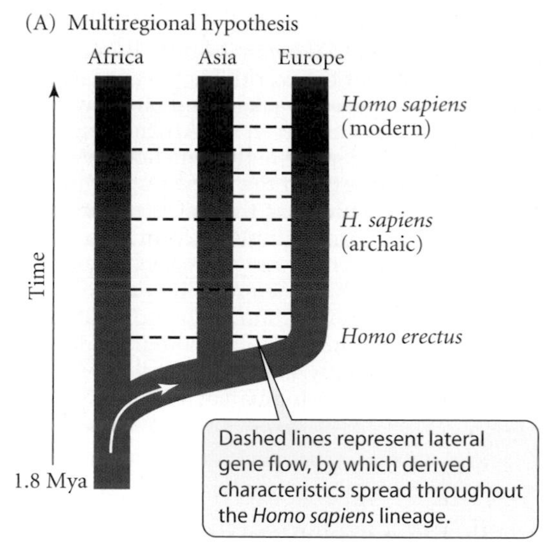

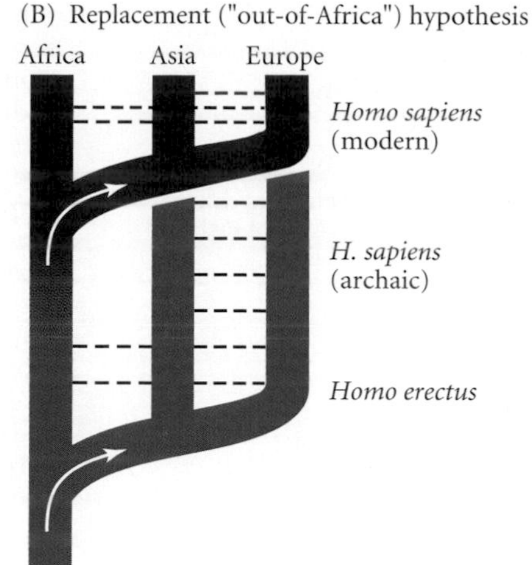

Figure 6.16 Two hypotheses on the origin of modern humans. (A) The multiregional hypothesis posits a single wave of expansion by *Homo erectus* from Africa to parts of Asia and Europe, and continuity of descent to the present day. (B) The replacement hypothesis proposes that populations of *H. erectus*, derived from African ancestors, gave rise to archaic *sapiens*, but that Asian and European populations of archaic *sapiens* became extinct when modern *sapiens* expanded out of Africa in a second wave of colonization.

modern Africans, Europeans, and Asians that trace back to the genetic differences that developed among populations of *erectus* and archaic *sapiens* nearly a million years ago. In contrast, the REPLACEMENT HYPOTHESIS, or OUT-OF-AFRICA HYPOTHESIS, holds that after archaic *sapiens* spread from Africa to Asia and Europe, modern *sapiens* evolved from archaic *sapiens* in Africa, spread throughout the world in a second expansion, and replaced the populations of archaic *sapiens* without interbreeding with them to any substantial extent (Figure 6.16B). That is, the modern *sapiens* that evolved from archaic *sapiens* in Africa was reproductively isolated from Eurasian populations of archaic *sapiens*—it was a distinct biological species. According to this hypothesis, most of the world's populations of archaic *sapiens* became extinct due to competition, and most genes in contemporary populations are descended from those carried by the population that spread from Africa.

Although this question is still subject to some debate (Templeton 2002), many genetic studies support the replacement hypothesis (Nei 1995; Jorde et al. 1998; Underhill et al. 2001). The first such studies employed sequence diversity in mitochondrial DNA (mtDNA) (Cann et al. 1987; Vigilant et al. 1991). A more extensive study of mtDNA used the complete mitochondrial sequence of 53 humans of diverse geographic origin, using a chimpanzee as an outgroup (Ingman et al. 2000). The phylogenetic analysis showed several basal clades of African haplotypes and a derived clade that includes not only several African haplotypes, but also all the non-African populations from throughout the world (Figure 6.17). Moreover, the non-African haplotypes vary less in nucleotide sequence than those found in Africa. These observations strongly support the replacement hypothesis. If, as in the multiregional hypothesis, some contemporary Asian populations were descended from indigenous populations of archaic *Homo sapiens* (and from indigenous *H. erectus*), and thus had a separate ancestry extending back a million years, we would expect some of their genes to have accumulated far more mutational differences than are observed. Indeed, mtDNA sequences from Neanderthal fossils are markedly divergent from modern human sequences (Ovchinnikov et al. 2000). It therefore appears likely that modern humans evolved from archaic *Homo sapiens* in Africa, and then colonized the rest of the world only about 200,000 to 30,000 years ago (see Chapter 10), replacing archaic *sapiens* without interbreeding (Klein 2003). This is a conclusion of the greatest importance, for it means that such genetic differences as exist among geographic populations of humans arose very recently, and that the human species is genetically much the same throughout the world.

Genetic similarities and differences among human populations have also been used to trace later movements. For example, sequence variation in a cluster of genes on the Y chromosome, which is carried only by men, has been studied on a worldwide basis (Underhill et al. 2001). Populations differ in both their proportions of different haplotypes and in

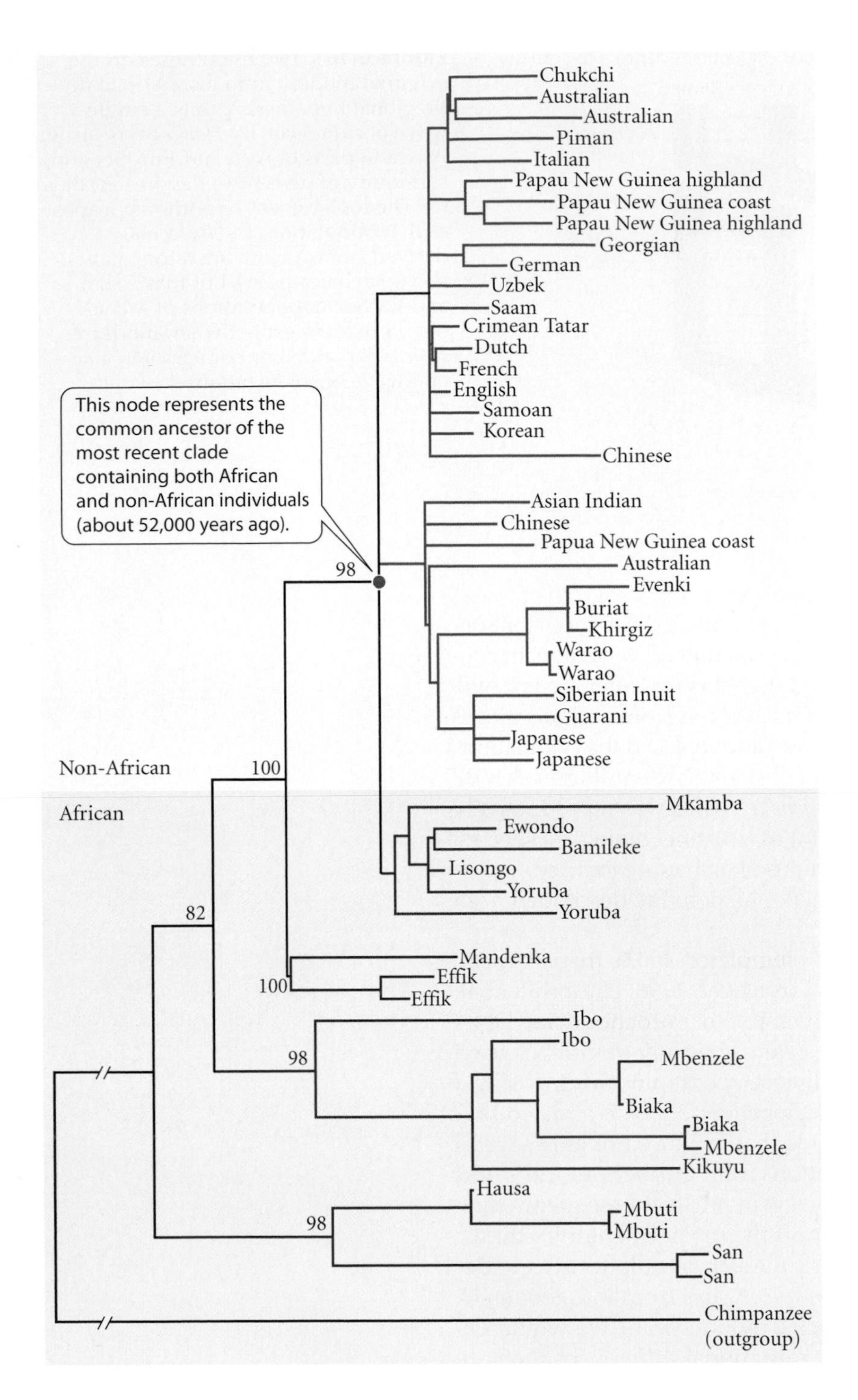

Figure 6.17 A gene tree based on complete sequences of mitochondrial genomes from human populations throughout the world. Haplotypes from individuals in Africa (green background) are phylogenetically basal, as expected given the African origin of the human species, and show high sequence diversity (represented by the lengths of the branches). Haplotypes taken from individuals in the rest of the world (yellow background) form a single clade of very similar haplotypes (denoted by short branches), as expected if these populations had been recently derived from a small ancestral population. Some populations (e.g., Australian) are represented by more than one individual. Numerals represent bootstrap values (see Box B in Chapter 2.) (After Ingman et al. 2000.)

how greatly those haplotypes differ in sequence from one another. The interpretation of such data can be difficult, partly because movements of people among populations over the course of time can obscure the genetic patterns that may have developed from the original course of colonization. Nonetheless, two groups of Y chromosome haplotypes that are basally located in the gene genealogy (groups I and II in Figure 6.18A) are restricted to Africa, supporting the replacement hypothesis. Non-African populations are characterized by haplotypes in the rest of the gene tree, consisting of several groups that are each more prevalent in some regions than in others. For example, group V is found in aboriginal Australians, whose ancestors arrived in Australia about 50,000 years ago, at about the same time that other humans were spreading throughout Eurasia (Figure 6.18B). Other groups of haplotypes differentiated in various parts of Europe and Asia, including Siberia, and were spread from one region to another by subsequent population movements. Group X haplotypes, descended from haplotypes found in Siberian populations, have a high frequency in Native American populations in both South and North America. Starting perhaps 15,000 to 12,000 years ago, several populations in northeastern Asia may have dispersed into North America at different times (Santos et al. 1999). The history suggested by Y chromosomes (which is considerably more complex than this brief description) supports inferences that had been previously drawn from other kinds of genetic data. The genetic relationships among populations parallel their linguistic relationships to some extent, suggesting that both genes and languages have a common history of divergence in isolation (Cavalli-Sforza et al. 1994).

Ecological Approaches to Biogeography

Whereas systematists often look first to evolutionary history in order to understand the reasons for a taxon's distribution, ecologists tend to look to factors operating now or in the very recent past. Whether a historical or an ecological perspective is most suitable may

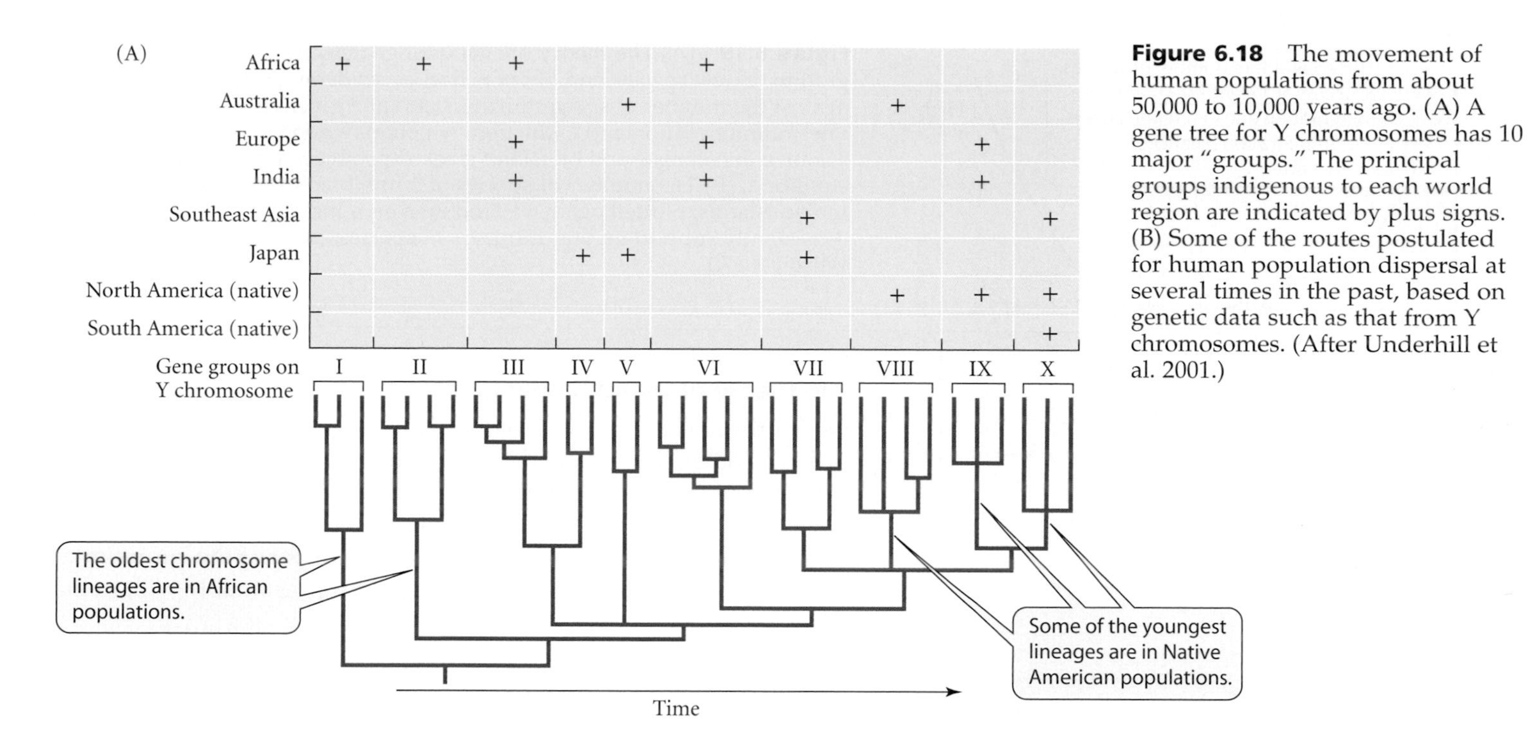

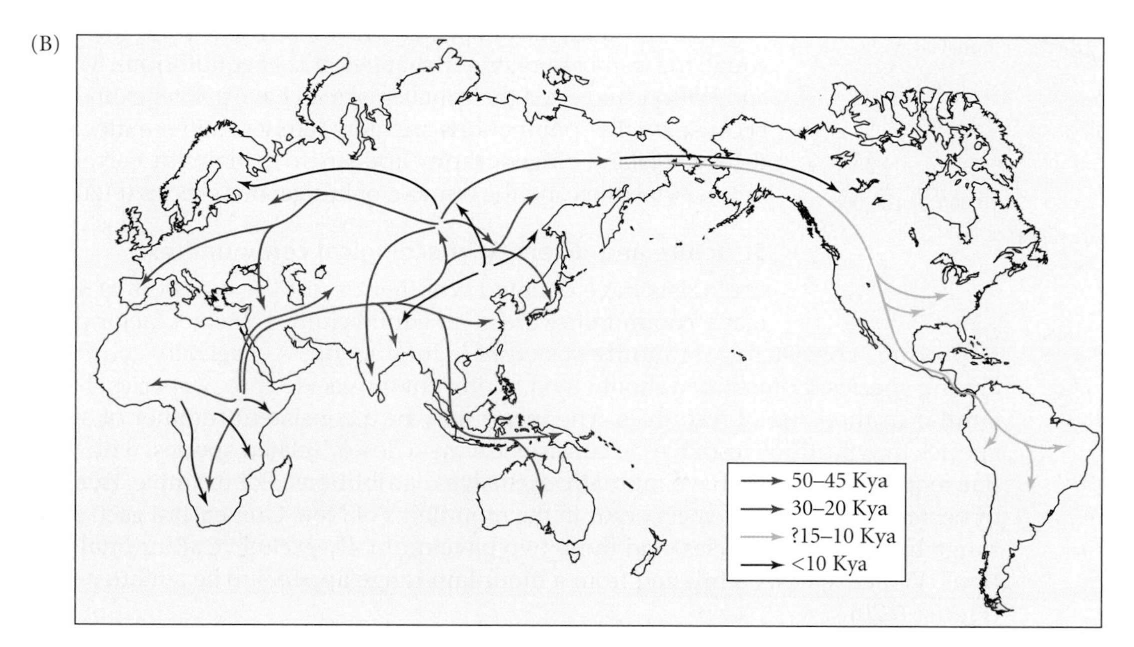

Figure 6.18 The movement of human populations from about 50,000 to 10,000 years ago. (A) A gene tree for Y chromosomes has 10 major "groups." The principal groups indigenous to each world region are indicated by plus signs. (B) Some of the routes postulated for human population dispersal at several times in the past, based on genetic data such as that from Y chromosomes. (After Underhill et al. 2001.)

depend on the particular questions posed and the spatial scale of the distributions under study. For example, phylogenetic history is likely to explain why cacti are native only to the Americas, but to explain why the saguaro cactus (*Carnegiea gigantea*) is restricted to certain parts of the Sonoran Desert, we would have to look toward ecological factors, such as the species' tolerance for rainfall and temperature, or perhaps the effects of competitors, herbivores, or pathogens. We might then assume that the species' range is at equilibrium (i.e., that it is not changing). Alternatively, we might entertain a NONEQUILIBRIUM HYPOTHESIS, such as the proposition that the species is still expanding from a glacial refuge. Although a species' range limit may have reached a short-term equilibrium determined by its present physiological tolerance, it might not have achieved an evolutionary equilibrium if its tolerance is still evolving.

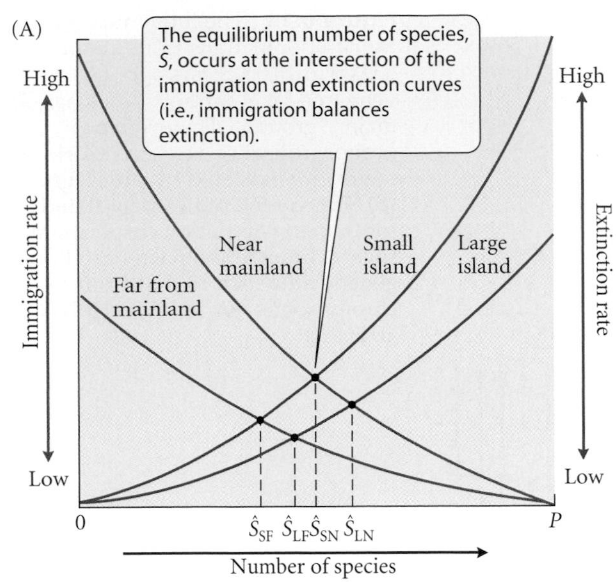

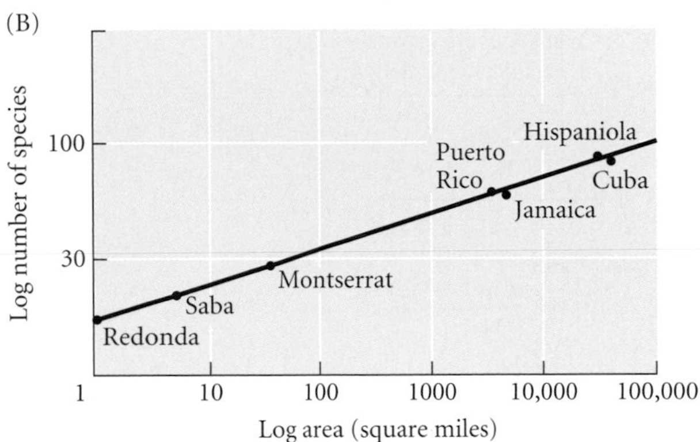

Figure 6.19 (A) The theory of island biogeography. The rates of immigration of new species and of extinction of resident species are plotted against the number of species on an island at a given time. Differences in rates of immigration and extinction, which may depend on distance from a source of colonists and on island size, respectively, result in different equilibria. (B) The number of species of amphibians and reptiles on West Indian islands, plotted against island area on a log-log plot. Larger islands consistently support greater numbers of species. (After MacArthur and Wilson 1967.)

The theory of island biogeography

One of the major topics in ecological biogeography is variation in the diversity of species among regions or habitats. For example, what determines the number of species on an island? Islands typically have fewer species than patches of the same size on continents. The traditional nonequilibrium hypothesis was that most of the continental species have not reached the islands yet (but presumably will, in the fullness of time).

Robert MacArthur and Edward O. Wilson (1967) proposed an equilibrium hypothesis instead (Figure 6.19A). The number of species on an island is increased by new colonizations, but decreased by extinctions. As long as the rate of new colonizations exceeds the rate of extinction, the number of species grows, but when the rates become equal, the number no longer changes; it is at equilibrium. MacArthur and Wilson suggested that smaller islands have greater extinction rates because smaller populations are more likely to suffer extinction. This theory of island biogeography appears to explain the correlation between island area and the number of indigenous species (Figure 6.19B).

Structure and diversity in ecological communities

Ecologists have debated whether or not the numbers of species in many communities are at an equilibrium. The chief factor presumed to produce consistent community structure is interactions—especially competition—among species. Competition should tend to prevent the coexistence of species that are too similar in their use of resources. The result may be a consistent number of sympatric species that partition resources in consistent ways. Closely related species, with very similar requirements, may have mutually exclusive distributions. For example, three species of nectar-feeding honeyeaters occur in the mountains of New Guinea, but each mountain range has only two species, and those two have mutually exclusive altitudinal distributions. Which species is missing from a mountain range appears to be a matter of chance (Figure 6.20).

Community convergence

Many examples of convergent evolution of individual taxa are known. For example, desert plants have independently evolved similar morphological features in many parts of the world (see Figure 6.1), and several groups of birds have independently evolved features suitable for feeding on nectar, such as a long, slender bill (see Figure 3.8). The question arises, are these individual instances part of a larger pattern of convergence of whole communities? If two regions present a similar array of habitats and resources, will species evolve to utilize and partition them in the same way? If so, it would suggest that communities have achieved an evolutionary equilibrium.

A striking example of community-level convergence has been described in the anoles (*Anolis*) of the West Indies (Williams 1972; Losos 1990, 1992; Losos et al. 1998). Anoles are a species-rich group of insectivorous, mostly arboreal Neotropical lizards (Figure 6.21). Different species are known to compete for food, and this competition has influenced the

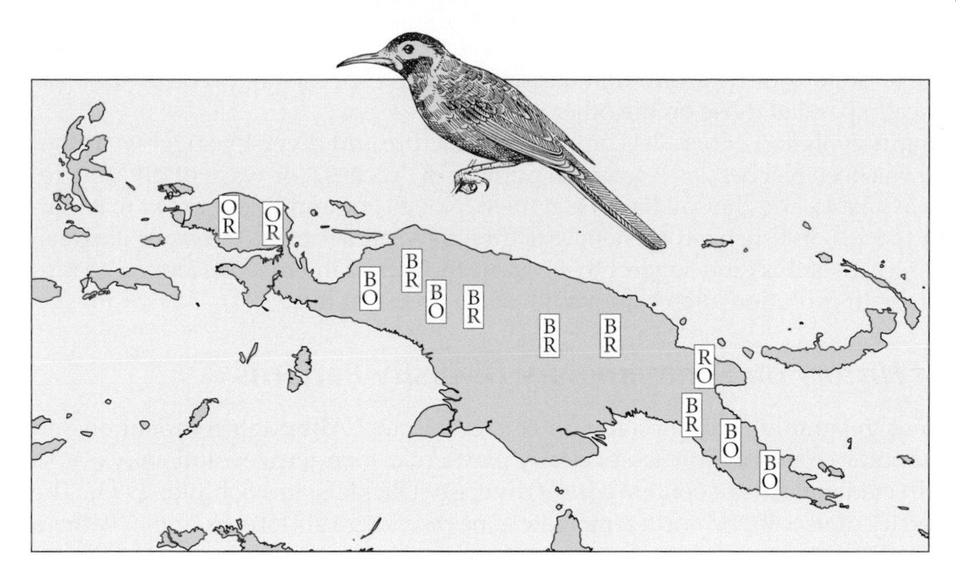

Figure 6.20 A "checkerboard" distribution in which species replace each other haphazardly. Among the various mountain ranges in New Guinea, three species of honeyeaters (*Melidectes*), denoted by letters O, R, and B, are distributed in pairs. Each pair has mutually exclusive altitudinal ranges, as shown by the stacked letters. The three species do not all coexist in any mountain range. (After Diamond 1975.)

structure of anole communities. Each of the small islands in the Lesser Antilles has either a single (solitary) species or two species. Solitary species are generally moderate in size, whereas larger islands have a small and a large species that can coexist because they take insect prey of different sizes and also differ in their microhabitats. The small species of the various islands are a monophyletic group, and so are the large species. Thus it appears that each island has a pair of species assembled from the small-sized and the large-sized clades.

The large islands of the Greater Antilles (Cuba, Hispaniola, Jamaica, Puerto Rico) harbor greater numbers of species. These anoles occupy certain microhabitats, such as tree crown, twig, and trunk, that are filled by different species on each island. The occupants of different microhabitats, called ECOMORPHS, have consistent, adaptive morphologies (see Figure 6.21). These ecomorphs have evolved repeatedly, for the species on each of the

Figure 6.21 Convergent morphologies, or "ecomorphs," of *Anolis* lizards in the West Indies. (A) *Anolis lineatopus* from Jamaica. (B) *A. strahmi* from Hispaniola. Both species have independently evolved the stout head and body, long hind legs, and short tail associated with living on lower tree trunks and on the ground. (C) *Anolis valencienni* from Jamaica. (D) *A. insolitus* from Hispaniola. Both are twig-living anoles that have convergently evolved a more slender head and body, shorter legs, and long tail. (Photographs by K. DeQueiroz and R. Glor, courtesy of J. Losos.)

islands form a monophyletic group that has radiated into species that ecologically and morphologically parallel those on the other islands.

Such extreme evolution of parallel community structure and diversity suggests that an equilibrium has been reached, as if a certain number of "niches," or ways of dividing resources, are available, and they all have been filled. Not all communities appear to be saturated with species, and such a consistent structure as the anoles present may be unusual. Nevertheless, cases of this kind suggest that basic principles of interactions among species may provide both evolution and ecology with some predictability.

Effects of History on Contemporary Diversity Patterns

What explains geographic variation in numbers of species? Although competition and other contemporary ecological processes clearly play a role, long-term evolutionary events have also affected patterns of contemporary diversity (Ricklefs and Schluter 1993). The species diversity of trees in the north temperate zone provides a striking example (Latham and Ricklefs 1993). Moist temperate forests are found primarily in Europe, eastern North America, and eastern Asia. The ratio of the number of tree species in these areas is 1:2:6; Asia has by far the greatest number of species. These differences in species diversity are paralleled by the diversity at higher taxonomic levels. In Asia, a greater proportion of taxa belong to primarily tropical groups than in Europe or America. These differences are not correlated with contemporary patterns of climate.

For about the first 40 million years of the Cenozoic, the Earth was warmer than it is today. Forests were spread across northern America and Eurasia, and many genera were distributed more broadly than they are today. The temperate flora of North America was separated from the tropical American flora by a broad seaway, and the temperate flora of Europe was disjunct from the African flora, but the northern Asian flora graded into the tropical flora, as it does today, from Siberia to the Malay Peninsula (Figure 6.22). Thus, in

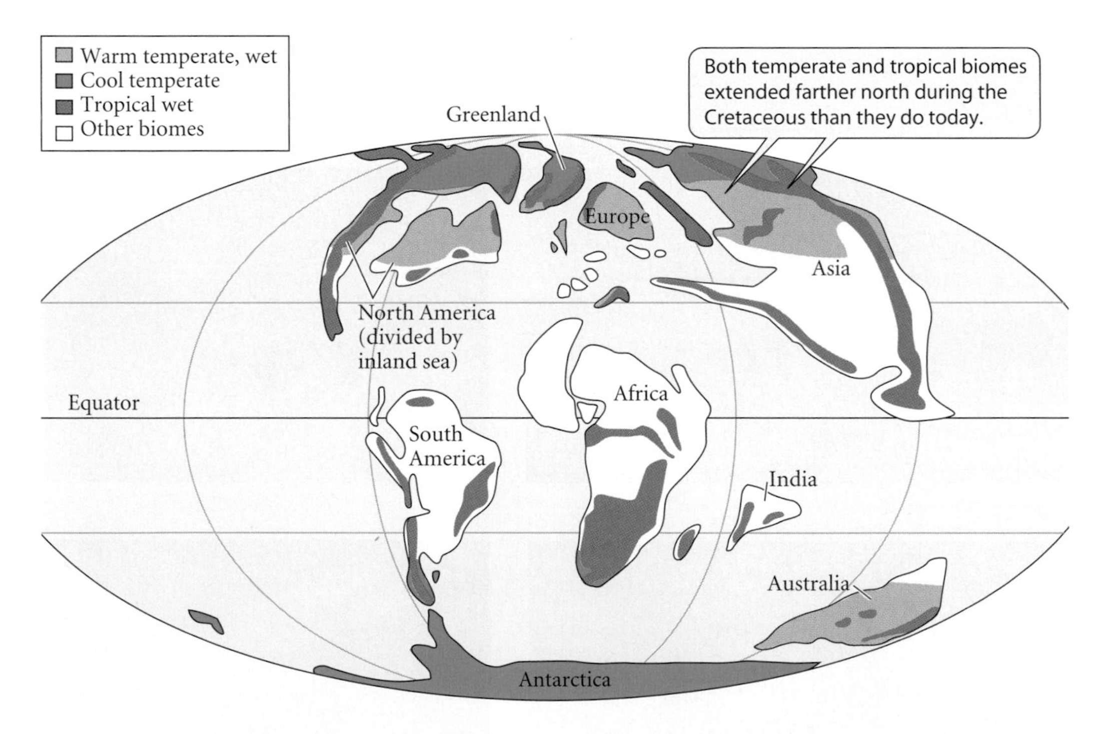

Figure 6.22 The distribution of warm temperate, cool temperate, and wet tropical biomes (vegetations types) at the end of the Cretaceous. A corridor of wet tropical vegetation extended farther south in eastern Asia than in Europe or eastern North America, which were separated from the major tropical areas. (After Latham and Ricklefs 1993.)

Asia, there was greater opportunity for tropical lineages to spread into and adapt to more temperate climates. Probably for this reason, eastern Asia in the Tertiary had more genera of trees than either Europe or eastern North America.

In the late Tertiary and the Quaternary, global cooling culminated in the Pleistocene glaciations, which extended farther south in Europe and eastern North America than in Asia. These glaciations devastated the flora of North America and especially of Europe, where its southward movement was blocked by the Alps, the Mediterranean Sea, and deserts. The continuous corridor to the Asian tropics, however, provided refuge for the Asian flora. A far greater proportion of genera became extinct in Europe and North America than in Asia. Thus contemporary differences in diversity among these regions appear to have been caused by two factors: a long Cenozoic history of differences in opportunities for dispersal, adaptation, and diversification and a recent history of differential extinction.

Summary

1. The geographic distributions of organisms provided Darwin and Wallace with some of their strongest evidence for the reality of evolution.
2. Biogeography, the study of organisms' geographic distributions, has both historical and ecological components. Certain distributions are the consequence of long-term evolutionary history; others are the result of contemporary ecological factors.
3. The historical processes that affect the distribution of a higher taxon include extinction, dispersal, and vicariance (fragmentation of a continuous distribution by the emergence of a barrier). These processes may be affected or accompanied by environmental change, adaptation, and speciation.
4. Histories of dispersal or vicariance can often be inferred from phylogenetic data. When a pattern of phylogenetic relationships among species in different areas is repeated for many taxa, a common history of vicariance is likely.
5. Disjunct distributions are attributable in some instances to vicariance and in others to dispersal.
6. Genetic patterns within species, especially phylogenetic relationships among genes that characterize different geographic populations, can provide information on historical changes in a species' distribution.
7. The local distribution of species is affected by ecological factors, including both abiotic aspects of the environment and biotic features such as competitors and predators.
8. The diversity of species in a local region may or may not be at an equilibrium. Interspecific interactions, especially competition, may limit species diversity and may result in different communities with a similar structure. In some cases, sets of species have independently evolved to partition resources in similar ways.
9. The species diversity of a higher taxon in a particular region is often a result both of current ecological factors and of long-term evolutionary factors.

Terms and Concepts

allochthonous
autochthonous
biogeographic realm
biogeography (phytogeography, zoogeography)
disjunct distribution
dispersal
ecological biogeography
endemic
historical biogeography
phylogeography
vicariance

Suggestions for Further Reading

J. H. Brown and M. V. Lomolino, *Biogeography* (Second Ed., Sinauer Associates, Sunderland, MA., 1998) is a comprehensive textbook of biogeography. A shorter textbook is C. B. Cox and P. D. Moore's *Biogeography: An ecological and evolutionary approach* (Blackwell Scientific Publications, Oxford, 1993).

R. E. Ricklefs and D. Schluter are the editors of *Species diversity in ecological communities: Historical and geographical perspectives* (University of Chicago Press, Chicago, 1993), a multi-authored collection of papers that includes both ecological and historical approaches to understanding species diversity.

Phylogeography is treated in depth by J. C. Avise in *Phylogeography* (Harvard University Press, Cambridge, Mass., 2000), and human phylogeography is included in J. Klein and N. Takahata, *Where do we come from? The molecular evidence for human descent* (Springer-Verlag, New York, 2002).

Problems and Discussion Topics

1. Until recently, the plant family Dipterocarpaceae was thought to be restricted to tropical Asia, where many species are ecologically dominant trees. Recently, a new species of tree in this family was discovered in the rain forest of Colombia, in northern South America. What hypotheses can account for its presence in South America, and how could you test those hypotheses?
2. As described in the text, the deepest split in cichlid phylogeny appears to be less than 60 million years old (see Figure 6.12). The most basal lineage of cichlids is restricted to Madagascar and India, where there are few other cichlids. If this "primitive" group's distribution is not due to ancient separation by the breakup of Gondwanaland, why should it be restricted to those areas? Why is this lineage not also found in Africa or South America? What evidence might bear on your hypotheses?
3. In their analysis of ratite biogeography, Haddrath and Baker argued that the distribution of the ostrich and kiwis is not attributable to rafting on fragments of Gondwanaland. Formulate a set of alternative hypotheses that could account for the distributions of these birds, specify what kind of evidence might support or refute each hypothesis, and then compare your analysis with these authors'.
4. Some biogeographers, subscribing to the "cladistic vicariance" school of thought (Humphries and Parenti 1986), hold that vicariance should always be the preferred hypothesis, and dispersal should be invoked only when necessary, because the vicariance hypothesis can be falsified (if it is false), whereas dispersal can account for any pattern and therefore is not falsifiable. What are the pros and cons of this position? (See Endler 1983.)
5. In some cases, it can be shown that species are physiologically incapable of surviving temperatures that prevail beyond the borders of their range. Do such observations prove that cold regions have low species diversity because of their harsh physical conditions?
6. The species diversity of plants, birds, mammals, and many other taxa declines from tropical regions toward the poles. What hypotheses account for this latitudinal gradient? What evidence is there for and against these hypotheses? (See Willig et al. 2003.)

7

The Evolution of Biodiversity

Biological diversity, or **biodiversity**, poses some of the most interesting questions in biology. Why are there more species of rodents than of primates, of flowering plants than of ferns? Why do some regions, such as the tropics, have more species than others, such as temperate areas? Why has the diversity of species changed over evolutionary time? Does diversity increase steadily, or has it reached a limit? Because so many factors can influence the diversity of species, these questions are both interesting and challenging.

Modern biodiversity. Tropical coral reefs are the most species-rich marine environments, harboring hundreds of species. Visible in the Philippine reef shown here are diverse corals and anemones (Cnidaria), feathery crinoids (Echinodermata), and fish (Vertebrata). (Photo © Photo library.com)

Biodiversity can be studied from the complementary perspectives of ecology and evolutionary history (Ricklefs and Schluter 1993). Ecologists focus primarily on factors that operate over short time scales to influence diversity within local habitats or regions. But factors that operate on longer time scales, such as climate change and evolution, also affect diversity. Within the last 50 years, global warming—almost certainly caused by humans burning fossil fuels—has noticeably altered the geographic distributions of many species (Parmesan et al. 1999; Root et al. 2003). On a scale of millions of years, extinction, adaptation, speciation, climate change, and geological change create the potential for entirely different assemblages of species. In this chapter, we will examine long-term patterns of change in diversity, caused by originations and extinctions of taxa on a scale of millions of years. Ecological and evolutionary studies of contemporary species are often useful for interpreting these paleontological patterns.

Estimating Changes in Taxonomic Diversity

Data on diversity in the fossil record, like all data in science, are incomplete and can be misleading unless steps are taken to account for their limitations. Paleobiologists continue to study the effects of sampling on their estimates of diversity (Alroy et al. 2001).

Estimates of diversity

The simplest expression of taxonomic diversity is a simple count of species (SPECIES RICHNESS). Over long time scales or large areas, species diversity is often estimated by compiling records, such as the publications or museum specimens that have been accumulated by many investigators, into faunal or floral lists of species. Most paleontological studies of diversity employ counts of higher taxa, such as families and genera, because they generally provide a more complete fossil record than individual species do.

Counts of taxa in the fossil record can be imprecise and even misleading, requiring that correction factors be included in data analysis (Raup 1972; Signor 1985; Foote 2000a). For example, records of fossil taxa are often described at the level of the stages into which each geological period is divided. Most such stages are about 5 to 6 million years (My) long, so the recorded duration of a taxon (its first and last occurrences in the record) is imprecise. Geological periods and stages vary in duration, and more recent geological times are represented by greater volumes and areas of fossiliferous rock. Therefore, it may be necessary to adjust the count of taxa by the amount of time and rock volume represented.

Because fossils are a small sample of the organisms that actually lived, a taxon is often recorded from several separated time horizons, but not from those in between. We therefore can deduce that in the same way, the actual times of origination and extinction of a taxon may have occurred before and after its earliest and latest records in the rocks. It follows that if many taxa actually became extinct in the same time interval, their last recorded occurrences—their *apparent* times of extinction—may be spread over several earlier intervals as well. Conversely, if many taxa actually originated at the same time, some of them may appear to have originated at later times.

Since our count of living (Recent) species is much more complete than our count of past species, taxa that are still alive today have apparently longer durations and lower extinction rates than they would if they had been recorded only as fossils. That is, we can list a living taxon as present throughout the last 10 My, let us say, even if its only fossil occurrence was 10 Mya. Because the more ancient a taxon is, the less likely it is to still be extant, diversity will seem to increase as we approach the present, a bias called the **pull of the Recent**. This bias can be reduced by counting only fossil occurrences of each taxon and not listing it for time intervals between its last fossil occurrence and the Recent. In practice, the pull of the Recent seems not to affect estimates of the history of diversity very much (Foote 2000a).

Because of unusually favorable preservation conditions or other chance events, a taxon may be recorded from only a single geological stage, even though it lived longer than that. Such "singletons" make up a higher proportion of taxa as the completeness of sampling decreases and therefore bias the sample; moreover, they can create a spurious correlation between rates of origination and extinction because they appear to originate and become extinct in the same time interval. Diversity may be more accurately estimated by ignoring such singletons and counting only those taxa that cross the border from one stage to another.

Rates

The rate of change in diversity depends on the rates at which taxa originate and become extinct. Several expressions for such rates have been used in the paleobiological literature. One is simply the number of taxa originating (or becoming extinct) per My (or other time unit). The most useful measure is the number of originations (or extinctions) per taxon per unit time (Foote 2000a).

The number of taxa (N) changes over time by origination (due to branching of lineages) and extinction. These events are analogous to the births and deaths of individual organisms in a population, so models of population growth can be adapted to describe changes in taxonomic diversity. For each time interval Δt, suppose S new taxa originate per taxon present at the start of the interval, and suppose E is the number of taxa that become extinct, per original taxon, during the interval. Then ΔN, the change in N, equals the number of "births," SN, minus the number of "deaths," EN, and the rate of change in diversity (the **diversification rate**), $\Delta N/\Delta t$, is

$$\frac{\Delta N}{\Delta t} = SN - EN \quad \text{or} \quad \frac{\Delta N}{\Delta t} = RN$$

where $R = (S - E)$ and is the PER CAPITA RATE OF INCREASE. The growth in number of taxa is positive if $R > 1$.

Between the beginning (time t_0) and end (time t_1) of the time interval, the "population" grows by multiplying its original size by the per capita rate of increase: $N_1 = N_0R$. If the rates S and E remain constant, then after the next interval Δt, the population will be $N_2 = N_1R = N_0R^2$. In general, after t time intervals, the number of taxa will be

$$N_t = N_0R^t$$

as long as the per capita origination and extinction rates remain the same. This equation describes **exponential growth** of the number of taxa (or of a population of organisms) (Figure 7.1A). For continuous growth, rather than growth in discrete intervals, the equivalent expressions are

$$\frac{dN}{dt} = rN \quad \text{and} \quad N_t = N_0e^{rt}$$

where r is the INSTANTANEOUS PER CAPITA RATE OF INCREASE.

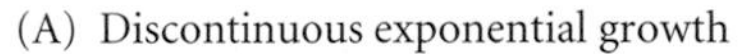
(A) Discontinuous exponential growth

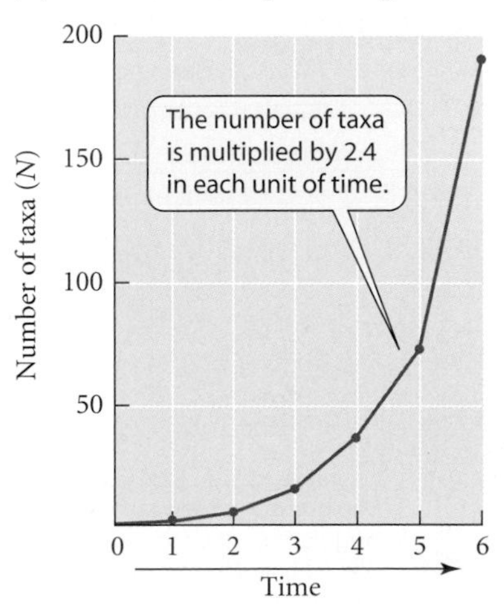
200
150
100
50
Number of taxa (N)
The number of taxa is multiplied by 2.4 in each unit of time.
0 1 2 3 4 5 6
Time

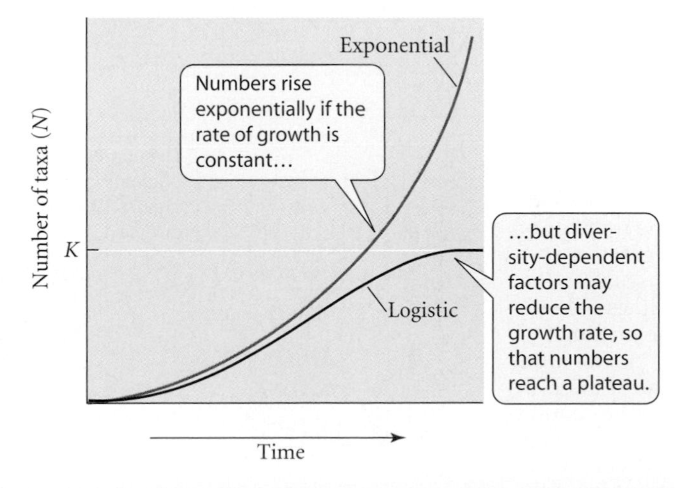
(B) Continuous exponential and logistic growth
Number of taxa (N)
K
Exponential
Numbers rise exponentially if the rate of growth is constant...
...but diversity-dependent factors may reduce the growth rate, so that numbers reach a plateau.
Logistic
Time

Figure 7.1 (A) Theoretical increase in the number of taxa (N), according to the equation $N_t = N_0R^t$, where R is the rate of increase in each discrete time interval (in this example, $R = 2.4$). (B) The increase in the number of taxa when change is continuous. The number of taxa grows exponentially if the rate of increase (r) is constant, but levels off if growth is logistic (i.e., if the rate of increase is negatively diversity-dependent).

Let us pursue the analogy between the number of taxa and the number of organisms in a population (in which, as we know, birth and death rates are not constant). Factors such as severe weather may alter these rates by proportions that are unrelated to the density of the population. In contrast, **density-dependent factors** may cause the per capita birth rate to decline, and/or the death rate to increase, with population density, so that population growth slows down and the population density may reach a stable equilibrium (Figure 7.1B). Such density-dependent factors include competition for food or space. Density-dependent population growth is modeled in ecology by the equation for **logistic growth**:

$$\frac{dN}{dt} = \frac{rN(K-N)}{K}$$

where K is the "carrying capacity"—the equilibrium population density, at which population increase has declined to zero. Paleobiologists have suggested that changes in the number of species or higher taxa may similarly be affected by **diversity-dependent factors** that might reduce origination rates or increase extinction rates as the number of species increases. A factor that might have this effect is competition among species for resources.

Timothy Walker and James Valentine (1984) proposed a modified logistic model, in which an ecosystem may have a stable equilibrium number of species that is less than the maximal possible number. Walker and Valentine's model (Figure 7.2) assumes that the per species rate of extinction, E, is diversity-independent, but that the per species rate of origination of new species, S, declines as the number of species (N) approaches the maximum (N_{max}) that the environment can support. Thus

$$S = S_0\left(1-\frac{N}{N_{max}}\right)$$

where S_0 is the highest possible rate of speciation. The resulting equation for the rate of change in the number of species is

$$\frac{\Delta N}{\Delta t} = (S-E)N = \left[S_0\left(1-\frac{N}{N_{max}}\right)-E\right]N$$

which, when $\Delta N/\Delta t = 0$, yields the equilibrium number of species (N^*):

$$N^* = \left(1-\frac{E}{S_0}\right)N_{max}$$

Thus the extent to which the equilibrium species number falls short of the maximal number of species that could coexist depends on how high the extinction rate is, relative to the speciation rate.

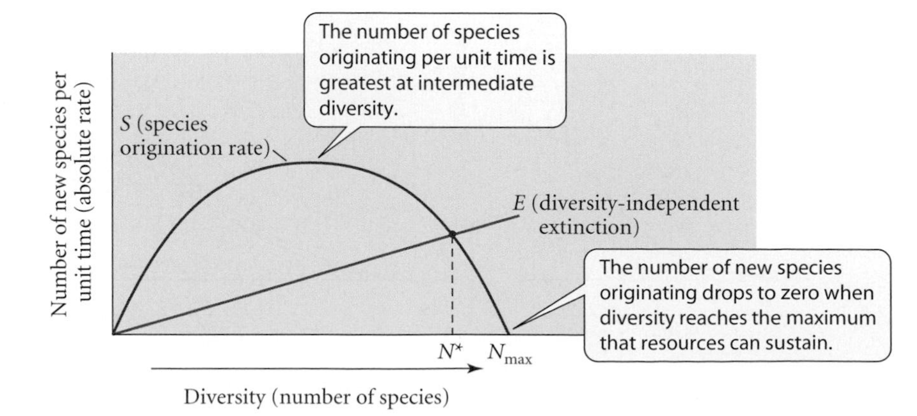

Figure 7.2 A model for the number of species at equilibrium over evolutionary time. The number of species originating per unit time (S) is greatest at intermediate diversity because it increases with the number of potential ancestors, but declines at high diversity levels, dropping to zero at the maximal number that resources can sustain (N_{max}). The number of extinctions per unit time is greater if there are more species. The number of species reaches equilibrium (N^*) when $S = E$—that is, when speciation and extinction rates are equal. Equilibrium diversity would be lower if line E, the extinction rate, were steeper. (After Walker and Valentine 1984.)

Taxonomic Diversity through the Phanerozoic

The most complete fossil record has been left by marine animals with hard parts (shells or skeletons). From the paleontological literature, Jack Sepkoski (1984, 1993) compiled data on the stratigraphic ranges of more than 4000 marine skeletonized families and 20,000 genera throughout the 542 million years of the Phanerozoic. His plot of the diversity of families (Figure 7.3A) shows, overall, a rapid increase in the Cambrian and Ordovician, a plateau throughout the rest of the Paleozoic, and a steady increase throughout the Mesozoic and Cenozoic, with diversity reaching a peak in the late Tertiary. This pattern is interrupted by decreases in diversity due to mass extinction events. The overall pattern is the same if the data are corrected for the biasing effects of taxa recorded from single stages and of Recent taxa, as described above (Figure 7.3B). Whether or not other biases have been fully taken into account is uncertain, and some authors have suggested that diversity in the Paleozoic may have been more similar to post-Paleozoic levels than the data seem to indicate (Alroy et al. 2001). This question continues to be a subject of vigorous research.

On land, diversity has also increased. The number of families of insects shows a steady increase since the Permian (Figure 7.4A), whereas the diversification of flowering plants

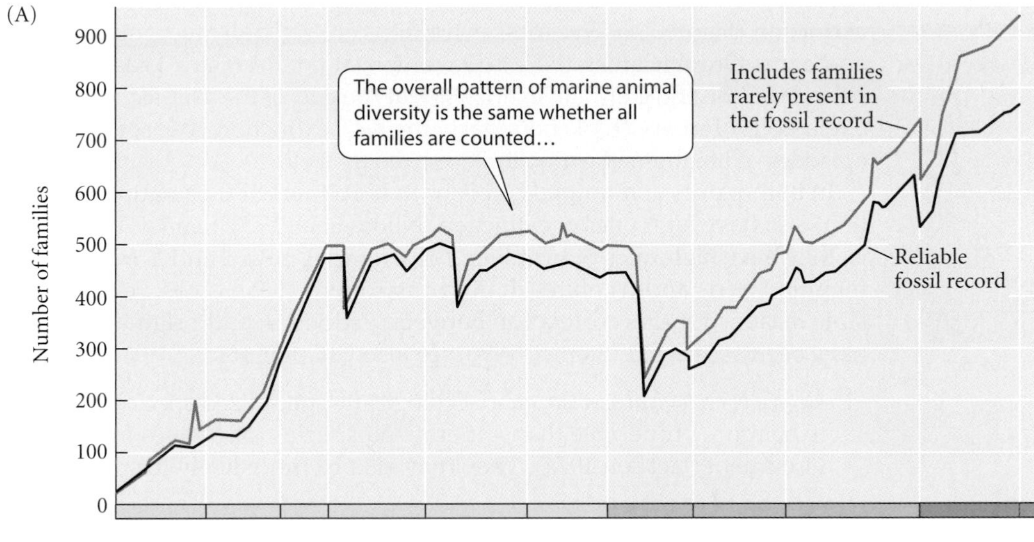

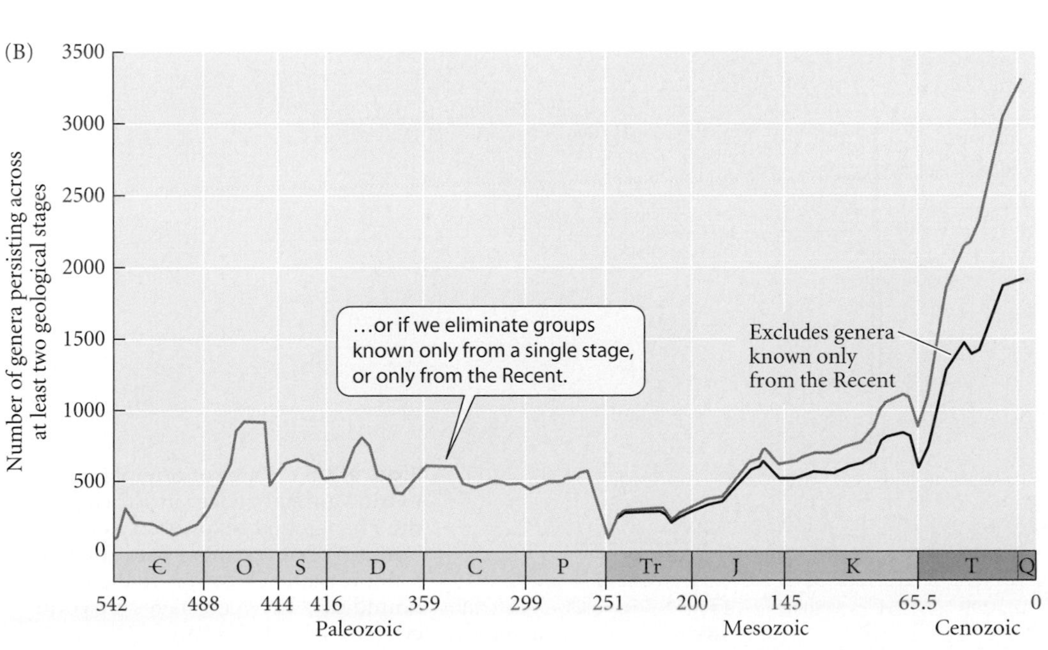

Figure 7.3 (A) Taxonomic diversity of skeletonized marine animal families during the Phanerozoic. The number of taxa entered for each geological stage (subdivisions of the geological periods; most stages represent 5–6 My.) includes all those whose known temporal extent includes that stage. The blue curve includes families that are rarely preserved; the black curve represents only families that have a more reliable fossil record. There are approximately 1900 marine animal families alive today, including those rarely preserved as fossils. (B) Taxonomic diversity of 25,049 skeletonized marine animal genera, counting only those crossing the boundary between two or more stages. Genera that existed for only a single stage are excluded. The black curve shows the diversity of genera only as represented in the fossil record (i.e., excluding their known occurrences in the Recent). (A after Sepkoski 1984; B after Foote 2000a.)

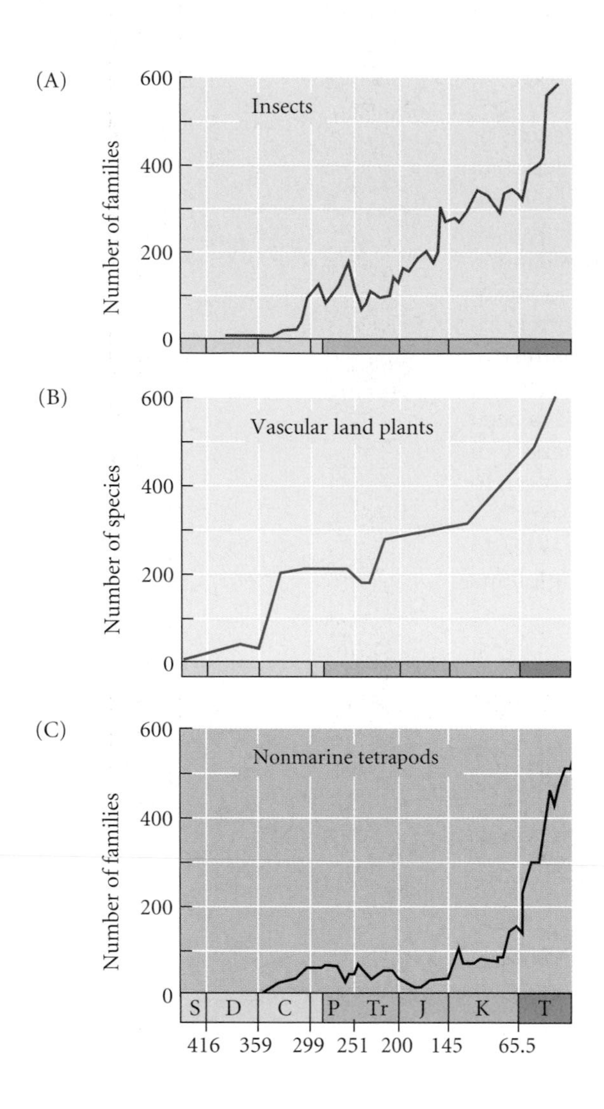

Figure 7.4 Changes in the number of known (A) families of insects, (B) species of vascular land plants, and (C) families of nonmarine tetrapod vertebrates. (A after Labandeira and Sepkoski 1993; B and C after Benton 1990.)

and of birds and mammals account for dramatic increases in the diversity of vascular plants (Figure 7.4B) and terrestrial vertebrates (Figure 7.4C) after the mid-Cretaceous.

Rates of origination and extinction

Since the Triassic, the rate of origination of marine animal taxa has been greater than the rate of extinction, resulting in the increase in diversity during the Mesozoic and Cenozoic, but both rates have fluctuated throughout Phanerozoic history. The rate of origination of new families was highest early in recorded animal evolution, in the Cambrian and Ordovician, and in the early Triassic, after the great end-Permian extinction (Figure 7.5). Extinction rates have varied dramatically. A distinction is often made between episodes during which exceptionally high numbers of taxa became extinct, the so-called **mass extinctions**, and periods of so-called "normal" or **background extinction** (Figure 7.6). Five mass extinctions are generally recognized: at the end of the Ordovician, in the late Devonian, at the Permian/Triassic (P/Tr) boundary (the end-Permian extinction), at the end of the Triassic, and at the Cretaceous/Tertiary (K/T) boundary (the K/T extinction), but several other episodes of heightened extinction occurred as well.

In both plants and animals, taxa with high rates of origination (speciation) also have high rates of extinction (Niklas et al. 1983; Stanley 1990). That is, they have high rates of **turnover**. For example, both *S* and *E* were higher in ammonoids and trilobites than in gastropods or bivalves. Several possible reasons for this correlation between extinction and origination rates have been suggested (Stanley 1990; see also Chapter 16):

1. *Degree of ecological specialization.* Ecologically specialized species are likely to be more vulnerable than generalized species to changes in their environment (Jackson 1974). They may also be more likely to speciate be-

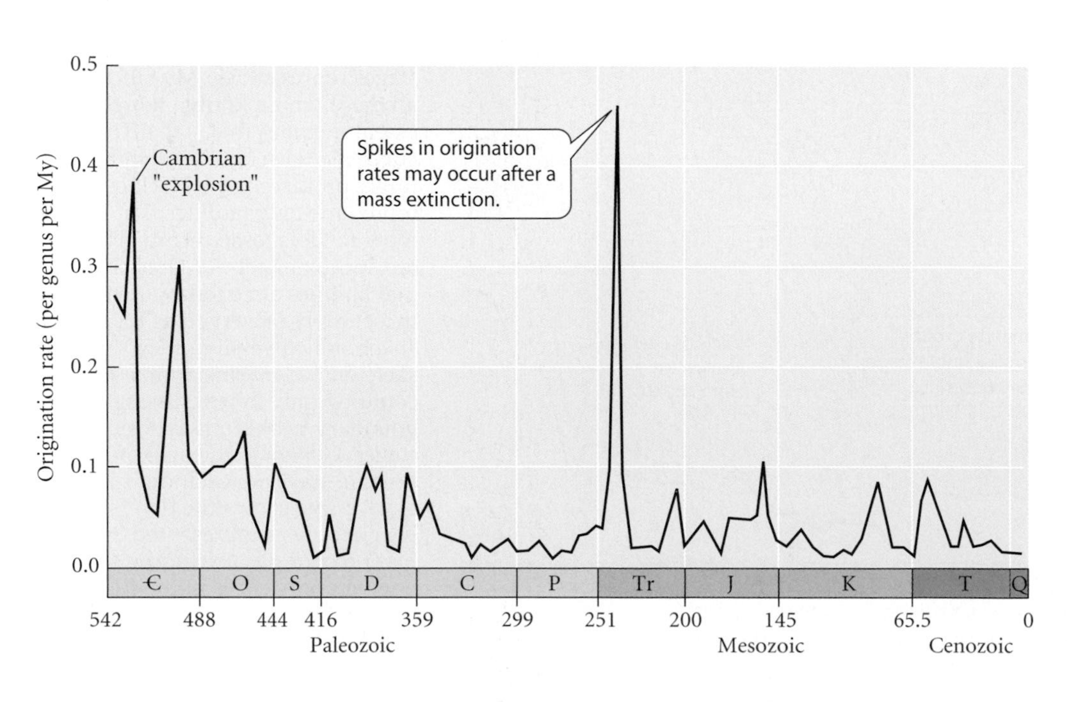

Figure 7.5 Rates of origination of marine animal genera in 107 stages of the Phanerozoic, expressed as the number of new genera per capita per million years. Only taxa that cross boundaries between stages are counted. (After Foote 2000a.)

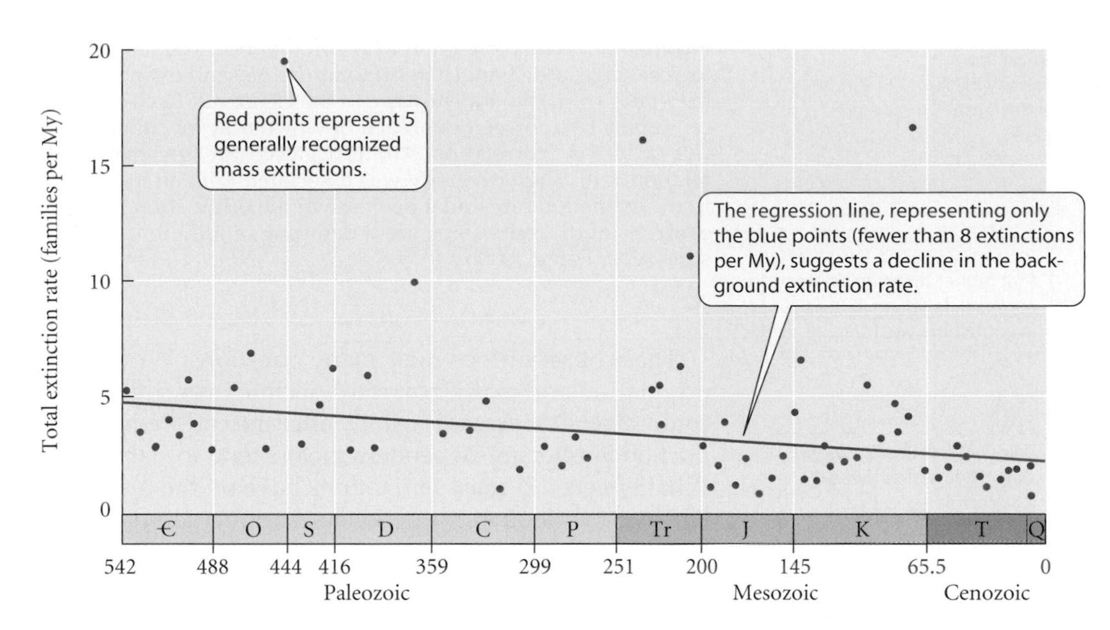

Figure 7.6 Extinction rates of marine animal families during the Phanerozoic, expressed as the number of families per million years. The solid regression line fits the blue points, which represent fewer than 8 extinctions per million years. The red points, which deviate significantly from the background cluster, mark the five major mass extinction events at the ends of the (1) Ordovician, (2) Devonian, (3) Permian (the end-Permian event), (4) Triassic, and (5) Cretaceous (the K/T extinction). (After Raup and Sepkoski 1982.)

cause of their more patchy distribution , and newly formed species may be more likely to persist by specializing on different resources and thus avoiding competition with other species.

2. *Population dynamics.* Species with low or fluctuating population sizes are especially susceptible to extinction. Some authors believe that speciation is also enhanced by small or fluctuating population sizes, although this hypothesis is controversial.
3. *Geographic range.* Species with broad geographic ranges tend to have a lower risk of extinction because they are not extinguished by local environmental changes. They also have lower rates of speciation (Jablonski and Roy 2003), probably because they have a high capacity for dispersal.

Taxa whose component species have high rates of origination and extinction are quite "volatile," fluctuating greatly in diversity—and are prone to extinction. The early extinction of many such groups may contribute to an observation that has not been fully explained: a long-term decline in origination and extinction rates throughout the Phanerozoic (Foote 2000b; see Figures 7.5 and 7.6).

Origination and extinction rates are one source of evidence that interactions among species have tended to stabilize diversity. Mike Foote (2000b) examined *changes* in marine animal diversity within each of 107 stages of the Phanerozoic record in relation to the *change* in per capita rates of origination (S) and extinction (E) from one stage to the next. He found that diversity increased more when S increased, and decreased more when E increased. More interestingly, extinction had a stronger effect on changes in diversity than did origination during the Paleozoic, but origination had a stronger effect during the Mesozoic and Cenozoic.

Foote then examined the correlation between the diversity at the start of each time interval and the change in the origination and extinction rates between that interval and the previous one. If changes in these rates were diversity-dependent, one would expect diversity to be negatively correlated with origination rate and positively correlated with extinction rate. The expected relationships were found, strongly supporting the hypothesis that diversity-dependent factors, such as competition among species, tend to stabilize diversity around an equilibrium (Figure 7.7).

In the same vein, Sepkoski (1984) statistically distinguished three major, taxonomically different "evolutionary faunas" that dominated the seas during the Phanerozoic (Figure 7.8). By modeling competition among these faunas as if they were three species populations, he found that the rise and fall of family diversity in each of the three faunas could be explained by diversity-dependent competition.

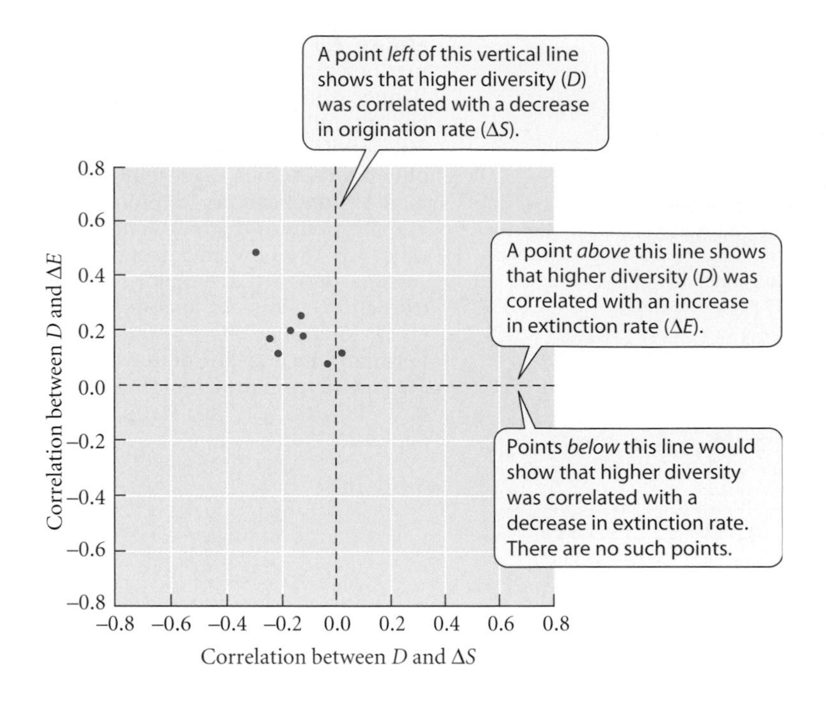

Figure 7.7 The correlation between diversity (D) in a geological stage and changes in origination (ΔS) and extinction (ΔE) rates from the mid-Jurassic to the Cenozoic. Each point represents two correlations, each of which may be either positive (a "direct" correlation, > 0) or negative (an "inverse" correlation, < 0). High diversity was associated with an increase in the extinction rate and a decrease in the origination rate—evidence of diversity-dependent damping of growth in diversity. (After Foote 2000b.)

These observations raise many questions: Why might origination and extinction rates decline over time? How could diversity increase steadily after the end-Permian extinction if diversity-dependent factors tend to stabilize it? What effects did mass extinctions have on the history of life? A closer look at extinction and origination may help to answer these questions.

Causes of extinction

Extinction has been the fate of almost all the species that have ever lived, but little is known of its specific causes. Biologists agree that extinction is caused by failure to adapt to changes in the environment. Ecological studies of contemporary populations and species point to habitat destruction as the most frequent cause of extinction by far, and some cases of extinction due to introduced predators, diseases, and competitors have been documented (Lawton and May 1995).

When a species' environment deteriorates, some populations may become extinct, and the geographic range of the species contracts, unless formerly unsuitable sites become suitable for colonists to establish new populations. If environmental changes cause populations to decline, the survival of those populations—and perhaps of the entire species—depends on adaptive genetic change. Whether or not this suffices to prevent extinction depends on how rapidly the environment (and hence the optimum phenotype) changes relative to the rate at which a character evolves. The rate of evolution may depend on the rate at which mutation supplies genetic variation and on population size, because smaller populations experience fewer mutations. Thus an environmental change that reduces population size also reduces the chance of adapting to it (Lynch and Lande 1993). Because a change in one environmental factor, such as temperature, may bring about changes in other factors, such as the species composition of a community, the survival of a species may require evolutionary change in several or many features.

Both abiotic and biotic changes have doubtless caused extinction. For example, during the Pliocene, the rate of extinction of bivalves and gastropods increased, chiefly in northern seas. This increase coincided with a decrease in temperature, a likely cause of the extinctions (Sepkoski 1996b). We will discuss the role of competition in extinction later in this chapter.

Declining extinction rates

It would seem reasonable to expect lineages of organisms to become more resistant to extinction over the course of time as they become better adapted. Evolutionary theory does not necessarily predict this, however, because natural selection, having no foresight, cannot prepare species for changes in the environment. If the environmental changes that threaten extinction are numerous in kind, we should not expect much carryover of "extinction resistance" from one change to the next. Consequently, we should expect that at any time t, the probability of extinction of a species (or higher taxon) would be the same, whether it is old (i.e., arose long before time t) or young (i.e., arose shortly before time t).

Extinctions of taxa in the fossil record can be analyzed by plotting the fraction that survive for different lengths of time (i.e., their age at extinction). If the probability of extinc-

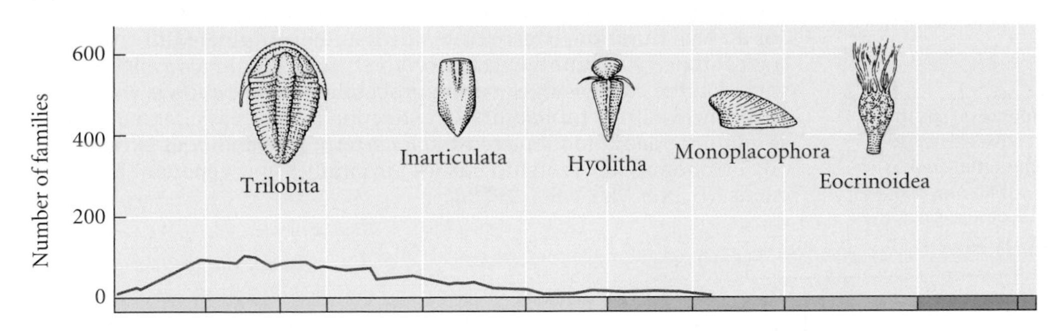

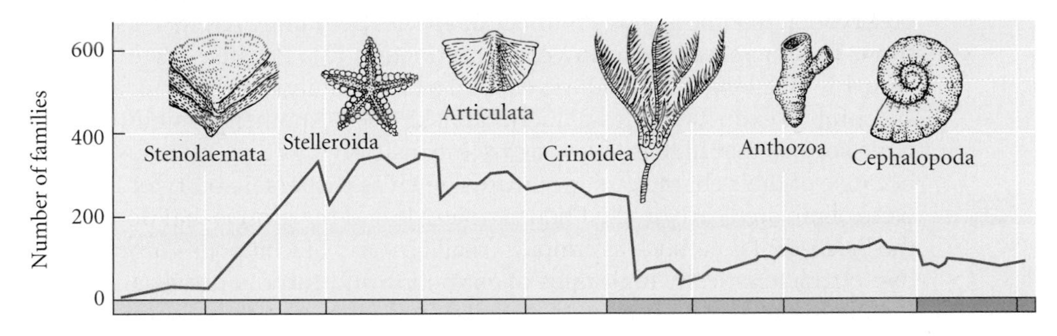

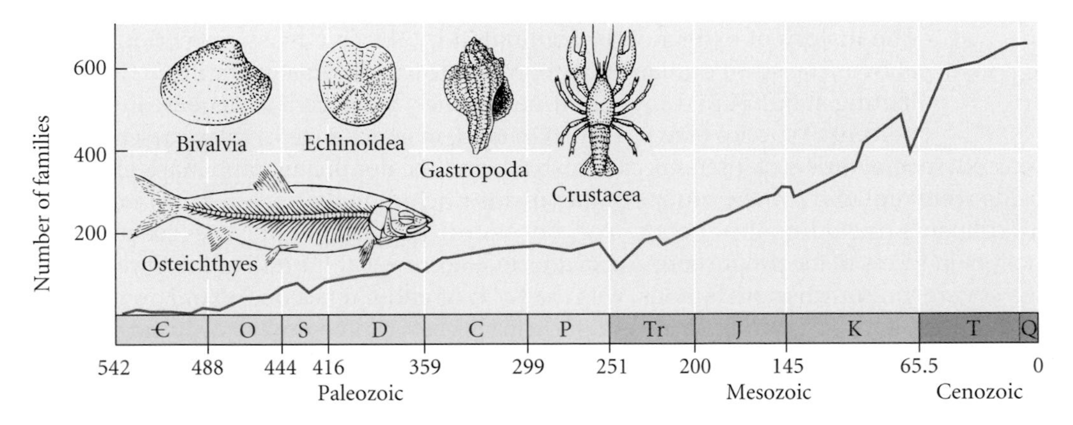

Figure 7.8 The history of diversity of the three "evolutionary faunas" in the marine fossil record, with illustrations of major forms. (The three diversity profiles add up to yield the overall diversity profile of Figure 7.3A.) (After Sepkoski 1984.)

tion is independent of age, the proportion of taxa surviving to increasingly greater ages should decline exponentially (just like the proportion of "surviving" parent atoms in radioactive decay; see Figure 4.2). Plotted logarithmically, the curve would become a straight line. If taxa evolve increasing resistance to causes of extinction as they age, the logarithmic plot should be concave upward, with a long tail (Figure 7.9A).

When Leigh Van Valen (1973) analyzed taxon survivorship in this way, he obtained rather straight curves, suggesting that the probability of extinction is roughly constant (Figure 7.9B). This is what we would expect if organisms are continually assaulted by new environmental changes, each carrying a risk of extinction. One possibility, Van Valen suggested, is that the environment of a taxon is continually deteriorating because of the evolution of other taxa. He proposed the **Red Queen hypothesis**, which states that, like the Red Queen in Lewis Carroll's *Through the Looking-Glass*, each species has to run (i.e., evolve) as fast as possible just to stay in the same place (i.e., survive), because its competitors, predators, and parasites also continue to evolve. There is always a roughly constant chance that it will fail to do so.

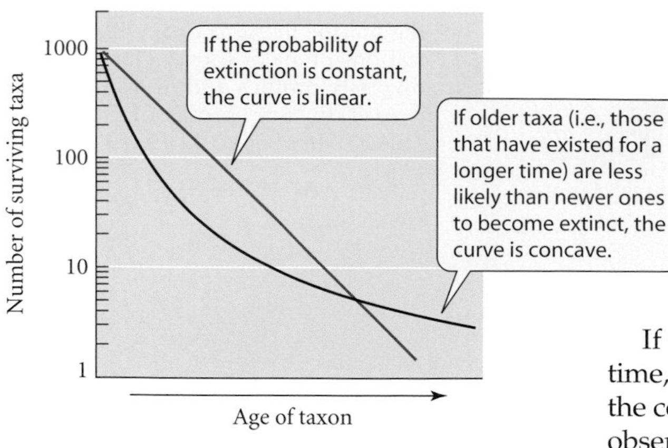
(A) Hypothetical curves
1000
100
10
1
Number of surviving taxa
Age of taxon
If the probability of extinction is constant, the curve is linear.
If older taxa (i.e., those that have existed for a longer time) are less likely than newer ones to become extinct, the curve is concave.

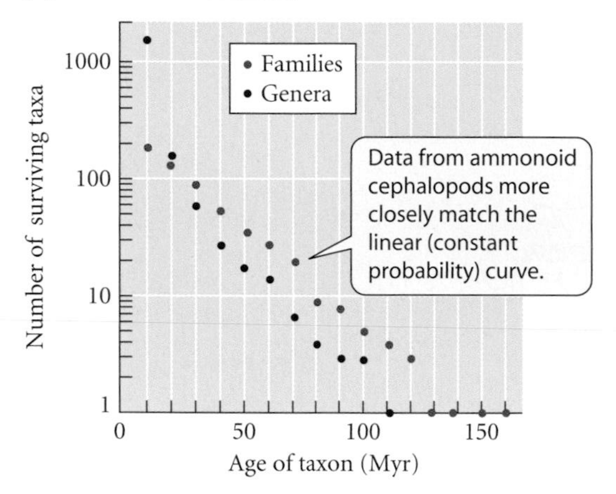
(B) Data for Ammonoidea
1000
100
10
1
0
50
100
150
Number of surviving taxa
Age of taxon (Myr)
• Families
• Genera
Data from ammonoid cephalopods more closely match the linear (constant probability) curve.

Figure 7.9 Taxonomic survivorship curves. Each curve or series of points represents the number of taxa that persisted in the fossil record for a given duration, irrespective of when they originated during geological time. (A) Hypothetical survivorship curves. In a semilogarithmic plot, the curve is linear if the probability of extinction is constant. It is concave if the probability of extinction declines as a taxon ages, as it might if adaptation lowered the long-term probability of extinction. (B) Taxonomic survivorship curves for families and genera of ammonoids. (B after Van Valen 1973.)

If we do not expect resistance to extinction to evolve progressively over time, what might explain the decline in the rate of background extinction over the course of the Phanerozoic (see Figure 7.6)? One hypothesis is based on the observation that the average number of species per family has increased over time. This increase would lower the extinction rate because it would presumably take longer for all the species of a large family than a small family to succumb to extinction (Flessa and Jablonski 1985). Another possibility, as we have seen, is that higher taxa that are intrinsically more prone to extinction because of their characteristic features (such as dispersal ability or habitat) were eliminated early in the Phanerozoic. Most of the groups that dominated the Paleozoic fauna, such as crinoids (sea lilies) and brachiopods (lamp shells), had characteristically high rates of extinction and turnover compared with the bivalves (clams), gastropods (snails), and other taxa that dominated the post-Paleozoic (Erwin et al. 1987).

Mass extinctions

The history of extinction is dominated by the five mass extinctions listed above. The end-Permian extinction was the most drastic (Figure 7.10), eliminating about 54 percent of marine families, 84 percent of genera, and 80–90 percent of species (Erwin 1993). On land, major changes in plant assemblages occurred, several orders of insects became extinct, and the dominant amphibians and therapsids were replaced by new groups of therapsids (including the ancestors of mammals) and diapsids (including the ancestors of dinosaurs). The second most severe mass extinction, in terms of the proportion of taxa affected, occurred at the end of the Ordovician. Less severe, but much more famous, was the K/T, or end-Cretaceous, extinction, which marked the demise of many marine and terrestrial plants and animals, including the dinosaurs (except for birds).

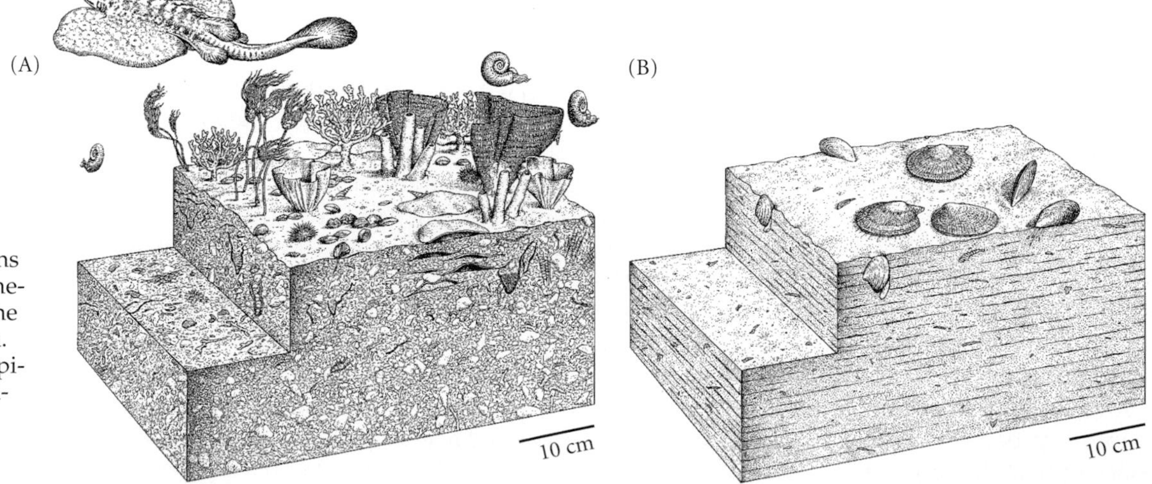
(A)
(B)
10 cm
10 cm

Figure 7.10 Reconstructions of an ancient seabed (A) immediately before and (B) after the end-Permian mass extinction. A rich fauna of burrowing, epifaunal, and swimming organisms was almost completely extinguished. (Artwork © J. Sibbick.)

CAUSES OF MASS EXTINCTIONS. The K/T extinction is famous because of the truly dramatic hypothesis, first suggested by Walter Alvarez and his colleagues (1980), that the dinosaurs were extinguished by the impact of an extraterrestrial body—an asteroid or large meteorite. Alvarez et al. postulated that this object struck the Earth with a force great enough to throw a pall of dust into the atmosphere, darkening the sky and lowering temperatures, thus reducing photosynthesis. Geologists now agree that such an impact occurred; it's site, the Chicxulub crater, has been discovered off the coast of the Yucatán Peninsula of Mexico. Most paleontologists agree that this impact caused the mass extinction at the K/T boundary, although some argue that extinctions of various taxa were too spread out in time to have all been caused by this catastrophe, and that the impact was only one of several environmental changes that interacted to cause the K/T extinction (MacLeod 1996).

The most drastic mass extinction by far was the end-Permian extinction. Many possible causes have been suggested. Recently, the hypothesis that massive volcanic eruption was responsible has become popular (Benton and Twitchett 2003). The end-Permian extinction was almost instantaneous, and it coincides with volcanic eruptions that produced enough lava to cover an area of eastern Russia equivalent to all of Europe (the formation called the Siberian traps). It is postulated that the global warming caused by these eruptions altered oceanic circulation, resulting in almost complete loss of oxygen in the deeper waters. Global warming may also have caused the release of vast quantities of methane, which further enhanced warming in a positive feedback spiral that brought life on Earth "close to complete annihilation 251 Mya" (Benton and Twitchett 2003).

VICTIMS, SURVIVORS, AND CONSEQUENCES. Mass extinctions were "selective" in that some taxa were more likely than others to survive. Survival of gastropods through the end-Permian extinction was greater for species with wide geographic and ecological distributions and for genera consisting of many species (Erwin 1993). Extinction appears to have been random with respect to other characteristics, such as mode of feeding. The pattern of selectivity was much the same as during periods of background extinction, when gastropods and other taxa with broad geographic distributions have had lower rates of extinction than narrowly distributed taxa (Boucot 1975). Patterns of survival through the end-Cretaceous mass extinction, however, differed from those during "normal" times (Jablonski 1995). During times of background extinction, survivorship of late Cretaceous bivalves and gastropods was greater for taxa with planktonic development (larvae dispersed by currents) and for genera consisting of numerous species, especially if those genera had broad geographic ranges. In contrast, during the end-Cretaceous mass extinction, planktonic and nonplanktonic taxa had the same extinction rates, and the survival of genera, although enhanced by broad distribution, was not influenced by their species richness. Thus the characteristics that were correlated with survival seem to have differed from those during "normal" times.

During mass extinction events, taxa with otherwise superb adaptive qualities succumbed because they happened not to have some critical feature that might have saved them from extinction under those circumstances. Evolutionary trends initiated in "normal" times were cut off at an early stage. For example, the ability to drill through bivalve shells and feed on the animals inside evolved in a Triassic gastropod lineage, but was lost when this lineage became extinct in the late Triassic mass extinction (Fürsich and Jablonski 1984). The same feature evolved again 120 My later, in a different lineage that gave rise to diverse oyster drills. A new adaptation that might have led to a major adaptive radiation in the Triassic was strangled in its cradle, so to speak.

Both physical and biotic environmental conditions were probably very different after mass extinctions than before. Perhaps for this reason, many taxa continued to dwindle long after the main extinction events (Jablonski 2002), while others, often members of previously subdominant groups, diversified. Full recovery of diversity took millions of years—as much as 100 million years after the end-Permian disaster.

The mass extinction events, especially the end-Permian and K/T extinctions, had an enormous effect on the subsequent history of life because, to a great extent, they wiped the slate clean. Stephen Jay Gould (1985) suggested, in fact, that there are "tiers" of evolutionary change, each of which must be understood in order to comprehend the full history of evolution. The first tier is microevolutionary change *within populations and species.* The second tier is "species selection," the *differential proliferation and extinction of species* during "normal" geological times, which affects the relative diversity of lineages with different characteristics (see Chapter 11). The third tier is the *shaping of the biota by mass extinctions,* which can extinguish diverse taxa and reset the stage for new evolutionary radiations, initiating evolutionary histories that are largely decoupled from earlier ones.

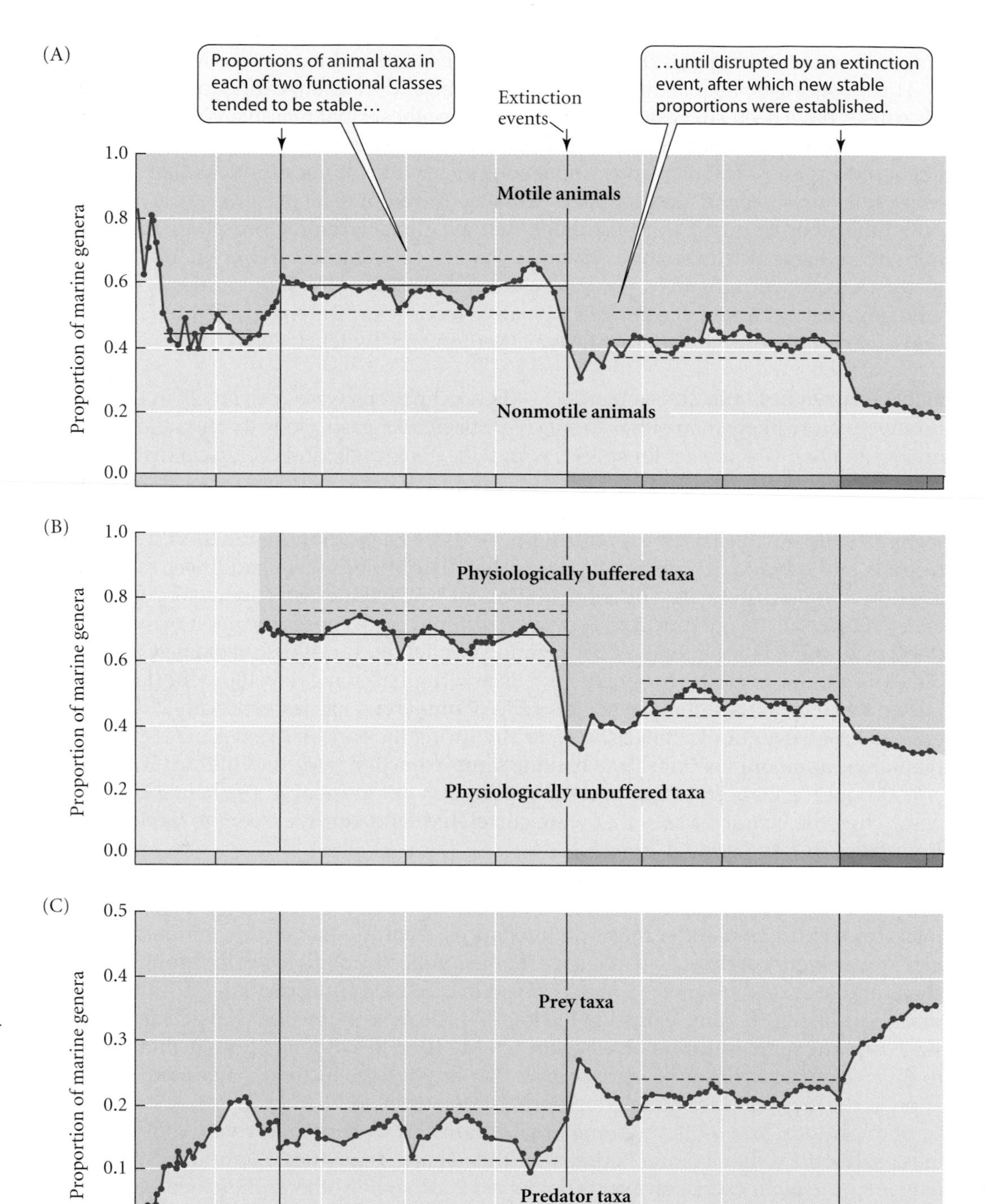

Figure 7.11 Changes in the proportions of marine animal genera classified by three functional criteria during the Phanerozoic. The proportions were roughly stable (dashed lines) between mass extinctions, but changed abruptly to a new stable state after mass extinction events at the end of the Ordovician, Permian, and Cretaceous (black lines). (A) Motile versus nonmotile animals. (B) Physiologically buffered versus unbuffered animals. "Buffered" taxa are those whose physiological systems (e.g., gills and circulatory systems) allow greater homeostatic control. (C) Predators versus nonpredators . (After Bambach et al. 2002.)

The results of a study by Bambach et al. (2002) lent some support to Gould's idea. Bambach and colleagues classified Phanerozoic marine animal genera by three functional criteria: whether they were passive (nonmotile, such as barnacles) or active (motile), whether they were physiologically "buffered" (with well-developed gills and circulatory system, such as crustaceans) or not (such as echinoderms), and whether or not they were predatory. With respect to all three kinds of functional groupings, the proportions of taxa with alternative characteristics remained stable over intervals as long as 200 My, even though the total diversity and the taxonomic composition of the marine fauna changed greatly (Figure 7.11). Shifts from one stable configuration to another are associated with the mass extinctions at the end of the Ordovician, Permian, and Cretaceous, suggesting that the extinction of long-prevalent (incumbent) taxa permitted the emergence of new community structures.

This observation—that extinction of one group permitted the efflorescence of others—describes one of the most important effects of mass extinction events and is a major theme in analyses of origination and diversification.

Origination and diversification

We turn now to the question of why increases in diversity have been greater in some lineages than in others and at some times than at others, and why diversity has tended to increase ever since the end-Permian extinction. Among the major factors that have fostered diversification are release from competition, ecological divergence, coevolution, and provinciality (Signor 1990; Benton 1990).

RELEASE FROM COMPETITION. Studies of both living and extinct organisms have shown that lineages often have diversified most rapidly when presented with ecological opportunity: what is often called "ecological space" or "vacant niches" not occupied by other species. In many isolated islands and bodies of water, descendants of just a few original colonizing species have diversified and filled ecological niches that are occupied in other places by unrelated organisms. Such adaptive radiations include the cichlid fishes in the Great Lakes of eastern Africa, the honeycreepers in the Hawaiian Islands, and Darwin's finches in the Galápagos Islands (see Figure 3.22). Islands and other habitats with taxonomically depauperate biotas typically harbor organisms that have evolved unusual new ways of life. For example, the larvae of almost all moths and butterflies are herbivorous, but in the Hawaiian Islands, the larvae of the moth genus *Eupithecia* are specialized for predation (Figure 7.12; Montgomery 1982). Probably such unusual forms are more prevalent where species diversity is reduced because they are not faced with as many predators or superior competitors in their early, relatively inefficient, stages of adaptation to new ways of life.

The fossil record provides many instances in which the reduction or extinction of one group of organisms has been followed or accompanied by the proliferation of an ecologically similar group. For example, conifers and other gymnosperms declined as angiosperms (flowering plants) diversified, and mammals radiated after the late Cretaceous extinction of the nonavian dinosaurs.

Several hypotheses can account for these patterns (Benton 1996; Sepkoski 1996a). Two of these hypotheses involve competition between species in the two clades. On one hand, the later group may have *caused* the extinction of the earlier group

Figure 7.12 A predatory moth caterpillar (*Eupithecia*) in the Hawaiian islands, holding a *Drosophila* that it has captured with its unusually long legs. Predatory behavior is extremely unusual in the order Lepidoptera. (Photo by W. P. Mull, courtesy of W. P. Mull and S. L. Montgomery.)

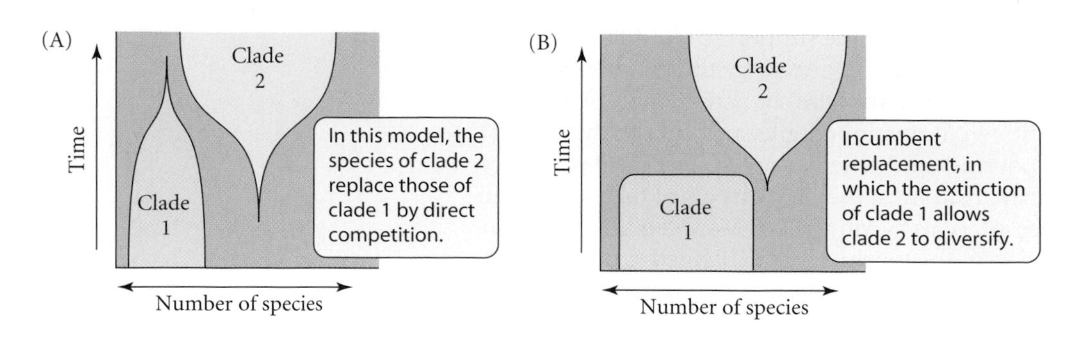

Figure 7.13 Models of competitive displacement and replacement. In each diagram, the width of a "spindle" represents the number of species. (A) Competitive displacement, in which the increasing diversity of clade 2 causes a decline in clade 1 by direct competitive exclusion. (B) Incumbent replacement, in which the extinction of clade 1 enables clade 2 to diversify. (After Sepkoski 1996a.)

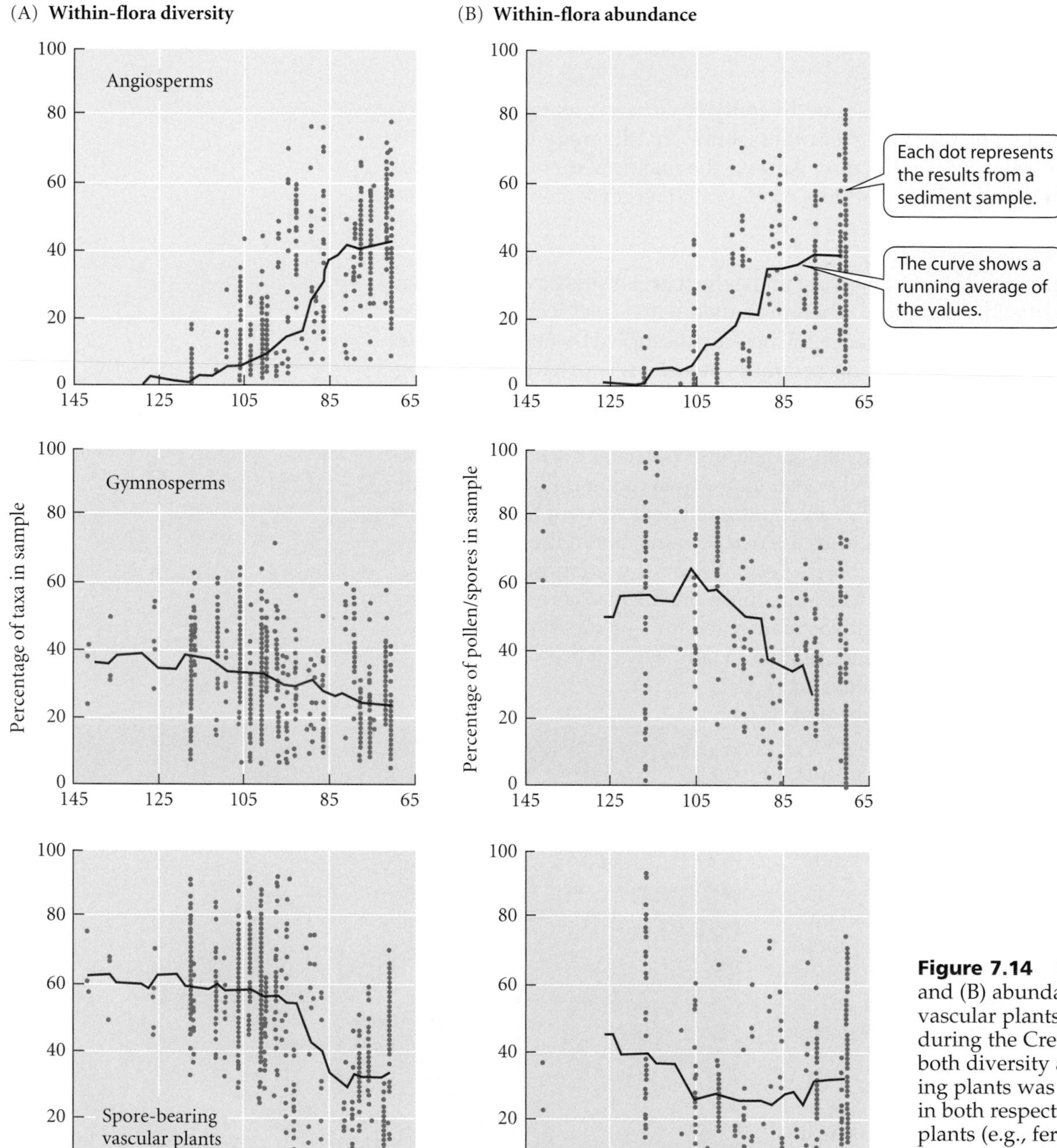

Figure 7.14 Changes in (A) diversity and (B) abundance of major groups of vascular plants within fossil samples during the Cretaceous. The increase in both diversity and abundance of flowering plants was mirrored by the decline, in both respects, of the spore-bearing plants (e.g., ferns) and the decline in abundance of gymnosperms (e.g., conifers). This pattern is consistent with competitive displacement. (After Lupia et al. 1999.)

by competition, a process called **competitive displacement** (Figure 7.13A). On the other hand, an incumbent taxon may have *prevented* an ecologically similar taxon from diversifying. Extinction of the incumbent taxon may then have vacated ecological "niche space," permitting the second taxon to radiate (Figure 7.13B). This process has been called **incumbent replacement** by Rosenzweig and McCord (1991), who argued that the second taxon may have had superior adaptive features, but nevertheless could not have displaced the earlier taxon by competition. Sepkoski (1996a) developed a mathematical model in which the number of species in two clades is affected by competition in a way similar to the number of individuals in populations of two competing species. In this model of logistic "population growth," changes in the species diversity of two clades reflect both displacement and incumbent replacement.

There are good reasons to believe that competition among species has affected changes in diversity (Sepkoski 1996a). The rate of increase in diversity has been much greater at times when diversity was unusually low—namely, during the Cambrian and after mass extinction events—than at other times, and there have been intervals of more than 200 million years during which diversity was quite stable (e.g., the late Paleozoic). But how competition has affected changes in diversity is controversial.

Competitive displacement may have occurred relatively rarely (Benton 1996). A pattern of replacement is consistent with competitive displacement if the earlier and later taxa lived in the same place at the same time, if they used the same resources, if the earlier taxon was not decimated by a mass extinction event, and if the diversity and abundance of the later taxon increased as the earlier one declined (Lupia et al 1999). Vascular plants, which certainly compete for space and light, showed exactly this pattern during the Cretaceous, when flowering plants increased in diversity and abundance at the expense of nonflowering plants, especially spore-bearing plants such as ferns (Figure 7.14). In another likely example of competitive displacement, the great increase in diversity of bivalves, especially after the end-Permian mass extinction, was accompanied by a decline in diversity of ecologically similar brachiopods; the pattern fits the theoretical model of competitive displacement (Figure 7.15; Sepkoski 1996a).

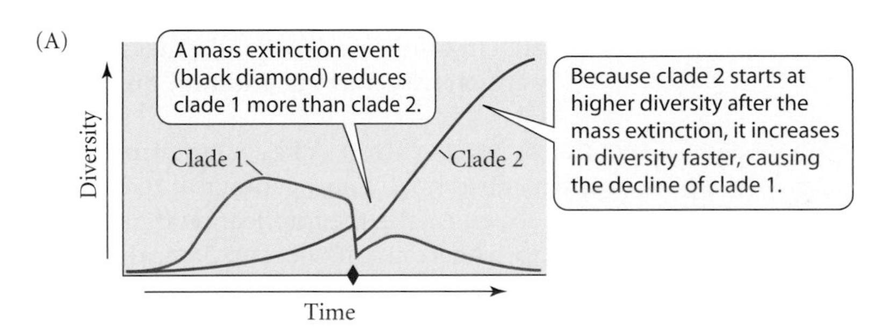

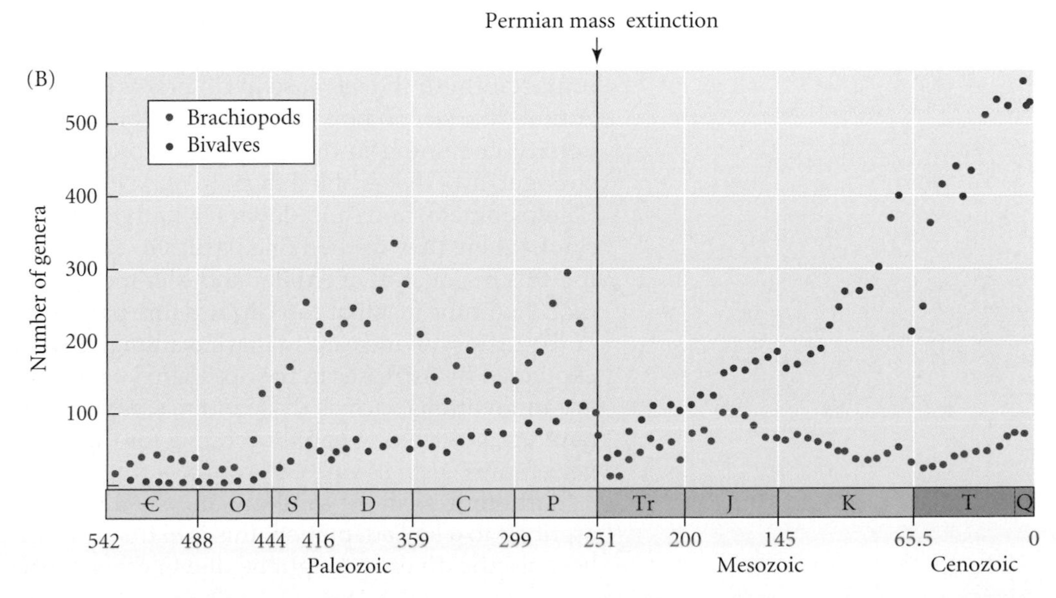

Figure 7.15 Probable competitive replacement of brachiopods by bivalves. (A) A model in which the numbers of species in two competing clades grow logistically until a severe extinction event reduces the diversity of one clade more than the other. The increase of clade 2 after the extinction event causes the decline of clade 1. (B) The number of genera of brachiopods and bivalves in each geological stage throughout the Phanerozoic. Bivalves increased at a higher rate after the diversity of brachiopods had been reduced by the end-Permian extinction event; brachiopods never regained their former diversity. This pattern fits the curves of the model, suggesting that the changes are consistent with competitive displacement. (After Sepkoski 1996a.)

(A)

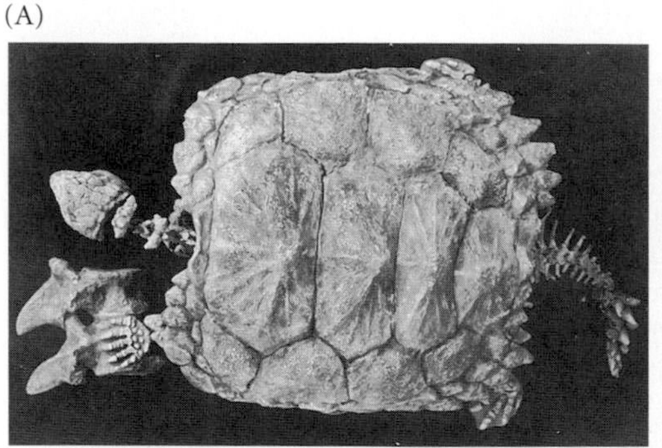

(B)

(C)

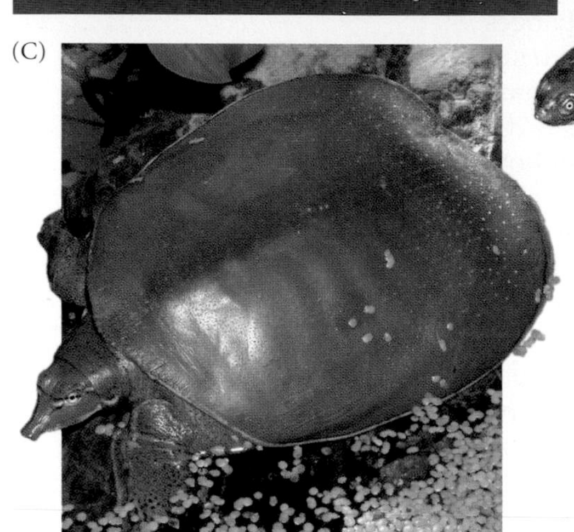

Figure 7.16 Pleurodiran and cryptodiran turtles replaced incumbent amphichelydian turtles, which became entirely extinct. (A) Amphichelydians, represented by the reconstructed skeleton of the earliest known turtle (*Proganochelys quenstedti*, Upper Triassic), could not retract their heads for protection. (B) The New Guinea snakeneck turtle (*Chelodina novaguineae*), a pleurodiran turtle that flexes its neck laterally. (C) The spiny softshell turtle (*Apalone spinifera*), a cryptodiran turtle that retracts the head by flexing its neck vertically. (A courtesy of E. Gaffney, American Museum of Natural History; B © Cliff and Dawn Frith/ANTPhoto.com; C © William Flaxington.)

Incumbent replacement has probably been more common than competitive displacement. The great radiation of placental mammals in the early Cenozoic is often credited to the K/T extinction of the last nonavian dinosaurs and other large "reptiles," which may have suppressed mammals by both competition and predation. Perhaps better evidence is supplied by replicated replacements. For instance, amphichelydians, the "stem group" of turtles, could not retract their heads and necks into their shells (Figure 7.16). Two groups of modern turtles, which protect themselves by bending the neck under or within the shell, replaced the amphichelydians in different parts of the world four or five times, especially during the K/T extinction event. The modern groups evidently could not radiate until the amphichelydians had become extinct. That this replacement occurred in parallel in different places and times makes it a likely example of release from competition (Rosenzweig and McCord 1991).

ECOLOGICAL DIVERGENCE. A **key adaptation** is an adaptation that enables an organism to occupy a substantially new ecological niche, often by using a novel resource or habitat. The term often carries the implication that the adaptation has enabled the diversification of a group. The group may occupy an **adaptive zone**, a set of similar ecological niches. For example, the many species of insectivorous bats and fruit-eating bats, which are nocturnal, occupy two adaptive zones that differ from those of diurnal insect- and fruit-eating birds. The term **ecological space** is roughly equivalent to a set of adaptive zones.

The evolution of the ability to use new resources or habitats has certainly contributed importantly to the increase in diversity over time (Niklas et al. 1983; Bambach 1985). Among the sea urchins (Echinoidea), for example, three orders increased greatly in diversity beginning in the early Mesozoic (Figure 7.17). The order Echinacea evolved stronger jaws that enabled them to use a greater variety of foods, while the heart urchins (Atelostomata) and sand dollars (Gnathostomata) became specialized for burrowing in sand, where they feed on fine particles of organic sediment. The key adaptations allowing this major shift of habitat and diet include a flattened form and a variety of highly modified tube feet that can capture fine particles and transfer them to the mouth. Much of the history of increase in marine animal diversity throughout the Phanerozoic can be explained by increases in the occupancy of ecological space accomplished by evolutionary innovations such as those of the sand dollars (Bambach 1985). Expansion into new habitats and feeding habits accounts for the diversification of most families of tetrapod vertebrates, such as various groups of frogs, snakes, and birds (Benton 1996).

Although, as in the sand dollars, the diversification of a clade can often be plausibly ascribed to a key adaptation, this is extremely difficult to demonstrate from a single case because the diversity might be due to other causes. Stronger evidence is provided if the

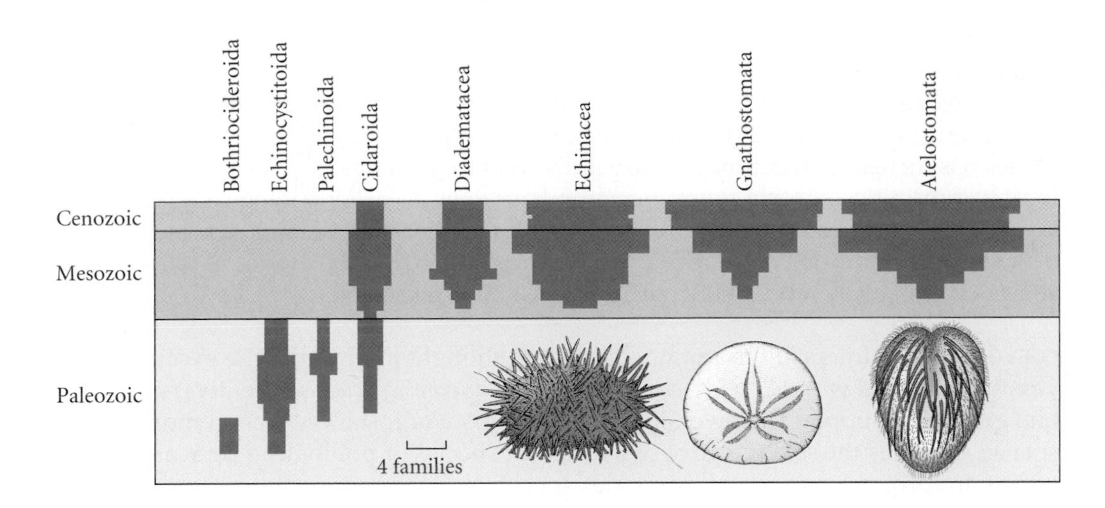

Figure 7.17 Echinoid diversity increased during the Mesozoic and Cenozoic, probably due to key adaptations described in the text. Diversification of sea urchins (order Echinacea), sand dollars (Gnathostomata), and heart urchins (Atelostomata) greatly increased the diversity of the echinoids. The width of the symmetrical profile of each group represents the number of families in that group at successive times. (After Bambach 1985.)

rate of diversification is correlated with a particular kind of character that has evolved independently in a number of different clades. Such tests have been applied mostly to living organisms. The diversity of a number of clades with a novel character can be compared with the diversity of their sister groups that retain the ancestral character state. Since sister taxa are equally old, the difference between them in number of species must be due to a difference in rate of diversification, not to age. If the convergently evolved character is consistently associated with high diversity, we would have support for the hypothesis that it has caused a higher rate of diversification.

Charles Mitter and colleagues (Mitter et al. 1988; Farrell et al. 1991) applied this method, called REPLICATED SISTER-GROUP COMPARISON, to herbivorous insects and plants. The habit of feeding on the vegetative tissues of green plants has evolved at least 50 times in insects, usually from predatory or detritus-feeding ancestors. Phylogenetic studies have identified the nonherbivorous sister group of 13 herbivorous clades. In 11 of these cases, the herbivorous lineage has more species than its sister group (Figure 7.18). This significant correlation supports the hypothesis that entry into the herbivorous adaptive zone has promoted diversification. These researchers then examined the species diversity of 16 clades of plants that have evolved rubbery latex (as in milkweeds) or resin (as in pines), both of which deter attack by herbivorous insects. Thirteen of these clades have more species than their sister clades, which lack latex or resin. These defensive features may have fostered diversification.

Figure 7.18 Two replicated sister-group comparisons of herbivorous clades of insects with their sister clades that feed on animals, fungi, or detritus. (Data from Mitter et al. 1988.)

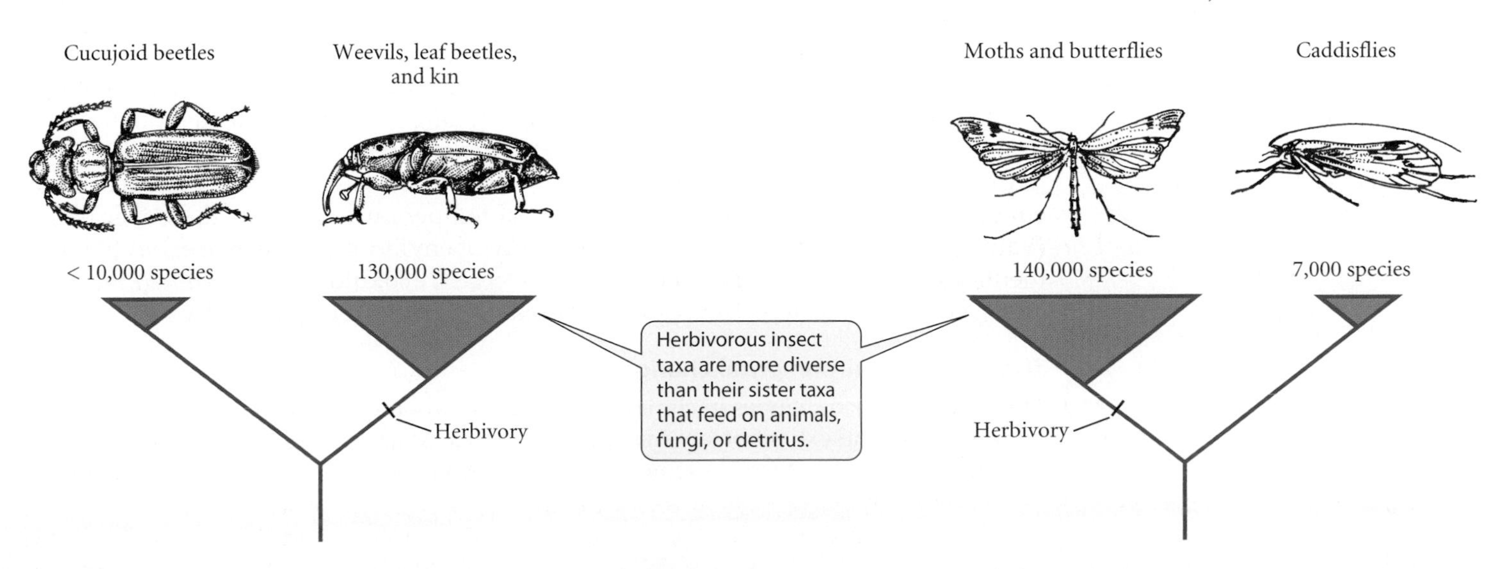

From studies of modern organisms, we know that much diversity resides in the great numbers of related species that reduce competition with one another by subtle differences in resource use. Anoles living together on an island, for example, forage in different microhabitats (see Figure 6.21). Some paleontologists have suggested that subdivision of niches has increased over time because individual fossil deposits, which generally record local communities, contain more species in later than in earlier geological periods (Benton 1990). The increase in species number seems greater than the increase in the variety of major growth forms or adaptive zones, suggesting that the ever greater number of species coexisted by more finely partitioning similar resources.

COEVOLUTION. Interactions among species are thought to promote the evolution of diversity in several ways. *Species serve as resources for other species*, so the diversification of one group can support the diversification of others. For instance, each of more than 700 species of figs is the sole resource of a different species of pollinating fig wasp, and the wasps, in turn, are parasitized by species-specific nematodes.

Coevolution of predators and their prey might also increase diversity. During the Mesozoic marine revolution, there occurred a great increase in the taxonomic diversity and morphological variety of both predators and prey (Vermeij 1987). Crustaceans and fishes evolved many ways of crushing or tearing mollusc shells, and molluscs evolved many defenses against the new predators, such as spines or thicker shells (see Figure 17.10).

PROVINCIALITY. The degree to which the world's biota is partitioned among geographic regions is called **provinciality**. A faunal or floral province is a region containing high numbers of distinctive, localized taxa (see Chapter 6). The fauna and flora of the contemporary world are divided into more provinces than ever before in the history of life. A trend from a cosmopolitan distribution of taxa to more localized distributions has persisted throughout much of the Mesozoic and Cenozoic and is thought by many paleontologists to be one of the most important causes of the increase in global diversity during this time (Valentine et al. 1978; Signor 1990).

Among marine animals, the number of faunal provinces was relatively low throughout much of the Paleozoic, and it dropped to an all-time low in the early Triassic, when most of the higher taxa that had survived the end-Permian mass extinction were so highly cosmopolitan that paleontologists recognize only a single, worldwide province during that time. During the Jurassic and Cretaceous, and especially the Tertiary, marine animals were distributed among an increasing number of latitudinally arranged provinces in both the Atlantic and Pacific regions. Similarly, among terrestrial vertebrates, a distinct fauna developed on each major land mass during the later Mesozoic and the Cenozoic, and the broad latitudinal distributions of many dinosaurs and other Mesozoic groups gave way to the much narrower latitudinal ranges of today's vertebrates.

Changes in the distribution of land masses due to plate tectonic processes are the fundamental cause of this trend. After the breakup of Pangaea in the Triassic, land masses ultimately became arrayed almost from pole to pole along a wider latitudinal span than ever before in Earth's history. This deployment of the continents created two increasingly disjunct ocean systems, the Indian-Pacific and the Atlantic, and established a pattern of ocean circulation that created a stronger latitudinal temperature gradient than ever before (Valentine et al. 1978). Not only did the variety of environments increase, but the fragmentation of land masses also allowed for divergent evolution and prevented the interchange of species that, by competition or predation, might lower diversity.

The role of environmental change

The ways in which changes in climate and other environmental factors have affected rates of origination and extinction are many and complex. Major changes in climate have been associated with increases in extinction and with changes in the distribution of habitats

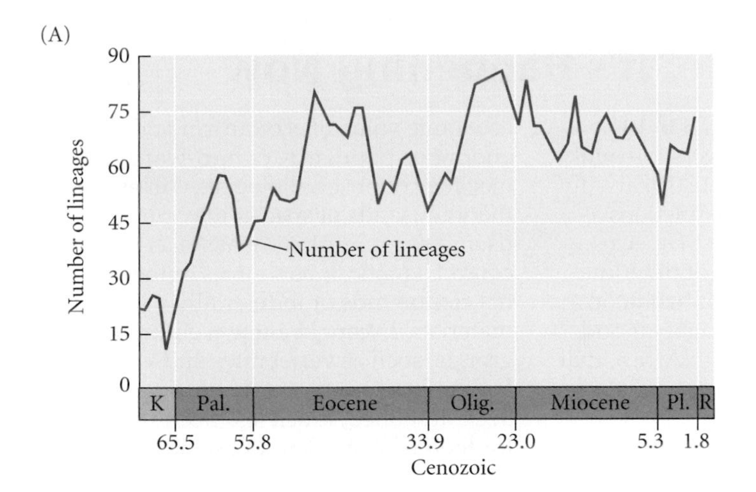

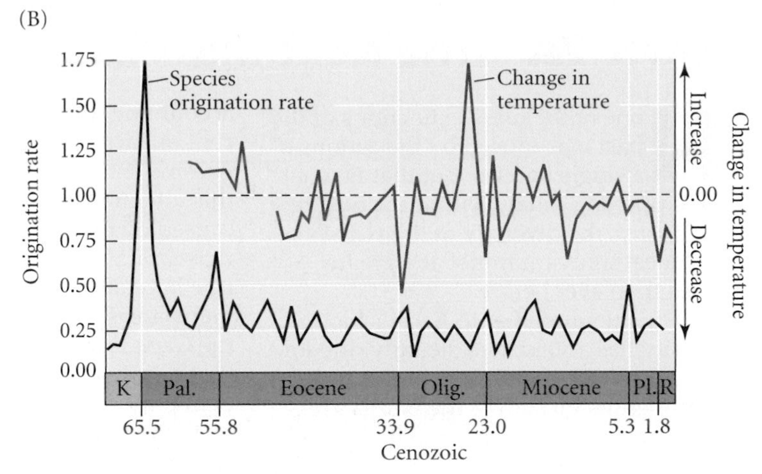

Figure 7.19 Changes in taxonomic diversity of Cenozoic mammals in North America. (A) The estimated history of diversity. (B) Fluctuations in the rate of origination of new lineages (black curve), compared with concurrent changes in temperature (red curve). The changes in temperature seem unrelated to origination rate, although some of them match declines in diversity that are evident in (A). (After Alroy et al. 2000.)

and vegetation types, which in turn have facilitated major changes in the distributions of taxa, often leading to diversification (Rothschild and Lister 2003). The origination of new taxa at such times, with new adaptations, may result more from these biotic changes than from the climate changes themselves. For example, in the mid-Eocene, about 50 to 40 Mya, the climate became cooler and drier, subtropical forests were widely replaced by savannahs in much of the temperate zone, and the diversity of primates and other arboreal mammals declined while that of large herbivores increased (Janis 1993). On the whole, however, changes in the rates of origination and extinction of mammals during the Tertiary are not closely correlated with changes in temperature (Figure 7.19). The importance of climate change in the evolution of diversity, relative to other factors such as key innovations and biotic interactions, remains to be elucidated (Alroy et al. 2000).

The Future of Biodiversity

As we have seen, there is evidence that diversity-dependent factors such as competition may tend to produce an equilibrium level of diversity. Indeed, diversity seems to have reached a long-sustained plateau, or equilibrium, at several times in Earth's history. Nevertheless, both marine and terrestrial diversity has increased spectacularly and rather steadily from the Triassic to the Pliocene. Although these observations may seem contradictory, they need not be. A system may shift from one equilibrium state to another when conditions change. At least three kinds of changes have altered conditions for organisms. First, changes have occurred in the physical environment, including those that increased provinciality. Second, the taxa that became dominant after mass extinctions were different from those that prevailed before, and would be expected to attain a different equilibrium level of diversity due to new patterns of competition and other interactions. There is no reason, for example, to expect the diversity of large herbivores and carnivores to be the same when they are mammals as when they are dinosaurs. Third, taxa have evolved to use new resources and habitats, changing the overall number of species that the planet's resources can support.

Has the diversity of organisms ever reached its greatest possible value? Probably not. The diversity of marine animals, which provide the best record, was greater in the Pliocene than ever before (Sepkoski 1996b), and the time since then has been too short to say whether or not diversity has attained a plateau. But whatever the future course of biological diversity might have been, it has been altered for the foreseeable future by human domination of Earth, and altered for the worse. We have initiated the next mass extinction (Box A), and unless we take strong, rapid action, much of the glorious variety of the living world will be extinguished as quickly as if another asteroid had smashed into the planet and again cast over it a pall of death.

BOX 7A The Next Mass Extinction: It's Happening Now

None of the mass extinctions of the past was caused by the actions of a single species. But that is what is happening today. Within the next few centuries, the diversity of life will almost certainly plummet at a greater pace than ever before.

The human threat to Earth's biodiversity has accelerated steadily due to ever more powerful technology and the exponential growth of the world's human population, which reached 6.4 billion in mid 2004 and is increasing by an estimated 76 million persons a year. The per capita rate of population growth is greatest in the developing countries, which are chiefly tropical and subtropical, but the per capita impact on the world's environment is greatest in the most highly industrialized countries. An average American, for example, has perhaps 140 times the environmental impact of an average Kenyan, because the United States is so profligate a consumer of resources (harvested throughout the world) and energy (with impacts ranging from strip mines and oil spills to insecticides and production of the "greenhouse gases" that cause global warming).

Some species are threatened by hunting or overfishing, and others by species that humans have introduced into new regions. But by far the greatest cause of extinction is the destruction of habitat. It is largely for this reason that 29 percent of North American freshwater fishes are endangered or already extinct, and that about 10 percent of the world's bird species are considered endangered by the International Council for Bird Preservation.

The numbers of species likely to be lost are highest in tropical forests, which are being destroyed at a phenomenal and accelerating rate. As E. O. Wilson (1992) said, "in 1989 the surviving rain forests occupied an area about that of the contiguous forty-eight states of the United States, and they were being reduced by an amount equivalent to the size of Florida each year." Several authors have estimated that 10 to 25 percent of tropical rain forest species—accounting for as much as 5 to 10 percent of Earth's species diversity—will become extinct in the next 30 years. To this toll must be added extinctions caused by the destruction of species-rich coral reefs, pollution of other marine habitats, and losses of habitat in many areas, such as Madagascar and the Cape Province of South Africa, that harbor unusually high numbers of endemic species.

In the long run, an even greater threat to biodiversity may be global warming caused by high and increasing consumption of fossil fuels and production of CO_2 and other "greenhouse" gases. Earth's climate has become 0.6°C warmer, on average, during the last century, and the rate of warming is accelerating. Although climate change will vary in different regions (some will actually suffer a cooling trend), snow cover, glaciers, and polar ice caps are rapidly diminishing, and some tropical areas are becoming much drier. These changes are much faster than most of the climate changes that have occurred in the past. Some species may adapt by genetic change, but there is already evidence that many species will shift their ranges. Such shifts, however, are difficult or impossible for many mountaintop and Arctic spcies, and for many others because their habitats and the habitat "corridors" along which they might disperse have been destroyed. Computer simulations, based on various scenarios of warming rate and species' capacity for dispersal, suggest that within the next 50 years, between 18 and 35 percent of species will become "committed to extinction"—that is, they will have passed the point of no return (Thomas et al. 2004).

If mass extinctions have happened naturally in the past, why should we be so concerned? Different people have different answers, ranging from utilitarian to aesthetic to spiritual. Some point to the many thousands of species that are used by humans today, ranging from familiar foods to fiber, herbal medicines, and spices used by peoples throughout the world. Others cite the economic value of ecotourism and the enormous popularity of bird-watching in some countries. Biologists will argue that thousands of species may prove useful (as many already have) as pest control agents, or as sources of medicinal compounds or industrially valuable materials. Except in a few well-known groups, such as vertebrates and vascular plants, most species have not even been described, much less been studied for their ecological and possible social value.

The rationale for conserving biodiversity is only partly utilitarian, however. Many people (including this author) cannot bear to think that future generations will be deprived of tigers, sea turtles, and macaws. They share with millions of others a deep renewal of spirit in the presence of unspoiled nature. Still others feel that it is in some sense cosmically unjust to extinguish, forever, the species with which we share the Earth.

Conservation is an exceedingly complicated topic; it requires not only a concern for other species, but compassion and understanding of the very real needs of people whose lives depend on clearing forests and making other uses of the environment. It requires that we understand not only biology, but also global and local economics, politics, and social issues ranging from the status of women to the reactions of the world's peoples and their governments to what may seem like elitist Western ideas. Anyone who undertakes work in conservation must deal with some of these complexities. But everyone can play a helpful role, however small. We can try to waste less; influence people about the need to reduce population growth (surely the most pressing problem of all); support conservation organizations; patronize environment-conscious businesses; stay aware of current environmental issues; and communicate our concerns to elected officials at every level of government. Few actions of an enlightened citizen of the world can be more important.

Summary

1. Analyses of diversity in the fossil record require procedures to correct for biases caused by the incompleteness of the record.
2. The record of skeletonized marine animals indicates a very rapid increase in diversity at the beginning of the Cambrian, a slower increase thereafter to an approximate equilibrium that lasted for almost two-thirds of the Paleozoic, a mass extinction at the end of the Permian, and an increase (with interruptions) since the beginning of the Mesozoic, accelerating in the Cenozoic. Terrestrial plants and vertebrates show a similar pattern, except that their diversity was relatively stable for much of the Mesozoic. By the Pliocene, the diversity of families and lower taxa was apparently higher than ever before in the history of life.
3. The "background" rate of extinction (in between mass extinctions) appears to have declined during the Phanerozoic, perhaps because higher taxa that were particularly prone to extinction became extinct early.
4. Five major mass extinctions (at the ends of the Ordovician, Devonian, Permian, Triassic, and Cretaceous), as well as several less pronounced episodes of heightened extinction rates, have occurred. The impact of a large extraterrestrial body at the end of the Cretaceous may have caused the extinction of many taxa, including the last of the nonavian dinosaurs. Although the cause of the incomparably devastating end-Permian extinction is unknown, it may have been the result of a rapid episode of severe global warming initiated by the volcanic release of vast quantities of lava.
5. Broad geographic and ecological distributions, rather than adaptation to "normal" conditions, enhanced the likelihood that taxa would survive mass extinctions. The diversification of many of the surviving lineages was probably released by extinction of other taxa that had occupied similar adaptive zones. Newly diversifying groups have sometimes displaced other taxa by direct competitive exclusion, but more often they have replaced incumbent taxa after these became extinct.
6. The increase in diversity over time appears to have been caused mostly by adaptation to vacant or underutilized ecological niches ("ecological space"), often as a consequence of the evolution of key adaptations, and by increasing provinciality (differentiation of the biota in different geographic regions) owing to the separation of land masses in the Mesozoic and Cenozoic and the consequent development of greater latitudinal variation in climate. Coevolution has probably also contributed to the growth in diversity.
7. The rates of both extinction and origination of taxa have been diversity-dependent. Such observations imply that diversity tends toward an equilibrium. However, an equilibrium can change over geological time because of changes in climates and the configuration of continents, and because organisms evolve new ways of using habitats and resources.
8. The next mass extinction, at a perhaps unprecedented rate, has already begun.

Terms and Concepts

adaptive zone
background extinction
biodiversity
competitive displacement
density-dependent factor
diversification rate
diversity-dependent factor
ecological space
exponental growth
incumbent replacement
key adaptation
logistic growth
mass extinction
provinciality
pull of the Recent
Red Queen hypothesis
turnover

Suggestions for Further Reading

Useful, succinct reviews of the history of diversity include journal articles by P. W. Signor ("The geological history of diversity," *Annual Review of Ecology and Systematics* 21: 509–539, 1990) and M. J. Benton ("Diversification and extinction in the history of life," *Science* 268: 52–58, 1995). Book-length summaries include J. W. Valentine (ed.), *Phanerozoic diversity patterns: Profiles in macroevolution* (Princeton University Press, 1985), and D. Jablonski et al. (eds.), *Evolutionary paleobiology* (University of Chicago Press, 1996).

The diversification of many groups of organisms is treated in P. D. Taylor and G. P. Larwood (Eds.), *Major evolutionary radiations* (Clarendon Press, Oxford, 1990). Mass extinctions are the topic of A. Hallam and P. B. Wignall's *Mass extinctions and their aftermath*, Oxford University Press, 1996) and M. J. Benton's *When life nearly died: The greatest mass extinction of all time* (Thames & Hudson, New York, 2003).

A thoughtful and informed work about the future of biodiversity is found in E. O. Wilson's *The diversity of life* (W.W. Norton, New York, 1999). The same subject is discussed in a more academic manner in *Principles of conservation biology*, edited by G. K. Meffe and C. R. Carroll (Sinauer Associates, Sunderland, MA., 1997).

Problems and Discussion Topics

1. Distinguish between the rate of speciation in a higher taxon and its rate of diversification. What are the possible relationships between the present number of species in a taxon, its rate of speciation, and its rate of diversification?
2. What factors might account for differences among taxa in their numbers of extant species? Suggest methods for determining which factor might actually account for an observed difference.
3. Ehrlich and Raven (1964) suggested that coevolution with plants was a major cause of the great diversity of herbivorous insects, and Mitter et al. (1988) presented evidence that the evolution of herbivory was associated with increased rates of insect diversification. However, the increase in the number of insect families in the fossil record was not accelerated by the explosive diversification of flowering plants (Labandeira and Sepkoski 1993). Suggest some hypotheses to account for this apparent conflict, and some ways to test them.
4. A factor that might contribute to increasing species numbers over time is the evolution of increased specialization in resource use, whereby more species coexist by more finely partitioning resources. Discuss ways in which, using either fossil or extant organisms, one might test the hypothesis that a clade is composed of increasingly specialized species over the course of evolutionary time.
5. In several phyla of marine invertebrates, lineages classified as new orders appear first in the fossil record in shallow-water environments and are recorded from deep-water environments only later in their history (Jablonski and Bottjer 1990). What might explain this observation? (Note: No one has offered a definitive explanation so far, so use your imagination.)
6. What effects is global warming likely to have on biological diversity? Consider changes in the climate of different geographic regions, in sea level, and in the geographic ranges of species, the likelihood that species will adapt to environmental changes, and possible causes of extinction. (See Walther et al. 2002; Parmesan and Yohe 2003.)

8

The Origin of Genetic Variation

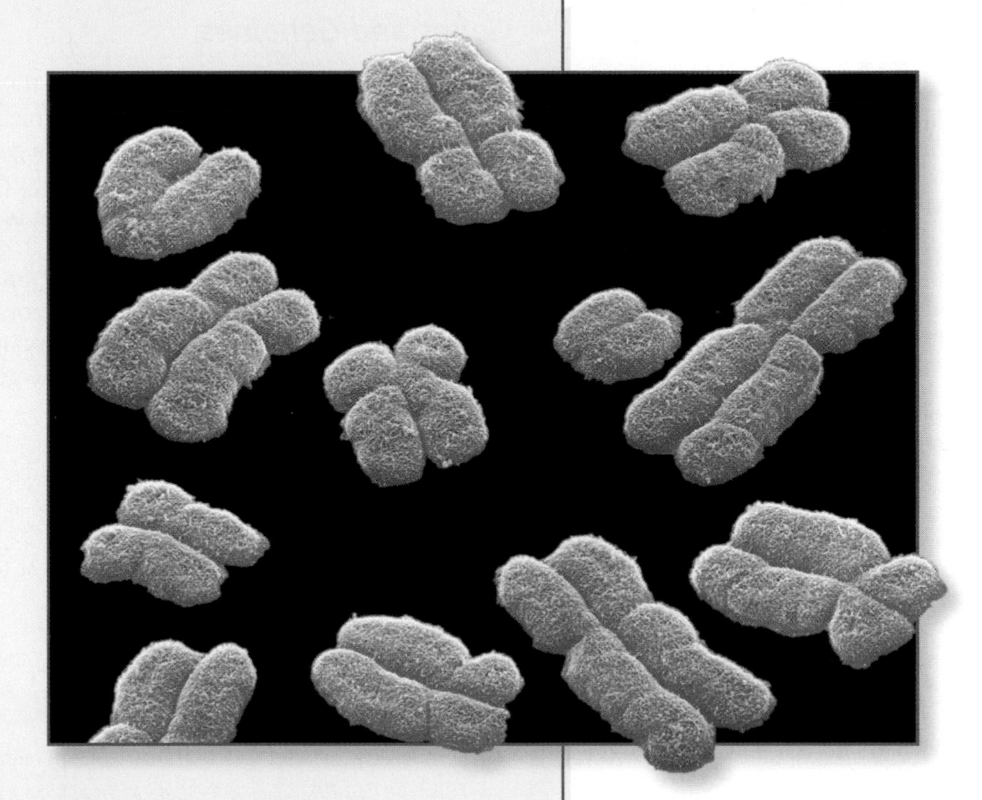

Human chromosomes. This scanning electron micrograph shows human chromosomes after replication, with identical pairs of chromatids joined at the centromeres. Mutation and recombination of the thousands of genes on each chromosome are the foundation of evolutionary change. (Photo © Andrew Syred/Photo Researchers, Inc.)

One of the landmarks in the history of science was the simultaneous publication in February 2001 of two "drafts" of the complete sequence of the human genome, one by the International Human Genome Sequencing Consortium (2001) and the other by a private company (Venter et al. 2001). There are still many gaps in the draft sequence, and the vast majority of apparent genes in the sequence have unknown functions, but the preliminary analyses nevertheless provide deep insights into the human genome and its evolutionary history.

The complete DNA sequence has been determined for more than a hundred species' genomes. These species include *Arabidopsis thaliana* (sometimes called mustard weed), the fruit fly *Drosophila melanogaster*, the nematode worm *Caenorhabditis elegans*, the fungus *Saccharomyces cerevisiae* (yeast), and the bacterium *Escherichia coli*, as well as a number of other bacteria, archaea, and viruses. Comparisons among these sequences will provide unprecedented volumes of information about the processes and history of evolution. To understand how the differences among the sequences arose, and therefore how phenotypic differences among organisms evolved, we must begin with the process of mutation.

Each of us was born with at least 300 new mutations that make our DNA different from that of our parents. At least one or two of these mutations are potentially harmful, especially if they are homozygous (present on both copies of the gene). At least 4500 different human genes have been described that, in their mutated form, cause inherited diseases or defects. Thousands of other such genes will be described in the near future.

This is the dark side of mutation. It is one we will revisit, but this chapter is more concerned with the positive role that mutation has played in evolution. Every gene, every variation in DNA, every characteristic of a species, every species itself, owes its existence to processes of mutation. Mutation is not the cause of evolution, any more than fuel in a car's tank is the cause of the car's movement. But it is the sine qua non, the necessary ingredient of evolution, just as fuel is necessary—though not sufficient—for traveling down a highway. The fundamental role of mutation makes it a logical starting point for our analysis of the causes of evolution.

Genes and Genomes

We will begin by describing mutation at the molecular level, but that requires a short review of the genetic material and its organization.

Except in certain viruses, in which the genetic material is RNA (ribonucleic acid), organisms' genomes consist of DNA (deoxyribonucleic acid), made up of a series of nucleotide **base pairs** (bp), each consisting of a purine (adenine, A, or guanine, G) and a pyrimidine (thymine, T, or cytosine, C). A haploid (gametic) genome of *Drosophila melanogaster* has about 1.5×10^8 bp, and that of a human about 3.2×10^9 bp (3.2 billion). However, DNA content varies greatly among organisms: it differs more than a hundredfold, for example, among species of salamanders, some of which have more than a hundred times as much DNA as humans. The genome of the single-celled protist *Amoeba dubia* is 200 times the size of a human's!

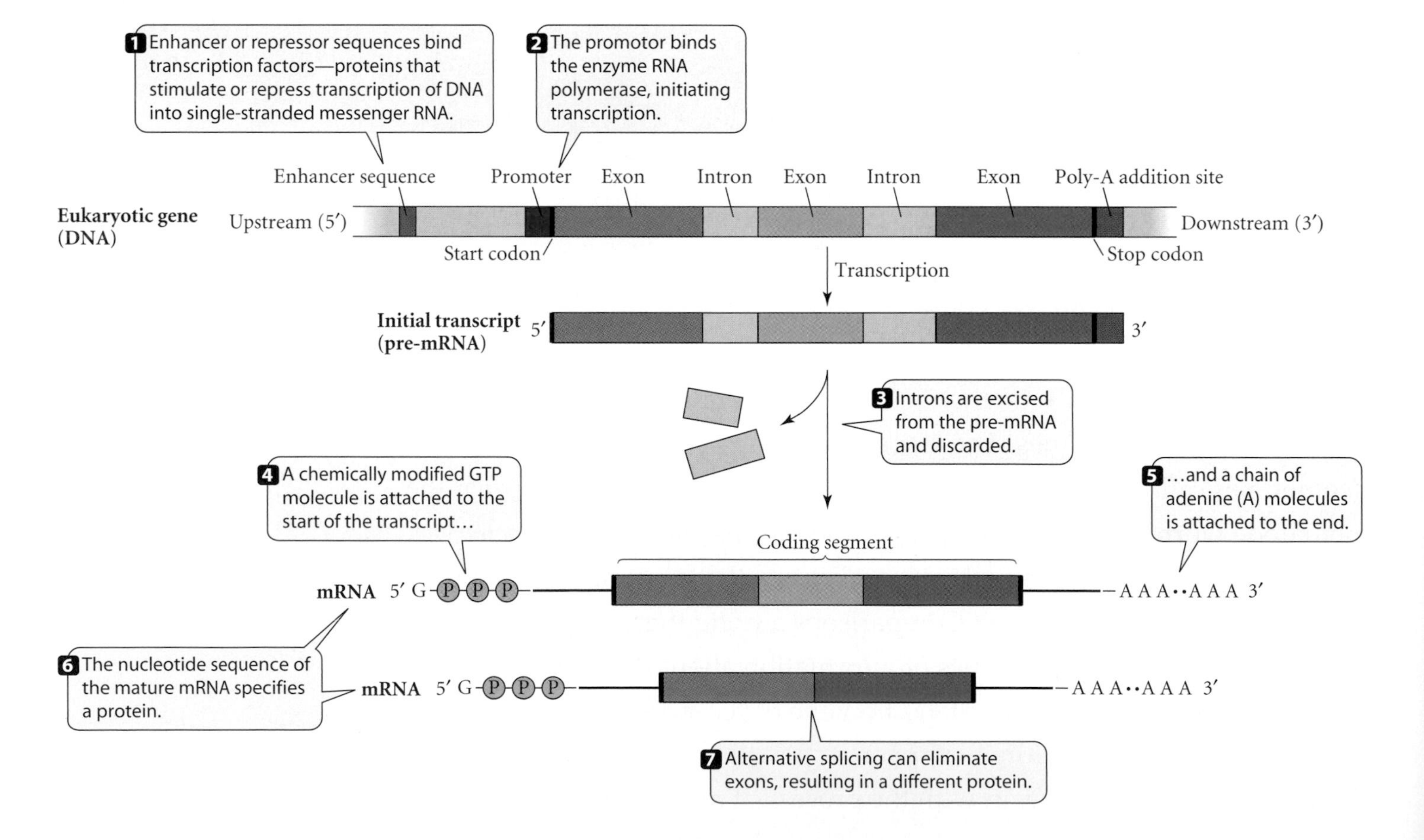

Figure 8.1 Diagram of a eukaryotic gene, its initial transcript (pre-mRNA), and the mature mRNA transcript. Proteins that regulate transcription bind to enhancer and repressor sequences, shown here upstream of the coding sequences. Transcription proceeds in the 5′ to 3′ direction. Introns are transcribed, but are spliced out of the pre-mRNA and discarded. The coding segment of the mature mRNA corresponds to the gene's exons. In some genes, alternative splicing may yield several different mRNAs.

Each chromosome carries a single long, often tightly coiled, DNA molecule, different portions of which constitute different genes. The term **gene** usually refers to a sequence of DNA that is transcribed into RNA, together with untranscribed regions that play roles in regulating its transcription. The word **locus** technically refers to the chromosome site occupied by a particular gene, but it is often used to refer to the gene itself. Thousands of genes in the human genome encode ribosomal and transfer RNAs (rRNAs and tRNAs) that are not translated into proteins. The number of protein-encoding genes is about 26,000 in *Arabidopsis*, 6000 in *Saccharomyces*, 13,000 in *Drosophila*, and 30,000 in mice and humans.

One strand of a protein-encoding gene is transcribed into RNA, a process regulated by **control regions**: untranscribed sequences (ENHANCERS and REPRESSORS) to which regulatory proteins produced by other genes bind. A gene may have many different enhancer sequences. In eukaryotes, the transcribed sequence of a gene consists of coding regions (**exons**) separated by noncoding regions (**introns**) (Figure 8.1). The average human gene has 1340 bp of coding sequence (encoding 445 amino acids) divided into 8.8 exons, which are separated by a total of 3365 bp of introns. After a gene is transcribed, the portions transcribed from exons are processed into a messenger RNA (mRNA) by splicing out the portions that were transcribed from introns. ALTERNATIVE SPLICING, whereby the mature mRNA corresponds to a variable number of the exons, can result in several proteins' being encoded by a single gene. At least 35 percent of human genes appear to be subject to alternative splicing, so the number of possible proteins may greatly exceed the number of genes.

Through the action of ribosomes, enzymes, and tRNAs, messenger RNA is translated into a polypeptide or protein on the basis of the **genetic code**, whereby a triplet of bases (a **codon**) specifies a particular amino acid in the growing polypeptide chain. The RNA code (Figure 8.2), which is complementary to the DNA code, consists of $4^3 = 64$ codons, which, however, encode only 20 amino acids. The amino acids are listed in Table 8.1, grouped by certain biochemical properties that affect the way the amino acid behaves in the process of protein folding. These properties have implications for protein evolution, which we will discuss in Chapter 19.

Most of the amino acids are encoded by two or more synonymous codons. The third position in a codon is the most "degenerate"; for example, all four CC– codons (CCU, CCC, CCA, CCG) specify proline. The second position is least degenerate—a substitution of one base for another in the second position usually results in an amino acid substitution in a protein. Three of the 64 codons are "stop" ("chain end") signals that terminate translation. Substitution in the third position of four other codons can therefore produce a "stop" codon, which may result in an incomplete, often nonfunctional protein.

It is a profoundly wonderful fact that the genetic code is nearly universal, from viruses and bacteria to pineapples and mammals. Moreover, the machinery of transcription and translation is remarkably uniform, so that sea urchin DNA or mRNA can be translated into a protein if injected into a frog. This uniformity is the basis of genetic engineering of crop plants with bacterial genes that encode natural insecticides, to give just one example.

In eukaryotes, the vast majority of DNA has no apparent function. Only 28 percent of the human genome

Second nucleotide

First nucleotide	U		C		A		G		Third nucleotide
U	UUU	Phe	UCU	Ser	UAU	Tyr	UGU	Cys	U
	UUC		UCC		UAC		UGC		C
	UUA	Leu	UCA		UAA	Stop	UGA	Stop	A
	UUG		UCG		UAG	Stop	UGG	Trp	G
C	CUU	Leu	CCU	Pro	CAU	His	CGU	Arg	U
	CUC		CCC		CAC		CGC		C
	CUA		CCA		CAA	Gln	CGA		A
	CUG		CCG		CAG		CGG		G
A	AUU	Ile	ACU	Thr	AAU	Asn	AGU	Ser	U
	AUC		ACC		AAC		AGC		C
	AUA		ACA		AAA	Lys	AGA	Arg	A
	AUG	Met	ACG		AAG		AGG		G
G	GUU	Val	GCU	Ala	GAU	Asp	GGU	Gly	U
	GUC		GCC		GAC		GGC		C
	GUA		GCA		GAA	Glu	GGA		A
	GUG		GCG		GAG		GGG		G

Figure 8.2 The genetic code, as expressed in messenger RNA. Three of the 64 codons are "stop" (chain end) signals; the other codons encode the 20 amino acids found in proteins. Note that many codons, especially those differing only in the third position, are synonymous. (The three-letter abbreviations for the amino acids are given in Table 8.1, which also explains the color-coding of the triplets.)

TABLE 8.1 *The amino acids*

Amino acid	One-letter abbreviation	Three-letter abbreviation	Biochemical properties (affect protein folding)
Polar, charged			
Positive charge (basic)			
Arginine	R	Arg	Electrically charged side chains attract water (are hydrophilic) and oppositely charged ions.
Histidine	H	His	
Lysine	K	Lys	
Negative charge (acidic)			
Aspartic acid	D	Asp	
Glutamic acid	E	Glu	
Polar, uncharged			
Serine	S	Ser	Uncharged polar side chains tend to form weak hydrogen bonds with water and with other polar or charged substances; mostly hydrophilic.
Threonine	T	Thr	
Asparagine	N	Asn	
Glutamine	Q	Gln	
Tyrosine	Y	Tyr	
Nonpolar			
Alanine	A	Ala	Nonpolar hydrocarbon side chains (hydrophobic) cluster toward center of protein, away from aqueous environment of cell cytoplasm.
Phenylalanine	F	Phe	
Leucine	L	Leu	
Isoleucine	I	Ile	
Methionine	M	Met	
Tryptophan	W	Trp	
Valine	V	Val	
Special cases			
Glycine	G	Gly	Smallest amino acid; side chain a single hydrogen atom.
Proline	P	Pro	Modified amino group "ring" limits hydrogen-bonding and rotational abilities.
Cysteine	C	Cys	Can form a disulfide bond with another cysteine.

is thought to be transcribed, and much of this consists of introns, so less than 5 percent (a generous estimate) of the genome encodes proteins. At least 45 percent of the human genome consists of repeated sequences, amounting to as many as 4.3 million repetitive elements that include **repeated sequences** of a few bp each. These sequences are sometimes referred to as **microsatellites**; **tandem repeats** (including clusters of short sequences that may number more than 2 billion repeats in some species); short interspersed repeats (SINEs) of 100–400 bp; long interspersed repeats (LINEs) that are more than 5 kilobases (kb) in length; and DNA transposons. Elements in the last three categories either are capable of undergoing or have been formed by the process of **transposition**: the production of copies that become inserted into new positions in the genome. DNA sequences that are capable of transposition are called **transposable elements** (**TEs**).

Many protein-encoding genes (probably at least 40 percent of human genes) are members of **gene families**: groups of genes that are similar in sequence and often have related functions. For example, the human hemoglobin gene family includes two subfamilies, α and β, located on different chromosomes (Figure 8.3). Different α and β polypeptides are combined to form different types of hemoglobin before and after birth. Both the α and β subfamilies also include hemoglobin **pseudogenes**: sequences that resemble the functional genes, but which differ at a number of base pair sites and are not transcribed because they have internal "stop" codons. Gene families are examples of repeated sequences that have diversified over time. Some gene families, such as that for mammalian olfactory receptors, have more than a thousand functional members.

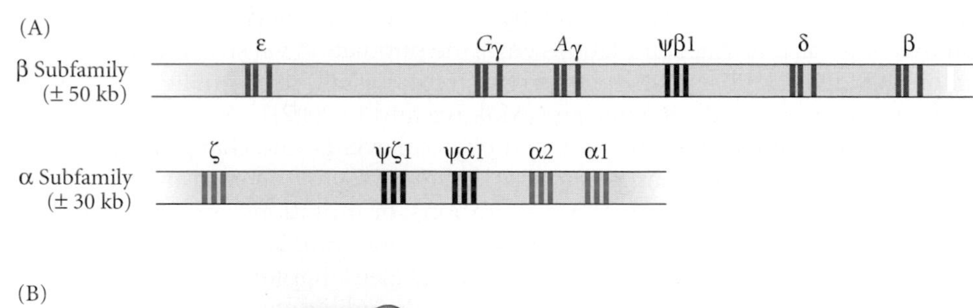

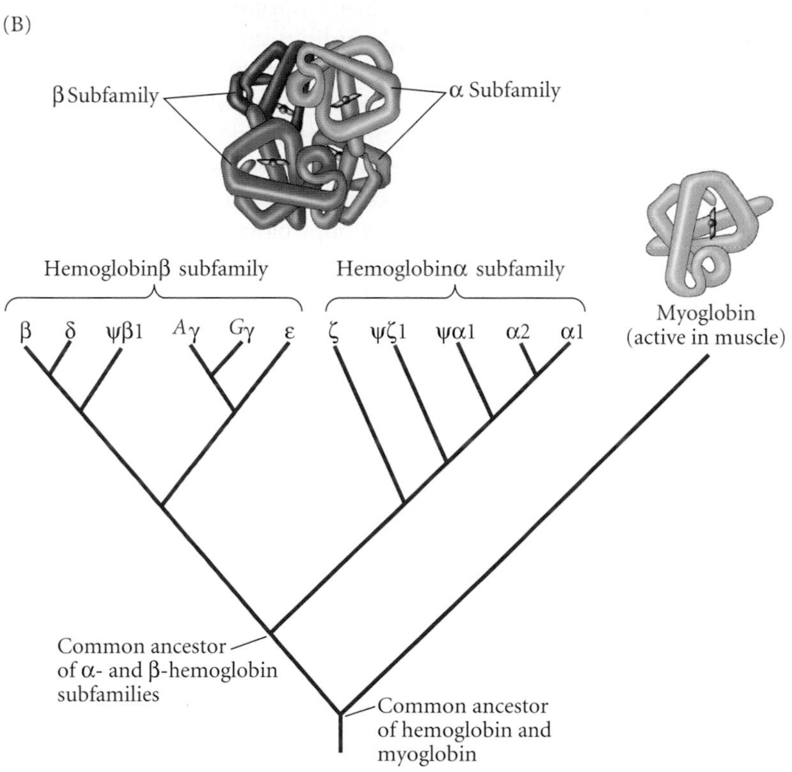

Figure 8.3 The human hemoglobin gene family has two subfamilies, α (green) and β (blue), located on different chromosomes. (A) Each functional gene is indicated by three lines, representing its three exons. Pseudogenes are denoted by ψ. (B) A gene tree for the hemoglobin gene family, with myoglobin as the outgroup. Myoglobin consists of a single protein unit. The hemoglobin protein (which carries oxygen in red blood cells) consists of four subunits, two each from the α and β subfamilies. (A after Lewin 1985; B after Hartwell et al. 2000.)

Gene Mutations

The word **mutation** refers both to the process of alteration of a gene or chromosome and to its product, the altered state of a gene or chromosome. It is usually clear from the context which is meant.

Before the development of molecular genetics, a mutation was identified by its effect on a phenotypic character. That is, a mutation was a newly arisen change in morphology, survival, behavior, or some other property that was inherited and could be mapped (at least in principle) to a specific locus on a chromosome. In practice, many mutations are still discovered, characterized, and named by their phenotypic effects. Thus we will frequently use the term "mutation" to refer to an alteration of a gene from one form, or **allele**, to another, the alleles being distinguished by phenotypic effects. In a molecular context, however, a gene mutation is an alteration of a DNA sequence, independent of whether or not it has any phenotypic effect. A particular DNA sequence that differs by one or more mutations from homologous sequences is called a **haplotype**. We will often refer to **genetic markers**, which are detectable mutations that geneticists use to recognize specific regions of chromosomes or genes.

Mutations have evolutionary consequences only if they are transmitted to succeeding generations. Mutations that occur in somatic cells may be inherited in certain animals and plants in which the reproductive structures arise from somatic meristems, but in those animals in which the germ line is segregated from the soma early in development, a mutation is inherited only if it occurs in a germ line cell. Mutations are thought to occur mostly during DNA replication, which usually occurs during cell division. In humans, more than five times as many new mutations enter the population via sperm than via eggs because more cell divisions have transpired in the germ line before spermatogenesis than before oogenesis in individuals of equal age (Makova and Li 2002).

DNA is frequently damaged by chemical and physical events, and changes in base pair sequence can result. Many such changes are repaired by DNA polymerase and by "proofreading" enzymes, but some are not. These alterations, or mutations, are considered by most evolutionary biologists to be *errors*. That is, *the process of mutation is thought to be not an adaptation*, but a consequence of unrepaired damage (see Chapter 17).

A particular mutation occurs in a single cell of a single individual organism. If that cell is in the germ line, it may give rise to a single gamete, or quite often, to a number of gametes that carry the mutation, and so may be inherited by a number of offspring. The occurrence of such "clusters" of mutations may have important evolutionary consequences (Woodruff et al. 1996). Initially, the mutation is carried by a very small percentage of individuals in the species population. If, because of natural selection or genetic drift, it ultimately becomes fixed (i.e., is carried by nearly the entire population), the mutation may be referred to as a **substitution**. The distinction between a mutation, which is simply an altered form of a gene, compared to some standard sequence, and a substitution is important. Most mutations do not become substitutions. Consequently, *mutation is not equivalent to evolution.*

Kinds of mutations

POINT MUTATIONS. Mutational changes of DNA sequences are of many kinds. The simplest mutation is a **base pair substitution** (Figure 8.4). In classic genetics, a mutation that maps to a single gene locus is called a **point mutation**; in modern usage, this term is often restricted to single base pair substitutions. A **transition** is a substitution of a purine for a purine (A ↔ G) or a pyrimidine for a pyrimidine (C ↔ T). **Transversions**, of eight possible kinds, are substitutions of purines for pyrimidines or vice versa (A or G ↔ C or T).

Mutations may have a phenotypic effect if they occur in genes that encode ribosomal and transfer RNA, nontranslated regulatory sequences such as enhancers, or protein-coding regions. Because of the redundancy of the genetic code, many mutations in coding regions are **synonymous mutations**, which have no effect on the amino acid sequence of the polypeptide or protein. **Nonsynonymous mutations**, in contrast, result in amino acid substitutions; they may have little or no effect on the functional properties of the polypeptide or protein, and thus no effect on the phenotype, or they may have substantial consequences. For example, a change from the (RNA) triplet GAA to GUA causes the amino acid valine to be incorporated instead of glutamic acid. This is the mutational event that in humans caused the abnormal β-chain in sickle-cell hemoglobin, which has many phenotypic consequences and is usually lethal in homozygotes.

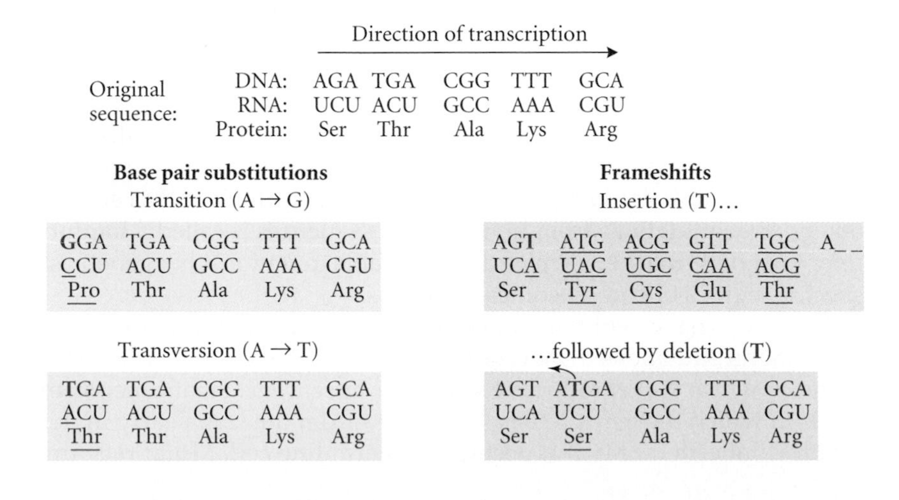

Figure 8.4 Examples of point mutations and their consequences for messenger RNA and amino acid sequences. (Only the transcribed, "sense" strand of the DNA is shown.) The boxes at the left show two kinds of base pair substitutions: a transition and a transversion at the first base position. The boxes at the right show two kinds of frameshift mutations.

If a single base pair becomes inserted into or deleted from a DNA sequence, the triplet reading frame is shifted by one nucleotide, so that downstream triplets are read as different codons and translated into different amino acids (Figure 8.4). Such insertions or deletions result in **frameshift mutations**. The greatly altered gene product is usually nonfunctional, although exceptions to this generalization are known.

A somewhat similar phenomenon is REPLICATION SLIPPAGE, which alters the number of short repeats in microsatellites. The growing (3′) end of a DNA strand that is being formed during replication may become dissociated from the template strand and form a loop, so that the next repeat to be copied from the template is one that had already been copied. Thus extra repeats will be formed in the growing strand. Microsatellite "alleles" that differ in copy number arise by replication slippage at a high rate.

SEQUENCE CHANGES ARISING FROM RECOMBINATION. Recombination typically is based on precise alignment of the DNA sequences on homologous chromosomes in meiosis. When homologous DNA sequences differ at two or more base pairs, **intragenic recombination** between them can generate new DNA sequences, just as crossing over between genes generates new gene combinations. DNA sequencing has revealed many examples of variant sequences that apparently arose by intragenic recombination.

Recombination appears to be the cause of a peculiar mutational phenomenon called GENE CONVERSION, which has been studied most extensively in fungi. The gametes of a heterozygote should carry its two alleles (A_1, A_2) in a 1:1 ratio. Occasionally, though, they occur in different ratios, such as 1:3. In these cases, an A_1 allele has been replaced specifically by an A_2 allele, rather than by any of the many other alleles to which it might have mutated: it seems to have been converted into A_2. This seems to occur when a damaged DNA strand of one chromosome is repaired by enzymes that insert bases complementary to the sequence on the undamaged homologous chromosome.

Unequal crossing over (unequal exchange) can occur between two homologous sequences or chromosomes that are not perfectly aligned. Recombination then results in a TANDEM DUPLICATION on one recombination product and a deletion on the other (Figure 8.5). The length of the affected region may range from a single base pair to a large block of loci (a segmental duplication), depending on the amount of displacement of the two misaligned chromosomes. Most unequal crossing over occurs between sequences that already include tandem repeats (e.g., ABBC) because the duplicate regions can pair out of register:

$$\begin{pmatrix} \text{ABBC...} \\ \text{...ABBC} \end{pmatrix}$$

This pairing generates further duplications (ABBBC). Unequal crossing over is one of the processes that have generated the extremely high number of copies of nonfunctional sequences that constitute much of the DNA in most eukaryotes. It has given rise to many gene families, and it has been extremely important in the evolution of greater numbers of functional genes and of total DNA (see Chapter 19).

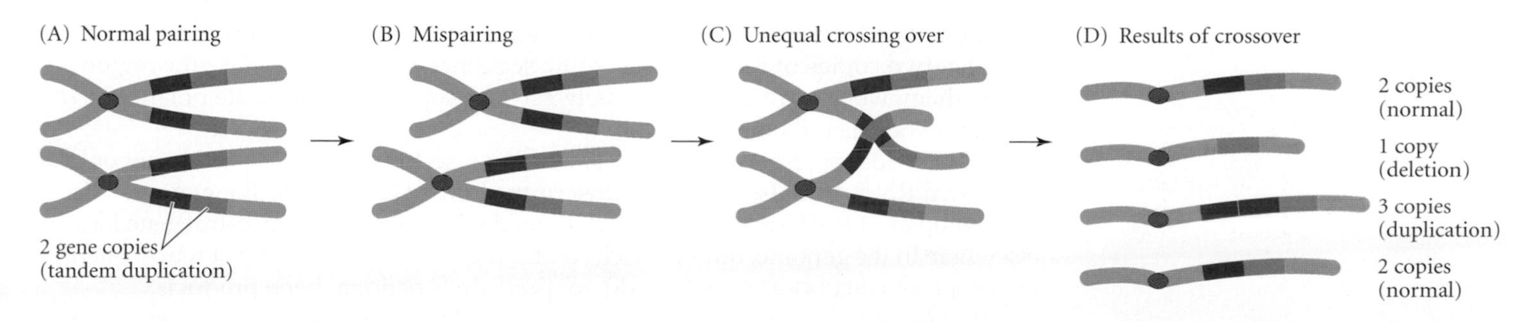

Figure 8.5 Unequal crossing over occurs most commonly when two repeated genes or sequences mispair with their homologues. Crossing over then yields a deletion of one of the sequences on one chromatid, and a duplication of the sequences (and hence three copies of that sequence) on the other chromatid. The duplicated sequence may encode part of a gene, or it may encode one or more complete genes. (After Hartl and Jones 2001.)

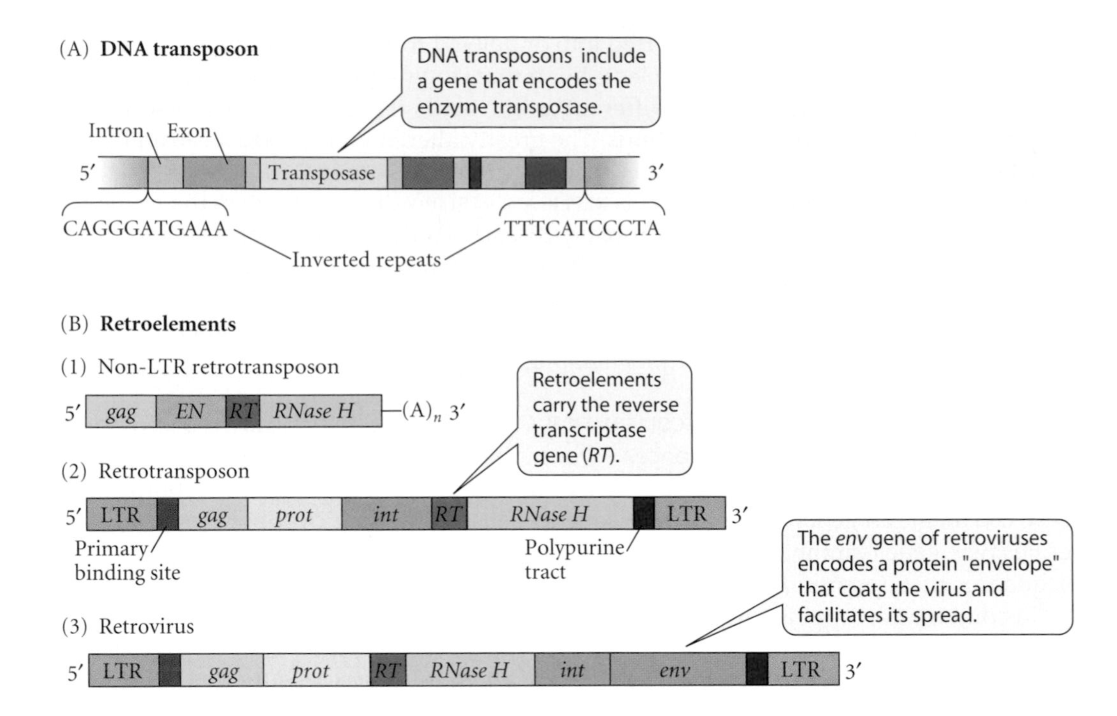

Figure 8.6 Some different kinds of transposable elements. (A) DNA transposable element are flanked by inverted repeats; these elements encode several genes in addition to a gene that encodes the enzyme (transposase) required for transposition. (B) Retroelements include a gene (*RT*) that encodes reverse transciptase, which transcribes and RNA copy into DNA, which is then inserted in the genome. Three kinds of retroelements are shown. (1) Non-LTR retrotransposons, which include LINEs, lack long terminal repeats (LTRs). (2) Retrotransposons, such as *copia* in *Drosophila*, are similar but have LTRs. (3) Retroviruses, the most complex retroelements, include the gene (*env*) that encodes a protein envelope.

CHANGES CAUSED BY TRANSPOSABLE ELEMENTS. Most transposable elements produce copies that can move to any of many places in the genome, and sometimes they carry with them other genes near which they had been located. These DNA sequences include genes that encode enzymes that accomplish the transposition (movement). The several kinds of TEs (Figure 8.6) include INSERTION SEQUENCES, which encode only enzymes that cause transposition, TRANSPOSONS, which encode other functional genes as well, and RETROELEMENTS, which carry a gene for the enzyme reverse transcriptase. Retroelements are first transcribed into RNA, which then is reverse-transcribed into a DNA copy (cDNA) that is inserted into the genome. Some retroelements are retroviruses (including the HIV virus that causes AIDS) whose RNA copies can cross cell boundaries. Retrotransposons, which include some LINE elements, are retroelements that act similarly, except that they do not cross cell boundaries and are copied only by cell division in the host.

Transposable elements often become excised from a site into which they had previously been inserted, but leave behind sequence fragments that tell of their former presence. From such cases and from more direct evidence, transposable elements are known to have many effects on genomes (Kazazian 2004; Bennetzen 2000):

- When inserted into a coding region, they alter, and usually destroy, the function of the protein, often by causing a frameshift or altering splicing patterns.
- When inserted into or near control regions, they can interfere with or alter gene expression (e.g., the timing or amount of transcription).
- They are known to increase mutation rates of host genes.
- They can cause rearrangements in the host genome, resulting from recombination between two copies of a TE located at different sites (Figure 8.7). Just as unequal crossing over between members of a gene family can generate duplications and deletions, so can recombination between copies of a TE located at nonhomologous sites. Recombination between two copies of a TE with the same sequence polarity can *delete* the region between them, whereas recombination between two copies with opposite polarity *inverts* the region between them.
- TEs that encode reverse transcriptase sometimes insert DNA copies (cDNA) not only of their own RNA, but also of RNA transcripts of other genes, into the genome. These cDNA copies of RNA (**retrosequences**) resemble the exons of an ancestral gene located elsewhere in the genome, but they lack control regions and introns. Most retrosequences are PROCESSED PSEUDOGENES, which do not produce functional gene products.

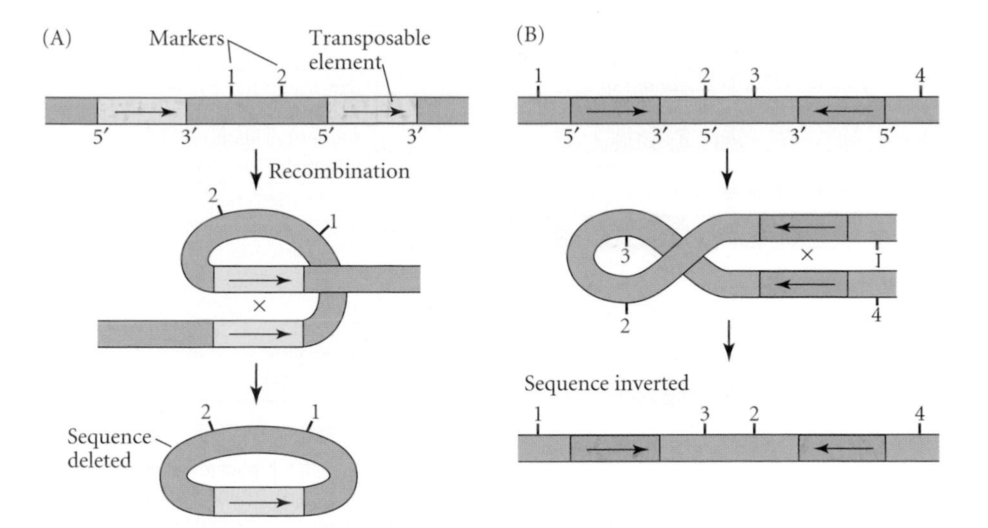

Figure 8.7 Recombination between copies of a transposable element can result in deletions and inversions. The boxes containing arrows represent transposable elements, with the polarity of base pair sequence indicated by the arrows. The numerals represent genetic markers. (A) Recombination (×) between two direct repeats (i.e., with the same polarity) excises one repeat and deletes the sequence between the two copies. (B) Recombination between two inverted repeats (with opposite polarity) inverts the sequence between them. (After Lewin 1985.)

- By transposition and unequal crossing over, TEs can increase in number, and so increase the size of the genome.

Most or all of these transposable element-induced effects have been observed within experimental populations of organisms such as maize (corn, *Zea mays*) and *Drosophila melanogaster*. The transposition rate of various retroelements ranges from about 10^{-5} to 10^{-3} per copy in inbred lines of *Drosophila*, resulting in an appreciable rate of mutation (Nuzhdin and Mackay 1994). All the kinds of changes engendered by transposable elements can be found by comparing genes and genomes of organisms of the same or different species. For instance, L1 retrotransposon insertions are associated with many disease-causing mutations in both mice and humans (Kazazian 2004), and a difference in flower color between two species of *Petunia* has been caused by the insertion and subsequent incomplete excision of a transposon that disabled a gene that controls anthocyanin pigment production (Quattrocchio et al. 1999).

Examples of mutations

Geneticists have learned an enormous amount about the nature and causes of mutations by studying model organisms such as *Drosophila* and *E. coli*. Moreover, many human mutations have been characterized because of their effects on health. Human mutations are usually rather rare variants that can be compared with normal forms of the gene; in some instances, newly arisen mutations have been found that are lacking in both of a patient's parents.

Single base pair substitutions are responsible for conditions such as sickle-cell anemia, described earlier, and for precocious puberty, in which a single amino acid change in the receptor for luteinizing hormone causes a boy to show signs of puberty when about 4 years old. Because many different alterations of a protein can diminish its function, the same phenotypic condition can be caused by many different mutations of a gene. For example, cystic fibrosis, a fatal condition afflicting one in 2500 live births in northern Europe, is caused by mutations in the gene encoding a sodium channel protein. The most common such mutation is a 3 bp deletion that deletes a single amino acid from the protein; another converts a codon for arginine into a "stop" codon; another alters splicing so that an exon is missing from the mRNA; and many of the more than 500 other base pair substitutions recorded in this gene are also thought to cause the disease (Zielenski and Tsui 1995). Mutations in any of the many different genes that contribute to the normal development of some characteristics can also result in similar phenotypes. For example, retinitis pigmentosa, a degeneration of the retina, can be caused by mutations in genes on 8 of the 23 chromosomes in the (haploid) human genome (Avise 1998).

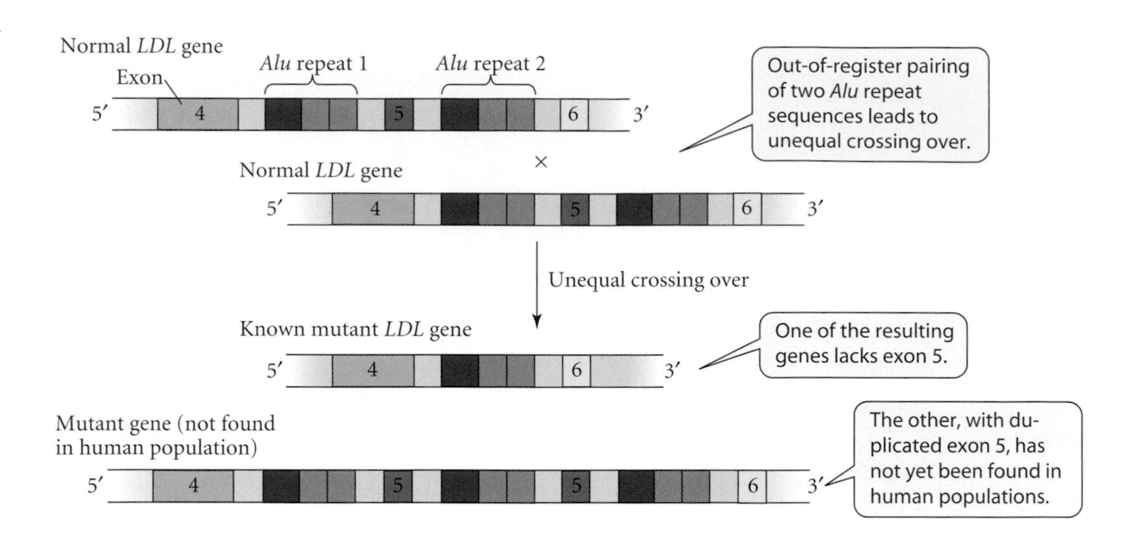

Figure 8.8 A mutated low-density lipoprotein (*LDL*) gene in humans lacks exon 5. It is believed to have arisen by unequal crossing over between two normal gene copies, due to out-of-register pairing between two of the repeated sequences (*Alu*, shown as blue boxes) in the introns. The numbered boxes are exons. (After Hobbs et al. 1986.)

Hemophilia can be caused by mutations in two different genes that encode blood-clotting proteins. In both genes, many different base pair substitutions, as well as small deletions and duplications that cause frameshifts, are known to cause the disease, and about 20 percent of cases of hemophilia-A are caused by an inversion of a long sequence within one of the genes (Green et al. 1995). Huntington disease, a fatal neurological disorder that strikes in midlife, is caused by an excessive number of repeats of the sequence CAG: the normal gene has 10 to 30 repeats, the mutant gene more than 75. Unequal crossing over between the two tandemly arranged genes for α-hemoglobin (see Figure 8.3) has given rise to variants with three tandem copies (duplication) and with one (deletion). The deletion of one of the loci causes α-thalassemia, a severe anemia. Another case of deletion, which results in high cholesterol levels, is the lack of exon 5 in a low-density lipoprotein gene. This deletion has been attributed to unequal crossing over, facilitated by a short, highly repeated sequence called *Alu* that is located in the introns of this gene and in many other sites in the genome (Figure 8.8).

These examples might make it seem as if mutations are nothing but bad news. While this is close to the truth—far more mutations are harmful than helpful—these mutations represent a biased sample. Many advantageous mutations have become incorporated into species' genomes (fixed) and thus represent the current **wild-type**, or normal, genes. For example, most genes that have arisen by reverse transcription from mRNA are nonfunctional pseudogenes, but at least one has been found that is a fully functional member of the human genome. Phosphoglycerate kinase is encoded by two genes. One, on the X chromosome, has a normal structure of 11 exons and 10 introns. The other, on an autosome, lacks introns, and clearly arose from the X-linked gene by reverse transcription. It is expressed only in the testes, a novel pattern of tissue expression that suggests that the gene plays a new functional role (see Li 1997).

When biologists seek those genes that have been involved in the evolution of a specific characteristic, they often use rare deleterious mutations of the kind described here as indicators of CANDIDATE GENES, those that may be among the genes they seek. For example, a rare mutation in the human *FOXP2* gene (*forkhead box 2*, which encodes a transcription factor) causes severe speech and language disorders. Two research groups, led by Jianzhi Zhang (Zhang et al. 2002) and Svante Pääbo (Enard et al. 2002), independently found that this gene has undergone two nonsynonymous (amino acid-changing) substitutions in the human lineage since the divergence of the human and chimpanzee lineages less than 7 Mya. This is a much higher rate of protein evolution than would be expected, considering that only one other such substitution has occurred between these species and the mouse, which diverged almost 90 Mya (Figure 8.9). Both research groups propose that these substitutions occurred in the human lineage less than 200,000 years ago and that they are among the important steps in the evolution of human language and speech.

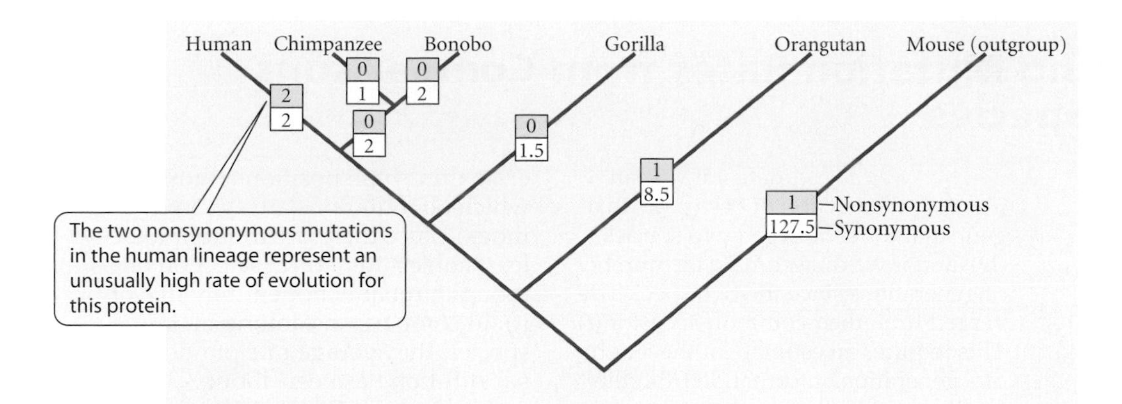

Figure 8.9 A phylogeny of the Hominoidea and its divergence from an outgroup, the mouse. Each box shows the number of nonsynonymous (yellow boxes) and synonymous (white boxes) substitutions in the *FOXP2* gene. The two nonsynonymous substitutions in the human lineage represent an unusually high rate of evolution of the FOXP2 protein, and may represent mutations that have been important in the evolution of language and speech. (After Zhang et al. 2002.)

Rates of mutation

Recurrent mutation refers to the repeated origin of a particular mutation, and the *rate* at which a particular mutation occurs is typically measured in terms of recurrent mutation: the number of independent origins per gene copy (e.g., per gamete) per generation or per unit time (e.g., per year). Mutation rates are estimates, not absolutes, and these estimates depend on the method used to detect mutations. In classical genetics, a mutation was detected by its phenotypic effects, such as white versus red eyes in *Drosophila*. Such a mutation, however, might be caused by the alteration of any of many sites within a locus; moreover, many base pair changes have no phenotypic effect. Thus phenotypically detected rates of mutation underestimate the rate at which all mutations occur at a locus. With modern molecular methods, mutated DNA sequences can be detected directly, so mutation rates can be expressed per base pair.

ESTIMATING MUTATION RATES. Rates of mutation are estimated in several ways (Drake et al. 1998). A relatively direct method is to count the number of mutations arising in a laboratory stock (which is usually initially homozygous), scoring mutations either by their phenotypic effects or by molecular methods. An indirect method (Box A) is based on the number of base pair differences between homologous genes in different species, relative to the number of generations that have elapsed since they diverged from their common ancestor. This method depends on the neutral theory of molecular evolution, which is described in Chapter 10.

Mutation rates vary among genes and even among regions within genes, but on average, as measured by phenotypic effects, a locus mutates at a rate of about 10^{-6} to 10^{-5} mutations per gamete per generation (Table 8.2). The average mutation rate per base pair, based mostly on the indirect method of comparing DNA sequences of different species, has been estimated at about 10^{-11} to 10^{-10} per replication in prokaryotes (see Table 8.3), or about 10^{-9} per sexual generation in eukaryotes. The mutation rate in the human genome has been estimated at about 4.8×10^{-9} per base pair per generation (Lynch et al. 1999).

Back mutation is mutation of a "mutant" allele back to the allele (usually the wild type) from which it arose. Back mutations are ordinarily detected by their phenotypic effects. They usually occur at a much lower rate than "forward" mutations (from wild type to mutant), presumably because many more substitutions can impair gene function than can restore it. At the molecular level, most phenotypically detected back mutations are not restorations of the original sequence, but instead result from a second amino

TABLE 8.2 ***Spontaneous mutation rates of specific genes, detected by phenotypic effects***

Species and locus	Mutations per 100,000 cells or gametes
Escherichia coli	
Streptomycin resistance	0.00004
Resistance to T1 phage	0.003
Arginine independence	0.0004
Salmonella typhimurium	
Tryptophan independence	0.005
Neurospora crassa	
Adenine independence	0.0008–0.029
Drosophila melanogaster	
Yellow body	12
Brown eyes	3
Eyeless	6
Homo sapiens	
Retinoblastinoma	1.2–2.3
Achondroplasia	4.2–14.3
Huntington's chorea	0.5

Source: After Dobzhansky 1970.

BOX 8A Estimating Mutation Rates from Comparisons among Species

In Chapter 10, we will describe the *neutral theory of molecular evolution*. This theory describes the fate of purely neutral mutations—that is, those mutations that neither enhance nor lower fitness. One possible fate is that a mutation will become fixed—that is, attain a frequency of 1.0—entirely by chance. The probability that this event will occur equals u, the rate at which neutral mutations arise. In each generation, therefore, the probability is u that a mutation that occurred at some time in the past will become fixed. After the passage of t generations, the fraction of mutations that will have become fixed is therefore ut.

If two species diverged from a common ancestor t generations ago, the expected fraction of fixed mutations in both species is $D = 2ut$, since various mutations have become fixed in both lineages. If the mutations in question are base pair changes, a fraction $D = 2ut$ of the base pairs of a gene should differ between the species, assuming that all base pairs are equally likely to mutate. Thus the average mutation rate per base pair per generation is $u = D/2t$.

Thus we can estimate u if we can measure the fraction of base pairs in a gene that differ between two species (D), and if we can estimate the number of generations since the two species diverged from their common ancestor (t). This requires an estimate of the length of a generation, information from the fossil record on the absolute time at which the common ancestor existed, and an understanding of the phylogenetic relationships among the living and fossilized taxa.

In applying this method to DNA sequence data, it is necessary to assume that most base pair substitutions are neutral and to correct for the possibility that earlier substitutions at some sites in the gene have been replaced by later substitutions ("multiple hits"). Uncertainty about the time since divergence from the common ancestor is usually the greatest source of error in estimates obtained by this method.

The best estimates of mutation rates at the molecular level have been obtained from interspecific comparisons of pseudogenes, other nontranslated sequences, and fourfold-degenerate third-base positions (those in which all mutations are synonymous), since these are thought to be least subject to natural selection (although probably not entirely free of it). In comparisons among mammal species, the average rate of nucleotide substitution has been about 3.3–3.5 per nucleotide site per 10^9 years, for a mutation rate of $3.3–3.5 \times 10^{-9}$ per site per year (Li and Graur 1991). If the average generation time were 2 years during the history of the lineages studied, the average rate of mutation per site would be about 1.7×10^{-9} per generation. Comparison of human and chimpanzee sequences yielded an estimate of 1.3×10^{-9} per site per year, assuming divergence 7 Mya. If the average generation time in these lineages has been 15–20 years, then the mutation rate is about 2×10^{-8} per generation. The human diploid genome has 6×10^9 nucleotide pairs, so this implies at least 120 new mutations per genome per generation—an astonishingly high number (Crow 1993).

acid substitution, either in the same or a different protein, that restores the function that had been altered by the first substitution. Advantageous mutations arose and compensated for severely deleterious mutations within 200 generations in experimental populations of *E. coli* (Moore et al. 2000).

EVOLUTIONARY IMPLICATIONS OF MUTATION RATES. With such a low mutation rate per locus, it might seem that mutations occur so rarely that they cannot be important. However, summed over all genes, the input of variation by mutation is considerable. If the haploid human genome has 3.2×10^9 bp and the mutation rate is 4.8×10^{-9} per bp per generation, an average zygote will carry about 317 new mutations. If only 2.5 percent of the genome consists of functional, transcribed sequences, 7 of these new mutations will be expressed and will have the potential to affect phenotypic characters (Lynch et al. 1999). Other authors have estimated the number of new base pair changes per zygote in the functional part of the genome as 0.14 in *Drosophila*, 0.9 in mice, and 1.6 in humans (Table 8.3). So, in a population of 500,000 humans, at least 800,000 new mutations arise every generation. If even a tiny fraction of these mutations were advantageous, the amount of new "raw material" for adaptation would be substantial, especially over the course of thousands or millions of years.

Experiments on *Drosophila* have confirmed that the total mutation rate per gamete is quite high. For example, Terumi Mukai and colleagues (1972), in a heroically large experiment, counted more than 1.7 million flies in order to estimate the rate at which the

TABLE 8.3 ***Estimates of spontaneous mutation rates per base pair and per genome***

	Base pairs		Mutation rate			
Organism	**in haploid genome**	**in effective genome**[a]	**per base pair per replication**	**per replication per haploid genome**	**per replication per effective genome**[a]	**per sexual generation per effective genome**[b]
T2, T4 phage	1.7×10^5	—	2.4×10^{-8}	0.0040	—	—
Escherichia coli	4.6×10^6	—	5.4×10^{-10}	0.0025	—	—
Saccharomyces cerevisiae (yeast)	1.2×10^7	—	2.2×10^{-10}	0.0027	—	—
Neurospora crassa (bread mold)	4.2×10^7	—	7.2×10^{-11}	0.0030	—	—
Caenorhabditis elegans	8.0×10^7	1.8×10^7	2.3×10^{-10}	0.018	0.004	0.036
Drosophila melanogaster	1.7×10^8	1.6×10^7	3.4×10^{-10}	0.058	0.005	0.14
Mouse	2.7×10^9	8.0×10^7	1.8×10^{-10}	0.49	0.014	0.9
Human	3.2×10^9	8.0×10^7	5.0×10^{-11}	0.16	0.004	1.6

Source: After Drake et al. 1998.
[a] The effective genome is the number of base pairs in functional sequences that could potentially undergo mutations that reduce fitness.
[b] Calculated for multicellular organisms in which multiple DNA replication events occur in development between zygote and gametogenesis.

chromosome 2 accumulates mutations that affect egg-to-adult survival (VIABILITY). They used crosses (see Figure 9.7) in which copies of the wild-type chromosome 2 were carried in a heterozygous condition so that deleterious recessive mutations could persist without being eliminated by natural selection. Every 10 generations, they performed crosses that made large numbers of these chromosomes homozygous and measured the proportion of those chromosomes that reduced viability. The mean viability declined, and the variation (variance) among chromosomes increased steadily (Figure 8.10). From the changes in the mean and variance, Mukai et al. calculated a mutation rate of about 0.15 per chromosome 2 per gamete. This is the sum, over all loci on the chromosome, of mutations that affect viability. Because chromosome 2 carries about a third of the *Drosophila* genome, the total mutation rate is about 0.50 per gamete. Thus almost every zygote carries at least one new mutation that reduces viability. Subsequent studies have indicated that the mutation rate for *Drosophila* is at least this high, and that it reduces viability by 1 to 2 percent per generation (Lynch et al. 1999). Indirect estimates indicate that humans likewise suffer about 1.6 new mutations per zygote that reduce survival or reproduction (Eyre-Walker and Keightley 1999).

Mutation rates vary among genes and chromosome regions, and they are also affected by environmental factors. MUTAGENS (mutation-causing agents) include ultraviolet light, X-rays, and a great array of chemicals, many of which are environmental pollutants. For example, mutation rates in birds and mice are elevated in industrial areas, and mice exposed to particulate air pollution in an urban-industrial site showed higher rates of mutation in repetitive elements than mice exposed only to filtered air at that site, or mice placed in a rural location (Figure 8.11).

Figure 8.12 shows that the variation in a typical phenotypic character is **polygenic**: it is based on several or many different genetic loci. It is very difficult to single out any of these loci to study the mutation rate per locus, but it is easy to estimate the **mutational variance** of the character—the increased variation in a population caused by new mutations in each generation. Studies of traits such as bristle number in *Drosophila* have shown that the mutational variance is high enough that an initially homozygous population would take only about 500 generations to achieve the level of genetic variation generally found in a natural population. The magnitude of the mutational variance varies somewhat among characters and species (Lynch 1988).

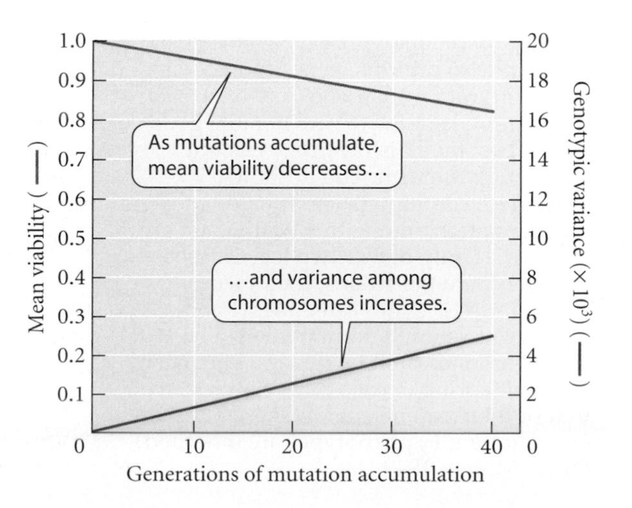

Figure 8.10 Effects of the accumulation of spontaneous mutations on the egg-to-adult survival of *Drosophila melanogaster*. The mean viability of flies made homozygous for chromosome 2 carrying new recessive mutations decreased, and the variation (variance) among those chromosomes increased. The rate of mutation was estimated from these data. (After Mukai et al. 1972.)

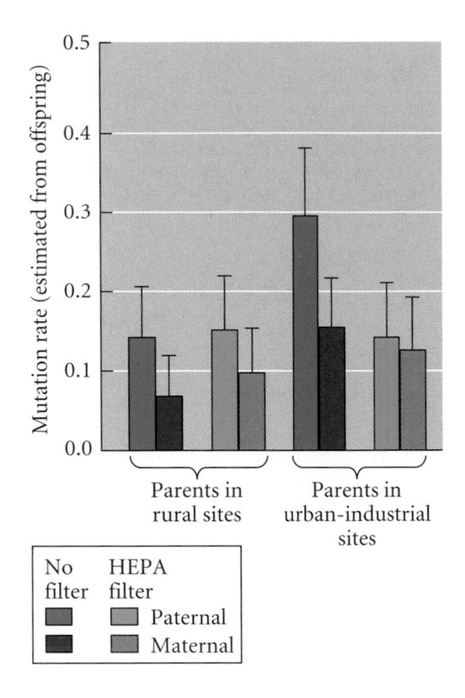

Figure 8.11 Mutation rates in mice, estimated from DNA sequences of two loci in their offspring. The mice were placed for 10 weeks in rural sites or in urban-industrial sites near steel mills and a major highway, where they were exposed either directly to the air or to air passed through a HEPA filter. Exposure to unfiltered urban-industrial air increased the mutation rate. (After Somers et al. 2004.)

In summary, although any given mutation is a rare event, the rate of origination of new genetic variation in the genome as a whole, and for individual polygenic characters, is appreciable. However, mutation alone does not cause a character to evolve from one state to another, because its rate is too low. Suppose that alleles A_1 and A_2 determine alternative phenotypes (e.g., red versus purple) of a haploid species, that half the individuals carry A_1, and that the rate of recurrent mutation from A_1 to A_2 is 10^{-5} per gene per generation. In one generation, the proportion of A_2 genes will increase from 0.5 to $[0.5 + (0.5)(10^{-5})] = 0.50000495$. At this rate, it will take about 70,000 generations before A_2 constitutes 75 percent of the population, and another 70,000 generations before it reaches 87 percent. This rate is so slow that, as we will see, factors other than recurrent mutation usually have a much stronger influence on allele frequencies, and thus are responsible for whatever evolutionary change occurs.

Phenotypic effects of mutations

A mutation may alter one or more phenotypic characters, such as size, coloration, or the amount or activity of an enzyme. Alterations in such features may affect survival and/or reproduction, the major components of FITNESS (see Chapter 11). It is often con-

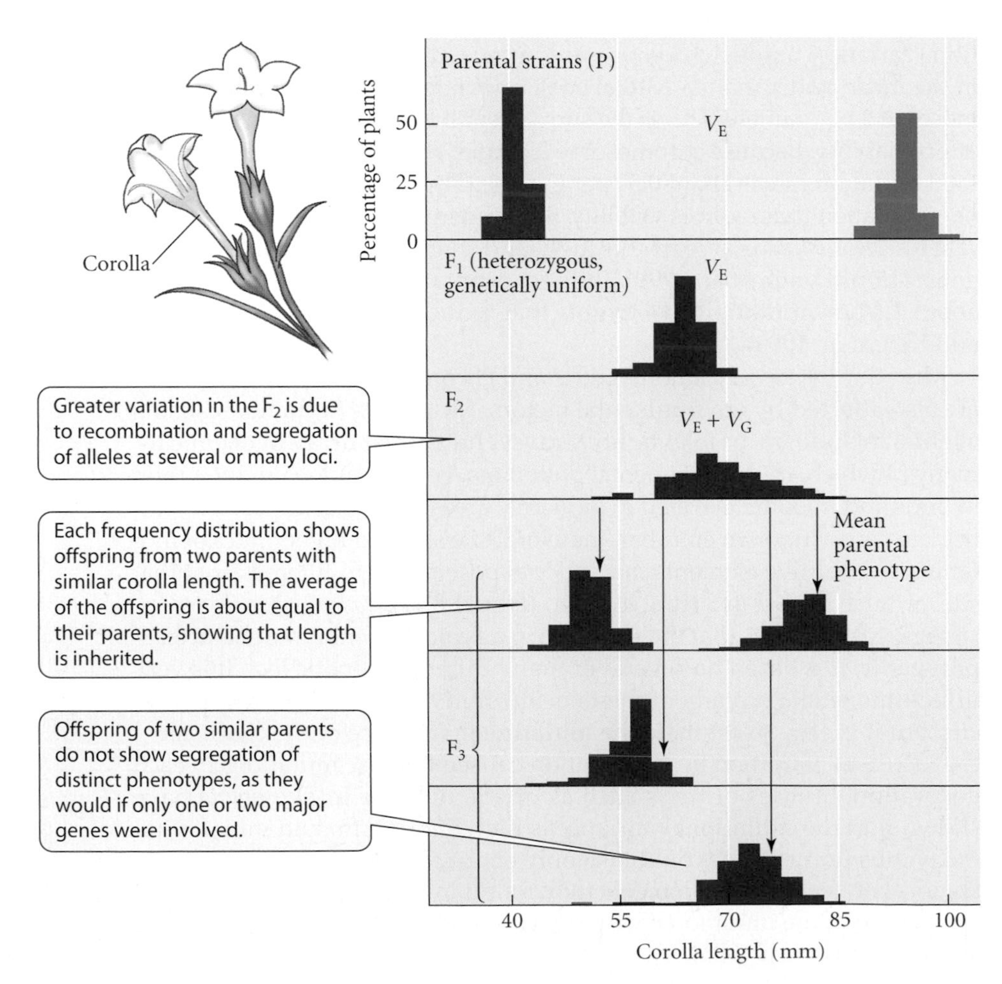

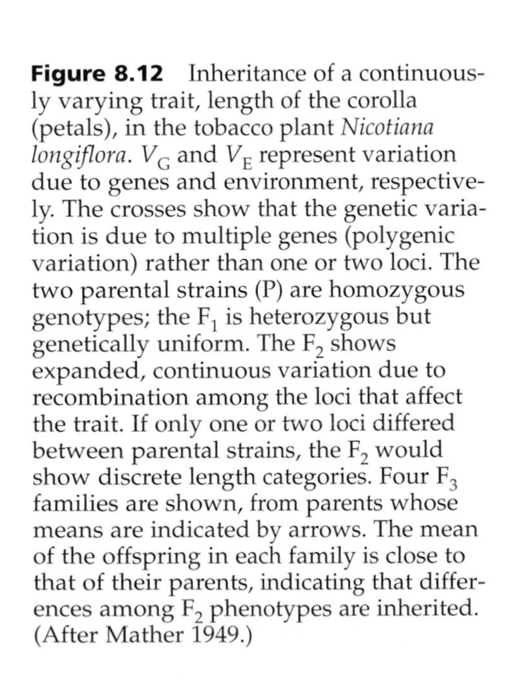

Figure 8.12 Inheritance of a continuously varying trait, length of the corolla (petals), in the tobacco plant *Nicotiana longiflora*. V_G and V_E represent variation due to genes and environment, respectively. The crosses show that the genetic variation is due to multiple genes (polygenic variation) rather than one or two loci. The two parental strains (P) are homozygous genotypes; the F_1 is heterozygous but genetically uniform. The F_2 shows expanded, continuous variation due to recombination among the loci that affect the trait. If only one or two loci differed between parental strains, the F_2 would show discrete length categories. Four F_3 families are shown, from parents whose means are indicated by arrows. The mean of the offspring in each family is close to that of their parents, indicating that differences among F_2 phenotypes are inherited. (After Mather 1949.)

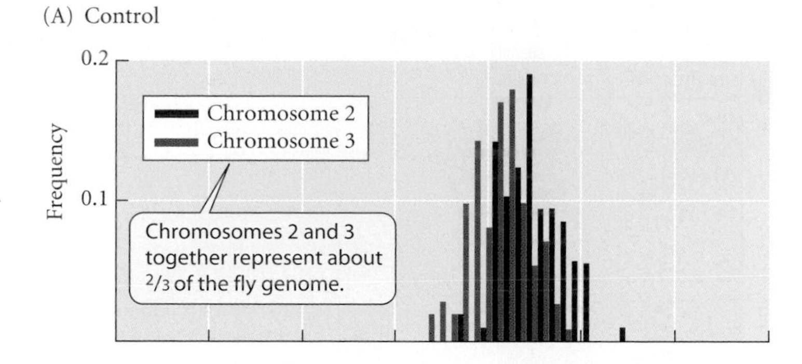
(A) Control
0.2
0.1
Frequency
Chromosome 2
Chromosome 3
Chromosomes 2 and 3 together represent about 2/3 of the fly genome.

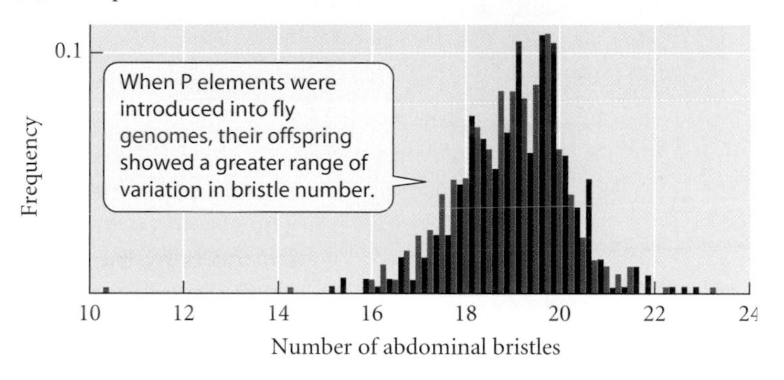
(B) Transposable elements introduced
0.1
Frequency
When P elements were introduced into fly genomes, their offspring showed a greater range of variation in bristle number.
10
12
14
16
18
20
22
24
Number of abdominal bristles

Figure 8.13 The frequency distribution of the number of abdominal bristles in (A) 392 homozygous control lines of *Drosophila melanogaster* and (B) 1094 homozygous experimental lines in which researchers used transposable elements (P elements) to induce mutations in chromosome 2 or chromosome 3. The mutations both increased and decreased bristle numbers compared with the control lines. (From Lyman et al. 1996.)

venient to distinguish between a mutation's effects on fitness and on other characters, even though they are connected.

The phenotypic effects of mutational changes in DNA sequence range from none to drastic. At one extreme, synonymous base pair changes are expected to have no evident phenotypic effect, and this is also apparently true of many amino acid substitutions, which seem not to affect protein function. The phenotypic effects of mutations that contribute to polygenic traits, such as bristle number in *Drosophila*, range from slight to substantial; in one study, mutations induced by insertion of transposable elements altered the number of abdominal bristles by about 0.9 bristle on average (Figure 8.13).

Among the most fascinating mutations are those in the "master control genes" that regulate the expression of other genes in developmental pathways. (We will discuss these genes in detail in Chapter 20.) The HOMEOTIC SELECTOR GENES, for example, determine the basic body plan of an organism, conferring a distinct identity on each segment of the developing body by producing DNA-binding proteins that regulate other genes that determine the features of each such segment. These genes derive their name from **homeotic mutations** in *Drosophila*, which redirect the development of one body segment into that of another. Mutations in the Antennapedia gene complex, for example, cause legs to develop in place of antennae (Figure 8.14). Another master control gene, *Pax6*, switches on about 2500 other genes required for eye development in mammals, insects, and many other animals (Gehring and Ikeo 1999). Mutations in this gene cause malformation or loss of eyes.

Dominance describes the effect of an allele on a phenotypic character when it is paired with another allele in the heterozygous condition. A fully dominant allele (say, A_1) produces nearly the same phenotype when heterozygous (A_1A_2) as when homozygous (A_1A_1), and its partner allele (A_2) in that instance is fully **recessive**. All degrees of INCOMPLETE DOMINANCE, measured by the degree to which the heterozygote resembles one or the other homozygote, may occur. Inheritance is said to be **additive** if the heterozygote's phenotype is precisely intermediate between those of the homozygotes. For example, A_1A_1,

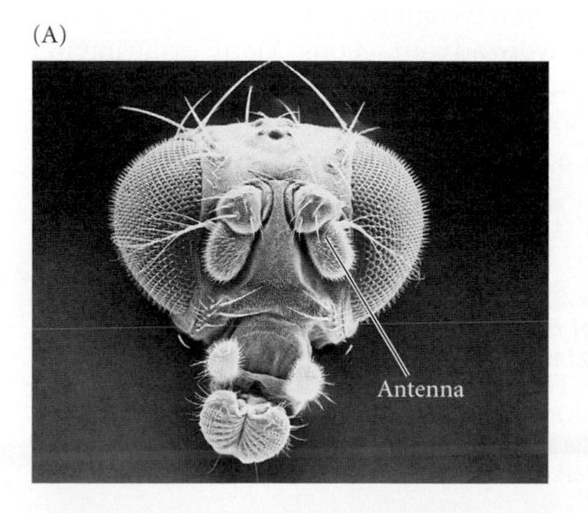
(A)
Antenna

(B)
Leg where antenna should be

Figure 8.14 The drastic phenotypic effect of homeotic mutations that switch development from one pathway to another. (A) Frontal view of the head of a wild-type *Drosophila melanogaster*, showing normal antennae and mouthparts. (B) Head of a fly carrying the *Antennapedia* mutation, which converts antennae into legs. The *Antennapedia* gene is part of a large complex of Hox genes that confer identity on segments of the body (see Chapter 20). (Photographs courtesy of F. R. Turner.)

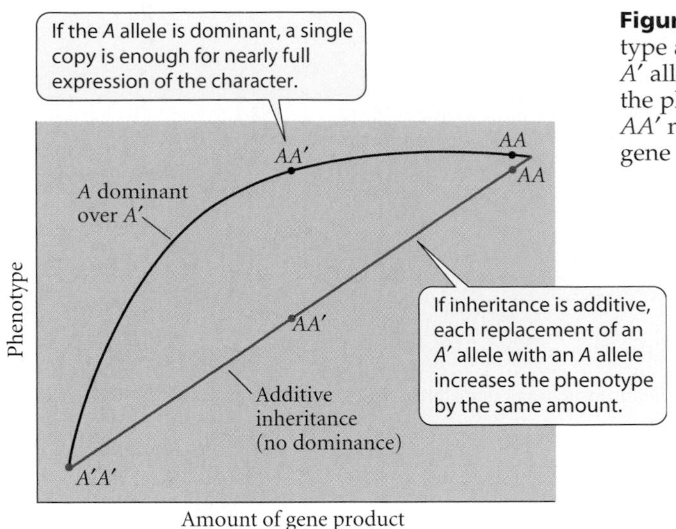
If the A allele is dominant, a single copy is enough for nearly full expression of the character.
AA′
AA
AA
A dominant over A′
Phenotype
AA′
If inheritance is additive, each replacement of an A′ allele with an A allele increases the phenotype by the same amount.
Additive inheritance (no dominance)
A′A′
Amount of gene product

Figure 8.15 Two of the possible relationships between phenotype and genotype at a single locus with two alleles. If inheritance is additive, replacing each *A′* allele with an *A* allele steadily increases the amount of gene product, and the phenotype changes accordingly. If *A* is dominant over *A′*, the phenotype of *AA′* nearly equals that of *AA* because the single dose of *A* produces enough gene product for full expression of the character.

A_1A_2, and A_2A_2 may have phenotypes 3, 2, and 1, respectively; the effects of replacing each A_2 with an A_1 simply add up. LOSS-OF-FUNCTION mutations, in which the activity of a gene product is reduced, are often at least partly recessive, while dominant mutations often have enhanced gene product activity (Figure 8.15).

Effects of mutations on fitness

The effects of new mutations on fitness may range from highly advantageous to highly disadvantageous. Undoubtedly, many mutations are neutral, or nearly so, having very slight effects on fitness (see Chapter 10). The *average*, or net, effect of those that do affect fitness is deleterious. This was shown, for example, by the decline in mean viability in Mukai's *Drosophila* experiment, described above (see Figure 8.10), and by the fitness effects of single mutations isolated in experimental populations of *E. coli* and yeast (Figure 8.16). A few mutations slightly enhanced fitness, some greatly decreased it, and the majority had small deleterious effects. Under some circumstances, slightly deleterious mutations may act as if they are nearly neutral and accumulate in populations. They may therefore have more harmful collective effects on a population than do strongly deleterious mutations, which are more rapidly expunged by natural selection.

The frequency distribution of mutational effects is not fixed, for the fitness consequences of many mutations depend on the population's environment and even on its existing genetic constitution. For example, the decline of fitness due to new mutations in some experimental *Drosophila* populations was more than 10 times greater if the flies were assayed under crowded, competitive conditions than under noncompetitive conditions (Shabalina et al. 1997).

Most mutations are **pleiotropic**—that is, they affect more than one character. For example, the *yellow* mutation in *Drosophila* affects not only body color, but also several components of male courtship behavior. In some cases, the basis of deleterious pleiotropic effects is understood; for example, some mutations that affect *Drosophila* bristle number also disrupt the development of the nervous system and reduce the viability of larvae, which do not have bristles (Mackay et al. 1992).

Evolution would not occur unless some mutations were advantageous. Many experiments that demonstrate advantageous mutations have been done with microorganisms such as phage, bacteria, and yeast because of their short genera-

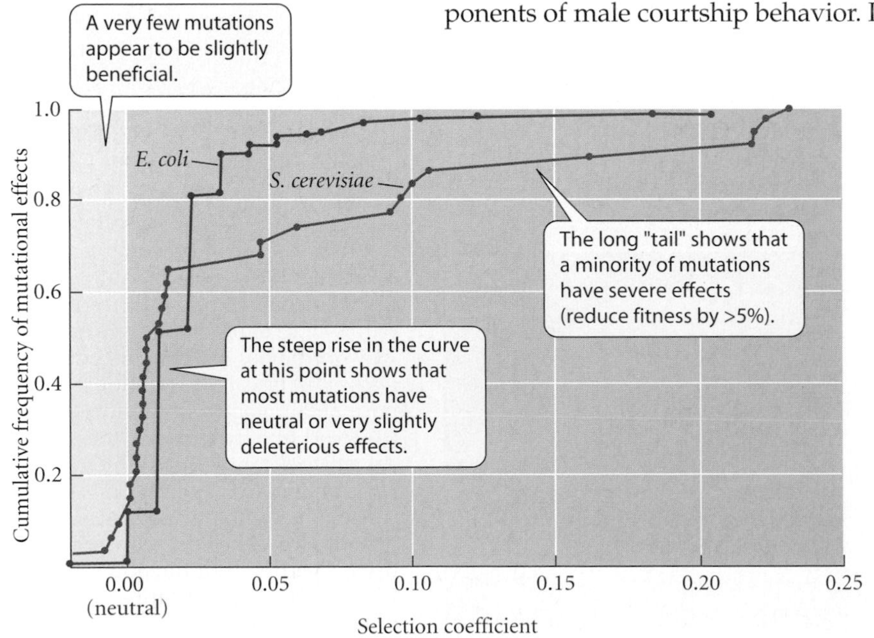
A very few mutations appear to be slightly beneficial.
E. coli
S. cerevisiae
The long "tail" shows that a minority of mutations have severe effects (reduce fitness by >5%).
The steep rise in the curve at this point shows that most mutations have neutral or very slightly deleterious effects.
Cumulative frequency of mutational effects
1.0
0.8
0.6
0.4
0.2
0.00 (neutral)
0.05
0.10
0.15
0.20
0.25
Selection coefficient

Figure 8.16 The cumulative frequency distributions of the effects of new mutations on fitness in the bacterium *Escherichia coli* and the yeast *Saccharomyces cerevisiae*. The higher the selection coefficient, the more the mutation reduces fitness. Beneficial effects are indicated by values to the left of 0.0 (neutral). (After Lynch et al. 1999.)

tion times and the ease with which huge populations can be cultured (Dykhuizen 1990; Elena and Lenski 2003).

Because bacteria can be frozen (during which time they undergo no genetic change) and later revived, samples taken at different times from an evolving population can be stored, and their fitness can later be directly compared. The fitness of a bacterial genotype is defined as its rate of increase in numbers relative to that of another genotype with which it competes in the same culture, but which bears a genetic marker and so can be distinguished from it. Suppose, for example, that a culture is begun with equal numbers of genotypes A and B, and that after 24 hours B is twice as abundant as A. If the bacteria have grown for x generations, each initial cell has produced 2^x descendants. Thus, if genotypes A and B have grown at the respective rates of 2^5 and 2^6 (i.e., B has produced one more generation per 24 hours), their relative numbers are 32:64, or 1:2. The relative fitnesses of the genotypes—that is, their relative growth rates—are measured by their rates of cell division per day: namely, 5:6, or 1.0:1.2. If the genotypes had the same growth rates—say, 2^5—both would increase in number, but their fitnesses would be equal.

Richard Lenski and colleagues used this method to trace the increase of fitness in populations of *E. coli* for an astonishing 20,000 generations. Each population was initiated with a single individual and was therefore genetically uniform at the start. Nevertheless, fitness increased substantially—rapidly at first, but at a decelerating rate later (Figure 8.17A). In a similar experiment (Bennett et al. 1992), *E. coli* populations adapted rapidly to several different temperatures (Figure 8.17B,C,D).

Bacteria can be screened for mutations that affect their biochemical capacities by placing them on a medium on which that bacterial strain cannot grow, such as a medium that lacks an essential amino acid or other nutrient. Whatever colonies do appear on the medium must have grown from the few cells in which mutations occurred that conferred a new biochemical ability. For example, Barry Hall (1982) studied a strain of *E. coli* that lacks the *lacZ* gene, which encodes β-galactosidase, the enzyme that enables *E. coli* to metabolize the sugar lactose as a source of carbon and energy. Hall screened populations for the ability to grow on lactose and recovered several mutations. A mutation in a different gene (*ebg*) altered an enzyme that normally performs another function so that it could break down lactose. Another mutation altered regulation of the *ebg* gene, and a third mutation altered the ebg enzyme so that it metabolized lactose into lactulose, which increased the cell's uptake of lactose from the environment. The three mutations together restored the metabolic capacities that had been lost

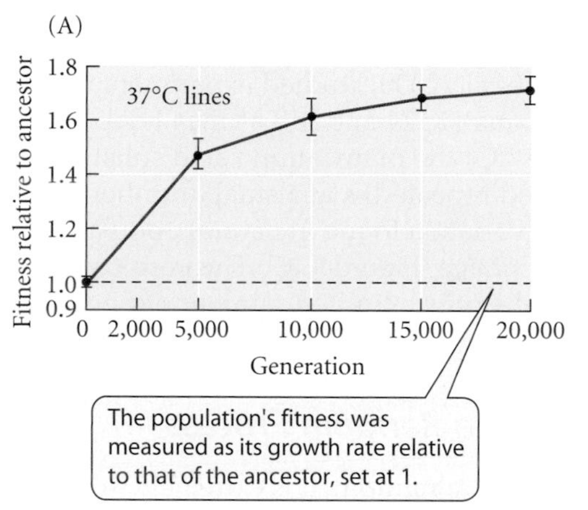

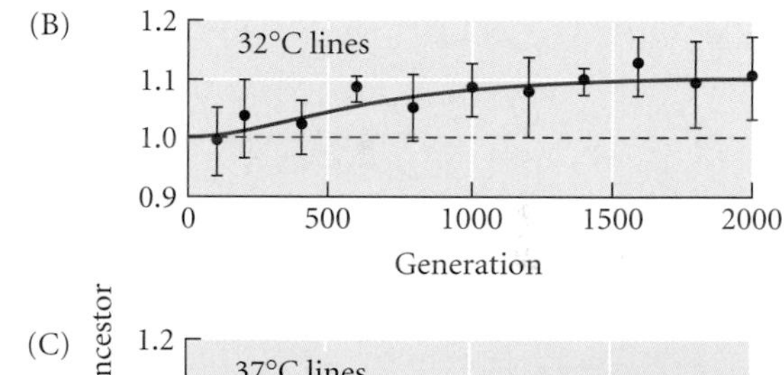

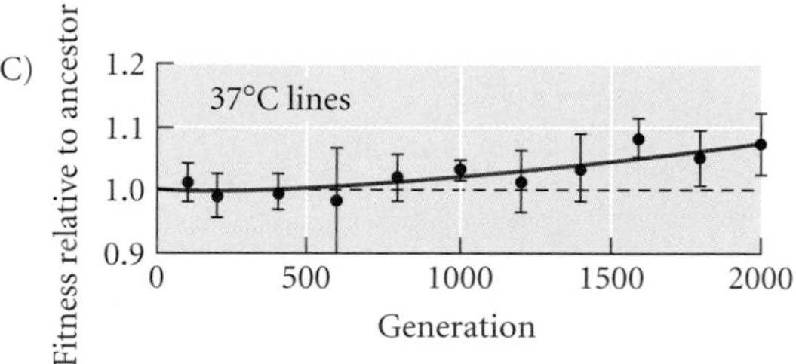

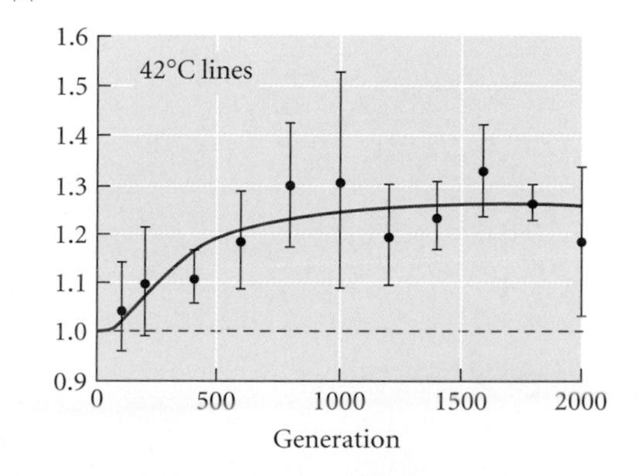

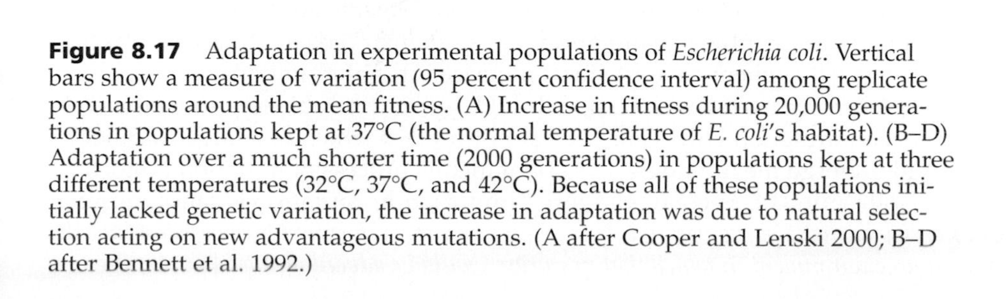

Figure 8.17 Adaptation in experimental populations of *Escherichia coli*. Vertical bars show a measure of variation (95 percent confidence interval) among replicate populations around the mean fitness. (A) Increase in fitness during 20,000 generations in populations kept at 37°C (the normal temperature of *E. coli*'s habitat). (B–D) Adaptation over a much shorter time (2000 generations) in populations kept at three different temperatures (32°C, 37°C, and 42°C). Because all of these populations initially lacked genetic variation, the increase in adaptation was due to natural selection acting on new advantageous mutations. (A after Cooper and Lenski 2000; B–D after Bennett et al. 1992.)

by the deletion of the original *lacZ* gene. Thus mutation and selection in concert can give rise to complex adaptations.

The limits of mutation

It cannot be stressed too strongly that even the most drastic mutations cause alterations of one or more *pre-existing* traits. Mutations with phenotypic effects alter developmental processes, but they cannot alter developmental foundations that do not exist. We may conceive of winged horses and angels, but no mutant horses or humans will ever sprout wings from their shoulders, for the developmental foundations for such wings are lacking.

The direction of evolution may be constrained if some conceivable mutations are more likely to arise and contribute to evolution than others. In laboratory stocks of the green alga *Volvox carteri*, for example, new mutations affecting the relationship between the size and number of germ cells correspond to the typical state of these characters in other species of *Volvox* (Koufopanou and Bell 1991).

Mutation may not constrain the rate or direction of evolution very much if many different mutations can generate a particular phenotype, as may be the case when several or many loci affect the trait (polygeny). For example, when different copper-tolerant populations of the monkeyflower *Mimulus guttatus* are crossed, the variation in copper tolerance is greater in the F_2 generation than within either parental population, indicating that the populations differ in the loci that confer tolerance (Cohan 1984).

Nevertheless, certain advantageous phenotypes can apparently be produced by mutation at only a very few loci, or perhaps only one. In such instances, the supply of rare mutations might limit the capacity of species for adaptation. The rarity of necessary mutations may help to explain why species have not become adapted to a broader range of environments, or why, in general, species are not more adaptable than they are (Bradshaw 1991). For instance, resistance to the insecticide dieldrin in different populations of *Drosophila melanogaster* is based on repeated occurrences of the same mutation, which, moreover, is thought to represent the same gene that confers dieldrin resistance in flies that belong to two other families (ffrench-Constant et al. 1990). This may mean that very few genes—perhaps only this one—undergo mutations that can confer dieldrin resistance, and that such a mutation is a very rare event.

Wichman et al. (2000) studied experimental populations of the closely related bacteriophage strains ϕX174 and S13 as they adapted to two species of host bacteria at high temperatures. Of the many amino acid substitutions that occurred in the populations, most occurred repeatedly, at a small number of sites, and many of those substitutions matched the variation found in natural populations and even the differences between the two kinds of phage (Figure 8.18). This result suggests that the natural evolution of these phage can take only a limited number of pathways, constrained by the possible kinds of advantageous mutations (Wichman et al. 2000).

Mutation as a Random Process

Mutations occur at random. It is extremely important to understand what this statement does and does not mean. It does not mean that all conceivable mutations are equally likely to occur, because, as we have noted, the developmental foundations for some imaginable transformations do not exist. It does not mean that all loci, or all regions within a locus, are equally mutable, for geneticists have described differences in mutation rates, at both the phenotypic and molecular levels, among and within loci. It does not mean that environmental factors cannot influence mutation rates: as remarked earlier, ultraviolet and other radiation, as well as chemical mutagens, increase rates of mutation.

Mutation *is* random in two senses. First, although we may be able to predict the *probability* that a certain mutation will occur, we cannot predict which of a large number of gene copies will undergo the mutation. The spontaneous process of mutation is stochastic rather than deterministic. Second, and more importantly, mutation is random in the sense that *the chance that a particular mutation will occur is not influenced by whether or not the organism is in an environment in which that mutation would be advantageous*. That is, the en-

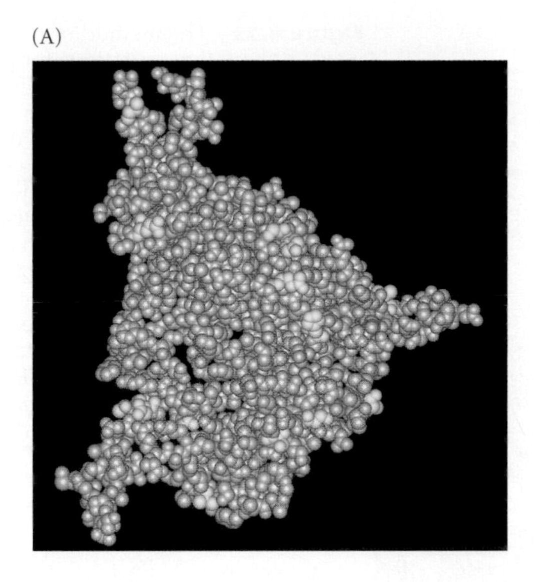

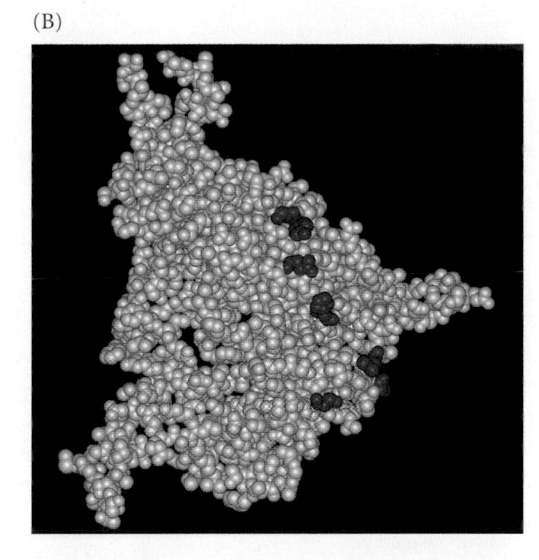

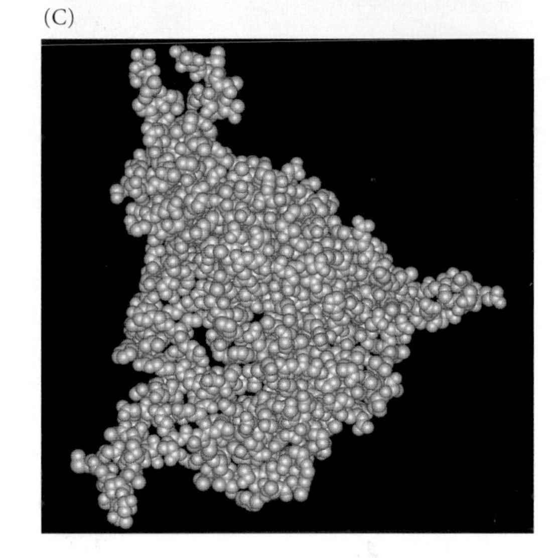

Figure 8.18 The surface of the major capsid protein (gpF) of phage strains ϕX174 and S13, showing parallel amino acid substitutions. (A) Amino acids that underwent substitution in the experimental lines are shown in yellow. (B) Amino acids shown in red are known to affect the fitness of wild phage in either of two bacterial host species. (C) Amino acids that represent differences between the two original strains, ϕX174 and S13, are shown in blue. Note that all the differences that have naturally evolved between these phage strains also occurred in the experimental lines. (After Wichman et al. 2000.)

vironment does not induce adaptive mutations. Indeed, it is hard to imagine a mechanism whereby most environmental factors could direct the mutation process by dictating that just the right base pair changes should occur.

The argument that adaptively directed mutation does not occur is one of the fundamental tenets of modern evolutionary theory. If it did occur, it would introduce a Lamarckian element into evolution, for organisms would then acquire adaptive hereditary characteristics in response to their environment. Such "neo-Lamarckian" ideas were expunged in the 1940s and 1950s by experiments with bacteria in which spontaneous, random mutation followed by natural selection, rather than mutation directed by the environment, explained adaptation.

One of these experiments was performed by Joshua and Esther Lederberg (1952), who showed that advantageous mutations occur without exposure to the environment in which they would be advantageous to the organism. The Lederbergs used the technique of REPLICA PLATING (Figure 8.19). Using a culture of *E. coli* derived from a single cell, the Lederbergs spread cells onto a "master" agar plate, without penicillin. Each cell gave rise to a distinct colony. They pressed a velvet cloth against the plate, and then touched the cloth to a new plate with medium containing the antibiotic penicillin, thereby transferring some cells from each colony to the replica plate, in the same spatial relationships as the colonies from which they had been taken. A few colonies appeared on the replica plate, having grown from penicillin-resistant mutant cells. When all the colonies on the master plate were tested for penicillin resistance, those colonies (and only those colonies) that had been the source of penicillin-resistant cells on the replica plate displayed resistance, showing that the mutations had occurred before the bacteria were exposed to penicillin.

Because of such experiments, biologists have generally accepted that mutation is adaptively random rather than directed. However. several investigators have reported results, again with *E. coli*, that at face value seem to suggest that some advantageous mutations might be directed by the environment. Their interpretations have been challenged by other investigators, and there does not appear to be convincing evidence for directed mutation (Sniegowski and Lenski 1995; Brisson 2003).

Recombination and Variation

All genetic variation owes its origin ultimately to mutation, but in the short term, a great deal of the genetic variation within populations arises through recombination. In sexually reproducing eukaryotes, genetic variation arises from two processes: the union of ge-

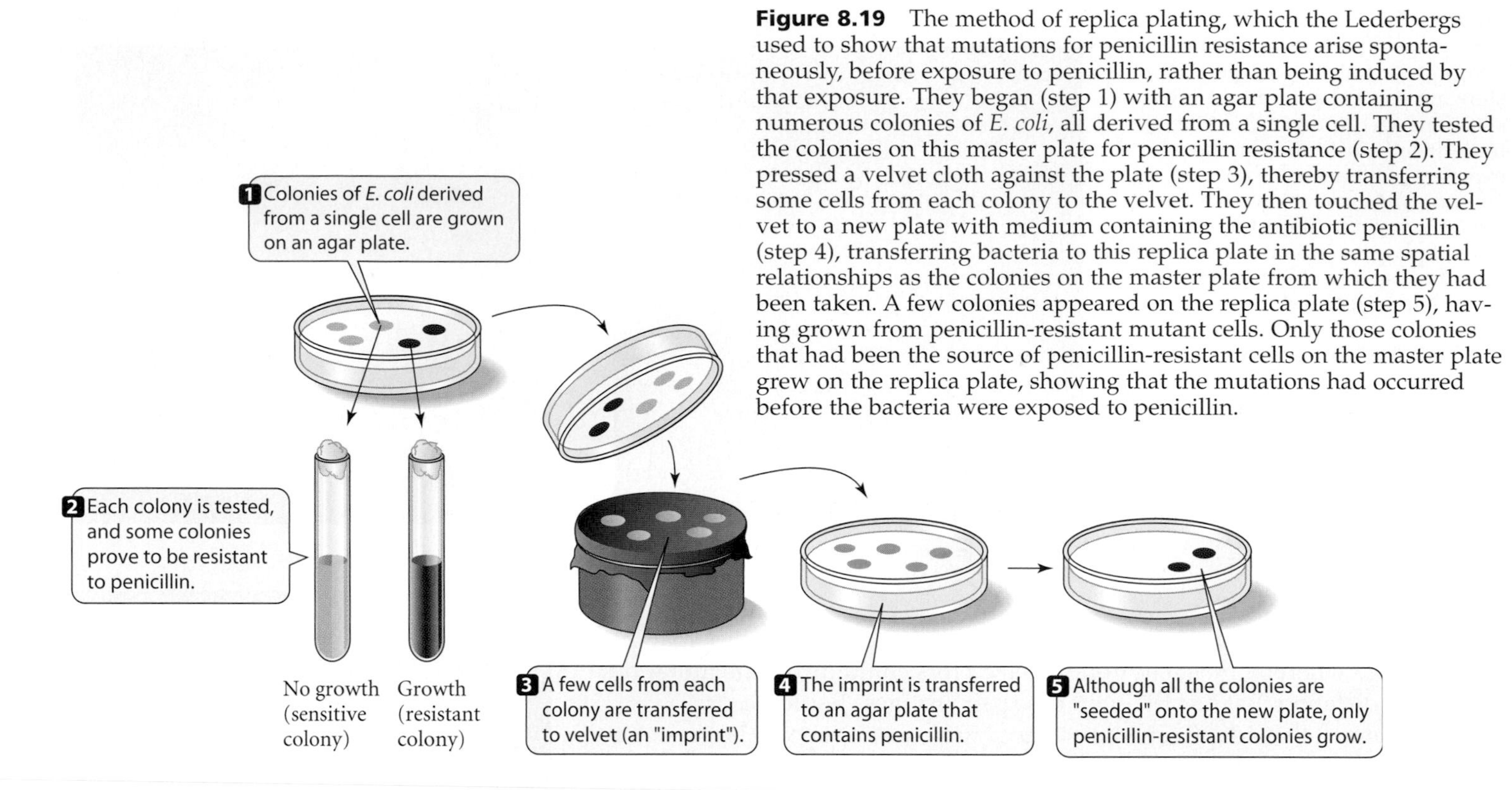

1 Colonies of E. coli derived from a single cell are grown on an agar plate.
2 Each colony is tested, and some colonies prove to be resistant to penicillin.
No growth (sensitive colony)
Growth (resistant colony)
3 A few cells from each colony are transferred to velvet (an "imprint").
4 The imprint is transferred to an agar plate that contains penicillin.
5 Although all the colonies are "seeded" onto the new plate, only penicillin-resistant colonies grow.

Figure 8.19 The method of replica plating, which the Lederbergs used to show that mutations for penicillin resistance arise spontaneously, before exposure to penicillin, rather than being induced by that exposure. They began (step 1) with an agar plate containing numerous colonies of *E. coli*, all derived from a single cell. They tested the colonies on this master plate for penicillin resistance (step 2). They pressed a velvet cloth against the plate (step 3), thereby transferring some cells from each colony to the velvet. They then touched the velvet to a new plate with medium containing the antibiotic penicillin (step 4), transferring bacteria to this replica plate in the same spatial relationships as the colonies on the master plate from which they had been taken. A few colonies appeared on the replica plate (step 5), having grown from penicillin-resistant mutant cells. Only those colonies that had been the source of penicillin-resistant cells on the master plate grew on the replica plate, showing that the mutations had occurred before the bacteria were exposed to penicillin.

netically different gametes, and the formation of gametes with different combinations of alleles, owing to independent segregation of nonhomologous chromosomes and to crossing over between homologous chromosomes.

The potential genetic variation that can be released by recombination is enormous. To cite a modest example: if an individual is heterozygous for only one locus on each of five pairs of chromosomes, independent segregation alone generates $2^5 = 32$ allele combinations among its gametes, and mating between two such individuals could give rise to $3^5 = 243$ genotypes among their progeny. If each locus affects a different feature, this represents a great variety of character combinations. If all five loci have equal and additive effects on a single polygenic character, such as size, the range of variation among the offspring can greatly exceed the difference between the parents. For instance, if each substitution of + and – alleles in a genotype adds or subtracts one unit of phenotype, two quintuply heterozygous parents (both +–+–+/–+–+–), both of size 20, could have offspring ranging in size from 15 (–/– at all five loci) to 25 (+/+ at all five loci). (Compare the F_1 and F_2 distribution of corolla lengths in Figure 8.12.)

In order to judge how much variation is released by recombination, a team led by the great population geneticist Theodosius Dobzhansky studied the effects of chromosomes they had "extracted" (by a series of crosses; see Figure 9.8) from a wild population of *Drosophila pseudoobscura* (Spassky et al. 1958). Homozygous chromosomes from natural populations of this species show tremendous variation in their effects on survival from egg to adult (see Figure 9.9). However, the Dobzhansky team chose 10 homologous chromosomes that conferred almost the same, nearly normal viability when homozygous and made all possible crosses between flies bearing those chromosomes. From the F_1 female offspring, in which crossing over had occurred, they then extracted recombinant chromosomes and measured their effect on viability when homozygous. Even though the original 10 chromosomes had differed little in their effect on viability, the variance in viability among the recombinant chromosomes was more than 40 percent of the variance found among homozygotes for much larger samples of chromosomes from natural populations. Thus a single episode of recombination among just 10 chromosomes generates a large fraction of a wild population's variability. Some of the recombinant chromosomes were "synthetic lethals," meaning that recombination between two chromosomes that yield normal viability produced chromosomes that were lethal when they were made homozygous. This finding implies that each of the original chromosomes carried an allele that did not lower viability on its own, but did cause death when combined with another allele, at another locus, on the other chromosome.

Recombination can both increase and decrease genetic variation. In sexually reproducing populations, genes are transmitted to the next generation, but genotypes are not: they end with organisms' deaths, and are reassembled anew in each generation. Thus an unusual, favorable gene combination may occasionally arise through recombination, but if the individuals bearing it mate with other members of the population, it will be lost immediately by the same process. Likewise, if some individuals have mostly + alleles that increase body size (+++/+++) and others have mostly – alleles that decrease it (– – –/– – –), the population will display considerable variation in body size. But, given recombination, most offspring will inherit mixtures of + and – alleles (e.g., +–+/– –+), and will have fairly similar intermediate sizes. (Compare the P generation with F_1 and F_2 in Figure 8.12.)

Recombination, therefore, has complicated effects on variation: it both retards adaptation by breaking down favorable gene combinations and enhances adaptation by providing natural selection with multitudinous combinations of alleles that have arisen by mutation.

Alterations of the Karyotype

An organism's **karyotype** is the description of its complement of chromosomes: their number, size, shape, and internal arrangement. In considering alterations of the karyotype, it is important to bear in mind that the loss of a whole chromosome, or a major part of a chromosome, usually reduces the viability of a gamete or an organism because of the loss of genes. Furthermore, a gamete or organism often is inviable or fails to develop properly if it has an **aneuploid**, or "unbalanced," chromosome complement—for example, if a normally diploid organism has three copies of one of its chromosomes. (For instance, humans with three copies of chromosome 21, a condition known as Down syndrome or trisomy-21, have brain and other defects.)

As we have seen, chromosome structure may be altered by duplications and deletions, which change the amount of genetic material (see Figure 8.5). Other alterations of the karyotype are changes in the number of whole sets of chromosomes (**polyploidy**) and rearrangements of one or more chromosomes.

Polyploidy

A diploid organism has two entire sets of homologous chromosomes (2*N*); a polyploid organism has more than two. (In discussing chromosomes, *N* refers to the number of different chromosomes in the gametic, or haploid, set, and the numeral refers to the number of representatives of each autosome.) Polyploids can be formed in several ways, especially when failure of the reduction division in meiosis produces diploid, or unreduced, gametes (Ramsey and Schemske 1998). The union of an unreduced gamete (with 2*N* chromosomes) and a reduced gamete (with *N* chromosomes) yields a TRIPLOID (3*N*) zygote. Triploids produce few offspring because most of their gametes have aneuploid chromosome complements. At segregation, each daughter cell may receive one copy of certain chromosomes and two copies of certain others (Figure 8.20A). However, tetraploid (4*N*) offspring can be formed if an unreduced (3*N*) gamete of a triploid unites with a normal gamete (*N*) of a diploid—or if two diploid gametes, whether from triploid or diploid parents, unite. Other such unions can form hexaploids (6*N*), octoploids (8*N*), or genotypes of even higher ploidy.

Each set of four homologous chromosomes of a tetraploid may be aligned during meiosis into a quartet (quadrivalent), and then may segregate in a balanced (two by two) or unbalanced (one by three) fashion (Figure 8.20B). In some such polyploids, aneuploid gametes may result and fertility may be greatly reduced. In other cases, the four chromosomes align not as a quartet, but as two pairs that segregate normally, resulting in bal-

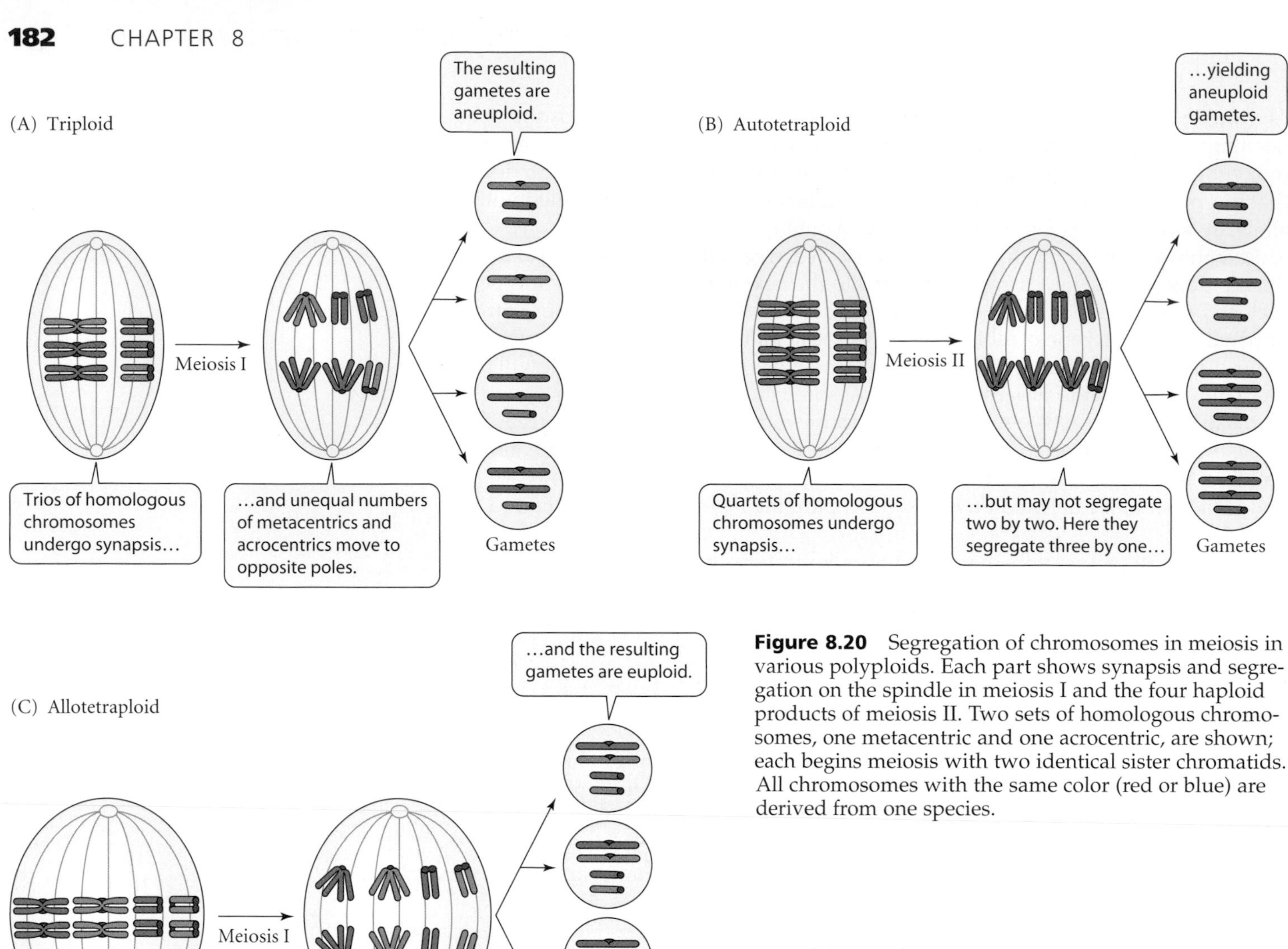
(A) Triploid
The resulting gametes are aneuploid.
Meiosis I
Trios of homologous chromosomes undergo synapsis...
...and unequal numbers of metacentrics and acrocentrics move to opposite poles.
Gametes
(B) Autotetraploid
...yielding aneuploid gametes.
Meiosis II
Quartets of homologous chromosomes undergo synapsis...
...but may not segregate two by two. Here they segregate three by one...
Gametes
(C) Allotetraploid
...and the resulting gametes are euploid.
Meiosis I
Each chromosome pairs with a single homologue from the same parental species.
Segregation is then normal...
Gametes

Figure 8.20 Segregation of chromosomes in meiosis in various polyploids. Each part shows synapsis and segregation on the spindle in meiosis I and the four haploid products of meiosis II. Two sets of homologous chromosomes, one metacentric and one acrocentric, are shown; each begins meiosis with two identical sister chromatids. All chromosomes with the same color (red or blue) are derived from one species.

anced (**euploid**), viable gametes, so that fertility is normal, or nearly so. This would seem to require that the chromosomes be differentiated so that each can recognize and pair with a single homologue rather than with three others.

Many species of plants and a few species of trout, tree frogs, and other animals have arisen by polyploidy (see Chapter 16). Estimates of the proportion of angiosperms that are polyploid range up to 50–70 percent (Stace 1989). Some recently arisen polyploids have arisen by the union of unreduced gametes of the same species; these organisms are known as **autopolyploids**. But the majority are **allopolyploids**, which have arisen by hybridization between closely related species. In allopolyploids, most of the parental species' chromosomes are different enough for the chromosomes from each parent to recognize and pair with each other, so that meiosis in an allotetraploid, for example, involves normal segregation of pairs rather than quartets of chromosomes (Figure 8.20C).

Chromosome rearrangements

Changes in the structure of chromosomes constitute another class of karyotypic alterations. These changes are caused by breaks in chromosomes, followed by rejoining of the pieces in new configurations. Some such changes can affect the pattern of segregation in meiosis, and therefore affect the proportion of viable gametes. Although most chromosome rearrangements seem not to have direct effects on morphological or other phenotypic features, an alteration of gene sequence sometimes brings certain genes under the influence of the control regions of other genes, and so alters their expression. It is not certain that such "position effects" have contributed to evolutionary change. Individual organisms may be homozygous or heterozygous for a rearranged chromosome, and are sometimes referred to as **homokaryotypes** or **heterokaryotypes** respectively.

INVERSIONS. Consider a segment of a chromosome in which ABCDE denotes a sequence of markers such as genes. If a loop is formed, and breakage and reunion occur at the point of overlap, a new sequence, such as A<u>DCB</u>E, may be formed. (The inverted sequence is underlined.) Such an **inversion**, with a rearranged gene order, is PERICENTRIC if it includes the centromere and PARACENTRIC if it does not.

During meiotic synapsis in an inversion heterozygote, alignment of the genes on the normal and inverted chromosomes requires the formation of a loop, which can sometimes be observed under the microscope (Figure 8.21A). Now suppose that in a paracentric inversion, crossing over occurs between loci such as B and C (Figure 8.21B). Two of the four strands are affected. One strand lacks certain gene regions (A), and also lacks a centromere; it will not migrate to either pole, and is lost. The other affected strand not only lacks some genetic material, but also has two centromeres, so the chromosome breaks when these centromeres are pulled to opposite poles. The resulting daughter cells lack certain gene regions, and will not form viable gametes. Consequently, in inversion heterokaryotypes (but not in homokaryotypes), *fertility is reduced* because many gametes are inviable, and *recombination is effectively suppressed* because gametes carrying the recombinant chromosomes, which lack some genetic material, are inviable.

In *Drosophila* and some other flies (Diptera), however, the incomplete recombinant chromosomes enter the polar bodies during meiosis, so that female fecundity is not reduced. It is particularly easy to study inversions in *Drosophila* and some other flies because the larval salivary glands contain giant (polytene) chromosomes that remain in a state of permanent synapsis (so that inversion loops are easily seen), and because these chromosomes display bands, each of which apparently corresponds to a single gene. The banding patterns are as distinct as the bar codes on supermarket products, so an experienced investigator can identify different sequences. INVERSION POLYMORPHISMS are common in *Drosophila*—more than 20 different arrangements of chromosome 3 have been described for *Drosophila pseudoobscura*, for example—and they have been extensively studied from both population genetic and phylogenetic points of view.

TRANSLOCATIONS. By breakage and reunion, two nonhomologous chromosomes may exchange segments, resulting in a **reciprocal translocation** (Figure 8.22). Meiosis in a translocation heterokaryotype often results in a high proportion of aneuploid gametes, so the fertility of translocation heterokaryotypes is often reduced by 50 percent or more. Consequently, polymorphism for translocations is rare in natural populations. Nevertheless, related species sometimes differ by translocations, which have the effect of moving groups of genes from one chromosome to another. The Y chromosome of the male *Drosophila miranda*, for ex-

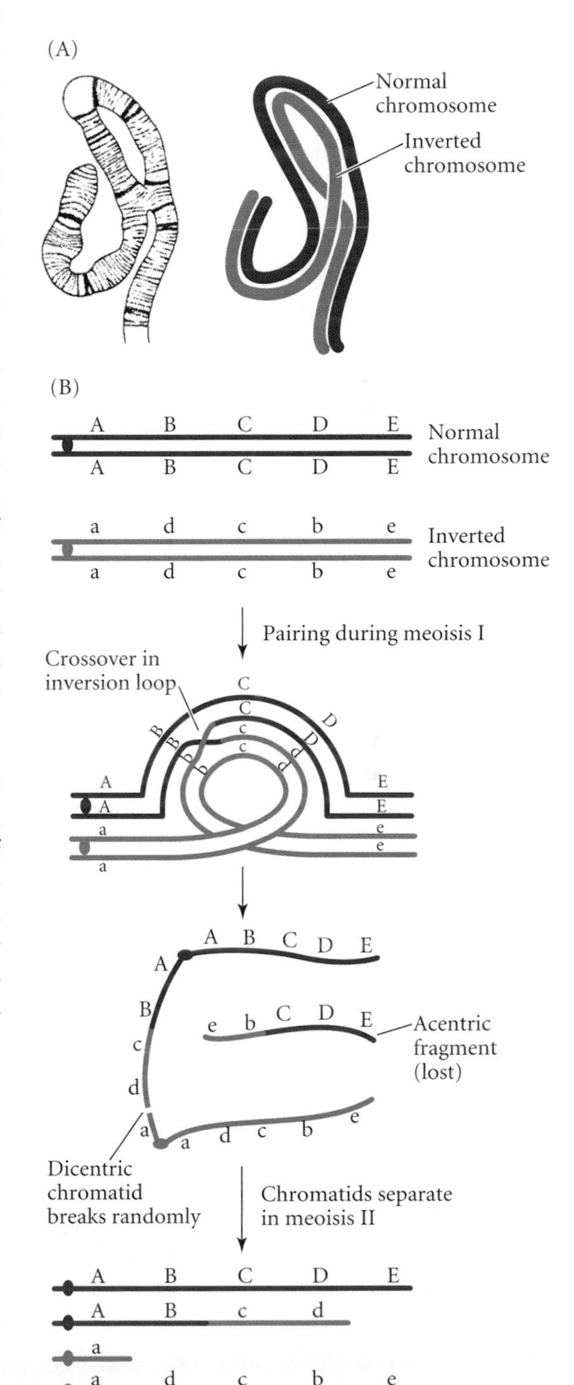

Figure 8.21 Chromosome inversions. (A) Synapsed chromosomes in a salivary gland cell of a larval *Drosophila pseudoobscura* heterozygous for *Standard* and *Arrowhead* arrangements. The two homologous chromosomes are so tightly synapsed that they look like a single chromosome. The "bridge" forms the loop shown in the diagram. Similar synapsis occurs in germ line cells undergoing meiosis. (B) Two homologous chromosomes differing by an inversion of the region B–D, and their configuration in synapsis. Crossing over between two chromatids (between B and C) yields products that lack a centromere or substantial blocks of genes. Because these products do not form viable gametes, crossing over appears to be suppressed. Only cells that receive the two chromatids that do not cross over become viable gametes. (After Strickberger 1968.)

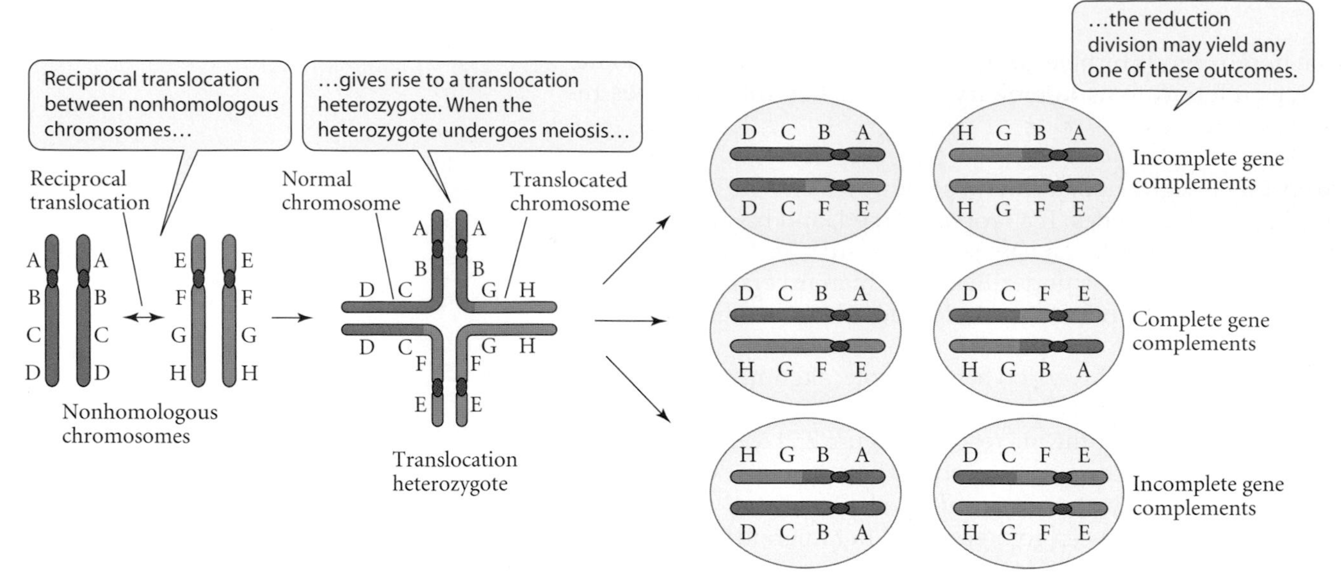

Figure 8.22 Reciprocal translocation between two nonhomologous chromosomes gives rise to a translocation heterozygote. When the heterozygote undergoes meiosis, many of the products will have incomplete gene complements.

ample, includes a segment that is homologous to part of one of the autosomes of closely related species.

FISSIONS AND FUSIONS. It is useful to distinguish ACROCENTRIC chromosomes, in which the centromere is near one end, from METACENTRIC chromosomes, in which the centromere is somewhere in the middle and separates the chromosome into two arms. In the simplest form of chromosome FUSION, two nonhomologous acrocentric chromosomes undergo reciprocal translocation near the centromeres so that they are joined into a metacentric chromosome (Figure 8.23A). More rarely, a metacentric chromosome may undergo **fission**. A simple fusion heterokaryotype has a metacentric, which we may refer to as AB, with arms that are homologous to two acrocentrics A and B. AB, A, and B together synapse as a "trivalent" (Figure 8.23B). Viable gametes and zygotes are often formed, but the frequency of aneuploid gametes can be quite high, especially for more complex patterns of fusion (see Chapter 16). Differences in chromosome number due to fusions often distinguish related species or geographic populations of the same species.

CHANGES IN CHROMOSOME NUMBER. Summarizing what we have covered so far, polyploidy (especially in plants), translocations, and fusions or fissions of chromosomes are the mu-

Figure 8.23 (A) A simple fusion of two acrocentric chromosomes to form arms A and B. (B) Segregation in meiosis of a fusion heterozygote can yield euploid (balanced) or aneuploid (unbalanced) complements of genetic material.

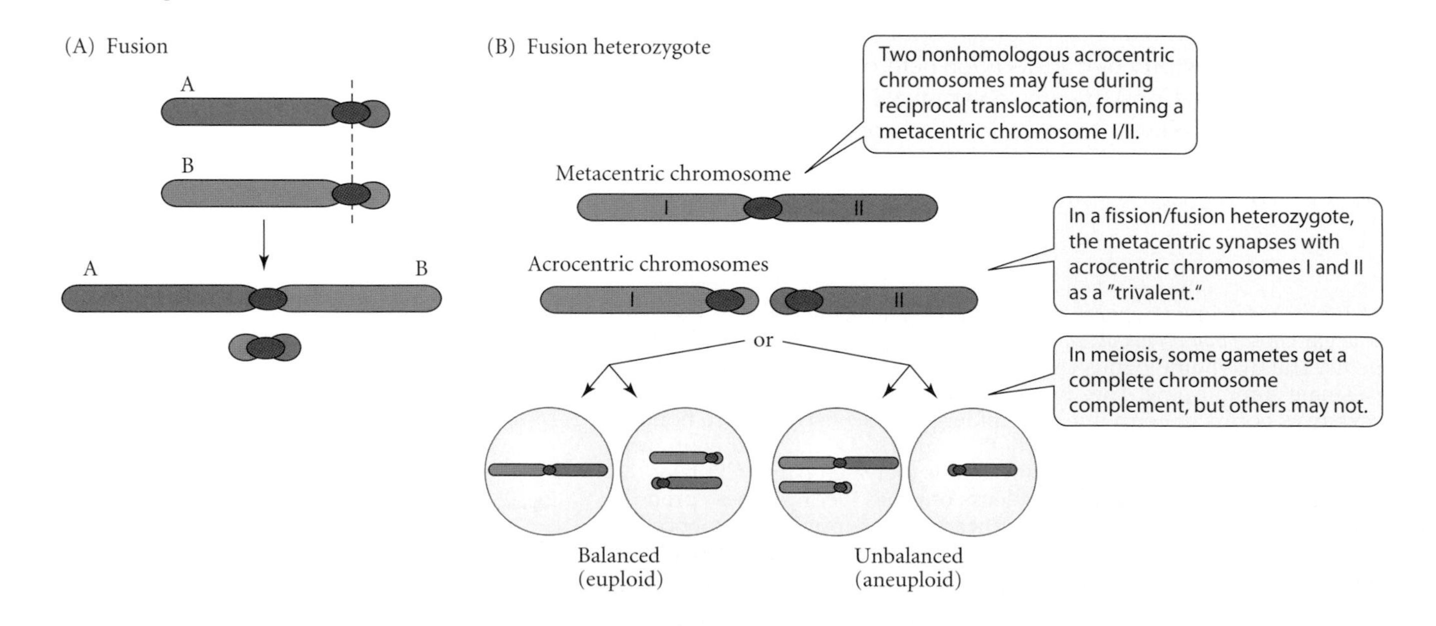

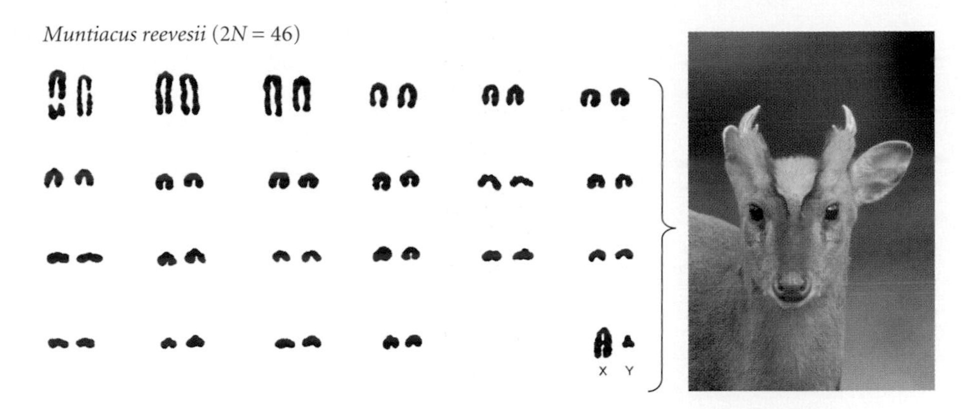

Figure 8.24 The diploid chromosome complements (taken from micrographs) of two closely related species of muntjacs (barking deer) represent one of the most extreme differences in karyotype known among closely related species. Despite the difference in karyotype, the species are phenotypically very similar. (*M. reevesii* photo © Mike Lane/Alamy Images; *M. muntiacus* © OSF/photolibrary.com)

tational foundations for the evolution of chromosome number. For example, the haploid chromosome number varies among mammals from 3 to 42 (Lande 1979), and among insects from 1 in an ant species to about 220 in some butterflies (the highest number known in animals). Related species sometimes differ strikingly in karyotype: in one of the most extreme examples, two very similar species of barking deer, *Muntiacus reevesii* and *M. muntiacus*, have haploid chromosome numbers of 23 and 3 or 4 (in different populations), respectively (Figure 8.24). Like that of all characteristics, evolution of the karyotype requires not only mutation, but other processes as well (see Chapter 16).

Muntiacus muntiacus (2*N* = 8)

The spontaneous rate of origin of any given class of chromosome rearrangement (e.g., reciprocal translocation) is quite high: about 10^{-4} to 10^{-3} per gamete (Lande 1979). However, a rearrangement involving breakage at any particular site(s) rarely arises, and is usually considered to be unique.

Summary

1. Mutations of chromosomes or genes are alterations that are subsequently replicated. They ordinarily do not constitute new species, but rather variant chromosomes or genes (alleles, haplotypes) within a species.
2. At the molecular level, mutations of genes include base pair substitutions, frameshifts, duplications and deletions of one or more base pairs (or of longer sequences that may include entire genes), and changes caused by insertion of various kinds of transposable elements. New DNA sequences also arise by intragenic recombination.
3. The rate at which any particular mutation arises is quite low: on average, about 10^{-6} to 10^{-5} per gamete for mutations detected by their phenotypic effects, and about 10^{-9} per base pair. The mutation rate, by itself, is too low to cause substantial changes of allele frequencies. However, the total input of genetic variation by mutation, for the genome as a whole or for individual polygenic characters, is appreciable.
4. The magnitude of change in morphological or physiological features caused by a mutation can range from none to drastic. In part because most mutations have pleiotropic effects, the *average* effect of mutations on fitness is deleterious, but some mutations are advantageous.
5. Mutations alter pre-existing biochemical or developmental pathways, so not all conceivable mutational changes are possible. Some adaptive changes may not be possible without just the right mutation of just the right gene. For these reasons, the rate and direction of evolution may in some instances be affected by the availability of mutations.
6. Mutations appear to be random, in the sense that their probability of occurrence is not directed by the environment in favorable directions, and in the sense that specific mutations cannot be predicted. The likelihood that a mutation will occur does not depend on whether or not it would be advantageous.
7. Recombination of alleles can potentially give rise to astronomical numbers of gene combinations, and in sexually reproducing organisms generates far more genetic variation per

generation than mutation alone. However, recombination also breaks apart favorable gene combinations and constrains the amount of variation displayed by polygenic characters.

8. Mutations of the karyotype (chromosome complement) include polyploidy and structural rearrangements (e.g., inversions, translocations, fissions, fusions). Many such rearrangements reduce fertility in the heterozygous condition.

Terms and Concepts

additive inheritance
allele
allopolyploid
aneuploid
autopolyploid
back mutation
base pair substitution
base pairs (bp)
codon
control regions (enhancers and repressors)
dominance
euploid
exon
fission (of chromosomes)
frameshift mutation
fusion (of chromosomes)
gene
gene family
genetic code
genetic marker
haplotype
heterokaryotype
homeotic mutation
homokaryotype
intragenic recombination
intron
inversion
karyotype
microsatellite
mutation
mutational variance
nonsynonymous mutation
pleiotropy
point mutation
polygenic
polyploidy
pseudogene
recessive
reciprocal translocation
recurrent mutation
repeated sequence
substitution
synonymous mutation
tandem repeat
transition
transposable element (TE)
transposition
transversion
unequal crossing over
wild-type

Suggestions for Further Reading

W.-H. Li, *Molecular Evolution* (Sinauer Associates, Sunderland, MA, 1997), treats evolutionary aspects of mutation at the molecular level. Reviews of mutation rates include J. W. Drake, B. Charlesworth, D. Charlesworth, and J. F. Crow, "Rates of spontaneous mutation"(1998, *Genetics* 148:1667–1686) and M. Lynch et al., "Perspective: Spontaneous deleterious mutation" (1999, *Evolution* 53: 645–663).

Problems and Discussion Topics

1. Consider two possible studies. (*a*) In one, you capture 3000 wild male *Drosophila melanogaster*, mate each with laboratory females heterozygous for the autosomal recessive allele *vg* (*vestigial*, which causes miniature wings when homozygous), and examine each male's offspring. You find that half the offspring of each of three males have miniature wings and have genotype *vgvg*. (*b*) In another study, you determine the nucleotide sequence of 1000 base pairs for 20 copies of the cytochrome *b* gene, taken from 20 wild mallard ducks. You find that at each of 30 nucleotide sites, one or another gene copy has a different base pair from all others. From these data, can you estimate the rate of mutation from the wild type to the *vg* allele (case *a*) or from one base pair to another (case *b*)? Why or why not?
2. From a laboratory stock of *Drosophila* that you believe to be homozygous wild type (++) at the *vestigial* locus, you obtain 10,000 offspring, mate each of them with homozygous *vg vg*

flies, and examine a total of 1 million progeny. Two of these are *vgvg*. Estimate the rate of mutation from + to *vg* per gamete. What assumptions must you make?

3. The following DNA sequence represents the beginning of the coding region of the alcohol dehydrogenase (*Adh*) gene of *Drosophila simulans* (Bodmer and Ashburner 1984), arranged into codons:

   CCC ACG ACA GAA CAG TAT TTA AGG AGC TGC GAA GGT

   (*a*) Find the corresponding mRNA sequence, and use Figure 8.2 to find the amino acid sequence. (*b*) Again using Figure 8.2, determine for each site how many possible mutations (changes of individual nucleotides) would cause an amino acid change, and how many would not. For the entire sequence, what proportion of possible mutations are synonymous versus nonsynonymous? What proportion of the possible mutations at first, second, and third base positions within codons are synonymous? (*c*) What would be the effect on the amino acid sequence of inserting a single base, G, between sites 10 and 11 in the DNA sequence? (*d*) What would be the effect of deleting nucleotide 16? (*e*) For the first 15 (or more) sites, classify each possible mutation as a transition or transversion, and determine whether or not the mutation would change the amino acid. Does the proportion of synonymous mutations differ between transitions and transversions?

4. A genus of Antarctic fishes, *Channichthys*, lacks hemoglobin. In its relative, *Trematomus*, hemoglobin serves its usual functions. Assuming that the gene encoding hemoglobin in *Channichthys* has no function, and is not transcribed, how might you expect the nucleotide sequence of this gene to differ between these two genera?

5. Ultraviolet light (UV) can induce mutations in organisms such as *Drosophila*. Because it damages DNA, and therefore essential physiological functions, it can also reduce survival. Suppose you expose a large number of *Drosophila* to UV, screen their adult offspring for new mutations, and discover that a few offspring carry mutations that increase the amount of black pigment, which can protect internal organs from UV. The progeny of an equal number of control flies, not exposed to UV, show fewer or no mutations that increase pigmentation. Can you conclude that the process of mutation responds to organisms' need for adaptation to the environment?

6. Researchers have used artificial selection (see Chapter 9) to alter many traits in *Drosophila melanogaster*, such as phototactic behavior and wing length. No one has selected *Drosophila* (about 2 mm long) to be as large (ca. 30 mm) as bumblebees (although I'm not sure if anyone has tried). Do you suppose this could be done? How would you attempt it? If your attempts were unsuccessful, what hypotheses could explain your lack of success? What role might mutation play in your experiment?

9

Variation

The processes of mutation and recombination described in the previous chapter give rise to genetic differences in many characteristics among the members of a population or species. This genetic variation is the foundation of evolution, for the great changes in organisms that have transpired over time and the differences that have developed among species as they diverged from their common ancestors all originated as genetic variation within species. Understanding the processes of evolution thus requires us to understand genetic variation and the ways it which it is transmuted into evolutionary change. Understanding genetic variation, moreover, provides insight into questions—ranging from the significance of "intelligence tests" to the meaning of "race"—that deeply affect human society. Of all the biological sciences, evolutionary biology is most dedicated to analyzing and understanding variation, a profoundly important feature of all biological systems.

Individuals of the same species often look different. Two alleles at a single locus are responsible for whether the flowers of *Linanthus parryae*, a desert plant of southern California, are blue or white. Evolutionary biologists started studying this variation more than 70 years ago. (Photo © Larry Blakely.)

Because genetic variation will be discussed throughout the rest of this book, we will need a brief review of its vocabulary:

- *Phenotype* refers to a characteristic in an individual organism, or in a group of individuals that are alike in this respect. Phenotypic variation is largely the result of genetic differences among individuals, but can also be the result of the direct effects of environmental variation on development. Figure 9.1 shows instances of both types of phenotypic variation.
- *Genotype* is the genetic constitution of an individual organism, or of a group of organisms that are alike in this respect, at one or more loci singled out for discussion.
- A *locus* (plural: *loci*) is a site on a chromosome, or more usually, the gene that occupies a site.
- An *allele* is a particular form of a gene, usually distinguished from other alleles by its effects on the phenotype.
- A *haplotype* is one of the sequences of a gene or DNA segment that can be distinguished from homologous sequences by molecular methods such as DNA sequencing.
- We will use the term *gene copy* when counting the number of representatives of a gene. In a diploid population (such as humans), each individual carries two copies of each autosomal gene, so a sample of 100 individuals represents 200 gene copies. The term *gene copy* is used without distinguishing allelic or sequence differences among the copies. For that purpose, we will sometimes refer to *allele copies*. Thus a heterozygous individual, A_1A_2, has two copies of the A gene: one copy of allele A_1, and one copy of allele A_2.

(A) Genetic variation

(B) Environmental variation

Figure 9.1 Phenotypic variation caused by genetic differences and by the environment. (A) White and "blue" forms of the snow goose *Chen caerulescens* are caused by two alleles at a single locus. The two forms occur within the same population and freely interbreed. (B) The varied plumages of the willow ptarmigan *Lagopus lagopus* are not due to different genotypes, but are a response to the environment. At each molt, the bird develops either white winter plumage or colored summer plumage, depending on the season in which the molt occurs. (A © Jim Zipp and John Bova/Photo Researchers, Inc.; B © Painet, Inc.)

Variation comes in many forms. In the simplest cases, two alleles account for most of the variation at a given locus (e.g., flower color in *Linanthus parryae*, or the color forms of the snow goose in Figure 9.1). In some cases, three or more alleles exist within a population. For instance, several forms of the African swallowtail butterfly *Papilio dardanus* are palatable mimics of different species of butterflies that birds find unpalatable (Figure 9.2A). In this case, the several different color patterns are inherited as if they were multiple alleles at one locus. In other instances, one or more phenotypic features vary because of allelic variation at two or more separate loci, such as those that control coloration and the pattern of bands on the shells of the land snail *Cepaea nemoralis* (Figure 9.2B). Multiple loci contribute to the continuous variation typical of most phenotypic features, such as the familiar variation in human hair and skin color (Figure 9.2C).

Distinguishing Sources of Phenotypic Variation

Because evolution consists of genetic changes in populations over time, evolutionary biologists are most interested in those variations that have a genetic basis. However, individuals may differ in phenotype because of genetic differences, environmental differences, or both. Any differences that are not caused by genetic differences are said to be environmental. Individual differences in behavior resulting from learning, for example, are environmental differences. A special class of such differences is referred to as **maternal effects**. This term refers to the effects of a mother on her offspring that are due not to the genes they inherit from her, but rather to nongenetic in-

Figure 9.2 Multiple alleles underlie some genetic variation. (A) Mimetic variation in the African swallowtail, the palatable species *Papilio dardanus*. Males are non-mimetic and all appear similar (top individual). The three individuals below are females, each of which has evolved a different color pattern that mimics a distantly related toxic species. This female-limited variation in *P. dardanus* is inherited as if it were due to multiple alleles at one locus, although it is actually caused by several closely linked genes. (B) Although a single locus is responsible for much of the variation in background shell color of the European land snail *Cepaea nemoralis*, several different loci contribute to variation in the number and width of dark bands. The factors that maintain this variation within snail populations have been studied extensively, but are not yet fully understood. (C) In *Homo sapiens*, alleles at several or many different loci contribute to variation in "quantitative characters" such as skin pigmentation, hair color and texture, and facial features. Such variation among humans, like that among the butterflies and snails shown here, does not affect the organisms' ability to interbreed. (A courtesy of Fred Nijhout; B, photo by the author; C © Painet, Inc.)

fluences, such as the amount or composition of yolk in her eggs, the amount and kind of maternal care she provides, or her physiological condition while carrying eggs or embryos. Nongenetic "paternal effects" have also been described in a few organisms.

CONGENITAL differences among individuals—those present at birth—may be caused by genes, by nongenetic maternal effects, or by environmental factors that act on an embryo before birth or hatching. Consumption of alcohol, tobacco, and many other drugs by pregnant women, for example, increases the risk of nongenetic birth defects in their babies.

To determine whether variation in a characteristic is genetic, environmental, or both, several methods can be used:

1. Phenotypes can be experimentally crossed to produce F_1, F_2, and backcross progeny. Mendelian ratios among the phenotypes of the progeny (e.g., 3:1 or 1:2:1) are taken as evidence of simple genetic control.

2. Correlation between the average phenotype of offspring and that of their parents, or greater resemblance among siblings than among unrelated individuals, suggests that genetic variation contributes to phenotypic variation. However, we must take maternal effects into account, and we must also be sure that siblings (or relatives in general) do not share more similar environments than nonrelatives. For example, human geneticists rely strongly on studies of adopted children to determine whether behavioral or other similarities among siblings are due to shared genes rather than shared environments.

3. The offspring of phenotypically different parents can be reared together in a uniform environment, often referred to as a **common garden**. Differences among offspring from different parents that persist in such circumstances are likely to have a genetic basis. Propagation in the common garden for at least two generations is advisable to distinguish genetic from maternal effects.

Fundamental Principles of Genetic Variation in Populations

We embark now on our study of genetic variation and the factors that influence it—that is, the factors that cause evolution within species. *The definitions, concepts, and principles introduced here are absolutely essential for understanding evolutionary theory.* We begin with a short verbal description of these ideas, followed by an explanation of a very important formal model.

At any given gene locus, a population may contain two or more alleles that have arisen over time by mutation. Sometimes one allele is by far the most common (and may be called the WILD TYPE), and the others are very rare; sometimes two or more of the alleles are each quite common. The relative commonness or rarity of an allele—its proportion of all gene copies in the population—is called the **allele frequency** (sometimes imprecisely referred to as the "gene frequency"). In sexually reproducing populations, the alleles, carried in eggs and sperm, become combined into homozygous (two copies of the same allele) and heterozygous (one copy each of two different alleles) genotypes. The **genotype frequency** is the proportion of a population that has a certain genotype (Figure 9.3). The proportions of the different genotypes are related to the allele frequencies in simple but important ways, as we will soon see. If the genotypes differ in a phenotypic character, the amount of variation in that character will depend not only on how different the genotypes are from each other, but also on the relative abundance (the frequencies) of the genotypes—which in turn depends on the allele frequencies.

Any alteration of the genotype frequencies in one generation will alter the frequencies of the alleles carried by the population's gametes when reproduction occurs, so the genotype frequencies of the following generation will be altered in turn. *Such alteration, from*

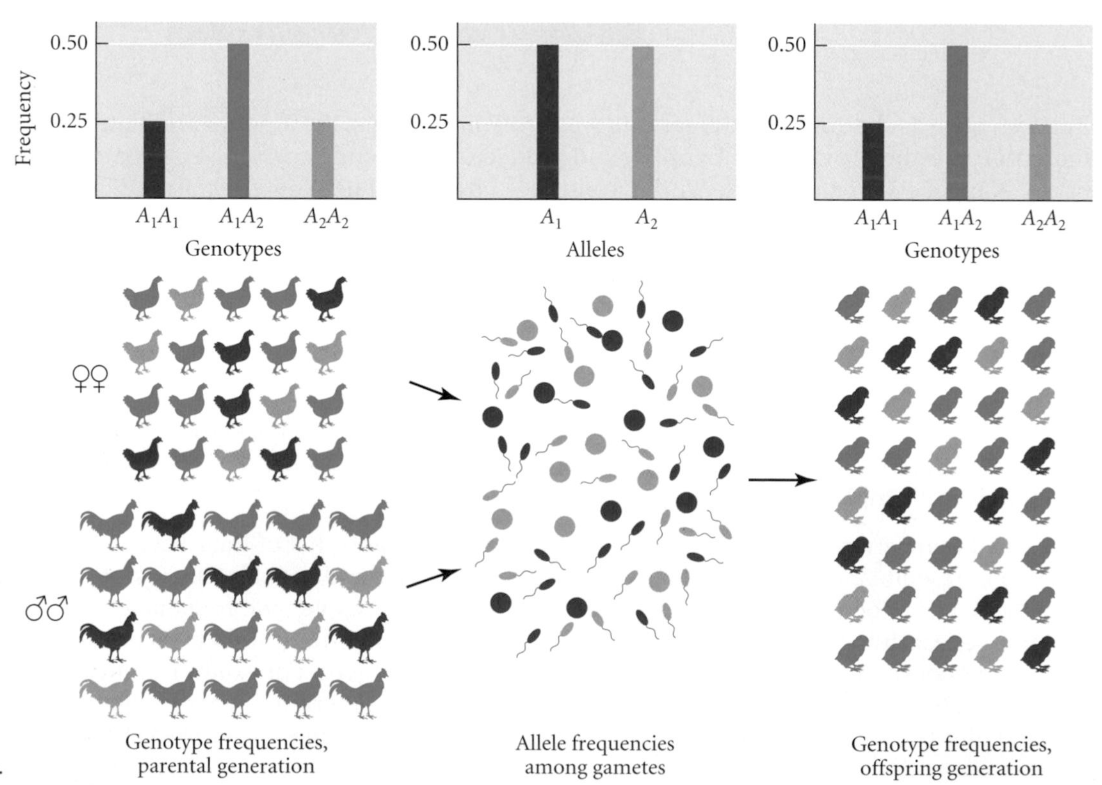

Figure 9.3 For a locus with two alleles (A_1, A_2), this diagram shows the frequency of three genotypes among females and males in one generation, the allele frequencies among their eggs and sperm, and the genotype frequencies among the resulting offspring.

generation to generation, is the central process of evolutionary change. However, the genotype and allele frequencies do not change on their own; something has to change them. *The factors that can cause the frequencies to change are the causes of evolution.*

Frequencies of alleles and genotypes: The Hardy-Weinberg principle

Imagine that a diploid population has 1000 individuals, and that for a particular gene locus in that population, there exist only two alleles, A_1 and A_2. Thus there are three possible genotypes for this locus: A_1A_1, A_2A_2 (both homozygous), and A_1A_2 (heterozygous).

Let us say 400 individuals have the genotype A_1A_1, 400 are A_1A_2, and 200 are A_2A_2. Let the frequencies of genotypes A_1A_1, A_1A_2, and A_2A_2 be represented by D, H, and R respectively, and let the frequencies of the alleles A_1 and A_2 be represented by p and q. Thus the genotype frequencies are $D = 0.4$, $H = 0.4$, and $R = 0.2$, and the allele frequencies are $p = 0.6$ and $q = 0.4$ (Figure 9.4A,B). The frequency of allele A_1 is the sum of all the alleles carried by A_1A_1 homozygotes, plus one-half of those carried by heterozygotes. Hence $p = D + H/2 = 0.6$. Likewise, the frequency of allele A_2 is calculated as $q = R + H/2 = 0.6$.

(A) Parental genotype frequencies (not in equilibrium)

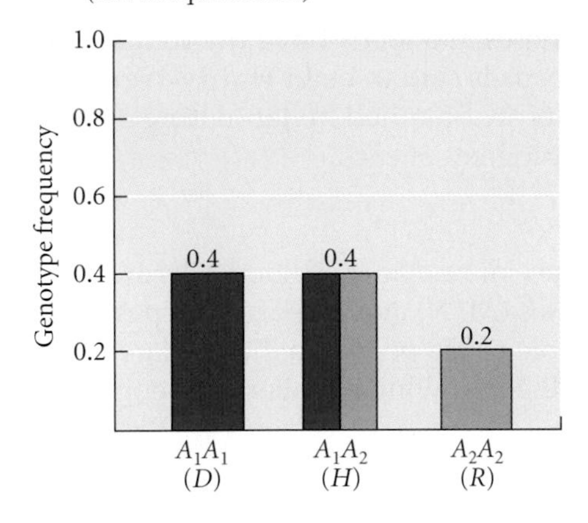

(B) Parental allele frequencies

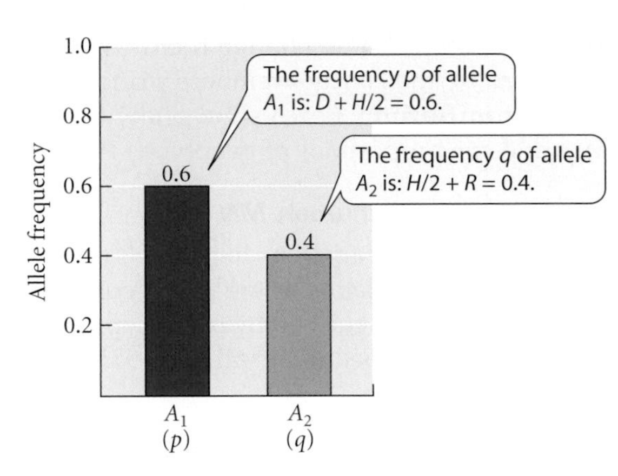

(C)

Offspring	Probability of a given mating producing the genotype		
A_1A_1	$\Pr[A_1 \text{ egg}] \times \Pr[A_1 \text{ sperm}] = p \times p = p^2$	0.6^2	$= 0.36$
A_1A_2	$\Pr[A_1 \text{ egg}] \times \Pr[A_2 \text{ sperm}] = p \times q = pq$	$0.6 \times 0.4 = 0.24$	$= 0.48$
	$\Pr[A_2 \text{ egg}] \times \Pr[A_1 \text{ sperm}] = q \times p = pq$	$0.4 \times 0.6 = 0.24$	
A_2A_2	$\Pr[A_2 \text{ egg}] \times \Pr[A_2 \text{ sperm}] = q \times q = q^2$	0.4^2	$= 0.16$

(D) Offspring genotype frequencies

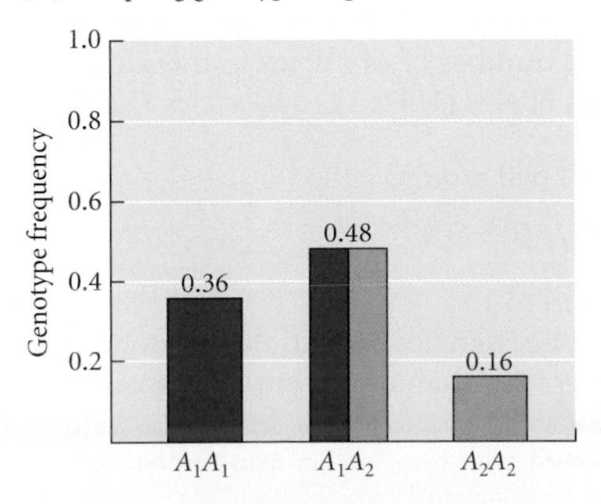

(E) Offspring allele frequencies

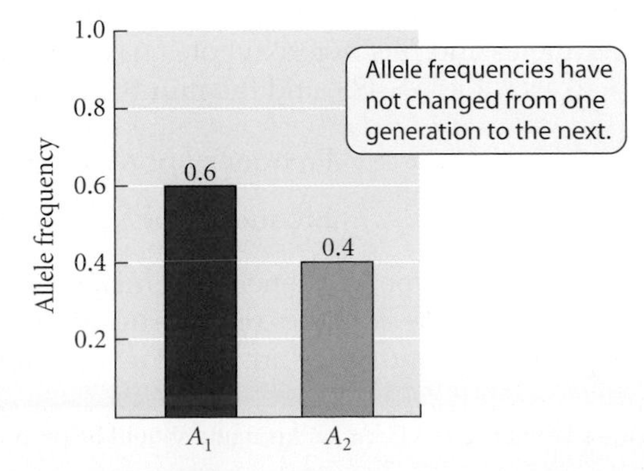

Figure 9.4 A hypothetical example illustrating attainment of Hardy-Weinberg genotype frequencies after one generation of random mating. (A) Genotype frequencies in the parental population. (B) Allele frequencies in the parental population. (C) Calculation of genotype frequencies among the offspring. (D) Genotype frequencies among the offspring, if all assumptions of the Hardy-Weinberg principle hold. (E) Allele frequencies among the offspring.

Now let us assume that each genotype is equally represented in females and males, and that *individuals mate entirely at random*. (This is an entirely different situation from the crosses encountered in elementary genetics exercises, in which familiar Mendelian ratios are produced by nonrandomly crossing females of *one* genotype with males of *one* genotype.) Random mating is conceptually the same as mixing eggs and sperm at random, as might occur, for instance, in those aquatic animals that release gametes into the water. A proportion p of eggs carry allele A_1, q carry A_2, and likewise for sperm. The probabilities (Pr) for all possible allele combinations are shown in Figure 9.4C, and the resulting offspring genotype and allele frequencies are shown in Figure 9.4D and E.

Notice that in the new generation, the allele frequencies are still $p = D + H/2 = 0.6$ and $q = H/2 + R = 0.4$. *The allele frequencies have not changed from one generation to the next*, although the alleles have become distributed among the three genotypes in new proportions. These proportions, denoted here as p^2, $2pq$, and q^2, constitute the HARDY-WEINBERG DISTRIBUTION of genotype frequencies.

The principle that genotypes will be formed in these frequencies as a result of random mating is named after G. H. Hardy and W. Weinberg,* who independently calculated these results in 1908. The Hardy-Weinberg principle, broadly stated, holds that whatever the initial genotype frequencies for two alleles may be, *after one generation of random mating, the genotype frequencies will be* $p^2{:}2pq{:}q^2$. Moreover, both these genotype frequencies and the allele frequencies *will remain constant in succeeding generations* unless factors not yet considered should change them. When genotypes at a locus have the frequencies predicted by the Hardy-Weinberg principle, the locus is said to be in **Hardy-Weinberg (H-W) equilibrium**. Box A shows how these results are obtained by tabulating all matings and the proportions of genotypes among the progeny of each.

An example: The human *MN* locus

Among the many variations in proteins on the surface of human red blood cells are those resulting from variation at the *MN* locus. Two alleles (*M*, *N*) and three genotypes (*MM*, *MN*, *NN*) are distinguished by blood typing. A sample of 320 people in the Sicilian village of Desulo (R. Ceppellini, in Allison 1955) yielded the following numbers of people carrying each genotype:

MM	*MN*	*NN*
187	114	19

We can now estimate the *frequency* of each genotype as the proportion of the total sample (320) that carries that genotype. Thus,

$$\text{Frequency of } MM = D = 187/320 = 0.584$$

$$\text{Frequency of } MN = H = 114/320 = 0.356$$

$$\text{Frequency of } NN = R = 19/320 = 0.059$$

Note that these frequencies, or proportions, must sum to 1.

Now we can calculate the allele frequencies. Each person carries two gene copies, so the total sample represents $320 \times 2 = 640$ gene copies. Because *MM* homozygotes each have two *M* alleles and *MN* heterozygotes have one, the number of *M* alleles in the sample is $(187 \times 2) + (114 \times 1) = 488$, and the number of copies of *N* is $(114 \times 1) + (19 \times 2) = 152$. Thus

$$\text{Frequency of } M = p = 488/640 = 0.763$$

$$\text{Frequency of } N = q = 152/640 = 0.237$$

As with the genotype frequencies, p and q must sum to 1.

These figures are *estimates* of the true genotype frequencies and allele frequencies in the population because they are based on a sample, rather than on a complete census. The

*Because Weinberg was German, his name should be pronounced "vine-berg," rather than "wine-berg."

BOX 9A Derivation of the Hardy-Weinberg Distribution

Let the frequencies of genotypes A_1A_1, A_1A_2, and A_2A_2 be D, H, and R, respectively, and let the frequencies of alleles A_1 and A_2 be p and q. The genotype frequencies sum to 1, as do the allele frequencies. The probability of a mating between a female of any one genotype and a male of any one genotype equals the product of the genotype frequencies. For reciprocal crosses between two different genotypes, we take the sum of the probabilities. The mating frequencies and offspring produced are:

Mating	Probability of mating*	Offspring genotype		
		A_1A_1	A_1A_2	A_2A_2
$A_1A_1 \times A_1A_1$	D^2	D^2		
$A_1A_1 \times A_1A_2$	$2DH$	DH	DH	
$A_1A_1 \times A_2A_2$	$2DR$		$2DR$	
$A_1A_2 \times A_1A_2$	H^2	$H^2/4$	$H^2/2$	$H^2/4$
$A_1A_2 \times A_2A_2$	$2HR$		HR	HR
$A_2A_2 \times A_2A_2$	R^2			R^2

*Note that there are a total of 9 mating possibilities: 3 female genotypes × 3 male genotypes. Reciprocal matings between two different genotypes count as 2.

Recalling that $p = D + H/2$ and $q = H/2 + R$, the frequency of each genotype among the offspring is:

A_1A_1: $D^2 + DH + H^2/4 = (D + H/2)^2 = p^2$

A_1A_2: $DH + 2DR + H^2/2 + HR = 2[(D + H/2)(H/2 + R)] = 2pq$

A_2A_2: $H^2/4 + HR + R^2 = (H/2 + R)^2 = q^2$

This result may also be obtained by recognizing that if genotypes mate at random, gametes, and therefore genes, also unite at random to form zygotes. Because the probability that an egg carries allele A_1 is p, and the same is true for sperm, the probability of an A_1A_1 offspring is p^2. A Punnett square shows the probability of each gametic union:

		Sperm	
		A_1 (p)	A_2 (q)
Eggs	A_1 (p)	A_1A_1 (p^2)	A_1A_2 (pq)
	A_2 (q)	A_1A_2 (pq)	A_2A_2 (q^2)

The Hardy-Weinberg principle can be extended to more complicated patterns of inheritance, such as multiple alleles. For example, if there are k alleles ($A_1, A_2, \ldots A_k$), the H-W frequency of homozygotes for any allele A_i is p_i^2, and that of heterozygotes for any two alleles A_i and A_j is $2p_ip_j$, where p_i and p_j are the frequencies of the two alleles in question. The total frequency of all heterozygotes combined (H) is sometimes expressed as the complement of the summed frequency of all homozygous genotypes, or $H = 1 - \Sigma p_i^2$.

For example, if there are three alleles A_1, A_2, and A_3 with frequencies p_1, p_2, and p_3, the H-W frequencies of the three possible homozygotes (A_1A_1, A_2A_2, A_3A_3) are p_1^2, p_2^2, and p_3^2; and the H-W frequencies of the heterozygotes (A_1A_2, A_1A_3, A_2A_3) are $2p_1p_2$, $2p_1p_3$, and $2p_2p_3$. The total frequency of heterozygotes is

$$H = 1 - (p_1^2 + p_2^2 + p_3^2)$$

larger the sample, the more confident we can be that we are obtaining accurate estimates of the true values. (This assumes that the sample is a *random* one, i.e., that our likelihood of collecting a particular type is equal to the true frequency of that type in the population.)

Our hypothetical example (see Figure 9.4) showed that for a given set of allele frequencies, a population might or might not have H-W genotype frequencies. (The frequencies 0.36, 0.48, and 0.16 in the offspring generation were in H-W equilibrium, but the frequencies 0.40, 0.40, and 0.20 in the parental generation were not.) Our real example shows a close fit to the H-W frequency distribution. The allele frequencies of M and N were $p = 0.763$ and $q = 0.237$. If we calculate the *expected genotype frequencies* under the H-W principle and multiply them by the sample size (320), we obtain the *expected* number of individuals with each genotype. These values in fact closely fit the *observed* numbers:

	Genotype		
	MM	***MN***	***NN***
	p^2	$2pq$	q^2
Expected H-W frequency	0.582	0.362	0.056
Expected number (H-W frequency × sample size)	186	116	18
Observed number	187	114	19

The significance of the Hardy-Weinberg principle: Factors in evolution

The Hardy-Weinberg principle is the foundation on which almost all of the theory of population genetics of sexually reproducing organisms—which is to say, most of the genetic theory of evolution—rests. Its importance cannot be overemphasized. We will encounter it in the theory of natural selection and other causes of evolution. It has two important implications: First, genotype frequencies attain their H-W values after a single generation of random mating. If some factor in the past had caused genotype frequencies to deviate from H-W values, a single generation of random mating would erase the imprint of that history. Second, according to the H-W principle, not only genotype frequencies, but also allele frequencies, remain unchanged from generation to generation. A new mutation, for example, will remain at its initial very low allele frequency indefinitely.

Like any mathematical formulation, the Hardy-Weinberg principle holds only under certain assumptions. Since allele frequencies and genotype frequencies often do change (i.e., evolution occurs), the assumptions of the Hardy-Weinberg formulation must not always hold. Therefore *the study of genetic evolution consists of asking what happens when one or more of the assumptions are relaxed.*

The most important assumptions of the Hardy-Weinberg principle are the following:

1. *Mating is random.* If a population is not **panmictic**—that is, if members of the population do not mate at random—the genotype frequencies may depart from the ratios $p^2{:}2pq{:}q^2$.
2. *The population is infinitely large* (or so large that it can be treated as if it were infinite). The calculations are made in terms of probabilities. If the number of events is finite, the actual outcome is likely to deviate, purely by chance, from the predicted outcome. If we toss an infinite number of unbiased coins, probability theory says that half will come up heads, but if we toss only 100 coins, we are likely not to obtain exactly 50 heads, purely by chance. Similarly, among a finite number of offspring, both the genotype frequencies and the allele frequencies may differ from those in the previous generation, *purely by chance*. Such random changes are called **random genetic drift**.
3. *Genes are not added from outside the population.* Immigrants from other populations may carry different frequencies of A_1 and A_2; if they interbreed with residents, this will alter allele frequencies and, consequently, genotype frequencies. Mating among individuals from different populations is termed **gene flow** or **migration**. We may restate this assumption as: There is no gene flow.
4. *Genes do not mutate from one allelic state to another.* Mutation, as we have seen (in Chapter 8), can change allele frequencies, although usually very slowly. The Hardy-Weinberg principle assumes no mutation.
5. *All individuals have equal probabilities of survival and of reproduction.* If these probabilities differ among genotypes (i.e., if there is a consistent difference in the genotypes' rates of survival or reproduction), then the frequencies of alleles and/or genotypes may be altered from one generation to the next. Thus the Hardy-Weinberg principle assumes that there is *no natural selection* affecting the locus.

Inasmuch as nonrandom mating, chance, gene flow, mutation, and selection can alter the frequencies of alleles and genotypes, these are the major factors that cause evolutionary change within populations.

Certain subsidiary assumptions of the Hardy-Weinberg principle can be important in some contexts. First, the principle, as presented, applies to autosomal loci; it can also be modified for sex-linked loci (which have two copies in one sex and one in the other). Second, the principle assumes that alleles segregate into a heterozygote's gametes in a 1:1 ratio. Deviations from this ratio are known and are called SEGREGATION DISTORTION or MEIOTIC DRIVE.

If the assumptions we have listed hold true for a particular locus, that locus will display Hardy-Weinberg genotype frequencies. But if we observe that a locus fits the Hardy-Weinberg frequency distribution, we cannot conclude that the assumptions hold true! For example, mutation or selection may be occurring, but at such a low rate that we cannot detect a deviation of the genotype frequencies from the expected values. Or, under some forms of natural selection, we might observe deviations from Hardy-Weinberg equilibrium if we measure genotype frequencies at one stage in the life history, but not at other stages.

Frequencies of alleles, genotypes, and phenotypes

At Hardy-Weinberg equilibrium, the frequency of heterozygotes is greatest when the alleles have equal frequencies (Figure 9.5). When an allele is very rare, almost all its carriers are heterozygous. If one allele (say, A_1) is dominant, masking a recessive allele (A_2) in heterozygotes, the dominant phenotype (genotypes A_1A_1 and A_1A_2) makes up $p^2 + 2pq$ of the population. Because almost all copies of a rare recessive allele are carried by heterozygotes, the allele may not be detected readily. If, for example, the frequency of a recessive allele is $q = 0.01$, only $(0.01)^2 = 0.0001$ of the population displays this recessive allele in its phenotype. Thus populations can carry **concealed genetic variation**. However, *a dominant allele may well be less common than a recessive allele*. "Dominance" refers to an allele's phenotypic effect in the heterozygous condition, not to its numerical prevalence. In the British moth *Cleora repandata*, for example, black coloration is inherited as a dominant allele, and "normal" gray coloration is recessive. In a certain forest, 10 percent of the moths were found to be black (Ford 1971).

Inbreeding

The Hardy-Weinberg principle assumes that a population is panmictic. What if mating does not occur at random? One form of nonrandom mating is **inbreeding**, which occurs when individuals are more likely to mate with relatives than with nonrelatives, or, more generally, when the gene copies in uniting gametes are more likely to be identical by descent than if they joined at random. Gene copies are said to be **identical by descent** if they have descended, by replication, from a common ancestor, relative to other gene copies in the population.

Box B shows that as inbreeding proceeds, the frequency of each homozygous genotype increases and the frequency of heterozygotes decreases by the same amount. The frequency of heterozygotes is $H = H_0(1 - F)$, where H_0 is the heterozygote frequency expected if the locus were in H-W equilibrium, and F is the **inbreeding coefficient**.

The ways in which genotype frequencies change are easily seen if we envision mating of individuals only with their closest relatives—namely, themselves—by self-fertilization. (You may have trouble imagining this if you think about people, but think about plants and you will see that it is not only plausible, but quite common.) The homozygous genotype A_1A_1 can produce only A_1 eggs and A_1 sperm, and thus only A_1A_1 offspring; likewise, A_2A_2 individuals produce only A_2A_2 offspring. Heterozygotes produce A_1 and A_2 eggs in equal proportions, and likewise for sperm. When these eggs and sperm join at random, ¼ of the offspring are A_1A_1, ½ are A_1A_2, and ¼ are A_2A_2. (The two allele copies carried by the homozygous offspring are identical by descent.) Thus the frequency of heterozygotes is halved in each generation, and eventually reaches zero. Conversely, F increases as inbreeding continues; in fact, F can be estimated by the deficiency of heterozygotes relative to the H-W equilibrium value (see Box B). If CONSANGUINEOUS mating (mating among relatives) is a consistent feature of a population, F will increase over generations at a rate that depends on how closely related the average pair of mates is.

The most extreme form of inbreeding, **self-fertilization** or **selfing**, occurs in many species of plants and a few animals. The wild oat *Avena fatua*, for example, reproduces mostly by selfing, and it has a low frequency of heterozygotes at the sev-

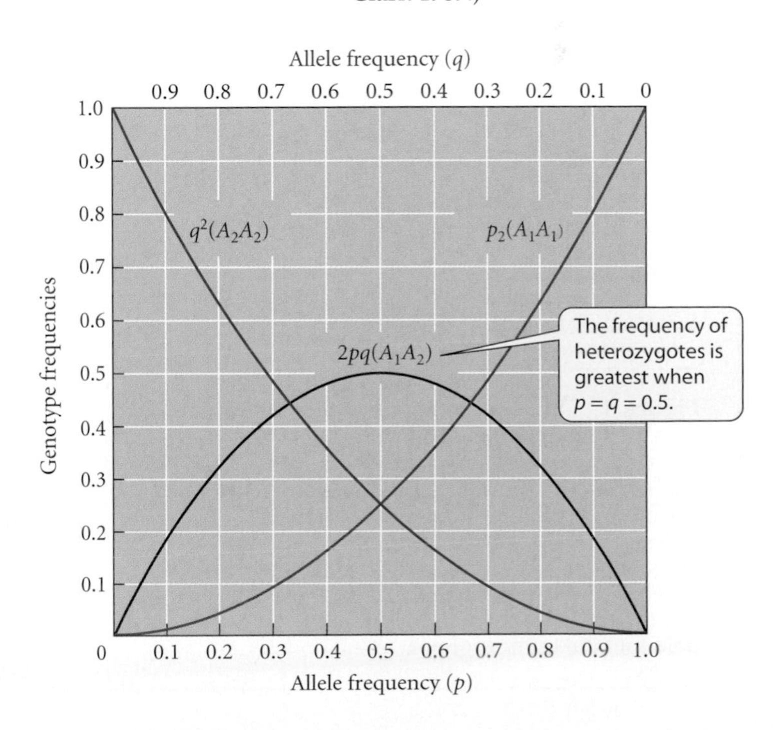

Figure 9.5 Hardy-Weinberg genotype frequencies as a function of allele frequencies at a locus with two alleles. Heterozygotes are the most common genotype in the population if the allele frequencies are between ⅓ and ⅔. (After Hartl and Clark 1989.)

BOX 9B Change of Genotype Frequencies by Inbreeding

Suppose we could label every gene copy at a locus and trace the descendants of each gene copy through subsequent generations. We would then be tracing alleles that are identical by descent. Some of the gene copies are allele A_1, and others are A_2. If we label one of the A_1 copies A_1*, then after one generation, a sister and brother may both inherit replicates of A_1* from a parent (with probability $0.5^2 = 0.25$). Among the progeny of a mating between sister and brother, both heterozygous for A_1*, one-fourth will be A_1*A_1*. These individuals carry two gene copies that are not only the same allele (A_1), but are also identical by descent (A_1*) (Figure 1). The A_1*A_1* individuals are said to be not only homozygous, but AUTOZYGOUS. ALLOZYGOUS individuals, on the other hand, may be either heterozygous or homozygous (if the two copies of the same allele are not identical by descent).

The inbreeding coefficient, denoted F, is the probability that an individual taken at random from the population will be autozygous. In a population that is not at all inbred, $F = 0$. In a fully inbred population, $F = 1$: all individuals are autozygous.

In a population that is inbred to some extent, F is the fraction of the population that is autozygous, and $1 - F$ is the fraction that is allozygous. If two alleles, A_1 and A_2, have frequencies p and q, the probability that an individual is allozygous *and* that it is A_1A_1 is $(1 - F) \times p^2$. Likewise, the fraction of the population that is allozygous and heterozygous is $(1 - F) \times 2pq$, and the fraction that is allozygous and A_2A_2 is $(1 - F) \times q^2$.

Turning our attention now to the fraction, F, of the population that is autozygous, we note that none of these individuals is heterozygous, because a heterozygote's alleles are not identical by descent. If an individual is autozygous, the probability that it is autozygous for A_1 is p, the frequency of A_1. Thus the fraction of the population that is autozygous and A_1A_1 is $F \times p$. Likewise, $F \times q$ is the fraction that is autozygous and A_2A_2.

Thus, taking into account the allozygous and autozygous fractions of the population, the genotype frequencies (Figure 2) are

	Allozygous		Autozygous		Genotype frequency
A_1A_1	$p^2(1 - F)$	+	pF	=	$p^2 + Fpq = D$
A_1A_2	$2pq(1 - F)$			=	$2pq\,(1 - F) = H$
A_2A_2	$q^2(1 - F)$	+	qF	=	$q^2 + Fpq = R$

Therefore, the consequence of inbreeding is that the frequency of homozygotes is higher, and the frequency of heterozygotes is lower, than in a population that is in Hardy-Weinberg equilibrium. Note that H, the frequency of heterozygotes in the inbred population, equals $(1 - F)$ multiplied by the frequency of heterozygotes we expect to find in a randomly mating population ($2pq$). Denoting $2pq$ as H_0, we have

$$H = H_0(1 - F) \quad \text{or} \quad F = (H_0 - H)/H_0$$

Thus we can estimate the inbreeding coefficient by two measurable quantities, the observed frequency of heterozygotes, H, and the "expected" frequency, $2pq$, which we can calculate from data on the allele frequencies p and q. In practice, then, the inbreeding coefficient F is a measure of the reduction in heterozygosity compared with a panmictic population with the same allele frequencies.

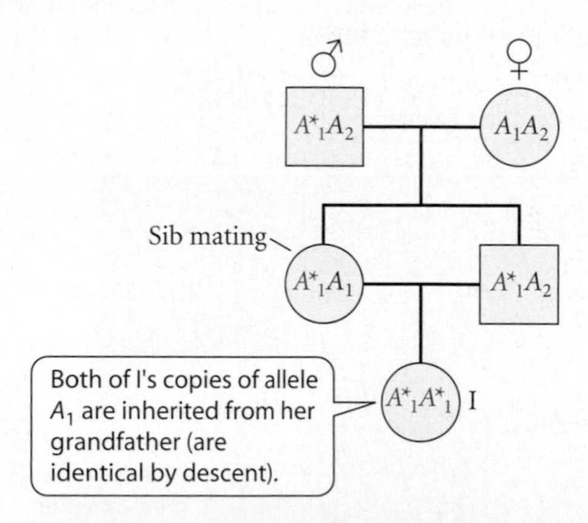

Figure 1 A pedigree showing inbreeding due to mating between siblings. Squares represent males, circles females. Copies of an A_1 allele, A_1* (red type), are traced through three generations. Individual I possesses two copies of A_1* that are identical by descent (she is autozygous). I's mother is homozygous for A_1, but the two copies are not identical by descent (she is allozygous).

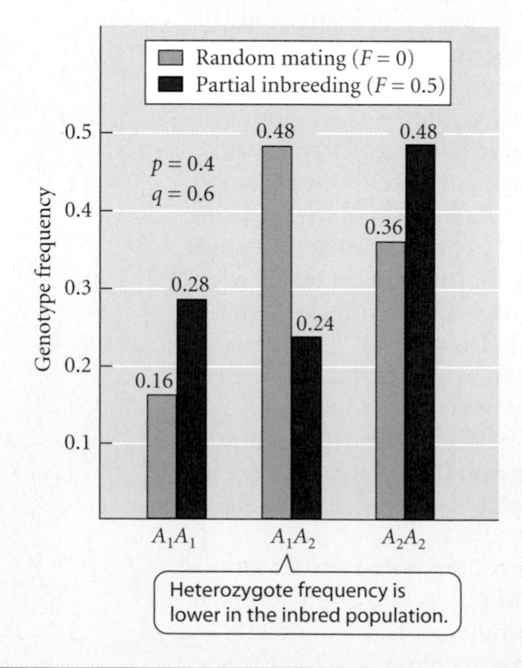

Figure 2 Genotype frequencies at a locus with allele frequencies $p = 0.4$ and $q = 0.6$ when mating is random ($F = 0$) and when the population is partially inbred ($F = 0.5$).

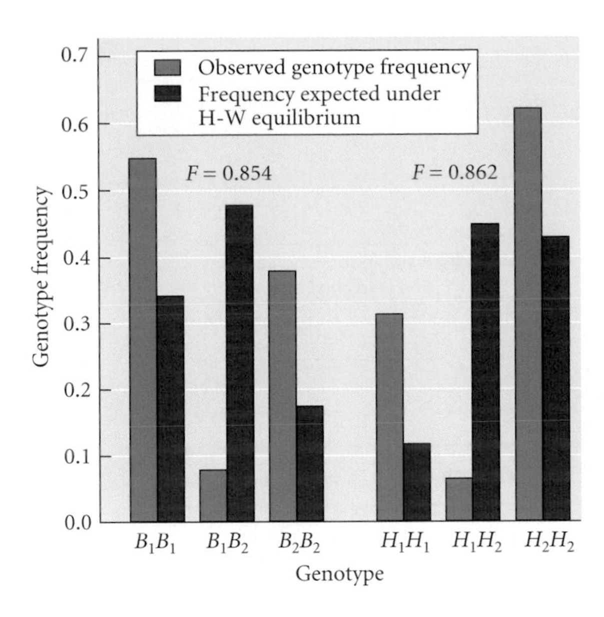

Figure 9.6 Genotype frequencies observed at two loci in a population of the self-fertilizing wild oat *Avena fatua* compared with those expected under Hardy-Weinberg equilibrium. Note that heterozygotes are deficient at both loci, and that calculated values of *F* are nearly the same for the two loci. (Data from Jain and Marshall 1967.)

eral loci studied (Figure 9.6). The inbreeding coefficients estimated from data on all the loci were nearly equal, which is as it should be, because inbreeding affects all loci in the same way.

Genetic Variation in Natural Populations

Polymorphism

Genetic **polymorphism** (*poly*, "many," *morph*, "form") is the presence in a population of two or more variants (alleles or haplotypes). The term is usually qualified to mean that the rarer allele has a frequency greater than 0.01. A locus or character that is not polymorphic is **monomorphic**. Polymorphism originally referred to genetically determined phenotypes, but the term has been extended to include variation at the molecular level.

Phenotypic polymorphisms caused by allelic differences at a single locus have been described in many species of plants and animals. In animals, single-locus polymorphisms have been described for characteristics such as color pattern (see Figure 9.1A) and behavior. In the fruit fly *Drosophila melanogaster*, for example, flies with different genotypes at the *per* (*period*) locus differ in whether their daily activity rhythm follows a 24-hour cycle or a cycle of slightly different length. Another interesting example is the polymorphism in *Layia glandulosa*, a relative of the sunflowers in the family Asteraceae (Figure 9.7). The presence or absence of ray florets is controlled by two loci. This difference in the inflorescence was the traditional distinction between two tribes of the Asteraceae. ("Tribe" is a taxonomic level between family and genus.) Thus characters that distinguish higher taxa can sometimes be found varying within species.

Genetic variation in viability

Beginning in the 1920s, geneticists found that the same mutations of *Drosophila* that had been studied in laboratory populations—mutations affecting coloration, bristles, wing shape, and the like—also existed in wild populations. In order to find and study these recessive alleles, it is necessary to "extract" chromosomes—that is, to create flies that are homozygous for a particular chromosome. This is accomplished by a series of crosses between a wild-type population and a laboratory stock that carries a dominant marker allele and an inversion that suppresses crossing over (Figure 9.8). A chromosome that has been made homozygous in this way may carry not only re-

Figure 9.7 A classic example of variation within a species, illustrating that the distinctive characters of higher taxa can sometimes have a simple genetic basis. The plant that produced the flowers labeled P_2 is conspecific with the more common form of *Layia glandulosa*, which produced the flowers labeled P_1, but lacks the ray florets with their long, white petals. When these two phenotypes were crossed, segregation of phenotypes in the F_2 generation, illustrated by the 12 flowers at the bottom, showed that the distinguishing characteristics are determined by only two loci. These anatomical differences distinguish two of the tribes into which the plant family Asteraceae was traditionally divided. (From Clausen et al. 1947.)

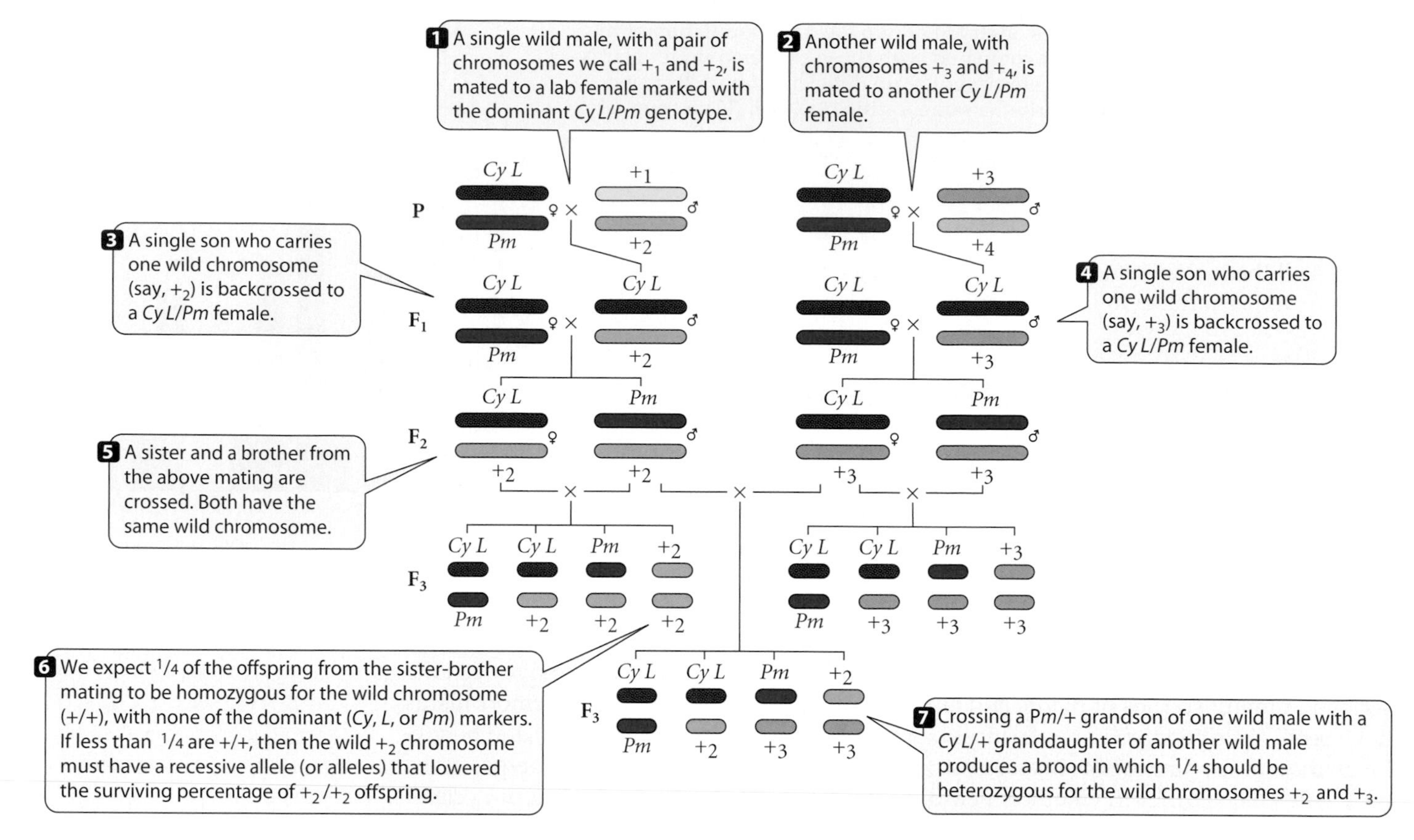

Figure 9.8 Crossing technique for "extracting" a chromosome from a wild-type male *Drosophila melanogaster* and making it homozygous in order to detect recessive alleles. In this figure, plus signs denote wild-type chromosomes, and the subscript number identifies a particular chromosome. The process is shown for two wild-type males (with chromosome pairs $+_1$,$+_2$ and $+_3$,$+_4$), each of which is crossed to a laboratory stock that has dominant mutant markers on the homologous chromosomes: *Cy* (curly wing) and *L* (lobed wing) on one and *Pm* (plum eye color) on the other. Each of these chromosomes also has inversions (see Chapter 8) that prevent crossing over. Consequently, one of each male's wild-type chromosomes (say, $+_2$), is transmitted intact to the F_3 generation. When crosses are performed as shown, the F_3 generation consists, in principle, of equal numbers of four genotypes, including a +/+ homozygote and the *Cy L*/+ and *Pm*/+ heterozygotes. The viability of these genotypes is measured by their proportion in the F_3 generation, relative to the expected 1:1:1:1 ratio. The lowermost family illustrates how flies heterozygous for two wild-type chromosomes ($+_2$ and $+_3$) can be produced. Their viability is also measured by deviation from the 1:1:1:1 ratio expected in this cross. (After Dobzhansky 1970.)

cessive alleles that affect morphological features, but also alleles that affect traits such as viability (survival from egg to adult). The crosses produce an F_3 generation of flies, of which one-fourth are expected to be homozygous for the wild chromosome. If no wild-type adult offspring (i.e., offspring without any of the dominant markers) appear, then the wild chromosome must carry at least one recessive **lethal allele**—that is, one that causes death before the flies reach the adult stage. Performing such crosses with many different wild flies makes it easy to determine what fraction of wild chromosomes cause complete lethality or reduce the likelihood of survival.

Theodosius Dobzhansky and his collaborators examined hundreds of copies of chromosome 2* sampled from wild *Drosophila pseudoobscura*. They discovered that about 10 percent of those copies were lethal when made homozygous (Figure 9.9). About half of

**Drosophila melanogaster*, *D. pseudoobscura*, and many other species of *Drosophila* have four pairs of chromosome, denoted X/Y, 2, 3, and 4. Chromosome 4 is very small and has few genes. The X chromosome and the two major autosomes, chromosomes 2 and 3, carry roughly similar numbers of genes.

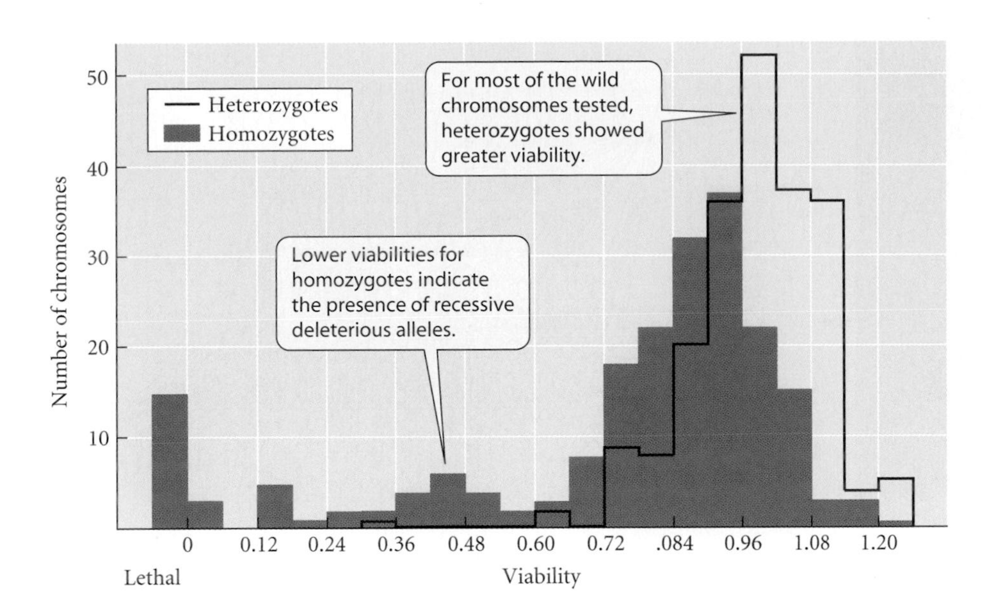

Figure 9.9 The frequency distribution of relative viabilities of chromosomes extracted from a wild population of *Drosophila pseudoobscura* by the method illustrated in Figure 9.8. A viability value of 1.00 indicates conformity to the expected ratios of laboratory and wild-type genotypes in the F_3 generation, as explained in Figure 9.8. The colored distribution shows the viability (relative survival from egg to adult) of homozygotes for 195 wild chromosomes. The great majority of chromosomes lower viability when made homozygous, indicating that they carry deleterious recessive alleles. Heterozygotes for various wild chromosomes, shown by the black line, have higher average viability. (After Lewontin 1974.)

the remaining chromosomes reduced survival to at least some extent. The other chromosomes in the genome yielded similar results, leading to the conclusion that almost every wild fly carries at least one chromosome that, if homozygous, substantially reduces the likelihood of survival. Moreover, many chromosomes cause sterility as well. Using a very different analytical method based on the mortality of children from marriages between relatives, Morton, Crow, and Muller (1956) concluded that humans are just like flies: "the average person carries heterozygously the equivalent of 3–5 recessive lethals [lethal alleles] acting between late fetal and early adult stages."

Such data, which have been confirmed by later work, point to a staggering incidence of life-threatening genetic defects. They imply, moreover, that *natural populations carry an enormous amount of concealed genetic variation* that is manifested only when individuals are homozygous. However, when flies carrying two different lethal chromosomes (i.e., derived from two wild flies) are crossed, the heterozygous progeny generally have nearly normal viability (Figure 9.9). Hence the two chromosomes must carry recessive lethal alleles at different loci: one lethal homozygote, for example, is *aaBB*, the other is *AAbb*, and the heterozygote, *AaBb*, has normal viability because each recessive lethal is masked by a dominant "normal" allele. From such data, it can be determined that the lethal allele at any one locus is very rare ($q < 0.01$ or so), and that the high proportion of lethal chromosomes is caused by the summation of rare lethal alleles at many loci.

Inbreeding depression

Because populations of humans and other diploid species harbor recessive alleles that have deleterious effects, and because inbreeding increases the proportion of homozygotes, populations in which many matings are consanguineous often manifest a decline in components of fitness, such as survival and fecundity. Such a decline is called **inbreeding depression**. This effect has long been known in human populations (Figure 9.10). For example, a very high proportion (27 to 53 percent) of individuals with Tay-Sachs disease, a lethal neurodegenerative disease caused by a recessive allele that is especially prevalent in Ashkenazic Jewish populations, are children of marriages between first cousins (Stern 1973).

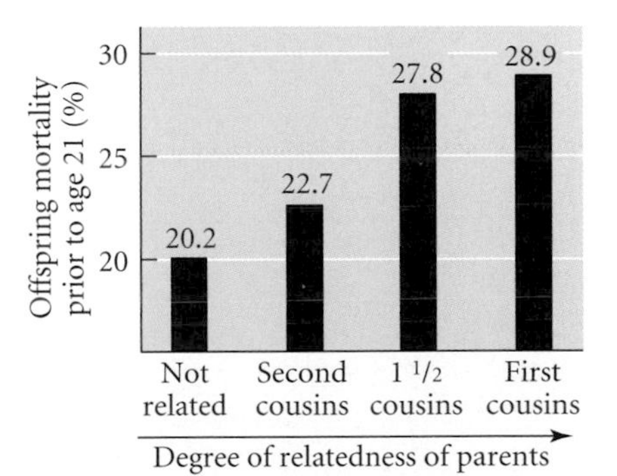

Figure 9.10 Inbreeding depression in humans: the more closely related the parents, the higher the mortality rate among their offspring. The data are for offspring up to 21 years of age from marriages registered in 1903–1907 in Italian populations. (After Stern 1973.)

Inbreeding depression is a well-known problem in the small, captive populations of endangered species that are propagated in zoos for reestablishing wild populations. Special breeding designs are required to minimize inbreeding in these populations (Frankham et al. 2002; Figure 9.11). Inbreeding also may increase the risk of extinction of small populations in nature. Thomas Madsen and colleagues (1995, 1999) studied an isolated Swedish population of a small poisonous snake, the adder (*Vipera berus*), that

(A)

(B)

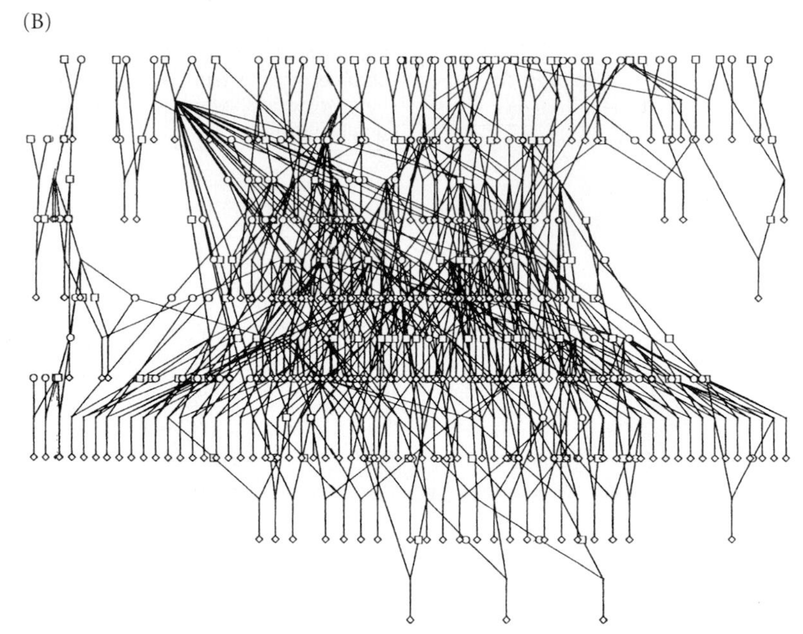

Figure 9.11 (A) The golden lion tamarin (*Leontopithecus rosalia*) is a small, highly endangered Brazilian monkey. Offspring from a captive breeding program in 140 zoos, which collectively maintain about 500 individuals, are now being reintroduced into natural preserves. (B) Inbreeding is reduced by an elaborate outcrossing scheme, illustrated by this pedigree. Lines show offspring of individual females (circles) and males (squares). (A © Tom & Pat Leeson/Photo Researchers, Inc.; B from Frankham et al. 2002.)

consisted of fewer than 40 individuals. The snakes were found to be highly homozygous (we will see shortly how this can be determined), the females had small litter sizes (compared with outbred adders in other populations), and many of the offspring were deformed or stillborn. The authors introduced 20 adult male adders from other populations, left them there for four mating seasons, and then removed them. Soon thereafter, the population increased dramatically (Figure 9.12), owing to the improved survival of the outbred offspring.

Genetic variation in proteins

Evolution would be very slow if populations were genetically uniform, and if only occasional mutations arose and replaced pre-existing genotypes. In order to know what the potential is for rapid evolutionary change, it would be useful to know how much genetic variation natural populations contain.

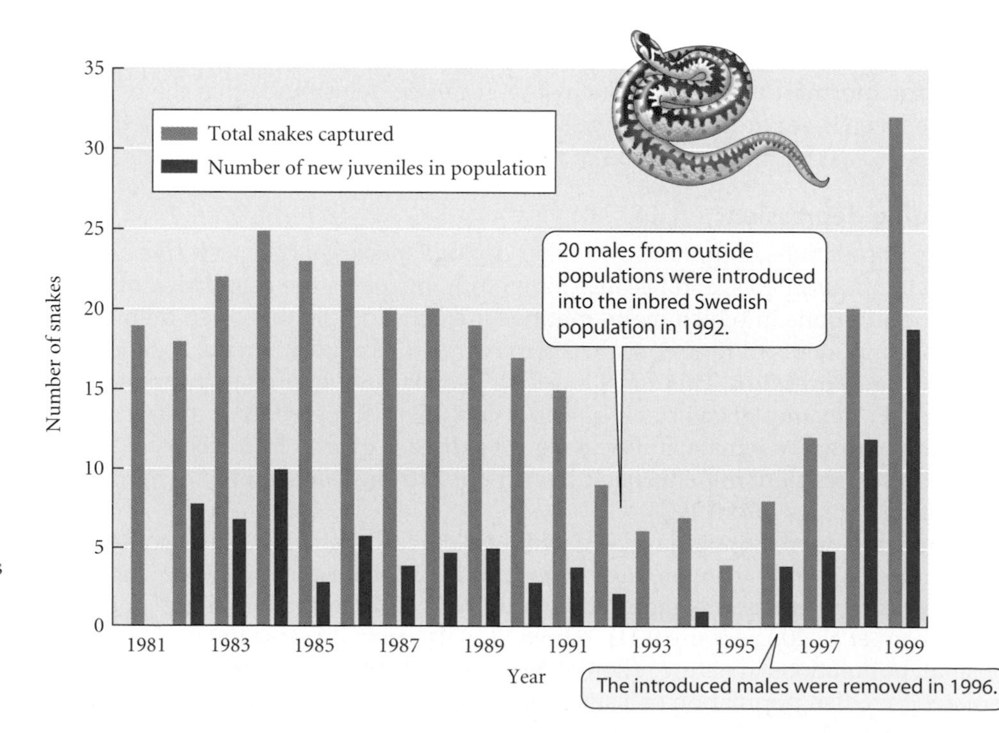

Figure 9.12 Population decline and increase in an inbred population of adders in Sweden. The gold bars represent the total number of males found in the population each year; the blue bars show the number of juveniles recruited into the population. (After Madsen et al. 1999.)

Answering this question requires that we know what fraction of the loci in a population are polymorphic, how many alleles are present at each locus, and what their frequencies are. To know this, we need to count both the monomorphic (invariant) and polymorphic loci in a random sample of loci. Ordinary phenotypic characters cannot provide this information because we cannot count how many genes contribute to phenotypically uniform traits.

In 1966, Richard Lewontin and John Hubby, working with *Drosophila pseudoobscura*, addressed this question in a landmark paper. They reasoned that because most loci code for proteins (including enzymes), an invariant enzyme should signal a monomorphic locus and a variable enzyme, a polymorphic locus. Biochemists had already devised techniques for visualizing certain proteins. In **electrophoresis**, a tissue extract (or a homogenate of a whole animal such as *Drosophila*) is placed in a starch gel or some other medium through which proteins can slowly move. An electrical current is applied to the gel, and the proteins move at a rate that depends on the molecules' size and net electrical charge. Certain amino acid substitutions alter the net charge, so some variants of the same protein, encoded by different alleles, can be distinguished by their mobility. The position of a particular enzyme in the gel can be visualized by letting the enzyme react with a substrate, then making the product visible as a colored blot by subjecting it to further reactions. If the locus is monomorphic, samples from all individuals display the same electrophoretic mobility; if it is polymorphic, mobility varies, and the various homozygotes and heterozygotes are distinguishable (Figure 9.13). A number of different enzymes and other proteins, each representing a different locus, can be investigated in this way, thus providing an estimate of the fraction of polymorphic loci. The amount of genetic variation is often underestimated by this technique, however, because not all amino acid substitutions alter electrophoretic mobility. Electrophoretically distinguishable forms of an enzyme that are encoded by different alleles are called **allozymes**.

Lewontin and Hubby examined 18 loci in populations of *Drosophila pseudoobscura*. In every population, about a third of the loci were polymorphic, represented by two to six different alleles, and these alleles segregated at high frequencies. The proportion of heterozygotes is a good measure of how nearly equal in frequency the alleles are (see Figure 9.5 and Box A). Assuming Hardy-Weinberg equilibrium, the frequency of heterozygotes (H) at a locus is $1 - \Sigma p_i^2$, where p_i is the frequency of the ith allele and p_i^2 is the frequency of homozygote A_iA_i. Averaged over all 18 loci (including the monomorphic ones), the frequency of heterozygotes was about 0.12 in each of Lewontin and Hubby's populations. This is equivalent to saying that an average individual is heterozygous at 12 percent of its loci. (This calculation is called the AVERAGE HETEROZYGOSITY, $\bar{H}$.) In a similar study of a human population, Harris and Hopkinson (1972), expanding an earlier study (Harris 1966), found that 28 percent of 71 loci were polymorphic, and that the average heterozygosity ($\bar{H}$) was 0.07. Since these pioneering studies, other investigators have examined hundreds of other species, most of which proved to have similarly high levels of genetic variation.

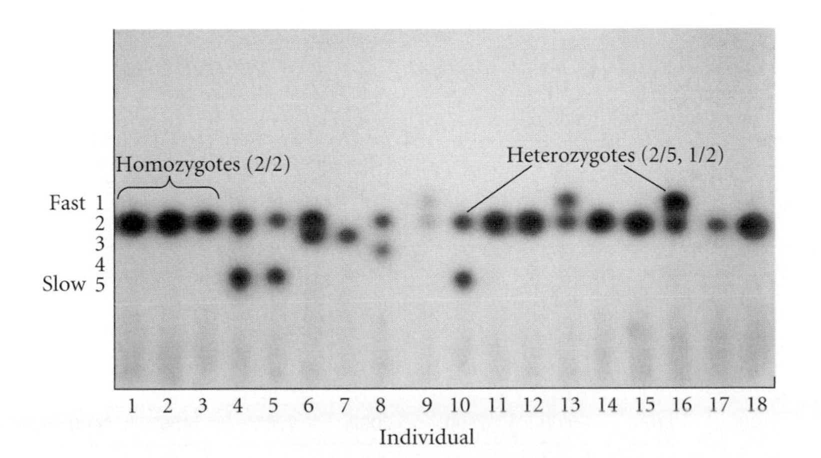

Figure 9.13 An electrophoretic gel, showing genetic variation in the enzyme phosphoglucomutase among 18 individual killifishes (*Fundulus zebrinus*). Five allozymes (alleles) can be distinguished by differences in mobility. The fastest, at top, is allele 1; the slowest, at bottom, is allele 5. Homozygotes display a single band, and heterozygotes two bands. From left to right, the genotypes are 2/2, 2/2, 2/2, 2/5, 2/5, 2/3, 3/3, 2/4, 1/2, 2/5, 2/2, 2/2, 1/2, 2/2, 2/2, 1/2, 2/2, 2/2. (Courtesy of J. B. Mitton.)

Lewontin and Hubby's paper (see also Lewontin 1974) had a great impact on evolutionary biology. Their data, and Harris's, confirmed that *almost every individual in a sexually reproducing species is genetically unique* (even with only two alleles each, 3000 polymorphic loci—the estimate for humans—could generate $3^{3000} = 10^{1431}$ genotypes, an unimaginably large number). Populations are far more genetically diverse than almost anyone had previously imagined. Lewontin and Hubby asked what factors could be responsible for so much variation, and their answers set in motion a research agenda that has kept population geneticists busy ever since. Their central question was, "Do forces of natural selection maintain this variation, or is it neutral, subject only to the operation of random genetic drift?"

Electrophoresis provided a tool for investigating many other questions in addition to this one. Before 1966, genes in nature could be studied only with simple phenotypic polymorphisms such as the coloration of the snow goose (see Figure 9.1A). But relatively few species have such polymorphisms, and their Mendelian basis often could not be documented in species that cannot be bred in captivity. Proteins provided, in almost every species, abundant polymorphisms with a clear genetic basis. These polymorphisms could be studied in their own right (e.g., to study natural selection), or they could be used simply as genetic markers to determine, for example, which individuals mate with each other. They have been widely used to determine how much related species or populations differ from one another genetically. And they were a major step toward the ever-increasing use of molecular information in evolutionary biology.

Variation at the DNA level

The first study of genetic variation using complete DNA sequencing was carried out by Martin Kreitman (1983), who sequenced 11 copies of a 2721-base-pair (bp) region in *Drosophila melanogaster* that includes the locus coding for the enzyme alcohol dehydrogenase (ADH). Throughout the world, populations of this species are polymorphic for two common electrophoretic alleles, "fast" (*Adh*^{F}) and "slow" (*Adh*^{S}). Kreitman sequenced only 11 gene copies because DNA sequencing at that time was still extremely laborious. Because of technical advances, studies today routinely include considerably larger samples.

Kreitman sequenced the four exons and three introns of the *Adh* gene, as well as noncoding flanking regions on either side (Figure 9.14). Among only 11 gene copies, he found 43 variable base pair sites, as well as six insertion/deletion polymorphisms (presence or absence of short runs of base pairs). Fewer sites were variable in the exons (1.8 percent) than in the introns (2.4 percent). The most striking discovery was that all but one of the 14 variations in the coding region were synonymous substitutions, the exception being a single nucleotide change responsible for the single amino acid difference between the *Adh*^{S} and *Adh*^{F} alleles.

Considerable sequence variation, especially synonymous variation, has been found in most of the genes and organisms that have been examined since Kreitman's pioneering study. A common measure of variation, analogous to the average heterozygosity ($\bar{H}$) used

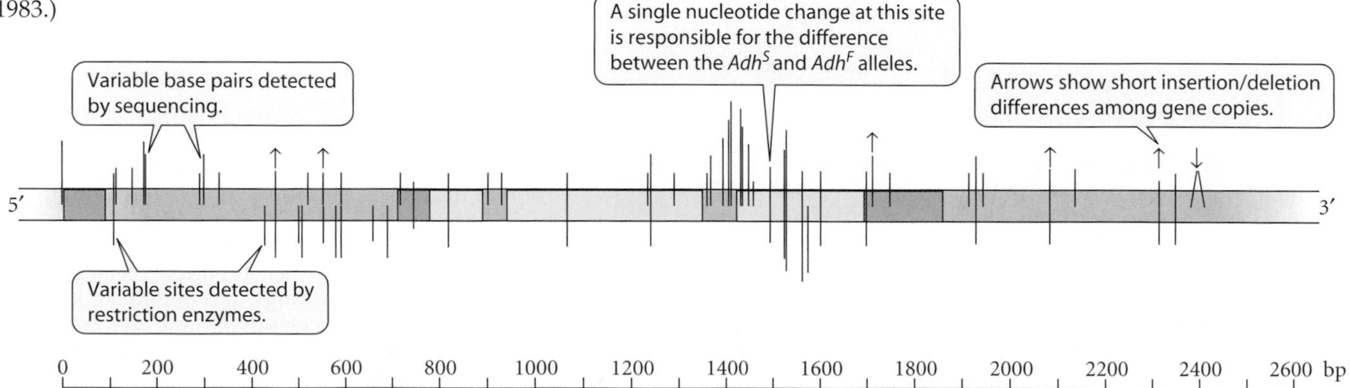
A single nucleotide change at this site is responsible for the difference between the AdhS and AdhF alleles.
Variable base pairs detected by sequencing.
Arrows show short insertion/deletion differences among gene copies.
5′
3′
Variable sites detected by restriction enzymes.
0
200
400
600
800
1000
1200
1400
1600
1800
2000
2200
2400
2600 bp

Figure 9.14 Nucleotide variation at the *Adh* locus in *Drosophila melanogaster*. Four exons (colored blocks) are separated by introns (gray blocks) are diagrammed. The yellow blocks represent the coding regions of the exons. The lines intersecting the diagram from above show the positions of 43 variable base pairs and 6 insertions or deletions (indicated by arrows) detected by sequencing 11 copies of this gene. The lines intersecting the diagram from below show the positions of 27 variable sites detected by another method (restriction enzymes) among 87 gene copies. The scale below shows base pair position along the sequence. (After Kreitman 1983.)

to quantify allozyme variation, is the average NUCLEOTIDE DIVERSITY PER SITE (π), the proportion of nucleotide sites at which two gene copies ("sequences") randomly taken from a population differ. For Kreitman's *Adh* sample, $\pi = 0.0065$, although over the genome as a whole, *D. melanogaster* is much more variable ($\pi = 0.05$) (Li 1997). In human populations, the average nucleotide diversity per site across loci is $\pi = 0.0008$.

Multiple loci and the effects of linkage

Each gene is LINKED to certain other genes, meaning that they are *physically associated* on the same chromosome. (In genetics and evolutionary biology, LINKAGE refers *only* to such a physical association between loci, not to the functional relationship or any other relationship among genes.) This linkage has important consequences. For example, under some circumstances, changes in allele frequencies at one locus cause correlated changes at other loci with which that locus is linked. In other words, if chromosomes with A_1 alleles tend to carry B_1 rather than B_2 alleles at another locus, increasing the frequency of A_1 would cause B_1 to increase in frequency.

Suppose two populations, one consisting only of genotype $A_1A_1B_1B_1$ and the other only of $A_2A_2B_2B_2$, are mixed together and mate at random. In the next generation, there will be three genotypes in the population, A_1B_1/A_1B_1, A_1B_1/A_2B_2, and A_2B_2/A_2B_2 (the slash separates the allele combinations an individual receives from its two parents). In this generation, there is a perfect association, or correlation, between specific alleles at the two loci: A_1 with B_1 and A_2 with B_2. This association is referred to as **linkage disequilibrium**. If there is no such association, the loci are in **linkage equilibrium**. (These two terms are slightly misleading because they may not necessarily refer to physical linkage; alleles at different loci can be associated even if they are not physically linked, or the loci may be physically linked, yet their alleles may not be associated.)

Recombination during meiosis reduces the level of linkage disequilibrium and brings the loci toward linkage equilibrium. If there were no recombination in our example, gametes would carry only the allele combinations A_1B_1 and A_2B_2, which, when united, could generate only the same three genotypes as before. However, recombination in the double heterozygote (A_1B_1/A_2B_2) gives rise to some A_1B_2 and A_2B_1 gametes, which, by uniting with A_1B_1 or A_2B_2 gametes, produce genotypes such as A_1B_1/A_1B_2. As this process continues, the "deficient" allele combinations (A_1B_2, A_2B_1) and genotypes increase in frequency slowly from generation to generation, until the alleles at each locus are randomized with respect to alleles at the other locus and there is no correlation between them (Figure 9.15). The tighter the linkage between the loci, the slower this process is.

When the loci are in linkage equilibrium, knowing which *A* allele an egg or sperm carried would not help us predict its *B* allele. In this case, the probability (frequency) of a gamete's carrying any one of the four possible allele combinations is the product of the probabilities of the alleles considered separately. These probabilities are the allele frequencies. If we denote the frequencies of alleles A_1 and A_2 as p_A and q_A, ($p_A + q_A = 1$), and of B_1 and B_2 as p_B and q_B, ($p_B + q_B = 1$), the frequency of the A_1B_1 combination among the eggs (or sperm) produced by the population as a whole is then p_Ap_B, and the frequency (proportion) of genotype $A_1A_1B_1B_1$ (A_1B_1/A_1B_1) in the next generation is $(p_Ap_B)^2$, or $p_A^2p_B^2$. If the observed frequencies of all the genotypes conform to the expected frequencies thus calculated, we can conclude that the loci are in a state of linkage equilibrium.Whether loci are in linkage equilibrium or linkage disequilibrium, the genotype frequencies at each locus, viewed individually, conform to Hardy-Weinberg frequencies.

In panmictic sexually reproducing populations, pairs of polymorphic loci often are found to be in linkage equilibrium, or nearly so. There are some interesting exceptions, however. For example, the European primrose *Primula vulgaris* is HETEROSTYLOUS, meaning that plants within a population differ in the lengths of stamens and style (pistil). Almost all plants have either the "pin" phenotype, with long style and short stamens, or the "thrum" phenotype, with short style and long stamens (Figure 9.16). In most experimental crosses, this difference is inherited as if it were due to a single pair of alleles, with *thrum* dominant over *pin*. Rarely, however, "homostylous" progeny are produced, in which the female and male structures are equal in length (either short or long). Thus style and stamen length are actually de-

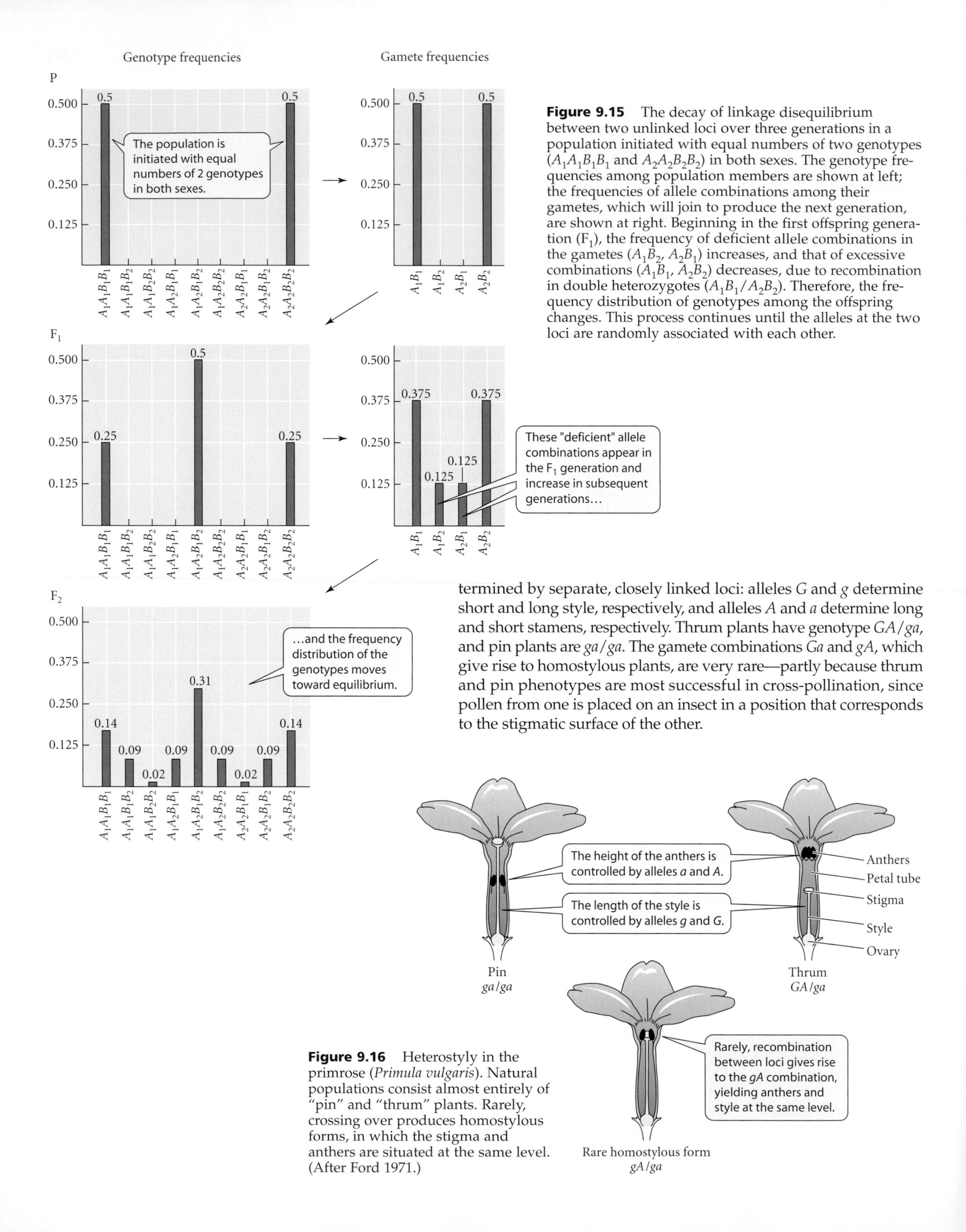

Figure 9.15 The decay of linkage disequilibrium between two unlinked loci over three generations in a population initiated with equal numbers of two genotypes ($A_1A_1B_1B_1$ and $A_2A_2B_2B_2$) in both sexes. The genotype frequencies among population members are shown at left; the frequencies of allele combinations among their gametes, which will join to produce the next generation, are shown at right. Beginning in the first offspring generation (F_1), the frequency of deficient allele combinations in the gametes (A_1B_2, A_2B_1) increases, and that of excessive combinations (A_1B_1, A_2B_2) decreases, due to recombination in double heterozygotes (A_1B_1/A_2B_2). Therefore, the frequency distribution of genotypes among the offspring changes. This process continues until the alleles at the two loci are randomly associated with each other.

termined by separate, closely linked loci: alleles *G* and *g* determine short and long style, respectively, and alleles *A* and *a* determine long and short stamens, respectively. Thrum plants have genotype *GA*/*ga*, and pin plants are *ga*/*ga*. The gamete combinations *Ga* and *gA*, which give rise to homostylous plants, are very rare—partly because thrum and pin phenotypes are most successful in cross-pollination, since pollen from one is placed on an insect in a position that corresponds to the stigmatic surface of the other.

Figure 9.16 Heterostyly in the primrose (*Primula vulgaris*). Natural populations consist almost entirely of "pin" and "thrum" plants. Rarely, crossing over produces homostylous forms, in which the stigma and anthers are situated at the same level. (After Ford 1971.)

Linkage disequilibrium is common in asexual populations (see Chapter 17) because they undergo little recombination. It is also found among very closely situated molecular markers, such as sites within genes (see Chapter 17). Researchers in human genetics can therefore use molecular markers, in a procedure called LINKAGE DISEQUILIBRIUM MAPPING, to find nearby mutations that cause genetic diseases. A great deal of such research, based on the recently sequenced human genome, is under way, and evolutionary geneticists are playing a major role in developing methods for analyzing such data.

Variation in quantitative traits

SOURCES OF VARIATION. Discrete genetic polymorphisms in phenotypic traits, such as pin or thrum flowers, are much less common than slight differences among individuals, such as variation in the number of bristles on the abdomen of *Drosophila* or in weight or nose shape among humans. Such variation, called **quantitative**, or CONTINUOUS, or METRIC variation, often approximately fits a **normal distribution** (Figure 9.17). The genetic component of such variation is often **polygenic**: that is, it is due to variation at several or many loci, each of which contributes to the variation in phenotype.

A simple model of the relation between genotype and phenotype in a case of quantitative variation, in which we envision only two variable loci, might be:

	A_1A_1	A_1A_2	A_2A_2
B_1B_1	5	6	7
B_1B_2	7	8	9
B_2B_2	9	10	11

In this example, relative to the genotype $A_1A_1B_1B_1$, each A_2 allele adds one unit, on average, and each B_2 allele adds two units to the phenotype. This is a model of purely **additive** allele effects.

Quantitative characters often vary both because of genes and because of nongenetic environmental factors, and by "developmental noise." The latter term refers to ineradicable variations in developmental processes that produce variation among individuals, as well as variation within individuals (e.g., asymmetry between the two sides of the same animal, which obviously have shared both the same genotype and the same environment).

If variation stems from both genetic and environmental sources, it is useless to ask whether a characteristic is "genetic" *or* "environmental," as if it had to be exclusively one or the other. Moreover, the relative amounts of genetic and environmental variation can differ with different circumstances, even in the same population. This is especially true if genotypes differ in their **norm of reaction**, the variety of different phenotypic states that can be produced by a single genotype under different environmental conditions. For example, Gupta and Lewontin (1982) measured the mean number of abdominal bristles in

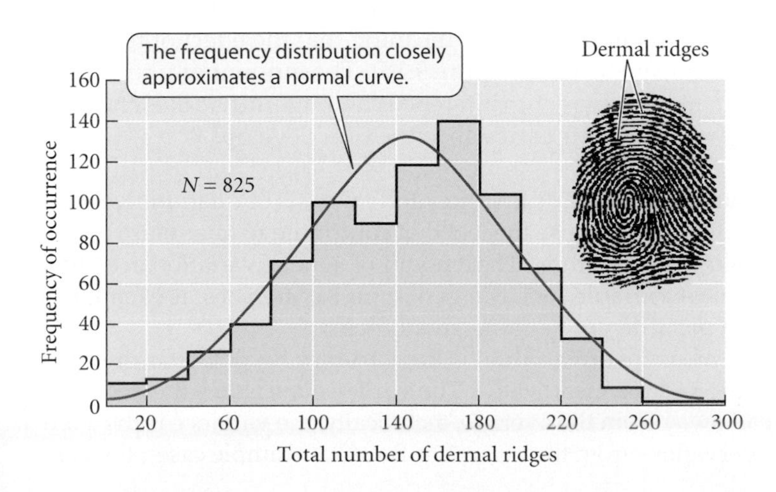

Figure 9.17 The frequency distribution of the number of dermal ridges, summed over all ten fingertips, in a sample of 825 British men. The distribution nearly fits a normal curve (red line). The number of dermal ridges is an additively inherited polygenic character, with a heritability of about 0.95. (After Holt 1955.)

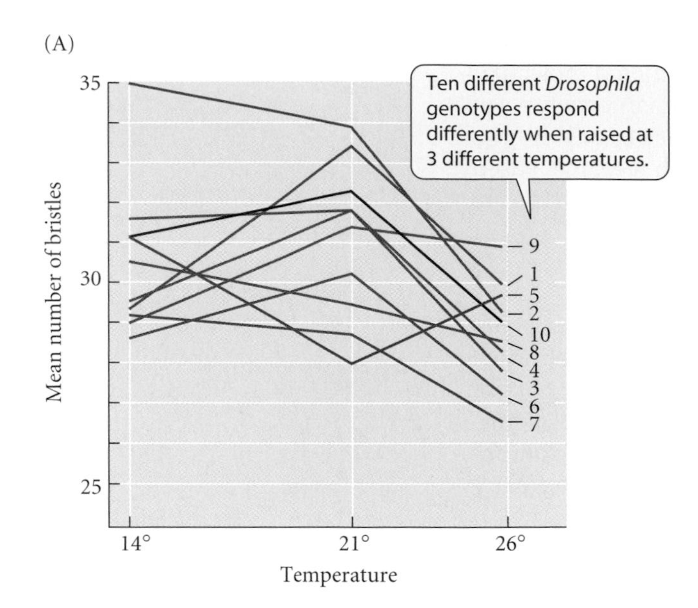
(A)
Ten different *Drosophila* genotypes respond differently when raised at 3 different temperatures.
Mean number of bristles
35
30
25
14°
21°
26°
Temperature
9
1
5
2
10
8
4
3
6
7

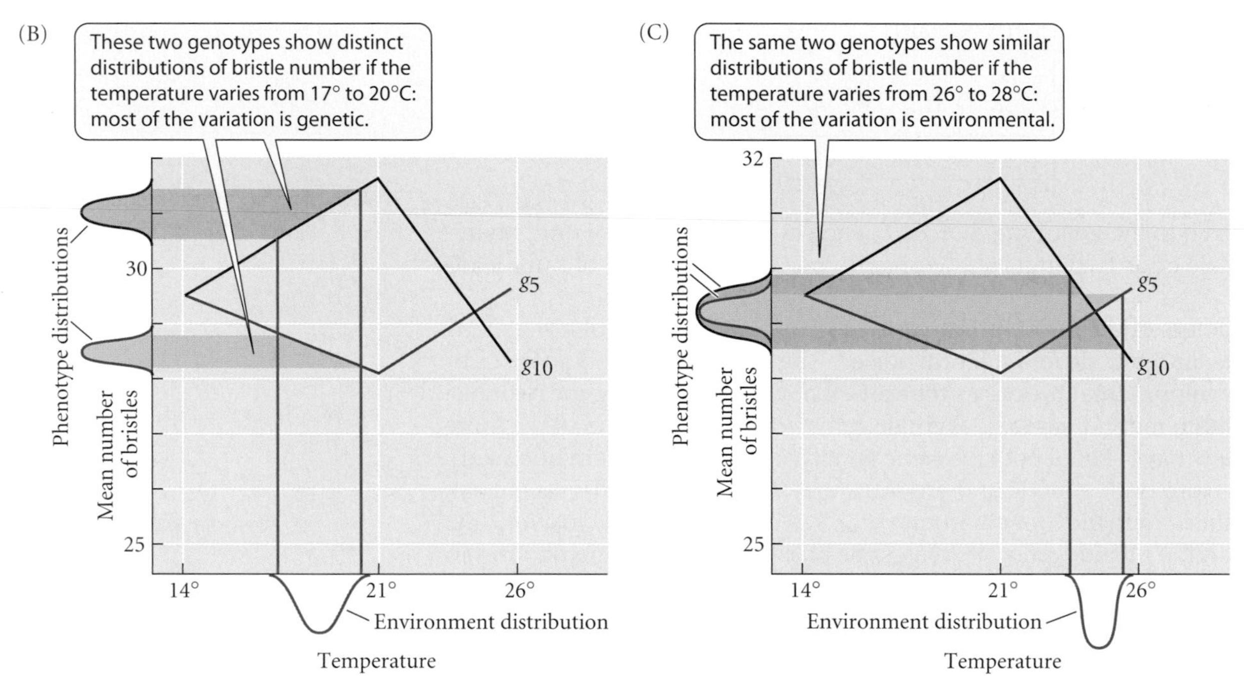
(B)
These two genotypes show distinct distributions of bristle number if the temperature varies from 17° to 20°C: most of the variation is genetic.
Phenotype distributions
Mean number of bristles
30
25
g5
g10
14°
21°
26°
Environment distribution
Temperature
(C)
The same two genotypes show similar distributions of bristle number if the temperature varies from 26° to 28°C: most of the variation is environmental.
32
25
Phenotype distributions
Mean number of bristles
g5
g10
14°
21°
26°
Environment distribution
Temperature

Figure 9.18 An example of genotype × environment interaction and how it can affect the relative contributions of genetic and environmental variation to phenotypic variation. (A) The number of bristles on the abdomen of male *Drosophila pseudoobscura* of 10 genotypes, each reared at three different temperatures. Temperature affects the development of bristle number differently for each genotype, so there is an interaction between genotype and environment. (B) If a population of flies of genotypes 5 and 10 from part A developed at a normally distributed range of temperatures between about 17°C and 20°C, the distribution of phenotypes would be bimodal, and most of the variation would be genetic, because at these temperatures the two genotypes differ greatly in phenotype. (Scan up along the vertical lines, which mark 17°C and 20°C, find their intersection with a genotype's norm of reaction, and then read across to the *y*-axis to find the number of bristles expected of flies that developed in that temperature range.) Flies of each genotype are expected to vary in bristle number since they develop at a variety of temperatures. (C) If the same two genotypes developed at temperatures varying between about 23°C and 25°C, most of the phenotypic variation would be environmental because the two genotypes would respond similarly to these temperatures. (After Gupta and Lewontin 1982.)

each of ten genotypes of *Drosophila pseudoobscura* raised at three different temperatures. They found a **genotype × environment interaction**, meaning that the effect of temperature on phenotype differed among genotypes (Figure 9.18). The amount of phenotypic variation in a population that is due to genetic differences among individuals thus depends on the particular range of temperatures at which the flies develop.

ESTIMATING COMPONENTS OF VARIATION. The description and analysis of quantitative variation are based on statistical measures because the loci that contribute to quantitative variation generally cannot be singled out for study. The amount of genetic variation in a character depends on the number of variable loci, the genotype frequencies at each locus (Figure 9.19), and the phenotypic difference among genotypes.

The most useful statistical measure of variation is the **variance**, which quantifies the spread of individual values around the mean value. The variance measures the degree to which individuals are spread away from the average; technically, the variance is the average squared deviation of observations from the mean (see Box C). In simple cases, the vari-

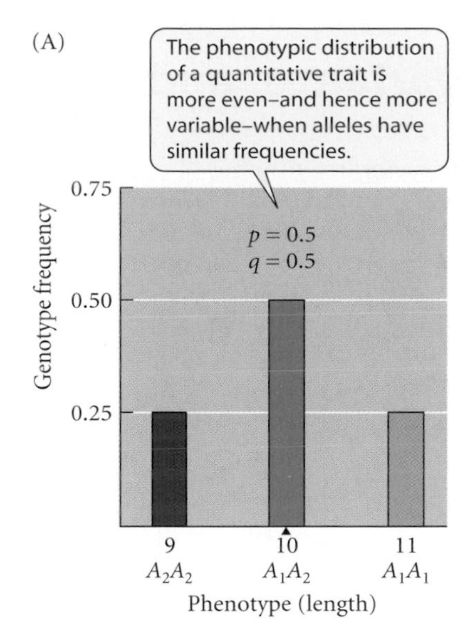

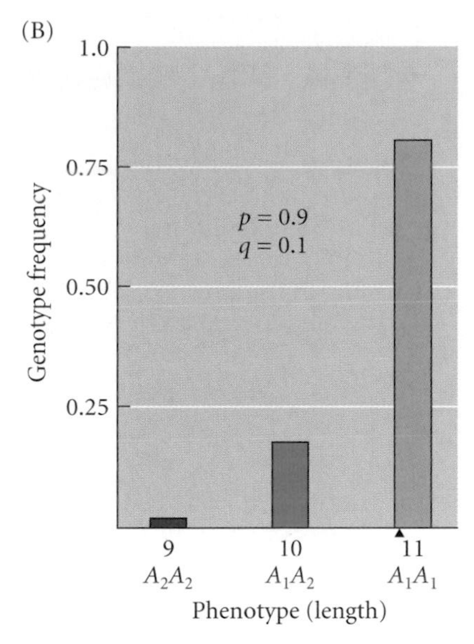

Figure 9.19 Variation in a quantitative trait, such as body length, due to alleles with frequencies (A) $p = 0.5$, $q = 0.5$ or (B) $p = 0.9$, $q = 0.1$. The black triangle denotes the mean. The distribution of lengths is more even, hence more variable, when the alleles have the same frequencies. (In the more variable population B, it would be harder to guess the phenotype of a randomly chosen individual.) The genetic variance, V_G, equals 0.500 in A and 0.472 in B. Hardy-Weinberg equilibrium is assumed in both cases.

ance in a phenotypic character (V_P) is the sum of **genetic variance** (V_G) and **environmental variance** (V_E): $V_P = V_G + V_E$. Oversimplifying, we can imagine that each genotype in a population has an average phenotypic value (of, say, body length), but that individuals with that genotype vary in their phenotypes due to environmental effects or developmental noise. The amount of variation among the averages of the different genotypes is the genetic variance, V_G, and the average amount of variation among individuals with the same genotype (at the relevant loci) is the environmental variance, V_E. The proportion of the phenotypic variance that is genetic variance is the **heritability** of a trait, denoted h^2. Thus,

$$h^2 = V_G/(V_G + V_E)$$

One way of detecting a genetic component of variation, and of estimating V_G and h^2, is to measure correlations* between parents and offspring, or between other relatives. For example, suppose that in a population, the mean value of a character in the members of each brood of offspring was exactly equal to the value of that character averaged between their two parents (the MIDPARENT MEAN) (Figure 9.20A). So perfect a correlation clearly would imply a strong genetic basis for the trait. In fact, in this instance,

*More properly, the *regression* of offspring mean on the mean of the two parents. The regression coefficient measures the slope of the relationship, and is conceptually related to the correlation.

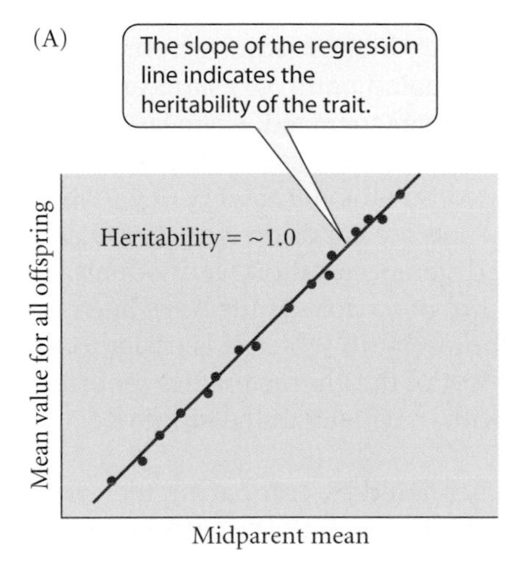

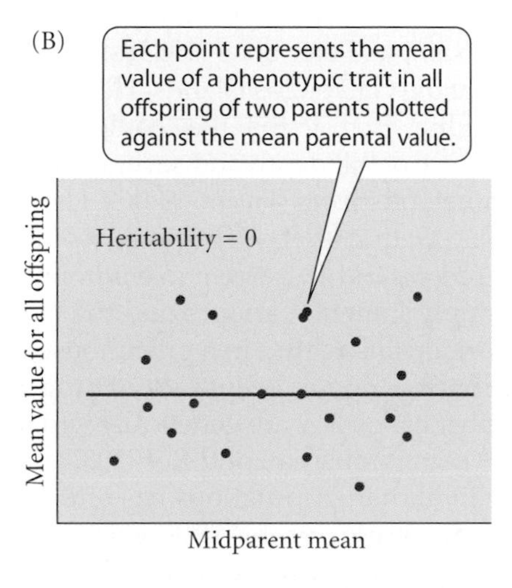

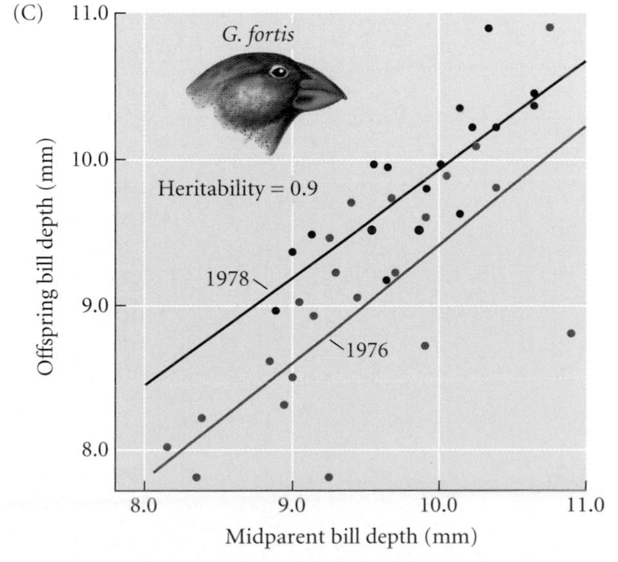

Figure 9.20 The relationship between the phenotypes of offspring and parents. Each point represents the mean of a brood of offspring, plotted against the mean of their two parents (midparent mean). (A) A hypothetical case in which offspring means are nearly identical to midparent means. The heritability is nearly 1.00. (B) A hypothetical case in which offspring and midparent means are not correlated. The slope of the relationship and the heritability are approximately 0.00. (C) Bill depth in the ground finch *Geospiza fortis* in 1976 and in 1978. Although offspring were larger in 1978, the slope of the relationship between offspring and midparent was nearly the same in both years. The heritability, estimated from the slope, was 0.90. (C after Grant 1986, based on Boag 1983.)

BOX 9C Mean, Variance, and Standard Deviation

Suppose we have measured a character in a number of specimens. The character may vary continuously, such as body length, or discontinuously, such as the number of fin rays in a certain fin of a fish. Let X_i be the value of the variable in the ith specimen (e.g., $X_3 = 10$ cm in fish number 3). If we have measured n specimens, the sum of the values is $X_1 + X_2 + \cdots + X_n$, or

$$\sum_{i=1}^{k} X_i$$

(or simply ΣX_i). The **arithmetic mean** (or average) is

$$\bar{x} = \frac{\sum x_i}{n}$$

If the variable is discontinuous (e.g., fin rays), we may have n_1 individuals with value X_1, n_2 with value X_2, and so on for k different values. The arithmetic mean is then

$$\bar{x} = \frac{n_1X_1 + n_2X_2 + \cdots + n_k}{n_1 + n_2 + \cdots + n_k}$$

The sum of n_i equals n, so we may write this as

$$\bar{x} = \frac{n_1X_1}{n} + \frac{n_2X_2}{n} + \cdots + \frac{n_kX}{n}$$

If we denote $n_i/n = f_i$, the frequency of individuals with value X_i, this becomes

$$\bar{x} = \sum_{i=1} (f_iX_i)$$

For example, in a sample of $n = 100$ fish, we may have $n_1 = 16$ fish with 9 fin rays ($X_1 = 9$), $n_2 = 48$ fish with 10 rays ($X_2 = 10$), and $n_3 = 36$ fish with 11 rays ($X_3 = 11$). There are thus three phenotypic classes ($k = 3$). The mean is

$$\bar{x} = \sum_{i=1} (f_iX_i) = (0.16)(9) + (0.48)(10) + (0.36)(11) = 10.2$$

Because we have only a sample from the fish population, this sample mean is an *estimate* of the true (parametric) mean of the population, which we can know only by measuring every fish in the population.

How shall we measure the amount of variation? We might measure the *range* (the difference between the two most extreme values), but this measure is very sensitive to sample size. A larger sample might reveal, for example, rare individual fish with 5 or 15 fin rays. These rare individuals do not contribute to our impression of the degree of variation. For this and other reasons, the most commonly used measures of variation are the variance and its close relative, the standard deviation. The true (parametric) variance is estimated by the mean value of the square of an observation's variation from the arithmetic mean:

$$V = \frac{(X_1 - \bar{x})^2 + (X_2 - \bar{x})^2 + \cdots + (X_n - \bar{x})^2}{n_1 - 1}$$

$$= \frac{1}{n-1} \sum_{i=1} n_i (Xi - \bar{x})^2$$

For statistical reasons, the denominator of a sample variance is $n - 1$ rather than n. In our hypothetical data on fin ray

V_G/V_P (i.e., the heritability, h^2) would equal 1.0: all the phenotypic variation would be accounted for by genetic variation. If the correlation were lower, with some of the phenotypic variation perhaps due to environmental variation, the heritability would be lower.

In a real-life example, Peter Boag (1983) studied heritability in the medium ground finch (*Geospiza fortis*) of the Galápagos Islands. He kept track of mated pairs of *G. fortis* and their offspring by banding them so that they could be individually recognized. He measured the phenotypic variance of bill depth and several other features, and correlated the phenotypes of parents and their offspring (Figure 9.20C). He estimated that the heritability of variation in bill depth was 0.9; that is, about 90 percent of the phenotypic variation was attributable to genetic differences, and 10 percent to environmental differences, among individuals.

More commonly, genetic variance and the heritability of various traits have been estimated from organisms reared in a greenhouse or laboratory, in which it is easier to set up controlled matings and to keep track of progeny. Most of the characteristics reported, for a great number of species, are genetically variable, with h^2 usually in the range of about 0.1 to about 0.9 (e.g., Mousseau and Roff 1987).

Heritability in human populations has often been estimated by comparing the correlation between dizygotic ("fraternal," or "nonidentical") twins with the correlation between monozygotic ("identical") twins, which are expected to have a higher correlation because they are genetically identical. Physical characteristics such as height, finger length,

BOX 9C (continued)

counts,

$$V = \frac{16(9-10.2)^2 + 48(10-10.2)^2 + 36(11-10.2)^2}{99} = 0.485$$

The variance is a very useful statistical measure, but it is hard to visualize because it is expressed in squared units. It is easier to visualize its square root, the **standard deviation** (see also Figure 4.22):

$$s = \sqrt{V}$$

For our hypothetical data, $s = \sqrt{0.485} = 0.696$. The meaning of this number can perhaps be best understood by contrast with a sample of 1 fish with 9 rays, 18 with 10 rays, and 81 with 11 rays—an intuitively less variable sample. Then $\bar{x} = 10.8$, $V = 0.149$, and $s = 0.387$. V and s are smaller in this sample than in the previous sample because more of the individuals are closer to the mean.

A continuous variable, such as body length, often has a bell-shaped, or normal, frequency distribution (Figure 1). In the mathematically idealized form of this distribution (which many real samples approximate quite well), about 68 percent of the observations fall within one standard deviation on either side of the mean, 96 percent within two standard deviations, and 99.7 percent within three. If, for example, body lengths in a sample of fish are normally distributed, then if $\bar{x} = 100$ cm and $s = 5$ cm (hence $V = 25$ cm^2), 68 percent of the fish are expected to range between 95 and 105 cm, and 96 percent to range between 90 and 110 cm. If the standard deviation is greater—say, $s = 10$ cm ($V = 100$ cm^2)—then, for the same mean, the range limits embracing 68 percent of the sample are broader: 90 cm and 100 cm.

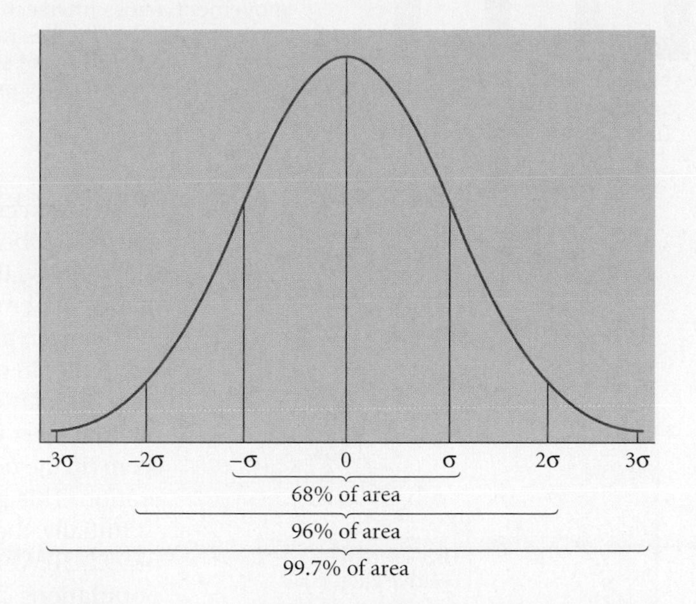

Figure 1 The normal distribution curve, with the mean taken as a zero-reference point, showing how the variable represented on the x-axis can be measured in standard deviations (σ). The bracketed areas show the fraction of the area under the curve (that is, the proportion of observations) embraced by one, two, and three standard deviations on either side of the mean. The true (parametric) value of the standard deviation is denoted by σ; the estimate of σ based on a sample is denoted s in this book

and head breadth are highly heritable (ca. 0.84–0.94) within populations, and the variation in dermatoglyphic traits (fingerprints) is nearly entirely genetically based (0.96) (Lynch and Walsh 1998).

RESPONSES TO ARTIFICIAL SELECTION. Because a character can be altered by selection only if it is genetically variable, **artificial selection** can be used to detect genetic variation in a character. To do this, an investigator (or a plant or animal breeder) breeds only those individuals that possess a particular trait (or combination of traits) of interest. Artificial selection may grade into natural selection, but the conceptual difference is that under artificial selection, the reproductive success of individuals is determined largely by a single characteristic chosen by the investigator, rather than by their overall capacity (based on all characteristics) for survival and reproduction.

Hundreds of such experiments have been performed. For example, Theodosius Dobzhansky and Boris Spassky (1969) bred the offspring of 20 wild female *Drosophila pseudoobscura* flies to form a base population, and from this population drew flies to establish several selected populations, each maintained in a large cage with food. Two of these populations were selected for positive phototaxis (a tendency to move toward light) and two for negative phototaxis (a tendency to move away from light). To select for phototaxis, the investigators put males and virgin females in a maze in which the flies had to

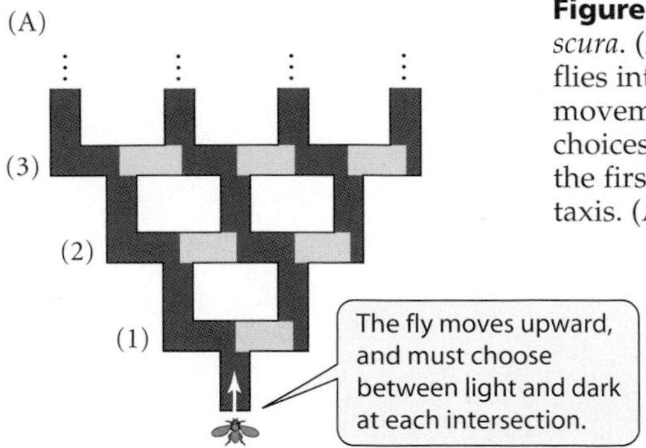

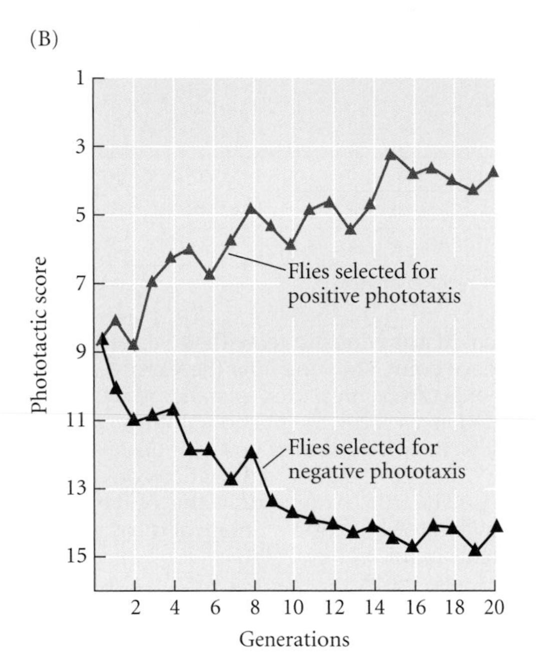

Figure 9.21 Selection for movement in response to light (phototaxis) in *Drosophila pseudoobscura*. (A) Diagram of part of a maze. The maze is oriented vertically below a light source, and flies introduced at the bottom move upward, choosing light or dark at each intersection. Lateral movement across intersections is prevented by barriers. The diagram shows the first 3 of the 15 choices made by the flies in the selection experiment. (B) The mean phototactic scores of flies in the first 20 generations of selection in two populations selected for positive and negative phototaxis. (After Dobzhansky and Spassky 1969.)

make 15 successive choices between light and dark pathways, ending up in one of 16 tubes (Figure 9.21A). Flies that made 15 turns toward light arrived at tube 1, those that made 15 turns toward dark arrived at tube 16, and those that made equal numbers of turns toward light and dark ended up in tubes 8 and 9. The mean and variance of phototactic scores were estimated from the number of flies in each of the 16 tubes. In each generation, 300 flies of each sex, from each population, were released into the maze, and the 25 flies of each sex that had the most extreme high score (in the positively selected population) or low score (in the negatively selected population) were saved to initiate the next generation. This procedure was repeated for 20 generations.

Initially, the flies were neutral, on average: the mean scores were 8 to 9 (Figure 9.21B). Very quickly, however, both the positively and negatively selected populations diverged, in opposite directions, from the initial mean. We may therefore infer that variation among flies in their response to light is partly hereditary. From the rapidity of the change, Dobzhansky and Spassky calculated that the heritability of phototaxis is about 0.09.

Experiments of this kind have shown that *Drosophila* species are genetically variable for almost every trait, including features of behavior (e.g., mating speed), morphology, life history (e.g., longevity), physiology (e.g., resistance to insecticides), and even features of the genetic system (e.g., the rate of crossing over). Artificial selection has been the major tool of breeders who have produced agricultural varieties of corn, tomatoes, pigs, chickens, and every other domesticated species, which often differ extremely in numerous characteristics. Such experiences reinforce the conclusion that *species contain genetic variation that could serve as the foundation for the evolution of almost all of their characteristics*. Moreover, the amount of genetic variation is so great that *most characters should be able to evolve quite rapidly*—far more quickly than Darwin ever imagined.

Variation among Populations

A few species consist of a single, panmictic population. The Devil's Hole pupfish (*Cyprinodon diabolis*), for example, is restricted to a single sinkhole near Death Valley, Nevada, and all the eels (*Anguilla rostrata*) from rivers throughout eastern North America and western Europe are thought to migrate to a single area near Bermuda to breed. The vast majority of species, however, are subdivided into several or many separate populations, with most mating taking place between members of the same population. Such populations of a single species often differ in genetic composition. Studies of differences among populations in different geographic areas, or **geographic variation**, have provided many insights into the mechanisms of evolution.

Patterns of geographic variation

If distinct forms or populations have overlapping geographic distributions, such that they occupy the same area and can frequently encounter each other, they are said to be **sympatric** (from the Greek *syn*, "together," and *patra*, "fatherland"). Populations with adjacent but nonoverlapping geographic ranges that come into contact are **parapatric** (Greek *para*, "beside"). Populations with separated distributions are **allopatric** (Greek *allos*,

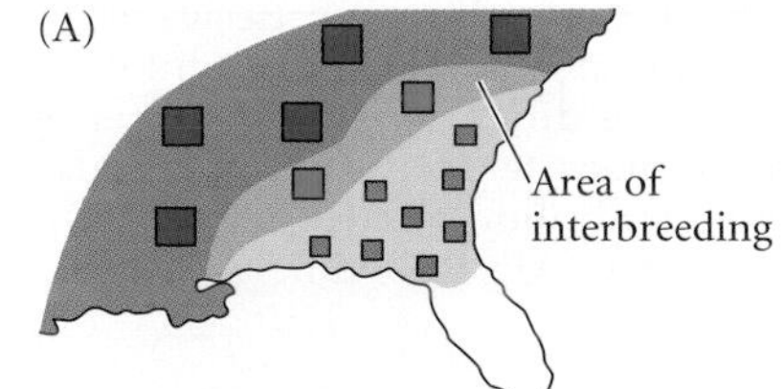

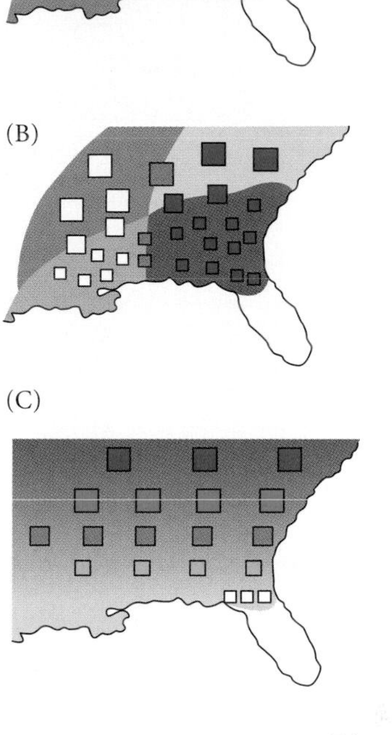

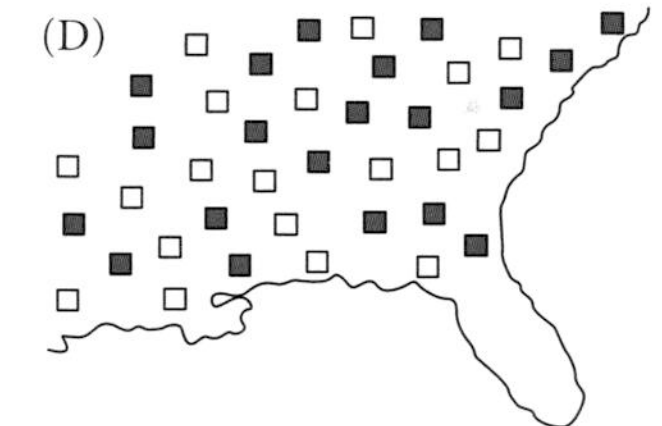

Figure 9.22 Highly diagrammatic representations of some common patterns of geographic variation within species. (A) Two classic subspecies that interbreed along a narrow border region. Size and color show concordant patterns of geographic variation. (B) Abrupt transition in two discordantly varying characters (size varies from north to south, color from west to east). (C) Concordant clines (gradual geographic change) in size and color. (D) A mosaic distribution of two phenotypes, as might be observed if they were associated with mosaic habitats (e.g., wet and dry).

"other"). Several patterns of geographic variation are frequently encountered (Figure 9.22) and can be the basis for the formation of new species, as discussed in Chapters 15 and 16.

A **subspecies**, or GEOGRAPHIC RACE, in zoological taxonomy means a recognizably distinct population, or group of populations, that occupies a different geographic area from other populations of the same species. (In botanical taxonomy, subspecies names are sometimes given to sympatric, interbreeding forms.) In some instances, subspecies differ in a number of features with concordant patterns of geographic variation (Figure 9.22A). For example, in the northern flicker (*Colaptes auratus*), the subspecies *auratus* and *cafer*, distributed in eastern and western North America, respectively, differ in the color of the wing feathers, in the crown and mustache marks, in the presence or absence of several other plumage marks, and in size (Short 1965; Moore and Price 1993). Despite these differences, the two populations interbreed in the Great Plains (Figure 9.23), forming a wide **hybrid zone** (a region in which genetically distinct parapatric forms interbreed).

Different characters often have discordant patterns of geographic variation (Figure 9.22B). For example, in the rat snake *Elaphe obsoleta* (Figure 9.24), some subspecies are distinguished by color and others by the pattern of stripes or blotches. These features, in turn, are not well correlated with the geographic distribution of mitochondrial DNA haplotypes, which distinguish western, central, and eastern groups of populations (Burbrink et al. 2000). Such discordance among different characters is also seen among human populations (see Figure 9.32). Many systematists opine that populations that differ by such discordant patterns should not be named as subspecies.

A gradual change in a character or in allele frequencies over geographic distance is called a **cline** (Figure 9.22C). For instance, body size in the white-tailed deer (*Odocoileus virginianus*) increases with increasing latitude over much of North America. This positive relationship between body size and latitude is so common in mammals and birds that it has been named BERGMANN'S RULE. This pattern, which is too consistent to be attributed to

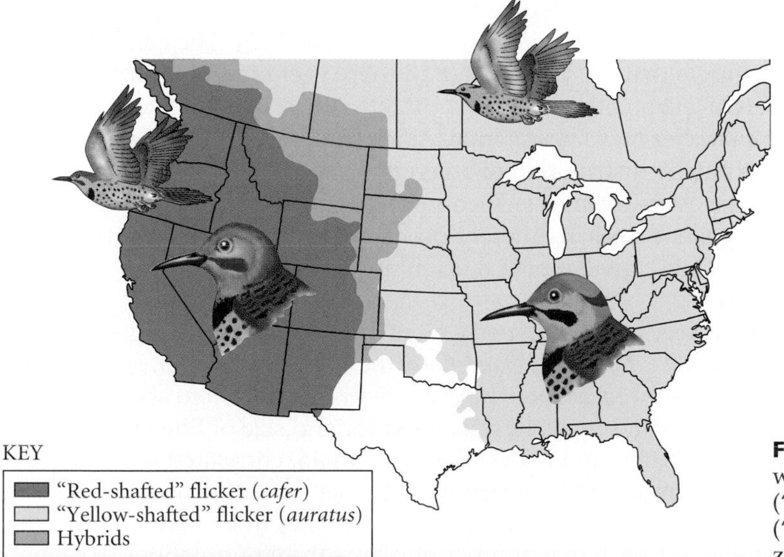

Figure 9.23 Two subspecies of a common North American woodpecker, the northern flicker (*Colaptes auratus*). The eastern ("yellow-shafted") subspecies (*C. a. auratus*) and the western ("red-shafted") subspecies (*C. a. cafer*) form a broad hybrid zone. (After Moore and Price 1993.)

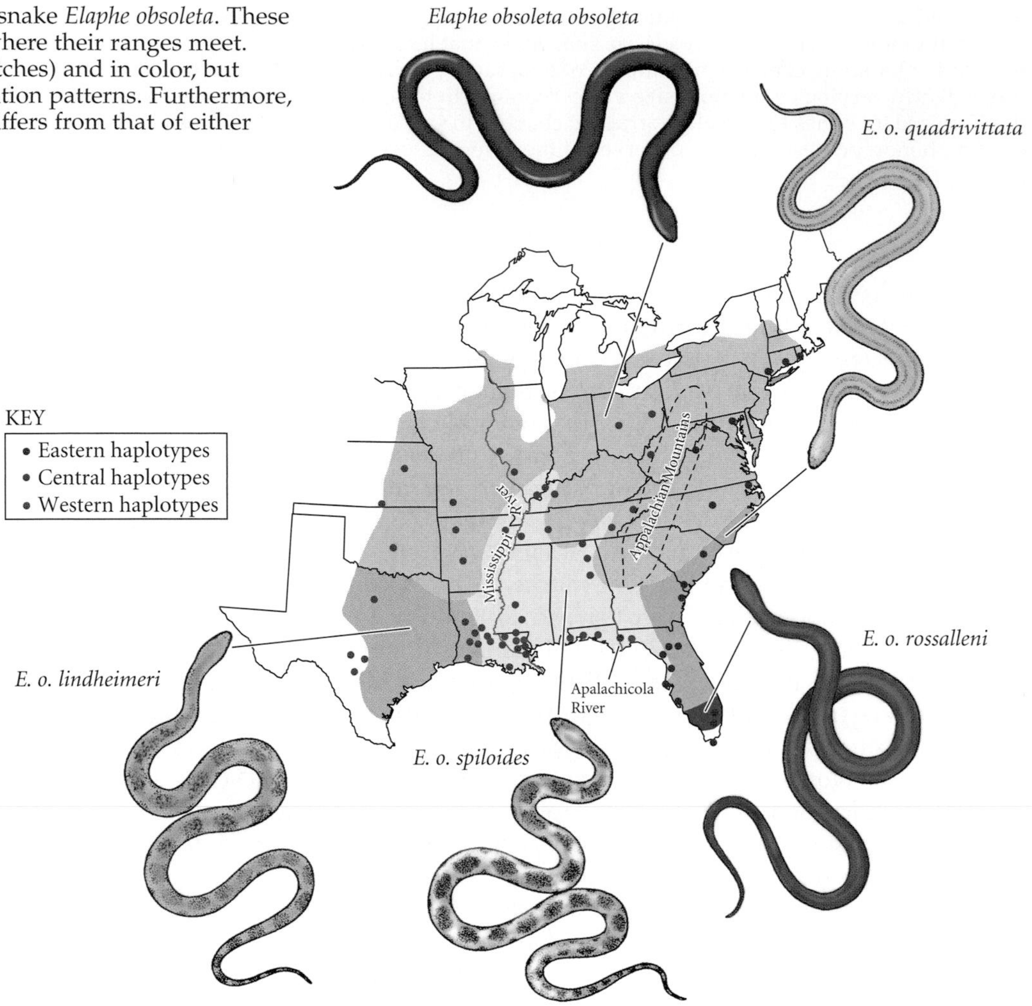

Figure 9.24 Five subspecies of the rat snake *Elaphe obsoleta*. These parapatric geographic races interbreed where their ranges meet. The races differ in pattern (stripes or blotches) and in color, but these characters have discordant distribution patterns. Furthermore, the distribution of mtDNA haplotypes differs from that of either character. (After Conant 1958.)

chance, provides important evidence of *adaptive geographic variation* due to natural selection. Larger body size is thought to be advantageous for homeotherms in colder climates because it reduces the surface area, relative to body mass, over which body heat is lost.

The alcohol dehydrogenase polymorphism in *Drosophila melanogaster* (see Figure 9.14) shows a cline in allele frequencies, with the Adh^F allele gradually increasing from low to high frequency as one moves toward higher latitudes (Figure 9.25). The consistency of this pattern strongly suggests that it is adaptive, and there is evidence that the trend is correlated with rainfall (Oakeshott et al. 1982).

In some instances, a character may show a more or less mosaic pattern (Figure 9.22D), often correlated with the patchy distribution of an environmental factor or habitat. Such habitat-associated phenotypes are often called **ecotypes**. In some instances, the same ecotypic feature differs in its genetic basis in different populations of a species, implying that it evolved independently. Thus convergent evolution may occur among different populations of the same species.

One of the most detailed studies of ecotypes was carried out by Jens Clausen, David Keck, and William Hiesey (1940). *Potentilla glandulosa*, the sticky cinquefoil, is a member of the rose family, distributed in western North America from sea level to above the timberline. Many characteristics, including plant size, the color and size of flowers, leaf morphology, and flowering time, differ among populations and are correlated with altitude.

Clausen and his collaborators set out to determine whether the differences among ecotypes are genetically based or are directly caused by differences in the environment. To do this, they divided, or cloned, each of a number of plants from populations of several different ecotypes from sites at different altitudes and grew them in common gardens at

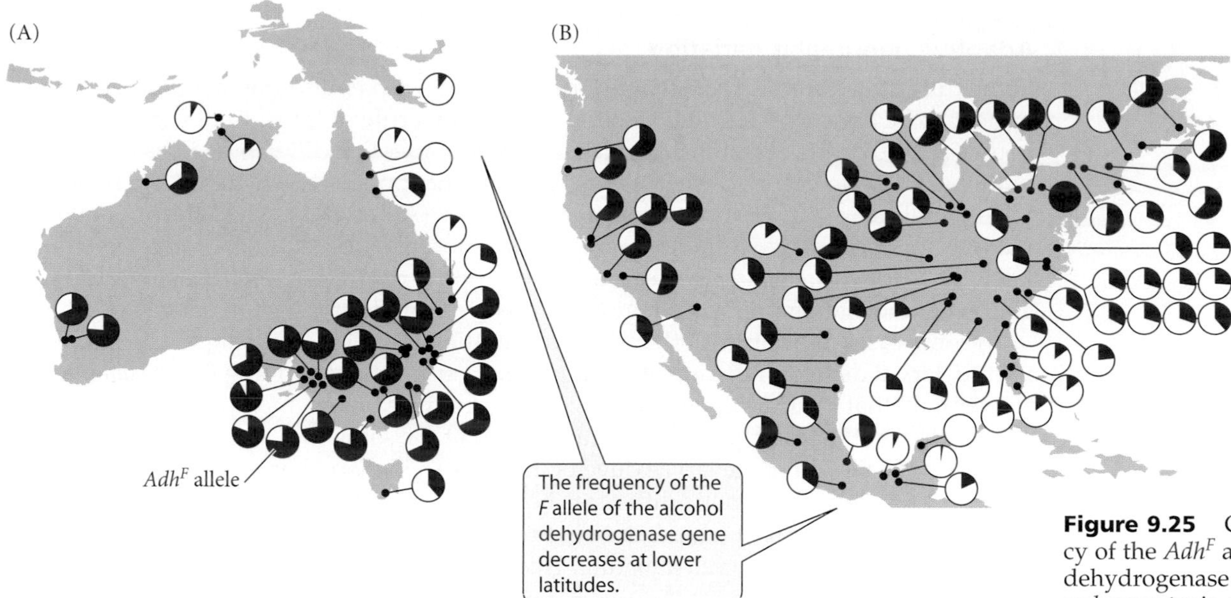

Figure 9.25 Clines in the frequency of the Adh^F allele at the alcohol dehydrogenase locus of *Drosophila melanogaster* in (A) Australia and (B) North America. The colored area of each "pie" diagram represents the frequency of Adh^F, which increases at higher latitudes on both continents. (After Oakeshott et al. 1982.)

three different altitudes in California. The differences between ecotypes in some features, such as flower color, remained unchanged, irrespective of altitude. Clausen et al. concluded that these features differ genetically and are not substantially affected by the environment. Other features, such as plant height, varied among the three gardens, showing that they were affected by the environment, but the ecotypes nevertheless differed one from another in each garden, implying that these features were also influenced by genetic differences (Figure 9.26). In further studies, Clausen et al. showed that the genetic differences among populations in these features are polygenic.

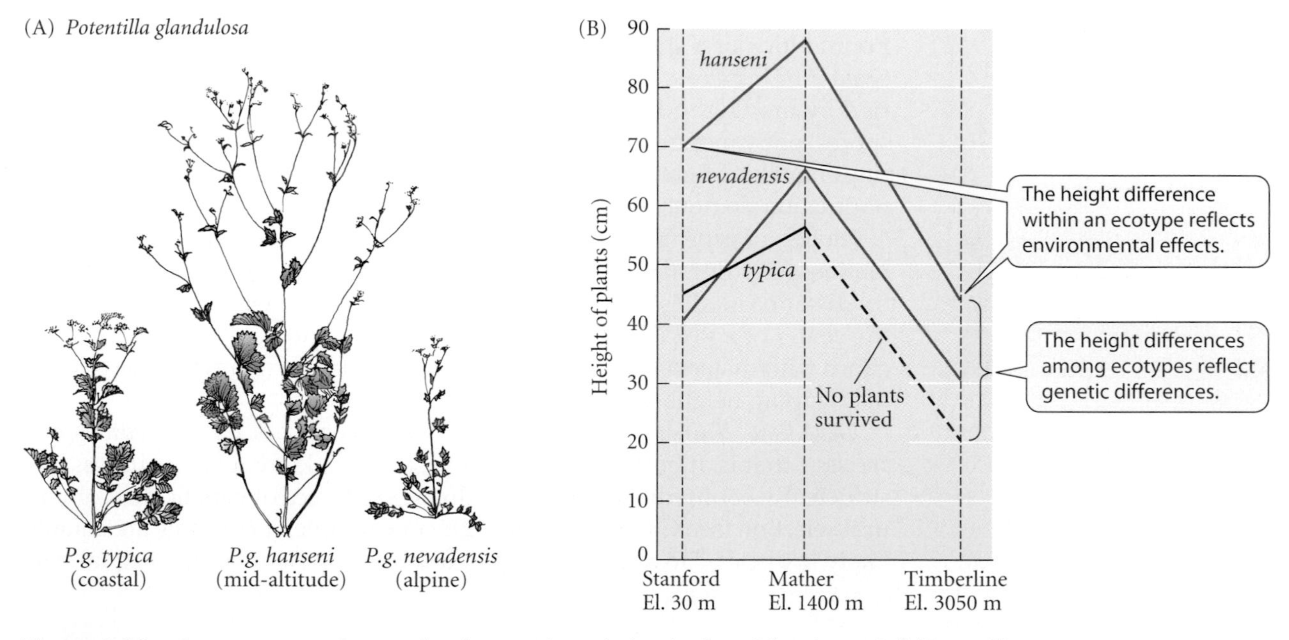

Figure 9.26 A common-garden study of ecotypic variation in the sticky cinquefoil (*Potentilla glandulosa*). (A) Representative specimens of coastal, mid-altitude, and alpine ecotypes, grown together at a site 1400 m above sea level. (B) The mean heights of the three ecotypes when grown in common gardens at three elevations. The differences among ecotypes at the same elevation reflect genetic differences, while the differences within each ecotype at different elevations reflect environmental effects. (After Clausen et al. 1940.)

Adaptive geographic variation

Some differences among populations are correlated with environmental differences and appear to be adaptive. Earlier, we cited Bergmann's rule of latitudinal variation in the body size of birds and mammals, which may reflect adaptive differences in heat exchange with the environment. Two other such "rules" are that populations of birds and mammals in colder climates tend to have shorter appendages (ALLEN'S RULE) and that animal populations in more arid environments are paler in color (GLOGER'S RULE). Shorter appendages have less surface area and lose heat more slowly. Pale coloration, matching the pale soils and vegetation common in arid regions, probably provides protection against predators.

In some cases, the nature of the adaptive variation is at first surprising. For example, in nature, larval development in montane populations of the green frog (*Rana clamitans*) is slower than that in lowland populations, simply because of lower temperatures at high altitudes. When the two types of larvae are reared at the same low temperature in the laboratory, however, montane larvae develop faster than lowland larvae (Berven et al. 1979). Thus the genetic difference between populations compensates, in part, for the effects of temperature in the field and runs counter to the phenotypic difference observed in nature. This pattern is termed **countergradient variation**.

In some cases, sympatric populations of two different species differ more than allopatric populations in one or more features. Such a pattern is termed **character displacement**. The most likely cause of character displacement is natural selection for features that reduce ecological competition between the species, or which reduce the chance that they will hybridize (see Chapters 16 and 18). Ecological character displacement is illustrated by the ground finches of the Galápagos Islands (Grant 1986). For example, *Geospiza fortis* and *G. fuliginosa* are the only ground finches on the islands of Daphne and Los Hermanos, respectively. These two populations have almost the same bill size and feed on the same kinds of seeds. On islands on which these same species co-occur, however, *G. fortis* has a substantially larger bill than *G. fuliginosa* and feeds on larger, harder seeds (Figure 9.27).

Gene flow

Populations of a species typically exchange genes with one another to a greater or lesser extent. This process is called gene flow. Genes can be carried by moving individuals (such as most animals, as well as seeds and spores) or by moving gametes (such as pollen and the gametes of many marine animals). Migrants that do not succeed in reproducing within their new population do not contribute to gene flow.

Models of gene flow treat organisms as if they formed either discrete populations (e.g., on islands) or continuously distributed populations (**isolation by distance** models). In an isolation by distance model, each individual is the center of a NEIGHBORHOOD, within which the probability of mating declines with distance from the center. The population as a whole consists of overlapping neighborhoods.

Gene flow, if unopposed by other factors, homogenizes the populations of a species—that is, it brings them all to the same allele frequencies unless it is sufficiently counterbalanced by the divergent forces of genetic drift or natural selection (see Chapters 10 and 12). For example, if migrants are equally likely to disperse to any of a group of discrete populations, all of equal size,

(A)

G. fuliginosa *G. fortis*

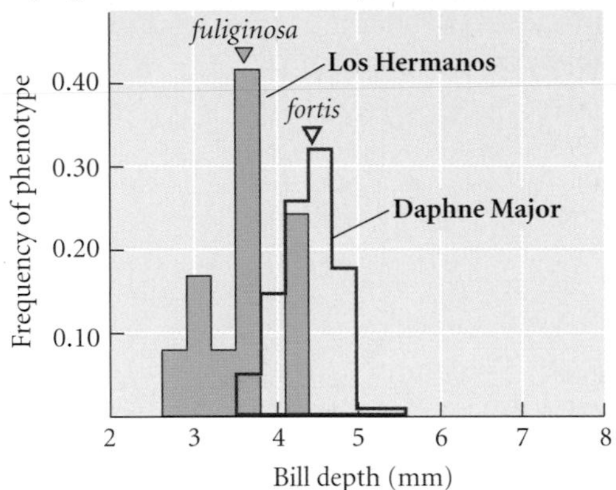

(B)

G. fuliginosa *G. fortis*

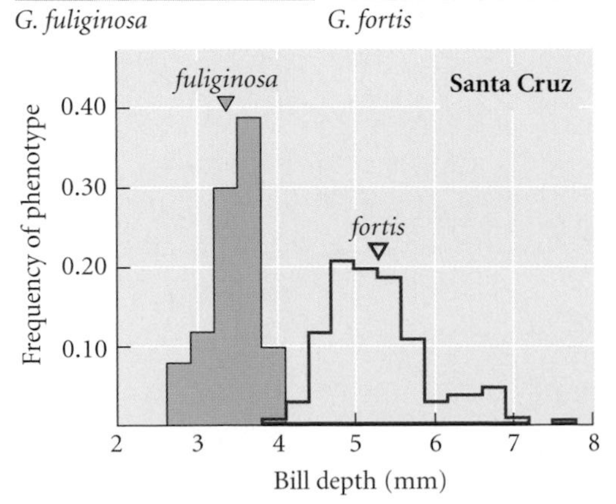

Figure 9.27 Character displacement in bill size in seed-eating ground finches of the Galápagos Islands. Bill depth is correlated with the size and hardness of seeds most used by each population; arrows show average bill depths. (A) Only *G. fortis* occurs on Daphne Major, and only *G. fuliginosa* occurs on Los Hermanos. (B) The two species co-occur on Santa Cruz, where they differ more in bill depth. (After Grant 1986; photos courtesy of Peter R. Grant.)

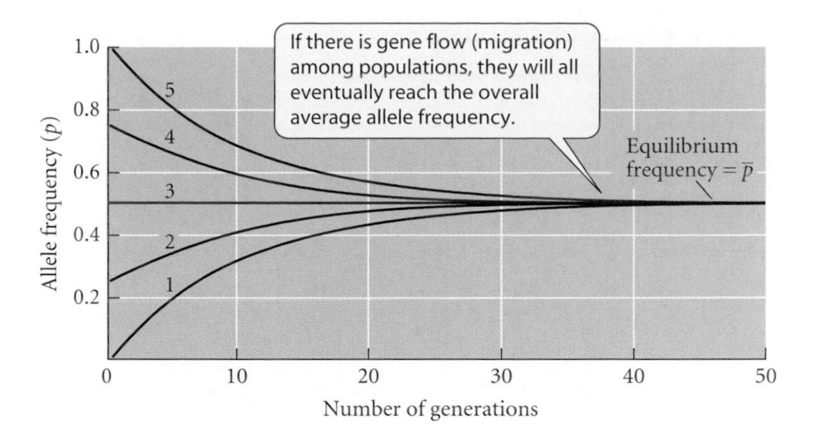

Figure 9.28 Gene flow causes populations to converge in allele frequencies. This model shows the changes over time in the frequency of one allele in five populations that exchange genes equally at the rate $m = 0.1$ per generation. (After Hartl and Clark 1989.)

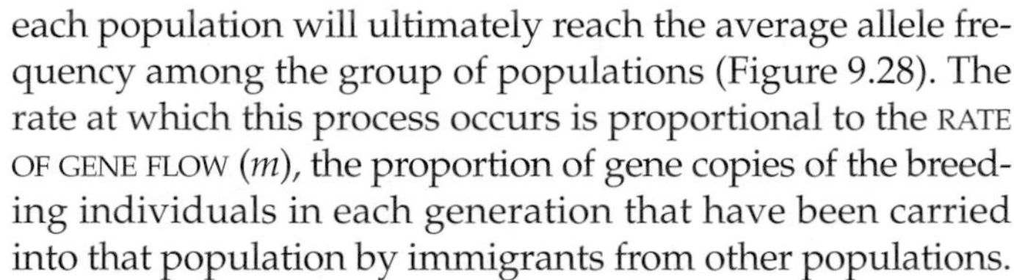

each population will ultimately reach the average allele frequency among the group of populations (Figure 9.28). The rate at which this process occurs is proportional to the RATE OF GENE FLOW (m), the proportion of gene copies of the breeding individuals in each generation that have been carried into that population by immigrants from other populations.

In this model, there is a fairly constant rate of migration among established populations. In some cases, another kind of gene flow may well be more important (McCauley 1993). If local populations at some sites become extinct, and those sites are then colonized by individuals from several other populations, the allele frequencies in the new colonies are a mixture of those among the source populations. Similarly, different populations will be genetically similar if they have all been recently founded by colonists from the same source population.

The characteristics of a species greatly affect its capacity for dispersal and gene flow. For example, animals such as land snails, salamanders, and wingless insects generally move little, and they are divided into relatively small, genetically distinct populations. Gene flow is greater among more mobile organisms, such as far-flying monarch butterflies (*Danaus plexippus*) and the many mussels and other marine invertebrates with planktonic larvae that are carried long distances by currents. However, even seemingly mobile species often display remarkably restricted dispersal. Despite their capacity for long-distance movement, many migratory species of salmon and birds breed near their birthplace, forming genetically distinct populations.

Rates of gene flow among natural populations can be estimated indirectly, from genetic differences among populations (see Chapter 10), or by direct methods, such as following the dispersal of individuals or their gametes. This can be done by releasing individuals bearing a marker—such as a genetic marker—that distinguishes them from others. In a pioneering study of isolation by distance, Dobzhansky and Wright (1943) released about 4800 *Drosophila pseudoobscura* flies, which carried an orange eye mutation to distinguish them from the native population, in a California woodland at the center of an array of banana-baited traps. From the distances at which the orange-eyed flies were recaptured over the next several days (the average time between emergence and mating), the neighborhood radius could be estimated as about 250 meters. The natural density of flies at the time of the study, in early summer, was about 50 per 100,000 square meters. Hence an average fly might encounter 500–1000 other flies. In a later study of the same species, flies released at an oasis in the desert environment of Death Valley, California, were recaptured at other oases almost 15 kilometers away (Coyne et al. 1982).

Bateman (1947) studied gene flow by pollen in maize (corn, *Zea mays*), by counting the number of heterozygous progeny of homozygous recessive plants ("seed parents") situated at various distances from a stand of homozygous dominant plants ("pollen parents"). At only 40 to 50 feet away, less than 1 percent of the progeny were fathered by the dominant plants (Figure 9.29). Similar studies have shown that gene flow by both pollen and seed dispersal is very restricted in many species of plants (Levin 1984).

Allele frequency differences among populations

Variation in allele frequencies among populations can be quantified in several ways. A commonly used measure, for a locus with two alleles, is

$$F_{ST} = \frac{V_q}{(\bar{q})(1-\bar{q})}$$

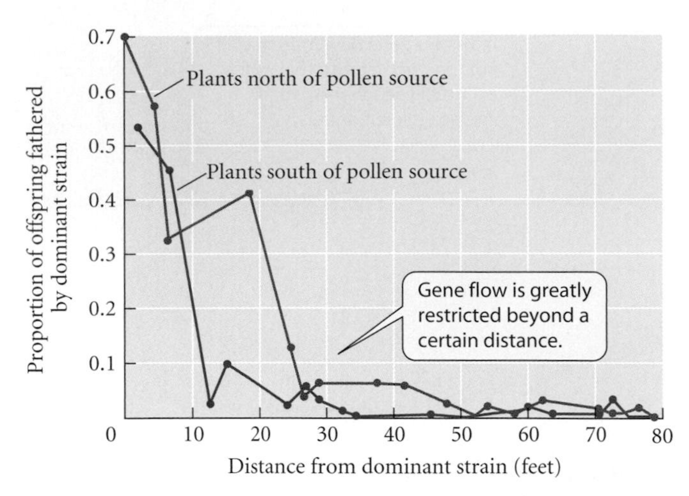

Figure 9.29 Gene flow in maize (corn), a wind-pollinated plant. The y-axis gives the proportion of offspring of recessive plants, grown at different distances from a strain of plants with a dominant allele, that were fathered by that dominant strain. The curves for plants situated north and south of the pollen source differ because of the effect of the prevailing wind on the dispersal of pollen. Most pollen is dispersed only a short distance. (After Bateman 1947.)

where $\bar{q}$ is the mean frequency of one of the alleles and V_q is the variance among populations in its frequency. A comparable measure, G_{ST}, may be calculated for a locus with more than two alleles. Both F_{ST} and G_{ST} range from 0 (no variation among populations) to 1 (populations are fixed for different alleles).

Armbruster et al. (1998) estimated G_{ST}, averaged over five allozyme loci, in samples of the mosquito *Wyeomyia smithii*, the larvae of which develop only in the water-holding leaves of pitcher plants. Samples were taken from southern sites in North America (in New Jersey, North Carolina, and the Gulf Coast) and from northern localities (in Manitoba, Ontario, Maine, and Pennsylvania) that had been covered by the most recent Pleistocene glacier. In both the southern and northern regions, more distant populations differed more in allele frequency (i.e., had a greater G_{ST}), an illustration of isolation by distance (Figure 9.30A). However, northern populations were genetically more similar than southern populations that were separated by comparable distances. The most likely interpretation is that there has not been enough time for northern populations to become substantially differentiated since the species colonized the formerly glaciated region.

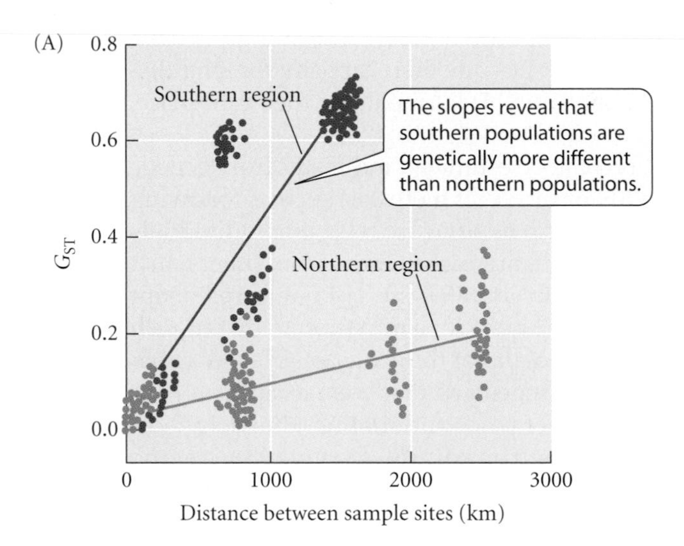

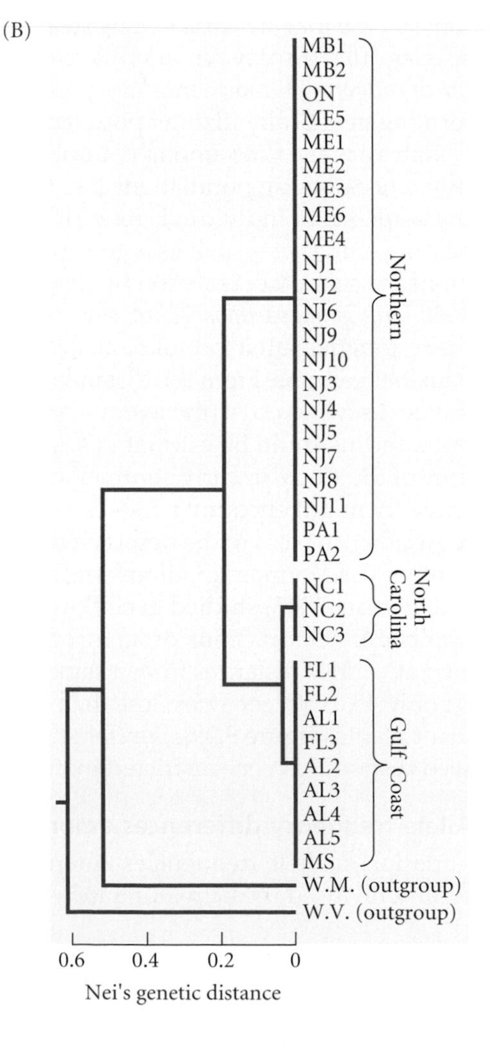

Figure 9.30 Genetic differentiation among populations of the North American pitcher-plant mosquito, *Wyeomyia smithii*. (A) G_{ST}, a measure of allele frequency differences, averaged over five loci, between southern samples (red points) and between northern samples (green points), in relation to the distance between sample localities. Each point represents a comparison of samples from two localities. (B) A phenogram, in which genetically more dissimilar populations, as measured by Nei's genetic distance, are joined at deeper levels than more similar populations. Populations of *Wyeomyia* from Canada (MB, ON), Maine (ME), New Jersey (NJ), and Pennsylvania (PA) are all more similar to one another than to those from North Carolina (NC) and the Gulf Coast (FL, AL, MS). W.M. and W.V. are closely related species of *Wyeomyia*. (After Armbruster et al. 1998.)

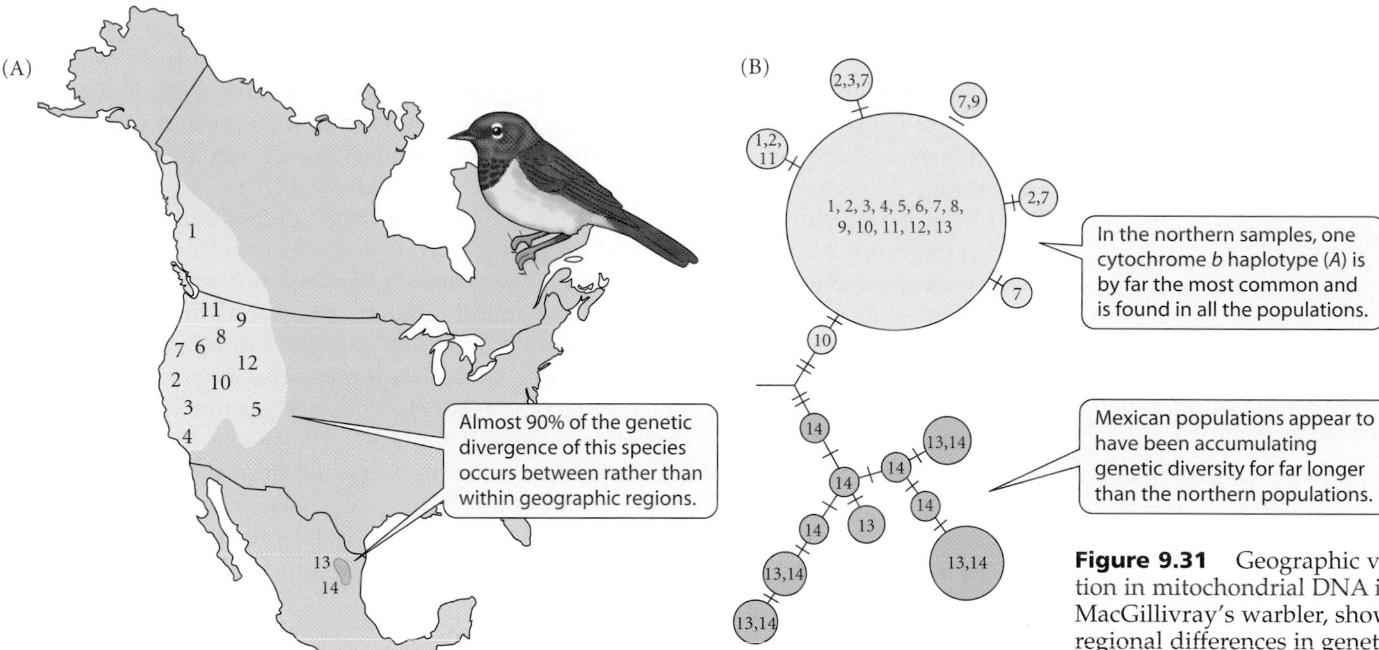
(A)
1
11
9
7
6
8
12
2
10
3
5
4
13
14
Almost 90% of the genetic divergence of this species occurs between rather than within geographic regions.
(B)
2,3,7
7,9
1,2, 11
1, 2, 3, 4, 5, 6, 7, 8, 9, 10, 11, 12, 13
2,7
7
10
14
13,14
14
14
14
14
13
13,14
13,14
13,14
In the northern samples, one cytochrome b haplotype (A) is by far the most common and is found in all the populations.
Mexican populations appear to have been accumulating genetic diversity for far longer than the northern populations.

Figure 9.31 Geographic variation in mitochondrial DNA in MacGillivray's warbler, showing regional differences in genetic diversity. (A) Sample localities in the western United States and Canada (1–12) and in Mexico (13, 14). (B) An evolutionary tree of the 17 cytochrome *b* haplotypes found. Each haplotype is represented by a circle, labeled with the localities in which it was found. The size of each circle is proportional to the total frequency of the haplotype it represents. Haplotypes are connected to one another by sequence similarity, with bars across branches indicating single nucleotide changes. Some haplotypes are placed at branch points, implying that they are the ancestors of other haplotypes in the sample. The diversity and degree of sequence divergence among haplotypes is greater in Mexican samples than in those from northern North America (After Milá et al. 2000.)

Another measure of the genetic difference between two populations is Nei's index of **genetic distance** (Nei 1987), which measures how likely it is that gene copies taken from two populations will be different alleles, given data on allele frequencies.* This index is often used in concert with a "clustering algorithm" to construct diagrams (**phenograms**) that portray the relative difference (or similarity) among populations (or species). A phenogram, for example, can be used to illustrate how similar the northern populations of pitcher-plant mosquitoes are compared with the difference between North Carolina and Gulf Coast populations (Figure 9.30B). Phenograms look like phylogenetic trees, but they represent a phylogeny only if the rate of divergence among the populations or species has been constant.

DNA sequences provide information not only on allele frequency differences among populations, but also on the genealogical (phylogenetic) relationships among alleles, which may cast light on the history of the populations (see Chapters 2 and 6). For example, populations of MacGillivray's warbler (*Oporornis tolmiei*) from temperate western North America all share one common haplotype (*A*) at the mitochondrial cytochrome *b* locus (Figure 9.31A,B). Rare variants, differing from haplotype *A* by single mutations, were found in some populations. In contrast, samples from northern Mexico carry a variety of haplotypes, differing from one another by as many as five mutations (Figure 9.31B). This pattern indicates that the Mexican population has been stable, accumulating genetic diversity for a long time. The temperate North American populations, on the other hand, probably stem from recent (postglacial) colonization by a few original founders that carried haplotype *A*. The rare haplotypes have only recently arisen from haplotype *A* by new mutations (Milá et al. 2000).

Almost all MacGillivray's warblers in Mexico have different haplotypes from those in temperate North America, so that almost 90 percent of the total genetic variation is *between* the two regions, and only 10 percent is *within* populations. The situation is almost the opposite if we compare human populations.

Geographic variation among humans

HUMAN "RACES." *Homo sapiens* is a single biological species. There exist no biological barriers to interbreeding among human populations, and even the cultural barriers that do exist often break down. For instance, despite racist barriers in the United States, a blood

*Nei's distance is calculated as $D = -\log\left(\frac{\Sigma p_{i1}p_{i2}}{\sqrt{\Sigma p_{i1}p_{i2}}}\right)$, where p_{i1} and p_{i2} are the frequencies of allele i in populations 1 and 2.

group allele, Fy^a, that is fairly abundant in European populations but virtually absent in African populations was found to have a frequency of 0.11 in the African-American population of Detroit, Michigan, from which it was calculated that 26 percent of the population's genes have been derived from the white population, at a rate of gene flow (m) of about 1 percent per generation (Cavalli-Sforza and Bodmer 1971).

Physical characteristics such as skin color, hair texture, shape of the incisors, and stature vary geographically in humans, and they have been used by various authors to define anywhere from 3 to more than 60 "races." The number of races is arbitrary, for each supposed racial group can be subdivided into an indefinite number of distinct populations. Among Africans, for example, Congo pygmies are the shortest of humans, and Masai are among the tallest. Variation in allozyme allele frequencies among villages of the Yanomama tribe in Venezuela is as great (F_{ST} = 0.077) as it is among the Mongoloid, Negroid, and Caucasoid "races" taken as a whole (F_{ST} = 0.069) (Hartl and Clark 1989).

The pattern of overall genetic variation among human populations, determined from proteins and other molecular markers, differs substantially from traditional racial divisions (Figure 9.32). Genetic differences among human populations consist of allele frequency differences only: at no known loci are "races" or other regional populations fixed for different alleles (Cavalli-Sforza et al. 1994). Early studies of allozyme variation showed that about 85 percent of the genetic variation in the human species is among individuals within populations, and only about 8 percent is among the major "races" (Nei and Roychoudhury 1982). Thus, "if everyone on earth became extinct except for the Kikuyu of East Africa, about 85 percent of all human variability would still be present in the reconstituted species" (Lewontin et al. 1984). A later analysis of 377 microsatellite loci in 52 indigenous populations came to a more extreme conclusion: differences among members of the same population account for 93 to 95 percent of genetic diversity, while differences among populations account for only about 5 percent (Rosenberg et al. 2002). Based on the slight allele frequency differences among populations, however, five major geographic clusters could be distinguished: sub-Saharan Africa, Europe and Central Asia, East Asia, Oceania, and native America. (No native Australians were included in this study.)

In view of the overwhelming genetic similarity of human populations, there is little reason to presume, a priori, that most traits will vary substantially among so-called "races," and many human geneticists feel that the concept of distinct human races has little scientific justification. This is not to deny, how-

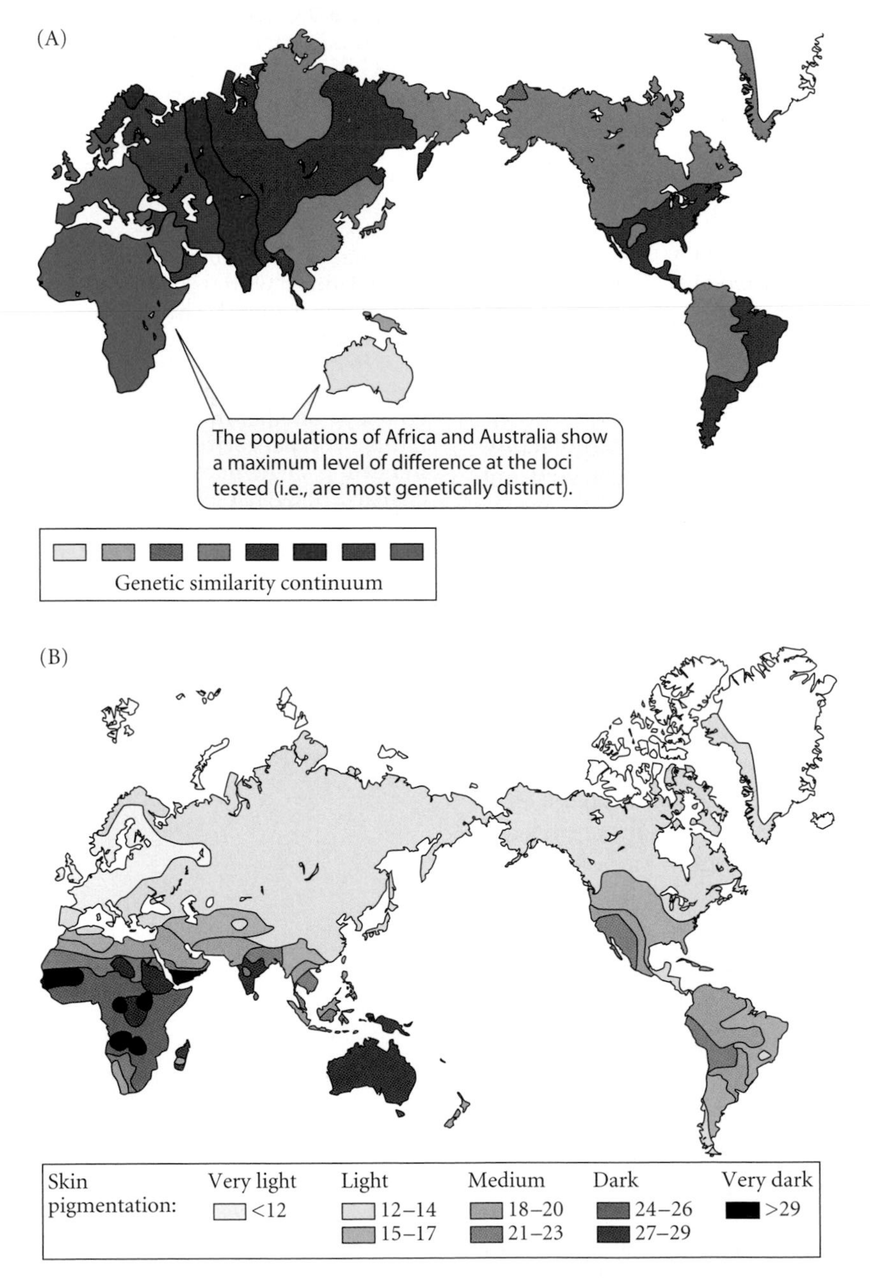
(A)
The populations of Africa and Australia show a maximum level of difference at the loci tested (i.e., are most genetically distinct).
Genetic similarity continuum
(B)
Skin pigmentation:
Very light
<12
Light
12–14
15–17
Medium
18–20
21–23
Dark
24–26
27–29
Very dark
>29

Figure 9.32 (A) A division of the world's human populations into eight classes of genetic similarity, based on numerous enzyme and blood group loci. The eight classes represented in the key are arrayed in order of increasing difference. (B) The geographic distribution of skin color, classified in eight grades of pigmentation intensity. Some considerable differences between the two maps are evident. (After Cavalli-Sforza et al. 1994.)

ever, that some loci show strong patterns of geographic variation: the sickle-cell hemoglobin allele is most frequent in parts of Africa, for example, and cystic fibrosis mutations are most prevalent in northern Europe.

VARIATION IN COGNITIVE ABILITIES. No topics in evolutionary biology are more controversial than those concerning the evolution and genetics of human behavioral characteristics, including cognitive abilities described as "intelligence." Although these abilities have evolved, and must therefore have a genetic foundation, much of the variation in their manifestation could well be nongenetic, especially in view of the enormous effects of social conditioning and learning. For example, almost everyone agrees that most of the cultural differences among groups are not genetic, for pronounced cultural differences exist among geographically neighboring peoples that are genetically indistinguishable and interbreed.

Determining the heritability of human behavioral traits is difficult because family members typically share not only genes but also environments—and humans cannot be deliberately reared in different experimentally altered environments. For this reason, studies of people adopted as children are critically important. The genetic component of variation is estimated by correlations between twins or other siblings reared apart, or by adoptees' correlation with their biological parents. In a variant of this method, the correlation of adopted children with their adoptive parents is contrasted with that of the biological (nonadopted) children of the same parents, which is expected to be higher if the variation has a genetic component. There is a risk that heritability will be overestimated in such studies, however, because adoption agencies often place children in homes that are similar (in factors such as religion and socioeconomic status) to those of their siblings. Some modern studies try to measure such environmental correlations and take them into account. Twins have played an important role in human genetic studies, since monozygotic twins should be more similar than dizygotic twins if variation has a genetic component. The genetic correlation between dizygotic twins should be no greater than between non-twin siblings.

It is exceedingly important to bear in mind that a genetic basis for a characteristic does not mean that the trait is fixed or unalterable (see Figure 9.26 for an example). A heritability value, therefore, may hold only for the particular population and the particular environment in which it was measured, and cannot be reliably extrapolated to other populations. By the same token, high heritability of variation *within* populations does not mean that differences *among* populations have a genetic basis. A character may display high heritability within a population, yet be altered dramatically by changing the environment. Twin studies have suggested that the heritability of human height is 0.8 or more, yet in many industrial nations, mean height has increased considerably within one or two generations due to nutritional and other improvements.

To a greater extent than in almost any other context, the "nature versus nurture" debate has centered on variations in the cognitive abilities collectively called "intelligence"—or, more properly, on IQ ("intelligence quotient") scores. IQ tests are supposed to be "culture-free," but they have been strongly criticized as favoring white, middle-class individuals. IQ testing has a long, sordid history of social abuse (Gould 1981). Even in the recent past, some individuals have wrongly argued that IQ cannot be improved by compensatory education because it is highly heritable (Jensen 1973), and others have suggested that African-Americans, who on average score lower than European-Americans on IQ tests, have genetically lower intelligence (Herrnstein and Murray 1994). (For counterarguments, see Fraser 1995; Fischer et al. 1996.)

Recent studies of the heritability of cognitive abilities, based on twins reared apart and on adopted compared with nonadopted children, have corrected many of the flaws of earlier studies; for example, they have assessed the similarity of the family environments in which separated twins were reared (Bouchard et al. 1990; Plomin et al. 1997; McClearn et al. 1997). "General cognitive ability" (IQ) appears to be substantially heritable, and, surprisingly, the heritability increases with age, from about 0.40 in childhood to 0.50 in adolescence to about 0.60 in adulthood and old age. The genetic component of specific cognitive abilities, such as verbal comprehension, spatial visualization, perceptual speed, and accuracy, is somewhat lower, and these features are only partly correlated with one another.

Despite the substantial heritability of IQ, there is abundant evidence that education and an enriched environment can substantially increase IQ scores (Fraser 1995). For example, one study found that children who remained in their parents' homes had an average IQ score of 107, those adopted into different homes an average of 116, and those who were returned to their biological mothers after a period of adoption, only 101 (Tizard 1973).

Probably the most incendiary question about IQ is whether genetic differences account for differences in average IQ scores among "racial" or ethnic groups, such as the 15-point difference (one standard deviation) between the average scores of European-Americans and African-Americans. Virtually all the evidence indicates that the average difference between blacks and whites is due to their very different social, economic, and educational environments (Nisbett 1995). In studies of black children adopted into white homes or reared in the same residential institution with white children, the two "racial" groups had similar scores. One study correlated individuals' IQ with their degree of admixture of alleles for several blood groups that differ in frequency between European and African populations. The correlation between IQ and degree of European ancestry was nearly zero. Finally, the average IQ of German children fathered by white American soldiers during World War II was nearly identical to the IQ of those with black American fathers.

Summary

1. Evolution occurs by the replacement of some genotypes by others. Hence evolution requires genetic variation.
2. The all-important concepts of allele frequency and genotype frequency are central to the Hardy-Weinberg principle, which states that in the absence of perturbing factors, allele and genotype frequencies remain constant over generations, with genotype frequencies at a ratio of $p^2{:}2pq{:}q^2$.
3. The potential causes of allele frequency changes at a single locus are those factors that can cause deviations from the Hardy-Weinberg equilibrium. These factors are (*a*) nonrandom mating; (*b*) finite population size, resulting in random changes in allele frequencies (genetic drift); (*c*) incursion of genes from other populations (gene flow); (d) mutation; and (*e*) consistent differences among genes or genotypes in reproductive success (natural selection).
4. Inbreeding occurs when related individuals mate and have offspring. Inbreeding increases the frequency of homozygous genotypes and decreases the frequency of heterozygotes. Most diploid populations contain rare recessive deleterious alleles at many loci, so inbreeding causes a decline in components of fitness (inbreeding depression).
5. Populations of most species contain a great deal of genetic variation. This variation includes rare alleles at many loci, which usually appear to be deleterious. But it also includes many common alleles, so that many loci—perhaps up to a third of them—are polymorphic, as revealed by enzyme electrophoresis. Most genes are variable when analyzed at the level of DNA sequence.
6. Many phenotypic traits, including morphological, physiological, and behavioral features, exhibit polygenic variation.
7. Alleles at different loci, affecting the same or different traits, sometimes are nonrandomly associated within a population, a condition called linkage disequilibrium.
8. Variation in most phenotypic traits includes both a genetic component and a nongenetic ("environmental") component. The proportion of the phenotypic variance that is due to genetic variation (genetic variance) is the heritability of the trait. The genetic variance and heritability can be estimated by breeding experiments and by artificial selection. Most characters appear to be so genetically variable that they should be able to evolve quite rapidly.
9. Some genetic differences among populations appear to be adaptive.
10. Genetic differences among different populations of a species take the form of differences in the frequencies of alleles that may also be polymorphic within populations. Unless countered by natural selection or genetic drift, gene flow among populations will cause them to become homogeneous.

11. Patterns of allele frequency differences and the phylogeny of alleles or haplotypes can shed light on the history that gave rise to geographic variation.

Terms and Concepts

additive effects (of alleles)
allele frequency
allopatric
allozyme
arithmetic mean
artificial selection
character displacement
cline
common garden
concealed genetic variation
countergradient variation
ecotype
electrophoresis
environmental variance
gene flow
genetic distance
genetic variance
genotype frequency
genotype × environment interaction
geographic variation
Hardy-Weinberg equilibrium
heritability
heterozygosity
hybrid zone
identical by descent
inbreeding
inbreeding coefficient
inbreeding depression
isolation by distance
lethal allele
linkage equilibrium, disequilibrium
maternal effect
migration
monomorphism
norm of reaction
normal distribution
panmictic
parapatric
phenogram
polygenic variation
polymorphism
quantitative variation
random genetic drift
selfing (self-fertilization)
standard deviation
subspecies
sympatric
variance

Suggestions for Further Reading

The study of variation builds on a rich history, much of which is summarized in three classic works by major contributors to evolutionary biology. Theodosius Dobzhansky's *Genetics of the evolutionary process* (Columbia University Press, New York, 1970) and its predecessors, the several editions of *Genetics and the origin of species*, are among the most influential books on evolution from the viewpoint of population genetics. Ernst Mayr's *Animal species and evolution* (Harvard University Press, Cambridge, MA, 1963) is a classic work on evolution that contains a wealth of information on geographic variation and the nature of species in animals, with important interpretations of speciation and other evolutionary phenomena. Richard C. Lewontin's *The genetic basis of evolutionary change* (Columbia University Press, New York, 1974) is an insightful analysis of the study of genetic variation by classic methods and by electrophoresis.

Excellent contemporary treatments of variation include *Principles of population genetics* by D. L. Hartl and A. G. Clark (Third Edition, Sinauer Associates, Sunderland, MA, 1997), which includes extensive treatment of much of the material in this and the next four chapters, and *Molecular markers, natural history, and evolution* by J. C. Avise (Second Edition, Sinauer Associates, Sunderland, MA, 2004), an outstanding description of the ways in which molecular variation is used to study topics such as mating patterns, kinship and genealogy within species, speciation, hybridization, phylogeny, and conservation.

Problems and Discussion Topics

1. In an electrophoretic study of enzyme variation in a species of grasshopper, you find 62 A_1A_1, 49 A_1A_2, and 9 A_2A_2 individuals in a sample of 120. Show that $p = 0.72$ and $q = 0.28$ (where p and q are the frequencies of alleles A_1 and A_2), and that the genotype frequencies of A_1A_1, A_1A_2, and A_2A_2 are approximately 0.51, 0.41, and 0.08, respectively. Demonstrate that the genotype frequencies are in Hardy-Weinberg equilibrium.

2. In a sample from a different population of this grasshopper species, you find four alleles at this locus. The frequencies of A_1, A_2, A_3, and A_4 are $p_1 = 0.50$, $p_2 = 0.30$, $p_3 = 0.15$, $p_4 = 0.05$. Assuming Hardy-Weinberg equilibrium, calculate the expected proportion of each of the 10 possible genotypes (e.g., that of A_2A_3 should be 0.09). Show that heterozygotes, of all kinds, should constitute 63.9 percent of the population. In a sample of 100 specimens, how many would you expect to be heterozygous for allele A_4? How many would you expect to be homozygous A_4A_4?
3. In the peppered moth (*Biston betularia*), black individuals may be either homozygous (A_1A_1) or heterozygous (A_1A_2), whereas pale gray moths are homozygous (A_2A_2). Suppose that in a sample of 250 moths from one locality, 108 are black and 142 are gray. (*a*) Which allele is dominant? (*b*) Assuming that the locus is in Hardy-Weinberg equilibrium, what are the allele frequencies? (*c*) Under this assumption, what *proportion* of the sample is heterozygous? What is the *number* of heterozygotes? (*d*) Under the same assumption, what proportion of black moths is heterozygous? (Answer: approximately 0.85.) (*e*) Why is it necessary to assume Hardy-Weinberg genotype frequencies in order to answer parts *b–d* ? (*f*) For a sample from another area consisting of 287 black and 13 gray moths, answer all the preceding questions.
4. In an experimental population of *Drosophila*, a sample of males and virgin females includes 66 A_1A_1, 86 A_1A_2, and 28 A_2A_2 flies. Each genotype is represented equally in both sexes, and each can be distinguished by eye color. Determine the allele and genotype frequencies, and whether or not the locus is in Hardy-Weinberg equilibrium. Now suppose you discard half the A_1A_1 flies and breed from the remainder of the sample. Assuming the flies mate at random, what will be the genotype frequencies among their offspring? (Hint: The proportion of A_2A_2 should be approximately 0.23.) Now suppose you discarded half of the A_1A_2 flies instead of A_1A_1. What will be the allele and genotype frequencies in the next generation? Why is the outcome so different in this case?
5. In an electrophoretic study of a species of pine, you can distinguish heterozygotes and both homozygotes for each of two genetically variable enzymes, each with two alleles (A_1, A_2 and B_1, B_2). A sample from a natural population yields the following numbers of each genotype: 8 $A_1A_1B_1B_1$, 19 $A_1A_2B_1B_1$, 10 $A_2A_2B_1B_1$, 42 $A_1A_1B_1B_2$, 83 $A_1A_2B_1B_2$, 44 $A_2A_2B_1B_2$, 48 $A_1A_1B_2B_2$, 97 $A_1A_2B_2B_2$, 49 $A_2A_2B_2B_2$. (*a*) Determine the frequencies of alleles A_1 and A_2 (p_A, q_A) and B_1 and B_2 (p_B, q_B). (*b*) Determine whether locus *A* is in Hardy-Weinberg equilibrium. Do the same for locus *B*. (*c*) *Assuming* linkage equilibrium, calculate the expected *frequency* of each of the nine genotypes. (Hint: The expected frequency of $A_1A_1B_1B_2$ is 0.106.) (*d*) From the results of part *c*, calculate the expected *number* of each genotype in the sample, and determine whether or not the loci actually are in linkage equilibrium. (*e*) From these calculations, can you say whether or not the loci are linked?
6. Until a few decades ago, most population geneticists believed that populations are genetically uniform, except for rare deleterious mutations. We now know that most populations are genetically very variable. Contrast the implications of these different views for evolutionary processes.
7. Different characters may vary more or less independently among geographic populations of a species (as in the rat snake example in Figure 9.24), or may vary concordantly (as in the flicker example in Figure 9.23). Suggest processes that could produce each pattern.
8. Suppose two alleles additively affect variation in a trait such as finger length. Contrast two populations with the same allele frequencies: mating is random in one ($F = 0$), whereas the other is undergoing inbreeding (perhaps $F = 0.25$). Explain why the variance in finger length is *greater* in the inbreeding population. As a challenge, algebraically determine the ratio of the variance in the inbreeding population to that in the panmictic population.
9. Several researchers have shown that groups of gypsy moths (*Lymantria dispar*) reared on more nutritious foliage lay eggs that hatch into larger larvae than those laid by groups reared on less nutritious foliage. Pose three hypotheses to account for this observation, and describe experiments to determine which is valid.

10

Genetic Drift: Evolution at Random

One of the first and most important lessons a student of science learns is that many words have very different meanings in a scientific context than in everyday speech. The word "chance" is a good example. Many nonscientists think that evolution occurs "by chance." What they mean is that evolution occurs without purpose or goal. But by this token, everything in the natural world—chemical reactions, weather, planetary movements, earthquakes—happens by chance, for none of these phenomena have purposes. In fact, scientists consider purposes or goals to be unique to human thought, and they do not view any natural phenomena as purposeful. But scientists don't view chemical reactions or planetary movements as chance events, either—because in science, "chance" has a very different meaning.

Polymorphism in snails. Populations of the European land snail *Cepaea nemoralis* are genetically polymorphic for background color and for the number and width of the dark bands on their shells. Extensive research has shown that both genetic drift and natural selection affect the allele frequencies for these traits. (Photo by D. McIntyre.)

Although the meaning of "chance" is a complex philosophical issue, scientists use chance, or **randomness**, to mean that when physical causes can result in any of several outcomes, we cannot predict what the outcome will be in any particular case. Nonetheless, we may be able to specify the *probability*, and thus the *frequency*, of one or another outcome. Although we cannot predict the sex of someone's next child, we can say with considerable certainty that there is a probability of 0.5 that it will be a daughter.

Almost all phenomena are affected simultaneously by both chance (unpredictable) and nonrandom, or DETERMINISTIC (predictable), factors. Any of us may experience a car accident due to the unpredictable behavior of other drivers, but we are predictably more likely to do so if we drive after drinking. So it is with evolution. As we will see in the next chapter, natural selection is a deterministic, nonrandom process. But at the same time, there are important random processes in evolution, including mutation (as discussed in Chapter 8) and random fluctuations in the frequencies of alleles or haplotypes: the process of **random genetic drift**.

Genetic drift and natural selection are the two most important causes of allele substitution—that is, of evolutionary change—in populations. Genetic drift occurs in all natural populations because, unlike ideal populations at Hardy-Weinberg equilibrium, natural populations are finite in size. Random fluctuations in allele frequencies can result in the replacement of old alleles by new ones, resulting in **nonadaptive evolution**. That is, while natural selection results in adaptation, genetic drift does not—so this process is not responsible for those anatomical, physiological, and behavioral features of organisms that equip them for reproduction and survival. Genetic drift nevertheless has many important consequences, especially at the molecular genetic level: it appears to account for much of the difference in DNA sequences among species.

Because all populations are finite, alleles at all loci are potentially subject to random genetic drift—but all are not necessarily subject to natural selection. For this reason, and because the expected effects of genetic drift can be mathematically described with some precision, some evolutionary geneticists hold the opinion that genetic drift should be the "null hypothesis" used to explain an evolutionary observation unless there is positive evidence of natural selection or some other factor. This perspective is analogous to the "null hypothesis" in statistics: the hypothesis that the data do not depart from those expected on the basis of chance alone.* According to this view, we should not assume that a characteristic, or a difference between populations or species, is adaptive or has evolved by natural selection unless there is evidence for this conclusion.

The theory of genetic drift, much of which was developed by the American geneticist Sewall Wright starting in the 1930s, and by the Japanese geneticist Motoo Kimura starting in the 1950s, includes some of the most highly refined mathematical models in biology. (But fear not! We shall skirt around almost all the math.) We will first explore the theory and then see how it explains data from real organisms. In our discussion of the theory of genetic drift, we will describe random fluctuations in the frequencies (proportions) of two or more kinds of self-reproducing entities that do not differ *on average* (or differ very little) in reproductive success (fitness). For the purposes of this chapter, those entities are alleles. But the theory applies to any other self-replicating entities, such as chromosomes, asexually reproducing genotypes, or even species.

The Theory of Genetic Drift

Genetic drift as sampling error

That chance should affect allele frequencies is readily understandable. Imagine, for example, that a single mutation, A_2, appears in a large population that is otherwise A_1. If the population size is stable, each mating pair leaves an average of two progeny that survive to reproductive age. From the single mating $A_1A_1 \times A_1A_2$ (for there is only one copy of A_2), the probability that one surviving offspring will be A_1A_1 is ½; therefore, the probability

*For example, if we measure height in several samples of people, the null hypothesis is that the observed means differ from one another only because of random sampling, and that the parametric means of the populations from which the samples were drawn do not differ. A statistical test, such as a *t*-test or analysis of variance, is designed to show whether or not the null hypothesis can be rejected. It will be rejected if the sample means differ more than would be expected if samples had been randomly drawn from a single population.

that two surviving progeny will both be A_1A_1 is ½ × ½ = ¼—which is the probability that the A_2 allele will be immediately lost from the population. We may assume that mating pairs vary at random, around the mean, in the number of surviving offspring they leave (0, 1, 2, 3 ...). In that case, as the pioneering population geneticist Ronald Fisher calculated, the probability that A_2 will be lost, averaged over the population, is 0.368. He went on to calculate that after the passage of 127 generations, the cumulative probability that the allele will be lost is 0.985. This probability, he found, is not greatly different if the new mutation confers a slight advantage: as long as it is rare, it is likely to be lost, just by chance.

In this example, the frequency of an allele can change (in this instance, to zero from a frequency very near zero) because the one or few copies of the A_2 allele may happen not to be included in those gametes that unite into zygotes, or may happen not to be carried by the offspring that survive to reproductive age. The genes included in any generation, whether in newly formed zygotes or in offspring that survive to reproduce, are a *sample* of the genes carried by the previous generation. Any sample is subject to random variation, or **sampling error**. In other words, the proportions of different kinds of items (in this case, A_1 and A_2 alleles) in a sample are likely to differ, by chance, from the proportions in the set of items from which the sample is drawn.

Imagine, for example, a population of land snails (*Cepaea nemoralis;* see the photograph that opens this chapter) in which (for the sake of argument) offspring inherit exactly the brown or yellow color of their mothers. Suppose 50 snails of each color inhabit a cow pasture. (The proportion of yellow snails is $p = 0.50$.) If 2 yellow and 4 brown snails are stepped on by cows, p will change to 0.511. Since it is unlikely that a snail's color affects the chance of its being squashed by cows, the change might just as well have been the reverse, and indeed, it may well be the reverse in another pasture, or in this pasture in the next generation. In this random process, the chances of increase or decrease in the proportion of yellow snails are equal in each generation, so the proportion will fluctuate. But an increase of, say, 1 percent in one generation need not be compensated by an equal decrease in a later generation—in fact, since this process is random, it is very unlikely that it will be. Therefore the proportion of yellow snails will wander over time, eventually ending up near, and finally at, one of the two possible limits: 0 and 1.0. It seems reasonable, too, that if the population should start out with, say, 80 percent brown and 20 percent yellow snails, it is more likely that the proportion of yellow will wander to zero than to 100 percent. In fact, the probability of yellow being lost from the population is exactly 0.20. Conversely, the probability that brown will reach 100 percent—that is, that it will be **fixed**—is 0.80.

Coalescence

The concept of random genetic drift is so important that we will take two tacks in developing the idea. Figure 10.1 shows a hypothetical, but realistic, history of gene lineages. First, imagine the figure as depicting lineages of individual asexual organisms, such as bacteria, rather than genes. We know from our own experience that not all members of our parents' or grandparents' generations had equal numbers of descendants; some had none. Figure 10.1 diagrams this familiar fact. We note that the individuals in generation t (at the right of the figure) are the progeny of only some of those that existed in the previous generation ($t - 1$): purely by chance, some individuals in generation $t - 1$ failed to leave descendants. Likewise, the population at generation $t - 1$ stems from only some of those individuals that existed in generation $t - 2$, and similarly back to the original population at time 0.

Now think of the objects in Figure 10.1 as copies of genes at a locus, in either a sexual or an asexual population. Figure 10.1 shows that as time goes on, more and more of the original gene lineages become extinct, so that the population consists of descendants of fewer and fewer of the original gene copies. In fact, if we look backward rather than forward in time, *all the gene copies* in the population ultimately *are descended from a single ancestral gene copy*, because given long enough, all other original gene lineages become extinct. The genealogy of the genes in the present population is said to **coalesce** back to a single common ancestor. Because that ancestor represents one of the several original alleles, the population's genes, descended entirely from that ancestral gene copy, must even-

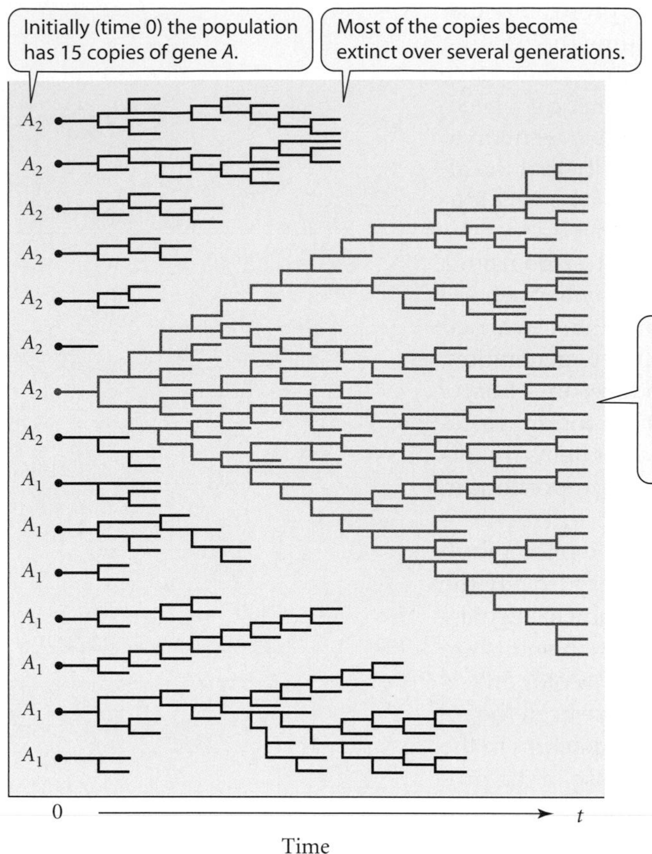
Initially (time 0) the population has 15 copies of gene A.
Most of the copies become extinct over several generations.
By time t, all copies of the gene present in the population are descended from (coalesce to) a single ancestral gene copy.
A_2
A_2
A_2
A_2
A_2
A_2
A_2
A_2
A_1
A_1
A_1
A_1
A_1
A_1
A_1
0
t
Time

Figure 10.1 A possible history of descent of gene copies in a population that begins (at time 0, at left) with 15 copies, representing two alleles. Each gene copy has 0, 1, or 2 descendants in the next generation. The gene copies present at time t (at right) are all descended from (coalesce to) a single ancestral copy, which happens to be an A_2 allele (the lineage shown in red). Gene lineages descended from all other gene copies have become extinct. If the failure of gene copies to leave descendants is random, then the gene copies at time t could equally likely have descended from any of the original gene copies present at time 0. (After Hartl and Clark 1989.)

tually become monomorphic: one or the other of the original alleles becomes fixed (reaches a frequency of 1.00). The smaller the population, the more rapidly all gene copies in the current population coalesce back to a single ancestral copy, since it takes longer for many than for few gene lineages to become extinct by chance.

In our example, all gene copies have descended from a copy of an A_2 allele, but because this is a random process, A_1 might well have been the "lucky" allele if the sequence of random events had been different. If, in the generation that included the single common ancestor of all of today's gene copies, A_1 and A_2 had been equally frequent ($p = q = 0.5$), then it is equally likely that the ancestral gene copy would have been A_1 or A_2; but if A_1 had had a frequency of 0.9 in that generation, then the probability is 0.9 that the ancestral gene would have been an A_1 allele. *Our analysis therefore shows that by chance, a population will eventually become monomorphic for one allele or the other, and that the probability that allele A_1 will be fixed, rather than another allele, equals the initial frequency of A_1.*

According to this analysis, for example, all the mitochondria of the entire human population are descended from the mitochondria carried by a single woman, who has been called "mitochondrial Eve," at some time in the past. (Mitochondria are transmitted only through eggs.) This does *not* mean, however, that the population had only one woman at that time: "mitochondrial Eve" happened to be the one among many women to whom all mitochondria trace their ancestry (in a pattern like that seen in Figure 10.1). Various nuclear genes likewise are descended from single gene copies in the past that were carried by many different members of the ancestral human population.

If this process occurs in a large number of independent, non-interbreeding populations, each with the same initial number of copies of each of two alleles at, say, locus A, then we would expect a fraction p of the populations to become fixed for A_1 and a fraction $1 - p$ to become fixed for A_2. Thus the genetic composition of the populations would diverge by chance. If the original populations had each contained three (or more) different alleles, rather than two, each of those alleles would become fixed in some of the populations, with a probability equal to its initial frequency (say, p_i).

As allele frequencies in a population change by genetic drift, so do the genotype frequencies, which conform to Hardy-Weinberg equilibrium among the new zygotes in each generation. If, for example, the frequencies p and q (that is, p and $1 - p$) of alleles A_1 and A_2 change from 0.5: 0.5 to 0.45: 0.55, then the frequencies of genotypes A_1A_1, A_1A_2, and A_2A_2 change from 0.25:0.50:0.25 to 0.2025:0.4950:0.3025. As was described in Chapter 9, the frequency of heterozygotes, H, declines as one of the allele frequencies shifts closer to 1 (and the other moves toward 0):

$$H = 2p(1 - p)$$

Bear in mind that this model, as developed so far, includes only the effects of random genetic drift. It assumes that other evolutionary processes—namely, mutation, gene flow, and natural selection—do not operate. Thus the model does not describe the evolution of adaptive traits—those that evolve by natural selection. We will incorporate natural selection in the following chapters.

Random fluctuations in allele frequencies

Let us take another, more traditional, approach to the concept of random genetic drift. Assume that the frequencies of alleles A_1 and A_2 are p and q in each of many independent populations, each with N breeding individuals (representing $2N$ gene copies in a diploid species). Small independent populations are sometimes called **demes**, and an ensemble of such populations may be termed a **metapopulation**. As before, we assume that the genotypes do not differ, *on average*, in survival or reproductive success—that is, the alleles are **neutral** with respect to fitness.

In each generation, the large number of newborn zygotes is reduced to N individuals by the time the next generation breeds, by mortality that is random with respect to genotype. By sampling error, the proportion of A_1 (p) among the survivors may change. The new p (call it p') could take on any possible value from 0 to 1.0, just as the proportion of heads among N tossed coins could, in principle, range from all heads to all tails. The probability of each possible value—whether it be the proportion of heads or the proportion of A_1 allele copies—can be calculated from the binomial theorem, generating a PROBABILITY DISTRIBUTION. Among a large number of demes, the new allele frequency (p') will vary, by chance, around a mean—namely, the original frequency, p.

Now if we trace one of the demes, in which p has changed from 0.5 to, say, 0.47, we see that in the following generation, it will change again from 0.47 to some other value, *either higher or lower with equal probability*. This process of random fluctuation continues over time. Since no stabilizing force returns the allele frequency toward 0.5, p will eventually wander (drift) either to 0 or to 1: *the allele is either lost or fixed*. (Once the frequency of an allele has reached either 0 or 1, it cannot change unless another allele is introduced into the population, either by mutation or by gene flow from another population.) The allele frequency describes a **random walk**, analogous to a New Year's Eve reveler staggering along a very long train platform with a railroad track on either side: if he is so drunk that he doesn't compensate steps toward one side with steps toward the other, he will eventually fall off the edge of the platform onto one of the two tracks, if the platform is long enough (Figure 10.2).

Just as an allele's frequency may increase by chance in some demes from one generation to the next, it may decrease in other demes. As a result, allele frequencies may vary among the demes. The *variance* in allele frequency among the demes continues to increase from generation to generation (Figure 10.3). Some demes reach $p = 0$ or $p = 1$ and can no longer change. Among those in which fixation of one or the other allele has not yet occurred, the allele frequencies continue to spread out, with all frequencies between 0 and 1 eventually becoming equally likely (Figure 10.4). Those that approach 0 or 1 tend to "fall over the edge," so the number of populations fixed for one or another allele continues to increase, until all demes in the metapopulation have become fixed. Thus *demes that initially are genetically identical evolve by chance to have different genetic constitutions*. (Remember, though, that we are assuming that the alleles have identical effects on fitness—that is, that they are neutral.)

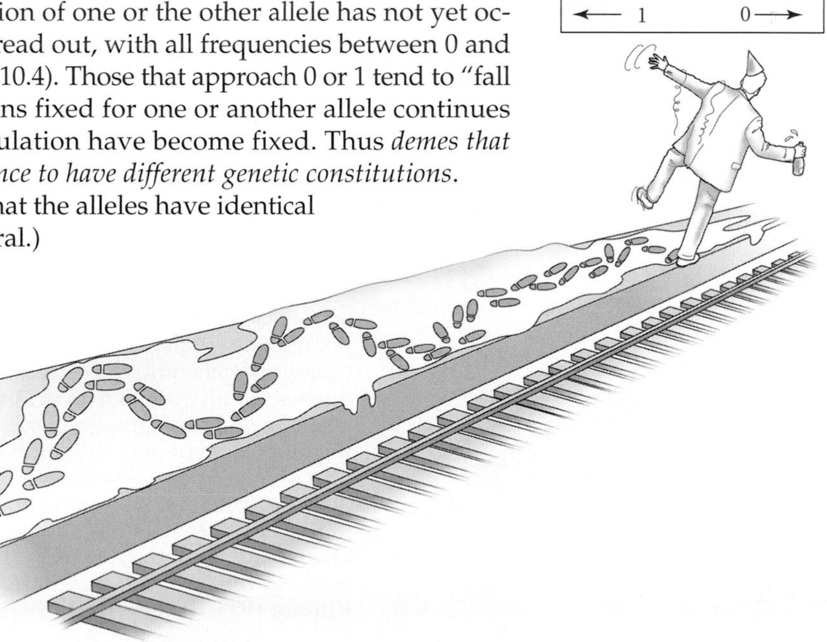

Figure 10.2 A "random walk" (or "drunkard's walk"). The reveler eventually falls off the platform if he is too far gone to steer a course toward the middle. The edges of the platform ("0" and "1") represent loss and fixation of an allele, respectively.

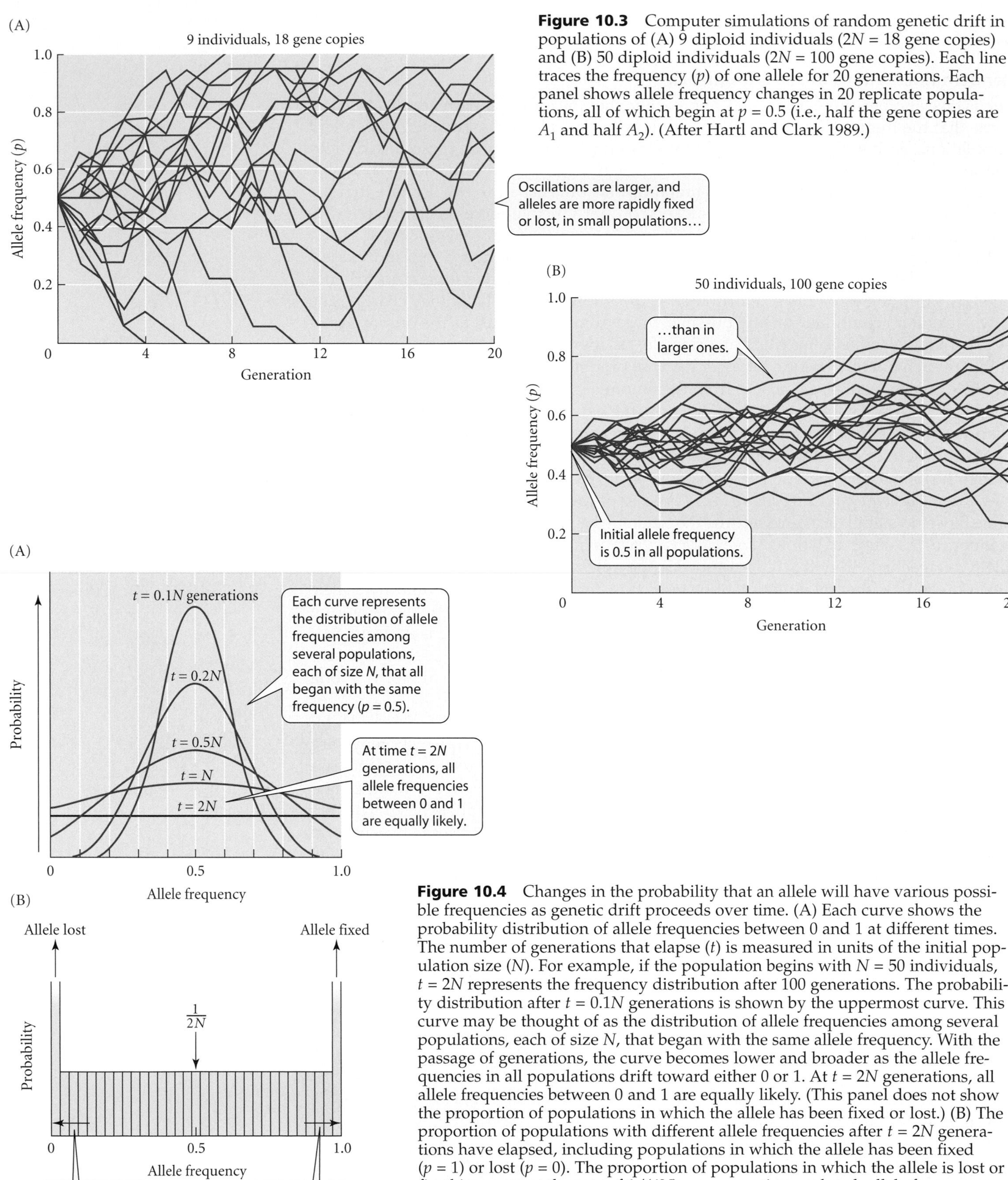
(A)
9 individuals, 18 gene copies
Allele frequency (p)
Generation
Oscillations are larger, and alleles are more rapidly fixed or lost, in small populations…
(B)
50 individuals, 100 gene copies
…than in larger ones.
Initial allele frequency is 0.5 in all populations.
Allele frequency (p)
Generation
(A)
Probability
t = 0.1N generations
t = 0.2N
t = 0.5N
t = N
t = 2N
Each curve represents the distribution of allele frequencies among several populations, each of size N, that all began with the same frequency (p = 0.5).
At time t = 2N generations, all allele frequencies between 0 and 1 are equally likely.
Allele frequency
(B)
Allele lost
Allele fixed
Probability
1/2N
Allele frequency
Allele frequencies drift downward to 0…
…or upward to 1.

Figure 10.3 Computer simulations of random genetic drift in populations of (A) 9 diploid individuals ($2N$ = 18 gene copies) and (B) 50 diploid individuals ($2N$ = 100 gene copies). Each line traces the frequency (p) of one allele for 20 generations. Each panel shows allele frequency changes in 20 replicate populations, all of which begin at $p = 0.5$ (i.e., half the gene copies are A_1 and half A_2). (After Hartl and Clark 1989.)

Figure 10.4 Changes in the probability that an allele will have various possible frequencies as genetic drift proceeds over time. (A) Each curve shows the probability distribution of allele frequencies between 0 and 1 at different times. The number of generations that elapse (t) is measured in units of the initial population size (N). For example, if the population begins with $N = 50$ individuals, $t = 2N$ represents the frequency distribution after 100 generations. The probability distribution after $t = 0.1N$ generations is shown by the uppermost curve. This curve may be thought of as the distribution of allele frequencies among several populations, each of size N, that began with the same allele frequency. With the passage of generations, the curve becomes lower and broader as the allele frequencies in all populations drift toward either 0 or 1. At $t = 2N$ generations, all allele frequencies between 0 and 1 are equally likely. (This panel does not show the proportion of populations in which the allele has been fixed or lost.) (B) The proportion of populations with different allele frequencies after $t = 2N$ generations have elapsed, including populations in which the allele has been fixed ($p = 1$) or lost ($p = 0$). The proportion of populations in which the allele is lost or fixed increases at the rate of $1/(4N)$ per generation, and each allele frequency class between 0 and 1 decreases at the rate of $1/(2N)$ per generation. (A after Kimura 1955; B after Wright 1931.)

Evolution by Genetic Drift

The following points, which follow from the previous discussion, are some of the most important aspects of evolution by genetic drift:

1. Allele (or haplotype) frequencies fluctuate at random within a population, and eventually one or another allele becomes fixed.
2. Therefore, the genetic variation at a locus declines and is eventually lost. As the frequency of one of the alleles approaches 1.0, the frequency of heterozygotes, $H = 2p(1-p)$, declines. The rate of decline in heterozygosity is often used as a measure of the rate of genetic drift within a population.
3. At any time, an allele's probability of fixation equals its frequency at that time, and is not affected or predicted by its previous history of change in frequency.
4. Therefore, populations with the same initial allele frequency (p) diverge, and a proportion p of the populations is expected to become fixed for that allele. A proportion $1-p$ of the populations becomes fixed for alternative alleles.
5. If an allele has just arisen by mutation, and is represented by only one among the $2N$ gene copies in the population, its frequency is

   $$p_t = \frac{1}{2N}$$

   and this is its likelihood of reaching $p = 1$. Clearly, it is more likely to be fixed in a small than in a large population. Moreover, if the same mutation arises in each of many demes, each of size N, the mutation should eventually be fixed in a proportion $1/(2N)$ of the demes. Similarly, of all the new mutations (at all loci) that arise in a population, a proportion $1/(2N)$ should eventually be fixed.
6. Evolution by genetic drift proceeds faster in small than in large populations. In a diploid population, the average time to fixation of a newly arisen neutral allele that does become fixed is $4N$ generations, on average. That is a long time if the population size (N) is large.
7. Among a number of initially identical demes in a metapopulation, the average allele frequency ($\bar{p}$) does not change, but since the allele frequency in each deme does change, eventually becoming 0 or 1, the frequency of heterozygotes (H) declines to zero in each deme and in the metapopulation as a whole.

Effective population size

The theory presented so far assumes highly idealized populations of N breeding adults. If we measure the actual number (N) of adults in real populations, however, the number we count (the CENSUS SIZE) may be greater than the number that actually contribute genes to the next generation. Among elephant seals, for example, a few dominant males mate with all the females in a population, so the alleles those males happen to carry contribute disproportionately to following generations; from a genetic point of view, the unsuccessful subdominant males might as well not exist (Figure 10.5). Thus the rate of genetic drift of allele frequencies, and of loss of heterozygosity, will be greater than expected from the population's census size, corresponding to what we expect of a smaller population. In other words, the population is *effectively* smaller than it seems. The **effective size** (denoted N_e) of an actual population is the number of individuals in an ideal population (in which every adult reproduces) in which the rate of genetic drift (measured by the rate of decline in heterozygosity) would be the same as it is in the actual population. For instance, if we count 10,000 adults in a population, but only 1000 of them successfully breed, genetic drift proceeds at the same rate as if the population size were 1000, and that is the effective size, N_e.

The effective population size can be smaller than the census size for several reasons:

1. *Variation in the number of progeny* produced by females, males, or both reduces N_e. The elephant seal represents an extreme example.

Figure 10.5 The effective population size among northern elephant seals (*Mirounga angustirostris*) is much lower than the census size because only a few of the large males compete successfully for the smaller females. The winner of the contest here will father the offspring of an entire "harem" of females. (Photo © Richard Hansen/Photo Researchers.)

2. Similarly, a *sex ratio* different from 1:1 lowers the effective population size.
3. *Natural selection* can lower N_e by increasing variation in progeny number; for instance, if larger individuals have more offspring than smaller ones, the rate of genetic drift may be increased at all neutral loci because small individuals contribute fewer gene copies to subsequent generations.
4. If *generations overlap*, offspring may mate with their parents, and since these pairs carry identical copies of the same genes, the effective number of genes propagated is reduced.
5. Perhaps most importantly, *fluctuations in population size* reduce N_e, which is more strongly affected by the smaller than by the larger sizes. For example, if the number of breeding adults in five successive generations is 100, 150, 25, 150, and 125, N_e is approximately 70 (the harmonic mean*) rather than the arithmetic mean, 110.

Founder effects

Restrictions in size through which populations may pass are called **bottlenecks**. A particularly interesting bottleneck occurs when a new population is established by a small number of colonists, or founders—sometimes as few as a single mating pair (or a single inseminated female, as in insects in which females store sperm). The random genetic drift that ensues is often called a **founder effect**. If the new population rapidly grows to a large size, allele frequencies (and therefore heterozygosity) will probably not be greatly altered from those in the source population, although some rare alleles will not have been carried by the founders. If the colony remains small, however, genetic drift will alter allele frequencies and erode genetic variation. If the colony persists and grows, new mutations eventually restore heterozygosity to higher levels (Figure 10.6).

Genetic drift in real populations

LABORATORY POPULATIONS. Peter Buri (1956) described genetic drift in an experiment with *Drosophila melanogaster*. He initiated 107 experimental populations of flies, each with 8 males and 8 females, all heterozygous for two alleles (*bw* and bw^{75}) that affect eye color (by which all three genotypes are recognizable). Thus the initial frequency of bw^{75} was 0.5 in all populations. He propagated each population for 19 generations by drawing 8 flies of each sex at random and transferring them to a vial of fresh food. (Thus each generation

*The HARMONIC MEAN is the reciprocal of the average of a set of reciprocals. If the number of breeding individuals in a series of t generations is $N_0, N_1, \ldots N_t$, N_e is calculated from $1/N_e = (1/t)(1/N_0 + 1/N_1 + \ldots + 1/N_t)$.

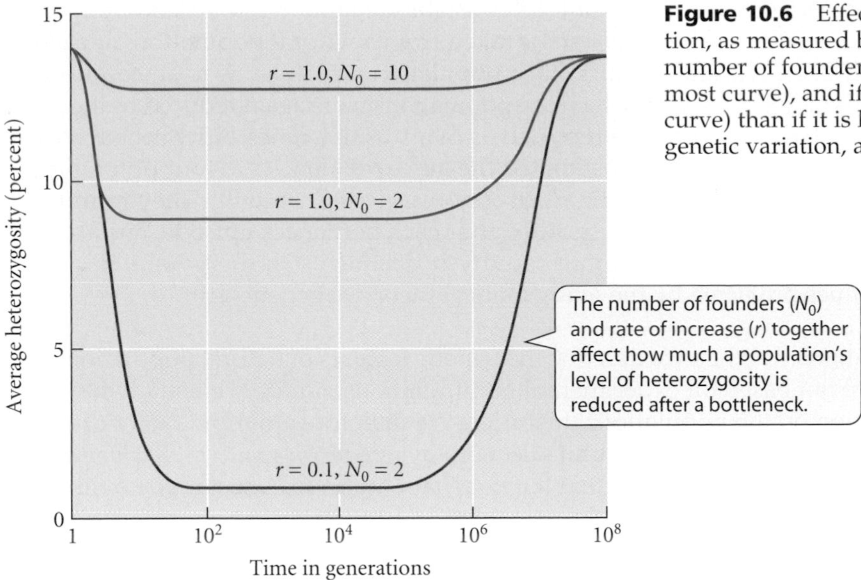
15
10
5
0
Average heterozygosity (percent)
$r = 1.0, N_0 = 10$
$r = 1.0, N_0 = 2$
$r = 0.1, N_0 = 2$
1
10^2
10^4
10^6
10^8
Time in generations
The number of founders (N_0) and rate of increase (r) together affect how much a population's level of heterozygosity is reduced after a bottleneck.

Figure 10.6 Effects of a bottleneck in population size on genetic variation, as measured by heterozygosity. Heterozygosity is reduced more if the number of founders is lower ($N_0 = 2$) than if it is higher ($N_0 = 10$; uppermost curve), and if the rate of population increase is lower ($r = 0.1$, lowest curve) than if it is higher ($r = 1.0$). Eventually, mutation supplies new genetic variation, and heterozygosity increases. (After Nei et al. 1975.)

was initiated with 16 flies × 2 gene copies = 32 gene copies.) The frequency of bw^{75} rapidly spread out among the populations (Figure 10.7); after one generation, the number of bw^{75} copies ranged from 7 ($q = 7/32 = 0.22$) to 22 ($q = 0.69$). By generation 19, 30 populations had lost the bw^{75} allele, and 28 had become fixed for it; among the unfixed populations, intermediate allele frequencies were quite evenly distributed. The results nicely matched those expected from genetic drift theory (see Figure 10.4).

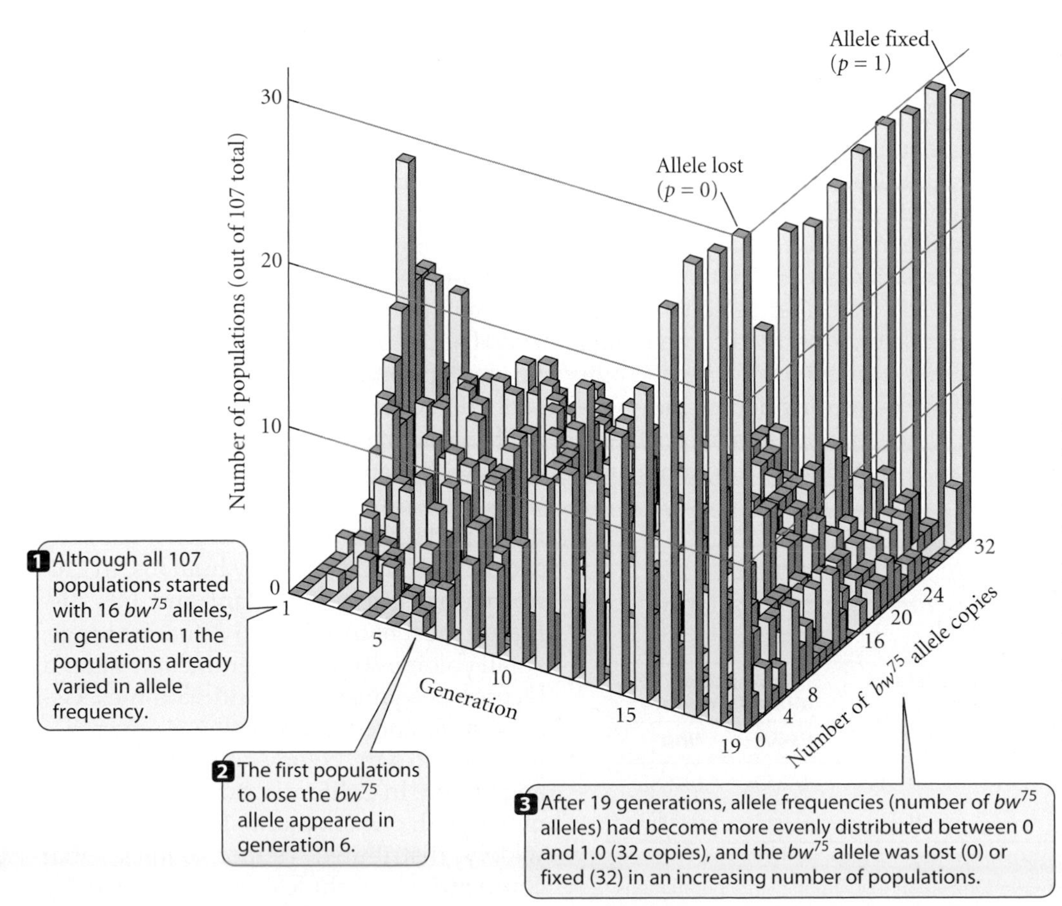
Allele fixed ($p = 1$)
Allele lost ($p = 0$)
30
20
10
0
Number of populations (out of 107 total)
1
5
10
15
19
Generation
0
4
8
16
20
24
32
Number of bw^{75} allele copies
1 Although all 107 populations started with 16 bw^{75} alleles, in generation 1 the populations already varied in allele frequency.
2 The first populations to lose the bw^{75} allele appeared in generation 6.
3 After 19 generations, allele frequencies (number of bw^{75} alleles) had become more evenly distributed between 0 and 1.0 (32 copies), and the bw^{75} allele was lost (0) or fixed (32) in an increasing number of populations.

Figure 10.7 Random genetic drift in 107 experimental populations of *Drosophila melanogaster*, each founded with 16 bw^{75}/bw heterozygotes, and each propagated by 16 flies (8 males and 8 females) per generation. The frequency distribution of the number of bw^{75} copies is read from front to back, and the generations of offspring proceed from left to right. The number of bw^{75} alleles, which began at 16 copies in the parental populations (i.e., a frequency of 0.5) became more evenly distributed between 0 and 32 copies with the passage of generations, and the bw^{75} allele was lost (0 copies) or fixed (32 copies) in an increasing number of populations. (After Hartl and Clark 1989.)

More recently, McCommas and Bryant (1990) established four replicate laboratory populations, using houseflies (*Musca domestica*) taken from a natural population, at each of three bottleneck sizes: 1, 4, and 16 pairs. Each population rapidly grew to an equilibrium size of about a thousand flies, after which the populations were again reduced to the same bottleneck sizes. This procedure was repeated as many as five times. After each recovery from a bottleneck, the investigators estimated the allele frequencies at four polymorphic enzyme loci for each population, using electrophoresis (see Chapter 9). They found that average heterozygosity ($\overline{H}$) declined steadily after each bottleneck episode, and that the smaller the bottlenecks were, the more rapidly it declined. On the whole, $\overline{H}$ closely matched the values predicted by the mathematical theory of genetic drift.

NATURAL POPULATIONS. When we describe the genetic features of natural populations, the data usually are not based on experimental manipulations, nor do we usually have detailed information on the populations' histories. We therefore attempt to *infer causes* of evolution (such as genetic drift or natural selection) by *interpreting patterns. Such inferences are possible only on the basis of theories* that tell us what pattern to expect if one or another cause has been most important.

Patterns of molecular genetic variation in natural populations often correspond to what we would expect if the loci were affected by genetic drift. For example, Robert Selander (1970) studied allozyme variation at two loci in house mice (*Mus musculus*) from widely scattered barns in central Texas. Selander considered each barn to harbor an independent population because mice rather seldom disperse to new barns, and those that do are often excluded by the residents. Having estimated the population size in each barn, Selander found that although small and large populations had much the same *mean* allele frequencies, the *variation* (variance) in allele frequency was much greater among the small populations, as we would expect from random genetic drift (Table 10.1).

Occasionally, we can check the validity of our inferences using independent information, such as historical data. For example, a survey of electrophoretic variation in the northern elephant seal (*Mirounga angustirostris;* see Figure 10.5) revealed *no* variation at any of 24 enzyme-encoding loci (Bonnell and Selander 1974)—a highly unusual observation, since most natural populations are highly polymorphic (see Chapter 9). However, although the population of this species now numbers about 30,000, it was reduced by hunting to about 20 animals in the 1890s. Moreover, the effective size was probably even lower, because less than 20 percent of males typically succeed in mating. The hypothesis that genetic drift was responsible for the monomorphism—a likely hypothesis according to the model we have just described—is supported by the historical data.

The reduced levels of genetic variation in populations that have experienced bottlenecks, such as the northern elephant seal, may have important consequences. Fixation of deleterious alleles, for example, can reduce survival and reproduction, increasing the risk of population extinction. Reduced viability in a small population of European adders—an instance of inbreeding depression—was described in Chapter 9. In rare cases, however, reduction of genetic variation may actually benefit a population. The Argentine ant (*Linepithema humile*) is relatively uncommon and coexists with many other ant species in its native Argentina, but it is highly invasive in many parts of the world to which it has been accidentally transported by humans. In California, it is very abundant and has displaced most native ants. In its native range, small colonies of the Argentine ant defend territories against conspecific colonies. Genetic differences among colonies gives rise to differences in "colony odor," which elicit aggression. In California, however, colonies merge with one another to form very large, widely distributed "supercolonies" that competitively exclude other ant species by

TABLE 10.1 ***Frequency of alleles at two loci relative to population size of house mice***

Estimated population size	Number of populations sampled	Mean allele frequency		Variance of allele frequency[a]	
		Es-3b	*Hbb*	*Es-3b*	*Hbb*
Small (median = 10)	29	0.418	0.849	0.0506	0.1883
Large (median = 200)	13	0.372	0.843	0.0125	0.0083

Source: After Selander 1970.
[a]Note that the variance of allele frequency is greater among small than among large populations.

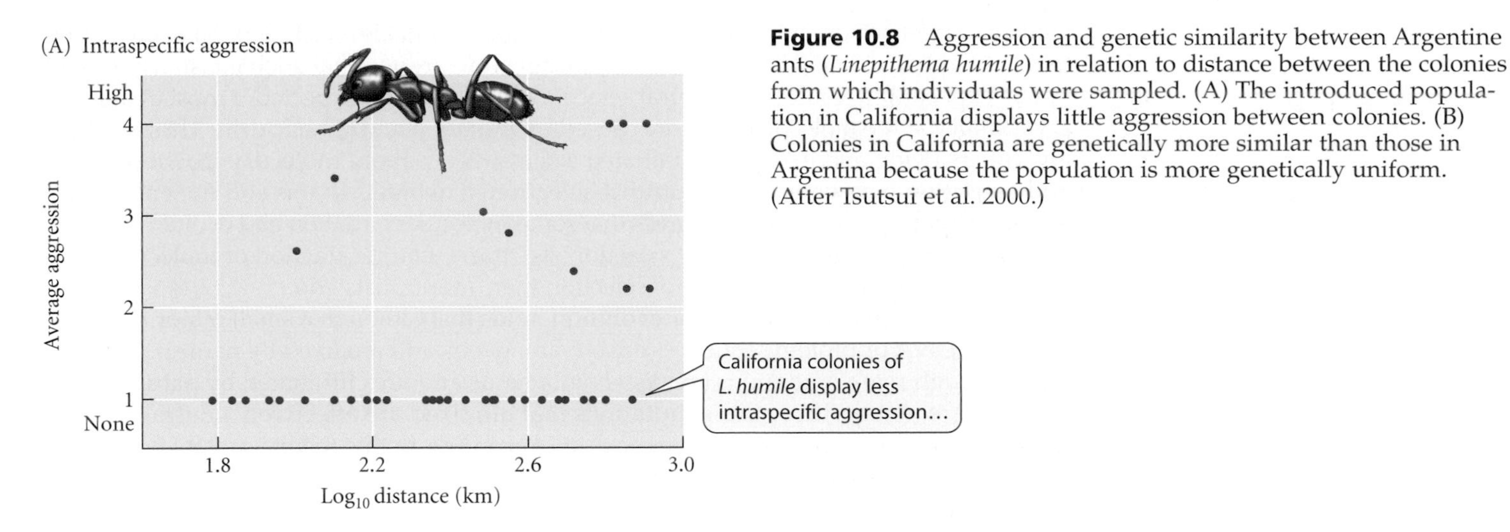

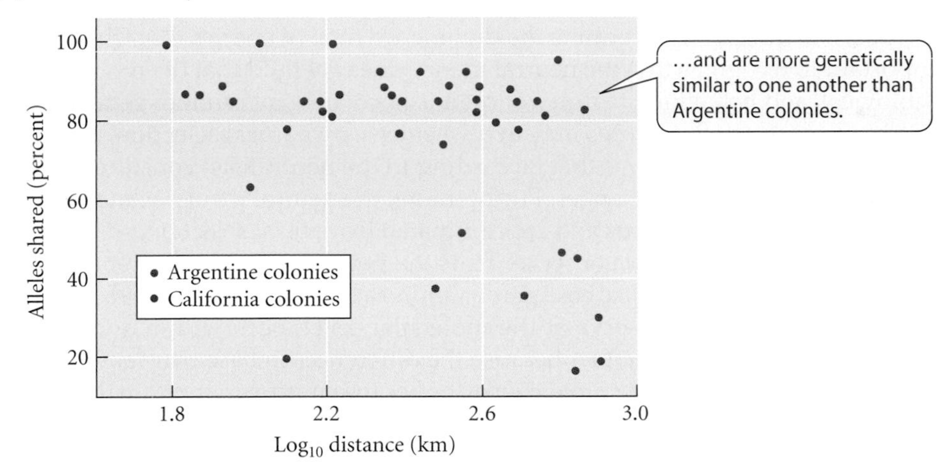

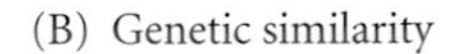

Figure 10.8 Aggression and genetic similarity between Argentine ants (*Linepithema humile*) in relation to distance between the colonies from which individuals were sampled. (A) The introduced population in California displays little aggression between colonies. (B) Colonies in California are genetically more similar than those in Argentina because the population is more genetically uniform. (After Tsutsui et al. 2000.)

their numerical superiority. Californian colonies are genetically very similar to one another, owing to a founder effect that greatly reduced genetic variation. The colonies therefore differ little in odor, are not aggressive to one another, and can therefore fuse into supercolonies (Tsutsui et al. 2000; Figure 10.8).

The Neutral Theory of Molecular Evolution

Whether or not random genetic drift has played an important role in the evolution of many of the morphological and other phenotypic features of organisms is a subject of considerable debate. There is no question, however, that at the levels of DNA and protein sequences, genetic drift is a major factor in evolution.

From the evolutionary synthesis of the late 1930s until the mid-1960s, most evolutionary biologists believed that almost all alleles differed in their effects on organisms' fitness, so that their frequencies were affected chiefly by natural selection. This belief was based on numerous studies of genes with morphological or physiological effects. But in the 1960s, the theory of evolution by random genetic drift of selectively neutral alleles became important as of two kinds of molecular data became available. In 1966, Lewontin and Hubby showed that a high proportion of enzyme loci are polymorphic. They argued that natural selection could not actively maintain so much genetic variation, and suggested that much of it might be selectively neutral. At about the same time, Motoo Kimura (1968) calculated the rates of evolution of amino acid sequences of several proteins, using the phylogenetic approach described in Chapter 2 (see Figure 2.14). He concluded that a given protein evolved at a similar rate in different lineages. He argued that such constancy would not be expected to result from natural selection, but would be expected if most evolutionary changes at the molecular level are caused by mutation and genetic drift. These authors and others (King and Jukes 1969) initiated a controversy about molecular polymorphism and evolution, known as the "neutralist-selectionist debate," that is still not entirely resolved. Although everyone now agrees that some molecular variation and evolution is neutral (i.e., a result of genetic drift), "selectionists" think a larger fraction of molecular evolutionary changes are due to natural selection than "neutralists" do.

The **neutral theory of molecular evolution** holds that although a small minority of mutations in DNA or protein sequences are advantageous and are fixed by natural selection, and although many mutations are disadvantageous and are eliminated by natural selection, the great majority of those mutations that are fixed are effectively neutral with respect to fitness and are fixed by genetic drift. According to this theory, most genetic variation at the molecular level—whether revealed by DNA sequencing or by enzyme electrophoresis—is selectively neutral and lacks adaptive significance. This theory, moreover, holds that evolutionary substitutions at the molecular level proceed at a roughly constant rate, so that the degree of sequence difference between species can serve as a MOLECULAR CLOCK, enabling us to determine the divergence time of species (see Chapter 2).

It is important to recognize that the neutral theory does *not* hold that the morphological, physiological, and behavioral features of organisms evolve by random genetic drift. Many—perhaps most—such features may evolve chiefly by natural selection, and they are based on base pair substitutions that (according to the neutralists) constitute a very small fraction of DNA sequence changes. Furthermore, the neutral theory acknowledges that many mutations are deleterious and are eliminated by natural selection, so that they contribute little to the variation we observe. Thus the neutral theory does not deny the operation of natural selection on *some* base pair or amino acid differences. It holds, though, that *most* of the variation we observe at the molecular level, both within and among species, has little effect on fitness, either because the differences in base pair sequence are not translated into differences at the protein level, or because most variations in the amino acid sequence of a protein have little effect on the organism's physiology.

Principles of the neutral theory

Let us assume that mutations occur in a gene at a constant rate of u_T per gamete per generation, and that because of the great number of mutable sites, every mutation constitutes a new DNA sequence (or allele, or haplotype). Of all such mutations, some fraction (f_0) are effectively neutral, so the **neutral mutation rate**, $u_0 = f_0 u_T$, is less than the total mutation rate, u_T. By **effectively neutral**, we mean that the mutant allele is so similar to other alleles in its effect on survival and reproduction (i.e., fitness) that changes in its frequency are governed by genetic drift alone, not by natural selection. (It is, of course, possible that the mutation does affect fitness to some slight extent, as we will see in Chapter 12. Then natural selection and genetic drift operate simultaneously, but because genetic drift is stronger in small than in large populations, the changes in the mutant allele's frequency will be governed almost entirely by genetic drift if the population is small enough. Therefore a particular allele may be effectively neutral, relative to another allele, when the population is small, but not when the population is large.)

The rate of origin of effectively neutral alleles by mutation, u_0, depends on the gene's function. If many of the amino acids in the protein it encodes cannot be altered without seriously affecting an important function—perhaps because they affect the shape of a protein that binds to DNA or to other proteins—then a majority of mutations in the gene will be deleterious rather than neutral, and u_0 will be much lower than the total mutation rate, u_T. Such a locus is said to have many **functional constraints**. On the other hand, if the protein can function well despite any of many amino acid changes (i.e., it is less constrained), u_0 will be higher. Within regions of DNA that code for proteins, we would expect the neutral mutation rate to be highest at third-base-pair positions in codons and lowest at second-base-pair positions because those positions have the highest and lowest redundancy,

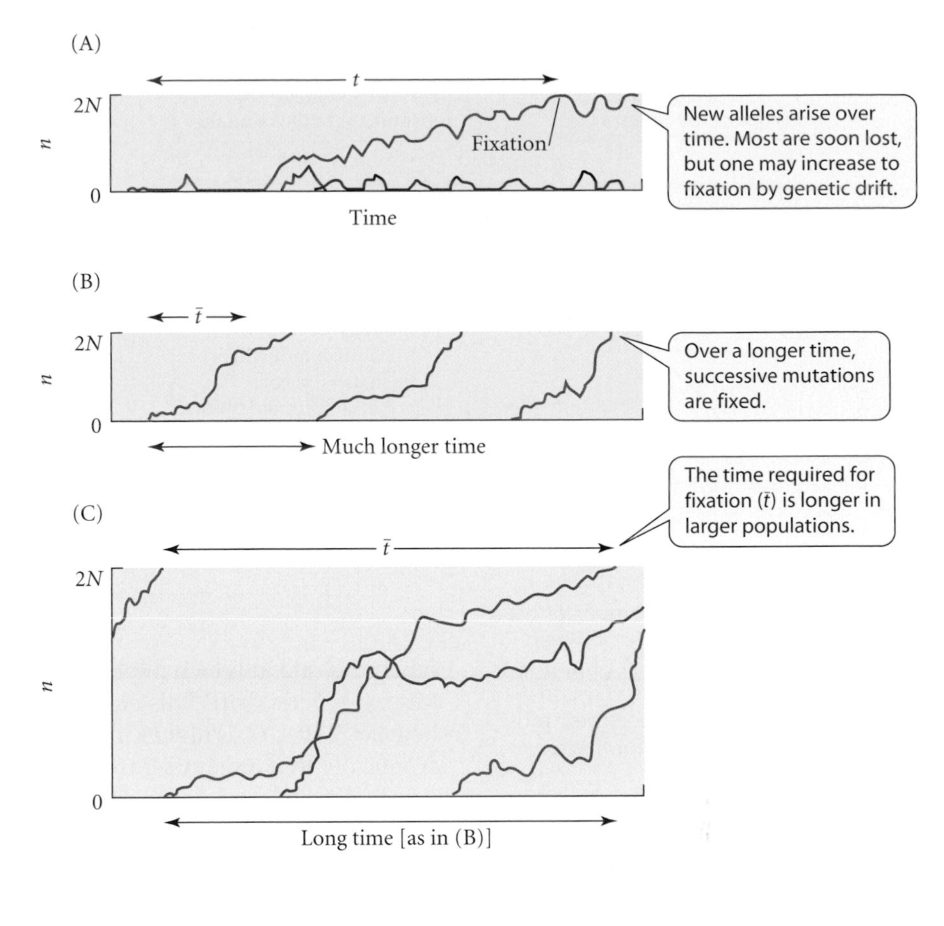
(A)
t
2N
n
Fixation
0
Time
New alleles arise over time. Most are soon lost, but one may increase to fixation by genetic drift.
(B)
t̄
2N
n
0
Much longer time
Over a longer time, successive mutations are fixed.
The time required for fixation (t̄) is longer in larger populations.
(C)
t̄
2N
n
0
Long time [as in (B)]

Figure 10.9 Evolution by genetic drift. Each graph plots the number of copies (n) of a mutation in a diploid population of N individuals ($2N$ gene copies) against time. (A) Most new mutations are lost soon after they arise, but occasionally an allele increases toward fixation by genetic drift. The average time required for such fixation is t. (B) Over a longer time, successive mutations become fixed at this locus. (C) The time required for fixation is longer in larger populations. Hence at any time there will be more neutral alleles in a larger population. (After Crow and Kimura 1970.)

respectively (see Figure 8.2). We would expect constraints to be least, or even nonexistent, and the neutral mutation rate to be greatest, for DNA sequences that are not transcribed and have no known function, such as introns and pseudogenes.

Now consider a population of effective size N_e in which the rate of neutral mutation at a locus is u_0 per gamete per generation (Figure 10.9). The *number* of new mutations is, on average, $u_0 \times 2N_e$, since there are $2N_e$ gene copies that could mutate. From genetic drift theory, we have learned that the probability that a mutation will be fixed by genetic drift is its frequency, p, which equals $1/(2N_e)$ for a newly arisen mutation. Therefore the number of neutral mutations that arise in any generation *and will someday be fixed* is

$$2N_e u_0 \times 1/(2N_e) = u_0$$

Since, on average, it will take $4N_e$ generations for such mutations to reach fixation, about the same number of neutral mutations should be fixed in every generation: *the rate of fixation of mutations is theoretically constant, and equals the neutral mutation rate*. This is the theoretical basis of the molecular clock (see Chapter 2). Notice that, surprisingly, the rate of substitution does not depend on the population size: each mutation drifts toward fixation more slowly if the population is large, but this is compensated for by the greater number of mutations that arise.

If two species diverged from their common ancestor t generations ago, and if each species has experienced u_0 substitutions per generation (relative to the allele in the common ancestor), then the number of base pair differences (D) between the two species should be $D = 2u_0t$, because each of the two lineages has accumulated u_0t substitutions. Therefore, if we have an estimate of the number of generations that have passed (see Box A in Chapter 8), the neutral mutation rate can be estimated as

$$u_0 = D/2t$$

This formula requires qualification, however. Over a sufficiently long time, some sites experience repeated base substitutions: a particular site may undergo substitution from, say, A to C and then from C to T or even back to A. Thus the observed number of differences between species will be less than the number of substitutions that have transpired. As the time since divergence becomes greater, the number of differences begins to plateau, as is evident from the number of differences per base pair between the mitochondrial DNA of different mammalian taxa (Figure 10.10; see also Figure 2.15).

In Figure 10.10, each point represents a pair of taxa for which the age of the common ancestor has been estimated from the fossil record. The number of base pair differences increases linearly for about 5–10 million years, then begins to level off; after about 40 mil-

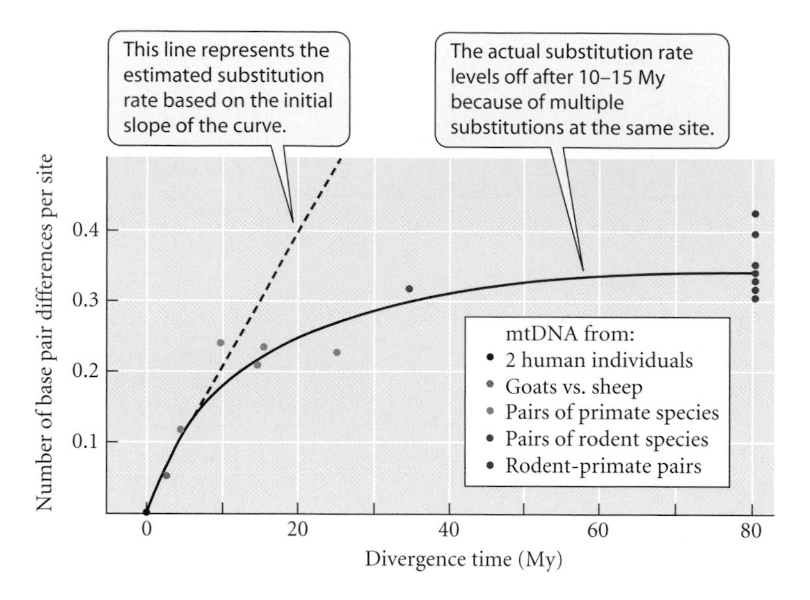

Figure 10.10 The number of base pair differences per site between the mitochondrial DNA of pairs of mammalian taxa, plotted against the estimated time since their most recent common ancestor. (After Brown et al. 1979.)

lion years, little further divergence is evident because of "multiple hits"—that is, successive substitutions. From the linear part of the curve, the mutation rate can be readily calculated, assuming that all the base pair differences represent neutral substitutions (in Figure 10.10, it is about 0.01 mutations per base pair per lineage per million years, or about 10^{-8} per year). As the curve begins to level off, however, the rate can be estimated only by making corrections for multiple substitutions (Li 1997). Data from taxa on the plateau cannot be used to estimate the substitution rate.

Within a population, there is turnover, or flux, of alleles or haplotypes (see Figure 10.9). As one or another allele approaches fixation (about every $4N$ generations, on average), other alleles are lost. But new neutral alleles arise continually by mutation, and although many are immediately lost by genetic drift, others drift to higher frequency and persist for some time in a polymorphic state before they are lost or fixed. Although the identity of the several or many alleles present in the population changes over time, the level of variation reaches an equilibrium when the rate at which alleles arise by mutation is balanced by the rate at which they are lost by genetic drift. This equilibrium level of variation, represented by the frequency of heterozygotes, H, is higher in a large population than in a small one. It can be shown mathematically that at equilibrium,

$$H = \frac{4N_e u_0}{4N_e u_0 + 1}$$

(Figure 10.11). For example, given the observed mutation rate for allozymes, 10^{-6} per gamete (Voelker et al. 1980), the equilibrium frequency of heterozygotes would be 0.004 if the effective population size N_e were 1000, but would be 0.50 if N_e were 250,000.

Variation within and among species

According to the neutral theory, the rate of allele substitution over time and the equilibrium level of heterozygosity are both proportional to the neutral mutation rate, u_0. If, because of differences in constraint or other factors, various kinds of DNA sequences or base pair sites differ in their rate of neutral mutation, those sequences or sites that differ more *between* related species should also display greater levels of variation *within* species. That is, *there should be a positive correlation between the heterozygosity at a locus and its rate of evolution.*

Figure 10.11 The equilibrium level of heterozygosity at a locus increases as a function of the product of the effective population size (N_e) and the neutral mutation rate (u_0). (After Hartl and Clark 1989.)

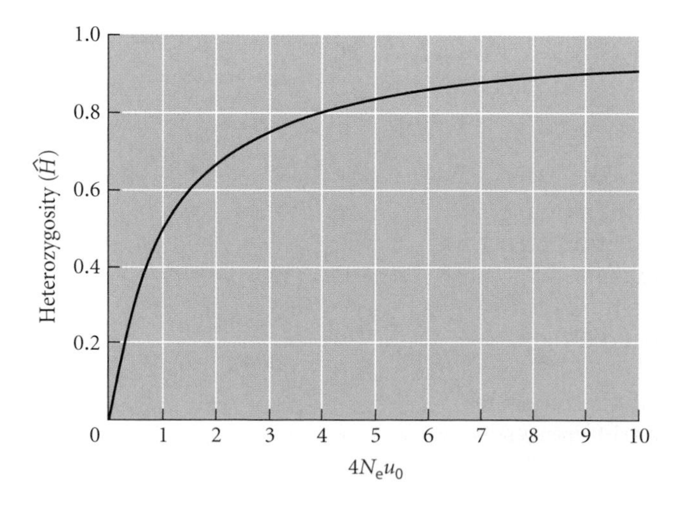

John McDonald and Martin Kreitman (1991) applied this principle in their analysis of DNA sequences of 6 to 12 copies of the coding region of the *Adh* (alcohol dehydrogenase) gene (see Figure 9.14) in each of three closely related species of *Drosophila*. Polymorphic sites (differences within species) and substitutions (differences between species) were classified as either synonymous or amino acid-replacing (nonsynonymous). If the neutral mutation rate is u_R for replacement changes and u_S for synonymous changes, then, according to the neutral theory, the ratio of replacement to synonymous differences should be the same—u_R:u_S—for both polymorphisms and substitutions, if indeed the replacement changes are subject only to genetic drift. The data (Table 10.2) showed, however, that only 5 percent of the polymorphisms, but fully 29 percent of the substitutions that distinguish species, are replacement changes. McDonald and Kreitman considered this result to be evidence that the evolution of amino acid-replacing substitutions is an adaptive process governed by natural selection. If most replacement substitutions are advantageous rather than neutral, they will increase in frequency and be fixed more rapidly than by genetic drift alone. They will therefore spend less time in a polymorphic state than selectively neutral synonymous changes do, and will thus contribute less to polymorphic variation within species.

TABLE 10.2 *Replacement (nonsynonymous) and synonymous substitutions and polymorphisms within and among three* Drosophila *species*[a]

	Polymorphisms	Substitutions
Replacement	2	7
Synonymous	42	17
Percent replacement	4.5	29.2

Source: Data from McDonald and Kreitman 1991.
[a]*D. melanogaster, D. simulans,* and *D. yakuba.*

Do comparisons among species support the neutral theory?

Recall from our discussion of the uses of molecular data in phylogenetic inference (see Chapter 2) that the rate of nucleotide or amino acid substitution can be estimated from the numbers of sequence differences among species in two ways. First, an *absolute* rate can be estimated by a calibration based on fossil evidence of the time since two or more taxa diverged from their common ancestor (see Figure 2.13). Second, the *relative* rates of evolution among different lineages can be estimated simply from the number of differences that have accumulated in each member of a monophyletic group, relative to an outgroup (see Figure 2.14).

Sequencing of DNA has provided a wealth of data on rates of molecular evolution. These data have provided evidence that most—although not all—DNA sequence evolution has been neutral. First, the rate of synonymous substitutions is generally greater than the rate of replacement substitutions, as in various genes of humans versus rodents (Table 10.3). That is, substitutions occur most frequently at third-base positions in codons and least frequently in second-base positions. Second, rates of substitution are higher in introns than in coding regions of the same gene, and even higher in pseudogenes, the nonfunctional genes related in sequence to functional genes (Figure 10.12). Third, some genes, such as histone genes, evolve much more slowly than others (Table 10.3). The genes that evolve most slowly are those thought to be most strongly constrained by their precise function. A striking example is that of the peptide hormone insulin, which is formed by the splicing of two segments of a proinsulin chain, the third segment of which (the C peptide) is removed, apparently playing no role other than in the formation of the mature insulin chain. Among mammals, the average rate of amino acid substitution has been 6 times greater in the C peptide locus than in the loci coding for the other portions of proinsulin (Kimura 1983). These classes of evidence all indicate that the *rate of evolution is greatest at DNA positions that, when altered, are least likely to affect function*, and therefore least likely to alter the organism's fitness. This conclusion provides strong support for the neutral theory.

Support for the neutral theory's prediction that rates of sequence evolution should be constant among phyletic lineages is more equivocal: some rates have been constant, and others have not. For example, rates of synonymous substitution have been about the same in the lineages leading to rodents, primates, and artiodactyls (even-toed hoofed

TABLE 10.3 *Rates of synonymous and replacement (nonsynonymous) substitutions in some protein-coding genes, calculated from the divergence between humans and several rodent species*

Gene	Number of base pairs compared	Replacement rate[a]	Synonymous rate[a]
Histone 3	135	0.00 ± 0.00	4.52 ± 0.87
Histone 4	102	0.00 ± 0.00	3.94 ± 0.81
Ribosomal protein S17	134	0.06 ± 0.04	2.69 ± 0.53
Actin α	376	0.01 ± 0.04	2.92 ± 0.34
Insulin	51	0.20 ± 0.10	3.03 ± 1.02
Insulin C peptide	31	1.07 ± 0.37	4.78 ± 2.14
α-globin	141	0.56 ± 0.11	4.38 ± 0.77
β-globin	146	0.78 ± 0.14	2.58 ± 0.49
Immunoglobulin κ	106	2.03 ± 0.30	5.56 ± 1.18
Interferon γ	136	3.06 ± 0.37	5.50 ± 1.45
Glyceraldehyde-3-phosphate dehydrogenase	332	0.20 ± 0.04	2.30 ± 0.30
Lactate dehydrogenase A	331	0.19 ± 0.04	4.06 ± 0.49

Source: From Li 1997.
[a]The rate is the number of substitutions per base pair per 10^9 years. A divergence time of 80 million (8×10^7) years between humans and rodents is assumed. Note that replacement rates vary far more than synonymous rates.

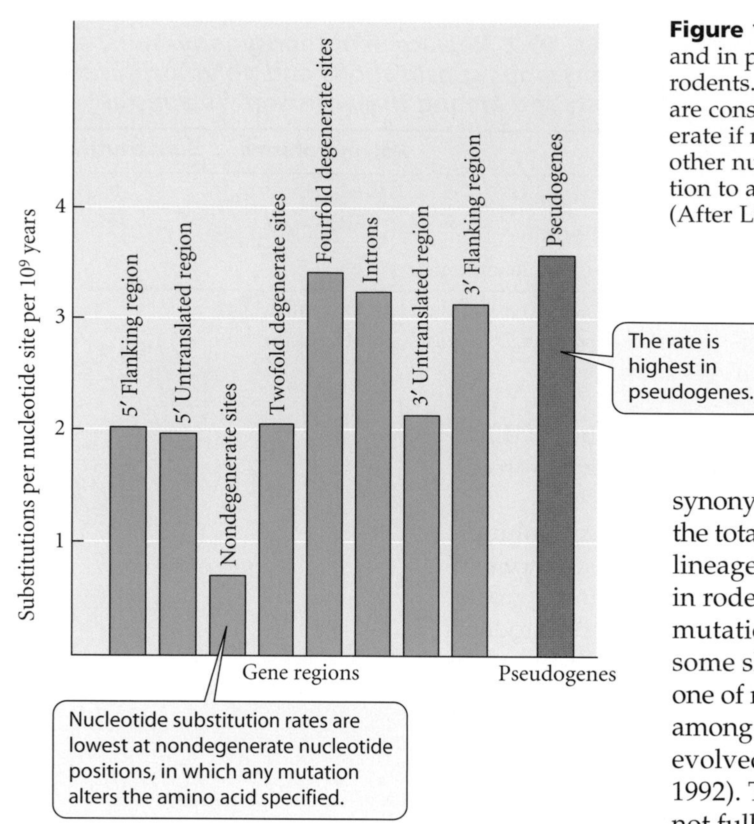

Figure 10.12 Average rates of substitution, in different parts of genes and in pseudogenes, estimated from comparisons between humans and rodents. Differences in the rate of molecular evolution among these classes are consistent with the neutral theory. Nucleotide positions are nondegenerate if no mutation would be synonymous; twofold degenerate if one other nucleotide would be synonymous; and fourfold degenerate if mutation to any of the other three nucleotides is synonymous (see Figure 8.2) (After Li 1997.)

mammals such as sheep and pigs), based on the relative rate test (Figure 10.13A). However, nonsynonymous substitutions have occurred at a significantly lower rate in primates than in the other two orders, and rodents have shown the highest rate (Figure 10.13B). The constancy of synonymous substitutions, which are presumably neutral, implies that the total rate of mutation (u_T) has been the same, per unit time, in these lineages. If so, then the higher rate of nonsynonymous substitutions in rodents could be due to either positive selection of advantageous mutations or lower effective population sizes, which would render some slightly deleterious mutations effectively neutral. This is only one of many examples of differences in the rate of sequence evolution among higher taxa. For instance, mitochondrial DNA sequences have evolved more slowly in turtles than in other vertebrates (Avise et al. 1992). The causes of variation in rates of sequence evolution are still not fully understood.

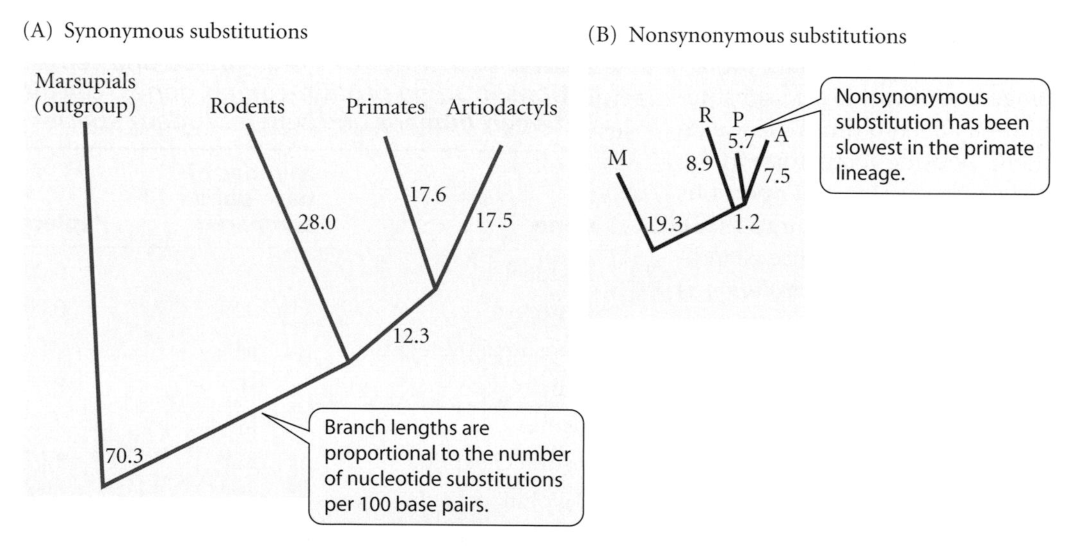

Figure 10.13 Constancy and inconstancy in the rate of sequence evolution. The number of nucleotide substitutions in 14 nuclear genes among marsupials (M), rodents (R), primates (P), and artiodactyls (A) is represented by the branch lengths, which are proportional to the number of substitutions per 100 base pairs. (A) Synonymous substitutions have occurred no faster in the rodent lineage (28.0) than in the lineages leading to either primates (12.3 + 17.6 = 29.9) or artiodactyls (12.3 + 17.5 = 29.8). (B) Nonsynonymous substitutions have occurred at a higher rate in rodents (8.9) than in the lineage from the rodent-primate ancestor to modern primates (1.2 + 5.7 = 6.9). The rate has been lower in primates (5.7) than in artiodactyls (7.5). In both diagrams, marsupials are an outgroup, and the substitutions along this branch cannot be partitioned into those between the lineage leading to marsupials and the lineage leading to placental mammals. (After Easteal and Collett 1994.)

Gene Flow and Genetic Drift

A measure of the variation in allele frequency among populations is F_{ST} (see Chapter 9). The rate at which populations drift toward fixation of one allele or another is inversely proportional to the effective population size, N_e (or N, for simplicity). However, the drift toward fixation is counteracted by gene flow from other populations, at rate m. These factors strike a balance, or equilibrium, at which the fixation index (F_{ST}) is approximately

$$\hat{F}_{ST} = \frac{1}{4Nm+1}$$

The quantity Nm is the *number* of immigrants per generation. If $m = 1/(N)$ (i.e., only one breeding individual per population is an immigrant, per generation), then $Nm = 1$, and $\hat{F}_{ST} = 0.20$. That is, even a little gene flow keeps all the demes fairly similar in allele frequency, and heterozygosity remains high. Recasting this equation as

$$Nm = \frac{(1/F_{ST})-1}{4}$$

enables us to indirectly estimate rates of gene flow among natural populations, since F_{ST} can be estimated from the variation of allele frequencies. In fact, such indirect estimates may be better than direct estimates of gene flow such as those described in Chapter 9, because direct observations are usually insufficient to detect long-distance migration, rare episodes of massive gene flow, and the perhaps rare (but nevertheless important) processes of population extinction and recolonization (Slatkin 1985). We must assume that the alleles for which we calculate F_{ST} are selectively neutral. (F_{ST} would underestimate gene flow if natural selection favored different alleles in different areas, and it would overestimate gene flow if selection favored the same allele everywhere.) This assumption can be evaluated by the degree of consistency among different loci for which F_{ST} is estimated. Genetic drift and gene flow affect all loci the same way, whereas natural selection affects different loci more or less independently. Therefore, if each of a number of polymorphic loci yields about the same value of F_{ST}, it is likely that selection is not strong. It is also necessary to assume that allele frequencies have reached an equilibrium between gene flow and genetic drift. This might not be the case if, for example, the sampled sites have only recently been colonized and the populations have not yet had time to differentiate by genetic drift. Their genetic similarity would then lead us to overestimate the rate of gene flow.

The pocket gopher *Thomomys bottae* is a burrowing rodent that seldom emerges from the soil. This species is famous for its localized variation in coloration and other morphological features, which has led taxonomists to name more than 150 subspecies. Moreover, local populations differ more in chromosome configuration than in any other known mammalian species. Such geographic variation suggests that gene flow might be relatively low. Indeed, 21 polymorphic enzyme loci in 825 specimens from 50 localities in the southwestern United States and Mexico showed extreme geographic differentiation (Figure 10.14). Among all 50 populations, the average F_{ST} was 0.412 (which might imply $Nm = 0.36$); among localities in Arizona, it was 0.198 (implying $Nm = 1.01$). The genetically most different populations were most geographically distant or segregated by expanses of unsuitable habitat—both factors that would reduce gene flow; however, even populations located close together differed considerably (Patton and Yang 1977).

Gene trees and population history

Earlier in this chapter, we introduced the principle of genetic drift by showing that because gene lineages within a population become extinct by chance over the course of time, all gene copies in a population today are descended from one gene copy that existed at some time in the past. The genealogical history of genes in populations is the basis of COALESCENT THEORY. This theory, applied to DNA sequence data, provides inferences about the structure and the effective size (N_e) of species populations (Hudson 1990). Recall that if the number of breeding individuals changes over time, N_e is approximately equal to the harmonic mean, which is much closer to the smallest number the population has experi-

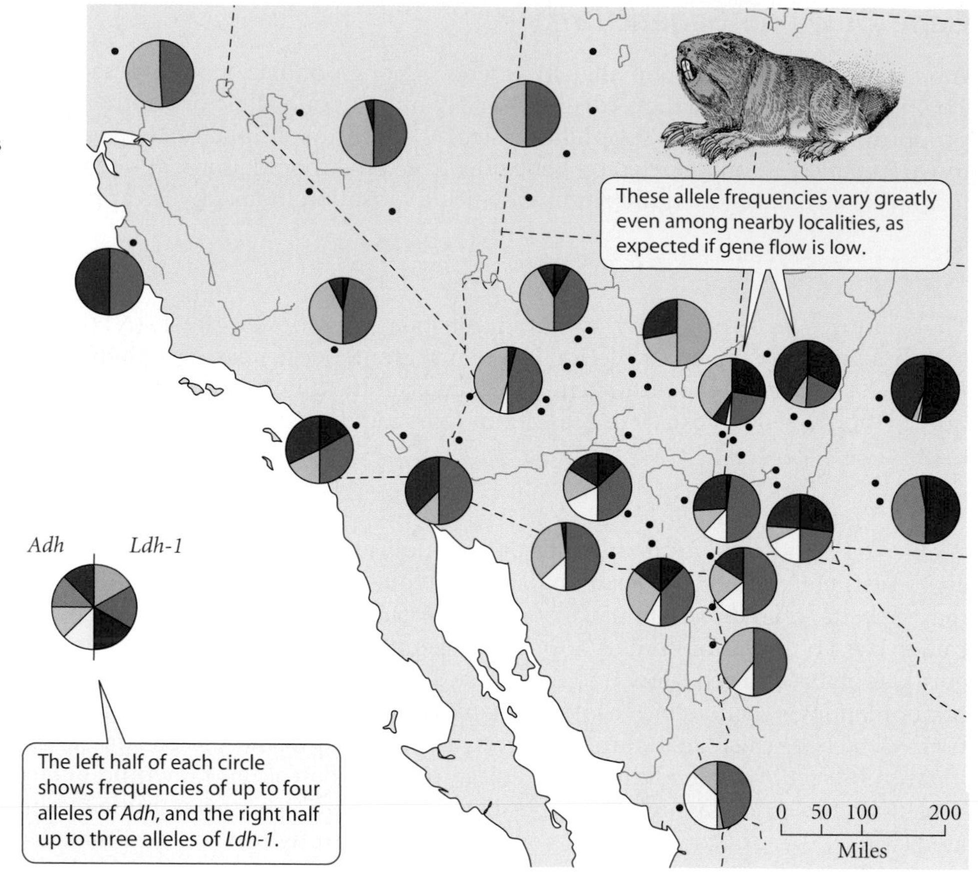
These allele frequencies vary greatly even among nearby localities, as expected if gene flow is low.
Adh
Ldh-1
The left half of each circle shows frequencies of up to four alleles of Adh, and the right half up to three alleles of Ldh-1.
0 50 100 200
Miles

Figure 10.14 Geographic variation in allele frequencies at two electrophoretic loci in the pocket gopher *Thomomys bottae*. The left half of each circle shows frequencies of up to four alleles of *Adh*, and the right half, those of up to three alleles of *Ldh-1*. The allele frequencies vary greatly even among localities located close to one another, as expected if gene flow is low. (After Patton and Yang 1977.)

enced than to the arithmetic mean. If, for example, the population has rapidly grown to its present large size from a historically much smaller size, N_e is close to the latter size, and this value can be estimated from coalescent theory.

Because the smaller the effective size (N_e) of a population, the more rapidly genetic drift transpires, the existing gene copies in a small population must stem from a more recent common ancestor than the gene copies in a large population (compare parts A and B in Figure 10.15). That is, if we look back in time from the present, it takes longer for the present genes in the larger population to coalesce to their common ancestor. Mathematical models show that in a haploid population of N_e individuals, the average time back to the common ancestor of all the gene copies (t_{CA}) is $2N_e$ generations, and in a diploid population, $t_{CA} = 4N_e$ generations. In a diploid population, the common ancestor of a random pair of gene copies occurred $2N_e$ generations ago. (For mitochondrial genes, carried in an effectively haploid state and transmitted only by females, $t_{CA} = N_e$ generations.)

If two randomly sampled gene copies had a common ancestor t generations ago, and each lineage experiences on average u mutations per generation, then each will have accumulated $u \times t$ mutations since the common ancestor. Therefore, the expected number of base pair differences between them (π) will be $2ut$, because there are two gene lineages. Since $t = 2N_e$, $\pi = 4N_e u$. We therefore *expect the average number of base pair differences between gene copies to be greater in large than in small populations*. (This difference is illustrated by the tick marks, representing mutations, on the gene trees in Figure 10.15.) In fact, if we have an estimate of the mutation rate per base pair (u), and if we measure the average proportion of sites that differ between random pairs of gene copies (π), *we can estimate the effective population size* as

$$N_e = \pi/4u$$

(For mitochondrial genes, the estimate of N_e is $N_e = \pi/u$.)

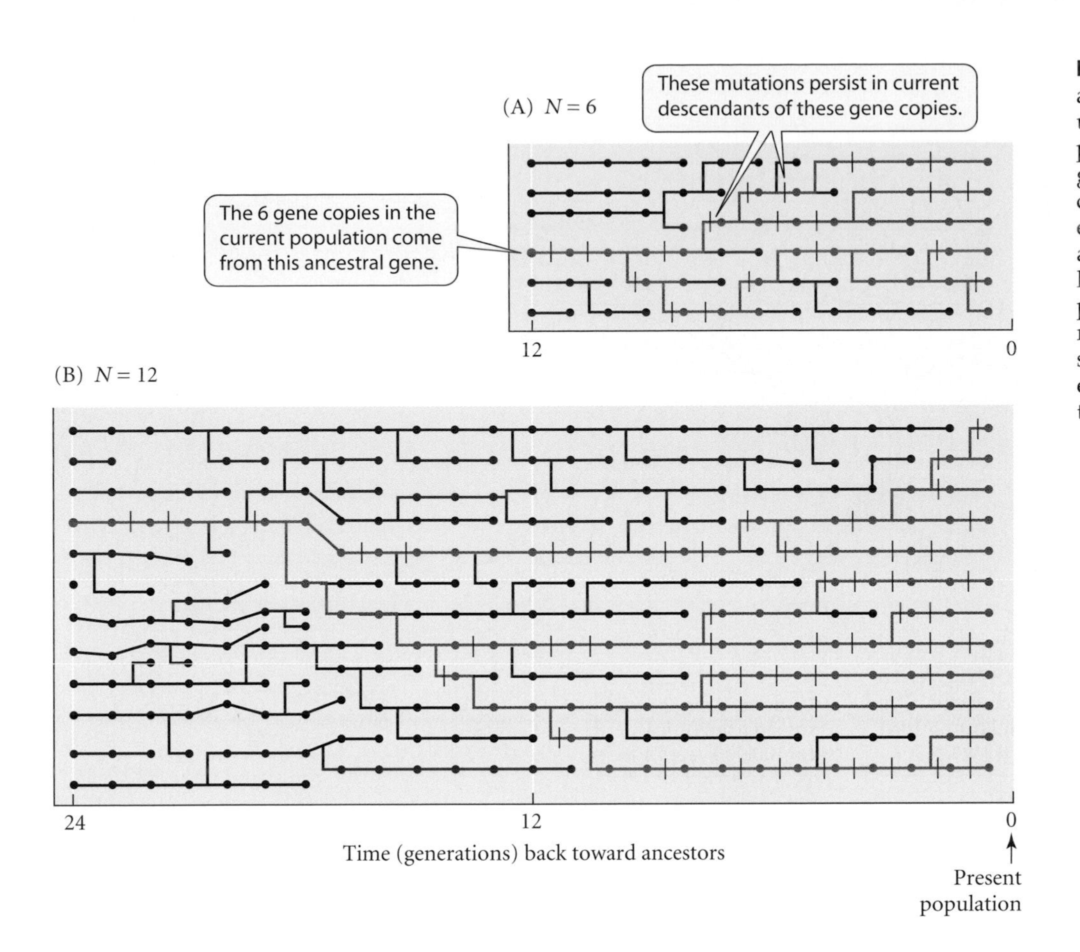

Figure 10.15 Coalescence time in (A) a small population and (B) a large population. Gene copies in the present-day population are shown in red, and their genealogy is shown back as far as their coalescence (common ancestry). The expected time back to the common ancestor of all gene copies in the population is $2N$ generations for a haploid population of N gene copies. Tick marks represent mutations, each at a unique site in the gene. More mutational differences among gene copies are expected if the coalescence time has been longer.

The origin of modern *Homo sapiens* revisited

The gene tree for human mitochondrial DNA, sampled from populations throughout the world, is rooted between most African mtDNA lineages and a clade of lineages that includes some African haplotypes and all non-African haplotypes (see Figure 6.17). Moreover, there are far fewer nucleotide differences among non-African haplotypes, on average, than among African haplotypes. We noted in Chapter 6 that these and other such data tend to support the "replacement" or "out-of-Africa" hypothesis, according to which the world's human population outside of Africa is descended from a relatively small population that spread from Africa rather recently, replacing Eurasian populations of archaic *Homo sapiens* without interbreeding with them to any significant extent.

The theory described above has been applied to human data of this kind. Several studies of mtDNA have concluded that all human mitochondrial genes, both African and non-African, are descended from a common ancestral gene that existed at t_{CA} = 156,000 to 250,000 years ago (Vigilant et al. 1991; Horai et al. 1995; Ingman et al. 2000). Similar conclusions were reached in analyses of Y chromosome sequences (carried by males) and autosomal microsatellite loci (Hammer 1995; Goldstein et al. 1995). Of course, this does not mean that the human population at that time consisted of one woman and one man; it means only that all other mitochondria and Y chromosomes in the population at that time have failed to leave descendants.

The most recent common ancestor of all mitochondrial or Y-linked genes existed before the various modern populations diverged from each other. The number of mutations distinguishing non-African from African sequences provides an estimate of when those populations diverged. Such estimates, from mtDNA, vary from 40,000 to 143,000 years ago. These estimates are much more recent than they would be if the major regional populations of humans were descended from the archaic *Homo sapiens* that inhabited those regions, but they conform to fossil and archaeological evidence that anatomically modern *H. sapiens* first occurred outside Africa about 100,000 years ago (Figure 10.16).

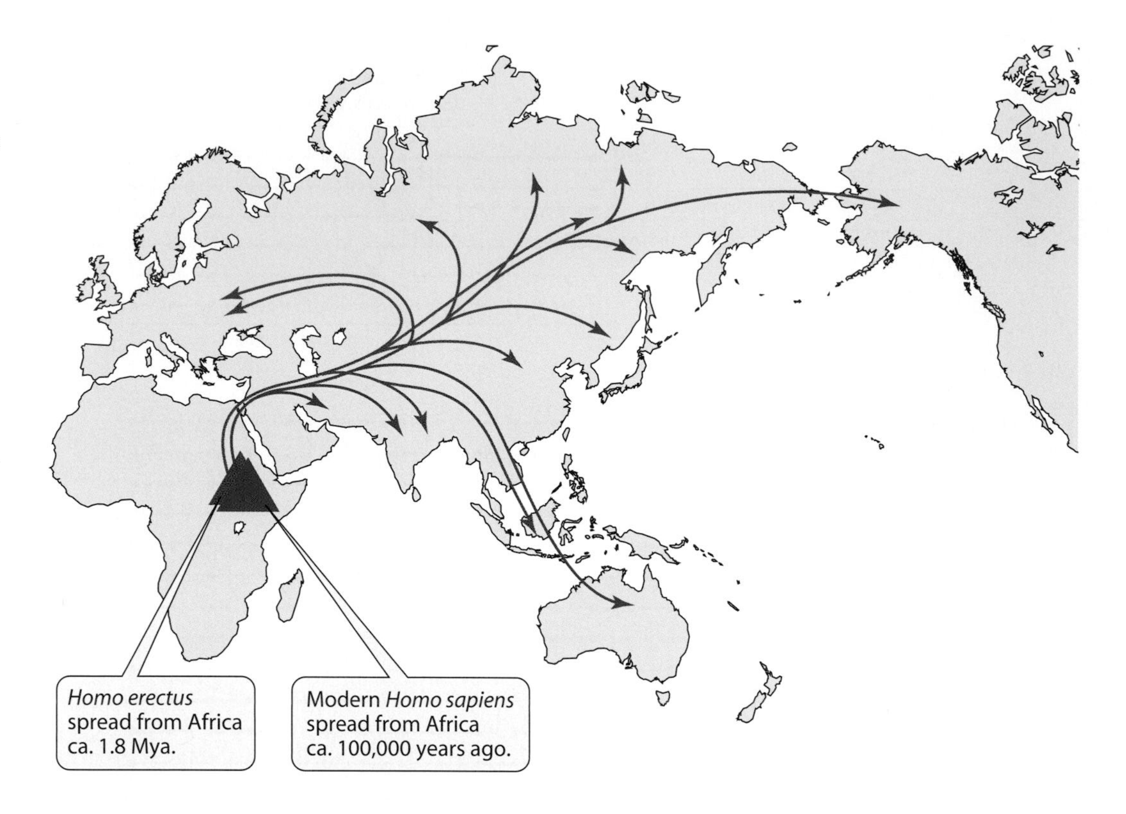
Homo erectus spread from Africa ca. 1.8 Mya.
Modern Homo sapiens spread from Africa ca. 100,000 years ago.

Figure 10.16 *Homo erectus* (red arrows) spread from Africa to Europe and Asia about 1.8 Mya, and evolved into *H. neanderthalensis* and "archaic *H. sapiens*." There is considerable evidence that modern *H. sapiens*, a separate species, evolved in Africa and spread from there (blue arrows) about 100,000 years ago. This colonization was associated with a small effective population size—perhaps smaller than 12,000—that has left its imprint on the low level of genetic variation within and among contemporary human populations outside Africa.

From data on variation in sequences (π), together with estimates of mutation rates, several studies have estimated the historical effective size of the human population (see Hammer 1995; Rogers 1995) and concluded that the world's human population is descended from an effective breeding population of only 4600 to 11,200 people! These ancestors must have formed a fairly localized population in Africa, for if they were distributed across Europe and Asia as well, the population density would have been so low that the demes could not have formed a cohesive species. The lower level of variation among non-African than among African sequences, moreover, indicates that the non-African population has grown from a smaller effective size than has the African population (Rogers and Harpending 1992). The replacement hypothesis appears to be supported by most genetic data.

Summary

1. The frequencies of alleles that differ little or not at all in their effect on organisms' fitness (neutral alleles) fluctuate at random. This process, called random genetic drift, reduces genetic variation and leads eventually to the random fixation of one allele and the loss of others, unless it is countered by other processes, such as gene flow or mutation.
2. Different alleles are fixed by chance in different populations.
3. The probability, at any time, that a particular allele will be fixed in the future equals the frequency of the allele at that time. For example, if a newly arisen mutation is represented by one copy in a diploid population of N individuals, the probability that it will increase to fixation is $1/(2N)$.
4. The smaller the effective size of a population, the more rapidly random genetic drift operates. The effective size is often much smaller than the apparent population size for a number of reasons.
5. Patterns of allele frequencies at some loci in both experimental and natural populations conform to predictions from the theory of genetic drift.
6. The theory of genetic drift has been applied especially to variation at the molecular level. The neutral theory of molecular evolution holds that, although many mutations are deleterious, and a few are advantageous, most molecular variation within and among species is selectively neutral. The fraction of mutations that are neutral varies: it is high for proteins that lack strong functional constraints and for sequences that are not transcribed. Likewise,

it is higher for synonymous than for nonsynonymous (amino acid-replacing) nucleotide substitutions.

7. As the neutral theory predicts, synonymous mutations and mutations in less constrained genes are fixed more rapidly than those that are more likely to affect function. The neutral theory also predicts that over long spans of time, substitutions will occur at an approximately constant rate for a given gene (providing a basis for the "molecular clock"). The rate of molecular evolution, as measured by differences among species, appears to be more nearly constant for synonymous than for nonsynonymous substitutions.
8. For neutrally evolving loci, the number of nucleotide differences among sequences increases over time due to new mutations, but genetic drift, leading to the loss of gene lineages, reduces genetic variation. When these factors balance, the level of sequence variation reaches equilibrium. Thus, given an estimate of the mutation rate, the level of sequence variation provides a basis for estimating the historical effective size (N_e) of a population.
9. Applying the above principles to human genes supports the hypothesis that the human population has descended from an African population of about 10,000 or fewer breeding members, from which colonists migrated into Europe and Asia less than 150,000 years ago.

Terms and Concepts

bottleneck
coalescence
deme
effective population size
effectively neutral mutation
fixation
founder effect
functional constraint
genetic drift (= random genetic drift)
metapopulation
neutral allele (neutral mutation)
neutral mutation rate
neutral theory of molecular evolution
nonadaptive evolution
random walk
randomness
sampling error

Suggestions for Further Reading

Fundamentals of molecular evolution by D. Graur and W.-H. Li (Sinauer Associates, Sunderland, MA, 2000) is a leading textbook on the topics discussed in this chapter, and covers many other topics as well. *The neutral theory of molecular evolution* by M. Kimura (Cambridge University Press, Cambridge, 1983) is a comprehensive (although somewhat dated) discussion of the neutral theory by its foremost architect.

Problems and Discussion Topics

1. For a diploid species, assume one set of 100 demes, each with a constant size of 50 individuals, and another set of 100 demes, each with 100 individuals. (*a*) If in each deme the frequencies of neutral alleles A_1 and A_2 are 0.4 and 0.6 respectively, what fraction of demes in each set are likely to become fixed for allele A_1 versus A_2? (*b*) Assume that a neutral mutation arises in each deme. Calculate the probability that it will become fixed in a population of each size. In what *number* of demes do you expect it to become fixed? (*c*) If a fixation occurs, how many generations do you expect it to take?
2. Assuming that the average rate of neutral mutation is 10^{-9} per base pair per gamete, how many generations would it take, on average, for 20 base pair substitutions to be fixed in a gene with 2000 base pairs? Suppose that the number of base pair differences in this gene between species A and B is 92, between A and C is 49, and between B and C is 91. Assuming that no repeated replacements have occurred at any site in any lineage, draw the phylogenetic tree, estimate the number of fixations that have occurred along each branch, and estimate the number of generations since each of the two speciation events.
3. Some evolutionary biologists have argued that the neutral theory should be taken as the null hypothesis to explain genetic variation within species or populations and genetic differences among them. In this view, adaptation and natural selection should be the preferred explanation only if genetic drift cannot explain the data. Others might argue that, since there is so much evidence that natural selection has shaped species' characteristics, selection

should be the explanation of choice, and that the burden of proof should fall on advocates of the neutral theory. Why might one of these points of view be more convincing than the other?

4. Some critics of Darwin's theory of evolution by natural selection have claimed that the concept of natural selection is tautologous (i.e., it is a "vicious circle"). They say, "Natural selection is the principle of the survival of the fittest. But the fittest are defined as those that survive, so there is no way to prove or disprove the theory." Argue against this statement, based on the contents of this chapter.
5. How can gene trees be used to estimate rates of gene flow among geographic populations of a species? What assumptions would have to be made? (See Slatkin and Maddison 1989.)
6. Several investigators want to use genetic markers such as allozymes to estimate gene flow (the average number of migrants per generation) among populations of several species. One wants to study movement of howler monkeys among forest patches left by land clearing in Brazil; another plans to study movement among populations of mink frogs in lakes in Ontario, north of Lake Superior; a third intends to study the warbler finch on the 17 major islands of the Galápagos archipelago, near the Equator. For which of these species is this approach most likely, or least likely, to yield a valid estimate of gene flow? Why?

Natural Selection and Adaptation

11

The theory of natural selection is the centerpiece of *The Origin of Species* and of evolutionary theory. It is this theory that accounts for the adaptations of organisms, those innumerable features that so wonderfully equip them for survival and reproduction; it is this theory that accounts for the divergence of species from common ancestors and thus for the endless diversity of life. Natural selection is a simple concept, but it nevertheless works in many and sometimes subtle ways. Although it is merely a statement about rates of reproduction and mortality, the theory of natural selection is perhaps the most important idea in biology. It is also one of the most important ideas in the history of human thought—"Darwin's dangerous idea," as the philosopher Daniel Dennett (1995) has called it—for it explains the apparent design of the living world without recourse to a supernatural, omnipotent designer.

Adapting to an adaptation. Male túngara frogs (*Physalaemus pustulosus*) have evolved a distinctive vocalization for signaling to female frogs; this adaptation helps them attract mates. Unfortunately for the frog, the predatory bat *Trachops cirrhosus* uses the frog's mating signals to find a meal. (Photo © Merlin Tuttle/BCI/Photo Researchers, Inc.)

An **adaptation** is a characteristic that enhances the survival or reproduction of organisms that bear it, relative to alternative character states (especially the ancestral condition in the population in which the adaptation evolved). Natural selection is the only mechanism known to cause the evolution of adaptations, so many biologists would simply define an adaptation as a characteristic that has evolved by natural selection. The word "adaptation" also refers to the process whereby the members of a population become better suited to some feature of their environment through change in a characteristic that affects their survival or reproduction. These definitions, however, do not fully incorporate the complex issue of just how adaptations (or the process of adaptation) should be defined or measured. We will touch on some of these complexities later in this chapter.

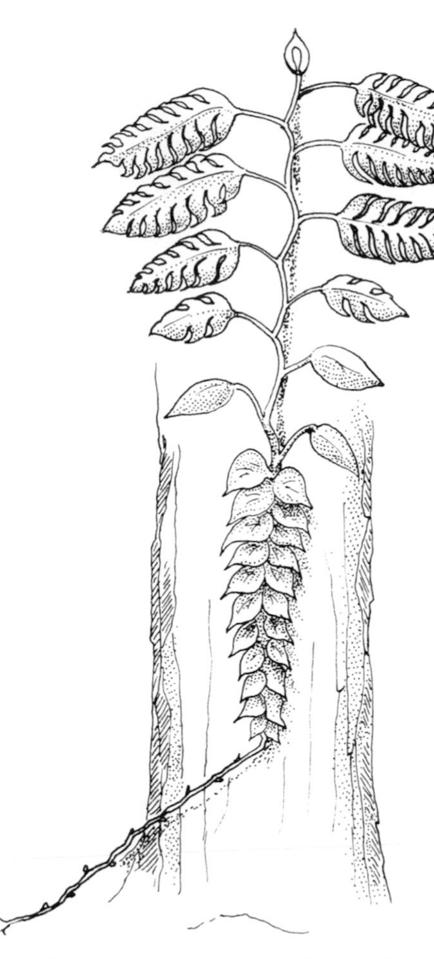

Figure 11.1 Different forms of leaves on the aroid vine *Monstera tenuis* (Araceae) in tropical American forests. The small "shingle leaves," appressed to a tree trunk, give way via transitional leaves to "adult leaves" far above the ground. The adult leaves are much larger, are deeply lobed, and are borne on long petioles. (After Lee and Richards 1991.)

Adaptations in Action: Some Examples

We can establish a few important points about adaptations by looking at four striking examples.

- *Philodendron* and other aroids (members of the family Araceae) are familiar household ornamental plants, most of which grow naturally in tropical forests. Many are vines. In some aroids, the young plant grows toward dark regions in its vicinity (rather than toward light, as do the growing tips of most plants). This strategy often brings it to a tree trunk, up which it grows (see Figure 3.19). The leaves produced along the clinging stem are flat and are closely appressed to the trunk, but the leaves produced in lighter regions high above the ground are entirely different in shape and are borne on long petioles so that they capture more light (Figure 11.1). Thus the developmental process is adaptive: it reacts to the environment by producing different phenotypes, suitable for different conditions, at different stages. This example also clearly demonstrates that adaptations are found among plants as well as animals. For Darwin, this was an important point, because Lamarck's theory, according to which animals inherit characteristics altered by their parents' behavior, could not explain the adaptations of plants.
- Among the 18,000 to 25,000 species of orchids, many have extraordinary modifications of flower structure and astonishing mechanisms of pollination. Figure 11.2 illustrates one of the more remarkable examples: pseudocopulatory pollination. Part of the flower is modified to look somewhat like a female insect, and the flower emits a scent that mimics the attractive sex pheromone (scent) of a female bee, fly, or thynnine wasp, depending on the orchid species. As a male insect "mates" with the flower, pollen is deposited precisely on that part of the insect's body that will contact the stigma of the next flower visited. Two points are noteworthy: First, the floral form and scent are adaptations to promote reproduction, rather than survival. Second, the plant achieves reproduction by deceiving, or exploiting, another organism; the insect gains nothing from its interaction with the flower.

(A)

(B)

Figure 11.2 Pseudocopulatory pollination. (A) An Australian orchid, *Chiloglottis formicifera*, attracts male thynnine wasps by scent and by the dark, somewhat insect-like structure on the flower's labellum. (B) A male wasp pollinates the flower as he attempts to copulate with it. (Photographs by W. P. Stoutamire, courtesy of W. P. Stoutamire and R. L. Dressler.)

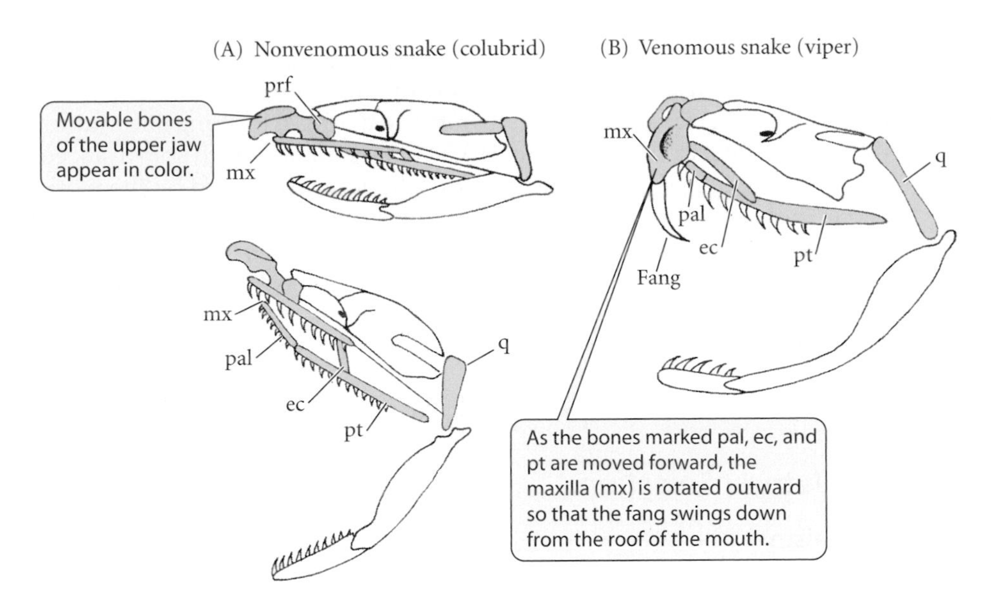
(A) Nonvenomous snake (colubrid)
(B) Venomous snake (viper)
Movable bones of the upper jaw appear in color.
prf
mx
pal
ec
pt
q
Fang
As the bones marked pal, ec, and pt are moved forward, the maxilla (mx) is rotated outward so that the fang swings down from the roof of the mouth.

(C)

Figure 11.3 The kinetic skull of snakes. The movable bones of the upper jaw are shown in gold. (A) The skull of a nonvenomous snake with jaws closed (top) and open (bottom). (B) A viper's skull. (C) The head of a red diamondback rattlesnake (*Crotalus ruber*) in strike mode. (A, B after Porter 1972; C © Tom McHugh/Photo Researchers, Inc.)

- In most terrestrial vertebrates, the skull bones are rather rigidly attached to one another, but in snakes, they are loosely joined. Most snakes can swallow prey much larger than their heads, manipulating them with astonishing versatility. They accomplish this by drawing the prey into the gullet with recurved teeth mounted on a number of freely moving bones that act as levers and fulcrums, operated by complex muscles. The lower jaw bones (mandibles) articulate to a long, movable quadrate bone that can be rotated downward so that the mandibles drop away from the skull, thus greatly increasing the mouth opening (Figure 11.3A). The front ends of the two mandibles are not fused (as they are in almost all other vertebrates), but are joined by a stretchable ligament. The mandibles are independently moved forward to engage the prey, then pulled back to bring it farther into the throat. Similarly, the tooth-bearing maxillary bones, which are suspended from the skull, are moved forward and backward by a series of other bones and muscles. This system is elaborated still further in rattlesnakes and other vipers, in which the maxilla is short and bears only a long, hollow fang, to which a duct leads from the massive poison gland (a modified salivary gland). The fang lies against the roof of the mouth when the mouth is closed. When the snake opens its mouth, the same lever system that moves the maxilla in nonvenomous snakes rotates the maxilla 90° (Figure 11.3B), so that the fang is fully erected and protrudes well beyond the margin of the mouth. Snakes' skulls, then, are complex mechanisms, "designed" in ways that an engineer can readily analyze. Their features have been achieved by modifications of the same bones that are found in other reptiles.
- Many species of animals engage in cooperative behavior, but it reaches extremes in some social insects. An ant colony, for example, includes one or more inseminated queens and a number of sterile females, the workers. Australian arboreal weaver ants (genus *Oecophylla*) construct nests of living leaves by the intricately coordinated action of numerous workers, groups of which draw together the edges of leaves by grasping one leaf in their mandibles while clinging to another (Figure 11.4). Sometimes several ants form a chain to

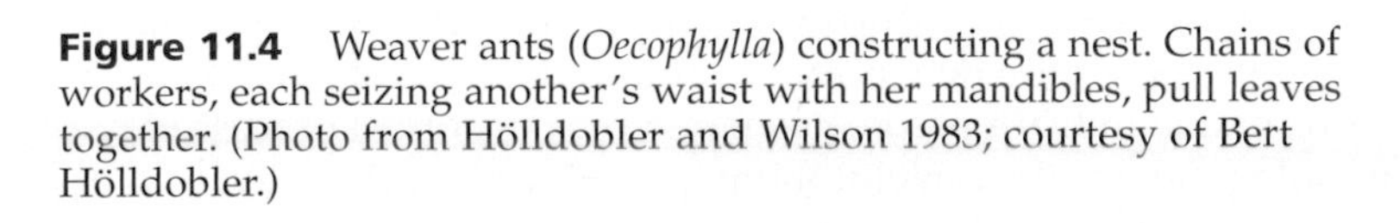

Figure 11.4 Weaver ants (*Oecophylla*) constructing a nest. Chains of workers, each seizing another's waist with her mandibles, pull leaves together. (Photo from Hölldobler and Wilson 1983; courtesy of Bert Hölldobler.)

collectively draw together distant leaf edges. The leaves are attached to one another by the action of other workers carrying larvae, which, when stimulated by the workers' antennae, emit silk from their labial glands. (The adult ants cannot produce silk.) The workers move the larvae back and forth between the leaf edges, forming silk strands that hold the leaves together. In contrast to the larvae of other ants, which spin a silk cocoon in which to pupate, *Oecophylla* larvae produce silk only when used by the workers in this fashion. These genetically determined behaviors are adaptations that enhance the reproductive success not of the worker ants that perform them, since the workers do not reproduce, but rather of their mother, the queen, whose offspring include both workers and reproductive daughters and sons. In some species, then, individuals have features that benefit certain other members of the same species. How such features evolve is a topic of special interest.

The Nature of Natural Selection

Design and mechanism

Most adaptations are *complex*, and most have the appearance of *design*—that is, they are constructed or arranged so as to accomplish some *function*, such as growth, feeding, or pollination, that appears likely to promote survival or reproduction. In the photograph that opens this chapter, for instance, the bat's wings are "designed" for flight. In inanimate nature, we see nothing comparable—we would not be inclined to think of erosion, for example, as a process designed to shape mountains.

The complexity and evident function of organisms' adaptations cannot conceivably arise from the random action of physical forces. For hundreds of years, it seemed that adaptive design could be explained only by an intelligent designer; in fact, this "argument from design" was considered one of the strongest proofs of the existence of God. For example, the Reverend William Paley wrote in *Natural Theology* (1802) that, just as the intricacy of a watch implies an intelligent, purposeful watchmaker, so every aspect of living nature, such as the human eye, displays "every indication of contrivance, every manifestation of design, which exists in the watch," and must, likewise, have had a Designer.

Supernatural processes cannot be the subject of science, so when Darwin offered a purely natural, materialistic alternative to the argument from design, he not only shook the foundations of theology and philosophy, but brought every aspect of the study of life into the realm of science. His alternative to intelligent design was design by the completely mindless process of natural selection, according to which organisms possessing variations that enhance survival or reproduction replace those less suitably endowed, which therefore survive or reproduce in lesser degree. This process cannot have a goal, any more than erosion has the goal of forming canyons, for *the future cannot cause material events in the present*. Thus the concepts of goals or purposes have no place in biology (nor in any other of the natural sciences). According to Darwin and contemporary evolutionary theory, the weaver ant's behavior has the appearance of design because among many random genetic variations (mutations) governing the behavior of an ancestral ant species, those displayed by *Oecophylla* enhanced survival and reproduction under its particular ecological circumstances.

Adaptive biological processes *appear* to have goals: weaver ants act as if they have the goal of constructing a nest; an aroid's leaf develops toward a suitable shape and stops developing when that shape is attained. We may loosely describe such features by TELEOLOGICAL statements, which express goals (e.g., "She studied *in order* to pass the exam"). But no conscious anticipation of the future resides in the cell divisions that shape an aroid's leaf or, as far as we can tell, in the behavior of weaver ants. Rather, the apparent goal-directedness is caused by the operation of a program—coded or prearranged information, residing in DNA sequences—that controls a process (Mayr 1988). A program likewise resides in a computer chip, but whereas that program has been shaped by an intelligent designer, the information in DNA has been shaped by a historical process of natural selection. Modern biology views the development, physiology, and behavior of organisms as the results of purely mechanical processes, resulting from interactions between programmed instructions and environmental conditions or triggers.

Definitions of natural selection

Many definitions of natural selection have been proposed (Endler 1986). For our purposes, we will define natural selection as *any consistent difference in fitness among phenotypically different classes of biological entities*. Let us explore this definition in more detail.

The **fitness**—often called the **reproductive success**—of a biological entity is its average per capita rate of increase in numbers. When we speak of natural selection among genotypes or organisms, the components of fitness generally consist of (1) the probability of survival to the various reproductive ages, (2) the average number of offspring (e.g., eggs, seeds) produced via female function, and (3) the average number of offspring produced via male function. "Reproductive success" has the same components, since survival is a prerequisite for reproduction.

Variation in the number of offspring produced as a consequence of competition for mates is often referred to as **sexual selection**, which some authors distinguish from natural selection. We will follow the more common practice of regarding sexual selection as a kind of natural selection.

Because the *probability* of survival and the *average* number of offspring enter into the definition of fitness, and because these concepts apply only to *groups* of events or objects, fitness is defined for a *set* of like entities, such as all the individuals with a particular genotype. That is, natural selection exists if there is an average (i.e., statistically consistent) difference in reproductive success. It is not meaningful to refer to the fitness of a single individual, since its history of reproduction and survival may have been affected by chance to an unknown degree, as we will see shortly.

Differences in survival and reproduction obviously exist among individual organisms, but they also exist below the organismal level, among genes, and above the organismal level, among populations and species. In other words, different kinds of biological entities may vary in fitness, resulting in different **levels of selection**. The most commonly discussed levels of selection are genes, individual organisms that differ in genotype or phenotype, populations within species, and species. Of these, selection among individual organisms (**individual selection**) and among genes (**genic selection**) are by far the most important.

Natural selection can exist only if different classes of entities differ in one or more features, or traits, that affect the components of fitness. Evolutionary biologists differ on whether or not the definition of natural selection should require that these differences be inherited (i.e., have a genetic basis). We will adopt the position taken by those (e.g., Lande and Arnold 1983) who define selection among individual organisms as a consistent difference in fitness among phenotypes. Whether or not this variation in fitness alters the frequencies of genotypes in subsequent generations depends on whether and how the phenotypes are inherited—but that determines the *response to selection*, not the process of selection itself. Although we adopt the phenotypic perspective, we will almost always discuss natural selection among heritable phenotypes because selection seldom has a lasting evolutionary effect unless there is inheritance.

Notice, finally, that according to our definition, natural selection exists whenever there is variation in fitness. Natural selection is not an external force or agent, and certainly not a purposeful one. It is a name for statistical differences in reproductive success among genes, organisms, or populations, and nothing more.

Natural selection and chance

If one neutral allele replaces another in a population by random genetic drift (see Chapter 10), then the bearers of that allele in that population have had a greater rate of increase than the bearers of the other. However, natural selection has not occurred, because the genotypes do not differ *consistently* in fitness: the alternative allele could just as well have increased. There is no *average* difference between the alleles, no *bias* toward the increase of one relative to the other. Fitness differences, in contrast, are *average* differences, *biases*, differences in the *probability* of reproductive success. This does not mean, of course, that every individual of a fitter genotype (or phenotype) survives and reproduces prolifically, while every individual of an inferior genotype perishes; some variation in survival and reproduction occurs independent of—that is, at random with respect to—phenotypic differences. But natural selection resides in the difference in rates of increase among biological entities that is *not* due to chance. *Natural selection is the antithesis of chance.*

If fitness and natural selection are defined by consistent, or average, differences, then we cannot tell whether a difference in reproductive success between two *individuals* is due to chance or to a difference in fitness. We cannot say that one identical twin had lower fitness than the other because she was struck by lightning (Sober 1984), or that the genotype of the Russian composer Tchaikovsky, who had no children, was less fit than the genotype of Johann Sebastian Bach, who had many. We can ascribe genetic changes to natural selection rather than random genetic drift only if we observe consistent, nonrandom changes in replicate populations, or measure numerous individuals of each phenotype and find an average difference in reproductive success.

Selection *of* and selection *for*

In the child's "selection toy" pictured in Figure 11.5, balls of several sizes, when placed in the top compartment, fall through holes in partitions, the holes in each partition being smaller than in the one above. If the smallest balls in the toy are all red, and the larger ones are all other colors, the toy will select the small, red balls. Thus we must distinguish *selection of objects* from *selection for properties* (Sober 1984). Balls are selected *for* the property of small size—that is, *because of* their small size. They are not selected for their color, or because of their color; nonetheless, here there is selection *of* red balls. Natural selection may similarly be considered a sieve that selects *for* a certain body size, mating behavior, or other feature. There may be incidental selection *of* other features that are correlated with those features.

The importance of this semantic point is that when we speak of the *function* of a feature, we imply that there has been natural selection *of* organisms with that feature and *of* genes that program it, but *for* the feature itself. We suppose that the feature *caused* its bearers to have higher fitness. The feature may, however, have other *effects*, or consequences, that were not its function, and *for* which there was no selection. For instance, there was selection for an opposable thumb and digital dexterity in early hominins, with the incidental effect, millions of years later, that we can play the piano. Similarly, a fish species may be selected for coloration that makes it less conspicuous to predators. The *function* of the coloration, then, is predator avoidance. An *effect* of this evolutionary change might well be a lower likelihood that the population will become extinct, but *avoidance of extinction is not a cause of evolution* of the coloration.

Figure 11.5 A child's toy that selects small balls, which drop through smaller and smaller holes from top to bottom. In this case there is selection *of* red balls, which happen to be the smallest, but selection is *for* small size. (After Sober 1984.)

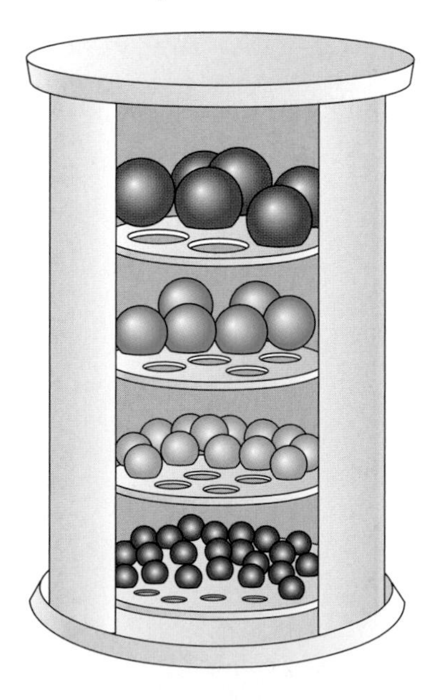

Experimental Studies of Natural Selection

We can illustrate the foregoing rather abstract points by presenting several studies of natural selection. These examples also show how natural selection can be studied by controlled experiments.

Bacterial populations

Bacteria and other microbes are useful for experimental evolutionary studies because of their very rapid population growth. Anthony Dean and colleagues (1986) studied competition between a wild-type strain of *Escherichia coli* and each of several strains that differed from the wild type only by mutations of the gene that codes for β-galactosidase, the enzyme that breaks down lactose. Pairs of genotypes, each consisting of the wild type and a mutant, were cultured together in vessels with lactose as their sole source of energy. The populations were so large that changes in genotype frequencies due to genetic drift alone would be almost undetectably slow. Indeed, in certain populations, the ratio of mutant to wild type did not change for many generations, indicating that these mutations were selectively neutral. One mutant strain, however, decreased in frequency, and so had lower

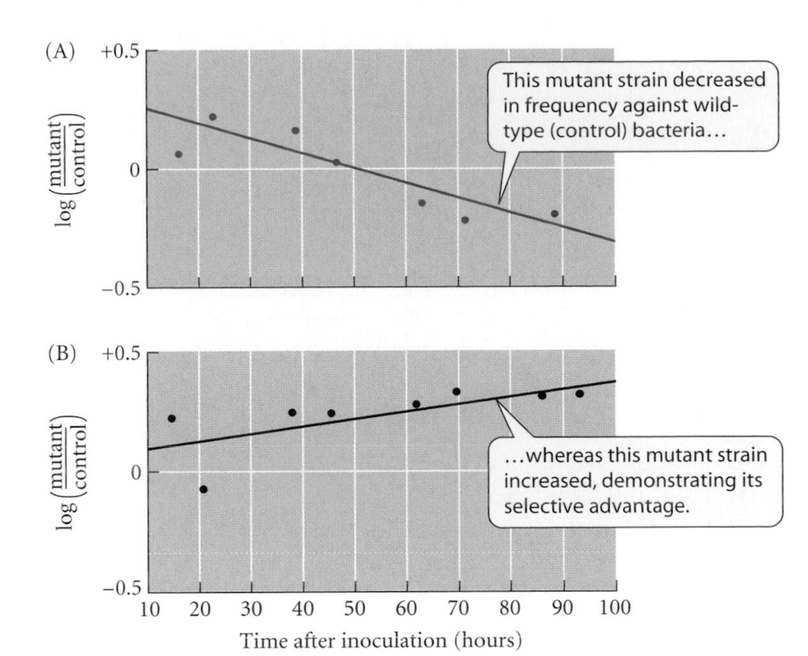
(A)
+0.5
0
−0.5
log (mutant/control)
This mutant strain decreased in frequency against wild-type (control) bacteria...
(B)
+0.5
0
−0.5
log (mutant/control)
...whereas this mutant strain increased, demonstrating its selective advantage.
10 20 30 40 50 60 70 80 90 100
Time after inoculation (hours)

Figure 11.6 Natural selection on mutations in the β-galactosidase gene of *Escherichia coli* in laboratory populations maintained on lactose. In each case, a strain bearing a mutation competed with a control strain bearing the wild-type allele. Populations were initiated with equal numbers of cells of each genotype; i.e., with log (ratio of mutant/control) initially equal to zero. Without selection, no change in the log ratio would be expected. (A) One mutant strain decreased in frequency, showing a selective disadvantage. (B) Another mutant strain increased in frequency, demonstrating its selective (adaptive) advantage. (After Dean et al. 1986.)

fitness than the wild type, apparently due to its lower enzyme activity. Another mutant strain, with higher enzyme activity, increased in frequency, displaying a greater rate of increase than the wild type (Figure 11.6).

This experiment conveys the essence of natural selection: it is a completely mindless process without forethought or goal. Adaptation—evolution of a bacterial population with a higher average ability to metabolize lactose—resulted from a difference in the rates of reproduction of different genotypes caused by a phenotypic difference (enzyme activity).

Another experiment with bacteria illustrates the distinction between "selection of" and "selection for." In *E. coli*, the wild-type allele *his*$^+$ codes for an enzyme that synthesizes histidine, an essential amino acid, whereas *his*$^-$ alleles are nonfunctional. These alleles are selectively neutral if histidine is supplied so that cells with the mutant allele can grow. Atwood et al. (1951) observed, to their surprise, that every few hundred generations, the allele frequencies changed rapidly and drastically in experimental cultures that were supplied with histidine (Figure 11.7). The experimenters showed that the *his* alleles were **hitchhiking** with advantageous mutations at other loci—a phenomenon readily observed in bacteria because their rate of recombination is extremely low. Occasionally, a genotype (say, *his*$^-$) would increase rapidly in frequency because of linkage to an advantageous mutation that had occurred at another locus. Subsequently, the alternative allele (*his*$^+$) might increase because of linkage to a new advantageous mutation at another locus altogether. Thus there was selection *for* new advantageous mutations in these bacterial populations, and selection *of* neutral alleles at the linked *his* locus.

Figure 11.7 Allele frequency fluctuates due to hitchhiking in a laboratory population of *Escherichia coli*. The *y*-axis represents the frequency of the selectively neutral *his*$^-$ allele compared with that of the wild-type *his*$^+$ allele. The frequency of the *his*$^-$ allele increases if a cell bearing it experiences an advantageous mutation at another locus, then decreases if a different, more advantageous mutation occurs in a wild-type cell. (After Nestmann and Hill 1973.)

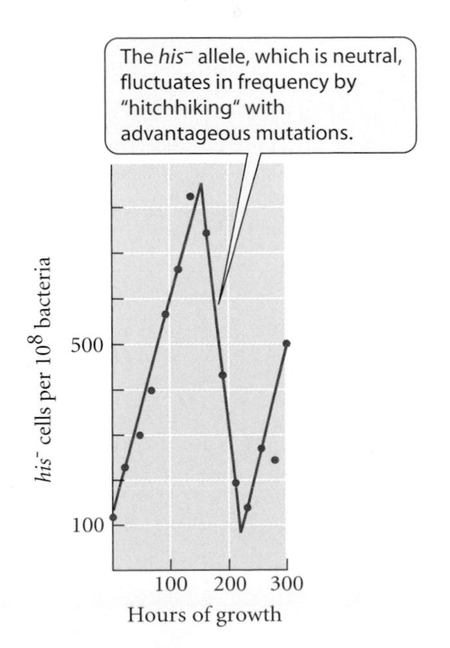
The his⁻ allele, which is neutral, fluctuates in frequency by "hitchhiking" with advantageous mutations.
his⁻ cells per 10⁸ bacteria
500
100
100 200 300
Hours of growth

Inversion polymorphism in *Drosophila*

Natural populations of *Drosophila pseudoobscura* are highly polymorphic for inversions—that is, differences in the arrangement of genes along a chromosome due to 180° reversals in the orientation of chromosome segments (see Chapter 8). Theodosius Dobzhansky suspected that these inversions affected fitness when he observed that the frequencies of several such arrangements within natural populations displayed a regular seasonal cycle (Figure 11.8A). This pattern implied changes in their relative fitnesses as a consequence of environmental changes, perhaps in temperature. He followed these observations with several experiments using POPULATION CAGES: boxes in which populations of several thousand flies are maintained by periodically providing cups of food in which larvae develop. In one such experiment, Dobzhansky (1948) used flies with two inversions, called Standard (ST) and Arrowhead (AR), that can be distinguished under the microscope by their banding patterns. One cage was initiated with 1119 ST and 485 AR chromosome copies (i.e., frequencies of 0.70 and 0.30, respectively). A second cage was initiated with ST and AR frequencies of 0.19 and 0.81, respectively. Within about 15 generations, the frequency of ST had dropped to and leveled off at about 0.54 in the first cage, and it had risen to almost the same frequency (0.50) in the second cage (Figure 11.8B). Later experiments showed that such changes occurred consistently in replicate populations.

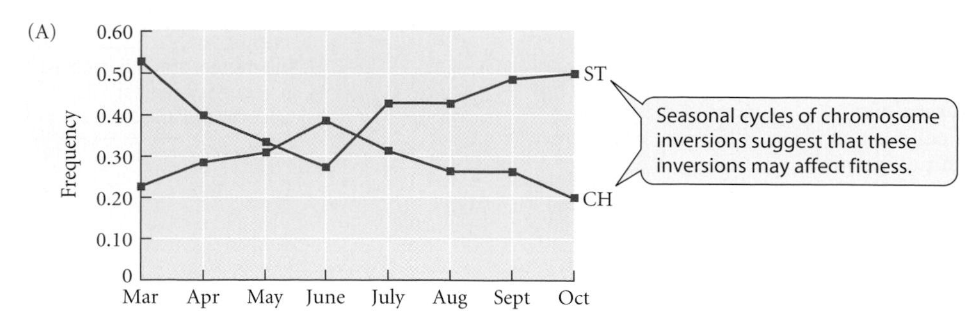

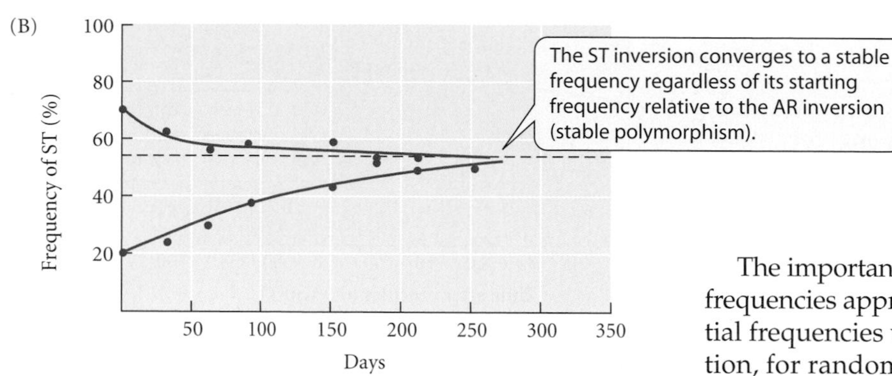

Figure 11.8 Changes in the frequencies of chromosome inversions in natural and laboratory populations of *Drosophila pseudoobscura*. (A) Seasonal fluctuations in the frequencies of inversions ST and CH in a natural population. The consistency of such cycles suggested to Dobzhansky that the inversions affect fitness. (B) The frequency of ST inversions in experimental populations. The frequency of ST arrived at about the same equilibrium level irrespective of starting frequencies of ST and AR. The convergence of the populations toward the same frequency shows that the ST and AR inversions affect fitness, and that natural selection maintains both in a population in a stable, or balanced, polymorphism. (A after Dobzhansky 1970; B after Dobzhansky 1948.)

The important feature of this experiment is that the chromosome frequencies approached a *stable equilibrium*, no matter what the initial frequencies were. This result can only be due to natural selection, for random genetic drift would not show such consistency. Moreover, natural selection must be acting in such a way as to *maintain variation* (polymorphism); it does not necessarily cause fixation of a single best genotype. When the genotype frequencies reach equilibrium, natural selection continues to occur, but evolutionary change does not.

Male reproductive success

The courting males of many species of animals have elaborate morphological features and engage in conspicuous displays; roosters provide a familiar example. Some such features appear to have evolved through female choice of males with conspicuous features, which therefore enjoy higher reproductive success than less elaborate males (see Chapter 14). For example, male long-tailed widowbirds (*Euplectes progne*) have extremely long tail feathers. Malte Andersson (1982) shortened the tail feathers of some wild males and attached the clippings to the tail feathers of others, thus elongating them well beyond the natural length. He then observed that males with shortened tails mated with fewer females than did normal males, and that males with elongated tails mated with more females (Figure 11.9).

Male guppies (*Poecilia reticulata*) have a very variable pattern of colorful spots. In Trinidad, males have smaller, less contrasting spots in streams inhabited by their major predator, the fish *Crenicichla*, than in streams without this predator. John Endler (1980) moved 200 guppies from a *Crenicichla*-inhabited stream to a site that lacked the predator. About 2 years (15 generations) later, he found that the newly established population had larger spots and a greater diversity of color patterns, so that the population now resembled those that naturally inhabit *Crenicichla*-free streams. Endler also set up populations in large artificial ponds in a greenhouse. After 6 months of population growth, he introduced *Crenicichla* into four ponds, released a less dangerous predatory fish (*Rivulus*) into four others, and maintained two control populations free of predators. In censuses after 4 and 10 generations, the number and brightness of spots per fish had increased in the ponds without *Crenicichla* and had declined in those with it (Figure 11.10). Males with more and brighter spots have greater mating success, but they are also more susceptible to being seen and captured by *Crenicichla*.

These experiments show that natural selection may sometimes lie only in differences in reproductive rate, not survival. Differences in mating success, which Darwin called *sexual selection*, result in adaptations for obtaining mates, rather than adaptations for survival. The guppy experiments also show that a feature may be subjected to *conflicting selection*

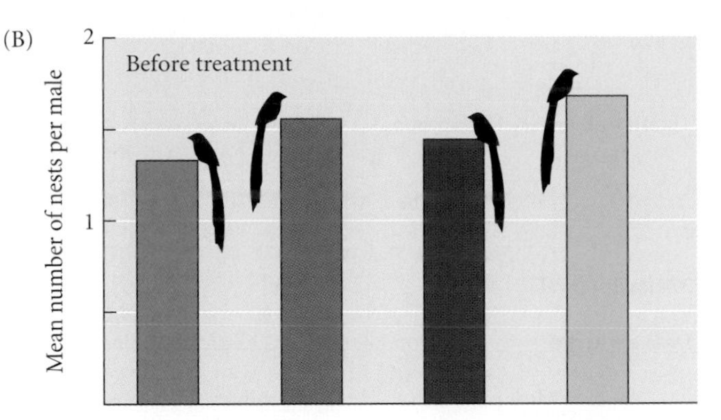

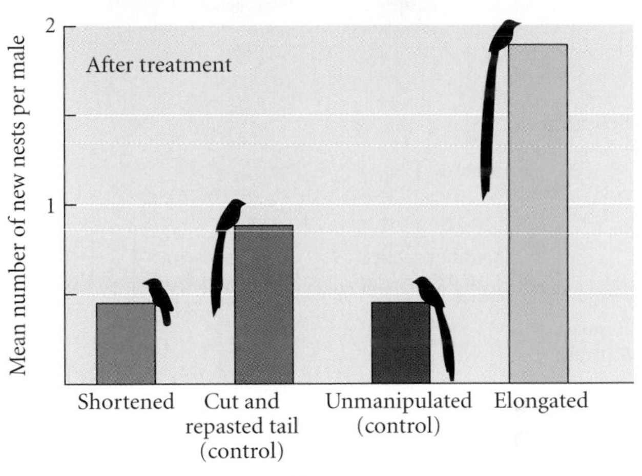

Figure 11.9 (A) A male long-tailed widowbird in flight. (B) Effects of experimental alterations of tail length on males' mating success, measured by the number of nests in the male's territory. Nine birds were chosen for each of four treatments: shortening or elongating the tail feathers, or controls of two types: one in which the tail feathers were cut and repasted, and one in which the tail was not manipulated. (After Andersson 1982; photo courtesy of Malte Andersson.)

pressures (such as sexual selection and predation), and that the direction of evolution may then depend on which is stronger. Many advantageous characters, in fact, carry corresponding disadvantages, often called COSTS or TRADE-OFFS: the evolution of male coloration in guppies is governed by a trade-off between mating success and avoidance of predation.

Population size in flour beetles

The small beetle *Tribolium castaneum* breeds in stored grains and can be reared in containers of flour. Larvae and adults feed on flour, but also eat (cannibalize) eggs and pupae. Michael Wade (1977, 1979) set up 48 experimental populations under each of three treatments. Each population was propagated from 16 adult beetles each generation. The control populations (treatment C) were propagated simply by moving beetles to a new vial of flour: each population in one generation gave rise to one population in the next. In treatment A, Wade deliberately selected for high population size by initiating each generation's 48 populations with sets of 16 beetles taken only from those few populations (out of the 48) in which the greatest number of beetles had developed. In treatment B, low population size was selected in the same way, by propagating beetles only from the smallest populations (Figure 11.11A).

Over the course of 9 generations, the average population size declined in all three treatments, most markedly in treatment B and least in treatment A (Figure 11.11B). The net reproductive rate also declined. In treatment C, these declines must have been due to evolution *within* each population, due to natural selection among the genotypes of individual beetles within each population. This process is *individual selection*, of the same kind we have assumed to operate on, say, the color patterns of guppies. But in treatments A and B, Wade imposed another level of selection by allowing some populations, or groups, but not others, to persist based on a phenotypic character-

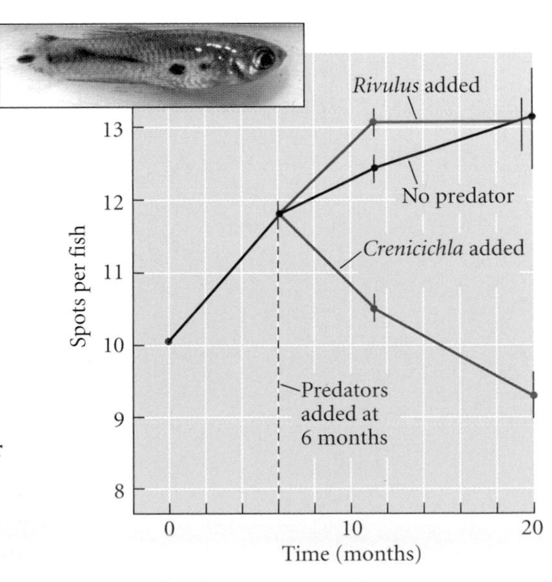

Figure 11.10 Evolution of male color pattern in experimental populations of guppies. Six months after the populations were established, some were exposed to a major predator of adult guppies (*Crenicichla*), some to a less effective predator that feeds mainly on juvenile guppies (*Rivulus*), and some were left free of predators (controls). Numbers of spots were counted after 4 and 10 generations. The vertical bars measure the variation among males. (After Endler 1980; photo courtesy of Anne Houde.)

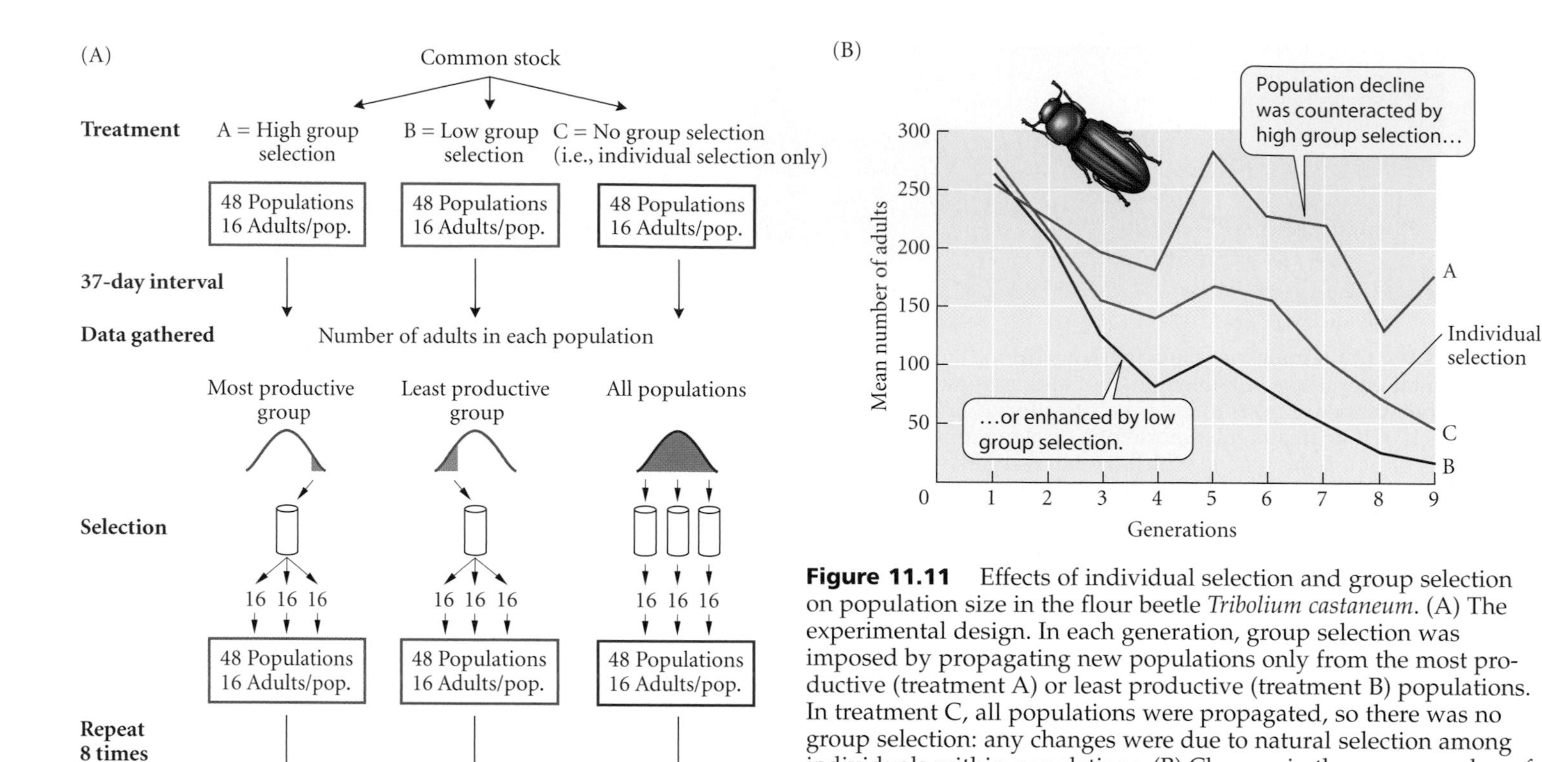

Figure 11.11 Effects of individual selection and group selection on population size in the flour beetle *Tribolium castaneum*. (A) The experimental design. In each generation, group selection was imposed by propagating new populations only from the most productive (treatment A) or least productive (treatment B) populations. In treatment C, all populations were propagated, so there was no group selection: any changes were due to natural selection among individuals within populations. (B) Changes in the mean number of adult beetles in the three treatments. (After Wade 1977.)

istic of each *group*—namely, its size. This process, called **group selection** or **interdemic selection**, operates *in addition to* individual selection among genotypes within populations. We must distinguish selection *within* populations from selection *among* populations.

The decline of population size in the control (C) populations seems like the very antithesis of adaptation. Wade discovered, however, that compared with the foundation stock from which the experimental populations had been derived, adults in the C populations had become more likely to cannibalize pupae, and females were prone to lay fewer eggs when confined with other beetles. For an individual beetle, cannibalism is an advantageous way of obtaining protein, and it may be advantageous for a female not to lay eggs if she perceives the presence of other beetles that may eat them. But although these features are advantageous to the individual, they are disadvantageous to the population, whose reproductive rate declines.

By selecting groups for low population size (treatment B), Wade *reinforced* these same tendencies. In treatment A, on the other hand, selection at the group level for large population size *opposed* the consequences of individual selection within populations. Compared with the C populations, beetles from treatment A had higher fecundity in the presence of other beetles, and they were less likely to cannibalize eggs and pupae. Thus selection among groups had affected the course of evolution.

This experiment shows that the size or growth rate of a population may decline due to natural selection even as individual organisms become fitter. It also illustrates that selection might operate at two levels: among individuals and among populations.

Selfish genetic elements

In many species of animals and plants, there exist "selfish" genetic elements, which are transmitted at a higher rate than the rest of an individual's genome and are detrimental (or at least not advantageous) to the organism (Hurst and Werren 2001). Many of these elements exhibit **segregation distortion**, or **meiotic drive**, meaning that the element is carried by more than half of the gametes of a heterozygote. For example, the *t* locus of the house mouse (*Mus musculus*) has several alleles that, in a male heterozygous for one of these alleles and for the normal allele T, are carried by more than 90 percent of the sperm. In the homozygous condition, certain of the t alleles are lethal, and others cause males to be sterile. Despite these disadvantages to the individual, the meiotic drive of the t alleles is so great that they reach high frequency in many populations. Another selfish element is a small chromosome called *psr* (which stands for "paternal sex ratio") in the parasitic wasp *Nasonia vitripennis*. It is transmitted mostly through sperm rather than eggs. When an egg is fertilized by a sperm containing this genetic element, it causes the destruction of all the other paternal chromosomes, leaving only the maternal set. In *Nasonia*, as in all Hymenoptera, diploid eggs become females and haploid eggs become males. The *psr* element thus converts female eggs into male eggs, thereby ensuring its own future propagation through sperm, even though this could possibly so skew the sex ratio of a population as to threaten its survival.

Selfish genetic elements forcefully illustrate the nature of natural selection: it is nothing more than differential reproductive success (of genes in this case), which need not result in adaptation or improvement in any sense. These elements also exemplify different levels of selection: in these cases, genic selection acts in opposition to individual selection. Selection among genes may not only be harmful to individual organisms, but might also cause the extinction of populations or species.

Levels of Selection

Selection of organisms and groups

It is common to read statements to the effect that oysters have a high reproductive rate "to ensure the survival of the species," or that antelopes with sharp horns refrain from physical combat because combat would lead to the species' extinction. These naive statements betray a misunderstanding of natural selection. If traits evolve by individual selection—by the replacement of less fit by more fit individuals, generation by generation—then the possibility of future extinction cannot possibly affect the course of evolution. Moreover, an **altruistic trait**—a feature that reduces the fitness of an individual that bears it for the benefit of the population or species—cannot evolve by individual selection. An altruistic genotype amid other genotypes that were not altruistic would necessarily decline in frequency, simply because it would leave fewer offspring per capita than the others. Likewise, if a population were to consist of altruistic genotypes, a selfish mutant—a "cheater"—would increase to fixation, even if a population of such selfish organisms had a higher risk of extinction (Figure 11.12).

There is a way, however, in which traits that benefit the population at a cost to the individual might evolve: by group selection. Populations made up of selfish genotypes, such as those with high reproductive rates that exhaust their food supply, might have a higher extinction rate than populations made up of altruistic genotypes. If so, then the species as a whole might evolve altruism through the greater survival of groups of altruistic individuals, even though individual selection within each group would act in the opposite direction (Figure 11.13A).

The hypothesis of group selection was criticized by George Williams (1966) in an influential book, *Adaptation and Natural Selection*. Williams argued that supposed adaptations that benefit the population or species, rather than the individual, do not exist: either the feature in

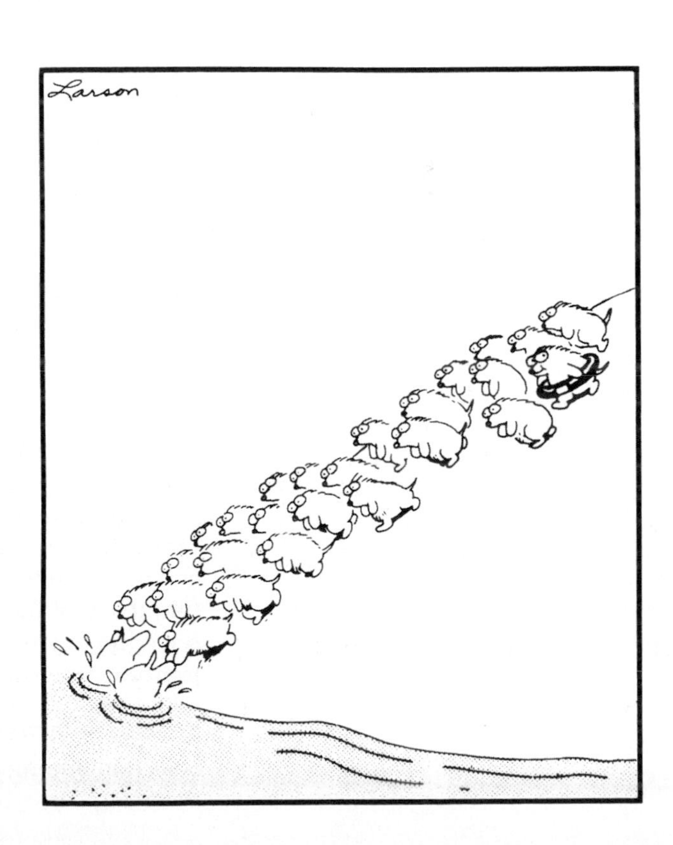

Figure 11.12 The mythical self-sacrificial behavior of lemmings, which (according to popular belief) rush *en masse* into the sea to prevent overpopulation. Cartoonist Gary Larson, in *The Far Side*, illustrates the "cheater" principle, and why such altruistic behavior would not be expected to evolve. (Reprinted with permission of Chronicle Features, San Francisco.)

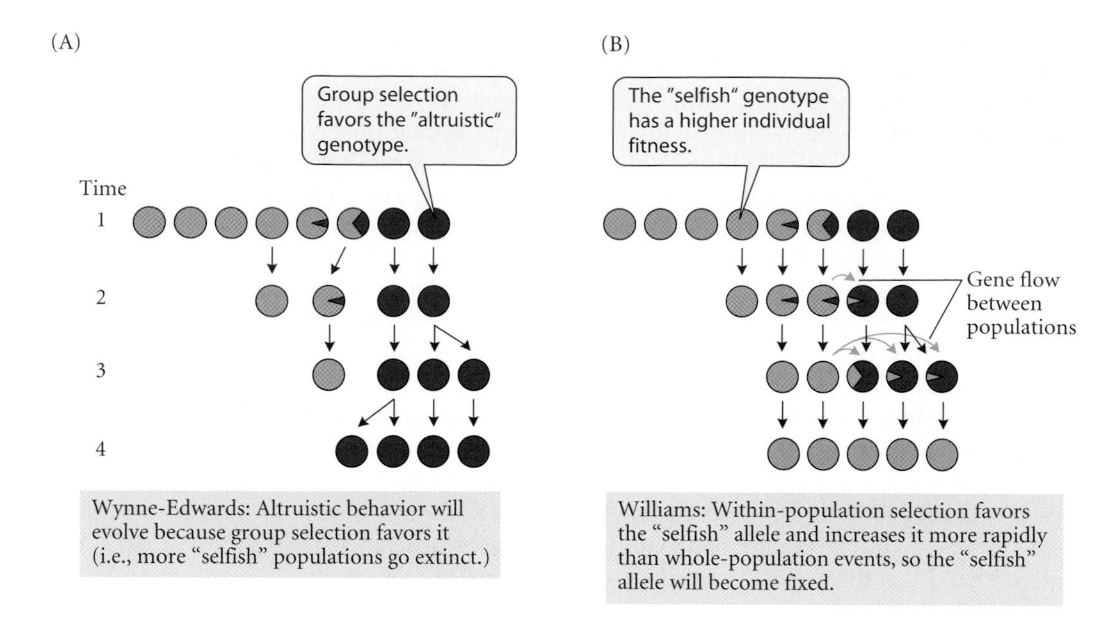
(A)
(B)
Group selection favors the "altruistic" genotype.
The "selfish" genotype has a higher individual fitness.
Time
1
2
3
4
Gene flow between populations
Wynne-Edwards: Altruistic behavior will evolve because group selection favors it (i.e., more "selfish" populations go extinct.)
Williams: Within-population selection favors the "selfish" allele and increases it more rapidly than whole-population events, so the "selfish" allele will become fixed.

Figure 11.13 Conflict between group and individual selection. Each circle represents a population of a species, traced through four time periods. Some new populations are founded by colonists from established populations, and some populations become extinct. The proportions of pink and blue areas in each circle represent the proportions of an "altruistic" and a "selfish" genotype in the population, the selfish genotype having a higher reproductive rate (individual fitness). Lateral arrows indicate gene flow between populations. (A) An altruistic trait may evolve by group selection if the rate of extinction of populations of the selfish genotype is very high. (B) Williams's argument: because individual selection operates so much more rapidly than group selection, the selfish genotype increases rapidly within populations and may spread by gene flow into populations of altruists. Thus the selfish genotype becomes fixed, even if it increases the rate of population extinction.

question is not an adaptation at all, or it can be plausibly explained by benefit to the individual or the individual's genes. For example, females of many species lay fewer eggs when population densities are high, but not to ensure a sufficient food supply for the good of the species. At high densities, when food is scarce, a female simply cannot form as many eggs, so her reduced fecundity may be a physiological necessity, not an adaptation. Moreover, an individual female may indeed be more fit if she forms fewer eggs in these circumstances and waits until food becomes more abundant, so that her subsequent offspring will have a greater chance of survival.

Williams based his opposition to group selection on a simple argument. Individual organisms are much more numerous than the populations into which they are aggregated, and they turn over—are born and die—much more rapidly than populations, which are born (formed by colonization) and die (become extinct) at relatively low rates. Selection at either level requires differences—among individuals or among populations—in rates of birth or death. Thus the rate of replacement of less fit by more fit individuals is potentially much greater than the rate of replacement of less fit by more fit populations, so individual selection will generally prevail over group selection (Figure 11.13B). Although some evolutionary biologists have argued that group selection is important in evolution (e.g., Wilson 1983), the majority view is that *few characteristics have evolved because they benefit the population or species*.

If adaptations that benefit the population are so rare, how do we explain worker ants that labor for the colony and do not reproduce, or birds that emit a warning cry when they see a predator approaching the flock? William Hamilton (1964) posited that such seemingly altruistic behaviors have evolved by **kin selection**, which may be viewed as selection at the level of the gene (see Chapter 14). An allele for altruistic behavior can increase in frequency within a population if the beneficiaries of the behavior are usually related to the individual performing it. Since the altruist's relatives are more likely to carry copies of the altruistic allele than are members of the population at random, when the altruist enhances the fitness of its relatives, even at some cost to its own fitness, it can increase the frequency of the allele (Figure 11.14). We may therefore define kin selection as a form of selection in which alleles differ in fitness by influencing the effect of their bearers on the reproductive success of individuals (kin) who carry the same allele by common descent.

Species selection

Selection among groups of organisms is called **species selection** or **taxon selection** when the groups involved are species or higher taxa (Stanley 1979; Williams 1992b). Consider

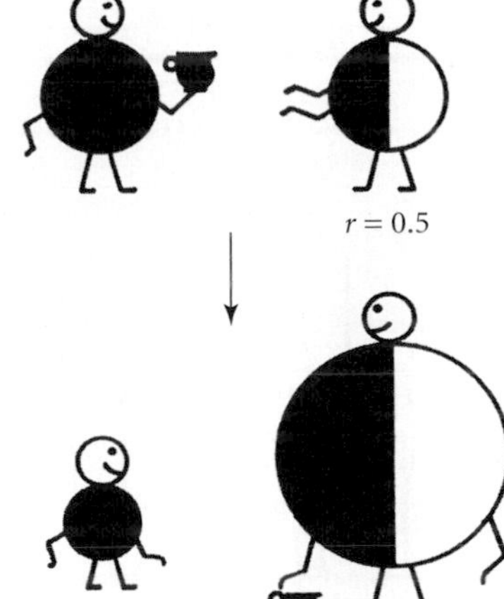

(A) Altruism
The altruist provides benefits to his brother, who shares 1/2 of his genes.
r = 0.5
Despite the altruist's decreased individual fitness, more of his genes are transmitted to the next generation.

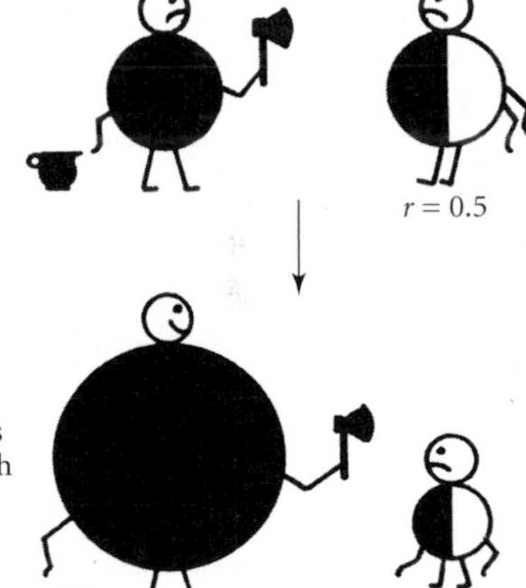

(B) Selfishness
The acts of the selfish individual harm his brother.
r = 0.5
The harmful act decreases the brothers fitness while the selfish individual's fitness increases.

Figure 11.14 The evolution of altruism and selfishness by kin selection. A family is represented by an altruistic or selfish individual and his brother; the fraction of the individual's gene copies that are shared by the brother (0.5) is indicated by the shaded part of the body. (A) The altruist provides some benefit (the vessel) to his brother, thereby increasing his brother's fitness (represented by size) to the extent that the number of shared alleles transmitted to the next generation more than compensates for the altruist's own reduction of genetic fitness. (B) The selfish individual, by acting harmfully (axe), reduces his brother's fitness, but enhances his own fitness even more, so that selfish behavior enhances the transmission of their shared alleles. (From Wilson 1975.)

two related clades that differ in several characters. For example, the orchids (Orchidaceae) generally have highly modified floral characters (scents, flower forms) that promote pollination by specialized insects that differ among orchid species, and have flower petioles that are twisted (Figure 11.15). Their relatives, the irises (Iridaceae), have straight petioles and are usually pollinated by less specialized insects. Suppose (as is likely to be the case) that changes in flower form were more likely to cause reproductive isolation in orchids than in irises, since any change in an orchid flower would be likely to attract a different specialized pollinator. Then the rate of speciation would be greater in orchids than in irises, and the number of species of orchids would grow more rapidly. (The Orchidaceae, with about 19,500 species, is in fact the largest family of plants; the Iridaceae has a mere 1750 species.) Considering both families together, we would expect, over evolutionary time, a change in the proportion of species with modified flowers and twisted petioles. The average state of species overall would change because of this difference in the "birth rate" of new species with one or another feature (speciation rate), analogously to a change in the proportions of different phenotypes within a population that differ in reproductive rate (Figure 11.16).

Differences in extinction rates among sets of species with different characteristics can also change average phenotypes. A good example is the prevalence of sexual species compared with closely related asexual forms.

Figure 11.15 Flowers of (A) the calypso orchid (*Calypso bulbosa*) and (B) the bowl-tubed iris (*Iris macrosiphon*). The highly modified flower of the orchid is anatomically upside down because the petiole that connects it to the stem is twisted 180 degrees. The very specialized pollination systems of many orchids may be the reason that they are the single largest family of plants. (A, photo © Painet, Inc.; B, photo by Gerald and Buff Corsi, California Academy of Sciences.)

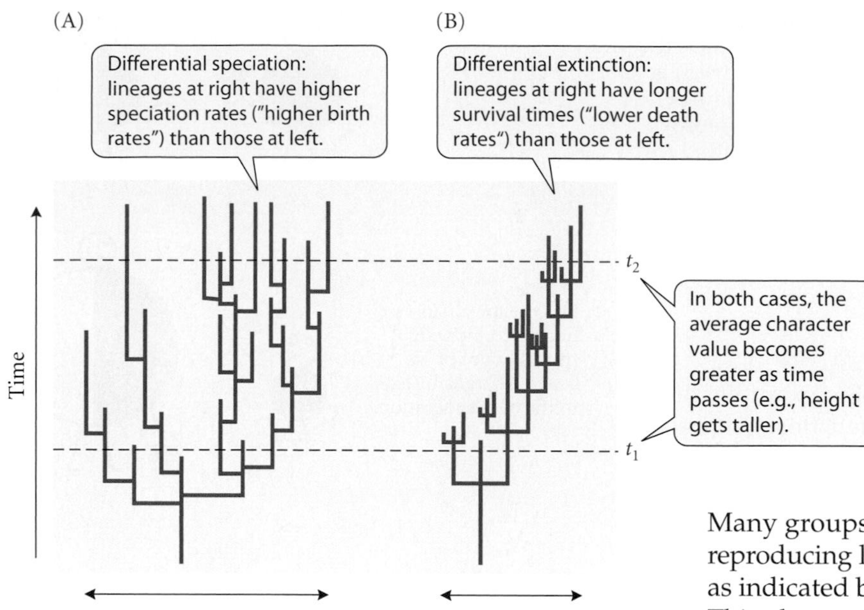

Figure 11.16 Species selection (differential proliferation of species with different character states). The x-axis represents a morphological character, such as body size. (A) Differential speciation: lineages with higher character values (toward the right of the phylogeny) have higher rates of speciation—analogous to higher birth rates of individual organisms—than those with lower values. (B) Differential extinction and survival: lineages with higher character values have longer survival times—analogous to higher survival rates of individual organisms. In both cases, the character value, averaged across species, is greater at time t_2 (upper dashed line) than at time t_1 (lower dashed line). (After Gould 1982.)

Many groups of plants and animals have given rise to asexually reproducing lineages, but almost all such lineages are very young, as indicated by their very close genetic similarity to sexual forms. This observation implies that asexual forms have a higher rate of extinction than sexual populations, since asexuals that arose long ago have not persisted (Normark et al. 2003).

In the orchid/iris example, there is species selection *for* specialized pollination (i.e., specialized pollination causes a higher speciation rate). Because of the correlation between petiole structure and mode of pollination, there has also been selection *of* (but not selection *for*) twisted petioles. The increasing incidence of twisted petioles among these plant species is an *effect* of a fortuitous association with speciation rate.

The Nature of Adaptations

Definitions of adaptation

All biologists agree that an adaptive trait is one that enhances fitness compared with at least some alternative traits. However, some authors include a historical perspective in their definition of adaptations, and others do not.

An ahistorical definition was provided by Reeve and Sherman (1993): "An adaptation is a phenotypic variant that results in the highest fitness among a specified set of variants in a given environment." This definition refers only to the current effects of the trait on reproductive success, compared with those of other variants. At the other extreme, Harvey and Pagel (1991) hold that "for a character to be regarded as an adaptation, it must be a derived character that evolved in response to a specific selective agent." This history-based definition requires that we compare a character's effects on fitness with those of a specific variant; namely, the ancestral character state from which it evolved. Phylogenetic or paleontological data may provide information about the ancestral state.

One reason for this emphasis on history is that a character state may be a simple consequence of phylogenetic history, rather than an adaptation. Darwin saw clearly that a feature might be beneficial, yet not have evolved for the function it serves today, or for any function at all: "The sutures in the skulls of young mammals have been advanced as a beautiful adaptation for aiding parturition [birth], and no doubt they facilitate, or may be indispensable for this act; but as sutures occur in the skulls of young birds and reptiles, which have only to escape from a broken egg, we may infer that this structure has arisen from the laws of growth, and has been taken advantage of in the parturition of the higher animals" (*The Origin of Species*, chapter 6). Whether or not we should postulate that a trait is an adaptation depends on such insights. For example, we know that beak length varies within and among species of Galápagos finches (see Figure 3.22), so it makes sense to ask whether there is an adaptive reason for the average beak length to be 11.5 mm in one

(A)

(B)

Figure 11.17 Exaptation and adaptation. (A) The wing might be called an exaptation for underwater "flight" in members of the auk family, such as this guillemot (*Uria aalge*). (B) The modifications of the wing for efficient underwater locomotion in penguins (here a Magellan penguin, *Spheniscus magellanicus*) may be considered adaptations. (A, photo © Allan Doug/photolibrary.com; B, photo © Pete Oxford/naturepl.com)

species and 8.1 mm in another species (Grant 1986). But it is not sensible to ask whether it is adaptive for the finch to have four toes rather than five, like the Galápagos iguanas, because the ancestor of birds lost the fifth toe and it has never been regained in any bird since. Five toes are probably not an option for birds because of genetic developmental constraints. Thus, if we ask why a species has one feature rather than another, the answer may be adaptation, or it may be phylogenetic history.

A **preadaptation** is a feature that fortuitously serves a new function. For instance, parrots have strong, sharp beaks, used for feeding on fruits and seeds. When domesticated sheep were introduced into New Zealand, some were attacked by an indigenous parrot, the kea (*Nestor notabilis*), which pierced the skin and fed on the sheep's fat. The kea's beak was fortuitously suitable for a new function and may be viewed as a preadaptation for slicing skin.

Preadaptations that have actually been co-opted to serve a new function have been termed **exaptations** (Gould and Vrba 1982). For example, the wings of alcid birds such as guillemots may be considered exaptations for swimming: these birds "fly" under water as well as in air (Figure 11.17A). An exaptation may be further modified by selection so that the modifications are adaptations for the feature's new function: the wings of penguins have been modified into flippers and cannot support flight in air (Figure 11.17B).

Recognizing adaptations

Not all the traits of organisms are adaptations. There are several other possible explanations of organisms' characteristics. First, a trait may be a necessary consequence of physics or chemistry. Hemoglobin gives blood a red color, but there is no reason to think that redness is an adaptation: it is a by-product of the structure of hemoglobin. Second, the trait may have evolved by random genetic drift rather than by natural selection. Third, the feature may have evolved not because it conferred an adaptive advantage, but because it was correlated with another feature that did. Genetic hitchhiking, as exemplified in the bacterial experiment by Atwood et al. described above, is one cause of such correlation. Pleiotropy—the phenotypic effect of a gene on multiple characters—is another. Fourth, as we saw in the previous section, a character state may be a consequence of phylogenetic history. For instance, it may be an ancestral character state, as Darwin recognized in his analysis of skull sutures.

Because there are so many alternative hypotheses, many authors believe that we should not assume that a feature is an adaptation unless the evidence favors this interpretation (Williams 1966). This is not to deny that a great many of an organism's features, perhaps the majority, are adaptations. Several methods are used to infer that a feature is an adaptation for some particular function. We shall note these methods only briefly and incompletely at this point, exemplifying them more extensively in later chapters.

COMPLEXITY. Even if we cannot immediately guess the function of a feature, *we often suspect it has an adaptive function if it is complex*, for complexity cannot evolve except by natural selection. For example, a peculiar, highly vascularized structure called a pecten projects in front of the retina in the eyes of birds (Figure 11.18). Only recently has evidence been developed to show that the pecten supplies oxygen to the retina, but it has always been assumed to play some important functional role because of its complexity and because it is ubiquitous among bird species.

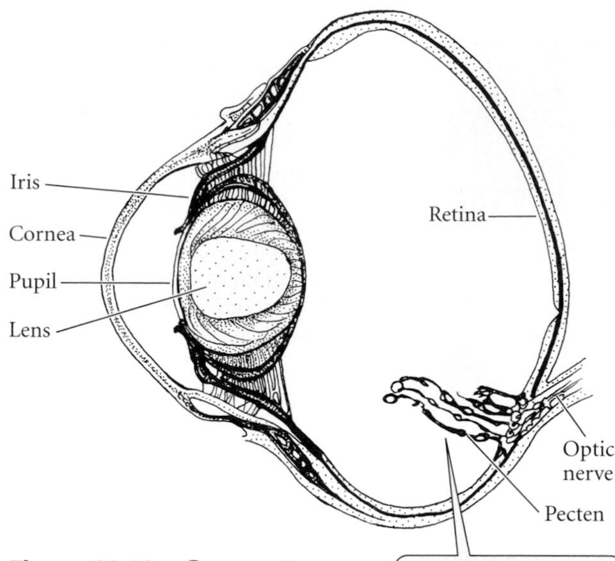
Iris
Cornea
Pupil
Lens
Retina
Optic nerve
Pecten
Among about 30 hypotheses that have been proposed for the function of the pecten, the most popular is that it supplies oxygen to the retina.

Figure 11.18 Cross section of a bird's eye, showing the pecten. Among the 30 or so hypotheses that have been proposed for the function of the pecten, the most likely is that it supplies oxygen to the retina. (After Gill 1995.)

DESIGN. The function of a character is often inferred from its *correspondence with the design* an engineer might use to accomplish some task, or with the *predictions of a model* about its function. For instance, many plants that grow in hot environments have leaves that are finely divided into leaflets, or which tear along fracture lines (Figure 11.19). These features conform to a model in which the thin, hot "boundary layer" of air at the surface of a leaf is more readily dissipated by wind passing over a small than over a large surface, so that the leaf's temperature is more effectively reduced. The fields of functional morphology and ecological physiology are concerned with analyses of this kind.

EXPERIMENTS. Experiments may show that a feature enhances survival or reproduction, or enhances performance (e.g., locomotion or defense) in a way that is likely to increase fitness, relative to individuals in which the feature is modified or absent. Andersson's (1982) alteration of the tail length of male widowbirds (see Figure 11.9) illustrates how artificially created variation may be used to demonstrate a feature's adaptive function—in this case, its role in mating success.

THE COMPARATIVE METHOD. A powerful means of inferring the adaptive significance of a feature is the **comparative method**, which consists of *comparing sets of species to pose or test hypotheses on adaptation and other evolutionary phenomena*. This method takes advantage of "natural evolutionary experiments" provided by convergent evolution. If a feature evolves independently in many lineages because of a similar selection pressure, we can often infer the function of that feature by determining the ecological or other selective factor with

(A)

(B)

Figure 11.19 Functional morphological analyses have shown that small surfaces shed the hot boundary layer of air that forms around them more readily than large surfaces. Many tropical and desert-dwelling plants have large leaves, but they are broken up into leaflets, as in the acacia (A), or split into small sections, as in the banana (B). The form of these leaves is therefore believed to be an adaptation for reducing leaf temperature. (A, photo © Premaphotos/naturepl.com; B, photo © Mireille Vautier/Alamy Images.)

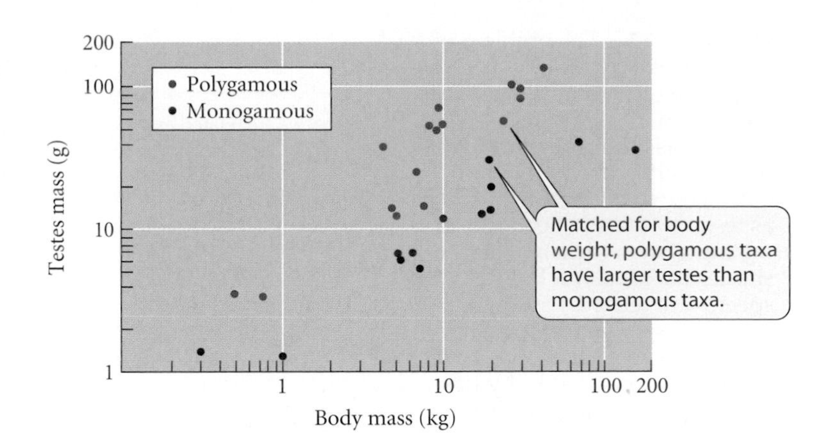

Figure 11.20 The relationship between weight of the testes and body weight among polygamous and monogamous primate taxa. (After Harvey and Pagel 1991.)

which it is correlated. For instance, a long, slender beak has evolved in at least six lineages of birds that feed on nectar (see Figure 3.8). Likewise, the independent evolution of large body size in northern populations of many mammal species implies that large size is an adaptation to low temperatures; the way in which it confers such an advantage can be deduced from physiological and physical principles, as we saw in Chapter 9.

Conversely, we can predict the correlation by postulating, perhaps on the basis of a model, the adaptive features we would expect to evolve repeatedly in response to a given selective factor. For example, in species in which a female mates with multiple males, the several males' sperm compete to fertilize eggs. Males that produce more abundant sperm should therefore have a reproductive advantage. In primates, the quantity of sperm produced is correlated with the size of the testes, so large testes should be expected to provide a greater reproductive advantage in polygamous than in monogamous species. Paul Harvey and collaborators compiled data from prior publications on the mating behavior and testis size of various primates and confirmed that, as predicted, the weight of the testes, relative to body weight, is significantly higher among polygamous than monogamous taxa (Figure 11.20).

This example raises several important points. First, although all the data needed to test this hypothesis already existed, the relationship between the two variables was not known until Harvey and collaborators compiled the data, because no one had had any reason to do so until an adaptive hypothesis had been formulated. Hypotheses about adaptation can be fruitful because they suggest investigations that would not otherwise occur to us.

Second, because the consistent relationship between testis size and mating system was not known a priori, the hypothesis generated a prediction. The predictions made by evolutionary theory, as in many other scientific disciplines, are usually predictions of what we will find when we collect data. Prediction in evolutionary theory does *not* usually mean that we predict the future course of evolution of a species. Predictions of what we will find, deduced from hypotheses, constitute the **hypothetico-deductive method**, of which Darwin was one of the first effective exponents (Ghiselin 1969; Ruse 1979).

Third, the hypothesis was supported by demonstrating that the average testis sizes of polygamous and monogamous taxa show a STATISTICALLY SIGNIFICANT difference. To do this, it is necessary to have a sufficient number of data points—that is, a large enough sample size. For a statistical test to be valid, each data point must be INDEPENDENT of all others. Harvey et al. could have had a larger sample size if they had included, say, 30 species of marmosets and tamarins (Callithricidae) as separate data points, rather than using only one. All the members of this family are monogamous. That suggests that monogamy evolved only once, and has been retained by all callithricids for unknown reasons: perhaps monogamy is advantageous for all the species, or perhaps an internal constraint of some kind prevents the evolution of polygamy even if it would be adaptive. Because our hypothesis is that testis size *evolves* in response to the mating system, we must suspect that the different species of callithricids represent only one evolutionary change, and so provide only one data point (Figure 11.21). Therefore, if we use convergent evolution (i.e., the comparative method) to test hypotheses of adaptation, we should count the *number of independent convergent evolutionary events* by which a character state evolved in the presence of one selective factor versus another (Ridley 1983; Felsenstein 1985; Harvey and Pagel 1991). Consequently, phylogenetic information is essential for proper use of the comparative method. (However, some biologists argue that counting only phylogenetically independent character changes may not be necessary if the characters are genetically vari-

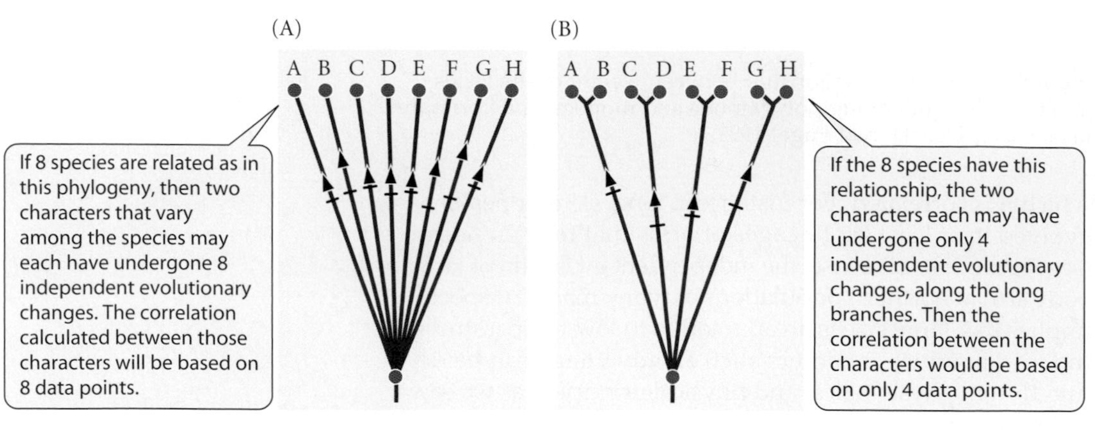

Figure 11.21 The problem of phylogenetic correlation in employing the comparative method. Suppose we test a hypothesis about adaptation by calculating the correlation between two characters, such as testis size (arrowheads) and mating system (ticks), in eight species (A–H). (A) If the species are related as shown in this phylogenetic tree, the character states in each species have evolved independently, by ticks and arrowheads and we have a sample of eight. (B) If the species are related as shown in this phylogenetic tree, the states of both characters may be similar in each pair of closely related species due to their common ancestry, rather than to independent adaptive evolution. Some authors maintain that the two species in each pair are not independent tests of the hypothesis; we would have four samples in this case. (After Felsenstein 1985.)

able, because the characters should be unconstrained and able to evolve to a different adaptive optimum quite rapidly [Reeve and Sherman 1993; Westoby et al. 1997].)

What Not to Expect of Natural Selection and Adaptation

We conclude this discussion of the general properties of natural selection and adaptation by considering a few common misconceptions of, and misguided inferences from, the theory of adaptive evolution.

The necessity of adaptation

It is naïve to think that if a species' environment changes, the species must adapt or else become extinct. Not all environmental changes reduce population size. Nonetheless, an environmental change that does not threaten extinction may set up selection for change in some characteristics. Thus white fur in polar bears may be advantageous, but not necessary for survival (Williams 1966). Just as a changed environment need not set in motion selection for new adaptations, new adaptations may evolve in an unchanging environment if new mutations arise that are superior to any pre-existing genetic variations. We have already stressed that the probability of extinction of a population or species does not in itself constitute selection on individual organisms, and so cannot cause the evolution of adaptations.

Perfection

Darwin noted that "natural selection will not produce absolute perfection, nor do we always meet, as far as we can judge, with this high standard in nature" (*The Origin of Species*, chapter 6). Selection may fix only those genetic variants with a higher fitness than other genetic variants in that population at that time. It cannot fix the best of all conceivable variants if they do not arise, or have not yet arisen, and the best possible variants often fall short of perfection because of various kinds of constraints. For example, with a fixed amount of available energy or nutrients, a plant might evolve higher seed numbers, but only by reducing the size of its seeds or some other part of its structure (see Chapter 17).

Progress

Whether or not evolution is "progressive" is a complicated question (Nitecki 1988; Ruse 1996). The word "progress" has the connotation of a goal, and as we have seen, evolution does not have goals. But even if we strip away this connotation and hold only that progress means "betterment," the possible criteria for "better" depend on the kind of organism. Better learning ability or greater brain complexity has no more evident adaptive advantage for most animals—for example, rattlesnakes—than an effective poison delivery system would have for humans. Measurements of "improvement" or "efficiency" must be relevant to each species' special niche or task. There are, of course, many examples of adaptive trends, each of which might be viewed as progressive within its special context. We will consider this topic in depth in Chapter 21.

Harmony and the balance of nature

As we have seen, selection at the level of genes and individual organisms is inherently "selfish": the gene or genotype with the highest rate of increase increases at the expense of other individuals. The variety of selfish behaviors that organisms inflict on conspecific individuals, ranging from territory defense to parasitism and infanticide, is truly stunning. Indeed, cooperation among organisms requires special explanations. For example, a parent that forages for food for her offspring, at the risk of exposing herself to predators, is cooperative, but for an obvious reason: her own genes, including those coding for this parental behavior, are carried by her offspring, and the genes of individuals that do not forage for their offspring are less likely to survive than the genes of individuals that do. This is an example of kin selection, an important basis for the evolution of cooperation within species (see Chapter 14).

Because the principle of kin selection cannot operate across species, "natural selection cannot possibly produce any modification in a species exclusively for the good of another species" (Darwin, *The Origin of Species*, chapter 6). If a species exhibits behavior that benefits another species, either the behavior is profitable to the individuals performing it (as in bees that obtain food from the flowers they pollinate), or they have been duped or manipulated by the species that profits (as are insects that copulate with orchids). Most mutualistic interactions between species consist of reciprocal exploitation (see Chapter 18).

The equilibrium we may observe in ecological communities—the so-called "balance of nature"—likewise does not reflect any striving for harmony. We observe coexistence of predators and prey not because predators restrain themselves, but because prey species are well enough defended to persist, or because the abundance of predators is limited by some factor other than food supply. Nitrogen and mineral nutrients are rapidly and "efficiently" recycled within tropical wet forests not because ecosystems are selected for or aim for efficiency, but because under competition for sparse nutrients, microorganisms have evolved to decompose litter rapidly, and plants have evolved to capture nutrients released by decomposition. Selection of individual organisms for their ability to capture nutrients has the *effect*, in aggregate, of a dynamic that we measure as ecosystem "efficiency." There is no scientific foundation for the notion that ecosystems evolve toward harmony and balance (Williams 1992a).

Morality and ethics

Natural selection is just a name for differences among organisms or genes in reproductive success. Therefore, it cannot be described as moral or immoral, just or unjust, kind or cruel, any more than wind, erosion, or entropy can be. Hence it cannot be used as a justification or model for human morality or ethics. Nevertheless, evolutionary theory has often been misused in just this way. Darwin expressed distress over an article "showing that I have proved 'might is right,' and therefore that Napoleon is right, and every cheating tradesman is also right." In the late nineteenth and early twentieth century, Social Darwinism, promulgated by the philosopher Herbert Spencer, considered natural selection to be a beneficent law of nature that would produce social progress as a result of untrammeled struggle among individuals, races, and nations. Evolutionary theory has likewise been used to justify eugenics and racism, most perniciously by the Nazis. But neither evolutionary theory nor any other field of science can speak of or find evidence of morality or immorality. These precepts do not exist in nonhuman nature, and science describes only what *is*, not what *ought to be*. The **naturalistic fallacy**, the supposition that what is "natural" is necessarily "good," has no legitimate philosophical foundation.

Summary

1. A feature is an adaptation for a particular function if it has evolved by natural selection for that function by enhancing the relative rate of increase—the fitness—of biological entities with that feature.
2. Natural selection is a consistent difference in fitness among phenotypically different biological entities, and is the antithesis of chance. Natural selection may occur among genes, individual organisms, and groups such as populations or species.

3. Selection at the level of genes or organisms is likely to be the most important because the numbers and turnover rates of these entities are greater than those of populations or species. Therefore, most features are unlikely to have evolved by group selection, the one form of selection that could in theory promote the evolution of features that benefit the species even though they are disadvantageous to the individual organism.
4. Not all features are adaptations. Methods for identifying and elucidating adaptations include studies of function and design, experimental studies of the correspondence between fitness and variation within species, and correlations between the traits of species and environmental or other features (the comparative method). Phylogenetic information may be necessary for proper use of the comparative method.
5. Natural selection does not necessarily produce anything that we can justly call evolutionary progress. It need not promote harmony or balance in nature, and, utterly lacking any moral content, it provides no foundation for morality or ethics in human behavior.

Terms and Concepts

adaptation
altruistic trait
comparative method
exaptation
fitness
function (vs. effect)
genic selection
group selection
hitchhiking
hypothetico-deductive method
individual selection
interdemic selection
kin selection
levels of selection
meiotic drive (segregation distortion)
naturalistic fallacy
preadaptation
reproductive success
selfish genetic elements
sexual selection
species selection
taxon selection

Suggestions for Further Reading

Adaptation and natural selection, by G. C. Williams (Princeton University Press, Princeton, NJ, 1966) is a classic: a clear, insightful, and influential essay on the nature of individual and group selection. See also the same author's *Natural selection: Domains, levels, and challenges* (Oxford University Press, New York, 1992).

Two books by R. Dawkins, *The selfish gene* (Oxford University Press, Oxford and San Francisco, 1989) and *The blind watchmaker* (Norton, New York, 1986), explore the nature of natural selection in depth, as well as treating many other topics in a vivid style for general audiences. More technical works include *The nature of selection: Evolutionary theory in philosophical focus* by E. Sober (MIT Press, Cambridge, MA, 1984), who treats evolution from the perspective of a philosopher of science; and *The comparative method in evolutionary biology* by P. H. Harvey and M. D. Pagel (Oxford University Press, Oxford, 1991), on the use and phylogenetic foundations of the comparative method.

Problems and Discussion Topics

1. Discuss criteria or measurements by which you might conclude that a population is better adapted after a certain evolutionary change than before.
2. Consider the first copy of an allele for insecticide resistance that arises by mutation in a population of insects exposed to an insecticide. Is this mutation an adaptation? If, after some generations, we find that most of the population is resistant, is the resistance an adaptation? If we discover genetic variation for insecticide resistance in a population that has had no experience of insecticides, is the variation an adaptation? If an insect population is polymorphic for two alleles, each of which confers resistance against one of two pesticides that are alternately applied, is the variation an adaptation? Or is each of the two resistance traits an adaptation?
3. Adaptations are features that have evolved because they enhance the fitness of their carriers. It has sometimes been claimed that fitness is a tautological, and hence meaningless, concept. According to this argument, adaptation arises from the "survival of the fittest,"

and the fittest are recognized as those that survive; consequently there is no independent measure of fitness or adaptiveness. Evaluate this claim. (See Sober 1984.)

4. It is often proposed that a feature that is advantageous to individual organisms is the reason for the great number of species in certain clades. For example, wings have been postulated to be a cause of the great diversity of winged insects compared with the few species of primitively wingless insects. How could an individually advantageous feature cause greater species diversity? How can one test a hypothesis that a certain feature has caused the great diversity of certain groups of organisms?
5. Provide an adaptive and a nonadaptive hypothesis for the evolutionary loss of useless organs, such as eyes in many cave-dwelling animals. How might these hypotheses be tested?
6. List the possible criteria by which evolution by natural selection might be supposed to result in "progress," and search the biological literature for evidence bearing on one or more of these criteria.

12

The Genetical Theory of Natural Selection

Natural selection is the most important concept in the theory of evolutionary processes. It is surely the explanation for most of the characteristics of organisms that we find most interesting, ranging from the origin of DNA as the genetic material to the complexities of the human brain. In its elementary form, natural selection—differential reproductive success—is a very simple concept. But its explanatory power is much greater if we take into account the many ways in which it can act and the ways in which its outcome is affected, especially in sexual organisms, by genetic factors such as recombination and the relationship between phenotype and genotype. When we take into account these complexities, we can begin to address a great variety of questions: Why are some characteristics, but not others, variable within species? How great a difference can we expect to see among different populations of a species, such as our own? Do populations of a species always evolve the same adaptation to a particular environmental challenge? How do cooperative and selfish behaviors evolve? Why do some species reproduce sexually and others asexually? How can we explain the extraordinary display feathers of the peacock, the immense fecundity of elms and oysters, the brevity of a mayfly's life, the pregnancy of the male seahorse, the abundance of transposable elements in our own genome?

Maximizing reproduction. The copious fruits of the rowan tree (*Sorbus aucuparia*) will be eaten by birds, which will then disperse the undigested seeds. Only a small percentage of the seeds will take root and grow to reproductive age. (Photo © Geoff Dore/naturepl.com.)

Darwin fully realized that a truly complete theory of evolutionary change would require understanding the mechanism of inheritance. That understanding began to develop only in 1900, when Mendel's publication was discovered. Modern evolutionary theory started to develop as the growing understanding of Mendelian genetics was synthesized with Darwin's theory of selection. The "genetical theory of natural selection" (as the pioneering population geneticist R. A. Fisher entitled his seminal 1930 book) is the keystone of contemporary evolutionary theory, on which our understanding of adaptive evolution depends.

As we delve into the genetical theory of natural selection, we should keep the following important points about natural selection in mind:

- *Natural selection is not the same as evolution*. Evolution is a two-step process: the origin of genetic variation by mutation or recombination, followed by changes in the frequencies of alleles and genotypes, caused chiefly by genetic drift or natural selection. Neither natural selection nor genetic drift accounts for the origin of variation.
- *Natural selection is different from evolution by natural selection*. In some instances, selection occurs—that is, in each generation, genotypes differ in survival or fecundity—yet the proportions of genotypes and alleles stay the same from one generation to another.
- Although natural selection may be said to exist whenever different phenotypes vary in average reproductive success, *natural selection can have no evolutionary effect unless phenotypes differ in genotype*. For instance, selection among genetically identical members of a clone, even though they differ in phenotype, can have no evolutionary consequences. Therefore, it is useful to describe the reproductive success, or fitness, of genotypes, even though genotypes differ in fitness only because of differences in phenotype.
- Because natural selection is variation in average reproductive success (which includes survival), a feature cannot evolve by natural selection unless it makes a positive contribution to the reproduction or survival of individuals that bear it. The long-haired tail of a horse, used as a fly-switch, could not have evolved merely because it increases horses' comfort; it must have resulted in increased reproductive success, perhaps by lowering mortality caused by fly-borne diseases.

Unlike genetic drift, inbreeding, and gene flow, which act at the same rate on all loci in a genome, the allele frequency changes caused by natural selection in a sexually reproducing species proceed largely independently at different loci. Moreover, different characteristics of a species evolve at different rates (mosaic evolution), as we would expect if natural selection brings about changes in certain features while holding others constant (see Chapter 2). Thus we are justified in beginning our analysis of natural selection with a single variable locus that alters a phenotypic character.

Fitness

Unless otherwise specified, the subsequent discussion of natural selection concerns selection at the level of individual organisms within populations. The consequences of natural selection depend on (1) the relationship between phenotype and fitness, and (2) the relationship between phenotype and genotype. These relationships, then, yield (3) a relationship between fitness and genotype, which determines (4) whether or not evolutionary change occurs.

Modes of selection

The relationship between phenotype and fitness can often be described as one of three MODES OF SELECTION (Figure 12.1). For a quantitative (continuously varying) trait, such as size, selection is **directional** if one extreme phenotype is fittest, **stabilizing** (NORMALIZING) if an intermediate phenotype is fittest, or **diversifying** (**disruptive**) if two or more phenotypes are fitter than the intermediates between them. Which *genotype* has the highest fitness under a given selection regime depends on the relationship between phenotype and genotype. For example, under directional selection for large size, genotype A_1A_1 would be most fit if it were largest, but A_1A_2 would be favored if it were larger than either homozygote. As we will soon see, this difference would have important evolution-

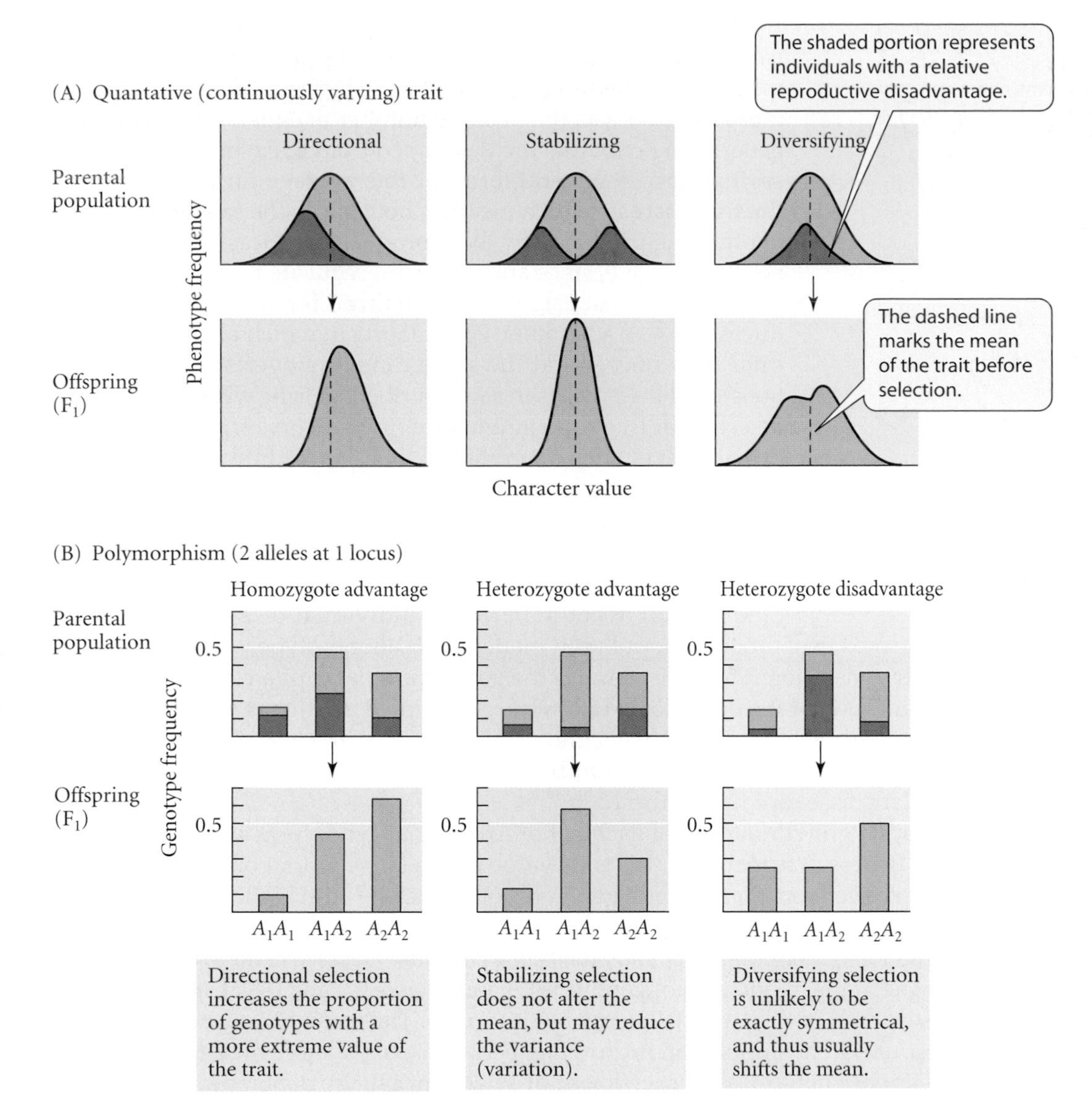

Figure 12.1 Modes of selection on (A) a heritable quantitative (continuously varying) character and (B) a polymorphism consisting of two alleles at one locus. The upper graphs in both (A) and (B) show the distribution in the parental generation, before selection occurs. The shaded portions represent individuals with a relative disadvantage (lower fitness). The dashed line in A represents the mean character value in the parental generation. The lower graphs in both (A) and (B) show the frequency distribution in the F_1 generation, after selection has occurred. (After Endler 1986.)

ary results: the population would become fixed for the largest phenotype if the homozygote were largest, but not if the heterozygote were largest.

The relationship between phenotype and fitness can depend on the environment, since different environmental conditions can favor different phenotypes. It also depends on how the mean and variation in a character are distributed relative to the fitness/phenotype relationship. Thus, if the mean body size is below the optimum, it will be directionally selected until it corresponds to the optimum (at least approximately); after that, it is subject to stabilizing selection.

Defining fitness

Because we are concerned with only those effects of selection that depend on inheritance, we will use models in which an average fitness value is assigned to each genotype. A genotype is likely to have different phenotypic expressions as a result of environmental influences on development, so the fitness of a genotype is the mean of the fitnesses of its several phenotypes, weighted by their frequencies. For example, a particular genotype of *Drosophila pseudoobscura* has a variable number of bristles, depending on the temperature at which the fly develops (see Figure 9.18). Thus, if fitness depended on bristle number, the fitness of a given genotype would depend on the proportions of flies that developed at each temperature.

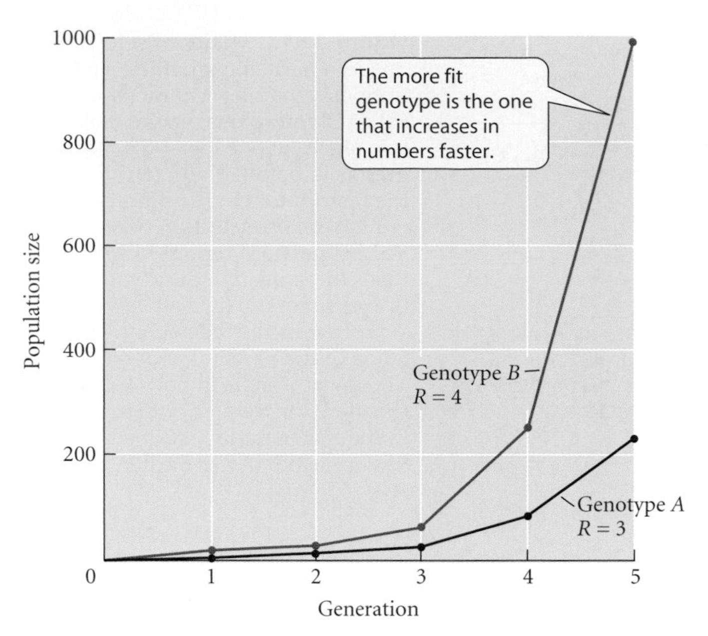

Figure 12.2 The growth of two genotypes with different per capita growth rates in an asexually reproducing population with nonoverlapping generations.

The fitness of a genotype is the average lifetime contribution of individuals of that genotype to the population after one or more generations, measured at the same stage in the life history. Often it suffices to measure fitness as the average number of eggs or offspring one generation hence that are descended from the average egg or offspring born. A general term for this average number is **reproductive success**, which includes not simply the average number of offspring produced by the reproductive process, but the number that survive, since survival is prerequisite to reproduction.

Fitness is most easily conceptualized for an asexually reproducing (PARTHENOGENETIC) population in which all adults reproduce only once, all at the same time (nonoverlapping generations), and then die, as in some parthenogenetic weevils and other insects that live for a single growing season. Suppose that in a population of such an organism, consisting only of females, the proportion of eggs of genotype A that survive to reproductive age is 0.05, and that each reproductive adult lays an average of 60 eggs (her FECUNDITY). Then the fitness of A is (proportion surviving) $\times$ (average fecundity) = $0.05 \times 60 = 3$. This is the number of offspring an average newborn individual of genotype A contributes to the next generation. A population made up of this genotype would grow by a factor of 3 per generation. Thus this value is the genotype's per capita replacement rate, or population growth rate, denoted R. Likewise, genotype B might have a survival rate of 0.10 and an average fecundity of 40, yielding a fitness of 4. With these per capita growth rates, both genotypes will increase in number, but the proportion of the genotype with higher R will increase rapidly (Figure 12.2).

The per capita growth rate, R_i, of each genotype i is that genotype's **absolute fitness**. The **relative fitness** of a genotype, W_i, is its value of R relative to that of some reference genotype. By convention, the reference genotype, often the one with highest R, is assigned a relative fitness of 1.0. Thus, in our example, $W_A = 3/4 = 0.75$ and $W_B = 1.0$. The **mean fitness**, $\bar{w}$, is then *the average fitness of individuals in a population relative to the fittest genotype*. In our example, if the frequencies of genotypes A and B were 0.2 and 0.8, respectively, the mean fitness would be $\bar{w} = (0.2)(0.75) + (0.8)(1.0) = 0.95$. The mean fitness does not indicate whether or not the population is growing, because it is a relative measure.

Another important term is the **coefficient of selection**, usually denoted s, which is the amount by which the fitness of one genotype is reduced relative to the reference genotype. In our example, $W_A = 0.75$, so $s = 0.25$. The coefficient of selection measures the **selective advantage** of the fitter genotype, or the intensity of selection against the less fit genotype.

It is easy to show mathematically that *the rate of genetic change under selection depends on the relative*, not the absolute, *fitnesses of genotypes*. The rate at which a population would become dominated by genotype B in our hypothetical example would be the same, whether genotypes A and B had R values of 0.6 versus 0.8, or 15 versus 20, or 300 versus 400.

Components of fitness

Survival and female fecundity are only two of the possible components of fitness. The components of fitness are more complex if a species reproduces sexually and if it reproduces repeatedly during the individual's lifetime. When generations overlap, as in humans and many other species that reproduce repeatedly, the absolute fitness of a genotype may be measured in large part by its per capita rate of population increase per unit time, r (see Chapter 17). This rate of increase depends on the proportion of individuals surviving to each age class and on the fecundity of each age class. Moreover, r is strongly affected by the age at which females have offspring, not just by their number; that is, genotypes may differ in the length of a generation. If females of genotypes A and B have the same number of offspring when they are 6 months and 12 months old, respectively, the rate of increase (the fitness) of A is about twice that of B, because A will have two generations of descendants by the time B has produced one generation of descendants. Often, differences among males in reproductive success also contribute to differences in fitness.

In sexually reproducing species, genotypes do not simply make copies of themselves; instead, they transmit haploid gametes. Therefore genotype frequencies depend on the allele frequencies among uniting gametes. These allele frequencies are affected by several components of selection at the "zygotic" (organismal) stage, and sometimes by selection at the gametic stage as well (Figure 12.3; Christiansen 1984). Table 12.1 summarizes the components of selection in a sexual species.

Evolution by natural selection depends on the way in which changes in allele frequencies are determined by the components of fitness of each zygotic and each gametic genotype. These components of fitness are combined (usually by multiplying them) into the overall fitness of each genotype. For instance, the fitness of the genotypes in the simple example above was found by multiplying the survival and fecundity of each genotype. In that example, one genotype had superior fecundity and the other had superior survival: a genotype may be superior to another in certain components of fitness and inferior in others, but its *overall fitness determines the outcome of natural selection.*

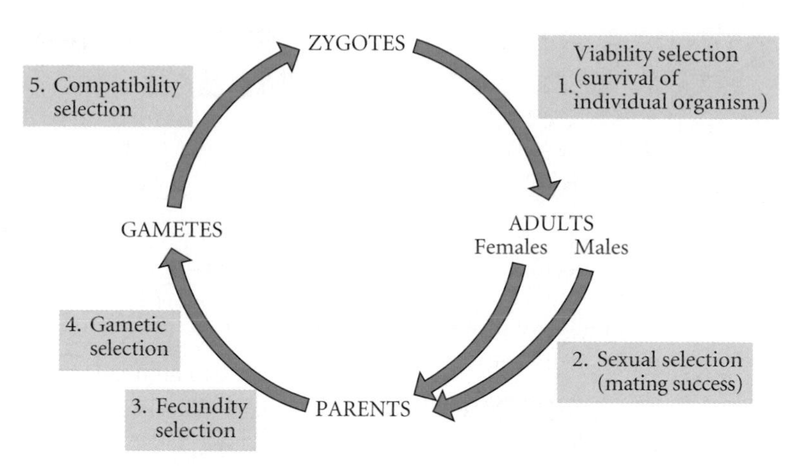

Figure 12.3 Components of natural selection that may affect the fitness of a sexually reproducing organism over the life cycle. Beginning with newly formed zygotes, (1) genotypes may differ in survival to adulthood; (2) they may differ in the numbers of mates they obtain, especially males; (3) those that become parents may differ in fecundity (number of gametes produced, especially eggs); (4) selection may occur among the haploid genotypes of gametes, as in differential gamete viability or meiotic drive; and (5) union of some combinations of gametic genotypes may be more compatible than others. (After Christiansen 1984.)

Models of Selection

In the following discussion, we make the simplifying assumptions that the population is very large, so genetic drift may be ignored; that mating occurs at random; that mutation and gene flow do not occur; and that selection at other loci does not affect the locus we are considering. We will later consider the consequences of changing these unrealistic assumptions. We also assume, for the sake of simplicity, that selection acts through differential survival among genotypes in a species with discrete generations. The principles are much the same for other components of selection and for species with overlapping generations, although these factors introduce complications when data from real populations are analyzed.

If a locus has two alleles (A_1, A_2) with frequencies p and q, the change in frequency from one generation to the next is expressed by Δp, which is positive if the allele is increasing in frequency, negative if it is decreasing, and 0 if p is at an equilibrium. In any model of selection, the change in allele frequencies depends on the relative fitnesses of the different genotypes and on the allele frequencies themselves. Box A provides a mathematical framework for several models of selection.

TABLE 12.1 ***Components of selection in sexually reproducing organisms***

I. Zygotic selection

A. *Viability.* The probability of survival of the genotype through each of the ages at which reproduction can occur. After the age of last reproduction, the length or probability of survival does not usually affect the genotype's contribution to subsequent generations, and so does not usually affect fitness.

B. *Mating success.* The number of mates obtained by an individual. Mating success is a component of fitness if the number of mates affects the individual's number of progeny, as is often the case for males, but less often for females, all of whose eggs may be fertilized by a single male. Variation in mating success is the basis of sexual selection.

C. *Fecundity.* The average number of viable offspring per female. In species with repeated reproduction, the contribution of each offspring to fitness depends on the age at which it is produced (see Chapter 17). The fertility of a mating may depend only on the maternal genotype (e.g., number of eggs or ovules), or it may depend on the genotypes of both mates (e.g., if they display some reproductive incompatibility).

II. Gametic selection

D. *Segregation advantage* (meiotic drive or segregation distortion). An allele has an advantage if it segregates into more than half the gametes of a heterozygote.

E. *Gamete viability.* Dependence of a gamete's viability on the allele it carries.

F. *Fertilization success.* An allele may affect the gamete's ability to fertilize an ovum (e.g., if there is variation in the rate at which a pollen tube grows down a style).

Directional selection

THEORY. The replacement of relatively disadvantageous alleles by more advantageous alleles is the fundamental basis of adaptive evolution. This replacement occurs

BOX 12A Selection Models with Constant Fitnesses

We first present a general model of allele frequency change under natural selection (Hartl and Clark 1997) and then modify it for specific cases. Suppose three genotypes at a locus with two alleles differ in relative fitness due to differences in survival:

	A_1A_1	A_1A_2	A_2A_2
Frequency at birth	p^2	$2pq$	q^2
Relative fitness	w_{11}	w_{12}	w_{22}

The ratio of A_1A_1: A_1A_2: A_2A_2 among surviving adults is

$$p^2w_{11}:2pqw_{12}:q^2w_{22}$$

and the ratio of the alleles (A_1: A_2) among their gametes is

$$[p^2w_{11} + \tfrac{1}{2}(2pqw_{12})]:[\tfrac{1}{2}(2pqw_{12}) + q^2w_{22}]$$

which simplifies to

$$p(pw_{11} + qw_{12}):q(pw_{12} + qw_{22})$$

The gamete frequencies, which are the allele frequencies among the next generation of offspring, are found by dividing each term by the sum of the gametes, which is

$$p(pw_{11} + qw_{12}) + q(pw_{12} + qw_{22})$$

$$= p_2w_{11} + 2pqw_{12} + q_2w_{22}$$

$$= \overline{w}$$

Thus the allele frequencies after selection (p', q') are the gamete frequencies, or

$$p' = \frac{p(pw_{11} + qw_{12})}{\overline{w}}$$

$$q' = \frac{q(pw_{12} + qw_{22})}{\overline{w}}$$

The change in allele frequency between generations is $\Delta p = p' - p$, or

$$\Delta p = \frac{p(pw_{11} + qw_{12}) - p\overline{w}}{\overline{w}}$$

Substituting for $\overline{w}$ and doing the algebra yields

$$\Delta p = \frac{pq[p(w_{11} - w_{12}) + q(w_{12} - w_{22})]}{\overline{w}} \quad \text{(A1)}$$

We can analyze various cases of selection by entering explicit fitness values for the $\overline{w}$'s. A few important cases are the following:

1. Advantageous dominant allele, disadvantageous recessive allele ($w_{11} = w_{12} > w_{22}$).

For w_{11}, w_{12}, and w_{22}, substitute 1, 1, and $1 - s$ respectively in Equation A1. The mean fitness is $p^2(1) + 2pq(1) + q^2(1 - s) = 1 - sq^2$ (bearing in mind that $p^2 + 2pq + q^2 = 1$). The equation for allele frequency change is

$$\Delta p = \frac{spq^2}{1 - sq^2}$$

or, equivalently,

$$\Delta q = \frac{-spq^2}{1 - sq^2} \quad \text{(A2)}$$

2. Advantageous allele partially dominant, disadvantageous allele partially recessive ($w_{11} > w_{12} > w_{22}$).

Let h, lying between 0 and 1, measure the degree of dominance for fitness, and substitute 1, $1 - hs$, and $1 - s$ for w_{11}, w_{12}, and w_{22}. (If $h = 0$, allele A_2 is fully recessive.) After sufficient algebra, we find that

$$\Delta p = \frac{-spq[h(1 - 2q) + sq]}{1 - 2pqhs - sq^2} \quad \text{(A3)}$$

which is positive for all $q > 0$, so allele A_1 increases to fixation. If $h = \tfrac{1}{2}$, Equation A3 reduces to $\Delta p = spq/[2(1 - sq)]$.

$$\Delta p = \frac{spq}{[2(1 - sq)]}$$

3. Fitness of heterozygote is greater than that of either homozygote ($w_{11} < w_{12} > w_{22}$).

Using s and t as selection coefficients, let the fitnesses of A_1A_1, A_1A_2, and A_2A_2 be $1 - s$, 1, and $1 - t$ respectively. Substituting these in Equation A1, we obtain

$$\Delta p = \frac{pq(-sp + tq)}{1 - sp^2 - tq^2} \quad \text{(A4)}$$

There is a stable "internal equilibrium" that can be found by setting $\Delta p = 0$; then $sp = tq$. Substituting $1 - p$ for q, the equilibrium frequency p is $t/(s + t)$. Thus the frequency of A_1 is proportional to the relative strength of selection against A_2A_2.

4. Fitness of heterozygote is less than that of either homozygote (that is, $w_{11} > w_{12} < w_{22}$).

As this is the reverse of the preceding case, let $1 + s$, 1, and $1 + t$ be the fitnesses of A_1A_1, A_1A_2, and A_2A_2. The equation for allele frequency change is

$$\Delta p = \frac{pq(sp - tq)}{1 + sp^2 + tq^2} \quad \text{(A5)}$$

Δp is positive if $sp > tq$, and negative if $sp < tq$. Setting $\Delta p = 0$ and solving for p, we find an internal equilibrium at $p = t/(s + t)$, but this is an unstable equilibrium. For example, if $s = t$, the unstable equilibrium is $p = 0.5$, but then Δp is positive if $p > q$ (i.e., if $p > 0.5$), and negative if $p < q$. The allele frequency therefore arrives at either of two stable equilibria, $p = 1$, or $p = 0$.

---

when the homozygote for an advantageous allele has a fitness equal to or greater than that of the heterozygote or of any other genotype in the population.

An advantageous allele may initially be fairly common if under previous environmental circumstances it was selectively neutral or was maintained by one of several forms of balancing selection (see page 280). However, an advantageous allele is likely to be initially very rare if it is a newly arisen mutation or if it was disadvantageous before an environmental change made it advantageous.

An advantageous allele that increases from a very low frequency is often said to INVADE a population. *Unless an allele can increase in frequency when it is very rare, it is unlikely to be-*

Figure 12.4 Warning (aposematic) coloration in a western coral snake (*Micrurus euryxanthus*), from the North American desert Southwest. If a population of dangerous or unpalatable organisms has a high frequency of such a color pattern, predators may rapidly learn to avoid the aposematically colored organisms, or may evolve to avoid them. It is less obvious how a new, rare mutation for such coloration, if it makes the organisms conspicuous to naive predators, can increase in frequency. (Photo © John Cancalosi/naturepl.com.)

come fixed in the population. According to this principle, some conceivable adaptations are unlikely to evolve because they could not increase if they were initially very rare. For instance, venomous coral snakes (*Micrurus*) are brilliantly patterned in red, yellow, and black (Figure 12.4). This pattern is presumed to be APOSEMATIC (warning) coloration, which is beneficial because predators associate the colors with danger and avoid attacking such snakes. How this coloration initially evolved has long been a puzzle, however, since the first few mutant snakes with brilliant colors would presumably have been easily seen and killed by naive predators. Given that all coral snakes are aposematically colored, it is understandable that predators might evolve an aversion to them (and, indeed, some predatory birds seem to have an innate aversion to coral snakes)—but how the snake's adaptation "got off the ground" is uncertain. (One possibility is that predators generalized from other brilliantly colored unpalatable or dangerous organisms, such as wasps, and avoided aposematic snakes from the beginning.)

A simple example of directional selection occurs if the fitness of the heterozygote is precisely intermediate between that of the two homozygotes (i.e., neither allele is dominant with respect to fitness). The frequencies and fitnesses of the three genotypes may be denoted as follows:

Genotype	A_1A_1	A_1A_2	A_2A_2
Frequency	p^2	$2pq$	q^2
Fitness	1	$1-(s/2)$	$1-s$

These fitnesses may be entered into Equation A1 in Box A, which, when solved, shows that the advantageous allele A_1 increases in frequency, per generation, by the amount

$$\Delta p = \frac{\frac{1}{2}spq}{1-sq} \quad (12.1)$$

where $(1 - sq)$ equals the mean fitness, $\overline{w}$.

Equation 12.1 tells us that Δp is positive whenever p and q are greater than zero. Therefore allele A_1 increases to fixation ($p = 1$), and $p = 1$ is a *stable equilibrium*. The rate of increase (the magnitude of Δp) is proportional to both the coefficient of selection s and the allele frequencies p and q, which appear in the numerator. Therefore the rate of evolutionary change increases as the variation at the locus increases. (It is approximately proportional to $2pq$, the frequency of heterozygotes, when selection is weak.)

Another important aspect of Equation 12.1 is that Δp is positive as long as s is greater than zero, even if it is very small. Therefore, as long as no other evolutionary factors intervene, *a character state with even a minuscule advantage will be fixed by natural selection*. Hence

Figure 12.5 A cryptic katydid (*Mimetica crenulata*) from Costa Rica. The irregular outline, twisted form, and conspicuous vein of the wing provide extraordinary resemblance to a dry, chewed, dead leaf. (Courtesy of P. Naskrecki.)

even very slight differences among species, in seemingly trivial characters such as the distribution of hairs on a fly or veins on a leaf, could conceivably have evolved as adaptations. This principle explains the extraordinary apparent "perfection" of some features. Some katydids, for example, resemble dead leaves to an astonishing degree, with transparent "windows" in the wings that resemble holes and blotches that resemble spots of fungi or algae (Figure 12.5). One might suppose that a less detailed resemblance would provide sufficient protection against predators, and some species are indeed less elaborately cryptic; but if an extra blotch increases the likelihood of survival by even the slightest amount, it may be fixed by selection (providing, we repeat, that no other factors intervene).

The same equations that describe the increase of an advantageous allele describe the fate of a disadvantageous allele: If A_1 and A_2 are advantageous and disadvantageous alleles, respectively, with frequencies p and q, and if $p + q = 1$, then $\Delta p = -\Delta q$. Selection that reduces the frequency of a deleterious mutation or eliminates it is referred to as **purifying selection**, which is simply directional selection in favor of the prevalent, advantageous homozygous genotype.

The number of generations required for an advantageous allele to replace one that is disadvantageous depends on the initial allele frequencies, the selection coefficient, and the degree of dominance (Figure 12.6). An advantageous allele can increase from low frequency more rapidly if it is dominant than if it is recessive because it is expressed in the heterozygous state, and until it reaches a fairly high frequency, it is carried almost entirely by heterozygotes. After a dominant advantageous allele attains high frequency, the deleterious recessive allele is eliminated very slowly, because a rare recessive allele occurs mostly in heterozygous form, and is thus shielded from selection.

One more theoretical conclusion can be drawn from Equation 12.1. The denominator is the mean relative fitness of individuals in the population, $\overline{w}$, which increases as the frequency (q) of the deleterious allele decreases. The mean fitness therefore increases as natural selection proceeds. In a graphical representation of this relationship (Figure 12.7A), we may think of the population as climbing up a "hillside" of increasing mean fitness until it arrives at the summit.

Equation 12.1, finally, can be used to draw an interesting inference from data. If we have data on the frequencies of genotypes at a locus (and therefore also have estimates of the frequencies of alleles, p and q), and if we also have data on allele frequencies in successive generations (i.e., an estimate of Δp), we can solve for s in the equation. This is one way of estimating the strength of natural selection. Several other methods are used to estimate selection coefficients,

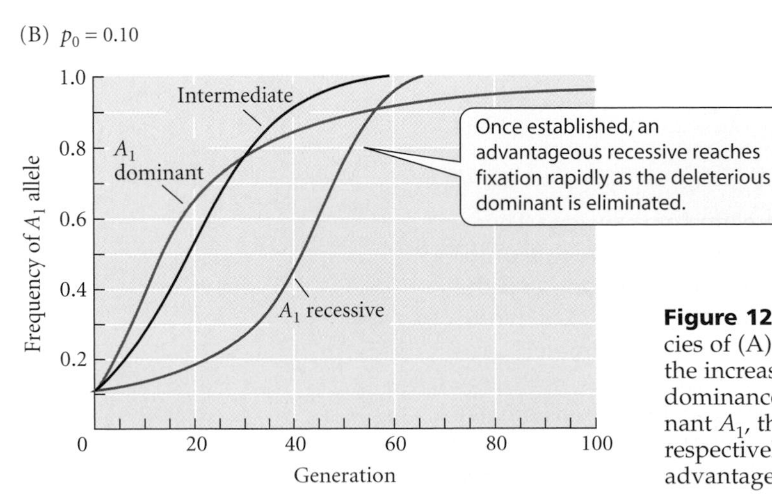

Figure 12.6 Increase of an advantageous allele (A_1) from initial frequencies of (A) $p_0 = 0.01$ and (B) $p_0 = 0.10$. The three curves in each graph show the increase of a fully dominant allele (green), an allele with intermediate dominance (black), and a recessive allele (red). For the advantageous dominant A_1, the fitnesses of genotypes A_1A_1, A_1A_2, and A_2A_2 are 1.0, 1.0, and 0.8 respectively; for the "intermediate" case they are 1.0, 0.9, and 0.8; for the advantageous recessive A_1, they are 1.0, 0.8, and 0.8.

Figure 12.10 Geographic variation aris
gene flow when genotype fitnesses differ
graphic gradient in an environmental fact
nesses W_{11}, W_{12}, and W_{22} of genotypes A_1
plotted for populations along an east–wes
higher fitness than A_2A_2 in the "west," an
(B) Frequency of the A_2 allele in an array
gradient. Each curve represents a differen
ranging from 0 to 100 percent (g = 1). The
flow, the steeper the cline in allele frequen

quency q of an allele A_2 will be 1 in cert
others. Gene flow among populations
into populations in which it is deleter
quency in each population thus arrives
by the balance between the incursion of
their elimination by selection. Thus gen
genetic variation within populations
greater than selection, a population inh
a distinctive environment will not beco
tiated from surrounding populations (I

If local populations are distributed
tal gradient over which the fitness
changes (Figure 12.10A), then even if th
gradually, we should expect an abrupt
cies (a STEP CLINE) in the absence of ger
tions along the gradient (Figure 12.10I
other has highest fitness in each popul
may be established if there is gene flow an
of the cline (the distance over which q c
where V measures the distance that g
against them (the coefficient of selectior
steep clines in allele frequencies will resu

Alleles at the aminopeptidase I locus
in Long Island Sound, a body of water
creases from the west to the east, where
of several allozymes, ap^{94}, increases fron
(Figure 12.11). The aminopeptidase I en

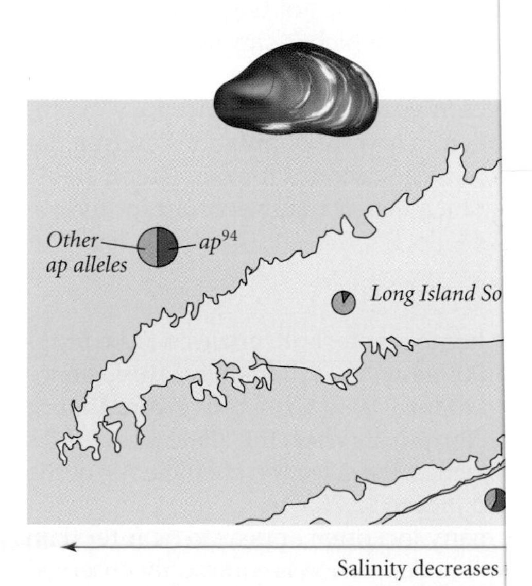

(A) Directional selection
$w_{11} = 1$, $w_{12} = 0.8$, $w_{22} = 0.2$
$\hat{p} = 1$
A_1A_1 homozygote fixed

(B) Directional selection
$w_{11} = 0.2$, $w_{12} = 0.8$, $w_{22} = 1$
$\hat{p} = 0$
A_1A_1 homozygote lost

(C) Overdominance (heterozygote advantage)
$w_{11} = 0.6$, $w_{12} = 1$, $w_{22} = 0.2$
$\hat{p} = 0.666$
Stable equilibrium of polymorphism

(D) Underdominance (heterozygote disadvantage)
$w_{11} = 1$, $w_{12} = 0.4$, $w_{22} = 0.8$
$\hat{p} = 0.4$
Unstable equilibrium; allele frequency changes toward either $p = 0$ or $p = 1$

Mean fitness ($\bar{w}$) — Frequency (p) of allele A_1

Figure 12.7 Plots of mean fitness ($\bar{w}$) against allele frequency (p) for one locus with two alleles when genotypes differ in survival. Each of these plots represents an "adaptive landscape," and may be thought of as a surface, or hillside, over which the population moves. From any given frequency of allele A_1 (p), the allele frequency moves in a direction that increases mean fitness ($\bar{w}$). The arrowheads show the direction of allele frequency change. (A) Directional selection. Here, A_1A_1 is the favored genotype. The equilibrium ($\hat{p} = 1$) is stable: the allele frequency returns to $p = 1$ if displaced. (B) Directional selection, in which the relative fitnesses are reversed compared with graph A, perhaps because of changed environmental conditions. A_2A_2 is now the favored genotype. (C) Overdominance (heterozygote advantage). From any starting point, the population arrives at a stable polymorphic equilibrium ($\hat{p}$). (D) Underdominance (heterozygote disadvantage). The interior equilibrium ($\hat{p} = 0.4$ in this example) is unstable because even a slight displacement initiates a change in allele frequency toward one of two stable equilibria: $\hat{p} = 0$ (loss of A_1) or $\hat{p} = 1$ (fixation of A_1). Therefore this curve represents an adaptive landscape with two peaks. (After Hartl and Clark 1989.)

such as estimating the survival rates (or other components of fitness) of different genotypes in natural populations (Endler 1986).

EXAMPLES OF DIRECTIONAL SELECTION. If a locus has experienced consistent directional selection for a long time, the advantageous allele should be near equilibrium—that is, near fixation. Thus the dynamics of directional selection are best studied in recently altered environments, such as those altered by human activities. Many examples of rapid evolution under such circumstances have been observed. Many are changes in polygenic traits, described in the next chapter.

An example of rapid evolution of a single-locus trait is the case of warfarin resistance in brown rats (*Rattus norvegicus*) (Bishop 1981). Warfarin is an anticoagulant: it inhibits an enzyme responsible for the regeneration of vitamin K, a necessary cofactor in the production of blood-clotting factors. Susceptible rats poisoned with warfarin often bleed to death from slight wounds. A mutation confers resistance by altering the enzyme to a form that is less sensitive to warfarin, but also less efficient in regenerating vitamin K, so that a higher dietary intake of the vitamin is necessary.

Warfarin has been used as a rat poison in Britain since 1953, and by 1958 resistance was reported in certain rat populations. Under exposure to warfarin, resistant rats have a strong survival advantage, and the frequency of the mutation has been known to increase rapidly to nearly 1.0 (Figure 12.8). Resistant rats suffer a strong disadvantage compared with susceptible rats, however, because of their greater need for vitamin K, and the frequency of the resistance allele drops rapidly if the poison is not administered.

Resistance to insecticides has evolved in many insects and mites (Metcalf and Luckmann 1994; Roush and Tabashnik 1990). Resistance appeared in many species in the 1940s, when synthetic pesticides came into wide use. By 1990, populations of more than 500 species were known to be resistant to one or more insecticides (Figure 12.9). Some species,

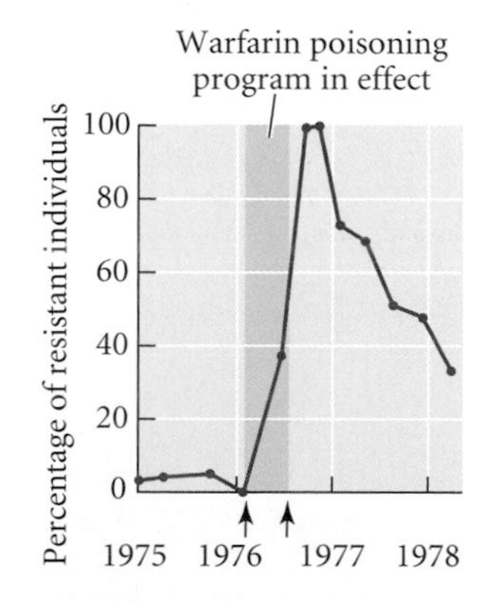

Figure 12.8 The proportion of warfarin-resistant individuals in a population of rats in Wales. The proportion increased when warfarin poisoning was practiced in 1976, but decreased after the poisoning program ended. (After Bishop 1981.)

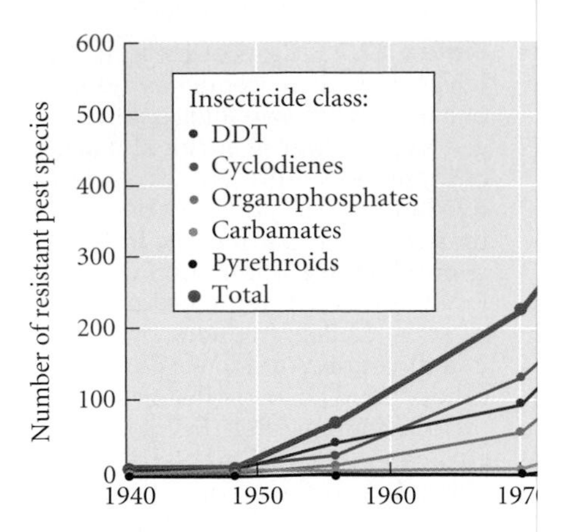

Figure 12.9 The cumulative numbers of arthropod pest species known to have evolved resistance to five classes of insecticides. The upper curve provides the total number of insecticide-resistant species. (After Metcalf and Luckmann 1994.)

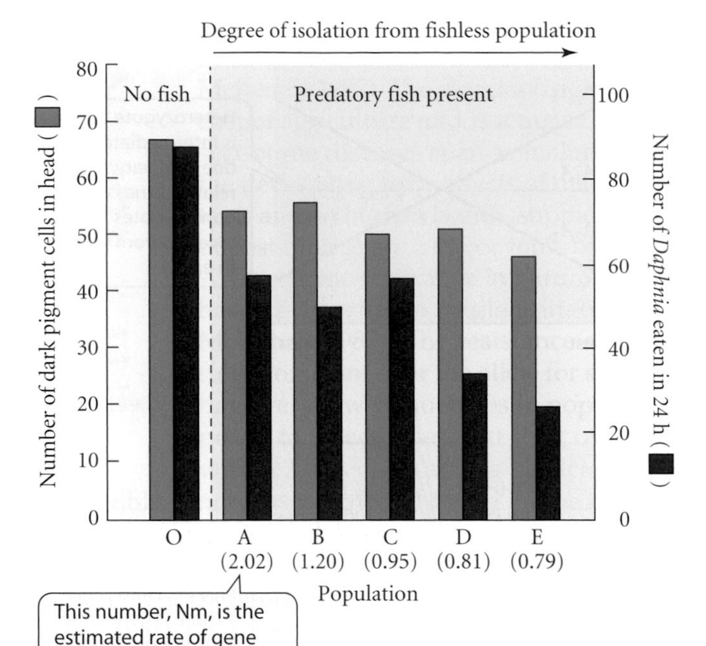

Figure 12.12 Gene flow reduces adaptation to fish predation among larvae of the salamander *Ambystoma barbouri*. Both graphs show mean character values in a "fishless" population (O) and in five populations that coexist with fish and are increasingly isolated from the fishless population. *Nm* estimates gene flow into populations A through E from fishless populations, based on F_{ST} values calculated from allozyme allele frequencies (see Chapter 10). Darkness of coloration (green bars) is estimated from the density of dark pigment cells on the head; paler coloration is favored in streams with fish. The red bars represent feeding rates when salamander larvae in experimental containers were provided with *Daphnia* crustaceans as food, but were then exposed to chemical cues (odorants) from fish. A low feeding rate indicates larval populations that were more immobile in the presence of fish odor, and thus more likely to escape the notice of predatory fish. (After Storfer and Sih 1998; Storfer et al. 1999.)

increasing the intracellular concentration of free amino acids and thereby helping to maintain osmotic balance in salt water. Richard Koehn and Jerry Hilbish (1987) and their colleagues found that aminopeptidase I activity is higher in mussels carrying the ap^{94} allele than in other genotypes. Because ap^{94} genotypes have higher intracellular amino acid levels, the ap^{94} allele is favored at oceanic salinity. In less saline waters, however, a lower concentration of amino acids suffices for osmotic balance, and high aminopeptidase I activity is disadvantageous because the breakdown of protein is costly in terms of both energy and nitrogen, and must be compensated by feeding. The investigators found that within Long Island Sound, ap^{94} has a high frequency among recently settled young mussels, but that its frequency decreases as they age, indicating that ap^{94} mussels have a higher mortality rate than others. The persistence of ap^{94} within the Sound therefore seems to be due to incursions of larvae from the ocean in each generation.

Gene flow can reduce the level of adaptation of populations to their local environment. For example, aquatic larvae of the salamander *Ambystoma barbouri* that occupy streams with predatory fish have paler coloration than populations in sites without fish; the pale coloration matches the background and thus reduces predation. Salamander populations from fish-inhabited streams also have a heightened response to the odor of fish: they seek shelter and become inactive, even though this reduces their feeding activity. However, neither of these adaptations is as well developed in populations that have high levels of gene flow from "fishless" populations (Figure 12.12; Storfer and Sih 1998; Storfer et al. 1999).

Polymorphism Maintained by Balancing Selection

Until the 1940s, the prevalent, or classic, view had been that at each locus, a best allele (the "wild type") should be nearly fixed by natural selection, so that the only variation should consist of rare deleterious alleles, recently arisen by mutation and fated to be eliminated by purifying selection. As we saw in Chapter 9, studies of natural populations revealed instead a wealth of variation. The factors that might be responsible for this variation are: (1) recurrent mutation producing deleterious alleles, subject to only weak selection; (2) gene flow of locally deleterious alleles from other populations in which they are favored by selection; (3) selective neutrality (i.e., genetic drift); and (4) maintenance of polymorphism by natural selection. The last of these hypotheses was championed by British ecological geneticists led by E. B. Ford and American population geneticists influenced by Theodosius Dobzhansky. They represented the BALANCE SCHOOL, holding that a great deal of genetic variation is maintained by **balancing selection** (which is simply selection that maintains polymorphism).

These contrasting points of view are still represented among contemporary evolutionary geneticists, and the causes of genetic variation in natural populations are still not well understood. Several models of natural selection can account for persistent, stable polymorphism, but we do not know the extent to which they actually account for the observed genetic variation within populations.

Heterozygote advantage

If the heterozygote has a higher fitness than either homozygote, both alleles will be propagated in successive generations, in which union of gametes will yield all three genotypes among the zygotes. Such **heterozygote advantage** is also termed **overdominance** or SINGLE-LOCUS HETEROSIS. It results in a stable equilibrium at which the allele frequencies depend on the balance between the fitness values (hence, the selection coefficients) of the two homozygotes (Figure 12.13; see also Box A and Figure 12.7C).

Genotypes that are heterozygous at several or many loci often appear to be fitter than more homozygous genotypes. For example, inbreeding depression is commonly observed when organisms become more homozygous under inbreeding (see Chapter 9). However, this phenomenon can be explained by dominance rather than overdominance: inbred lines become homozygous for deleterious recessive alleles. Likewise, it can be very difficult to show that the heterozygote (*Aa*) at a particular locus has higher fitness than both homozygotes (*AA*, *aa*). Suppose most of the chromosomes in a population were *Ab* and *aB*, where *A* and *B* are favorable dominant alleles and *a* and *b* are unfavorable recessive alleles at two closely linked loci. The homozygotes *Ab*/*Ab* and *aB*/*aB*, each expressing an unfavorable recessive, would have lower fitness than the heterozygote *Ab*/*aB*. If we were aware of the phenotypic effect of only one of the loci, it would appear to display overdominance for fitness. Such apparent but spurious superiority of the heterozygote at the observed locus is called **associative overdominance**.

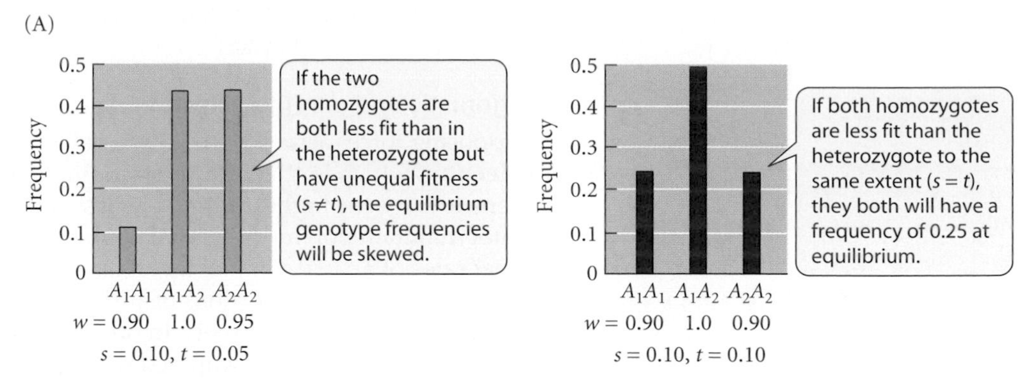

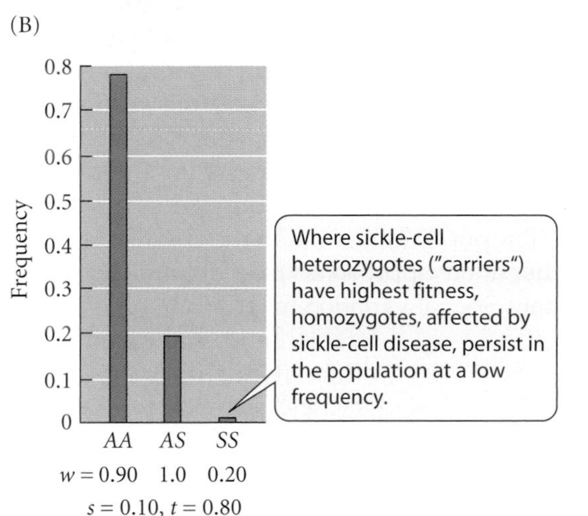

(B)
Frequency
Where sickle-cell heterozygotes ("carriers") have highest fitness, homozygotes, affected by sickle-cell disease, persist in the population at a low frequency.
AA *AS* *SS*
w = 0.90 1.0 0.20
s = 0.10, t = 0.80

Figure 12.13 (A) Genotype frequencies among newborn zygotes at a locus with heterozygote advantage (overdominance) for several relative fitnesses of the two homozygotes. The fitnesses of A_1A_1, A_1A_2, and A_2A_2 are $1 - s$, 1, and $1 - t$; these values, and those of s and t (the coefficients of selection), are shown below each graph. (B) Expected frequencies for the sickle-cell hemoglobin polymorphism, using fitness values estimated for an African population exposed to malaria. *AA* is the normal homozygote, *AS* carries the sickle-cell trait, and *SS* expresses sickle-cell anemia.

Few cases of overdominance for fitness have been well documented. Homozygotes have lower survival than heterozygotes at a locus in the plant *Arabidopsis thaliana*, but the reasons are not known (Mitchell-Olds 1995). The best-understood case of heterozygote advantage is the β-hemoglobin locus in some African and Mediterranean human populations (Cavalli-Sforza and Bodmer 1971). One allele at this locus encodes sickle-cell hemoglobin (*S*), which is distinguished by a single amino acid substitution from normal hemoglobin (encoded by the *A* allele). At low oxygen concentrations, *S* hemoglobin forms elongate crystals, which carry oxygen less effectively, causing the red blood cells to adopt a sickle shape and to be broken down more rapidly. Heterozygotes (*AS*) suffer slight anemia; homozygotes (*SS*) suffer severe anemia (sickle-cell disease) and usually die before reproducing. "Normal" homozygotes (*AA*) suffer higher mortality from malaria than heterozygotes (*AS*), however, because the protozoan that causes malaria (*Plasmodium falciparum*) develops in red blood cells. Since the red blood cells of heterozygotes are broken down more rapidly, the growth of the protozoan is curtailed. Thus heterozygotes survive at a higher rate than either homozygote (Figure 12.13B), and the frequency of *S* is quite high (about $q = 0.13$), in some parts of Africa with a high incidence of malaria. (The relative fitnesses have been estimated as $W_{AA} = 0.89$, $W_{AS} = 1.0$, $W_{SS} = 0.20$.) The heterozygote advantage therefore arises from a balance of OPPOSING SELECTIVE FACTORS: anemia and malaria. In the absence of malaria, balancing selection gives way to directional selection, because then the *AA* genotype has the highest fitness. In the African-American population, which is not subject to malaria, the frequency of *S* is about 0.05, and is presumably declining due to mortality.

Sickle-cell hemoglobin is one of several polymorphisms associated with resistance to malaria, which causes 2 million deaths annually. Thalassemia, characterized by partial or complete suppression of synthesis of the α- or β-hemoglobin chain, is caused by many different point mutations and large deletions that reach high frequencies in many tropical and subtropical human populations. Thalassemia is associated with resistance to malaria, but it is not yet certain that heterozygotes have higher fitness than homozygotes

(Clegg and Weatherall 1999). A malaria-associated polymorphism at the *G6PD* locus is described later in this chapter.

Antagonistic and varying selection

The opposing forces acting on the sickle-cell polymorphism are an example of **antagonistic selection**, which in this instance maintains polymorphism because the heterozygote happens to have the highest fitness. Unless this is the case, antagonistic selection usually does not maintain polymorphism (Curtsinger et al. 1994). Suppose, for example, that the survival rates of an insect in the larval stage are 0.5 for genotypes A_1A_1 and A_1A_2 and 0.4 for A_2A_2, whereas the proportions of surviving larvae that then survive through the pupal stage are 0.6 and 0.9, respectively. In other words, the A_1 allele improves larval survival, but reduces pupal survival, compared with the A_2 allele. However, for genotypes A_1A_1 and A_1A_2, the proportion surviving to reproductive adulthood is (0.5)(0.6) = 0.30, and for A_2A_2 it is (0.4)(0.9) = 0.36. Thus, A_2A_2 has a net selective advantage, and allele A_2 will be fixed.

Within a single breeding population, a fluctuating environment may favor different genotypes in different generations (TEMPORAL FLUCTUATION), or different genotypes may be best adapted to different microhabitats or resources (SPATIAL VARIATION). Like antagonistic selection, variation in the environment *does not necessarily maintain genetic variation*, although under some circumstances it may do so (Felsenstein 1976; Hedrick 1986).

Temporal fluctuation in the environment may slow down the rate at which one or another allele approaches fixation, but usually it does not preserve multiple alleles indefinitely. Spatial variation in the environment is most likely to maintain polymorphism if different homozygotes within a single population are best adapted to different microhabitats or resources—that is, if they have different "niches." (This phenomenon is sometimes called **multiple-niche polymorphism**.) A stable multiple-niche polymorphism is more likely if each individual organism usually experiences only one of the environments. It also is more likely if selection is "soft" than if it is "hard." SOFT SELECTION occurs when the number of survivors in a patch of a particular microenvironment is determined by competition for a limiting factor, such as space or food, and the *relatively* superior genotype has a higher probability of survival. Then selection determines the genotype frequencies among the surviving adults, but not their total number. HARD SELECTION occurs when the likelihood of survival of an individual in a microenvironment depends on its *absolute* fitness, not on the density of competitors. Selection then determines not only the genotype frequencies, but also the total number of survivors. (For example, selection for insecticide resistance may be hard, since the survival of an individual insect may depend only on whether it has a resistant or susceptible genotype. In contrast, selection for long mouthparts that enable a bee to obtain nectar from flowers more rapidly than shorter-tongued bees would be soft selection if the shorter-tongued bees could obtain nectar, and survive and reproduce prolifically, in the absence of superior longer-tongued competitors.)

The black-bellied seedcracker (*Pyrenestes ostrinus*) provides a nice example of multiple-niche polymorphism (Smith 1993). Populations of this African finch have a bimodal distribution of bill width (Figure 12.14A). The difference between wide and narrow bills seems to be due to a single allele difference. The two morphs differ in the efficiency with which they process seeds of different species of sedges, their major food: wide-

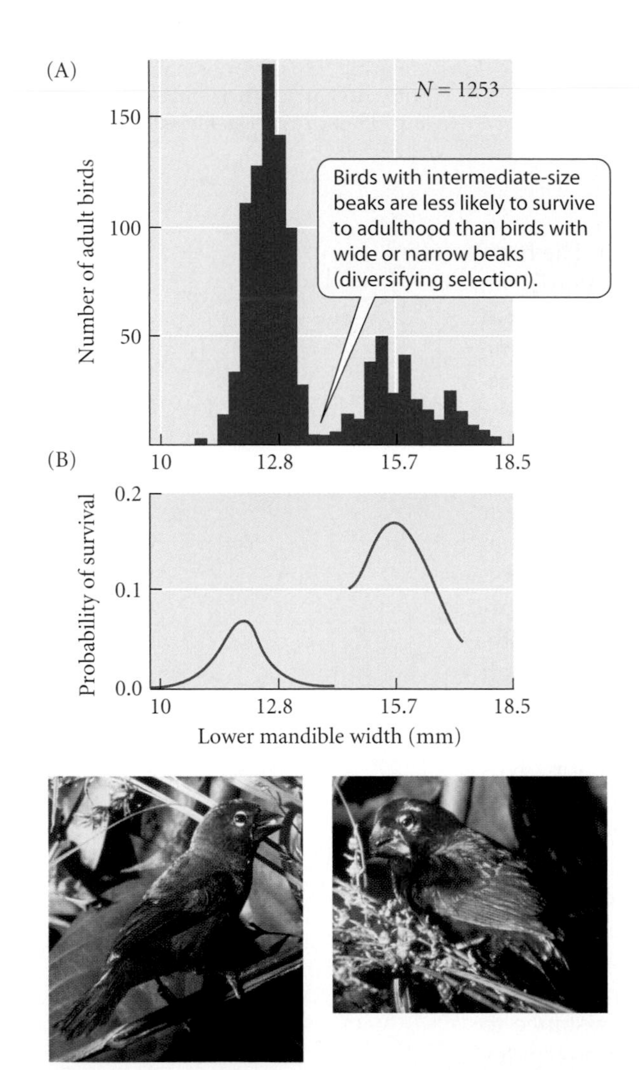
(A)
N = 1253
150
100
50
Number of adult birds
Birds with intermediate-size beaks are less likely to survive to adulthood than birds with wide or narrow beaks (diversifying selection).
(B)
10
12.8
15.7
18.5
0.2
0.1
0.0
Probability of survival
Lower mandible width (mm)

Figure 12.14 Multiple-niche polymorphism in the black-bellied seedcracker. (A) Probability of survival to adulthood of banded juvenile birds in relation to their lower mandible width, a measure of bill size. (B) The distribution of lower mandible width among adults is bimodal. The small- and large-billed morphs are shown at left and right, respectively. (After Smith 1993; photographs courtesy of Thomas B. Smith.)

billed birds process hard seeds, and narrow-billed birds soft seeds, more efficiently. Smith banded more than 2700 juvenile birds, and found that survival to adulthood was lower for birds with intermediate bills than for either wide- or narrow-billed birds (Figure 12.14B). Thus diversifying selection, arising from the superior fitness of different genotypes on different resources, appears to maintain the polymorphism.

Frequency-dependent selection

In the models we have considered so far, the fitness of each genotype is assumed to be constant within a given environment. Very often, however, *the fitness of a genotype depends on the genotype frequencies in the population*. The population then undergoes **frequency-dependent selection**. This kind of balancing selection appears to maintain many polymorphisms, and it has many other important consequences, especially for the evolution of animal behavior, as we will see in Chapter 14.

INVERSE FREQUENCY-DEPENDENT SELECTION. In **inverse frequency-dependent selection**, the rarer a phenotype is in the population, the greater its fitness (Figure 12.15A). For example, the per capita rate of survival and reproduction of a dominant phenotype (with genotype A_1A_1 or A_1A_2) may be greatest when it is very rare and decrease as it becomes more common; the same may be true of the recessive phenotype (with genotype A_2A_2). Thus, when A_2 is at high frequency, it declines because A_2A_2 has lower fitness than A_1A_1 and A_1A_2, and likewise for A_1. Whatever the initial allele frequencies may be, they shift toward a stable equilibrium value, which in this case occurs when the frequencies of the two phenotypes are equal (i.e., when $q^2 = 0.5$). At this point, the mean fitnesses of both phenotypes are the same: neither has an advantage over the other.

Many biological phenomena can give rise to inverse frequency-dependent selection. The self-incompatibility alleles of many plant species are a striking example. These alleles enforce outcrossing because pollen carrying any one allele cannot effectively grow down the stigma of the plant that produced it or of any other plant that carries the same allele. Thus, if there are three alleles, S_1, S_2, and S_3, in a plant population, pollen of type S_1 can grow on stigmas, and fertilize ovules, of genotype S_2S_3, but not on S_1S_2 or S_1S_3; S_2 pollen can fertilize only S_1S_3 plants, and S_3 pollen can fertilize only S_1S_2 plants. (Notice that plants cannot be homozygous at this locus.) New S alleles that arise by mutation can increase in frequency, since they are rare at first, until they reach a frequency of about $1/k$, where k is the number of S alleles in the population. We should expect to find a large number of self-incompatibility alleles, all about equal in frequency, at equilibrium. In some plant species, indeed, hundreds of such alleles exist.

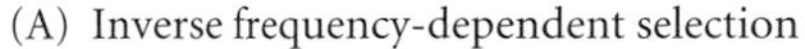

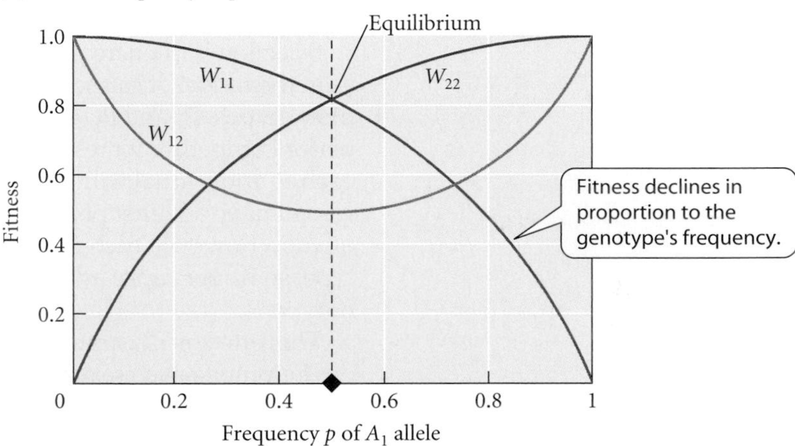

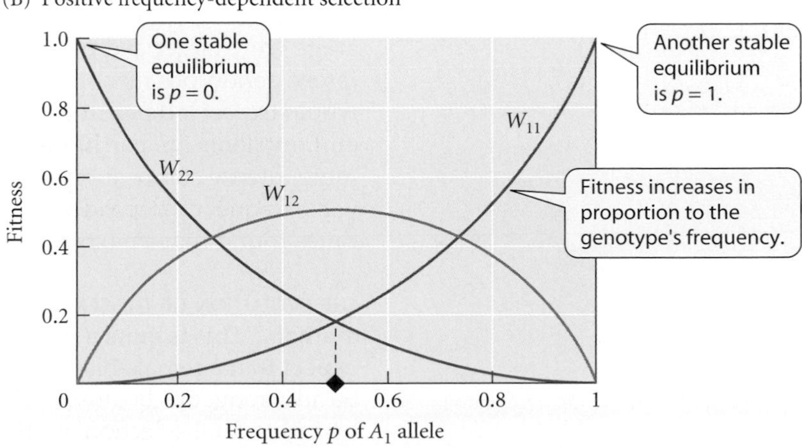

Figure 12.15 Two forms of frequency-dependent selection, in which the fitness of each genotype depends on its frequency in a population. (A) Inverse frequency-dependent selection. A genotype's fitness declines with its frequency, which depends on allele frequencies (p, the frequency of A_1). The fitnesses of A_1A_1, A_1A_2, and A_2A_2 are $W_{11} = 1 - sp^2$, $W_{12} = 1 - spq$, and $W_{22} = 1 - sq^2$ in this model; the curves are calculated for $s = 1$. A stable equilibrium exists at $p = q = 0.5$. (B) Positive frequency-dependent selection, in which a genotype's fitness increases with its frequency. The fitnesses of A_1A_1, A_1A_2, and A_2A_2 are $1 + sp^2$, $1 + 2spq$, and $1 + sq^2$. (After Hartl and Clark 1989.)

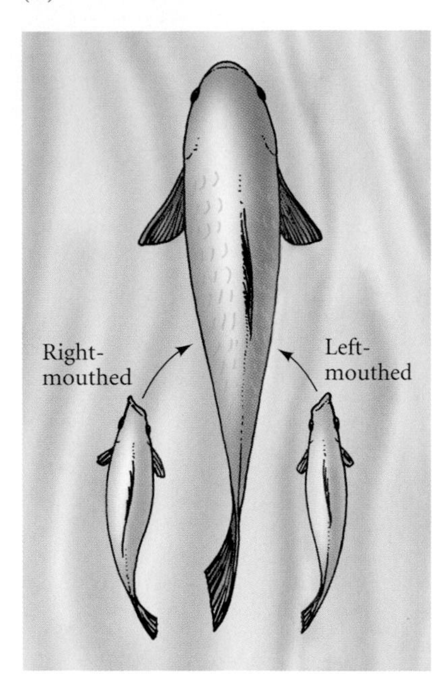

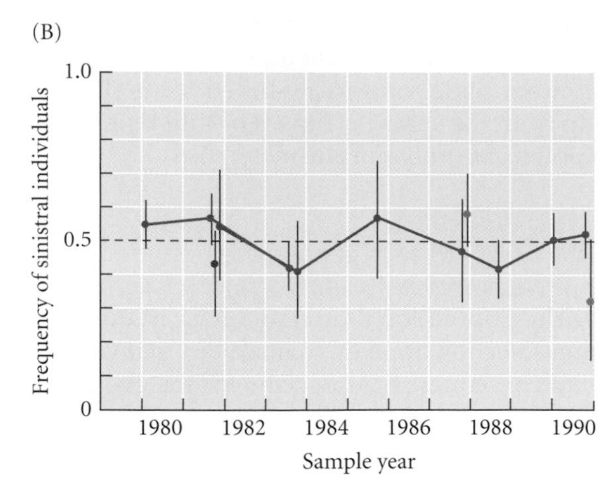

Figure 12.16 An inverse frequency-dependent polymorphism in the scale-eating cichlid *Perissodus microlepis*. (A) Dextral, or right-mouthed, and sinistral, or left-mouthed, *P. microlepis* attack prey from opposite sides. (B) Fluctuations in the frequency of the sinistral form at two nearby locations along the shore of Lake Tanganyika. (B after Hori 1993.)

Michio Hori (1993) analyzed a fascinating example of frequency-dependent selection in the cichlid fish *Perissodus microlepis* in Lake Tanganyika, which feeds by approaching other cichlids from behind and snatching a mouthful of scales from the prey's flank. This cichlid's mouth is twisted to the right (dextral) or left (sinistral) (Figure 12.16A). Dextral individuals always approach the left side of their prey, and sinistral individuals the right. The two morphs, which appear to be based on two alleles at a single locus, fluctuate in frequency around 0.5 (Figure 12.16B). The frequencies are stabilized by the prey's escape behavior: when the dextral form is more abundant, the prey are more wary of approaches on their left side, and consequently are attacked more frequently on the right, by the rarer sinistral form. Conversely, the dextral form has greater feeding success when the sinistral form is more common.

Selection is often frequency-dependent when genotypes compete for limiting resources, as in the model of soft selection described above. Suppose that two genetically determined phenotypes, P_1 and P_2, can survive on either resource 1 or resource 2, but that P_1 is a superior competitor for resource 1 and P_2 is a superior competitor for resource 2. If P_1 is rare, each P_1 individual will compete for resource 1 chiefly with inferior P_2 competitors, and so will have a higher per capita rate of increase than P_2 individuals. As P_1 increases in frequency, however, each P_1 individual competes with more individuals of the same genotype, so its per capita advantage, relative to P_2, declines. The same pattern applies to phenotype P_2.

This effect is illustrated in an experiment on the grass *Anthoxanthum odoratum*, which can be propagated asexually by vegetative cuttings. In a natural environment, Norman Ellstrand and Janis Antonovics (1984) planted small "focal" cuttings and surrounded each with competitors. The surrounding plants were either the same genotype as the focal individual, taken as vegetative shoots from the same parent plant, or different genotypes, obtained from sexually produced seed. After a season's growth, focal individuals on average were larger and produced more seed (i.e., had higher fitness) if surrounded by different genotypes than if surrounded by the same genotype. This pattern is just what we would expect if different genotypes use somewhat different resources (perhaps certain mineral nutrients are partitioned differently by adjacent plants that differ in genotype). The more intense competition among individuals of the same genotype would then impose inverse frequency-dependent selection. In the same vein, agricultural researchers have sometimes found that a mixture of crop varieties yields more than fields of a single variety.

THE EVOLUTION OF THE SEX RATIO. Why is the sex ratio about even (1:1) in many species of animals? This is quite a puzzle, because from a group-selectionist perspective, we might expect that a female-biased sex ratio (i.e., production of more females than males) would be advantageous because such a population could grow more rapidly. If sex ratio evolves by individual selection, on the other hand, and if all females have the same number of progeny, why should a genotype producing an even sex ratio have an advantage over any other?

The solution to the puzzle was provided by the great population geneticist R. A. Fisher (1930), who realized that because every individual has both a mother and a father, females and males must contribute equally to the ancestry of subsequent generations, and must therefore have the same average fitness. Therefore individuals that vary in the sex ratio of their progeny can differ in the number of their grandchildren (and later descendants), and thus differ in fitness if this is measured over two or more generations.

To see why this is so, let us define the sex ratio as the proportion of males, and distinguish the sex ratio in a population (the POPULATION SEX RATIO, *S*) from that in the progeny of an individual female (an INDIVIDUAL SEX RATIO, *s*). In a large, randomly mating population, the fitness of a genotype that produces a given individual sex ratio depends on the population sex ratio, which in turn depends on the frequencies of the various individual-sex-ratio genotypes. Because the average per capita reproductive success of the minority sex is greater than that of the majority sex, selection favors genotypes whose individual sex ratios are biased toward that sex that is in the minority in the population as a whole.

For example, suppose the population sex ratio is 0.25 (1 male to 3 females) because it consists of a genotype with this individual sex ratio. Suppose each female has 4 offspring. The average number of progeny per female is 4, but since every offspring has a father, the average number of progeny sired by a male is 12 (since each male mates with 3 females on average). Thus a female has 4 grandchildren through each of her daughters and 12 through each son, for a total of $(3 \times 4) + (1 \times 12) = 24$ grandchildren. Now suppose a rare genotype with an individual sex ratio of 0.50 (2 daughters and 2 sons) enters the population. Each such individual has $2 \times 4 = 8$ grandchildren through her daughters and $2 \times 12 = 24$ grandchildren through her sons, for a total of 32. Since this is a greater number of descendants than the mean number per individual of the prevalent female-biased genotype, any allele that causes a more male-biased progeny in this female-biased population will increase in frequency. Likewise, if the population sex ratio were male-biased, an allele for female-biased individual sex ratios would spread. By this reasoning, a genotype that produces an even sex ratio (0.5) has highest fitness and cannot be replaced by any other genotype (Figure 12.17). (A genotype that produces a sex ratio of 0.5 represents an EVOLUTIONARILY STABLE STRATEGY, or ESS, as we will see Chapter 14.)

Alexandra Basolo (1994) tested this theory using the platyfish *Xiphophorus maculatus*, which has three kinds of sex chromosomes, W, X, and Y. Females are XX, WX, or WY, and males are XY or YY. Of the six possible crosses, four yield a sex ratio of 0.5, but XX mated with YY yields all sons, and WX mated with XY yields 0.25 sons, so there is variation among genotypes in individual sex ratios. Basolo set up experimental populations with different frequencies of these chromosomes, which carried different color-pattern alleles so that chromosome frequencies could be followed. Two populations were initiated with sex ratios (proportion of males) of 0.25 percent and 0.78 percent. Within only two generations, the sex ratio in both populations evolved nearly to 0.5, as Fisher's theory predicts (Figure 12.18).

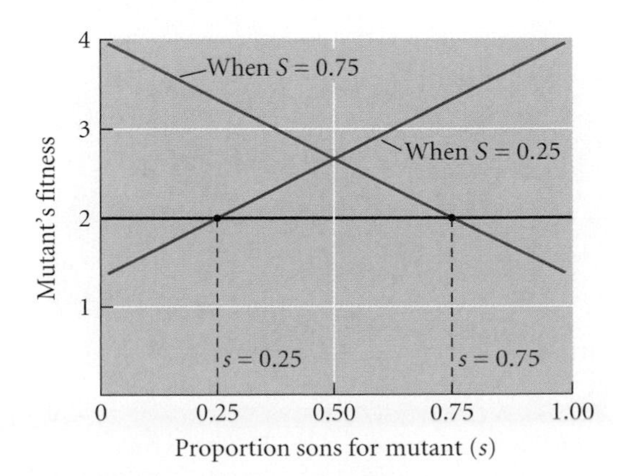

Figure 12.17 Frequency-dependent selection on sex ratio. Mutants may arise that vary in individual sex ratio (*s*, the proportion of sons among a female's progeny). The fitness of each such mutant, based on its average number of grandchildren, depends on the sex ratio in the population (*S*). The average fitness of individuals in the population equals 2. When $S = 0.25$—i.e., when 25 percent of the population is male,—the fitness of a mutant is directly proportional to the proportion of sons among its offspring, and it is greater than the mean fitness of the prevalent genotype if the mutant's *s* exceeds 0.25. Any such mutant will therefore increase in frequency. Conversely, if $S = 0.75$, the fitness of a mutant is inversely proportional to its individual sex ratio, and its fitness increases in frequency if its *s* is less than 0.75. (After Charnov 1982.)

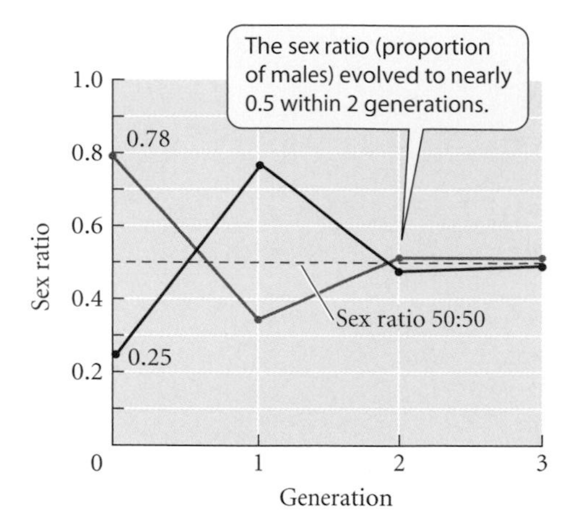

Figure 12.18 Changes in sex ratio (proportion of males) in two experimental populations of platyfish, initiated with sex ratios (proportions of males) of 0.25 and 0.78. In both populations, the sex ratio evolved nearly to 0.5 within only two generations. (After Basolo 1994.)

Multiple Outcomes of Evolutionary Change

One of the most important principles in evolution is that initial genetic conditions often determine which of several paths, or trajectories, of genetic change a population will follow. Thus *the evolution of a population often depends on its previous evolutionary history*. Positive frequency-dependent selection and heterozygote disadvantage are two important factors that can give rise to multiple outcomes—**multiple stable equilibria**.

Positive frequency-dependent selection

In **positive frequency-dependent selection**, the fitness of a genotype is greater the more frequent it is in a population. As a result, *whichever allele is initially more frequent will be fixed* (see Figure 12.15B).

For example, the unpalatable tropical butterfly *Heliconius erato* has many different geographic races that differ markedly in color pattern. Each race is monomorphic. Adjacent geographic races interbreed at zones only a few kilometers wide. The geographic variation in color pattern in this species is closely paralleled by that in another unpalatable species of *Heliconius*—a spectacular example of Müllerian mimicry (Figure 12.19). Predators that have had an unpleasant experience with one or a few individuals of either species tend thereafter to avoid similar butterflies, of either species.

James Mallet and Nicholas Barton (1989) showed that within *Heliconius erato*, gene flow from one geographic race to another is countered by positive frequency-dependent selection: immigrant butterflies that deviate from the locally prevalent color pattern are selected against because predators have not learned to avoid attacking butterflies with unusual color patterns. On either side of a contact zone between two geographic races in Peru, Mallet released *H. erato* of the other race and, as a control, butterflies of the same race from a different locality. The butterflies, marked so they could be recognized, were repeatedly recaptured for some time thereafter. Compared with control butterflies, with the same color pattern as the population into which they were released, far fewer of those with the allopatric color pattern were captured. From bill marks left on the wings of butterflies that had escaped from birds, the authors concluded that the missing butterflies were lost to bird predation, and calculated an average selection coefficient of 0.52 against the "wrong" color pattern in either population. This amounts to a selection coefficient of about $s = 0.17$ at each of the three major loci that control the differences in color pattern between the races—very strong selection indeed.

Heterozygote disadvantage

The case in which the heterozygote has lower fitness than either homozygote is called **heterozygote disadvantage** or **underdominance**. If a population is initially monomorphic for A_1A_1, and A_2 then enters at low frequency by mutation or gene flow, almost all A_2 alleles are carried by heterozygotes (A_1A_2); since their fitness is lower than that of A_1A_1, selection reduces the frequency (q) of A_2 to zero. Likewise, A_1 is eliminated if it enters a monomorphic population of A_2A_2, which also has a greater fitness than A_1A_2. Monomorphism for either A_1A_1 or

Figure 12.19 An extraordinary case of geographic variation and Müllerian mimicry. The butterflies on the right represent geographic races of a single species, *Heliconius erato*, from regions in Central and South America. The butterflies on the left are races of a rather distantly related species, *Heliconius melpomene*, from the corresponding regions. Both species are unpalatable to predators. Races of *H. erato* interbreed with one another to some extent where they meet, but the hybrid zone between them is very narrow; the same is true of *H. melpomene*. (From Cornell University Insect Collection, courtesy of Andrew Brower.)

A_2A_2 is therefore a stable equilibrium, and the initially more frequent allele is fixed by selection. This model applies to some chromosome rearrangements, such as inversions and translocations, for which heterozygotes may have reduced fertility due to improper segregation in meiosis (see Figures 8.21, 8.22, and the discussion below).

If the homozygotes' fitnesses are different, but both greater than that of the heterozygote, the mean fitness in a population that has become fixed for the less fit homozygote is less than if it were fixed for the other homozygote, but *selection cannot move the population from the less fit to the more fit condition*. Thus *a population is not necessarily driven by natural selection to the most adaptive possible genetic constitution.*

Adaptive landscapes

Recall that we can calculate the mean fitness ($\bar{w}$) of individuals in a population with any conceivable allele frequency (p) and plot a curve showing $\bar{w}$ as a function of p (see Figure 12.7A,B). When fitnesses are constant, natural selection changes allele frequencies in such a way that mean fitness ($\bar{w}$) increases, so that the population moves up the slope of this curve. The current location of the population on this slope is then a simple guide to how allele frequencies will change under selection: simply see which direction of allele frequency change will increase $\bar{w}$. For an underdominant locus, the curve dips in the middle and slopes upward to $p = 0$ and $p = 1$ (see Figure 12.7D). Thus natural selection decreases or increases p depending on whether a population begins to the left or the right of the minimum of the $\bar{w}$ curve.

Curves such as those in Figure 12.7 are often referred to as **adaptive landscapes**, or ADAPTIVE TOPOGRAPHIES. The curve in Figure 12.7D represents two **adaptive peaks** separated by an **adaptive valley**. This metaphor, introduced by Sewall Wright, is widely used in evolutionary biology, and we will refer to it in Chapter 16. Each point on the curve (the adaptive landscape) represents the *average* fitness of individuals in a hypothetical population made up of (in this case) three genotypes with frequencies p^2, $2pq$, and q^2. Each possible value of p—each possible hypothetical population—yields a different value on the x-axis, and therefore a different point on the landscape. When, as in Figure 12.7D, the relationship of $\bar{w}$ to p has two or more maxima, two genetically different populations (e.g., with $p = 0$ or $p = 1$ in this case) can (but need not) have the same mean fitness ($\bar{w}$) *under the same environmental conditions*. Different environments, which might alter the fitnesses of the genotypes relative to one another, are represented not by different points on any one landscape, but rather by *different landscapes*—different relationships between $\bar{w}$ and p.

Interaction of selection and genetic drift

In developing the theory of selection so far, we have assumed an effectively infinite population size. However, *in a finite population, allele frequencies are simultaneously affected by both selection and chance*. As the movement of an airborne dust particle is affected both by the deterministic force of gravity and by random collisions with gas molecules (Brownian movement), so the effective size (N_e) of a population and the strength of selection (s) both affect changes in allele frequencies. The effect of random genetic drift is negligible if selection on a locus is strong relative to the population size—that is, if s is much greater than $1/(4N_e)$. Conversely, if s is much less than $1/(4N_e)$, selection is so weak that the allele frequencies change mostly by genetic drift: the alleles are *nearly neutral*.

The effect of population size on the efficacy of selection has several important consequences. First, a population may not attain exactly the equilibrium allele frequency predicted from its genotypes' fitnesses; instead, it is likely to wander by genetic drift in the vicinity of the equilibrium frequency. Second, a slightly advantageous mutation is less likely to be fixed by selection if the population is small than if it is large, because it is more likely to be lost simply by chance. Conversely, deleterious mutations can become fixed by genetic drift, especially if selection is weak and the population is small. Third, population bottlenecks provide temporary conditions under which genetic drift may counteract selection so that a deleterious allele may increase in frequency. For example, slightly deleterious mutations might be fixed, contributing to divergence among populations at the molecular level.

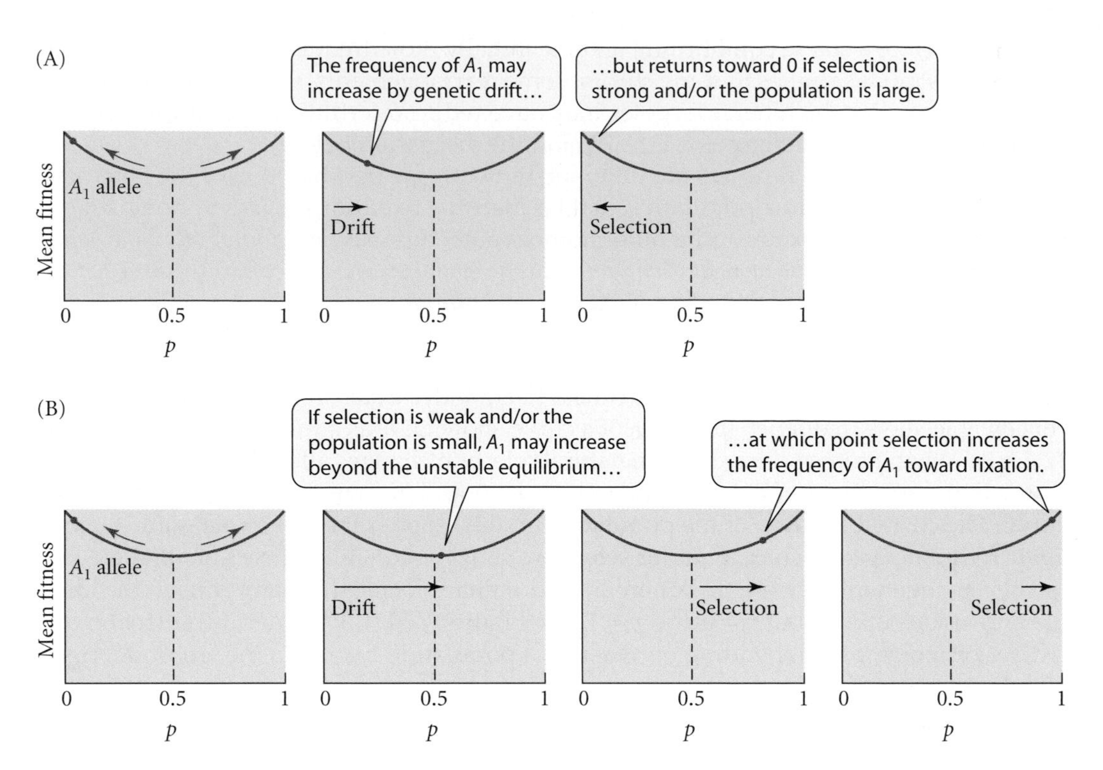

Figure 12.20 A peak shift due to the joint action of genetic drift and natural selection. For two alleles or chromosome rearrangements that lower the fitness of the heterozygote, the fitnesses of A_1A_1, A_1A_2, and A_2A_2 are 1, $1 - s$, and 1 respectively. The curve shows the mean fitness ($\overline{w}$) for each possible value of p, the frequency of A_1, and may be considered an adaptive landscape with two peaks (at $p = 0$ and $p = 1$). An unstable equilibrium exists at $p = 0.5$, indicated by the dashed line. The value of p at a particular time is shown by a colored dot. (A) The frequency of A_1, initially near 0, may increase by genetic drift, but returns toward 0 if the population is large. (B) If the population becomes small, however, p may increase beyond the unstable equilibrium frequency (0.5) due to genetic drift. If this occurs, selection increases the frequency of A_1 toward fixation.

The principle that genetic drift could prevail over selection might be especially important if heterozygotes are inferior in fitness, so that the adaptive landscape has two peaks (see Figure 12.7D). Selection alone cannot move a population down the slope of one peak and across a valley to the slope of another peak, even if the second peak is higher: a population does not first become poorly adapted so that it can then become better adapted (Figure 12.20A). But during episodes of very low population size, allele frequencies may fluctuate so far by genetic drift that they cross the adaptive valley—after which selection can move the population "uphill" to the other peak (Figure 12.20B). The probability that such a **peak shift** will occur (Barton and Charlesworth 1984) depends on the population size and on the difference in height (mean fitness) between the valley and the initially occupied peak.

Thus, when there are multiple stable equilibria, *genetic drift and selection may act in concert to accomplish what selection alone cannot*, moving a population from one adaptive peak to another. This theory explains how populations can come to differ in underdominant chromosome rearrangements such as translocations and pericentric inversions. Some chromosome rearrangements are thought to conform to the model of underdominance because heterozygotes have lower fertility than either homozygote (see Chapter 8). For example, local populations of the Australian grasshopper *Vandiemenella viatica* are monomorphic for different chromosome fusions and pericentric inversions. Heterozygotes for these chromosomes have many aneuploid gametes. Any such chromosome, introduced by gene flow into a population monomorphic for a different arrangement, is reduced in frequency by natural selection, so no two "chromosome races" are sympatric; instead, they meet in "tension zones" only 200–300 meters wide (White 1978). Because these grasshoppers are flightless and quite sedentary, local populations are small, providing the opportunity for genetic drift to occasionally initiate a peak shift whereby a new chromosome arrangement is fixed.

Molecular Signatures of Natural Selection

Theoretical expectations

Variation in DNA sequences can provide evidence of the action of natural selection if the pattern of variation differs from patterns expected under the neutral theory of molecular evolution. We saw in Chapter 10, for example, that at equilibrium between mutation and genetic drift, the expected amount of sequence variation in a diploid population, as expressed by the frequency of heterozygotes per nucleotide site, is

$$\frac{4N_e u_0}{4N_e u_0 + 1}$$

where N_e is the effective population size and u_0 is the rate of neutral mutation. Furthermore, the nucleotides at different variable sites in a DNA sequence should not be correlated with one another at equilibrium because recombination, even between closely linked sites, should eventually lead to linkage equilibrium (see Chapter 9).

Now suppose that natural selection acts at a particular base pair site within a gene, and consider the *effects of this selection on neutral variation at sites that are closely linked to the selected site*. POSITIVE DIRECTIONAL SELECTION (directional selection for an advantageous mutation) reduces variation at closely linked sites. If an advantageous mutation occurs in a gene for which neutral variation exists in the population, and if this mutation is fixed by selection, then all the copies of the gene in the population will be descended from the single copy in which the mutation occurred. The neutral variant sites linked to the mutation will also be fixed, by hitchhiking. Thus all neutral variation in the gene is eliminated by a **selective sweep**, and variation is only slowly reconstituted as new neutral mutations occur among the copies of the advantageous gene. Selective sweeps eliminate variation over a broader length of DNA in regions of the genome that have low rather than high rates of recombination. If, furthermore, we observe the population when the advantageous mutation has reached high frequency, but before it is fixed, we will observe linkage disequilibrium among various neutral polymorphic sites, because the advantageous mutation has remained associated with particular nucleotides at linked sites, which therefore are correlated with one another.

The effects of a selective sweep are evident in gene genealogies. Consider two unlinked loci, one that has been evolving solely by genetic drift (Figure 12.21A) and one that has experienced a selec-

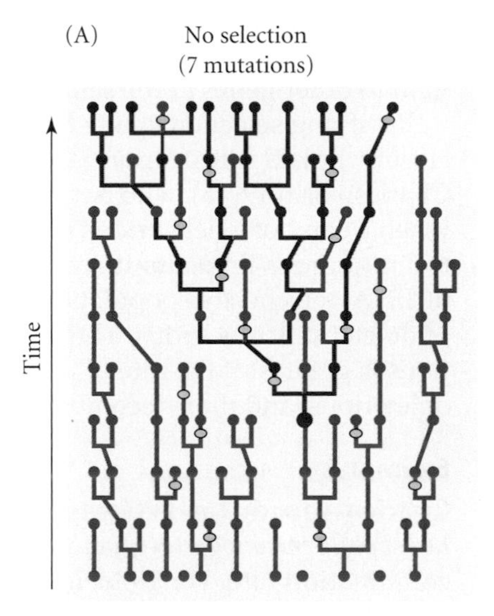

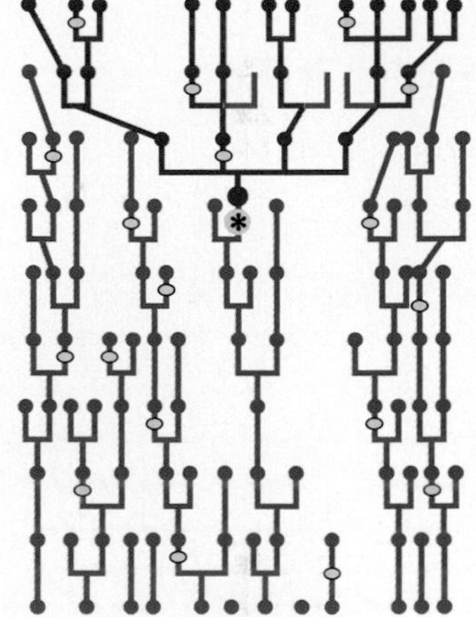

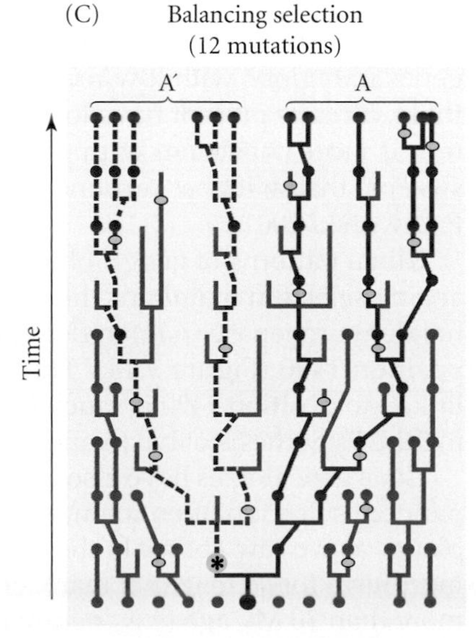

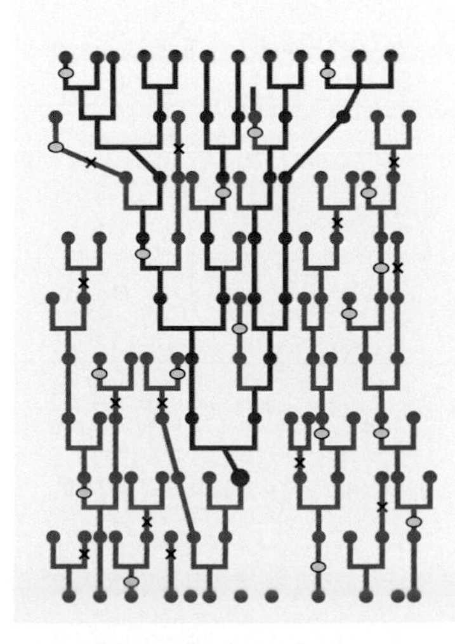

Figure 12.21 Schematic genealogies of gene copies in a population, showing the effects of three modes of selection on nucleotide diversity, compared with the neutral model (compare with Figure 10.15). The contemporary population (represented by 12 gene copies) is at the top of each diagram, and the ancestry of contemporary gene copies is marked by the red gene trees. In each diagram, some gene lineages (blue lines) become extinct by chance. These diagrams assume that no recombination occurs within the gene. Ovals represent selectively neutral mutations, each at a different site within the gene. (A) Neutral mutations only. Contemporary gene copies vary by 7 mutations. (B) Positive selection of an advantageous mutation, marked by an asterisk. In a selective sweep, this gene lineage replaces all others, so contemporary copies differ only in the 5 mutations that have occurred since the advantageous mutation. (C) Balancing selection of alleles *A* and *A′*, the latter lineage (dashed lines) arising by the mutation marked with an asterisk. The two gene lineages persist for a long time, and so accumulate more mutations than in the neutral case. (D) Background selection. Deleterious mutations, marked by ×, eliminate some gene copies, hence reducing the number of surviving neutral mutations.

tive sweep (Figure 12.21B). Compared with the neutrally evolving gene, the copies of the gene that was fixed by selection are descended from a more recent common ancestor (the one in which the favorable mutation occurred); they have had less time to accumulate different neutral mutations, and so are more similar in sequence. A selective sweep resembles a bottleneck in population size in that it reduces variation and increases the genealogical relatedness among gene copies, but a bottleneck will affect the entire genome, not just the portion surrounding an advantageous mutation.

The effects of balancing selection (e.g., heterozygote advantage or frequency-dependent selection) are opposite to those of positive directional selection. Assume that two variants are maintained at a polymorphic site, and again, assume that recombination is low in the vicinity of that site. The gene copies in the population are all descended from two ancestral copies (bearing the original, selectively advantageous alternative nucleotides), each of which was the progenitor of a lineage of genes that have accumulated neutral mutations in the vicinity of the selected site (Figure 12.21C). Thus, compared with a gene with solely neutral variation, a gene subjected to balancing selection will display elevated variation in the vicinity of the selected site (Strobeck 1983). In a genealogy of sequences sampled from a population, the common ancestor of all the sequences may be older than if they had been evolving solely by genetic drift because selection has maintained two gene lineages longer. In fact, the polymorphism may have been maintained by selection for so long that speciation has occurred in the interim. In that case, both lineages of genes may have been inherited by two (or more) species, and some gene copies in each species may be genealogically more closely related to genes in the other species than to other genes in the same species.

Purifying selection against deleterious mutations reduces neutral polymorphism at closely linked sites. Brian Charlesworth and colleagues (Charlesworth et al. 1993; Charlesworth 1994a), who have termed this effect **background selection**, pointed out that when a copy of a deleterious mutation is eliminated from a population, selectively neutral mutations linked to it are eliminated as well (Figure 12.21D). Thus, for this region of DNA, the effective population size is reduced to the proportion of gametes that are free of deleterious mutations. The reduction in the level of heterozygosity for neutral mutations is greatest if the rate of deleterious mutation is high, if the mutations are strongly deleterious, and if the recombination rate is very low.

Examples

Charles Aquadro and colleagues (1994) surveyed sequence variation within genes of *Drosophila melanogaster* that are known to lie in chromosome regions that differ in recombination rate. For 19 loci, they found a strong correlation between the recombination rate and ≠ (the proportion of base pairs that differ between random pairs of gene copies; see Chapter 10). In other words, sequence variation was lowest in parts of the genome with the lowest recombination rate. Other researchers have found not only that genes in regions with low recombination rates have fewer sequence variants, but that those variants present have low frequencies. Theoretical analyses indicate that this pattern is more consistent with positive selection for advantageous mutations (selective sweeps) than with background selection against deleterious mutations (Andalfatto and Przeworski 2001).

Clinal patterns of geographic variation (see Figure 9.25) show that some kind of balancing selection maintains the polymorphism for "fast" and "slow" allozymes of alcohol dehydrogenase (*Adh*) in *Drosophila melanogaster*, which is due to a single mutation at position 1490 (Figure 9.14). The level of synonymous polymorphism at sites closely linked to position 1490 is much higher than elsewhere in the *Adh* gene region, supporting the hypothesis of balancing selection (Figure 12.22).

Gene genealogies have also provided evidence for balancing selection. Frequency-dependent selection often maintains polymorphisms in the self-incompatibility alleles of plants, as we saw above. In the family Solanaceae, selection has maintained such a polymorphism for so long that many of the alleles in different genera of plants that diverged more than 30 My ago (e.g., petunia and tobacco) are genealogically more closely related

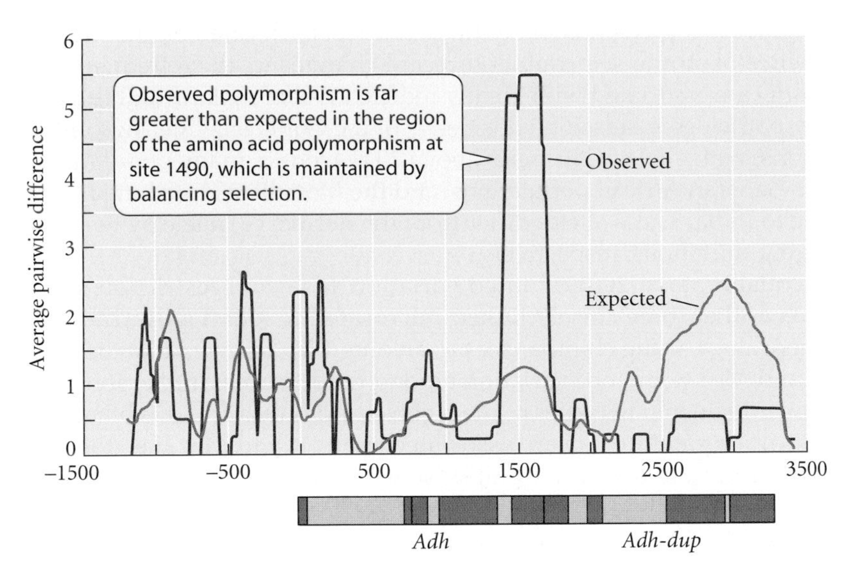

Figure 12.22 A plot of the level of nucleotide polymorphism in the *Adh* region in *Drosophila melanogaster*. The region includes the *Adh* locus and the "*Adh-dup*" gene; the boxes below the graph represent the exons of these genes. Numbers on the *x*-axis represent nucleotide position, with the beginning of the *Adh* gene set at 0. An untranslated region lies upstream of *Adh*. The expected level of variation is based on a neutral model and the level of divergence in each region between *D. melanogaster* and a closely related species. The observed level of synonymous variation is very high at sites in the vicinity of the nonsynonymous polymorphism at position 1490, as expected if this polymorphism has been maintained by balancing selection. (After Kreitman and Hudson 1991.)

to each other than to other alleles in the same species (Figure 12.23A). Similarly, allele lineages of certain of the major histocompatibility (MHC) genes are older than the divergence between humans and chimpanzees (Figure 12.23B). The proteins encoded by these genes bind foreign peptides (antigens), a key step in the immune response. Because variant MHC proteins may differ in specificity for different antigens, heterozygotes may have a broader spectrum of responses and may therefore have higher fitness than homozygotes (Nei and Hughes 1991).

(A)

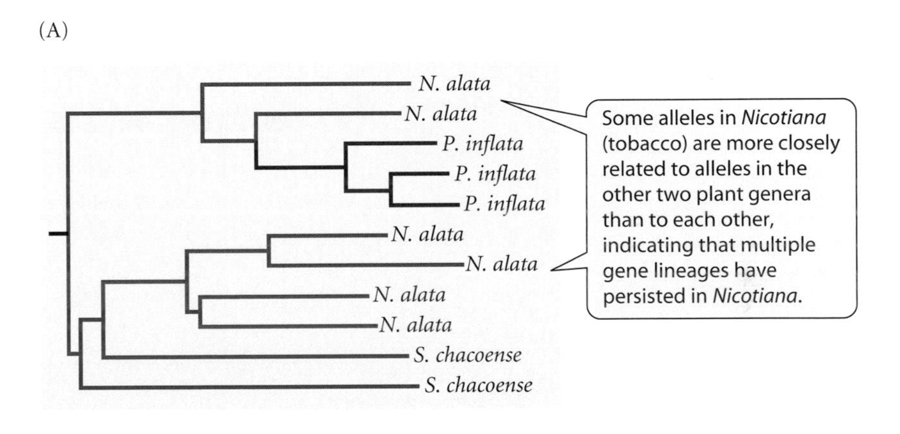

(B)

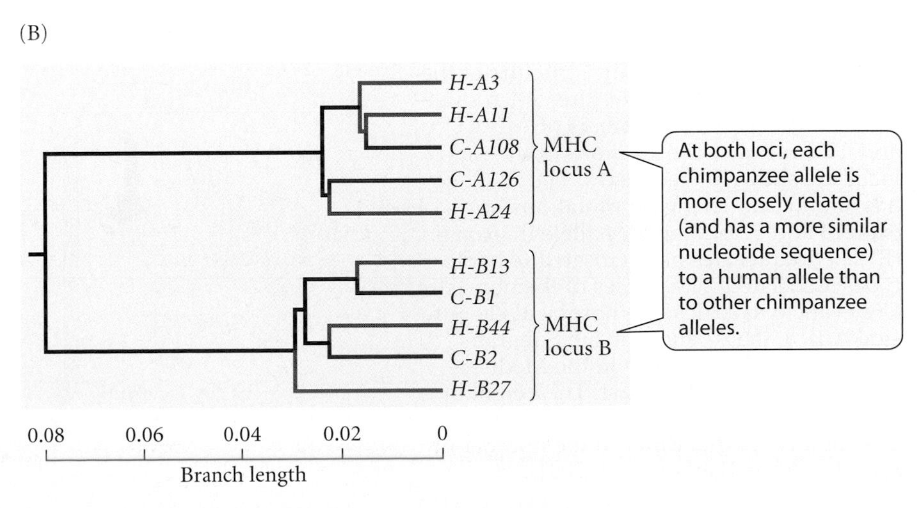

Figure 12.23 Due to long-lasting balancing selection, a polymorphism may be inherited by two or more species from their common ancestor, and certain haplotypes in each species may be most closely related to haplotypes in the other species. These trees show relationships among gene sequences (haplotypes) from two or more species; each species is indicated by a different color. (A) Alleles at the self-incompatibility locus in the potato family, Solanaceae. The common ancestor of the six alleles sequenced in the tobacco *Nicotiana alata* is older than the common ancestor of the three genera *Nicotiana*, *Petunia*, and *Solanum*.(B) The phylogenetic relationships among six alleles in humans and four alleles in chimpanzees at the major histocompatibility (MHC) loci A and B. Both species have loci A and B, which form monophyletic clusters, indicating that the two loci arose by gene duplication before speciation gave rise to the human and chimpanzee lineages. At both loci, each chimpanzee allele is more closely related (and has more similar nucleotide sequence) to a human allele than to other chimpanzee alleles. Thus polymorphism at each locus in the common ancestor has been carried over into both descendant species. These polymorphisms are at least 5 million years old. (A after Ioerger et al. 1990; B after Nei and Hughes 1991.)

The human *G6PD* (glucose-6-phosphate dehydrogenase) locus, located on the X chromosome, shows evidence of strong selection and recent changes in allele frequencies. Many mutations of *G6PD* are associated with greatly reduced risk of malaria, but this advantage is balanced by pathologies caused by deficiency of the G6PD enzyme. Two of the most common mutations, with allele frequencies above 20 percent in many populations, are the A^- allele, in sub-Saharan African populations, and the *Med* allele, distributed from Mediterranean Europe to India. These alleles differ from the normal (*B*) allele by two and one amino acid-changing mutations, respectively.

Sara Tishkoff and collaborators (2001) studied variation at several restriction sites* within the *G6PD* locus and at three closely linked microsatellite loci (Figure 12.24A). Analysis of the gene genealogy, using chimpanzees as an outgroup, showed that the *B* allele is ancestral to A^- and *Med*. Sequence variation among the A^- copies and among the *Med* copies is much lower than among copies of the *B* allele, and only a few combinations of variants at the three microsatellite loci are associated with A^- and *Med* (i.e., there is strong linkage disequilibrium). In fact, almost all A^- sequences from throughout Africa are the same, and *Med* copies likewise almost all have the same sequence across a broad geographic range (Figure 12.24B).

The greatly reduced variation and high linkage disequilibrium are consistent with the hypothesis that the A^- and *Med* mutations have increased rapidly and recently due to natural selection. If the increase in allele frequency had happened long ago, new microsatellite mutations would have replenished variation; moreover, recombination would have

*A restriction site is a sequence of several base pairs that is recognized and cleaved by restriction enzyme. Polymorphism in the sequence is detected by variation in whether or not the enzyme cleaves it.

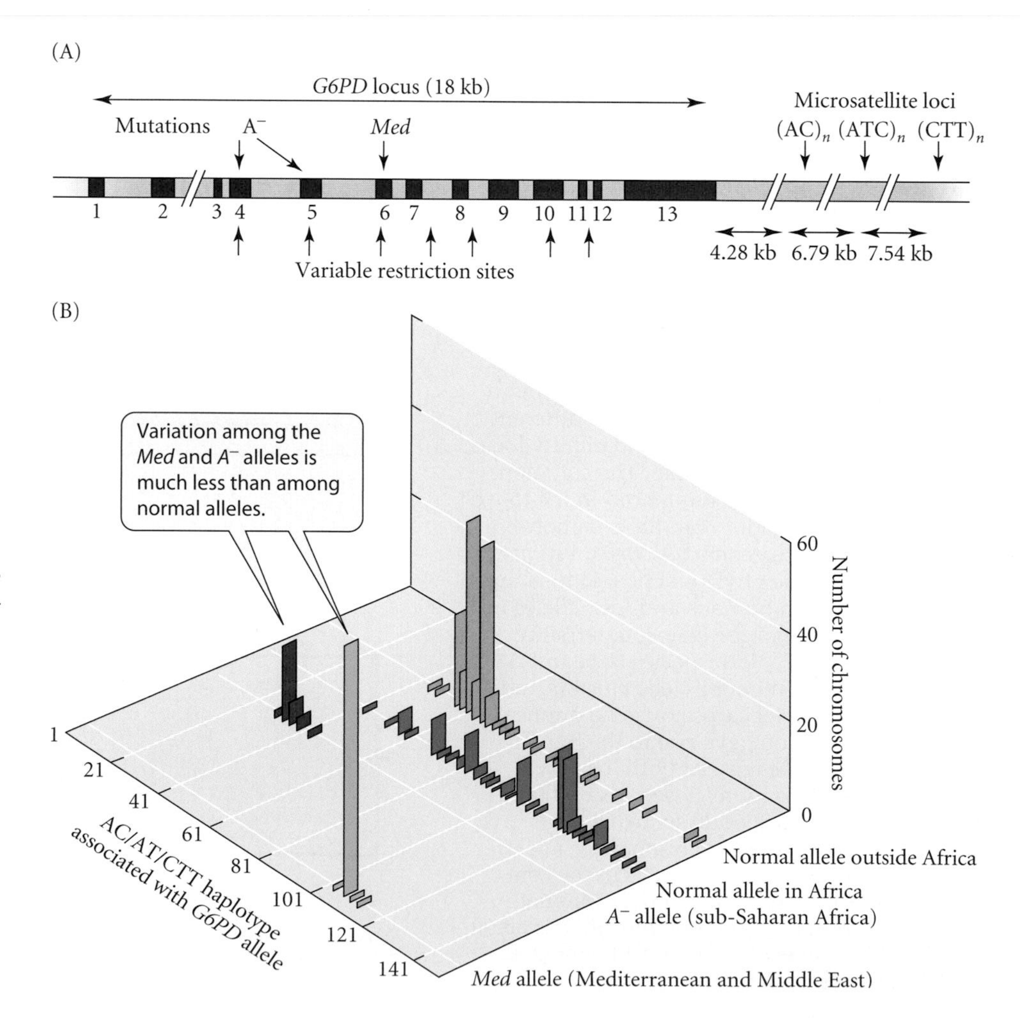

Figure 12.24 Variation at and near the human *G6PD* locus. (A) The structure of the locus and its distance from three microsatellite loci with variable numbers of AC, AT, and CTT repeats. The 13 exons of the *G6PD* locus are shown as red boxes. The A^- allele has amino acid-changing mutations in exons 4 and 5, and the *Med* allele in exon 6. (B) The frequencies of microsatellite haplotypes associated with copies of the normal *G6PD* allele in Africa, the normal allele outside Africa, the A^- allele in sub-Saharan Africa, and the *Med* allele in the Mediterranean and the Middle East. The height of a column shows the frequency of one of the 149 haplotypes that differ in the number of repeats. (After Tishkoff et al. 2001.)

broken down the strong association between an advantageous *G6PD* mutation and the microsatellite variants with which it was linked at first. When Tishkoff and collaborators performed computer simulations of various possible histories of change, they found that in simulations involving genetic drift, but not selection, levels of microsatellite variation were much higher that those observed, and levels of linkage disequilibrium were lower. The observed data on variation among A^- copies are best predicted by a simulation model in which the A^- allele has a selective advantage (s) of 0.044, and in which this allele has increased rapidly within the last 6357 years (with a range from 3840 to 11,760 years ago). The *Med* allele, likewise, has had a selective advantage of about 0.034, and increased within the last 3330 (range 1600–6640) years. Furthermore, low levels of synonymous nucleotide variation in *Plasmodium falciparum*, the malaria-causing protozoan, imply that its population size has increased greatly within the last 10,000 years (Volkman et al. 2001; Joy et al. 2003). The genetic data from both human and *Plasmodium* populations are consistent with archaeological and historical evidence that malaria has become a significant source of human mortality only within the last 10,000 years. The probable cause of this change is the spread of slash-and-burn agriculture and clearing of forests, land use practices that increase breeding habitats for the anopheline mosquitoes that carry *Plasmodium*.

The Strength of Natural Selection

Until the 1930s, most evolutionary biologists followed Darwin in assuming that the intensity of natural selection is usually very slight. By the 1930s, however, examples of very strong selection came to light. One of the first examples was INDUSTRIAL MELANISM in the peppered moth (*Biston betularia*). In some parts of England, a black form of the moth carrying a dominant allele increased in frequency after the onset of the Industrial Revolution. Museum collections dating since the mid-nineteenth century in England, after the onset of the Industrial Revolution, show that in less than a century, the "typical" pale gray form declined and the black (melanic) form increased from about 1 percent to more than 90 percent in some areas. The rate of change is so great that it implies a very substantial selective advantage, possibly as high as 50 percent, for the melanic form (Haldane 1932). There is considerable evidence, obtained by several independent researchers, that birds attack a greater proportion of gray than black moths where tree trunks, due to air pollution, lack the pale lichens that would otherwise cover them (Figure 12.25A); however, other factors also appear to affect the allele frequencies (Majerus 1998).

As air pollution has become regulated, conditions have reverted to favor the typical gray phenotype, and the frequency of the melanic form has declined very rapidly in Great Britain, Europe, and the United States (Figure 12.25B; Grant and Wiseman 2002; Cook 2003). By applying the appropriate equation for allele frequency change (see Box A) to the

(A)

(B)

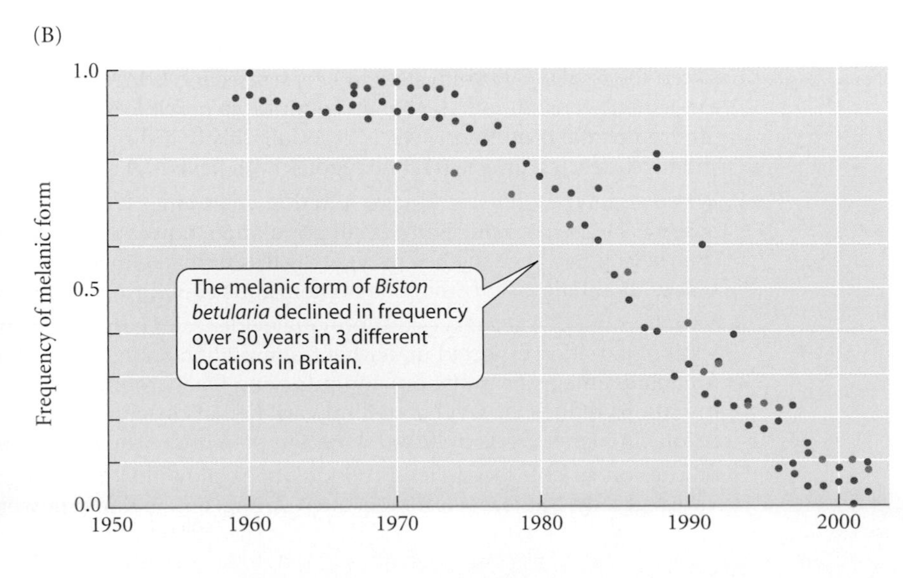
1.0
0.5
0.0
Frequency of melanic form
1950
1960
1970
1980
1990
2000
The melanic form of Biston betularia declined in frequency over 50 years in 3 different locations in Britain.

(C)

Figure 12.25 (A) The pale "typical" form and the dark melanic form of the peppered moth (*Biston betularia*) on a dark tree trunk. The British biologist H. B. D. Kettlewell pinned these and other freshly killed specimens to both dark and pale trunks, and determined that birds took a higher proportion of the more conspicuous phenotype in each situation. (B) The decline in the frequency of the melanic form in three British localities, indicated by different symbols. (C) The "typical" and melanic forms of the moth on a pale tree trunk covered with lichens, which grow in areas with clean air. (A, C, © photolibrary.com; B after Cook 2003.)

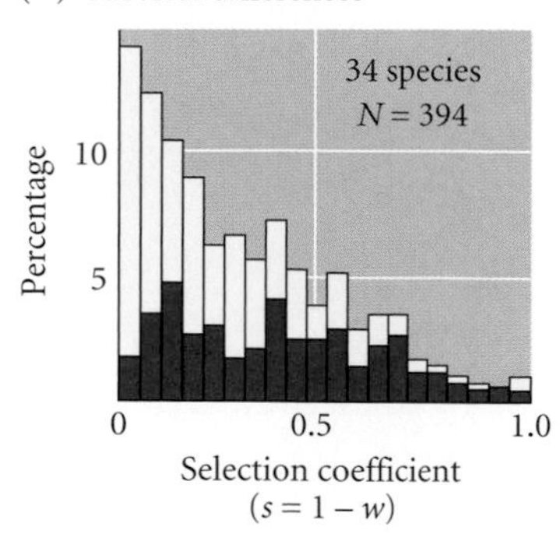

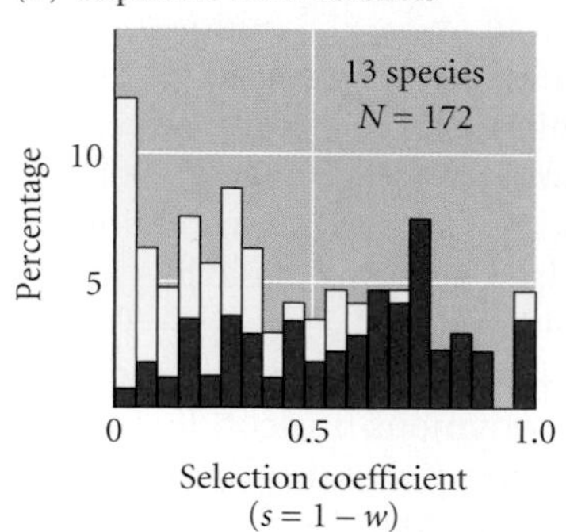

Figure 12.26 A compilation of selection coefficients (s) reported in the literature for discrete, genetically polymorphic traits in natural populations of various species. The total height of each bar represents the percentage of all reported values in each interval, and the red portion represents the percentage shown by statistical analysis to be significantly different from zero. N is the number of values reported, based on one or more traits from each of a number of species. (A) Selection based on differences in survival. (B) Selection based on differences in reproduction (fecundity, fertility, and sexual selection). (After Endler 1986.)

data, Lawrence Cook has estimated selection coefficients (s) against the melanic form of 0.05 to more than 0.20 in various British sites.

In the last few decades, components of fitness have been quantified for polymorphic traits in many species. Although cases in which s is close to zero are probably underrepresented because investigators may not report studies that yielded no evidence of selection, the selection coefficients estimated in natural populations range from low to very high (Figure 12.26). Moreover, selection acting through both survival and reproduction can be very strong. Thus natural selection is a powerful factor in evolution and is often far stronger than Darwin ever would have imagined.

Summary

1. Even at a single locus, the diverse genetic effects of natural selection cannot be summarized by the slogan "survival of the fittest." Selection may indeed fix the fittest genotype, or it may maintain a population in a state of stable polymorphism, in which inferior genotypes persist.
2. The absolute fitness of a genotype is measured by its rate of increase, the major components of which are survival, female and male mating success, and fecundity. In sexual species, differences among gametic (haploid) genotypes may also contribute to selection among alleles.
3. Rates of change in the frequencies of alleles and genotypes are determined by differences in their relative fitness, and are also affected by genotype frequencies and the degree of dominance at a locus.
4. Much of adaptive evolution by natural selection consists of replacement of previously prevalent genotypes by a superior homozygote (directional selection). However, genetic variation at a locus often persists in a stable equilibrium condition, owing to a balance between selection and recurrent mutation, between selection and gene flow, or because of any of several forms of balancing selection.
5. The kinds of balancing selection that maintain polymorphism include heterozygote advantage, inverse frequency-dependent selection, and variable selection arising from variation in the environment.
6. Often the final equilibrium state to which selection brings a population depends on its initial genetic constitution: there may be multiple possible outcomes, even under the same environmental conditions. This is especially likely if the genotypes' fitnesses depend on their frequencies, or if two homozygotes both have higher fitness than the heterozygote.
7. When genotypes differ in fitness, selection determines the outcome of evolution if the population is large; in a sufficiently small population, however, genetic drift is more powerful than selection. When the heterozygote is less fit than either homozygote, genetic drift is necessary to initiate a shift from one homozygous equilibrium state to the other.
8. Variation in DNA sequences can provide evidence of natural selection. Compared with the level of variation expected under neutral mutation and genetic drift alone, positive selection (of an advantageous mutation) and purifying (background) selection against deleterious mutations reduce the level of neutral variation at closely linked sites. Balancing selection results in higher levels of linked variation than under the neutral theory.
9. Studies of variable loci in natural populations show that the strength of natural selection varies greatly, but that selection is often strong, and is thus a powerful force of evolution.

Terms and Concepts

absolute fitness
adaptive landscape
adaptive peak/valley
antagonistic selection
associative overdominance
background selection
balancing selection
coefficient of selection
components of fitness
cost of adaptation (trade-off)
directional selection
disruptive selection
diversifying selection
frequency-dependent selection (inverse or positive)
heterozygote advantage/disadvantage
mean (average) fitness
multiple stable equilibria
multiple-niche polymorphism
overdominance
peak shift
purifying selection
relative fitness
reproductive success
selective advantage
selective sweep
stabilizing selection
underdominance

Suggestions for Further Reading

Natural selection in the wild, by J. A. Endler (Princeton University Press, Princeton, NJ, 1986), analyzes methods of detecting and measuring natural selection and reviews studies of selection in natural populations. Textbooks of population genetics, such as D. L. Hartl and A. G. Clark's *Principles of population genetics* (third edition, Sinauer Associates, Sunderland, MA,1997) and P. W. Hedrick's *Genetics of populations* (Jones and Bartlett, Sudbury, MA, 2000), present the mathematical theory of selection in depth.

Problems and Discussion Topics

1. If a recessive lethal allele has a frequency of 0.050 in newly formed zygotes in one generation, and the locus is in Hardy-Weinberg equilibrium, what will be the allele frequency and the genotype frequencies at this locus at the beginning of the next generation? (Answer: q = 0.048; $p^2 = 0.9071$, $2pq = 0.0907$; $q^2 = 0.0023$.) Calculate these values for the succeeding generation. If the lethal allele arises by mutation at a rate of 10^{-6} per gamete, what will be its frequency at equilibrium?
2. Suppose the egg-to-adult survival of A_1A_1 is 80 percent as great as that of A_1A_2, and the survival of A_2A_2 is 95 percent as great. What is the frequency (p) of A_1 at equilibrium? What are the genotype frequencies among zygotes at equilibrium? Now suppose the population has reached this equilibrium, but that the environment then changes so that the relative survival rates of A_1A_1, A_1A_2, and A_2A_2 become 1.0, 0.95, and 0.90. What will the frequency of A_1 be after one generation in the new environment? (Answer: 0.208.)
3. If the egg-to-adult survival rates of genotypes A_1A_1, A_1A_2, and A_2A_2 are 90, 85, and 75 percent, respectively, and their fecundity values are 50, 55, and 70 eggs per female, what are the approximate absolute fitnesses (R) and relative fitnesses of these genotypes? What are the allele frequencies at equilibrium? Suppose the species has two generations per year, that the genotypes do not differ in survival, and that the fecundity values are 50, 55, and 70 in the spring generation and 70, 65, and 55 in the fall generation. Will polymorphism persist, or will one allele become fixed? What if the fecundity values are 55, 65, 75 in the spring and 75, 65, 55 in the fall?
4. In pines, mussels, and other organisms, investigators have often found that components of fitness such as growth rate and survival are positively correlated with the number of allozyme loci at which individuals are heterozygous rather than homozygous (Mitton and Grant 1984; Zouros 1987). The interpretation of such data has been controversial (see references in Avise 2004). Provide two hypotheses to explain the data, and discuss how they might be distinguished.
5. Considering the principles of mutation, genetic drift, and natural selection discussed in this and previous chapters, do you expect *adaptive* evolution to occur more rapidly in small or in large populations? Why? Answer the same question with respect to nonadaptive (*neutral*) evolution.

6. Discuss whether or not natural selection would be expected to (*a*) increase the abundance (population size) of populations or species; (*b*) increase the rate at which new species evolve from ancestral species, thus increasing the number of species.
7. Both creationists (e.g., Wells 2000) and a science writer (Hooper 2002) have charged that H. B. D. Kettlewell's famous evidence that predatory birds exert natural selection on the color forms of the peppered moth was deeply flawed and possibly deceitful. Grant (2002), Cook (2003), and others have rebutted this claim, and provide an entrée into the extensive research literature on this topic. Discuss whether or not this charge, if true, would weaken the case for evolution by natural selection. Read some of the relevant literature and write an essay on whether or not there is evidence for Kettlewell's claim.

13

Evolution of Phenotypic Traits

If you consider human characteristics, or those of any other species you are familiar with, you will probably be hard pressed to think of a phenotypic feature that displays a single-locus genetic polymorphism with two or three discrete states. Even human eye color doesn't come just in brown and blue, despite what some elementary biology textbooks might lead you to believe. Instead, variation in most phenotypic characters, such as height or finger length or life span, is continuous, or "quantitative," and it is based on the effects of several or many variable gene loci, as well as those of the environment. At first surmise, then, it is a bit difficult to see how the one-locus models of evolution described in the previous chapter apply to most of phenotypic evolution. The field of **quantitative genetics** was developed to analyze quantitative characters, and its methods are used by biologists who study the evolution of morphology, life history characteristics, behavior, and other phenotypic traits.

Artificial selection. Darwin drew on the great changes wrought by artificial selection, such as those between a domesticated basset hound and its ancestor, the wolf, to show that selection can alter characters far beyond the range of variation seen within a species. (Wolf © Painet, Inc.; basset © Lynn M. Stone/naturepl.com.)

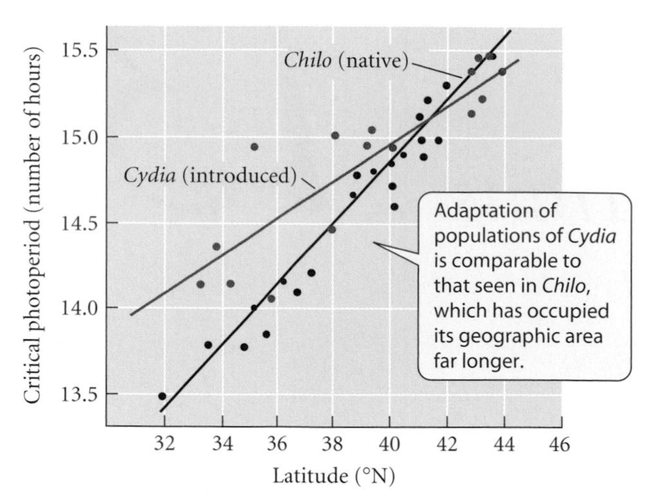

Figure 13.1 Geographic variation in the critical photoperiod for entering diapause in two species of moths at various latitudes. *Chilo suppressalis* is native to Japan. *Cydia pomonella* is an introduced pest in North America, where it has spread within the last 250 years. The two species now occupy the same latitudinal range. (After Tauber et al. 1986.)

Evolution Observed

In the previous chapter, we encountered a few examples of adaptive change in allele frequencies at a single locus—such as those affecting warfarin resistance in rats—that have been observed in natural populations. However, most observed instances of rapid evolution by natural selection involve quantitative characters that have (or probably have) a polygenic basis. Rapid adaptation, at rates far greater than the average evolutionary rates documented in the fossil record, is most often seen when a species is introduced into a new region or when humans alter features of its environment (Endler 1986; Taylor et al. 1991). The following examples show that rapid evolution has occurred in morphology, physiology, and behavior.

In many insects, the cue for entering diapause, a state of low metabolic activity that is necessary for surviving the winter, is a critical photoperiod (day length). Northern populations are typically genetically programmed to enter diapause at longer day lengths than southern populations because winter arrives at northern latitudes sooner, when days are still relatively long. The codling moth (*Cydia pomonella*), a major pest of apples, is a European species that was first recorded in New England in 1750, and has since expanded its range over 12 degrees of latitude. Populations have diverged genetically so that they display the same adaptive cline in critical photoperiod as other moth species that have occupied the same latitudinal span for a much longer time (Figure 13.1).

Until the 1950s, German populations of the blackcap (*Sylvia atricapilla*), a European songbird, migrated only to the western Mediterranean region for the winter. Since that time, more and more German blackcaps have overwintered in Britain, migrating northwest rather than southwest to the Mediterranean. When feeling the urge to migrate, a caged bird flutters in the right direction for its migration if it can see the night sky. (Most small birds migrate at night and use star patterns to determine direction.) Peter Berthold and colleagues (1992) used this behavior to show that offspring of blackcaps from populations that overwinter in Britain (who have never made the migratory journey themselves) orient toward the northwest, while offspring from populations that overwinter in the Mediterranean orient toward the southwest. The difference between the populations is genetically based and has evolved in a few decades. The authors suggest that the selective advantage of this change in behavior lies in improved winter weather and other conditions in Britain and in earlier spring return from Britain than from the Mediterranean.

Copper, zinc, and other heavy metals are toxic to plants, but in several species of grasses and other plants, metal-tolerant populations have evolved where soils have been contaminated by mine works that range from over 700 to less than 100 years old (Figure 13.2). In some cases, tolerance has evolved within decades on a microgeographic scale, such as in the vicinity of a zinc fence. Tolerance is based on a variable number of genes, depending on the species and population. When tolerant and nontolerant genotypes of a

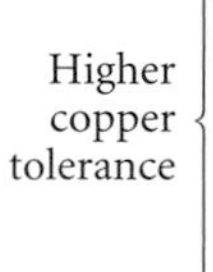

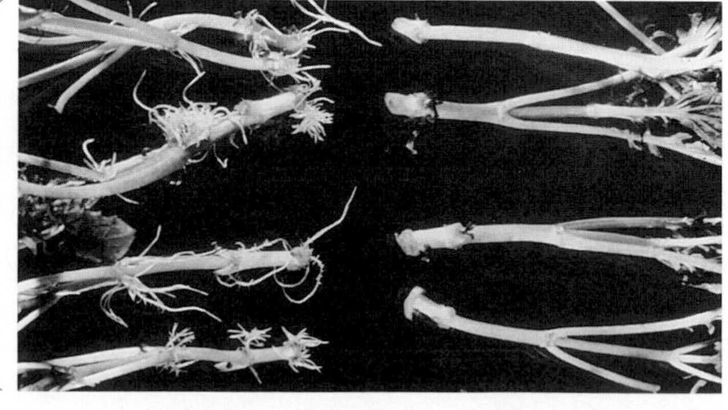

Figure 13.2 Variation in root growth of the monkeyflower *Mimulus guttatus* grown in a copper solution. The plants with the longest roots (left) are more tolerant of copper than those at the right. This variation has a genetic basis, which has enabled some populations of this and other plant species to adapt rapidly to high soil concentrations of copper. (From Macnair 1981; courtesy of M. Macnair.)

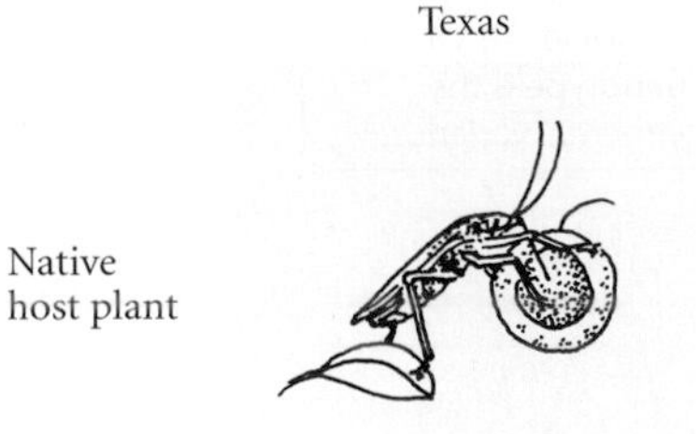

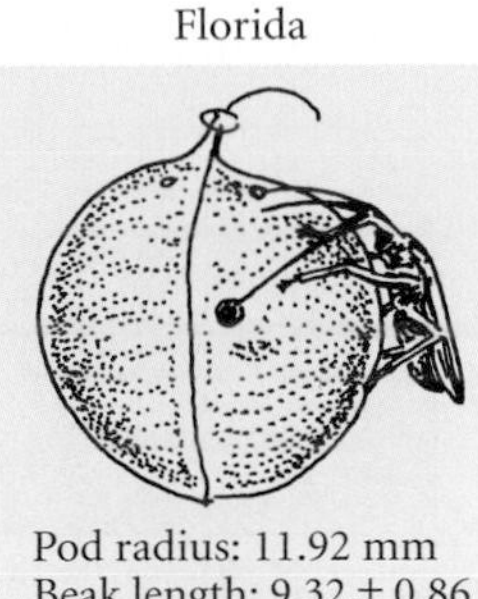

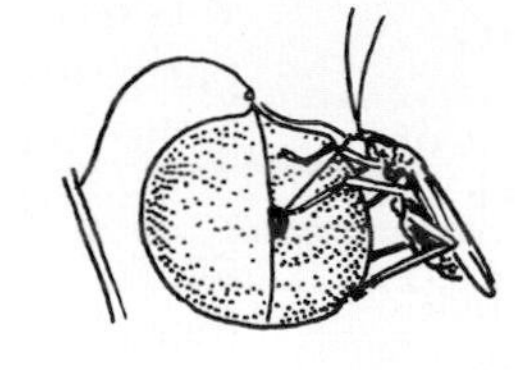

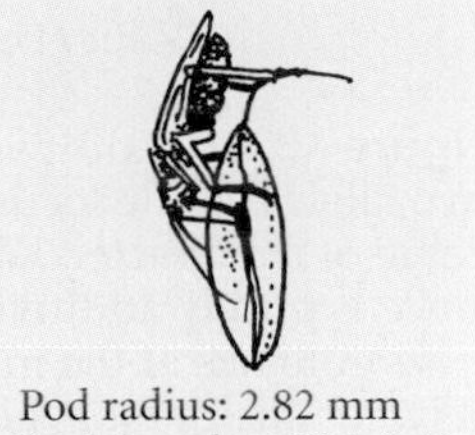

Figure 13.3 Soapberry bugs (*Jadera haematoloma*) and their native and introduced host plants in Texas and Florida, drawn to scale. The bug's beak is the needle-like organ projecting from the head at right angles to the body. The average pod radius of each host species and the average (with standard deviation) beak length of associated *Jadera* populations are given. Beak length has evolved rapidly as an adaptation to the new host plant. (After Carroll and Boyd 1992.)

species are grown in competition with other plant species in the absence of the metal, the relative fitness of the tolerant genotypes is often much lower than that of the nontolerant genotypes, implying a cost of adaptation (Antonovics et al. 1971; Macnair 1981).

Several species of insects, such as the soapberry bug (*Jadera haematoloma*) (Carroll and Boyd 1992; Carroll et al. 1997), have adapted rapidly to new food plants. This insect feeds on seeds of plants in the soapberry family (Sapindaceae) by piercing the enveloping seed pod with its slender beak (Figure 13.3). In Texas, its natural host plant is the soapberry tree, which has a small pod, whereas in Florida it feeds naturally on balloon vine, in which a large spherical capsule envelops the seed. In both regions, some bug populations now feed mostly on introduced plants that have become common only within the last 20 to 50 years. In Texas, the major contemporary host plant is the Asian round-podded golden rain tree (*Koelreuteria paniculata*), which has a larger pod than the native host. In Florida, most bugs feed on the flat-podded rain tree (*Koelreuteria elegans*), also from Asia, which has a flatter, smaller seed pod than the native host. The bugs feed most efficiently if the beak is the right length for reaching seeds. Measurements on museum specimens collected at different times in the past show that the mean beak length of the bugs has changed steadily since the foreign host plants were introduced. Today, the mean beak length of Texan populations that feed on *K. paniculata* is 8 percent longer than that of populations that feed on the native host, and in Florida, beaks are 25 percent shorter where soapberry bugs feed on *K. elegans* than where they still feed on the native balloon vine. These differences have a genetic basis. Adaptation to an altered environment—to new food plants—has resulted in large, rapid changes in morphology.

Components of Phenotypic Variation

In order to discuss the evolution of quantitative traits, we must recall and expand on some concepts introduced in Chapter 9. An important measure of variation is the *variance* (V), defined as the average of the squared deviations of observations from the *arithmetic mean* of a sample. The square root of the variance is the *standard deviation* ($s = \sqrt{V}$), measured in the same units as the observations. If a variable has a normal (bell-shaped) frequency distribution, about 68 percent of the observations lie within one standard deviation on either side of the mean, 96 percent within two standard deviations, and 99.7 percent within three (see Box C in Chapter 9).

The **phenotypic variance** (V_P) in a phenotypic trait is the sum of the variance due to differences among genotypes (the **genetic variance**, V_G) and the variance due to direct effects of the environment and developmental noise (the **environmental variance**, V_E). Thus $V_P = V_G + V_E$. Considering for the moment the effect of only one locus on the phenotype, we may take the midpoint between the two homozygotes' means as a point of reference (Figure 13.4). Then the mean phenotype of A_1A_1 individuals deviates from that midpoint by $+a$, and that of A_2A_2 by $-a$. The quantity a, the ADDITIVE EFFECT of an allele, measures how greatly the phenotype is affected by the genotype at this locus. If the inheritance of the phenotype is entirely additive, the heterozygote's phenotype is exactly in between that of the two homozygotes. Such additive effects are responsible for a critically important component of genetic variance, the **additive genetic variance**, denoted V_A. (The genetic variance

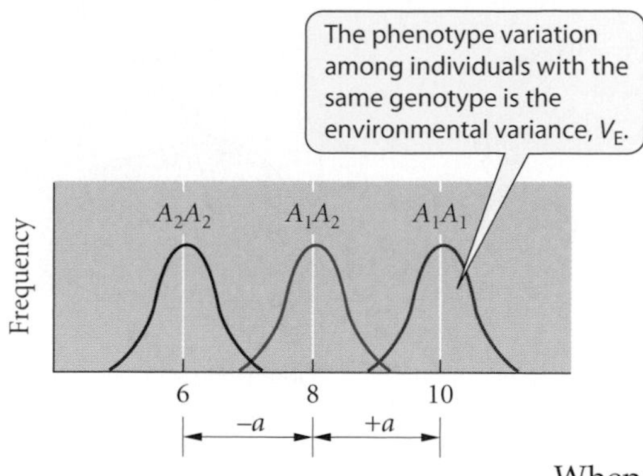
The phenotype variation among individuals with the same genotype is the environmental variance, V_E.
A_2A_2
A_1A_2
A_1A_1
Frequency
6
8
10
$-a$
$+a$
Value of character z

Figure 13.4 Additive effects of two alleles at one locus on the value of a character. When inheritance is purely additive, the heterozygote lies at the midpoint between the homozygotes. The additive effect of substituting an A_1 or an A_2 allele in the genotype is a (in this case, $a = 2$). The magnitude of a affects that of the additive genetic variance, V_A. The phenotypic variation within each genotype is measured by the environmental variance, V_E.

V_G may also include nonadditive components, due to dominance and epistatic gene interactions, that we will not concern ourselves with here.)

The additive genetic variance depends both on the magnitude of the additive effects of alleles on the phenotype and on the genotype frequencies. If one genotype is by far the most common, most individuals are close to the average phenotype, so the variance is lower than if the several genotypes have more equitable frequencies, as they will if the allele frequencies are more nearly equal. When, as in Figure 13.4, two alleles (with frequencies p and q) have purely additive effects, V_A at a single locus is

$$V_A = 2pqa^2$$

When several loci contribute additively to the phenotype, the average phenotype of any particular genotype is the sum of the phenotypic values of each of the loci. Likewise, V_A is the sum of the additive genetic variance contributed by each of the loci.

The additive genetic variance plays a key role in evolutionary theory because *the additive effects of alleles are responsible for the degree of similarity between parents and offspring* and therefore *are the basis for response to selection within populations*. When alleles have additive effects, the expected average phenotype of a brood of offspring equals the average of their parents' phenotypes.* Evolution by natural selection requires that selection among phenotypically different parents be reflected in the mean phenotype of the next generation. Therefore, V_A enables a **response to selection**—a change in the mean character state of one generation as a result of selection in the previous generation.

The proportion of phenotypic variance that is due to additive genetic differences among individuals is referred to as the character's **heritability in the narrow sense**, h^2_N. The heritability is determined by the additive genetic variance (V_A), which depends on allele frequencies; and by the environmental variance (V_E), which depends in part on how variable the environmental factors are that affect the development or expression of the character. That is,

$$h_N^2 = V_A/V_P$$

where $V_P = V_G + V_E$ and V_G is the sum of V_A and nonadditive genetic components. Because allele frequencies and environmental conditions may vary among populations, an estimate of heritability is strictly valid only for the population in which it was measured, and only in the environment in which it was measured. Moreover, it is wrong to think that if a character has a heritability of 0.75, the feature is ¾ "genetic" and ¼ "environmental," as if a character is formed by mixing genes and environment the way one would mix paints to achieve a desired color. It is the *variation* in the character that is statistically partitioned, and even this partitioning is a property of the particular population, not a fixed property of the feature. For evolutionary studies, moreover, the additive genetic variance is often more informative than the heritability (Houle 1992).

The additive genetic variance is estimated from resemblances among relatives. In order to achieve accurate estimates, it is important that relatives not develop in more similar environments than nonrelatives; otherwise, it may not be possible to distinguish similarity due to shared genes from similarity due to shared environments.

The narrow-sense heritability (h^2_N) of a trait equals the slope (b) of the *regression* of offspring phenotypes (y) on the average of the two parents of each brood of offspring (x) (see Figure 13.8 below; see also Figure 9.20). The slope of this relationship is calculated so as to minimize the sum of the squared deviations of all the values of y from the line. As the correspondence between offspring and parents is reduced, the slope of the regression declines (compare parts A and B of Figure 13.8). Therefore, anything that reduces the similarity between parents and their offspring (such as environmental effects on phenotype, or dominance) reduces heritability. There are also ways of estimating V_A and other components of genetic variance from similarities among other relatives, such as siblings ("full sibs") or half sibs (individuals that have only one parent in common).

*The correlation between parent and offspring phenotypes is lower if there is dominance at a locus.

Before proceeding with our analysis of quantitative characters, we must understand their genetic foundations: the number of loci involved, the ways they affect characters, and the role of linkage among loci.

How Polygenic Are Polygenic Characters?

The number of loci that contribute to variation in a character may be less than the number that actually contribute to its development. However, only variable loci can be detected—and detecting those that have small phenotypic effects is not easy. The principal method of detection relies on the association of phenotypic differences with genetic markers. Suppose, as a simple hypothetical example, that we have two inbred lines of a species of fly that has two pairs of chromosomes, that the lines differ in wing length, and that they also carry different alleles at one marker locus on each chromosome (Figure 13.5). Line 1 is homozygous for alleles X_1 and Y_1 on chromosomes 1 and 2, respectively, while line 2 is homozygous for alleles X_2 and Y_2. A backcross between the F_1 and line 2 ($X_1X_2Y_1Y_2 \times X_2X_2Y_2Y_2$) yields progeny with four combinations of markers. Some progeny, such as $X_1X_2Y_2Y_2$ and $X_2X_2Y_2Y_2$, differ only with respect to chromosome 1. If they also differ in wing length, then chromosome 1 must carry *at least one* locus, linked to locus X, that contributes to variation. (Compare genotypes 3 and 4, or 1 and 2, in Figure 13.5.) Likewise, if genotypes distinguished by the marker Y, such as $X_2X_2Y_1Y_2$ and $X_2X_2Y_2Y_2$, differ in wing length, then chromosome 2 must also carry at least one gene that affects this character. (Compare genotypes 2 and 4 or 1 and 3 in Figure 13.5.)

Suppose, further, that two marker loci (Y and Z) on chromosome 2 also distinguish the inbred lines. Then some progeny differ at locus Y only, or at locus Z only, due to crossing over in the F_1 parent. If wing length differs between flies that differ only at marker Y, and also differs between flies that differ only at marker Z, then each of the chromosome regions marked by these genes must have at least one gene that affects wing length. Such results would imply that at least three loci affect the character. (Compare genotypes 4 and 5, and also genotypes 4 and 6, in Figure 13.5; note too that these two contrasts indicate that these chromosome regions differ in the magnitude of their effects on the phenotype.)

The more markers there are, distributed along all the organism's chromosomes, the greater the number of chromosome regions that can be compared, and the greater the potential number of trait-affecting loci that may be detected. (Of course, some markers may not be associated with any difference in the trait studied, and may provide no evidence for a closely linked gene affecting the character.) Traditionally, geneticists used morphological mutations as markers for studies

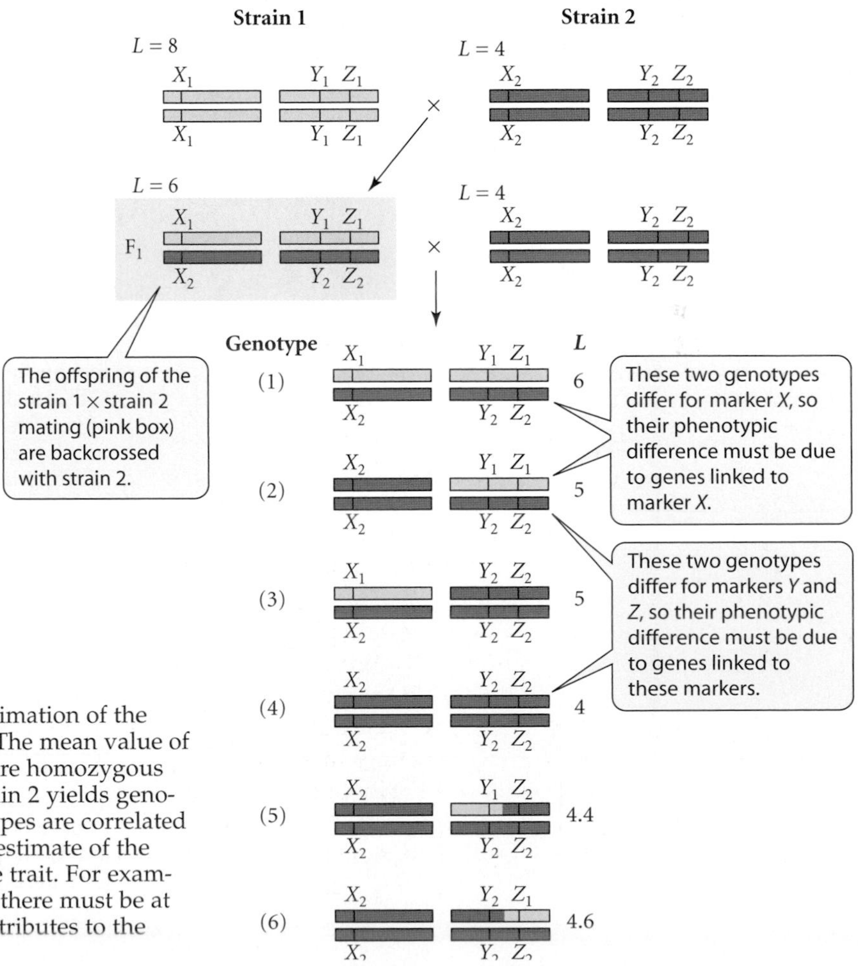

Figure 13.5 Illustration of the principle underlying estimation of the number of loci affecting variation in a quantitative trait. The mean value of the trait (L) is shown for each genotype. Strains 1 and 2 are homozygous for three marker loci (X, Y, Z). Backcrossing the F_1 to strain 2 yields genotypes 1–6 (among others). Differences in L among genotypes are correlated with differences in the markers they carry, providing an estimate of the minimal number of genes contributing to variation in the trait. For example, contrasting genotype 1 with either 2 or 3 shows that there must be at least one locus on each of the two chromosomes that contributes to the character difference between strains 1 and 2.

of this kind, but they now increasingly use molecular markers (such as polymorphic restriction sites or transposable elements), because many more such markers can be found in well-studied organisms. This method is called **QTL mapping** ("QTL" stands for "quantitative trait locus" or "loci"). A QTL is not necessarily a single locus, but rather a chromosome interval that may include many genes, of which one or more affect the character under study. In order to detect and locate many potential loci by the use of many molecular markers, very large numbers of progeny must be compared, and the data are subjected to special statistical analyses.

Trudy Mackay and her colleagues have performed especially intensive studies of the genetics of bristle number in *Drosophila melanogaster* (e.g., Nuzhdin et al. 1999; Dilda and Mackay 2002). The bristles they studied, on the abdomen and the sternopleuron (part of the side of the thorax), have a sensory function and are part of the peripheral nervous system. The investigators selected for high and low bristle numbers in laboratory populations founded by large numbers of wild flies (Long et al. 1995). They used transposable elements on the X chromosome and chromosome 3 as markers to analyze the genetic differences between the high- and low-selected strains. They detected 53 QTL, of which 33 affected sternopleural bristle number, 31 affected abdominal bristle number, and 11 affected both characters (Figure 13.6). Some QTL alleles had rather large effects, adding or subtracting several bristles, but many had smaller effects. Many QTL displayed both additive effects on the phenotype and strong EPISTATIC INTERACTIONS, meaning that the joint effect of some pairs of loci differed from the sum of their individual effects. The phenotypic effects of many of the QTL differed strongly between the sexes, and the effects of some QTL depended on the temperature at which the flies were raised (that is, these QTL displayed genotype × environment interactions, as described in Chapter 9; see Figure 9.18).

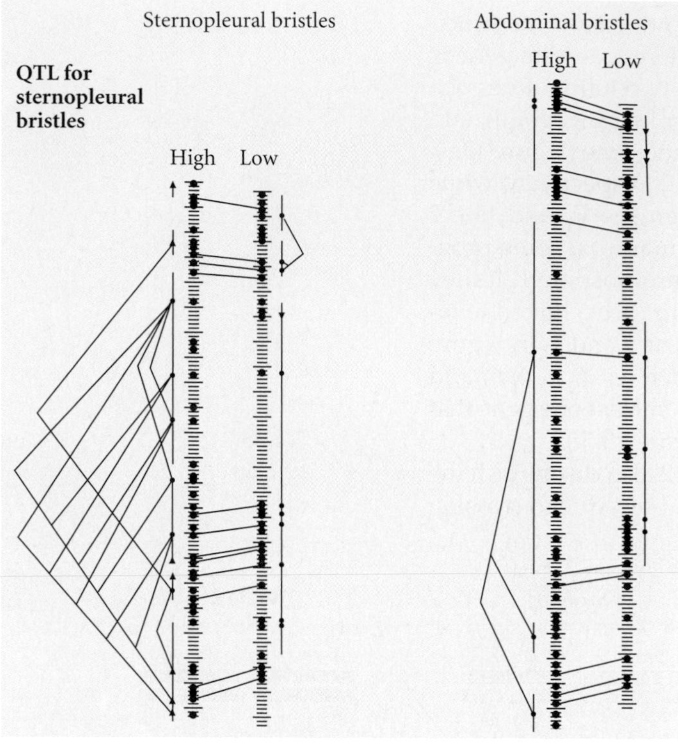

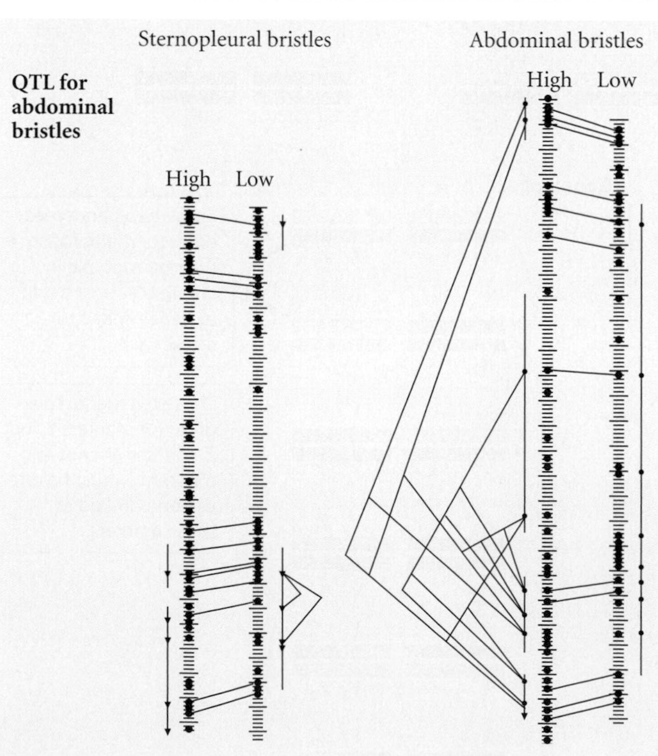

Figure 13.6 This diagram shows that even on a single chromosome, many loci may contribute to variation in a trait, and that variation in two traits may be due partly to shared (pleiotropic) loci and partly to unshared loci. The figure shows the locations of QTL on chromosome 3 that contribute to differences in the number of bristles between populations of *Drosophila melanogaster* selected for changes in the number of sternopleural bristles (left) or for changes in the number of abdominal bristles (right). In both selection regimes, different populations were selected for more ("High") or fewer ("Low") bristle number. Both kinds of bristles were analyzed in all the selected populations, so the locations of QTL for both sternopleural bristles (top) and abdominal bristles (bottom) were estimated for all selected lines. Thus the upper left "ladder" shows chromosomal locations of QTL for sternopleurals in the populations selected for sternopleural change, and the lower left "ladder" shows the QTL for abdominal bristles in the same populations. Black circles represent transposable elements used as genetic markers. Lines between chromosomes join markers found in both the "High" and "Low" populations. The location of each QTL is marked by a circle or triangle, with a thin line showing the interval within which the actual gene may be located. Angled lines join QTL that interact to affect bristle number nonadditively. Some QTL are the same in both the upper and lower diagrams—that is, they affect both kinds of bristles. (From Dilda and Mackay 2002.)

Most of the QTL mapped to chromosome sites where CANDIDATE LOCI were located: genes known from past studies to affect bristle morphology or development of the peripheral nervous system. DNA sequence variation at some of these loci has been shown to correlate with bristle variation, and alleles of these loci distinguished by DNA sequence segregate at high frequencies in natural populations of *Drosophila* (Lai et al. 1994). The variation at these loci is therefore not due to mutation–selection balance, but is either selectively neutral or is maintained by balancing selection (Chapter 12).

These and other such studies have important implications for the study of quantitative traits (Mackay 2001; Dilda and Mackay 2002). Models of quantitative traits generally assume that variation in a trait is due to many loci, each with alleles that have small effects. However, some of the variation may in fact be due to loci with large effects. The statistical models have often assumed that variation stems from the sum of additively acting alleles at functionally interchangeable loci. Variation in bristle number, however, is due to loci that are not interchangeable: they have different developmental roles, and they interact with one another in producing the phenotype. However, models that take into account such complex gene interactions—although important for researchers in this field—are beyond the scope of this book.

Linkage Disequilibrium

When alleles at two loci are randomly associated with each other, they are said to be in a state of LINKAGE EQUILIBRIUM; when they are nonrandomly associated, they are in LINKAGE DISEQUILIBRIUM (see Chapter 9).

These concepts, which are important in many contexts, are best thought of in terms of the frequencies of different combinations of alleles among all the gametes produced in the population. Let the frequencies of alleles A_1 and A_2 be p_A and q_A, and those of alleles B_1 and B_2 be p_B and q_B. There are four gamete types (A_1B_1, A_1B_2, A_2B_1, A_2B_2), with frequencies that we might denote g_{11}, g_{12}, g_{21}, and g_{22} respectively. If the alleles at the two loci A and B are randomly associated, then the frequency of any gamete type will be the simple product of the allele frequencies. (For instance, the expected frequency of A_1B_2 gametes is $g_{12} = p_Aq_B$.) Under this condition of linkage equilibrium, the loci could be closely linked, but the probability that a copy of the A_1 allele would have a B_1 as its neighbor on the same chromosome would be p_B, its frequency in the population at large.

If the loci are in linkage disequilibrium, however, the gamete frequencies depart from their expected values; they cannot be predicted from the allele frequencies, but must be measured instead. A COEFFICIENT OF LINKAGE DISEQUILIBRIUM may be defined as $D = (g_{11} \times g_{22}) - (g_{12} \times g_{21})$. If $D > 0$, then gametes A_1B_1 and A_2B_2, as well as genotypes formed by their union (such as $A_1A_1B_1B_1$), will be more frequent than expected.

If both loci contribute additively to a particular character, and if the genotypes $A_1A_1B_1B_1$ and $A_2A_2B_2B_2$ have the most extreme phenotypes (e.g., largest and smallest size), then, because linkage disequilibrium elevates the frequency of the extreme phenotypes, *positive linkage disequilibrium increases the phenotypic (and genetic) variance*. Conversely, if $D < 0$, these genotypes are less frequent than expected, and the phenotypic variance is correspondingly reduced. Recombination, which tends to bring the frequencies toward linkage equilibrium, would therefore either increase or decrease variation, depending on whether D is negative or positive. If the two loci affect different characters, a positive association between them ($D > 0$) can create a correlation between the two features, as we shall see below.

With the passage of generations, *the level of linkage disequilibrium declines due to recombination in the double heterozygotes* (Figure 13.7; see also Figure 9.15), at a rate proportional to the recombination rate between the loci. Why, then, might we ever find two loci, or two sites within a gene, in linkage disequilibrium? There are several possible reasons.

1. Nonrandom mating can maintain linkage disequilibrium. In an extreme case, the sample of organisms that display linkage disequilibrium may actually include two reproductively isolated species.

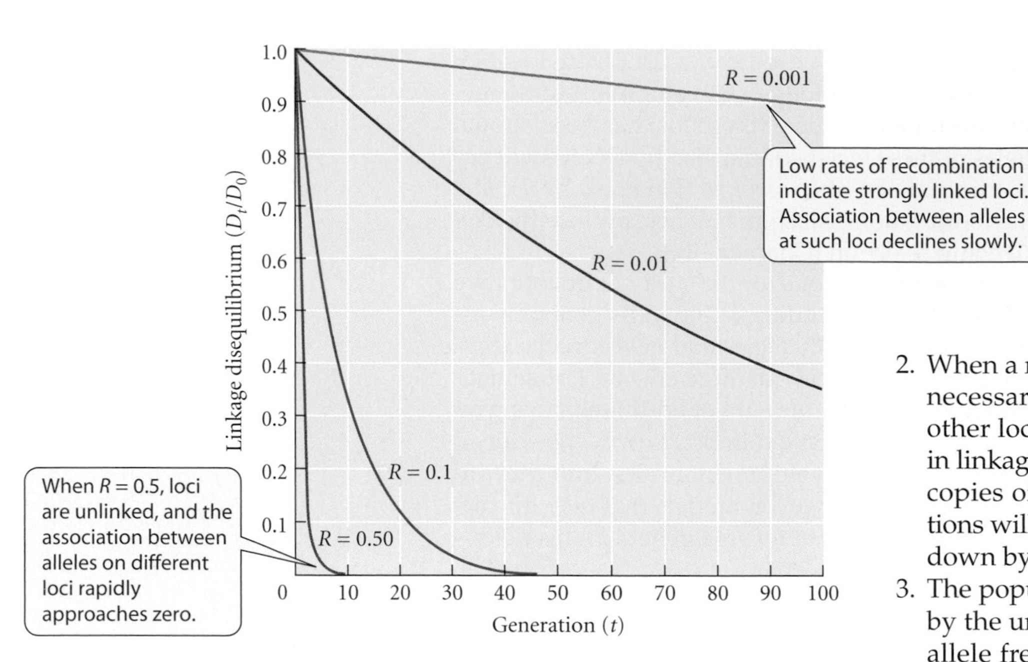

Linkage disequilibrium (D_t/D_0)
1.0
0.9
0.8
0.7
0.6
0.5
0.4
0.3
0.2
0.1
0
10 20 30 40 50 60 70 80 90 100
Generation (t)
$R = 0.001$
$R = 0.01$
$R = 0.1$
$R = 0.50$
Low rates of recombination indicate strongly linked loci. Association between alleles at such loci declines slowly.
When $R = 0.5$, loci are unlinked, and the association between alleles on different loci rapidly approaches zero.

Figure 13.7 The decrease in linkage disequilibrium (D) over time, relative to its initial value (D_0), for pairs of loci with different recombination rates (R). $R = 0.50$ if loci are unlinked. (After Hartl and Clark 1989.)

2. When a new mutation arises, the single copy is necessarily associated with specific alleles at other loci on the chromosome, and therefore is in linkage disequilibrium with those alleles. The copies of this mutation in subsequent generations will retain this association until it is broken down by recombination.
3. The population may have been formed recently by the union of two populations with different allele frequencies, and linkage disequilibrium has not yet decayed.
4. Recombination may be very low or nonexistent. Chromosome inversions and parthenogenesis (asexual reproduction) have this effect.
5. Linkage disequilibrium may be caused by genetic drift. If the recombination rate is very low, the four gamete types in the example above may be thought of as if they were four alleles at one locus. One of these "alleles" may drift to high frequency by chance, creating an excess of that combination relative to others.
6. Natural selection may cause linkage disequilibrium if two or more gene combinations are much fitter than recombinant genotypes.

Evolution of Quantitative Characters

Genetic variance in natural populations

Heritable variation has been reported for the great majority of traits in which it has been sought, in diverse species (Lynch and Walsh 1998). Not all characters are equally variable, however. For example, characters strongly correlated with fitness (such as fecundity) tend to have lower heritability than characters that seem unlikely to affect fitness as strongly (Mousseau and Roff 1987). However, the low h^2_N of fitness components arises from the greater magnitude of other variance components, especially V_E, in the denominator of the expression $h^2_N = V_A/V_P$. The additive genetic variance (V_A) of components of fitness is actually higher than that of morphological and other traits, probably because many physiological and morphological characteristics affect fitness (Houle et al. 1996).

In some cases, traits do not appear to be genetically variable at all. The paucity of genetic variation would then be a genetic constraint that could affect the direction of evolution (for example, an insect might adapt to some species of plants rather than to others) or prevent adaptation altogether. For instance, in sites near mines where the soil concentration of copper or zinc is high, a few species of grasses have evolved tolerance for these toxic metals (see Figure 13.2), but most species of plants have not. Populations of various species of grasses growing on normal soils were screened for copper tolerance by sowing large numbers of seeds in copper-impregnated soil. Small numbers of tolerant seedlings were found in those species that had evolved tolerance in other locations, but no tolerant seedlings were found in most of the species that had not formed tolerant populations (Macnair 1981; Bradshaw 1991). Genetic variation that might enable the evolution of copper tolerance seems to be rare or absent in those species.

Response to selection

In the simplest model of the effect of selection on a quantitative trait z, such as the tail length of rats, we assume that z has a normal frequency distribution in a population. (A roughly normal distribution is expected if a large number of loci, all with relatively small effects on the character, freely recombine.) Suppose an experimenter imposes selection for greater tail length by breeding only those rats in a captive population with tails longer than a certain value. This form of selection is called **truncation selection**. The mean tail length of the selected parents differs from that of the population from which they were taken ($\bar{z}$) by an amount S, the **selection differential** (Figure 13.8A). The average tail length ($\bar{z}'$) among the offspring of the selected parents differs from that of the parental generation as a whole ($\bar{z}$) by an amount R, the **response to selection** (Figure 13.8A, right-hand graph). The magnitude of R is proportional to the heritability of the trait (compare graphs

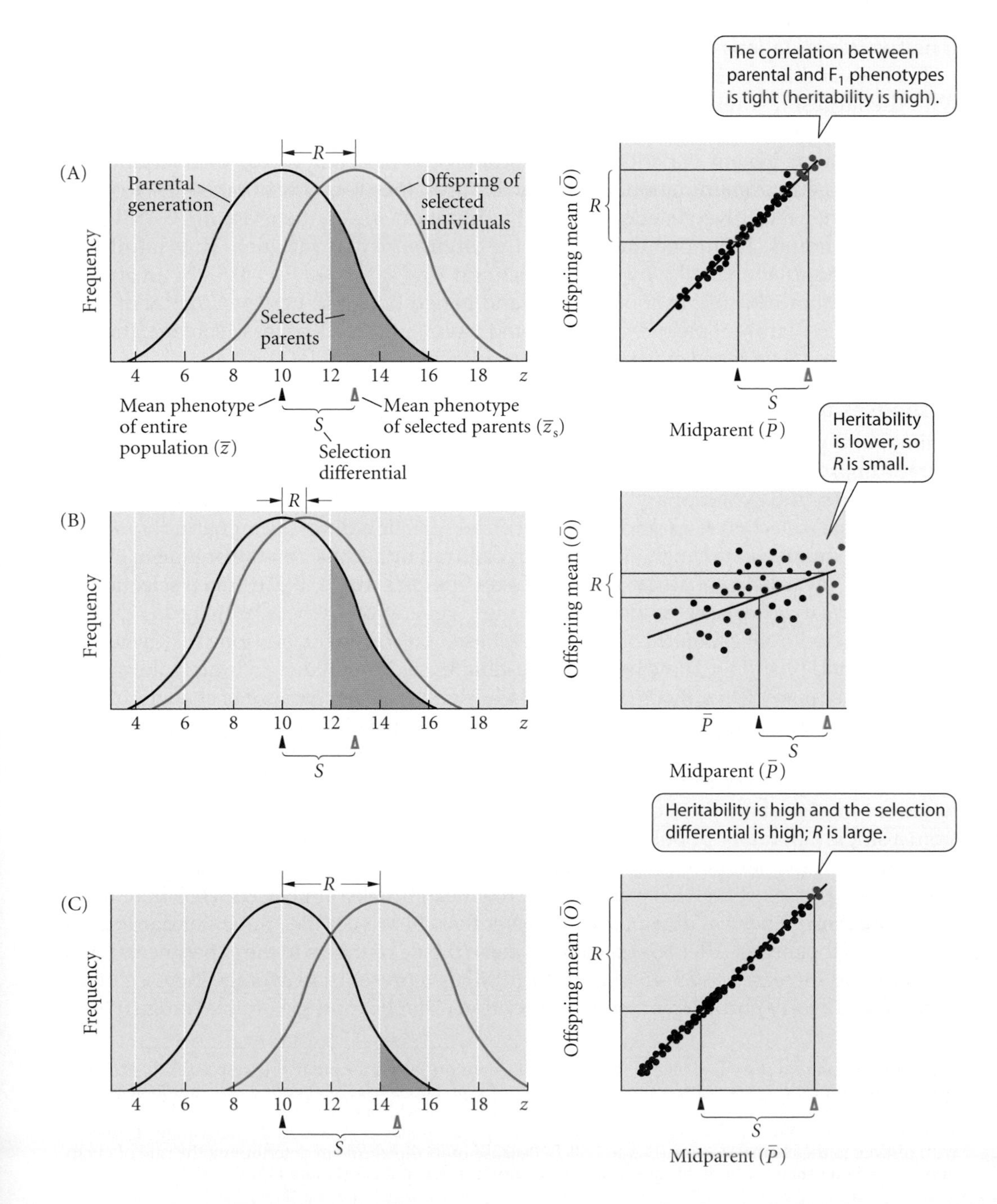

Figure 13.8 The response of a quantitative character to selection depends on the heritability of the character and the selection differential. (A) The black curve shows a normally distributed trait, z, with an initial mean of 10. Truncation selection is imposed, such that individuals with $z > 12$ reproduce. The graph at the right shows a strong correspondence between the average phenotype of pairs of selected parents ($\bar{P}$) and that of their broods of offspring ($\bar{O}$), i.e., high heritability. Blue circles represent the selected parents and offspring, black circles the rest of the population (were it to have bred). Because heritability is high, the selection differential S results in a large response to selection (R). The blue curve is the distribution of z in the next generation, whose mean lies R units to the right of the mean of the parental generation. (B) The circumstances are the same, but the relationship between phenotypes of parents and offspring is more variable and has a lower slope in the graph at the right, i.e., lower heritability. Consequently, the selection differential S translates into a smaller response (R). The frequency distribution of offspring (blue curve) shifts only slightly to the right. (C) Here the heritability is very high, as in part A, but the selected parents have $z > 14$. Thus the selection differential S is larger, resulting in a greater response to selection (R).

A and B in Figure 13.8) and to the selection differential S (compare graphs A and C).* In fact, the change in mean phenotype in offspring, R, due to selection of parents, S, can be read directly from the regression line as

$$R = h^2_N S$$

Since this equation can be rearranged as $h^2_N = R/S$, heritability can be estimated from a selection experiment in which S (which is under the experimenter's control) and R are measured. Such an estimate of h^2_N is called the **realized heritability**. This is how (as you may recall from Chapter 9) Dobzhansky and Spassky estimated that the heritability of phototaxis in *Drosophila pseudoobscura* was about 0.09 (see Figure 9.21).

As selection proceeds, it increases the frequencies of those alleles that produce phenotypes closer to the optimum value. As those frequencies increase, multilocus genotypes (combinations of alleles at different loci) that had been extremely rare become more common, so phenotypes arise that had been effectively absent before. *Thus the mean of a polygenic character shifts beyond the original range of variation as directional selection proceeds*, even if no further mutations occur.

If alleles at different loci differ in the magnitude of their effects on the phenotype, those with the largest favorable effects are likely to be fixed first (Orr 1998). In the absence of complicating factors, prolonged directional selection should ultimately fix all favored alleles, eliminating genetic variation. Further response to selection would then require new variation, arising from mutation. For many features, the MUTATIONAL VARIANCE, V_m—the infusion of new additive genetic variance by mutation—is on the order of $10^{-3} \times V_E$ per generation; that is, about one-thousandth the environmental variance. Thus a fully homozygous population could, by mutation, attain $V_A/(V_A + V_E) = h^2 = 0.5$, for an entirely additively inherited trait, in about a thousand generations. (If, however, many of these mutations have harmful pleiotropic effects and have a net selective disadvantage, the "usable" mutational variance may be much less; Hansen and Houle 2004.)

Responses to artificial selection

Animal and plant breeders have used artificial selection to alter domesticated species in extraordinary ways (Figure 13.9). Darwin opened *The Origin of Species* with an analysis of such changes, and evolutionary biologists have drawn useful inferences about evolution from artificial selection ever since then. Artificial selection differs from natural selection because the human experimenter focuses on one trait rather than on the organism's overall fitness. Nevertheless, natural selection often operates much like artificial selection.

Responses to artificial selection over just a few generations generally are rather close to those predicted from estimates of heritability based on correlations among relatives, such as parents and offspring. These heritability estimates seldom predict accurately the change in a trait over many generations of artificial selection, however, because of changes in linkage disequilibrium and genetic variance, input of new genetic variation by mutation, and the action of natural selection, which often opposes artificial selection (Hill and Caballero 1992). Such effects were found, for example, in an experiment by B. H. Yoo (1980), who selected for increased numbers of abdominal bristles in lines of *Drosophila melanogaster* taken from a single laboratory population. For 86 generations, Yoo scored bristle numbers on 250 flies, and bred the next generation from the top 50 flies, of each sex. In the base population from which the selection lines were drawn, the mean bristle number was 9.35 in females, and more than 99 percent of females had fewer than 14 bristles (i.e., three standard deviations above the mean). After 86 generations, mean bristle numbers in the experimental populations had increased to 35–45 (Figure 13.10). This represents an average increase of 316 percent, or 12 to 19 phenotypic standard deviations—far beyond the original range of vari-

*The simple equation presented here is the "breeders' equation," used for predicting responses to artificial selection in domesticated species. A conceptually related equation is used more often by evolutionary biologists: $\Delta\bar{z} = V_A/\bar{w} \times d\bar{w}/d\bar{z}$, where z is a character state, V_A is the additive genetic variance, and w is fitness. Thus the rate of evolution of the character mean ($\Delta\bar{z}$) is proportional to V_A and to the derivative of fitness with respect to the character value (Lande 1976a). Because of its direct role in determining the rate of evolution, V_A is better than the heritability, h^2_N, as a measure of "evolvability" (Houle 1992).

Figure 13.9 Results of artificial selection: domesticated breeds of pigeons, all developed from the wild rock pigeon by selective breeding. Each breed has features quite unlike the wild ancestor, such as (A) the long bill and large featherless region surrounding the eye of the English carrier; (B) the greatly inflated esophagus, erect stance, elongated body and legs, and feathered toes of the English pouter; and (C) the tail of the fantail with its 32 (or more) feathers. The fantail is remarkable because the tails of all species in the pigeon family (and in most other families of birds) normally have 12 feathers. These drawings are from Darwin's book *The Variation of Animals and Plants under Domestication* (1868). Darwin kept pigeons, studied pigeon breeds extensively, and used information from pigeons and other domesticated species in developing his argument for evolution.

ation. In a very short time, selection had accomplished an enormous evolutionary change, at a rate far higher than is usually observed in the fossil record.

This progress was not constant, however. Some of the experimental populations showed periods of little change, followed by episodes of rapid increase. Such irregularities in response are partly due to the origin and fixation of new mutations with rather large effects

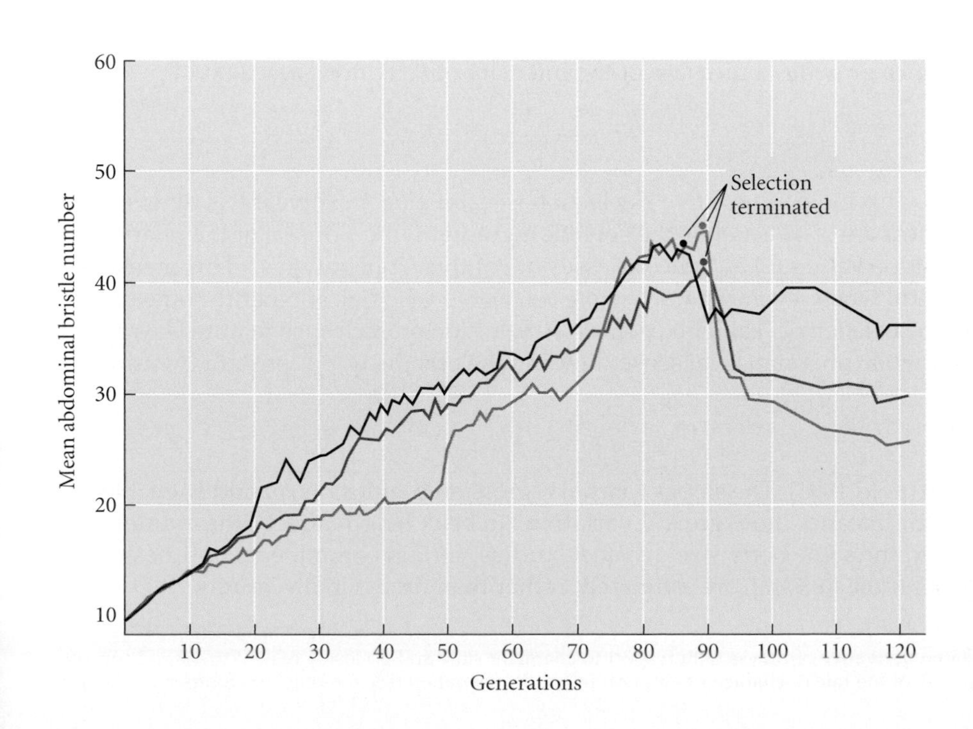

Figure 13.10 Responses to artificial selection for increased numbers of abdominal bristles in three laboratory populations of *Drosophila melanogaster*. After about 86 generations, the means had increased greatly. Selection was terminated at the points indicated by colored circles and bristle number declined thereafter, indicating that genotypes with fewer bristles had higher fitness. (After Yoo 1980.)

(Mackay et al. 1994). Several populations eventually stopped responding: they reached a **selection plateau**. This cessation of response to selection was not caused by loss of genetic variation, because when Yoo terminated ("relaxed") selection after 86 generations, mean bristle number declined, proving that genetic variation was still present.

A selection plateau and a decline when selection is relaxed are commonly observed in selection experiments. These patterns are caused by natural selection, which opposes artificial selection: genotypes with extreme values of the selected trait have low fitness. The changes in fitness are due to both hitchhiking of deleterious alleles (linkage disequilibrium) and pleiotropy. Yoo found, for example, that lethal alleles had increased in frequency in the selected populations because they were closely linked to alleles that increased bristle number. Other investigators have shown that some alleles affecting bristle number have pleiotropic effects that reduce viability (Kearsey and Barnes 1970; Mackay et al. 1992).

Several investigators have found that artificially selected traits change faster in large than in small experimental populations (e.g., Weber and Diggins 1990; López and López-Fanjul 1993). This is because more genetic variation is introduced by mutation in large than in small populations, large populations lose variation by genetic drift more slowly, and selection is more efficient in large populations. (Recall from Chapter 12 that whether allele frequency change is affected more by selection or by genetic drift depends on the relationship between the coefficient of selection and the population size.)

Selection in Natural Populations

Measuring natural selection on quantitative characters

In studies of natural populations, several measures of the strength of natural selection on quantitative traits have been used. The simplest indices of selection can be used if the mean ($\bar{z}$) and variance (V) of a trait are measured within a single generation before ($\bar{z}_b$, V_b) and then again after ($\bar{z}_a$, V_a) selection has occurred. (For instance, these measurements may be made on juveniles and then on those individuals that successfully survive to adulthood and reproduce.) Then, if selection is directional, an index of the **intensity of selection** is

$$i = \frac{\bar{z}_a - \bar{z}_b}{\sqrt{V_P}}$$

where V_P is the phenotypic variance.

If selection is stabilizing or diversifying, the change within a generation in the phenotypic variance provides a measure of the intensity of SELECTION ON VARIANCE:

$$j = \frac{V_a - V_b}{V_b}$$

(Endler 1986). This index is negative if selection is stabilizing, positive if it is diversifying.

Another measure of selection is the **selection gradient**: the slope, b, of the relation between phenotype values (z) and the fitnesses (w) of those phenotypes.* This measure is especially useful when several characters are correlated with one another to some degree, such as, say, beak length (z_1) and body size (z_2). Selection on each such feature (say, z_1) can be estimated while (in a statistical sense) holding the other (say, z_2) constant by using the equation

$$w = a + b_1 z_1 + b_2 z_2$$

(Lande and Arnold 1983). The slopes (partial regression coefficients) b_1 and b_2 enable one to estimate, for instance, how greatly variations in beak length affect fitness among individuals with the same body size. In most studies, certain components of fitness, such as juvenile-to-adult survival, are estimated, rather than fitness in its entirety.

*The slope is the derivative of fitness with respect to character state and is closely related to $d\bar{w}/d\bar{z}$, one of the determinants of the rate of character evolution in Lande's equation (see the previous footnote, p. 306).

Examples of selection on quantitative characters

DARWIN'S FINCHES. Peter and Rosemary Grant (1986, 1989) and their colleagues have carried out long-term studies on some of the species of Darwin's finches on certain of the Galápagos Islands. They have shown that birds with larger (especially deeper) bills feed more efficiently on large, hard seeds, whereas there is some evidence that small, soft seeds are more efficiently utilized by birds with smaller bills. When the islands suffered a severe drought in 1977, seeds, especially small ones, became sparse, medium ground finches (*Geospiza fortis*) did not reproduce, and their population size declined greatly due to mortality. Compared with the pre-drought population, the survivors were larger and had larger bills (Figure 13.11). From the differences in morphology between the survivors ($\bar{z}_a$) and the pre-drought population ($\bar{z}_b$), the intensity of selection i and the selection gradient b were calculated for three characters:

Character	*i*	*b*
Weight	0.28	0.23
Bill length	0.21	–0.17
Bill depth	0.30	0.43

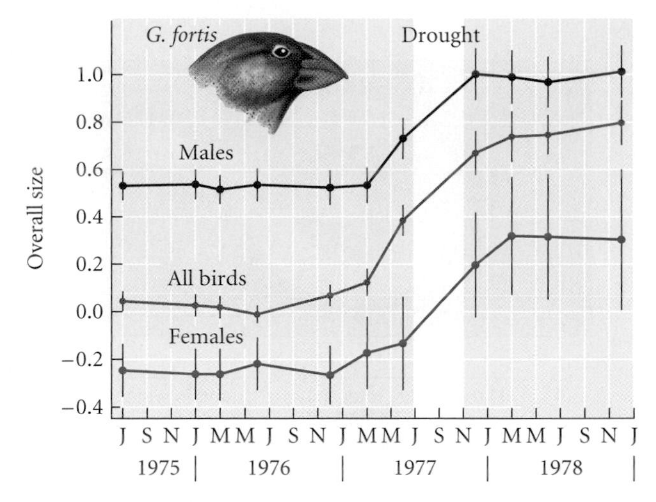

Figure 13.11 Changes in the mean size of the ground finch *Geospiza fortis* on Daphne Island in the Galápagos archipelago due to mortality during a drought in 1977. Changes occurring in 1977 and 1978 are all the result of mortality; no reproduction occurred during this period. "Overall size" is a composite of measurements of several characters. (After Grant 1986.)

The values of i show that each character increased by about 0.2 to 0.3 standard deviations, a very considerable change to have occurred in one generation. The values of b show the strength of the relationship between survival and each character while holding the other characters constant. Selection strongly favored birds that were larger and had deeper bills because they could more effectively feed on large, hard seeds, virtually the only available food. The negative b values show that selection favored shorter bills. Nevertheless, bill length increased, in opposition to the direction of selection, because bill length is correlated with bill depth. Thus a feature can evolve in a direction opposite to the direction of selection if it is strongly correlated with another trait that is more strongly selected. (We will soon return to this theme.)

Why don't these finches evolve ever larger bills? The Grants found that during normal years, birds with smaller bills survive better in their first year of life, probably because they feed more efficiently on abundant small seeds; in addition, small females tend to breed earlier in life than large ones. Thus conflicting selection pressures create stabilizing selection that, on average, favors an intermediate bill size.

EVIDENCE OF STABILIZING SELECTION. Many traits are subject to stabilizing selection, so the mean changes little, if at all. For example, human infants have lower rates of mortality if they are near the population mean for birth weight than if they are lighter or heavier (Figure 13.12; Karn and Penrose 1951).

Figure 13.12 Stabilizing selection for birth weight in humans. The rate of infant mortality is shown by the points and the line fitting them. The histogram shows the distribution of birth weights in the population. (After Cavalli-Sforza and Bodmer 1971.)

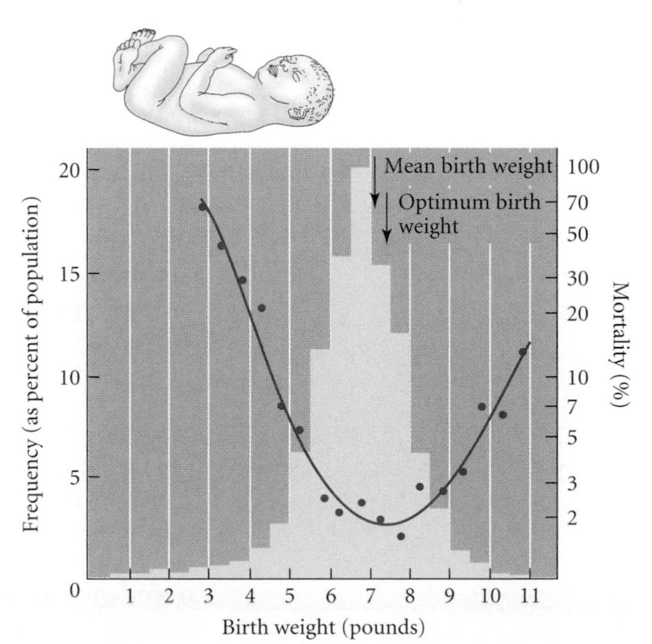

Stabilizing selection often occurs because of **trade-offs**, antagonistic agents of selection (Travis 1989). Arthur Weis and colleagues (1992) found that different natural enemies impose conflicting selection on the size of galls induced by the goldenrod gall fly (*Eurosta solidaginis*). The larva of this fly induces a globular growth (gall) on the stem of its host plant, goldenrod. Much of the variation in gall size is due to genetic variation in the fly, which feeds and pupates within the gall, emerging in spring. During the summer, parasitoid wasps, *Eurytoma gigantea* and *E. obtusiventris*, lay eggs on the fly larvae by inserting their ovipositors

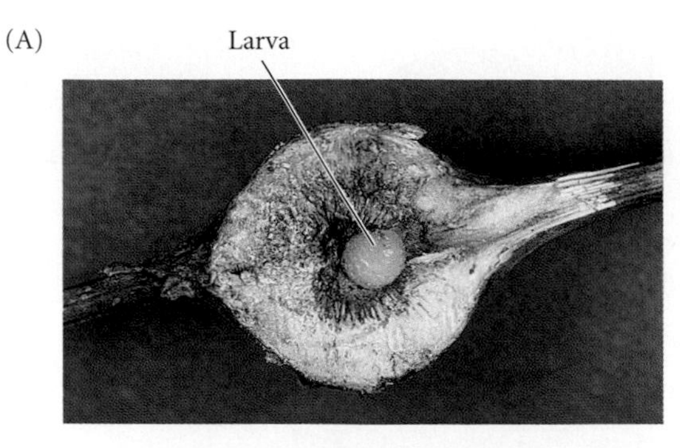
(A)
Larva

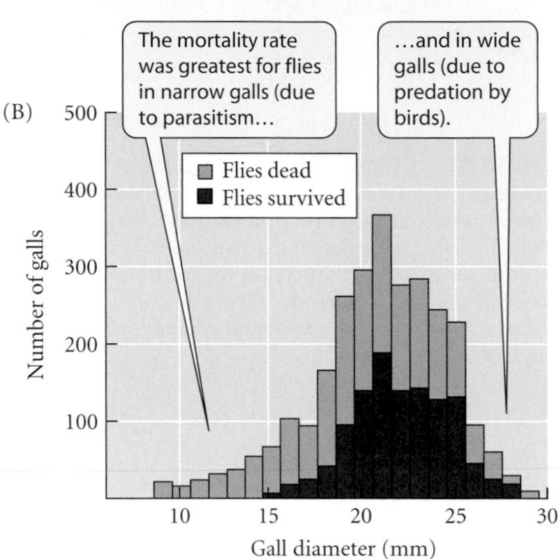
(B)
The mortality rate was greatest for flies in narrow galls (due to parasitism…
…and in wide galls (due to predation by birds).
Flies dead
Flies survived
Number of galls
500
400
300
200
100
10
15
20
25
30
Gall diameter (mm)

Figure 13.13 (A) Larva of the gall fly (*Eurosta solidaginis*), shown inside the dissected stem gall of a goldenrod plant. (B) Stabilizing selection on the size of galls made by *E. solidaginis*. The height of each bar shows the proportion of each plant gall size in the population, and the blue portion shows the proportion of fly larvae that survived. Flies in intermediate-sized galls had the highest survivorship. (A © Scott Camazine/Photo Researchers, Inc.; B after Weis et al. 1992.)

through the gall wall. During winter, woodpeckers and chickadees open galls and feed on the fly pupae. Mortality caused by parasitoids and birds can be determined by examining galls in the spring.

The researchers found that the parasitoid *E. gigantea* very consistently selected for wide galls (i.e., mortality was greatest for fly larvae in narrow galls) because the wasp's ovipositor cannot penetrate thick gall tissue. *Eurytoma obtusiventris* generally selected for intermediate-sized galls, whereas birds most frequently attacked wide galls, selecting for narrower gall diameter. Taken together, these enemies imposed rather strong stabilizing selection ($j = -0.30$), but because selection by parasitoids was weaker than selection by birds, a directional component ($i = 0.34$) was detected as well (Figure 13.13).

VESTIGIAL FEATURES IN CAVE ORGANISMS. A long-standing puzzle is why, over the course of evolution, features that have no function become vestigial and are ultimately lost. For instance, snakes have vestigial legs or none at all; many flowers have only female or male function and possess vestigial stamens or pistil.

The Lamarckian explanation, whereby organs are maintained or lost as a result of use or disuse, has long since been rejected. Neo-Darwinian theory offers two possible explanations: either mutations that cause degeneration of an unused character become fixed by genetic drift because variations in the character are selectively neutral, or there is selection against an unused organ, perhaps because it interferes with some important function or requires energy and materials that could better be used for other purposes. In addition, selection could indirectly reduce an unused organ if, due to pleiotropy, it were negatively correlated with another feature that increased due to selection (Fong et al. 1995).

Cave-dwelling populations of the amphipod crustacean *Gammarus minus*, like many animals in caves, have eyes that are highly reduced compared with those of their surface-living relatives. The heritability of eye size is high. Male amphipods remain mounted on their mates for a week or two after mating, guarding them against other males. Jones and Culver (1989) compared the eye sizes of paired amphipods with those of unpaired amphipods, which were assumed to have lower reproductive success on average. For 2 years in succession, mating individuals of both sexes had smaller eyes, on average, than unmated individuals. The investigators estimated that the selection gradient (b) was about -0.30, indicating that selection for small eyes was quite strong. Why amphipods with reduced eyes should have higher mating success is not clear, but the authors speculated, based on neurobiological studies by other researchers, that reduction of the unused visual system might free more of the central nervous system to process nonvisual sensory input.

THE STRENGTH OF NATURAL SELECTION. The strength of selection has been estimated in many studies of quantitative traits in natural populations (Kingsolver et al. 2001). The strength of selection is commonly quite modest, although strong selection (b greater than, say, 0.25) has often been recorded (Figure 13.14A). Stabilizing selection and diversifying selection (i.e., selection on variance) appear to be about equally common (Figure 13.14B). There is a tendency for the strength of selection due to variation in mating success and female fecundity to be greater, on average, than that of selection due to differences in survival.

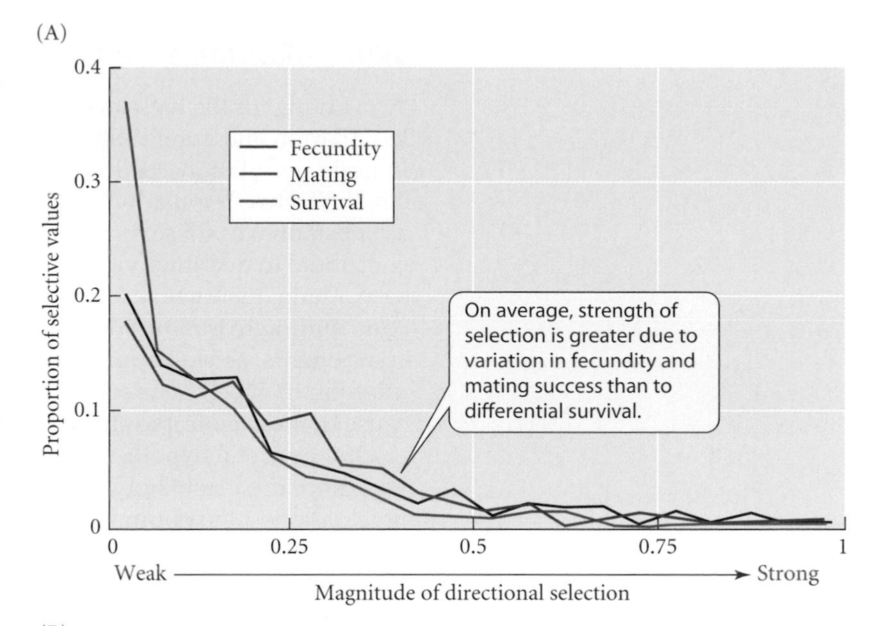

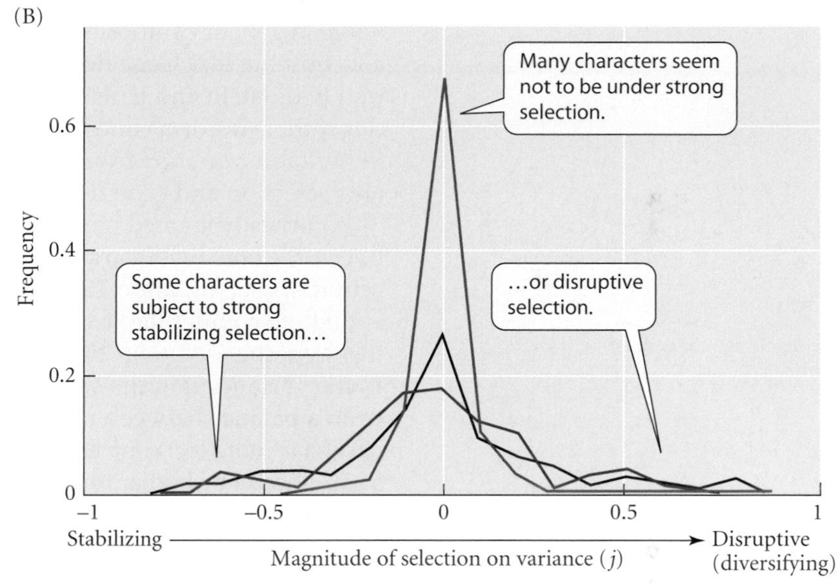

Figure 13.14 The frequency distribution of the strength of selection on traits in natural populations, as compiled from data from numerous studies. (A) The three curves show the strength of directional selection based on variation in fecundity, mating success, and survival. (B) The frequency distribution of selection on variance—either stabilizing selection (negative selection values) or diversifying selection (positive selection values). Stabilizing and diversifying selection appear to be about equally common. (After Kingsolver et al. 2001.)

A Neutral Model of the Evolution of Quantitative Characters

If alleles that contribute to variation in a polygenic trait are selectively neutral, variation and evolution of the trait are affected only by mutation (which increases variation) and genetic drift (which erodes it). The variance that arises per generation by mutation, V_m, is proportional to the number of mutating loci, the mutation rate per locus, and the average phenotypic effect of a mutation. As we have seen, V_m is often about 0.001 times V_E (Lynch 1988).

At equilibrium, when mutation is balanced by genetic drift, the genetic variance and heritability should, theoretically, reach a stable value, which should be quite high if the effective population size is large. Although the genetic variance stays fairly constant, the mean does not. As mutations that affect the character arise and are fixed by genetic drift, the mean will fluctuate at random. If a number of isolated populations are derived from an initially uniform ancestral population, mutation and genetic drift can cause genetic divergence among them in a polygenic character, just as they do at a single locus (see Chapter 10). The rate of divergence among the populations, measured by the variance among their character means, depends only on the mutation rate, not on the population size. (Recall that this is also true of the rate at which populations diverge by substitution of neutral mutations at a single locus; see Chapter 10.) Such constancy is expected only if selection does not affect the character.

Evolutionary geneticists have compared rates of morphological evolution over millions of years with the rates expected if evolution transpired by genetic drift. Michael Lynch (1990) studied the rates of evolution of skeletal features of several groups of mammals, based on fossils (e.g., horse and hominid lineages) as well as on differences between living species (such as lion and cheetah) for which there are good estimates of the time since the common ancestor. Lynch assumed that the mutation rate (V_m) for such features is about the same as has been estimated for skeletal features of mice and other living species of mammals. He found that almost all the features had evolved at much *lower* rates than expected under mutation and genetic drift. Only the cranial capacity of *Homo sapiens* has evolved at rates that may be higher than expected from the neutral model. The very low rate at which most characters seem to have evolved suggests that stabilizing selection has maintained them at roughly constant values for long periods. Furthermore, fossils show that many features fluctuate rather rapidly, but show little net change (see Figure 4.21), so that the rate of evolution over the long term is much less than it is over shorter intervals of time.

What Maintains Genetic Variation in Quantitative Characters?

Accounting for the high levels of genetic variation in quantitative traits is a difficult problem (Barton and Turelli 1989). Although the neutral theory predicts high genetic variance in large populations, estimates of h^2_N for characters of species that are thought to have small effective population sizes (e.g., Darwin's finches, salamanders, mice) are often just as substantial as those for more populous species (Houle 1989). Furthermore, alleles that contribute to quantitative variation are probably seldom selectively neutral. As we have seen, many quantitative characters are subject to fairly intense selection. Moreover, genes contributing to quantitative traits have pleiotropic effects on survival and other fitness components, as we know from studies of *Drosophila* bristles. Thus even characters that might in themselves be selectively "trivial" are probably subject to indirect selection because of the pleiotropic effects of the underlying genes (Dobzhansky 1956).

Of the several hypotheses that have been advanced to account for quantitative genetic variation, the most likely may be VARIABLE SELECTION and MUTATION-SELECTION BALANCE. Fluctuation in the optimal phenotype from one generation to another can delay the loss of genetic variation, although it does not prevent it indefinitely. Gene flow between populations with different optimal phenotypes can maintain variation in each population. Moreover, populations in which stabilizing selection favors the same phenotype can diverge in genetic composition as mutation and genetic drift create turnover in alleles at the contributing loci. Gene flow among such populations can help to maintain genetic variation (Goldstein and Holsinger 1992). However, laboratory populations, maintained under rather uniform conditions and isolated from gene flow, do not differ substantially in heritable variation from natural populations, casting doubt on the importance of variable selection and gene flow (Bürger et al. 1989).

A currently favored hypothesis is that levels of polygenic variation reflect a balance between the erosion of variation by stabilizing selection and the input of new variation by mutation (V_m) (Lande 1976b; Houle et al.1996). There is some doubt that V_m is high enough to counter the strong stabilizing selection that acts on many traits, a point against this hypothesis (Turelli 1984). Moreover, some of the alleles that contribute substantially to the variance in traits such as bristle number have higher frequencies than predicted from a balance between mutation and purifying selection (Lai et al. 1994). On the other hand, far more loci contribute to a fitness-related trait than to a single morphological trait, so V_m should be higher for fitness-related traits, and should maintain higher genetic variance (V_A). In fact, V_A is considerably greater for fitness-related traits than for morphological traits, as predicted by the mutation–selection balance hypothesis (Houle et al. 1996).

Correlated Evolution of Quantitative Traits

Evolutionary change in one character is often correlated with change in other features. For example, species of animals that differ in body size differ predictably in many individual features, such as the length of their legs or intestines. Correlated evolution can have two causes: correlated selection and genetic correlation.

Correlated selection

In **correlated selection**, there is independent genetic variation in two or more characters, but selection favors some combination of character states over others, usually because the characters are functionally related.

Edmund Brodie (1992) found evidence of correlated selection on color pattern and escape behavior in the garter snake *Thamnophis ordinoides* (Figure 13.15). This snake can have a uniform color, spots, or lengthwise stripes. Brodie found that when chased down an experimental runway, some snakes fled in a straight line, while others repeatedly reversed their course. Resemblance among siblings indicated that both coloration and escape behavior are highly heritable. Other investigators had noted that among different species of snakes, spotted species have irregular flight patterns or tend to be sedentary, relying on their cryptic patterns to avoid predation. Striped species tend to flee rapidly in a single direction, probably because visually hunting predators find it difficult to judge the speed and position of a moving stripe.

Brodie scored color pattern and propensity to reverse course in 646 newborn snakes born to 126 pregnant females. He marked each snake so it could be individually recognized, released the snakes in a suitable habitat, and periodically sought them thereafter. Brodie had reason to believe that most of the snakes that were not recaptured were eaten by crows and other predators. He found that the survival rate was greatest for those that had both strong striping and a low reversal propensity, and for those that had both a non-striped pattern and a high reversal propensity (Figure 13.15). Other phenotypes, such as striped snakes that reversed course when chased, had lower survival rates. Thus there was correlated selection on color pattern and escape behavior in the direction that had been predicted from comparisons among species of snakes and from the theory of visual perception.

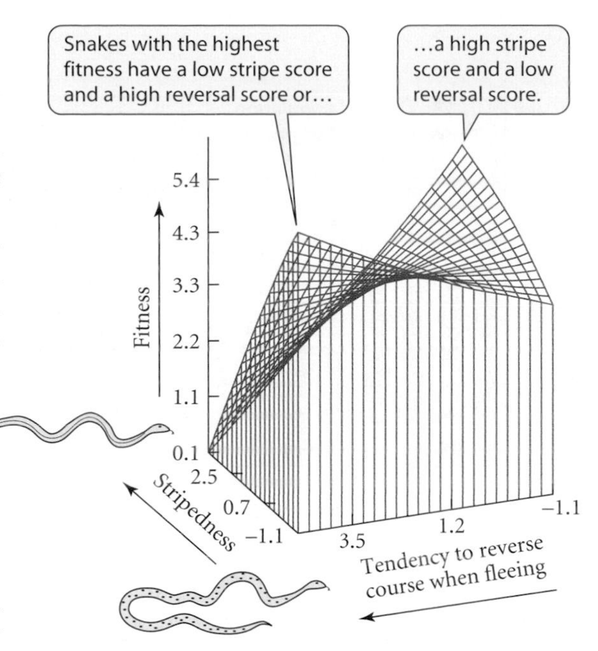

Figure 13.15 A fitness surface for combinations of two traits in the garter snake *Thamnophis ordinoides*, based on survival in the field. The height of a point on the surface represents the fitness (relative survival) of individuals with particular values of stripedness and reversal (tendency to reverse course when fleeing). (After Brodie 1992.)

Genetic correlation

We often observe that characters are correlated within a species: tall people, for example, tend to have long arms. The magnitude of a correlation between two characters—the degree to which they vary in concert—is expressed by the CORRELATION COEFFICIENT (r), which ranges from +1.0 (for a perfect correlation in which both features increase or decrease together) to –1.0 (when one feature decreases exactly in proportion to the other's increase). For uncorrelated characters, $r = 0$.

The **phenotypic correlation**, r_P, between, say, body size and fecundity is simply what we measure in a random sample from a population. Just as the phenotypic variance may have both genetic and environmental components, so too may the phenotypic correlation. Two features of individuals with the same genotype may vary together because both are affected by environmental factors, such as nutrition. Such features display an **environmental correlation**, r_E. In genetically variable populations, the correlated variation may also be caused by genetic differences that affect both characters, causing a **genetic correlation**, r_G.

Genetic correlations can have two causes. One cause is linkage disequilibrium among the genes that independently affect each character. The other cause is pleiotropy—the influence of the same genes on different characters. A genetic correlation due to pleiotropy will be perfect ($r_G = 1.0$ or –1.0) if the alleles that increase one character all increase (or all decrease) the other character. If some genes affect only one of the characters, or if some alter both characters in the same direction (+,+ or –,–) while others have opposite effects on the two characters (+,– or –,+), the genetic correlation will be imperfect.

Both linkage disequilibrium and pleiotropy can change over time, so genetic correlations need not be constant, but may evolve (Turelli 1988). A genetic correlation caused by linkage disequilibrium, such as the correlation between pistil length and stamen height in the primrose *Primula* (see Figure 9.16), will decline due to recombination unless selection for the adaptive gene combinations maintains it. Correlations caused by pleiotropy may also change, although more slowly than those caused by linkage disequilibrium. Some alleles may become fixed, so that the loci contribute no variation and therefore no correlation, while other loci, perhaps affecting only one character, remain variable.

Another cause of change in genetic correlation is natural selection, which may favor MODIFIER ALLELES that alter the pleiotropic effects of other loci. For example, natural populations of Australian blowflies (*Lucilia cuprina*) exposed to the insecticide diazinon rapidly evolved resistance due to an increase in the frequency of a resistance allele, *R*. At first, resistant flies had reduced viability (when tested in the absence of diazinon) and a high incidence of bilateral asymmetry (which is thought to reflect disruptions of development). These features were pleiotropic effects of the *R* allele. Over the course of several years, however, viability increased and the incidence of asymmetry decreased. These changes were not due to change at the *R* locus, because the *R* allele still evinced pleiotropic effects when it was backcrossed from the resistant field population into non-

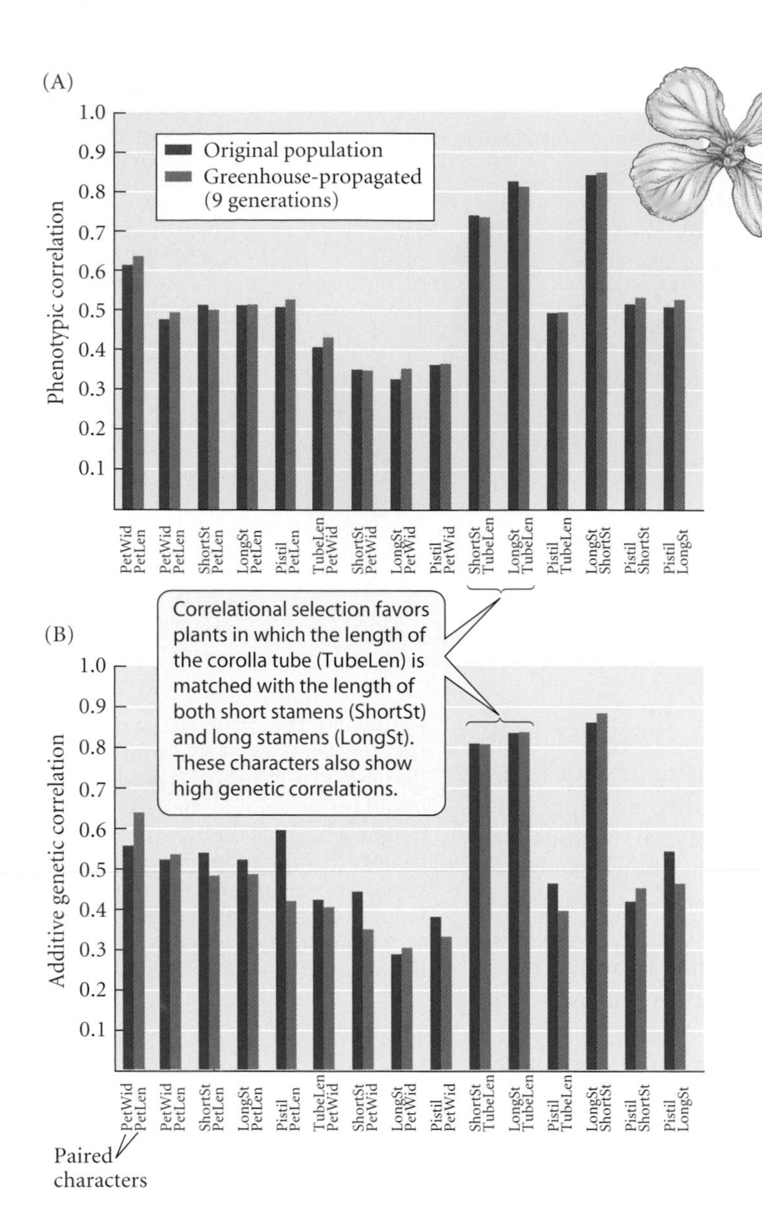

Figure 13.16 Phenotypic (A) and genetic (B) correlations among six measurements of flower parts in the wild radish. Correlations caused by linkage disequilibrium would be expected to decline over time, but the correlations after nine generations of random mating are similar to those in the original population, suggesting that they are a result of pleiotropy. The characters are petal length and width (PetLen, PetWid), corolla tube length (TubeLen), length of short and long stamens (LongSt, ShortSt), and pistil length (Pistil). The flower has four long and two short stamens. (After Conner 2002.)

resistant laboratory stocks. Rather, natural selection had increased the frequency of modifier alleles at other loci, which altered and mitigated the deleterious effects of the *R* allele (McKenzie and Clarke 1988).

Examples of genetic correlation

Genetic correlation between traits can be estimated from correlations among relatives in the same way that genetic variance can. For example, when Brodie (1993) scored escape behavior and color pattern in newborn garter snakes, he found that both traits were more similar among offspring of the same wild-caught mother than among offspring of different mothers, providing evidence of heritable variation. Moreover, in one population, there was a significant genetic correlation ($r_G = -0.17$) between color pattern and escape behavior: members of families with a striped pattern made few reversals when chased, and nonstriped individuals generally changed direction after a short distance. In another geographic population, these traits were not genetically correlated. This difference between populations is consistent with the hypothesis that linkage disequilibrium, rather than pleiotropy, is the cause of the genetic correlation. Notice that the genetic correlation between the traits parallels the correlation in their effects on fitness, mentioned earlier: the traits are COADAPTED, displaying an adaptive correlation.

Jeffrey Conner (2002) estimated phenotypic and genetic correlations among six flower characters of the wild radish (*Raphanus raphanistrum*), both among offspring of field-collected plants and in a sample of plants that were propagated in a greenhouse for nine generations. Stamen and corolla tube lengths, which are under correlational selection in wild populations, showed some of the strongest correlations (Figure 13.16). In this case, the genetic correlations are probably caused by pleiotropy, because they were not diminished by nine generations of recombination.

How genetic correlation affects evolution

Genetic correlations among characters can cause them to evolve in concert. They can also either enhance or retard the rate of adaptive evolution, depending on the circumstances. In extreme cases, they may severely constrain adaptation.

If two characters, z_1 and z_2, are genetically correlated, the rate and direction of evolution of z_1 depend on both direct selection (on z_1 itself) and selection on z_2. If selection on z_2 is much stronger than on z_1, z_1 may change mostly due to its correlation with z_2, and the change may not be in an adaptive direction. For example, Arnold (1981) estimated a genetic correlation of 0.89 between the feeding reaction of newborn garter snakes to slugs (z_1) and to leeches (z_2) (Figure 13.17). If both slugs and leeches are pres-

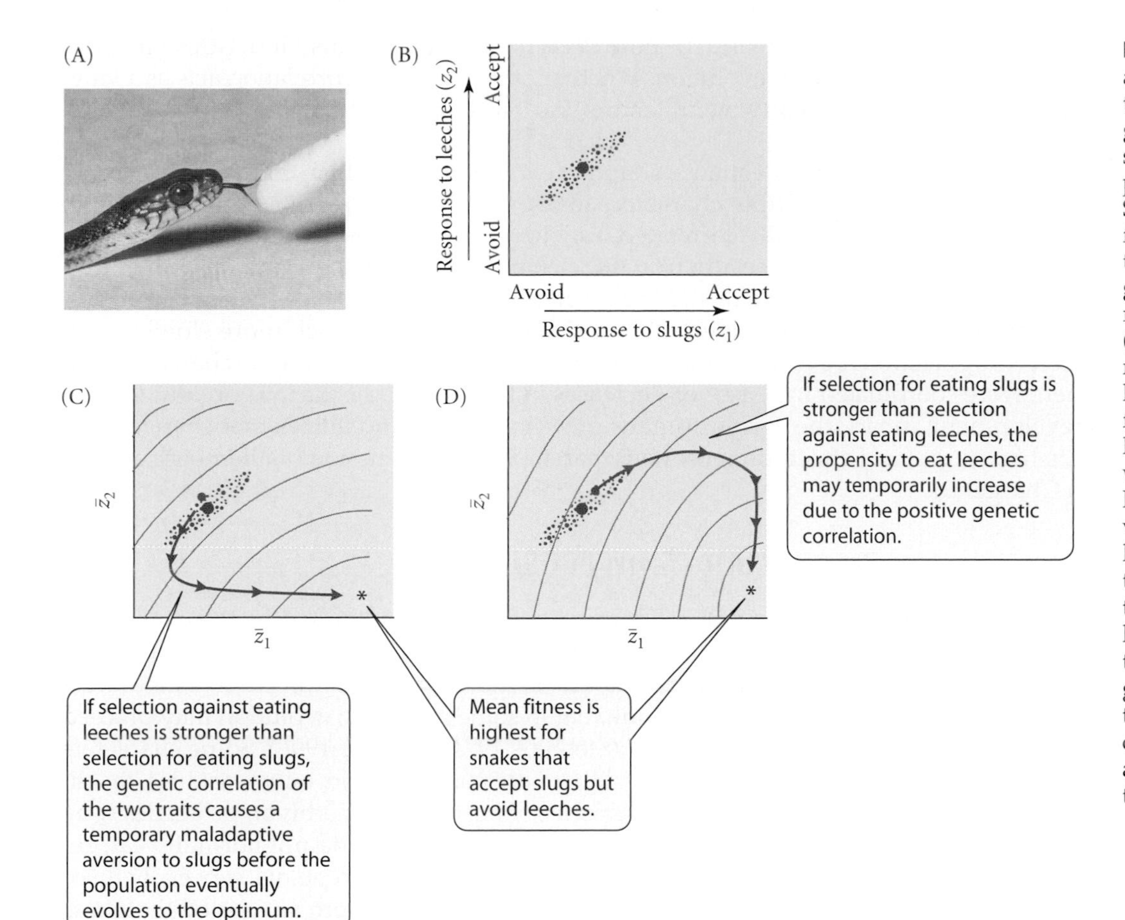
(A)
(B)
Response to leeches (z_2)
Accept
Avoid
Avoid
Accept
Response to slugs (z_1)
(C)
$\bar{z}_2$
$\bar{z}_1$
(D)
$\bar{z}_2$
$\bar{z}_1$
If selection for eating slugs is stronger than selection against eating leeches, the propensity to eat leeches may temporarily increase due to the positive genetic correlation.
If selection against eating leeches is stronger than selection for eating slugs, the genetic correlation of the two traits causes a temporary maladaptive aversion to slugs before the population eventually evolves to the optimum.
Mean fitness is highest for snakes that accept slugs but avoid leeches.

Figure 13.17 Possible evolutionary implications of a genetic correlation between feeding responses in garter snakes. (A) A newborn garter snake investigating the odor of a potential prey item on a cotton swab. Snakes show a higher response to favored prey types by flicking the tongue more often. (B) The positive genetic correlation (r_G) between response to slugs (z_1) and to leeches (z_2). The "cloud" of small points represents individual genotypes; the large point is the population mean for both traits. (C, D) Possible fitness landscapes for a snake population in which slugs and leeches both occur. Mean fitness, shown by contours for various mean responses to slugs and leeches, is highest (*) for populations that accept slugs but avoid leeches. If the traits were not genetically correlated, they would evolve directly to the optimum (*). Because of the genetic correlation, the joint evolution of these two traits takes a curved path, and may involve maladaptation in one trait. (Photo courtesy of S. J. Arnold.)

ent in the environment, there may be direct selection in favor of slug-feeding (increased z_1), but stronger selection against leech-feeding (decreased z_2), since a leech can kill a snake by biting its digestive tract after being swallowed. If adaptive aversion to leeches evolves (decreased z_2), maladaptive aversion to slugs (decreased z_1) may also evolve, at least temporarily, as a correlated effect (Figure 13.17C). (We assume that other kinds of food are available, so that the population can persist even if it evolves aversion to both leeches and slugs.) Conversely, if leeches are rare and slugs are the most abundant food, selection for feeding on slugs may be stronger, the population may evolve a slug-feeding habit, and it may also evolve the maladaptive habit of occasionally eating leeches (Figure 13.17D). After the more strongly selected trait approaches its optimum (e.g., z_2, strong aversion to leeches, in Figure 13.17C), the weakly selected trait can evolve to its optimum (z_1, positive feeding response to slugs). This change is based mostly on those genes that affect only z_1.

A conflict may therefore exist between the genetic correlation of characters and directional selection on those characters. When such a conflict exists, the two characters may evolve to their optimal states only slowly, and may even evolve temporarily in a maladaptive direction. (We have already seen that selection for a deeper bill in the Galápagos finch *Geospiza fortis* caused average bill length to increase, even though selection favored a shorter bill.) In some cases, a genetic correlation may be so strong that one or both traits cannot reach their optimal states. For example, there is a necessary trade-off between the number and the size of eggs (or seeds) that an organism can produce because the resources that it can allocate to reproduction are limited. This trade-off creates a negative genetic correlation, with some genotypes producing more but smaller eggs and others producing fewer but larger ones (see Chapter 17). Although selection might favor both more and larger eggs, increasing both is possible to only a very limited extent.

Thus genetic correlations, owing in some cases to trade-offs of this kind, can sometimes act as genetic constraints on evolution. Whether or not a genetic correlation acts as a long-term constraint depends on several factors, such as how readily the genetic correlation changes.

Genetic correlations may enhance adaptive evolution, rather than constrain it, in some cases (Wagner 1988). Multiple characters may evolve as an integrated ensemble more rapidly if they are genetically correlated, as when they are subject to the same developmental controls. This is especially true if the characters are functionally related. For instance, the size of each organ (e.g., lungs, gut, bones) must match the overall body size if an animal is to function properly. Body size would evolve much more slowly in response to selection if every organ had to undergo independent genetic change than if there were coordinated increases or decreases in the sizes of the various organs. During development, in fact, the various organs grow in concert, and alleles that change body size have correlated effects on most body parts. (See the discussion of allometric growth in Chapter 3).

Can Genetics Predict Long-Term Evolution?

If the response to selection in natural populations were never limited by the availability of genetic variation in single characters or combinations of characters, the rate and direction of adaptive evolution would depend only on the strength and direction of natural selection. There is reason, however, to think that in some instances, evolution may proceed along "genetic lines of least resistance" (Stebbins 1974; Schluter 1996). Some characters may have very little genetic variation, and therefore may constrain, or at least bias, the direction of evolution. For example, when species of host-specific herbivorous beetles were screened for their propensity to feed on novel plants that they do not normally eat, genetic variation was found in the feeding responses to only certain plants, especially those most closely related to the insects' normal host plants. This pattern suggested that these beetles might more readily adapt to feed on closely related plants than on distantly related plants—which is exactly what has occurred in the evolution of these beetles and many other groups of herbivorous insects (Futuyma et al. 1995).

If two characters are genetically correlated, the greatest genetic variation lies along the long axis of the ellipse formed by a plot of individual character values (Figure 13.18A). Dolph Schluter (2000) referred to this axis as the "maximal genetic variation" (g_{max}), and predicted that, at least in the short term, adaptive evolution should be greatest along this axis because of the constraining effect of genetic correlation between characters under directional selection. Over time, however, g_{max} should have less of an effect , because characters that are not perfectly genetically correlated can eventually evolve to their optimum values (as described earlier). For several adaptive morphological characters in stickleback fishes, sparrows, and other vertebrates, Schluter determined the direction of divergence of a species from a closely related species and compared it to g_{max}, the direction of evolution predicted by the genetic correlation between characters (Figure 13.18B). As predicted, the deviation between these two directions was least for species that had diverged only recently, and it increased with time since common ancestry. The rather close initial correspondence between the actual and genetically predicted directions suggests that patterns of genetic variation and correlation can persist for long periods of time and can influence the direction of evolution. Schluter estimated that this influence may last for up to 4 million years.

The pattern Schluter describes suggests that genetic correlations among characters may remain consistent for a long time. Whether or not this is generally true is one of the most important, most poorly understood aspects of phenotypic evolution. Some investigators have found that the genetic correlations among certain characters are fairly similar among geographic populations of a species or among closely related species, but other studies have found lower similarities, suggesting that the strength of genetic correlations can evolve rather rapidly (Steppan et al. 2002).

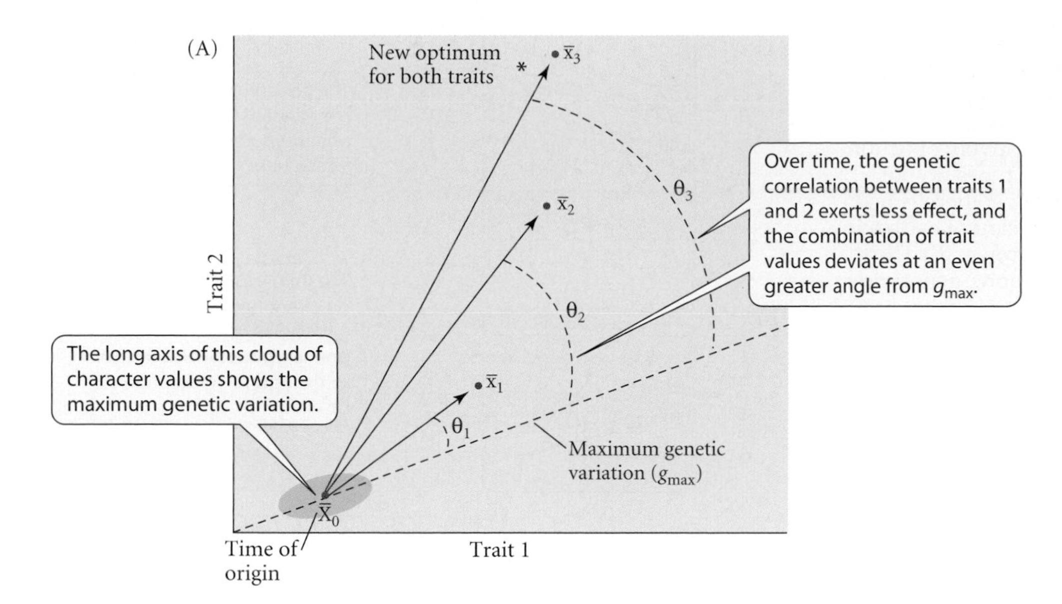

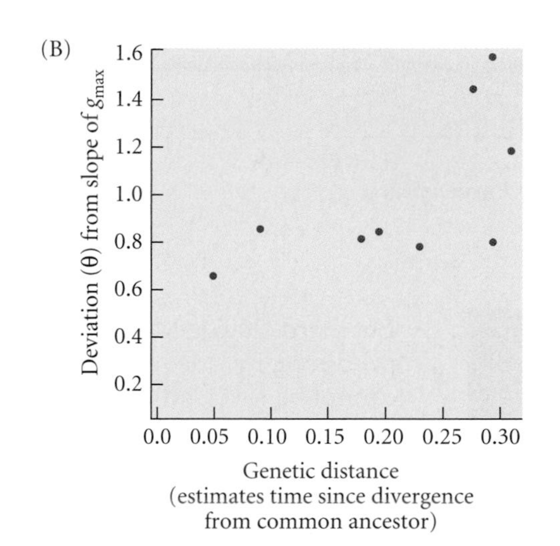

Figure 13.18 Evolution along genetic lines of least resistance. (A) The ellipse shows the distribution of values of the genetically correlated characters 1 and 2. The maximal genetic variation (g_{max}) is for combinations of trait values along the long axis of the ellipse. If the ancestral population mean is at point $\bar{x}_0$ and it experiences directional selection toward a new optimum (*), the genetic correlation will cause evolution toward $\bar{x}_1$ at first. The direction of evolution (arrow) deviates by angle θ_1 from g_{max}. Over time, the genetic correlation will exert a lesser effect, so that the line between $\bar{x}_0$ and $\bar{x}_2$ deviates more from g_{max} (angle θ_2). The deviation is still greater (θ_3) at a later time, when the population approaches the optimum for both traits. (B) Genetic correlations among several adaptive traits (e.g., bill dimensions) were estimated in song sparrows and used to determine g_{max}. Points show the deviation (angle θ in panel A) between g_{max} and the difference in these traits between the song sparrow and nine other species of sparrows, in relation to the molecular genetic distance between these species and the song sparrow. The low deviation of closely related species (those at low genetic distance) from g_{max} shows that g_{max} in the song sparrow predicts the initial direction of evolution fairly well. The theory of genetic correlation predicts that the characters of more distantly related species should display a larger deviation from g_{max}, as is observed in these data. (After Schluter 1996.)

Norms of Reaction

The correspondence between genotypic differences and phenotypic differences depends on developmental processes. In some cases, these processes may reduce the phenotypic expression of genetic differences; in other instances, a single genotype may produce radically different phenotypes in response to environmental stimuli. The **norm of reaction** of a genotype is the set of phenotypes it expresses in different environments (Figure 13.19; also see Figure 9.18). The norm of reaction can be visualized by plotting the genotype's phenotypic value in each of two or more environments.

When the effect of environmental differences on the phenotype differs from one genotype to another in a population, the reaction norms of the genotypes are not parallel, and the phenotypic variance includes a variance component (referred to as $V_{G \times E}$) due to **genotype × environment** (G × E) **interaction** (Figure 13.19B,C). If all the genotypes have parallel reaction norms (Figure 13.19A), there is no G × E interaction ($V_{G \times E} = 0$).

Phenotypic plasticity

In many species, adaptive **phenotypic plasticity** has evolved: a genotype has the capacity to produce different phenotypes, suitable for different environmental conditions (West-

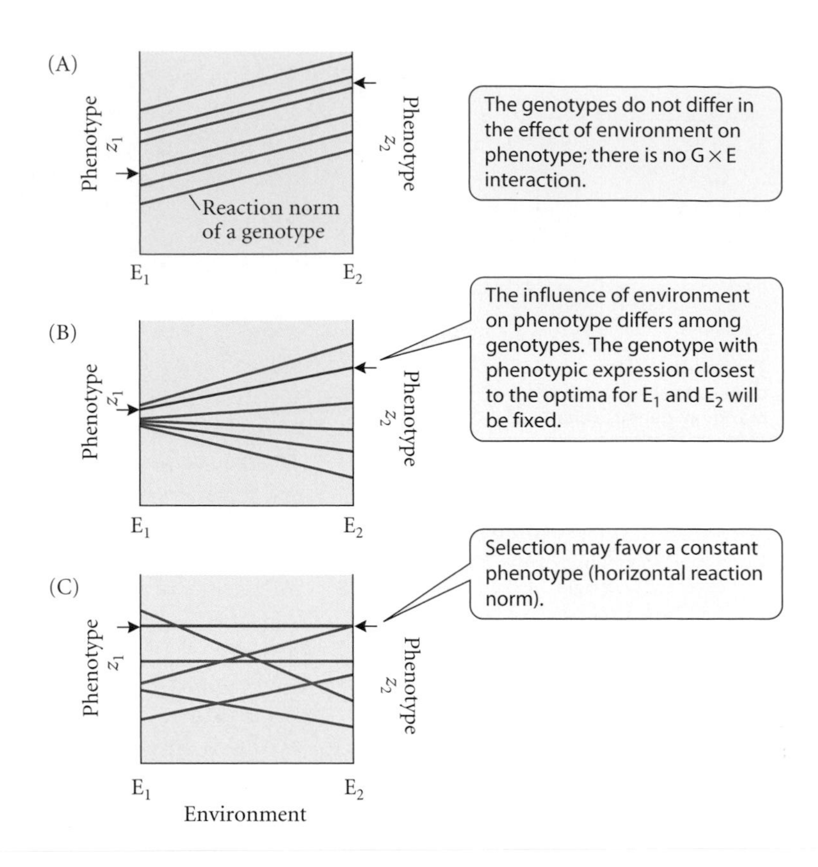
(A)
Phenotype z_1
Phenotype z_2
Reaction norm of a genotype
E_1
E_2
The genotypes do not differ in the effect of environment on phenotype; there is no G × E interaction.
(B)
Phenotype z_1
Phenotype z_2
E_1
E_2
The influence of environment on phenotype differs among genotypes. The genotype with phenotypic expression closest to the optima for E_1 and E_2 will be fixed.
(C)
Phenotype z_1
Phenotype z_2
E_1
E_2
Environment
Selection may favor a constant phenotype (horizontal reaction norm).

Figure 13.19 Genotype × environment interaction and the evolution of reaction norms. Each line represents the reaction norm of a genotype—its expression of a phenotypic character in environments E_1 and E_2. The character states expressed are labeled z_1 and z_2. The arrows indicate the adaptively optimum phenotype in each environment. (A) The genotypes do not differ in the effect of environment on phenotype; there is no G × E interaction. The optimal norm of reaction cannot evolve in this case, because no genotype matches the arrows. (B) The effect of environment on phenotype differs among genotypes; G × E interaction exists. The genotype with the norm of reaction closest to the optima in E_1 and E_2 (red line) will be fixed. New mutations that bring the phenotype closer to the optimum for each environment may be fixed thereafter. (C) Selection may favor a constant phenotype, irrespective of environment. A genotype with a horizontal reaction norm (red line) may be optimal.

Eberhard 2003). Many such differences, produced by "developmental switches," cannot be reversed during the organism's lifetime (Figure 13.20A). In some semiaquatic plants, for instance, the form of a leaf depends on whether it develops below or above water (Figure 13.20B). Adaptive phenotypic plasticity of this kind has evolved by natural selection for those genotypes with norms of reaction that most nearly yield the optimal phenotype for the various environments the organism commonly encounters (Schlichting and Pigliucci 1998).

(A)
Catkins
Caterpillars

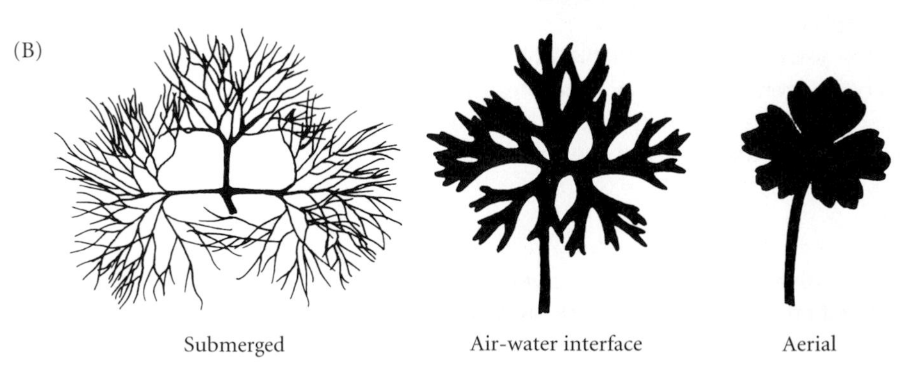
(B)
Submerged
Air-water interface
Aerial

Figure 13.20 Examples of phenotypic plasticity. (A) Larvae of the geometrid moth *Nemoria arizonaria* that hatch in the spring (left) resemble the oak flowers (catkins) on which they feed. Those that hatch in the summer (right) feed on oak leaves, and resemble twigs. (B) The form of a leaf of the water-crowfoot *Ranunculus aquatilis* depends on whether it is submerged, aerial, or situated at the air-water interface during development. (A, photos courtesy of Erick Greene; B from Cook 1968.)

Canalization

In other cases, the most adaptive norm of reaction may be a constant phenotype, buffered against alteration by the environment (Figure 13.19C). It may be advantageous, for example, for an animal to attain a fixed body size at maturity or metamorphosis, despite variations in nutrition or temperature that affect the rate of growth. The developmental system underlying the character may then evolve so that it resists environmental influences on the phenotype (Scharloo 1991). Conrad Waddington, one of the first biologists to integrate developmental biology and evolutionary biology, referred to this phenomenon as **canalization**.

Waddington (1953) used the concept of canalization to interpret some curious experimental results. A crossvein in the wing of *Drosophila* sometimes fails to develop if the fly is subjected to heat shock as a pupa. By selecting and propagating flies that developed a crossveinless condition in response to heat shock, Waddington bred a population in which most individuals were crossveinless when treated with heat. But after further selection, a considerable portion of the population was crossveinless even without heat shock, and the crossveinless condition was heritable. A character state that initially developed in response to the environment had become genetically determined, a phenomenon that Waddington called **genetic assimilation**.

Although this result is reminiscent of the discredited theory of inheritance of acquired characteristics, it has a simple genetic interpretation. Genotypes of flies differ in their susceptibility to the influence of the environment (heat shock)—that is, they differ in the degree of canalization, so that some are less easily deflected into an aberrant developmental pattern. Selection for this pattern favors alleles that canalize development into the newly favored pathway. As such alleles accumulate, less environmental stimulus is required to produce the new phenotype. The finding that genetic assimilation does not occur in inbred populations that lack genetic variation supports this interpretation (Scharloo 1991).

Evolution of variability

Whereas VARIATION refers to the differences actually present in a sample or a species, the word VARIABILITY, in the strict sense, refers to the ability, or potential, to vary (Wagner et al. 1997). For example, in insects, the number of compound eyes (two, or in a few species, none) seems able to vary much less than the number of units (ommatidia) that compose each compound eye. In mammals, because of the developmental correlation between the size of their bodies and the size of their brains and intestines, some conceivable variations—large bodies with tiny brains, for instance—are seldom or never seen. Developmental processes therefore affect variability, the extent to which genetic variation can be expressed as phenotypic variation. Does variability depend solely on immutable "laws" of development, or does it evolve by natural selection? This question applies to both variability in individual characters and correlations among characters.

The variability of individual characters is affected by the evolution of canalization. A character that is insensitive to alteration by environmental factors is ENVIRONMENTALLY CANALIZED. A character may also become GENETICALLY CANALIZED; that is, it may acquire low sensitivity to the effects of mutations. In such instances, the phenotype may remain unchanged even if the genes underlying its development vary considerably.

Threshold traits, for example, are expressed as discrete alternatives, but are controlled by polygenic variation rather than by single loci. The polygenic variation is not expressed phenotypically unless development is perturbed substantially (beyond a threshold) by a large enough genetic or environmental change. For example, in natural populations of *Drosophila melanogaster* and related species, there is almost no variation in the number of bristles (four) on the scutellum (part of the thorax). In homozygotes for the *scute* (*sc*) mutation, however, bristle number is variable, due to the expression of polygenic variation at other loci (Figure 13.21A). Thus this mutation breaks down canalization, and, conversely, the normal allele at the *scute* locus may be considered to exert genetic canalization. It has an *epistatic* effect on the quantitative loci, overseeing their phenotypic expression. Using an experimental population that was homozygous for *scute* and was therefore variable in bristle number, James Rendel and colleagues (1966) imposed selec-

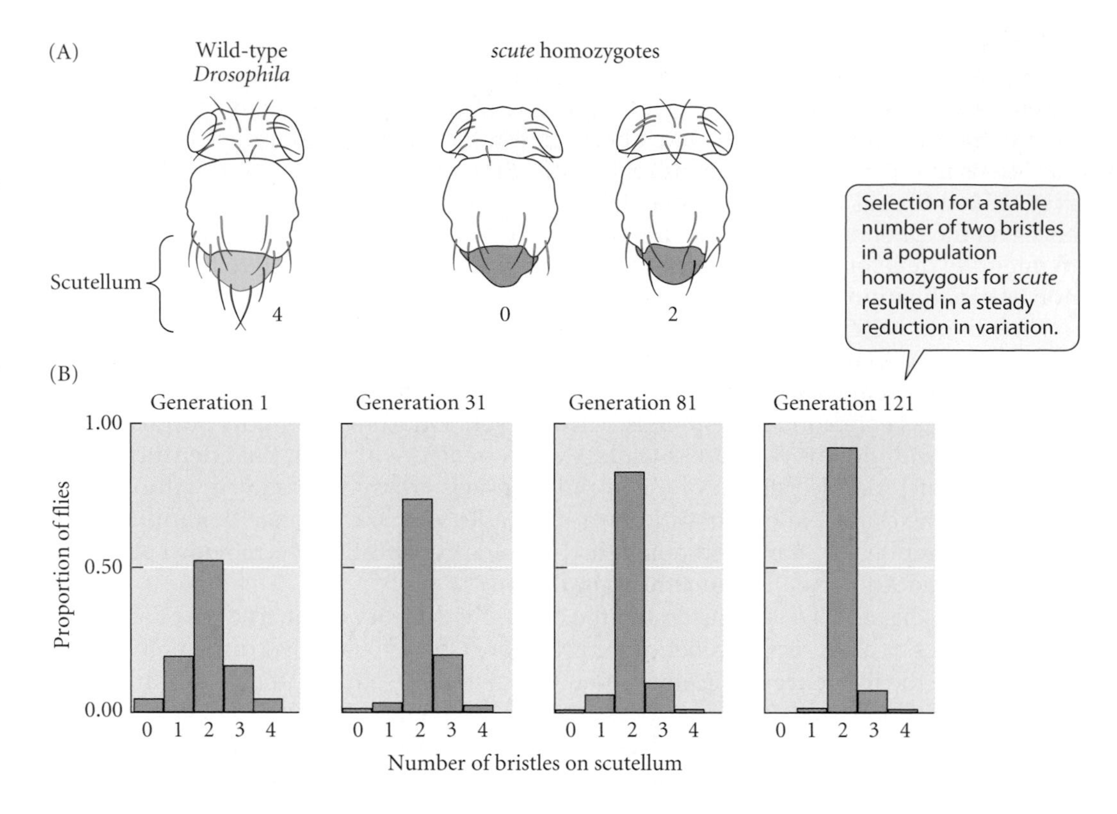

Figure 13.21 Canalization by artificial selection. (A) The top of the head and thorax of *Drosophila melanogaster*. The posterior part of the thorax, the scutellum, bears four bristles in wild-type flies, but a variable number (e.g., zero or two) in *scute* homozygotes. (B) Selection for a stable number of two bristles in a population homozygous for *scute* resulted in a steady reduction of variation, shown here at 1, 31, 81, and 121 generations of selection. (Based on data from Rendel et al. 1966.)

tion against variation by breeding from the least variable families (i.e., those that *most consistently* had two bristles). Within about 30 generations, the phenotypic variation became greatly reduced (Figure 13.21B). The investigators apparently had selected for genes that canalized development into a new pathway.

Can natural selection produce the same result? Wagner et al. (1997) and Kawecki (2000) have explored the evolution of canalization mathematically. According to their studies, alleles for environmental canalization should increase if there is prolonged stabilizing selection against deviations from an optimal phenotype. The evolution of genetic canalization, however, would be expected only under rather restricted conditions. If directional selection fluctuates rapidly in direction, canalization of the phenotype may be advantageous because it prevents a response to selection in one generation that is maladaptive a few generations later. Under consistent, long-term stabilizing selection, canalization evolves only if selection is neither too weak nor too strong. If stabilizing selection is weak, alleles that prevent the phenotypic expression of mutations have too slight a selective advantage to increase. If stabilizing selection is strong, it eliminates new mutations so fast that few individuals deviate from the optimum, so there is little selection for alleles that prevent the phenotypic expression of the mutations.

The theory of canalization may explain why some characters—such as some synapomorphies of higher taxa—have remained unchanged for vast periods of time. For example, the earliest known Devonian amphibians had a variable number of about eight or nine toes (see Chapter 4). Soon afterward, however, amphibians "settled on" five-digit (pentadactyl) limbs, and almost no tetrapod vertebrates since then have had more than five digits. Therefore, there was no inescapable rule that feet could have no more than five toes, but the developmental processes evolved so that the maximum digit number came to be constrained. We return to the problem of phylogenetically "conservative" features in Chapter 21.

The hypothesis of morphological integration (Olson and Miller 1958), or more generally, **phenotypic integration** (Pigliucci and Preston 2004), holds that functionally related characteristics should be genetically correlated with one another; thus the characters should remain coordinated even as they vary within species and as they evolve. Günter

Wagner and Lee Altenberg (1996) have shown theoretically that prolonged directional selection favors modifier alleles that enhance a pleiotropic correlation between functionally related traits along an axis pointing toward the optimum for the characters (marked with an asterisk in Figure 13.18A). For example, if it were functionally important for the upper and lower mandibles of a bird's bill to be the same length, then selection for a longer bill would include selection for alleles that coordinate the development of the two mandibles, creating a pleiotropic correlation between them.

Much of the evidence bearing on this hypothesis comes from studies of correlations among the various parts of flowers, since the match between flower structure and the size and position of a pollinating animal is thought to affect plants' reproductive success (Armbruster et al. 2000, 2004). There is more information from phenotypic than genetic correlations, but flower structure usually shows a strong correspondence between them. In some cases, the hypothesis of phenotypic integration appears to be upheld. For example, the sexual structures of the amazing Australian triggerplants (*Stylidium*) form a column that snaps forward to hit an insect on the back when the insect is properly positioned on the landing platform formed by the lower petals and touches a sensitive point on the column (Figure 13.22). The phenotypic correlation between the length of the column and of the landing platform is very high (0.81 in one population studied). In general, floral structures are more highly correlated with one another than with vegetative structures. On the whole, however, the evidence that genetic correlations reflect selection for proper function, rather than developmental pathways that may or may not be adapted to the species' special ecological situation, is equivocal (Armbruster et al. 1999; Herrera et al. 2002). Just how prevalent adaptive phenotypic integration is remains to be seen.

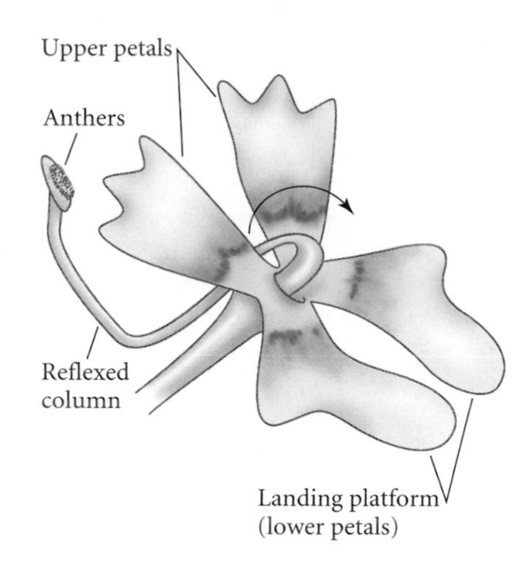

Figure 13.22 The flower of the triggerplant *Stylidium bicolor*. The column consists of fused stamens and pistil. When an insect lands on the platform formed by the lower petals and touches the base of the column, the column snaps forward. Thus the precision of pollen placement (and pickup of pollen from other flowers) depends on the dimensions of the column and the landing platform. The phenotypic correlation between these characters is quite high (0.81). (From Armbruster et al. 2004.)

Summary

1. Variation (variance) in a phenotypic trait (V_P) may include genetic variance (V_G) and variance due to the environment (V_E). Genetic variance may include both additive genetic variance (V_A) due to the additive effects of alleles and nonadditive genetic variance due to dominance and epistasis. Only the additive variance creates a correlation between parents and offspring (and can be measured by this correlation). Thus only V_A enables response to selection.
2. The ratio V_A/V_P is the heritability (h^2_N) of a trait. Heritability is not fixed, but depends on allele frequencies and on the amount of phenotypic variation induced by environmental variation. The short-term effect of selection ("response" to selection) on a character can be predicted from the heritability and the strength of selection.
3. Most, although not all, characters show substantial genetic variance in natural populations and can therefore evolve rapidly if selection pressures change. Many examples of rapid evolution, within a century or less, have been described.
4. Quantitative trait loci (QTL) can be mapped using molecular or other markers. The variation in many traits is due to variation at several or many loci, some with large and others with small effects. For certain characters, some of the genes have been identified and their function is known.
5. Linkage disequilibrium among loci can affect the variance of a character. Linkage disequilibrium is reduced by recombination.
6. Artificial selection experiments show that traits can often be made to evolve far beyond the initial range of variation. The response to selection is based on both genetic variation in the original population and new mutations that occur during the experiments.
7. Stabilizing selection is common in natural populations, either because the character is nearly at its optimum or because conflicting selection pressures or negative pleiotropic effects prevent further change. Diversifying selection also appears to be common.
8. The causes of high levels of genetic variation (high V_A and h^2_N) in natural populations are uncertain, but input by mutation may balance losses due to selection and genetic drift.
9. Linkage disequilibrium and especially pleiotropy cause genetic correlations among characters, which, together with correlations caused by environmental factors, give rise to phenotypic correlations. The evolution of a trait is governed both by selection on that trait and by selection on other traits with which it is genetically correlated. The effect of a genetic corre-

lation depends on its strength and degree of permanence. Genetic correlations can enhance the rate of adaptation (if functionally interdependent features show adaptive correlation), can cause a trait to evolve in a maladaptive direction (if selection on a correlated trait is strong enough), or may reduce the rate at which characters evolve toward their optimal states. It is not certain whether genetic correlations are especially strong among characters that are functionally integrated (the hypothesis of phenotypic integration).

10. The norm of reaction—the expression of the phenotype under different environmental conditions—can evolve if genotypes vary in the degree to which the phenotype is altered by the environment in which an individual develops. Some characters exhibit adaptive phenotypic plasticity, whereas selection in other cases favors constancy of phenotype despite differences in environment.
11. Canalization is the buffering of development against alteration by environmental or genetic variation. Canalized characters include threshold characters, in which underlying polygenic variation is not phenotypically expressed unless a drastic mutation or environmental perturbation breaks down canalization. Canalization can evolve under some circumstances. The evolution of canalization may explain the constancy of some characters over vast periods of evolutionary time.

Terms and Concepts

additive genetic variance
canalization
correlated selection
environmental correlation
environmental variance
genetic assimilation
genetic correlation
genetic variance
genotype × environment interaction
heritability
intensity of selection
norm of reaction
phenotypic correlation
phenotypic integration
phenotypic plasticity
phenotypic variance
QTL mapping
quantitative genetics
realized heritability
response to selection
selection differential
selection gradient
selection plateau
threshold trait
trade-off
truncation selection

Suggestions for Further Reading

Introduction to quantitative genetics, by D. S. Falconer and T. F. C. Mackay (fourth edition, Longman Group Ltd., Harlow, U.K., 1996) is a widely read, clear introduction to the subject. D. A. Roff, in *Evolutionary quantitative genetics* (Chapman and Hall, New York, 1997), provides a comprehensive treatment of the evolution of quantitative traits. An advanced treatment is *Genetics and analysis of quantitative traits*, by M. Lynch and J. B. Walsh (Sinauer Associates, Sunderland, MA, 1998).

An important review of some recent research is T. F. C. Mackay's "The genetic architecture of quantitative traits" (2001, *Annual Review of Genetics* 35: 303–339). Phenotypic plasticity, canalization, and related topics are the subject of C. D. Schlichting and M. Pigliucci's *Phenotypic evolution: A reaction norm perspective* (Sinauer Associates, Sunderland, MA, 1998).

Problems and Discussion Topics

1. Under artificial selection for increased body weight, what will be the response to selection (R), after one generation, for the following values of phenotypic variance (V_P), additive genetic variance (V_A), environmental variance (V_E), and selection differential (S)? (*a*) V_P = 2.0 grams^2, $V_A = 1.25$ g^2, $V_E = 0.75$ g^2, $S = 1.33$ g; (*b*) $V_P = 2.0$ g^2, $V_A = 0.95$ g^2, $V_E = 1.05$ g^2, $S = 1.33$ g; (*c*) $V_P = 2.0$ g^2, $V_A = 1.25$ g^2, $V_E = 0.75$ g^2, $S = 2.67$ g. (Answer for part *a*: mean weight will increase by about 0.83 g.) If the parameters remain the same for successive generations of selection, and the initial mean weight is 10 grams, what is the expected mean after two generations of selection in each case?

2. If most quantitative genetic variation within populations were maintained by a balance between the origin of new mutations and selection against them, then most mutations might be eliminated before environments could change and favor them. If the "residence times" of most mutations were short enough, the alleles that distinguish different populations or species would generally not be those found segregating within populations (Houle et al. 1996). How might one determine whether the alleles that contribute to among-population differences in the means of quantitative traits are also polymorphic within populations?
3. It has been suggested that genetic correlation between the expressions of a trait in the two sexes is responsible for some apparently nonadaptive traits, such as nipples in men and the muted presence in many female birds of the bright colors used by males in their displays (Lande 1980). Suggest ways of testing this hypothesis. What other traits might have evolved because they are genetically correlated with adaptive traits, rather than being adaptive themselves?
4. Debate the proposition that paucity of genetic variation and genetic correlations do not generally constrain the rate or direction of evolution.
5. Consider a character that is typical of the species in a higher taxon and may indeed be an important synapomorphic character for the clade. (For example, the number of petals is such a character in many genera and families of plants.) How might you decide whether this consistency is a result of intrinsic, unchangeable developmental "rules" or of a history of selection for canalization?
6. Traditional quantitative genetics is based on a theory of multiple anonymous loci, the functional roles of which are unknown. Using the "candidate loci" approach, the sequence and function of some of these loci is now being discovered. In what ways is this important for understanding the evolution of phenotypic traits?

14

Conflict and Cooperation

Darwin first conceived of natural selection when he read the economist Thomas Malthus's theory that population growth must inevitably cause competition for food and other resources. Malthus's *Essay on Population* inspired Darwin's realization that in all species, only a fraction of those born survive to reproduce, and that the survivors must often be those best equipped to compete for limiting resources. Thus conflict has been inherent in the idea of natural selection from the start. Darwin soon realized, however, that not all of natural selection stems from overt struggle among members of a species We also observe cooperation in nature: among the cells and organs of an individual organism, between mates in the act of sexual reproduction, among individuals in many social species of animals, and even between some mutualistic pairs of species. Yet conflict is so ubiquitous, even inevitable, that the question of how evolutionary theory can account for cooperation has occupied evolutionary biologists ever since Darwin.

Cooperation and antagonism. Among female wasps (*Polistes gallica*) that start a nest together, one becomes the queen and prevents the others from reproducing. The queen's nonreproductive daughters rear their sisters and brothers. (Photo © Bartomeu Borrell/ AGE Fotostock.)

Members of a species may struggle for more than food or space, as will shall see, and social interactions among individuals, even between mates or between parents and their offspring, may entail surprising elements of conflict. More surprisingly, evolutionary battles may rage even within individual organisms, between different genes. Contest, conflict, and cooperation are often overtly expressed in the behavior of animals, but they subtly pervade the lives of all organisms and may be manifested in many other ways. Conflict and cooperation are fundamental for understanding the evolution of a vast variety of biological phenomena, ranging from mating displays to sterile social insects, from cannibalism to some most peculiar features of human pregnancy.

Characteristics that benefit the population or species, but not the individual, can evolve only by group selection, as we saw in Chapter 11. Because group selection is generally a weak agent of evolutionary change, most evolutionary biologists seek explanations of the evolution of cooperation and conflict based on selection at the level of the individual organism or the gene. The models we will describe in this chapter ask how a character state would increase the reproductive success of its bearer, or of the allele that determines that character state.

A Framework for Conflict and Cooperation

Levels of organization

In this chapter, we will often be concerned with entities at one level that interact to form higher-level entities. For instance, individual cells may interact and form a multicellular organism; female and male organisms may form mated pairs, at least briefly; individual organisms may be organized into flocks or colonies. Both cooperation and conflict may be, and usually are, inherent in such interactions. Kern Reeve and Laurent Keller (1999) have pointed out that a cooperative group of lower-level units is stable only if evolutionary "attractive forces"—the advantages that each lower-level unit derives from the interaction—exceed "repulsive" and "centrifugal" forces (Figure 14.1). Repulsive forces refer to the harm one unit may inflict on another, and centrifugal forces are factors that might increase the fitness of a lower-level unit if it left the cooperative group and struck out on its own. For instance, female wasps that build a nest together benefit from mutual defense against predators; however, one female aggressively dominates the others and reduces their egg production. Whether or not it would pay a subordinate female to leave and build her own nest depends on ecological factors that affect the likelihood that she would succeed.

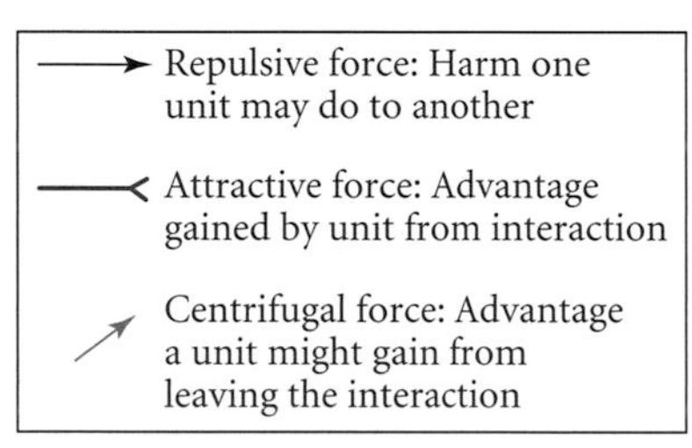

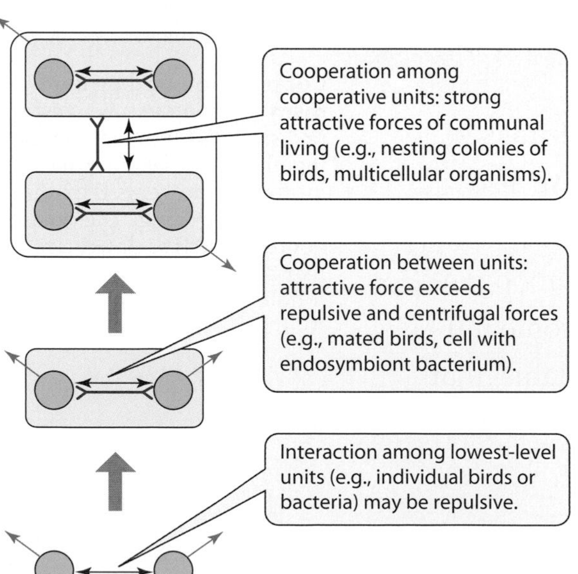

Cooperation among cooperative units: strong attractive forces of communal living (e.g., nesting colonies of birds, multicellular organisms).
Cooperation between units: attractive force exceeds repulsive and centrifugal forces (e.g., mated birds, cell with endosymbiont bacterium).
Interaction among lowest-level units (e.g., individual birds or bacteria) may be repulsive.

Figure 14.1 Higher-level cooperative units (rectangles) may be formed from lower-level units (circles) if an attractive fitness force overcomes repulsive and centrifugal forces. (After Reeve and Keller 1999.)

Inclusive fitness and kin selection

The advantage and disadvantage each lower-level unit derives from a cooperative interaction must be measured by its fitness—that is, its contribution of genes to subsequent generations. Here we must introduce a concept that is fundamental to understanding whether or not cooperation evolves; namely, **inclusive fitness**. It is often useful in this context to think of selection at the level of the gene. Bear in mind, then, that one allele replaces another in a population if it leaves more copies of itself in successive generations by whatever effect it may have. The inclusive fitness of an allele is its effect on both the fitness of the individual bearing it (DIRECT FITNESS) and the fitness of other individuals that carry copies of the same allele (INDIRECT FITNESS). We may likewise think of the inclusive fitness of an individual organism. At this level, selection based on inclusive fitness is called **kin selection** because these other individuals are the bearer's relatives, or kin.

Kin selection is one of the most important explanations for cooperation (Hamilton 1964; Michod 1982). Let us suppose that an individual performs an act that benefits another individual, but incurs a cost to itself: a reduction in its own (direct) fitness. The fundamental principle of kin selection is that an allele for such an ALTRUISTIC trait can increase in frequency only if the number of extra copies of the allele passed on by the altruist's beneficiary (or beneficiaries) to the next generation as a result of the altruistic interaction is greater, on average, than the number of allele copies lost by the altruist. This principle is formalized in **Hamilton's rule**, which states that *an altruistic trait can increase in frequency if the benefit* (b) *received by the donor's relatives, weighted by their relationship* (r) *to the donor, exceeds the cost* (c) *of the trait to the donor's fitness*. That is, altruism spreads if

$$rb > c$$

The COEFFICIENT OF RELATIONSHIP, r, is the fraction of the donor's genes that are identical by descent with any of the recipient's genes (Grafen 1991). (See Chapter 9 for a discussion of identity by descent.) For example, at an autosomal locus in a diploid species, an offspring inherits one of its mother's two gene copies, so $r = 0.5$ for mother and offspring. For two full siblings, $r = 0.5$ also, because the probability is 0.25 that both siblings inherit copies of the same gene from their mother, and likewise from their father.

The simplest example of a trait that has evolved by kin selection is parental care. If females with allele A enhance the survival of their offspring by caring for them, whereas females lacking this allele do not, then if parental care results in more than two extra surviving offspring, A will increase in frequency, even if parental care should cost the mother her life. If $c = 1$ (death of mother) and $b = 1$ (survival of an extra offspring), then since a mother's A allele has a probability of $r = 0.5$ of being carried by her offspring, Hamilton's rule is satisfied by the survival of more than two extra offspring, relative to noncaring mothers.

Interactions among other ("collateral") relatives also follow Hamilton's rule. For instance, the relationship between an individual and her niece or nephew is $r = 0.25$, so alleles that cause aunts to care for nieces and nephews will spread only if they increase the fitness benefit by more than fourfold the cost of care. The more distantly related the beneficiaries are to the donor, the greater the benefit to them must be for the allele for an altruistic trait to spread.

Parental care illustrates why indiscriminate altruism cannot evolve by individual selection. If allele A caused a female to dispense care to young individuals in the population at random, it could not increase in frequency because, on average, the fitness of all genotypes in the population, whether they carried A or not, would be equally enhanced. Thus the only difference in fitness among genotypes would be the reduction in fitness associated with dispensing care.

Frequency-dependent selection on interactions

Interactions among conspecific organisms often are affected by frequency-dependent selection (see Chapter 12). Consider competition for resources between two haploid genotypes, A and A', which prefer to eat different kinds of food. If A' arises as a rare mutant in a population of A individuals, its fitness (per capita rate of survival and reproduction) will be greater than that of A because it is rare and suffers little competition for food. As A' increases in frequency, however, competition among A' individuals increases, so fitness declines. When the intensity of competition experienced by an average individual of the two genotypes is equal, their fitnesses are equal, and a stable allele frequency is attained.

Evolutionarily stable strategies

One way of modeling the evolution of frequency-dependent traits has been adopted from the mathematical theory of games. The central concept of this approach is the **evolutionarily stable strategy** (**ESS**), which John Maynard Smith (1982) defined as "*a strategy such that, if all the members of a population adopt it, then no mutant strategy could invade under the influence of natural selection.*" That is, it is a phenotype that cannot be replaced by any other phenotype under the prevailing conditions. A strategy may be PURE, meaning that an individual always has the same phenotype, or MIXED, meaning that an individual's phenotype varies over time, as is often the case with behavior. ESS models frequently describe

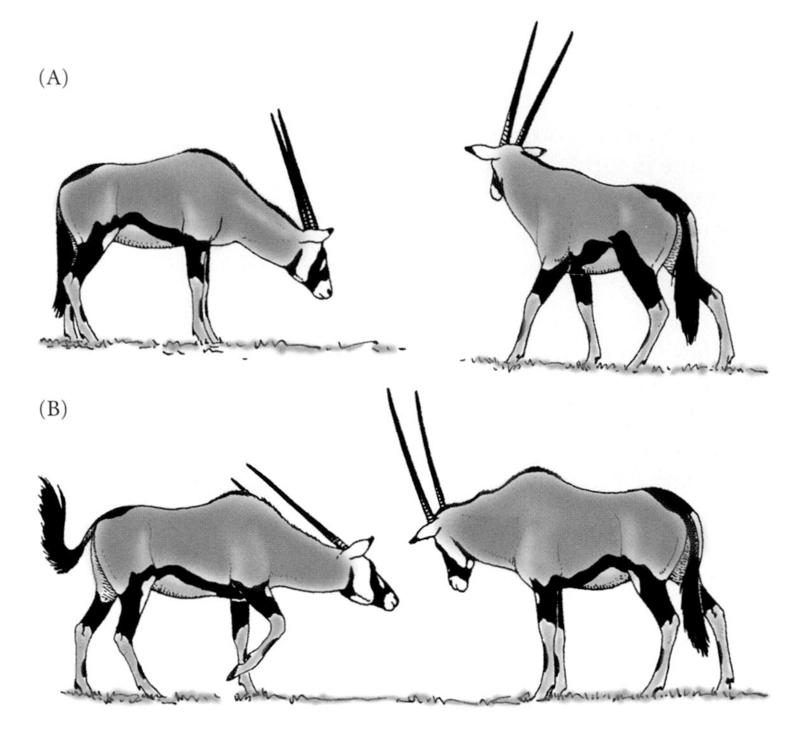

Figure 14.2 Ritualized aggressive behavior in the East African oryx (*Oryx beisa*). (A) A subordinate male (left) responds to a dominant male by lowering his head. (B) If the dominant male intensifies the threat by rotating his horns forward, the subordinate assumes a still more submissive attitude (note the laid-back horns) that leaves him defenseless against attack. The dominant male is then unlikely to attack him. (After Walther 1984.)

interactions between two individuals, each of which has one of two or more phenotypes (strategies). For each possible pairwise combination of strategies, an individual has a different payoff: the increment or decrement of fitness that it receives. The payoff depends not only on the individual's own phenotype, but also on that of the individual it interacts with.

One of the earliest problems to which Maynard Smith and others applied ESS theory was conflict between animals, such as between males competing for a territory or a mate. In many species, aggressive encounters consist largely of "ritualized" displays (Figure 14.2), seldom escalating into physical conflicts that could cause injury or death. The question is whether it is individually advantageous not to escalate conflicts.

Suppose, first, that there are two possible pure strategies: "Hawk," which escalates the conflict either until it is injured or until the opponent retreats, and "Dove," which retreats as soon as its opponent escalates the conflict (Box A). Is it better to be a Hawk, more likely to win the prize, or a Dove, more likely to escape injury and live to compete another day? Dove is never an ESS, because a Hawk genotype can always increase in a population of

BOX 14A ESS Analysis of Animal Conflict

Assume that the Hawk (H) phenotype escalates a conflict until it is injured (with a cost in fitness C) or until it gains the contested resource (with a gain in fitness V) when its opponent retreats. The Dove (D) phenotype retreats when threatened by H, neither gaining nor losing fitness. If two Doves meet, one gains the resource, so the average gain to an individual Dove in such encounters is $V/2$. A Hawk that encounters another Hawk wins with probability ½ and loses with probability ½, so its average payoff is $V/2 - C/2$. Thus we have a "payoff matrix:"

	Opponent	
Payoff to	**H**	**D**
H	$\frac{1}{2}(V-C)$	V
D	0	$V/2$

If $E(x,y)$ is the payoff to individual x in a conflict with y, then for any two strategies I and J, the fitnesses (w) of the phenotypes are

$$w(\text{I}) = w_0 + (1 - p)E(\text{I,I}) + pE(\text{I,J})$$

$$w(\text{J}) = w_0 + (1 - p)E(\text{J,I}) + pE(\text{J,J})$$

where p and $1 - p$ are the frequencies of J and I in the population and w_0 is a "baseline" fitness. If J is a rare mutant and I is an ESS, $w(\text{I}) > w(\text{J})$ by definition. From the expressions above, this will be true only if $E(\text{I,I}) > E(\text{J,I})$ or if $E(\text{I,I}) = E(\text{J,I})$ and $E(\text{I,J}) > E(\text{J,J})$.

If Dove (D) were an ESS, I = D, J = H; but since $E(\text{D,D}) < E(\text{H,D})$ (that is, $V/2 < V$), D is not an ESS. Hawk (H) is an ESS if $V > C$, because $E(\text{H,H}) = (V - C)/2$ is greater than $E(\text{D,H}) = 0$ if $V > C$. If, however, $V < C$, neither H nor D is an ESS.

Suppose, however, $V < C$, and there is a strategy F that entails playing H with probability P and D with probability $1 - P$. If genotypes vary in P, the fittest will be the one that plays each role often enough for its advantage to offset its disadvantage. That is, it must get an equal expected payoff from each randomly played role. From the payoff matrix, the expected payoffs from playing H and D with probabilities P and $1 - P$ are respectively $P(V - C)/2 + (1 - P)V$ and $P(0) + (1 - P)(V/2)$. Equating these and solving for P, $P = V/C$. Thus playing "Hawk" with probability P is an ESS (Maynard Smith 1982).

(A)

(B)

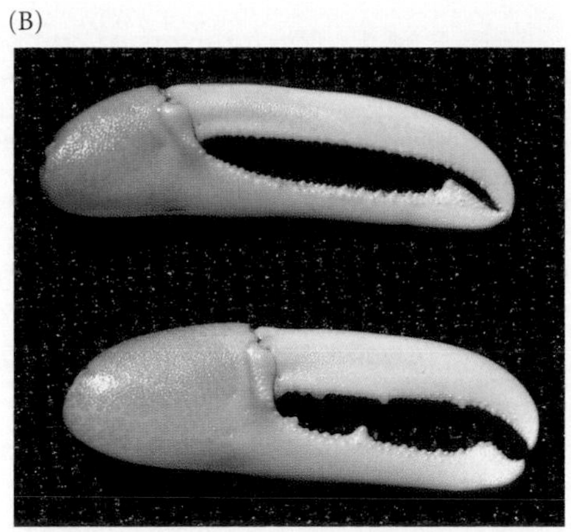

Figure 14.3 (A) A male fiddler crab (*Uca* sp.) signaling with its greatly enlarged claw. (B) A slender regenerated claw and a robust original claw of *Uca annulipes*. The regenerated claw is an effective bluff, even though it does not have the strength needed to win a fight with an intact male. (Photos courtesy of Patricia Backwell; B from Backwell et al. 2000.)

Doves. Hawk is an ESS if the fitness gained in a successful contest (V), even with another Hawk, is greater than the cost of injury (C). If, however, $C > V$, ESS analysis shows that a pure Hawk phenotype can be replaced by a "mixed strategy," such that an individual will adopt Hawk behavior with probability $P = V/C$. Thus this model predicts that the optimal behavior is variable and is contingent on conditions such as the value of the resource.

We can also postulate an Assessor strategy, in which the individual escalates the conflict if it judges the opponent to be smaller or weaker, but retreats if it judges the opponent to be larger or stronger. In most theoretical cases, the Assessor strategy is an ESS. In accord with these models, many animals react aggressively or not depending on their opponent's size or correlated features. For example, male toads (*Bufo bufo*) that clasp females before the eggs are laid and fertilized are often aggressively displaced by larger males, but not by smaller ones. A male is unlikely to try to displace a mounted male that is larger than himself, or one that emits a deeper-pitched croak when touched—the pitch being correlated with body size (Davies and Halliday 1978).

Characters such as the toad's croak might be HONEST SIGNALS of the individual's fighting ability or resource-holding potential, or they might be deceptive signals, indicating greater fighting ability than the individual actually has. Theoretically, deceptive signals should be unstable in evolutionary time because selection would favor genotypes that ignored the signals, which, having then lost their utility, would be lost in subsequent evolution. Thus, many existing signals of resource-holding potential are probably honest assessment signals (Grafen 1990; Johnstone and Norris 1993). However, dishonest signals are not uncommon (Bradbury and Vehrencamp 1998). In male fiddler crabs, for example, one claw is greatly enlarged and is used both to fight other males and to attract females. In the species *Uca annulipes*, males that lose their large claw regenerate one that is almost the same size, but has less muscle and is much weaker. Although such males lose physical fights with intact males, they effectively bluff and deter potential opponents, and they are apparently just as successful in attracting females (Backwell et al. 2000; Figure 14.3).

Sexual Selection

The concept of sexual selection

Sexual reproduction is the epitome of cooperation, but it also entails conflict, both among members of the same sex and between the sexes. Both of the mating partners have a "genetic interest" in the production and welfare of their offspring, who carry genes from each parent. However, if the mates are unrelated, neither has any genetic interest in the other's survival or reproductive success with other partners.

Darwin introduced the concept of sexual selection to describe differences among individuals of a sex in the number or reproductive capacity of mates they obtain. Sexual selection was Darwin's solution to the problem of why conspicuous traits such as the bright

TABLE 14.1 *Mechanisms of competition for mates and characters likely to be favored*

Mechanism	Characters favored
Same-sex contests	Traits improving success in confrontation (e.g., large size, strength, weapons, threat signals); avoidance of contests with superior rivals
Mate preference by opposite sex	Attractive and stimulatory features; offering of food, territory, or other resources that improve mate's reproductive success
Scrambles	Early search and rapid location of mates; well-developed sensory and locomotory organs
Endurance rivalry	Ability to remain reproductively active during much of season
Sperm competition	Ability to displace rival sperm; production of abundant sperm; mate guarding or other ways of preventing rivals from copulating with mate
Coercion	Adaptations for forced copulation and other coercive behavior
Infanticide	Similar traits as for same-sex contests
Antagonistic coevolution	Ability to counteract the other sex's resistance to mating (by, e.g., hyperstimulation); egg's resistance to sperm entry

Source: After Andersson 1994; Andersson and Iwasa 1996.

colors, horns, and displays of males of many species have evolved. He proposed two forms of sexual selection: contests between males for access to females and female choice (or "preference") of some male phenotypes over others. Several other bases for sexual selection have been recognized (Table 14.1).

Sexual selection exists because females produce relatively few, large gametes (eggs) and males produce many small gametes (sperm). *This difference creates an automatic conflict* between the reproductive strategies of the sexes: a male can mate with many females, and he often suffers little reduction in fitness if he should mate with an inappropriate female, whereas all a female's eggs can potentially be fertilized by a single male, and her fitness can be significantly lowered by inappropriate matings. Thus encounters between males and females often entail conflicts of reproductive interest (Trivers 1972). Furthermore, the OPERATIONAL SEX RATIO, the relative numbers of males and females in the mating pool at any time, is often male-biased because males mate more frequently. Commonly, then, *females are a limiting resource for males*, which compete for mates, but males are not a limiting resource for females. Thus *variation in mating success is generally greater among males than among females* (Figure 14.4), and indeed, is a measure of the intensity of sexual selection (Wade and Arnold 1980). In some species, however, the tables are turned. In cases of so-called SEX ROLE REVERSAL, as in phalaropes (Figure 14.5) and seahorses (see Figure 14.19A), males care for the young and can tend fewer offspring than a female can produce. In such instances, females may compete for males.

Figure 14.4 Among elephant seals, a few dominant males defend harems of females against other males (see Figure 10.5). A plot of the number of offspring produced by each of 140 elephant seals over the course of several breeding seasons shows much greater variation in reproductive success among males than among females. (After Gould and Gould 1989, based on data of B. J. LeBoeuf and J. Reiter.)

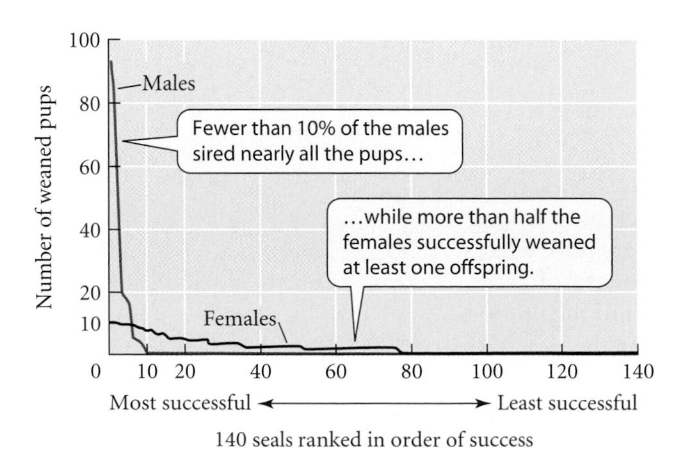

Contests between males and between sperm

Male animals often compete for mating opportunities through visual displays of bright colors or other ornaments, many of which make a male look larger. The males of some species fight outright and possess weapons, such as horns or tusks, that can inflict injury (Figure 14.6). Plumage patterns and songs are used to establish dominance by many birds, such as the red-winged blackbird (*Agelaius phoeniceus*), males of which have a bright red shoulder patch (Andersson 1994). Male blackbirds that were experimentally silenced were likely to lose their territories to intruders; however, intruders were deterred by tape recordings of male song. When territorial

Figure 14.5 A pair of red phalaropes (*Phalaropus fulicarius*) on their breeding grounds in Alaskan tundra. In contrast to most birds, the female phalarope (on the left) is more brightly colored than the male. Female phalaropes court males, which care for the eggs and young. (Photo © K. Karlson.)

males were removed and replaced with stuffed and mounted specimens whose shoulder patches had been painted over to varying extents, the specimens with the largest red patches were avoided by potential trespassers.

In sexual selection by male contest, directional selection for greater size, weaponry, or display features can cause an "arms race" that results in evolution of ever more extreme traits. Such "escalation" becomes limited by opposing ecological selection (i.e., selection imposed by ecological factors) if the cost of larger size or weaponry becomes sufficiently great (West-Eberhard 1983). The equilibrium value of the trait is likely to be greater than it would be if only ecological selection were operating. As Darwin noted, the duller coloration and lack of exaggerated display features in females and nonbreeding males of many species implies that these features of breeding males are ecologically disadvantageous.

Closely related to contests among males for mating opportunities are the numerous ways in which males reduce the likelihood that other males' sperm will fertilize a female's eggs (Thornhill and Alcock 1983; Birkhead and Møller 1992). Males of many species of birds defend territories, keeping other males away from their mates (although studies of DNA markers show that females of many such species nevertheless engage in high rates of extra-pair copulation). The males of many frogs, crustaceans, and insects clasp the female, guarding her against other males for as long as she produces fertilizable eggs. In some species of *Drosophila*, snakes, and other animals with internal fertilization, the seminal fluid of a mating male reduces the sexual attractiveness of the female to other males, reduces her receptivity to further mating, or forms a copulatory plug in the vagina (Partridge and Hurst 1998; see the discussion below on "chase-away" sexual selection).

Sexual competition between males may continue during and after copulation. In many damselflies, the male's genitalia are adapted to remove the sperm of previous mates from the female's reproductive tract (Figure 14.7). In many animals, **sperm competition** occurs when the sperm of two or more males have the opportunity to fertilize a female's eggs (Parker 1970; Birkhead 2000). In some such cases, a male can achieve greater reproductive success than other males simply by producing more sperm. This explains why polygamous

(A)

(B)

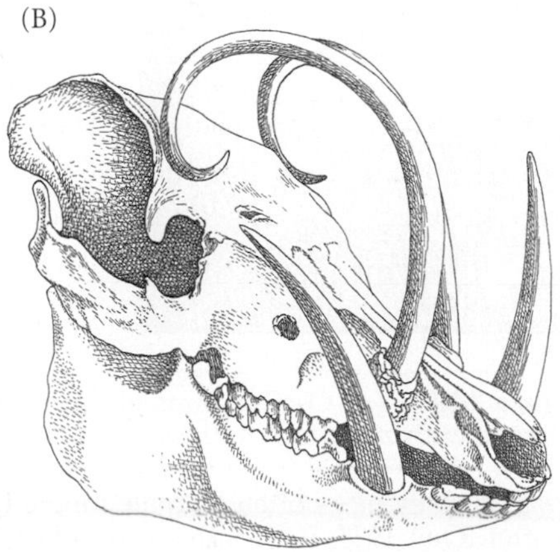

Figure 14.6 Some armaments of male mammals used in contests over mates: (A) The antlers of red deer (elk; *Cervus elaphus*). (B) The extraordinary upper and lower canine teeth of the babirusa (*Babyrousa babyrussa*), a tropical Asian pig. This historical drawing is from *The Malay Archipelago* (1869) by Alfred Russel Wallace. (A photo by J.H. Robinson/Photo Researchers, Inc.)

(A)

(B)

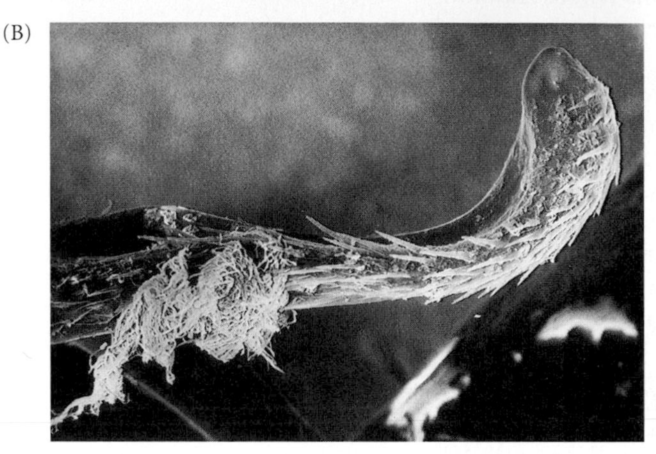

Figure 14.7 An elaborate mechanism for improving a male's likelihood of paternity: the penis of the black-winged damselfly *Calopteryx*. (A) The end of the penis includes a lateral horn at the top. (B) Close-up of the lateral horn, showing spinelike hairs and a clump of a rival's sperm. (Courtesy of J. Waage.)

species of primates tend to have larger testes than monogamous species (see Figure 11.20). SPERM PRECEDENCE, whereby most of a female's eggs are fertilized by the sperm of only one of the males with which she has mated, occurs in many insects and other species. In *Drosophila melanogaster*, the degree of sperm precedence is affected by genetic variation among females and by genetic variation among males in the ability of their sperm to displace other males' sperm and to resist displacement (Clark et al. 1994; Clark and Begun 1998).

Sexual selection by mate choice

In many species of animals, individuals of one sex (usually the male) compete to be chosen by the other. The evolution of sexually selected traits by mate choice presents some of the most intriguing problems in evolutionary biology and is the subject of intense research (Andersson 1994; Johnstone 1995; Andersson and Iwasa 1996; Houle and Kondrashov 2002).

Females of many species of animals mate preferentially with males that have larger, more intense, or more exaggerated characters such as color patterns, ornaments, vocalizations, or display behaviors. For example, female peafowl, widowbirds, barn swallows, and other species prefer males with longer tail feathers (see Figure 11.9). The preferred male characters are often ecologically disadvantageous. For instance, females of the Central American túngara frog (*Physalaemus pustulosus*) prefer male calls that include not only a "whine," but also a terminal "chuck" sound—but this part of the call also attracts predatory fringe-lipped bats (*Trachops cirrhosus*) (Figure 14.8). Female choosiness may likewise have costs: the time spent searching for acceptable males has been shown to reduce reproductive output in several species (Andersson 1994).

Sexually selected traits are often extraordinarily diverse among related species. Closely related species of hummingbirds, birds of paradise, cichlid fishes, and many other groups show astonishing differences in colors, ornaments, and displays that are thought to have evolved by sexual selection (see Figure 15.6).

Subject to limits imposed by ecological selection, male traits will obviously evolve to exaggerated states if they enhance mating success. But why should females have a preference for these traits, especially for features that seem so arbitrary—and even dangerous for the males that bear them? The several hypotheses that have been proposed include (1) *direct* and (2) *indirect benefits* to

(A)

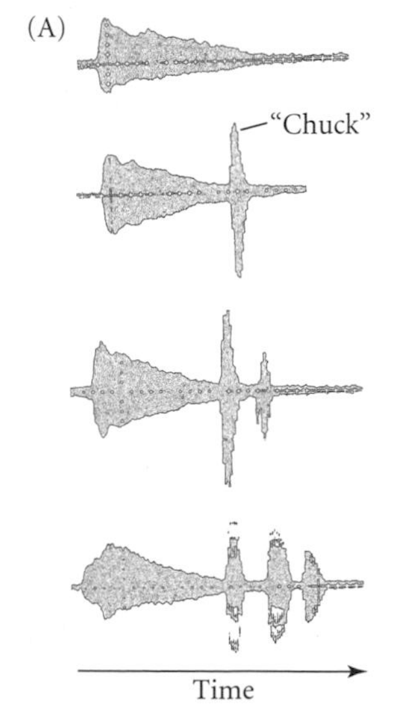

(B)

Figure 14.8 (A) Oscillograms of four calls of male túngara frogs (*Physalaemus pustulosus*), showing calls with zero (top) to three (bottom) "chucks." Each oscillogram displays a plot of amplitude against time. Female frogs are most attracted to calls with "chucks." (B) Frog-eating bats (*Trachops cirrhosus*) prey more often on male túngara frogs that emit "chucks" than on those that do not. (A after Ryan 1985; B © Merlin Tuttle/BCI/Photo Researchers, Inc.)

choosy females, (3) *sensory bias*, and (4) *antagonistic coevolution*. The category of indirect benefits includes two principal hypotheses.

Direct benefits of mate choice

The least controversial hypothesis applies to species in which the male provides a direct benefit to the female or her offspring, such as nutrition, a superior territory with resources for rearing offspring, or parental care. Under these circumstances there is selection pressure on females to recognize males that are superior providers by some feature that is correlated with their ability to provide. Once this capacity has evolved in females, their preference selects for males with the distinctive, correlated character. The coloration of male house finches (*Carpodacus mexicanus*), for example, varies from orange to bright red. Females prefer to mate with bright males, and brighter males bring food to nestlings at a higher rate (Hill 1991).

Sensory bias

In some cases, female preference may evolve before the preferred male trait does (Ryan 1998). For example, certain traits may be intrinsically stimulating and evoke a greater response simply because of the organization of the sensory system. The occurrence of such **sensory bias** is well supported by studies of sensory physiology and animal behavior. Animals frequently show greater responses to SUPERNORMAL STIMULI that are outside the usual range of stimulus intensity, and even artificial neural networks trained to recognize patterns show biases for exaggerated stimuli.

Phylogenetic studies have shown that female preferences sometimes evolve before the preferred male trait (Figure 14.9A). For instance, in some species of the fish genus *Xiphophorus* (swordtails), part of the male's tail is elongated into a "sword" (Figure 14.9B). Alexandra Basolo (1995, 1998) found that females preferred males fitted with plastic swords not only in sword-bearing species of *Xiphophorus*, but also in a species of *Xiphophorus* that lacks the sword and in the swordless genus *Priapella*, the sister group of *Xiphophorus*. In fact, female *Priapella* had a stronger preference for swords than females of a sword-bearing species of *Xiphophorus*. The female preference evidently evolved in the common ancestor of these genera, and thus provided selection for a male sword when the mutations for this feature arose.

Indirect benefits of mate choice

The most difficult problem in accounting for the evolution of female preferences is presented by species in which the male provides no direct benefit to either the female or her offspring, but contributes only his genes. In this case, alleles affecting female mate choice increase or decrease in frequency depending on the fitness of the females' offspring. Thus

Figure 14.9 Evidence supporting sexual selection due to sensory bias in female mate choice. (A) A hypothetical phylogeny of three species that differ in presence (+) or absence (–) of a male trait (T) and in female preference (P) for that trait. Both T and P are absent in the outgroup species, evidence that their presence is derived. That P is present in both species 1 and 2, whereas T is present in 2 only, is evidence that a female preference, or sensory bias, for the trait T evolved before the trait did. Real species corresponding to species 1 and 2 include (B) poeciliid fishes that lack (top) or possess (bottom) a swordlike tail ornament in the male and (C) wolf spiders that lack (top) or possess (bottom) tufts of bristles on the front legs. (A after Ryan 1998; B, photos courtesy of Alexandra Basolo; C photos courtesy of George Uetz.)

(A)

Species 1 P^+/T^-

Species 2 P^+/T^+

Outgroup P^-/T^-

Trait arises in males

Female preference (sensory bias for trait)

Ancestral state P^-/T^-

(B)

(C)

females may benefit indirectly from their choice of mates (Kirkpatrick and Barton 1997; Kokko et al. 2002).

The two prevalent models of such indirect benefits are **runaway sexual selection** (sometimes called the "sexy son" hypothesis), in which the sons of females that choose a male trait have improved mating success because they inherit the trait that made their fathers appealing to their mothers, and **good genes models**, in which the preferred male trait indicates high viability, which is inherited by the offspring of females who choose such males. Good genes models are sometimes called **handicap models** because the male trait indicates high viability despite the ecological handicap that it poses to survival (Zahavi 1975).

RUNAWAY SEXUAL SELECTION. In runaway sexual selection, as proposed by R. A. Fisher (1930), the evolution of a male trait and a female preference, once initiated, becomes a self-reinforcing, snowballing or "runaway" process (Lande 1981; Kirkpatrick 1982; Pomiankowski and Iwasa 1998). This process is often referred to as the "Fisherian model" of sexual selection (even though Fisher discussed both kinds of indirect benefits). In the simplest form of the model, haploid males of genotypes T_1 and T_2 have frequencies of t_1 and t_2, respectively. T_2 has a more exaggerated trait, such as a longer tail, that carries an ecological disadvantage, such as increasing the risk of predation. Females of genotype P_2 (with frequency p_2) prefer males of type T_2, whereas P_1 females exhibit little preference (or prefer T_1). It is assumed that alleles P_1 and P_2 do not affect survival or fecundity, and thus are selectively neutral.

Although the expression of genes P and T is sex-limited, both sexes carry both genes and transmit them to offspring. Because P_2 females and T_2 males tend to mate with each other, linkage disequilibrium will develop: their offspring of both sexes tend to inherit both the P_2 and T_2 alleles. The male trait and the female preference thus become genetically correlated (see Chapter 13), so that any increase in the frequency of the male trait is accompanied by an increase in the frequency of the female preference through hitchhiking (Figure 14.10).

Perhaps T_2 males have a slight mating advantage over T_1 because they are both acceptable to P_1 females and preferred by the still rare P_2 females. Whether for this or another reason, suppose t_2 increases slightly. Because of the genetic correlation between the loci, an increase in t_2 is accompanied by an increase in p_2. That is, T_2 males have more progeny, and their daughters tend to inherit the P_2 allele, so P_2 also increases in frequency. As P_2 increases, T_2 males have a still greater mating advantage because they are preferred by more females; thus the association of the P_2 allele with the increasing T_2 allele can increase P_2 still further.

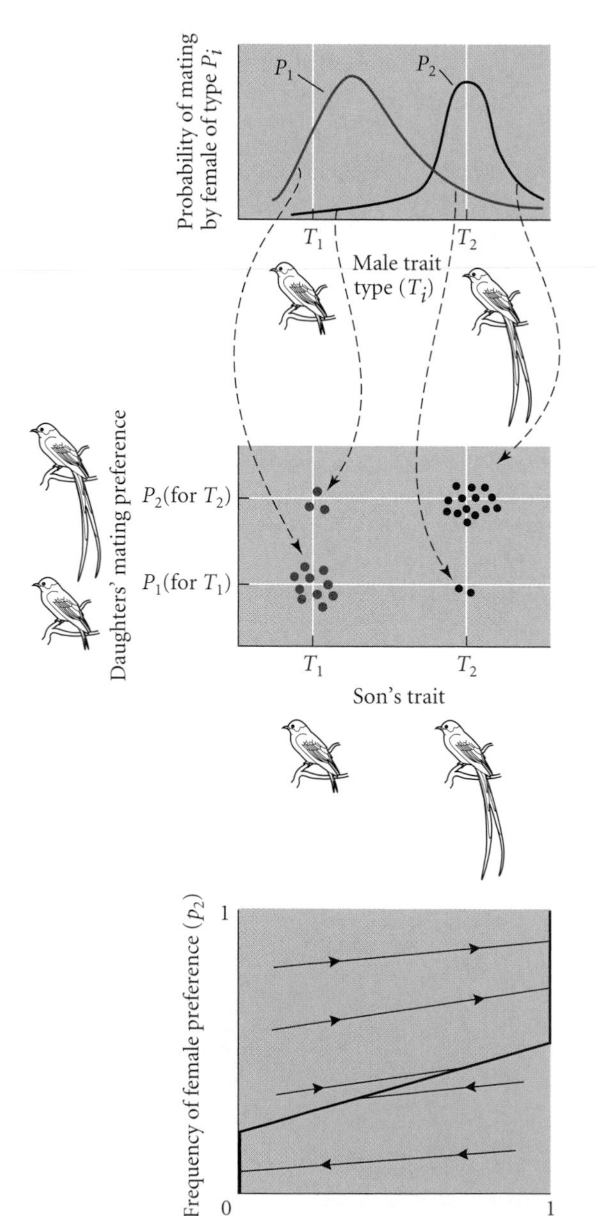

Figure 14.10 A model of runaway sexual selection by female choice. T_1 and T_2 represent male genotypes that differ in some trait, such as tail length. P_1 and P_2 females have different preferences for T_1 versus T_2 males, as shown in the upper graph. The resulting pattern of mating creates a correlation in the next generation between the tail length of sons and the mate preference of daughters (middle graph). Thus a genetic correlation is established in the population, in which alleles P_2 and T_2 are associated to some extent. Any change in the frequency (t_2) of the allele T_2 thus causes a corresponding change in p_2. In the lower graph, each point in the space represents a possible population with some pair of frequencies p_2 and t_2, and the vectors show the direction of evolution. When p_2 is low, t_2 declines due to ecological selection, so p_2 declines through hitchhiking. When p_2 is high, sexual selection for T_2 males outweighs ecological selection, so t_2 increases, and p_2 also increases through hitchhiking. Along the solid line, allele frequencies are not changed by selection, but may change by genetic drift. (Lower graph after Pomiankowski 1988.)

If both the male trait and the female preference are polygenically inherited, the same principle holds, but new genetic variation due to mutation can enable indefinite evolution of both characters, and under some theoretical conditions, the evolutionary process "runs away" toward preference for an ever more exaggerated male trait.

As we have seen, many exaggerated sexually selected traits carry ecological costs for the males that bear them. Female choice may also carry an ecological cost; for instance, rejecting a male in search of a more acceptable one may delay reproduction or entail other risks. Such costs may prevent the runaway process from occurring, or they may lead to an equilibrium at which the male trait and the female preference are less extreme than if there were no cost. Moreover, too extreme a preference may become so disadvantageous that it evolves in the opposite direction, resulting in CYCLICAL EVOLUTION: both preference and trait may evolve first in one direction (toward, say, longer tails), then in the other, and then back again over the course of many generations. This theory may account for many cases in which, as phylogenetic analyses show, species have lost sexually selected male traits (Wiens 2001).

If females have genetically variable responses to each of several or many male traits, different traits or combinations of traits may evolve, depending on initial genetic conditions (Pomiankowski and Iwasa 1998). Thus runaway sexual selection can follow different paths in different populations, so that *populations may diverge in mate choice and become reproductively isolated*. Sexual selection is therefore a powerful potential cause of speciation (see Chapter 16). Runaway sexual selection of this kind could explain the extraordinary variety of male ornaments among different species of hummingbirds and many other kinds of animals.

INDICATORS OF GENETIC QUALITY. Because females risk substantial losses of fitness if their offspring do not survive or reproduce, an appealing hypothesis is that females should evolve to choose males with high genetic quality, so that their offspring will inherit "good genes" and so have a superior prospect of survival and reproduction. Any male trait that is correlated with genetic quality—any INDICATOR of "good genes"—could be used by females as a guide to advantageous matings, so selection would favor a genetic propensity in females to choose mates on this basis. Female preference for male indicator traits should be most likely to evolve if the trait is a **condition-dependent indicator** of fitness—as many male display traits indeed are. In this case, males with allele T can develop the indicator trait to a fuller extent if they are in good physiological condition because they also carry "good genes"; say, allele B at another locus. Because females with a "preference allele" (P) mate with males that carry both T and B, their offspring inherit all three alleles. Thus linkage disequilibrium develops among the three loci, and both P and T increase in frequency due to their association with the advantageous allele B. (Single-gene control of the three traits is assumed here for simplicity, but any of the traits could be polygenic.)

In this model, the strength of indirect selection on female preference is proportional to the genetic variance in fitness in the population. However, natural selection should reduce variance in fitness. Several authors have addressed the question of how genetic variance could be maintained and therefore provide a foundation for female preference. For instance, it has been proposed that continual change in the genetic composition of parasite populations may maintain genetic variation in resistance traits in populations of their hosts, and that sexually selected male traits may be indicators of resistance to parasites (Hamilton and Zuk 1982).

A simpler hypothesis is that genetic variation for fitness components is replenished by recurrent mutation. The mutational variance for fitness components is quite high, as we saw in Chapter 13, and indeed, these traits are variable in natural populations. Locke Rowe and David Houle (1996) developed a model in which a male's condition, and thus the size of the display ornament he can develop, is inversely proportional to the number of deleterious mutations he carries. The high mutational variance in condition will sustain genetic variance in the display trait and enable long-sustained evolution of both more extreme ornaments and more extreme female preferences. The assumptions of this model are realistic. A high courtship rate in male dung beetles, for example, is a condition-de-

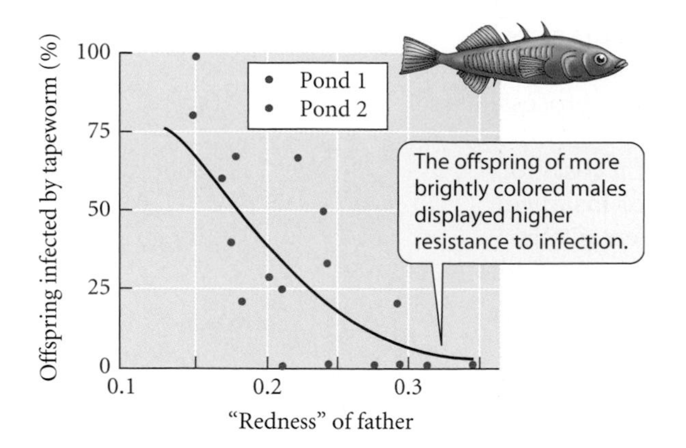

Figure 14.11 Evidence for the "good genes" hypothesis of female choice. The percentage of young sticklebacks that became infected when exposed to tapeworm larvae was correlated with the coloration of their fathers. (After Barber et al. 2001.)

pendent trait that is preferred by females; both courtship rate and condition are genetically variable, as the model predicts (Kotiaho et al. 2001).

EVIDENCE ON THE ROLE OF INDIRECT EFFECTS. There is evidence for both models of indirect effects. For instance, Allison Welch and coworkers (1998) studied the growth of gray tree frog tadpoles (*Hyla versicolor*) in broods consisting of eggs from the same mother but fertilized by two different fathers, one of which emitted longer courtship calls than the other. Females strongly prefer long calls. Averaged over 25 broods, the offspring of males with long calls had higher growth rates and a larger size at metamorphosis, especially when they were reared on a low food supply. Since male tree frogs provide no direct benefits to females or offspring, and since the tadpoles presumably differed only in their paternal genes, the courtship call preferred by females appears to be correlated with male genes that improve offspring growth. In other words, the call duration seems to be an honest signal of male genetic quality. In a very similar experiment, Barber et al. (2001) studied a character preferred by female sticklebacks (*Gasterosteus aculeatus*): the intensity of red coloration of the male's belly. They found that young sticklebacks with bright red fathers were more resistant to infection by tapeworms than their half-sibs who had dull red fathers (Figure 14.11). The red coloration is based on carotenoid pigments, which are obtained from food and which appear to enhance development of an effective immune system. Quite a few other studies have provided evidence for the good genes hypothesis, although the correlation between the male trait and offspring survival is usually very low (Møller and Alatalo 1999).

There is evidence, too, for the Fisherian runaway process. For example, male sandflies (*Lutzomyia longipalpis*) aggregate in LEKS (areas where males gather and compete for females that visit in order to mate). Females typically reject several males before choosing one with whom to mate. T. M. Jones and coworkers (1998), using experimental groups of males, determined which were most often, and which least often, chosen by females. Females who mated with the most "attractive" versus the least "attractive" males did not differ in survival or in the number of eggs they laid; that is, there was no evidence of a direct benefit of mating with attractive males. Nor was the father's attractiveness correlated with his offspring's survival or with his daughters' fecundity; that is, there was no evidence that attractive fathers transmitted good genes for survival or egg production (Figure 14.12A). However, when the investigators provided virgin females with trios of sons—one from a highly attractive father, one

(A)

Mean survival (days)

Low / Average / High

Sons · Non-reproducing daughters · Reproducing daughters

The survival of sandflies was not correlated with their fathers' attractiveness to females...

(B)

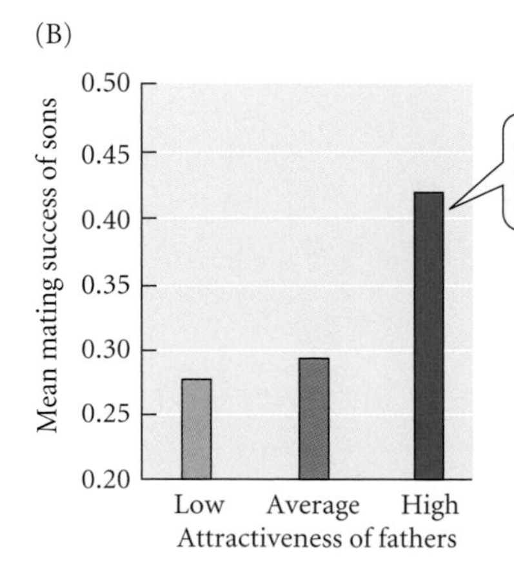

Figure 14.12 Evidence for Fisher's "sexy son" hypothesis. (A) Survival is shown for the sons and for both the reproducing and nonreproducing daughters of three classes of fathers that differed in attractiveness to females. There is no indication that attractive fathers contribute "good genes" to their offspring. (B) The *sons* of highly attractive male sandflies, however, show strikingly high mating success when females choose between them and the sons of less attractive fathers. Thus the indirect benefit to females of mating with an attractive male is that her sons will enjoy greater reproductive success. (After Jones et al. 1998.)

from an "average" father, and one from an unattractive father—the sons of attractive fathers were by far the most successful in mating, as predicted by the Fisherian "sexy son" hypothesis (Figure 14.12B). In this case, we do not know what male characteristic female sandflies prefer. In a similar study of the spotted cucumber beetle (*Diabrotica undecimpunctata howardi*), however, Tallamy et al. (2003) found that females prefer males that stroke them more frequently with their antennae. The male's stroke frequency does not affect his mate's reproduction or his offspring's viability or fecundity, but fast-stroking fathers have fast-stroking sons who are preferred by females. Thus this study, too, provides evidence for the Fisherian, but not for the good genes, model.

Antagonistic coevolution

Although the sexes must cooperate in order to produce offspring, conflict between the sexes is also pervasive (Parker 1979; Chapman et al. 2003). For example, unless a species is strictly monogamous, a male profits if he can cause a female to produce as many of his offspring as possible, even if this reduces her subsequent ability to survive and reproduce. Thus genes that govern male versus female characters may conflict, resulting in **antagonistic coevolution** (Rice and Holland 1997; Holland and Rice 1998). Such evolution may consist of a protracted "arms race" in which change in a male character is parried or neutralized by evolution of a female character, which in turn selects for change in the male character, and so on in a chain reaction.

Consider interactions between gametes in externally fertilizing species such as abalones (large gastropods, *Haliotis*). Sperm compete to fertilize eggs, so selection on sperm always favors a greater ability to penetrate eggs rapidly. But selection on eggs should favor features that slow sperm entry, or else POLYSPERMY (entry by multiple sperm) may result. Polyspermy disrupts development, and indeed, eggs have elaborate mechanisms to prevent it. The conflict between the interests of egg and sperm may result in continual antagonistic coevolution of greater penetration ability by sperm and countermeasures by eggs. Such coevolution might be the cause of the extraordinarily rapid evolution of the amino acid sequence of the sperm lysin protein of abalones, a protein released by sperm that bores a hole through the vitelline envelope surrounding the egg. Nonsynonymous differences between the lysin genes of different species of abalones have evolved much faster than synonymous differences, a sure sign of natural selection (Vacquier 1998; see also Palumbi 1998 and Chapter 19).

Genetic conflict between the sexes has been studied in greatest detail in *Drosophila melanogaster*, in which mating reduces female survival, both because females are harassed by courting males and because the male seminal fluid contains toxic proteins (Chapman et al. 1995). A study of genetic variation showed that males whose sperm best resist displacement by a second male's sperm cause the greatest reduction of their mates' life span: a clear case of sexual conflict (Civetta and Clark 2000). In a series of clever experiments, William Rice and Brett Holland have shown that male and female *Drosophila* are locked in a coevolutionary "arms race" (Rice 1992, 1996; Holland and Rice 1999). In one experiment, Rice crossed flies for 29 generations so that genetically marked segments of two autosomes were inherited only through females. Thus alleles in these chromosome regions that enhanced female fitness could increase in frequency, irrespective of their possible effect on males. Rice observed that the proportion of females among adult flies increased in the course of the experiment, due to both improved survival of females and diminished survival of males—just as predicted by the antagonistic coevolution hypothesis.

In a second experiment, Rice constructed experimental populations of flies in which males could evolve, but females could not. (In each generation, the males were mated with females from a separate, non-evolving stock with chromosome rearrangements that prevented recombination with the males' chromosomes, and only sons were saved for the next generation.) After 30 generations, the fitness of these males had increased, compared with control populations, but females that mated with these males suffered greater mortality, perhaps because of enhanced semen toxicity (Figure 14.13). This experiment suggests that males and females are continually evolving, but in balance, so that we cannot see the change unless evolution in one sex is prevented.

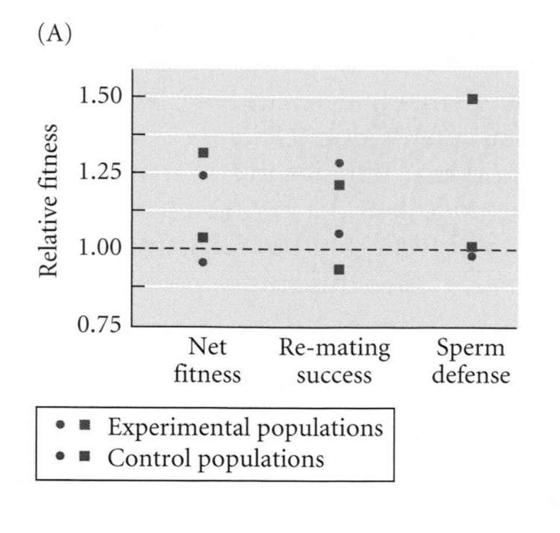

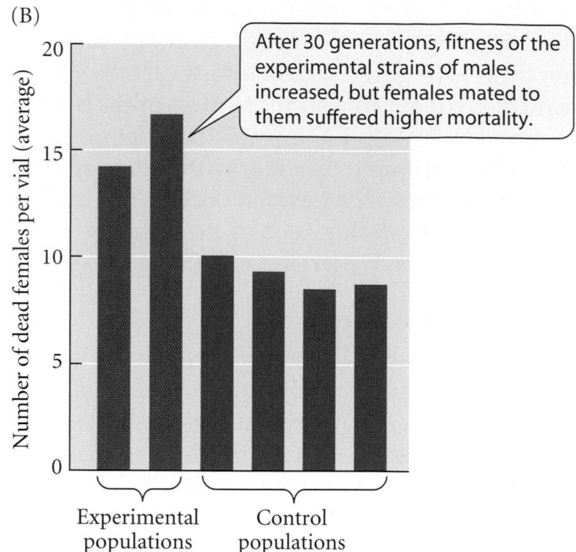

Figure 14.13 Experimental evidence of genetic conflict between the sexes. (A) Measures of fitness of males from two experimental populations in which only males could evolve, relative to that of males from control populations. The measures are net fitness (number of sons per male), the rate at which the males re-mated with previously mated females, and sperm defense (the ability of a male's sperm to fertilize eggs of a female that was given the opportunity to mate with a second male). (B) When males from these experimental populations were mated with females from another stock, they caused higher female mortality than did males from four control populations. (After Rice 1996)

Semen toxicity is only one of many examples of harm that males inflict on their mates. "Forced copulation" by groups of male mallard ducks has been known to drown females, and female bedbugs suffer reduced survival and reproduction from repeated "traumatic insemination." (The male bedbug pierces his mate's abdominal wall with his genital structure, rather than inserting it into her genital opening: Stutt and Siva-Jothy 2001.) Thus females evolve resistance to males' inducements to mate, and their resistance selects for features that enable males to overcome the females' reluctance—a dynamic that has been termed CHASE-AWAY SEXUAL SELECTION (Holland and Rice 1998). Among species of water striders, for example, in which males force females to copulate, the evolution of features that enable males to grip females has been accompanied by changes that enable females to break the males' grip (Figure 14.14).

Under chase-away selection, males may evolve increasingly strong stimuli, such as brighter colors or more elaborate song, to induce reluctant females to mate. In such cases, more elaborate male traits evolve not because females evolve to prefer them, but because females are selected to resist less elaborate traits. If this hypothesis is true, then males may

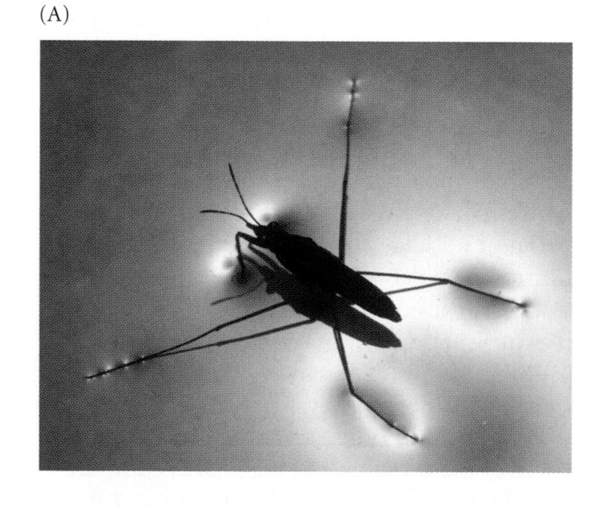

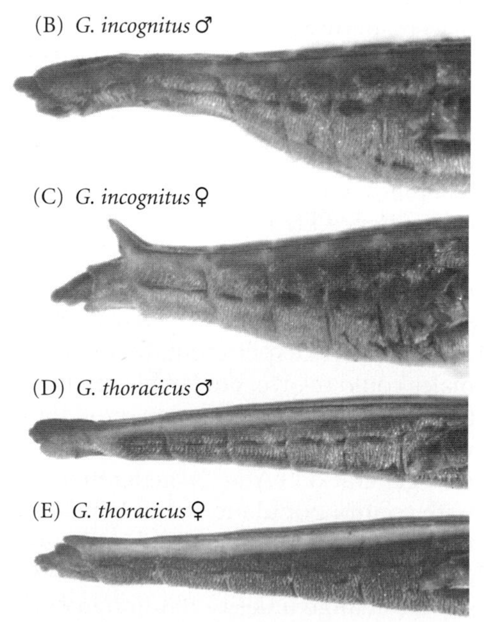

Figure 14.14 (A) Among water striders (genus *Gerris*), which live on the surface of water, males forcibly copulate with females. (B, C) In *G. incognitus*, the abdomen of the male has exaggerated grasping adaptations (extended, flattened genital segments), and that of the female has features (prolonged, erect abdominal spines; curved abdomen tip) that obstruct the male's grip during premating struggles. (D, E) The closely related species *G. thoracicus* shows the ancestral condition, in which these adaptations have not evolved. (A © photolibrary.com; B–E from Arnqvist and Rowe 2002.)

accumulate characteristics that were once sufficient to induce mating, but are sufficient no longer. Recall, for example, the surprising observation that the swordtail's sword stimulates conspecific females less than female *Priapella*, the swordless relative. This is just what we would expect if the ancestral attraction to swords, still evident in *Priapella*, had been replaced by female resistance after the sword evolved. Similarly, tufts of leg bristles in the wolf spider *Schizocosa ocreata* are a derived male character, relative to their absence in a closely related species, *S. rovneri* (see Figure 14.9C). Female spiders were presented with video images manipulated to show courting males with added, enlarged, or removed tufts. These manipulations did not affect the mating receptiveness of female *S. ocreata*, but adding tufts greatly enhanced the receptiveness of females of the tuftless species (McClintock and Uetz 1996). These results are consistent with the hypothesis that male tufts evolved in *S. ocreata* to stimulate females, but female resistance evolved thereafter.

Social Interactions and the Evolution of Cooperation

Darwin's theory of natural selection, as we have seen, is based on *individual* advantage: traits that enhance an individual's fitness relative to that of other members of the population—"selfish" traits—increase in frequency if they are heritable. Thus cooperative interactions seem antithetical to evolution by natural selection and require explanation. Until the 1960s, it was common for biologists to assume that cooperation—especially altruism—had evolved because it benefited the population or species—that is, by group selection. The modern study of cooperation, which for the most part rejects group selection, issues largely from William Hamilton's (1964) theory of kin selection and from the realization, articulated most forcefully by George Williams (1966), that group selection is a weaker process than individual selection (see Chapter 11).

Theories of cooperation and altruism

Traits that appear harmful to the possessor but beneficial to other individuals have been explained by four major classes of individual selection hypotheses: *manipulation*, *individual advantage*, *reciprocation*, and *kin selection*.

MANIPULATION. A donor may dispense aid to a recipient not because it is adaptive, but simply because the donor is being manipulated or coerced. For example, brood-parasitic birds such as cowbirds and some cuckoos lay their eggs in other birds' nests. The eggs or nestlings of the brood-parasitic birds often mimic those of their hosts. The parasitic nestling successfully solicits care from the foster parents, even though it often destroys the legitimate eggs or nestlings (see Chapter 18). Brood parasitism occurs both between species and within species: the pair of birds seen feeding a brood of nestlings often are tending not only to their own offspring, but to some that have a different father (the result of extra-pair copulation or "cuckoldry," as mentioned above) or a different mother (who stealthily laid an egg or two in another female's nest). Presumably the hosts cannot distinguish their own eggs or young from those of the interlopers.

INDIVIDUAL ADVANTAGE. Cooperative behavior often evolves simply because it is advantageous to the individual. Joining a flock or herd, for example, provides protection from predation owing to safety in numbers. It is often advantageous to each individual to be as close as possible to the center of the group, thus using other group members as shields against approaching predators. The effect of this behavior is to increase the compactness and cohesion of the group (Hamilton 1971). Unmated males of some birds, such as pied kingfishers (Reyer 1990), help unrelated pairs to rear offspring, evidently because their best chance of obtaining a mate is to replace the mated male in the pair if he dies (Figure 14.15). Among unrelated female paper wasps (*Polistes dominulus*) who build a nest together, one aggressively achieves dominance over the others and prevents them from reproducing; but dominant females may and do get killed, in which case a subordinate female takes over the nest and achieves reproductive success (Queller et al. 2000). Thus apparent altruism may actually have a delayed benefit.

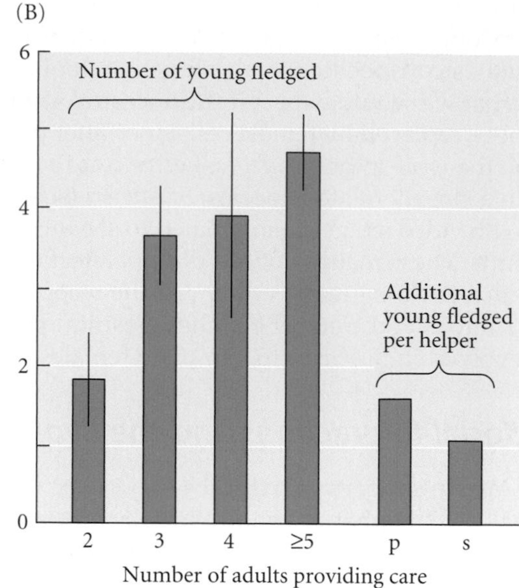

Figure 14.15 (A) The pied kingfisher, *Ceryle rudis*. (B) Survival of young pied kingfishers as a function of the number of adults providing care. Groups of more than two adults include male helpers. The number of surviving young per helper, over and above the number reared by parents without helpers, is shown at the right for primary (p) and secondary (s) helpers. (Photo © Tony Heald/naturepl.com; B after Reyer 1990.)

In *The Defiant Ones* (1958), Tony Curtis and Sidney Poitier portrayed two escaped convicts who, because they are chained to each other, must cooperate even though they dislike each other. When the fitness of each member of a group depends on the fitness of the other members—of the group as a whole—cooperation is clearly in every individual's interest. This principle is important in many biological contexts; perhaps the simplest examples are cooperation among the genes in a cell and among the cells in an organism. If the cell dies, so do all the included genes; if the organism dies, so do its cells. Selection at the higher level—cell or organism—thus eliminates outlaw genes or renegade cells that selfishly diminish the survival of the group. It also favors mechanisms that suppress or destroy such outlaws when they arise. We will see clear examples of this principle later in this chapter.

RECIPROCATION. Delayed benefits also figure in the hypothesis of **reciprocal altruism**: it can be advantageous for an individual to help another if the recipient will provide reciprocal aid in the future (Trivers 1971). For example, vampire bats (*Desmodus rotundus*), which feed on mammalian blood, form roosting groups, in which members that have fed successfully sometimes feed regurgitated blood to other group members that have been unsuccessful in foraging. The recipients reciprocate at other times (Wilkinson 1988). Such cooperation will evolve only under certain conditions, since cheating is often possible (Axelrod and Hamilton 1981).

The theory of reciprocal altruism is part of a more general TRANSACTIONAL MODEL OF REPRODUCTIVE SKEW, first developed by Sandra Vehrencamp (1983). The basic idea is that dominant individuals gain from the assistance of subordinate helpers, and "pay" those helpers by allowing them to reproduce just a little more than they could if they left the group and reproduced on their own. Considerable evidence supports this idea (Keller and Reeve 1994, 1999). For example, a dominant dwarf mongoose (a weasel-like African carnivore) suppresses reproduction by older subordinates less than younger subordinates. As subordinates grow older, they are better able to disperse and breed in another group, so the model predicts that dominants should offer them a greater share of the group's reproductive output as an incentive to stay (Creel and Waser 1991). Counterintuitively, the model predicts that the degree of reproductive skew (the inequality of reproductive success among group members) should increase with the coefficient of relationship between dominants and subordinates because closely related subordinates are already "rewarded" by the inclusive fitness they gain from helping dominant relatives reproduce. Species as different as wasps and lions fit this prediction.

Interactions among related individuals

At the start of this chapter, we introduced the concept of kin selection, which is one of the most important explanations of cooperation among conspecific individuals. For kin selection to operate, individuals must dispense benefits more often to kin than to nonkin. This can happen in two ways. First, many animals can distinguish at least some kin from nonkin and behave differently toward them (Sherman et al. 1997). Second, the population may be structured, at least at that point in the life cycle when the altruistic behavior occurs, so that interacting individuals are more likely to be related than unrelated. A nestling bird, for example, is more often associated with siblings, on average, than with unrelated individuals. Such a population structure requires that individuals have not become randomly mixed before the time of dispersal.

KIN RECOGNITION AND CANNIBALISM. Many species of animals are cannibalistic, preying on smaller individuals of the same species or on relatively defenseless life history stages, such as eggs (Pfennig 1997). Many such species discriminate kin from nonkin and are less likely to eat related individuals. For example, tadpoles of the spadefoot toad *Scaphiopus bombifrons* develop into detritus- and plant-feeding omnivores if they eat these materials early in life, or into cannibalistic carnivores, with large horny beaks and large mouth cavities, if they eat animal prey. Omnivores associate more with their siblings than with nonrelatives, whereas carnivores do the opposite. Given equal opportunity, carnivores eat siblings much less frequently than unrelated individuals.

COOPERATIVE BREEDING. In many species of birds and in several mammals and fishes, young are reared not only by their parents, but also by other individuals that are physiologically able to reproduce, but do not. In most such species of birds, the helpers are young males that are prevented by competition from acquiring mates or territories. Several factors may explain cooperative breeding (Emlen 1991; Cockburn 1998), including kin selection, since in many species, the helpers aid their parents in rearing their siblings.

Hans-Ulrich Reyer (1990) studied pied kingfishers (*Ceryle rudis*) by individually marking birds so that their relationships and behavior could be monitored. Brood care is often provided not only by the breeding pair, but also by one or more young, unmated males, which include both "primary helpers" that help their parents and "secondary helpers" that are unrelated to the pair they help. Helping increases the direct fitness of a secondary helper because if a breeding male dies, the helper is likely to become the widow's mate in the next breeding season. Primary helpers gain indirect fitness because they increase the survival of their siblings (see Figure 14.15).

SOCIAL INSECTS. The most extreme altruism is found in EUSOCIAL animals: those in which nearly or completely sterile individuals (workers) rear the offspring of reproductive individuals, usually their parent (or parents). Eusociality is known in one species of mammal (the naked mole-rat *Heterocephalus glaber*), in all species of termites (Isoptera), in many Hymenoptera, and in a few other kinds of insects (Wilson 1971; Crozier and Pamilo 1996; Bourke and Franks 1995; Keller 1993).

Eusociality has evolved independently many times among the wasps, bees, and ants in the order Hymenoptera. In all eusocial Hymenoptera, the workers are females that do not mate. These insects have been especially important in studies of social interactions and the role of kin selection because they are HAPLODIPLOID: females develop from fertilized eggs and are diploid, but males develop from unfertilized eggs and are haploid. Consequently, **coefficients of relationship** (r) among relatives differ from those in diploid species (Figure 14.16). Whereas r in diploid species is 0.5 both between parent and offspring and between full siblings, a female hymenopteran is more closely related to her sister ($r = 0.75$) than to her daughter ($r = 0.5$), and is less closely related to her brother ($r = 0.25$). In a colony with a single, singly mated queen, the inclusive fitness of a female (a worker) may be greater if she devotes her energy to rearing reproductive sisters (future queens) than to rearing daughters. Thus kin selection is thought to have shaped the interactions among members of a hymenopteran colony, and it has been hypothesized that

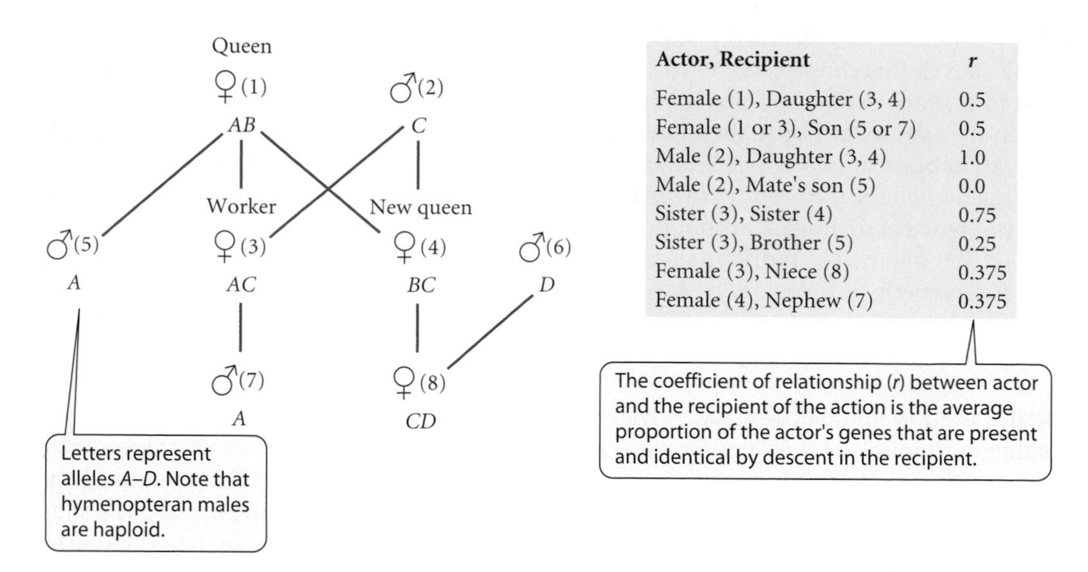

Actor, Recipient	r
Female (1), Daughter (3, 4)	0.5
Female (1 or 3), Son (5 or 7)	0.5
Male (2), Daughter (3, 4)	1.0
Male (2), Mate's son (5)	0.0
Sister (3), Sister (4)	0.75
Sister (3), Brother (5)	0.25
Female (3), Niece (8)	0.375
Female (4), Nephew (7)	0.375

Figure 14.16 Some coefficients of relationship among relatives in a hymenopteran species, such as an ant, assuming a single, singly mated queen per colony. Individuals in the pedigree are numbered as in the table, and one of the possible genotypes of each is labeled with letters representing alleles A–D. The coefficient of relationship (r) is calculated by finding the proportion (1.0 or 0.5) of an actor's genes that are inherited from one of its parents, multiplying by the probability that a copy of the same gene has also been inherited by the recipient, and summing these products over the actor's one or two parents. Notice that sisters are more closely related than mother and daughter or son. (After Bourke and Franks 1995.)

the haplodiploid condition in Hymenoptera is one of the factors that has predisposed these insects to evolve eusociality (Hamilton 1964).

The role of kin selection has been tested by a theory of *conflict between queens and workers*, proposed by Trivers and Hare (1976). The queen's fitness would be maximized by investing equally in reproductive daughters and sons, because she is equally related (r = 0.5) to both (see Chapter 12 on the evolution of the sex ratio). But the inclusive fitness of workers would be maximized if three-fourths of the reproductive siblings they reared were sisters (r = 0.75) and one-fourth were brothers (r = 0.25). A queen can control the sex ratio among newly laid eggs by determining which eggs are fertilized with sperm stored in her reproductive tract. Workers can control whether a female larva develops into a queen or a worker by the nutrition they provide to the larva. Thus workers could control the sex ratio among their adult reproductive siblings by withholding care from male larvae and by altering the proportion of female larvae that develop into queens versus workers. If kin selection has shaped the behavior of workers, we would expect the sex ratio of reproductive offspring to be biased toward 0.75 (3:1 female:male) in colonies with a single queen. In colonies with multiple queens, workers would have a coefficient of relationship lower than 0.75 to many of the offspring they rear (since they are not full sisters), so the sex ratio should be closer to 0.5 (1:1).

Many data support these predictions (Bourke and Franks 1995; Crozier and Pamilo 1996). The average sex ratio is closer to 3:1 in single-queen than in multiple-queen species of ants, and it is almost exactly 1:1 in slave-making species of ants, in which the workers that rear the slave-maker's brood are members of other species, captured by the slave-maker's own workers. The slaves have no genetic interest whatever in the captor's reproductive success, and so are not expected to alter the sex ratio by preferential treatment.

Moreover, there is direct evidence in the wood ant (*Formica exsecta*) that workers manipulate the sex ratio as kin selection theory predicts (Sundström et al. 1996). Although all colonies have about the same sex ratio among eggs, the sex ratio among pupae becomes shifted toward females in single-queen colonies, but toward males in multiple-queen colonies, which would be advantageous for the queens (Figure 14.17). Kin selection may not be the only possible explanation of these patterns (Mehdiabadi et al. 2003), but so far it seems the most likely.

Figure 14.17 Evidence that workers in hymenopteran colonies manipulate the sex ratio to increase their inclusive fitness, as predicted by kin selection theory. The bars show the change in the proportion of females between the egg and pupal stage in colonies of wood ants with single queens and in colonies with multiple queens, in which each worker is unrelated to many of the offspring. (After Sundström et al. 1996.)

A Genetic Battleground: The Nuclear Family

At first surmise, it would seem that relationships within families should be the epitome of cooperation, since the parents' fitness depends on producing surviving offspring. However, evolutionary biologists have come to understand that these interactions are pervaded with potential conflict, and that much of the diversity of reproductive behavior and life histories among organisms stems from the balance between conflict and cooperation. (Incidentally, the ways in which some animal species behave toward family members starkly show that natural selection utterly lacks morality, as we pointed out in Chapter 11.)

Mating systems and parental care

Whether or not one or both parents care for offspring varies greatly among animal species and partly determines the MATING SYSTEM, the pattern of how many mates individuals have and whether or not they form pair bonds (Clutton-Brock 1991; Davies 1991). Providing care (such as guarding eggs against predators, or feeding offspring) increases offspring survival, which enhances the fitness of both parents (and of the offspring). But parental care is also likely to have a cost. It entails risk, and it requires the expenditure of time and energy that the parent might instead allocate to further reproduction: a female might lay more eggs, and a male might find more mates.

In most animals, neither parent provides care to the eggs or offspring after eggs are laid, and one or both sexes, especially in long-lived species, may mate with multiple partners (PROMISCUOUS MATING). In many birds and mammals, females provide care, but males do not, and males may mate with multiple females (POLYGYNY). In some fishes and frogs and a few species of birds, only males guard eggs or care for offspring, and in some such species, the female may mate and lay eggs with different males (POLYANDRY). (These terms stem from the Greek *polys*, "many"; *gyne*, "woman"; *andros*, "man.") In many species of birds, some mammals, and a few insects such as dung beetles, a female and male form a "socially monogamous" pair bond and contribute biparental care of the offspring. (However, as we saw earlier in this chapter, many such birds engage in frequent "extra-pair copulation" and are not sexually monogamous. Females may also increase their reproductive success by laying eggs in the nests of unwitting foster parents.)

Each parent in a mated pair will maximize her or his fitness by some combination of investing care in current offspring and attempting to produce still more offspring. In a species with biparental care, each parent would profit by leaving as much care as possible to the other partner, as long as any resulting loss of her (or his) fitness due to the death of her (or his) current offspring were more than compensated by the offspring she (or he) would have from additional matings. If offspring survival were almost as great with uniparental care as with biparental care, selection would favor females that defected, abandoning the brood to the care of the male—or vice versa (Figure 14.18). Thus a conflict between mates arises as to which will evolve a promiscuous habit and which will care for the eggs or young. Selection favors defection more strongly in the sex for which parental care is more costly (in terms of lost opportunities for further reproduction). On the other hand, if the survival of offspring depends very strongly on care, the fitness of both parents may be maximized by forming a socially monogamous pair bond and cooperating in offspring care.

This theory may explain why in birds and mammals, parental care is generally provided by females or by both mates, while in fishes and frogs, it is usually provided by males (Clutton-Brock 1991; Figure 14.19). Fishes and frogs guard the eggs or young, but do not feed them. Males can often mate with multiple females and guard all their eggs in a single nest; thus they pay a lesser cost than females, whose subsequent reproduction depends on replenishing the massive resources they expend in egg production. In birds and mammals that must feed their young, parental care is more costly for males than for females, since males could potentially obtain many matings in the time they must spend rearing one brood exclusively.

All else being equal, the strength of natural selection for parental care is proportional to the probability that individuals are actually caring for their own offspring (carrying the

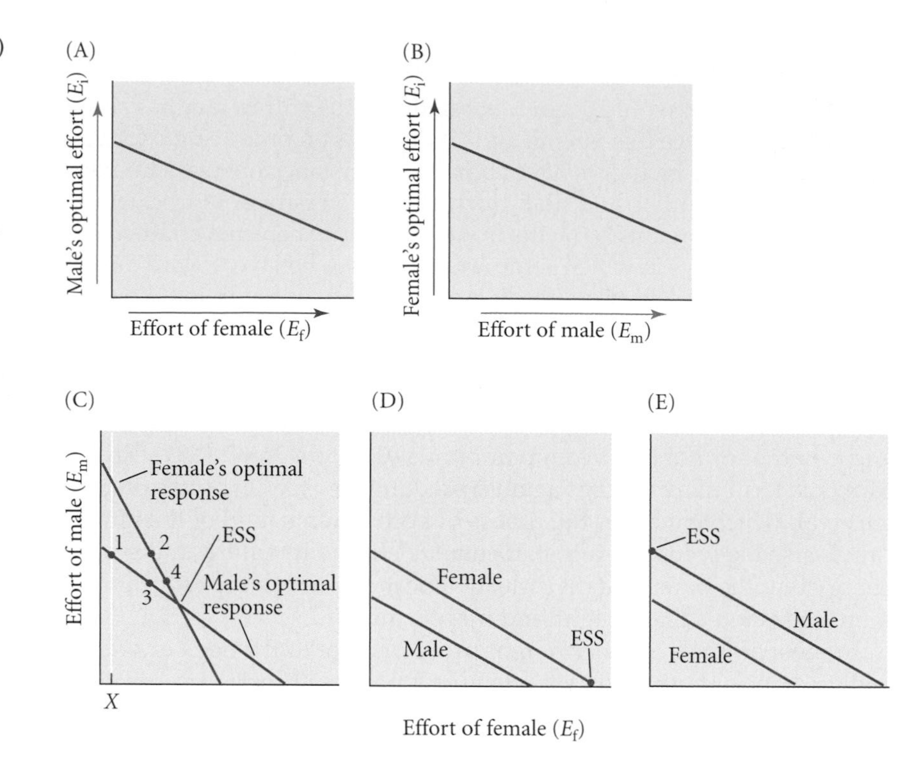

Figure 14.18 An ESS model of parental care. (A, B) The optimal parental effort expended by each sex declines, the more effort its partner expends. (C) Curves for males and females plotted together. Their intersection marks the ESS, the evolutionarily stable strategy. If, for example, the population starts with female effort (E_f) equal to X, male effort (E_m) evolves to point 1; but then the optimal E_f is at point 2 on the female's optimality line. When E_f evolves to point 2, E_m evolves to point 3; but then E_f evolves to point 4. Eventually, E_m and E_f evolve to the intersection (the ESS), no matter what the initial conditions are. (D, E) Conditions can be envisioned in which the optimal curves for the sexes do not intersect and the ESS is care by only the female (D) or the male (E). (After Clutton-Brock and Godfray 1991.)

carer's genes). Male parental care, for example, is expected to be proportional to the males' "confidence of paternity"—which may be quite low in species with frequent extra-pair copulation (Whittingham et al. 1992). The dunnock (*Prunella modularis*) is a songbird with a variable mating system: some individuals form monogamous pairs, some polygynous trios (two females and one male), and some polyandrous trios (two males and one female). Although male reproductive success is maximized by polygyny, females in polyandrous trios have greatest reproductive success, which may account for the persistent variation in mating system. Polyandrous males each provide parental care only in proportion to the amount of time they spent with the female during her fertile period, presumably using this as an estimate of their paternity (Davies 1992). In some other species, males simply evict or kill offspring that are unlikely to be their own. In other words, they practice INFANTICIDE, our next topic.

(A)

(B)

Figure 14.19 Parental care. (A) A male Australian seahorse (*Hippocampus breviceps*) giving birth. Males of this species carry and nurture developing young in their pouch. (B) A female red-necked wallaby (*Macropus rufogriseus*) carries her single offspring in her pouch. (A © Paul A. Zahl/ Photo Researchers, Inc.; B © John Cancalosi/ naturepl.com)

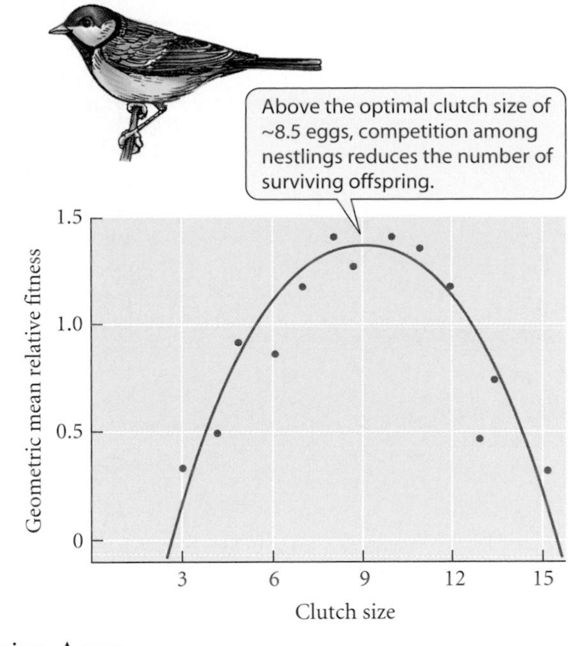

Figure 14.20 An estimate of fitness as a function of the number of eggs per clutch in a population of great tits (*Parus major*) in Oxford, England, from 1960 to 1982. Fitness is estimated by the geometric mean number of offspring that survived to the next breeding season from a given clutch size. (The geometric mean takes year-to-year variation into account; for example, in years of poor food supply, mortality is disproportionately high in large broods.) (After Boyce and Perrins 1987.)

Infanticide, abortion, and siblicide

There are a number of circumstances in which an individual's fitness may be enhanced by killing young members of the same species (Hausfater and Hrdy 1984; Clutton-Brock 1991; Borries et al. 1999). For example, infanticide may be sexually selected. In many species of mammals, including lions and some primates and rodents, males that replace a mated male kill the offspring of his mates. The infanticidal male can then father his own offspring faster, because females become fertile and sexually receptive sooner if they are not nursing young. DNA analyses of wild langur monkeys (*Presbytis entellus*) showed that males that kill infants are the fathers of the infants that are subsequently born to the same mothers (Borries et al. 1999).

Parents sometimes kill offspring as a way of adaptively regulating brood size. A parent's fitness is proportional (all else being equal) to the number of surviving offspring, which equals (size of egg clutch or brood) × (per capita probability of survival). In species with parental care, the probability of survival may decrease as brood or clutch size increases due to competition among offspring for resources such as food. Thus, for example, in birds, the clutch size that maximizes the parent's fitness is often less than the greatest clutch size that the female can lay (Figure 14.20). Moreover, the parental care expended on an excessively large brood can reduce the parent's survival and subsequent reproduction, reducing its lifetime fitness. Thus it can be adaptive for a parent to reduce the number of offspring to an optimal number (possibly even zero). For example, females of some species of mice kill some or all of their young if food is scarce or the number born is too high. Likewise, shortage of food or disturbance by predators can induce many birds to desert their eggs or young, especially early in the breeding season (when the parents have the opportunity to renest elsewhere). In the same vein, plants typically abort the development of many of their offspring (seeds), allocating limited resources to fewer but larger seeds with a greater chance of survival.

When the offspring in a brood (sibs) share an environment, such as young birds in a nest or larval parasitic wasps within a caterpillar, they may actively fight for resources, and larger individuals may kill smaller siblings (SIBLICIDE). Siblicide is the norm in some species of eagles and boobies, in which the female lays two eggs, but one of the nestlings always kills the other. (The second egg may be the female's "insurance" in case the first is inviable.) In other species, such as owls and herons, the number of survivors of sibling competition depends on the food supply (Figure 14.21). The young of some shark species eat their siblings while they share their mother's uterus.

Figure 14.21 Siblicide in the brown booby (*Sula leucogaster*). The parent is sheltering a large chick that has forced its sibling out of the nest. The parent ignores the dying chick. (Courtesy of J. Alcock.)

Parent-offspring conflict

Parents and offspring typically differ with respect to the optimal level of parental care (Trivers 1974; Godfray 1999). A parent's investment of energy in caring for one offspring may reduce her production of other (e.g., future) offspring. Production of other offspring increases the inclusive fitness of both the parent and each current offspring. However, the parent is equally related to all her offspring ($r = 0.5$), whereas each offspring values parental investment in itself twice as much as investment in a full sib ($r = 0.5$) and four times as much as investment in a half sib ($r = 0.25$). Therefore offspring are expected to try to obtain more resources from a parent than it is optimal for the parent to give, resulting in **parent-offspring conflict**. Animals may evolve features that enable offspring to extract extra resources from their parents; for example, baby coots (*Fulica americana*) have unusual reddish plumes that stimulate their parents to feed them (Lyon et al. 1994).

From this perspective, siblicide is more advantageous to a sib-killing offspring than to its parents, and so it is surprising that parent birds usually stand by without interfering while one of the offspring slaughters the others (see Mock 2004). The theory of parent-offspring conflict, however, does explain why young birds and mammals often beg for more food than the parents are "willing" to give, and why adults must aggressively wean their offspring from further care. The best-documented example of parent-offspring conflict, perhaps, is one we have already discussed: the conflict between queens and workers of social insects over the sex ratio of reproductive offspring.

Genetic Conflicts

Thyme (*Thymus vulgaris*), a familiar food seasoning, is a member of the mint family. Like most mints, some thyme plants have hermaphroditic flowers, with both male (stamens) and female (pistil) parts. Many thyme plants, however, lack anthers; they are "male-sterile," or female. Male sterility is caused by a mitochondrial gene. The interesting point is that all thyme plants carry this "cytoplasmic male sterility" (CMS) factor. Hermaphrodites have male function only because they carry a "restorer" gene, inherited on a nuclear chromosome, that counteracts CMS. Remarkably, there exist hermaphroditic plant species in which all individuals carry both a CMS gene and a nuclear restorer. Why should a gene exist, only to be counteracted by another gene?

The answer appears to be that there exist conflicts between different genes as an outcome of selection at the gene level. Such conflicts can arise whenever a gene has a transmission advantage over other genes, perhaps by a segregation advantage during meiosis or perhaps by not following the rules of meiosis at all. In Chapter 11, we encountered "selfish genes" that increase in frequency not because they provide an advantage to the organism that carries them, but simply because they have an advantage in transmission from one generation to the next. One example is the *t* allele in mice, which causes sterility or death in homozygotes, but increases in frequency due to meiotic drive: it is transmitted to more than 90 percent of a heterozygous male's sperm. Such "outlaw genes," which promote their own spread at a faster rate than other parts of the genome, can create a context in which there is selection for genes at other loci to suppress their effects. When this is the case, a **genetic conflict** is said to exist (Hurst et al. 1996).

Cytoplasmic male sterility in plants illustrates how such conflicts arise (Figure 14.22). Since mitochondria are maternally inherited, any mitochondrial allele that can increase the number of female relative to male offspring will increase in frequency relative to an allele that does not alter the sex ratio from 1:1. Male-sterile plants produce more seeds than hermaphroditic plants because protein and other resources are diverted from pollen to seed production. Thus mitochondrially inherited cms^+ alleles increase in frequency. In contrast, recall (from Chapter 12) that because every individual has a mother and a father, nuclear alleles that produce an even sex ratio are advantageous. Thus production of excess females is disadvantageous to nuclear genes, since they are not transmitted by pollen from male-sterile plants. A mutation of a nuclear gene that nullifies the effect of a cms^+ allele may therefore increase in frequency. The only function of such a gene, then, is to counteract the effect of a selfish gene at another locus.

Mitochondrial and other cytoplasmically inherited genes commonly conflict with nuclear genes, as cytoplasmic male sterility illustrates. For example, they may exaggerate sexual conflict, since they benefit from enhancing female, but not male, fitness (Zeh 2004). Mitochondrial genes that reduce sperm competitive ability, for instance, could be fixed if they increase female fitness.

Because of their effects on sex ratio, genes on the sex chromosomes of animals may similarly initiate genetic conflicts. In many species, the sex ratio among the progeny of a single mating depends on the proportion of eggs fertilized by X-bearing versus Y-bearing

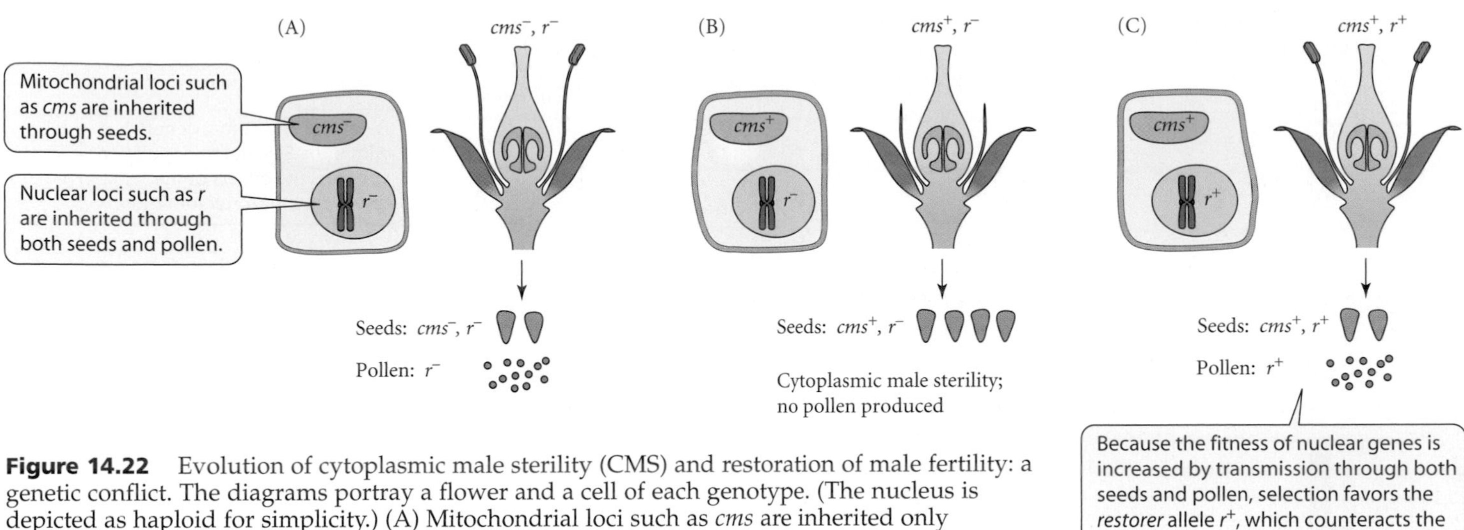

Figure 14.22 Evolution of cytoplasmic male sterility (CMS) and restoration of male fertility: a genetic conflict. The diagrams portray a flower and a cell of each genotype. (The nucleus is depicted as haploid for simplicity.) (A) Mitochondrial loci such as *cms* are inherited only through eggs, and nuclear loci such as *r* are inherited through both eggs and pollen. (B) A cms^+ mutation eliminates pollen production, and may increase seed number due to the plant's reallocation of energy and resources from pollen to eggs. Thus cms^+ increases in frequency. (C) Because the fitness of nuclear genes is increased by transmission through both eggs and pollen, selection favors the restorer allele r^+, which counteracts the effects of cms^+. The population may become fixed for antagonistic alleles cms^+ and r^+, but may be phenotypically indistinguishable from the ancestral population, cms^-r^-.

sperm, yielding daughters and sons respectively. X-linked genes that can increase the proportion of successful X-bearing sperm (by meiotic drive), distorting the sex ratio of the progeny, have a transmission advantage and may increase in frequency. If they do, there is selection for autosomal genes that achieve highest fitness by restoring the 1:1 sex ratio.

In matings of *Drosophila simulans* from the Seychelles Islands, the progeny sex ratio is about 1:1, but the F_1 males from a cross between Seychelles females and males from elsewhere in the world produce a great excess of female offspring (Figure 14.23). This skewed sex ratio is the result of a *segregation-distorter* gene on the X chromosomes of Seychelles flies that causes degeneration of the male's Y-bearing sperm. The effect of this chromosome is not ordinarily seen because the Seychelles population also has recessive autosomal genes that suppress the distortion. Other populations of *D. simulans* lack both the X-linked *distorter* and the suppressor alleles, so the effects of *distorter* become evident when it is crossed into their genetic background. In the Seychelles and other areas where the *distorter* gene has increased in frequency, there has been selection for autosomal suppression of its effects (Atlan et al. 1997).

Our final example of genetic conflicts may be a surprise: it concerns pregnancy in humans and other mammals (Haig 1993, 1997). Here, the cooperation one might expect between mother, father, and offspring is compromised by parent-offspring conflict and a genetic conflict between the parents that is played out in the offspring.

We have seen that an offspring is expected to try to get more resources from its mother than the mother should willingly give, since she profits more than the offspring does from reserving some resources for subsequent offspring. Some of the interactions between the human fetus and the mother follow this prediction. For example, mothers increase their production of insulin (which causes cells to remove glucose from the blood) during pregnancy—the very time you might expect them to decrease insulin levels in order to provide sugar to the fetus. However, the fetus produces extremely high levels of a hormone (hPL) that counteracts insulin, so that the net result is no alteration of blood glucose concentration. This escalation of opposing hormones seems to serve no purpose, but is just what one might expect of a parent-offspring conflict.

Even more remarkable is the expression of certain genes that are IMPRINTED so that whether or not they are transcribed in the embryo depends on whether they were inher-

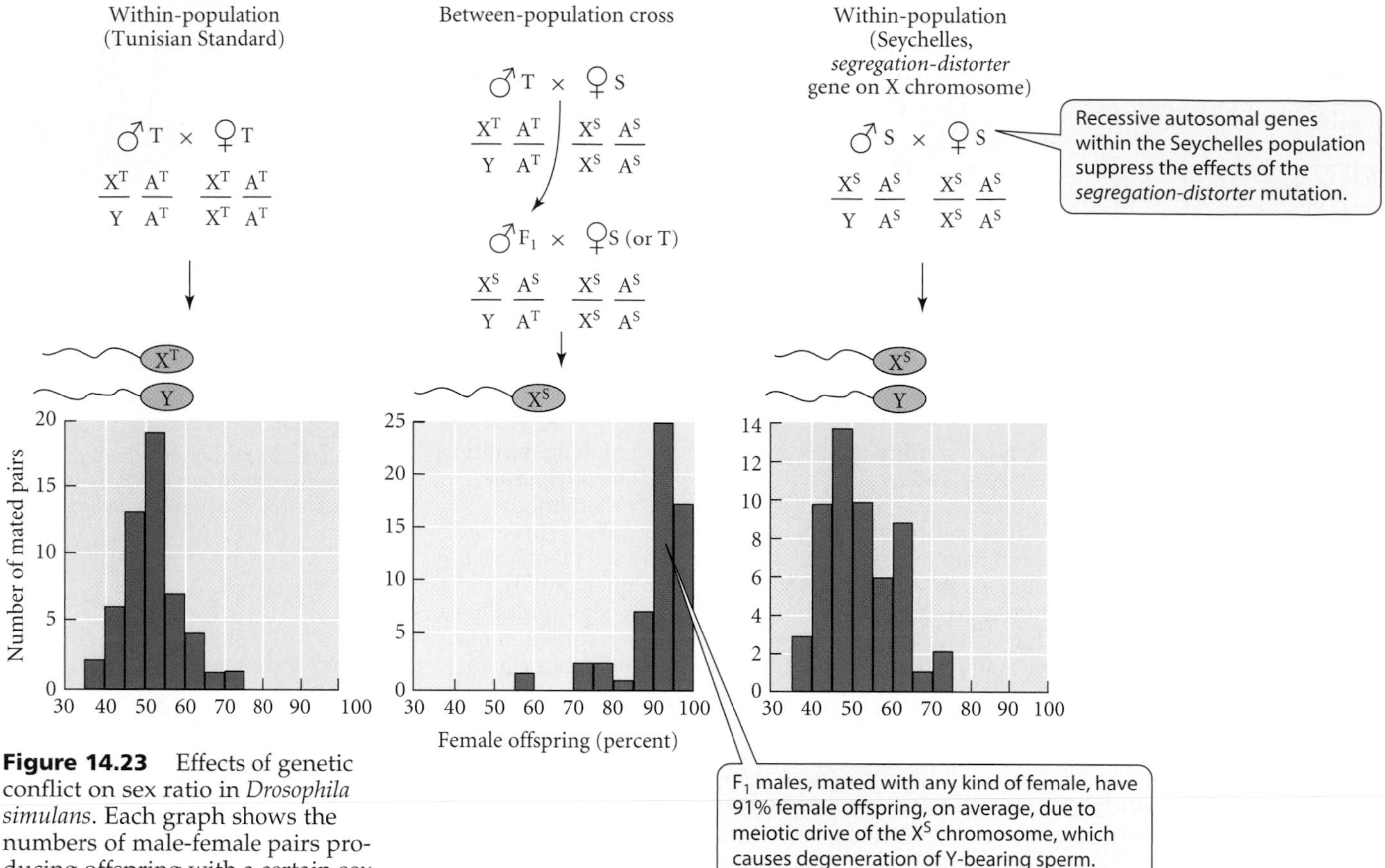

Figure 14.23 Effects of genetic conflict on sex ratio in *Drosophila simulans*. Each graph shows the numbers of male-female pairs producing offspring with a certain sex ratio (percentage of females). Results are shown for crosses within a "standard" population from Tunis (T, at left), crosses within the Seychelles population (S, at right), and offspring of F_1 males from crosses between these two populations (center). The parents' genotypes are symbolized by X, Y, and A for X chromosomes, Y chromosomes, and autosomes respectively. (After Atlan et al. 1997.)

ited from the mother or the father. (For example, the maternal copy of a gene may be methylated, in which case it is not expressed in the offspring.) In terms of selection at the level of genes, the fitness of paternally inherited genes will be greatest if the embryo survives, since copies of those genes will not be carried by the mother's subsequent offspring if she mates with a different male. In contrast, maternally transmitted genes will also be carried by the mother's subsequent offspring, so their inclusive fitness is greatest if they prevent the mother from nurturing the current embryo exclusively, to the detriment of later offspring. Therefore, paternally inherited genes should enhance the embryo's ability to obtain nourishment from the mother, whereas maternally inherited genes should temper the embryo's ability to do so. In mice, a pair of conflicting genes shows just this pattern. Only the paternal copy of the gene for IGF-2 (insulin-like growth factor 2), which promotes the early embryo's ability to obtain nutrition from the uterus, is expressed in the embryo. IGF-2 is degraded by another protein (IGF-2R, insulin-like growth factor 2 receptor); only the maternal allele of the gene for this protein is expressed. Imprinting of such genes is likely if an allele has an inclusive fitness benefit when it is derived from one parent, but an inclusive fitness cost when it is derived from the other (Mochizuki et al. 1996; Haig 1997).

Parasitism, mutualism, and the evolution of individuals

The concept of genetic conflict bears on many topics in evolutionary biology (Hurst et al. 1996). Consider, for example, a genetic element—a gene or set of genes—that can replicate faster than the rest of the genome with which it is associated. It might be, for example, a bacterium that lives within a host organism's cells. (Such an organism is called an ENDOSYMBIONT.) If the population of endosymbionts within a single host is genetically variable, selection within that population favors (by definition) the genotype that increases in numbers faster than others. Because growth and reproduction of the endosymbionts depend on resources obtained from the host, excessive growth in the number of symbionts may reduce the host's fitness.

The fitness of an endosymbiont genotype is measured by the number of new hosts it infects per generation. If the endosymbionts are transmitted **horizontally**—that is, laterally among members of the host population (Figure 14.24A), the number of new hosts infected may be proportional to the number of symbiont progeny released from each old host. If symbionts escape to new hosts before the old host dies, their fitness does not depend strongly on the reproductive success of the individual host in which the parent symbionts reside. Therefore, selection favors symbiont genotypes with a high reproductive rate, even if they kill the host. In other words, selection favors evolution of a PARASITE that may be highly virulent.

Suppose, in contrast, that symbionts are transmitted mostly **vertically**, from a mother host to daughter hosts (Figure 14.24B). Symbiont and host are now chained to each other, and the survival and reproductive success of the symbionts now depend entirely on the fitness of the host. Selection for high proliferation *within* the symbiont populations occupying each host is opposed by selection *among* the populations of symbionts that occupy different hosts. On balance, selection at the group level favors genotypes with restrained reproduction—those that do not extract so many resources from the host as to cause its death before it can transmit the endosymbionts to its progeny. Selection may even favor alleles in the symbiont that enhance the host's fitness, since that also enhances the fitness of the symbionts carried by that host. Furthermore, selection favors host alleles that control or inhibit the symbiont (an instance of genetic conflict). Evolution in both the symbiont and its host may therefore result in **mutualism** (an interaction in which two genetic entities enhance each other's fitness). (We return to the evolution of parasitism and mutualism in Chapter 18.)

In the extreme case, the symbiont may become an integral, essential part of the host. Many eukaryotes harbor vertically transmitted intracellular bacteria that play indispensable biochemical roles, such as those that reside in special cells in aphids and synthesize essential amino acids (Figure 14.24C). Mitochondria and chloroplasts are considered organelles in eukaryotic cells, but they were originally endosymbiotic bacteria. In these cases, the host and the symbiont have highly correlated reproductive interests: any advantage to one party provides an advantage to the other.

The strong correlation between the reproductive interests of associated elements—such as the organelle and nuclear genomes of eukaryotes—leads us to understand why indi-

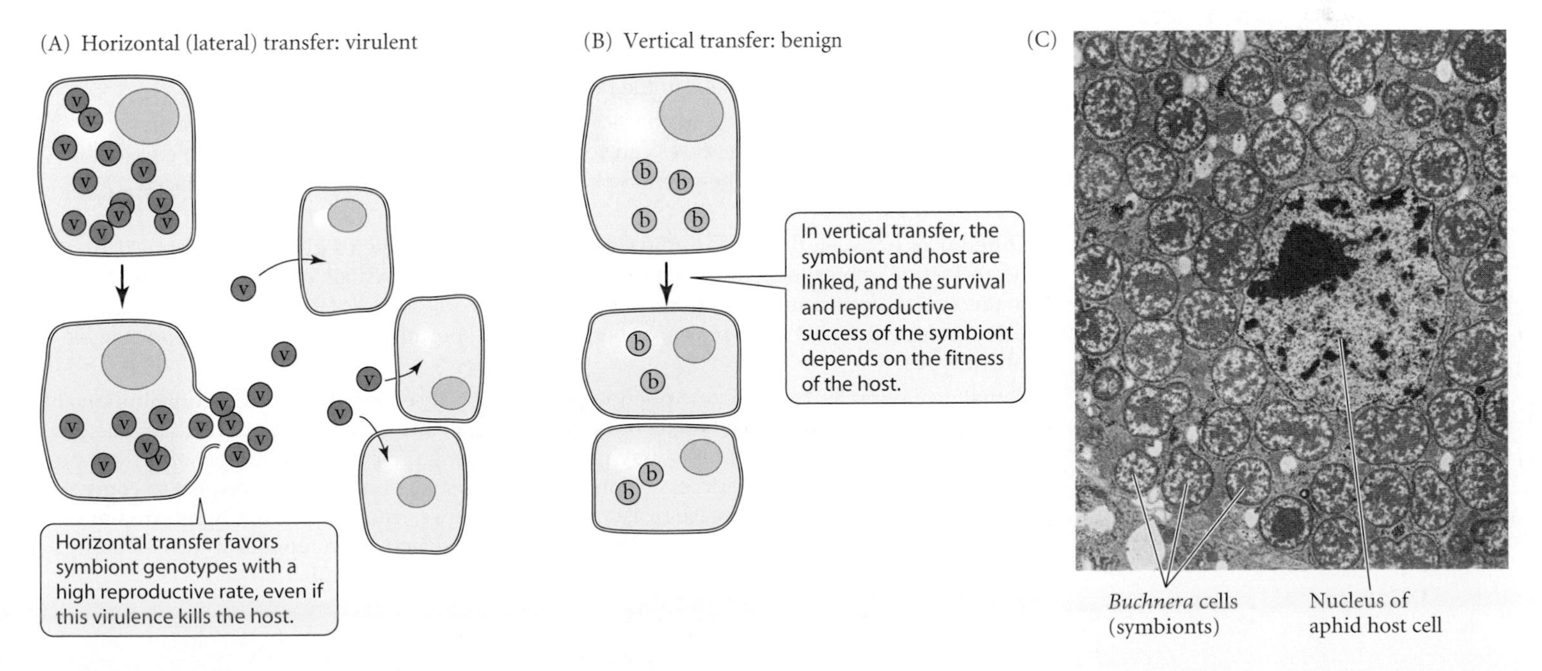
(A) Horizontal (lateral) transfer: virulent
Horizontal transfer favors symbiont genotypes with a high reproductive rate, even if this virulence kills the host.
(B) Vertical transfer: benign
In vertical transfer, the symbiont and host are linked, and the survival and reproductive success of the symbiont depends on the fitness of the host.
(C)
Buchnera cells (symbionts)
Nucleus of aphid host cell

Figure 14.24 (A) Horizontal transmission of endosymbiotic elements (v) from one host to unrelated hosts selects for a high level of virulence. (B) Vertical transmission of endosymbiotic elements from a host to its descendants favors relatively benign endosymbionts (b) with a lower reproductive rate. (C) In the extreme case, a vertically transmitted symbiont may become an integral part of the host. These intracellular bacteria (*Buchnera*) in specialized cells (bacteriocytes) of an aphid supply essential amino acids to the host. (C, photo courtesy of N. Moran and J. White.)

vidual organisms exist at all (Buss 1987; Maynard Smith and Szathmáry 1995; Michod 1997; Frank 1997). An organism is more than a group of cells; for example, dividing bacteria that remain loosely attached, but physiologically independent of one another, do not constitute an organism. Rather, the cells of a multicellular organism cooperate and play different roles—including the distinction between cells that give rise to gametes (germ cells) and those that do not (the soma). Why should unicellular ancestors, in which each cell had a prospect of reproduction, have given rise to multicellular descendants in which some cells sacrifice this prospect?

The fundamental answer is kin selection: if the cell lineages in a multicellular organism arise by mitosis from a single-celled egg or zygote, the genes of cooperative cells that sacrifice reproduction for the good of the cell "colony" are propagated by closely related reproductive cells. However, the coefficient of relationship is reduced to the extent that mutational differences arise among cells. A mutation that increases the rate of cell division has a selective advantage *within* the colony, and may have a further advantage if it increases the chance that those cells that carry it will be included in the reproductive tissues and give rise to gametes. But unregulated cell division—as in cancer—usually harms the organism. Selection at the level of whole colonies of cells—organisms—therefore opposes selection among cells within colonies. It has favored mechanisms of "policing" that regulate cell division and prevent renegade cell genotypes from disrupting the integrated function of the organism. In animals, it has resulted in the evolution of a germ line that is segregated early in development from the soma, thereby excluding most disruptive mutations from the gametes and reducing their possible fitness advantage. Selection for organismal integration may be responsible for the familiar but remarkable fact that almost all organisms begin life as a single cell, rather than as a group of cells. This feature increases the kinship among all the cells of the developing organism, reducing genetic variation and competition within the organism and increasing the heritability of fitness. The result, then, has been the emergence of the "individual," and with it, the level of organization at which much of natural selection and evolution take place.

Summary

1. Many biological phenomena result from conflict or cooperation among organisms or among genes. Although group selection may sometimes play a role in the evolution of cooperation, the evolution of most interactions can be explained best by selection at the level of individual organisms or genes.
2. Cooperation often evolves by kin selection, based on differences among alleles or genotypes in inclusive fitness: the combination of an allele's direct effects (on its carrier's fitness) and its indirect effects (on other copies of the allele, borne by the carrier's kin). Hamilton's rule describes the condition for increase of an allele for an altruistic trait in terms of the coefficient of relationship, the benefit to the beneficiary, and the cost to the donor.
3. Characteristics that contribute to conflict and cooperation often evolve by frequency-dependent selection. Such features can sometimes be modeled by calculating the evolutionarily stable strategy (ESS): the phenotype that, once established, cannot be replaced by mutant phenotypes.
4. Differences between the sexes in the size and number of gametes give rise to conflicts of reproductive interest and to sexual selection, in which individuals of one sex compete for mates (or for opportunities to fertilize eggs). The several forms of sexual selection include direct competition between males, or between their sperm, and female choice among male phenotypes.
5. Females may prefer certain male phenotypes because of sensory bias, direct contributions of the male to the fitness of the female or her offspring, or indirect contributions to female or offspring fitness. Indirect benefits may include fathering offspring that are genetically superior with respect to mating success ("runaway sexual selection") or with respect to components of viability, such as disease resistance ("good genes" models). Sexually selected male features may also evolve by antagonistic coevolution: selection on females to resist mating, and on males to overcome female resistance with irresistible stimuli.

6. The leading explanations for cooperation or apparent altruism (features that benefit other individuals, to the detriment of the possessor) are manipulation, individual advantage, reciprocation, and kin selection.
7. The genetic benefit of caring for offspring is an increase in the number of current offspring that survive. Its cost is the number of additional offspring that the parent could expect to have if she/he abandoned the offspring and reproduced again. Parental care is expected to evolve only if its genetic benefit exceeds its genetic cost. Whether or not one or both parents evolve to provide care can depend on the male's confidence of paternity and on the relative cost/benefit ratio for each parent.
8. Conflicts between parents and offspring may arise because a parent's fitness may be increased by allocating some resources to its own survival and future reproduction, thus providing fewer resources to current offspring than would be optimal from the offspring's point of view. This principle may be one of several reasons that in many species, parents may reduce their brood size by aborting some embryos, killing some offspring, or allowing siblicide among their offspring.
9. Conflicts may exist among different genes in a species' genome. For example, a gene may spread at a faster rate than other parts of the genome, engendering selection for other genes to prevent it from doing so. At loci that are transmitted through only one sex, gene-level selection favors alleles that alter the sex ratio in favor of that sex. Such alteration creates selection at other loci for suppressors that restore the 1:1 sex ratio.
10. Other phenomena explained by genetic conflicts include genomic imprinting, which affects the expression of maternally and paternally derived alleles in mammalian embryos, and the evolution of integration among cells—the very essence of multicellular organisms.

Terms and Concepts

antagonistic coevolution
coefficient of relationship
condition-dependent indicator (of fitness)
ESS (evolutionarily stable strategy)
genetic conflict
good genes model (of female choice)
Hamilton's rule
handicap model (of female choice)
horizontal transmission
inclusive fitness
kin selection
mutualism
parent-offspring conflict
reciprocal altruism
runaway sexual selection
sensory bias
sperm competition
vertical transmission

Suggestions for Further Reading

An outstanding, easily readable introduction to the study of behavior, emphasizing evolutionary explanations, is *Animal behavior: An evolutionary approach*, by J. Alcock (seventh edition, Sinauer Associates, Sunderland, MA, 2001). Sexual selection is comprehensively reviewed by M. Andersson in *Sexual selection* (Princeton University Press, Princeton, NJ, 1994). An excellent set of essays on many aspects of cooperation and conflict is *Levels of selection*, edited by L. Keller (Princeton University Press, Princeton, NJ, 1999). Genetic conflict in particular is reviewed in L. D. Hurst, A. Atlan, and B. O. Bengtsson, "Genetic conflicts" (1996, Q. Rev. Biol. 71: 317–364). J. Maynard Smith and E. Szathmáry provide an outstanding analysis of the origin of levels of organization, from cells to societies, in *The major transitions in evolution* (Oxford University Press, 1995).

Problems and Discussion Topics

1. In many socially monogamous species of parrots and other birds, both sexes are brilliantly colored or highly ornamented. Is sexual selection likely to be responsible for the coloration and ornaments of both sexes? How can there be sexual selection in pair-bonding species with a 1:1 sex ratio, since every individual presumably obtains a mate? Which of the models of sexual selection described in this chapter might account for these species' characteristics? (See Jones and Hunter 1999 for an experimental study of this topic.)

2. Describe the contexts in which animals might evolve signals used in interactions among members of the same species or in interactions with different species. Would you expect each kind of signal to be honest or dishonest? Why?
3. Many species of albatrosses and other seabirds nest in colonies on islands. The adults search for food at sea, sometimes over great distances. Some ecologists who study seabirds turn to their favorite topic of conversation in a bar one evening, and one says, "I can imagine an albatross genotype that destroys the eggs or nestlings of its colonymates when the parents are away foraging, and leaves them dead without eating them. Do you think such a behavior would evolve?" One of his companions says, "Yes, because it would make more food available for the albatross and its offspring." Another says, "No, because it would threaten the survival of the population." The fourth says, "You're both wrong. I have a different explanation for why albatrosses don't kill others' chicks." What is the fourth ecologist's explanation, and why does she think her companions are wrong?
4. Although the worker females in eusocial Hymenoptera do not mate, some of them can lay haploid eggs that develop into sons. In what ways would doing so affect the inclusive fitness of a worker? How would it affect the fitness of her mother, the queen? How would other workers be expected to react to these male larvae? Is there a genetic conflict of interest within a colony of social insects? What is known about egg laying by workers? (See Keller 1999.)
5. Speculate about why some genetic elements (e.g., mitochondria and chloroplasts) are usually inherited through only one parent, rather than biparentally, as are most nuclear genes. (See Hurst et al. 1996.)
6. Kin selection explains why organisms may provide benefits to relatives. Is there a conflict between the principle of kin selection and the evolution of siblicide and abortion?
7. Some models described in this chapter predict phenomena that seem familiar from contemporary human societies, such as male aggressiveness and competition for mates. There is an enormous and very controversial literature on whether these and other behavioral features in humans have a genetic basis and have evolved for reasons postulated in evolutionary models, or whether they are culturally formed. Analyze reasons for and against the proposition that men are more aggressive than women because of sexual selection for competitiveness. (See, for example, Daly and Wilson 1983 versus Kitcher 1985.)
8. What kind of evidence can be used to tell whether a common human behavior is an evolved adaptation (with a genetic basis) or a result of learning and culture? What difference does it make?

15

Species

Speciation forms the bridge between the evolution of populations and the evolution of taxonomic diversity. The diversity of organisms is the consequence of CLADOGENESIS, the branching or multiplication of lineages, each of which then evolves (by ANAGENESIS, or evolution within species) along its own path. Each branching point in the great phylogenetic tree of life marks a speciation event: the origin of two species from one.

In speciation lies the origin of diversity, and the study of speciation bridges microevolution and macroevolution.

Many events in the history of evolution are revealed to us only by virtue of speciation. If a single lineage evolves great changes but does not branch, the record of all steps toward its present form is erased, unless they can be found in the fossil record. But if the lineage branches frequently, and if intermediate stages of a character are retained in some branches that survive to the present, then the history of evolution of the feature may be represented, at least in part, among living species. This fact is used routinely to infer phylogenetic relationships among living taxa and to trace the evolution of characteristics on phylogenetic trees (see Figure 3.3).

Similar but separate. The fire-bellied toad (*Bombina bombina*) and the yellow-bellied toad (*B. variegata*), both found in Europe, are thought to have diverged during the Pliocene. Where their ranges meet, hybrids are found but are less viable than the nonhybrid individuals of either species. (Photo Researchers, Inc.; *B. bombina* © Stephen Dalton, *B. variegata* © Steinhart Aquarium.)

The most important consequence of speciation is that different species undergo independent divergence, maintaining separate identities, evolutionary tendencies, and fates (Wiley 1978). Some authors have also suggested that speciation may facilitate the evolution of new morphological and other phenotypic characters—that is, that a characteristic that would not evolve in a single, unbranched lineage may be able to do so if the lineage branches. (This view, however, is not widely accepted.)

Some steps toward speciation may occur fast enough for us to study directly, but the full history of the process is usually too prolonged for one generation, or even a few generations, of scientists to observe. Conversely, speciation is often too fast to be fully documented in the fossil record, and even an ideal fossil record could not document some of the genetic processes in speciation that are still inadequately understood. Thus the study of speciation is based largely on inferences from living species.

What Are Species?

Many definitions of "species"—which is Latin for "kind"—have been proposed (Table 15.1). It is important to bear in mind that a definition is not true or false, because the definition of a word is a convention. Still, if a conventional definition of a word has been established, one can apply it in error. Although a rose by any other name would smell as sweet, we would be wrong to call a rose a skunk cabbage. A definition can be more or less useful, and it can be more or less successful in accurately characterizing a concept or an object of discussion. Probably no definition of "species" suffices for all the contexts in which a species-like concept is used. Jerry Coyne and Allen Orr (2004), the authors of a recent comprehensive book on speciation, note that species can be defined in a way that (1) enables us to classify organisms systematically, (2) corresponds to discrete groups of similar organisms, (3) helps us understand how discrete clusters of organisms arise in nature, (4) represents products of evolutionary history, and/or (5) applies to the largest possible variety of organisms. It turns out that no two of these possible goals always coincide, and, as Coyne and Orr point out, it is unlikely that any one species concept will serve most of these purposes.

Linnaeus and other early taxonomists held what Ernst Mayr (1942, 1963) has called a TYPOLOGICAL or ESSENTIALIST notion of species. Individuals were members of a given species if they sufficiently conformed to that "type," or ideal, in certain morphological characters that were "essential" fixed properties—a concept descended from Plato's "ideas" (see Chapter 1). Thus a bird specimen was a member of the species *Corvus corone*, the carrion

TABLE 15.1 *Some species concepts*

Biological species concept Species are groups of actually or potentially interbreeding natural populations that are reproductively isolated from other such groups (Mayr 1942).

Evolutionary species concept A species is a single lineage (an ancestor-descendant sequence) of populations or organisms that maintains an identity separate from other such lineages and which has its own evolutionary tendencies and historical fate (Wiley 1978).

Phylogenetic species concepts (1) A phylogenetic species is an irreducible (basal) cluster of organisms that is diagnosably distinct from other such clusters, and within which there is a parental pattern of ancestry and descent (Cracraft 1989). (2)A species is the smallest monophyletic group of common ancestry (de Queiroz and Donoghue 1990).

Genealogical species concept Species are "exclusive" groups of organisms, where an exclusive group is one whose members are all more closely related to one another than to any organism outside the group (Baum and Shaw 1995).

Recognition species concept A species is the most inclusive population of individual biparental organisms that share a common fertilization system (Paterson 1985).

Cohesion species concept A species is the most inclusive population of individuals having the potential for phenotypic cohesion through intrinsic cohesion mechanisms (Templeton 1989).

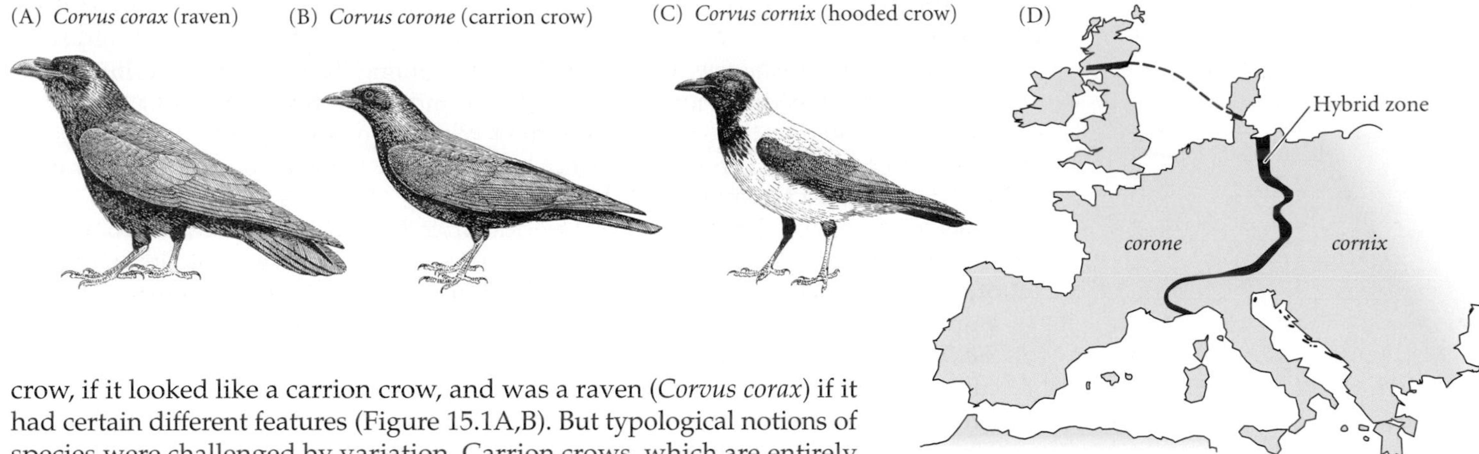

Figure 15.1 (A–C) Three closely related birds. (A) Ravens (*Corvus corax*) are distinguished from crows by their larger size, heavier bill, shaggy throat feathers, more pointed tail (when viewed from above or below), and voice. Even though (B) carrion crows (*Corvus corone*) and (C) hooded crows (*Corvus cornix*) are superficially more different from each other than from the raven, hybrids showing various mixtures of their plumage patterns are found in central Europe (D). The two forms of *Corvus* have been classified by some taxonomists as subspecies of a single species, but because they appear to exchange genes only to a very limited extent, they might better be considered species. (A–C from Goodwin 1986; D after Mayr 1963.)

crow, if it looked like a carrion crow, and was a raven (*Corvus corax*) if it had certain different features (Figure 15.1A,B). But typological notions of species were challenged by variation. Carrion crows, which are entirely black, are readily distinguished from hooded crows, which are black and gray (Figure 15.1C)—except in a narrow region in central Europe, where crows have various amounts of gray. Are there two species or one?

Variation among and within populations has engendered debates about the definition of species that are as active now as ever before (see Table 15.1). At present, the most commonly advocated species concepts are variations on the **phylogenetic species concept** and the **biological species concept**. These concepts differ chiefly in that phylogenetic species concepts emphasize species as the outcome of evolution, the *products* of a history of evolutionary divergence, whereas biological species concepts emphasize the *process* by which species arise and take a prospective view of the future status of populations (Harrison 1998). No matter what species concept is adopted, some populations of organisms will not be unambiguously assigned to one species or another. There are borderline cases, because the species properties advanced in every definition evolve gradually.

Phylogenetic species concepts

Phylogenetic species concepts (PSC), which are gaining acceptance especially among systematists, emphasize the phylogenetic history of organisms. One of several definitions of a phylogenetic species is "an irreducible (basal) cluster of organisms diagnosably different from other such clusters, and within which there is a parental pattern of ancestry and descent" (Cracraft 1989). This definition would presumably apply to both sexual and asexual organisms. According to this definition, speciation would occur whenever a population undergoes fixation of a genetic difference—even a single DNA base pair—that distinguishes it from related populations. The study of speciation, then, would be simply the study of divergence between populations.

The biological species concept

The biological species concept (BSC) has been and continues to be the species concept most frequently used by evolutionary biologists who are concerned with processes of evolution, and it is the concept used in this book. It was defined by Ernst Mayr (1942): "*Species are groups of actually or potentially interbreeding populations, which are reproductively isolated from other such groups*." "Reproductive isolation" means that any of several biological differences between the populations greatly reduce gene exchange between them, even when they are not geographically separated. These differences might or might not include mortality or sterility of hybrids between the two forms. Mayr (1942) and other advocates of the BSC have not insisted that populations must be 100 percent reproductively isolated in order to qualify as species; they recognize that there can be a little genetic "leakage" between some species.

The roots of the biological species concept are very old, for it has always been recognized that morphologically very different organisms (e.g., different sexes) might be born of the same parents and thus be **conspecific** (members of the same species). Nevertheless, the BSC arose from studies of variation that showed that morphological similarity and difference do not suffice to define species. Several critical observations contributed to its development:

1. *Variation within populations.* Characteristics vary among the members of a single population of interbreeding individuals. The white and blue forms of the snow goose (see Figure 9.1A), known to be born to the same mother, represent a genetic polymorphism, not different species. A mutation that causes a fruit fly to have four wings rather than two is just that: a mutation, not a new species.
2. *Geographic variation.* Populations of a species differ; there exists a spectrum from slight to great difference; and intermediate forms, providing evidence of interbreeding, are often found where such populations meet. Human populations are a conspicuous example.
3. *Sibling species.* **Sibling species** are reproductively isolated populations that are difficult or impossible to distinguish by morphological features, but which are often recognized by differences in ecology, behavior, chromosomes, or other such characters. The discovery that the European mosquito *Anopheles "maculipennis"* was actually a cluster of six sibling species had great practical importance because some transmit human malaria and others do not. (Box A provides definitions of sibling species and some other terms related to the biological species concept; later in the chapter Box B describes the diagnosis of a sibling species.)

BOX 15A Some Terms Encountered in the Literature on Species

Some of the following terms are frequently, and others infrequently, used in the literature on species. These definitions conform to usage by adherents to the biological species concept (e.g., Mayr 1963).

Geographic isolation Reduction or prevention of gene flow between populations by an extrinsic barrier to movement, such as topographic features or unfavorable habitat.

Reproductive isolation Reduction or prevention of gene flow between populations by genetically determined differences between them.

Allopatric populations Populations occupying separated geographic areas.

Parapatric populations Populations occupying adjacent geographic areas, meeting at the border.

Sympatric populations Populations occupying the same geographic area and capable of encountering one another.

Sibling species Reproductively isolated species that are difficult to distinguish by morphological characteristics.

Sister species Species that are thought, on the basis of phylogenetic analysis, to be each other's closest relatives, derived from an immediate common ancestor. (Cf. sister groups in phylogenetic systematics.)

Chronospecies Phenotypically distinguishable forms in an ancestor-descendant series in the fossil record that are given different names.

Subspecies Populations of a species that are distinguishable by one or more characteristics and are given subspecific names (see the *Elaphe* subspecies, Figure 9.24). In zoology, subspecies have different (allopatric or parapatric) geographic distributions and are equivalent to "geographic races;" in botany, they may be sympatric forms.

Race A vague term, sometimes equivalent to subspecies and sometimes to polymorphic genetic forms within a population.

Ecotype Used mostly in botany to designate a phenotypic variant of a species that is associated with a particular type of habitat; may be designated a subspecies.

Polytypic species A geographically variable species, often divided into subspecies. (Most species are polytypic, whether or not subspecies have been named.) The German term *Rassenkreis* (*Rasse*, "race"; *Kreis*, "circle," in the sense of "a circle of friends") is equivalent.

Hybrid zone A region where genetically distinct populations meet and interbreed to some extent, resulting in some individuals of mixed ancestry ("hybrids").

Introgression The movement, or incorporation, of genes from one genetically distinct population (usually considered a species or semispecies) into another.

Semispecies Usually, one of two or more parapatric, genetically differentiated groups of populations that are thought to be partially, but not fully, reproductively isolated; nearly, but not quite, different species.

Superspecies Usually, the aggregate of a group of semispecies. Sometimes designates a group of closely related allopatric or nearly allopatric forms that are designated different taxonomic species.

Domain and application of the biological species concept

All concepts have limitations. A concept may have a limited *domain of application* (for example, "matter" is ambiguous on a subatomic scale). A concept may inadequately describe BORDERLINE CASES. (The concept of an "individual organism," for instance, is ambiguous in a grove of aspen trees that have grown by vegetative propagation from a single seed.) There may also be *practical limitations* in applying the concept. (The "human population of the world" is a presumably valid concept, but technological and economic limitations prevent us from counting the population accurately.)

DOMAIN. The domain of the BSC is restricted to sexual, outcrossing organisms. It is also limited to short intervals of time, since it is meaningless to ask whether an ancestral population could have interbred with its descendants a million years later. To be sure, asexually reproducing organisms are given names, such as *Escherichia coli*, and putative ancestral and descendant populations in the fossil record may be distinguished by names such as the mid-Pleistocene *Homo erectus* and the later *Homo sapiens*. These cases illustrate that the word "species" has two overlapping but distinct meanings in biology. One meaning is embodied in the BSC. The other meaning of "species" is *a taxonomic category*, just like "genus" and "family." Some organisms that bear binomial names (such as *Escherichia coli*) are taxa in the species category, but are not biological species.

BORDERLINE CASES. Interbreeding versus reproductive isolation is not an either/or, all-or-none distinction. There exist graded levels of gene exchange among adjacent (parapatric) populations and sometimes between sympatric, more or less distinct populations. Several such situations are encountered frequently.

Narrow **hybrid zones** exist where genetically distinct populations meet and interbreed to a limited extent, but in which there exist partial barriers to gene exchange (Figure 15.1D). The hybridizing entities are often recognized as species, but may be called **semispecies**. A collection of semispecies is a **superspecies** (Mayr 1963).

Partial, but not completely free, gene exchange sometimes occurs between populations that are rather broadly sympatric (sympatric hybridization). Sympatric hybridization seems more common in plants than in animals, and it is one reason why many botanists have been reluctant to adopt the biological species concept. For example, different species of oaks (*Quercus*) often hybridize to some extent (Figure 15.2). However, advocates of the BSC recognize that a low level of gene exchange may occur between closely related species.

Geographic variation in status is encountered when genetically different populations appear to be conspecific in certain geographic regions, but to be different species elsewhere. For example, two sparrowlike birds with very different color patterns, the spotted towhee (*Pipilo maculatus*) and the collared towhee (*Pipilo ocai*), hybridize in some localities in Mexico, but coexist with little or no evidence of hybridization in other localities (Figure 15.3).

PRACTICAL DIFFICULTIES. The greatest practical limitation of the BSC lies in determining whether or not geographically segregated (allopatric) populations belong to the same

Figure 15.2 An example of sympatric hybridization. The gray oak (*Quercus grisea*) and Gambel's oak (*Q. gambelii*) have broadly overlapping ranges in the southwestern United States, including much of Texas, New Mexico, Arizona, and Colorado. Hybrids showing variation in leaf shape and other features are found in many of these places. (Photos courtesy of M. Cain.)

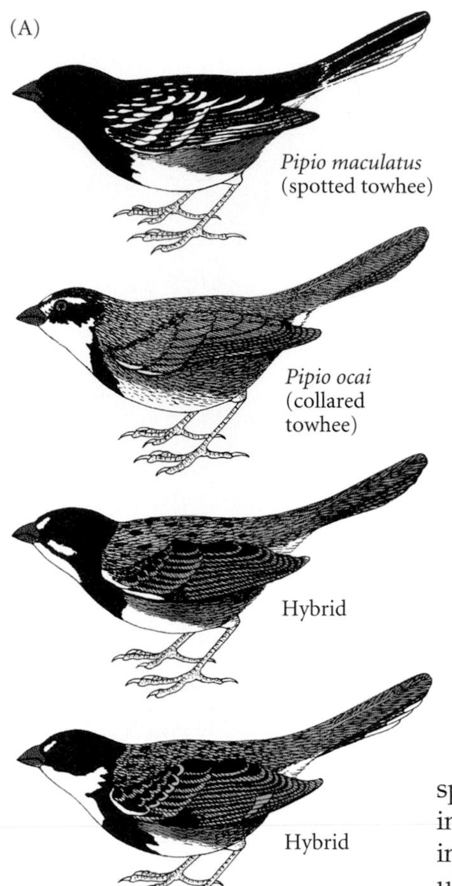

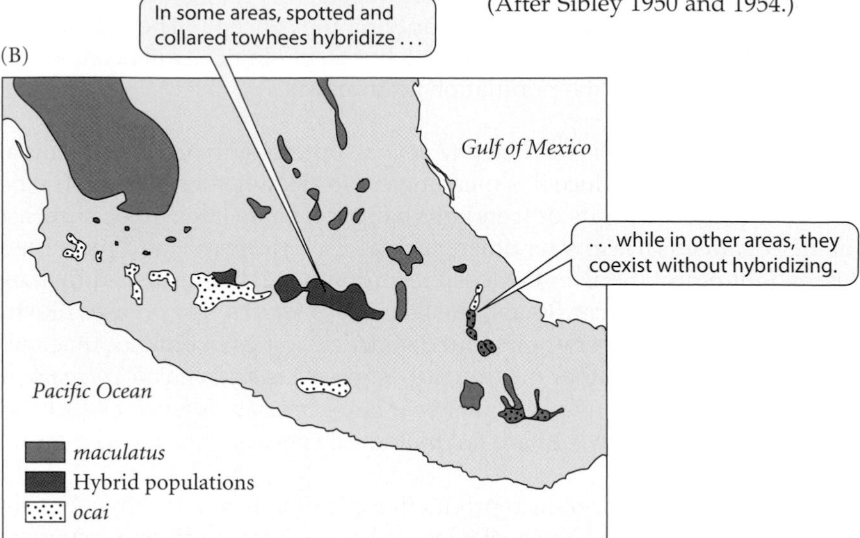

Figure 15.3 (A) The spotted towhee (*Pipilo maculatus*, top), the collared towhee (*P. ocai*, second from top), and two of the many plumage patterns of backcross hybrids. (B) The ranges of towhees in mountain areas of central Mexico. (After Sibley 1950 and 1954.)

species, for applying the BSC requires us to assess whether or not they would *potentially* interbreed if they should someday encounter each other. "Potential interbreeding" is an integral part of the BSC because the heart of this species concept lies in the idea that populations with intrinsic barriers to gene exchange can undergo independent evolutionary change, even if they should become sympatric. Moreover, it clearly would be absurd to distinguish as a species every population that is geographically, but perhaps very temporarily, isolated. (Would every laboratory stock of fruit flies be a different species?)

In many instances, range extension or colonization could well bring presently separate populations into contact, so the evolutionary future of the populations depends on whether or not they have evolved reproductive isolation. Humans have inadvertently or purposely introduced many species into new areas, some of which have hybridized with native populations (Abbott 1992), and many native species have extended their ranges considerably over the course of decades. After the Pleistocene glaciations, disjunct populations of a great many species expanded their ranges, met one another, and in many instances now interbreed. Future changes in climate will undoubtedly affect some currently disjunct populations in similar ways.

In principle, one could experimentally test for reproductive isolation between allopatric populations, and such tests have been performed with *Drosophila* and many other organisms by bringing them together in a laboratory or garden. For many organisms, however, such tests are not feasible (although the mere impracticality of such studies does not invalidate the concept of reproductive isolation). In practice, therefore, the classification (i.e., naming) of allopatric populations is often somewhat arbitrary. Commonly, allopatric populations have been classified as species if their differences in phenotype or in DNA sequence are as great as those usually displayed by sympatric species in the same group.

When species concepts conflict

Advocates of the BSC and the PSC are most likely to classify populations differently in two circumstances. First, allopatric populations that can be distinguished by fixed characters are species according to the PSC, but if the diagnostic differences are slight, advocates of the BSC may recognize the populations as geographic variants of a single species.

Second, in some cases, a local population of a widespread species evolves reproductive isolation from other populations, which remain reproductively compatible with one an-

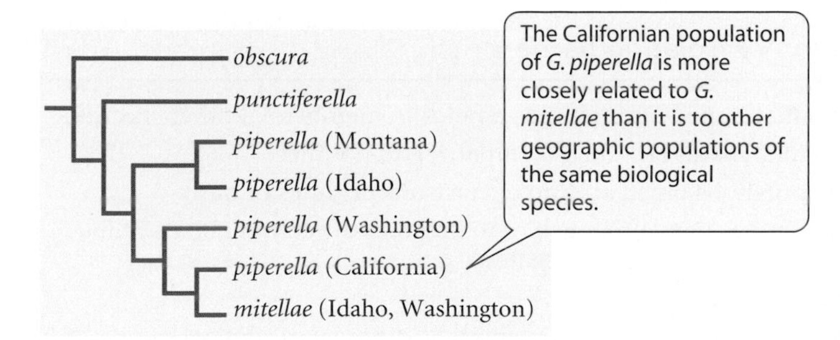
obscura
punctiferella
piperella (Montana)
piperella (Idaho)
piperella (Washington)
piperella (California)
mitellae (Idaho, Washington)
The Californian population of G. piperella is more closely related to G. mitellae than it is to other geographic populations of the same biological species.

Figure 15.4 The phylogeny of some species and populations in the moth genus *Greya*, based on mitochondrial DNA sequence data. This pattern would be expected if a local population of ancestral *G. piperella* evolved reproductive isolation and became the distinct biological species *G. mitellae*. According to the biological species concept, *G. mitellae* and *G. piperella* would be recognized as two species, one of which (*G. piperella*) is paraphyletic. According to the phylogenetic species concept, if *G. mitellae* is recognized as a species, the various populations of *G. piperella* should also be called species. (After Harrison 1998.)

other. Phylogenetic study may then show that the "new" species is more closely related to some populations of the "old" species than some of the "old" populations are to one another. Under the BSC, two species would be recognized, one of which is paraphyletic (Figure 15.4). (Recall from Chapter 3 that a paraphyletic taxon lacks one or more of the descendants of the common ancestor of the members of that taxon, and that paraphyletic taxa are not acceptable under the cladistic philosophy of classification.) Under the PSC, the various distinguishable populations of the paraphyletic group might be named as distinct species.

Barriers to Gene Flow

Gene flow between biological species is largely or entirely prevented by biological differences that have often been called ISOLATING MECHANISMS, but which we will term **isolating barriers**, or BARRIERS TO GENE FLOW. Under the BSC, therefore, *speciation—the origin of two species from a common ancestral species—consists of the evolution of biological barriers to gene flow*. As we noted above, mere physical isolation does not define populations as different species, although isolation by topographic or other barriers is considered to be instrumental in the formation of species.

It is an error to think that sterility of hybrids is the criterion of species as conceived in the BSC. There are many kinds of isolating barriers (Table 15.2). The most important distinction is between prezygotic and postzygotic barriers. Most prezygotic barriers are premating barriers, although some species are isolated by postmating, prezygotic barriers. Any of these barriers may be incomplete; for example, interspecific mating may occur at a low rate, or hybrid offspring may have reduced fertility (be "partially sterile").

Premating barriers

Premating barriers prevent (or reduce the likelihood of) transfer of gametes to members of other species.

ECOLOGICAL ISOLATION. Many species breed at different times of year (seasonal isolation). Two closely related field crickets (*Gryllus pennsylvanicus* and *G. veletis*), for example, reach reproductive age in the fall and spring, respectively, in the northeastern United States (Harrison 1979). Some species are isolated by habitat, so potential mates seldom meet. For instance, two Japanese species of herbivorous ladybird beetles (*Henosepilachna nipponica* and *H. yasutomii*) feed on thistles (*Cirsium*) and blue cohosh (*Caulophyllum*), respectively. Each species mates exclusively on its own host plant, and this ecological segregation appears to be the only barrier to gene exchange (Katakura and Hosogai 1994). Disturbance of habitat sometimes results in a breakdown of ecological isolation; for example, the wild irises *Iris fulva* and *I. hexagona* hybridize in Louisiana where their habitats (bayous and marshes, respectively) have been disturbed (Nason et al. 1992).

BEHAVIORAL ISOLATION. In animals, **behavioral isolation** (also called SEXUAL ISOLATION or ETHOLOGICAL ISOLATION) is an important barrier to gene flow among sympatric species that may frequently encounter each other, but simply do not mate. The **specific mate recognition system** of a species consists of signals and responses between potential mates

TABLE 15.2 *A classification of isolating barriers*

I. *Premating barriers:* Features that impede transfer of gametes to members of other species
- A. Ecological isolation: Potential mates (although sympatric) do not meet
  1. Temporal isolation (populations breed at different seasons or times of day)
  2. Habitat isolation (populations have propensities to breed in different habitats in the same general area, and so are spatially segregated)
- B. Potential mates meet but do not mate
  1. Behavioral (sexual or ethological) isolation (in animals, differences prevent populations from mating)
  2. Pollinator isolation (in plants, populations transfer pollen by different animal species or on different body parts of a single pollinator; may also be classified as ecological isolation)

II. *Postmating, prezygotic barriers:* Mating or gamete transfer occurs, but zygotes are not formed
- A. Mechanical isolation (copulation occurs, but no transfer of male gametes takes place because of failure of mechanical fit of reproductive structures)
- B. Copulatory behavioral isolation (failure of fertilization because of behavior during copulation or because genitalia fail to stimulate properly)
- C. Gametic isolation [failure of proper transfer of gametes or of fertilization, either due to intrinsic incompatibility or to competition between conspecific and heterospecific gametes (conspecific sperm precedence or pollen tube precedence)]

III. *Postzygotic barriers:* Hybrid zygotes are formed but have reduced fitness
- A. Extrinsic (hybrid fitness depends on context)
  1. Ecological inviability (hybrids do not have ecological niche in which they are competitively equal to parent species)
  2. Behavioral sterility (hybrids are less successful than parent species in obtaining mates)
- B. Intrinsic (hybrid fitness is low because of problems that are relatively independent of environmental context)
  1. Hybrid inviability (developmental problems cause reduced survival)
  2. Hybrid sterility (usually due to reduced ability to produce viable gametes; also "behavioral sterility," neurological incapacity to perform normal courtship)

Source: After Coyne and Orr 2004.

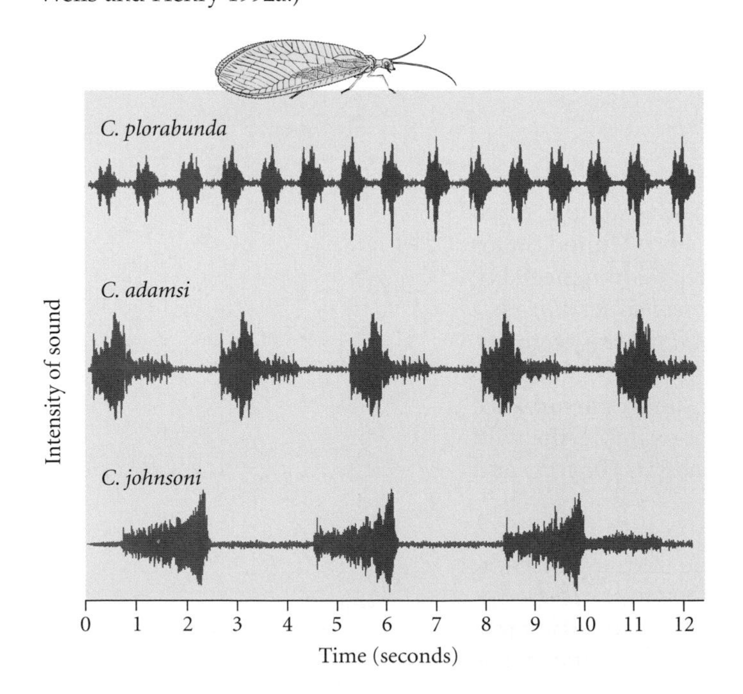

Figure 15.5 Oscillograms of the songs of three morphologically indistinguishable species of green lacewings (*Chrysoperla*). Each oscillogram displays a plot of amplitude against time. Two of the species, *C. adamsi* and *C. johnsoni*, were distinguished and named only recently, after study of songs and DNA differences made it clear that they are distinct species. (After Martínez Wells and Henry 1992a.)

(Paterson 1985), and one sex (often the female) will not respond to inappropriate signals. In three morphologically indistinguishable species of green lacewings (*Chrysoperla*), for example, a male and a female engage in a duet, initiated by the male, of low-frequency songs produced by vibrating the abdomen (Martínez Wells and Henry 1992b). Mating does not occur unless the female sings back to the male. The species produce very different songs (Figure 15.5), and females respond much more often to tape recordings of their own species than those of other species. Moreover, they discriminate against the intermediate songs produced by hybrids.

Sexual isolation in many animals (e.g., many mammals and insects) is based on differences in chemical mating signals (sex pheromones). Many other groups (e.g., many birds, fishes, jumping spiders) use visual signals, sometimes accompanied by acoustic or chemical signals, in their courtship displays (Figure 15.6). Differences in such signals often underlie sexual isolation. In many organisms, the courtship signals have not been identified, but it is nevertheless possible to measure sex-

Figure 15.6 Secondary sexual characteristics, such as bright color patterns and elaborate crests and tail feathers, vary greatly among male hummingbirds. Featured prominently in courtship displays, they undoubtedly contribute to reproductive isolation. Females of different species are much more similar to one another. (Left to right, above: *Sappho sparganura*, *Ocreatus underwoodii*, *Lophornis ornata*; below: *Stephanotis lalandi*, *Popelairia popelairii*, *Topaza pella*). (After illustrations by A. B. Singer in Skutch 1973.)

ual isolation by comparing the frequency of conspecific and heterospecific matings in experimental settings.

In plants, the nearest equivalent of behavioral isolation is pollination of different species by different pollinating animals that respond to differences in the color, form, or scent of flowers (Grant 1971). For example, the monkeyflower *Mimulus lewisii*, like most members of the genus, is pollinated by bees, and has pink flowers with a wide corolla and ridges of yellow hairs that probably guide bees to the nectar. Its close relative, *M. cardinalis*, has a narrow, red, tubular corolla and is pollinated by hummingbirds (Schemske and Bradshaw 1999; Figure 15.7). Some plant species differ in where they place their

Figure 15.7 (A) *Mimulus lewisii* has the broadly splayed petals characteristic of many bee-pollinated flowers. (B) An F_1 hybrid between *M. lewisii* and *M. cardinalis*. (C) *Mimulus cardinalis* has the red coloration and narrow, tubular form that has evolved independently in many bird-pollinated flowers. (D–F) Some F_2 hybrids, showing the variation that Bradshaw and Schemske used to analyze the genetic basis of differences between these two species. (From Schemske and Bradshaw 1999.)

(A)

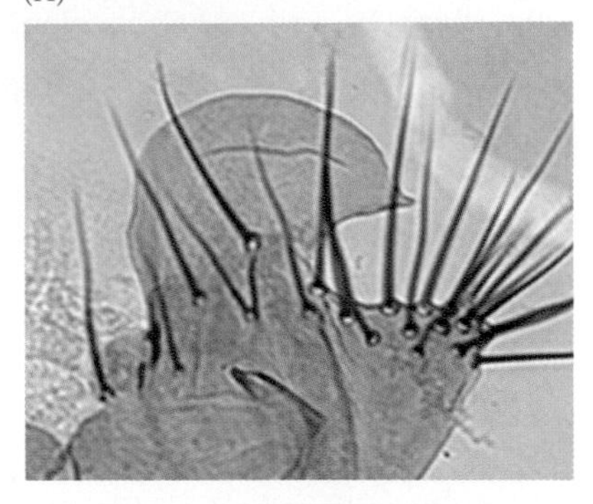

(B)

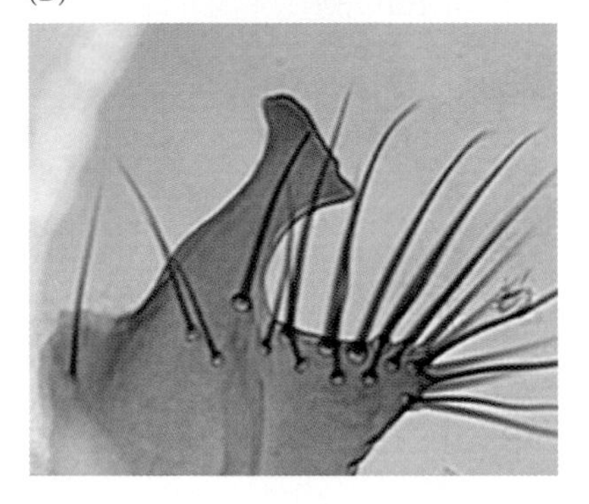

(C)

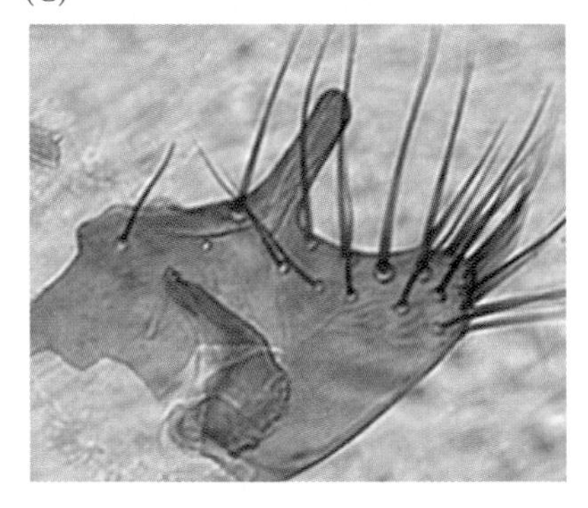

Figure 15.8 The posterior lobe of the genital arch in males of three closely related species of *Drosophila*: (A) *D. simulans*, (B) *D. sechellia*, and (C) *D. mauritiana*. This is almost the only morphological feature by which these species differ. Differences in genitalia can contribute to reproductive isolation between species if copulation between them occurs. (Photos courtesy of J. R. True.)

pollen on the body of a pollinator. For instance, the Swedish orchids *Platanthera bifolia* and *P. chlorantha* differ in the spacing between the pollinia (pollen masses), so that the pollinia of *P. bifolia* adhere to the proboscis of moths, and those of *P. chlorantha* adhere to the moths' eyes. Because of differences in flower form and scent, these plants also tend to attract different species of moths (Nilsson 1983).

Postmating, prezygotic barriers

On the border between premating and postzygotic barriers are features that prevent successful formation of hybrid zygotes even if mating takes place. In many groups of insects and some other taxa, the genitalia of related species differ in morphology, and it was suggested long ago that each species' male genitalia are a special "key" that can open only a conspecific female's "lock." Only a few studies support this hypothesis, but there is good reason to believe that females terminate mating, and prevent transfer of sperm, if a male's genitalia do not provide suitable tactile stimulation (Eberhard 1996; Figure 15.8).

Transfer of sperm is no guarantee that the sperm will fertilize a female's eggs. In some insects, such as ground crickets (*Nemobius*), fertilization by a heterospecific male's sperm will occur if the female mates only with that male, but if she also mates with a conspecific male, only the conspecific sperm will be successful in fertilizing her eggs. This phenomenon is known as CONSPECIFIC SPERM PRECEDENCE (Howard 1999). A similar effect is seen in some plants, in which heterospecific pollen cannot compete well against conspecific pollen in growing down through the style to reach the ovules.

GAMETIC ISOLATION occurs when gametes of different species fail to unite. This barrier is important in many externally fertilizing species of marine invertebrates that release eggs and sperm into the water. Because cell surface proteins determine whether or not sperm can adhere to and penetrate an egg, divergence in such proteins can result in gametic isolation (Palumbi 1998). Among species of abalones (large gastropods), the sperm protein lysin dissolves the vitelline membrane of only conspecific eggs, enabling the sperm to enter. The failure of heterospecific eggs and sperm to unite is related to the high rate of divergence between abalone species of the amino acid sequence of both lysin and the vitelline membrane protein with which it interacts (Galindo et al. 2003; see Chapter 19).

Postzygotic barriers

Postzygotic barriers consist of reduced survival or reproductive rates of hybrid zygotes that would otherwise backcross to the parent populations, introducing genes from each to the other. These barriers are sometimes classified as either "extrinsic" or intrinsic," depending on whether or not their effect depends on the environment.

HYBRID INVIABILITY. Hybrids between species often (though by no means always) have lower survival rates than nonhybrids. Quite often mortality is intrinsic, occurring in the embryonic stages due to failure of proper development, irrespective of the environment. Little is known about the malfunctions in development that cause mortality in hybrids. Especially in plants, hybrid inviability may be extrinsic, in that hybrids may have higher survival in intermediate or disturbed habitats than in those occupied by the parent species (Anderson 1949; Cruzan and Arnold 1993).

HYBRID STERILITY. The reduced fertility of many hybrids is an intrinsic barrier that can be caused by *structural differences between the chromosomes* that cause segregation of some ANEUPLOID gametes during meiosis (i.e., gametes with an unbalanced complement of chromosomes) or by *differences between the genes* from the two parents, which interact disharmoniously. These two causes are difficult to distinguish, and it is not clear how often differences in chromosome structure reduce fertility (King 1993; Rieseberg 2001).

As is sometimes true of hybrid inviability, hybrid sterility is often limited to the HETEROGAMETIC sex. (The heterogametic sex is the one with two different sex chromosomes, or with only one sex chromosome; the HOMOGAMETIC sex has two sex chromosomes of the same type. The male is heterogametic in mammals and most insects; the female is heterogametic in birds and butterflies.) This generalization is called **Haldane's rule**, and it appears to be one of the most consistent generalizations that can be made about speciation (Coyne and Orr 1989b).

Hybrid sterility and inviability may be manifested not only in F_1 hybrids, but also in F_2 and backcross offspring. This phenomenon has been observed in crosses both between species and among different geographic populations of the same species, in which case it is referred to as **F_2 breakdown** in fitness. For example, survival of F_2 larvae in a cross between *Drosophila pseudoobscura* from California and from Utah was lower than that in either "pure" population. This observation was interpreted to mean that recombination in the F_1 generated various combinations of alleles that were "disharmonious." In contrast, alleles at different loci within the same population have presumably been selected to form harmonious combinations. They are said to be COADAPTED, and each population is said to have a **coadapted gene pool** (Dobzhansky 1955).

How Species Are Diagnosed

Biological species are *defined* as reproductively isolated populations, but *diagnosing* species—distinguishing them in practice—is seldom done by directly testing their propensity to interbreed or their ability to produce fertile offspring. Morphological and other phenotypic characters are the usual evidence used for diagnosing species (Figure 15.9), even though species are not defined by their degree of phenotypic difference. Morphological and other phenotypic characters, judiciously interpreted, serve as *markers* for reproductive isolation or community among sympatric populations. If a sample of sympatric organisms falls into two discrete clusters that differ in two or more characters, it is likely to represent two species.

How can phenotypic differences of this kind indicate a barrier to gene exchange? From elementary population genetics (see Chapter 9), we know that a locus in a single population with random mating should conform fairly closely to Hardy-Weinberg genotype frequencies. Furthermore, two or more loci should be nearly at linkage equilibrium, unless very strong selection or suppression of recombination exists. If these loci affect a more or less additively inherited character, its variation will have a single-peaked, more or less normal distribution. If they affect different characters, variation in those characters is likely

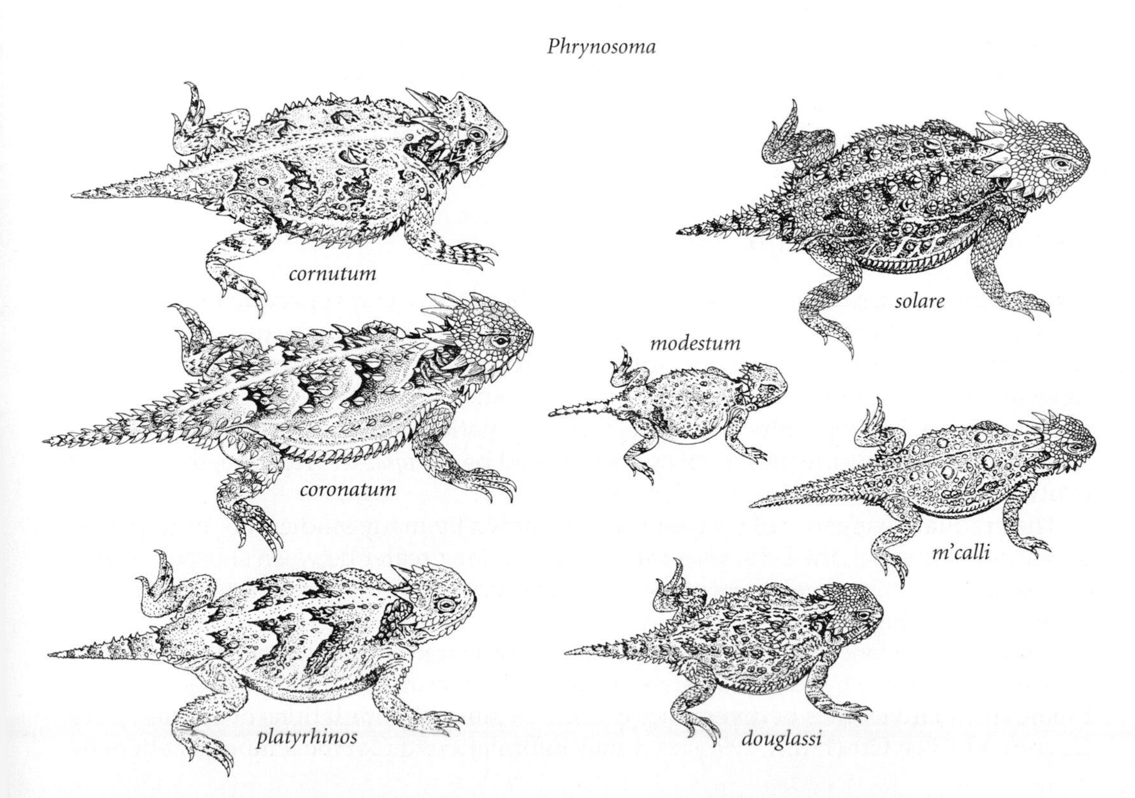

Figure 15.9 An example of species distinguished by morphological characters. These seven species of horned lizards (*Phrynosoma*) from western North America can be distinguished by differences in the number, size, and arrangement of horns and scales as well as body size and proportions, color pattern, and habitat. (From Stebbins 1954.)

BOX 15B Diagnosis of a New Species

Species in the leaf beetle genus *Ophraella* each feed on one species or a few related species of plants. *O. notulata*, for example, has been recorded only from two species of *Iva* along the eastern coast of the United States. This species is most readily distinguished from other species of *Ophraella* by the number and pattern of dark stripes on each wing cover.

Some leaf beetles found in Florida closely resembled *O. notulata*, but were collected on ragweed, *Ambrosia artemisiifolia*. This host association suggested the possibility that these beetles were a different species. In a broader study of the genus, I collected samples of beetles from both *Ambrosia* and *Iva* throughout Florida and examined them by enzyme electrophoresis (Futuyma 1991). I found consistent differences in allele frequencies between samples from *Iva* and *Ambrosia* at three loci, even in samples from both plants in the same locality. In the most extreme case, one allele had an overall frequency of 0.968 in *Ambrosia*-derived specimens, but was absent in *Iva*-derived specimens, in which a different allele had a frequency of 0.989. No specimens had heterozygous allele profiles that would suggest hybridization. Thus these genetic markers were evidence of two reproductively isolated gene pools.

A careful examination then revealed average differences between *Ambrosia*- and *Iva*-associated beetles in a few morphological characters, such as the shape of one of the mouthparts and the relative length of the legs. None of these morphological differences distinguished all individuals of one species from all individuals of the other. Later studies showed that adults and newly hatched larvae strongly prefer their natural host plant (*Ambrosia* or *Iva*) when given a choice, and that the beetles mate preferentially with their own species. In laboratory crosses, viable eggs were obtained by mating female *Ambrosia* beetles with males from *Iva*, but not the reverse. Few of the hybrid larvae survived to adulthood, and none laid viable eggs. Based on all of this evidence, I concluded that the *Ambrosia*-associated form, which I named *Ophraella slobodkini* in honor of the ecologist Lawrence Slobodkin, is a sibling species of *O. notulata*.

Ophraella slobodkini

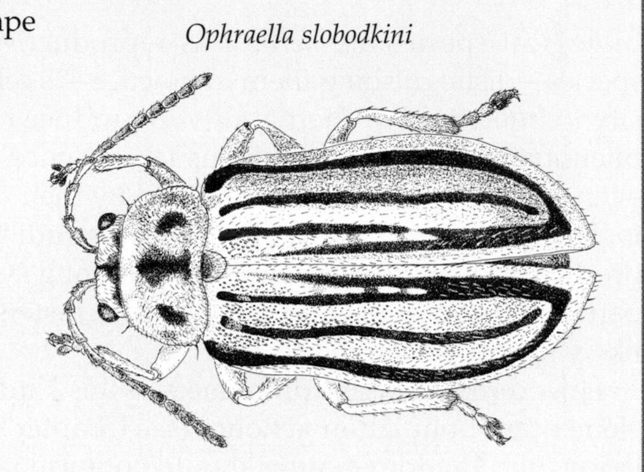

not to be strongly correlated. Conversely, if a sample includes two (or more) reproductively isolated populations with different allele frequencies, then a single locus should show a deficiency of heterozygotes compared with the Hardy-Weinberg expectation; variation in a polygenic character may have a bimodal distribution; and variation in different, genetically independent characters may be strongly correlated. (See Box B for an application of these principles.) On the basis of these principles, molecular markers at two or more loci can provide an even clearer indication of reproductive isolation than phenotypic traits.

Differences among Species

Some of the differences among species include those that are responsible for reproductive isolation. Others are adaptive differences related to ecological factors, such as temperature tolerance and habitat use; still others are presumably neutral differences that have arisen by mutation and genetic drift. Any such character difference may have evolved partly in geographically segregated populations *before* they became different species, partly *during* the process of speciation, and partly *after* the reproductive barriers evolved.

The gradual divergence of species has been shown by many studies. For example, allozyme differences in the *Drosophila willistoni* group are greater between subspecies than between geographic populations of the same subspecies, and they are greater still among sibling species (Figure 15.10).

The genetic distance (D) between populations, measured by the degree of difference in their allozyme allele frequencies, can be used as an approximate "molecular clock" to estimate the relative times of divergence of various pairs of populations or species. Jerry Coyne and Allen Orr (1989a, 1997) used such information to plot the temporal pattern by

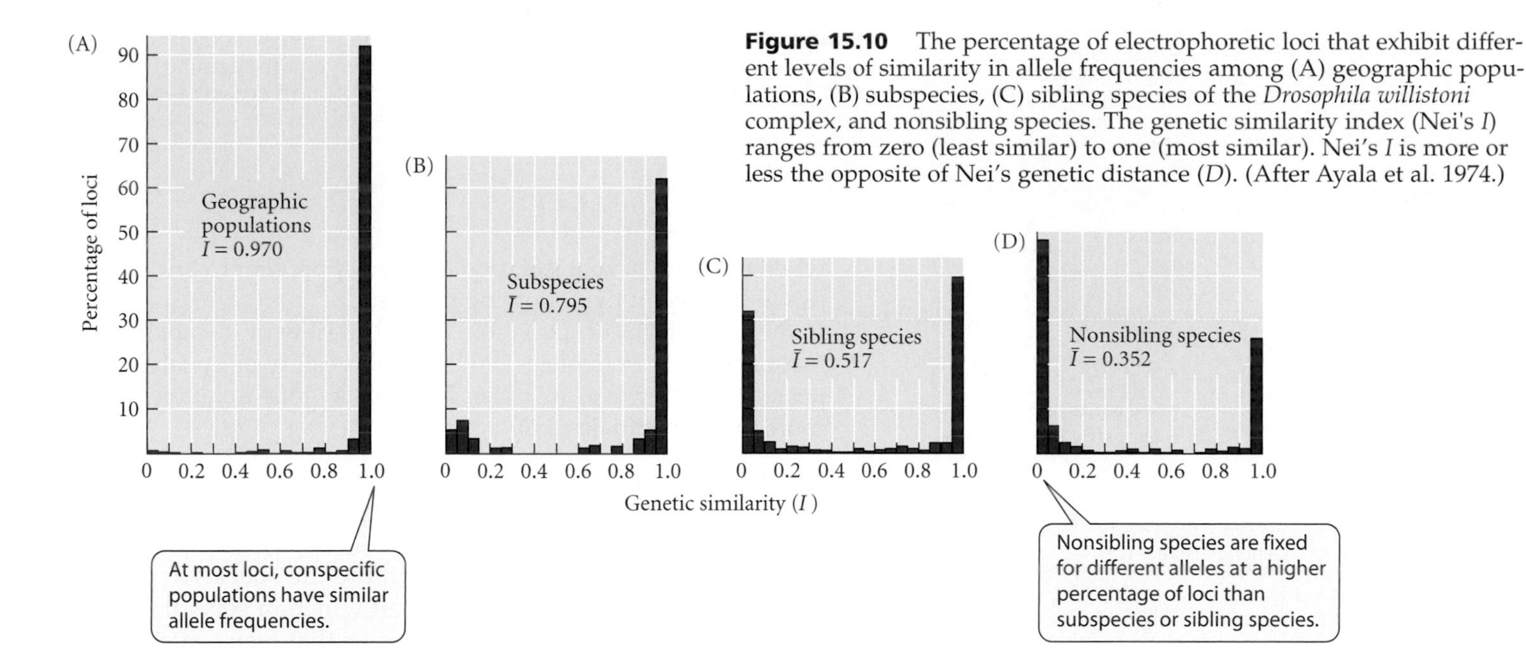

Figure 15.10 The percentage of electrophoretic loci that exhibit different levels of similarity in allele frequencies among (A) geographic populations, (B) subspecies, (C) sibling species of the *Drosophila willistoni* complex, and nonsibling species. The genetic similarity index (Nei's I) ranges from zero (least similar) to one (most similar). Nei's I is more or less the opposite of Nei's genetic distance (D). (After Ayala et al. 1974.)

which reproductive isolation evolves. They compiled data on reproductive isolation from reports, published over the previous 60 years, of experiments on numerous combinations of populations or species of *Drosophila*. They arrived at several important conclusions:

1. The strength of both prezygotic and postzygotic isolation increases gradually with the time since the separation of the populations (Figure 15.11). That is, speciation is a gradual process.
2. The time required for full reproductive isolation to evolve is very variable, but on average, it is achieved when D is about 0.30–0.53, which (based on a molecular clock calibrated from the few fossils of *Drosophila*) corresponds to about 1.5–3.5 million years. However, a considerable number of fully distinct species appear to have evolved in less than 1 million years.
3. Among recently diverged populations or species, premating isolation is, overall, a stronger barrier to gene exchange than postzygotic isolation (hybrid sterility or inviability). However, in Coyne and Orr's results, this effect was entirely due to sympatric taxa. Prezygotic isolation was stronger among sympatric than among allopatric pairs of taxa (Figure 15.12). This finding bears on a controversy about

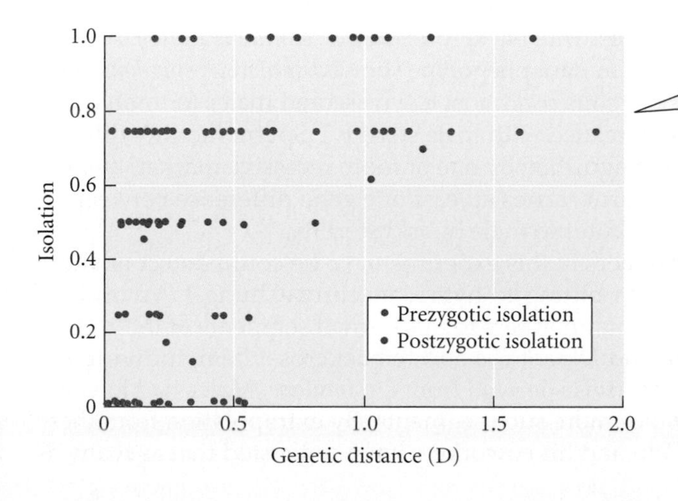

Figure 15.11 The level of prezygotic or postzygotic reproductive isolation between pairs of populations and species of *Drosophila*, plotted against genetic distance (D). Time since divergence can be inferred from genetic distance. Prezygotic isolation was measured by observing mating versus failure to mate between flies from different populations when confined together in the laboratory. Postzygotic isolation was based on survivial and fertility of hybrid individuals, measured in the laboratory. (After Coyne and Orr 1997.)

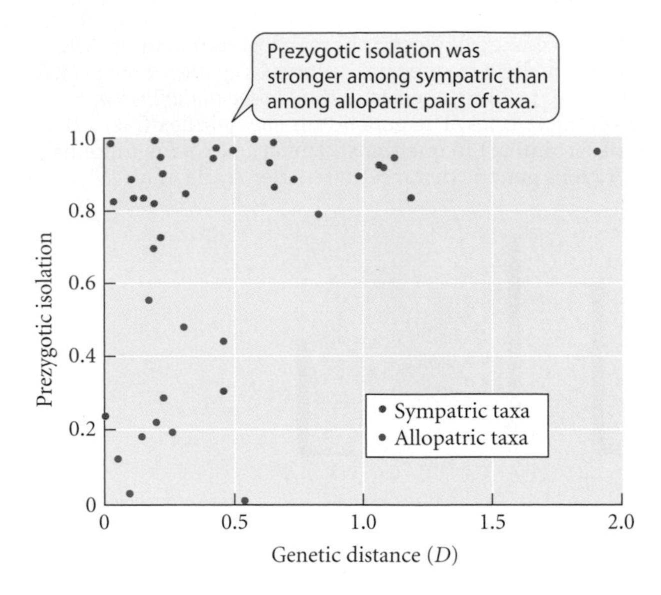

Figure 15.12 The level of prezygotic isolation between allopatric and sympatric pairs of *Drosophila* populations, plotted against genetic distance (*D*). At low genetic distances—interpreted as recent divergence—prezygotic isolation is greater between sympatric than between allopatric populations. (After Coyne and Orr 1997.)

whether or not sexual isolation evolves to prevent hybridization, as we will see in the following chapter.

4. In the early stages of speciation, hybrid sterility or inviability is almost always seen in males only; female sterility or inviability appears only when taxa are older. Thus postzygotic isolation evolves more rapidly in males than in females.

Because genetic differences continue to accumulate long after two species achieve complete reproductive isolation, some of the genes, and even some of the traits, that now confer reproductive isolation may not have been instrumental in forming the species in the first place. Thus it can be very difficult to tell which of several demonstrable isolating barriers, or which of the many gene differences that may confer hybrid sterility, was the cause of speciation. Such information can be obtained by studying populations that have achieved reproductive isolation only very recently (see below).

The Genetic Basis of Reproductive Barriers

In analyzing barriers to gene exchange, we wish to know whether the genetic differences required for speciation consist of few or many genes and how those genes act. Because some genetic differences accrue *after* speciation has occurred, we must compare populations that have speciated very recently, or are still in the process of doing so, in order to answer these questions.

Genes affecting reproductive isolation

The most extensive information on the genetics of reproductive barriers has been obtained for certain *Drosophila* species because so many genetic markers—the sine qua non of any genetic analysis—are available for those well-studied species. Theodosius Dobzhansky (1936, 1937), one of the most influential figures of the evolutionary synthesis, pioneered the use of genetic markers to study hybrid sterility. This method, effectively the same as QTL mapping (see Chapter 13), is still used by modern researchers.

Jerry Coyne (1984) analyzed hybrid sterility in crosses between *D. simulans* and its close relative *D. mauritiana*. F_1 hybrid males are sterile, but hybrid females are fertile. Coyne used a *D. simulans* stock with recessive visible mutations on the X chromosome and on each arm of both of the autosomes. Female *D. simulans* were crossed to *D. mauritiana* males, and the fertile F_1 hybrid females were backcrossed to *D. simulans* males. Among the resulting male progeny, a recessive mutant phenotype showed which *D. simulans* chromosome arms were carried in homozygous condition. Coyne scored males for motile versus immotile sperm (immotility is correlated with male sterility). Sperm motility differed for each pair of genotype classes distinguished by one or more recessive markers (Figure 15.13). Therefore, each chromosome arm carries at least one gene difference between *D. simulans* and *D. mauritiana* that contributes to male hybrid sterility.

Coyne used only five genetic markers in this experiment, so he could detect no more than five linked genetic factors contributing to hybrid sterility. Chung-I Wu and his coworkers have used multiple molecular markers to mark small segments of the X chromosomes of *Drosophila mauritiana* and *D. sechellia* and to backcross them, individually or in combination, into a genome otherwise derived from *D. simulans*. Males had lowered fertility whenever they carried two or more such segments. By extrapolation from these short segments of chromosomes, Wu and his coworkers have suggested that as many as

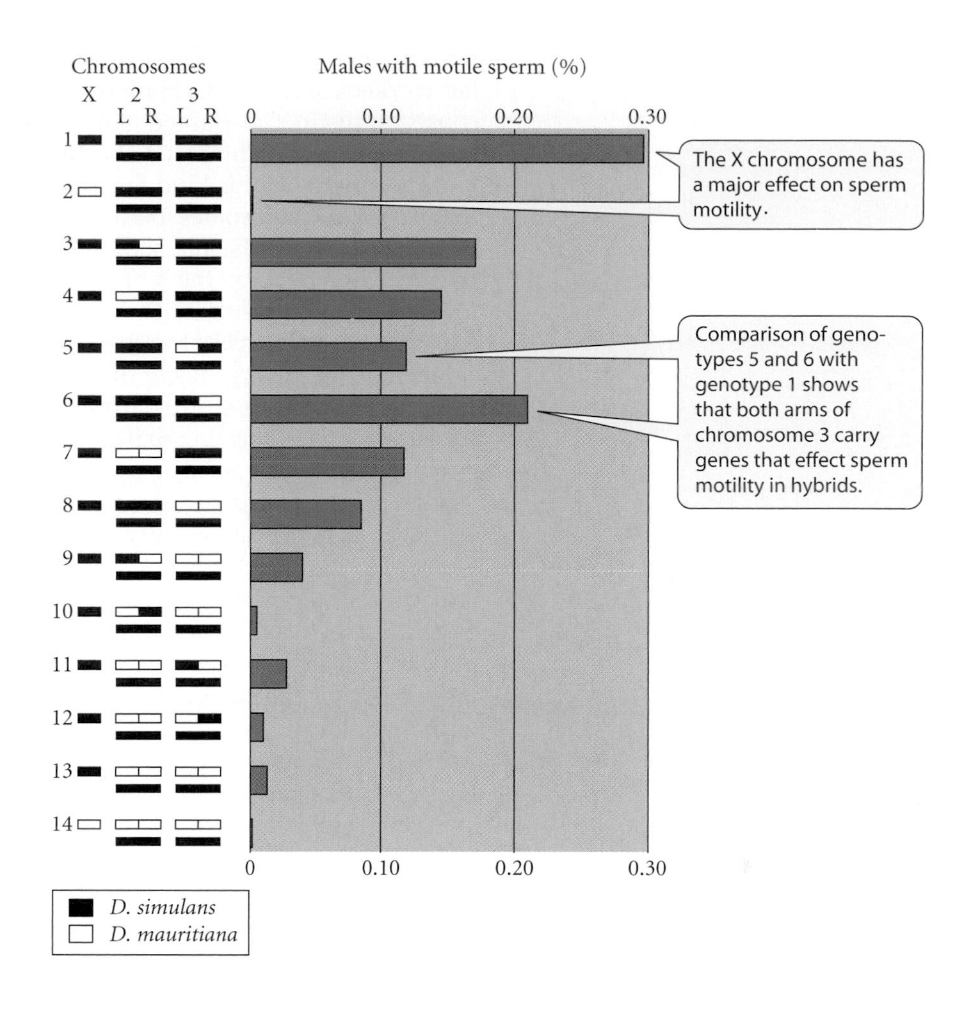

Figure 15.13 The proportion of males with motile sperm in nonhybrid *Drosophila simulans* and in backcross hybrids with various combinations of chromosome arms from *D. simulans* and *D. mauritiana*. All genotypes have a *D. simulans* Y chromosome (not shown). Note that every chromosome arm derived from *D. mauritiana* reduced sperm motility compared with that of the standard *D. simulans* genotype (1). (After Coyne 1984.)

40 gene differences on the X chromosome and 120 in the genome as a whole might cause hybrid male sterility among these closely related species (Wu and Hollocher 1998). A similar study of *Drosophila simulans* and *D. melanogaster* suggested that about 200 genes can contribute to hybrid inviability (Presgraves 2003).

Far fewer gene differences, however, are *sufficient* to confer postzygotic isolation. A subspecies of *D. pseudoobscura* found near Bogotá, Colombia, is thought to have diverged from the subspecies in the United States less than 200,000 years ago (more recently than *D. simulans* and its relatives evolved). Male hybrids from one of the two possible crosses between the two subspecies are sterile, and the sterility appears to be based on only about five gene regions, of which four are required for any sterility at all (Orr and Irving 2001). It therefore appears that the early acquisition of hybrid sterility requires few gene differences.

Genes that contribute to hybrid sterility or inviability do not have these effects in nonhybrid individuals; these effects must therefore stem from interactions between genes in the two different species. That is, *epistatic interactions* contribute to postzygotic isolation (Figure 15.14). In Wu and Hollocher's study, as already mentioned, crossing a combination of two or more marked chromosome segments from one species into another reduces

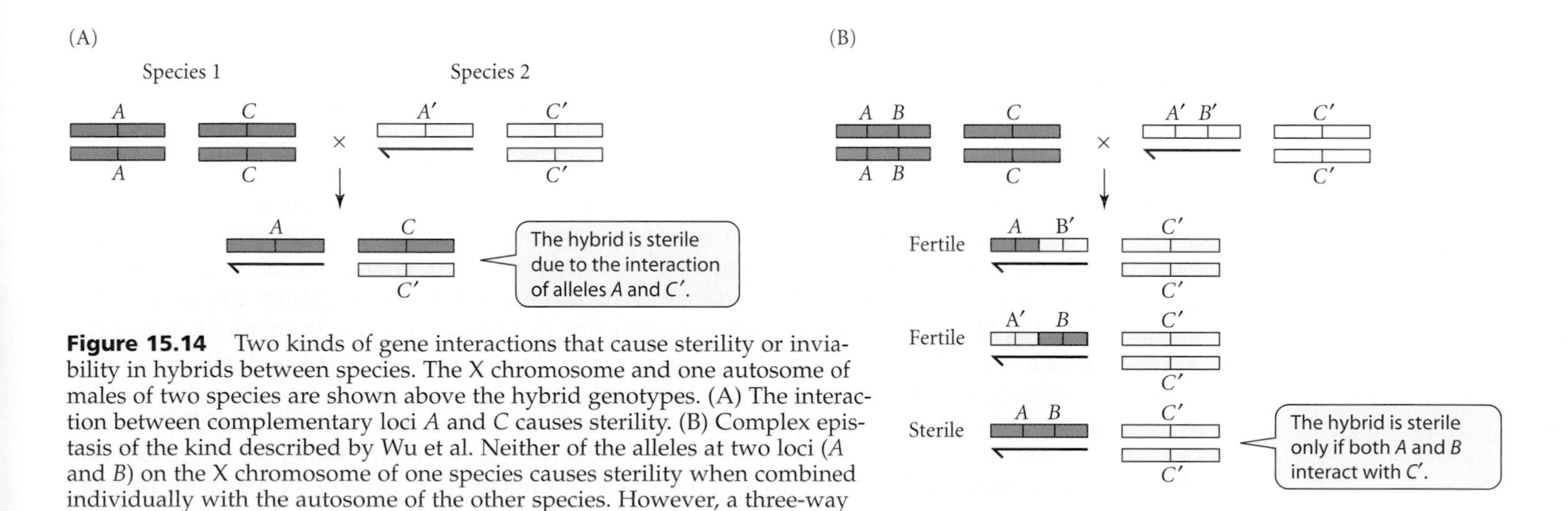

Figure 15.14 Two kinds of gene interactions that cause sterility or inviability in hybrids between species. The X chromosome and one autosome of males of two species are shown above the hybrid genotypes. (A) The interaction between complementary loci *A* and *C* causes sterility. (B) Complex epistasis of the kind described by Wu et al. Neither of the alleles at two loci (*A* and *B*) on the X chromosome of one species causes sterility when combined individually with the autosome of the other species. However, a three-way interaction between the pair of alleles (*A*, *B*) of one species and an autosomal allele (*C*′) of the other species causes sterility. The three hybrid genotypes illustrated here are produced by experimental crosses that are not shown.

fertility (Figure 15.14B). This evidence of COMPLEX EPISTASIS supports the convictions of earlier workers, such as S. C. Harland (1936) and Ernst Mayr (1963), who argued that species consist of distinct coadapted gene pools, or systems of genes that interact harmoniously within species, but interact disharmoniously if mixed together.

Experiments of this kind generally show that the X chromosome has a greater effect than any of the autosomes. Because the X chromosome does not cause sterility when combined with conspecific autosomes, the sterility effect in hybrids must arise from an epistatic interaction between the X of one species and autosomal genes of the other species (Figure 15.14A). This result sheds some light on the more rapid evolution of sterility in male than in female *Drosophila* hybrids, an instance of Haldane's rule. It has been proposed that X-linked genes diverge faster than autosomal genes because favorable X-linked recessive alleles are most exposed to natural selection (since males carry only one X). In addition, autosomal genes affecting male sterility have diverged faster than those affecting female sterility, possibly because of sexual selection.

Like postzygotic isolation, premating isolation is frequently based on polygenic traits, although in some cases only a few genes are involved (Ritchie and Phillips 1998). Differences in the shape of the male genitalia of *Drosophila simulans* and *D. mauritiana*, which may cause interspecific matings to terminate prematurely (see Figure 15.8), are affected by at least 19 loci (Zeng et al. 2000). Sexual isolation based on pheromones, in contrast, may be based on a few genes with major effects. For example, the European corn borer moth (*Ostrinia nubilalis*) includes two "pheromone races" (or sibling species) that differ in the ratio of two components (E- and Z-11 tetradecenyl acetate) of the sex pheromone emitted by females. A single autosomal gene difference accounts for most of the difference in the E:Z ratio. Both in the field and in wind tunnels, males of each race are almost exclusively attracted to the E:Z ratio produced by females of their own race. A single autosomal gene, not linked to the gene for the female's pheromone ratio, controls the difference in the physiological responses of the males' antennae to the two pheromone components, but the behavioral response of males to different E:Z ratios is sex-linked (Roelofs et al. 1987). This and similar studies show that the male and female components of communication that result in sexual isolation are usually genetically independent.

In a massive QTL study of F_2 progeny from crosses between the bee-pollinated monkeyflower *Mimulus lewisii* and its hummingbird-pollinated relative *M. cardinalis* (see Figure 15.7), Bradshaw et al. (1998) found that each of 12 characters that distinguish the species' flowers differed by between one and six loci. For most of these features, one locus accounted for at least 25 percent of the difference between the species. Schemske and Bradshaw then placed a full array of F_2 hybrids in an area where the two species are sympatric and observed pollination. At least 4 of the 12 floral traits affected the relative frequency of visitation by bees versus hummingbirds. Two QTL underlying these characters could be shown to have a significant effect on pollinator isolation. Thus allele substitutions of both large and small effect appear to contribute to reproductive isolation between these species.

Functions of genes that cause reproductive isolation

What are the nature and function of the genes that cause reproductive isolation? For most prezygotic isolating barriers, answering this question is a matter of understanding the genetic mechanisms of development of ordinary phenotypic characters, such as the features used in courtship or the behaviors involved in mate choice or habitat preference. The genes that underlie the breakdown of fertility or viability in hybrids are much more mysterious, and only in the last few years has there been progress toward understanding what these genes actually do.

In one of the best-understood cases, Daven Presgraves and collaborators (2003) showed that one of the *Drosophila simulans* genes that causes hybrid inviability when crossed into a *D. melanogaster* genetic background encodes a nucleoporin protein (Nup96), one of about 30 such proteins composing the nuclear pore complexes that regulate the passage of proteins and RNA between the cell nucleus and the cytoplasm. This gene interacts with at least one *D. melanogaster* gene that has not yet been characterized. Presgraves found that

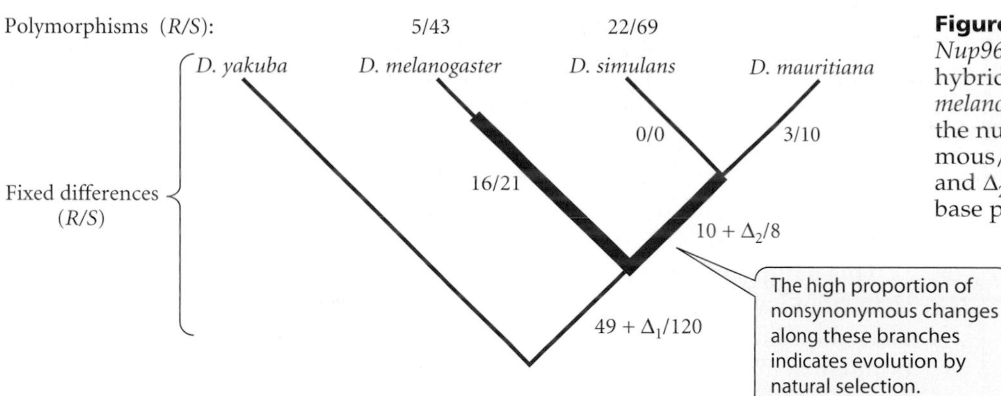

Figure 15.15 The evolutionary history of the *Nup96* gene, which contributes to inviability of hybrids between *Drosophila simulans* and *D. melanogaster*. On each branch of the phylogeny, the numbers indicate the number of nonsynonymous/synonymous base pair substitutions. Δ_1 and Δ_2 indicate insertion or deletion of several base pairs. (After Presgraves et al. 2003.)

in both species lineages, amino acid-replacing nucleotide substitutions have occurred in the *Nup96* gene at a high rate relative to synonymous substitutions, a clear indication that natural selection, rather than genetic drift, has driven divergence (Figure 15.15). Why selection favored changes in this protein is not yet known.

Chromosome differences and postzygotic isolation

Chromosome differences among species include alterations of chromosome structure (see Chapter 8) and differences in the number of chromosome sets (polyploidy, treated in Chapter 16). The role of structural alterations in postzygotic isolation and speciation is controversial (King 1993; Rieseberg 2001; Coyne and Orr 2004). An important question is whether heterozygosity for chromosome rearrangements causes reduced fertility (postzygotic isolation) in hybrids due to segregation of aneuploid gametes in meiosis.

A RECIPROCAL TRANSLOCATION is an exchange between two homologous chromosomes. Suppose, for example, that 1.2 and 3.4 represent two metacentric chromosomes in one population, with 1 and 2 representing the arms of one and 3 and 4 the arms of the other. A second population that is fixed for a translocation might have chromosomes 1.4 and 3.2. The F_1 hybrid would have all four chromosome types (1.2, 3.4, 1.4, 3.2). Only if the two parental combinations segregated (1.2 and 3.4 to one pole, 1.4 and 3.2 to the other) would balanced (euploid) gametes be formed (see Figure 8.22). Other patterns of segregation (e.g., 1.2 and 3.2 to one pole, 3.4 and 1.4 to the other) would yield unbalanced (aneuploid) gametes that lack considerable genetic material.

Some species differ by multiple translocations. For example (Dobzhansky 1951), the jimsonweeds *Datura stramonium* and *D. discolor* both have 12 pairs of chromosomes. In the F_1 hybrid, 7 pairs form normal synapsed pairs in meiosis. The other 5 pairs have undergone multiple translocations. If these chromosomes in *D. stramonium* are designated 1.2, 3.4, 5.6, 7.8, and 9.10, those of *D. discolor* represent 1.3, 2.7, 4.10, 5.9, and 6.8. In synapsis, a ring of ten chromosomes is formed, with the two arms of each *stramonium* chromosome aligned with an arm of each of two *discolor* chromosomes (Figure 15.16). The opportunities for aneuploid segregation are numerous.

Perhaps for this reason, such chromosome rearrangements are seldom found as polymorphisms within populations. More often, they are nearly or entirely monomorphic in different populations, except where those populations meet in narrow hybrid zones. For example, parapatric races of the burrowing mole-rat *Spalax ehrenbergi* in Israel differ in the number of chromosome

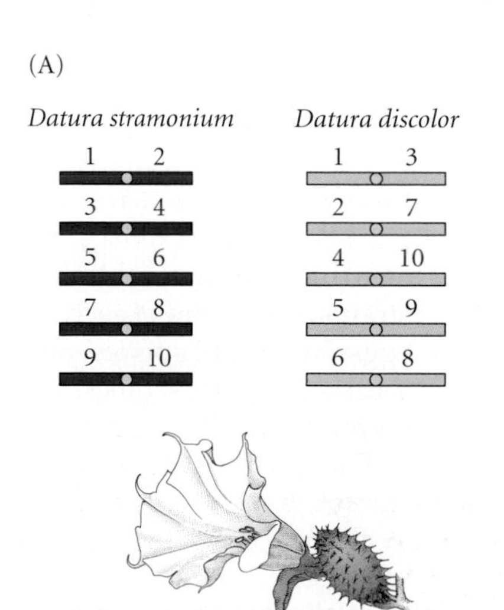

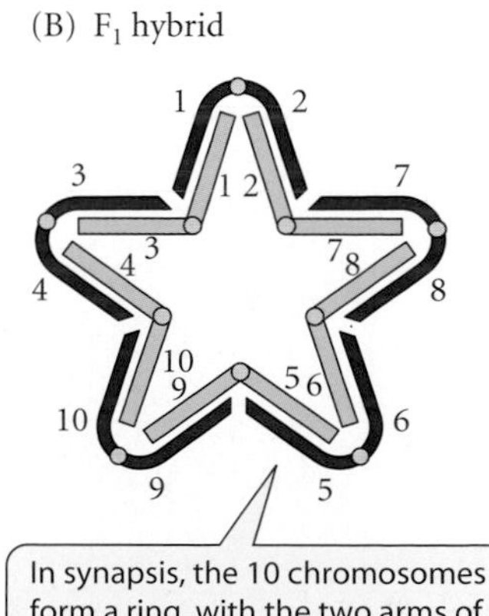

(B) F1 hybrid
In synapsis, the 10 chromosomes form a ring, with the two arms of each stramonium chromosome aligned with the corresponding arm from each of two discolor chromosomes.

Figure 15.16 (A) Five chromosomes of the jimsonweeds *Datura stramonium* and *D. discolor*, which differ by five reciprocal translocations. Homologous chromosome arms are correspondingly numbered. Only one member of each chromosome pair is shown. (B) A diagram of the possible arrangement of these chromosomes in synapsis in an F_1 hybrid.

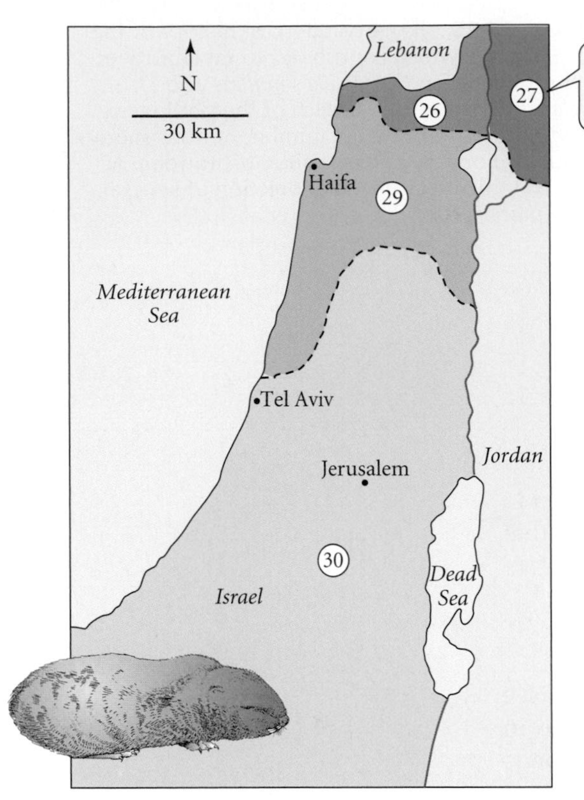

Figure 15.17 The distribution of four "races" of the mole-rat *Spalax ehrenbergi* with different chromosome numbers. Pairs of races meet at very narrow hybrid zones, indicated by the dashed lines. (After Nevo 1991.)

pairs due to chromosome fusions. Hybrids between these races are found in zones that range from 2.8 km to only 0.3 km wide (Figure 15.17).

This pattern is expected if chromosomal heterozygotes have lower fitness than homozygotes (are UNDERDOMINANT), perhaps due to reduced fertility caused by aneuploidy. If so, a chromosome introduced by gene flow from one population into another would seldom increase in frequency; because its initial frequency would be low, it would occur mostly in heterozygous condition, and it would probably be eliminated by selection (see Chapter 11). However, it is very difficult to tell whether the reduced fitness of hybrids is caused by the structural differences between the chromosomes or by differences between the genes of the parent populations. Certainly, chromosome rearrangements reduce gene exchange between populations. For example, the sunflowers *Helianthus annuus* and *H. petiolaris* differ by inversions and translocations that affect some chromosomes, but not others (Rieseberg et al. 1999). The rearranged chromosomes show a more abrupt transition in a hybrid zone between these species than do the chromosomes lacking rearrangements. But as we will see in our discussion of hybrid zones below, this pattern may be due to genetic incompatibility, not to production of aneuploid gametes. According to Coyne and Orr (2004), evidence that chromosome heterozygotes have reduced fertility because of meiotic irregularities is convincing for some plants and mammals, but not for other organisms, at least so far.

Cytoplasmic incompatibility

A possible cause of or contributor to speciation in insects is **cytoplasmic incompatibility**, caused by endosymbiotic bacteria in the genus *Wolbachia* that are inherited in egg cytoplasm, but are not transmitted by sperm (Werren 1998). Eggs that carry *Wolbachia* develop normally when fertilized by sperm from either infected or uninfected males. In the offspring of a cross between a *Wolbachia*-infected male and an uninfected female, however, the paternal chromosomes are destroyed or lost very early in development. It appears that *Wolbachia* in a male modifies his sperm in such a way that paternal chromosomes are destroyed in the zygote unless *Wolbachia* in the egg cytoplasm "cure" the modification. The loss of paternal chromosomes results in zygotic death in diploid species such as *Drosophila*. In haplodiploid insects, such as wasps and other Hymenoptera, fertilized (diploid) eggs normally develop into females, and unfertilized eggs develop into haploid males. Therefore, zygotes in which the paternal chromosomes are destroyed survive, but develop into nonhybrid males that carry the maternal set of chromosomes (Figure 15.18, cross 2).

Wolbachia typically sweeps through an uninfected population so that it becomes entirely infected. However, different strains of *Wolbachia* have independent incompatibility reactions: an egg must carry the same strain of *Wolbachia* as an infected male carries, or else the chromosome loss will occur. Thus if two populations carry different *Wolbachia* strains, the sperm of each population's males is incompatible with the other population's eggs. In diploid species, offspring from mixed matings ("hybrids") therefore die. In haplodiploid species, offspring from mixed matings lose their paternal chromosomes and therefore develop as males (not hybrids) carrying their mother's chromosomes.

Exactly this pattern has been found in crosses between the parasitic wasps *Nasonia vitripennis* and *N. giraulti*. Mating a male of one species with a female of the other results in all male offspring with the haploid egg's nuclear genome (Figure 15.18, crosses 3 and 4). When males are treated with radiation, however, so that their sperm are not modified by *Wolbachia*, the offspring of mixed matings include diploid females with chromosomes from both species (Figure 15.18, cross 5), showing that *Wolbachia* is responsible for the in-

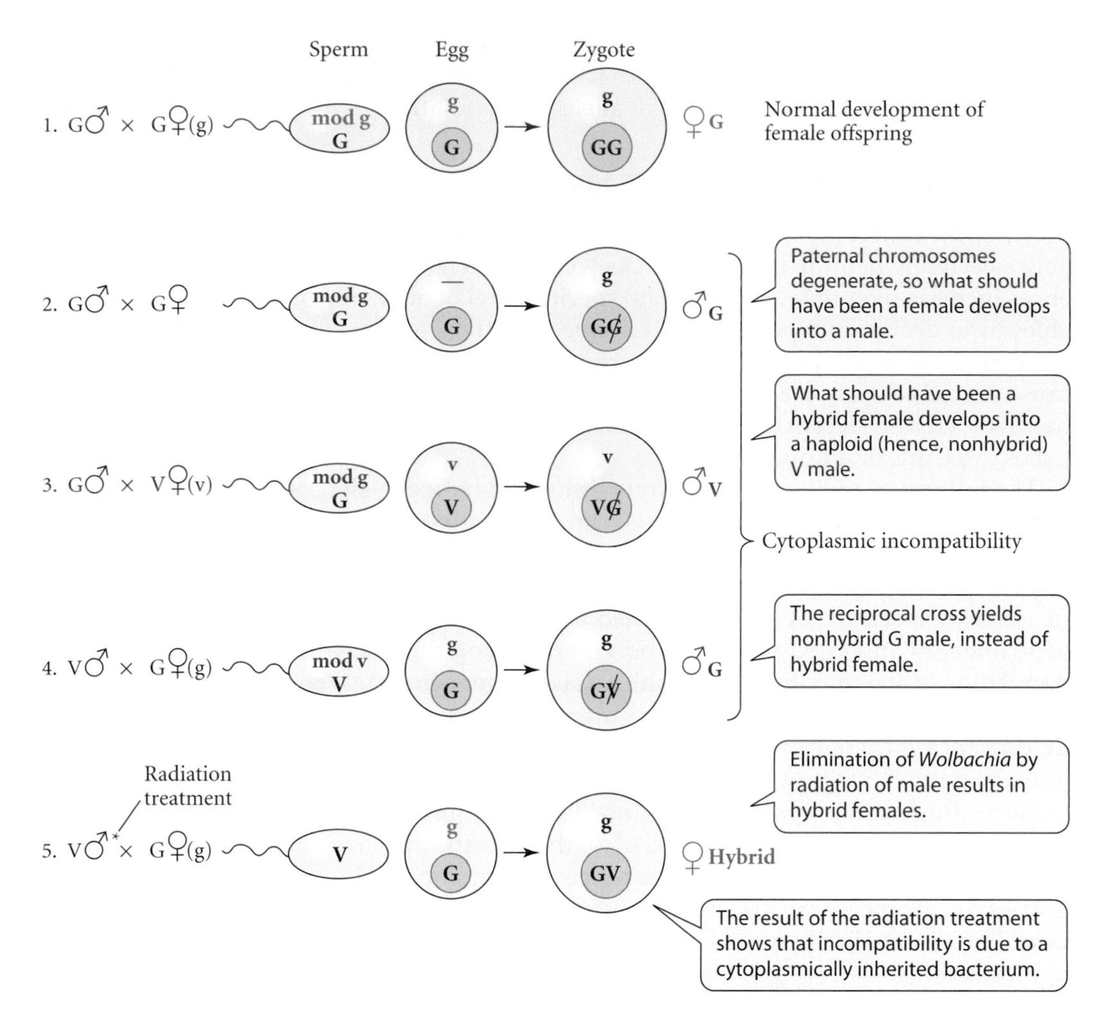

Figure 15.18 Cytoplasmic incompatibility in the wasps *Nasonia vitripennis* (V) and *N. giraulti* (G). Uppercase letters within the sperm, egg, and zygote in each cross indicate the source of chromosomes, and lowercase letters indicate *vitripennis*-specific (v) or *giraulti*-specific (g) strains of *Wolbachia*. Sperm modified by *Wolbachia* are labeled "mod v" or "mod g." (1) Diploid (female) offspring from a normal pairing between *N. giraulti* that both carry *Wolbachia* strain *g*. (2) Cytoplasmic incompatibility between infected male and uninfected female *N. giraulti*, yielding haploid (male) offspring due to loss of paternal chromosomes. (3) A cytoplasmically incompatible cross between the two species, yielding nonhybrid male *N. giraulti* offspring due to loss of paternal chromosomes. (4) The reciprocal cross between species, likewise yielding nonhybrid male offspring. (5) The same cross, but using irradiated, *Wolbachia*-free males, yields diploid female F_1 hybrid offspring.

compatibility. There are other reproductive barriers between these species, including sexual isolation, so *Wolbachia*-induced incompatibility may not have been the cause of speciation in this case. However, conditions under which bidirectional cytoplasmic incompatibility might initiate speciation can be envisioned (Werren 1998).

The significance of genetic studies of reproductive isolation

Both prezygotic and postzygotic isolation are generally due to differences between populations at several gene loci. A mismatch between genes affecting sexual communication, such as a male courtship signal and a female response, creates reproductive isolation. Similarly, a functional mismatch between genes gives rise to hybrid sterility or inviability. In both cases, reproductive isolation requires that populations diverge by at least two allele substitutions. Thus an $A_1A_1B_1B_1$ ancestral population may give rise to populations with genotypes $A_1A_1B_2B_2$ and $A_2A_2B_1B_1$, the incompatibility between A_2 and B_2 being the cause of reproductive isolation. This kind of epistatic incompatibility between loci is called a **Dobzhansky-Muller incompatibility** because Theodosius Dobzhansky, in 1934, and Hermann Muller, in 1940, postulated that such interactions underlie postzygotic isolation.

As we have seen, the number of gene differences that suffice for postzygotic isolation may be rather small, but more such differences accrue over time. Thus reproductive isolation eventually becomes irreversible, and the evolutionary lineages undergo independent genetic change thereafter.

Finally, genetic studies have shown that differences between species, including characters that confer reproductive isolation, have the same kinds of genetic foundations as variation within species. Thus there is no foundation for the opinion, held by some earlier biologists such as the geneticist Richard Goldschmidt (1940), that species and higher taxa arise through qualitatively new kinds of genetic and developmental repatterning. Moreover, reproductive isolation, like the divergence of any other character, usually evolves by the gradual substitution of alleles in populations.

Molecular Divergence among Species

Many of the differences between species in allozymes and DNA sequences are presumably selectively neutral, or nearly so. As we have seen (e.g., in Figure 15.11), such differences increase over time. However, no specific level of allozyme or DNA divergence enables us to declare that populations have become different species. Some reproductively isolated populations display little or no divergence in molecular markers, presumably because reproductive isolation, perhaps based on genetic change in one or a few characters, has evolved very recently. This is especially true in some groups of cichlids and other fishes (McCune and Lovejoy 1998).

The pattern of molecular variation within and between **sister species**—two species with an immediate common ancestor—can shed light on their history of divergence, especially when data from DNA sequences or from restriction site differences among gene copies are used to estimate the phylogeny of DNA sequences (the "gene tree"). Two populations (or species) that become isolated from each other at first share many of the same gene lineages, inherited from their polymorphic common ancestor (Figure 15.19). Each population is at first polyphyletic with respect to gene lineages (i.e., its genes are derived from several ancestral genes). Thus, with respect to some loci, individuals in each population are genealogically less closely related to one another than they are to some individuals in the other population (Figure 15.19, time t_1).

According to the COALESCENT THEORY described in Chapter 10, genetic drift in each species eventually results in the loss of all the ancestral lineages of DNA sequence variants except one; that is, coalescence to a common ancestral gene copy occurs in each species. (This process can also be caused by directional selection for a favorable mutation.) Gene lineages are lost by genetic drift at a rate inversely proportional to the effective population size. At some point, one population (population 1 in Figure 15.19) will become

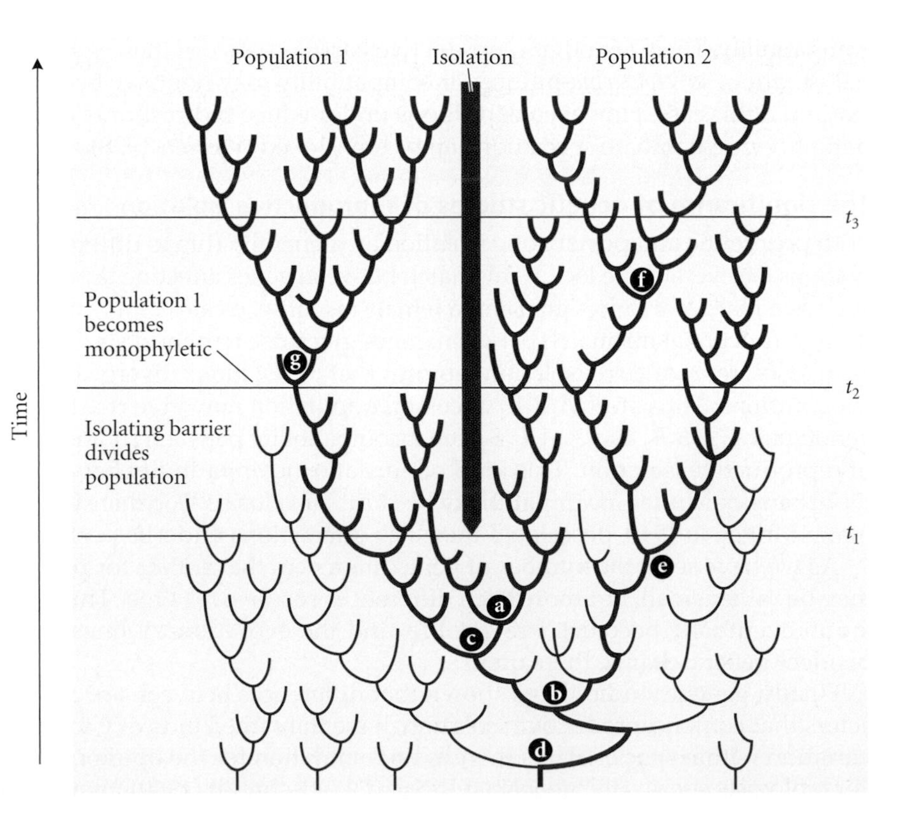

Figure 15.19 The transition from genetic polyphyly to paraphyly to monophyly in speciation. Colored branches show lineages of haplotypes at a single locus in a population that becomes divided into two populations at time t_1. Due to genetic drift or natural selection, all ancestral lineages except one are eventually lost, so that all the gene copies in the contemporary populations (at the top) coalesce to ancestral gene copy b. Population 1 loses gene lineages more rapidly than population 2, perhaps because it is smaller. Between times t_1 and t_2, both populations are polyphyletic for haplotype lineages, since some gene copies in each population, such as those derived from gene copy a, are more closely related to some gene copies in the other population than to some other gene copies (e.g., those derived from b) in their own population. Population 1 becomes monophyletic (coalescing to gene copy g) sooner—at time t_2—than population 2, perhaps because of its smaller size. Between times t_2 and t_3, population 2 is genetically paraphyletic, since its gene copies derived from ancestor c are more closely related to population 1's gene copies than they are to gene copies in population 2 that have been derived from ancestor e. At time t_3, population 2 also becomes monophyletic, with all gene copies derived from f. (After Avise and Ball 1990.)

monophyletic for a single gene lineage, while the other population, if it is larger, retains both this and other gene lineages (population 2 in Figure 15.19, at time t_2). At this time, the more genetically diverse population will be paraphyletic with respect to this gene, and some gene copies sampled from population 2 will be more closely related to population 1's gene copies than they are to other copies from population 2. Thus the phylogenetic relationships between genes from organisms in both populations will not correspond to the relationships among the individual organisms or the populations. Eventually, however, both populations will become monophyletic for gene lineages (Figure 15.19, time t_3), and the relationships among genes will reflect the relationships among populations. This process of sorting of gene lineages into species, or **lineage sorting**, is faster if effective population sizes (N_e) are small (Neigel and Avise 1986; Pamilo and Nei 1988).

Closely related species often share ancestral polymorphisms; in other words, lineage sorting has not gone to completion. For example, mitochondrial DNA haplotypes fall into two major lineages in 30 species of cichlid fishes from Lake Malawi in eastern Africa. These species, classified into 11 genera, are diverse in morphology and ecology, including fishes that feed on algae, molluscs, and the scales and fins of other fishes. Nevertheless, many of the species share haplotypes from both of the major mitochondrial clades (see Figure 2.19). The overall level of DNA sequence divergence among the species is very low, indicating that they have arisen very recently (Moran and Kornfield 1993). However, shared polymorphisms can persist for a long time if natural selection maintains the variation in both species. For example, humans and chimpanzees are each other's closest relatives, sharing several gene lineages at two loci in the major histocompatibility complex (MHC), retained from their common ancestor since divergence occurred about 5 million years ago. The polymorphism is thought to have been retained because of the role that MHC proteins may play in defense against pathogens.

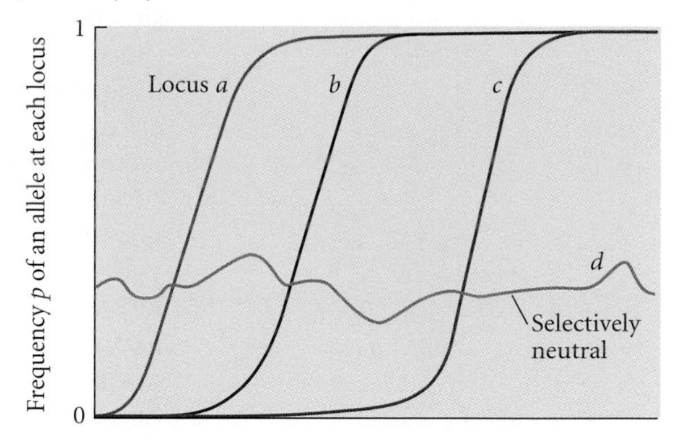

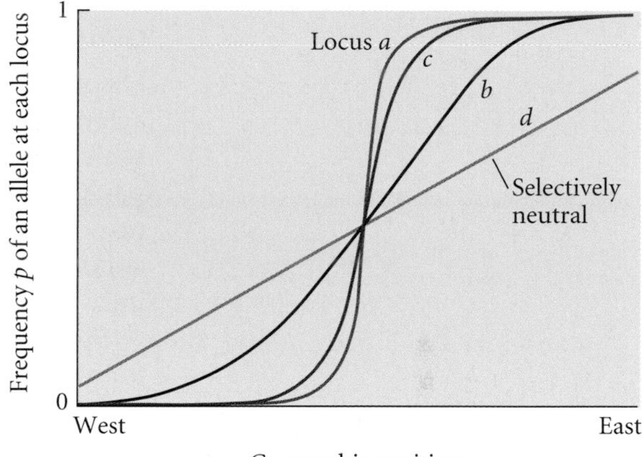

Figure 15.20 Expected patterns of variation in the frequency (p) of alleles or characters across a hybrid zone originating by (A) divergent selection along an environmental gradient (a primary zone) and (B) secondary contact. "Stepped clines" are shown for four loci. In (A), the loci are assumed to be affected differently by the environmental gradient, so the clines have different geographic positions. Allelic variation at locus d (orange line) is assumed to be nearly selectively neutral. (B) In a secondary hybrid zone, clines for all loci, including that for the nearly neutral alleles at locus d, are expected to have about the same location. The width of the cline depends on the strength of selection, relative to gene flow, at the locus or at closely linked loci. After a long enough time, the cline at locus d will come to resemble that in (A) due to gene flow. In some circumstances, the two causes of hybrid zones can result in indistinguishable patterns.

Hybridization

Hybridization occurs when offspring are produced by interbreeding between genetically distinct populations (Harrison 1990). Hybridization in nature interests evolutionary biologists because the hybridizing populations sometimes represent intermediate stages in the process of speciation. In some cases, hybridization may be the source of new adaptations or even of new species (Arnold 1997).

Primary and secondary hybrid zones

A **hybrid zone** is a region where genetically distinct populations meet and mate, resulting in at least some offspring of mixed ancestry (Harrison 1990, 1993). A character or locus that changes across a hybrid zone exhibits a CLINE that may be quite steep; for example, the alleles for the gray hood in hybridizing hooded and carrion crows would display a steep cline in frequency across the zone of hybridization between these forms (see Figure 15.1D).

Hybrid zones are thought to be caused by two processes. PRIMARY HYBRID ZONES originate in situ as natural selection alters allele frequencies in a series of more or less continuously distributed populations. Thus the position of the zone is likely to correspond to a sharp change in one or more environmental factors. SECONDARY HYBRID ZONES are formed when two formerly allopatric populations that have become genetically differentiated expand so that they meet and interbreed (**secondary contact**). It is not always easy to determine whether a hybrid zone is primary or secondary (Endler 1977). However, we might expect that in a primary hybrid zone, natural selection on different loci or characters would result in clines with different geographic positions, and that selectively neutral variation would not display a clinal pattern (Figure 15.20A). In contrast, the hybrids between populations that meet at secondary hybrid zones often have low intrinsic fitness due to heterozygote disadvantage or breakdown of coadapted gene complexes. (Such hybrid zones are often referred to as **tension zones**.) The clines in characters that differentiate the populations, therefore, need not match changes in the environment and are expected to be COINCIDENT (i.e., located in the same place) (Figure 15.21B). Clines in selectively neutral markers should be coincident with the others, although they may become broader over time due to gene flow.

The eastern European fire-bellied toad (*Bombina bombina*) and the western European yellow-bellied toad (*B. variegata*) meet in a long hybrid zone that is only about 6 kilometers wide (see the chapter-opening photo). The two species (or semispecies) differ in allozymes and in several morphological features (Figure 15.21), and hybrids have lower rates of survival than nonhybrids due to epistatic incompatibility. These taxa are thought to have arisen during the Pliocene, and to have formed a secondary hybrid zone after spreading from different refuges in southeastern and southwestern Europe that they occupied during the Pleistocene glacial periods (Szymura 1993).

Genetic dynamics in a hybrid zone

Dispersal, selection, and linkage all affect the distribution of alleles and phenotypic characters in hybrid zones. Let us consider how these factors affect clines in tension zones, in which hybrids have low fitness due to epistatic incompatibility or heterozygote disadvantage at certain loci (Barton and Gale 1993). Because the hybrids have low fitness irrespective of variation in environmental conditions, the geographic position of a tension zone is not determined by ecological factors. Its position may move if, for example, more individuals disperse across the hybrid zone in one direction than the other.

Suppose populations (semispecies) 1 and 2 have come into contact, that they are fixed for alleles A_1 and A_2, respectively, and that the fitness of A_1A_2 is reduced. Dispersal of in-

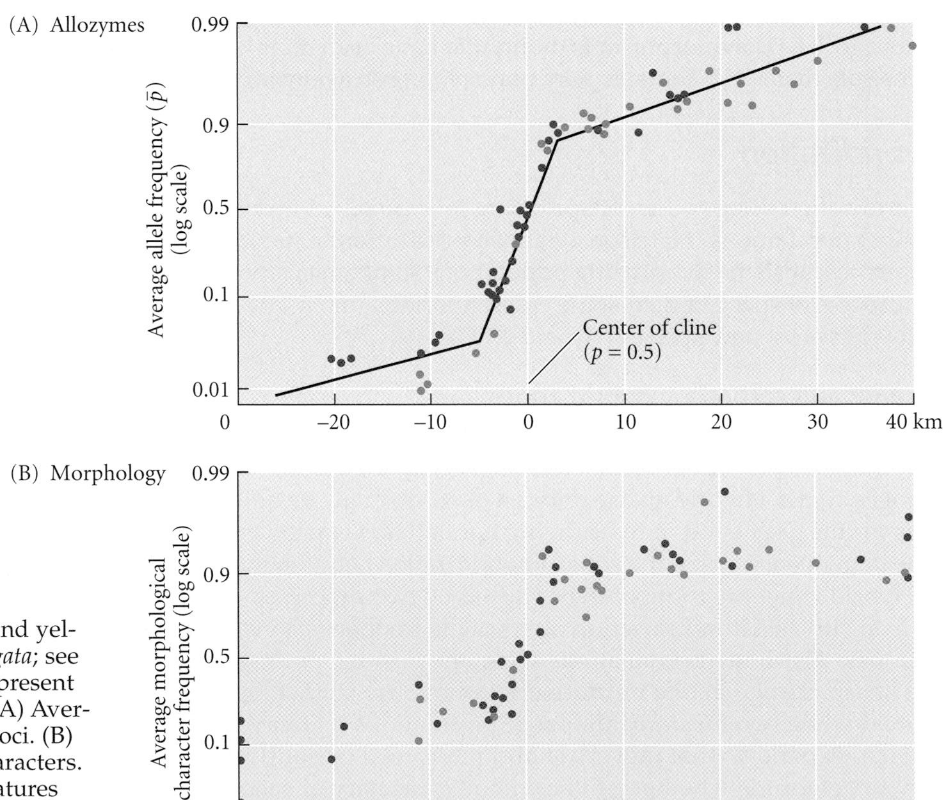

Figure 15.21 The hybrid zone between fire- and yellow-bellied toads (*Bombina bombina* and *B. variegata*; see chapter-opening photo). Red and blue points represent two different 60-kilometer transects in Poland. (A) Average allele frequencies at six diagnostic enzyme loci. (B) Average frequencies of seven morphological characters. The clines in enzyme loci and morphological features are coincident, suggesting that this hybrid zone was formed by contact between two formerly allopatric populations. (After Szymura 1993.)

dividuals of each semispecies into the range of the other, followed by random mating, constitutes gene flow that tends to make the cline in allele frequency broader and shallower. Gene flow continues if F_1 hybrids backcross with the parental (nonhybrid) genotypes. However, allele A_1 cannot increase in frequency within population 2, nor can A_2 within population 1, because of the heterozygote's disadvantage (see Chapter 12). The term "tension zone" is based on this "standoff": neither population's allele can invade the other population. The lower the fitness of the F_1 hybrid, the fewer A_1 or A_2 allele copies will be introduced into the parent populations. Thus selection against hybrids acts as a barrier to gene flow, and the steepness of the cline at the A locus depends on the strength of selection against hybrids relative to the magnitude of dispersal. Unless the rate of dispersal or interbreeding changes, this cline persists indefinitely.

Now consider another locus, with alleles B_1 and B_2 fixed in populations 1 and 2. If there is selection against B_1B_2 heterozygotes, the cline at this locus will parallel the cline at the A locus; in fact, they will reinforce each other, since the two loci both reduce the F_1 hybrid's fitness. Suppose, though, that the B_1 and B_2 alleles are selectively neutral. If they are very closely linked to the A locus, then they will hitchhike with the A alleles and form a cline coincident with that of the A locus in position and steepness. In fact, they will mark the existence and location of the locus (A) that contributes to low hybrid fitness. On the other hand, if the B locus is on another chromosome, lacking genes that reduce hybrid fitness, the B_1 and B_2 alleles will diffuse through the hybrid zone and spread into the other semispecies, forming a cline that becomes more shallow over time (see Figure 15.21B). The diffusion occurs because if F_1 hybrids reproduce at all, some of the offspring of the backcross $A_1A_2B_1B_2 \times A_1A_1B_1B_1$ (i.e., $F_1 \times$ population 1) will be $A_1A_1B_1B_2$. Thus some copies of the B_2 allele, dissociated by recombination from the deleterious heterozygous allele combination A_1A_2, will be introduced into population 1. Backcrosses to population 2 will similarly spread B_1 allele copies into that population. Thus some alleles or characters can spread from each semispecies into the other (a process called **introgression**), whereas others cannot.

A lowered fitness of hybrids at one locus therefore reduces the flow of neutral (or advantageous) alleles between populations, but the reduction is greater for alleles at closely linked loci than at loosely linked or unlinked loci. If a chromosome rearrangement that includes a selected locus reduces crossing over, then neutral alleles at all loci within the rearrangement, as well as the rearrangement itself, will display the same steep cline as the selected locus. This may account for the pattern seen in the sunflower hybrid zone mentioned on page 370, in which rearranged chromosomes have steeper clines than structurally similar chromosomes (Rieseberg 2001). In fact, Loren Rieseberg and collaborators (1999) showed that genes in these chromosome rearrangements reduce hybrid fertility.

Neutral alleles that are tightly linked to selected loci will have a cline that is steep in the middle of the hybrid zone, but shallow farther from the zone. Moreover, since some neutral genetic markers will be tightly linked to a locus that reduces hybrid fitness, whereas others may be unlinked, the number of markers with steep clines provides an estimate of the number of loci contributing to the low fitness of hybrids (i.e., to postzygotic reproductive isolation). For example, allozyme clines in the *Bombina* hybrid zone have been used to estimate that about 55 loci contribute to the reduction in the fitness of hybrids between fire-bellied and yellow-bellied toads (Szymura 1993; Barton and Gale 1993).

The fate of hybrid zones

As the last glaciers receded about 8000 years ago, many organisms rapidly colonized the newly exposed landscape. Thus many postglacial hybrid zones, such as that between the European carrion and hooded crows, are undoubtedly several thousand years old (Mayr 1963). Such apparent stability raises the question of whether or not hybridization will continue indefinitely.

Hybrid zones may have several fates:

1. A hybrid zone may persist indefinitely, with selection maintaining steep clines at some loci even while the clines in neutral alleles dissipate due to introgression. If the hybrid zone is a tension zone, it may move. It may become lodged in a region of low popu-

lation density, or may eventually move to the far edge of the range of one of the semispecies, resulting in its extinction. If, however, some of the character differences are favored by different environments, the positions of the clines in those characters will be stable.
2. Natural selection may favor alleles that enhance prezygotic isolation, resulting ultimately in full reproductive isolation.
3. Alleles that improve the fitness of hybrids may increase in frequency. In the extreme case, the postzygotic barrier to gene exchange may break down, and the semispecies may merge into one species.
4. Some hybrids may become reproductively isolated from the parent forms and become a third species.

These possibilities will be discussed in the following chapter as we examine the processes of speciation.

Summary

1. Many definitions of "species" have been proposed. The biological species concept is the one most widely used by evolutionary biologists. It defines species by reproductive discontinuity, not by phenotypic difference, although phenotypic differences may be indicators of reproductive discontinuity. Among other species concepts, the most widely adopted is the phylogenetic species concept, according to which species are sets of populations with character states that distinguish them.
2. The biological species concept (BSC) has a restricted domain; moreover, some populations cannot be readily classified as species or not because the origin of reproductive discontinuity is a gradual process. Although the BSC applies validly to allopatric populations, it is often difficult to determine whether or not such populations are distinct species.
3. The biological differences that constitute barriers to gene exchange are many in kind, the chief distinction being between prezygotic (e.g., ecological or sexual isolation) and postzygotic barriers (hybrid inviability or sterility). Some species are also isolated by postmating, prezygotic barriers (e.g., gametic isolation).
4. Among prezygotic barriers to gene exchange, sexual (ethological) isolation is important in animals. It entails a breakdown in communication between the courting and the courted sexes, and therefore, usually, genetic divergence in both the signal and the response. The differences in male and in female components are usually genetically independent.
5. Postzygotic isolation can be due to differences in nuclear genes, structural differences in chromosomes, or incompatibility between cytoplasmic factors such as endosymbiotic bacteria. Genic differences that yield hybrid sterility or inviability consist of differences at at least two (and usually considerably more) loci that interact disharmoniously in the hybrid. The role of chromosome differences in hybrid sterility is not well understood.
6. Reproductive barriers evolve gradually. Hybrid sterility or inviability of the heterogametic sex (often male) usually evolves before its manifestation in the homogametic sex (often female); this is known as Haldane's rule.
7. Levels of molecular divergence between closely related species vary greatly, and some recently formed species cannot be distinguished by molecular markers. Some species share ancestral molecular polymorphisms. In some cases, some gene copies in one species are more closely related to gene copies in another species than to other gene copies in the same species. In such cases, the phylogeny of genes may not match the phylogeny of the species that carry the genes.
8. Species (or semispecies) sometimes hybridize, often in hybrid zones, which in many cases are regions of contact between formerly allopatric populations. Alleles at some loci, but not others, may introgress between hybridizing populations, forming allele frequency clines of varying steepness. The steepness of such a cline depends on the rate of dispersal, the strength of selection, and linkage to selected loci.

Terms and Concepts

behavioral isolation
biological species concept (BSC)
coadapted gene pool
conspecific
cytoplasmic incompatibility
Dobzhansky-Muller incompatibility
F_2 breakdown
Haldane's rule
hybrid zone (primary, secondary)
introgression
isolating barriers
lineage sorting
phylogenetic species concept (PSC)
postzygotic barriers
prezygotic barriers
secondary contact
semispecies
sibling species
sister species
specific mate recognition system
superspecies
tension zone

Suggestions for Further Reading

A recent book by J. A. Coyne and H. A. Orr, *Speciation* (Sinauer Associates, Sunderland, MA, 2004) is the most comprehensive book on speciation in more than 40 years. The authors analyze hypotheses and data about speciation carefully and summarize a great amount of relevant literature. They provide an extensive discussion of species concepts and a justification of the biological species concept in particular.

Three books by Ernst Mayr—*Animal species and evolution* (Harvard University Press, Cambridge, MA, 1963), its abridged successor, *Populations, species, and evolution* (Harvard University Press, Cambridge, MA, 1970), and its predecessor, *Systematics and the origin of species* (Columbia University Press, New York, 1942)—are the classic works on the nature of animal species and speciation. For plants, an equally foundational work is *Variation and evolution in plants*, by G. L. Stebbins (Columbia University Press, New York, 1950). This topic is also treated by V. Grant in *Plant speciation* (Columbia University Press, New York, 1981).

Endless forms: Species and speciation, edited by D. J. Howard and S. H. Berlocher (Oxford University Press, Oxford and New York, 1998) is a stimulating collection of essays on species concepts and speciation. *Hybrid zones and the evolutionary process*, edited by R. G. Harrison (Oxford University Press, New York, 1993) compiles essays, including case studies of animals and plants, on this complex topic.

Problems and Discussion Topics

1. Some degree of genetic exchange occurs in bacteria, which reproduce mostly asexually. What evolutionary factors should be considered in debating whether or not the biological species concept can be applied to bacteria?
2. Suppose the phylogenetic species concept were adopted in place of the biological species concept. What would be the implications for (*a*) evolutionary discourse on the mechanisms of speciation; (*b*) studies of species diversity in ecological communities; (*c*) estimates of species diversity on a worldwide basis; (*d*) conservation practices under such legal frameworks as the U.S. Endangered Species Act?
3. Botanists have generally been more reluctant than zoologists to adopt the biological species concept. Read and discuss their arguments in, for example, papers by Raven (1976) and Levin (1979).
4. Identify two species of plants in the same genus that grow in your area. Propose a research program that would enable you to (*a*) judge whether or not there is any gene flow between the species and (*b*) determine what the reproductive barriers are that maintain the differences between them.
5. Studies of hybrid zones have shown that mitochondrial and chloroplast DNA markers frequently introgress farther, and have higher frequencies, than nuclear gene markers (Avise 1994, pp. 284–290). Thus, far from the hybrid zone, individuals that have no phenotypic indications of hybrid ancestry may have mitochondrial or chloroplast genomes of the other species (or semispecies). How would you account for this pattern?

6. Suppose you are interested in a poorly studied genus of leaf beetles that occurs in a distant land you can't afford to visit. A biologist friend who will be doing research at a field station in that region volunteers to collect some specimens and preserve them in alcohol for you. You can then study both their morphology and DNA sequences. How will you decide how many species are in the collection your friend brings back? Suppose your friend extracts mitochondrial DNA from each specimen and brings it back safely, but loses the specimens en route. How can you decide how many species are represented—or can you?
7. Suppose that two or more related taxa are polyphyletic for gene lineages, as in the cichlid fishes in Figure 2.19. What evidence might enable you to decide whether this is due to incomplete sorting (i.e., lack of coalescence) of ancestral polymorphism or to introgressive hybridization?

16 Speciation

In this chapter, we turn to the question of how species arise. If we considered species to be merely populations with distinguishing characteristics, the question of how they originate would be easily answered: natural selection or genetic drift can fix novel alleles or characteristics (see Chapters 10–13). But if the permanence of these distinctions depends on reproductive isolation, and if we consider reproductive isolation a defining feature of species, then the central question about speciation must be how genetically based barriers to gene exchange arise.

Speciation in sunflowers. The sunflower *Helianthus anomalus* thrives in sand dunes. One of several species that have arisen from hybrids between *H. annuus* and *H. petiolaris fallax* (see Figure 16.20), *H. anomalus* can survive in drier and harsher environments than either parent species. (Photo by Jason Rick, courtesy of Loren Rieseberg.)

The difficulty this question poses is most readily seen if we consider postzygotic reproductive barriers, such as hybrid inviability or sterility. If two populations are fixed for genotypes A_1A_1 and A_2A_2, but the heterozygote A_1A_2 has low reproductive success, how could these populations have diverged? Whatever allele the ancestral population may have carried (say, A_1), the low fitness of A_1A_2 would have prevented the alternative allele (A_2) from increasing in frequency and thus forming a reproductively incompatible population. Suppose, furthermore, that reproductive isolation between the populations is based on more than one locus.

TABLE 16.1 ***Modes of speciation***

I. Classified by geographic origin of reproductive barriers
- A. Allopatric speciation
  - 1. Vicariance
  - 2. Peripatric speciation
- B. Parapatric speciation
- C. Sympatric speciation

II. Classified by genetic and causal bases[a]
- A. Genetic divergence (allele substitutions)
  - 1. Genetic drift
  - 2. Peak shift (peripatric speciation)
  - 3. Natural selection
    - a. Ecological selection
      - i. for reproductive isolation
      - ii. of reproductive barriers
      - iii. of pleiotropic genes
    - b. Sexual selection
- B. Cytoplasmic incompatibility
- C. Cytological divergence
  - 1. Polyploidy
  - 2. Chromosome rearrangement
- D. Recombinational speciation

[a]Most of the genetic and causal bases might act in an allopatric, parapatric, or sympatric geographic context, and some of the causal bases listed under "Genetic divergence" apply also to cytoplasmic incompatibility, cytological divergence, and/or recombinational speciation.

The problem with a polygenic trait as a cause of reproductive isolation is that recombination generates intermediates. If several loci govern, for example, time of breeding, $A_1A_1B_1B_1C_1C_1$ and $A_2A_2B_2B_2C_2C_2$ might breed early and late in the season, respectively, and so be reproductively isolated, but the many other genotypes, with intermediate breeding seasons, would constitute a "bridge" for the flow of genes between the two extreme genotypes. *The problem of speciation, then, is how two different populations can be formed without intermediates*. This problem holds, whatever the character that confers prezygotic or postzygotic isolation may be.

Modes of Speciation

The many conceivable solutions to this problem are the MODES OF SPECIATION. The modes of speciation that have been hypothesized can be classified by several criteria (Table 16.1), including the *geographic origin* of the barriers to gene exchange, the *genetic bases* of the barriers, and the *causes of evolution* of the barriers.

Speciation may occur in three kinds of geographic settings that blend one into another (Figure 16.1). **Allopatric speciation** is the evolution of reproductive barriers in populations that are prevented by a geographic barrier from exchanging genes at more than a negligible rate. A distinction is often made between allopatric speciation by **vicariance** (divergence of two large populations; Figure 16.1A) and **peripatric speciation** (divergence of a small population from a widely distributed ancestral form; Figure 16.1B). In **parapatric speciation**, neighboring populations, between which there is modest gene flow, diverge and become reproductively isolated (Figure 16.1C). **Sympatric speciation** is the evolution of reproductive barriers within a single, initially randomly mating population (Figure 16.1D). Allopatric, parapatric, and sympatric speciation form a continuum, differing only in the degree to which the initial reduction of gene exchange is accomplished by a physical barrier extrinsic to the organisms (as in allopatric speciation) or by evolutionary change in the biological characteristics of the organisms themselves (as in sympatric speciation). Allopatric speciation is widely acknowledged

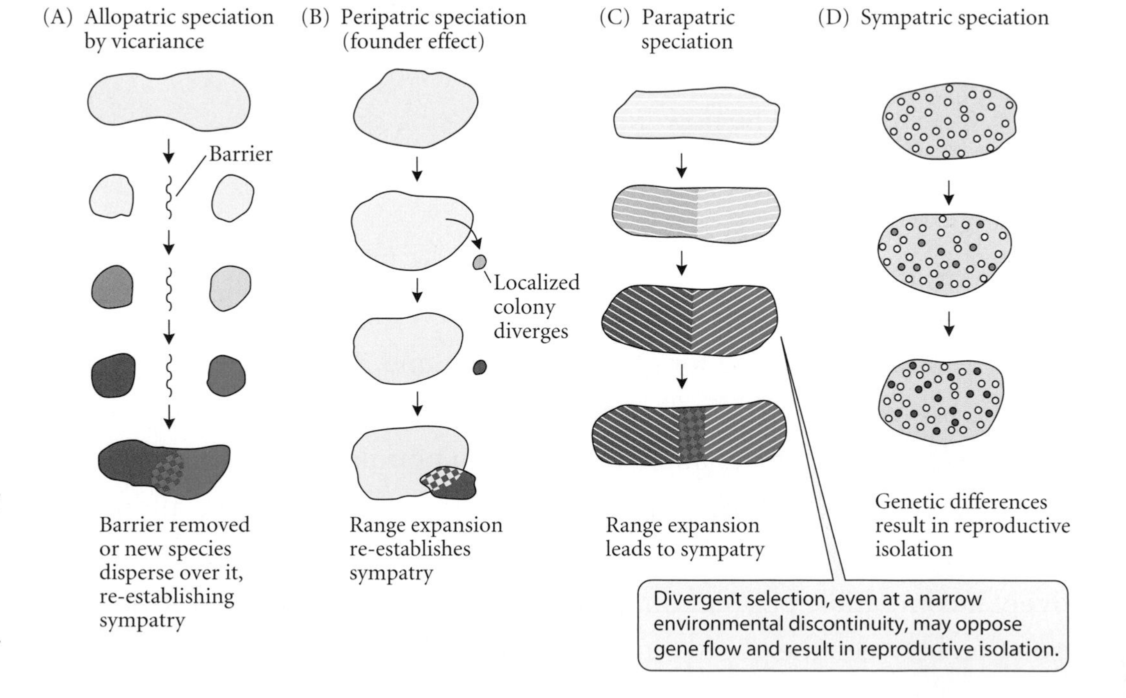
(A) Allopatric speciation by vicariance
(B) Peripatric speciation (founder effect)
(C) Parapatric speciation
(D) Sympatric speciation
Barrier
Localized colony diverges
Barrier removed or new species disperse over it, re-establishing sympatry
Range expansion re-establishes sympatry
Range expansion leads to sympatry
Genetic differences result in reproductive isolation
Divergent selection, even at a narrow environmental discontinuity, may oppose gene flow and result in reproductive isolation.

Figure 16.1 Diagrams of successive stages in each of four models of speciation differing in geographic setting. (A) Allopatric speciation by vicariance. (B) The peripatric, or founder effect, model of allopatric speciation. (C) Parapatric speciation. (D) Sympatric speciation.

to be a common mode of speciation; the incidence of parapatric and sympatric speciation is debated.

From a genetic point of view, the reproductive barriers that arise may be based on genetic divergence (allele differences at, usually, several or many loci), cytoplasmic incompatibility, or cytological divergence (polyploidy or structural rearrangement of chromosomes). We will devote most of this chapter to speciation by genetic divergence.

The causes of evolution of reproductive barriers, as of any characters, are genetic drift and natural selection of genetic alterations that have arisen by mutation. Peripatric speciation, a hypothetical form of speciation that is also referred to as TRANSILIENCE or SPECIATION BY PEAK SHIFT, requires both genetic drift and natural selection. Both sexual selection and ecological causes of natural selection may result in speciation. In some cases, there may be selection *for* reproductive isolation—that is, to prevent hybridization. (Recall the distinction between *selection for* and *selection of* characters, discussed in Chapter 11.) Alternatively, reproductive isolation may arise as a *by-product* of genetic changes that occur for other reasons (Muller 1940; Mayr 1963). In this case, there may be ADAPTIVE DIVERGENCE of the isolating character itself (e.g., climatic factors may favor breeding at two different seasons, with the effect that the populations do not interbreed), or the reproductive barrier may arise as a pleiotropic by-product of genes that are selected for their other functions.

Allopatric Speciation

Allopatric speciation is the evolution of genetic reproductive barriers between populations that are geographically separated by a physical barrier such as topography, water (or land), or unfavorable habitat. The physical barrier reduces gene flow sufficiently for genetic differences between the populations to evolve that prevent gene exchange if the populations should later come into contact (see Figure 16.1A). Allopatry is defined by a severe reduction of movement of individuals or their gametes, not by geographic distance. Thus, in species that disperse little or are faithful to a particular habitat, populations may be "microgeographically" isolated (e.g., among patches of a favored habitat within a lake). All evolutionary biologists agree that allopatric speciation occurs, and many hold that it is the prevalent mode of speciation, at least in animals (Mayr 1963; Coyne and Orr 2004).

Allopatric populations may expand their range and come into contact. (Consider, for example, the postglacial expansion portrayed in Figure 6.15.) If sufficiently strong isolating barriers have evolved during the period of allopatry, they may become sympatric without exchanging genes. If incomplete reproductive isolation has evolved, they will form a hybrid zone (see Chapter 15, where we described the possible fates of hybrid zones). Because range expansion may have occurred in the past, sympatric sister species that we observe today may well have speciated allopatrically, not sympatrically.

Evidence for allopatric speciation

Because both natural selection and genetic drift cause populations to diverge in genetic composition, it is probably inevitable that if separated long enough, geographically separated populations will become different species. Many species show incipient prezygotic and/or postzygotic reproductive isolation among geographic populations. For example, Stephen Tilley and colleagues (1990) examined sexual isolation among dusky salamanders (*Desmognathus ochrophaeus*) from various localities in the southern Appalachian Mountains of the eastern United States. They brought males and females from different populations (heterotypic pairs) and from the same population (homotypic pairs) together and scored the proportion of these pairs that mated. Among the various pairs of populations, an index of the strength of sexual isolation varied continuously, from almost no isolation to almost complete failure to mate (Figure 16.2). The more geographically distant the populations, the more genetically different they were (as measured by Nei's D, an index of genetic distance based on allozyme frequencies), and the less likely they were to mate.

Speciation can often be related to the geological history of barriers. For example, the emergence of the Isthmus of Panama in the Pliocene divided many marine organisms into

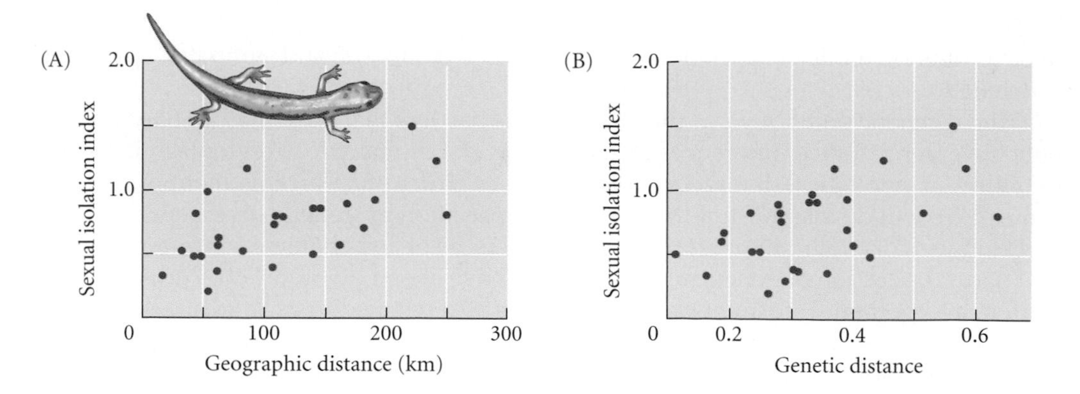

Figure 16.2 The degree of sexual isolation between populations of the salamander *Desmognathus ochrophaeus* is correlated with (A) the geographic distance between the populations as well as with (B) their genetic distance (Nei's *D*, which measures the difference in allozyme frequencies at several loci). (After Tilley et al. 1990.)

Pacific and Caribbean populations, some of which have diverged into distinct species. Among seven such species pairs of snapping shrimp, only about 1 percent of interspecific matings in the laboratory produced viable offspring (Knowlton et al. 1993). In some cases, CONTACT ZONES between differentiated forms mark the meeting of formerly allopatric populations. For example, Eldredge Bermingham and John Avise (1986; Avise 1994) analyzed the genealogy of mitochondrial DNA in samples of six fish species from rivers throughout the coastal plain of the southeastern United States. In all six species, DNA sequences form two distinct clades characterizing eastern and western populations, and the two clades make contact in the same region of western Florida (Figure 16.3). This pattern implies that gene flow between east and west was reduced at some time in the past. The amount of sequence divergence between the two DNA clades suggests that isolation occurred 3–4 million years ago. At that time, sea level was much higher than at present, forming a barrier to dispersal by freshwater fishes.

In many cases, related species replace each other geographically, reflecting their presumably allopatric origins. For example, the leopard frogs of North America, most of which were at one time considered geographic races of *Rana pipiens*, consist of about 27 species in two clades (Hillis 1988). The geographic ranges of the species in each clade overlap only slightly or not at all (Figure 16.4). These species are actually or potentially isolated by habitat, breeding season, mating call, and/or hybrid inviability.

Allopatric speciation is also supported by negative evidence. No pairs of sister species of birds occur together on any isolated island, implying that speciation does not occur on land masses that are too small to provide geographic isolation between populations (Coyne and Price 2000). Where two or more closely related species of birds do occur on an island, other islands or a continent can be identified as a source of invading species, and in all cases there is evidence that the several species invaded the island at separate times. For example, many of the islands in the Galápagos archipelago harbor two or more

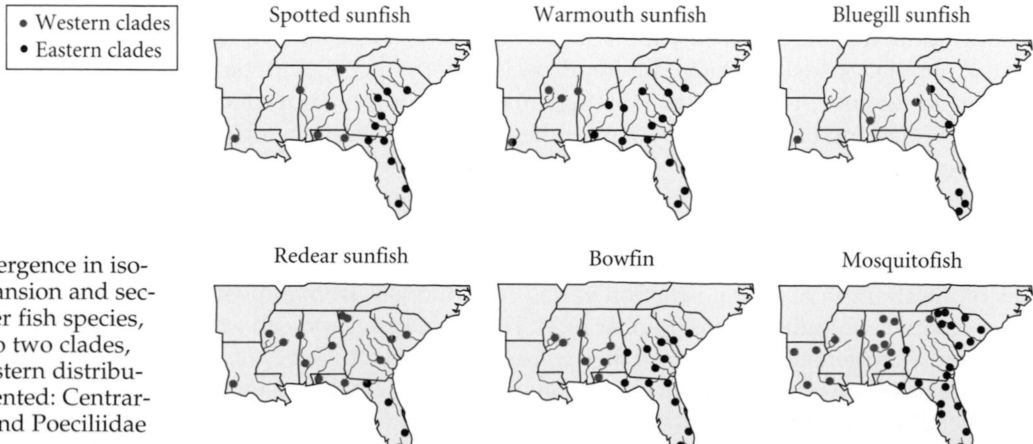

Figure 16.3 Evidence for genetic divergence in isolated "refuges," followed by range expansion and secondary contact. In each of six freshwater fish species, mitochondrial DNA haplotypes fall into two clades, one with a western and one with an eastern distribution. Three families of fishes are represented: Centrarchidae (sunfishes), Amiidae (bowfin), and Poeciliidae (mosquitofish). (After Avise 1994.)

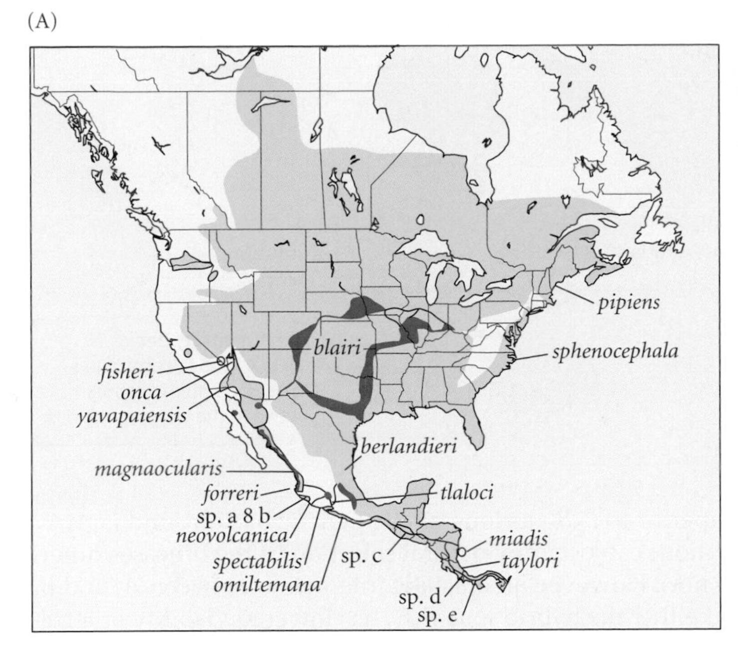

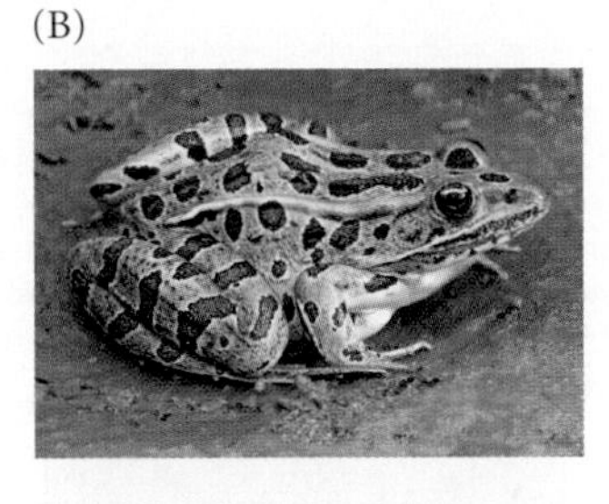

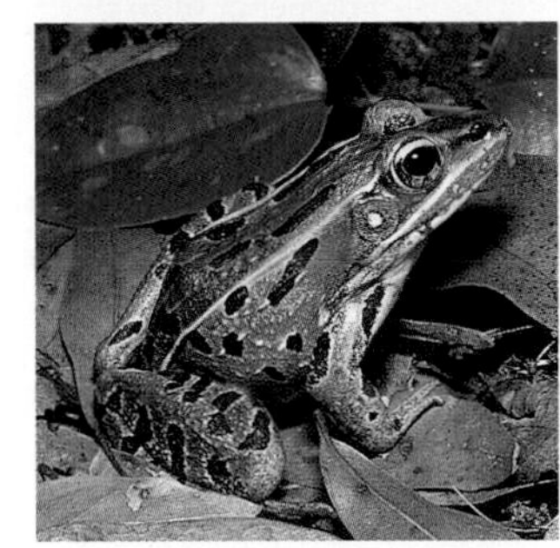

Figure 16.4 (A) The distribution of one clade of species in the *Rana pipiens* (leopard frog) complex. Only recently have studies of the overlap zones (dark) between some of these forms shown that they are reproductively isolated species. The largely allopatric or parapatric distributions of these species suggest that spatial separation has enabled them to become differentiated. (B) *Rana pipiens pipiens*, a northern species, and (C) the Florida leopard frog, *Rana sphenocephala sphenocephala*. (A after Hillis 1988; B, © Rod Planck/Photo Researchers Inc.; C, © Suzanne and Joseph Collins/Photo Researchers, Inc.)

species of Darwin's finches, which evolved on different islands and later became sympatric. But Cocos Island, isolated far to the northeast of the Galápagos, has only one species of finch, which occupies several of the ecological niches that its relatives in the Galápagos Islands fill (Werner and Sherry 1987; see Figure 3.22 and Chapter 18).

Tim Barraclough and Alfried Vogler (2000) reasoned that over time, the amount of overlap between the geographic ranges of species that have formed by allopatric speciation can only increase from zero, whereas overlap between species that originated by sympatric speciation should stay the same or decrease. For several clades of closely related birds, insects, and fishes, they plotted degree of range overlap against degree of molecular difference between species, which they used as an index of time since gene exchange was curtailed. Several groups showed increasing overlap with time, as expected from allopatric speciation (Figure 16.5A,B), whereas two groups of insects displayed a pattern consistent with the possibility of sympatric speciation (Figure 16.5C,D).

Mechanisms of vicariant allopatric speciation

Models of vicariant allopatric speciation have been proposed based on genetic drift, natural selection, and a combination of these two factors. The combination of genetic drift and selection is discussed later, in relation to peripatric speciation.

THE ORIGIN OF INCOMPATIBILITY. How can failure to interbreed, or inability of hybrids to reproduce, arise if they imply fixation of alleles that lower reproductive success? Such an increase to fixation, of course, would be counter to natural selection. Dobzhansky (1936) and Muller (1940) provided a theoretical solution to this problem that does not envision increasing an allele's frequency in opposition to selection. It requires that the reproductive barrier be based on differences at two or more loci that have complementary effects on fitness. (That is, fitness depends on the combined action of the "right" alleles at both loci.)

Suppose the ancestral genotype in two allopatric populations is $A_2A_2B_2B_2$ (Figure 16.6). For some reason, A_1 replaces A_2 in population 1 and B_1 replaces B_2 in population 2, yielding populations monomorphic for $A_1A_1B_2B_2$ and $A_2A_2B_1B_1$, respectively. Both A_1A_2 and A_1A_1 have fitness equal to or greater than A_2A_2 in population 1, *as long as the genetic background is* B_2B_2; likewise,

Figure 16.5 The degree of overlap in the geographic ranges of pairs of closely related species, plotted against the genetic divergence between them, which is an index of time since speciation. Overlap increases with time in fairy wrens and swordtail fish, as expected if speciation is allopatric in these groups. There is no correlation between overlap and time since divergence in tiger beetles or *Rhagoletis* fruit flies, a pattern that is consistent with sympatric speciation. (After Barraclough and Vogler 2000.)

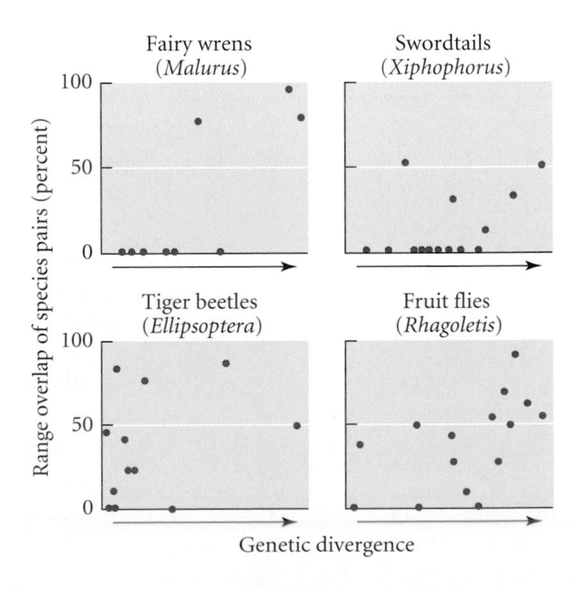

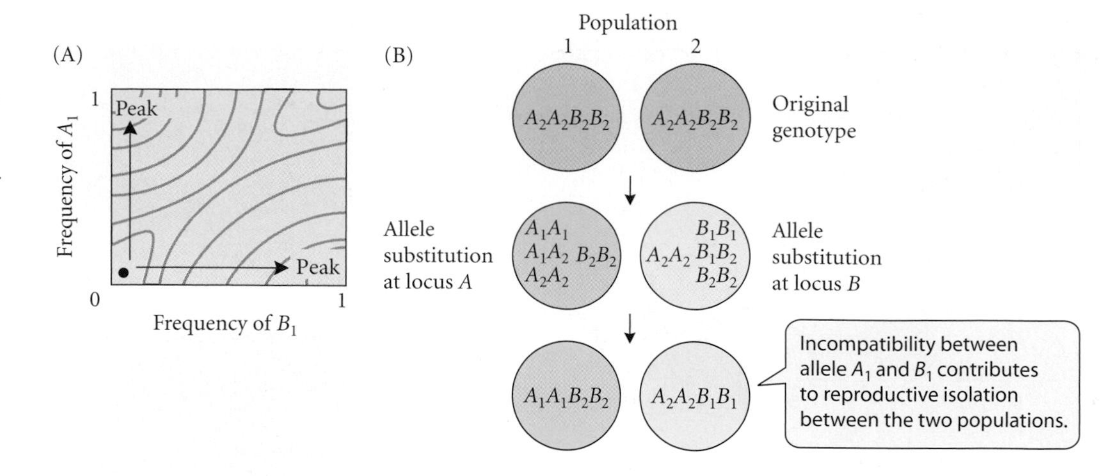

Figure 16.6 The Dobzhansky-Muller theory of allele substitution leading to reproductive isolation between two populations, each initially composed of genotype $A_2A_2B_2B_2$. (A) The adaptive landscape, in which contour lines represent mean fitness as a function of allele frequencies at both loci, shows how the two populations may move uphill toward different adaptive peaks. (B) Each population undergoes an allele substitution at a different locus (substituting either A_1 or B_1). The hybrid combination $A_1A_2B_1B_2$ has low fitness (as indicated by the "valley" in the center of the landscape) due to prezygotic or postzygotic incompatibility between A_1 and B_1.

B_1B_2 and B_1B_1 are equal or superior to B_2B_2, as long as the genetic background is A_2A_2. Therefore these allele substitutions can occur by natural selection (if the fitnesses differ) or by genetic drift (if they do not). However, an epistatic interaction between A_1 and B_1 causes incompatibility, so that either the hybrid $A_1A_2B_1B_2$ has lowered viability or fertility, or $A_1A_1B_2B_2$ and $A_2A_2B_1B_1$ are isolated by a prezygotic barrier, such as different sexual behavior. The important feature of this model is that *neither population has passed through a stage in which inferior heterozygotes existed*. Neither of the incompatible alleles has ever been "tested" against the other within the same population.

A mutation at a third locus, say C_1, may be fixed in the $A_1A_1B_1B_1$ population, but be incompatible with either A_2 or B_2, and further lower the hybrid's fitness. Likewise, mutation D_1 might be fixed, and be incompatible with A_2, B_2, or C_2. As more allelic differences accumulate, later mutations have more opportunities to cause incompatibility and lower the fitness of hybrids. Thus the degree of hybrid sterility or inviability may increase exponentially with the passage of time (Orr 1995).

This model is supported by genetic data showing that reproductive isolation is based on epistatic interactions (called Dobzhansky-Muller incompatibilities) among several or many loci (see Chapter 15). The allele substitutions could be caused by either genetic drift or natural selection. For the moment, we will leave open the possibility of speciation by random fixation of alleles, and consider the ways in which natural selection may contribute to the origin of species.

THE ROLE OF NATURAL SELECTION. The most widely held view of vicariant allopatric speciation is that it is caused by *natural selection, which causes the evolution of genetic differences that create prezygotic and/or postzygotic incompatibility*. Some—perhaps most—of the reproductive isolation evolves while the populations are allopatric, so that a substantial or complete barrier to gene exchange exists when the populations meet again if their ranges expand (Mayr 1963). Thus speciation is usually an *effect*—a by-product—of the divergent selection that occurred during allopatry. The divergent selection may be ecological selection or sexual selection.

A contrasting possibility is that natural selection favors prezygotic (e.g., sexual) reproductive barriers because of their isolating function—because they prevent their bearers from having unfit hybrid progeny. Selection would then result in *reinforcement of reproductive isolation*.

Ecological selection and speciation

Allopatric populations and species undergo both adaptive divergence and evolution of reproductive isolation, but showing that reproductive isolation is a *result* of adaptive divergence requires evidence that the two processes are genetically and causally related to each other. The most direct evidence comes from laboratory studies of *Drosophila* and houseflies, in which investigators have tested for reproductive isolation among subpopulations drawn from a single base population and subjected to divergent selection for various morphological, behavioral, or physiological characteristics (Rice and Hostert 1994). In many of these studies, partial sexual isolation or postzygotic isolation developed, demonstrating that substantial progress toward speciation can be observed in the laboratory, and that it can arise as a correlated response to divergent selection. That is, reproductive isolation was due to pleiotropic effects of genes for the divergently selected character, or to closely linked genes.

Several examples from natural populations support the hypothesis that divergent ecological adaptation causes reproductive isolation as a by-product (Coyne and Orr 2004). In the monkeyflower *Mimulus guttatus*, for example, hybrids between a copper-tolerant population and a nontolerant population had low viability. Genetic analysis showed that the gene for copper tolerance (or an allele at a very closely linked locus) was responsible for this effect (Macnair and Christie 1983). Daniel Funk (1998) predicted that if reproductive isolation arises as a by-product of ecological adaptation, then allopatric populations of an insect that are adapted to different host plant species should display stronger incipient isolation from each other than allopatric populations that feed on the same host plant. As he predicted, a willow-adapted population of the leaf beetle *Neochlamisus bebbianae* was more strongly sexually isolated from two maple-feeding populations than the latter were from each other (Figure 16.7).

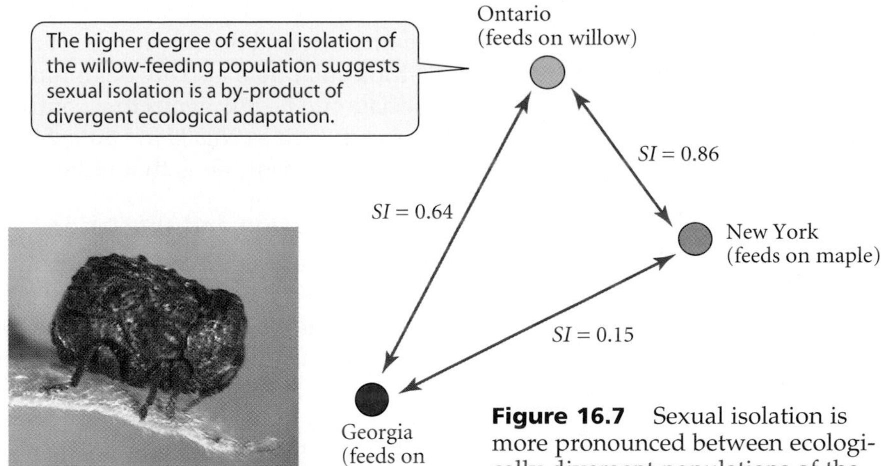

Figure 16.7 Sexual isolation is more pronounced between ecologically divergent populations of the leaf beetle *Neochlamisus bebbianae* than between populations that are separated by geographic distance alone. The numbers represent a sexual isolation index (SI) derived from laboratory mating trials. These results suggest that sexual isolation is a by-product of divergent ecological adaptation. (After Funk 1998; photo by Christopher Brown, courtesy of Daniel Funk.)

In contrast to these examples, in which pleiotropy or linkage may be responsible for an association between ecological adaptation and reproductive isolation, a character that contributes to reproductive isolation may diverge because of its ecological role. A good example is the case of the two monkeyflowers (*Mimulus*) described in Chapter 15, which have become adapted to different pollinators (see Figure 15.7). Three-spined sticklebacks (*Gasterosteus*) have undergone PARALLEL SPECIATION in several Canadian lakes, where a limnetic (open-water) "ecomorph" coexists with a benthic (bottom-feeding) "ecomorph" that is smaller and differs in shape. These ecomorphs are sexually isolated and have evolved independently in each lake; that is, speciation has occurred in parallel (Figure 16.8A). Parallel ecological divergence implies that ecological selection has shaped the differences between the ecomorphs. In laboratory trials, fish of the same ecomorph from different lakes mate almost as readily as those from the same lake, but different ecomorphs mate much less frequently (Figure 16.8B). Thus features associated with ecological divergence affect

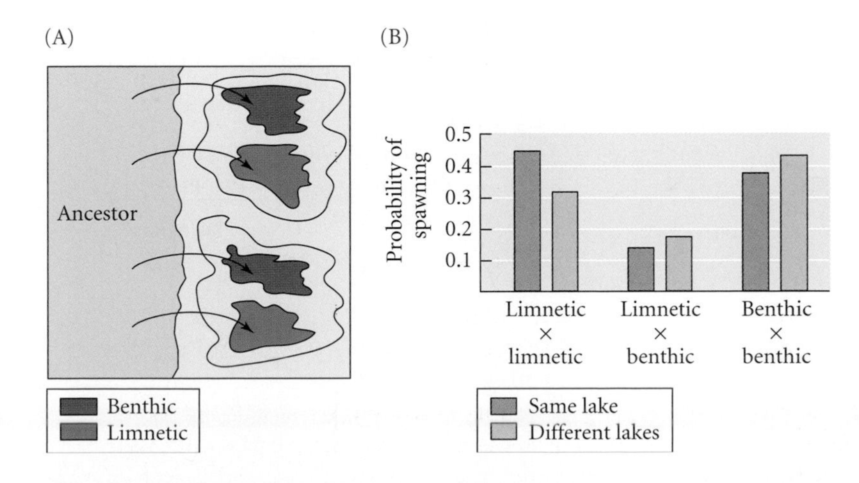

Figure 16.8 Parallel speciation in the three-spined stickleback (*Gasterosteus*). (A) Pairs of open-water (limnetic) and bottom-foraging (benthic) "ecomorphs" have arisen independently in different lakes. (B) Females mate preferentially with males on the basis of their morphology, whether they are from the same or different lakes. This isolating character is evidently adaptive, since it has repeatedly evolved in the same way. (A after Schluter and Nagel 1995; B after Rundle et al. 2000.)

reproductive isolation. One such feature is probably the difference in body size between the limnetic and benthic forms (Rundle et al. 2000).

Molecular data also are beginning to provide evidence of a role for natural selection in speciation. The few genes that contribute to reproductive isolation and that have been sequenced, such as *Nup96* in *Drosophila* (see Figure 15.15), show the high rate of amino acid-replacing substitutions that indicates directional selection.

Sexual selection and speciation

Models of sexual selection of male traits by female choice show that divergent traits and preferences can evolve in different populations of an ancestral species, resulting in speciation (Lande 1981; Pomiankowski and Iwasa 1998; Turelli et al. 2001). The expected result would be the diversity of different male traits that distinguish species of hummingbirds (see Figure 15.6) and many other groups of animals.

It is very likely that sexual selection has been an important cause of speciation, especially in highly diverse groups, such as African lake cichlids, Hawaiian *Drosophila*, pheasants, and birds of paradise, in which males are commonly highly (and diversely) colored or ornamented (Panhuis et al. 2001). Comparisons of the species diversity of sister groups of birds suggest that sexual selection has enhanced diversity (Figure 16.9). Groups of birds with promiscuous mating systems have higher diversity than sister clades in which pair bonds are formed and the variance in male mating success is presumably lower—resulting in weaker sexual selection (Mitra et al. 1996). Because sister clades, by definition, are equal in age, the difference in diversity implies a higher rate of speciation (or possibly a lower extinction rate) in clades that experience strong sexual selection. Diversity of species and subspecies has likewise been correlated with the evolution of sexually selected feather ornaments, such as crests and elongated tail feathers (Møller and Cuervo 1998).

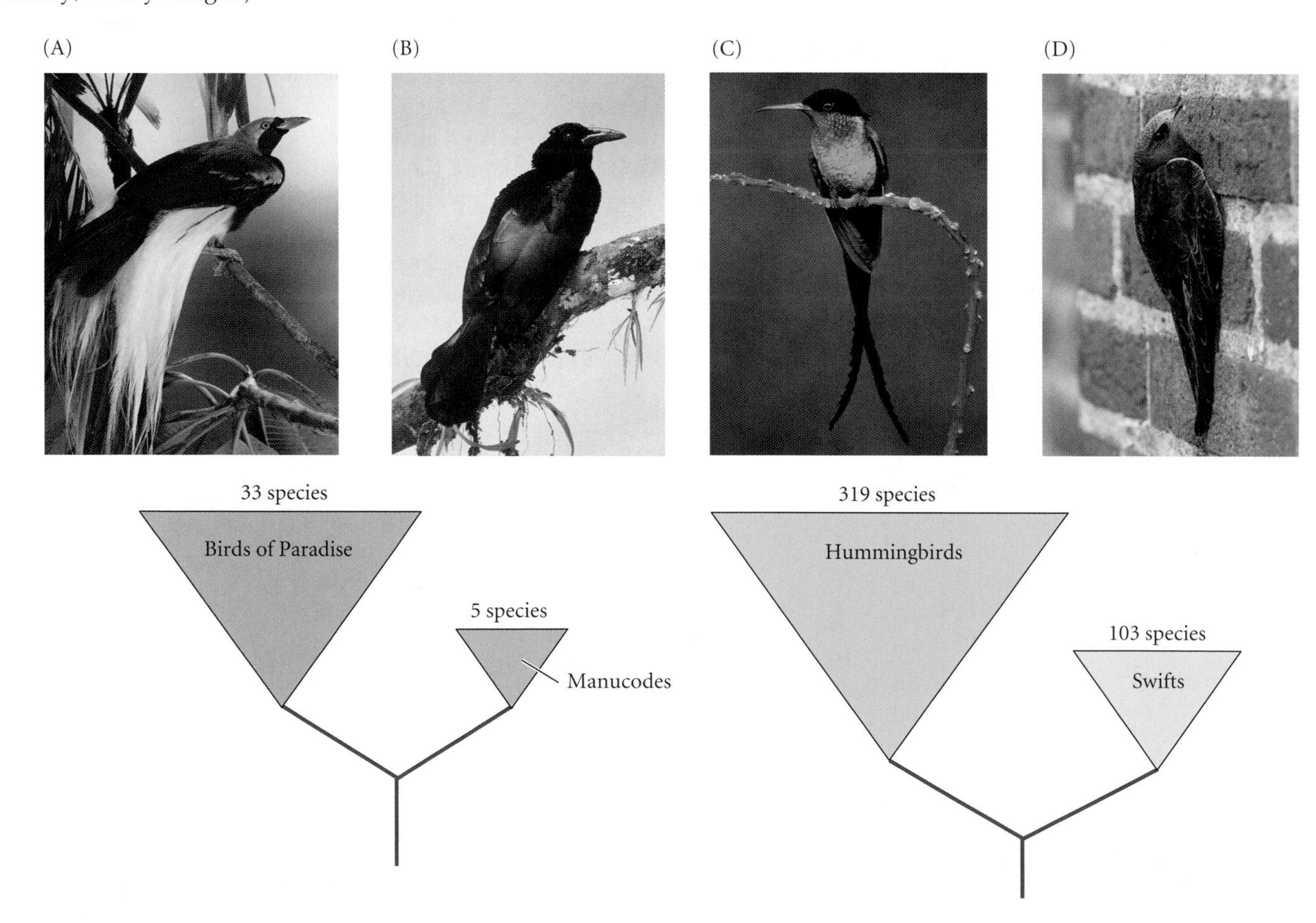

Figure 16.9 In sister clades of birds that differ in their mating system, those clades that mate promiscuously and do not form a pair bond (A, C) tend to have more species than nonpromiscuous clades (B, D) that do form pairs. The promiscuously mating clades are thought to experience stronger sexual selection. (A) The promiscuous male lesser bird of paradise (*Paradisaea minor*), and (B) a nonpromiscuous manucode (*Manucodia comrii*). (C) A male streamertail hummingbird (*Trochilus polytmus*) and (D) a common swift (*Apus apus*), member of a clade that forms pair bonds. (A, © Tony Tilford/photolibrary.com; B, © W. Peckover, VIREO; C, © RobertTyrrell/OSF/photolibrary.com; D, © National Trust Photolibrary/Alamy Images.)

Sexually selected characters often act as barriers to interbreeding. For example, characteristics of the male call in the túngara frog (*Physalaemus pustulosus*) are sexually selected by female preferences (see Chapter 14). Females of this species prefer calls of their own species to calls of other *Physalaemus* species, implying that the call differences act as reproductive barriers (Ryan and Rand 1993). Likewise, the male color pattern of some closely related African lake cichlids acts as a reproductive barrier between species and is sexually selected within species (McKaye et al. 1984; Seehausen et al. 1999). Sexual selection in these cichlids has probably contributed to their extraordinarily high rate of speciation.

Michael Ritchie (2000) has provided explicit evidence that sexual selection within populations results in reproductive isolation. The male song of Mediterranean populations of the bushcricket *Ephippiger ephippiger* has a single syllable, whereas males from the Pyrenees produce polysyllabic songs. Ritchie provided females with a choice between tape recordings of synthetic songs that varied in syllable number and observed which speaker the females approached. Mediterranean females showed increasing preference, the lower the syllable number, and should therefore exert directional selection on males for monosyllabic songs (Figure 16.10). In contrast, Pyrenees females responded most to songs with five syllables, which is actually more than most males in this polysyllabic population emit. Thus females exert directional sexual selection for greater syllable number in this population. Why don't Pyrenees males match the females' preference? Possibly females have evolved to resist male stimulation, as the "chase-away" model of sexual selection posits (Holland and Rice 1998; see Chapter 14).

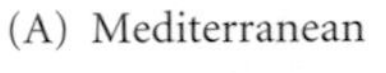

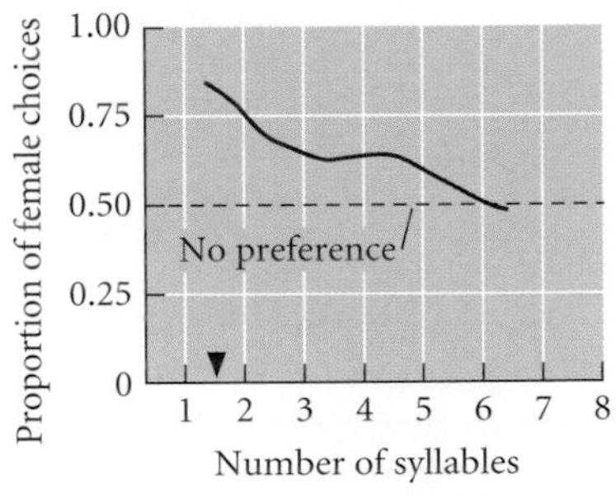

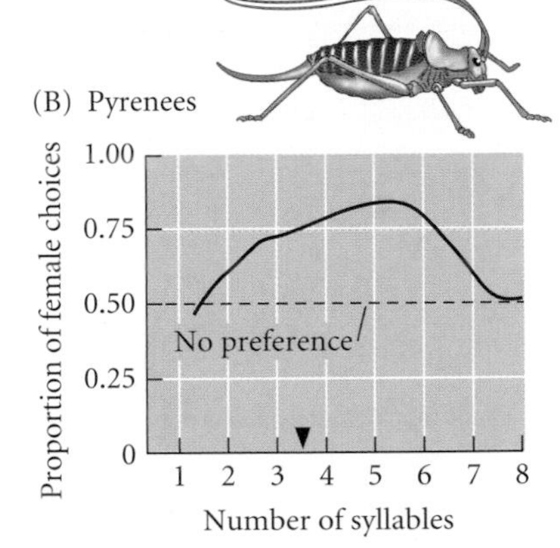

Figure 16.10 Divergent sexual selection for male song in the bushcricket *Ephippiger ephippiger*. The plots show the proportion of females that chose male songs with different numbers of syllables. Black triangles indicate the mean number of syllables in male song in each region. Females of a Mediterranean population (A) impose directional selection for few syllables, whereas Pyrenees females (B) impose selection for a greater number. Such differences in sexual selection can result in reproductive isolation. (After Ritchie 2000.)

Reinforcement of reproductive isolation

We have seen that reproductive isolation can arise as a *side effect* of genetic divergence due to natural selection. However, many persons have supposed that reproductive isolation evolves, at least in part, as an *adaptation to prevent the production of unfit hybrids*. The champion of this viewpoint was Theodosius Dobzhansky, who expressed the hypothesis this way:

> Assume that incipient species, A and B, are in contact in a certain territory. Mutations arise in either or in both species which make their carriers less likely to mate with the other species. The nonmutant individuals of A which cross to B will produce a progeny which is adaptively inferior to the pure species. Since the mutants breed only or mostly within the species, their progeny will be adaptively superior to that of the nonmutants. Consequently, natural selection will favor the spread and establishment of the mutant condition. (Dobzhansky 1951, 208)

Dobzhansky introduced the term "isolating mechanisms" to designate reproductive barriers, which he believed were indeed mechanisms designed to isolate. In contrast, Ernst Mayr (1963), among others, held that although natural selection might enhance reproductive isolation, reproductive barriers arise mostly as side effects of allopatric divergence, whatever its cause may be. Mayr cited several lines of evidence: sexual isolation exists among fully allopatric forms; it has not evolved in several hybrid zones that are thought to be thousands of years old; features that promote sexual isolation between species are usually not limited to regions where they are sympatric and face the "threat" of hybridization. It is now generally agreed that natural selection can enhance prezygotic reproductive isolation between hybridizing populations, but how often this process plays a role in speciation is not known (Howard 1993; Noor 1999; Turelli et al. 2001).

In most organisms, natural selection cannot strengthen *postzygotic* isolation between hybridizing populations because such a process would require that alleles that reduce fertility or survival increase in frequency, which would be precisely antithetical to the meaning of natural selection! (See Grant 1966 and Coyne 1974 for possible exceptions.) Postzygotic isolation arises, instead, by fixation of alleles at different loci in separate populations that happen to reduce fertility or viability when they are combined in hybrids. *Within* a population of hybrids, we would expect mutations that *improve* fitness to increase in frequency. Such an increase would reduce the advantage of genes that accentuate prezygotic isolation, and the likely result would be fusion of the populations. For example, populations of the common shrew (*Sorex araneus*) in England differ in complex chromosome rearrangements that should greatly reduce the fertility of heterozygotes. In contact zones between populations, however, the complex rearrangements have been replaced by simple chromosome arrangements that do not reduce hybrid fertility very much (Searle 1993).

The enhancement of prezygotic barriers that Dobzhansky envisioned is often called **reinforcement** of prezygotic isolation. This process has been cited as a cause of **character displacement**, meaning a *pattern* whereby characters differ more where two taxa are sympatric than where they are allopatric (Brown and Wilson 1956).*

Theoretical models have shown that the likelihood of reinforcement is reduced by some factors and enhanced by others. A simple model illustrates some of these factors (Felsenstein 1981; Sanderson 1989). Suppose alleles A_1 and A_2 are fixed in populations (semispecies) 1 and 2, respectively, and that in F_1 hybrids, the heterozygote A_1A_2 has a selective disadvantage (s) due to reduced fertility. Another locus, B, governs ASSORTATIVE MATING, such that B_1 carriers have a certain degree of preference for other B_1 carriers as mates, and B_2 carriers likewise prefer B_2 mates. (B_2 might alter the season of mating or a phenotypic character that mates use to recognize each other.) Initially, both populations are fixed for B_1, and hence mate randomly. The question is whether the new mutation B_2 can increase in frequency within one of the populations (say, population 2), establishing partial sexual isolation between the A_1 and the A_2 carriers where they meet at a hybrid zone.

If the fertility of A_1A_2 is zero ($s = 1$), the populations are fully reproductively isolated. If mutation B_2 arises in population 2, it is always associated with allele A_2. Therefore, B_2 carriers who mate with each other have fertile A_2A_2 offspring, whereas carriers of the combination A_1B_1 frequently mate with carriers of the combination A_2B_1, and produce sterile offspring. The mean fitness of B_2 is thus greater than that of B_1, so B_2 increases in frequency within population 2, and partial prezygotic isolation evolves, as Dobzhansky proposed.

However, we have assumed that mutation B_2 has no other effect on fitness. It might have disadvantageous pleiotropic effects, or sexual selection might disfavor it outside the hybrid zone. If so, B_2 might increase in frequency in and near the hybrid zone, but would be selected against elsewhere. Moreover, gene flow into the hybrid zone would increase the frequency of B_1, counteracting the increase of B_2 by selection. Hence the evolution of complete prezygotic isolation in the hybrid zone may be unlikely.

In this model, B_2 can increase, forming a partial prezygotic barrier between the populations, only because it is associated with A_2 alleles, and thus reduces the frequency of A_1A_2 offspring. Now suppose the F_1 hybrid (A_1A_2) has only partly reduced fertility, so that some recombinant offspring are produced by backcrossing of A_1A_2 with the parental types A_1A_1 and A_2A_2. Because of recombination, the association of B_2 with A_2 breaks down. A_1B_2-bearing backcross progeny mate with A_2B_2 individuals from population 2, producing A_1A_2 offspring with reduced fertility. Thus the advantage of B_2 is diminished, since it is advantageous only insofar as it prevents individuals from engaging in matings that result in A_1A_2 offspring. B_2 may therefore not increase in frequency, especially if gene flow or pleiotropic disadvantage tends to reduce its frequency. Thus recombination between loci that reduce the fitness of hybrids and loci that govern assortative mating is a powerful factor working against the reinforcement of prezygotic isolation. (It also reduces the likelihood of sympatric evolution of assortative mating, as we will see in the section below on sympatric speciation.)

Models based on different assumptions, however, show that reinforcement can occur under some circumstances (Liou and Price 1994; Kirkpatrick and Servedio 1999; Cain et al. 1999). For example, instead of assuming a single locus (B) for assortative mating, assume that one locus (P) governs female mate preference and another a male phenotypic

*Some authors follow Butlin (1989) in using *reproductive character displacement* to mean the process of evolution of accentuated prezygotic barriers to mating between fully developed species—i.e., populations that do not actually exchange genes. They may be fully isolated by postzygotic barriers (e.g., full hybrid sterility), or selection may favor greater mating discrimination or more different courtship signals for other reasons (e.g., to avoid wasting time dallying with members of the other species in unconsummated courtship). Butlin restricts the term *reinforcement* to the evolution of stronger mating barriers between taxa that can exchange genes through hybrids that have low but nonzero fitness. This text uses the traditional definitions (Howard 1993).

trait (T) that differs between the "semispecies." Then divergent sexual selection for different mate preference alleles (P_1, P_2) in the two populations may be reinforced by the low fitness of hybrids. Because sexual selection generally creates an association (linkage disequilibrium) between P-locus and T-locus alleles (see Chapter 15), and because the populations are assumed to differ already in the frequencies of the male trait alleles T_1 and T_2, associations between genes reducing hybrid fitness and genes affecting the mating system are less likely to be broken down by recombination than in the previous model.

Reinforcement of prezygotic isolation appears to occur fairly often (Howard 1993; Noor 1999). The geographic range of *Drosophila persimilis* is contained within that of the more broadly distributed *D. pseudoobscura*, and these species occasionally hybridize in nature, producing sterile male but fertile female hybrids. Female *D. pseudoobscura* from two sympatric populations, confined with male *D. persimilis*, mated less frequently than females from three allopatric populations, having evidently evolved greater discrimination (Noor 1995). Green tree frogs (*Hyla cinerea*) and barking tree frogs (*H. gratiosa*) sometimes hybridize where they co-occur, and the hybrids, although fertile, appear to have reduced mating success and survival. The mating call of male *H. cinerea* is slightly more different from the call of *H. gratiosa* in ponds where these species coexist than in allopatric *H. cinerea* populations (Figure 16.11A), and sympatric female *H. cinerea* show a stronger preference for conspecific males' calls when offered a choice between recordings of both species (Figure 16.11B).

Peripatric speciation

THE HYPOTHESIS. One of Ernst Mayr's most influential and controversial hypotheses was **founder effect speciation** (1954), which he later termed *peripatric speciation* (1982b). He based this hypothesis on the observation, in many birds and other animals, that isolated populations with restricted distributions, in locations peripheral to the distribution of a probable "parent" species, often are highly divergent, to the point of being classified as different species or even genera. For example, the small lizard *Uta stansburiana* exhibits only subtle geographic variation throughout western North America, but populations on different islands in the Gulf of California vary so greatly in body size, scalation, coloration, and ecological habits that some have been named separate species (Soulé 1966).

Mayr proposed that genetic change could be very rapid in localized populations founded by a few individuals and cut off from gene exchange with the main body of the species. He reasoned that allele frequencies at some loci would differ from those in the parent population because of accidents of sampling—i.e., genetic drift—simply because a small number of colonists would carry only some of the alleles from the source population, and at altered frequencies. (He termed this initial alteration of allele frequencies the FOUNDER EFFECT; see Chapter 10.) *Because epistatic interactions among genes affect fitness, this initial change in allele frequencies at some loci would alter the selective value of genotypes at other, interacting loci.* Hence selection would alter allele frequencies at these loci, and this in turn might select for changes at still other epistatically interacting loci. The "snowballing" genetic change that might result would incidentally yield reproductive isolation. As Mayr (1954) pointed out, this hypothesis implies that substantial evolution would occur so rapidly, and on so localized a geographic scale, that it would probably not be documented in the fossil record. If such a new species expanded its range, it would appear suddenly in the fossil record, without evidence of the intermediate phenotypic changes that had occurred. Thus this hypothesis, he said, might help to explain the rarity of fos-

(A)

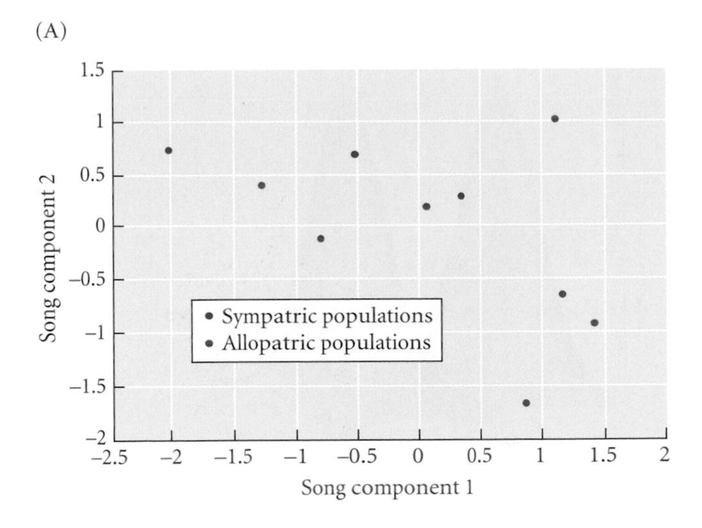

(B)

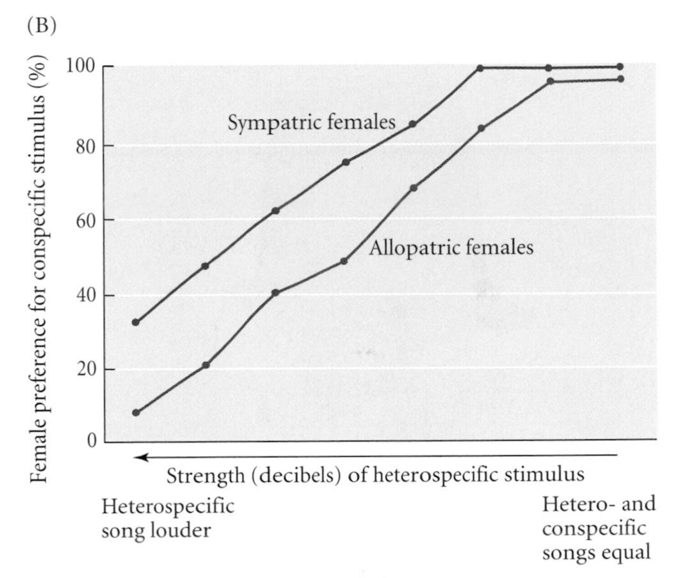

Figure 16.11 Evidence for reinforcement of reproductive isolation in the green tree frog (*Hyla cinerea*). (A) The male's song differs, especially in "component 1," between populations that are sympatric with the barking tree frog and those that are allopatric. Each song component is a statistical combination of several song characteristics. (B) When recordings of the songs of both species are played to female *Hyla cinerea*, females from sympatric populations show a greater preference for their own species' song than do females from allopatric populations. The difference between their responses is most pronounced when the heterospecific recording is louder than the conspecific recording. (After Höbel and Gerhardt 2001.)

(A) Peak shift across an adaptive valley

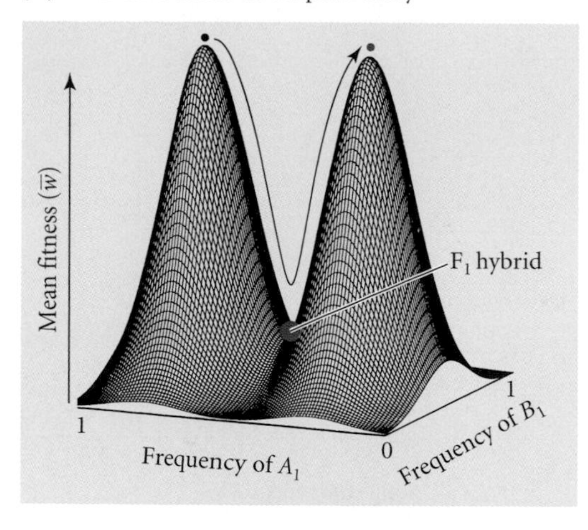

(B) Genetic drift along an adaptive ridge

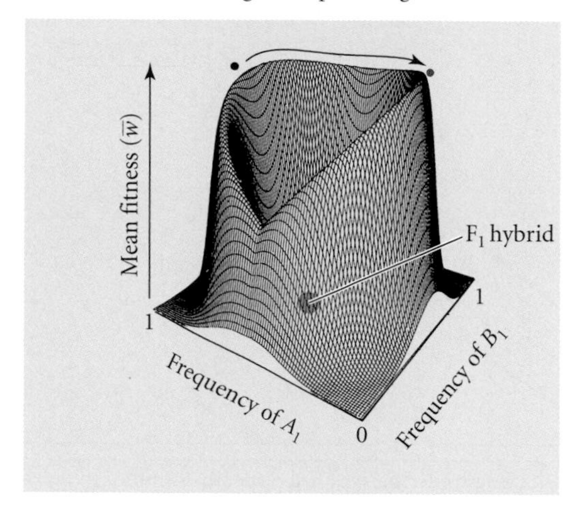

Figure 16.12 Two adaptive landscapes show how peripatric speciation might occur. The height of a point on the three-dimensional landscape represents the mean fitness of a population ($\bar{w}$). The mean fitness is a function of allele frequencies at loci *A* and *B*. (A) In a peak shift, a population evolves from one adaptive peak to another by moving downhill (lowering fitness) and then uphill. The F_1 hybrid of a cross between populations on the two peaks lies in the valley; i.e., it has low fitness, which causes some reproductive isolation between the populations. (B) Genetic drift along an adaptive ridge, in which genetic constitutions with the same fitness connect the beginning and end states of a population. The F_1 hybrid between populations with these genetic constitutions lies inside the crater. (After Gavrilets and Hastings 1996.)

silized transitional forms among species and genera. Mayr thus anticipated, and provided the theoretical foundation for, the idea of PUNCTUATED EQUILIBRIUM (see Chapters 4 and 21) advanced by Eldredge and Gould (1972).

Similar hypotheses have been advanced by Hampton Carson (1975), who suggested that genetic reorganization may be enhanced by repeated fluctuations in population size, and by Alan Templeton (1980), who emphasized that founder events may alter the frequency of alleles of large effect at only a few loci, affecting only certain characters such as courtship behavior. Selection might then alter polygenic modifier loci to bring about a new coadapted state of the character.

The usual interpretation of these hypotheses employs the metaphor of an adaptive landscape (see Figure 12.20). The colony undergoes a shift between two "adaptive peaks"—that is, from one adaptive genetic constitution (that of the parent population) through a less adaptive constitution (an "adaptive valley") to a new adaptive equilibrium (Figure 16.12A). The process begins when genetic drift in the small, newly founded population shifts allele frequencies from the vicinity of one adaptive peak to the slope of the other. This stage cannot be accomplished by natural selection, since selection cannot reduce mean fitness. However, selection can move the allele frequencies up the slope away from the valley toward the new peak. The most elementary peak shift would be substitution of one allele (or chromosome) for another when the heterozygote has lower fitness than either homozygote. Mayr, Carson, and Templeton envisioned a more complex shift between peaks represented by different "coadapted" gene combinations (such as $A_1A_1B_1B_1$ and $A_2A_2B_2B_2$).

THEORETICAL CONSIDERATIONS. Speciation by peak shift is considered unlikely by many theoretical population geneticists (Charlesworth and Rouhani 1988; Turelli et al. 2001). In their view, reproductive isolation is caused by the low fitness of heterozygous hybrids; that is, by a deep adaptive valley. If the adaptive valley is very deep (i.e., if there is strong selection against heterozygotes), genetic drift is unlikely to move allele frequencies across the valley from one peak to another, unless the founder population is very small. But in that case, genetic variation—including the rare alleles that might be supposed to initiate the evolutionary change—is likely to be lost. Thus, if the adaptive valley is shallow enough for a peak shift to be likely, the genetic difference between the populations will cause little reproductive isolation; if selection is strong and the valley is deep, the populations will be reproductively well isolated, but the shift to the new genetic composition is unlikely to occur.

Other models, however, show that peak shifts may be more likely if different assumptions are made (Price et al. 1993; Wagner et al. 1994; Gavrilets 2004). For example, a small population may move by genetic drift along an "adaptive ridge" to the other side of an adaptive valley from the parent population (Figure 16.12B). Hybrids between the two populations would have low fitness, even though mean fitness would not decrease within either population (see also Figure 16.6). When large numbers of epistatically interacting loci are considered, there are likely to be many such "ridges" of high fitness separated by valleys or "holes" of low fitness, and the evolution of reproductive isolation by genetic drift is quite probable (Gavrilets 2004).

EVIDENCE FROM NATURAL POPULATIONS. Many species do originate, as Mayr said, as localized "buds" from a widespread parent species. The best evidence consists of phylogenies of DNA sequences (gene genealogies) taken from various geographic populations of widespread species and closely related species with narrower, often peripheral, distributions. The localized species often prove to be more closely related to certain populations of the widespread species than the populations of the widespread species are to one another (Avise 1994). The genealogy of mitochondrial DNA in the moths *Greya piperella* and *Greya mitellae*, in Figure 15.4, is an example.

The environment of peripheral populations often differs substantially from that occupied by more central populations, both in abiotic factors such as climate and in the species composition of the community. Thus peripheral populations may often diverge simply because of natural selection, not because of founder effects. The hypothesis that a species arose from a population that was not only peripheral, but also small, would be supported by evidence that the population had lost most of the genetic variation present in its more populous ancestor. A likely example is *Drosophila sechellia*, which is restricted to the Seychelles Islands, a group of small islands in the Indian Ocean east of Africa (Kliman et al. 2000). This species is closely related to *D. mauritiana*, found on the island of Mauritius in the Indian Ocean, and to *D. simulans*, which is native to Africa and has become cosmopolitan due to inadvertent human transport. At many loci, the gene genealogy of *D. sechellia* is nested within that of *D. simulans*, as expected if it arose by colonization from Africa (Figure 16.13). The level of DNA sequence diversity within *D. sechellia* is much lower than in *D. mauritiana* or *D. simulans*. This observation clearly indicates that the effective population size of *D. sechellia* has been much smaller than those of the other species, but whether or not the reduced population size initiated the evolution of reproductive isolation is not known. The case of *Drosophila sechellia* is rather unusual, for most analyses of genetic variation in sister species have not provided any evidence of bottlenecks.

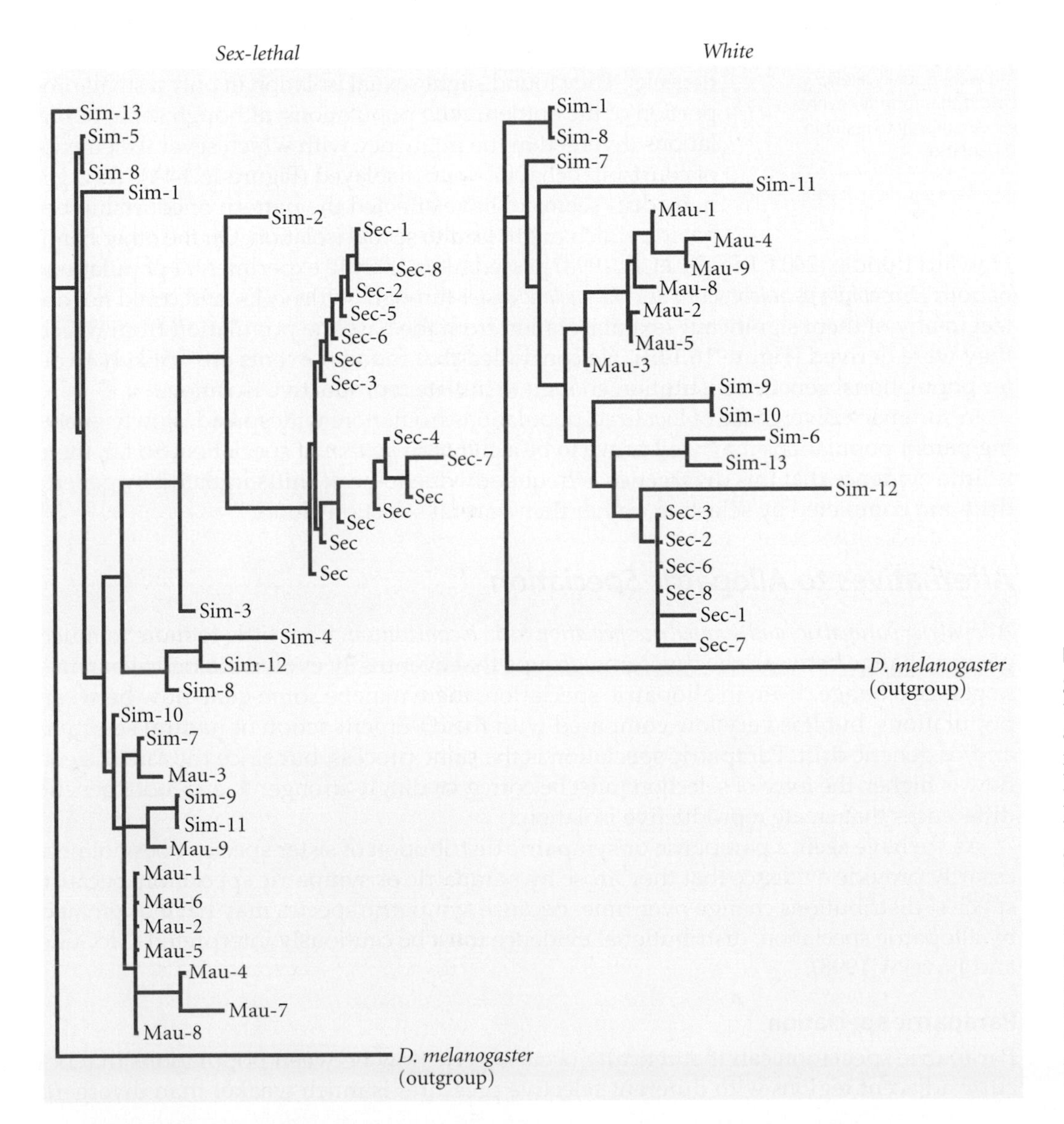

Figure 16.13 Gene trees for two genes (*Sex-lethal* and *white*) suggest that *Drosophila mauritiana* (Mau) and *D. sechellia* (Sec), which are restricted to small islands in the Indian Ocean, originated as small populations. Each numbered "twig" is a different haplotype. Since their common ancestry with the widespread species *D. simulans* (Sim), the smaller populations of the island species have coalesced faster, so that at both loci, Mau and Sec sequences are nested within the Sim gene tree. Probably a *simulans*-like ancestor colonized each island and evolved into a new species. (After Kliman et al. 2000.)

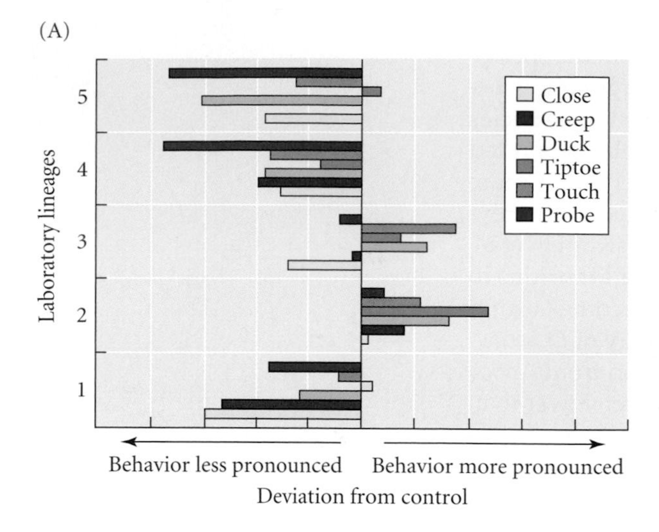

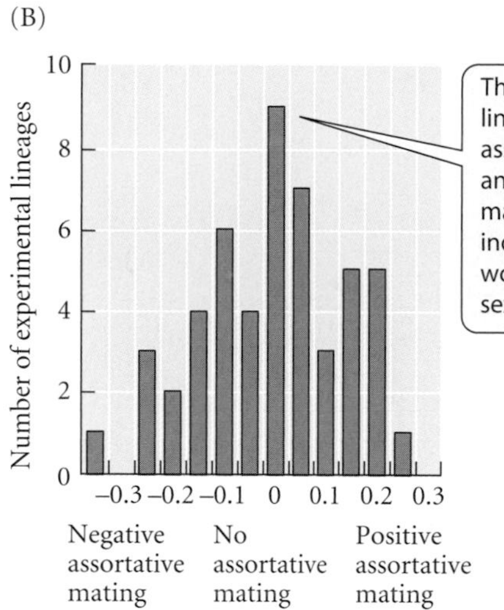

Figure 16.14 Tests of the peripatric speciation hypothesis in laboratory populations. (A) Courtship behavior among bottlenecked laboratory populations of houseflies (*Musca domestica*). For each of five experimental lineages that underwent several reductions to 1, 4, or 16 pairs, the percentage of courtships that included each of the designated behavioral elements and terminated in successful copulation is shown, expressed as a deviation from the values in a large (control) population. The data suggest that genetic drift has altered male display elements and female responses, either diminishing or enhancing the likelihood of successful mating. (B) The degree of assortative mating displayed by 50 bottlenecked laboratory lines of *Drosophila melanogaster*, when given the choice of mating with individuals from its own population (positive) or from the base population (negative assortative mating). Positive and negative assortative mating were equally frequent, and the average degree of assortative mating was nearly zero. (A after Meffert and Bryant 1991; B after Rundle et al. 1998.)

EXPERIMENTAL EVIDENCE. Several investigators have passed laboratory populations through repeated bottlenecks to see whether reproductive isolation can evolve in this way (summarized by Rice and Hostert 1993; Coyne and Orr 2004). In one such experiment, Lisa Meffert and Edwin Bryant (1991) passed populations of houseflies (*Musca domestica*) through several bottlenecks. They found slight sexual isolation in only a small proportion of the bottlenecked populations, although some populations diverged in the frequency with which several elements of courtship behavior were displayed (Figure 16.14A). Thus genetic drift seems to have affected the pattern of courtship behavior, which might lead to sexual isolation. On the other hand, Howard Rundle (2003; Rundle et al. 1998) passed more than 40 experimental populations of both *Drosophila pseudoobscura* and *D. melanogaster* through bottlenecks, and could not detect in any of them significant sexual isolation from the baseline population from which they were derived (Figure 16.14B). He concluded that founder events are unlikely to alter populations' genetic constitution enough to initiate reproductive isolation.

In summary, divergence of localized populations from more widespread, slowly evolving parent populations may well prove to be a common *pattern* of speciation. So far, there is little evidence that this divergence is frequently due to peak shifts initiated by genetic drift and completed by selection, rather than natural selection alone.

Alternatives to Allopatric Speciation

Allopatric, parapatric, and sympatric speciation form a continuum, from little to more to much gene exchange between the diverging groups that eventually evolve biological barriers to gene exchange. Even in allopatric speciation, there may be some gene flow between populations, but it is very low compared with the divergent action of natural selection and/or genetic drift. Parapatric speciation is the same process, but since the rate of gene flow is higher, the force of selection must be correspondingly stronger to engender genetic differences that create reproductive isolation.

As we have seen, a parapatric or sympatric distribution of sister species does not necessarily provide evidence that they arose by parapatric or sympatric speciation, because species' distributions change over time. Because sympatric species may have originated by allopatric speciation, distributional evidence must be cautiously interpreted (McCune and Lovejoy 1998).

Parapatric speciation

Parapatric speciation can theoretically occur if gene flow between populations that occupy adjacent regions with different selective pressures is much weaker than divergent selection for different gene combinations (Endler 1977). Strong selection at a sharp border between different habitats poses a barrier to gene exchange, caused by the reproductive failure of individuals with the "wrong" genotype or phenotype that migrate across the border. Consequently, clines at various loci may tend to develop at the same location, resulting in a primary hybrid zone that has developed in situ, but might look like a secondary hybrid zone (Endler 1977; Barton and Hewitt 1985). Steady genetic divergence may eventually result in complete reproductive isolation.

Another possibility is that populations isolated by distance can evolve reproductive incompatibility. Divergent features that arise at widely separated sites in the species' distribution may spread, supplanting ancestral features as they travel, and preventing gene exchange when they eventually meet. Russell Lande (1982) has theorized that prezygotic isolation could arise in this way due to divergent sexual selection.

Parapatric speciation undoubtedly occurs and may even be common, but it is very difficult to demonstrate that it provides a better explanation than allopatric speciation for real cases (Coyne and Orr 2004). Possibly the best-documented example of the parapatric origin of reproductive isolation is attributable not to these theories, but to selection for isolation—i.e., reinforcement. *Anthoxanthum odoratum* is one of several grasses that have evolved tolerance to heavy metals in the vicinity of mines within the last several centuries (see Chapter 13). Several populations, under very strong selection for heavy metal tolerance, have diverged from neighboring nontolerant populations (on uncontaminated soil) not only in tolerance, but in flowering time; moreover, they self-pollinate more frequently, having become more self-compatible (Figure 16.15). Both characteristics provide considerable reproductive isolation from adjacent nontolerant genotypes. In this case, the historical evidence indicates that the divergence is very recent, and the species is so ubiquitously distributed that parapatric divergence cannot be doubted.

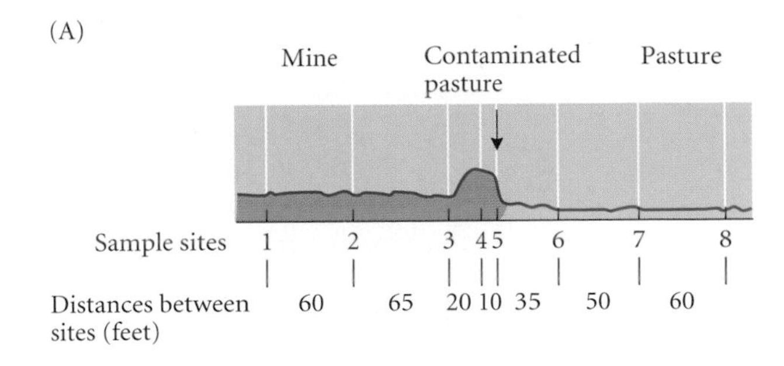

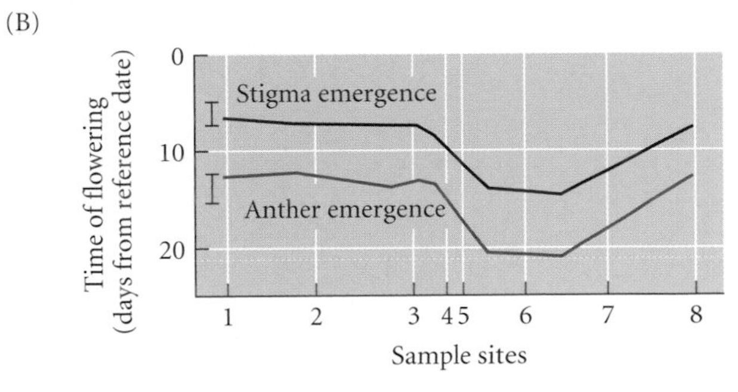

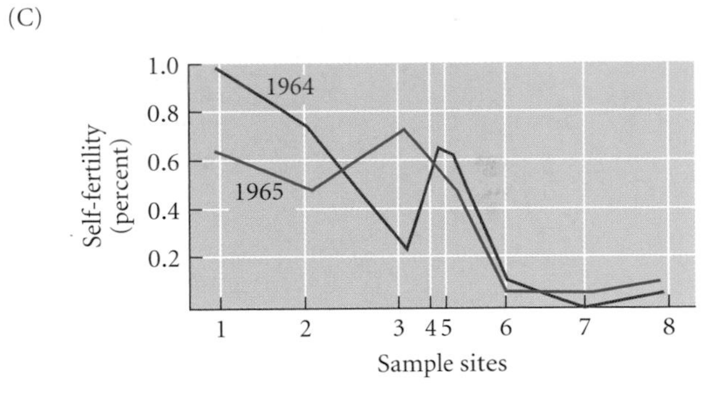

Figure 16.15 Parapatric evolution of reproductive isolation over a very short distance in the grass species *Anthoxanthum odoratum*. (A) Sample sites and distances along a transect from an area contaminated with lead and zinc (blue) into an area with uncontaminated soil. (B) Flowering time, in a common garden, of plants derived from different sample sites along the transect. (C) Seed set by plants with flowers bagged to prevent cross-pollination. The differences in both flowering time and capacity for self-fertilization are likely to reduce gene flow between plants from contaminated and uncontaminated environments. (After McNeilly and Antonovics 1968; Antonovics 1968.)

Sympatric speciation

THE CONTROVERSY. Sympatric speciation is a highly controversial subject. Speciation would be sympatric if a biological barrier to gene exchange arose within an initially randomly mating population *without any spatial segregation of the incipient species*—that is, if speciation occurred despite high initial gene flow. The difficulty any model of sympatric speciation must overcome is how to reduce the frequency of the intermediate genotypes that would act as a conduit of gene exchange between the incipient species.

Ernst Mayr (1942, 1963) was the most vigorous and influential critic of the sympatric speciation hypothesis, demonstrating that many supposed cases are unconvincing and that the hypothesis must overcome severe theoretical difficulties. Under certain special circumstances, however, these difficulties are not all that severe (Diehl and Bush 1989; Dieckmann and Doebeli 1999; see Turelli et al. 2001).

MODELS OF SYMPATRIC SPECIATION. Most models of sympatric speciation postulate disruptive (diversifying) selection (see Chapter 12) whereby certain homozygous genotypes have high fitness on one or the other of two resources (or in two microhabitats) and intermediate (heterozygous) phenotypes have lower fitness, perhaps because they are not as well adapted to either resource. Divergent adaptation to the resources might be based on one or on several loci. Selection might then favor alleles at one or more other loci that cause nonrandom mating, reducing the frequency of unfit heterozygous offspring. Thus the incipient species would come to differ at several loci, governing both mating behavior and adaptation to different resources. The problem, as Joseph Felsenstein (1981) pointed out, is that recombination would break apart these adaptive genetic packages.

Consider two loci in an herbivorous insect, A (for "adaptation") and B (for "behavior"). Genotypes A_1A_1 and A_2A_2 survive well on plant species 1 and 2, respectively, but the heterozygote has lower fitness on either plant than the better-adapted homozygote. Another locus, B, controls mate preference (we will shortly consider what happens if it controls host plant preference instead). Both sexes of the ancestral genotype B_1B_1 prefer to mate with each other rather than with B_2B_2 individuals, which also prefer to mate with each other, and which are initially rare. Allele B_2 increases in frequency, and partial assortative mating (sexual isolation) evolves, if the B_2 allele is associated (in linkage disequilibrium) with the A_2 allele, because B_2 carriers then produce fewer unfit A_1A_2 progeny than B_1 carriers. However, recombination breaks down the association between A_2 and B_2, and therefore reduces the selective advantage that would enable B_2 to increase. Thus, just as recombination can prevent reinforcement of sexual isolation in a hybrid zone, as we saw earlier in this chapter, it makes the sympatric evolution of assortative mating unlikely, unless selection against heterozygotes is very strong (Felsenstein 1981). The same principle makes it even more unlikely that additional alleles at other loci that would further enhance assortative mating could increase in frequency.

Sympatric speciation is somewhat more probable in several variant models. For example, suppose that insect genotypes A_1A_1 and A_2A_2 are best adapted to different host plants, and that locus B affects the insect's choice of host plant. Assume that the insects mate on the host plant chosen. Many herbivorous insects do exactly that; in fact, Guy Bush (1969) proposed this model based on his study of true fruit flies (Tephritidae). In such insects, a genetic difference in host preference (or, more generally, habitat preference), if it affects both sexes, automatically causes assortative mating. Speciation then occurs by sympatric evolution of ecological isolation (see Table 15.2), rather than by sexual isolation as such.

In this model, carriers of the allele B_1 prefer host 1 and carriers of B_2 prefer host 2. The optimal gene combinations are $A_1A_1B_1B_1$ (adapted to and attracted to host 1) and $A_2A_2B_2B_2$ (adapted to and attracted to host 2); individuals with other gene combinations (e.g., $A_2A_2B_1B_1$) are attracted to a host plant on which their offspring will survive poorly. Thus selection favors the divergent gene combinations and promotes linkage disequilibrium, so that the antagonism between selection and recombination is lower than in the model described previously. In computer simulations such as those by James Fry (2003), the frequencies of alleles such as A_2 and B_2 that confer adaptation to and preference for a new host may rapidly increase (Figure 16.16A). Gene flow may be strongly reduced, so that the population divides into two host-associated, ecologically isolated incipient species. However, if host preference is a continuous, polygenic trait, reproductive isolation will not evolve unless selection is strong (Figure 16.16B). Somewhat similar models describe sympatric speciation by adaptation to a continuously distributed resource, such as prey size (Dieckmann and Doebeli 1999; Kondrashov and Kondrashov 1999). Some authors have questioned how realistic these models are (Gavrilets 2004; Coyne and Orr 2004).

EVIDENCE ON SYMPATRIC SPECIATION. Because the conditions required for sympatric speciation to occur are theoretically more limited than those for allopatric speciation, and because there is so much evidence for allopatric speciation, sympatric speciation must be demonstrated rather than assumed for most groups of organisms. Such demonstration may be quite difficult. Nevertheless, many possible examples, supported by varying degrees of evidence, have been proposed.

Many experiments have been done in which laboratory populations of *Drosophila* have been subjected to disruptive selection and then tested for prezygotic isolation (Rice and Hostert 1993). In most, no sexual isolation developed. The exception was in experiments in which the disruptively selected character is one that automatically causes assortative mating as a correlated effect. In one experiment, for example, flies were disruptively selected for habitat preference and length of development. Within fewer than 30 generations, two subpopulations of flies had developed, strongly isolated by development time and to a lesser extent by habitat preference (Figure 16.17).

"Host races" of specialized herbivorous insects—partially reproductively isolated subpopulations that feed on different host plants—have often been proposed to represent

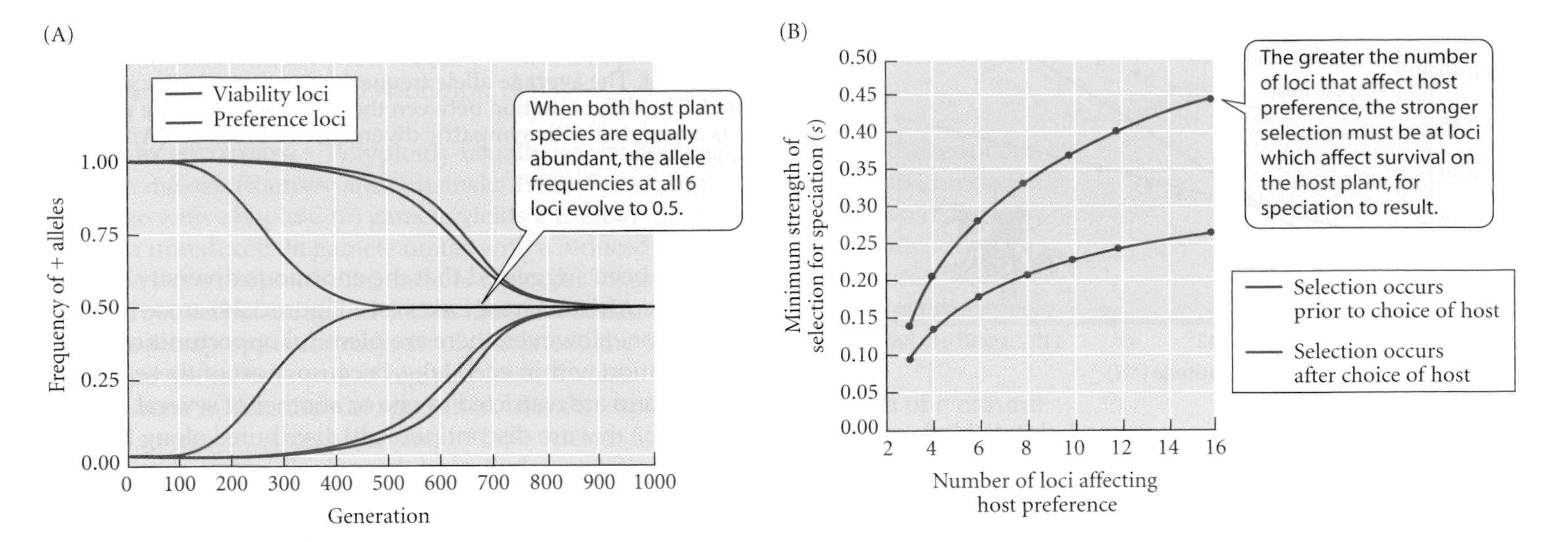

Figure 16.16 Some results of a computer simulation of sympatric speciation in an insect that mates on its host plant. (A) Alleles that enhance survival on or preference for one host plant species are referred to as "+ alleles"; those that have the complementary effect are called "– alleles." The simulation shows changes in the frequency of the + allele at two loci that affect survival (viability loci) and four loci that affect host preference (preference loci) when the + allele at each locus begins with a frequency near zero or one. Eventually, half the population prefers and survives better on one host and half on the other, representing progress toward reproductive isolation. (B) These curves show how strong selection at viability loci has to be to result in speciation when preference is controlled by multiple loci. The strength of selection at each viability locus is *s* (the coefficient of selection; see Chapter 12). The upper and lower curves model life histories in which selection occurs before and after the choice of host occurs. (After Fry 2003.)

sympatric speciation in progress. The apple maggot fly (*Rhagoletis pomonella*) is the most extensively studied case (Bush 1969; Feder 1998). Adult flies emerge from pupae in July and August and mate on the host plant; the larvae develop in ripe fruits, and in the autumn drop to the ground, spending the winter as pupae. The major ancestral host plants throughout eastern North America were hawthorns (*Crataegus*). About 150 years ago, *R. pomonella* was first recorded in the northeastern United States as a pest of cultivated apples (*Malus*), which are related to hawthorns. Subsequently, infestation of apples spread westward and southward. Allele frequencies at several loci differ significantly between apple- and hawthorn-derived flies, showing that gene exchange between them is limited (Figure 16.18). Gene exchange is reduced (to about 2 percent) by several factors, including a difference in preference for apples versus hawthorns and by a difference of about 3 weeks between mating activity on apple and on hawthorn. The earlier mating time on apples is correlated with an earlier time of emergence from the pupal state, which may have been advantageous because apple fruits ripen and are suitable for larval development earlier than hawthorn fruits. The genetic basis of the difference in development time may have evolved in hawthorn-feeding populations in Mexico, and was fortuitously advantageous for developing on apples (Feder et al. 2003). The critical step, divergence in host preference, presumably occurred in sympatry.

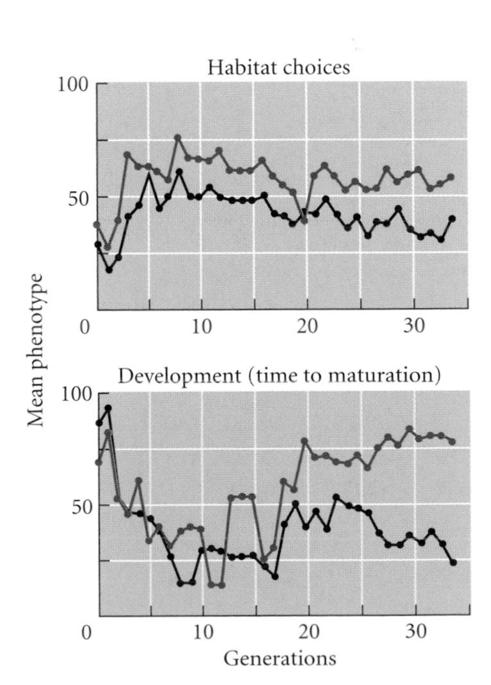

Figure 16.17 Sympatric divergence in experimental populations of *Drosophila melanogaster* subjected to disruptive selection. An experimental population was disruptively selected for three choices between habitats and for development time. In each generation, flies with similar phenotypes were mated with one another. Over 30 generations, two subpopulations (shown as red versus black) emerged and developed partial reproductive isolation. The lines in each figure show the mean phenotypes of progeny of females in each subpopulation. (After Rice and Salt 1990.)

(A)

(B)

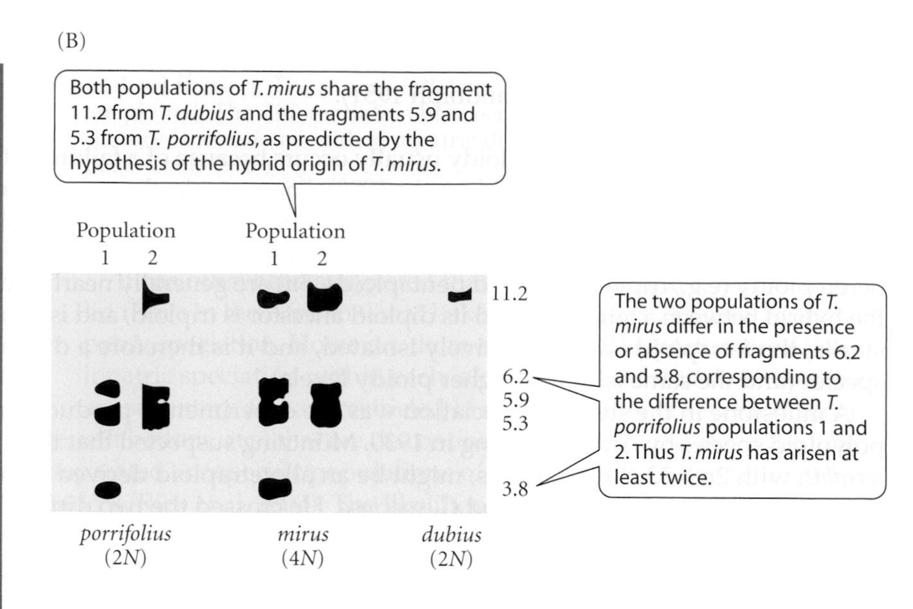
Both populations of T. mirus share the fragment 11.2 from T. dubius and the fragments 5.9 and 5.3 from T. porrifolius, as predicted by the hypothesis of the hybrid origin of T. mirus.
Population
1 2
Population
1 2
11.2
6.2
5.9
5.3
3.8
The two populations of T. mirus differ in the presence or absence of fragments 6.2 and 3.8, corresponding to the difference between T. porrifolius populations 1 and 2. Thus T. mirus has arisen at least twice.
porrifolius (2N)
mirus (4N)
dubius (2N)

Figure 16.19 Newly arisen allotetraploid species of goatsbeards (*Tragopogon*). (A) The flower heads of the diploid species *T. porrifolius* (1), *T. dubius* (2), and *T. pratensis* (3), and of the fertile tetraploid species *T. mirus* (4, from 1 × 2) and *T. micellus* (5, from 2 × 3). The unlabeled flower heads are the several diploid hybrids, which are mostly sterile. (B) Molecular evidence for the multiple origin of the hybrid species *T. mirus* (two populations) from *T. dubius* and *T. porrifolius* (two populations). These fragments of a ribosomal RNA-encoding DNA sequence, the result of cleavage by a restriction enzyme, were sorted by size in an electrophoretic gel. (A from Ownbey 1950; B after Soltis and Soltis 1991.)

Although polyploidy may confer new physiological and ecological capabilities, it does not confer major new morphological characteristics, such as differences in the structure of flowers or fruits. Thus polyploidy does not cause the evolution of new genera or other higher taxa (Stebbins 1950).

Recombinational speciation

Hybridization sometimes gives rise not only to polyploid species, but also to distinct species with the same ploidy as their parents. Among the great variety of recombinant offspring produced by F_1 hybrids between two species, certain genotypes may be fertile but reproductively isolated from the parent species. These genotypes may then increase in frequency, forming a distinct population (Rieseberg 1997). This process has been called **recombinational speciation** or HYBRID SPECIATION (Grant 1981).

Recombinational speciation seems to be rare in animals, but may be more common in plants (Rieseberg and Wendel 1993; Rieseberg 1997). Diploid species of hybrid origin have been identified by morphological, chromosomal, and molecular characters. For example, in a molecular phylogenetic analysis of part of the sunflower genus, *Helianthus*, Loren Rieseberg and coworkers found that hybridization between *Helianthus annuus* and *H. petiolaris* has given rise to three other distinct species (*H. anomalus*, *H. paradoxus*, *H. deserticola*) (Figure 16.20). Although F_1 hybrids between the parent species have low fertility, the derivative species are fully fertile and are genetically isolated from the parent species by postzygotic incompatibility. Because recombination breaks down the initial associations among genetic markers derived from the two parent species, the sizes of chromosomal blocks derived from each parent can be used to estimate how long it took for speciation to occur. On this basis, one of the hybrid species, *H. anomalus*, is estimated to have arisen within about 60 generations (Ungerer et al. 1998).

The recombinant species grow in very different (drier or saltier) habitats than either parent species, flower later, and have unique morphological and chemical features. *H. anomalus*, for example, has thicker, more succulent leaves and smaller flower heads than either parent species (see the photograph that opens this chapter). Such "extreme" traits transgress the range of variation between the two parent species. Rieseberg and coworkers (2003) crossed *H. annuus* and *H. petiolaris*, the parent species, and grew the backcross progeny in a greenhouse for several generations. Using genetic markers on all the chromosomes, Rieseberg et al. found that the experimental hybrids had combinations of *annuus* and *petiolaris* chromosome segments that matched those found in the three hybrid

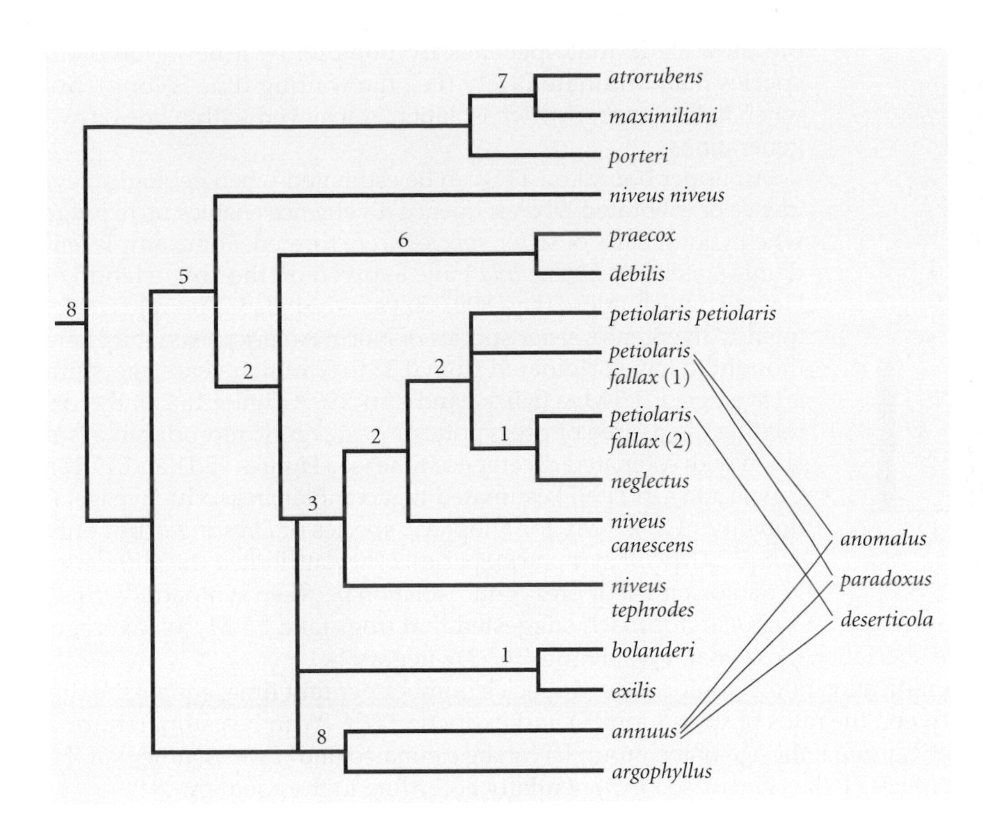

Figure 16.20 The hybrid origin of diploid species of sunflowers. The phylogeny, based on sequences of chloroplast DNA and nuclear ribosomal DNA, shows that *Helianthus anomalus*, *H. paradoxus*, and *H. deserticola* have arisen from hybrids between *H. annuus* and *H. petiolaris fallax*. The numbers of synapomorphic base pair substitutions are shown along the branches of the phylogenetic tree. Hybridization results in a netlike rather than strictly branching phylogenetic tree. (After Rieseberg and Wendel 1993.)

species—confirming that these species indeed arose from hybridization. Almost all the extreme, "transgressive" traits of *H. anomalus* and the other two hybrid species, such as small flower heads, occurred among the experimentally produced hybrids. To a considerable degree, then, the hybridization experiment replayed the origin of these species. Thus hybridization, by generating diverse gene combinations on which selection can act, can be a source of new species with novel morphological and ecological features.

How Fast Is Speciation?

The phrase "rate of speciation" has several meanings (Coyne and Orr 2004). One is the TRANSITION TIME or TIME FOR SPECIATION (TFS), the time required for (nearly) complete reproductive isolation to evolve, once the process has started (Figure 16.21A). Another is the BIOLOGICAL SPECIATION INTERVAL (BSI), the average time between the origin of a new species and when that species branches (speciates) again. The BSI includes not only the TFS, but also the "waiting time" before the process of speciation begins again. For exam-

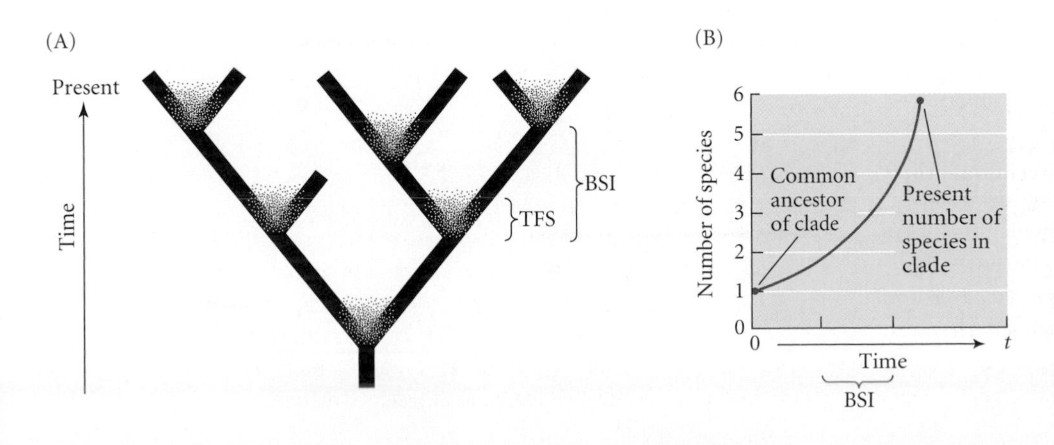

Figure 16.21 Two meanings of the "rate of speciation." (A) The time for speciation (TFS) is the time between the beginning and end of the evolution of reproductive isolation. The biological speciation interval (BSI) is the average time that has elapsed between two sequential forks in the phylogeny. (B) The BSI can be estimated from the amount of time between the present number of species and the common ancestor of the clade, assuming that the number has grown exponentially and that no extinctions have occurred. Extinctions (as in panel A) will cause us to overestimate BSI.

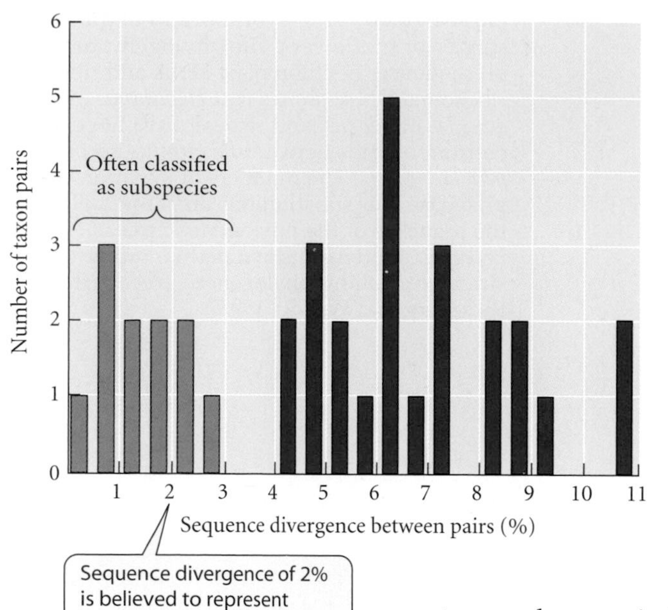

Figure 16.22 Sequence divergence of mitochondrial DNA between pairs of closely related North American songbirds. Most of the pairs with less than 4 percent sequence divergence have been classified as subspecies by some authors and as species by others. Most pairs of species appear to be more than 2 Myr old, suggesting that TFS is at least this great. (After Klicka and Zink 1997.)

ple, in a clade that speciates by polyploidy, a new polyploid species may originate rarely (i.e., the waiting time is long), but when it does, reproductive isolation is achieved within one or two generations.

An upper bound on TFS can be estimated when geological evidence or calibrated DNA sequence divergence enables us to judge when young pairs of sister species were formed. For example, endemic species of *Drosophila* have evolved on the "big island" of Hawaii, which is less than 800,000 years old. Based on rates of sequence divergence, sister species of North American songbirds are thought to have originated from 0.35 to 5 million years ago, with an average of 2.6 Mya (Klicka and Zink 1997; Figure 16.22). By correlating the degree of prezygotic or postzygotic reproductive isolation with estimated divergence time (see Figures 15.11 and 15.12), Coyne and Orr (1997) estimated that complete reproductive isolation takes 1.1–2.7 My for allopatric species of *Drosophila*, but only 0.08–0.20 My for sympatric species. (They attributed this difference to reinforcement of prezygotic isolation between sympatric forms.) A similar approach suggested that frogs take 1.5 My, on average, to complete speciation (Sasa et al. 1998; Figure 16.23, category I).

The diversification rate, R, or increase in species number per unit time, equals the difference between the rates of speciation (S) and extinction (E). R can be estimated for a monophyletic group if the age of the group (t) can be estimated and if we assume that the number of species (N) has increased exponentially according to the equation

$$N_t = e^{Rt}$$

(We encountered this approach in Chapter 7 when we considered long-term rates of diversification in the fossil record; see page 141.) The average time between branching events on the phylogeny is $1/R$, the reciprocal of the diversification rate (Figure 16.21B). This number estimates BSI, the average time between speciation events, if we assume there has been no extinction ($E = 0$). According to estimates made using this approach, BSI in animals ranges from less than 0.3 My (in the phenomenal adaptive radiation of cichlid fishes in the Great Lakes of Africa) to more than 10 My in various groups of mol-

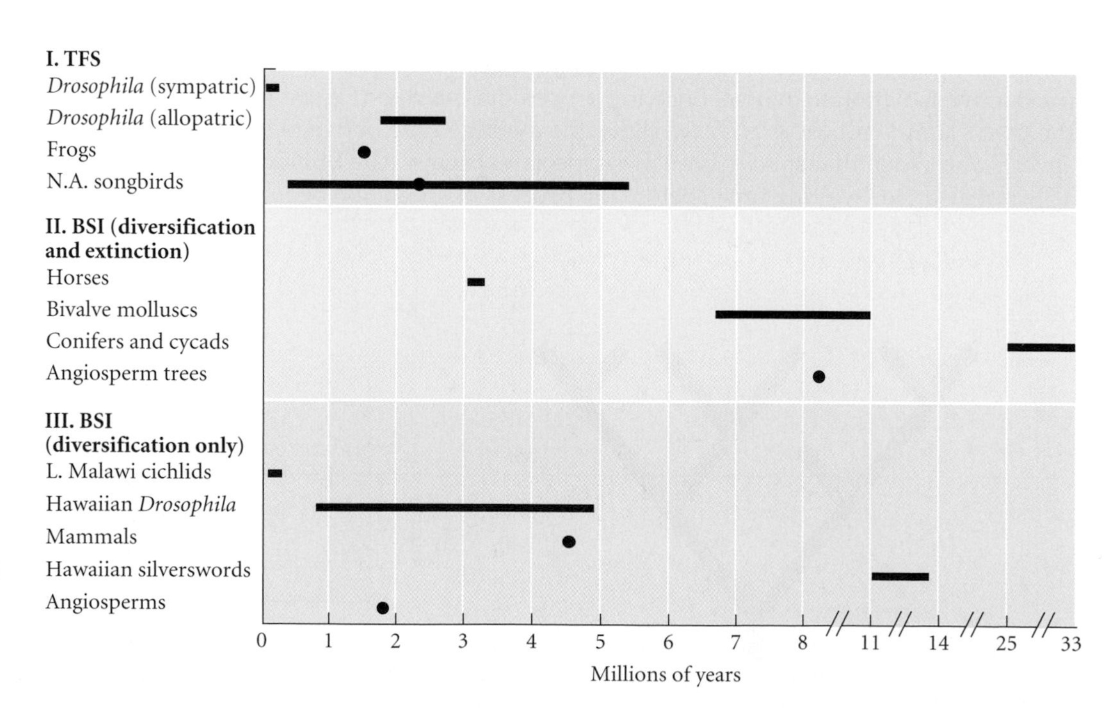

Figure 16.23 Estimates of time required for the speciation process in various groups of organisms. The average is provided for some groups, and a range of values is provided for others. In category I, estimates of TFS are based on the relation between degree of reproductive isolation and time-calibrated genetic distance. In category II, estimates of biological speciation interval (BSI) are based on data on diversification and extinction from the fossil record. In category III, estimates of BSI are based on net diversification rates, not accounting for extinction. Data from the fossil record suggest that BSI can vary greatly among clades. (Data from Coyne and Orr 2004.)

luscs. When estimates of *E* from the fossil record are taken into account, BSI is about 3 My for horses and is still very long for bivalve molluscs (6–11 My; Figure 16.23, category II). BSI for plants may be longer than for animals, on average; conifers (25 My) and cycads (33 My) appear to speciate especially slowly.

Whatever approach is taken, speciation rates clearly vary greatly—as we would expect from theories of speciation. We expect the process of speciation (TFS) to be excruciatingly slow if it proceeds by mutation and drift of neutral alleles; we expect it to be faster if it is driven by ecological or sexual selection, and to be accelerated if reinforcement plays a role. Some possible modes of speciation, such as polyploidy, recombinational speciation, sympatric speciation, and speciation by peak shifts, should be very rapid when they occur—although they may occur rarely, resulting in long intervals (BSI) between speciation events. As we have already seen, substantial reproductive isolation apparently evolved within about a century in the apple maggot fly *Rhagoletis pomonella* and the hybrid sunflower species *Helianthus anomalus;* on the other hand, some sister taxa of snapping shrimps (*Alpheus*) on opposite sides of the Isthmus of Panama have not achieved full reproductive incompatibility in the 3.5 My since the isthmus arose (Knowlton et al. 1993).

What characteristics favor high rates of speciation? The best way to determine whether a characteristic affects the rate of diversification is to compare the species diversity of replicated sister groups that differ in the characteristic of interest (a replicated sister group comparison; see Figures 7.18 and 16.9). Many features are correlated with diversification rate in various groups of organisms, but it is often hard to tell whether they enhance the speciation rate or diminish the extinction rate. Among characteristics studied so far, those that seem most likely to have increased speciation rate as such seem to be animal (rather than wind) pollination in plants and features that indicate intense sexual selection in animals (Coyne and Orr 2004). These observations suggest the intriguing possibility that diversification in some groups of animals owes more to the simple evolution of reproductive isolation (due to sexual selection) than to ecological diversification. This conclusion may call into question the belief that ecological divergence is the main engine of evolutionary radiation (Schluter 2000).

Consequences of Speciation

The most important consequence of speciation is that it is the sine qua non of diversity. For sexually reproducing organisms, every branch in the great phylogenetic tree of life represents a speciation event, in which populations became reproductively isolated and therefore capable of independent, divergent evolution, including, eventually, the acquisition of those differences that mark genera, families, and still higher taxa. Speciation, then, stands at the border between MICROEVOLUTION—the genetic changes within and among populations—and MACROEVOLUTION—the evolution of the higher taxa in all their glorious diversity.

In their hypothesis of punctuated equilibrium, Eldredge and Gould (1972) (see also Stanley 1979; Gould and Eldredge 1993) proposed that speciation may be required for morphological evolution to occur at all. From the observation that many fossil lineages change little over the course of millions of years (see Chapter 7), they proposed that in broadly distributed species, internal constraints may prevent adaptive evolution. They suggested, based on Mayr's (1954) proposal that founder events trigger rapid evolution from one genetic equilibrium to another, that most evolutionary changes in morphology are triggered by and associated with peripatric speciation.

Population geneticists generally reject this hypothesis; after all, morphological characters vary among populations of a species, just as they do among reproductively isolated species (Charlesworth et al. 1982). Thus, as Gould (2002) himself concluded, there is no reason to think that speciation (acquisition of reproductive isolation) triggers morphological evolution. Nevertheless, morphological change might be associated with speciation in the fossil record because reproductive isolation enables morphological differences between populations to persist in the long term (Futuyma 1987). Although different local

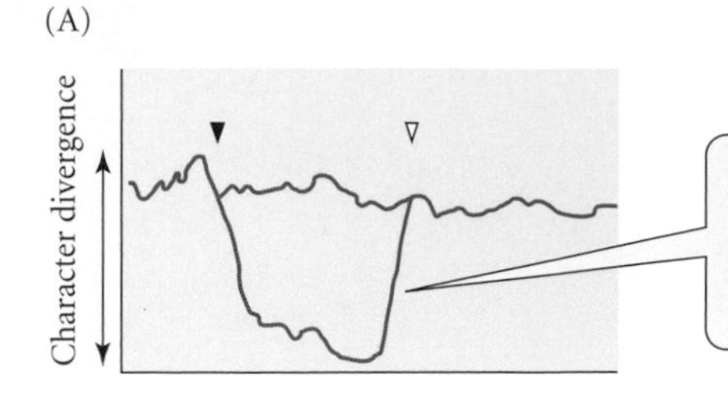
(A)
Character divergence
Geographically isolated populations diverge due to natural selection, but the divergence is lost by interbreeding if reproductive isolation has not evolved before the geographic barrier breaks down.

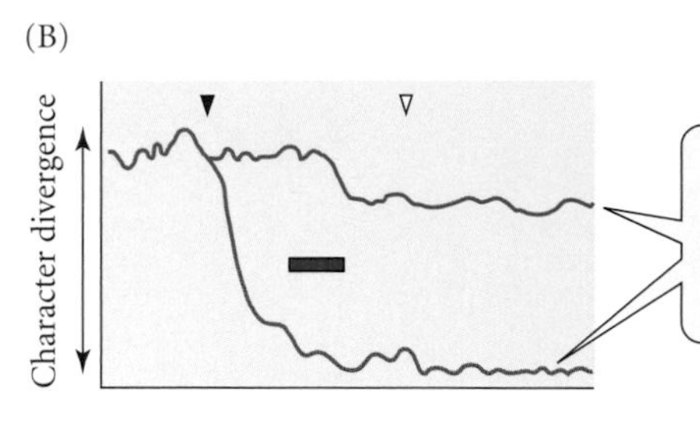
(B)
Character divergence
Divergence is permanent, and may continue to evolve after breakdown of the geographic barrier, if reproductive isolation has evolved.

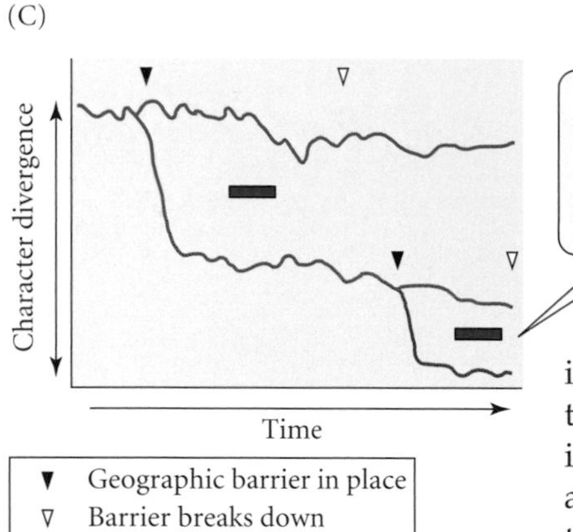
(C)
Character divergence
Repetition of the process in (B) can enable still further departures from the ancestral character state to become permanent.
Time
Geographic barrier in place
Barrier breaks down
Reproductive isolation

Figure 16.24 A model of how speciation might facilitate long-term evolutionary change in morphological and other phenotypic characters. Shifts in geographic ranges that bring divergent populations into contact may cause loss of their divergent features due to interbreeding (A), unless the populations evolve reproductive isolation while allopatric (B, C).

populations may diverge rapidly due to selection, local populations are ephemeral: as climate and other ecological circumstances change, divergent populations move about and come into contact sooner or later. Much of the divergence that has occurred will then be lost by interbreeding—unless reproductive isolation has evolved (Figure 16.24). A succession of speciation events, each "capturing" further change in a character, may result in a long-term trend.

This hypothesis predicts, perhaps counterintuitively, that directional evolutionary change should be more pronounced during times of environmental stability (when the geographic isolation of populations lasts long enough for reproductive isolation to evolve) than during times of environmental change (when the distribution of habitats and populations changes, promoting gene flow). Paleontologists have described just such a pattern (Jansson and Dynesius 2002). Despite the great environmental changes during the Pleistocene glaciations, for example, early Pleistocene fossil beetles are identical to living species. Their geographic distributions have changed radically, however, and it is to these incessant changes in distribution, resulting in gene flow, that Coope (1979) attributes their morphological stability. Likewise, gradual morphological evolution in trilobites was more pronounced in stable than in unstable environments (Sheldon 1990). Perhaps, then, as Ernst Mayr (1963, p. 621) said, "Speciation … is the method by which evolution advances. Without speciation, there would be no diversification of the organic world, no adaptive radiation, and very little evolutionary progress. The species, then, is the keystone of evolution."

Summary

1. Probably the most common mode of speciation is allopatric speciation, in which gene flow between populations is reduced by geographic or habitat barriers, allowing genetic divergence by natural selection and/or genetic drift.
2. In vicariant allopatric speciation, a widespread species becomes sundered by a geographic barrier, and one or both populations diverge from the ancestral state.
3. In a simple model of the evolution of reproductive isolation, complementary allele substitutions that do not reduce the fitness of heterozygotes occur at several loci in one or both populations. Epistatic interactions between alleles fixed in the two populations reduce the fitness of hybrids (postzygotic isolation). Likewise, genetic divergence may result in prezygotic isolation.
4. The genetic changes that cause the evolution of reproductive isolation in allopatric populations may be caused by genetic drift or by divergent ecological or sexual selection. In only a few cases have the effects of ecological selection been well documented. Sexual selection is probably an important cause of prezygotic isolation in animals.
5. Prezygotic isolation evolves mostly while populations are allopatric, but may be reinforced when the populations become parapatric or sympatric. Whether or not reinforcement occurs depends on the nature of the prezygotic barrier and on its genetic basis. Sexual selection may enhance the likelihood of reinforcement, whereas recombination may reduce it.

6. Peripatric speciation, or founder effect speciation, is a hypothetical form of allopatric speciation in which genetic drift in a small peripheral population initiates rapid evolution, and reproductive isolation is a by-product. The likelihood of this form of speciation differs greatly depending on the mathematical model used. Although the geographic pattern of speciation predicted by this hypothesis may be common, there is little evidence for the process of drift-induced speciation.
7. Sympatric speciation, the origin of reproductive isolation within an initially randomly mating population, may occur due to disruptive selection. However, the sympatric evolution of sexual isolation is unlikely, due to recombination among loci affecting mating and those affecting the disruptively selected character. Sympatric speciation may occur, however, if recombination does not oppose selection. For example, if disruptive selection favors preference for different habitats and if mating occurs within those habitats, prezygotic isolation may result. How often this occurs is debated.
8. Instantaneous speciation by polyploidy is common in plants. Allopolyploid species arise from hybrids between genetically divergent populations. Establishment of a polyploid population probably requires ecological or spatial segregation from the diploid ancestors because backcross offspring have low reproductive success. Polyploid species can have multiple origins.
9. In a few documented cases (in plants), some genotypes of diploid hybrids are fertile and are reproductively isolated from the parent species, and so give rise to new species (recombinational speciation).
10. The time required for speciation to proceed to completion is highly variable. It is shorter for some modes of speciation (polyploidy, recombinational speciation) than others (especially speciation by mutation and drift of neutral alleles that confer incompatibility). The process of speciation may require 2–3 million years, on average, for some groups of organisms; it is much longer in some cases and very much shorter in others. The rate at which new species arise, estimated from phylogenetic and paleontological data, is likewise very variable. Sexual selection in animals and animal pollination in plants appear to enhance rates of speciation.
11. Speciation is the source of the diversity of sexually reproducing organisms, and it is the event responsible for every branch in their phylogeny. It probably does not stimulate evolutionary change in morphological characters, as suggested by the hypothesis of punctuated equilibria, but it may contribute to long-term progressive divergence by preventing interbreeding between populations from undoing the changes wrought by natural selection.

Terms and Concepts

allopatric speciation
character displacement
founder effect speciation
parapatric speciation
peripatric speciation
recombinational speciation
reinforcement
sympatric speciation
vicariance

Suggestions for Further Reading

As noted in Chapter 15, *Speciation*, by J. A. Coyne and H. A. Orr, is the most comprehensive recent work on the subject. Those with a mathematical bent will enjoy the wide-ranging treatment of models of speciation by Sergei Gavrilets in *Fitness landscapes and the origin of species* (Princeton University Press, Princeton, NJ, 2004).

Problems and Discussion Topics

1. Why is it difficult to demonstrate that speciation has occurred parapatrically or sympatrically?
2. Coyne and Orr (1997) found that sexual isolation is more pronounced between sympatric populations than between allopatric populations of the same apparent age, and took this finding as evidence for reinforcement of sexual isolation. It might be argued, though, that any pairs of sympatric populations that were not strongly sexually isolated would have merged, and would have been unavailable for study. Thus the degree of sexual isolation in sympatric compared with allopatric populations might be biased. How might one rule out this possible bias? (Read Coyne and Orr after suggesting an answer.)

3. Suppose that full reproductive isolation between two populations has evolved. Can speciation in this case be reversed, so that the two forms merge into a single species? Under what conditions is this probable or improbable?
4. Referring to the discussion of parallel speciation in sticklebacks, can a single biological species arise more than once (i.e., polyphyletically)? How might this possibility depend on the nature of the reproductive barrier between such a species and its closest relative?
5. The heritability of an animal's preference for different habitats or host plants might be high or low. How might heritability affect the likelihood of sympatric speciation by divergence in habitat or host preference?
6. Biological species of sexually reproducing organisms usually differ in morphological or other phenotypic traits. The same is often true of taxonomic species of asexual organisms such as bacteria and apomictic plants. What factors might cause discrete phenotypic "clusters" of organisms in each case?
7. In many groups of plants, low levels of hybridization between related species are not uncommon, yet only a few cases of the origin of "hybrid species" by recombinational speciation have been documented. What factors make recombinational speciation likely versus unlikely?
8. Choose a topic from this chapter and discuss how its treatment would be altered if one adopted a phylogenetic species concept rather than the biological species concept.

How to Be Fit: Reproductive Success

17

A very long life history. A bristlecone pine (*Pinus aristata*) on an arid mountaintop in Nevada. Perhaps as much as 4600 years old, it is among the oldest known individual organisms. (Photo © D. Cavagnaro/Visuals Unlimited.)

Species differ from one another not just in their anatomy and physiology, but in many aspects of their life cycles. For example, sea anemones and corals can live for close to a century, bristlecone pine trees (*Pinus aristata*) have survived for 4600 years, and vegetatively propagating clones of quaking aspen (*Populus tremuloides*) can live for more than 10,000 years. In contrast, annual plants die less than a year after germinating, and many small animals, such as some rotifers, live for at most a few weeks.

Reproductive output likewise varies greatly: many bivalves and other marine invertebrates release thousands or millions of tiny eggs in each spawning, whereas a blue whale (*Balaenoptera musculus*) gives birth to a single offspring that weighs as much as an adult elephant, and a kiwi (*Apteryx*) lays a single egg that weighs 25 percent as much as its mother (Figure 17.1).

Patterns of development also differ greatly among species. Some insects develop very rapidly: a newly laid egg of *Drosophila melanogaster* may be a reproducing adult 10 days later, and a parthenogenetic aphid may carry an embryo even before she herself is born. In contrast, periodical cicadas (*Magicicada*) feed underground for 13 or 17 years before

Figure 17.1 An X-ray photograph of a kiwi (*Apteryx*), showing the relatively enormous egg, which weighs 25 percent of the female's body weight. (Photo courtesy Otorohanga Zoological Society.)

emerging to live as adults, which die less than a month later (Figure 17.2). Many species, such as humans, reproduce repeatedly, whereas others, such as century plants (*Agave*) and some species of salmon, reproduce only once, and then die.

Organisms also differ in how they reproduce. Many animals have separate sexes, but earthworms and other SIMULTANEOUS HERMAPHRODITES have both female and male functions at the same time, and certain sea basses and other SEQUENTIAL HERMAPHRODITES develop first as one sex and later switch to the other sex. Species of plants may consist of simultaneous or sequential hermaphrodites only, hermaphrodites and females, hermaphrodites and males, or separate sexes. Even hermaphroditic plants differ with respect to whether they outcross or self-pollinate, and so differ in degree of inbreeding. The process of reproduction often involves sex, accompanied by recombination, but many organisms reproduce asexually, often by parthenogenesis—development from an unfertilized egg.

What accounts for such extraordinary variation in species' survival and reproduction—the very features that we would expect to be most intimately related to their fitness? There are apparently many ways to achieve high fitness, or reproductive success. To make sense of all this diversity, evolutionary ecologists and evolutionary geneticists have developed theories of the evolution of life histories and breeding systems.

Individual Selection and Group Selection

Why do codfish produce hundreds of thousands of eggs? Is it to compensate for their high mortality and thus to ensure the survival of the species? Why do people die "of old age?" Is it to make room for the vigorous new generation that will propagate the species? Why do so many species reproduce sexually? Because sex and recombination enhance genetic variation, and therefore enable the species to adapt to environmental changes?

Even some professional biologists have been known to answer "yes" to these questions. But either they have assumed that these characteristics did not evolve by Darwinian natural selection (selection among individuals), or they have not realized that the good of the species does not affect the course of selection among individuals—they have not fully understood the meaning of natural selection.

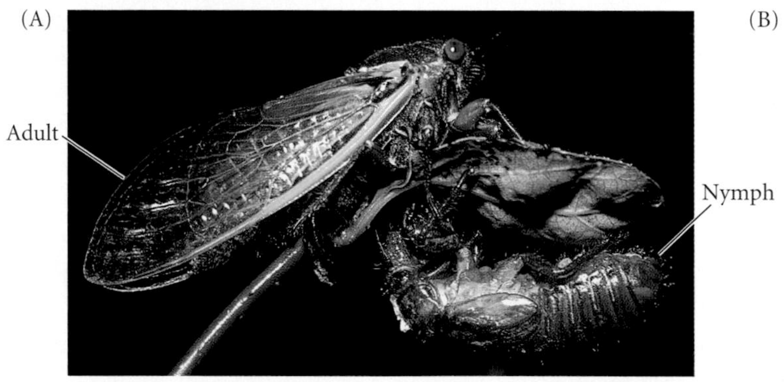

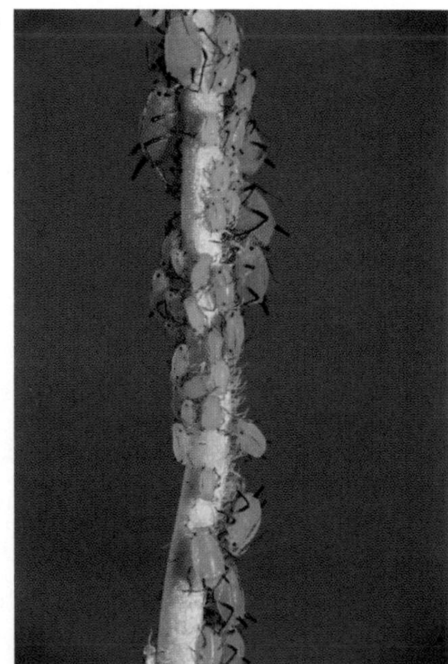

Figure 17.2 Two insects that differ in generation time and rate of increase due to a great difference in age at which reproduction begins. (A) Adult and nymph of a periodical cicada (*Magicada septendecim*). Emerging after 17 years spent underground as a nymph, feeding on sap in plant roots, the adult cicada is now ready to reproduce. Periodical cicadas have the longest generation time known among insects. (B) A cluster of milkweed aphids (*Aphis nerii*) belonging to several generations; they have descended by parthenogenetic reproduction from a single female. Aphids give birth to live young that, in some species, may start to develop offspring of their own even before they are born. (A © Jimm Zipp/Photo Researchers, Inc.; B © photolibrary.com.)

Because they are components of fitness, differences in fecundity and life span must have evolved at least partly by natural selection. Selection among populations—the only possible cause of evolution of a trait that is harmful to the individual but beneficial to the population or species—is generally a weaker force than selection among individuals, as we saw in Chapter 11. This must be especially true for life history traits, which are components of individuals' fitness.

The possibility of future extinction due to excessive population growth or inadequate reproduction is irrelevant to, and cannot affect, the course of natural selection among individuals. A mutation that increased the fecundity of humans (or any other species), for example, would increase individual fitness (if it had no other effects) and would therefore become fixed—even if overpopulation and mass starvation should ensue. Instead of supposing that a species' fecundity evolves to balance mortality, *we should consider the level of mortality to be the ecological consequence of the level of fecundity*, since most populations are regulated by density-dependent factors such as limited food and other resources (Williams 1966).

At first surmise, then, we should expect any species to evolve ever greater fecundity and an ever longer life span. The problem, therefore, is to understand what advantage low fecundity or a short life span—or the genes that underlie them—might provide to individual organisms, rather than to whole populations or species. By the same token, we must beware of supposing that sexual reproduction has evolved because it benefits species by enabling them to adapt to changes in the environment.

Life History Evolution

We will first focus on the evolution of life history traits, especially the potential life span, the ages at which reproduction begins and ends, and how many offspring females produce at each age (Stearns 1992; Charlesworth 1994b; Roff 2002). These traits affect the growth rates of populations.

Life history traits as components of fitness

In Chapter 12, we defined the fitness of a genotype for the simple case in which females reproduce once and then die (a **semelparous** life history). In this case, reproductive success—the number of descendants of an average female after one generation—is R, the product of the probability of a female's survival to reproductive age (L) and the average number of offspring per survivor (M):

$$R = LM$$

For **iteroparous** species—those in which females reproduce more than once—the calculation of fitness is more complex. The average number of offspring per female is the sum of the offspring an average female produces at each age, weighted by the probability that a female survives to that age. We use x to denote age, l_x to denote the probability of survival to age x (i.e., the proportion of eggs or newborns that survive to age x), and m_x to denote the average fecundity (number of eggs or newborns) at age x. (Figure 17.3 illustrates l_x and m_x for several populations of a species of *Drosophila*.) Suppose that at ages 1, 2, 3, and 4 years, females lay 0, 4, 8, and 0 eggs, respectively, and that their chances of surviving to those ages are 75%, 50%, 25%, and 10%, respectively. Then we can write a simple life table:

x	l_x	m_x	$l_x m_x$
0	1.00	0	0
1	0.75	0	0
2	0.50	4	2
3	0.25	8	2
4	0.10	0	0
5	0.00	0	0
$\Sigma = R$			4

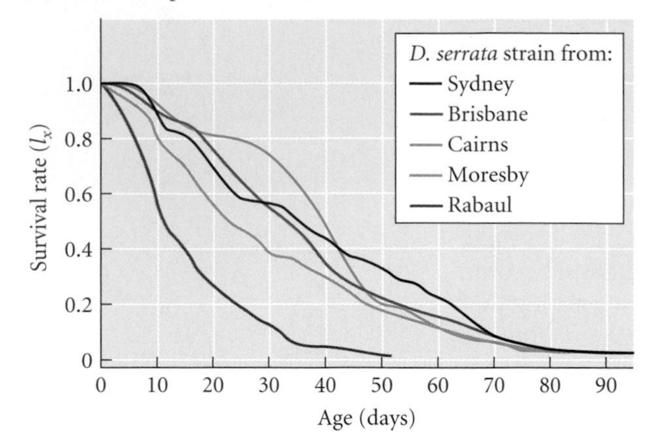

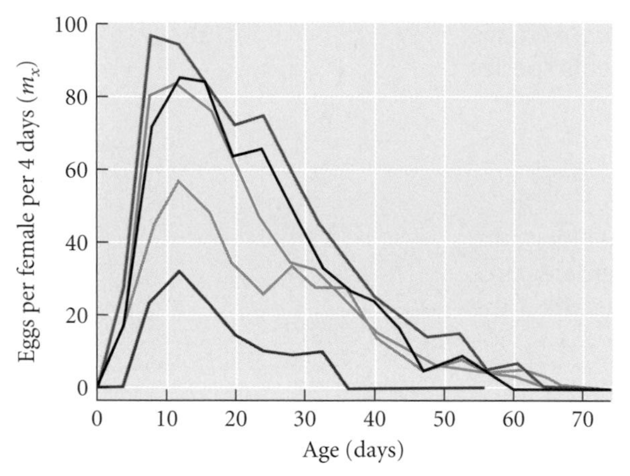

Figure 17.3 Genetic variation in life history characteristics. The graphs show (A) age-specific probability of survival, l_x, and (B) fecundity, m_x, in strains of the Australian fruit fly *Drosophila serrata* from five localities when raised in the laboratory. The survival curves show the fraction of newborns that survive to each age, and the fecundity curves show the average egg production per female at each age. Egg production peaks a few days after the flies transform from the pupal to the adult stage. These curves show only the adult (post-pupa) stage of the life history. (After Birch et al. 1963.)

R is calculated as

$$R = \sum l_x m_x$$

Thus each female is replaced, on average, by $R = 4$ offspring. (By convention, ecologists generally count just daughters in such analyses, and assume that sons are produced in equal numbers.) This sum is also the growth rate, per generation, of the genotype: if the population starts with N_0 individuals, its size after g generations will be

$$N = N_0 R^g$$

A genotype with higher R would increase faster in numbers, and thus have higher fitness.

If genotypes differ in the length of a generation, their fitness can be compared only by their increase per unit time, not per generation. The per capita rate of population increase per unit time is denoted r, which is related to R by $R = e^r$. If the population starts with N_0 individuals, its size after t time units will be

$$N = N_0 \, e^{rt}$$

Like R, r depends on the probability of survival and the fecundity at each age. Under most circumstances, r is a suitable measure of a genotype's fitness.

All else being equal, increasing l_x—survival to any age x—up to and including the reproductive ages will increase R (or r) and therefore increase fitness. If, as for human females (or in the hypothetical life table above), there is a postreproductive life span (when $m_x = 0$), changing the probability of survival to advanced postreproductive ages does not alter R. If, however, the reproductive period is extended into older ages, then increasing survival to those ages does increase R. Similarly, increasing m_x (fecundity at any age x) increases fitness, all else being equal.

Offspring produced at an early age increase fitness more because they contribute more to population growth (r) than the same number of offspring produced at a later age. That is, they have greater "value" in terms of fitness. For instance, suppose females reproduce either at age 2 or age 3, but have the same fecundity. Then 2-year-old females contribute more to the future population size than do 3-year-olds. Because fewer individuals survive to age 3 than to age 2, 2-year-olds collectively will leave more offspring. Moreover, population growth is like compound interest. Just as your bank account grows faster if you make a deposit now than if you wait, the offspring of 2-year-olds will themselves contribute offspring (i.e., "gain interest") before the offspring of 3-year-olds do so. Thus a genotype that reproduces earlier in life has a shorter generation time, and higher fitness (as measured by r), than a genotype that delays reproduction until a later age.

Trade-offs

Because traits evolve so as to maximize fitness, we might naively expect organisms to evolve ever greater fecundity, ever longer life, and ever earlier maturation. That all organisms are nevertheless limited in these respects may be attributed to various constraints.

Phylogenetic constraints arise from the history of evolution, which has bequeathed to each lineage certain features that constrain the evolution of life history traits and other characters. For instance, although many insects feed as adults, and so obtain energy and protein that enable them to form successive clutches of eggs, adult silkworm moths and some other insect groups lack functional mouthparts, so their fecundity is limited by the resources they stored when they fed as larvae. Most such insects lay only one batch of eggs and then die. In most groups of birds, the number of eggs per clutch varies within and among species, but all species in the order Procellariiformes (albatrosses, petrels, and relatives) lay only a single egg.

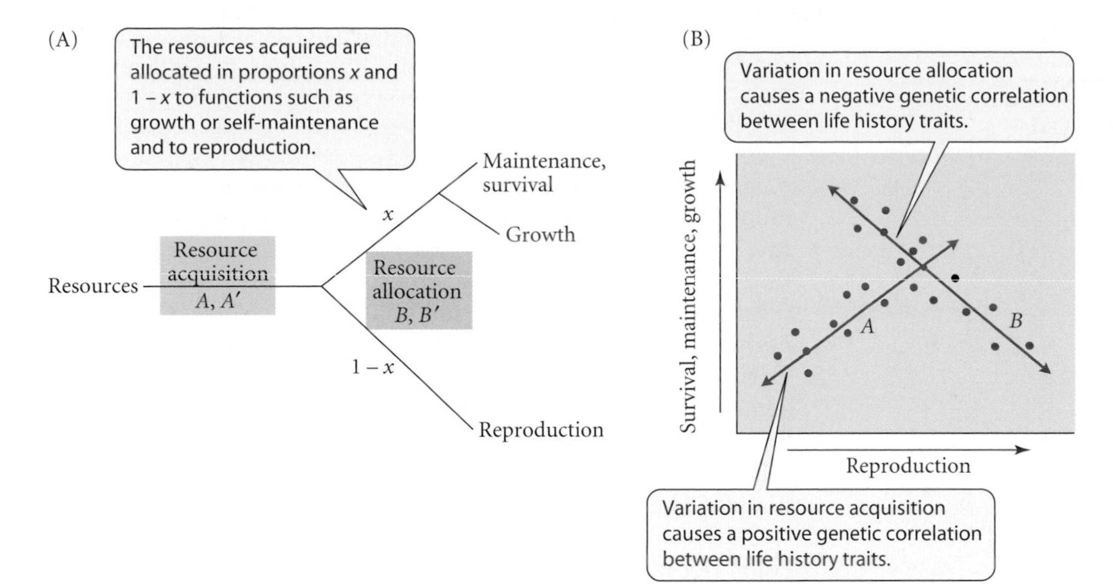

Figure 17.4 Factors giving rise to positive or negative genetic correlations between life history traits such as survival (or growth) and reproduction. (A) Variation at locus *A* affects the amount of energy or other resources that an individual acquires from the environment. Variation at locus *B* affects allocation of resources to functions such as growth or self-maintenance and to reproduction, in proportions x and $1 - x$. (B) Genotypes that differ at locus *A* are represented by green circles, those that differ at locus *B* by red circles. The overall genetic correlation between survival and reproduction depends on the relative magnitudes of variation in resource acquisition versus resource allocation.

Other constraints, termed PHYSIOLOGICAL or GENETIC CONSTRAINTS, are less well understood, but may be detected by comparisons among different genotypes or phenotypes. Some such constraints constitute **trade-offs**, whereby the advantage of a change in a character is correlated with a disadvantage in other respects. For example, the reproductive activities of animals often increase their risk of predation, so there is a trade-off between reproduction and survival: recall that the courtship calls of male túngara frogs (*Physalaemus pustulosus*) attract both female frogs and predatory bats (see Chapter 14).

Some physiological trade-offs may result in **antagonistic pleiotropy**, wherein genotypes manifest an inverse relationship between different components of fitness. For example, if genotypes differ in the amount of energy or nutrients they allocate to reproduction (often called **reproductive effort**) versus their own maintenance or growth, then increased fecundity may be correlated with decreased subsequent survival or growth. This ALLOCATION TRADE-OFF would be manifested as a *negative genetic correlation* between reproduction and survival. If there were also genetic variation in the amount of resources individuals acquired from the environment, however, this could give rise to a *positive correlation* between reproduction and survival (Figure 17.4) (van Noordwijk and deJong 1986; Bell and Koufopanou 1986). The allocation trade-off might still constrain evolution, but the trade-off might be difficult to detect in this case.

There are several ways to detect trade-offs (Reznick 1985):

1. Correlations between the means of two or more traits in different populations or species can strongly suggest a trade-off, although such correlations might result from other, unknown differences among the populations. We would expect, and can often document, an allocation trade-off between many small versus fewer large offspring if parents must allocate limited resources (Figure 17.5).
2. Phenotypic or, better, genetic correlations between traits within populations can be useful indicators of the extent to which enhancement of one component of fitness would be immediately accompanied by reduction of another (see Chapter 13). For instance, Law et al. (1979) grew individuals of each of

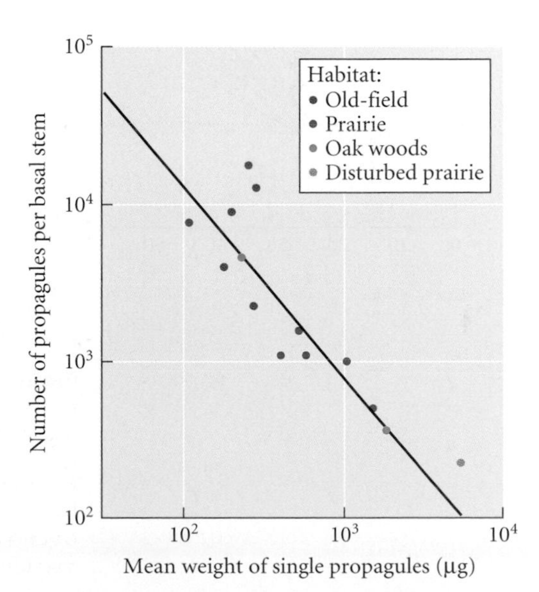

Figure 17.5 The relationship between number and weight of propagules (seeds) among species of goldenrods (*Solidago*) suggests an allocation trade-off. The colonizing species, growing in old-field habitats, tend to produce smaller seeds than species that grow in more stable prairies, where competition may be intense and favor larger offspring. (After Werner and Platt 1976.)

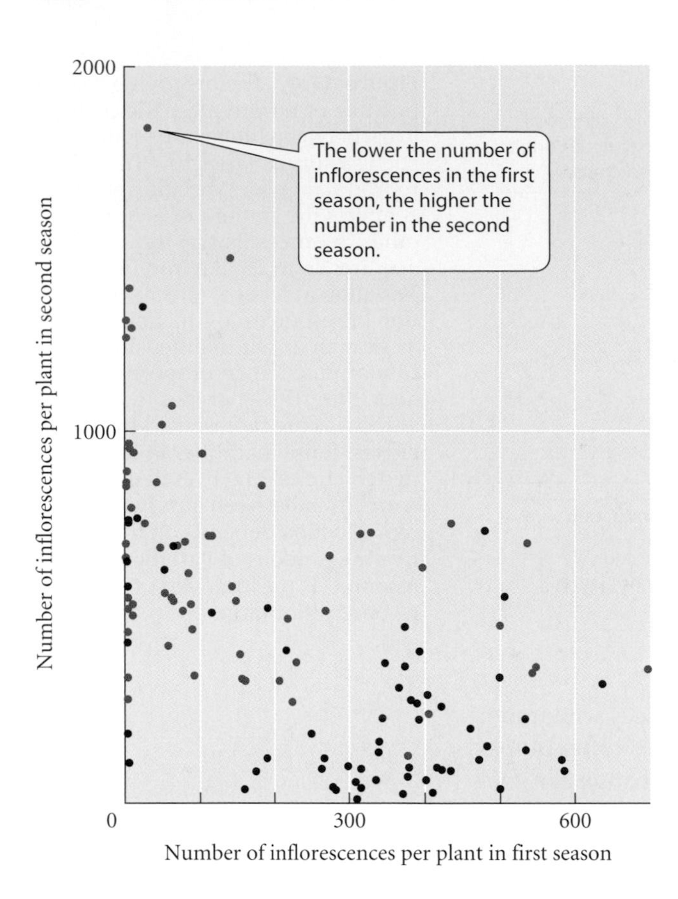

Figure 17.6 The number of inflorescences per plant in the grass *Poa annua* in the first and second seasons of life. Each point represents the mean of the individuals in a single family. Plants from two different habitats, indicated by different colors, were grown together in the same plot. (After Law 1979.)

many families of meadow grass (*Poa annua*) in a randomized array. They found that families that, on average, produced more inflorescences in their first season produced fewer in the second (Figure 17.6), and also achieved less vegetative growth. This experiment demonstrated a genetic basis for a cost of reproduction.

3. Correlated responses to artificial or natural selection provide some of the most consistent evidence of trade-offs (Reznick 1985; Stearns 1992). Linda Partridge and colleagues (1999) set up ten selection lines of *Drosophila melanogaster* from the same base population. They selected five "young" populations by rearing offspring from eggs laid by females less than one week old, and five "old" populations by propagating from eggs laid when females were 3 to 4 weeks old. After 19 generations, the mean life span of the "young" populations did not differ from that of the base population, but the longevity of the "old" populations had increased—as we would expect, since only flies that lived at least 3 weeks could contribute genes to subsequent generations (Figure 17.7A). However, the fecundity of 1-week-old females in the "old" populations decreased compared with that of the base or "young" populations (Figure 17.7B). Thus survival to greater age seems to have been achieved at the expense of reproduction early in life—an important result, as we will see shortly.
4. Experimental manipulation of one trait and observation of the effect on other traits often reveals trade-offs. For instance, Sgrò and Partridge (1999) followed the selection experiment on *Drosophila* longevity by experimentally sterilizing females from both "young" and "old" populations, either by gamma radiation or by inheritance of a dominant allele that causes female sterility. In both experimental treatments, the difference

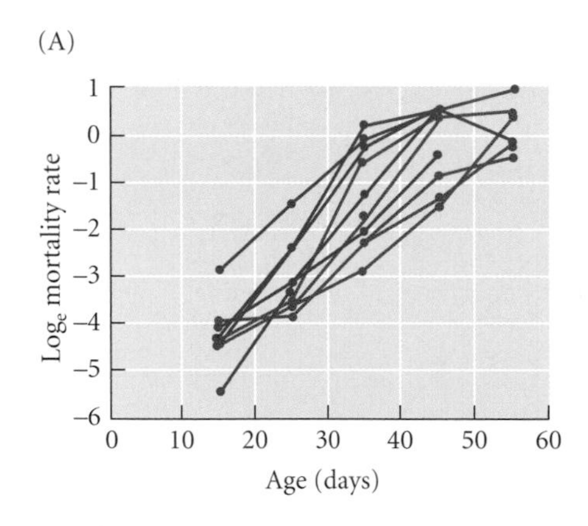

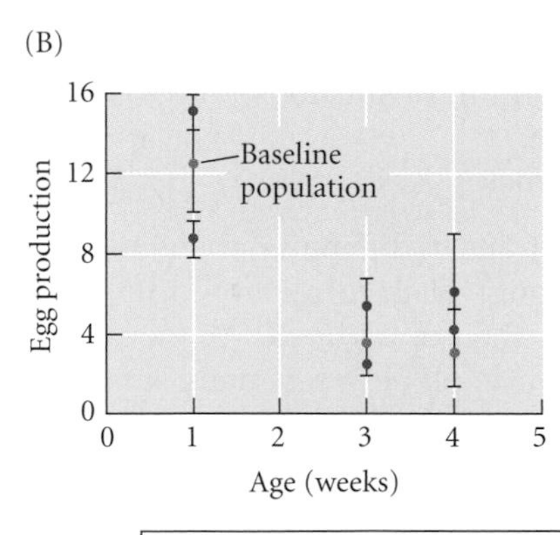

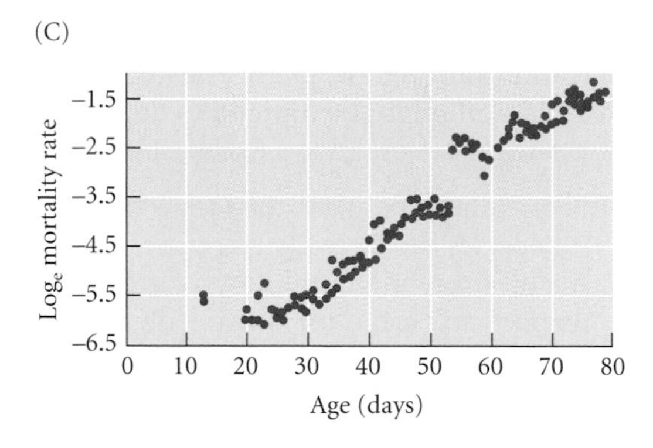

Figure 17.7 Results of selection of laboratory populations of *Drosophila* for age at reproduction. (A) The mortality rate, per 10-day interval, is lower at older ages for populations that were selected for reproduction at "old" age than for populations that were selected for reproduction at "young" age. (B) Relative to the "young" populations and the baseline population, the "old" populations had lower egg production when young. (C) The difference in mortality rate between "old" and "young" populations disappeared when a gene that prevents female reproduction was crossed into the populations, suggesting that the costs of reproduction increase mortality. (After Partridge et al. 1999 and Sgrò and Partridge 1999.)

in longevity between the "young" and "old" populations disappeared, proving that longevity is affected by a physiological cost of reproduction (Figure 17.7C). These results are consistent with other evidence that in *Drosophila* and many other insects, mating activity and egg production reduce the longevity of both sexes, and that virgins live longer than nonvirgins (Fowler and Partridge 1989; Bell and Koufopanou 1986).

The Theory of Life History Evolution

Because reproduction early in life contributes more to the rate of population growth than an equal production of offspring later in life, the SENSITIVITY of fitness to small changes in life history traits—the effect of a given magnitude of change in fecundity or survival on fitness (r)—depends on the age at which the change is expressed (Charlesworth 1994). For the hypothetical data in Table 17.1, for example, an increase in survival to the first age class (l_1) would obviously increase fitness. On the other hand, increasing survival from age 5 to age 6 (l_5) would not alter fitness at all, because this species does not reproduce beyond age 5 ($m_6 = 0$). Therefore, *natural selection usually does not favor postreproductive survival.* (Postreproductive survival may be advantageous, however, if postreproductive parents care for their offspring, as in humans.)

Furthermore, the selective advantage of a slight increase in survival or fecundity at an advanced age increases fitness (r) less than an equal increase at an early age, because the contribution of a cohort to population growth declines with age. (Note the decline in the "sensitivity values" $S_l(x)$ and $S_m(x)$ with age in Table 17.1.) A simple reason is that older females are less likely to be alive to reproduce. Another reason, as we have seen, is that offspring born to older females contribute less to population growth than those born to young females.

Life span and senescence

Most organisms in which germ cells are distinct from somatic tissues undergo physiological degeneration with age, a process called aging or **senescence**. Why does this occur? There are two related hypotheses on the evolution of senescence and limited life span (Rose 1991). Both rest on the principle that the selective advantage of an enhanced probability of survival declines with age.

TABLE 17.1 *A hypothetical example of a life table and of the sensitivity of fitness (r) to changes in age-specific survival and fecundity*[a]

Age class	Number of survivors	Fraction of survivors	Average fecundity	Survival × Fecundity			Sensitivity of r to m_x	Sensitivity of r to l_x
x		l_x	m_x	$l_x m_x$	e^{-rx}	$e^{rx} l_x m_x$	$S_m(x)$	$S_s(x)$
0	1000	1.000	0.00	0.00	1.000	0.000	0.335	0.334
1	750	0.750	0.00	0.000	0.796	0.000	0.200	0.334
2	600	0.600	1.20	0.720	0.634	0.456	0.128	0.182
3	480	0.480	1.40	0.672	0.505	0.339	0.081	0.068
4	360	0.360	1.03	0.396	0.402	0.159	0.049	0.018
5	180	0.180	0.96	0.144	0.320	0.046	0.019	0.018
6	100	0.100	0.00	0.000	0.255	0.000	0.011	—
Sums:					1.932 = R			1.000

Source: After Stearns 1992.

[a] The instantaneous rate of increase, r, is calculated by "trial and error" from the equation $1 = \sum_{x=a}^{x=z} e^{-rx} l_m m_x$ and is found to be 0.228.

The sensitivity coefficients $S_m(x)$ and $S_s(x)$ indicate the effect on r of a small change in fecundity (m_x) or survival (l_x), respectively, at age x.

They are calculated, respectively, as $S_m(x) = \frac{e^{-rx} l_m}{T}$ *and* $S_s(x) = \sum_{y=x}^{y=z} \frac{e^{-ry} l_y m_y}{T}$.

Peter Medawar (1952) proposed that deleterious mutations that affect later age classes accumulate in populations at a higher frequency than those that affect earlier age classes because selection against them is weak. If there are many such loci, then the causes of senescence should vary among individuals. This hypothesis predicts that genetic variation in fitness-related traits (such as those affecting survival) should be greater in late than in early age classes. The other hypothesis, proposed by George Williams (1957), postulates antagonistic pleiotropy, or genetic trade-offs. Because of the greater contribution of earlier age classes to fitness, an allele that is advantageous early in life, such as one that increases reproductive effort, has a selective advantage even if it is deleterious later in life (perhaps because it reduces allocation of energy and materials to maintenance, repair, and defense).

The evidence for Medawar's mutation accumulation hypothesis is mixed. It predicts that the additive genetic variance in fitness components should be greater in older age classes. In *Drosophila*, the predicted relationship was found for mortality rate and male mating success, but not for female fecundity (Charlesworth 1994b). On the other hand, Williams's hypothesis of antagonistic pleiotropy is supported by selection experiments like those by Linda Partridge's group (see Figure 17.7), which provide evidence of a negative relationship between early reproduction and both longevity and later reproduction. These experiments are among the most striking confirmations of evolutionary hypotheses that had been posed long before.

Age schedules of reproduction

If survival contributes to fitness only as long as reproduction continues, why don't organisms reproduce indefinitely? The answer is that all else being equal, there is always an advantage to reproducing earlier in life. Since early reproduction is correlated with lowered subsequent reproduction (as in Figure 17.6), we should expect organisms to be semelparous, allocating all their resources to a single early burst of reproduction rather than to maintaining themselves. We must therefore ask why so many species are nevertheless iteroparous.

Reproducing at an early age may increase the risk of death, decrease growth, or decrease subsequent fecundity so as to lower r compared with what it would be if reproduction were deferred. Fecundity, for example, is often correlated with body mass in species that grow throughout life, such as many plants and fishes. In such species, allocating resources to growth, self-maintenance, and self-defense rather than to immediate reproduction is an investment in the much greater fecundity that may be attained later in life. Accordingly, mathematical models show that *repeated reproduction is more likely to evolve if adults have high survival rates from one age class to the next, and if the rate of population increase is low*. These factors also favor later, rather than earlier, reproductive maturity in an iteroparous species. Since some resources are used for growth, maintenance, and defense, the effort devoted by an iteroparous species to reproduction will be less at each reproductive age than what a semelparous species devotes to reproduction in its single "big bang" reproductive episode. As individuals age, however, the benefit of withholding energy from reproduction declines because the intrinsic disadvantages of reproducing late in life become greater. Therefore, we would expect that at some point in life, *the proportion of energy or other resources devoted to reproduction by iteroparous species should increase with age* (Williams 1966; Charlesworth 1994b).

Figure 17.8 Reproductive effort—an index of the proportion of biomass allocated to inflorescences—in annual (semelparous) and perennial (iteroparous) species of British grasses. Allocation to reproduction is greater in the semelparous annual species. (After Wilson and Thompson 1989.)

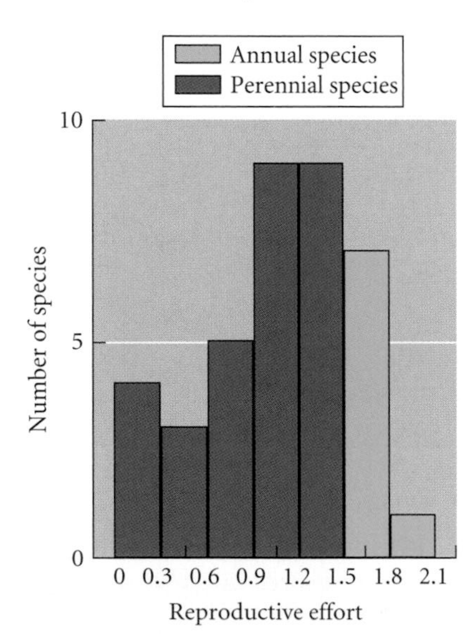

These theoretical predictions have been supported by many studies. Comparisons among species within several taxa support the prediction that reproductive effort, in each reproductive episode, should be lower in iteroparous than in semelparous organisms (Roff 2002). For example, inflorescences make up a lower proportion of plant weight in perennial than in annual species of grasses (Figure 17.8; Wilson and Thompson 1989). Likewise, the prediction that high adult survival rates favor delayed maturation and high reproductive effort later in life is upheld by studies of mammals, fishes, lizards and snakes, and other groups: species that have long life spans in nature also mature at a later age (Figure 17.9) (Promislow and Harvey 1991; Shine and Charnov 1992).

David Reznick and colleagues have studied guppies (*Poecilia reticulata*) in Trinidad (e.g., Reznick et al. 1990; Reznick and Travis 2002). In some streams, the cichlid fish *Crenicichla alta* preys heavily on large (mature) guppies (see Chapter 11). In other streams, or above waterfalls, *Crenicichla* is absent, and there is much less predation. Predation by *Crenicichla* should favor the evolution of maturity and reproduction early in life, and guppies from *Crenicichla*-dominated streams indeed mature faster and at smaller sizes, reproduce more frequently, have higher reproductive effort (measured as weight of embryos relative to weight of mother), and have more and smaller offspring than guppies from low-predation streams. In two streams, Reznick and colleagues moved guppies from below a waterfall, where they are preyed on by *Crenicichla*, to sites above the waterfall, where guppies and *Crenicichla* were absent. After several generations, the researchers took guppies from the sites of origin and introduction and reared their offspring in a common laboratory environment. As predicted by life history theory, the populations relieved of predation on large adults had evolved delayed maturation and larger adult size, and they tended toward fewer, larger offspring and lower reproductive effort (Figure 17.10)

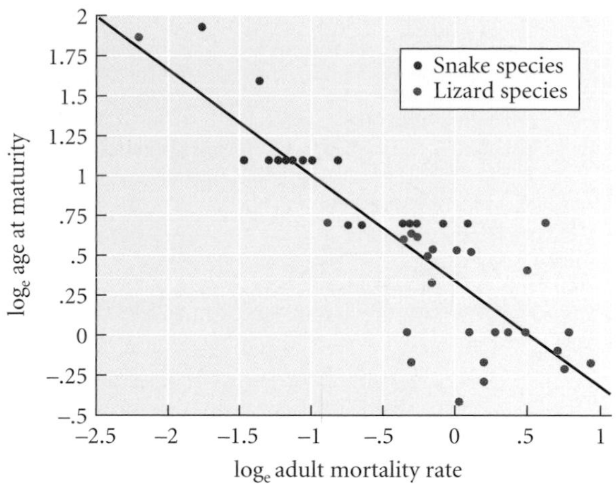

Figure 17.9 Among 16 species of snakes and lizards, the lower the annual mortality rate of adults, the later reproduction begins. This pattern conforms to the prediction that delayed onset of reproduction is most likely to evolve in species with high rates of adult survival. (After Shine and Charnov 1992.)

Number and size of offspring

All else being equal, a genotype with higher fecundity has higher fitness than one with lower fecundity. Why, then, do some species, such as humans, albatrosses, and kiwis, have so few offspring?

The British ecologist David Lack (1954) proposed that the *optimal* clutch size for a bird is the number of eggs that yields the greatest number of surviving offspring. The number of survivors from larger broods may be less than the number from more modest clutches because parents are unable to feed larger broods adequately. This decrease in offspring survival has proved to be one of several costs of large clutch size in birds (Stearns 1992). A long-term study of reproductive success in great tits (*Parus major*) showed that in good years, when caterpillars were abundant and offspring were well fed, it was advantageous to lay as many eggs as possible, but that in bad years, large clutches yielded fewer sur-

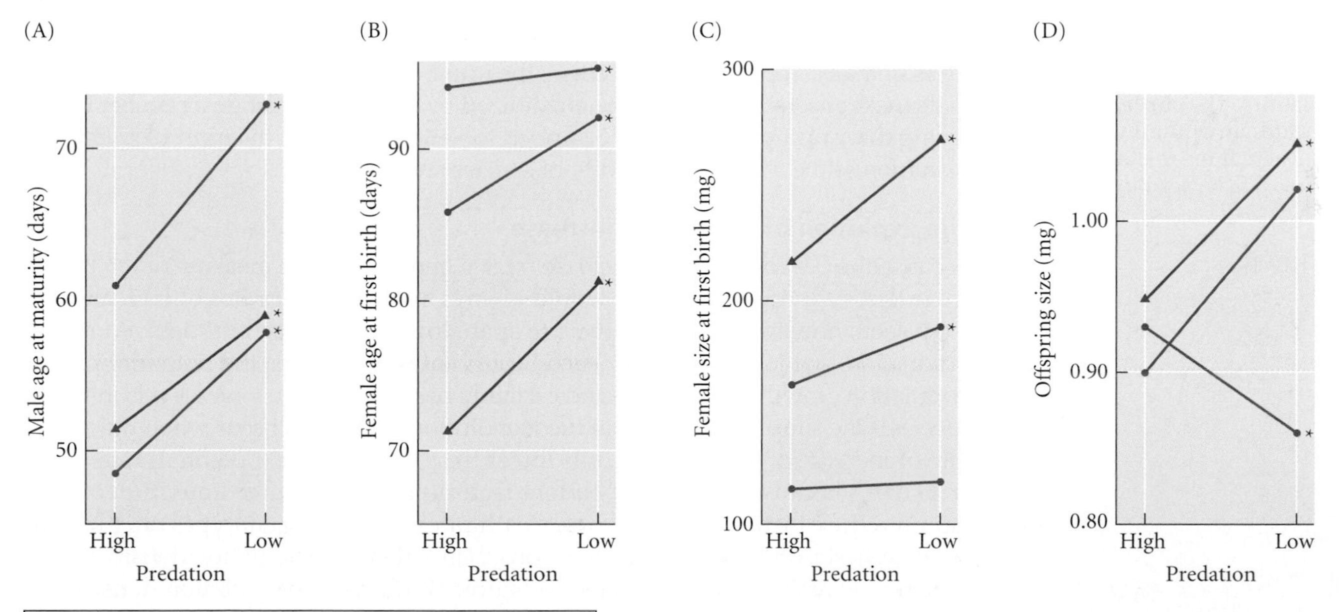
(A)
Male age at maturity (days)
70
60
50
High
Low
Predation
(B)
Female age at first birth (days)
90
80
70
(C)
Female size at first birth (mg)
300
200
100
(D)
Offspring size (mg)
1.00
0.90
0.80
▲ Contrasts the means of two natural populations in high-predation streams and two in low-predation streams.
• Contrasts an experimental population, isolated from predators for 7 generations, with a downstream control population that experiences high predation.
• Contrasts an experimental population isolated for about 18 generations, with a high-predation downstream control.

Figure 17.10 Differences between guppy populations in high-predation and low-predation environments, assayed in "common-garden" comparisons of second-generation laboratory-reared offspring of wild females. Asterisks indicate statistically significant differences. Low-predation populations tend to evolve (A) a later age at maturity in males and (B) in females; (C) a larger size at maturity in females; and (D) a larger offspring size at birth. (Data from Reznick and Travis 2002.)

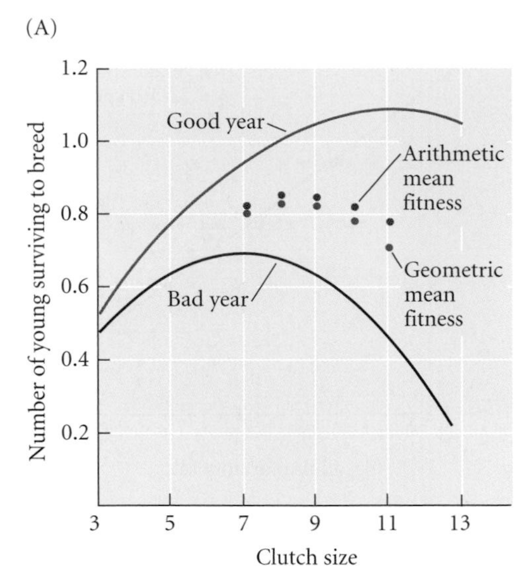

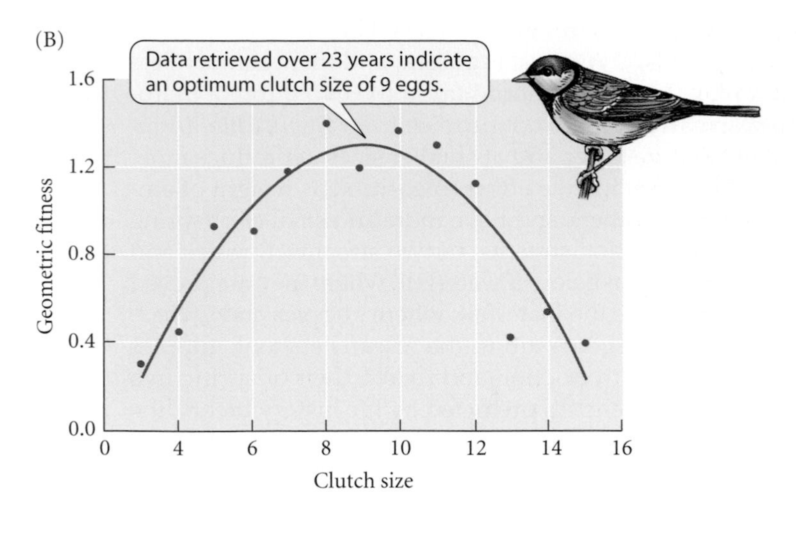

Figure 17.11 Optimal clutch size in the great tit. (A) The curves show the theoretical relationship (based on field data) between clutch size and the number of surviving young in a good year, when the optimum is 11 eggs, compared with a bad year, when food is scarce and the optimum is 7 eggs. The arithmetic and geometric mean fitness of birds that produce each clutch size if good and bad years are equally frequent is also shown. The geometric mean (the square root of the product of successive values) is appropriate to use when a phenotype's fitness varies across generations. (B) Calculation of the geometric mean fitness resulting from different clutch sizes, based on data on offspring survival in a 23-year study of a population of great tits. The curve fitted to the points indicates an optimum clutch size of 9 eggs. The actual mean clutch size in this population is 8.5. (After Boyce and Perrins 1987.)

viving offspring than smaller clutches (Figure 17.11A). Averaged over years, the optimal clutch size was about nine eggs (Figure 17.11B).

Great tit offspring from large clutches usually weigh less than those from smaller clutches. Whenever parents can provide only a limited amount of yolk, endosperm, nourishment, or other forms of parental care (collectively referred to as **parental investment**), there may be a negative correlation between number and size of offspring, and greater initial size usually enhances survival and growth rate. (For example, human twins have a lower birth weight, on average, than single-born infants, and infant survival of human quadruplets and quintuplets is notoriously precarious.) Barry Sinervo (1990) experimentally manipulated hatchling size in the lizard *Sceloporus occidentalis* by removing yolk from eggs with a hypodermic needle. Smaller hatchlings ran more slowly, which probably would reduce their survival in the wild. In such cases, females' reproductive success—the number of *surviving* offspring they leave to the next generation—should be maximized by producing a modest number of offspring that are larger and better equipped for survival. This is the most plausible explanation of the very low fecundity of species such as humans.

The evolution of the rate of increase

Because the per capita rate of increase (r) of a genotype is the measure of its fitness, we might suppose that species would always evolve higher rates of increase. We have seen, however, that a shorter life span, lower fecundity, and delayed maturation, all of which lower r, can each be advantageous. Thus the potential rate of population growth can evolve—and certainly has evolved—to lower levels in many species. One simple reason is that most evolution is likely to occur when the actual rate of increase of a population, r, is lower than the **intrinsic** (potential) **rate of increase** (r_m) because density-dependent factors such as resource limitation or predation reduce birth rates or increase death rates. Different genotypes are likely to have highest r under crowded conditions than when the population density is low (when r_m is realized), as illustrated in Figure 17.12. As the population density ap-

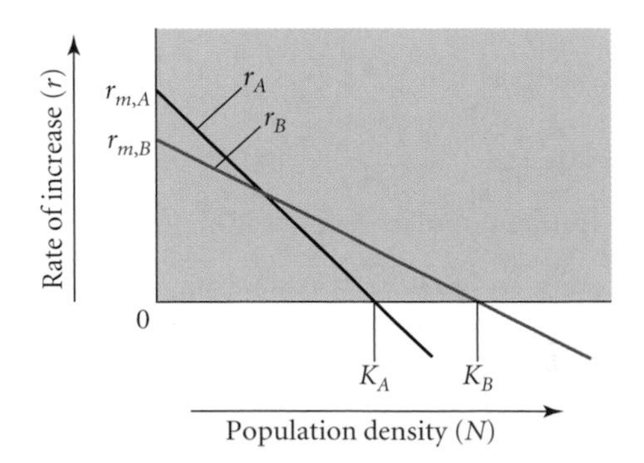

Figure 17.12 A model of density-dependent selection of rates of increase. The instantaneous per capita rate of increase, r, declines for genotypes A and B as population density (N) increases. The intrinsic rate of increase (r_m), the population growth rate at very low density, is lower for genotype B, but this genotype has a selective advantage at high density. (After Roughgarden 1971.)

proaches equilibrium (K), a more competitive genotype may sustain positive population growth while inferior competitors decline in density (have negative r).

In populations that are regulated by density-dependent factors (see Chapter 7) and occupy relatively stable environments, predation or competition for resources often causes heavier mortality among juveniles than adults. The mortality of seedling trees in a mature forest, for example, is exceedingly high, but if a tree does survive beyond the sapling stage—perhaps because a treefall has opened a light gap—it is likely to have a long life. As we have seen, these conditions favor the evolution of iteroparous reproduction late in life, and thus favor the evolution of long life spans. Moreover, under competitive conditions, juvenile survival can be enhanced by large size, so producing large eggs or offspring—and fewer of them—may be advantageous. Thus many authors have concluded that traits associated with a low intrinsic rate of increase—delayed maturation, production of few, large offspring, a long life span—are likely to evolve in species that occupy stable, competitive, or resource-poor environments. For example, species of beetles, fish, and other animals that inhabit caves generally develop very slowly and produce large eggs at an extraordinarily low rate (Culver 1982).

Male reproductive success

Much of the theory of life history evolution described for females applies to males as well. A substantial cost of reproduction, for example, may impose selection for delayed maturation and iteroparous reproduction. Competition for mates—the basis of sexual selection (see Chapter 14)—is often very costly (Andersson 1994). Larger males, or males that invest more energy in competing for females, are often more successful both in contests with other males and in attracting females (Figure 17.13). For these reasons, males of many species are larger than females and begin to breed at a later age.

The cost of reproduction underlies some interesting variations in male life histories. One such variation is ALTERNATIVE MATING STRATEGIES. In some species, large males display and/or defend territories to attract females, whereas small males do not, but rather "sneak" about, intercepting females and attempting to mate with them. In some instances, "sneaker" males have lower reproductive success than displaying males, so their behavior is probably not an adaptation; they are probably making the best of a bad situation, as they are unable to compete successfully. In other instances, however, the sneaker strategy appears to be an alternative adaptation, yielding the same fitness as the display strategy. For example, in the Pacific coho salmon (*Oncorhynchus kisutch*), large, red "hooknose" males develop hooked jaws and enlarged teeth, and fight over females, whereas "jack" males are smaller, resemble females, do not fight, and breed when they are only about a third as old as the hooknose males. Based on data on survival to breeding age and frequency of mating, the fitnesses of these two types of males appear to be nearly equal (Gross 1984).

Some plants, annelid worms, fishes, and other organisms undergo sex change (sequential hermaphroditism). In species that grow in size throughout reproductive life, a

(A)

(B)

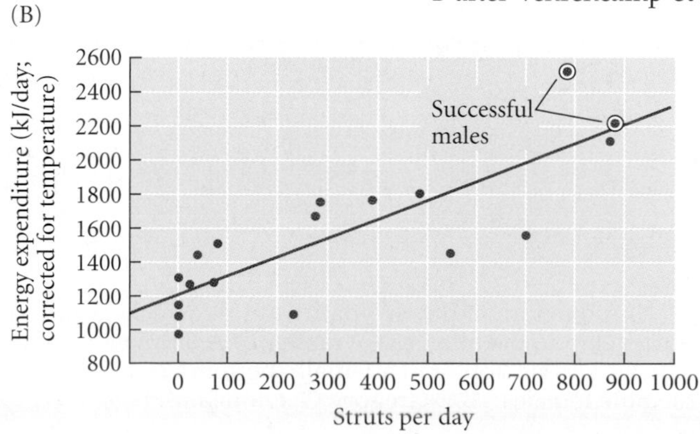

Figure 17.13 Male competition for females requires large expenditures of energy. (A) A male sage grouse (at left) displays for females on an arena, or lek. One or a few dominant males are likely to be chosen as mates by all the females that visit the lek. Note that the males of this species are larger than the females. (B) The energy expended by displaying male sage grouse increases with the number of displays (struts). In the group studied, only the two most active males were successful in mating. (A © Peter Arnold, Inc./Alamy Images; B after Vehrencamp et al. 1989.)

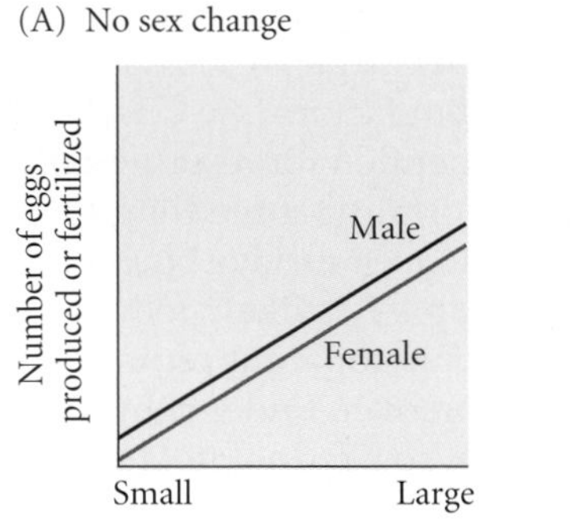

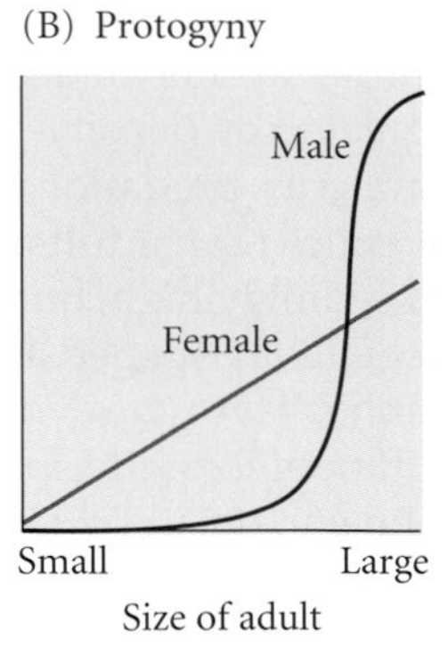

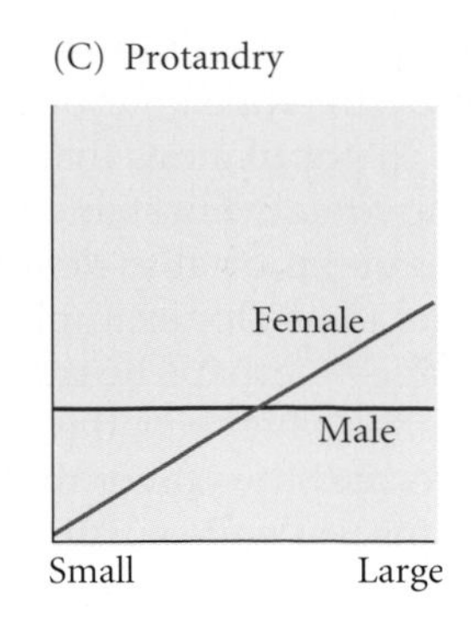

Figure 17.14 A model for the evolution of sex change in sequential hermaphrodites. (A) When reproductive success increases equally with body size in both sexes, there is no selection for sex change. (B) A switch from female to male (protogyny) is optimal if male reproductive success increases more steeply with size than female reproductive success. (C) The opposite relationship favors the evolution of protandry, in which males become females when they grow to a large size. (After Warner 1984.)

sex change can be advantageous if reproductive success increases with size to a greater extent in one sex than in the other (Figure 17.14). In the bluehead wrasse (*Thalassoma bifasciatum*), for instance, some individuals start life as females and later become brightly colored "terminal-phase males" that defend territories. Other individuals begin as "initial-phase males," which resemble females and spawn in groups (Figure 17.15). A female usually does not produce as many eggs as a large terminal-phase male typically fertilizes. Both females and initial-phase males become terminal-phase males at about the size at which this form achieves superior reproductive success (Warner 1984).

Modes of Reproduction

Organisms vary greatly in what is sometimes called their GENETIC SYSTEM: whether they reproduce sexually or asexually, self-fertilize or outcross, are hermaphroditic or have separate sexes. Discovering why and how each of these characters evolved poses some of the most challenging problems in evolutionary biology and is the subject of some of the most creative contemporary research on evolution.

The genetic system affects genetic variation, which of course is necessary for the long-term survival of species. This fact has been cited for more than a century as the reason for the existence of recombination and sexual reproduction. But, as we have seen, arguments that invoke benefits to the species are suspect because they rely on group selection, which is ordinarily a weak agent of evolution. The question, then, is whether or not natural selection *within* populations can account for features of the genetic system.

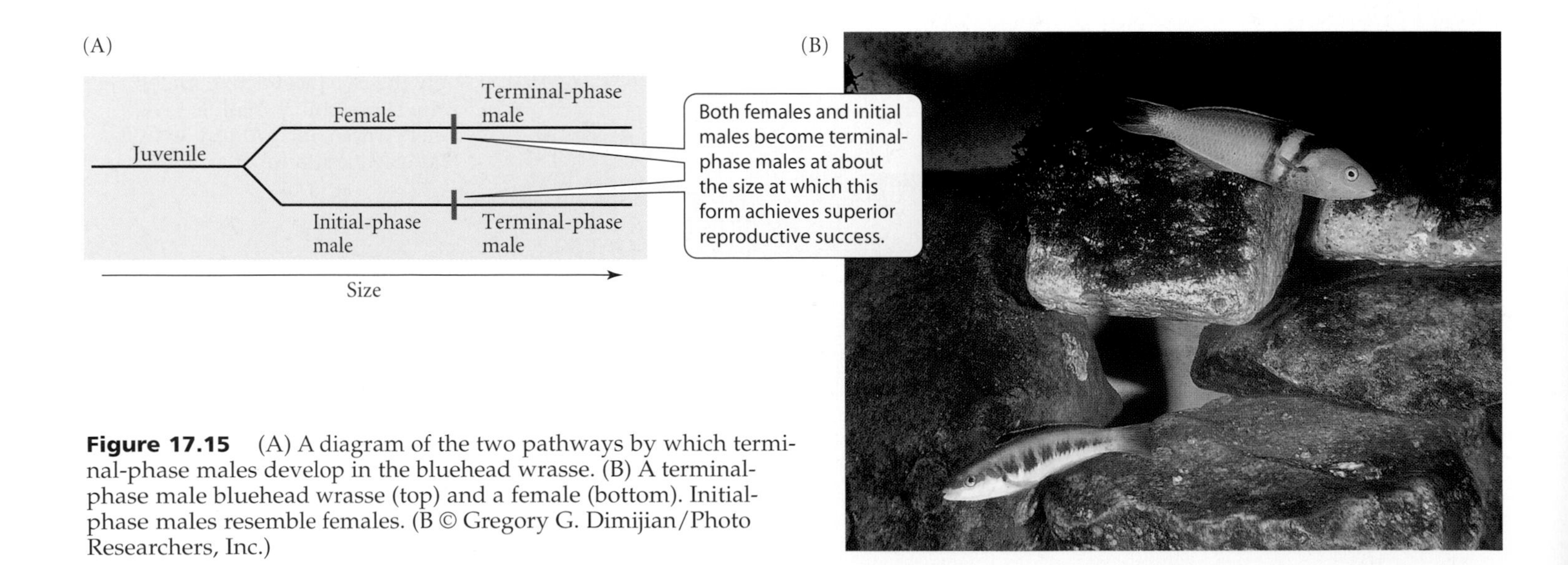

Figure 17.15 (A) A diagram of the two pathways by which terminal-phase males develop in the bluehead wrasse. (B) A terminal-phase male bluehead wrasse (top) and a female (bottom). Initial-phase males resemble females. (B © Gregory G. Dimijian/Photo Researchers, Inc.)

The evolution of mutation rates

The nature of this question can be appreciated if we first think about the evolution of mutation rates. Two hypotheses have been proposed: either the mutation rate has evolved to some optimal level, or it has evolved to the minimal possible level. Variation in factors such as the efficacy of DNA repair provides potential genetic variation in the genomic mutation rate. According to the hypothesis of optimal mutation rate, selection has favored somewhat inefficient repair enzymes. According to the hypothesis of minimal mutation rate, mutation exists only because the repair system is as efficient as it can be, or because selection is not strong enough to favor investment of energy in a more efficient repair system. According to this hypothesis, the process of mutation is not an adaptation.

Group selection would favor an optimal (greater than zero) mutation rate because genetically invariant species would become extinct, leaving only species that experience mutation. We do not know how fast this process would occur because the faster the environment changes, the higher the mutation rate must be to avert extinction (Lynch and Lande 1993).

Alternatively, we can ask how evolution *within* populations affects the mutation rate. This can be done by postulating a "mutator" locus that affects the mutation rate of other genes. Let us assume that mutator alleles that increase the mutation rate affect the fitness of their bearers only indirectly, via the mutations they cause. It turns out that the fate of such a mutator allele depends on the level of recombination. In an asexual population, the mutator allele is likely to decline in frequency because copies of the allele are permanently associated with the mutations they cause, and far more mutations reduce than increase fitness. But occasionally, a mutator allele causes a beneficial mutation. It will then increase in frequency by hitchhiking with the mutation it has caused.

In a sexual population, however, recombination will soon separate the mutator allele from a beneficial mutation it has caused, so it will not hitchhike to high frequency. Because deleterious mutations occur at so many loci, the mutator will usually be associated with one or another of them, and will therefore decline in frequency. Therefore, *natural selection within sexual populations will tend to eliminate any allele that increases mutation rates, and mutation rates should evolve toward the minimal achievable level*—even though mutation is necessary for the long-term survival of a species. Consequently, most evolutionary biologists believe that the existence of mutation is not an adaptation in most organisms, but rather an unavoidable effect of the physics and chemistry of DNA replication (Leigh 1973; Sniegowski et al. 2000).

Mutator alleles actually occur at considerable frequencies in some natural populations of the bacterium *Escherichia coli*. Because *E. coli* reproduces mostly asexually, it is understandable that mutator alleles sometimes increase to fixation in experimental populations (Figure 17.16). It has been shown that these increases are caused by hitchhiking with new beneficial mutations (Shaver et al. 2002).

Figure 17.16 A mutator allele that increases the mutation rate throughout the genome increased in frequency (solid line) and eventually became fixed in an experimental population of *Escherichia coli* by hitchhiking with one or more advantageous mutations. Samples of clones carrying the mutator allele (red circles) showed an increase in fitness over time, whereas the fitness of clones that lacked the allele (green circles) did not change. A mutator allele would not be expected to increase in frequency in a sexual population. (After Shaver et al. 2002.)

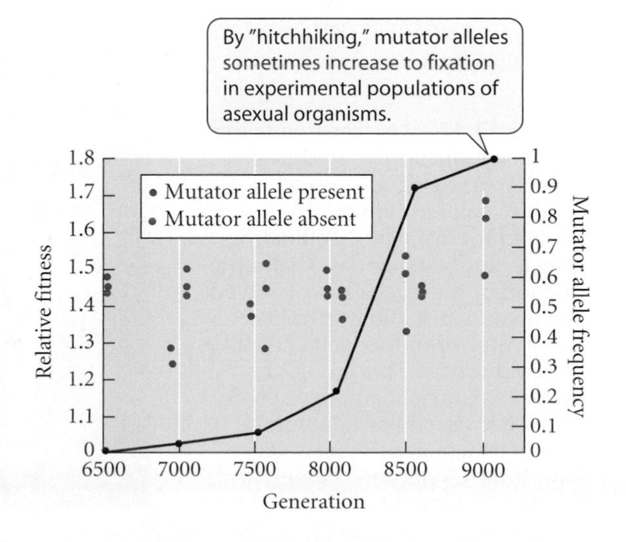

Sexual and asexual reproduction

Sex usually refers to the union (SYNGAMY) of two genomes, usually carried by gametes, followed at some later time by REDUCTION, ordinarily by the process of meiosis and gametogenesis. Sex often, but not always, involves outcrossing between two individuals, but it can occur by self-fertilization in some organisms. Sex almost always includes SEGREGATION of alleles and RECOMBINATION among loci, although the extent of recombination varies greatly. Most sexually reproducing species have distinct female and male sexes, which are defined by a difference in the size of their gametes (ANISOGAMY). In ISOGAMOUS organisms, such as *Chlamydomonas* and many other algae, the uniting cells are the same size; such species have MATING TYPES, but not distinct sexes. Species in which individuals are either female or male, such as willow trees and mammals, are termed **dioecious** or GONOCHORISTIC; species such as roses and earthworms, in which an individual can produce both kinds of gametes, are **hermaphroditic** or COSEXUAL.

Asexual reproduction may be carried out by **vegetative propagation**, in which an offspring arises from a group of cells, as in plants that spread by runners or stolons, or by **parthenogenesis**, in which the offspring develops from a single cell. The most common kind of parthenogenesis is **apomixis**, whereby meiosis is suppressed and an offspring develops from an unfertilized egg. The offspring is genetically identical to its mother, except for whatever new mutations may have arisen in the cell lineage from which the egg arose. A lineage of asexually produced, and thus genetically nearly identical, individuals may be called a CLONE.

In some taxa, recombination and the mode of reproduction can evolve rather rapidly. Using artificial selection in laboratory populations of *Drosophila*, investigators have altered the rate of crossing over between particular pairs of loci, and have even developed parthenogenetic strains from sexual ancestors (Carson 1967; Brooks 1988). Asexual populations are known in many otherwise sexual species of plants and animals, such as crustaceans and insects. However, many of the features required for sexual reproduction seem to degenerate rapidly in populations that have evolved asexual reproduction, so reversal from asexual to sexual reproduction becomes very unlikely (Normark et al. 2003).

The problem with sex

The traditional explanation of the existence of recombination and sex is that they increase the rate of adaptive evolution of a species, either in a constant or a changing environment, and thereby reduce the risk of extinction. That there is indeed a long-term advantage of sex is indicated by phylogenetic evidence that most asexual lineages of eukaryotes have arisen quite recently from sexual ancestors, and by the observation that such lineages often retain structures that once had a sexual function. A typical example is the dandelion *Taraxacum officinale*, which is completely apomictic, but is very similar to sexual species of *Taraxacum*, retaining nonfunctional stamens and the brightly colored petal-like structures that in its sexual relatives serve to attract cross-pollinating insects (Figure 17.17A). If a parthenogenetic lineage were able to persist for many millions of years, it should have diverged greatly from its sexual relatives and given rise to a morphologically and ecologically diverse clade. Such diversity, betokening persistence since an ancient origin of asexuality, exists in only a few eukaryote groups, such as the bdelloid rotifers (Figure 17.17B; Normark et al. 2003). Most asexual lineages that arose a very long time ago must have become extinct, since there are so few ancient asexual forms.

The recency of most parthenogenetic lineages suggests that sex reduces the risk of extinction. If this were the reason for its prevalence, sex might be one of the few characteristics of organisms that has evolved by group selection. But recombination and sex also have serious disadvantages. One is that *recombination destroys adaptive combinations of genes*. For example, recall (from Chapter 9) that in the primrose *Primula vulgaris*, plants with a

(A)

(B)

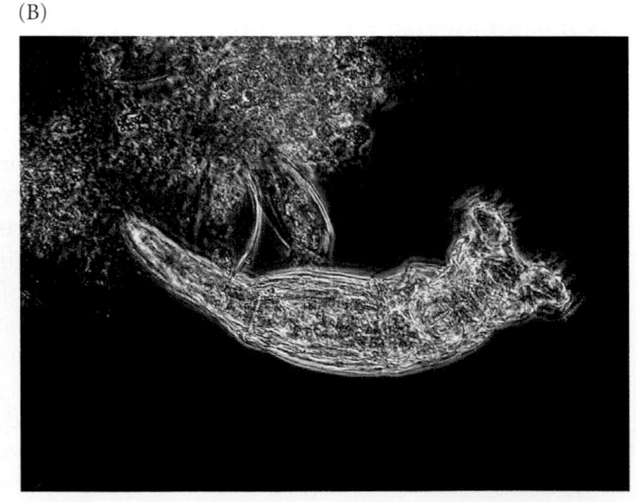

Figure 17.17 (A) The dandelion *Taraxacum officinale* reproduces entirely asexually, by apomixis. The brightly colored flower, which evolved for attracting pollinating insects, suggests that asexual reproduction in this species has evolved recently. In fact, some species of this genus reproduce sexually. (B) A bdelloid rotifer. This group of rotifers is unusual among metazoans because it has apparently been parthenogenetic for a very long time. (A photo © Painet, Inc.; B © Eric V. Gravé.)

short style and long stamens carry a chromosome with the gene combination *GA*, whereas those with a long style and short stamens are homozygous for the combination *ga*. These are adaptive combinations because each can successfully cross with the other; but recombination produces combinations such as *Ga*, which has both short stamens and a short style, and is less successful in pollination. In general, asexual reproduction preserves adaptive combinations of genes, whereas sexual reproduction breaks them down and reduces linkage disequilibrium between them. An allele that promotes recombination, perhaps by increasing the rate of crossing over, may decline in frequency if it is associated with gene combinations that reduce fitness (Figure 17.18).

Sexual reproduction has a second disadvantage that is great enough to make its existence one of the most difficult puzzles in biology. This disadvantage is the COST OF SEX. Imagine two genotypes of females with equal fecundity, one sexual and one asexual. In many sexual species, only half of all offspring are female. However, all the offspring of an asexual female are female (because they inherit their mother's sex-determining genes). If sexual and asexual females have the same fecundity, then a sexual female will have only half as many grandchildren as an asexual female (Figure 17.19). Therefore, the rate of increase of an asexual genotype is approximately twice as great as that of a sexual genotype (all else being equal), so an asexual mutant allele would very rapidly be fixed if it occurred in a sexual population. In the long term, of course, the evolution of asexuality might doom a population to extinction, but, given this twofold advantage of asexuality, it is doubtful that extinction occurs frequently enough to prevent the replacement of sexual with apomictic genotypes within populations. The problem, therefore, is to discover whether there are any *short-term advantages* of sex that can overcome its short-term disadvantages (Maynard Smith 1978; Charlesworth 1989; Kondrashov 1993).

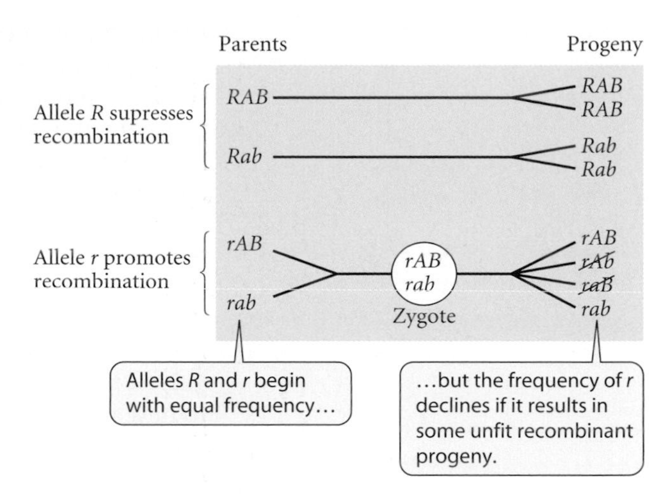

Figure 17.18 Selection against an allele (*r*) that promotes recombination, if allele combinations *Ab* and *aB* have lower fitnesses than *AB* and *ab*. The parental generation has two copies of *R* and two copies of *r*. Allele *R*, which suppresses sex and recombination, increases in frequency because of its association with the favored genotypes *AB* and *ab*. The lower diagram pictures an organism, such as some algae, in which a diploid zygote undergoes meiosis, and the dominant part of the life cycle is haploid. The same principle holds for organisms in which the diploid phase of the life cycle is dominant.

Hypotheses for the advantage of sex and recombination

Many explanations for the prevalence of recombination and sex have been proposed (Kondrashov 1993), but most are variations on the following few themes:

REPAIR OF DAMAGED DNA. A possible benefit of recombination is that breaks and other lesions in a DNA molecule can be repaired by copying from an intact sequence on a homologous chromosome (Bernstein and Bernstein 1991). According to this hypothesis, the formation of new gene combinations is a by-product of the molecular mechanism of DNA repair, not the *raison d'être* of recombination or sex. However, this hypothesis fails to explain the elaborate mechanisms of meiosis and syngamy, so most evolutionary biologists believe that the maintenance of sex in most species must be attributed to other causes involving variation and selection (Maynard Smith 1988; Barton and Charlesworth 1998).

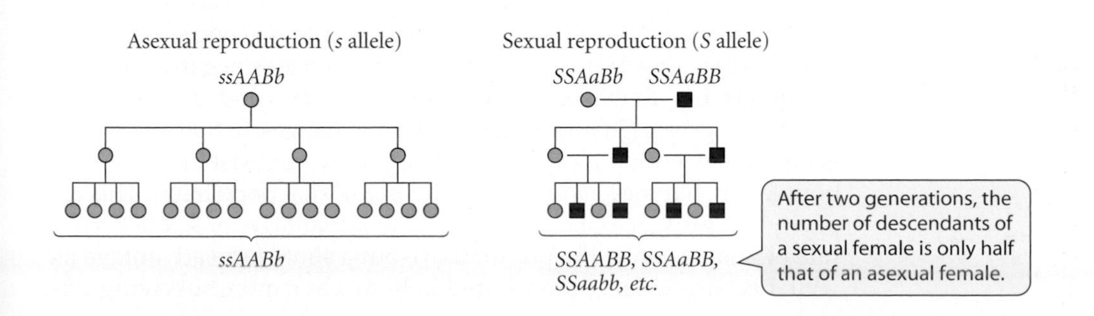

Figure 17.19 The cost of sex. The disadvantage of an allele *S*, which codes for sexual reproduction, compared with an allele *s*, coding for asexual reproduction. Circles represent females, squares males. Each of two females in the same population produces four equally fit offspring, but the frequency of the *S* allele drops rapidly from two-thirds in the first generation to one-third by the third generation due to the production of males. However, the *S* allele would increase if the environment were to change so that a recombinant genotype such as *aabb* had much a higher survival rate than other genotypes.

SIBLING COMPETITION. Suppose genotypes can differ in their use of limiting resources. If siblings compete for resources, then a patch of habitat can sustain more progeny from a sexual family than from an asexual family because genetically identical asexual siblings compete more intensely (Williams 1975; Bell 1982). However, sibling competition is far from universal, and thus probably does not explain the prevalence of sexual reproduction among species (Barton and Charlesworth 1998).

MULLER'S RATCHET: PREVENTING MUTATIONAL DETERIORATION. Herman Muller (1964), who won a Nobel Prize for his role in discovering the mutagenic effect of radiation, proposed a hypothesis that has been named MULLER'S RATCHET (Figure 17.20). In an asexual population, deleterious mutations at various loci create a spectrum of genotypes carrying 0, 1, 2, ... m mutations. Individuals can carry more mutations that their ancestors did (due to new mutations), but not fewer. Thus the zero-mutation class declines over time because its members experience new deleterious mutations. Moreover, due to genetic drift in a finite population, the zero class may be lost by chance, despite its superior fitness. (The smaller the population, the more likely this is to happen.) Thus all remaining genotypes have at least one deleterious mutation. Sooner or later, by the same process of drift, the one-mutation class is lost, and all remaining individuals carry at least two mutations. The accidental loss of superior genotypes continues, and is an irreversible process—a ratchet. This reduction of fitness is likely to lower population size, and this, in turn, increases the rate at which the least mutation-laden genotypes are lost by genetic drift. Thus there may be an accelerated decline of fitness—a "mutational meltdown"—leading to extinction (Lynch et al. 1993). In contrast, *recombination in a sexual population reconstitutes the least mutation-laden classes of genotypes* by generating progeny with new combinations of favorable alleles.

The decline in fitness caused by Muller's ratchet would be slow, so the advantage of sex under this hypothesis is unlikely to counterbalance the twofold cost of sex. However, this hypothesis may, in part, explain the longer survival of sexual than asexual lineages.

Figure 17.20 Muller's ratchet. The frequency of individuals with different numbers of deleterious mutations (0–10) is shown for an asexual population at three successive times. The class with the lowest mutation load (0 in top graph, 1 in middle graph) is lost over time, both by genetic drift and by its acquisition of new mutations. In a sexual population, class 0 can be reconstituted, since recombination between genomes in class 1 that bear different mutations can generate progeny with none. (After Maynard Smith 1988.)

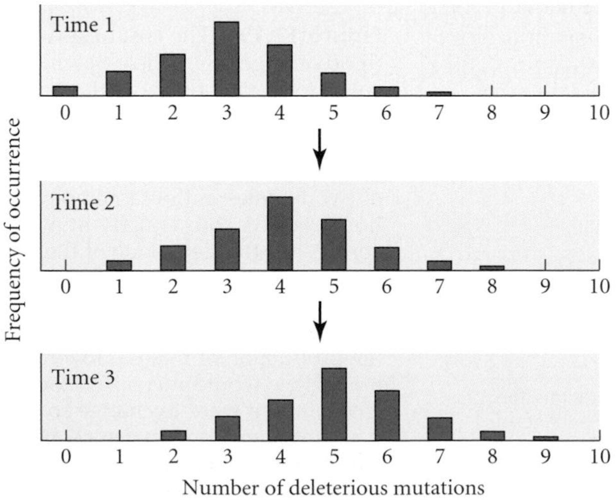

THE MUTATIONAL DETERMINISTIC PROCESS. Suppose deleterious mutations at different loci reduce fitness more when they are combined in a single genotype than would be expected from their separate effects—that is, they show "synergistic epistasis." Alexey Kondrashov (1988) pointed out that such mutations will be eliminated from a sexual population more rapidly than from an asexual population because recombination in the sexual population brings them together. Thus the mean fitness ($\overline{w}$) of a sexual population is likely to be maintained at a higher level than that of an asexual population. Kondrashov showed that this effect could counteract the twofold cost of sex if there is at least one new deleterious mutation per genome per generation. It is not known whether the rate of deleterious mutation is actually high enough to balance the cost of sex. Moreover, deleterious mutations in the bacterium *Escherichia coli* appear not to have the synergistic effects on fitness that the model requires (Elena and Lenski 1997).

ADAPTATION TO FLUCTUATING ENVIRONMENTS. Suppose a polygenic character is subject to stabilizing selection, but the optimum character state fluctuates due to a fluctuating environment (Maynard Smith 1980). Let us assume that alleles A, B, C, D ... additively increase a trait such as body size, and alleles a, b, c, d ... decrease it. Stabilizing selection for intermediate size reduces the variance and creates negative linkage disequilibrium, so that combinations such as *AbCd* and *aBcD* are present in excess (see Chapter 13). If selection changes so that larger size is favored, combinations such as *ABCD* may not exist in an asexual population, but they can arise rapidly in a sexual population. This capacity can provide not only a long-term advantage for sex (a higher rate of adaptation of the population), but a short-term advantage as well, because sexual parents are likely to leave more surviving offspring than asexual parents. For this hypothesis to work, the selection regime must fluctuate rather frequently, and some factor must maintain genetic variation, because a long-term regime of stabilizing selection for a constant optimal phenotype would fix a homozygous genotype (such as *AABBccdd*) (see Chapter 13).

One popular possibility is that genetic variation is maintained, and sex is favored, by parasites. As a resistant host genotype (e.g., *ABCD*) increases in frequency, a parasite might evolve to attack it. Parasite genotypes that can attack less common host genotypes (such as *abCD*) would then become rare, so the uncommon host genotypes would acquire higher fitness and increase in frequency. Continuing cycles of coevolution between host and parasite might select for sex, which would continually regenerate rare combinations of alleles. This role of parasites is an instance of the Red Queen hypothesis (see Chapter 7), according to which a species must evolve continually to keep up with evolutionary changes in its natural enemies or competitors.

The Red Queen hypothesis for the advantage of sex may hold true only under special conditions, such as very strong selection by parasites (Otto and Nuismer 2004), but it has some evidence in its favor (Jokela and Lively 1995; Lively and Dybdahl 2000). For example, the sexual form of the freshwater snail *Potamopyrgus antipodarum* is more abundant than the asexual form in sites where a trematode parasite is abundant (Figure 17.21A), and common asexual genotypes are more heavily infected by sympatric trematodes than rare clones are, as the Red Queen hypothesis predicts (Figure 17.21B). But it is by no means clear that parasite-host coevolution often occurs on the time scale that the hypothesis requires.

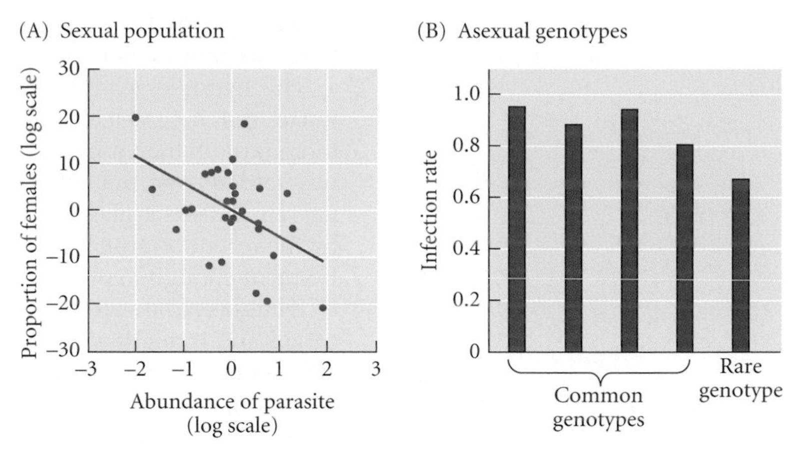

Figure 17.21 Evidence that selection by a parasitic trematode may favor sexual reproduction in a freshwater snail species that has both sexual and asexual genotypes. (A) The proportion of sexual genotypes is greater in local populations that are exposed to a high incidence of the parasite, as shown by the decreased proportion of females in such populations. (B) When snails of different asexual genotypes were exposed to trematodes, individuals with a rare genotype were less likely to become infected than were those with four common genotypes. (A after Jokela and Lively 1995; B after Lively and Dybdahl 2000.)

ENHANCED ADAPTATION UNDER DIRECTIONAL SELECTION. The most obvious apparent advantage of sex is that it enhances the rate of adaptation to new environmental conditions by combining new mutations or rare alleles. In an asexual population, beneficial mutations *A* and *B* would be combined only when a second mutation (*B*) occurs in a growing lineage that has already experienced mutation *A* (or vice versa). In a sexual population, *A* and *B* mutations that have occurred in different lineages could be combined more rapidly (Figure 17.22). However, this difference in evolutionary rate holds only for large populations. In small populations, mutations are so few that the first (*A*) is likely to be fixed by selection be-

Figure 17.22 Effects of recombination on the rate of evolution. *A*, *B*, and *C* are new mutations that are advantageous in concert. In asexual populations (1 and 3), combination *AB* (or *ABC*) is not formed until a second mutation (such as *B*) occurs in a lineage that already bears the first mutation (*A*). In a large sexual population (2), independent mutations can be brought together in a lineage more rapidly by recombination, so adaptation is more rapidly achieved. In small sexual populations (4), however, the interval between the occurrence of favorable mutations is so long that a sexual population does not adapt more rapidly than an asexual population. (After Crow and Kimura 1965.)

Population 1: Large, asexual

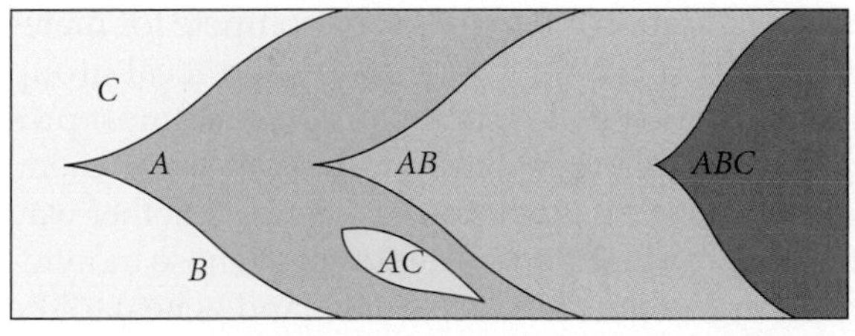

Population 2: Large, sexual

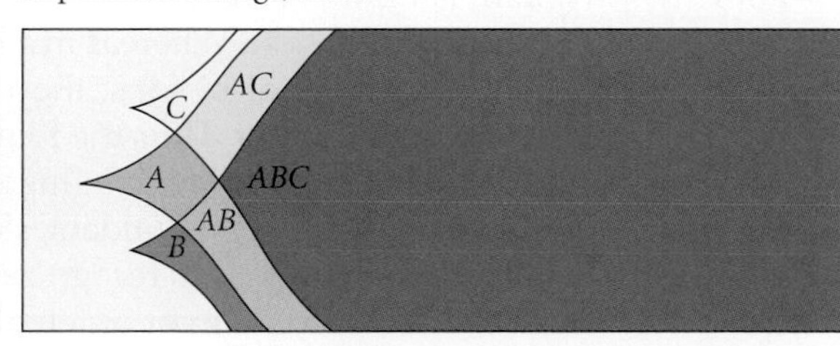

Population 3: Small, asexual

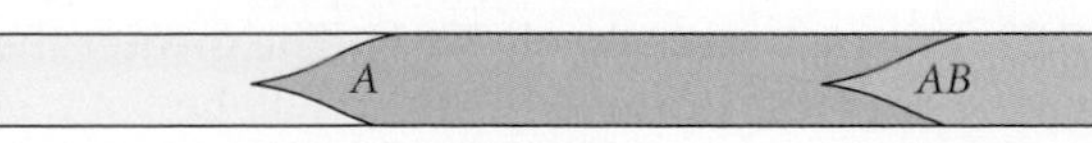

Population 4: Small, sexual

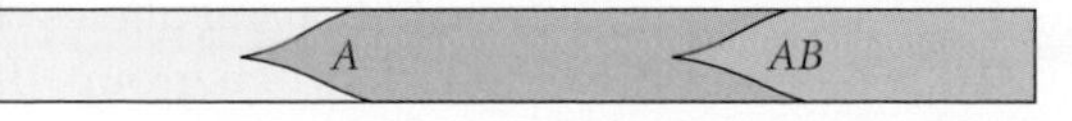

fore the second (B) arises, whether the population is asexual or sexual. Thus recombination may or may not speed up adaptive evolution.

This hypothesis has been experimentally confirmed. For example, sexual populations of the unicellular green alga *Chlamydomonas reinhardtii* evolved higher fitness in the laboratory more rapidly than asexual populations; furthermore, as the hypothesis predicts, this effect was seen in large, but not in small, populations (Colegrave 2002). Slower adaptation by asexual populations is likely to be a major reason for their high rate of extinction, as documented by the recent origin of most asexual eukaryotes. But directional selection is unlikely to be frequent enough to provide the short-term advantage needed to counter the cost of sex.

It has been difficult to demonstrate that any single hypothesis accounts for the prevalence of sex, and it might well be that a combination of factors is usually at play (see West et al. 1999 and associated commentaries). Answering the question of sex may require both better genetic data (e.g., on mutation rates) and more studies of selection in natural populations.

Sex Ratios, Sex Allocation, and Sex Determination

Large gametes (eggs) versus small gametes (sperm) define the two sexes. Why should there be distinct sexes? Models have shown that anisogamy evolves if one genotype is favored because the large size of its gametes enhances the survival of the offspring, and another genotype is favored because it can make many gametes. Individuals that produce a third, intermediate gamete size enjoy neither the advantage of size nor the advantage of numbers. Thus two types of gametes, of highly disparate size, are expected to evolve (Charlesworth 1978).

Given the two kinds of gametes, we would like to understand why some species are hermaphroditic and others are dioecious, and why the sex ratio in dioecious species is nearly 0.5 (1:1) in some species but not in others. The theory of sex allocation has been developed to explain such variation (Charnov 1982; Frank 1990).

The evolution of sex ratios

The **sex ratio** is defined as the proportion of males. As in Chapter 12, we distinguish the sex ratio in a population (the POPULATION SEX RATIO) from that in the progeny of an individual female (an INDIVIDUAL SEX RATIO). In Chapter 12, we saw that in a large, randomly mating population, a genotype with a male-biased individual sex ratio has an advantage if the population sex ratio is female-biased, and vice versa, because each individual of the minority sex has a higher number of offspring than each individual of the majority sex. A genotype with an individual sex ratio of 0.5 is an ESS (evolutionarily stable strategy) because it has, per capita, the greatest number of grandchildren (see Figure 12.18).

In many species, however, mating occurs not randomly among members of a large population, but within small local groups descended from one or a few founders. After one or a few generations, progeny emerge into the population at large, then colonize patches of habitat and repeat the cycle. In many species of parasitoid wasps, for example, the progeny of one or a few females emerge from a single host and almost immediately mate with each other; the daughters then disperse in search of new hosts. Such species often have female-biased sex ratios.

William Hamilton (1967) explained such "extraordinary sex ratios" by what he termed LOCAL MATE COMPETITION. Whereas in a large population a female's sons compete for mates with many other females' sons, they compete only with one another in a local group founded by their mother. Thus the founding female's genes can be propagated most prolifically by producing mostly daughters, with only enough sons to inseminate all of them. Additional sons would be redundant, since they all carry their mother's genes. Another way of viewing this situation is to recognize that groups founded by genotypes whose individual sex ratio is biased in favor of females contribute more individuals (and genes) to the population as a whole than groups founded by unbiased genotypes. The difference among local groups in their production of females increases the frequency, in the population as a whole, of female-biasing alleles (Figure 17.23; Wilson and Colwell 1981). The greater the number of founders of a group, the more nearly even the optimal sex ratio will be.

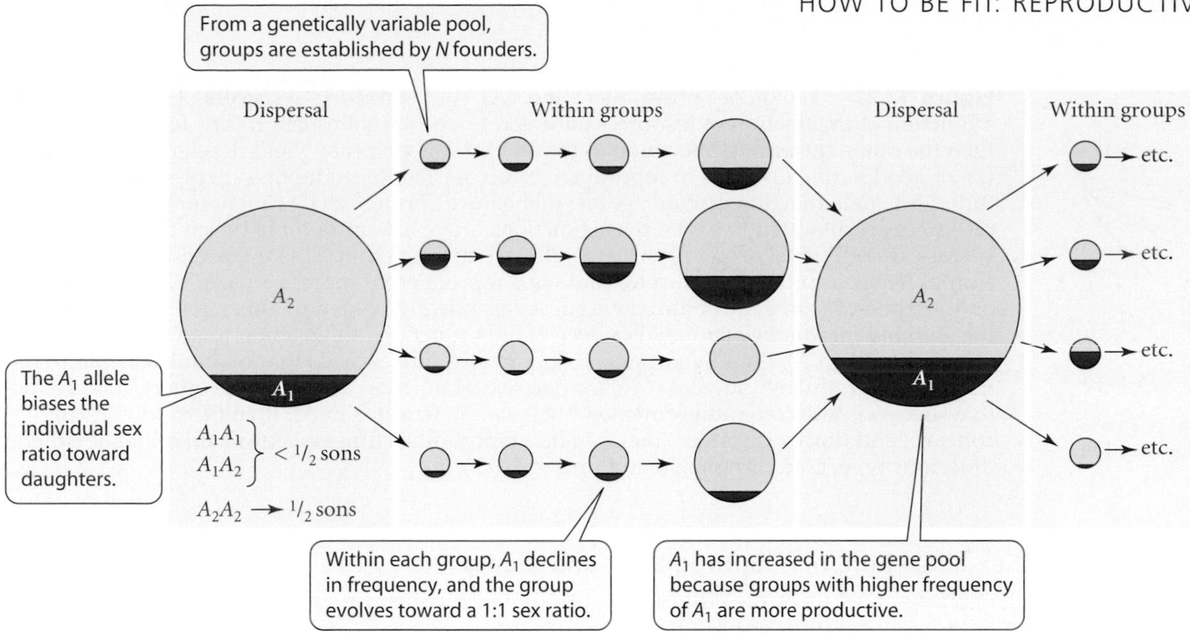
From a genetically variable pool, groups are established by N founders.
Dispersal
Within groups
Dispersal
Within groups
etc.
etc.
etc.
etc.
A_2
A_1
A_2
A_1
The A_1 allele biases the individual sex ratio toward daughters.
A_1A_1, A_1A_2 < 1/2 sons
A_2A_2 → 1/2 sons
Within each group, A_1 declines in frequency, and the group evolves toward a 1:1 sex ratio.
A_1 has increased in the gene pool because groups with higher frequency of A_1 are more productive.

Figure 17.23 A model of the evolution of a female-biased sex ratio in a population that is structured into local groups, but periodically forms a single pool of dispersers. The frequency of A_1, an allele that biases the individual sex ratio toward daughters, is indicated by the dark portion of each circle. The size of each circle represents the size of a group or population. From a genetically variable pool, groups are established by one or a few founders. The frequency of A_1 varies among groups by chance. Although A_1 declines in frequency within each group over the course of several generations, the growth in group size is greater the higher the frequency of A_1 (because of the greater production of daughters). When individuals emerge from the groups, they form a pool of dispersers, in which A_1 has increased in frequency since the previous dispersal episode. (After Wilson and Colwell 1981.)

Some of the best evidence for this theory is provided by the adaptive plasticity of the sex ratio in parasitoid wasps such as *Nasonia vitripennis*. The progeny of one or more females develop in a fly pupa and mate with each other immediately after emerging. Based on the theory described above, we would expect females to produce more daughters than sons, but we would expect the sex ratio to increase with the number of families developing in a host. Moreover, if the second wasp that lays eggs in a host can detect previous parasitization, we would expect her to adjust her individual sex ratio to a higher value than that of the first female. John Werren (1980) calculated the theoretically optimal individual sex ratio of a second female, then measured the individual sex ratios of second females by exposing fly pupae successively to strains of *Nasonia* that were distinguishable by an eye color mutation. On the whole, his data fit the theoretical prediction very well (Figure 17.24).

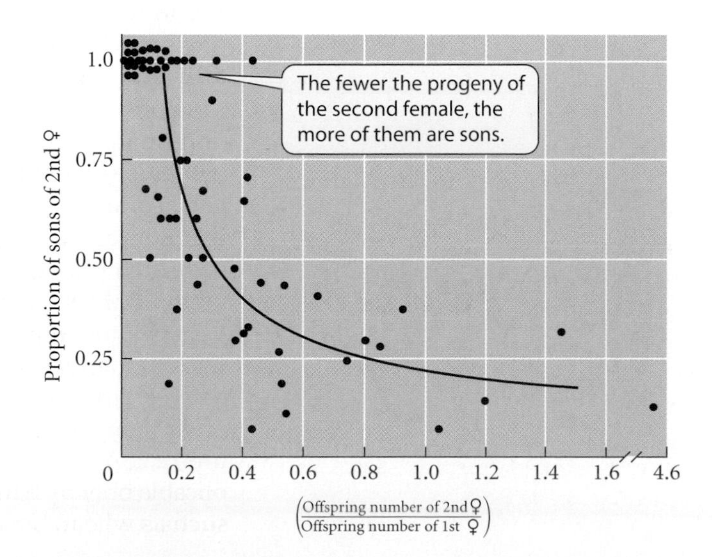
The fewer the progeny of the second female, the more of them are sons.
Proportion of sons of 2nd ♀
1.0
0.75
0.50
0.25
0
0.2
0.4
0.6
0.8
1.0
1.2
1.4
1.6
4.6
(Offspring number of 2nd ♀ / Offspring number of 1st ♀)

Figure 17.24 Adaptive adjustment of individual sex ratio by a parasitoid wasp in which females usually mate with males that emerge from the same host. The points show the relationship between the proportion of sons among the offspring of a "second" female (i.e., one that lays eggs in a fly pupa in which another female has already laid eggs) and the proportion of her offspring in the host. The curved line is the theoretically predicted individual sex ratio in second females' broods, as a function of the sex ratio in the first female's brood and the relative number of the two females' offspring. If the second female's offspring made up only a small fraction of the total, that female's optimum "strategy" should be to produce mostly sons, which could potentially inseminate many female offspring of the first female. As predicted, the fewer the progeny of the second female, the more of them are sons. (After Werren 1980.)

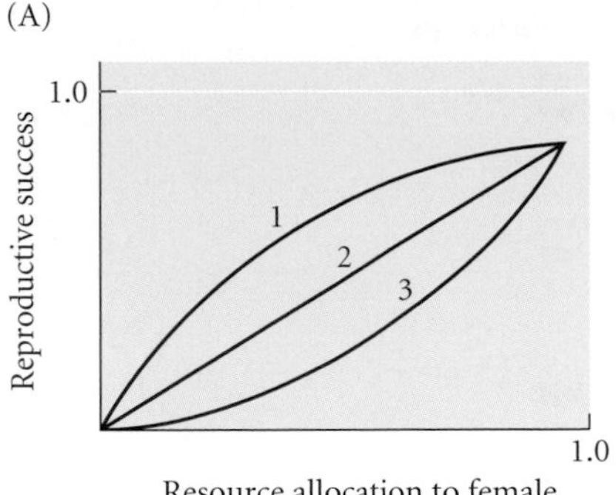

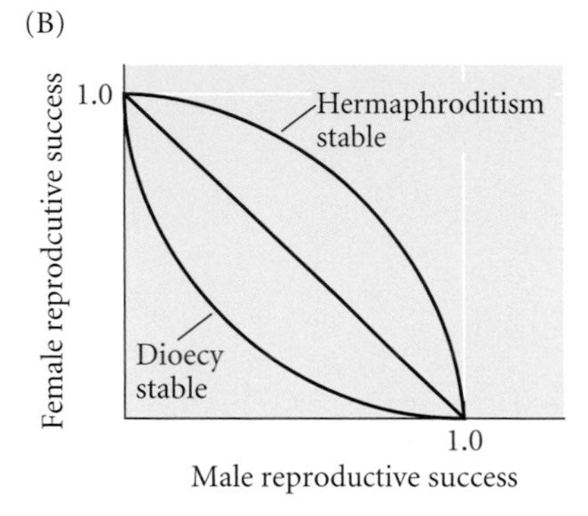

Figure 17.25 The theory of sex allocation. (A) The reproductive success of an individual as a function of the fraction of resources allocated to one sexual function (say, female) rather than the other. Increasing allocation to that sexual function may yield decelerating (1), linear (2), or accelerating (3) gains in reproductive success. (B) Reproductive success gained through female function plotted against that gained through male function. Because resources are allocated between these functions, there is a trade-off between the reproductive success an individual achieves through either sexual function. This trade-off is linear, and the sum of female reproductive success and male reproductive success equals 1.0 at each point on the trade-off curve, if reproductive success is linearly related to allocation (curve 2 in A). If the gain in reproductive success is a decelerating function of allocation to one sex or the other (curve 1 in A), then the fitness of a hermaphrodite exceeds that of a unisexual individual (with reproductive success = 1.0 for one sexual function and 0 for the other). If reproductive success is an accelerating function of allocation (curve 3 in A), then the trade-off curve is concave, and dioecy is stable—that is, a hermaphrodite's fitness is lower than that of either dioecious type. (After Thomson and Brunet 1990.)

Sex allocation, hermaphroditism, and dioecy

The energy and resources that a potentially hermaphroditic individual devotes to reproduction may be allocated to female functions (e.g., eggs, seeds) and to male functions (e.g., sperm, mate-seeking) in different proportions. Whatever is allocated to one sexual function cannot be allocated to the other, so there must be a trade-off between the reproductive success that can be realized through the two sexual functions. Dioecious species may be thought of as those in which individuals allocate all their reproductive energy to one sexual function or the other.

The reproductive success that individuals achieve might be a linear function of their proportional allocation to, say, male function (curve 2 in Figure 17.25A); it might be an accelerating function (curve 3); or it might decelerate, showing diminishing returns (curve 1). It can be shown that in the latter case, the theoretically optimal strategy is a hermaphroditic one (Figure 17.25B), whereas it is optimal to be one sex or the other if reproductive success increases more than linearly with increasing allocation. One factor that may affect the shape of the trade-off curve in Figure 17.25B is the cost of developing the structures required for the two sexual functions. If most structures associated with reproduction enhance both male and female function (e.g., petals), hermaphroditism may be favored, because with relatively little extra expenditure the individual can produce offspring via both sexual functions. If male and female functions require different structures, however, a pure male or female incurs the cost of developing only one set of structures, whereas a hermaphrodite incurs both costs and thus is likely to have lower fitness.

Inbreeding and Outcrossing

Self-fertilization (selfing), which results in inbreeding, occurs in some hermaphroditic animals (such as certain snails) and in a great many plants. Many plants can both self-fertilize and export pollen to the stigmas of other plants (**outcrossing**). Characteristics that promote outcrossing include dioecy (separate sexes), asynchronous male and female function (maturation and dispersal of pollen either before or after the stigma of the same flower is receptive), and several kinds of self-incompatibility (Matton et al. 1994). In the most common self-incompatibility system, the growth of a haploid pollen grain is inhibited if its allele at a self-incompatibility locus matches either allele in the diploid stigma on which it lands. The DNA sequences of self-incompatibility alleles show that these polymorphisms are stable and very old. Among genera of Solanaceae (potatoes, tobacco, petunias, and others), for example, there is as much as 40 percent difference in amino acid sequence among self-incompatibility alleles within a species (indicating an old divergence between these alleles), and some alleles within a species are less closely related to each other than to those from other genera that diverged up to 36 million years ago (see Figure 12.23A). That a polymorphism should have been retained for such a long time is explicable only by balancing selection (Clark 1993; see Chapter 12). In contrast, many plants, such as wheat, have evolved a strong tendency toward self-fertilization within flowers.

Many such species produce only a little pollen and have small, inconspicuous flowers that lack the scent and markings to which pollinators are attracted. In some cases, flowers remain budlike and do not open at all.

It is more difficult to identify adaptations to avoid inbreeding in animals than in plants. The major factors that may reduce inbreeding are prereproductive dispersal and INCEST AVOIDANCE: avoidance of mating with relatives (Thornhill 1993). Incest avoidance has been described in a few species of birds and mammals, such as chimpanzees. In several species of rodents, sibling pairs have a much reduced likelihood of mating, and house mice prefer mating with individuals that differ from them at the major histocompatibility (MHC) loci, which play an important role in the immune response (Potts et al. 1991).

Behavioral avoidance of inbreeding inevitably brings to mind the well-known "incest taboo," often codified in religious and civil law, in human societies. This is a highly controversial topic with respect to both the actual incidence of inbreeding ("incest") and the interpretation of the social taboo. Societies vary as to which kin matings are prohibited (Ralls et al. 1986). Moreover, the incidence of closely incestuous sexual activity is evidently far higher than society has generally been ready to recognize, especially in the form of sexual attention forced upon young women by fathers and uncles. These observations raise doubt about whether a strong, genetically based aversion to incest has evolved in our species (at least in males). Some anthropologists hold that outbreeding is a social device (not an evolved genetic trait) to establish coalitions between families or larger groups in order to reap economic and other benefits of cooperation.

Inbreeding increases homozygosity, and it is often accompanied by inbreeding depression (see Chapter 9). Inbreeding depression seems usually to be caused by homozygosity for deleterious recessive (or nearly recessive) alleles, but it may occasionally be caused by homozygosity at overdominant loci, at which heterozygotes have highest fitness (Uyenoyama et al. 1993). If inbreeding depression is caused by homozygous recessive alleles, selection may be expected to "purge" the population of those alleles as inbreeding continues, so genetic variation should decline and mean fitness should increase. The mean fitness of a long-inbred population might therefore equal or even exceed that of the initial outbreeding population (Lande and Schemske 1985).

Spencer Barrett and Deborah Charlesworth (1991) provided some evidence for this hypothesis, using a highly outcrossing Brazilian population and an almost exclusively self-fertilizing Jamaican population of the aquatic plant *Eichhornia paniculata*. For five generations, Barrett and Charlesworth self-pollinated plants in each population, then cross-pollinated the inbred lines. In the naturally outcrossing Brazilian lines, flower number declined as inbreeding proceeded (Figure 17.26A), but it increased dramatically in crosses between the inbred lines. In contrast, the naturally selfing Jamaican lines conformed to the purging theory: they showed neither inbreeding depression nor heterosis when crossed with each other (Figure 17.26B).

Advantages of inbreeding and outcrossing

R. A. Fisher (1941) first pointed out that a plant genotype that can both self-fertilize and disperse pollen (a PARTIAL SELFER) has a strong selective advantage over either an exclusive outcrosser or an exclusive selfer. Each partially selfing individual can transmit genes to the next generation in three ways: through its eggs (ovules), through its pollen by selfing, and through its pollen as a sire of outcross

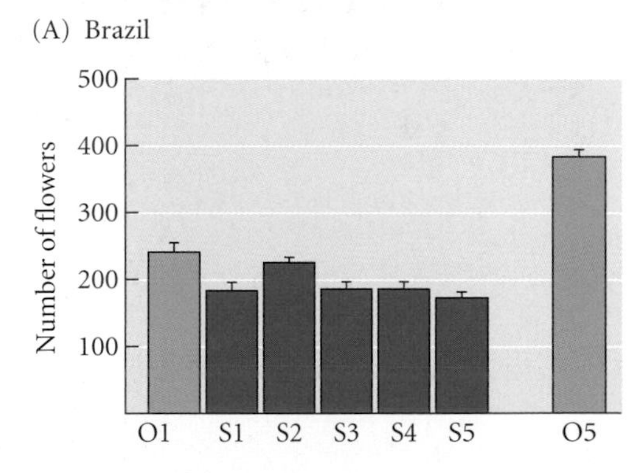

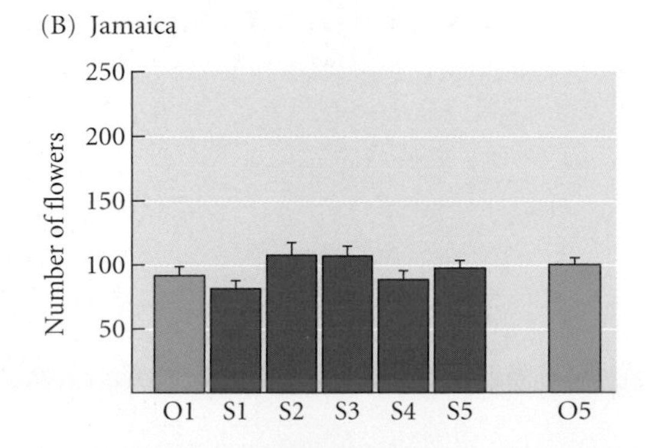

Figure 17.26 The mean number of flowers produced by *Eichhornia paniculata* plants from (A) a naturally outcrossing Brazilian population and (B) a naturally inbred Jamaican population in the first outcross generation (O1), in five generations of selfing (S1–S5), and in outcrosses between selfed plants after five generations (O5). The naturally outcrossing Brazilian population displayed inbreeding depression (compare O1 through S5) and heterosis in outcrossed plants (O5), but the naturally inbred Jamaican population did not. (After Barrett and Charlesworth 1991; photo courtesy of S.C.H Barrett.)

progeny. An exclusively outcrossing genotype, however, transmits genes only through its eggs and through its outcrossed pollen. On average, the ratio of genes transmitted by partial selfers and by outcrossers is 3:2, and partial selfing has a 50 percent advantage. Since exclusive selfers do not reproduce through outcrossed pollen, they too suffer a 50 percent disadvantage relative to partial selfers. Thus, for exclusive outcrossing to evolve, its advantage would have to be greater than the 50 percent disadvantage that Fisher described.

The most important advantage of characteristics such as self-incompatibility is thought to be avoidance of inbreeding depression (Lloyd 1992; Charlesworth and Charlesworth 1978). The evolution of obligate or predominant self-fertilization would thus have to overcome two obstacles: inbreeding depression and the loss of reproductive success through outcross pollen. Several possible advantages of selfing might outweigh these disadvantages. First, exclusive selfers may save energy and resources because they usually have small flowers and produce little pollen. However, a slight initial reduction of flower size seems unlikely to offset the severe disadvantages of selfing (Jarne and Charlesworth 1993).

Second, even if selfing is disadvantageous on average, it may occasionally produce a highly fit homozygous genotype that sweeps to fixation, carrying with it the alleles that increase the selfing rate (Holsinger 1991). A related possibility is that selfing may "protect" locally adapted genotypes from "outbreeding depression" due to gene flow and recombination. For example, Nikolas Waser and Mary Price (1989) crossed *Ipomopsis aggregata* plants separated by various geographic distances, planted the seeds under uniform conditions, and measured fitness-related characteristics in the progeny. Fitness was greatest in the progeny of moderately distant plants; it was lower in the progeny of plants situated close together (presumably an example of inbreeding depression), and in the progeny of distant plants (outbreeding depression). Waser and Price postulated that distant populations have different coadapted gene combinations, perhaps adapted to different environmental conditions, and that recombination between such gene combinations would be prevented by selfing.

Third, and perhaps most important, is REPRODUCTIVE ASSURANCE: a plant is almost certain to produce some seeds by selfing, even if scarcity of pollinators, low population density, or other adverse environmental conditions prevent cross-pollination. There is abundant support for this hypothesis (Jarne and Charlesworth 1993; Wyatt 1988). For example, adaptations for self-fertilization are especially common in plants that grow in harsh environments, where insect visitation is low or unpredictable, and on islands, where populations are sparse for some time after colonization.

Summary

1. The components of fitness (the per capita rate of increase of a genotype, r) are the age-specific values of survival and fecundity. Natural selection on morphological and other phenotypic characters results from their effects on these life history traits. Life history traits can be understood as the product of individual selection, not group selection.
2. Constraints, especially trade-offs between reproduction and survival and between the number and size of offspring, prevent organisms from evolving infinitely long life spans and infinite fecundity.
3. The effect of changes in survival (l_x) or fecundity (m_x) on fitness depends on the age at which such changes are expressed and declines with age. Hence selection for reproduction and survival at advanced ages is weak.
4. Consequently, senescence (physiological aging) evolves. Senescence may be a result, in part, of the negative pleiotropic effects on later age classes of genes that have advantageous effects on earlier age classes.
5. If reproduction is very costly (in terms of growth or survival), repeated (iteroparous) and/or delayed reproduction may evolve, provided that reproductive success at later ages more than compensates for the loss of fitness incurred by not reproducing earlier. Otherwise a semelparous life history, in which all of the organism's resources are allocated to a single reproductive effort, is optimal. Iteroparity is especially likely to evolve if juvenile mortality is high relative to adult mortality is low and if population density is stable.

6. The optimal clutch size is often less than the maximal potential clutch size because of trade-offs between the number of offspring and their size or survival.
7. The evolution of reproductive effort by males is governed by similar principles as in females. Delayed maturation may evolve if larger males are more successful in attracting or competing for mates. Similar principles explain phenomena such as sequential hermaphroditism (sex change with age) and alternative mating strategies.
8. The evolution of features of genetic systems, such as rates of mutation and recombination, sexual versus asexual reproduction, and rates of inbreeding, can usually be understood best as consequences of selection at the level of genes and individual organisms, rather than group selection.
9. Alleles that increase mutation rates are generally selected against because they are associated with the deleterious mutations they cause. Therefore, we would expect mutation rates to evolve to the minimal achievable level, even if this should reduce genetic variation and increase the possibility of a species' extinction.
10. Asexual populations have a high extinction rate, so sex has a group-level advantage in the long term. But this is unlikely to offset the short-term advantage of asexual reproduction.
11. In a constant environment, alleles that decrease the recombination rate are advantageous because they lower the proportion of offspring with unfit recombinant genotypes. In addition, asexual reproduction has approximately a twofold advantage over sexual reproduction because only half of the offspring of sexuals (i.e., the females) contribute to population growth, whereas all of the (all-female) offspring of asexuals do so. Therefore, the prevalence of recombination and sex requires explanation.
12. There are several hypotheses for the short-term advantage of sex: (*a*) recombination allows the repair of damaged DNA; (*b*) a sexual parent's genetically diverse progeny may partition resources more efficiently than an asexual parent's; (*c*) in small asexual populations, fitness declines because genotypes with few deleterious mutations, if lost by genetic drift, cannot be reconstituted, as they are in populations with recombination (Muller's ratchet); (*d*) deleterious mutations that interact to reduce fitness can be more effectively purged by natural selection in sexual than in asexual populations, thus maintaining higher mean fitness; (*e*) recombination enables the mean of a polygenic character to evolve to new, changing optima in a fluctuating environment; (*f*) the rate of adaptation, by fixing combinations of advantageous mutations, may be higher in sexual than asexual populations, if the populations are large.
13. In large, randomly mating populations, a 1:1 sex ratio is an evolutionarily stable strategy, because if the population sex ratio deviates from 1:1, a genotype that produces a greater proportion of the minority sex has higher fitness. If, however, populations are characteristically subdivided into small local groups whose offspring then colonize patches of habitat anew, a female-biased sex ratio can evolve because female-biased groups contribute a greater proportion of offspring to the population as a whole.
14. The evolution of hermaphroditism versus dioecy (separate sexes) depends on how reproductive success via female or male function is related to the allocation of an individual's energy or resources. Dioecy is advantageous if the reproductive "payoff" from one or the other sexual function increases disproportionately with allocation to that function.
15. All else being equal, a genotype that both self-fertilizes and outcrosses has a 50 percent advantage over an obligate outcrosser or an obligate selfer. However, outcrossing can be advantageous because it prevents inbreeding depression in an individual's progeny. Conversely, self-fertilization may evolve if fewer resources need be expended on reproduction, if an allele for selfing becomes associated with advantageous homozygous genotypes, or if selfing ensures reproduction despite low population density or scarcity of pollinators.

Terms and Concepts

antagonistic pleiotropy
apomixis
dioecious
hermaphroditic
intrinsic rate of increase
iteroparous
outcrossing
parental investment
parthenogenesis
reproductive effort

semelparous
senescence
sex
sex ratio
trade-offs
vegetative propagation

Suggestions for Further Reading

An informative and provocative book for a general audience by a prominent evolutionary biologist is J. Roughgarden's *Evolution's rainbow: Diversity, gender, and sexuality in nature and people* (University of California Press, Berkeley, 2004). *The evolution of life histories*, by S. C. Stearns (Oxford University Press, Oxford,1992) and *Life history evolution*, by D. A. Roff (Sinauer Associates, Sunderland, MA, 2002) are comprehensive treatments of that topic. *The evolution of sex*, edited by R E. Michod and B. R. Levin (Sinauer Associates, Sunderland, MA, 1988), and articles by Kondrashov (1993), Barton and Charlesworth (1998), Jarne and Charlesworth (1993), and Uyenoyama et al. (1993) treat the evolution of genetic systems.

Problems and Discussion Topics

1. Female parasitoid wasps search for insect hosts in which to lay eggs, and they can often discriminate among individual hosts that are more or less suitable for their offspring. Behavioral ecologists have asked whether or not the wasps' willingness to lay eggs in less suitable hosts varies with age. On the basis of life history theory, what pattern of change would you predict? Does life history theory make any other predictions about animal behavior?
2. Suppose that a mutation in a species of annual plant increases allocation to chemical defenses against herbivores, but decreases production of flowers and seeds (i.e., there is an allocation trade-off). What would you have to measure in a field study in order to predict whether or not the frequence of the mutation will increase?
3. Within many species of birds and mammals, clutch size is larger in populations at high latitudes than in populations at low latitudes. Species of lizards and snakes at high latitudes often have smaller clutches, and are more frequently viviparous (bear live young rather than laying eggs), than low-latitude species (see references in Stearns 1992). What selective factors might be responsible for these patterns?
4. An important life history characteristic, not discussed in this chapter, is dispersal between hatching and reproductive age. The extent of dispersal varies considerably among different organisms. What are the advantages versus disadvantages of dispersing? How might the evolution of dispersal be affected by group selection versus individual selection? (See Olivieri et al. 1995 and references therein.)
5. Populations of some species of fish, insects, and crustaceans consist of both sexually and obligately asexually reproducing individuals. Would you expect such populations to become entirely asexual or sexual? What factors might maintain both reproductive modes? How might studies shed light on the factors that maintain sexual reproduction?
6. Some mites, scale insects, and gall midges display "paternal genome loss" (or *pseudoarrhenotoky*, a fine word for parlor games). Males develop from diploid (fertilized) eggs, but the chromosomes inherited from the father become heterochromatinized and nonfunctional early in development, so that males are functionally haploid. How might this peculiar genetic system have evolved?
7. Many parthenogenetic "species" of plants and animals are known to be genetically highly variable. Determine from the literature whether this genetic variation is due to mutation within asexual lineages or to multiple origins of asexual genotypes from a sexually reproducing ancestor.
8. The number of chromosome pairs in the genome ranges from one (in an ant species) to several hundred (in some butterflies and ferns). Within a single genus of butterflies (*Lysandra*), the haploid number varies from 24 to about 220 (White 1978). Is it likely that chromosome number evolves by natural selection? Is there evidence for or against this hypothesis?

18

Coevolution: Evolving Interactions among Species

About 750 species of figs—trees and vines in the genus *Ficus*—grow in the tropical and subtropical regions of the world. Their unusual inflorescence consists of several hundred tiny flowers that line the inside wall of a hollow sphere called a syconium (Figure 18.1 and at right). In many species, the syconium, which eventually matures into a fleshy fruit that is important food for the many birds and mammals that distribute the fig's seeds, contains both female and male flowers. Almost every species of fig is pollinated exclusively by a single species of minute wasp of the family Agaonidae, and each such wasp depends on a single species of fig. The phylogeny of many of the wasp species matches the phylogeny of the figs they pollinate, suggesting that the wasps and figs have speciated in parallel—perhaps because they absolutely depend on each other.

An interdependent relationship. Almost every species of fig depends on pollination by a single species of wasp; the wasps in turn develop only in that fig species. Here wasps are shown inside the fig synconium, which has been cut open. Each of the plant's structures is a single fig flower. (Photo © Gregory Dimijian/Photo Researchers, Inc.)

The wasp's behavior is exquisitely adapted to the plant on which it depends. One or more female wasps enter a syconium of the right fig species through a narrow opening, bearing pollen in special pockets in

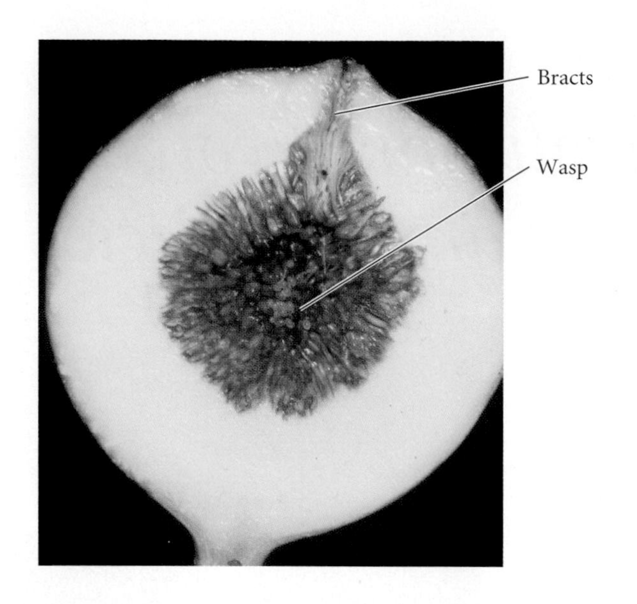

Figure 18.1 A section through the syconium of a fig, showing the many small flowers that line the chamber and the entry, with numerous bracts through which female pollinator wasps must force their way. (Photo © OSF/photolibrary.com)

their legs or thorax. The wasp carefully deposits pollen on the stigmas, deposits one egg in the ovary of each of many flowers, and then dies. Each larva feeds on the developing seed within a single flower, and as many as half of the female flowers may produce wasps rather than seeds. By the time the wasps develop and mate with each other within the syconium, the male flowers have matured. The young female wasps gather pollen, exit through holes in the syconium wall that the male wasps have chewed, and fly to other plants of the same species, to start their next generation.

This is a rather extreme example of reciprocal adaptation of species to each other, but almost all species have evolved adaptations for interacting with other species. Such adaptations, some of which are quite extraordinary, have enhanced the diversity of life, and have had profound effects on the structure of ecological communities.

In this chapter, we will consider interactions among species in terms of their effects on the fitness of individual organisms (not, as in some ecological theory, from the viewpoint of their effects on population growth). Most of the species with which an individual might interact can be classified as RESOURCES (used as nutrition or habitat), COMPETITORS (for resources such as food, space, or habitat), ENEMIES (species for which the focal species is a consumable resource), or COMMENSALS (species that profit from but have no effect on the focal species). In MUTUALISTIC interactions (such as the relation between a fig and its wasp), each species uses the other as a resource. Some interactions are more complex, often because they are mediated by a third species. For example, different unpalatable species of butterflies that resemble one another may profit from their resemblance because predators that have learned to avoid one may avoid the other as well (see Figures 12.19 and 18.24). Moreover, the nature and strength of an interaction may vary depending on environmental conditions, genotype, age, and other factors. There is genetic variation, for example, in virulence within species of parasites and in resistance within species of hosts. Some mycorrhizal fungi, associated with plant roots, enhance plant growth in infertile soil, but depress it in fertile soil. Thus the selection that species may exert on each other may differ among populations, resulting in a "geographic mosaic" of coevolution that differs from one place to another (Thompson 1999).

The Nature of Coevolution

The possibility that an evolutionary change in one species may evoke a reciprocal change in another species distinguishes selection in interspecific interactions from selection stemming from conditions in the physical environment. Reciprocal genetic change in interacting species, owing to natural selection imposed by each on the other, is **coevolution** in the narrow sense.

The term "coevolution" includes several concepts (Futuyma and Slatkin 1983; Thompson 1994). In its simplest form, two species evolve in response to each other (**specific coevolution**). For example, Darwin envisioned predatory mammals, such as wolves, and their prey, such as deer, evolving ever greater fleetness, each improvement in one causing selection for compensating improvement in the other, in an "evolutionary arms race" between prey and predator (Figure 18.2A). **Guild coevolution**, sometimes called **diffuse coevolution** (Figure 18.2B), occurs when several species are involved and their effects are

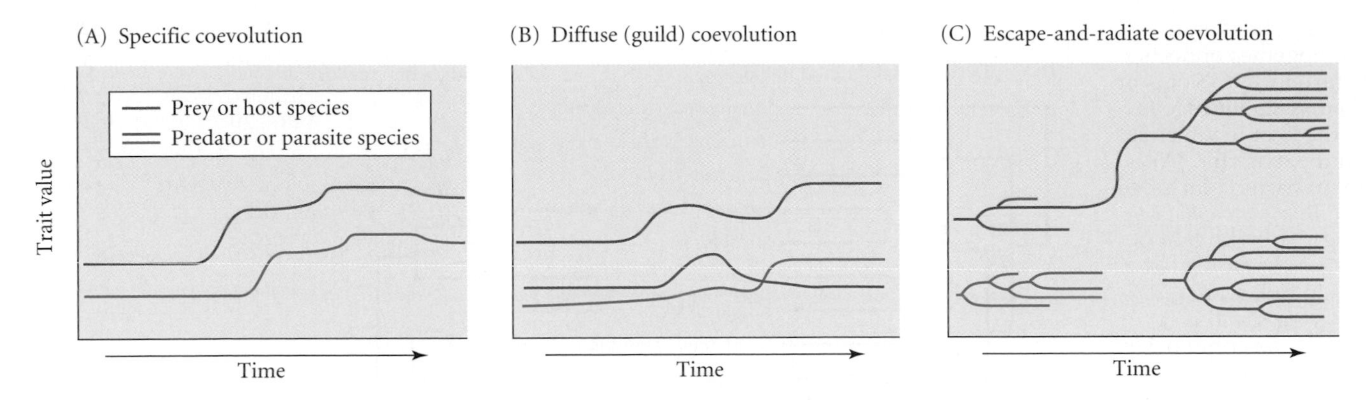

Figure 18.2 Three kinds of coevolution. In each graph, the horizontal axis represents evolutionary time and the vertical axis shows the state of a character in a species of prey or host and one or more species of predators or parasites. (A) Specific coevolution. (B) Guild, or diffuse, coevolution, in which a prey species interacts with two or more predators. (C) Escape-and-radiate coevolution. One of several prey or host species evolves a major new defense, escapes association with a predator or parasite, and diversifies. Later, a different predator or parasite adapts to the host clade and diversifies.

not independent. For example, genetic variation in the resistance of a host to two different species of parasites might be correlated (Hougen-Eitzman and Rausher 1994). In **escape-and-radiate coevolution** (Figure 18.2C), a species evolves a defense against enemies, and is thereby enabled to proliferate into a diverse clade. For example, Paul Ehrlich and Peter Raven (1964) proposed that species of plants that evolved effective chemical defenses were freed from predation by most herbivorous insects, and thus diversified, evolving into a chemically diverse array of food sources to which different insects later adapted and then diversified in turn.

Phylogenetic Aspects of Species Associations

The term "coevolution" has also been applied to a history of parallel diversification, as revealed by concordant phylogenies, of associated organisms such as hosts and their parasites or endosymbionts. Figs and their pollinators have largely concordant phylogenies, as do aphids and the endosymbiotic bacteria (*Buchnera*) that live within special cells and supply the essential amino acid tryptophan to their hosts (see Figure 14.24C). The phylogeny of these bacteria is completely concordant with that of their aphid hosts (Figure 18.3A). The simplest interpretation of this pattern is that the association between *Buchnera* and aphids dates from the origin of this insect family, that there has been little if any cross-infection between aphid lineages, and that the bacteria have diverged in concert with speciation of their hosts. However, phylogenetic correspondence is rarely this great. The phylogeny of the chewing lice that infest pocket gophers matches the host phylogeny fairly well, but there are some mismatches, probably caused by the horizontal transfer, or HOST SWITCHING, of lice from one gopher lineage to another (Figure 18.3B; Hafner et al. 2003). Discordance between phylogenies can arise from several other causes as well, such as extinction of parasite lineages (Page 2003).

If parasites disperse from one host to another through the environment (as do plant-feeding insects), they are more likely to shift between host species, and the phylogenies are rarely strongly concordant. Nevertheless, the phylogenies often provide evidence of ancient associations. For example, phylogenetically basal lineages of leaf beetles, long-horned beetles, and weevils all feed mostly on cycads or conifers—plant lineages that evolved before the angiosperms, with which the more phylogenetically "advanced" beetles are associated (Farrell 1998). These beetle lineages therefore are likely to have retained their association with cycads and conifers since the Jurassic. The fossil record also attests to the great age of some such associations; for example, late Cretaceous fossils of ginger

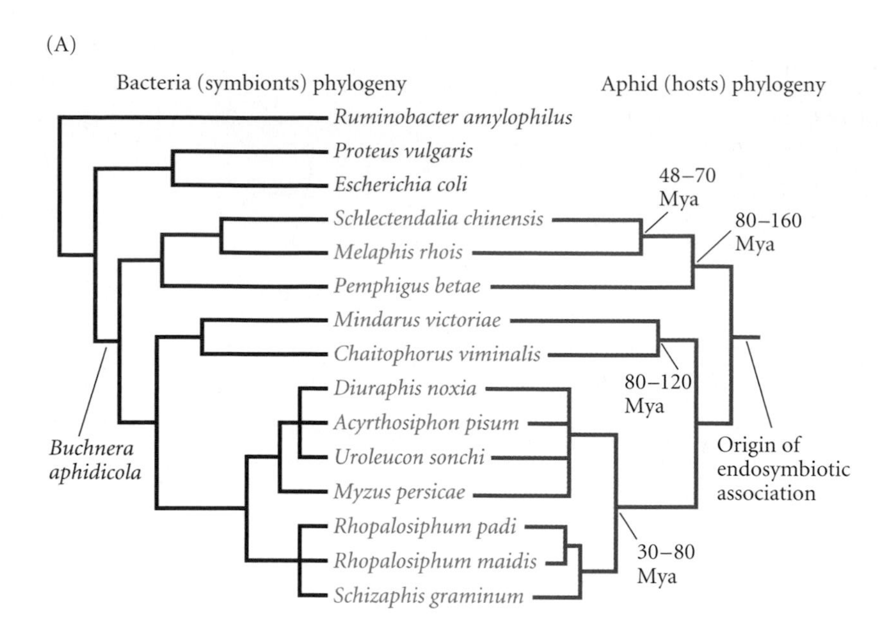

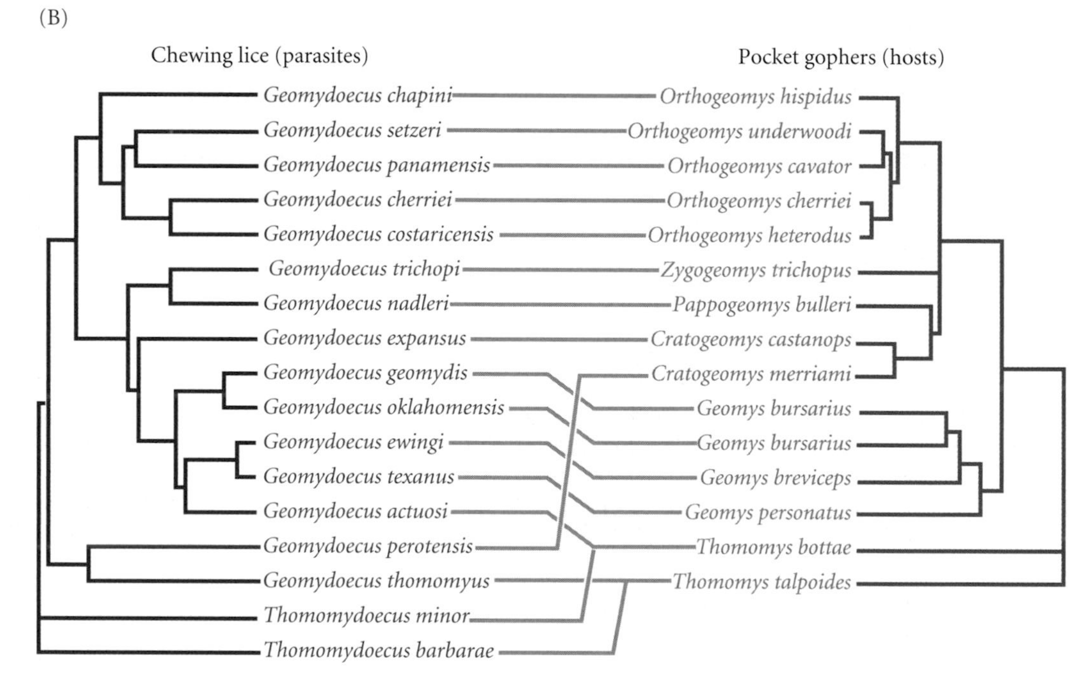

Figure 18.3 Congruent and incongruent phylogenies of hosts and host-specific endosymbionts or parasites. Each parasite lineage is specialized on the host to which it is connected in the diagram. (A) The phylogeny of bacteria included under the name *Buchnera aphidicola* is perfectly congruent with that of their aphid hosts. Several related bacteria (names in red) were included as outgroups in this analysis. Names of the aphid hosts of the *Buchnera* lineages are given in blue. The estimated ages of the aphid lineages are based on fossils and/or biogeography. (B) Phylogenies of pocket gophers and their chewing lice. Note areas of both congruence (e.g., the uppermost five gopher/louse pairs) and incongruence (e.g., the gopher *C. merriami* and its louse *G. perotensis*). (A after Moran and Baumann 1994; B after Hafner et al. 2003.)

plants show exactly the same distinctive damage that is inflicted on living gingers by certain leaf beetles (subfamily Hispinae) today (Labandeira 2002).

Coevolution of Enemies and Victims

In considering the *processes* of evolutionary change in interacting species, we will begin with interactions between enemies and victims: predators and their prey, parasites and their hosts, herbivores and their host plants. Predators and parasites have evolved some extraordinary adaptations for capturing, subduing, or infecting their victims (Figure 18.4). Defenses against predation and parasitism can be equally impressive, ranging from cryptic patterning (Figure 18.5A; see also Figure 12.5), to the highly toxic chemical defenses of both plants and animals (Figure 18.5B), to the most versatile of all defenses—the vertebrate immune system, which can generate antibodies against thousands of foreign com-

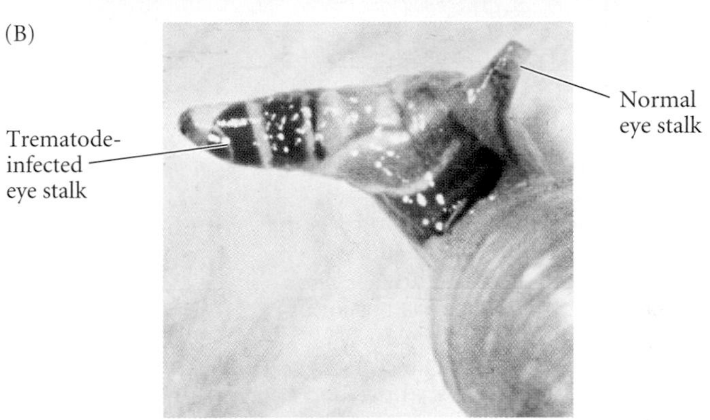

Figure 18.4 Predators and parasites have evolved many extraordinary adaptations to capture prey or infect hosts. (A) The dorsal fin spine of a deep-sea anglerfish (*Himantolophus*) is situated above the mouth and modified into a luminescent fishing lure. (B) The larva of a parasitic trematode (*Leucochloridium*) migrates to the eye stalk of its intermediate host, a land snail, and turns it a bright color to make the snail more visible to the next host in the parasite's life cycle, a snail-eating bird such as a thrush. (A, © David Shale/naturepl.com; B, photo by P. Lewis, courtesy of J. Moore.)

pounds (see Figure 19.9). Many such adaptations appear to be directed at a variety of different enemies or prey species, so although it is easy to demonstrate adaptations in a predator or a prey species, it is usually difficult to show how any one species has coevolved with another.

Theoretically, the coevolution of predator and prey might take any of several courses (Abrams 2000): it might continue indefinitely in an unending escalation of an evolutionary arms race (Dawkins and Krebs 1979); it might result in a stable genetic equilibrium; it might cause continual cycles (or irregular fluctuations) in the genetic composition of both species; or it might even lead to the extinction of one or both species.

An unending arms race is unlikely because adaptations that increase the offensive capacity of the predator or the defensive capacity of the prey entail allocations of energy and other costs that at some point outweigh their benefits. Consequently, a stable equilibrium may occur when costs equal benefits. For example, the toxic SECONDARY COMPOUNDS that plants use as defenses against herbivores, such as the tannins of oaks and the terpenes of pines, can account for more than 10 percent of a plant's energy budget. Such high levels of chemical defense are especially typical of slowly growing plant species, suggesting that they impose economic costs (Coley et al. 1985). Genetic lines of wild parsnip (*Pastinaca sativa*) containing high levels of toxic furanocoumarins suffered less attack from webworms, and matured more seeds, than lines with lower levels when grown outdoors; in the greenhouse, however, where they were free from insect attack, the lines with higher levels of furanocoumarins had lower seed production (Berenbaum and Zangerl 1988). Costs of this kind may explain why plants are not more strongly defended than they are, and thus why they are still subject to insect attack.

Another kind of cost arises if a defense against one enemy makes the prey more vulnerable to another. For example, terpenoid compounds called cucurbitacins enhance the resistance of cucumber plants (*Cucumis sativus*) to spider mites, but they attract certain cucumber-feeding leaf beetles (Dacosta and Jones 1971).

Figure 18.5 Examples of defenses against predation. (A) The cryptically colored leaf-tailed gecko (*Uroplatus phantasticus*) blends with the floor of its dry forest home in Madagascar. (B) The toxins in the brilliant blue skin of *Dendrobates azureus* have been put to human use, as its common name of "poison dart frog" implies. Its color warns potential predators away. (A © Nick Garbutt/naturepl.com; B © Barry Mansell/naturepl.com)

(A) Resistance locus (host)

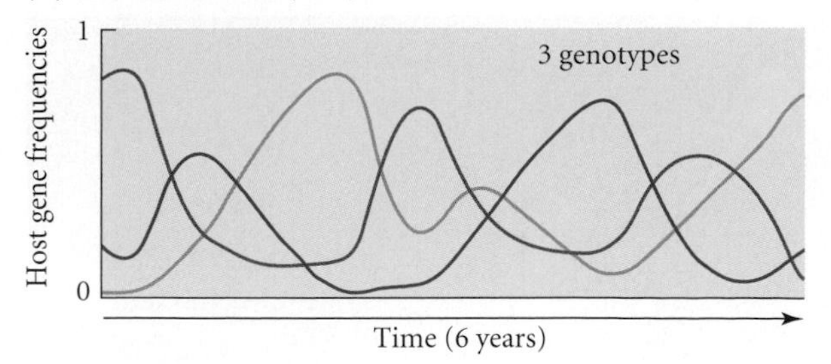

(B) Infectivity locus (parasite)

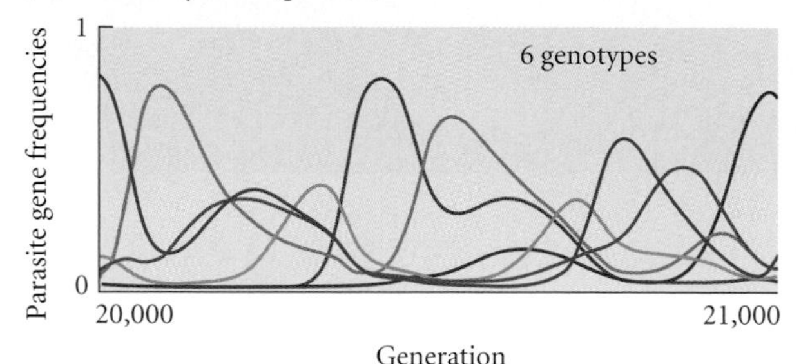

Figure 18.6 A computer simulation of genetic changes at (A) a resistance locus in a host and (B) an infectivity locus in a parasite. The host is diploid and has three resistance alleles; the parasite is haploid and has six infectivity alleles. Each parasite genotype can overcome the defenses of one of the six host genotypes (e.g., parasite P_1 can attack host H_1H_1). Both populations remain polymorphic and fluctuate irregularly in genetic composition. (After Seger 1992.)

Models of enemy-victim coevolution

GENE-FOR-GENE MODELS. Coevolution of enemies and victims has been modeled in several ways, appropriate to different kinds of characters. For example, models of evolution at one or a few loci are appropriate for **gene-for-gene interactions**, which were first described in cultivated flax (*Linum usitatissimum*) and flax rust (*Melampsora lini*), a basidiomycete fungus. Similar systems have been described or inferred in several dozen other pairs of plants and fungi, as well as in cultivated wheat (*Triticum*) and one of its major pests, the Hessian fly (*Mayetiola destructor*). In each such system, the host has several loci at which a dominant allele (R) confers resistance to the parasite. At each of several corresponding loci in the parasite, a recessive allele (v) confers infectivity—the ability to infect and grow in a host with a particular R allele (Table 18.1). If resistance has a cost, any particular resistance allele (R_i) will decline in frequency when the parasite's corresponding infectivity allele (v_i) has high frequency, because R_i is then ineffective. As a different R allele (R_j) increases in frequency in the host population, the corresponding infectivity allele v_j increases in the parasite population. According to computer simulations, such frequency-dependent selection can cause cycles or irregular fluctuations in allele frequencies (Figure 18.6). In wild populations of Australian flax, the frequencies of different rust genotypes fluctuated from year to year (Figure 18.7). On the whole, highly infective genotypes—those that could attack the greatest number of flax genotypes—occurred in highly resistant flax populations, and less infective rusts were found in less resistant flax populations (Thrall and Burdon 2003).

QUANTITATIVE TRAITS. Coevolutionary models of a defensive polygenic character (y) in a prey species and a corresponding polygenic character (x) in a predator are mathematically complex and include many variables that can affect the outcome (Abrams 2000). An important distinction is whether the capture rate of the prey by the predator increases as the difference ($x - y$) increases (e.g., when the predator's speed is greater than the prey's) or decreases (e.g., if it depends on a close match between the size of the prey and the size of the predator's mouth). In the former case, mathematical analyses suggest that both

TABLE 18.1 ***Gene-for-gene interactions between a parasite and its host***

Parasite genotype	Host genotype			
	R_1— R_2—	R_1— r_2r_2	r_1r_1 R_2—	r_1r_1 r_2r_2
V_1—V_2—	–	–	–	+
V_1—v_2v_2	–	–	+	+
v_1v_1 V_2—	–	+	–	+
v_1v_1 v_2v_2	+	+	+	+

Source: After Frank 1992.

Note: In each species, two loci, with dominant and recessive alleles at each locus, control resistance (of the host) and infectivity (of the parasite). A + sign indicates that the parasite genotype can grow on a host of a given genotype (i.e., the parasite is infective and the host is susceptible); the – signs indicate that the host genotype is resistant to the parasite genotype.

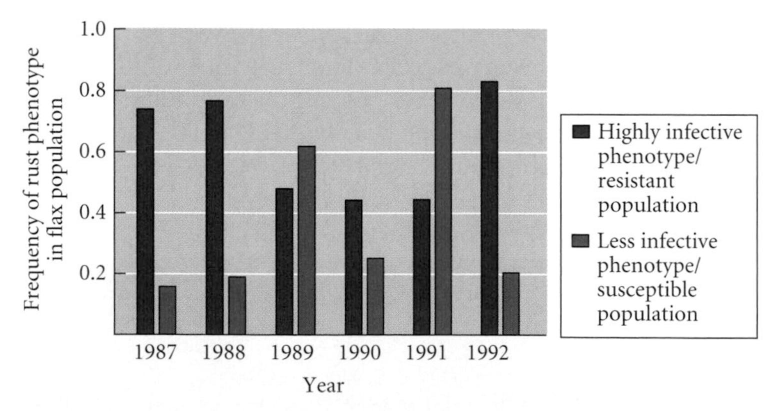

Figure 18.7 Changes in the frequencies of two phenotypes of flax rust over the course of 6 years in Australian populations of wild flax. A rust phenotype capable of infecting most resistant phenotypes of flax had high frequencies in a population of plants that were resistant to most other rust phenotypes. In a nearby flax population, in which 80 percent of plants were susceptible to the highly infective rusts, a less infective rust had fairly or very high frequencies. (After Thrall and Burdon 2003.)

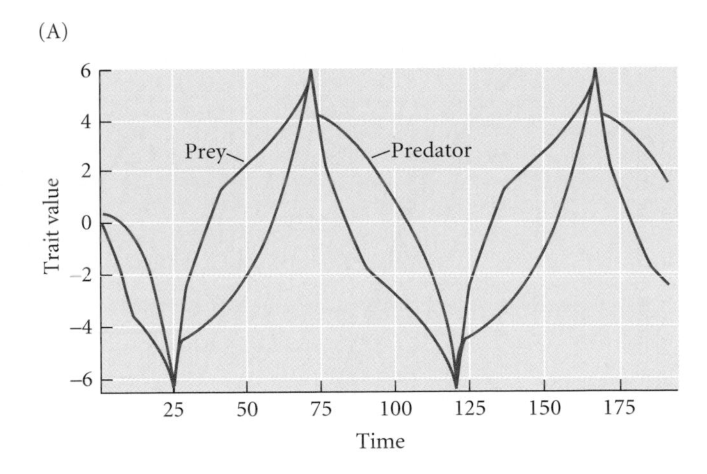

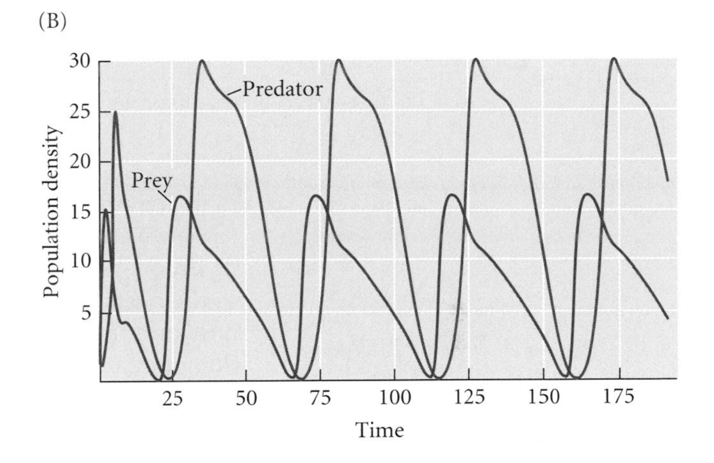

Figure 18.8 Computer simulation of coevolution between prey and predator in which the optimal predator phenotype (e.g., mouth size) matches a prey phenotype (e.g., size). (A) Evolution of character state means. As a character state diverges from a reference value, its fitness cost prevents it from evolving indefinitely in either direction. The evolution of the predator's character state lags behind the prey's. (B) Changes in character state means may be paralleled by cycles in population density, arising partly from changes in the match between the predator's character and the prey's. (After Abrams and Matsuda 1997.)

species will often evolve in the same direction (e.g., toward greater speed), arriving at an equilibrium point that is determined by physiological limits or excessive investment costs. However, suppose the capture rate depends on a close match between x and y, that deviation too greatly in either direction increases the cost of x (or y), and that $\bar{x} = \bar{y}$. Then either increasing or decreasing y will improve prey survival. In this case, y will evolve in one or the other direction, and x will evolve to track y. Eventually y may evolve in the opposite direction as its cost becomes too great, and x will evolve likewise. Continuing cycles of change in the characteristics of both species might result, and these genetic changes may contribute to cycles in population density (Figure 18.8).

Examples of predator-prey coevolution

It has not yet been possible to obtain data on long-term coevolution in natural populations, but there is plentiful indirect evidence that enemies and victims affect each others' evolution. For example, during the Mesozoic, new, highly effective predators of molluscs, such as shell-crushing fishes and crustaceans that could either crush or rip shells, evolved. The diversity of shell form in bivalves and gastropods then increased as various lineages evolved thicker shells, thicker margins of the shell aperture, or spines and other excrescences that could foil at least some of these predators (Figure 18.9; Vermeij 1987).

The rough-skinned newt (*Taricha granulosa*) of northwestern North America has one of the most potent known defenses against predation: the neurotoxin tetrodotoxin (TTX). Most populations have high levels of TTX in the skin (one newt has enough to kill 25,000 laboratory mice), but a few populations, such as the one on Vancouver Island, have almost none (Brodie and Brodie 1999; Brodie et al. 2002). Populations of the garter snake *Thamnophis sirtalis* from outside the range of this newt have almost no resistance to TTX. But populations that are sympatric with toxic newts feed on them, and can be as much as a hundred

Figure 18.9 Some features of living molluscs that, like those that evolved in the Mesozoic, provide protection against predators. Spines on the shells of bivalves of the genus *Arcinella* (A) and gastropods of the genus *Murex* (B) prevent some fishes from swallowing the animal and may reduce the effectiveness of crushing predators. (C) The narrow aperture of *Cypraea mauritiana* prevents predators from reaching the gastropod's body. (Photos by D. McIntyre.)

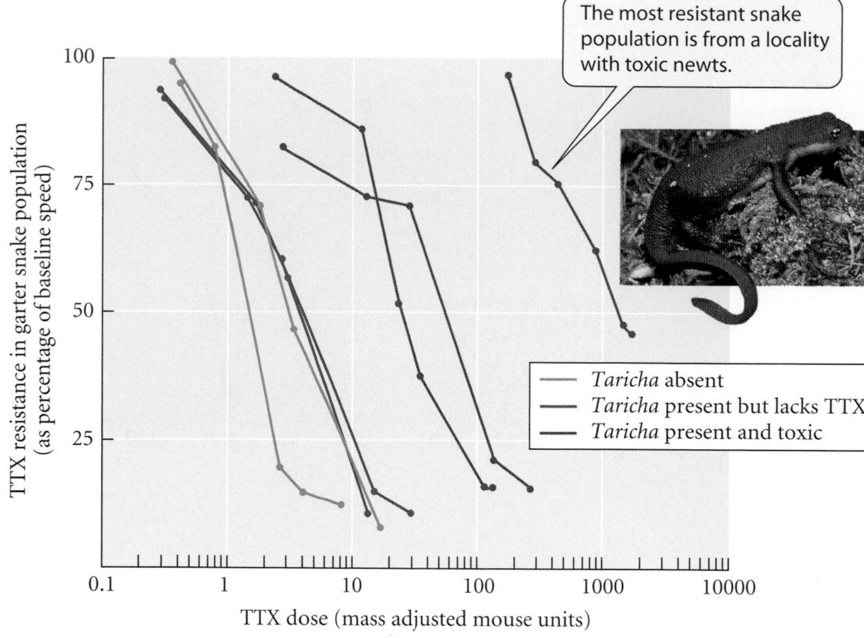

Figure 18.10 Variation in TTX resistance, measured by crawling speed after injection in relation to dose, in garter snakes from several localities. The least resistant population is from Maine, where the toxic rough-skinned newt (*Taricha granulosa*) does not occur. Two of the other nonresistant populations coexist with newt populations that lack TTX. The three most resistant populations are sympatric with toxic newt populations. (After Brodie and Brodie 1999; photo © Henk Wallays.)

times more resistant to TTX than allopatric populations (Figure 18.10). Similarly, many species of insects feed on plants that bear diverse chemical toxins and have evolved a variety of mechanisms of resistance to those toxins. Plants in the carrot family, for example, contain toxic furanocoumarins, but the high activity of a detoxifying enzyme enables larvae of the black swallowtail butterfly (*Papilio polyxenes*) to feed on such plants with impunity (Berenbaum 1983).

Brood-parasitic birds, such as cowbirds and some species of cuckoos, lay eggs only in the nests of certain other bird species. Cuckoo nestlings eject the host's eggs from the nest, and the host ends up rearing only the parasite (Figure 18.11A). Adults of host species do not treat parasite nestlings any differently from their own young, but some host species do recognize parasite eggs, and either eject them or desert their nest and start a new nest and clutch.

The most striking counteradaptation among brood parasites is egg mimicry (Rothstein and Robinson 1998). Each population of the European cuckoo (*Cuculus canorus*) contains several different genotypes that prefer different hosts and lay eggs closely resembling those of their preferred hosts (Figure 18.11B). Some other individuals lay nonmimetic eggs. Some host species accept cuckoo eggs, some frequently eject them, and others desert parasitized nests. By tracing the fate of artificial cuckoo eggs placed in the nests of various

(A)

(B)

Figure 18.11 (A) A fledgling European cuckoo (*Cuculus canorus*) being fed by its foster parent, a reed warbler (*Acrocephalus scirpaceus*). (B) Mimetic egg polymorphism in the cuckoo. The left column shows eggs of six species parasitized by the cuckoo (from top: robin, pied wagtail, dunnock, reed warbler, meadow pipit, great reed warbler). The second column shows a cuckoo egg laid in the corresponding host's nest. The match is quite close except for cuckoo eggs laid in dunnock nests. The right column shows artificial eggs used by researchers to test rejection responses. (A © David Kjaer/naturepl.com; B, photo by M. Brooke, courtesy of N. B. Davies.)

bird species, Nicholas Davies and Michael Brooke (1998) found that species that are not parasitized by cuckoos (due to unsuitable nest sites or feeding habits) tend not to eject cuckoo eggs, whereas among the cuckoos' preferred hosts, those species whose eggs are mimicked by cuckoos rejected artificial eggs more often than those whose eggs are not mimicked. These species have evidently adapted to brood parasitism. Moreover, populations of two host species in Iceland, where cuckoos are absent, accepted artificial cuckoo eggs, whereas in Britain, where those species are favored hosts, they rejected such eggs. Surprisingly, among suitable host species, those that are rarely parasitized by cuckoos did not differ in discriminatory behavior from those commonly parasitized. Davies and colleagues suspect that the rarely parasitized species were more commonly parasitized in the past, but that their ability to reject cuckoo eggs has selected against the cuckoo genotypes that parasitized these species.

Infectious disease and the evolution of parasite virulence

The two greatest challenges that a parasite faces are moving itself or its progeny from one host to another (transmission) and overcoming the host's defenses. Some parasites are transmitted **vertically**, from a host parent to her offspring, as in the case of *Wolbachia* bacteria, which are transmitted in insects' eggs (see Chapter 15). Other parasites are transmitted **horizontally** among hosts in a population via the external environment (e.g., human rhinoviruses, the cause of the common cold, are discharged by sneezing), via contact between hosts (e.g., the causes of venereal diseases, such as the gonorrhea bacterium), or via carriers (VECTORS, such as the mosquitoes that transmit the malaria-causing protist and the yellow fever virus).

The effects of parasites on their hosts vary greatly. Those that reduce the survival or reproduction of their hosts are considered **virulent**. We are concerned here with understanding the evolutionary factors that affect the degree of virulence. This topic has immense medical implications because the evolution of virulence can be rapid in "microparasites" such as viruses and bacteria (Ewald 1994; Bull 1994). The level of virulence depends on the evolution of both host and parasite. For example (Fenner and Ratcliffe 1965), after the European rabbit (*Oryctolagus cuniculus*) became a severe rangeland pest in Australia, the myxoma virus, from a South American rabbit, was introduced to control it. Periodically after the introduction, wild rabbits were tested for resistance to a standard strain of the virus (Figure 18.12A), and virus samples from wild rabbits were tested for virulence in a standard laboratory strain of rabbits (Figure 18.12B). Over time, the rabbits evolved greater resistance to the virus, and the virus evolved a lower level of virulence. Although some almost avirulent strains were detected, the virus population as a whole did not become avirulent.

THEORY OF THE EVOLUTION OF VIRULENCE. Many people imagine that parasites generally evolve to be benign (avirulent) because the parasite's survival depends on that of the host population. However, a parasite may evolve to be more benign or more virulent depending on many factors (May and Anderson 1983; Bull 1994; Frank 1996).

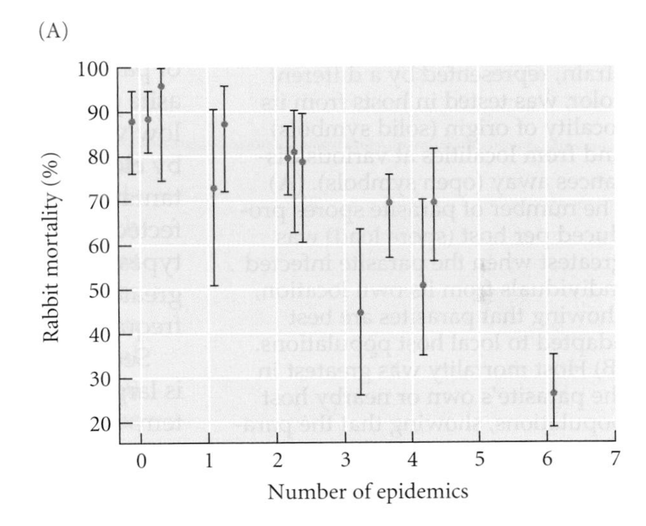

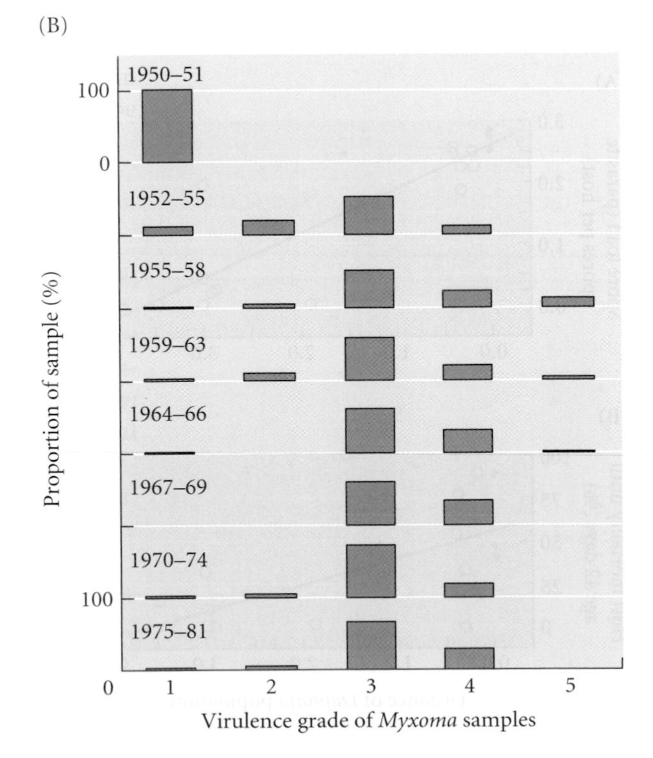

Figure 18.12 Coevolution in rabbits and myxoma virus after the virus was introduced into the rabbit population in Australia. (A) Mortality in field-collected rabbits exposed to a standard virus strain declined as the wild population experienced more epidemics. (B) Virus samples from the wild, tested on a standard rabbit stock, were graded from low (1) to high (5) virulence. Average virulence decreased over time, but stabilized at an intermediate level. (A after Fenner and Ratcliffe 1965; B after May and Anderson 1983.)

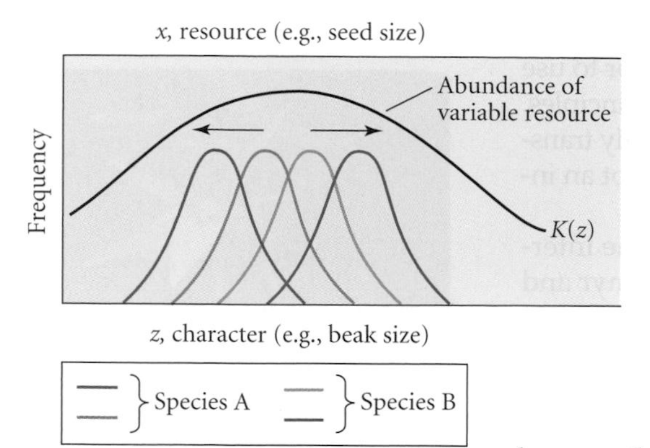

Figure 18.18 A model of evolutionary divergence in response to competition. The x-axis represents a quantitative phenotypic character (z), such as bill size, that is closely correlated with some quality of a resource, such as the average size of the food items eaten by that phenotype. The curve $K(z)$ represents the frequency distribution of food items that vary in size. Two variable species (orange and green) initially overlap greatly in z, and therefore in the food items they depend on. Those phenotypes in each species that overlap with the fewest members of the other species experience less competition, and so may have higher fitness. Divergent selection on the two species is expected to shift their character distributions (red, dark green) so that they overlap less.

because they are less abundant, and they will experience less interspecific competition because they tend not to use the same resources as the other species. Therefore, the most extreme genotypes will have higher fitness. Such density-dependent diversifying selection can result in the two species' evolving less overlap in their use of resources and in a shift of their phenotype distributions away from each other (Slatkin 1980; Taper and Case 1992). Divergence in response to competition between species is often called ecological character displacement (see below).

Because recombination among loci restricts the variance in a polygenic character that determines resource use (see Chapter 13), a broad spectrum of resources may not be fully utilized by just one or two species. In that case, one or more additional species, differing from the first two, may be able to invade the community. Both the invaders and the previous residents may then evolve further shifts in resource utilization that minimize competition. The species may also diverge in other respects that reduce competition, such as habitat use. For example, although some of the bumblebee species described in Figure 18.17 differ in proboscis length and thus in the flowers they use, others are similar in this respect, but those species occupy different habitats (i.e., altitudinal zones).

Brown and Wilson (1956) coined the term **character displacement** to describe a pattern of geographic variation wherein sympatric populations of two species differ more greatly in a characteristic than allopatric populations. One possible reason for such a pattern is that the characteristic is associated with the use of food or another resource, and that the species have evolved differences in resource use where they would otherwise compete with each other. (Hence, "character displacement" is often used to mean the process of divergence due to competition.) The kind of geographic pattern that Brown and Wilson described has provided some of the best evidence for evolutionary divergence in response to competition (Taper and Case 1992; Schluter 2000). For example, the Galápagos ground finches *Geospiza fortis* and *G. fuliginosa* differ more in bill size where they coexist than where they occur singly (see Figure 9.27). Differences in bill size are correlated with the efficiency with which the birds process seeds that differ in size and hardness, and the population size of these finch species is often food-limited, resulting in competition (Grant 1986). A rather similar example is the case of sticklebacks in the *Gasterosteus aculeatus* complex. In northwestern North America, several lakes each have two reproductively isolated forms, one benthic and one limnetic (see Figure 16.8), which differ in body shape, mouth morphology, and the number and length of the gill rakers. Other lakes have only a single form of stickleback, with intermediate morphology (Schluter and McPhail 1992).

Ecological release is another geographic pattern, wherein a species or population exhibits greater variation in resource use and in associated phenotypic characters if it occurs alone than if it coexists with competing species. Ecological release is most often characteristic of island populations. For example, the sole finch species on Cocos Island (in the Pacific Ocean northeast of the Galápagos Islands) has a much broader diet, and forages in more different ways, than do any of its relatives in the Galápagos Islands, where there are many more species (Werner and Sherry 1987). Similarly, the only species of woodpecker on the Caribbean island of Hispaniola exhibits greater sexual dimorphism in the length of the bill and tongue than do related continental species that coexist with other woodpeckers, and the sexes differ in where and how they forage (Figure 18.19; Selander 1966).

(A) *Melanerpes striatus*

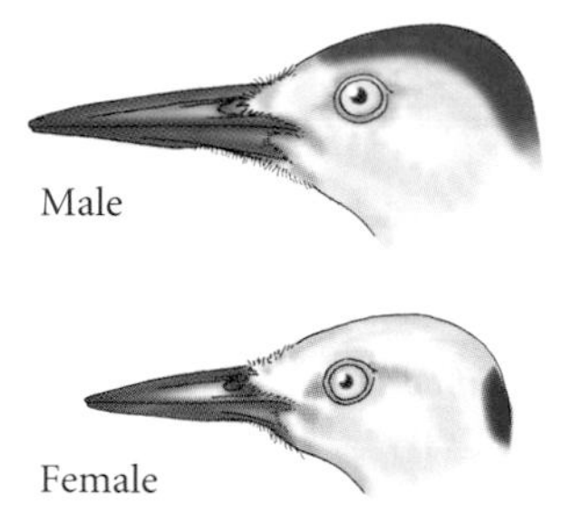

(B) *Melanerpes aurifrons*

Figure 18.19 Ecological release. The difference in bill size between the sexes is greater in *Melanerpes striatus* (A), the only species of woodpecker on the island of Hispaniola, than in continental species such as *M. aurifrons* (B), which is sympatric with other species of woodpeckers. Bill size is correlated with differences in feeding behavior, so greater sexual dimorphism results in broader resource use in *M. striatus*. (After Selander 1966.)

Community patterns

Of all interspecies interactions, competition has been most emphasized in ecologists' attempts to detect and explain repeatable patterns in ecological communities. Competition is often supposed to limit species diversity within contemporary assemblages, and it may have affected species diversity over long periods of evolutionary time (see Chapter 7).

To some extent, ecological interactions may guide the evolution of interacting species along predictable paths, resulting in convergent patterns. This doesn't always happen, however; for example, blood-drinking bats are restricted to tropical America, even though the abundant hoofed mammals in Africa would provide plenty of food for such species. Similarly, the species diversity of lizards that live in deserts in North and South America is lower on both of those continents than in Australia, where lizards are more diverse in deserts and also occupy wetlands, a habitat that very few American lizards use (Schluter and Ricklefs 1993).

Nevertheless, some surprisingly consistent patterns have resulted from convergent evolution. In Chapter 6, for example, we described the remarkable parallel evolution of *Anolis* lizards on different Caribbean islands, each of which has morphologically and ecologically corresponding species that typically seek food in different microhabitats (see Figure 6.21). Cuba, Hispaniola, Jamaica, and Puerto Rico all have clades with four ecomorphs that are adapted for foraging in four different forest microhabitats: on the crowns of trees, in the trunk-crown region, on twigs, and in the trunk-ground area (Figure 18.20). The most reasonable interpretation of this pattern is that as new species have arisen on each island, they have evolved in similar ways to avoid competition by adapting to the same kinds of previously unused microhabitats.

In a similar vein, species of forest-dwelling bird-eating hawks (*Accipiter*) that differ in body size differ correspondingly in the size of the prey species they usually take. Pairs of sympatric species of *Accipiter* consistently differ more in body size than if pairs of species were taken at random from the 47 species in the world (Figure 18.21; Schoener 1984). Such examples suggest that principles of ecological organization may confer some predictability on the course of evolutionary diversification.

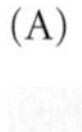

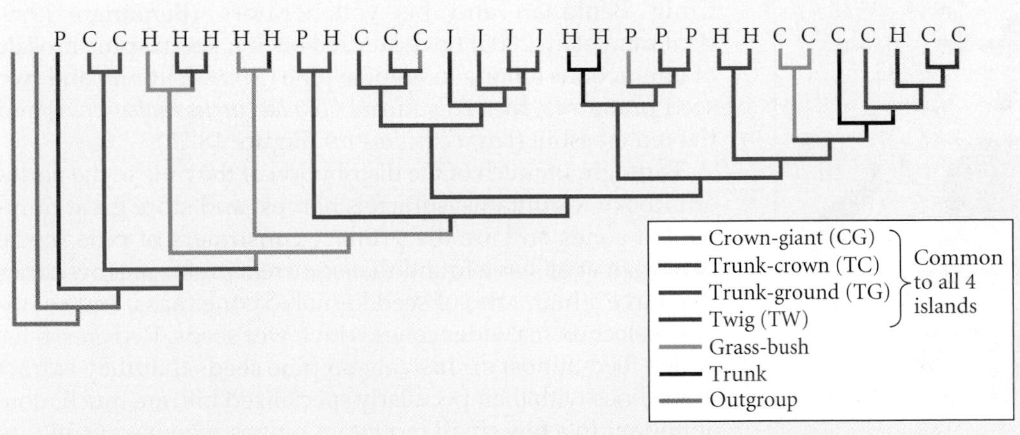

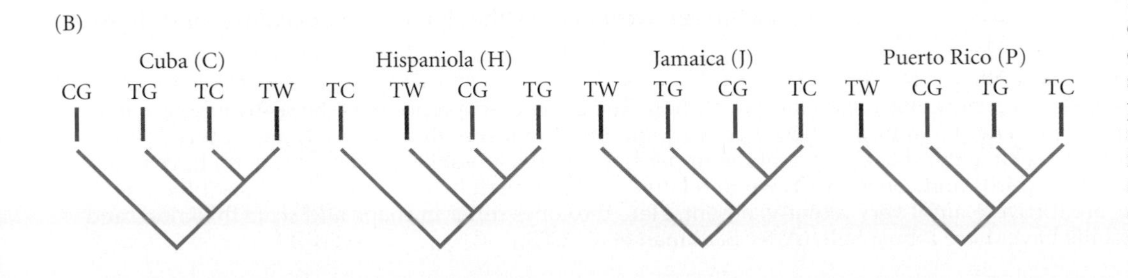

Figure 18.20 (A) A molecular phylogeny of *Anolis* species in the Greater Antilles indicates frequent transitions among the ecomorph classes. The letters at the top indicate the island on which each species occurs. (C, Cuba; H, Hispaniola; J, Jamaica; P, Puerto Rico). (B) A phylogenetic tree, for each island, of the four ecomorphs that are common to all the islands, extracted from the full phylogeny. (After Losos et al. 1998.)

Figure 18.21 Throughout the world, coexisting species of bird-eating hawks (*Accipiter*) differ more in body size than would be expected if species were selected at random. Hawks of different sizes feed on correspondingly different species of prey. These data imply either that coexisting species evolve differences in prey use to reduce competition or that only species that differ in prey use can coexist. (After Schoener 1984.)

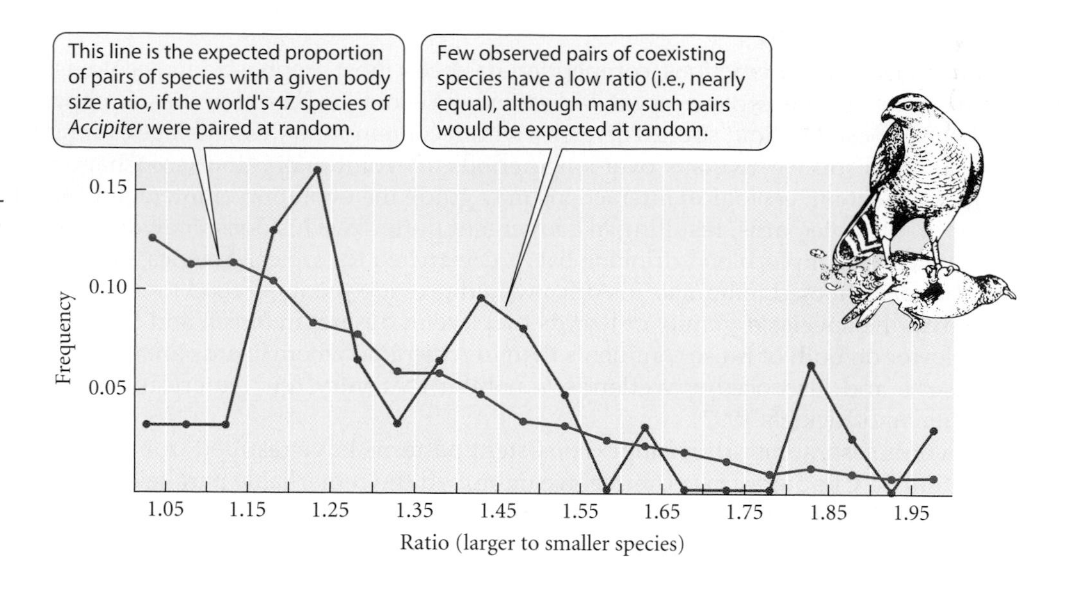
This line is the expected proportion of pairs of species with a given body size ratio, if the world's 47 species of *Accipiter* were paired at random.
Few observed pairs of coexisting species have a low ratio (i.e., nearly equal), although many such pairs would be expected at random.
Frequency
0.15
0.10
0.05
1.05
1.15
1.25
1.35
1.45
1.55
1.65
1.75
1.85
1.95
Ratio (larger to smaller species)

Multispecies interactions

Each species in an ecological community interacts with several or many other species. Consequently, its evolutionary response to interaction with one species may be influenced by the effect of another, in any of many possible ways. Two examples will illustrate this point.

A THREE-SPECIES SELECTION MOSAIC. John Thompson (1999) has observed that the selection imposed on a species by its interactions with other species may vary from one geographic population to another, resulting in a **geographic mosaic** of coevolution. Selection may be stronger in some places than in others, or even favor different characteristics, and gene flow among such populations may result in locally inadequate adaptation. Craig Benkman and his collaborators (Benkman 1999; Benkman et al. 2003) have studied such a geographic mosaic of interactions among lodgepole pine (*Pinus contorta*) and two seed predators, the red squirrel (*Tamiasciurus hudsonicus*) and the red crossbill (*Loxia curvirostra*) (Figure 18.22).

Throughout much of the distribution of the pine in the northern Rocky Mountains, squirrels harvest and store great numbers of cones and are the primary consumers of pine seeds. Benkman et al. have found that squirrels prefer narrow cones that have a high ratio of seed kernel to cone mass, and so impose selection for wider cones with fewer seeds. Red crossbills, which feed almost exclusively on pine seeds that they extract from cones with their peculiarly specialized bill, are much more abundant in a few small mountain ranges where squirrels are absent than where squirrels occur. Crossbills feed less effectively on larger, wider cones that have thicker scales—and these are

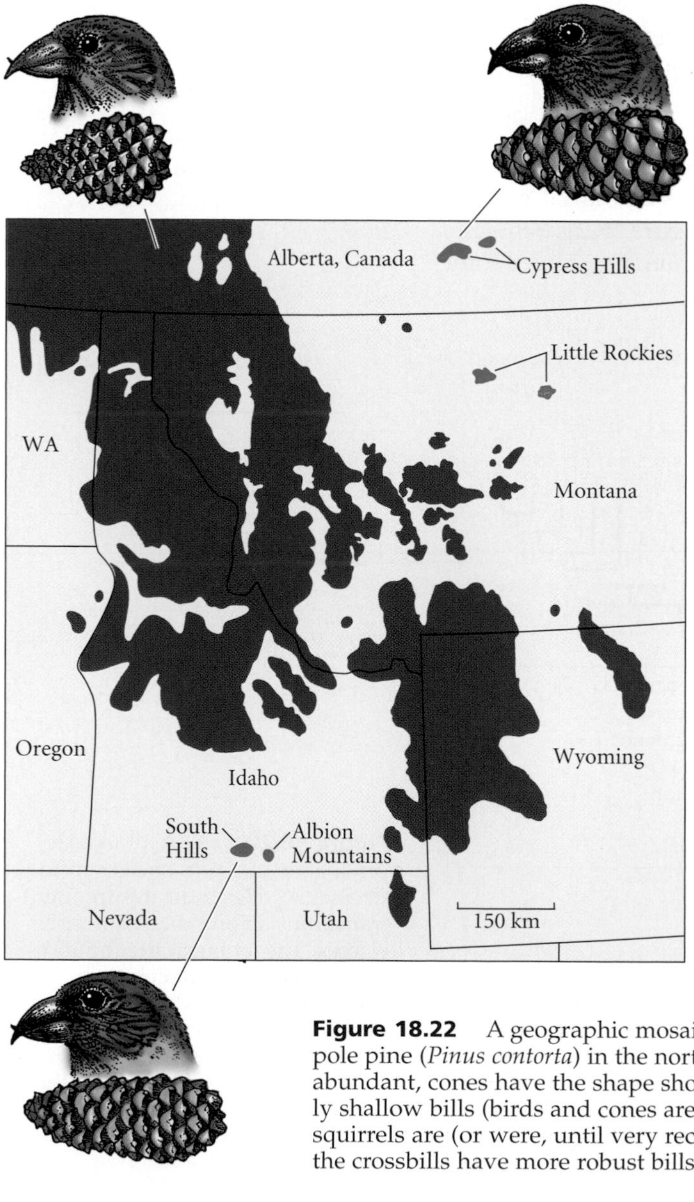

Figure 18.22 A geographic mosaic of coevolution. The colored area represents the distribution of lodgepole pine (*Pinus contorta*) in the northern Rocky Mountains. In most of this area (red), red squirrels are abundant, cones have the shape shown at the upper left, and red crossbills (*Loxia curvirostra*) have relatively shallow bills (birds and cones are drawn to relative scale). In peripheral mountain ranges (blue), red squirrels are (or were, until very recently) absent. Here the cones differ in shape and scale thickness, and the crossbills have more robust bills. (After Benkman et al. 2003.)

Figure 18.23 Batesian mimicry. The palatable red-spotted purple butterfly (*Limentis arthemis*; top) resembles the pipevine swallowtail (*Battus philenor*; below) which stores distasteful, poisonous chemicals that it obtains from the plant it eats when it is a larva. Predators that learn, from unpleasant experience, to avoid the model also will tend to avoid attacking the mimic. (*Limentis* © Michael Gadomski/Photo Researchers, Inc.; *Battus* © S. McKeever/Photo Researchers, Inc.)

precisely the characteristics that have evolved in pine populations that suffer seed predation only from crossbills. Correspondingly, crossbill populations in these locations have evolved longer, deeper bills than in regions where red squirrels occur. These bill characteristics have been found, in tests of caged crossbills, to enhance feeding efficiency on large, thick-scaled cones. Thus coevolution between pines and crossbills is apparent where these species interact strongly, but where squirrels are the dominant seed predator, they drive the evolution of cone characteristics, and crossbills adapt accordingly.

MIMICRY RINGS. Defensive mimicry, in which one or more species gain protection against predators from their resemblance to one another, provides model systems for studying many evolutionary phenomena (Mallet and Joron 1999; Joron and Mallet 1998; Turner 1977). Traditionally, two forms of defensive mimicry have been recognized (see Chapter 3). In Batesian mimicry, a palatable species (a mimic) resembles an unpalatable species (a model; Figure 18.23). In Müllerian mimicry, two or more unpalatable species are co-mimics (or co-models). In both cases, predators learn, from unpleasant experience, to avoid potential prey that look like the unpalatable species. (Such learning has been experimentally documented, especially with birds preying on butterflies and other insects.) Often, although not always, the models and mimics display conspicuous aposematic (warning) patterns.

Groups of species that benefit from defensive mimicry are known as MIMICRY RINGS. In many cases, mimicry rings include both strongly unpalatable and mildly unpalatable species; the latter may be "quasi-Batesian" mimics of the more unpalatable species. In many cases, several mimicry rings are found in the same region, each consisting of multiple species of similar Müllerian mimics and often including some palatable Batesian mimics (and/or mildly unpalatable quasi-Batesian mimics) as well (Figure 18.24). Espe-

Figure 18.24 A mimicry ring. *Heliconius melpomene* and *H. erato* have a very different color pattern in the Mayo and upper Huallaga rivers, in eastern Peru, than in the lower Huallaga drainage, where they join a mimicry ring with a "rayed" pattern. This ring of unpalatable species includes four other species of *Heliconius*, three other genera of butterflies (the top three species in the center column), and a moth (center column, bottom). (Courtesy of J. Mallet.)

(A) Batesian pair: Unpalatable model, palatable mimic

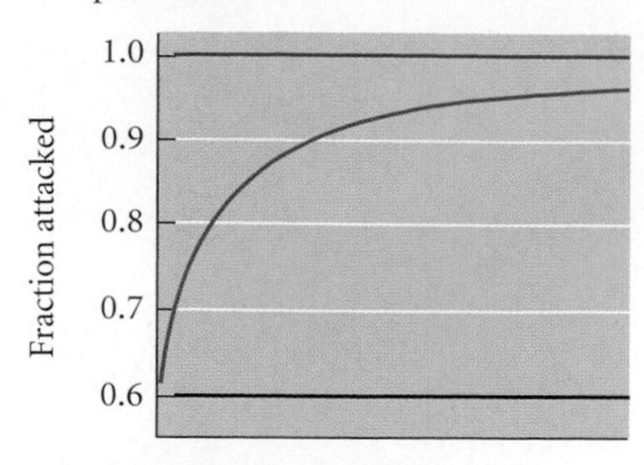

(B) Müllerian pair: Mimic and model close in palatability

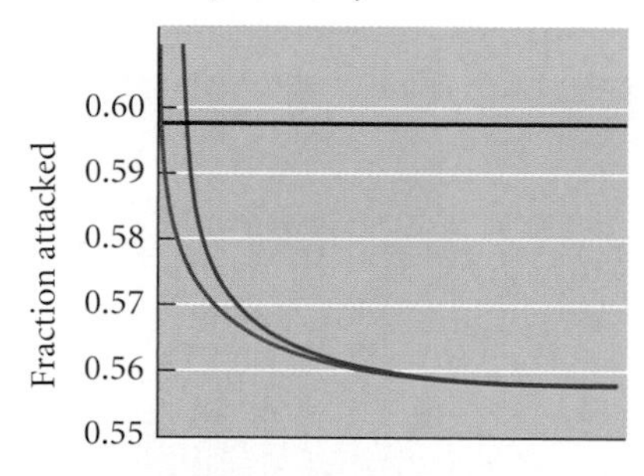

(C) Quasi-Batesian pair: Unpalatable model, mildly unpalatable mimic

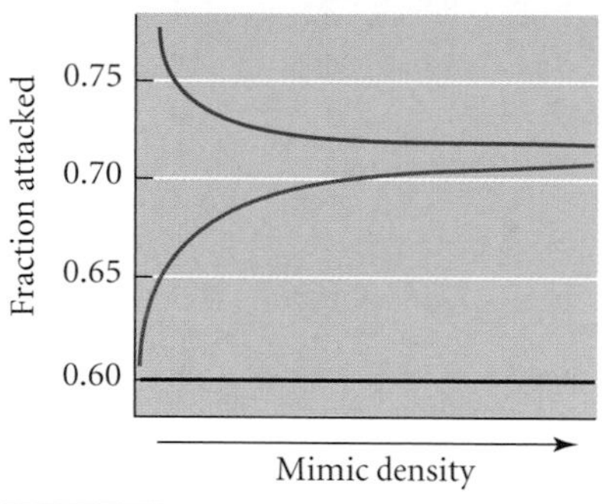

— Mimic alone
— Model and mimic
— Model alone (reference constant)

Figure 18.25 Computer simulations of the probability of predator attack on model-mimic pairs of prey species. In each graph, a constant probability of attack on the unpalatable model, when it is alone, is assumed, and the probability of attack is scaled relative to that level. (A) A palatable mimic experiences high predation when its model is absent ("mimic alone"), much lower predation when it is rare, and increasingly higher predation as its density increases relative to that of the model. (B) A Müllerian species pair in which one species ("mimic") is only slightly less unpalatable than the other ("model"). As the mimic's density increases, predators' aversion is reinforced at a higher rate, so predation on both species declines. (C) A mildly unpalatable "quasi-Batesian" mimic, if alone, suffers somewhat less predation as its density increases because predators learn to avoid it—but then soon forget and attack again, because they have had only a mild aversive experience. High densities of the mimic will result in higher predation on the model because predators are only seldom strongly deterred. Hence the aposematic color pattern tends to lose its advantage. (After Mallet and Joron 1999.)

cially among swallowtail butterflies, Batesian mimics are sometimes polymorphic within populations, with each morph resembling a different unpalatable model (e.g., the African swallowtail *Papilio dardanus*; see Figure 9.2A). Müllerian mimics are almost never polymorphic within populations, but different geographic races of certain species may have different aposematic color patterns and may belong to different mimicry rings (e.g., the geographic races of *Heliconius erato* and *H. melpomene*; see Figures 12.19 and 18.24).

Selection on a mimetic phenotype can depend on both its density, relative to that of a model species, and the degree of unpalatability of the model. A predator is more likely to avoid eating a butterfly that looks like an unpalatable model if it has had a recent reinforcing experience (e.g., swallowing a butterfly with that pattern, and then vomiting). If, however, it has recently swallowed a tasty butterfly with that phenotype, it will be more, not less, inclined to eat the next butterfly with that phenotype. Thus the rarer a palatable Batesian mimic is, relative to an unpalatable model, the more likely predators are to associate its color pattern with unpalatability, and so the greater the advantage of resembling the model will be (Figure 18.25A). (The degree of unpalatability of the model also affects the outcome because the more unpleasant the predator's experience has been, the longer its aversion to that color pattern is likely to last.) Mimetic polymorphism in Batesian mimics such as *Papilio dardanus* can therefore evolve by frequency-dependent selection: a rare new phenotype that mimics a different model species will have higher fitness that a common mimetic phenotype, simply because it is less common and predators will not have had an opportunity to learn that butterflies with that phenotype are palatable rather than unpalatable.

Since Müllerian mimics jointly reinforce aversion learning by predators, there is likely to be strong stabilizing selection for a common color pattern in all sympatric unpalatable species (Figure 18.25B). However, many apparent Müllerian mimics, such as *Heliconius melpomene*, may actually be mildly unpalatable quasi-Batesian mimics, lowering the effectiveness of the aposematic color pattern (Figure 18.25C). Mutant phenotypes in a quasi-Batesian species might have a selective advantage if they resembled a different mimicry ring with much more abundant or much more unpalatable model species (Mallet and Joron 1999). This hypothesis is likely to account for geographic variation in mimetic patterns.

Summary

1. Coevolution is reciprocal evolutionary change in two or more species resulting from the interaction between them. Species also display many adaptations to interspecific interactions that appear one-sided, rather than reciprocal.
2. Phylogenetic studies can provide information on the age of associations between species and on whether or not they have codiversified or acquired adaptations to each other. The phylogenies of certain symbionts and parasites are congruent with the phylogenies of their hosts, implying cospeciation, but in other cases such phylogenies are incongruent and imply shifts between host lineages.

3. Coevolution in predator-prey and parasite-host interactions can theoretically result in an ongoing evolutionary arms race, a stable genetic equilibrium, indefinite fluctuations in genetic composition, or even extinction.
4. Parasites (including pathogenic microorganisms) may evolve to be more or less virulent, depending on the correlation between virulence and the parasite's reproductive rate, vertical versus horizontal transmission between hosts, infection of hosts by single versus multiple parasite genotypes, and other factors. Parasites do not necessarily evolve to be benign.
5. Mutualism is best viewed as reciprocal exploitation. Selection favors genotypes that provide benefits to another species if this action yields benefits to the individual in return. Thus the conditions that favor low virulence in parasites, such as vertical transmission, can also favor the evolution of mutualisms. Mutualisms may be unstable, because "cheating" may be advantageous, or stable, if it is individually advantageous for each partner to provide a benefit to the other.
6. Evolutionary responses to competition among species may lead to divergence in resource use and sometimes in morphology (character displacement). These responses have fostered adaptive radiation. In some cases, adaptive diversification has occurred repeatedly, in parallel, in response to competition.
7. Coevolutionary interactions between species may be altered by a third species. Such interactions can have several consequences, including geographic variation in the intensity and direction of coevolutionary selection.

Terms and Concepts

character displacement
coevolution
diffuse coevolution
ecological release
escape-and-radiate coevolution
gene-for-gene interactions
geographic mosaic
guild coevolution
horizontal transmission
mutualism
specific coevolution
symbiotic
vertical transmission
virulence

Suggestions for Further Reading

J. N. Thompson, in *The coevolutionary process* (University of Chicago Press, Chicago, 1994), discusses the evolution and ecology of many interactions, especially among plants and their herbivores and pollinators. He develops one of that book's themes further in *The geographic mosaic of coevolution* (University of Chicago Press, 2004).

Plant-animal interactions are the focus of essays by prominent researchers in *Plant-animal interactions: An evolutionary approach*, edited by C. M. Herrera and O. Pellmyr (Blackwell Science, Oxford, 2002). "Models of parasite virulence," by S. A. Frank (1996, *Quarterly Review of Biology* 71: 37–78), is an excellent entry into this subject. *The ecology of adaptive radiation*, by D. Schluter (Oxford University Press, Oxford, 2000), includes extensive treatment of the evolution of ecological interactions and their role in diversification.

Problems and Discussion Topics

1. How might coevolution between a specialized parasite and a host that either is or is not attacked by numerous other species of parasites differ?
2. How might phylogenetic analyses of predators and prey, or of parasites and hosts, help to determine whether or not there has been a coevolutionary "arms race"?
3. The generation time of a tree species is likely to be 50 to 100 times longer than that of many species of herbivorous insects and parasitic fungi, so its potential rate of evolution should be slower. Why have trees, or other organisms with long generation times, not become extinct due to the potentially more rapid evolution of their natural enemies?
4. Design an experiment to determine whether greater virulence is advantageous in a horizontally transmitted parasite and in a vertically transmitted parasite.
5. Some authors have suggested that selection by predators may have favored host specialization in herbivorous insects (e.g., Bernays and Graham 1988). How might this occur? Com-

pare the pattern of niche differences among species that might diverge due to predation with the pattern that might evolve due to competition for resources.

6. Provide a hypothesis to account for the extremely long nectar spur of the orchid *Angraecum sesquipedale* (see Figure 18.15) and the long proboscis of its pollinator. How would you test your hypothesis?
7. In simple ecological models, two resource-limited species cannot coexist stably if they use the same resources. Hence coexisting species are expected to differ in resource use because of the extinction, by competition, of species that are too similar. Therefore, coexisting species could differ either because of this purely ecological process of "sorting" or because of evolutionary divergence in response to competition. How might one distinguish which process has caused an observed pattern? (See Losos 1992 for an example.)

Evolution of Genes and Genomes

19

Cytochrome *c* protein structure. The evolutionarily ancient enzyme cytochrome *c*, an essential component of the cellular respiratory chain, is present in all eukaryotic species. Its gene sequence has been determined for many species, and computer graphics allow us to visualize the protein's structure based these sequences. This structure—including α-helices (red) and a central heme group (yellow)—is extremely similar in species as varied as tuna (top) and rice (bottom).

From the time of Aristotle in the fourth century BC to Darwin, Huxley, and Owen in the nineteenth century, the study of the diversity and history of life focused on morphology and, to a lesser extent, behavior. Only in the second half of the twentieth century did it become possible to compare the genes and molecules of different species, and thus to understand more clearly both the evolutionary relationships among species and population processes such as gene flow and genetic drift. The impact of molecular biology on evolutionary biology has been so profound that it is hard to imagine that evolutionary biology could experience further methodological and conceptual shifts of similar magnitude. And yet, the tools of genomics (a suite of biotechnologies that can be described as molecular biology writ large) are causing just such an impact. Genomics is to twenty-first century evolutionary biology what protein electrophoresis and DNA sequencing were to the field in the twentieth century, and likewise promises to provide as many questions as answers.

Today it is routine for the complete nucleotide sequence of all the genes of an organism to be determined (Box A). As this chapter is being written, the list of microbial species whose genomes are completely sequenced is well over 200, and the genomes of five vertebrates—human, mouse, dog, chicken, and pufferfish—are completely known for all but the most technically difficult regions of the genome. Now that complete genomes can be characterized, biologists can study in detail the population genetics of hundreds of genes simultaneously. We can fully catalogue the diverse genes along a chromosome, and we have a basis for studying the interactions among them and the ways in which these interactions influence their evolutionary fates. We can also study the myriad links between genic and organismal evolution in exciting detail. For instance, gene families, such as the immunoglobulin superfamily or the family of calcium-dependent ion channels involved in brain function, provide the molecular foundation for many of the profound organismal adaptations that have occurred in the history of life.

Comparative genomics—the comparative study of whole genomes—began about 10 years ago for simpler genomes; information for complex eukaryotes has become available only in the last 5 years. The reach of comparative genomics will soon extend beyond model species such as mice and *Drosophila,* or species of immediate health concern, such as the malaria parasite *Plasmodium falciparum* and its mosquito host *Anopheles gambiae*, to other organisms of intrinsic evolutionary interest. Comparative genomics will allow us to refine in ever greater detail the Tree of Life; it will also enable us to track the travels of genes as they evolve within species and occasionally move between species.

The differences in timing, level, and location of expression of genes (**differential gene expression**; see Box 20A) has long been thought to contribute more to morphological differences between species than do point substitutions within protein-coding genes, espe-

BOX 19A How to sequence a genome

When it was first conceived in the mid-1980s, sequencing of a whole genome the size of the human genome seemed impossibly unrealistic. Today, at least one new vertebrate genome is sequenced each year, and a new microbial genome of about 2 million base pairs, is sequenced each week! What advances have enabled these amazing feats?

A critical advance was automation of nearly every step in DNA sequencing. Today, robots perform many routine tasks such as isolation of DNA, amplification by the polymerase chain reaction (PCR), and the actual sequencing. A single robot can do in one hour what would take a technician a day or more. Remarkably, many of the techniques used in genome sequencing are the same as those used to obtain DNA sequences for several decades. For example, dideoxy chain termination sequencing, invented in 1977, has not been modified except that DNA fragments generated during the sequencing reaction are visualized not by radioisotopes, but by colored fluorescent dyes that are read by lasers in automated sequencers.

Preparation of a genome for sequencing usually requires four steps: (1) Isolation of DNA; (2) shearing of chromosome-sized DNA fragments into manageable fragments, 2 kb to 200 kb in length; (3) cloning these fragments into vectors that can be propagated with the aid of a bacterium; and (4) sequencing of both ends of the clones. These clones are called BACs if they are very large, and are sequences by "shotgun sequencing": first shearing the larger clones into smaller fragments, then sequencing enough random clones to cover the original large fragment by chance. All of these laboratory-intensive steps are followed by a crucial computational phase in which the raw data are "assembled" into large "contigs" consisting of large groups of smaller sequences that overlap with one another to form longer sequences.

There was heated debate as to the most efficient means of sequencing the whole human genome. Researchers in the publicly funded Human Genome Project believed that it was necessary to meticulously determine the positions of individual DNA fragments along chromosomes before DNA sequencing. Researchers headed by Craig Ventner of Celera Genomics, a private biotechnology company, claimed that this laborious mapping process could be by passed and that sequence contiguity could be ensured simply by "shotgun sequencing" an entire genome: sequencing so many clones that they would eventually overlap and cover the genome. Both camps proved to be partly correct. It is now routine to bypass the mapping phase and even parts of the BAC cloning phase, and apply shotgun sequencing even to large vertebrate genomes. However, the accuracy of the resulting genome sequence is greatly increased by first cloning the genome into BACs and mapping their positions relative to one another, especially if the genome contains many repeated sequences.

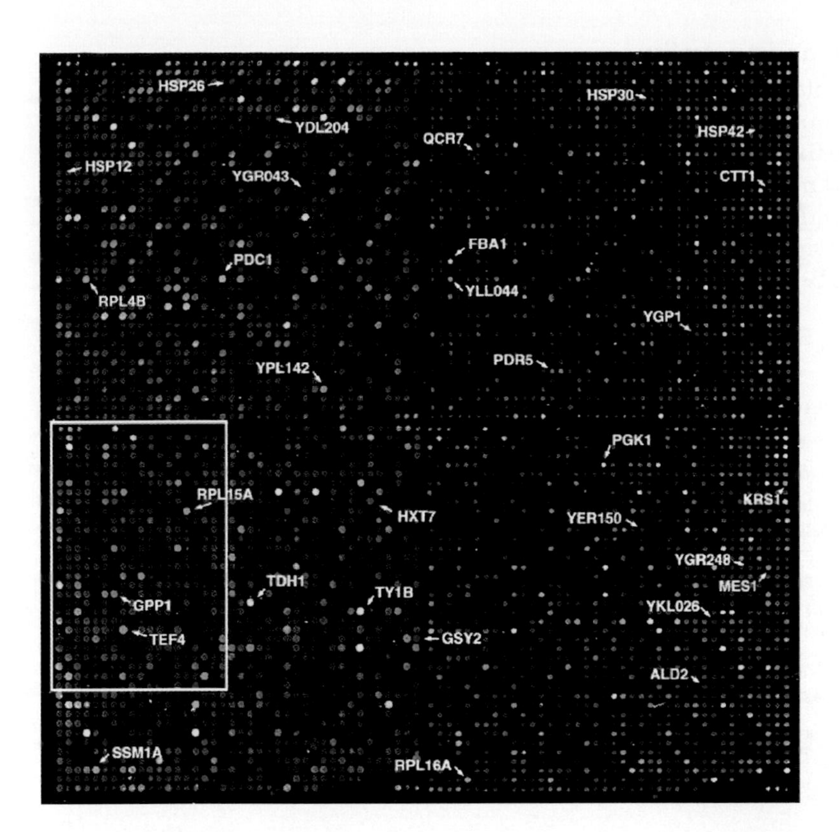

Figure 19.1 A microarray experiment. Microarrays are used to examine differential gene expression (for example, differences in the level of transcription) between different tissue types, developmental stages, and species. In this image, the expression differences, measured by mRNA levels, of thousands of yeast genes are indicated by dots on this microscope slide, comparing the two physiological stages of fermentation (anaerobic) and respiration (aerobic). The color of each spot indicates changes in the expression of a specific gene for the two physiological states. Genes whose expression increases during the shift to aerobic growth are red, whereas genes that are repressed during the shift to aerobic growth are in green. Genes whose expression levels remain unchanged are in yellow. Microarrays are one example of new genomic technologies that are transforming evolutionary biology in a variety of ways. (From DeRisi et al. 1997; © AAAS.)

cially among closely related species such as chimpanzees and humans. This hypothesis arose from the findings of Mary-Claire King and Allan Wilson (King and Wilson 1975) and others, indicating that despite their gross morphological and behavioral differences, the proteins of humans and chimpanzees differed by at most 1 percent in their nucleotide sequences—a figure that is consistent with the 1.44 percent nucleotide divergence between the species found in the first whole-chromosome comparisons (Watanabe et al. 2004). Technologies such as microarrays (Figure 19.1), in which the expression status of thousands of genes can be monitored for an organism in different physiological or behavioral states, enable us to study the evolution of gene expression. We can now correlate morphological differences between species with gene expression status across their entire genomes, rather than investigating species differences arduously on a gene-by-gene basis.

This chapter explores the evolution of genes and genomes, a field that mixes new technologies with older evolutionary principles. Despite the advent of rapidly changing technologies and the discovery of novel molecular processes, the principles by which genes evolve in populations and by which species diverge over time have been developing since the modern synthesis. For example, RNA editing is a recently discovered process by which the sequence of a gene's mRNA transcript can be altered, often in ways that compensate for premature stop codons or other defects in the gene. And we now know that horizontal gene transfer takes place much more frequently among microbial species than previously thought. These discoveries do not alter the fact that the genome as a whole is subject to the same principles of mutation, drift and adaptive evolution that have been studied for decades—although we can add a wealth of recent insights into chromosome evolution and genome structure to the evolutionary principles covered earlier in this book.

Evolution of Genes and Proteins

In Chapter 10, we introduced Motoo Kimura's neutral theory of molecular evolution: DNA sequences and gene evolution are usually considered to be evolving neutrally *unless* they display the signatures of adaptive evolution.

Adaptive evolution and neutrality

The neutral theory states that the vast majority of evolutionary change in genes and chromosomes occurs via mutation followed by random drift, rather than by adaptive mutations being driven to fixation by selection. Adaptive mutations do occur, but they are rare. To be sure, more and more evidence of adaptive molecular evolution is being found; for example, a recent study suggested that as many as 45 percent of all amino acid substitutions in *Drosophila simulans* and *D. yakuba* have been fixed by natural selection (Smith and Eyre-Walker 2002). Still, the neutral theory is the starting point in any analysis of DNA sequence evolution.

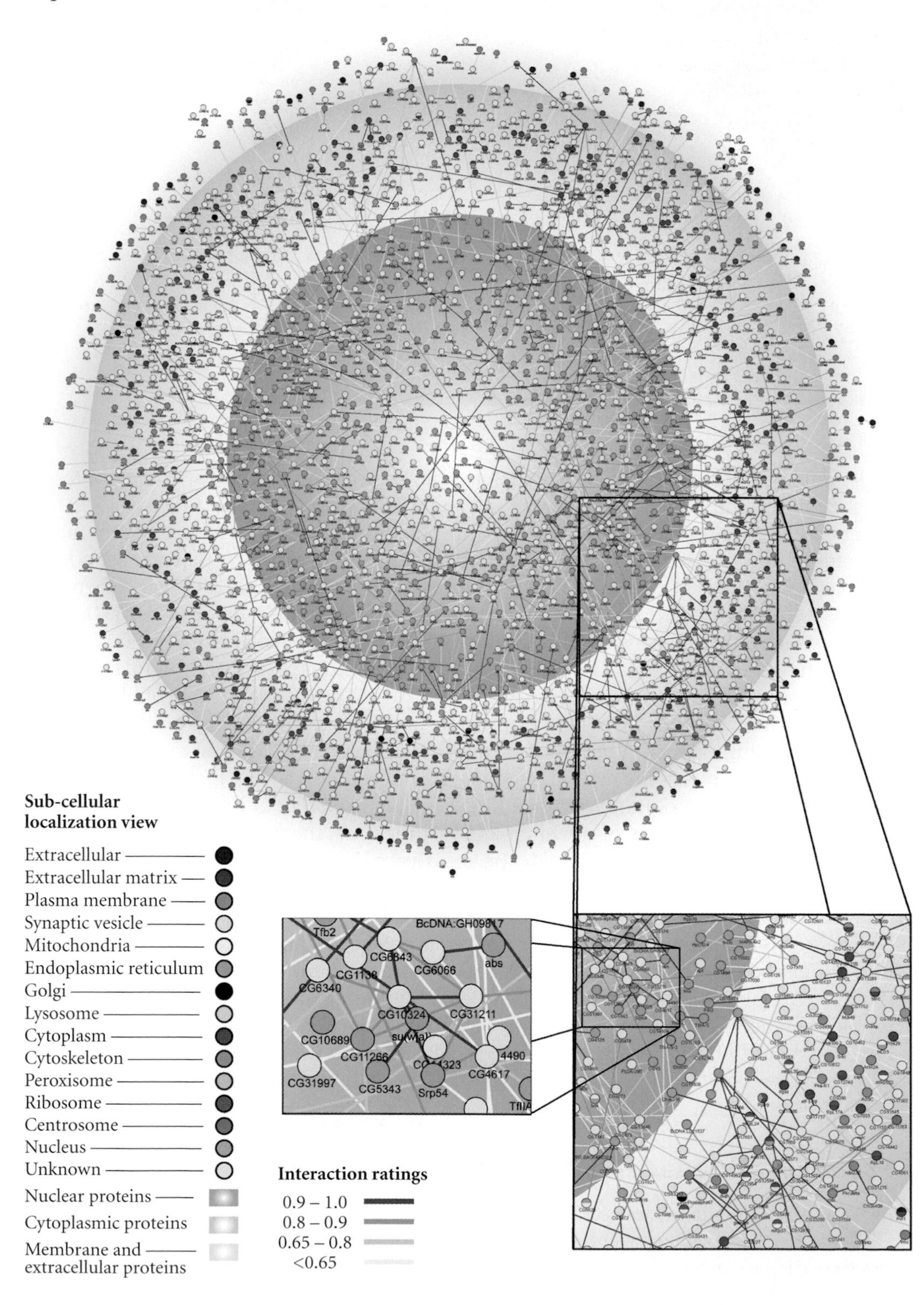

Figure 19.2 A *Drosophila* protein interaction network. The network describes interactions among 7,048 proteins for a total of 20,405 interactions. Each dot represents a *Drosophila* protein, and each line represents an interaction. Proteins within the yellow ring perform functions in the cell membrane and extracellularly; proteins within the blue sphere operate in the cytoplasm; and those within the green core operate in the nucleus. Statistical analysis of the network suggests two levels of protein interactions: a short-range level within multiprotein complexes, and a more global level, presumably corresponding to interactions between protein complexes. (From Giot et al. 2003; © AAAS.)

The neutral theory considers polymorphisms within species to be a transient state, one in which a new allele that has arisen by mutation is on its way to either fixation or loss by drift. It also predicts that most change in DNA sequences and proteins will take place in regions—whether coding or noncoding—that would not affect organismal fitness drastically if modified by mutations. A classic example involves fibrinopeptides—short peptides that are cleaved from their nascent protein, fibrinogen, during blood coagulation. The cleaving of fibrinogen produces fibrin, a threadlike protein essential for blood clotting, and the short fibrinopeptides, which are discarded. Fibrinopeptides have no apparent function, so most amino acid-changing mutations are inconsequential; they exhibit a remarkably high rate of evolution, as measured by the number of amino acid substitutions per million years (estimated by the method outlined in Chapters 2 and 10). By contrast, the fibrin that is produced by the cleaving process evolves at a much slower rate, because more mutations can affect function and are deleterious.

The rate of evolution k of a gene, or a site within a gene, can be expressed as

$$k = f_0 v$$

where v is the total rate of mutation and f_0 is the fraction of mutations that are neutral. This fraction is higher for a sequence such as fibrinopeptide (in which $f_o = 1$) than for proteins such as hemoglobin and cytochrome *c*, in which many amino acids have clear functions. These functions may involve interactions with other proteins, which may be diagrammed as **protein interaction networks**, in which lines connect pairs of proteins that are known to physically interact with each other. Such interaction networks have been constructed for yeast and *Drosophila* (Figure 19.2).

Fraser et al. (2002) compiled an interaction network for 2245 yeast proteins, and then chose a subset of 164 proteins for which they could reliably estimate k, the number of amino acid substitutions between homologous genes of yeast and the nematode *Caenorhabditis elegans*. As the neutral theory predicts, they found a significant negative correlation between the value of k for a given protein and the number of other proteins with which it interacts (Figure 19.3). Moreover, the effect of a given protein on organismal fitness, as measured by the reduction in the growth rate of yeast strains in which that protein had been experimentally deleted, is positively correlated with the number of proteins with which it interacts. Fraser and colleagues also showed that proteins in the same interaction clusters evolve at more similar rates than expected by chance, suggesting that interacting proteins coevolve.

Sequence evolution under purifying and positive selection

PURIFYING SELECTION (Chapter 12) occurs when new sequence variants are selected against, causing $f_0 < 1$ and $k < v$. Under strong purifying selection, most nonsynonymous mutations are selected against, but synonymous mutations can still accumulate, since changes at these sites do not change the protein's amino acid sequence. One index of purifying selection in a protein-coding gene, therefore, is a low ratio of nonsynonymous to synonymous substitutions. The numbers of nonsynonymous and synonymous substitutions are typically normalized by the number of sites in each category, resulting in the ratio ω of nonsynonymous substitutions per nonsynonymous site (d_N) divided by the number of synonymous substitutions per synonymous site (d_S). Protein-encoding genes under purifying selection are characterized by very low values for ω. Genes for histones, for example, are among the most highly constrained genes

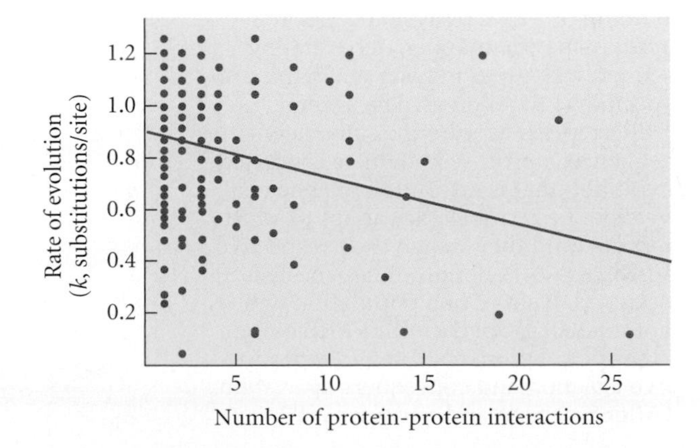

Figure 19.3 Inverse correlation between the rate of protein evolution and the number of protein-protein interactions in a network, based on 3541 interactions between 2245 different yeast proteins. This relationship suggests that such interactions impose functional constraints that reduce the rate of evolution. (After Fraser et al. 2002.)

in the genome, with virtually no nonsynonymous substitutions even in comparisons of very distant mammals (see Table 10.3).

POSITIVE SELECTION—substitution of a mutation that increases fitness—accelerates the accumulation of nonsynonymous mutations over and above the mutation rate (the rate of fixation of neutral mutations). If the number of advantageous substitutions in a gene exceeds the number of neutral substitutions (i.e., if $\omega > 1$), positive Darwinian selection has acted on the gene.

Rapid evolution at nonsynonymous sites has been documented in a variety of genes, including those involved in disease resistance (such as immunoglobulins and cytokines), immune evasion by parasites, and reproduction (Yang and Bielawski 2000). Swanson and Vacquier (2002) have described examples of rapid evolution in reproductive proteins, including those that mediate sperm-egg recognition in marine invertebrates, mammals, and fruit flies. For example, the VERL protein expressed by the egg envelope of abalones is thought to play a role in reproductive isolation among several sympatric abalone species (Chapter 15). VERL has a species-specific sequence that has evolved rapidly. The sperm protein lysin interacts with VERL during fertilization in a sequence-specific manner, and has coevolved with it (Figure 19.4). High values of ω indicate that there has been strong positive selection on lysin. Similarly, many of the proteins produced by the accessory glands of male *Drosophila*, which are transferred to females along with sperm during mating, undergo rapid sequence evolution. These proteins increase the competitiveness of the male's sperm, affect female behavior, and can decrease female lifespan (Chapter 14). Female proteins that interact with accessory gland proteins have evolved rapidly, perhaps to combat these adverse effects (Swanson and Vacquier 2002).

Adaptive molecular evolution in primates

Molecular evolutionary processes are understood better in primates than in any other mammalian lineage. The first good example of adaptive convergent evolution at the molecular level was in the enzyme lysozyme, which breaks down bacterial cell walls. Both ruminants (such as cows, deer, sheep and giraffes) and colobine monkeys (such as langurs) have evolved a modified foregut (the rumen), in which bacteria digest the cellulose from the plant

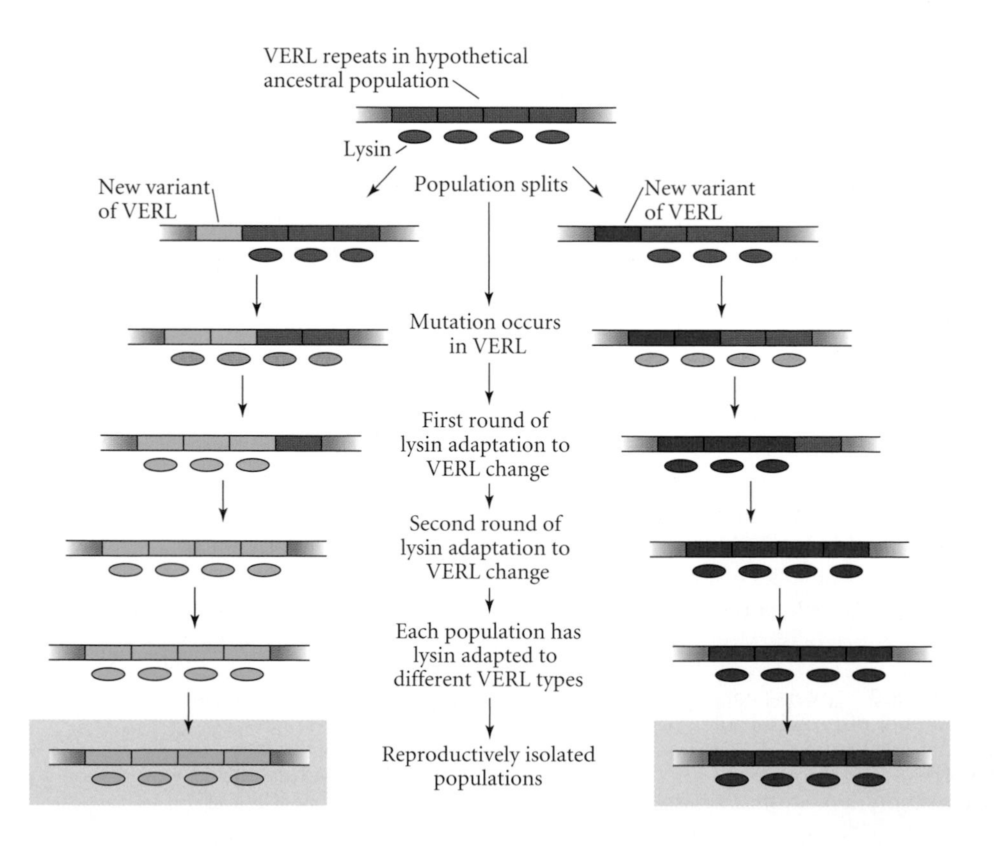

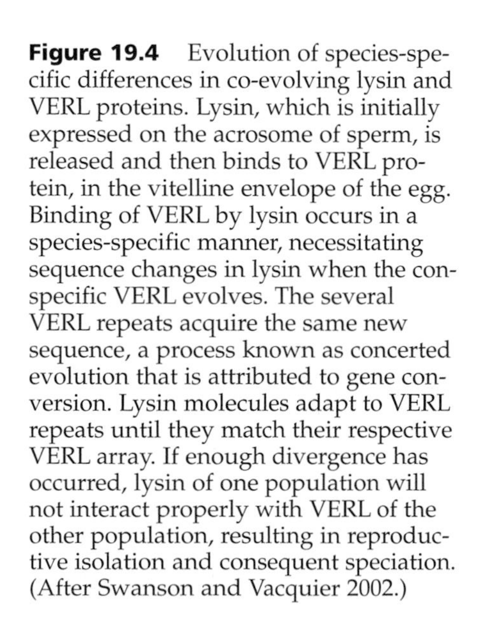

Figure 19.4 Evolution of species-specific differences in co-evolving lysin and VERL proteins. Lysin, which is initially expressed on the acrosome of sperm, is released and then binds to VERL protein, in the vitelline envelope of the egg. Binding of VERL by lysin occurs in a species-specific manner, necessitating sequence changes in lysin when the conspecific VERL evolves. The several VERL repeats acquire the same new sequence, a process known as concerted evolution that is attributed to gene conversion. Lysin molecules adapt to VERL repeats until they match their respective VERL array. If enough divergence has occurred, lysin of one population will not interact properly with VERL of the other population, resulting in reproductive isolation and consequent speciation. (After Swanson and Vacquier 2002.)

material that the animals eat. Lysozyme in the rumen enables the animal to digest bacterial cell contents—a major source of protein. Caro-Beth Stewart, Allan Wilson and colleagues (Stewart et al. 1987) found that, like cows, langur monkeys express high levels of lysozyme in the foregut. More remarkably, they found that the langur's protein has five amino acid substitutions (compared to other primates) that are also found in the lysozyme of cows. Consequently, the langur lysozyme is more similar to the lysozyme of the cow, not other primates, at specific amino acid positions (Figure 19.5). Messier and Stewart (1997) have since studied the DNA sequence evolution of the primate lysozyme gene, and have found that it underwent an episode of accelerated nonsynonymous substitution in the ancestral lineage leading to colobines, associated with the evolutionary change in diet from fruit to foliage. Adding to the history of convergence in this enzyme, Kornegay et al. (1994) found similar amino acid substitutions in the lysozyme of a leaf-eating bird, the hoatzin, which also has high levels of cellulose-digesting bacteria.

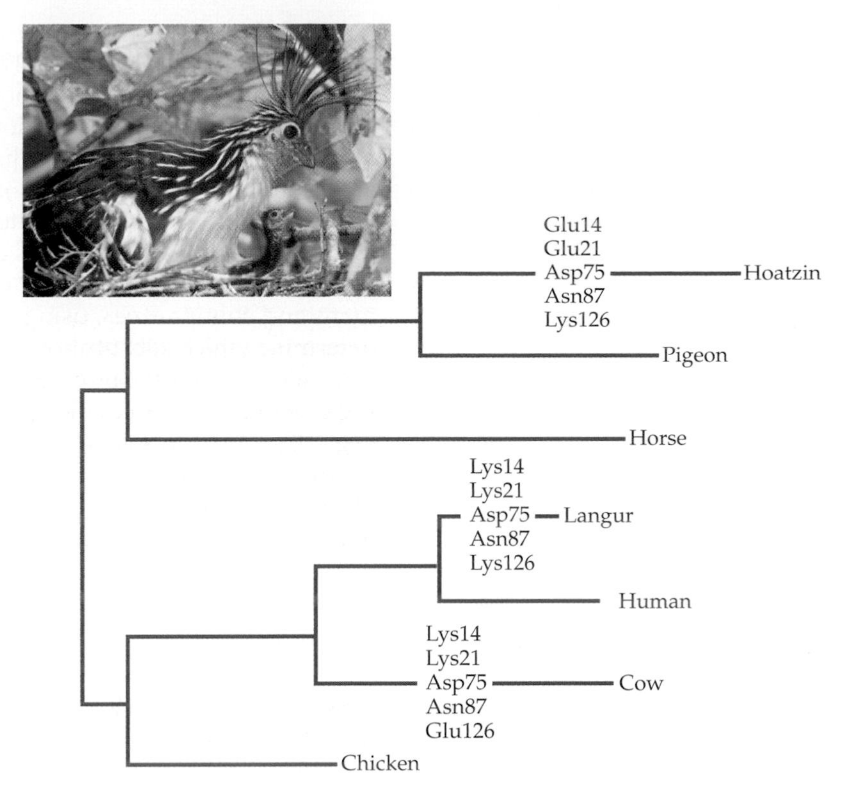

Figure 19.5 This gene tree for lysozyme provides phylogenetic evidence for molecular convergence in primate, ruminant, and avian lysozymes. Branch lengths are proportional to the total number of amino acid replacements along them. Convergent replacements occur at positions 14, 21, 75, 87, and 126 in the lineages leading to the lysozymes of three species that digest bacteria in the foregut—cow, langur, and the hoatzin (*Opisthocomus hoazin*; from the Amazonian region of South America, it is the only bird known to possess this trait). For example, position 75 changed to aspartic acid independently in all three foregut-fermenting lineages. The convergent amino acid replacements presumably represent adaptive biochemical changes (see Table 8.1) in the enzyme. (After Kornegay et al. 1994; photo © photolibrary.com.)

Genes that exhibit adaptive evolution on the branch leading exclusively to humans could be responsible for some of the traits, such as speech, that distinguish humans from other primates. At least two genes have been found that show evidence of adaptive change in the human lineage after it branched from the chimpanzee lineage. Mutations in a gene known as *forkhead box 2* (*FOXP2*) were found in families that have a high incidence of abnormalities in speech and grammar. Two groups of researchers, led by Jianzhi Zhang (Zhang et al. 2002) and Svante Pääbo (Enard et al. 2002), examined sequence evolution of this gene and found that although only one amino acid substitution had occurred in the 130 Myr separating mice and the ancestor of humans and chimpanzees, two substitutions have occurred in the human lineage during the 4–6 Myr since it separated from chimpanzees—a significant acceleration in the rate of amino acid substitution. Zhang's group showed that this gene is virtually invariant in another 28 orders of mammals that last shared a common ancestor around 100 million years ago.

Another gene, *sarcomeric myosin heavy chain* (*MYH*), is highly expressed in the chewing muscles of chimpanzees and is responsible for the large size of these muscles in this and other non-human hominins. In humans, *MYH* has been inactivated by a frameshift mutation that occurred about 2.4 Mya, after humans separated from chimpanzees (Stedman et al. 2004). The loss of *MYH* function is thought to be responsible for the reduced size of these muscles in humans. In this example, ω increased from less than 0.1 to approximately 1, as expected if constraints were relaxed following loss of function.

Olfactory perception plays an important role in many mammalian behaviors, but primates, especially humans, possess a much less sensitive sense of smell than most other mammals. The olfactory receptor (OR) genes present a molecular example of the tendency in organisms for "structures of little use" to degrade and degenerate, as Darwin noted in *The Origin of Species*. There are approximately 1200 functional OR genes in the mouse genome, but only about 550 in humans. Apparently selective pressures to maintain OR gene functionality are relaxed in humans, and mutations that reduce function and cause pseudogene formation have been fixed (Gilad et al. 2003). Zhang and Webb (2003) found the same tendency in five genes that are normally expressed in the vomeronasal organ in primates. Mammals use this organ to perceive pheromones, which are used extensively in social interactions. Two distinct families of these genes are functional and conserved in several lineages of Old World primates (catarrhine monkeys), but they have degenerated dramatically in hominins, having accumulated numerous stop codons since their separation from ancestral catarrhines approximately 23 million years ago.

Adaptive evolution across the genome

As the genomes of more and more organisms are completely sequenced, it becomes possible to move beyond descriptions of the evolutionary dynamics of single genes or gene families, and to examine the distribution of selective histories and evolutionary rates across the entire genome. For example, human population genetic studies now include surveys of polymorphism in literally hundreds of genes, using a variety of new molecular approaches (e.g., Stephens et al. 2001). Andrew Clark and collaborators (Clark et al. 2003) estimated ω values (the ratio d_N/d_S) for the coding regions of 7645 genes from humans and chimpanzees, using mouse genes as an outgroup. The study enabled them to determine which substitutions occurred in the human and which in the chimpanzee lineage. Using a statistical method that permitted identification of specific codons exhibiting $\omega > 1$, they found evidence of adaptive evolution in 873 genes along the human lineage. Certain functional groups of genes were especially prone to show adaptive evolution, such as those encoding olfactory receptors and amino acid catabolism. The authors suggested that these changes reflect the behavioral and dietary changes in the human lineage. Several adaptively evolving human genes play key roles in early development, pregnancy, and hearing. It is particularly interesting that many of the adaptively evolving genes are known to cause genetic diseases when mutated.

Genome Diversity and Evolution

Diversity of genome structure

The structures of genomes across the major branches of life differ widely. Viral and bacterial genomes are models of efficiency, maximizing the speed of genome replication and minimizing unnecessary genes. Eukaryotic genomes—particularly those of mammals, amphibians, and some plants—are by comparison large and lumbering, harboring vast regions of noncoding and repeated DNA seqeunces with unknown functions. Although much of this noncoding DNA is unlikely to be "junk" (as was postulated in the early 1970s), a typical mammalian genome is by any measure extravagant in its excesses and complexity compared to a bacterial genome.

Whereas many eukaryotic genes are interrupted by introns, the genomes of Bacteria and Archaea have few introns, and these are spliced out by a different mechanism—self-splicing—than that used by eukaryotes. Because of some similarities between prokaryotic and eukaryotic introns, Walter Gilbert, who discovered introns in the 1970s, suggested in 1987 that introns have been present since the common ancestor of all extant life, and have simply been lost in those genes and genomes that do not possess them today. This "introns early" hypothesis is still prevalent today (Poole et al. 1998), but the opposite view—that introns are structures that appeared in large numbers in the eukaryotic genome only recently, long after the prokaryotic-eukaryotic split (the "introns late" hypothesis)—is supported by the lack of introns in many phylogenetically basal eukaryotes (Palmer and Logsdon 1991). Also, most introns are restricted to specific clades of plants and animals, and can therefore be inferred to have entered eukaryotic genomes relatively recently (Figure 19.6).

Eukaryotic genomes contain far more noncoding DNA than prokaryotic genomes do. Only about 1.5 percent of the human genome, for example, is composed of protein-encoding sequences. Up to 95 percent of a typical human gene consists of introns. Moreover, there are vast regions of noncoding DNA, much of which may be "selfish DNA" that merely replicates itself and accumulates within genomes. However, more than 10 percent of noncoding DNA is highly conserved between long-diverged species, such as humans and mice, suggesting a function maintained by purifying selection (Shabalina et al. 2001). Moreover, many noncoding regions, including introns, are transcribed into RNA sequences such as "microRNAs" that are usually about 22 bp long. These diverse sequences, some represented by as many as 50,000 copies per genome, perform important functions in gene regulation (Bartel 2004).

Finally, many eukaryotic genes, unlike prokaryote genes, are subject to alternative splicing (AS), wherein many (rather than just one) mRNAs are encoded by a single gene (see

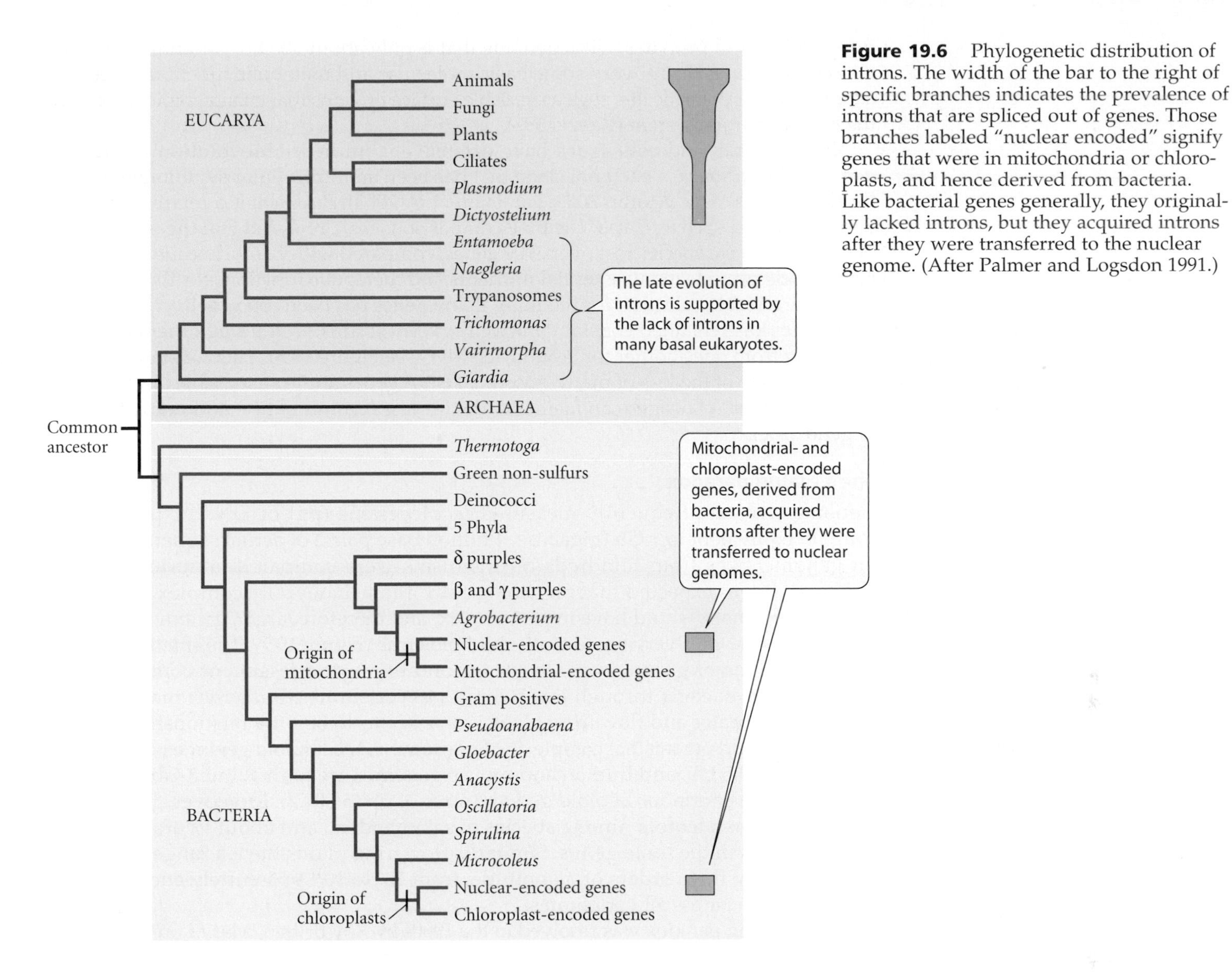

Figure 19.6 Phylogenetic distribution of introns. The width of the bar to the right of specific branches indicates the prevalence of introns that are spliced out of genes. Those branches labeled "nuclear encoded" signify genes that were in mitochondria or chloroplasts, and hence derived from bacteria. Like bacterial genes generally, they originally lacked introns, but they acquired introns after they were transferred to the nuclear genome. (After Palmer and Logsdon 1991.)

Figure 8.1). For example, the *CD44* gene, a cell-surface glycoprotein that modulates interactions between cells, contains 21 exons, at least 12 of which can undergo AS (Roberts and Smith 2002), potentially yielding thousands of splice variants. AS appears to be a major mechanism by which metazoans can increase the functional diversity of proteins from a limited set of genes. Almost nothing is known yet about how the pattern and control of alternative splicing vary among related taxa, or about how they have evolved.

Viral and microbial genomes: The smallest genomes

According to life history theory (Chapter 17), rapid growth and early reproduction are advantageous in organisms that frequently experience rapid population growth. Thus, in organisms such as viruses and bacteria, small genomes are advantageous because they can be copied faster—a key adaptation in situations where interspecific or host–parasite competition for resources is intense. Indeed, many viruses and bacteria have streamlined their genomes by doing away with many genes. Some of them, by exploiting the genomes of their hosts, make do with extremely small and focused genomes. For example, many viral genomes encode proteins for only three functions: replication of their genome, construction of their outer core, and integration into the host genome. The RNA genome of HIV, for instance, is only 9.8 Kb (kilobase, 1000 base pairs), encoding nine open reading frames (genes). The bacterium *Mycoplasma genitalium*, a parasite of the genitalia and respiratory tracts of primates, has a genome that is only about 580 kb, containing 468 protein-coding genes that govern some basic molecular and metabolic functions as well as adaptations for parasitic life, such as variable surface proteins that change rapidly to evade the host's immune system (Razin 1997).

Nancy Moran and colleagues have extensively studied the evolution of reduced genomes in *Buchnera*, a bacterial clade that has been an intracellular symbiont of aphids for the past 200 Myr (Moran 2003; van Ham et al. 2003). *Buchnera* is a relative of better-known bacteria such as *E. coli*. Genome comparisons have revealed that the common ancestor of *Buchnera* species lost over 2000 genes, compared with *E. coli*. Because *Buchnera*'s host aphids provide many essential nutrients and metabolic functions for the symbionts (see Figure 14.24), selection for retaining many genes has been relaxed. Intriguingly, the gene order has remained constant among *Buchnera* strains even though they have been diverging from one another for around 150 Myr (see Figure 18.3). This constancy may be a consequence of the loss of many ribosomal RNA genes and transposable elements that in other bacterial lineages can facilitate illegitimate recombination and rearrangement of gene order.

The C-value paradox

Genome sizes are frequently measured in picograms (pg) of DNA; 1 picogram is roughly equivalent to 1 Gb (gigabase, 1 billion base pairs) of actual sequence. As data on genome sizes from hundreds of organisms were compared, a curious pattern emerged. It was expected that physiologically and behaviorally complex organisms, such as mammals, would have more complex, and therefore larger, genomes than simpler organisms. On a very broad scale, this holds true (Figure 19.7). The smallest genome size in each major group seems to correspond to our impression of complexity, increasing from bacteria through invertebrates to vertebrates. But *within* major groups, such as vertebrates and flowering plants, there seems to be little relationship between genome size and organismal complexity. For example, the haploid genome of the pufferfish is about 0.5 Gb, and human and mouse genomes are both about 3 Gb—but salamanders have enormous haploid genome sizes of up to 50 Gb. Moreover, genome size varies more than tenfold among species of salamanders, and about twofold even between species in the same genus. Similarly, flowering plants have a range of genome sizes spanning three orders of magnitude, from 10^8 to 10^{11} bp, entirely encompassing the range seen across all vertebrates.

This seeming paradox was resolved in the 1960s by Roy Britten, Mary Lou Pardue and others. Long before DNA sequencing was widely used, genome size was estimated by a number of means, one of which yielded a measure of DNA content known as the C_0t value,* or C value. The C value is a comparative measure of *single-copy DNA content*, which contains the majority of functional genes, versus *total DNA content*, which usually

*C_0t stands for the concentration of DNA (in moles of nucleotides per liter) × time (in seconds).

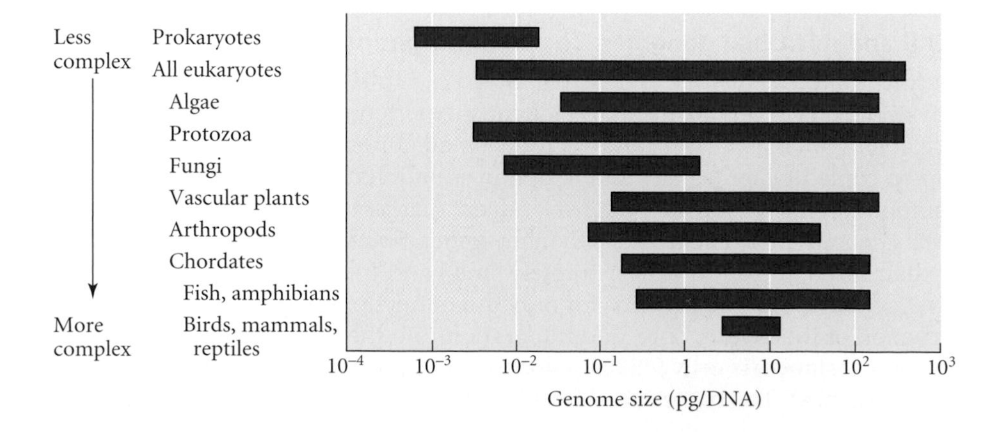

Figure 19.7 Genome size variation. The bars indicate the range of genome size for particular clades. Taxa are arranged top to bottom in order of increasing organismal complexity. Within eukaryotes, there is little relationship between maximum genome size and organismal complexity This "C-value paradox" is thought to arise because lineages vary greatly in the amount of repetitive (noncoding) DNA their genomes contain. (1 pg of DNA is approximately equivalent to 1 billion base pairs.) (After Gregory 2001.)

contains a substantial amount of repetitive sequences, many of which are not useful for the organism. The C value can be approximated biochemically by reassociation kinetics and construction of C_0t curves that plot the conversion of single-stranded DNA to double-stranded (reassociated) DNA over time. The faster the DNA reassociates, the higher the percentage of repetitive sequences. Thus the lack of correspondence between genome size and phenotypic complexity in eukaryotes was dubbed the **C-value paradox**, as researchers discovered that not all the DNA in a genome carries information that is used during the development and functioning of an organism. Genomes contain a great deal of noninformative, highly repetitive DNA that varies greatly in amount among species.

Repetitive sequences and transposable elements

A substantial fraction of the genome—nearly half the human genome, and about 34 percent of the *Drosophila* genome—consists of repeated DNA sequences that are referred to as low repetitive, middle repetitive, and highly repetitive DNA, depending on copy number. They are also termed SATELLITE DNA because of their location in a chemical gradient in a centrifuge. A major source of repeated DNA in human and other mammalian genomes is transposable elements (TEs), sequences that can copy and transpose themselves, or "jump," to other regions of the genome (Chapter 8). Most TEs are retroelements, of several kinds (see Figure 8.7), that are transposed by producing an RNA transcript followed by reintegration of new copies into the genome as DNA. The organism in which retroelements reside is often referred to as the host. The genes of most TEs do not contribute to development or function of the host organism; rather, they encode only proteins essential for replication and transposition of the retroelement itself. They are an example of a SELFISH GENETIC ELEMENT, or "SELFISH GENE."

The copies of retroelements that are transposed to new sites in the genome undergo point mutations, just as host genes do. They either continue to produce daughter elements or, more frequently, degenerate by mutation and become inactive elements that no longer transpose. Mutations in these elements can be used to determine relationships among the copies in a genome, and the age of a family of transposable elements can be estimated in the same way that divergence times of species can be estimated. This requires an estimate of the absolute rate of point substitution, usually obtained from rates estimated for the host genome itself. For example, *Alu* elements, which belong to a group of TEs called short interspersed nuclear elements (SINEs), are abundant in hominoid primates (*Homo, Pan, Gorilla, Pongo*, and *Hylobates*), to which they are restricted. Since they are not found in other Old World primates, *Alu* elements must have originated in an ancestor of the hominoid lineage after its divergence from the other primate lineages, about 25 Mya. There are over 500,000 *Alu* copies in the human genome, making up over 10 percent of human DNA. The number of point mutations observed among hundreds of *Alu* elements scattered across the human genome implies that they underwent an ancient proliferation about 40–50 Mya, and have since slowed down in their rate of transposition (Figure 19.8).

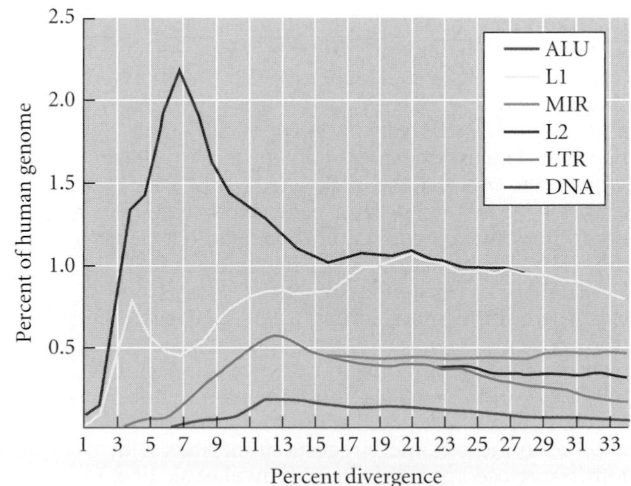

Figure 19.8 Age distribution of retroelements in the human genome. On the X axis, the percent divergence between two sampled copies of a retroelement is indicated, whereas the Y axis is a percent count of copies with that degree of divergence. Six different retroelements are indicated: *Alu* elements, L1 and L2 LINE elements, Mammalian interspersed repeats (MIR), long-terminal repeat (LTR) elements and DNA transposons. If percent substitution can be used as a rough proxy for time, the figure suggests that many *Alu* elements underwent a burst of proliferation about 40–50 Mya (7 percent divergence), with decreased rates of proliferation at earlier and later dates. By contrast, L1 lines seem to have had a protracted proliferative phase a very long time ago, with a second spike in numbers corresponding to roughly 4 percent divergence (~25 Mya). MIRs and L2s are uniformly ancient and on their way to degradation and loss from the genome. (After Deininger and Batzer 2002.)

The effect of a new transposition event on the fitness of the individual bearing it depends largely on where the transposition occurs. The production and insertion of new copies of transposable elements must often have little effect on the fitness of the host organism, but many transposition events are known to be deleterious, because they can affect the function of genes near which, or within which, they are inserted. TEs tend to occur in regions between genes and in introns, probably because those that occur within coding regions often cause deleterious mutations and are eliminated by purifying selection.

Transposition can have at least two kinds of genetic effects. First, they cause mutations; 10 percent of all genetic mutations in mice are thought to result from retroelement transpositions, many of them into coding regions or control regions. Second, the repeated copies of a transposable element in different parts of the genome can provide templates for illegitimate recombination, resulting in chromosome or gene rearrangements that often carry deletions of some genetic material (see Figure 8.8). Approximately 0.3 percent of all human genetic disorders, such as many hereditary leukemias, are thought to arise in this way (Deininger and Batzer 2002). The reduction of host fitness due to deleterious mutations and chromosome rearrangements is thought to be the chief reason that transposable elements are not even more abundant in the host genome than they are (Charlesworth and Langley 1989).

Occasionally, transposition can lead to adaptive evolution. Perhaps the most dramatic example involves the origin of the vertebrate immune system (Figure 19.9). Each of the huge number of possible antibody proteins is produced by a composite gene consisting of three sequences (called *V*, *D*, and *J*) that are physically separated in the genome but are joined together during an immune response by a mechanism known as *V-D-J* recombination. Each *V*, *D*, and *J* segment is flanked by 12-bp and 23-bp recombination signal sequences that precisely direct the joining of the different elements into a single functioning gene. David Schatz and colleagues (Agrawal et al. 1998) found that the proteins responsible for *V-D-J* joining, encoded by the genes *RAG1* and *RAG2*, lack introns—a com-

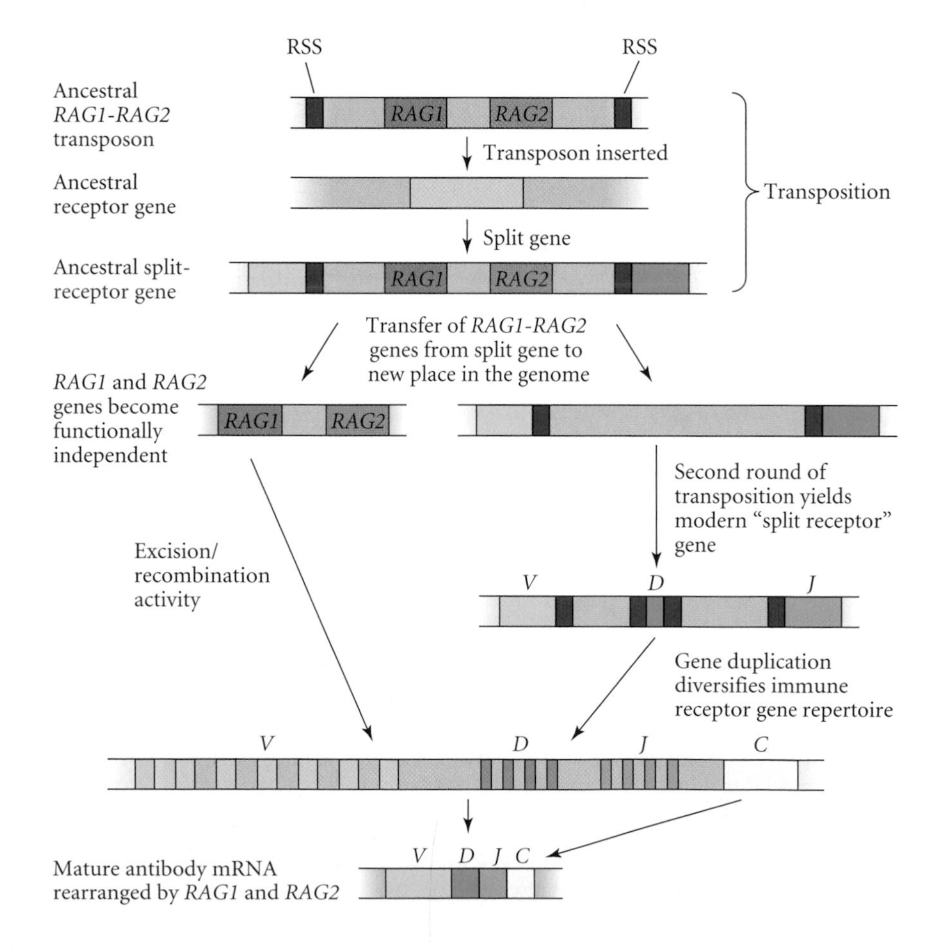

Figure 19.9 Birth of the vertebrate immune system via retroposition of *RAG* genes. The protein products of *RAG1* and *RAG2* perform the recombination reactions of the vertebrate immune system by binding to recombination signal sequences (RSS) adjacent to receptor gene coding elements (light green, dark green, and light blue rectangles). These events are very similar to those catalyzed by transposable elements during excision from the genome. The diagram displays one hypothesis for the origin of "split genes" in the immunoglobulin and T-cell receptor loci. The key evolutionary events were (top) the initial insertion of a ancestral *RAG*-containing transposon into a progenitor variable-region receptor gene, thereby splitting it into different elements with RSS flanking coding regions (the ancestral split-receptor gene). Subsequent to this transposition, *RAG1* and *2* genes are lost from split genes and become functionally and genomically independent. The transpostion and gene loss events are repeated to produce the *V*, *D*, *J* -type split genes found in vertebrate receptor genes today, and after several rounds of gene duplication, leading to the modern clusters of *V*, *D*, and *J* elements (bottom right). By recognizing the RSS of *V-D-J* genes and cleaving out the intervening spacer sequences, the *RAG* genes are essentially performing their ancestral role in transposition and excision. The fact that the *RAG1* and *2* genes are uninterrupted by introns in most genomes studied, and occur adjacent to one another in the genome, suggests similarities with retrotransposed genes. (After Agrawal et al. 1998.)

mon feature of genes derived by retrotransposition of an mRNA—and can cause transposition of specific sequences into specific targets of DNA. The products formed by these events contain short terminal duplications, reminiscent of the long terminal repeats (LTRs) produced by typical TEs. These authors postulate that the vertebrate immune system may have originated with the arrival of a TE that contained the ancestors of the *RAG1* and *RAG2* genes. This TE split an ancestral immunoglobulin-like gene into several components, the ancestors of *V*, *D*, and *J* sequences, each flanked by terminal repeats that evolved into the recombination signal sequences that flank these elements in modern genomes. The birth of the immune system—one of the traits that may have enabled ancestral vertebrates to expand into new niches and combat the myriad parasites that result from such events—may have been a "lucky strike" of a transposable element into an immunoglobulin gene.

Michael Lynch and John Conery (2003) have pointed out that a variety of genomic features that appear to have little fitness advantage for organisms—introns, transposable elements, large tracts of noncoding DNA—may be more prevalent in species with small effective population sizes. They have suggested that viruses and bacteria have extremely large population sizes that facilitate the sweep of advantageous mutations that enable genomic streamlining. By contrast, eukaryotes have smaller population sizes that facilitate the fixation of nonadaptive traits (Chapter 10). This is the best hypothesis advanced so far that would explain the diversity of genome sizes and structures.

The Origin of New Genes

It is obvious that the approximately 30,000 different functional genes in mammalian genomes must have evolved from a much lower number in the earliest ancestor of living organisms. Presumably, all genes in the human genome ultimately descend from a single gene or set of genes that provided the first programs for life on earth. Moreover, the number of functional genes differs among major groups of organisms. How do such genes arise, and what processes lead to the origin of novel genes?

Evolutionary biologists have described several mechanisms by which the genes in a species' genome have originated, either from pre-existing genes in the same genome or in the genome of a different species. These mechanisms include lateral gene transfer, exon shuffling, gene chimerism, retrotransposition, motif multiplication, and gene duplication (Long et al. 2003a).

Lateral gene transfer

In the 1970s, Carl Woese analyzed 16S rRNA sequences and suggested that the Tree of Life is divided into three major "empires" or "domains": Bacteria, Archaea, and Eucarya (see Figure 2.1), and that Archaea and Eucarya appear to be sister clades. Since Woese's original work, this phylogeny has been supported by many other gene sequences, but some gene sequences seem to indicate a sister-group relationship between Eucarya and Bacteria instead. As we saw in Chapter 2 (see Figure 2.21), when different genes provide strong support for different phylogenies, **lateral gene transfer** (LGT) between lineages is a likely hypothesis. It is now thought that genetic material was often transferred across quite different lineages early in the history of life, and some authors even favor a tree in which eukaryotes represent a fusion of sorts between Archaea and Bacteria (Doolittle 1999).

Lateral gene transfer (LGT, also called horizontal gene transfer) has also occurred more recently. For example, the eukaryote protist *Entamoeba histolytica,* which causes over 50 million cases of human dysentery annually, can live anaerobically in the human colon and in tissue abcesses due to fermentation enzymes that most other eukaryotes lack. A phylogenetic analysis (Figure 19.10) showed that several of these fermentation genes were obtained by lateral transfer from Archaea (Field et al. 2000). Perhaps 40 to 50 human genes have their origins in bacteria (Salzberg et al. 2001). New genome sequence data suggest that LGT may be especially frequent among prokaryotes, and that novel adaptive mechanisms, often born on chromosomally distinct plasmids, are especially likely to spread phylogenetically by LGT (Ochman et al. 2000).

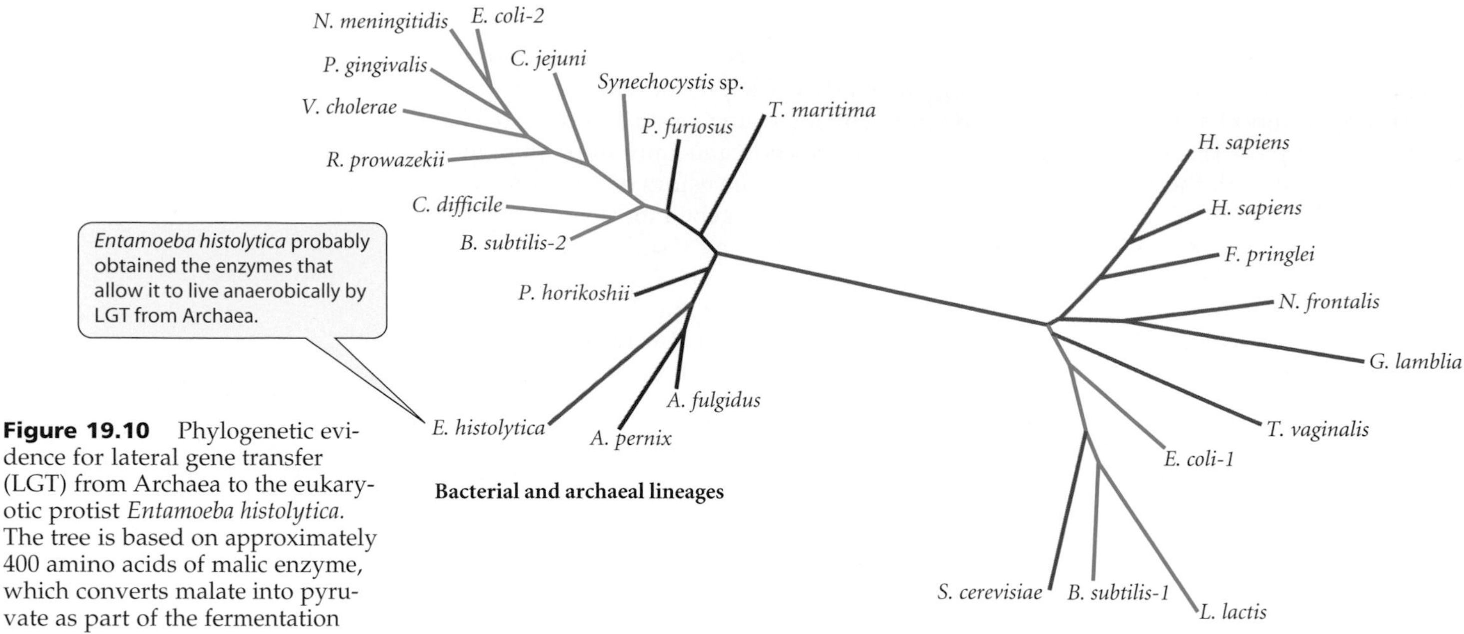

Figure 19.10 Phylogenetic evidence for lateral gene transfer (LGT) from Archaea to the eukaryotic protist *Entamoeba histolytica.* The tree is based on approximately 400 amino acids of malic enzyme, which converts malate into pyruvate as part of the fermentation pathway. Archaeal lineages are red, bacterial green, and eukaryotic lineages are shown in blue. A tree in which the *Entamoeba histolytica* enzyme clusters with other protist lineages is 46 steps longer than the most parsimonious tree, and therefore a worse explanation of the amino acid data. However, cell structure and other gene sequences clearly show that *Entamoeba* is a typical eukaryote. (After Field et al. 2000.)

Exon shuffling

Lateral gene transfer adds genes to the genome of a particular lineage, but the genes already exist in the source species. Several other mechanisms produce truly new genes. One of these mechanisms is exon shuffling.

Eukarote genes include introns and exons. There is often a close correspondence between the division of a gene into exons and the division of the protein into domains. A protein **domain** (or "module") is a small (~100 amino acids) segment that can fold into a specific three-dimensional structure independently of other domains. Protein domains frequently have specific functions, although usually they cannot perform these functions completely in the absence of other domains that would together make up a mature protein. For example, antibodies function primarily through their immunoglobulin domains, two of which together form the major cleft into which foreign proteins fit during an immune response (Figure 19.11).

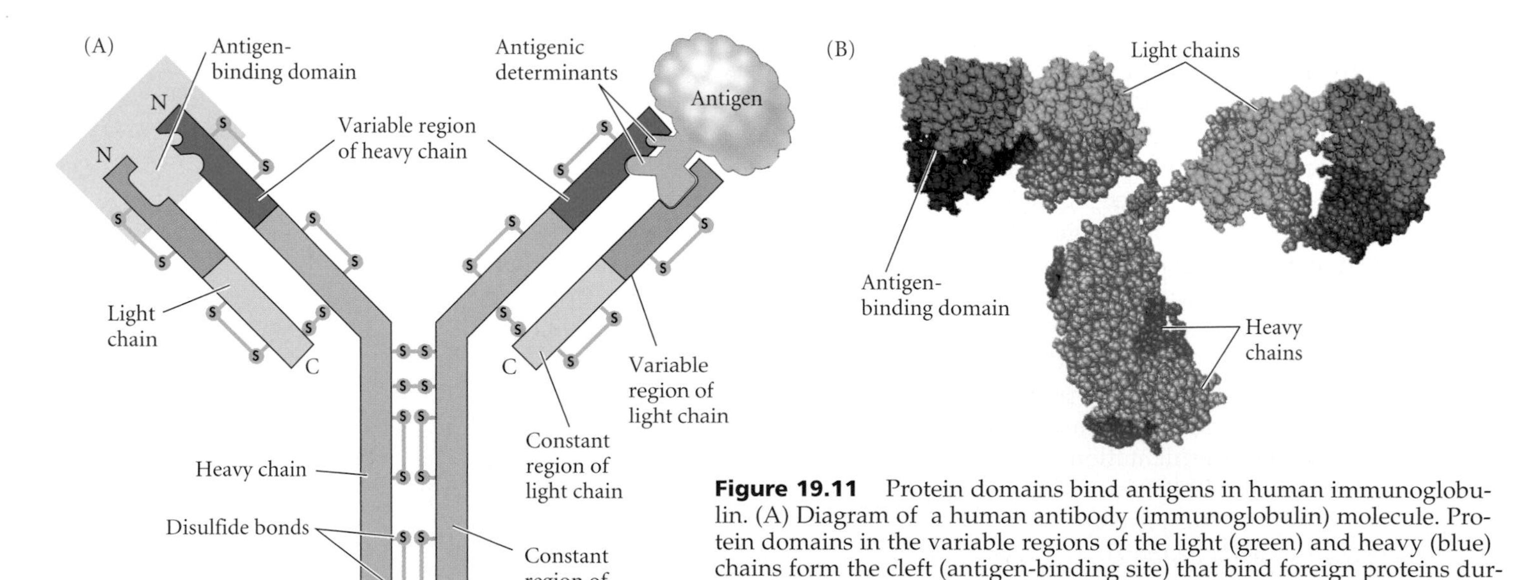

Figure 19.11 Protein domains bind antigens in human immunoglobulin. (A) Diagram of a human antibody (immunoglobulin) molecule. Protein domains in the variable regions of the light (green) and heavy (blue) chains form the cleft (antigen-binding site) that bind foreign proteins during the immune response. The immense variability of these domains is crucial to the antibody function of the immune system. (B) Three-dimensional molecular model in an orientation similar to (A).

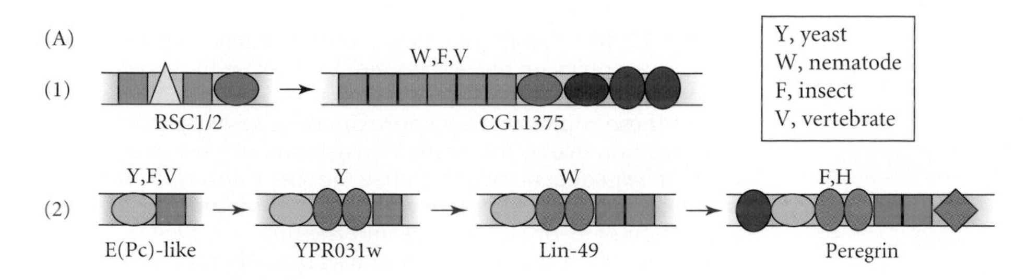

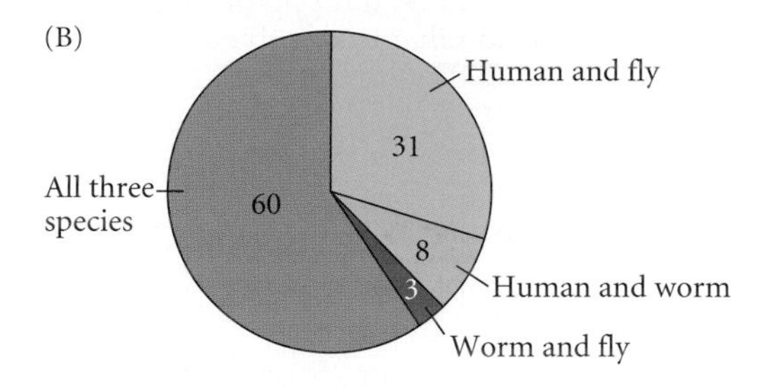

Figure 19.12 Evolution and conservation of domains in diverse proteins. (A) Diversification of chromatin proteins via exon shuffling and domain accretion. Each shape indicates a different protein domain (a bright red oval, for example, indicates a zinc finger domain). Arrows show how new proteins may have evolved by the terminal addition, intercalation, and shuffling of domains. In Group (2), for example, the protein YPR031w differs from the E(Pc)-like protein by the intercation of two PHD domains (orange) between two enhancers (Ep1 and Ep2; green). The intercalation may have occurred after the origin of separate genes by gene duplication. Next in this series, the Lin-49 protein can be derived from the YPR031w protein by terminal addition of a bromo domain (blue square); the further addition of an initial zinc finger domain (red) and a terminal BMB domain (brown). (B) Similarities in domain structure of different proteins such as those in (A) are called "conserved domain architecture." This diagram shows the proportion of chromatin proteins in which domain architecture is shared among three species with nearly fully sequenced genomes. (After International Human Genome Consortium 2001.)

Hundreds of different protein domains are known, and most proteins are mosaics consisting of several different domains. For example, various collagens have different sequences of up to five different types of domains, some of which may be repeated. Some domains, such as the nucleotide-binding domain and the heme-binding domain, occur in many different proteins.

Comparison of gene structures in the human genome and in genomes from *Drosophila* and yeast has shown that many genes evolve by **domain accretion**, whereby new genes are produced by the addition of domains to the beginnings or ends of ancestral genes. For example, many chromatin-associated genes have evolved by addition of a variety of domains, such as chromodomains, zinc-fingers, and helicase/ATPase domains (Figure 19.12).

The approximate correspondence between gene exons and protein domains in some proteins has led to the hypothesis of **exon shuffling**, which states that much of the diversity of genes has evolved as new combinations of exons have been produced by illegitimate (nonhomologous) recombination that occurs in the intervening introns (Figure 19.13). Long et al. (2003b) estimate that 19 percent of all exons in the eukaryotic genomes have arisen from pre-existing exons via exon shuffling. One of the first well-documented examples of an exon shuffling event in plants also illustrates how exon shuffling can confer new functions on genes. In potato (*Solanum tuberosum*), cytochrome *c1* has nine exons, the first three of which are responsible for situating the protein in mitochondria, the site of enzyme catalysis. In an extensive search of the protein sequence data bases, Long et al. (1996) found that the first three exons of cytochrome *c1* of potato show significant sequence similarity to the first three exons of an otherwise unrelated gene (*Gapdh*) that encodes glyceraldehyde-3-phosphate dehydrogenase in maize, *Arabidopsis*, and other plants. Thus, the organelle-targeting function in a leader sequence of cytochrome *c1* has been taken over by a sequence acquired from *Gapdh*.

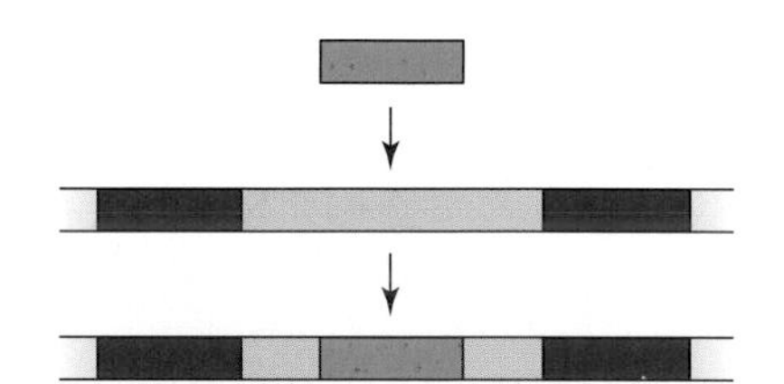

Figure 19.13 Origin of new genes via intron-mediated exon shuffling. Colored boxes are exons and the gray boxes connecting or flanking them are introns. A pre-existing exon (blue) is here shown to be inserted into an intronic region of a different gene. This process could be accomplished by illegitimate recombination between introns of the two participating genes. (After Long et al. 2003b.)

Gene chimerism and processed pseudogenes

A **chimeric gene** is one that consists of pieces derived from two or more different ancestral genes. We have seen that some chimeric genes arose by exon shuffling, in which case the exons are separated by introns, the sites at which genetic exchange occurred. Chimeric genes may also arise by retrotransposition, in which a mature mRNA is reverse-tran-

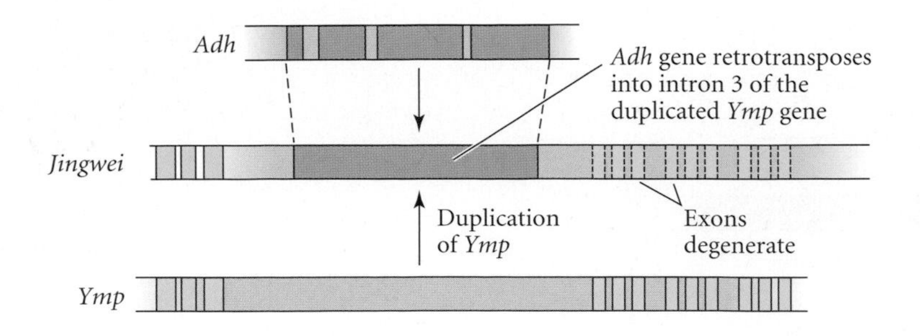

Figure 19.14 Origin of a new *Drosophila* gene, *jingwei* (blue box). Two copies of *Yellow emperor* (*Ymp*) arose by gene duplication. An ancestral *Adh* gene retrotransposed into intron 3 of one of these copies of *Ymp* approximately 2.5 Mya. After the retroposition event, the exons downstream of the novel *Adh* exon of *Ymp* degenerated because the *Adh* transcript provided a new stop codon. Although *Adh* is known to function in the regulation of alcohol tolerance and the function *Ymp* is largely unknown, the function of *jingwei* was recently found to be related to the metabolism of hormones and pheromones (Zhang et al. 2004), a new function that has evolved in the past 2–3 Myr. (After Long et al. 2003a.)

scribed into cDNA, and inserted into another gene. The result is a gene with contiguous sequence—groups of exons or entire genes—that is uninterrupted by introns, yet corresponds closely to sequences that in other genes are interrupted by introns.

The first discovered example of this process was the *jingwei* gene, found only in the species *Drosophila teissieri* and *D. yakuba*. The *jingwei* gene consists of four exons, the first three of which are homologous to the *Yellow-emperor* (*Ymp*) gene found in both these and other *Drosophila* species Figure 19.14). These three have a length typical for *Drosophila* exons, whereas the fourth exon is as long as an entire gene. The sequence of this fourth exon is more than 90 percent similar to the entire coding sequence of the well-studied gene for alcohol dehydrogenase (*Adh*). Although the *Adh* gene contains multiple introns, the fourth "exon" of *jingwei* is intronless: clearly, a retrotransposed copy of *Adh* landed in the middle of the third intron of the *Ymp* gene in the ancestor of *D. teissieri* and D. *yakuba*, producing a chimeric gene. The first three exons of *Ymp* modified the expression pattern of the retrotransposed *Adh* gene, because *jingwei* exhibits the precise testis-specific expression pattern shown by *Ymp* in both species, despite additional expression in other tissues and developmental stages in *D. yakuba*.

Even without corroborating evidence of a related gene possessing introns, any intronless gene is likely to have originated by retrotransposition, which often leads to intron loss. For example, most of the large and diverse family of G-protein coupled receptors (GPCRs) found in humans and other mammals are intronless, whereas many invertebrate GPCRs have introns (Gentles and Karlin 1999). It is likely that the mammalian gene family originated by retrotransposition of an ancestral gene that had introns.

Many such retrotransposition events result in new but nonfunctional "genes" called **processed pseudogenes**.* Although their DNA sequence may resemble that of the related functional genes from which they were originally copied, they commonly have deletions that destroy the reading frame of the gene and stop codons that occur before the correct termination point. Processed pseudogenes are common in the human and other eukaryotic genomes; in humans there are at least 8000 processed pseudogenes, and probably many more (Torrents et al. 2003; Zhang et al. 2003). Ultimately, the origin of processed pseudogenes, and even their identity as such, becomes unrecognizable because they accumulate mutations that erode signature sequences that would indicate a pseudogene's origin.

Motif multiplication and exon loss

The multiplication of specific motifs within genes can give rise to new genes with new functions. The notothenioid fishes of the Southern Ocean, around Antarctica, are famous for their ability to thrive at ocean temperatures at which most vertebrates' blood would freeze. These fishes have evolved a variety of antifreeze glycoprotein (*AFGP*) genes encoding short polypeptides that serve to break up ice crystals and prevent their blood from freezing. *AFGP* genes encode short, three-amino acid monomers (Threonine-Alanine-Alanine, or ThrAlaAla; Figure 19.15A) repeated over and over, comprising 234 amino acids

*A *pseudogene* is any nonfunctional DNA sequence that has been derived from a functional gene; a *processed pseudogene* is a pseudogene that has arisen via retrotransposition of mRNA into cDNA.

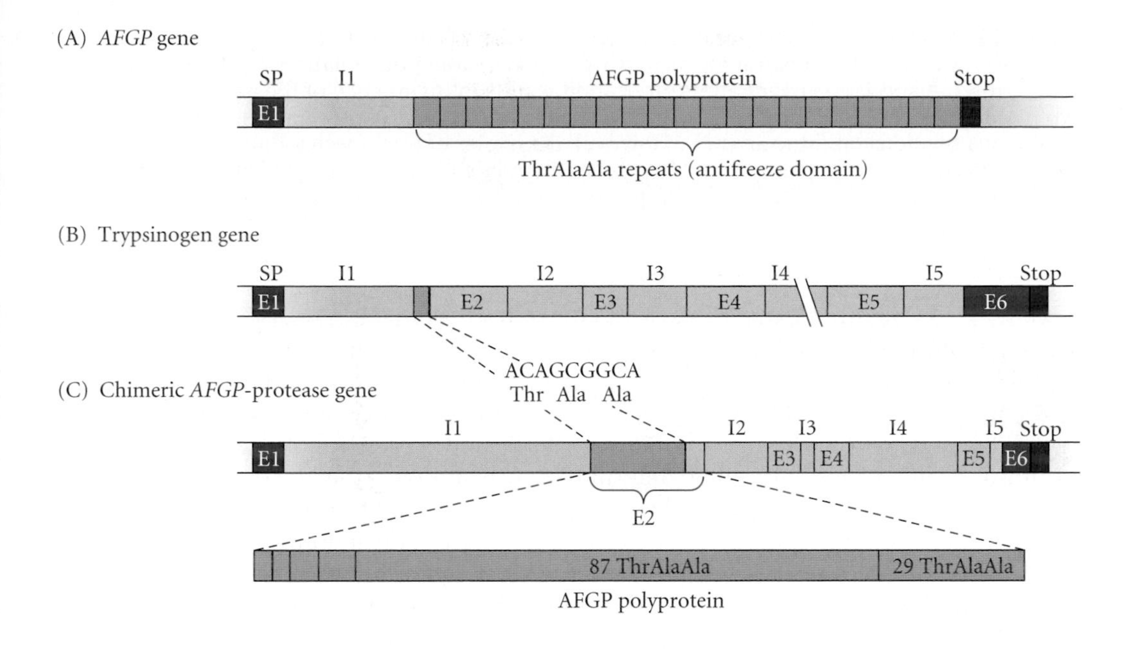

Figure 19.15 The evolution of *AFGP* genes of Antarctic notothenioid fishes. (A) A typical functional *AFGP* gene from an Antartctic notothenioid, showing the repeated ThrAlaAla motifs encoded in exon 2 (E2) that comprise the antifreeze domain. (B) A typical trypsinogen gene showing the single ThrAlaAla motif at the beginning of exon 2, as well as exons 1 and 6 (in red) that are also found in the *AFGP* gene. (C) A chimeric *AFGP* gene from a giant Antarctic toothfish (*Dissostichus mawsoni*). The exon in blue, encoding the ThrAlaAla repeats, is a manyfold expansion of the single motif located just upstream of exon 2 in the trypsinogen gene. The *Dissostichus AFGP* gene possesses many of the exons of a typical trypsinogen gene, but has a hybrid second exon (E2) consisting mostly of an expanded triplet motif antifreeze domain. (After Cheng and Chen 1999.)

in all. They share several features in common with the gene for trypsinogen, which has a completely different function as a protease.

The entire beginning of the trypsinogen gene, extending as far as the beginning of exon 2, resembles structures in the beginning of the *AFGP* gene, including a single ThrAlaAla motif at the start of trypsinogen exon 2 (Figure 19.15B). Additionally, the sequence of exon 6 and the 3′ end of the *AFGP* gene resembles that of trypsinogen. However, the *AFGP* gene lacks any equivalent of the central exons of the trypsinogen gene. The evidence suggests that the single triplet amino acid motif in exon 2 of an ancestor of the trypsinogen gene expanded to produce exon 2 of the *AFGP* gene. At some point during or after this event, exons 3–5 of the ancestral trypsinogen gene were lost. In fact, a chimeric *AFGP* gene has been found in one species, the giant Antarctic toothfish (*Dissostichus mawsoni*), that seems to represent an evolutionary intermediate in the transition from a trypsinogen-like ancestor and a modern *AFGP* gene (Figure 19.14C). Such genes are believed to have played a crucial function permitting the notothenioids to colonize and diversify in the extremely cold waters surrounding Antarctica.

Gene duplication and gene families

One of the most common ways in which new genes originate is by **gene duplication**, in which new genes arise as copies of pre-existing genes. Many genes are members of larger groups of genes, called **gene families**, that are related to one another by clear common ancestry, and which often have diverse functions that nonetheless have a common theme. For example, globin genes (see Figure 8.3) all encode proteins that have a heme-binding domain and can bind oxygen. In mammals, the ε-, ζ-, and γ-globin chains have higher oxygen affinities and are expressed in embryonic tissues, whereas the α- and β-globins function in the adult. The molecular mechanisms of gene duplication are still poorly understood; it occurs at the DNA level (since the members of gene families usually have similar intron-exon structure), and unequal crossing over (see Figure 8.5) is known to change copy number in gene families.

The relationships among members of a gene family can be analyzed phylogenetically, both among species and within a species. These two kinds of relationships among genes represent two forms of homology, and so warrant different terms: **orthology** and **paralogy**. Orthologous genes are found in different organisms (e.g., different species) and have diverged from a common ancestral gene by phylogenetic splitting at the organismal level. In contrast, the members of gene families are paralogous, originating from a common ancestral gene by gene duplication (Figure 19.16).

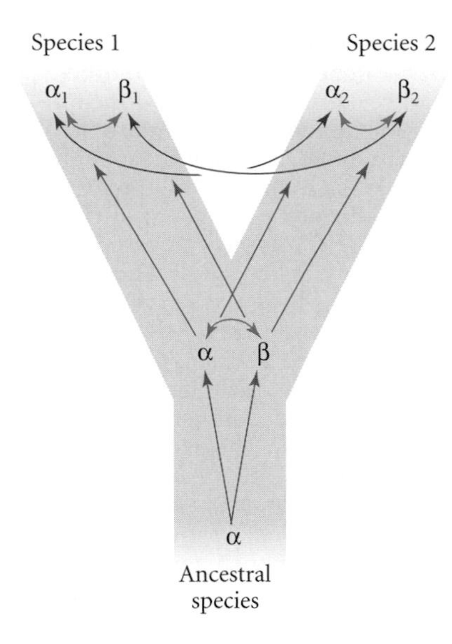

Figure 19.16 Orthology and paralogy in gene families. When an ancestral gene (α) undergoes duplication, the two resulting genes (α + β) have a paralogous relationship to one another (blue arrow). A speciation event after duplication results in divergence of the ancestral set of two paralogous genes. Within the genomes of the two diverged species, the α + β still have a paralogous relationship (blue arrows). However, the copies of α in species 1 and species 2 are orthologous (red arrows), because the two genes are related to one another via speciation, not duplication. Likewise, the copies of β in species 1 and 2 are orthologous.

Many genes get duplicated as parts of large chromosomal blocks, or even as parts of whole genomes (polyploidy; see Chapter 16). Evidence for such events comes from finding that many duplicated genes seem to have diverged at approximately the same time, based on comparisons of DNA sequences. Whether the entire genome was duplicated in one "big bang," or whether much of it was duplicated piecemeal over a long time (the "continuous mode" model) can be determined by examining the age distribution of divergences between many duplicated paralogs, making judicious use of molecular clocks. For example, it has been postulated that the entire vertebrate genome was duplicated in the common ancestor of jawed vertebrates and then again in the ancestor of the fishes (Prince and Pickett 2002). Gu et al. (2002) estimated the time of duplication for 1,739 gene duplication events inferred from a phylogenetic analysis of 749 different gene families in the human genome. The distribution of divergences between paralogs provided evidence for both the "big-bang" hypothesis (a major peak in the frequency distribution at about 500 Mya) and the "continuous mode" hypothesis (the continuous distribution of divergence times, including a peak near the base radiation of mammals) (Figure 19.17). Thus the large-scale structure of mammalian genomes may have been produced by both a large and many small duplications.

Large duplications (paralogous regions) often contain hundreds of genes. For example, interspersed among the many pathogen-resistance genes of the major histocompatibility complex (MHC) on human chromosome 6 are many housekeeping genes, such as collagen (*COL*) genes. Similar sets of genes, in the same order, have been found on chromosomes 1, 9, and 19 as well (Figure 19.18). For several of the housekeeping genes, the

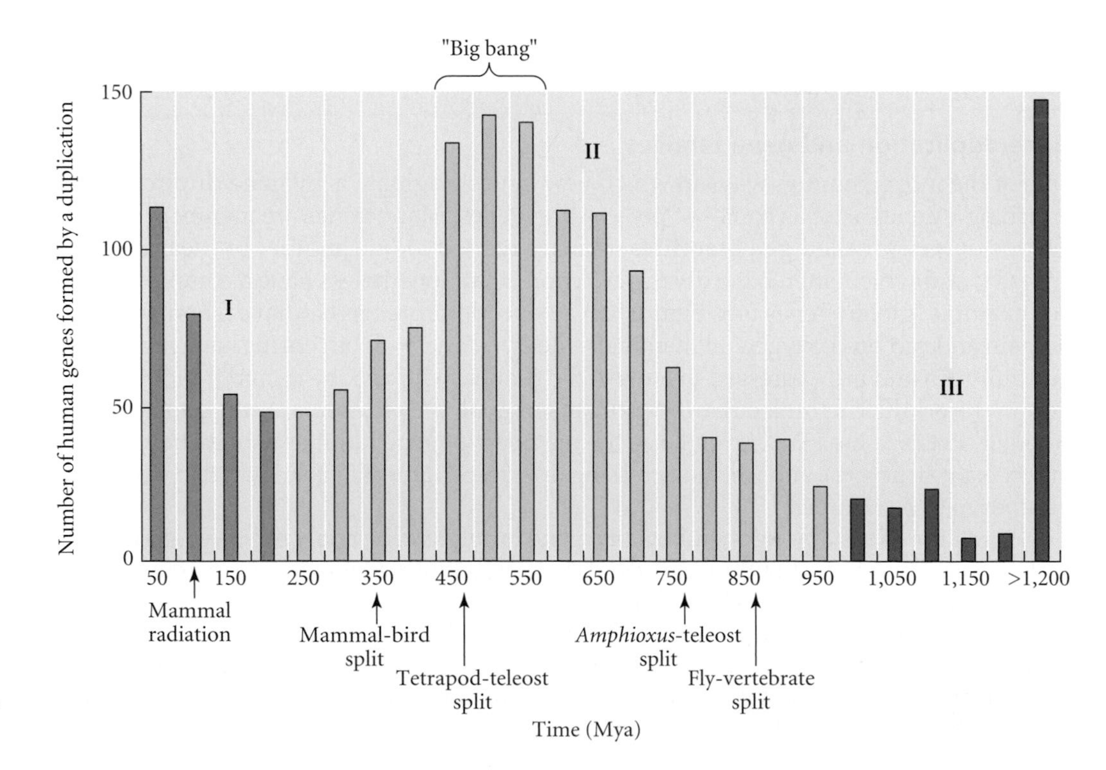

Figure 19.17 Use of age distribution of gene duplication events to infer whole-genome duplications. The X-axis shows the estimated time of divergence of gene duplicates in the human genome based on analysis of sequence differences. The Y-axis shows a count of gene duplicates in each age class. The figure shows three waves, I, II, and III. Wave I corresponds to gene duplications that occurred during the radiation of mammals and consists of certain large gene families, such as the immunoglobulin family in mammals. Wave II corresponds to duplications that took place during early vertebrate evolution and consists of tissue-specific isoforms and other developmental loci. Wave III consists of duplications that took place very early in metazoan evolution, when a number of novelties in signal transduction pathways are thought to have originated. (After Gu et al. 2002.)

Chromosome arm			
6p (MHC)	**19p**	**1q**	**9q**
RXRB	–	*RXRG*	*RXRA*
COL11A2	–	*COL11A1*	*COL5A1*
RING3	–	–	*RING3*-like
LMP2/LMP7	–	–	*PSMB7*
TAP1/TAP2	–	–	*ABC2*
NOTCH4 (INT3)	*NOTCH3*	*(NOTCH2)*	*NOTCH1*
PBX2	–	*PBX1*	*PBX3*
TNX	–	*TNR*	*HXB*
CYP21	*CYP2*	–	–
C4A/C4B	*C3*	–	*C5*
HSPA1	–	*(HSPA6/7)*	*GRP78*
HLA-A,-B,-C	–	*CD1*	–
–	*VAV1*	–	*VAV2*
–	*LMNB2*	*LMNA*	–
–	–	*SPTA*	*SPTAN1*
–	–	*ABL2*	*ABL1*

Paralogous *Notch* genes are present on all four chromosomes and occur upstream of a gene family known as PBX.

Figure 19.18 Block duplication. Paralogous genes belonging to a number of gene families are listed for four human chromosomes. Many of these paralogs occur in the same relative positions to one another on these chromosomes, suggesting duplication of an ancestral block. (After Kasahara 1997.)

existence of such paralogous regions has been detected even in invertebrates such as the nematode *C. elegans*. MHCs are only found in jawed vertebrates, but the chromosomal region containing them may have a very ancient origin.

As shown in Figure 19.19, gene families vary greatly in size, from just two members (the most common number in both yeast and human genomes; Dujon et al. 2004; Gu et al. 2002) to as many as 800 (e.g., the immunoglobulin superfamily) or 1000 (e.g., human ribosomal RNA genes). The size of gene families often coincides with specific adaptations. For example, mammals have hundreds of genes that encode olfactory receptor proteins, each of which binds one or a few odorant chemicals.

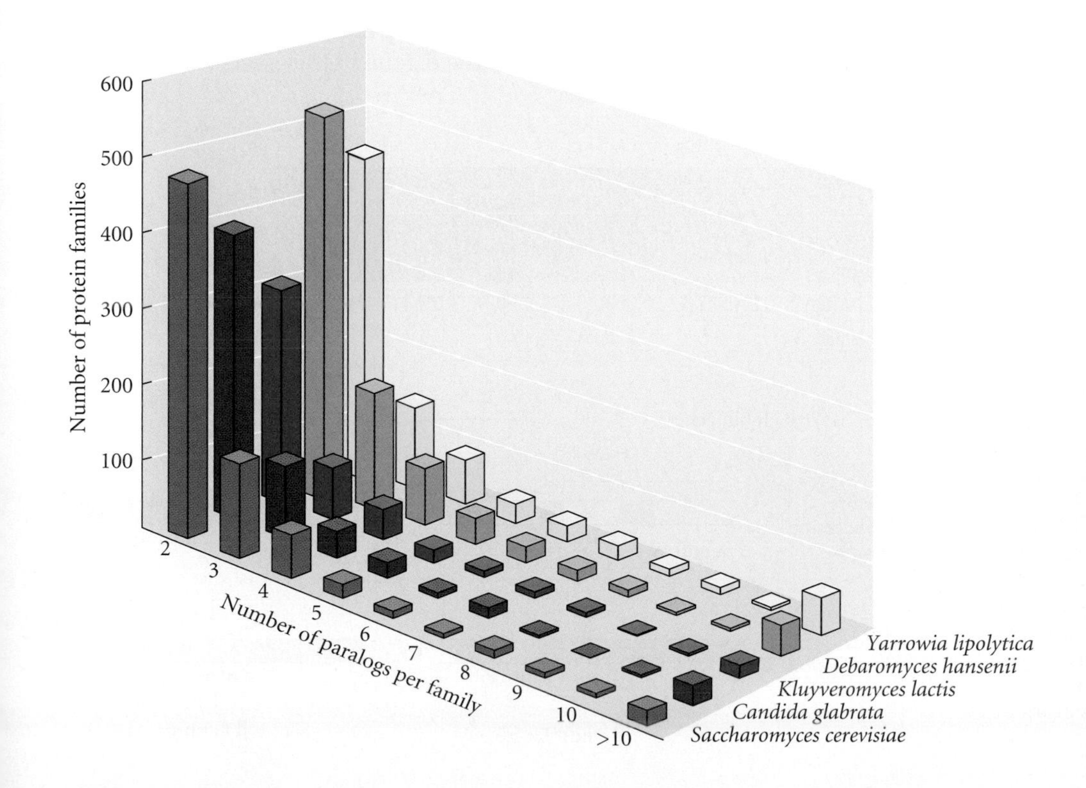

Figure 19.19 Distribution of number of paralogs in the complete genomes of five species of yeast. This distribution has a similar shape in many organisms in which it has been tested, except that the largest size classes of paralogs are larger in complex eukaryotes such as humans than in simpler eukaryotes such as yeast. The differences among these species suggest that differential amplification of genes can take place in different species, often as a result of adaptation to different environments. (After Dujon et al. 2004.)

Phylogenetic and Adaptive Diversification in Gene Families

Duplicate genes can have several possible fates. They may diverge in sequence and (usually) in function, in several different ways that we will describe in the next section. Alternatively, one copy may remain functional, while the other becomes a nonfunctional pseudogene. It is also possible for a locus to be deleted. Thus genes may undergo "birth" and "death," resulting in a turnover of the membership of a gene family.

Gene conversion

Duplicate genes may also undergo **gene conversion**, which occurs when sequence information from one locus is transferred unidirectionally to other members of the gene family, so that most or all acquire essentially the same sequence. The molecular process of gene conversion is poorly understood in most organisms, but its consequence, **concerted evolution** of the gene family, is well known. It results in production of the same gene product from multiple loci, which can be adaptive if large quantities of the product are needed. Ribosomal RNA, a major component of ribosomes, is an example, and is produced by a large gene family. It would be deleterious if different transcripts of rRNA had different sequences. Abalone VERL, cited early in this chapter, is another example of concerted evolution.

The globin gene family of many animals is known to undergo concerted evolution and frequent gene conversion, sometimes with observable consequences for gene expression patterns. For example, the δ-globin gene of many primates is typically expressed at a very low level, whereas the the β-globin gene is highly expressed. However, in galagos (a phylogenetically basal branch of African primates, related to the Madagascan lemurs), the δ-globin gene is highly expressed. To understand the basis of this pattern, Morris Goodman's group (Tagle et al. 1991) sequenced the δ- and β-globin genes of a galago and compared these to each other and to the pattern of divergence observed between these genes in humans. Whereas in humans the coding regions, upstream regulatory regions, and intron 1 are highly divergent between the δ- and β-globin genes, these same regions are virtually identical for the galago genes (Figure 19.20). By scrutinizing the sequence differences between the galago δ- and β-genes, it was determined that a conversion tract spanning the first one third of the gene had been transferred from the β- to the δ-globin gene. By applying a molecular clock to the number of noncoding differences observed in intron 2, the researchers estimated that the conversion event occurred 18–24 Mya, long after galagos and humans had a common ancestor. The fact that the 5′ regulatory regions of the galago genes were identical explained the high expression level of the δ gene in this species.

(A)

Globin pair	Percent sequence divergence: 5′ region	Coding regions	Intron 1	Intron 2
Human δ - Human β	14.8	23.0	12.4	—
Galago δ - Galago β	0.8	5.0	0.0	18.0

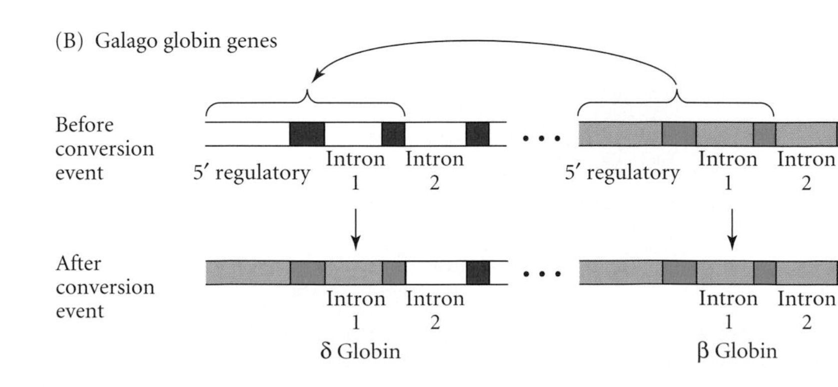

Figure 19.20 Evidence for localized gene conversion and concerted evolution in the primate globin gene family. (A) Percent sequence divergence between various regions of the δ and β globin genes in humans and in a galago. Note the high sequence similarity between the 5′ upstream region, the coding regions, and intron 1 of the galago genes. (B) Conversion of the first third of the galago δ gene by the galago β gene, in effect converting the δ gene into another β gene. The bracket above the genes before gene conversion indicates the region participating in the conversion event. Colors of the exons and gray and white denoting introns indicate the origin of various regions. The hybrid origin of the galago δ gene (bottom left) is indicated by the multiple colors . (After Tagle et al. 1991.)

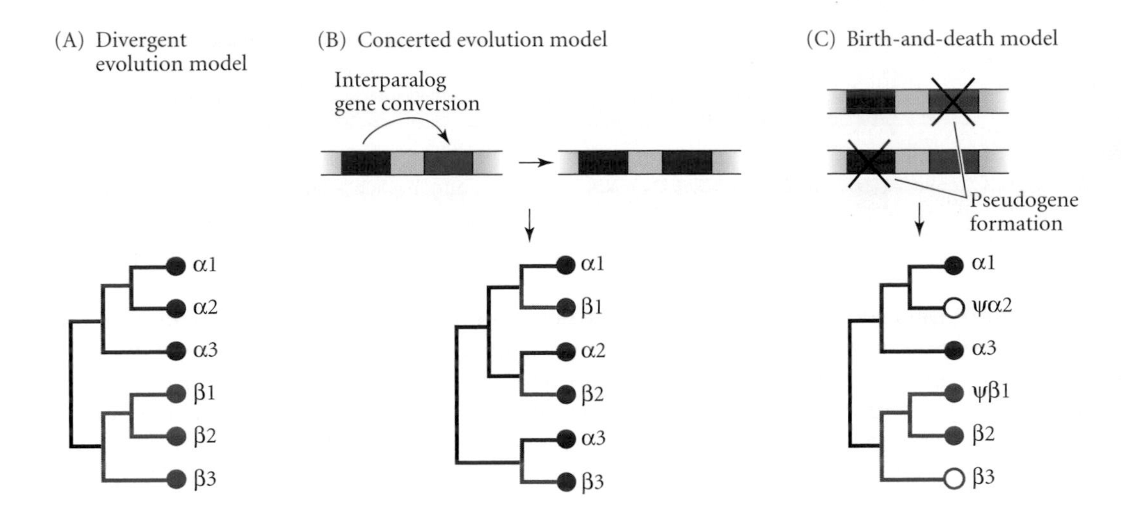

Figure 19.21 Phylogenetic consequences of duplication, speciation, and gene conversion in gene families. In each diagram, individual paralogs are indicated by Greek letters and a distinct color, and different species in which a given paralog is found are indicated by a number. Thus α1 is the copy of gene α in species 1; β2 is the copy of gene β in species 2. (A) In the divergent evolution model, gene duplication takes place before speciation. The result is a tree in which the paralogs from different species form separate clusters, within which the genes share the same phylogenetic relationships as the species do. (B) In the concerted evolution model, frequent interparalog gene conversion results in very closely related genes occurring in the same genome. Converted genes are colored purple because sometimes only portions of genes are converted, as in Figure 19.20. Such genes cluster closely with one another in a phylogenetic analysis and result in clusters of genes that are largely species-specific. (C) In the birth-and-death model, pseudogene formation balances gene duplication, the result being an equilibrium number of functional genes within a given genome. Pseudogene formation results in a patchwork of functional genes (colored circles) and pseudogenes (white circles) in the phylogeny.

Phylogenetic patterns following gene duplication

When duplication occurs in the common ancestor of two or more species, these several fates of duplicate genes result in different phylogenetic patterns (Ota and Nei 1994):

1. When gene duplication precedes speciation and the duplicates diverge in sequence both within and between species, the major clusters in the resulting phylogenetic tree will correspond to the different paralogs. Within each paralog, the phylogenetic relationships of the species sampled will be reflected (Figure 19.21A).
2. If loci have undergone concerted evolution, synapomorphic mutations that occurred in any ancestral species will be shared by all paralogs in all the species derived from that ancestor. Therefore, a tree based on these sequences will display the species' phylogeny, and the paralogs will cluster within each species (Figure 19.21B)
3. If some paralogs are lost in one or more species, or have not been characterized because their sequence has diverged so much that they have not been recognized as paralogs, the phylogenetic tree may display an intermingling of functional and nonfunctional genes (Figure 19.21C).

Selective fates of recently duplicated loci

Paralogous genes are initially redundant: when gene duplication occurs, two identical copies of a gene suddenly exist within a genome. This opens the possibility for functional diversification, which may occur according to two models.

NEOFUNCTIONALIZATION. Susumo Ohno (1970) first articulated the classical model of **neofunctionalization**, whereby one of the duplicates retains its original function and the other acquires a new function, due to fixation of certain new mutations. Evidence of this process is the rapid accumulation of nonsynonymous substitutions in only one of the recently duplicated copies. Neofunctionalization will occur only if advantageous new mutations occur *before* one of the duplicates loses function due to fixation of disabling mutations, thus becoming a pseudogene.

Two genes in the ribonuclease gene family in primates—eosinophil cationic protein (*ECP*) and eosinophil-derived neurotoxin (*EDN*)—provide an example. Zhang et al. (1998) found that the *ECP* gene experienced a large number of nonsynonymous substitutions after duplication from the *EDN* gene. Moreover, *ECP* possesses an anti-pathogen function not found in *EDN*, suggesting that functional diversity was acquired after duplication by rapid accumulation of amino acid substitutions.

SUBFUNCTIONALIZATION. Force et al. (1999) have proposed another mode of adaptive divergence, **subfunctionalization**, whereby each gene duplicate becomes specialized for

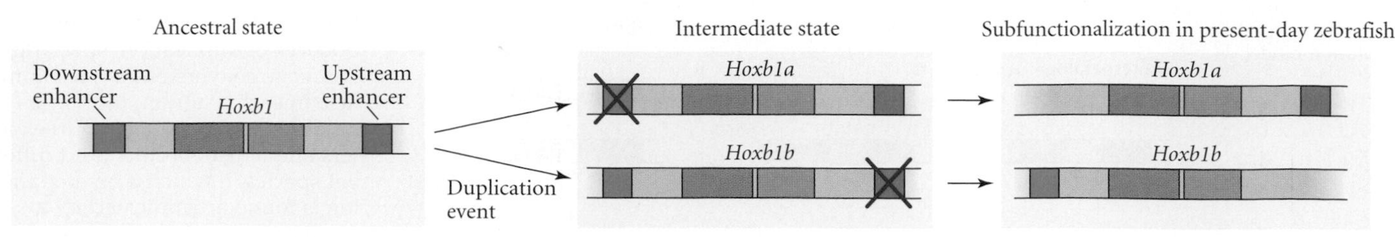

Figure 19.22 The DDC model of subfunctionalization, as illustrated by *Hox* genes. In an ancestral vertebrate, the *Hoxb1* gene had dual functions and was regulated by dual upstream and downstream enhancer domains. In an intermediate derived species, this gene underwent a duplication event, followed by loss of one of these functions and change of expression pattern in the two novel paralogs (center). This loss is indicated by the crossing out of either the blue or green enhancer, respectively. The derived state of the duplicated genes (right) consists of two paralogs each with a single function—a process known as subfunctionalization. (After Prince and Pickett 2002.)

a subset of the functions originally performed by the ancestral single-copy gene. This model, called the DDC model (for duplication-degeneration-complementation), hypothesizes that an ancestral gene had two or more functions, and that in each paralog, complementary mutations are fixed that reduce or eliminate a different function. The paralogs are therefore no longer redundant, so both are preserved by natural selection, and may later undergo further functional specialization and evolutionary change. The DDC model differs from the classical model in that both of the duplicate loci are expected to undergo changes in sequence and function compared to the ancestral gene. It also predicts a higher rate of retention of gene duplicates. In polyploid plants, over 15 percent of gene duplicates tend to be functionally retained, a much higher rate than predicted by the classical model (Prince and Pickett 2002).

In the zebrafish, the single *Hoxb1* gene found in other vertebrates (such as mice) has been duplicated into *Hoxb1a* and *Hoxb1b*. Whereas mouse *Hoxb1* is expressed continuously in the developing hindbrain, zebrafish *Hoxb1a* and *Hoxb1b* are expressed sequentially, with *Hoxb1a* terminating about 10 hours after fertilization, at which point *Hoxb1b* takes over (Prince 2002; Prince and Pickett 2002). *Hoxb1a* and *b* have each lost single regulatory sequences upstream and downstream of the gene, both of which are present and functional in mouse *Hoxb1*. The complementary expression profiles and complementary degenerate mutations of the zebrafish genes exemplify subfunctionalization (Figure 19.22).

Rates of gene duplication

The great diversity of large gene families indicates that gene duplication is a common and ubiquitous process. Recently, Michael Lynch and colleagues (2000) estimated that the rate of gene duplication in *C. elegans*, humans and other recently sequenced genomes is about 0.01 duplication per gene per million years—much higher than previously thought. However, Gao and Innan (2004) found an unexpectedly high rate of gene conversion between duplicated paralogs in yeast. These conversion events made the paralogs appear much more similar in sequence, and hence much younger, than Lynch and colleagues initially thought, and suggest a much lower overall rate of gene duplication in yeast.

Regardless of these new findings, and even though the vast majority of new duplicates simply degenerate and do not contribute to the functional genome, adaptive divergence of duplicate genes is clearly one of the primary modes of functional gene diversification in both eukaryotic and microbial genomes, and is one of the keystones of adaptive diversification at the molecular level. The study of gene duplication illustrates the rapid pace of comparative genomics, and how multiple types of data and inference are required to refine our picture of genome evolution.

Summary

1. New technologies used in the emerging field of genomics enable biologists to study the genome on hitherto unprecedented scales. The structures of thousands of genes and the content of many entire genomes can be compared to one another to search for patterns of ancestry and evolution. Microarray technology provides a means for investigating the expression status of thousands of genes simultaneously when the organism's environment or physiological state is perturbed.
2. The neutral theory of molecular evolution, developed by Motoo Kimura in the 1960s, provides the basis for understanding gene evolution. By contrast, a detailed "theory of

genomes" that can explain the diversity of genomic structures and sizes across life forms is only just now emerging, and relies heavily on older population-genetic principles.

3. The ratio of the number of nonsynonymous to synonymous substitutions per site in protein-coding genes provides evidence of the extent of adaptive evolution. Ratios (ω) of less than 1, signifying purifying selection, are commonly found, but many cases of adaptive evolution at the molecular level (ratios of greater than 1, indicating recent positive selection) have also been demonstrated.
4. Introns are ubiquitous in the genes of most eukaryotes. Their presence in the nuclear copies of many genes that originated in organelle genomes, and their absence in several phylogenetically basal eukaryotes, suggest that for most genes introns have been late evolutionary arrivals.
5. Genome size varies by several orders of magnitude across life forms. The "C-value paradox" refers to the discrepancy between genome size and organismal complexity in eukaryotes. It was resolved by noting that the coding portion of genomes may increase with organismal complexity, whereas the noncoding portion, made up of highly repetitive DNA, transposable elements, and other types of "selfish DNA," varies with features other than complexity, such as population size.
6. New genes arise in genomes through a variety of mechanisms. Lateral gene transfer occurs when a gene is transferred between completely unrelated genomes, presumably by viruses or other genomic vectors. New genes can also arise from preexisting genes by exon shuffling of protein domains. Such shuffling can create chimeric genes with novel functions.
7. Genes can also arise by retrotransposition. Such retrotransposed genes often recruit new exons and regulatory regions up- and downstream, thereby gaining functions and expression patterns that differ from the progenitor gene. Retrotransposition can also give rise to pseudogenes.
8. Genes can duplicate individually or as parts of large chromosomal regions and sometimes as part of whole genome duplications. Gene duplication is frequent in eukaryotic and prokaryotic genomes and provides an opportunity for the generation of novel genes with derived functions. Gene duplication is the major mode of growth of gene families, which are the most complex expression of coding region diversity in genomes.
9. Orthologous genes are genes that are homologous by descent: they have diverged from a common ancestral gene only as a consequence of speciation and divergence among the organisms harboring them. By contrast, paralogous genes are homologous as a result of gene duplication.
10. Duplicate genes sometimes undergo concerted evolution, wherein all or part of the DNA sequences of a gene are transferred unidirectionally to other members of the gene family. Like gene duplication, concerted evolution has dramatic consequences for the phylogenetic relationships of genes in gene familiies and can be inferred by phylogenetic analysis of orthologs and paralogs from multiple species.
11. Neofunctionalization is the process whereby a newly duplicated gene acquires a new function relative to its ancestral gene. In subfunctionalization, by contrast, new gene duplicates undergo complementary degenerative mutations that knock out one of several functions present in the ancestral gene. Thus both duplicates retain one of the ancestral gene's functions, and are preserved. Subfunctionalization is thought to be common among recent gene duplicates.

Terms and Concepts

chimeric gene
concerted evolution
C-value paradox
differential gene expression
domain
domain accretion
exon shuffling
gene conversion
gene duplication
gene family
lateral gene transfer
neofunctionalization
orthology
paralogy
positive selection
processed pseudogene
purifying selection
"selfish DNA"
subfunctionalization

Suggestions for Further Reading

Fundamentals of molecular evolution by D. Graur and W.-H. Li (Sinauer Associates, Sunderland, MA 2000) is a highly readable introduction to many aspects of sequence evolution. Many of these topics are treated in greater depth by A. L. Hughes in *Adaptive evolution of genes and genomes* (Oxford University Press, Oxford 2000). In *A primer of genome science* (second edition, Sinauer Associates, Sunderland, MA 2005), G. Gibson and S. V. Muse provide an introduction to the techniques and promise of this important new field.

Problems and Discussion Topics

1. What hypotheses could account for differences among closely lineages in the rate of synonymous substitutions in a gene? Of nonsynonymous substitutions? In each case, what data will enable you to distinguish among the hypotheses?
2. What are the major differences between viral, bacterial, and eukaryotic genomes? What demographic differences between species might account for some of the differences we observe at the level of whole genomes?
3. How can genomic data be used to address the neutralist/selectionist controversy? Specifically, how could such data determine the relative roles of selection and genetic drift as causes of (a) genetic variation within species, and (b) sequence differences between species?
4. How might you tell if a specific adaptive substitution that has been fixed in a lineage was caused by insertion of a transposable element?
5. What factors might determine the amount of highly repetitive DNA in a genome? How might you test your hypotheses?
6. "Codon bias" is a phenomenon whereby certain codons are substituted more often in phylogenetic lineages than other (synonymous) codons that encode the same amino acid. What might account for codon bias?
7. Distinguish the different hypotheses for how functionally different genes may evolve after gene duplication. What data can distinguish between these hypotheses?
8. What factors might determine whether duplicate genes undergo functional divergence, concerted evolution, or degeneration into pseudogenes?
9. Suppose you are discussing evidence for evolution with someone who does not believe in evolution. Describe three kinds of molecular data that provide evidence that different lineages (e.g., humans and chimpanzees) have evolved from a common ancestor.

Evolution and Development

20

The great morphological complexity and diversity that we see in multicellular organisms is produced by developmental processes that have evolved in response to natural selection. But how do these developmental processes evolve? Direct development in animals illustrates many of the issues involved in addressing this question. DIRECT DEVELOPMENT occurs when embryos develop directly into adultlike forms instead of progressing through a larval stage (INDIRECT DEVELOPMENT). This striking divergence in developmental mode has evolved independently in many animal lineages, including sea urchins, ascidians, frogs, and salamanders (Figure 20.1). The evolutionary forces and genetic mechanisms promoting such radical, and sometimes rapid, changes in development and life history have mystified biologists for over a century. Comparisons of embryogenesis and larval morphogenesis, especially among marine invertebrates, are central topics in both classical developmental biology and modern evolutionary developmental biology.

Gene expression during development molds morphology. Embryos of the garter snake (*Thamnophis*) do not develop limbs. The expression of certain genes during the course of the snake's development prevents the formation of tetrapod limb-forming regions and converts nearly all of the embryo's vertebrae into rib-bearing thoracic vertebrae. (Photo courtesy of Anne C. Burke.)

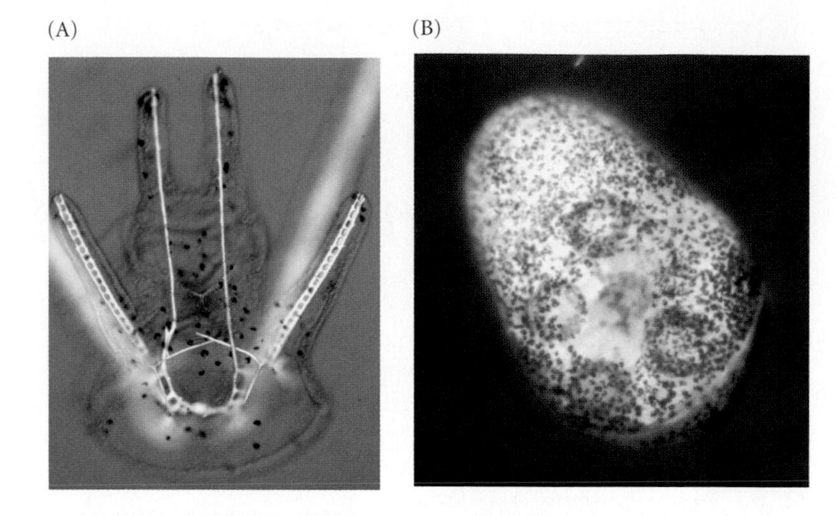

Figure 20.1 Direct versus indirect development. (A) A pluteus larva of the indirect-developing sea urchin *Heliocidaris tuberculata.* (B) A nonfeeding larva of the direct-developing congeneric species *H. erythrogramma*. In *H. etythrogramma*, the ancestral larval mode has been lost and embryos initiate the program for adult morphogenesis without an intervening pluteus stage. None of the complex morphological features of the pluteus are present in *H. erythrogramma* larvae, yet these two species are so closely related that they can be interbred in the laboratory. (Photos courtesy of R. Raff.)

These examples suggest several questions: What are the selective pressures that favor such a novel evolutionary trajectory? How could such a profound alteration of early development evolve so many times? And, perhaps most challenging, what genetic and developmental processes are involved in these evolutionary alterations? It is likely that selection for rapid development promotes the evolution of direct development. But even though some of the genes that underlie these alternative developmental trajectories are beginning to be uncovered, the developmental mechanisms involved—and more importantly, the reasons why these mechanisms are apparently more flexible in some groups of organisms than others—are still mysteries.

The field of **evolutionary developmental biology**, or **EDB** (often called "evo-devo"), seeks to understand the mechanisms by which development has evolved, both in terms of developmental processes (for example, what novel cell or tissue interactions are responsible for novel morphologies in certain taxa) and in terms of evolutionary processes (for example, what selection pressures promoted the evolution of these novel morphologies). Two of the main questions or themes that concern evolutionary developmental biologists are, first, *what role has developmental evolution played in the history of life on Earth?* and second, *do the developmental trajectories that produce phenotypes bias the production of variation or constrain trajectories of evolutionary change?* Natural selection acts on phenotypes produced by development, but ultimately we want to understand how the modes by which development produces those phenotypes affect evolutionary potentials and trajectories.

Hox Genes and the Dawn of Modern EDB

Biologists dating back to Geoffroy Saint-Hilaire (1772–1844), Karl Ernst von Baer (1792–1876), and Darwin himself were fascinated by the patterns of similarity and divergence in development among species. However, until quite recently, the fields of evolutionary biology and developmental biology proceeded along mostly separate paths, with seemingly distinct research programs and methodologies (Gould 1977; Depew and Weber 1994; Wilkins 2002). In the past three decades, however, burgeoning information about the genetic mechanisms of morphogenesis in model organisms, as well as the molecular genetic techniques developed to obtain that information, have been integrated with many strands of evolutionary research to form the highly interdisciplinary field now known as EDB.

The discovery and characterization of the Hox cluster of **homeobox genes** in animals in the 1970s and 1980s marks the dawn of modern EDB (Wilkins 2002). The Hox genes are the best-known class of homeotic selector genes, which control the patterning of specific body structures, as we saw in Chapter 8. Hox genes control the identity of segments along the anterior-posterior body axis of all metazoans. Mutations in the Hox genes often cause transformations of one type of segment into another. In *Drosophila melanogaster*, for example, a mutation of the *Ultrabithorax* (*Ubx*) gene transforms the third thoracic segment (T3), which normally bears the tiny halteres (the *Drosophila* homologue of the hindwing of four-winged insects), into a second thoracic segment (T2), which bears wings (Figure 20.2). A mutation in another Hox gene, *Antennapedia* (*Antp*), causes the misexpression of Antp protein in the cells that normally give rise to the antennae, resulting in the replacement of antennae with legs (see Figure 8.14). *Antp* is normally expressed only in the second thoracic segment (T2), where it controls the development of T2-specific body structures, including legs.

In *Drosophila*, the Hox genes occur in two complexes (clusters) of genes on chromosome 3, termed the Antennapedia complex and the bithorax complex. The pioneering genetic work on the bithorax complex was done between the 1940s and the 1970s by E. B. Lewis, and that on the Antennapedia complex in the 1970s and 1980s by Thomas Kaufman and his colleagues. These investigators found that the genes in both complexes control the anterior-posterior identity of segments corresponding to their order on the chromosome (Figure 20.3). They also discovered that the eight *Drosophila* Hox genes are members of a single gene family, and that the proteins they encode share a particular amino acid sequence that binds DNA, subsequently named the **homeobox** (in the gene) or the **homeodomain** (in the protein). This finding supported Lewis's idea, proposed in the 1960s, that the Hox genes regulate the transcription of other genes. Other researchers were stunned to discover that all other animal phyla also possess a set of Hox genes These genes have homeodomain sequences similar to those of their homologues in *Drosophila* and have the same gene order and orientation as in *Drosophila* (except that they form a single gene complex in most animals). Mammals have four Hox gene complexes (denoted *Hoxa*, *Hoxb*, *Hoxc*, and *Hoxd*) in different parts of the genome, and a total of 13 different Hox genes (as opposed to only 8 in *Drosophila*), although not all of the complexes have all 13 members (see Figure 20.5).

Staining for Hox proteins or mRNA (see Box A) showed that the anterior-posterior expression of the Hox genes corresponds to their mutant phenotypes. For example, as pre-

(A)

(B)

Figure 20.2 Effects of homeotic mutations. (A) A wild-type *Drosophila melanogaster* has a single pair of wings and a pair of small winglike structures called halteres. (B) This mutant fly was experimentally produced by combining several mutations in the regulatory region of the *Ultrabithorax* (*Ubx*) gene. The third thoracic segment has been transformed into another second thoracic segment, bearing wings instead of haltares. (Photos courtesy of E. B. Lewis.)

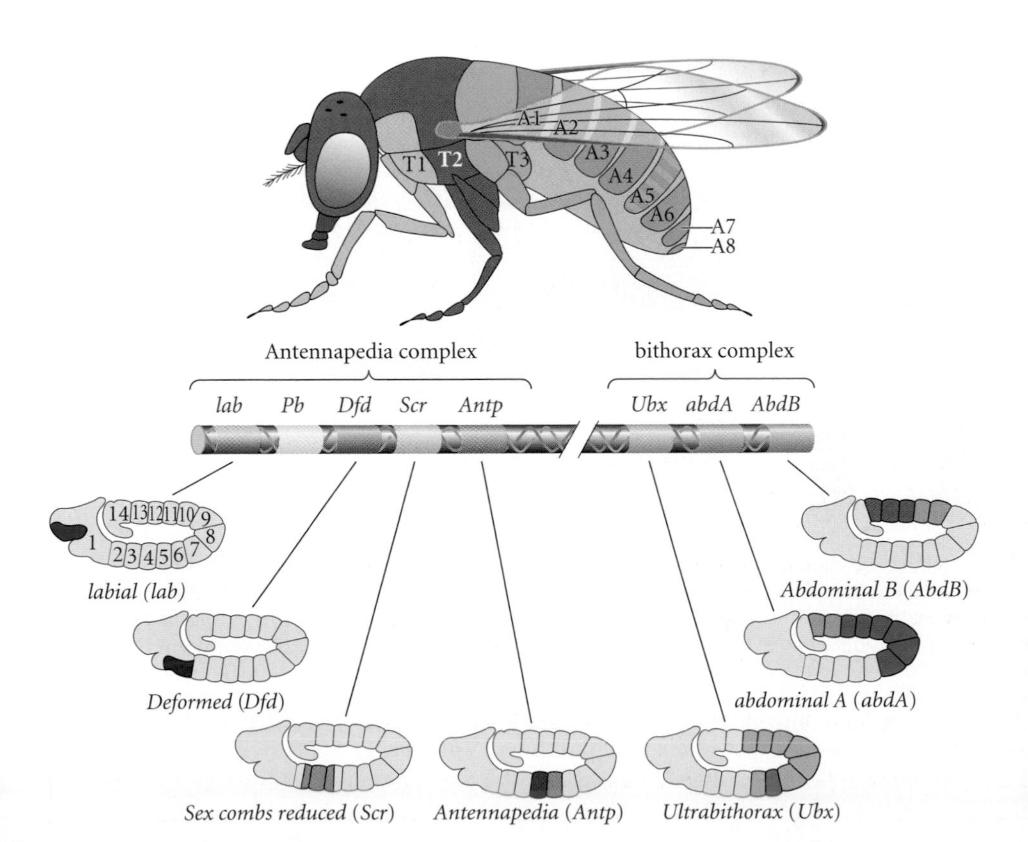

Figure 20.3 Hox gene expression in *Drosophila*. In the center is a map of the genes of the Antennapedia and bithorax complexes, with their functional domains shown in color. The regions of Hox gene expression are shown for the blastoderm of the *Drosophila* embryo (below) and the regions that form from them in the adult fly (above). Darker shaded areas represent those segments in which gene expression is highest (i.e., produces the most protein product). (After Dessain et al. 1992 and Kaufman et al. 1990.)

BOX 20A Characterizing Gene and Protein Expression during Development

Perhaps the most important type of data sought in developmental genetics and evolutionary developmental biology are the expression patterns of specific genes, and the proteins they encode, during development. These patterns are SPATIO-TEMPORAL, with a spatial component (referring to specific cells, tissues, segments, or structures) and a temporal component (referring to specific developmental stages). Gene expression patterns can be visualized by three methods, each requiring different tools and hence currently usable for certain species, but not others.

IN SITU HYBRIDIZATION subjects tissues or whole specimens to a chemical process designed to stabilize messenger RNA molecules in the cells in which they are produced. Then a species-specific, single-stranded RNA or DNA "probe" corresponding to the gene of interest is applied. It hybridizes by base-pairing with the mRNA of interest. The probe is either chemically modified so that it can be detected by a staining procedure or labeled with a radioisotope so that it can be detected by autoradiography (Figure A). Alternatively, extractions of mRNA from different tissues or developmental stages are run in separate lanes on an electrophoretic gel. The mRNA from the gel is then blotted onto a membrane, which is exposed to the same sort of radioactively or chemically labeled probe used in the in situ approach. This procedure is known as a NORTHERN BLOT.

Gene expression patterns can be analyzed at the protein level using antibodies. (The mRNA and protein expression patterns of a given gene may not be identical, due to regulation of translation.) Antibodies are produced by injecting a mammal (e.g., a rat) with the protein of interest (the antigen). The animal produces antibodies (immunoglobulin molecules) that bind specifically to that protein. These "primary" antibodies are collected by passing the animal's blood serum over a resin column containing the antigen and then eluting the antibodies from the column in concentrated form. Tissues or embryos are prepared in a similar way as for in situ staining and incubated with the primary antibody. A secondary antibody, an immunoglobulin that specifically binds to the primary antibody, is then applied to the specimen. The secondary antibody is modified so that it can be detected either by an enzymatic reaction producing a colored product or by fluorescence (Figure B). An alternative to staining fixed tissue is to prepare protein extracts from different tissues or developmental stages and run each extract as a separate lane on an electrophoretic gel. The protein from the gel is then blotted onto a membrane, which is then incubated with primary and secondary antibodies as described above. This procedure is known as a WESTERN BLOT.

A third method is to study the transcription patterns of particular fragments of putative *cis*-regulatory DNA by using REPORTER CONSTRUCTS in cultured cells or in transgenic (genetically engineered) individuals. Reporter constructs consist of the regulatory DNA of interest, spliced upstream of a "reporter gene" that encodes a protein whose expression can be easily visualized under the microscope. One such protein is β-galactosidase, a bacterial enzyme that processes a particular sugar into a blue product. Another is a protein from jellyfish (GFP) that fluoresces bright green when irradiated with light of a particular wavelength. Because reporter construct analysis requires the use of gene transfer technology, it can be undertaken only in certain well-studied model species, such as *Drosophila*, *Caenorhabditis elegans*, *Arabidopsis*, and mice. Figure C shows the nematode *Caenorhabditis briggsae* expressing a GFP reporter construct containing *cis*-regulatory DNA from the *myo-2* gene, which directs the reporter gene's expression in the pharynx.

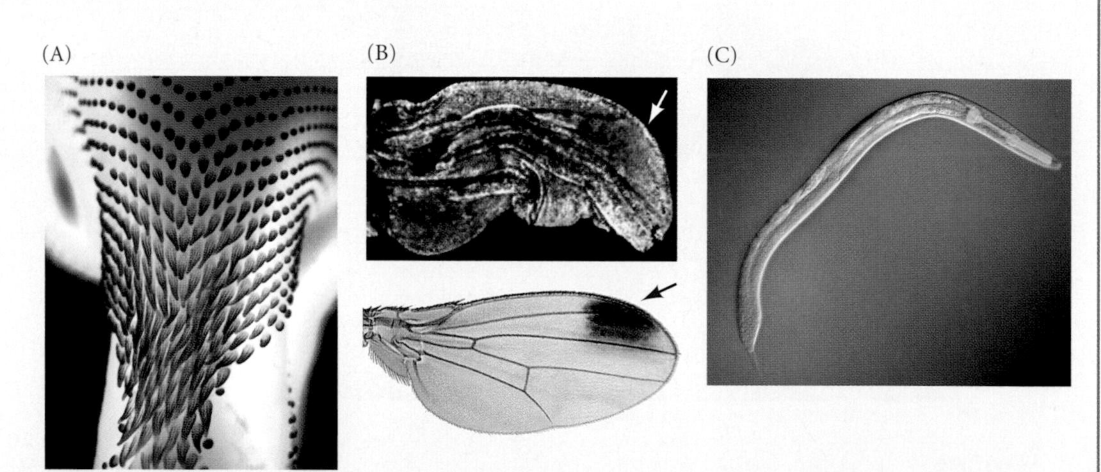

Methods of visualizing gene expression patterns in developing animal tissues. (A) In situ hybridization of *sonic hedgehog* mRNA in developing feathers on the neck region of a chicken. (B) Fluorescent antibody staining of the proteins Yellow (in green) and Ebony (in pink) in the pupal wing of male *Drosophila biarmipes*. The *Ebony* gene is expressed where the pigmented spot is located in the fully developed wing (below). (C) Green fluorescent protein (GFP) reporter gene expression of a transgenic construct containing *cis*-regulatory DNA from the gene *myo-2*, which directs expression in the pharynx of the nematode *Caenorhabditis briggsae* (visualized in bright green). (A, photo by Matthew Harris; B, photos by John True; C, photo by Eric Haag, used by permission of Takao Inoue and Eric Haag.)

Figure 20.4 Segment-specific patterning functions of Hox genes in the vertebrate hindbrain. This schematic diagram of a mouse embryo shows the hindbrain, consisting of a series of segments. The horizontal bars indicate segmental patterns of *Hoxb* gene expression in the mouse hindbrain and spinal cord, with darker color corresponding to areas of relatively high gene expression. The double-headed arrows connect the genes in the *Hoxb* cluster to the homologous Hox genes in *Drosophila*. (After McGinnis and Krumlauf 1992.)

dicted, *Ubx* is expressed in the T3 segment (as well as the anterior abdomen), where it was long known to be required for normal segment identity (see Figure 20.3). Vertebrate Hox expression patterns, although more complex, are also generally expressed in specific anterior-posterior patterns (Figure 20.4).

Mapping the presence and absence of Hox genes on the metazoan phylogenetic tree shows their evolutionary history (Figure 20.5). Two Hox genes have been found in radially symmetrical Cnidaria (jellyfishes, corals), which are the sister group of the Bilateria. Several novel Hox genes arose in the lineage leading to all Bilateria, representing new Hox classes (as evidenced by their homeodomain sequences) that presumably can define increasing degrees of anterior-posterior axis identity.

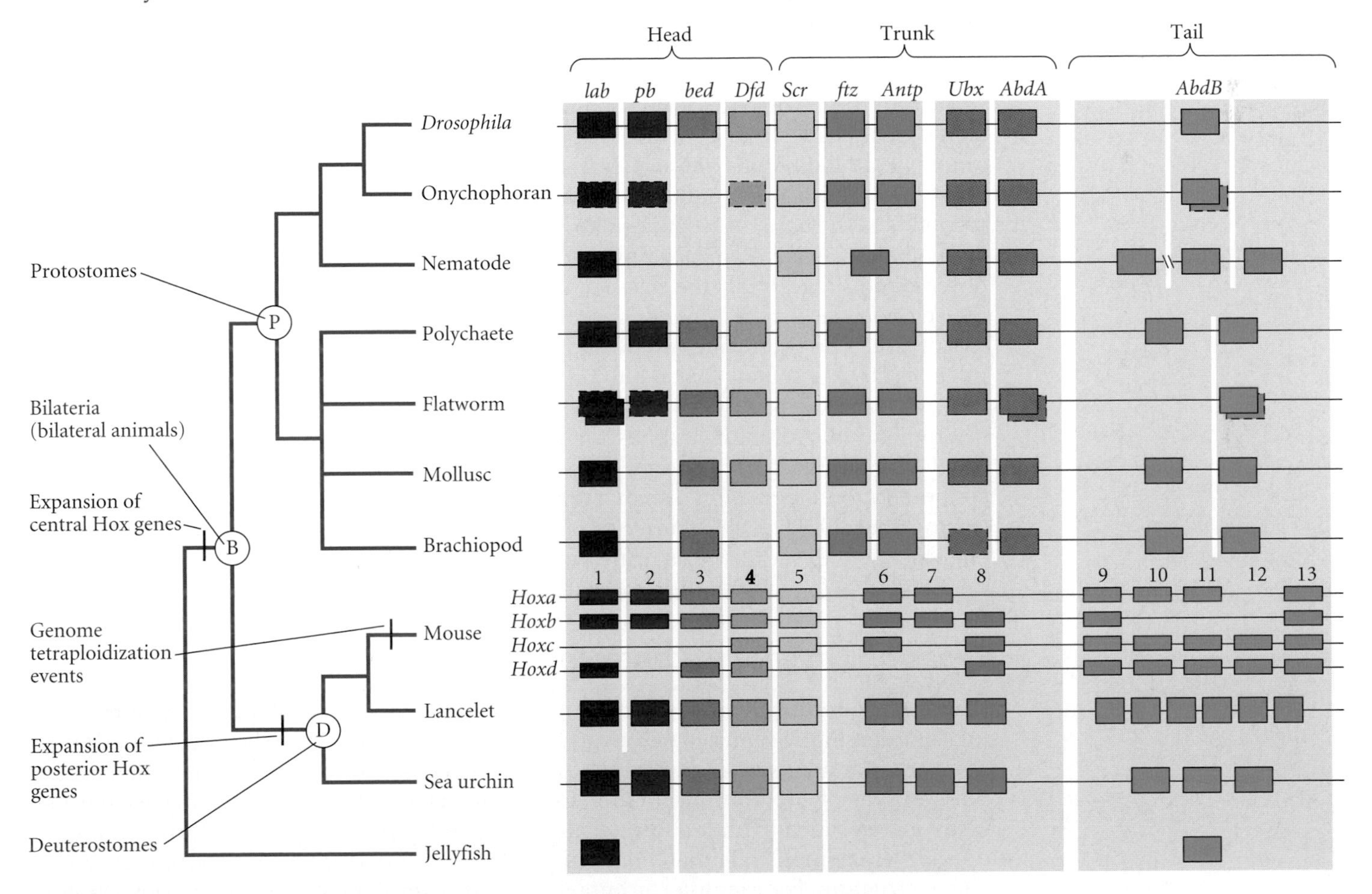

Figure 20.5 Probable evolution of the metazoan Hox gene complex. Vertical white lines delineate currently accepted groups of orthologous Hox genes. Important gene duplication events are indicated by the labeled tick marks. Genes with solid outlines indicate that complete homeobox sequences are known; dashed outlines indicate that only partial homeobox sequences are known. (After Carroll et al. 2001.)

We can hardly overstate the importance of the Hox gene discoveries for our understanding of how animal diversity evolved. For the first time, a common developmental genetic framework unified the ontogeny of all metazoans; before then, few biologists imagined that vertebrates and invertebrates would share such fundamental developmental genetic underpinnings. These discoveries set the stage for further investigations into the potential commonalities among animals in other aspects of development (e.g., in dorsal-ventral patterning and limb development), which have turned out to be plentiful (Carroll et al. 2001; Wilkins 2002). Ironically, such interest in morphological evolution was precipitated by the discovery of *conserved*, as opposed to *evolving*, features. New questions were immediately raised: What is the basis of body plan *differences* among animal taxa? How can seemingly conserved genetic factors play a role in these differences?

These questions are among the most actively investigated in current EDB. For example, the *Ubx* gene was used to test an early hypothesis that major structural differences, such as hindwing differences between dipteran flies and butterflies, might result from simply turning transcription of a particular Hox gene on or off. Sean Carroll and colleagues showed that differences between the tiny *Drosophila* haltere and the large butterfly hindwing were, in fact, *not* due to differences in Hox gene expression; *Ubx* is expressed in both (Warren et al. 1994). Therefore, divergence among taxa in hindwing morphology must be due to differences in the expression of other genes. The Hox genes encode proteins (**transcription factors**) that regulate transcription by binding to DNA control regions (called **promoters**, **enhancers**, or ***cis*-regulatory elements**) of "downstream" or **target genes** (see Box B). Therefore, it is likely that morphological divergence is caused by changes in the expression of genes that the Hox genes regulate. In fact, *Drosophila* and butterfly hindwings differ in the expression of several *Ubx* target genes (Carroll et al. 1995; Weatherbee et al. 1998).

A second type of investigation has shown that differences in segmental Hox expression domains are strikingly correlated with the evolution of animal body plans. For example, in all the groups of crustaceans except the basal branchiopods, which have only one type of thoracic appendage, the anterior margin of expression of *Ubx* and *abdA* corresponds to the boundary between thoracic segments bearing maxillipeds (small, leglike appendages specialized for feeding) and those bearing thoracic limbs (see Figure 3.7). This observation suggests that the change in spatial expression of these Hox genes has enabled the segments and their appendages to become different (to become INDIVIDUALIZED; see Chapter 3). Many correlations of this kind have been found throughout the Bilateria (Carroll et al. 2001), implying that evolutionary change in the expression patterns of Hox genes may underlie key body plan adaptations both within and among phyla.

Studies of the role of Hox genes in animal diversification established a new framework in which to think about morphological evolution, based on the idea that changes in the spatio-temporal regulation of a shared set of genes (sometimes referred to as a "toolkit"; Carroll et al. 2001) are the primary causes of morphological evolution. In the rest of this chapter, we will see how this framework of gene regulatory evolution can help us understand how organismal form evolves.

Types of Evidence in Contemporary EDB

Patterns of gene expression (see Box A) are now frequently used together with morphological, comparative embryological, and phylogenetic data to infer the developmental genetic origins and histories of morphological characters. However, developmental genetic data, such as phenotypic information from mutants or individuals that have been genetically manipulated to under- or overexpress a gene or protein of interest, are required to definitively demonstrate that a particular gene is required for the development of a tissue or structure. For example, recent work on the evolution of angiosperm flowers illustrates how information from mutant phenotypes can be used to form evolutionary hypotheses. In the flowers of dicots, three classes of transcription factors, designated A, B, and C, are required to pattern the four whorls, or concentric rings, of distinct structures: sepals, petals, stamens, and carpels (pistils) (Ma and DePamphilis 2000; Figure 20.6A).

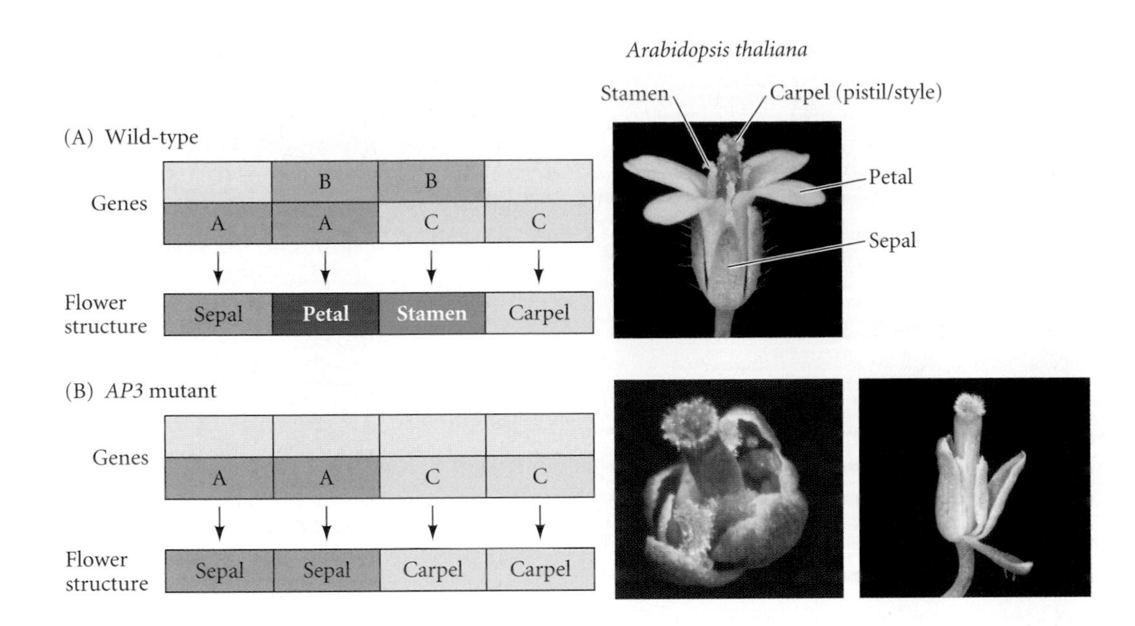

Figure 20.6 The ABC model of flower development. (A) Schematic of the four whorls of angiosperm floral organs with the expression patterns of the A, B, and C class genes. A wild-type *Arabidopsis* flower (right) illustrates the phenotypic result of normal gene expression. (B) In *Arabidosis* mutants lacking expression of the B-class gene *AP3*, petals are transformed into sepals and stamens are transformed into carpels (pistils). (Photos courtesy of J. Bowman.)

The sepals are determined by A gene expression alone, the petals by overlapping A and B gene expression, the stamens by overlapping B and C gene expression, and the carpels by C gene expression alone. Loss of expression of particular gene classes during flower development results in conversion of one structure into another. For example, in *Arabidopsis thaliana* (a dicot), mutations in the B gene *APETALA3* (*AP3*) convert petals to sepals and stamens to pistils (Figure 20.6B). Similar mutant phenotypes and expression patterns of homologues of the ABC genes in a monocot, maize (corn), indicate that the floral patterning system is ancient and have helped to confirm long-supposed homologies between floral structures in different plant taxa (Ambrose et al. 2000). Nucleotide sequence data suggest that the ABC system originated long before the appearance of angiosperms (Purugganan 1997), and presumptive homologues of B and C class genes have been shown to be expressed in similar patterns in nonflowering seed plants ("gymnosperms") (Rutledge et al. 1998; Winter et al. 1999), which diverged from flowering plants (angiosperms) at least 300 million years ago. The ancestral function of these genes may have been to pattern the male and female reproductive organs.

The Evolving Concept of Homology

Under the phylogenetic concept of homology, which is fundamental to all of comparative biology and systematics, homologous features are those that have been inherited, with more or less modification, from a common ancestor in which the feature first evolved. That is, homologous structures are synapomorphies (Donoghue 1992; G. P. Wagner 1989b). Homology may be suggested by a combination of similarity in position relative to other body structures, similarity in at least some structural features, and the presence of intermediate forms, either among species (fossil or extant) or during ontogeny. Once such observations are made, hypotheses of homology are evaluated by the congruence of the character with a phylogeny of the taxon derived from other characters (see Chapter 3).

We would therefore expect genetically and developmentally similar characters to be homologous and phylogenetically homologous structures to have similar genetic and developmental bases. However, many observations conflict with these expectations, leading several evolutionary biologists to propose an additional concept, the **biological homology concept** (Roth 1988, 1991; G. P. Wagner 1989a; P. J. Wagner 1996).

SERIALLY HOMOLOGOUS structures, such as arthropod legs or vertebrate teeth, share much of the same developmental genetic machinery in their ontogeny, but they are clearly not historically homologous within a species. A more profound conflict is posed by many evo-

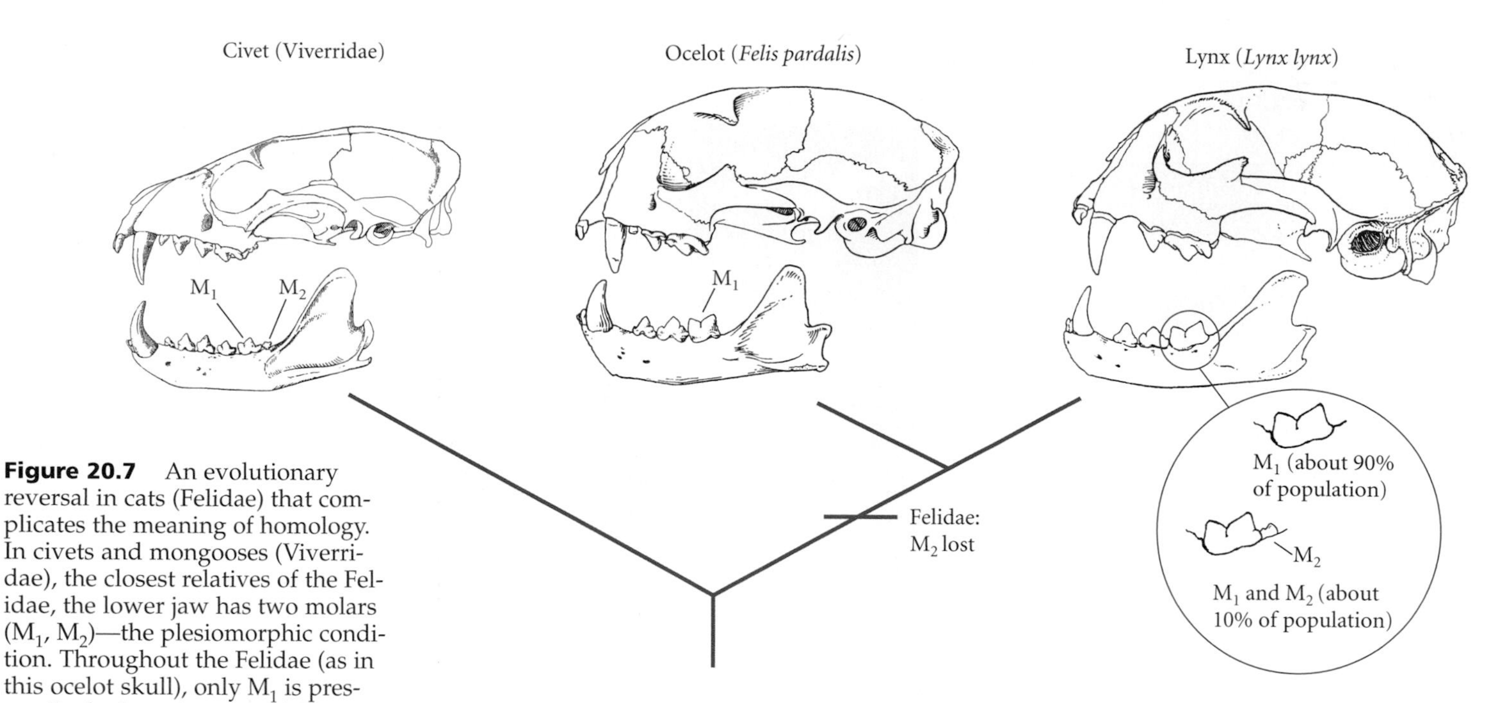

Figure 20.7 An evolutionary reversal in cats (Felidae) that complicates the meaning of homology. In civets and mongooses (Viverridae), the closest relatives of the Felidae, the lower jaw has two molars (M_1, M_2)—the plesiomorphic condition. Throughout the Felidae (as in this ocelot skull), only M_1 is present. In the lynx, however, M_2 reappears in about 10 percent of individuals. (After Hall and Kelson 1959; Kurtén 1963.)

lutionary reversals. For example, the second molar that is present in most carnivores was lost in the ancestor of the cat family (Felidae), but has reappeared in one cat, the lynx, in which it is found in about one in ten individuals (Kurtén 1963; Figure 20.7). The second molar of the lynx is not phylogenetically homologous to the second molar of other carnivores; it is homoplasious. However, from a developmental genetic standpoint, it may represent the "same" tooth, because the mechanisms producing it may be very similar or identical to those producing the structure in other carnivores (Raff 1996).

Another conflict between phylogenetic and biological homology is that phylogenetically homologous traits often have different developmental and genetic foundations. For example, digits differentiate sequentially from back (postaxial) to front (preaxial) in all tetrapods except salamanders, whose digits differentiate in the reverse order. As another example, animal eye lenses all contain various crystallin proteins, but the lens crystallins of different animal lineages have evolved from a wide variety of different ancestral proteins, as we will see below. Conversely, developmentally and functionally similar structures in different taxa may not be phylogenetically homologous. In perhaps the best example, animal eyes evolved independently in very diverse taxa, but in all of these taxa, a highly conserved transcription factor, Pax6, controls eye development. We will examine these examples in more detail later in this chapter.

The concept of biological homology suggests that a feature may be homologous among species at one level of organization (e.g., phenotypic), but not at another level (e.g., genetic or developmental). It emphasizes the idea that multicellular organisms are constructed from a set of more or less conserved tools. These tools may be individual genes and proteins or multiprotein circuits. Morphological evolution has consisted, in large part, of "tinkering" with this toolkit (Jacob 1977). One of the greatest challenges of current EDB studies is to understand how biologically homologous structural units are assembled from these tools.

Evolutionarily Conserved Developmental Pathways

The genes that regulate morphogenesis function in hierarchies or networks termed **developmental pathways** or **developmental circuits** (Box B). These genes encode signaling proteins that relay molecular signals between cells, transcription factors, which respond to signaling pathways by increasing (up-regulating) or decreasing (down-regulating) transcription at target genes, and structural genes, which encode the proteins that actually do the work of development and physiology (e.g., enzymes and cytoskeletal proteins). Sev-

BOX 20B Components of Developmental Pathways

Hox genes are examples of homeotic selector genes, which control cascades of gene expression (i.e., transcription) during the patterning and development of particular tissues, organs, or regions of the body. If a selector gene such as *Ultrabithorax* is not expressed properly, specific tissues or organs may not develop at all, or may be transformed into inappropriate structures (see Figure 20.2).

Transcription factors, including most homeotic selector genes, control the expression of many other genes, including the "structural" genes that encode the proteins, such as enzymes and cell structural components, that actually do the work of morphogenesis. The actions of transcription factors are often regulated in part by cell signaling pathways (Figure 1). These pathways rely on receptor proteins in the cell membrane that respond to extracellular signals such as hormones and short-range signaling proteins called MORPHOGENS. Receptor proteins relay these signals to the genes encoding transcription factors. Seven such cell signaling pathways (each named for a constituent protein, such as Hedgehog or Notch) have been found in animals; others are known in plants. All of these pathways are conserved between *Drosophila* and mammals, and all are involved in many aspects of morphogenesis and pattern formation throughout the developing body, having evolved many specific functions in particular animal lineages, often through gene duplication. Most cell signaling pathways are used multiple times during development, suggesting that morphogenetic novelty may often evolve by re-deploying these pathways in different tissues and at different developmental stages.

Cell signaling pathways and transcription factors are linked into developmental pathways (also called developmental circuits). Such circuits are involved, for example, in patterning the *Drosophila* wing, which takes place in the wing imaginal disc (Figure 2). The end result of developmental circuits are patterns of gene expression that guide

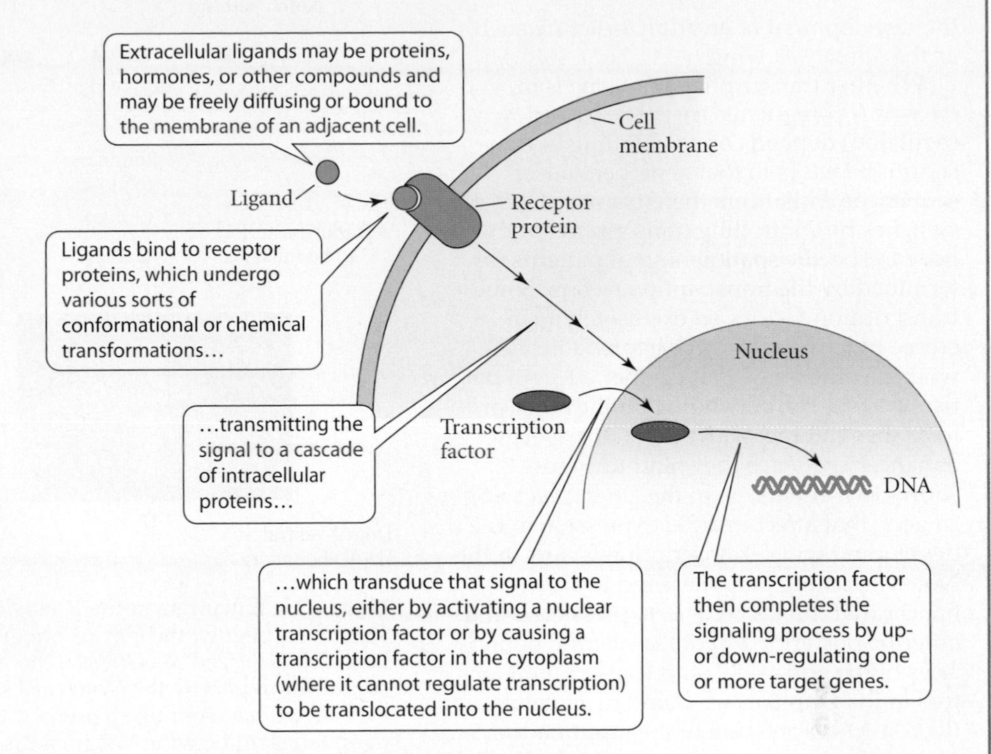

Figure 1 Transduction of intercellular signals through a cell signaling pathway.

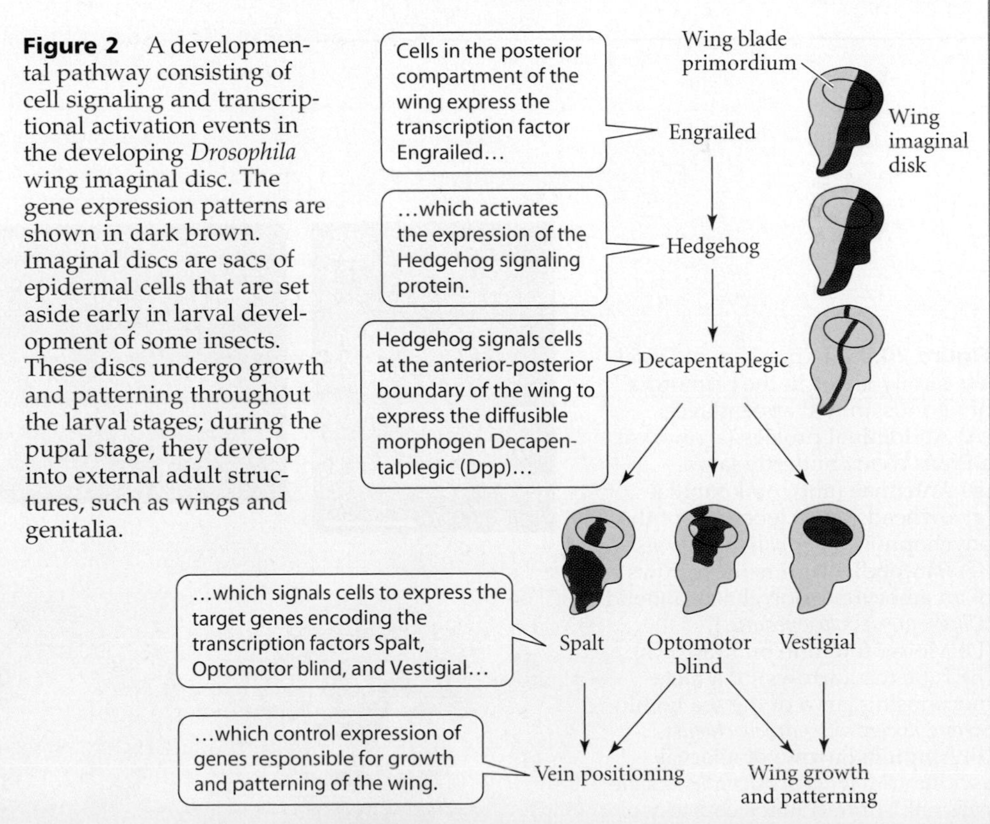

Figure 2 A developmental pathway consisting of cell signaling and transcriptional activation events in the developing *Drosophila* wing imaginal disc. The gene expression patterns are shown in dark brown. Imaginal discs are sacs of epidermal cells that are set aside early in larval development of some insects. These discs undergo growth and patterning throughout the larval stages; during the pupal stage, they develop into external adult structures, such as wings and genitalia.

BOX 20B (continued)

the development of an adult structure such as the *Drosophila* wing.

Whether transcription of a gene is increased (up-regulated) or decreased (down-regulated) depends on the binding of transcription factors to that gene's enhancer sequences. Enhancers therefore act as genetic switches by controlling transcription of the gene in specific spatiotemporal patterns determined by the transcription factors. Some transcription factors act exclusively to increase or to decrease gene expression, whereas others may have either effect in particular contexts, depending on the other proteins they interact with and on the specific enhancer sequence they bind to. Figure 3 shows two enhancers in the *Drosophila vestigial* gene that affect *vestigial* expression in the developing wing. Transcription factors in the Notch signaling pathway bind to an enhancer that directs *vestigial* expression at the anterior-posterior and dorsal-ventral boundaries of the wing field, and transcription factors in the Dpp pathway bind to an enhancer that directs expression in the four quadrants of the wing field that complement the boundary pattern. Normal wing development requires this pattern of expression of *vestigial*.

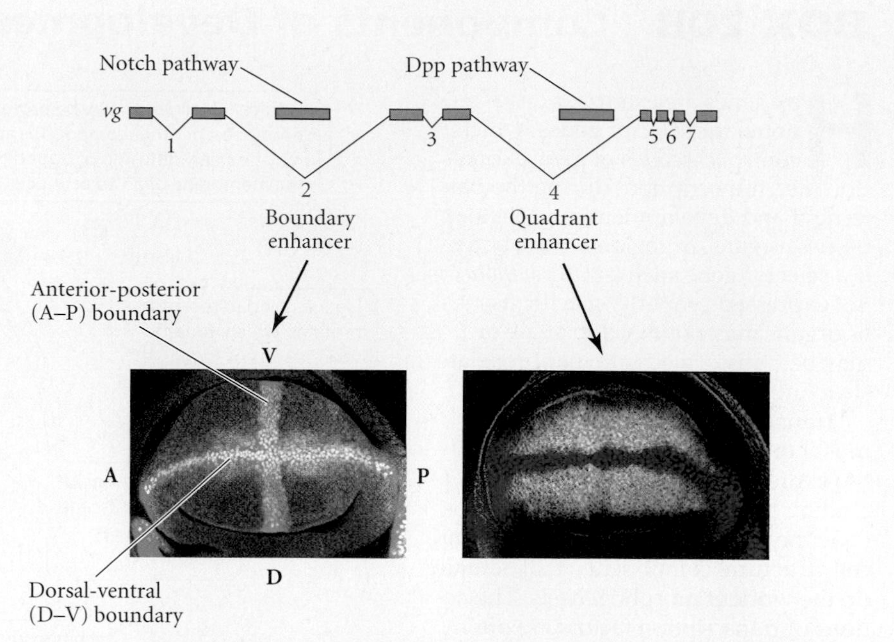

Figure 3 Enhancers in the *Drosophila vestigial* (*vg*) gene. Exons of *vestigial* are shown in gray; the introns are numbered. Two different enhancers have been characterized in *vestigial*. One, the "boundary enhancer" (in intron 2), is controlled in part by the Notch cell signaling pathway and activates *vestigial* expression in a cross-like pattern at the anterior-posterior and dorsal-ventral boundaries of the wing imaginal disc. The other, the "quadrant" enhancer (in intron 4), is controlled in part by the Dpp signaling pathway and activates *vestigial* expression in a complementary four-quadrant pattern. (Photos courtesy of Sean P. Carroll.)

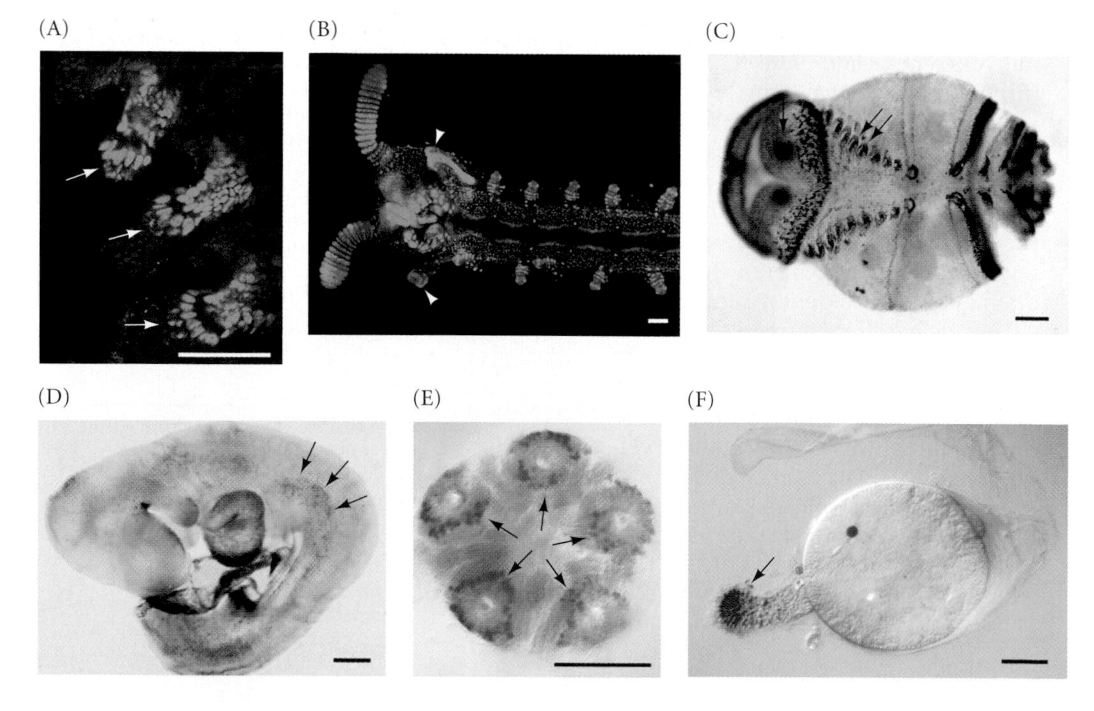

Figure 20.8 Expression of *Distal-less* family genes in the primordia of various animal appendages. (A) Abdominal prolegs (arrows) of a *Precis coenia* butterfly larva. (B) Antennae (ant), oral papillae (arrowheads), and lobopods of the onychophoran *Peripatopsis capensis*. (C) Parapodial rudiments (arrows) of an embryonic polychaete annelid (*Chaetopterus variopedatus*). (D) Mouse forelimb bud (arrows). (E) Tube feet (arrows) of a metamorphosing larva of the sea urchin *Strongylocentrotus droebachiensis*. (F) Ampulla (arrow) of a larval ascidian (*Molgula occidentalis*). Scale bars = 0.1 mm. (Photos courtesy of Grace Boekhoff-Falk.)

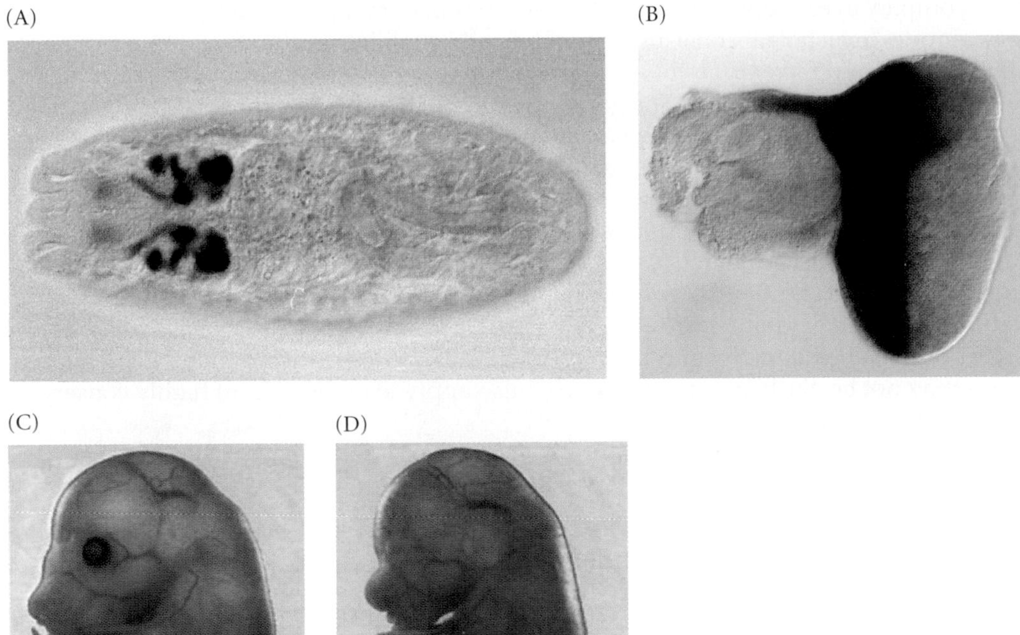

Figure 20.9 Expression of *Pax6/eyeless* genes in animal eye development, visualized by blue staining of a β-galactosidase reporter gene. (A, B) *Drosophila*. (A) Expression of *eyeless* in the larval eye precursors and other nearby tissues. (B) Close-up of eye imaginal disc (large lobe on right), a group of larval cells that develop into the adult eye. The *eyeless* gene is expressed in the cells anterior to a boundary-like feature called the morphogenetic furrow, which moves from posterior to anterior during eye development and delineates the boundary between differentiated and undifferentiated cells. The smaller lobe on the left is the antennal imaginal disc, which gives rise to the adult antenna. (C, D) Mouse. (C) The developing mouse eye can be seen as a circle of pigment. (D) The mouse mutant *Small eye* (*Sey*) lacks *Pax6* expression and has small or missing eyes. (A, B courtesy of Georg Halder; C, D courtesy of Robert Hill.)

eral developmental pathways that control the formation of major organs or appendages seem to be largely controlled by highly conserved transcription factors (reviewed in Carroll et al. 2001). The *Distalless* gene, for example, encodes a transcription factor that governs the development of body outgrowths that differentiate into very diverse structures in different phyla (Figure 20.8) (Panganiban et al. 1997).

A famous example of such a gene is *eyeless*, originally discovered in a classic *Drosophila* mutant with greatly reduced or missing eyes. Mutations in the mammalian homologue of *eyeless*, which is called *Pax6*, also cause reduction of the eyes. *Pax6/eyeless* activates the transcription of a hierarchy of regulatory proteins that control the development and differentiation of the eye. Expression of *Pax6/eyeless* is localized to the developing eye in embryos (Figure 20.9). Amazingly, when researchers genetically engineered *Drosophila* to express *eyeless* in various parts of the body where it is not normally expressed (**ectopic expression**), they discovered that this gene was sufficient to induce the development of ectopic eyes at these positions (Figure 20.10A,B). Even more astounding is the functional conservation of the *Pax6/eyeless* gene between vertebrates and invertebrates: mouse and squid *Pax6* genes can induce ectopic eyes when expressed in *Drosophila* (Figure 20.10C). Recent evidence suggests that at least two genes regulated by *Pax6* have conserved functions in *Drosophila* and mammalian eye development (Oliver et al. 1995; Xu et al. 1997), which helps to explain how *Pax6* homologues can function when placed in the genomes of such divergent animal species.

The question of how an organ as complex as the human eye evolved is one in which Darwin himself, and many theorists after him, have invested a great deal of thought. Darwin suggested that various intermediate stages, capable of photoreception, may have had adaptive value, leading to the evolution of

Figure 20.10 Ectopic eye formation in *Drosophila*. (A) A small ectopic eye on the base of the antenna (arrow), formed by ectopic expression of the *eyeless* gene. (B) A close-up of an ectopic eye, showing the same morphology as the large, normal compound eye nearby. (C) Ectopic expression of the human *Pax6* gene (driven by enhancers for the gene *dpp*) cause eyes to form in many locations (arrows). (A, B courtesy of W. G. Gehring; C courtesy of Nadean Brown.)

more complex eyes. Consistent with this idea, basal animal phyla such as Cnidaria have very simple "eyespots," consisting of a few cells containing photoreceptive pigments (rhodopsins), and many types of eyes, varying in complexity, are found among other invertebrates (see Figure 21.12). The great diversity of complex animal eye structures led many investigators to believe that eyes had evolved many times (Fernald 2000). However, the apparently universal role of *eyeless/Pax6* and of its target genes suggests a single origin. Currently, the most widely accepted explanation is that *Pax6* has a very ancient function in regulating the expression of universal components of photoreceptor organs, such as rhodopsin pigments, but that in various lineages of animals, different morphological features evolved independently, possibly due to the actions of different genes that became regulated by the *Pax6* pathway (see Wilkins 2002, pp. 148–155). Thus the capacity for photoreception may be homologous, but the structural features that carry out this function—eyes—may not be. Such an explanation may also apply to other cases of highly conserved regulators of organogenesis.

The Evolution of Gene Regulation: The Keystone of Developmental Evolution

The regulation of gene expression, whereby genes are activated or repressed at particular times and in particular tissues but unaffected in others, is achieved by discrete enhancers for each gene. These DNA sequences bind particular sets of transcription factors in specific cells or at specific developmental stages. For example, several genes expressed in the developing *Drosophila* wing are regulated by the transcription factors Scalloped and Vestigial, which activate genes required for wing development (Guss et al. 2001). The noncoding DNA (introns) of these genes contains one to several binding sites for Scalloped and Vestigial, each 8 to 9 nucleotides long (see Figure 3 in Box B). An enhancer consists of one or more such sites, which bind one or more transcription factors. Enhancers can be studied using reporter constructs in cultured cells or transgenic (genetically engineered) individuals (see Box A).

A particular gene often has a number of different enhancers. This **regulatory modularity** is thought to enable evolutionary changes in the development of specific tissues and body structures. That is, changes in the enhancers, rather than changes in the amino acid sequences of proteins, are thought to be responsible for many phenotypic adaptations. However, demonstrating the validity of this attractive idea is very difficult. The best examples thus far have been found in *Drosophila* and maize.

Using interspecies hybridization and gene mapping (see Chapter 13), Stern (1998) demonstrated that differences between *Drosophila melanogaster* and *D. simulans* in the pattern of epidermal cell hairs on the legs of the T3 segment map to the *Ubx* locus. Furthermore, he found that the *Ubx* gene is expressed in patterns that precisely correlate with the hair patterns (Figure 20.11). Because the two species do not differ in the amino acid sequence of the Ubx protein, Stern concluded that regulatory changes at the *Ubx* gene must be responsible for the differences between them. Using similar genetic mapping techniques, John Doebley and colleagues (Doebley et al. 1995; Hubbard et al. 2002) demonstrated that *teosinte branched 1*,* a gene in maize encoding a

*A mutation such as *teosinte branched 1* that causes a phenotype similar to that of another species is sometimes referred to as a PHYLETIC PHENOCOPY (Stebbins and Basile 1985; Stark et al. 1999.)

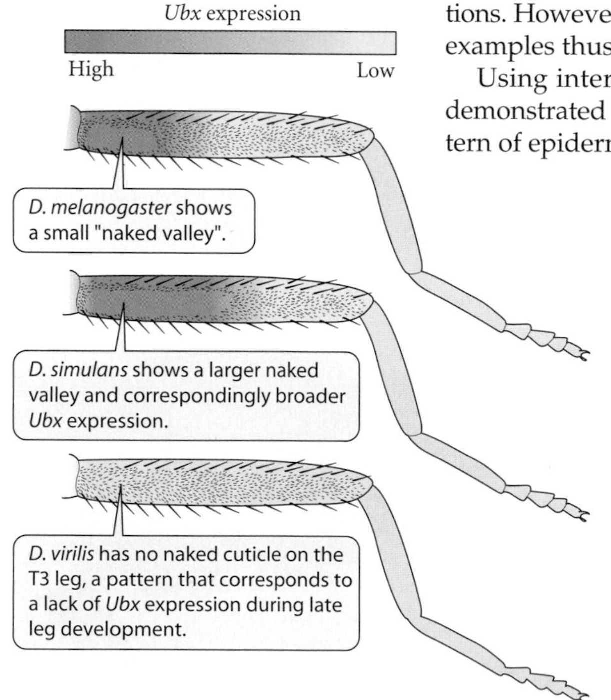
Ubx expression
High
Low
D. melanogaster shows a small "naked valley".
D. simulans shows a larger naked valley and correspondingly broader *Ubx* expression.
D. virilis has no naked cuticle on the T3 leg, a pattern that corresponds to a lack of *Ubx* expression during late leg development.

Figure 20.11 The role of *Ubx* in the evolution of the pattern of epidermal cell hairs in the third thoracic (T3) leg of *Drosophila*. Areas with *Ubx* expression during late leg development correspond with areas lacking epidermal cell hairs. This species difference was demonstrated to map genetically to the *Ubx* locus. (After Carroll et al. 2001.)

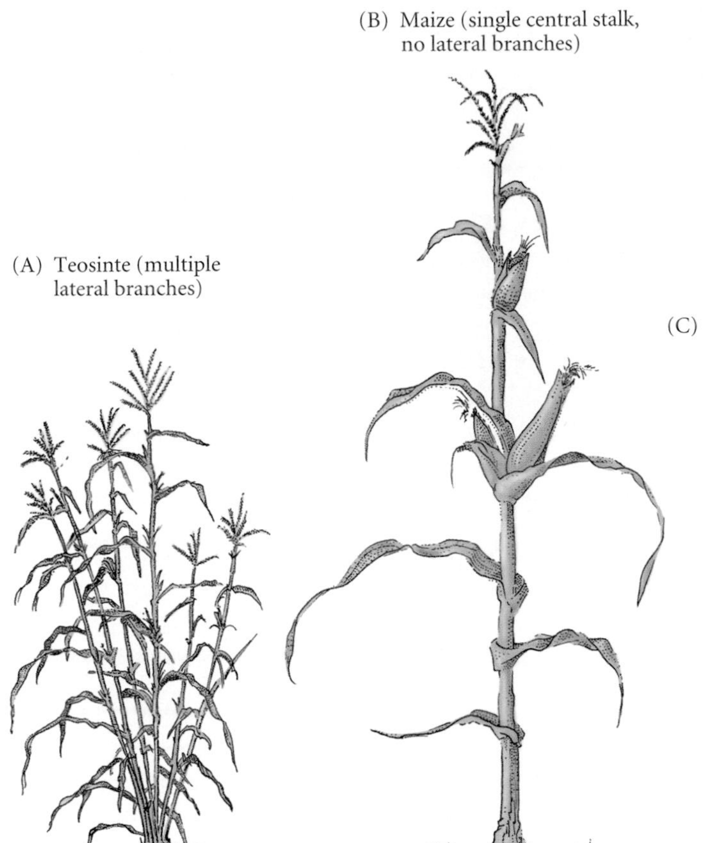

Figure 20.12 The *teosinte branched 1* (*tb1*) gene underlies a major difference in growth pattern between cultivated corn, or maize (*Zea mays* L. subspecies *mays*) and its wild ancestor, teosinte (*Zea mays* L. subspecies *parviglumis*). (A) Teosinte plants have multiple lateral branches. (B) Maize plants have a single central stalk and no lateral branches. (C) A plant from an F_2 population of a maize × teosinte cross that is homozygous for a genomic region containing the teosinte allele of *tb1*, shows the teosinte-like growth pattern. Plants from the same F_2 population that are homozygous for the maize allele of *tb1* show the maize-like growth pattern. (C courtesy of John Doebley.)

transcription factor required for the normal structure of the plant's body axis, controls differences in branching architecture between maize and its wild ancestor, teosinte (Figure 20.12).

An evolutionary change in the level of expression of a gene might be caused by an alteration of a closely linked (*cis*) regulatory sequence (such as an enhancer) that is affected differently by the same transcription factor—produced by another locus—in both species. Alternatively, the difference in gene expression between the species might be the result of a ***trans*-regulatory** difference: a difference in the transcription factor itself. Patricia Wittkopp and colleagues (2004) reasoned that if a gene expression difference between two related species is caused by the divergence of a *cis*-regulatory sequence, the same expression difference between their genes should also occur within an F_1 hybrid, because each species' gene would remain associated with its linked regulatory sequence (Figure 20.13; compare cases 1 and 2). However, if the expression difference etween the species is caused by divergent transcription factors (*trans*-regulators), both transcription factors would be produced in the F_1 hybrid, and both would diffuse through the nucleoplasm to the target genes of both species—so they should cause both target genes to be expressed at about the same level (Figure 20.13, case 3).

The investigators studied the expression of 29 genes in hybrids between *D. melanogaster* and *D. similans.* They found that about half the genes displayed only *cis*-regulatory divergence, while the other half showed signs of both *cis*- and *trans*-regulatory divergence. This seminal study suggests that changes in enhancers are likely to account for many evolutionary changes in gene expression.

Modularity in morphological evolution

The body plans of multicellular organisms are composed of discernible units, such as body segments, appendages, or the petals and sepals of angiosperm flowers. In animals, many

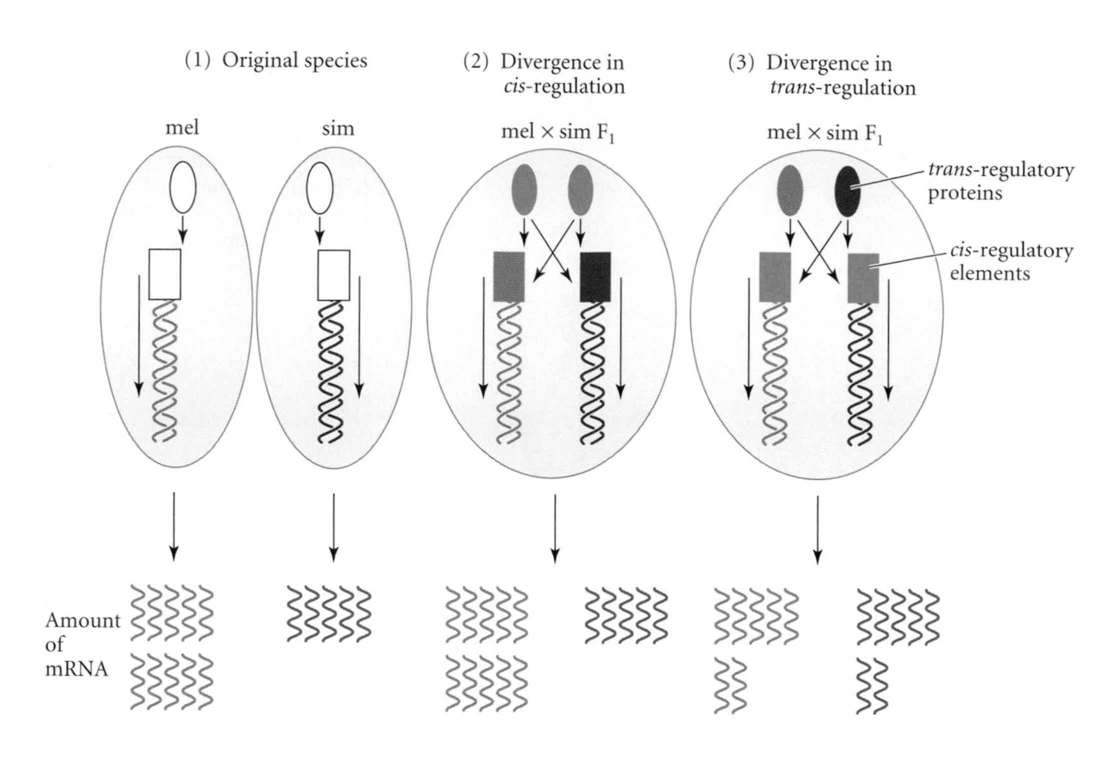

Figure 20.13 A test for divergence of gene expression due to divergence in *cis*-regulatory elements or *trans*-regulatory proteins. The *cis*-regulatory elements are represented by rectangles; *trans*-regulatory proteins (which bind to the *cis*-regulatory elements) are shown as ovals. (1) The expression of the gene of interest (measured by the production of mRNA) differs between *D. melanogaster* (mel) and *D. simulans* (sim); the ratio of mel/sim mRNA is 2:1. Whether this difference is due to differences in *cis*-regulatory elements or *trans*-regulatory proteins is unknown (open symbols). (2) In an F_1 hybrid, the same relative difference in gene expression is expected if the two species' *cis*-regulatory elements have diverged (different shades of blue), since neither element alters the expression of the homologous gene from the other species. (The *trans*-regulatory proteins are presumed to be the same in both species, indicated by gray.) (3) If the species difference is due to divergence in *trans*-regulatory proteins, however, equal expression of the two species' genes (1:1 mRNA ratio) is expected, since the homologous genes from both species are equally exposed to the *trans*-regulatory proteins of both parental species.

serially homologous structures have quite different morphologies (e.g., arthropod appendages). Similarly, elements such as teeth that repeat within a segment may vary greatly in form. Much of the spectacular diversity that we see in nature is the result of such changes in individual segments or structures. Thus the individual parts of the body plan show dissociation, or individualization—that is, they can develop and evolve in an independent manner. This phenomenon is an example of mosaic evolution (see Chapter 3).

The degree to which the development of different body structures is independent is referred to as MODULARITY, and the individual structures or units can be thought of as MODULES. How is modularity achieved by developmental pathways? Insights about gene regulation, of the kind described above, will be the key to answering this question.

Co-option and the evolution of novel characters

Many genes and signaling pathways have multiple developmental roles. For example, the transcription factor *Distalless* is required to organize the development of legs, wings, and antennae of all insects, but in some butterflies, it is also expressed later in specific positions on the developing wing, where it is involved in setting up the color patterns known as "eyespots" (see Figure 20.16A). Such cases suggest that, over the course of evolution, genes and pathways have been redeployed to serve new functions.

Change in the function of pre-existing features in adaptive evolution has been known ever since Darwin. Gould and Vrba (1982) coined the term **exaptations** to refer to novel uses of pre-existing morphological traits (see Chapter 11). Developmental biologists have used the terms **recruitment** (Wilkins 2002) and **co-option** (reviewed by True and Carroll 2002) to refer to the evolution of novel functions for pre-existing genes and developmental pathways.

Co-option of single genes for new functions may be common. The members of many gene families have diversified into different developmental and physiological roles (Chapter 19). In one of the most interesting such cases, the diverse crystallin proteins of animal eye lenses have been co-opted from a number of genes (Wistow and Piatigorsky 1988; Wistow 1993) (Figure 20.14). Two types of crystallins, α and β, which are common to all vertebrates, are derived from stress proteins, which help stabilize cellular functioning during environmental stresses, such as excess heat. Various lineages of animals have also derived taxon-specific crystallins from distinct enzymes, such as lactate dehydrogenase in

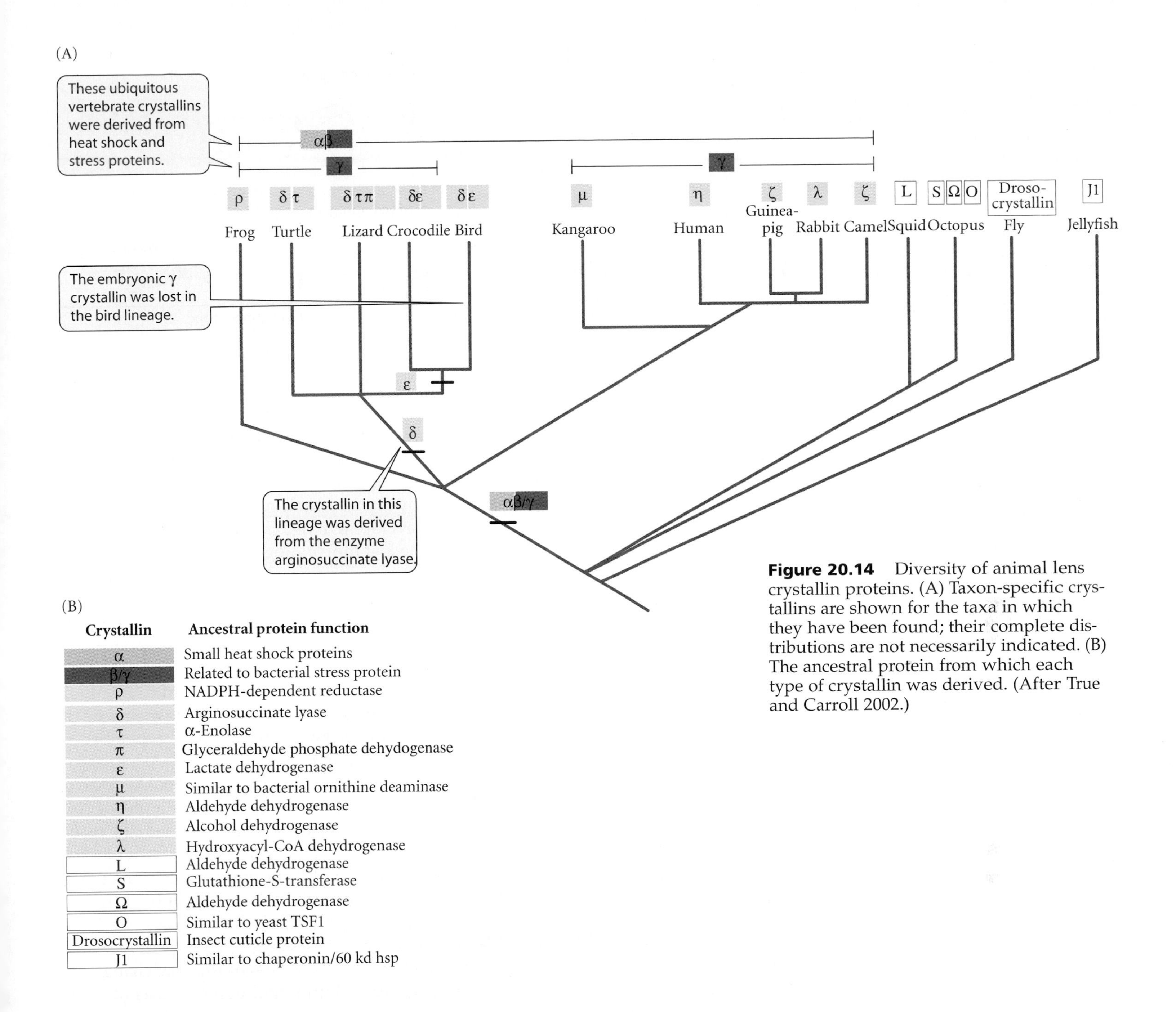

Crystallin	Ancestral protein function
α	Small heat shock proteins
β/γ	Related to bacterial stress protein
ρ	NADPH-dependent reductase
δ	Arginosuccinate lyase
τ	α-Enolase
π	Glyceraldehyde phosphate dehydogenase
ε	Lactate dehydrogenase
μ	Similar to bacterial ornithine deaminase
η	Aldehyde dehydrogenase
ζ	Alcohol dehydrogenase
λ	Hydroxyacyl-CoA dehydrogenase
L	Aldehyde dehydrogenase
S	Glutathione-S-transferase
Ω	Aldehyde dehydrogenase
O	Similar to yeast TSF1
Drosocrystallin	Insect cuticle protein
J1	Similar to chaperonin/60 kd hsp

Figure 20.14 Diversity of animal lens crystallin proteins. (A) Taxon-specific crystallins are shown for the taxa in which they have been found; their complete distributions are not necessarily indicated. (B) The ancestral protein from which each type of crystallin was derived. (After True and Carroll 2002.)

reptiles and glutathione-S-transferase in cephalopods. In the eye lens, crystallins are expressed at high levels and packed tightly into transparent matrices that are resistant to environmental stress and are designed to endure for the entire adult life of the animal. To achieve this function, many crystallins have undergone amino acid substitutions since they were co-opted from their ancestral function, but in most cases these proteins still share extensive homology with the ancestral proteins. These crystallins are derived from duplicated enzyme genes. In other cases, gene duplication has not occurred, and both the crystallin and the enzyme are encoded by the same gene (e.g., τ crystallin/α enolase in fishes, reptiles, and birds).

Figure 20.15 shows two ways in which a new structure might evolve by co-option of a developmental pathway. First, expression of a developmental regulatory protein may persist after the developmental stage in which it is needed. If additional target genes evolve the ability to be activated by the protein (e.g., by evolving a novel enhancer), then a new

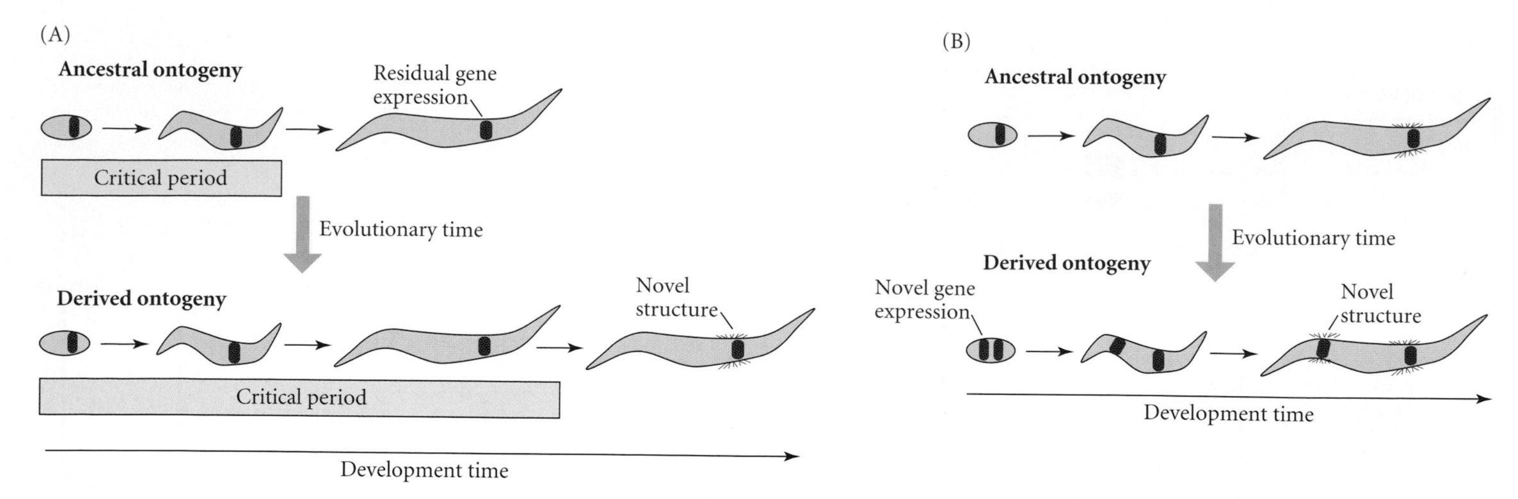

Figure 20.15 Two modes by which variation in the expression of a developmental regulatory gene may be co-opted to promote novel morphogenetic features. (A) Temporal co-option. Gene expression that persists after the critical period (a period during which the function of that gene is required) may be utilized during evolution to regulate novel morphogenetic processes. This utilization results in "terminal addition" of a novel feature and the lengthening of the critical period. (B) Spatial co-option. A novel, initially neutral, domain of expression may arise as a pleiotropic effect of evolutionary change in another developmental program. This novel expression may be co-opted to promote morphogenesis of an ancestral feature in the novel region of the body. (After True and Carroll 2002.)

morphogenetic process may arise later in development (Figure 20.15A). Alternatively, a developmental pathway originally expressed in one region of the embryo may become expressed in a different region, leading to a duplication of that structure in the new region (Figure 20.15B). Two cases of evolutionary novelty in which such events may have occurred are illustrated in Figure 20.16. During development of the "eyespots" on the wings of nymphalid butterflies, which are among the last morphological features to develop, several regulatory proteins are expressed that are required earlier for segmentation and early wing patterning (Figure 20.16A). Patterning of the tetrapod limb involves expression of the Hox genes in nested anterior-posterior patterns, similar to the way in which they are initially expressed during patterning of the anterior-posterior body axis (Figure 20.16B).

In plants, a developmental circuit has been co-opted in the evolution of divided, or compound, leaves, in which more or less separate leaflets develop along the main axis of the leaf (reviewed by Bharathan and Sinha 2001; Byrne et al. 2001). *KNOX1*, a member of a homeobox protein family, is expressed in the apical meristem of most plants, in association with the maintenance of growth of the stem. *KNOX1* expression must be deactivated in the leaf primordium to enable normal leaf development. However, in the developing compound leaves of the tomato plant, leaflet primordia form as bulges on the main leaf primordium, and *KNOX1* expression reappears in these regions. A growth control pathway has apparently been co-opted in the evolution of compound leaves, which has occurred many times in angiosperms (Goliber et al. 1999; Bharathan and Sinha 2001).

The developmental genetics of heterochrony

Much of morphological evolution has entailed HETEROCHRONY—evolutionary changes in the timing of development (see Chapter 3). The developmental genetic basis of heterochrony has been little studied in model organisms, but genetic approaches have been used to study neoteny in salamanders (Voss and Shaffer 2000; Voss et al. 2003). In tiger salamanders (*Ambystoma tigrinum*), the standard ontogenetic sequence involves loss of the tail fin and external gills in terrestrial adults. Several related species, including the axolotl (*A. mexicanum*), attain reproductive maturity while remaining fully aquatic, retaining the gills and other larval features (see Figure 3.14). Some species are capable of facultatively transforming into the typical terrestrial adult form if their habitats dry up, whereas others have lost this ability completely.

In salamanders, thyrotropin-releasing hormone (TRH) stimulates the pituitary gland to release another hormone, thyroid-stimulating hormone (TSH), which stimulates the thyroid gland to release a third hormone, thyroxin. Thyroxin triggers metamorphosis by inducing morphogenetic events in several different tissues. In *A. mexicanum*, the TRH cascade does not take place, but metamorphosis can be induced by injecting TRH into the animal (Shaffer and Voss 1996). This observation suggests that the evolution of neoteny

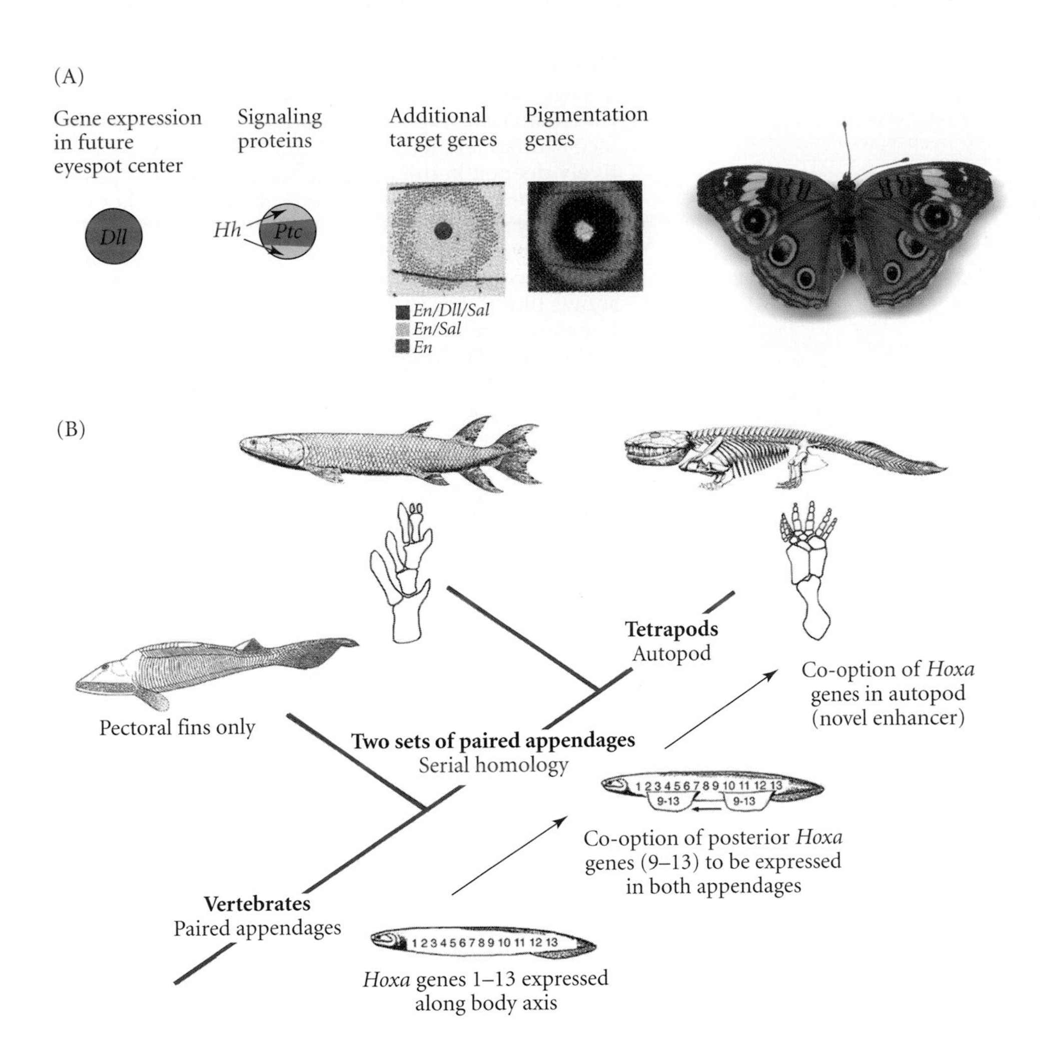

Figure 20.16 Co-option of developmental circuits in the evolution of novelties. (A) Butterfly "eyespots" are the developmental products not only of genes for pigmentation, but also of many co-opted genes and pathways that play important roles in establishing the body plan. These include the signaling proteins Hedgehog (Hh) and Patched (Ptc), and the transcription factors Distalless (Dll), Spalt (Sal), and Engrailed (En). (B) Co-option of the vertebrate *Hoxa* genes during evolution of the tetrapod appendage. Ancestrally, Hox genes were expressed only along the anterior-posterior axis of the developing body. The evolution of paired fore- and hindlimbs involved novel gene expression, presumably using novel enhancer sequences of *Hoxa* 9–13. The evolution of hands and feet (autopods) involved further novel expression patterns of the *Hoxa* genes. (From True and Carroll 2002.)

in *A. mexicanum* involved inactivation of the TRH cascade. Crosses between *A. tigrinum* and *A mexicanum* suggest that this difference is controlled by multiple genes. Interestingly, Voss et al. (2003) found evidence that neoteny has a different genetic basis in each of three species in which it has evolved independently. One of the genes may correspond to a thyroid receptor locus that is required in order for cells to respond to the TSH signal.

The evolution of allometry

ALLOMETRY refers to the differential growth rates of different parts of the body (see Chapter 3). Morphological evolution, including changes in shape, often involves alterations in such allometries, achieved by heterochronic changes in the growth of individual body parts. For example, in the horse lineage from *Hyracotherium* to *Equus* (see Chapter 4), the length of the face and depth of the lower jaw (mandible) are allometrically related to body size, and both increased along with body size during the evolution of browsing forms in the Eocene and Oligocene (Figure 20.17A). In the Miocene, *Merychippus* became adapted to grazing on grasses, and evolved a relatively deeper mandible, as well as more anteriorly situated molars relative to the eye socket (Figure 20.17B). These evolutionary changes are thought to have been necessary to accommodate the bases of the molars, which extend deep into the jaw bones. These teeth descend throughout life to compensate for abrasion of the grinding surfaces by grasses, which wear down teeth more rapidly than other plant foods.

Allometric relationships can evolve quite rapidly, as in the dung beetle *Onthophagus taurus*. Males of this beetle exhibit a very conspicuous POLYPHENISM: small males do not develop horns, whereas those that attain a threshold body size develop horns that are used

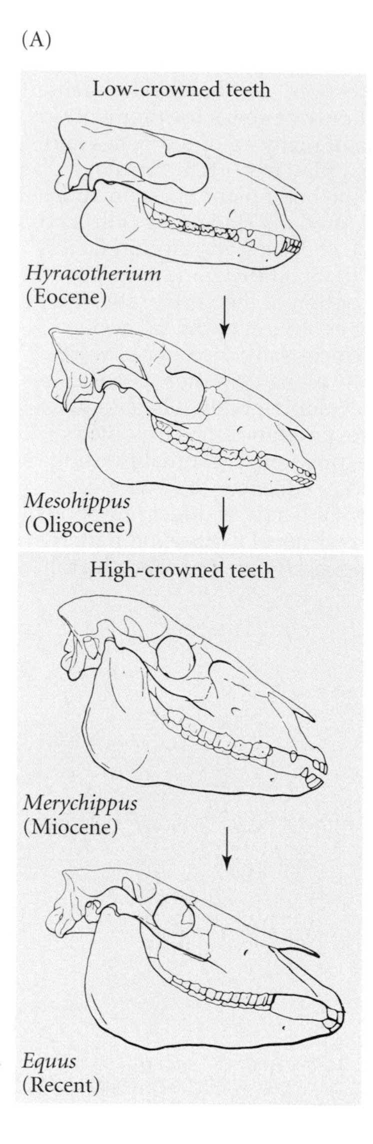

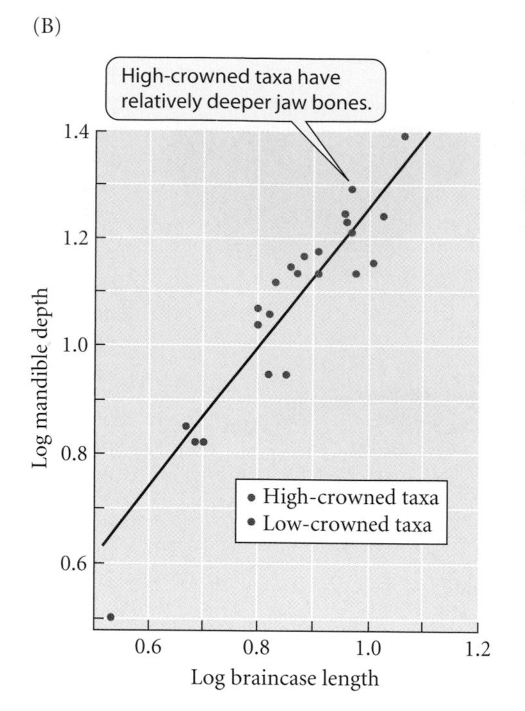

Figure 20.17 Evolution by allometry in the horse lineage. (A) Skulls of four equid taxa in the lineage from *Hyracotherium* to the modern horse, *Equus*. The face is relatively longer, and the lower jaw is relatively deeper, in the more recent taxa, and the posterior molars are shifted forward of the eye socket in *Merychippus* and *Equus*, which have high-crowned teeth that extend deep into the skull. (B) A log-log plot of two measurements in various fossil horses shows that lower jaw depth increased disproportionately in relation to body size (for which length of the braincase was used as an index). (After Radinsky 1984.)

in male-male combat. In the mid-twentieth century, this beetle was introduced from Europe to North America and Australia in order to reduce dung accumulation in pastures. The threshold size for horn development has diverged from the ancestral condition in both introduced populations, even though variation of this magnitude was not evident in the founding populations (Figure 20.18). As in the cases of heterochrony discussed above, these evolutionary changes are believed to be caused by changes in the response of specific tissues (i.e., the cells destined to give rise to the horns) to global signals such as hormones and perhaps nutritional supplies, which may be direct determinants of body size.

Developmental Constraints and Morphological Evolution

Traditional neo-Darwinian theory explains how natural selection, genetic drift, and gene flow, acting on the raw material of genetic variation, have produced the astonishing variety of organisms. But does it explain why organisms have *not* evolved certain features, or in certain directions? Does it explain why there are no live-bearing turtles, for instance, or why frogs have no more than four digits on their forelimbs? Such questions have led evolutionary biologists to ask what the **constraints** on evolution might be.

Several kinds of constraints on evolution have been distinguished. Some are universal, in that they affect all organisms; an example is the constant presence of gravity during mor-

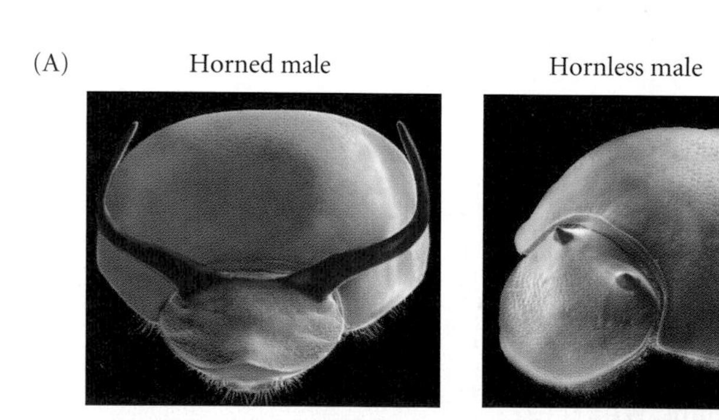

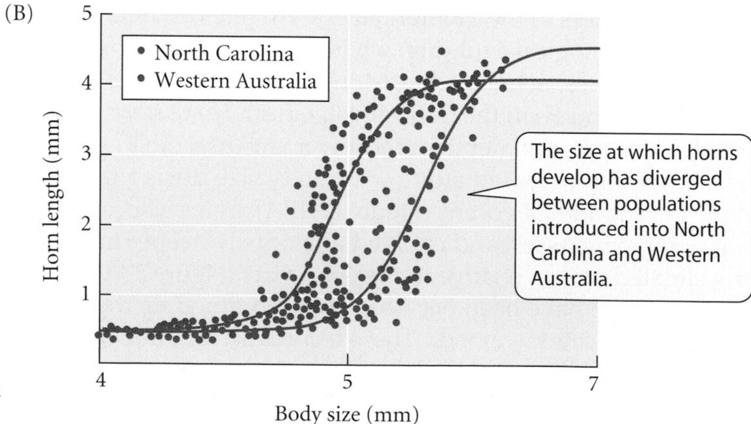

Figure 20.18 Rapid evolution of an allometric threshold in the dung beetle *Onthophagus taurus*. (A) Morphology of horned and hornless males at the same developmental stage; horns have been artificially emphasized with blue. (B) Horn length shows a steep allometric relationship to body size, and has diverged between populations introduced into Western Australia and North Carolina. (A, photos courtesy of Doug Emlen; B after Moczek et al. 2002.)

phogenesis. Others, referred to as PHYLOGENETIC CONSTRAINTS, are more local, affecting only a group of related organisms. There are several types of constraints on evolution:

1. *Physical constraints.* Some structures do not evolve because the properties of biological materials (e.g., bones, epidermis, DNA, RNA, etc.) do not permit them. Physical constraints can be phylogenetically local. Insects, for example, conduct oxygen and carbon dioxide by means of diffusion in narrow tubes, or tracheae, which branch throughout the body. Limits on diffusion rates are thought to set an upper limit on insect body size.
2. *Selective (or functional) constraints.* Some features do not appear in particular lineages because they are always disadvantageous, or because they might interfere with the function of an existing trait.
3. *Genetic constraints.* Genetic variation in a particular phenotype may not be present, as discussed in Chapter 13. Developmental pathways are expected to have varying degrees of tolerance for variation in their components, and their limits of tolerance may limit variation in the resulting traits. Moreover, if two traits share a common pathway of morphogenesis, the genes underlying that pathway will have strong pleiotropic effects, resulting in genetic correlations that limit the freedom with which those traits can vary relative to each other (see Chapter 13). Thus genetic constraints, such as paucity of variation and genetic correlation, are closely related to developmental constraints.
4. *Developmental constraints.* Maynard Smith et al. (1985) defined a developmental constraint as "a bias on the production of various phenotypes caused by the structure, character, composition, or dynamics of the developmental system." The two most common phenomena attributed to developmental constraint are absence or paucity of variation, including the absence of morphogenetic capacity (i.e., lack of cells, proteins, or genes required for the development of a structure), and strong correlations among characters, which may result from interaction between tissues during development or the involvement of the same genes or developmental pathways in multiple morphogenetic processes.

Developmental constraints can be revealed by embryological and genetic manipulations in the laboratory. In a classic experiment, Pere Alberch and Emily Gale (1985) used the mitosis-inhibiting chemical colchicine to inhibit digit development in the limb buds of salamanders (*Ambystoma*) and frogs (*Xenopus*) (Figure 20.19). The treatment consistently caused specific digits to be missing in each species, and the missing digits were the preaxial ones in frogs and the postaxial ones in salamanders. These results reflected the different order of digit differentiation in the two species (see above); the last digits to form tended to be the most sensitive to the colchicine treatment. Furthermore, the results strongly reflected evolutionary trends: salamanders have often lost postaxial digits, and frogs have repeatedly experienced preaxial digit reduction, during evolution. Although the digit number variation in this study was produced artificially, the re-

(A)

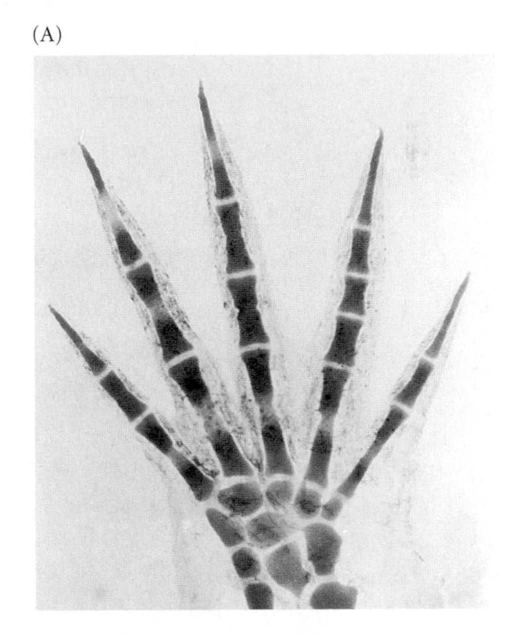

(B)

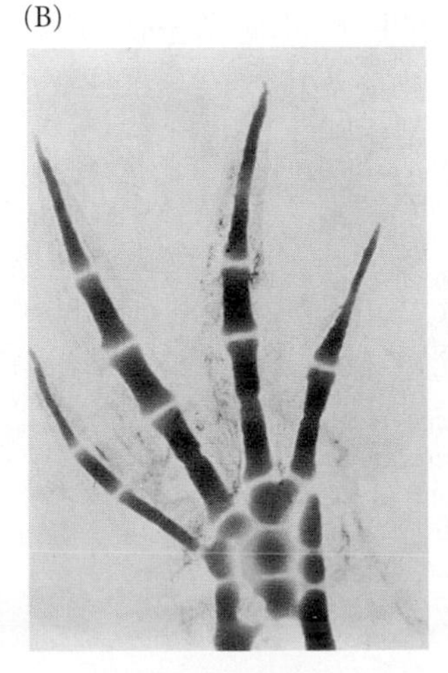

(C)

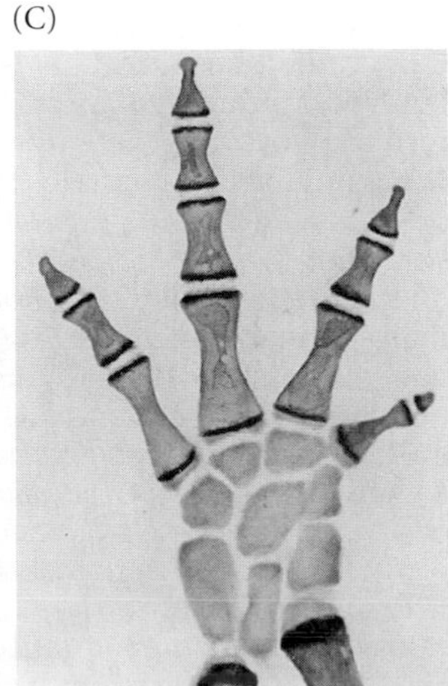

Figure 20.19 Evidence for developmental constraints. (A) X-ray of the right hind foot of an axolotl salamander (*Ambystoma mexicanum*), showing the normal five-toed condition. (B) The left hind foot of the same individual, which was treated by an inhibitor of mitosis during the limb bud stage. The foot lacks the postaxial toe and some toe segments, and is smaller than the control foot. (C) A normal left hind foot of the four-toed salamander (*Hemidactylium scutatum*) has the same features as the experimentally treated foot of the axolotl. (From Alberch and Gale 1985; photos courtesy of the late P. Alberch.)

sults suggest that naturally occurring variation in developmental systems may be constrained by intrinsic, species-specific developmental programs.

Although in practice it is very difficult to rule out selective constraints, developmental or genetic constraints might explain some common evolutionary patterns:

1. *The absence of features in certain lineages.* For example, although viviparity (giving birth rather than laying eggs) has evolved in lizards and snakes, it has not evolved in turtles, even though it might well be advantageous in sea turtles by freeing them from the need to come to land to lay eggs (Williams 1992b).
2. *Directional trends* (see Chapter 3), such as cumulative elaboration of a particular structure. The developmental system may impose a bias such that certain kinds of variation are produced and not others, enabling particular evolutionary trajectories to be taken. For example, if a morphogenetic novelty arises in the ancestor of a lineage, the descendant species may gradually evolve structural modifications that could not have evolved otherwise. Bony frontal horns have evolved in only a few groups of mammals, such as the Bovidae (antelopes, goats, sheep, and cattle), among which horn sizes and shapes vary greatly. In contrast to the unbranched horns of bovids, the evolution of branching in the antlers of the Cervidae (deer) has enabled a very different variety of forms to evolve (Figure 20.20).
3. *Parallel evolution of traits in independent lineages.* For example, similar patterns of abdominal pigmentation and cuticular cell hair morphology have evolved in several lineages of *Drosophila*. Gompel and Carroll (2003) demonstrated that changes in the expression of the transcription factor *bric-a-brac*, which controls sexually dimorphic pigmentation and cuticular patterning (see Figure 20.21), are correlated with interspecific differences in these traits. This finding suggests that only one genetic pathway

(A) Bovidae: Bony, unbranched horns

(B) Cervidae: Branched antlers

Figure 20.20 Bony horns are a novelty that have evolved in only a few mammalian lineages. (A) The frontal horns of the Bovidae (antelope, cattle, sheep, and goats) have evolved a stunning variety of forms, all of which are unbranched. Left to right: Cape buffalo (*Syncerus caffer*); greater kudu (*Tragelaphus strepsiceros*); and Nubian ibex (*Capra nubiana*). (B) The antlers borne by males of many deer species (Cervidae) also display wide diversity, but on a "branched template." Left to right: caribou (*Rangifer tarandus*); Indian muntjac (*Muntiacus muntiacus*); and the moose (*Alces alces*). (Buffalo, ibex, caribou, moose © Painet, Inc.; kudu © John Cancalosi/naturepl.com; muntjac © OSF/photolibrary.com.)

was readily available for evolution in these abdominal traits, possibly due to limitations in genetic variation at other pigmentation and cuticular development genes.

4. *"Standardization,"* or the reduction of morphological variation of traits in the fossil record following an initial phase of high variation. One of several possible explanations for this pattern is **canalization**: the evolution of modifications of the developmental system such that the most highly advantageous phenotype is more reliably produced (Waddington 1942; see Chapter 13). Several proteins that are present in virtually all eukaryotic cells, called "heat shock proteins" or chaperonins, have been suggested to play roles in this process, although what those roles may be is not yet clear. These proteins aid in the proper folding of all cellular proteins and are expressed at high levels when the cell experiences various environmental stresses, such as thermal shock or infection.
5. *Morphological stasis over long periods of evolutionary time.* The absence of evolutionary change has many possible explanations, of which developmental constraint is one (see Chapter 21).
6. *Similarities in embryological stages among higher taxa.* Such similarities might result from the need to conserve early developmental processes so as not to disturb the later events that depend on them (Riedl 1978). von Baer's law (see Figure 3.12) may be a consequence of developmental constraints. For example, the notochord, which persists throughout life in the most "primitive" vertebrates, is almost completely lost in the postembryonic forms of "advanced" vertebrates, but is needed in the embryo in order to induce the differentiation of central nervous system tissues.

The Developmental Genetic Basis of Short-Term Morphological Evolution

Morphological novelty and diversity originate in intraspecific variation and in differences between closely related species. Most of this variation is polygenic and can be studied by quantitative trait locus (QTL) mapping (see Chapter 13). The kinds of genes and the kinds of alleles involved in these differences (i.e., *cis*-regulatory or amino acid-coding sequences) are particularly interesting to EDB researchers. The few QTL studies that implicate specific genes suggest that developmental regulatory loci are commonly involved in morphological differences between species, and that these loci can include major developmental regulatory genes such as *Ubx*, as we saw above. Thus the many crucial developmental functions of these genes do not preclude their involvement in short-term evolutionary change. Changes in the regulatory regions of these genes have been implicated more often than changes in their amino acid-coding sequences, confirming the importance of variation in gene regulation in morphological evolution.

Within *Drosophila melanogaster*, naturally occurring variation in bristle number is caused partly by nucleotide substitutions at regulatory loci encoding both cell signaling proteins and transcription factors (Lai et al. 1994; Long et al. 1998, 2000). An example of regulatory differences among closely related species involves the gene *bric-a-brac* (*bab*), which encodes a transcription factor required for sexually dimorphic pigmentation and cuticle morphology in *Drosophila* and is associated with a QTL for abdominal pigmentation in *D. melanogaster* (Kopp et al. 2003). Sexually dimorphic abdominal pigmentation evolved once, in the lineage leading to the *melanogaster* species group after its divergence from the *obscura* species group. In the *melanogaster* group, *bab* is expressed in the posterior abdomen of females, in which it is required to prevent the development of male abdominal pigmentation, and it is repressed in males, enabling the development of pigmentation (Figure 20.21A–C). However, the species in the *montium* lineage, a subgroup of the *melanogaster* group, generally lack sex-specific pigmentation patterns, even though *bab* expression is down-regulated in males—in which it may regulate sexual dimorphism in abdominal cuticle patterns rather than in pigmentation (Figure 20.21D). Developmental genetic data from *D. melanogaster* indicate that *bab*'s role is to produce sexual dimorphism

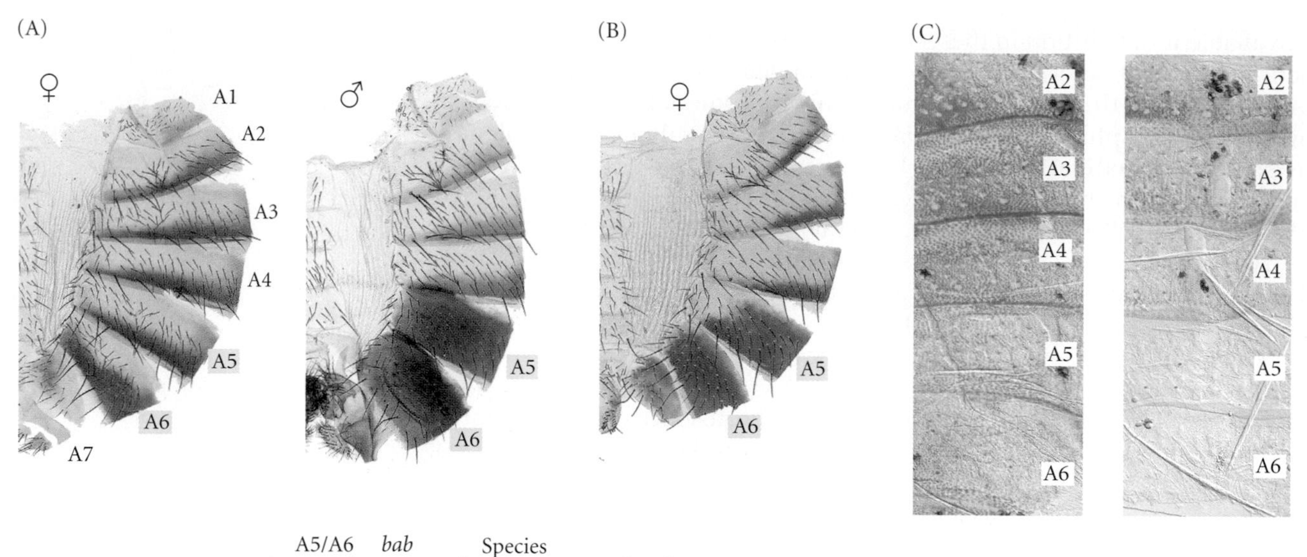

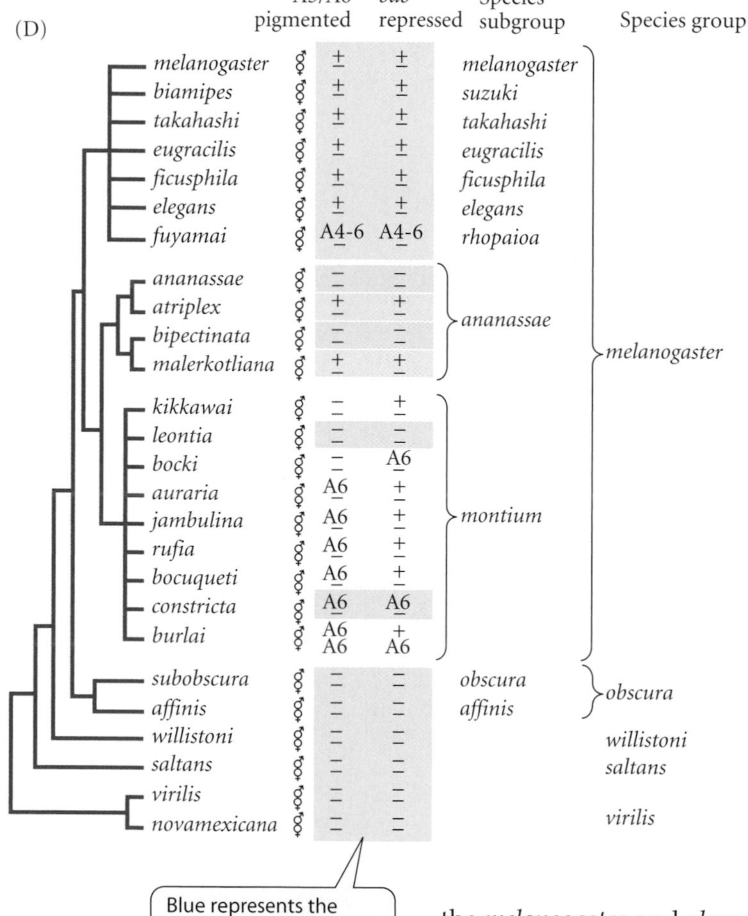

Figure 20.21 Role of *bric-a-brac* (*bab*) expression in sexually dimorphic abdominal pigmentation in *Drosophila*. (A) Abdominal cuticle of wild-type female and male *D. melanogaster*, showing differences in the pigmentation of the fifth (A5) and sixth (A6) abdominal segments. (B) A *bab* mutant female, showing male-like pigmentation in A5 and A6. (C) A reporter construct shows *bab* expression (blue) in developing female (left) and male (right) abdomens. Note the lack of *bab* expression in A5 and A6 of the male. (D) Sexually dimorphic *bab* expression is associated with sexually dimorphic abdominal pigmentation. The + and – signs indicate presence or absence of sexually dimorphic A5 and A6 pigmentation. In the right column, a + sign indicates in the repression of *bab* in A5 and A6 segments. Orange indicates species with correlated sexually dimorphic abdominal pigmentation and *bab* expression. Yellow indicates species with male-specific down-regulation of *bab*, but no male-specific pigmentation. Blue indicates species that are sexually monomorphic in both abdominal pigmentation and abdominal *bab* expression (the ancestral condition). (After Kopp et al. 2000. Photos courtesy of A. Kopp and S. Carroll.)

in specific abdominal segments by integrating information about anterior-posterior position in the abdomen, which is conferred by expression of the Hox gene *Abdominal B* (*AbdB*), and about sexual identity, which is determined by expression of the transcription factor *doublesex* (*dsx*). Kopp et al. (2000) have proposed that, following the divergence of the *melanogaster* and *obscura* lineages about 25 million years ago, transcriptional regulation of *bab* evolved to integrate the *AbdB* and *dsx* signals, presumably through changes in its *cis*-regulatory DNA.

The Molecular Genetic Basis of Gene Regulatory Evolution

Comparative studies of gene expression are showing that gene duplication and nucleotide substitution both contribute to the diversification of gene regulation. For example, the genes *snail* and *slug* in vertebrates are both descendants, via gene duplication, of an ancestral gene in the common invertebrate ancestor of all vertebrates, which is thought to be developmentally similar to the cephalochordate amphioxus (the living sister group of

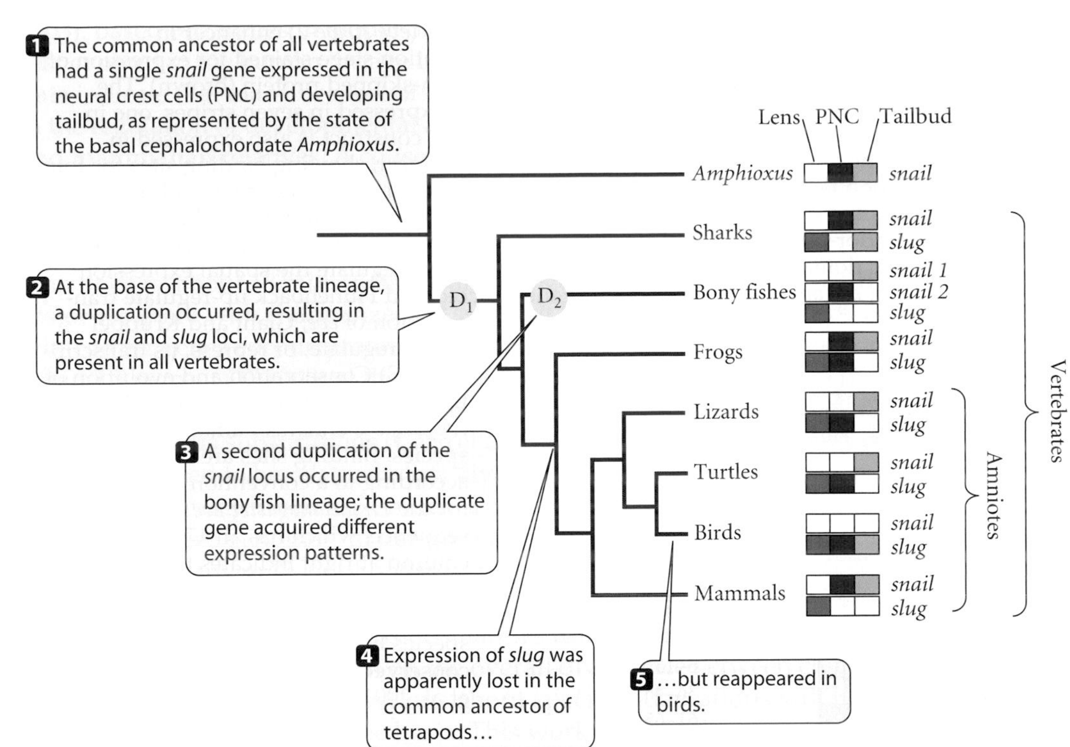

Figure 20.22 Evolution of the roles of the genes *snail* and *slug* in vertebrate development. Colored symbols indicate expression of *snail* and *slug* loci in the developing eye lens, premigratory neural crest cells (PNC), and tailbud. (After Locascio et al. 2002.)

vertebrates). In both invertebrates and vertebrates, proteins encoded by the *snail* gene family (which includes both *snail* and *slug*) have evolutionarily conserved functions in mesoderm formation and neural development during embryogenesis. In the lineage containing amphioxus and the vertebrates, *snail* and *slug* acquired expression and new developmental roles in the eye lens, the tailbud, and the premigratory neural crest cells (PNCs; these are epidermal cells that originate in the neural crest and migrate to different parts of the body, where they differentiate into a wide variety of cell types, including pigment cells and jaw and facial tissues). Recently, Locascio et al. (2002) have demonstrated that, although expression of *slug* in the developing lens is conserved across vertebrates, the particular roles of *snail* and *slug* in PNCs and tailbud cells have greatly diversified (Figure 20.22). For example, birds have lost *snail* expression in the tailbud, but have gained *slug* expression there, whereas mammals have retained *snail* expression in the PNCs and tailbud, and do not express *slug* in either of these tissues. This pattern of regulatory evolution does not conform to a simple duplication-diversification model (see Chapter 19), especially in the case of the chicken, which has apparently "re-activated" *slug* expression in the tailbud, a condition that disappeared much earlier in vertebrate history. Locascio and colleagues propose that a rare event such as chromosomal translocation may account for the reappearance of regulatory elements that direct *slug* expression in the tailbud of birds.

Molecular evolutionary studies in closely related species show that even when enhancer function is conserved, the binding sites within an enhancer can change in surprising ways, presumably due to neutral or nearly neutral nucleotide substitutions. In a groundbreaking research program, Michael Ludwig, Martin Kreitman, and colleagues (Ludwig and Kreitman 1995; Ludwig et al. 1998, 2000) examined sequence and functional differences among *Drosophila* species in a specific *cis*-regulatory sequence from the gene *even-skipped*, a transcription factor required for segmentation. In *Drosophila melanogaster*, this enhancer, called *eve stripe-2*, is a 670-base-pair segment of DNA that is necessary and sufficient to direct expression of *even-skipped* in the second of seven stripes in which the protein is expressed in wild-type embryos (Figure 20.23A). This enhancer contains 17 binding sites for transcription factors encoded by four different regulatory genes that are expressed in anterior-posterior gradients and that, by activation and repression, combinatorially determine the precise limits of the *even-skipped* stripes (Figure 20.23B). Kreitman's group se-

7. During evolution, genes and developmental pathways have often been co-opted, or recruited, for new functions that probably are responsible for the evolution of many novel morphological traits. This process results from evolutionary changes in gene regulation and diversification in the function and expression patterns of duplicated genes.
8. Many differences among species are due to heterochronic or allometric changes in the relative developmental rates of different body parts or in the rates or durations of different life history stages. The modularity of morphogenesis in different body parts and in different developmental stages facilitates such changes.
9. Several kinds of constraints on evolution may determine that certain evolutionary trajectories are followed and not others. Developmental systems are thought to impose some constraints on morphological evolution. Such constraints may be restricted to certain clades (i.e., may be phylogenetically "local").
10. The developmental genetic basis of short-term evolution (i.e., within species and among closely related species) is an active area of EDB research . Several studies suggest that changes in gene regulation are important in short-term morphological evolution.
11. The enhancers of genes may evolve either by gradual nucleotide substitutions, resulting in the gain and loss of individual transcription factor binding sites, or by gene duplications, transpositions, or other genome-level processes. Presently, the relative importance of these types of change is not known.

Terms and Concepts

biological homology concept
canalization
***cis*-regulatory elements**
constraints
co-option
developmental circuits
developmental pathways
ectopic expression
enhancers
evolutionary developmental biology (EDB)
exaptations
homeobox
homeobox genes
homeodomain
Hox genes
promoters
recruitment
regulatory modularity
target genes
transcription factors
***trans*-regulation**

Suggestions for Further Reading

Evolutionary biologists' interest in development was rekindled in the 1970s largely by Stephen Jay Gould, who also portrayed the early history of the subject in *Ontogeny and phylogeny* (Harvard University Press, Cambridge, MA, 1977). An excellent, very readable introduction to contemporary evolutionary developmental biology, emphasizing regulation of gene expression, is *From DNA to diversity: Molecular genetics and the evolution of animal design* by S. B. Carroll, J. K. Grenier, and S. D. Weatherbee (second edition, Blackwell Science, Malden, MA, 2005). A more extended treatment is *The evolution of developmental pathways* by A. S. Wilkins (Sinauer Associates, Sunderland, MA, 2002).

Rudolph A. Raff, one of the founders of EDB, portrayed the field somewhat earlier in his very influential book *The shape of life: Genes, development, and the evolution of animal form* (University of Chicago Press, Chicago, 1996). E. H. Davidson, a developmental biologist whose insights helped to shape the field, does much the same in *Genomic regulatory systems: Development and evolution* (Academic Press, San Diego, 2001).

Problems and Discussion Topics

1. If two allometrically related traits show a very strong correlation among species, does this provide evidence that the traits are developmentally constrained and could not change independently? How else might such a strong correlation be interpreted?

2. Formulate a hypothesis on a developmental genetic mechanism by which ancestrally identical serially homologous structures might begin to acquire differences—i.e., individual identity.
3. If mutations such as those of the *Ubx* gene can drastically change morphology in a single step, why should most evolutionary biologists maintain that evolution has generally proceeded by successive small steps?
4. How might modularity of growth control for a particular organ or body part evolve? What selective pressures might promote size changes in individual body parts versus size changes in the whole organism?
5. Would you predict that novel structures that require complex morphogenetic processes are more often gained or lost in evolution? How might you address this question given a group of organisms with known phylogenetic relationships that vary with respect to presence or absence of a complex structure, such as an eyespot or a pair of appendages modified for feeding?
6. How might the regulatory DNA sequences underlying a heterochronic change (e.g., an increase in the developmental rate of the larval stage of an insect, resulting in a shorter larval period) differ from those responsible for an evolutionary novelty in a particular body segment (e.g., a novel wing pigmentation spot)? What spatio-temporal components of the developmental system would you expect to be acting in these two cases, and what sorts of genes (encoding transcription factors, signaling proteins, hormones) would you look for as candidate genes underlying these two types of evolutionary change?
7. What developmental mechanisms might underlie an evolutionary change in shape? For example, what might cause the fingers of a bat to grow so much longer (relative to the body) than in other mammals? Can you find any research publications that bear on this question?
8. When a lineage loses an ancestral character (such as the loss of the second molar in the Felidae; see Figure 20.7), how do you think a derived lineage, such as the lynx, regains the lost trait? What ecological or population-level factors might act to cause the disappearance of the trait, and what factors might be involved in regaining or maintaining the developmental pathway needed for morphogenesis of that trait?

21

Macroevolution: Evolution above the Species Level

Structural changes accompany changes in function. The African elephant (*Loxodonta africana*) is so familiar, it is easy to forget how extraordinary it is that an animal's incisor teeth should have developed into tusks used for display and defense rather than feeding, and that a nose can evolve to function as an arm tipped with a prehensile hand. Fossils show that these features evolved gradually. (Photo © Mike Wilkes/naturepl.com)

The phenomena of evolution are often divided into **microevolution** (meaning, mostly, processes that occur within species) and **macroevolution**, which is often defined as "evolution above the species level." "Macroevolution" has slightly different meanings to different authors. To Stephen Jay Gould (2002, p. 38), it meant "evolutionary phenomenology from the origin of species on up." These phenomena include patterns of origination, extinction, and diversification of higher taxa, the subject of Chapter 7. To other authors, macroevolution is restricted to the evolution of great phenotypic changes, or the origin of characteristics that diagnose higher taxa (e.g., Levinton 2001). The subject matter of macroevolutionary studies, however defined, includes patterns that have developed over great periods of evolutionary time—patterns that are usually revealed by paleontological or comparative phylogenetic studies, even if their explanation lies in genetic and ecological processes that can be studied in living organisms. Thus we want to know how fast evolution happens and what determines its pace, whether the great differences that distinguish higher taxa have arisen gradually or discontinuously, what the mechanisms are by which novel features have come into existence, and whether or not there are grand trends, or progress of any kind, in the history of life.

Much of the modern study of macroevolution stems from themes and principles developed by the paleontologist George Gaylord Simpson (1947, 1953), who focused on rates and directions of evolution perceived in the fossil record, and Bernhard Rensch (1959), a zoologist who inferred patterns of evolution from comparative morphology. Contemporary macroevolutionary studies draw on the fossil record, on phylogenetic patterns of evolutionary change, on evolutionary developmental biology, and on our understanding of genetic and ecological processes.

Rates of Evolution

As we noted in Chapters 4 and 7, rates of evolution vary greatly. Simpson (1953), who pioneered the study of evolutionary rates, distinguished rates at which single characters or complexes of characters evolve (which he called PHYLOGENETIC RATES) from TAXONOMIC RATES, the rates at which taxa with different characteristics originate, become extinct, and replace one another. For example, Steven Stanley (1979) analyzed several clades that have increased exponentially in species diversity during the Cenozoic. Among the most rapidly radiating groups are murid rodents (mice and rats) and colubrid snakes, diverse groups that arose in the Miocene and took about 1.98 Myr and 1.24 Myr, respectively, to double in species number. These rates mean that, without extinction, each rodent species would speciate, on average, within about 2 Myr (assuming that each species bifurcates into two "daughter" species). This interval is roughly the same as, or even greater than, the time required for speciation that has been estimated from genetic differences between sister species of living organisms (see Chapter 16). Thus a duration of 1 or 2 million years per speciation event is more than enough to account for the evolution of great diversity, even in the most rapidly proliferating groups.

Rates of character evolution

Individual characters evolve at rates that differ greatly (see Chapters 4 and 13). Many features in fossilized lineages, such as the tooth dimensions of the early horse *Hyracotherium* (see Figure 4.23), show the pattern that Eldredge and Gould (1972) called **punctuated equilibrium**: long periods of little change (which they called **stasis**) interrupted by brief episodes of much more rapid change (Figure 21.1A, left panel). During these brief periods (of hundreds of thousands of years), the rate of change per generation is roughly the same as rates measured for characteristics that have been altered by novel selection pressures within the last few centuries (see Chapter 4). Over longer time intervals (of several millions of years), however, the average rate of evolution of most characters is much lower—low enough for known mutation rates for polygenic characters to supply the genetic variation required for long-term evolution. At these low rates, even genetic drift, to say nothing of natural selection, could explain the net change in the feature, if the rate of evolution were constant (see Chapter 13). However, the long-term average rate masks not only "punctuational" episodes of rapid evolution (right-hand panel of Figure 21.1A), but also rapid fluctuations that may transpire even during periods of apparent stasis (Figure 21.1B). Selection, rather than genetic drift, may well be the cause of these rapid changes. Thus rates of evolution of quantitative traits, determined from the fossil record or from comparisons among living species, are consistent with evolution by natural selection and/or genetic drift, with information on mutation and genetic variation, and with observations on short-term rates of evolution that have been inferred from responses to laboratory selection (see Chapter 13), responses of natural populations to environmental change (see Chapter 13), and divergence among populations and among closely related species (see Chapter 15).

Punctuated equilibrium, revisited

Recall that in the theoretical *model* of punctuated equilibrium that Eldredge and Gould proposed to explain the *pattern* of stasis and punctuation, rapid changes (the punctuational events) represent evolution in newly formed species that originate as small, local populations. That is, Eldredge and Gould applied Mayr's model of founder effect spe-

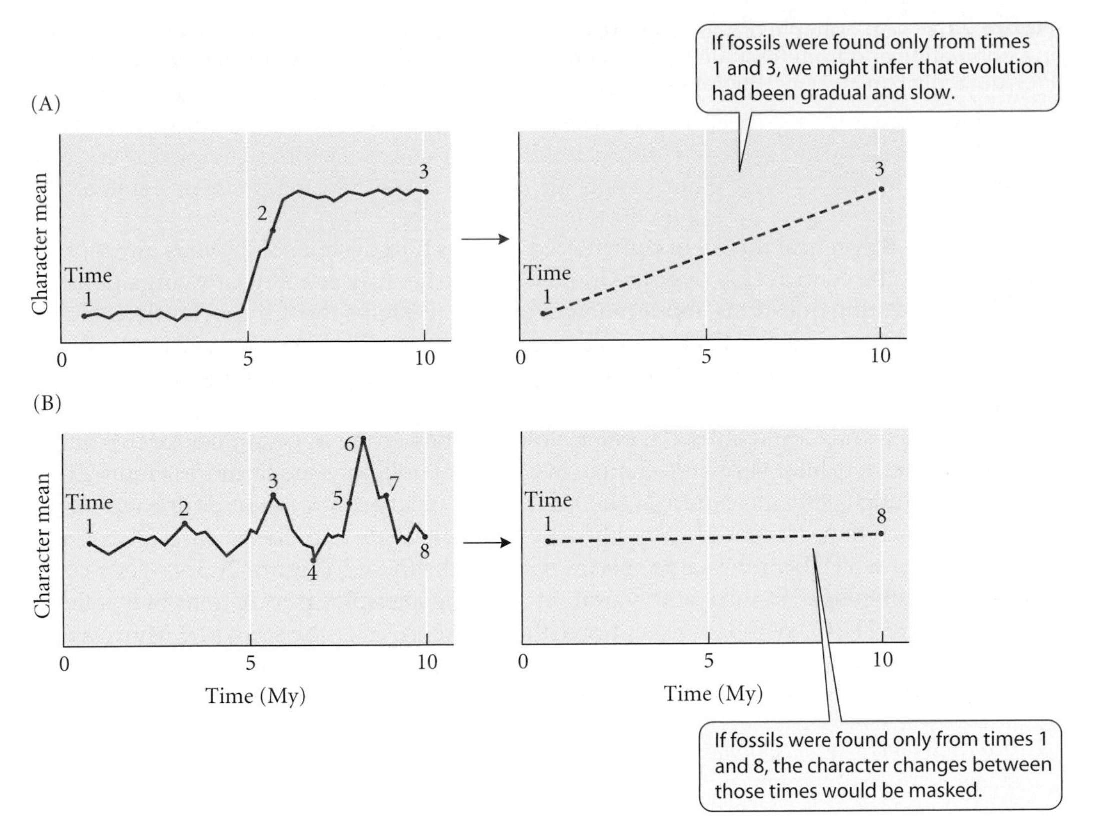

Figure 21.1 Averaged over long periods, the rate of evolution may be low, even though there are episodes of rapid evolution. (A) A punctuated pattern of shift from one rather static character mean to another. If we had fossils only from times 1 and 3, we would not know there had been periods of slow and rapid evolution. (B) A pattern of rapid fluctuations in the character mean, but with little net change over a long period. Given only fossils from times 1 and 8, we might think that little evolution had occurred.

ciation (peripatric speciation) to macroevolutionary change, proposing that most morphological characters cannot evolve (because of internal genetic constraints) except when genetic drift initiates a shift to a new adaptive equilibrium (see Chapter 16).

The fossil record would provide some support for Eldredge and Gould's theoretical model if morphological change were ordinarily accompanied by bifurcation of a lineage—i.e., true speciation. It can be difficult to distinguish change with bifurcation from "punctuated gradualism" (see Chapter 4), but there are some convincing examples of both gradual change without speciation (e.g., the rodent example in Figure 4.19) and morphological change associated with true speciation (e.g., the bryozoan genus *Metrarhabdotos*, in Figure 4.20). Obviously, we would expect to find some instances of coupled speciation and morphological divergence in the fossil record, since fossils of different species cannot be distinguished except by phenotype. But the punctuated equilibrium hypothesis requires that morphological evolution be almost inevitably accompanied by speciation, and it is not clear that the evidence supports this expectation.

Mayr's model of speciation requires that genetic drift (the founder effect) move a small population from the vicinity of one adaptive peak across an adaptive valley (i.e., opposing the action of natural selection) to the slope of a different adaptive peak. As we saw in Chapter 16, this is unlikely to occur unless selection is very weak or unless the population is so small that it is at risk of losing genetic variation altogether. Consequently, many population geneticists are skeptical that peripatric speciation is at all common, and so far there is little evidence that species are formed in this way. Moreover, geographic variation within species, as well as the rapid adaptive evolution of populations exposed to new selection pressures, show that speciation is not required for adaptive phenotypic change (Levinton 2001). Therefore, few evolutionary biologists espouse the theoretical model of Eldredge and Gould (1972), and even its authors have agreed that speciation is not a necessary trigger of adaptive, directional morphological evolution (Eldredge 1989; Gould 2002, p. 796). (Nonetheless, speciation may contribute to anagenetic evolution, as we will see in the following section.)

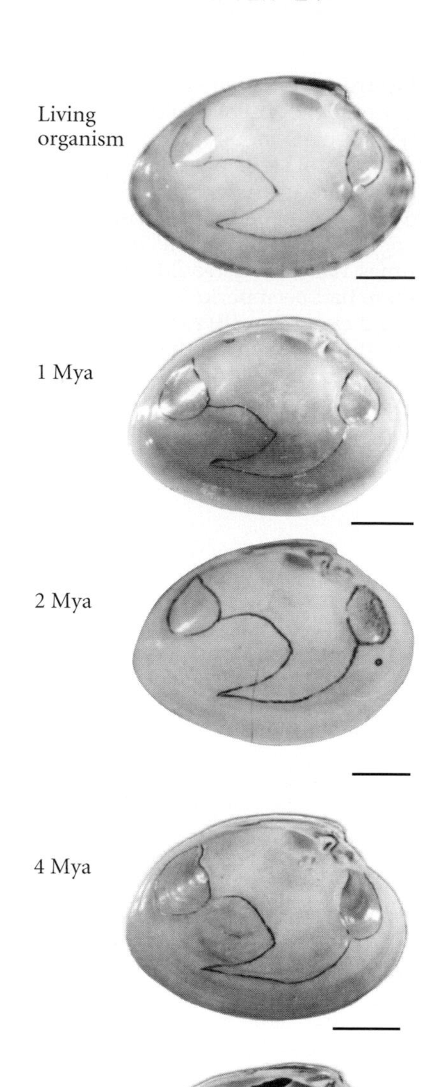

Figure 21.2 An example of stasis: specimens of the bivalve *Macrocallista maculata* from a living population and from fossil deposits dated at 1, 2, 4, and 17 million years ago. All are from Florida. Scale bars = 1 cm. (Photos courtesy of Steven M. Stanley.)

Stasis

Although the theoretical model of punctuated equilibrium described above is almost certainly wrong, the controversy over the model had the healthy effect of drawing attention to many interesting questions about macroevolution. Perhaps most importantly, it established "stasis" as an important and puzzling phenomenon. Rapid evolution is not a problem for evolutionary biology to explain. The problem, rather, is to explain why evolution is often so slow.

As Eldredge and Gould stressed, paleontologists have long known that species in the fossil record often exhibit very little change over several million years or more (Figure 21.2). Stanley and Yang (1987) measured 24 shell characters in large samples of 19 lineages of bivalves, in which they compared early Pliocene (4 Mya) fossils with their nearest living relatives (most of which bear the same species name as the fossils) (Figure 21.3A). They compared these differences, in turn, with variation among geographic populations of 8 of these species (Figure 21.3B). With few exceptions, the difference over the span of 4 Myr was no greater than the difference among contemporary conspecific populations.

Three major hypotheses have been proposed to account for stasis within species lineages:

1. *Internal genetic or developmental constraints.* Eldredge and Gould (1972) proposed that stasis is caused by internal genetic or developmental constraints, which would be manifested by lack of genetic variation or by genetic correlations too strong to permit characters to evolve independently to new optima. But although such constraints may indeed play a role in evolution, as we will see below, they cannot explain the constancy of size and shape of many quantitative characters, which are almost always genetically variable and only imperfectly correlated with one another (see Chapter 13).
2. *Stabilizing selection for a constant optimum phenotype.* It may seem unlikely that natural selection could favor the same character state over millions of years, during which both physical and biotic environmental factors would almost inevitably change. A dramatic example of such change is the succession of the many glacial and interglacial episodes of the Pleistocene, during which climates, the geographic ranges of species, and the associations among species changed drastically and repeatedly (see Chapters 5 and 6). However, a species' "effective environment" may be far more constant over time that we might expect due to **habitat tracking** (Eldredge 1989): the shifting of the geographic distributions of species in concert with the distribution of their typical habitat. The distribution of cold-climate plants such as spruce, for example, has shifted southward during glacial and northward during interglacial times; similarly, many groups of aquatic and semiaquatic insects are found in desert regions, but they live in such water as is available, and have not adapted to dry habitats. A variation on this theme is the proposal that many features are often undergoing adaptive change, but that they track an optimum that fluctuates within narrow limits—perhaps because habitat tracking keeps the species within narrow environmental bounds.
3. *Ephemeral local divergence.* It is possible that substantial adaptive changes occur in local or regional populations of a species, but that they leave no lasting imprint on the phenotype of the species, or even any fossil evidence of their occurrence, because they are too brief and localized (Futuyma 1987). Many adaptations to different food types or microhabitats arise in local populations as long as local selection is not overwhelmed by gene flow from surrounding populations with ancestral character states (Holt and Gaines 1992). Because a food item or microhabitat may have a wide, although patchy, distribution, such a new adaptive phenotype might spread over a broad geographic range—except that interbreeding of migrants with intervening ancestral-type populations would result in recombination and the loss of the distinctive

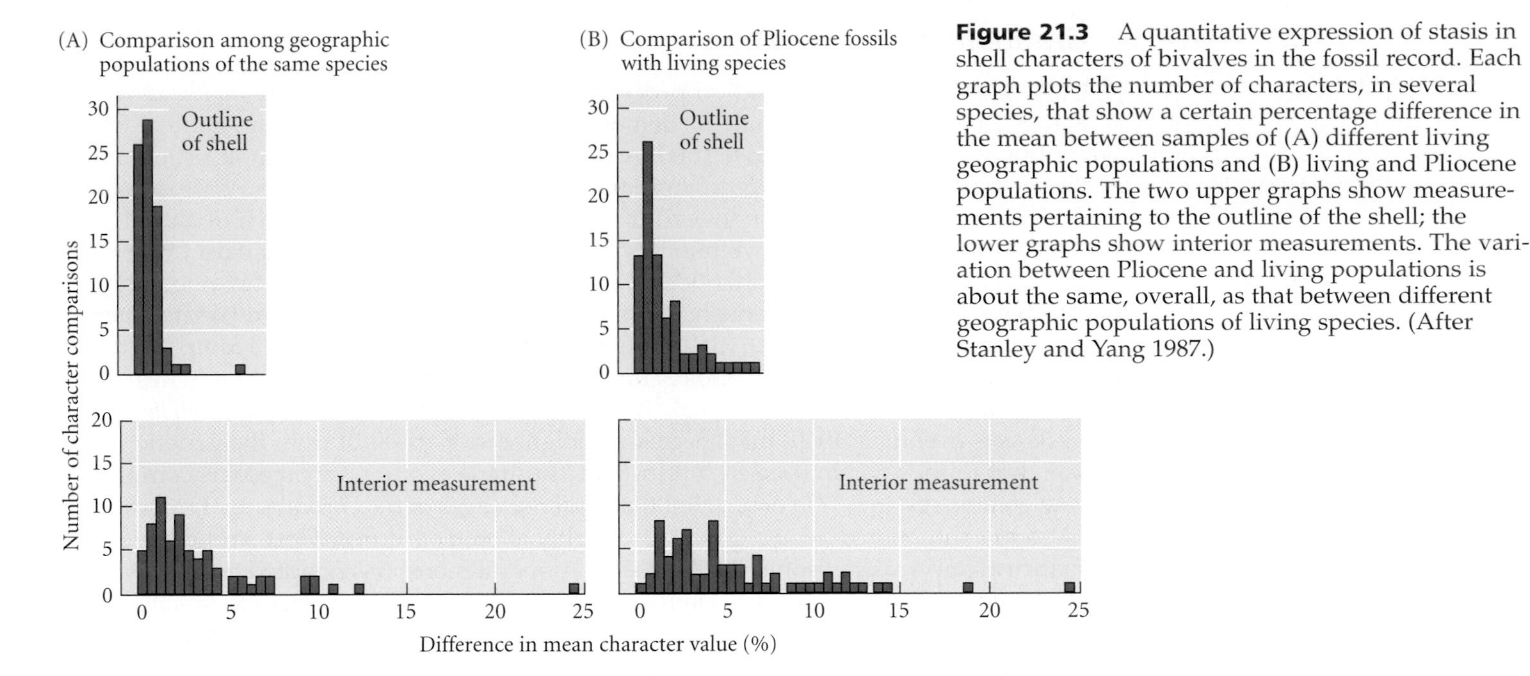

Figure 21.3 A quantitative expression of stasis in shell characters of bivalves in the fossil record. Each graph plots the number of characters, in several species, that show a certain percentage difference in the mean between samples of (A) different living geographic populations and (B) living and Pliocene populations. The two upper graphs show measurements pertaining to the outline of the shell; the lower graphs show interior measurements. The variation between Pliocene and living populations is about the same, overall, as that between different geographic populations of living species. (After Stanley and Yang 1987.)

phenotype (especially if it were polygenic; see Chapter 13). Moreover, if the geographic distribution of patches with the new resource shifted (perhaps due to climate change), the divergent character state would be lost due to interbreeding with the (probably more widespread) ancestral phenotype as individuals dispersed and colonized newly favorable sites and the populations from which they dispersed became extinct. Therefore, the existence of the new phenotype might be as brief as the intervals between climate-induced shifts in the geographic distribution of resources and species, and for this reason might not be registered in the fossil record. If, however, the phenotypically divergent local population became reproductively isolated from the ancestral phenotype (i.e., became a different species), it might track the geographic distribution of its habitat or resource without interbreeding with the ancestral phenotype, and so maintain a long-lasting, divergent identity (see Figure 16.24). Thus, although speciation may not cause anagenetic adaptive change, it may confer long life on such changes, leading to a possible association between speciation and morphological evolution (i.e., the *pattern* of punctuated equilibrium).

This last scenario leads us to the perhaps counterintuitive conclusion that long-term evolutionary change is more likely to occur in relatively stable than in frequently changing environments. Evidence of sustained evolution in stable environments has led some paleontologists to the same conclusion (Sheldon 1987; see Figure 4.3), and there is considerable evidence that the drastic climatic fluctuations in the Pleistocene inhibited both speciation and persistent adaptive phenotypic change (Jansson and Dynesius 2002). For example, even though many groups of beetles had undergone enormous diversification before the Pleistocene, G. R. Coope (1995) has found that species of Quaternary fossil beetles can be matched in precise morphological detail with living species. They "have remained constant both in their morphology and their environmental requirements throughout the whole of the Quaternary period. … From the British Isles alone there are now over two thousand fossil species known that are the precise match of their present-day equivalents." Coope attributes this constancy to habitat tracking, by which "populations continuously split up and reform as they progress to and fro across a complex landscape," during which "the gene pools were kept well stirred." Adaptive changes that are advantageous among most of the geographic populations of a species may be favored by widespread environmental changes, but otherwise such instability is likely to dampen evolutionary change.

Gradualism and Saltation

Darwin proposed that evolution proceeds gradually, by small steps. In *The Origin of Species*, he wrote that "if it could be demonstrated that any complex organ existed, which could not possibly have been formed by numerous, successive, slight modifications, my theory would absolutely break down." His ardent supporter Thomas Henry Huxley, however, cautioned that his theory of evolution would be just as valid even if evolution proceeded by leaps. Some later evolutionary biologists, such as the paleontologist Otto Schindewolf (1950), proposed just this. Schindewolf declared that the differences among higher taxa have "arisen discontinuously, by saltation," and that "the first bird [*Archaeopteryx*] hatched from a reptilian egg." (**Saltation**, from the Latin *saltus*, means "a jump.") The accomplished geneticist Richard Goldschmidt, one of the first to propose that genes act by controlling the rates of biochemical and developmental reactions, argued in *The Material Basis of Evolution* (1940) that species and higher taxa arise not from the genetic variation that resides within species, but instead "in single evolutionary steps as completely new genetic systems." He postulated that major changes of the chromosomal material, or "systemic mutations," would give rise to highly altered creatures. Most would have little chance of survival, but some few would be "hopeful monsters" adapted to new ways of life. He pointed, for example, to a mutation that transforms the halteres of *Drosophila* into more winglike structures that resemble, he said, the rudimentary wings of a dipteran fly (*Termitoxenia*) that inhabits termite nests (Figure 21.4).

If, between even the most extremely different organisms, there existed a full panoply of all possible intermediate forms, each differing from similar forms ever so slightly, we would have little doubt that evolution is a history of very slight changes. Such is often the case when we examine differences among individuals in a population, among populations of a species, and among closely related species, such as those in the same genus. Moreover, quite different species are often connected by intermediate forms, so that it becomes arbitrary whether the complex is classified as two genera (or subfamilies, or families) or as one (see Chapter 3). Nonetheless, there exist many conspicuous gaps among phenotypically similar clusters of species, especially among those classified as higher taxa such as orders and classes. No living species bridge the gap between cetaceans (dolphins and whales) and other mammals, for example.

The most obvious explanation of phenotypic gaps among living species is extinction of intermediate forms that once existed—as the cetaceans themselves illustrate. Of course, the common ancestor of two quite different forms need not have appeared precisely intermediate between them, because the two phyletic lines may have undergone quite different modifications (Figure 21.5). For instance, DNA sequences imply that, among living animals, whales are most closely related to hippopotamuses, but early fossil cetaceans do not have a hippo-like appearance in the slightest (see Figure 4.11). Certainly, the fossil record provides many examples of the gradual evolution of higher taxa (see Chapter 4). Nevertheless, many higher taxa appear in the fossil record without intermediate antecedents.

One of the most enduring controversies in evolutionary biology has been whether such phenotypic gaps simply represent an inadequate fossil record—that is, evolution was gradual, but we simply lack the data to prove it—or whether evolution really proceeded by saltation. A saltation would result from the fixation of a single mutation of large effect. This concept is entirely different from the hypothesis of punctuated equilibrium, which allows that evolutionary changes in morphology might (or might not) have been continuous, passing through many intermediate

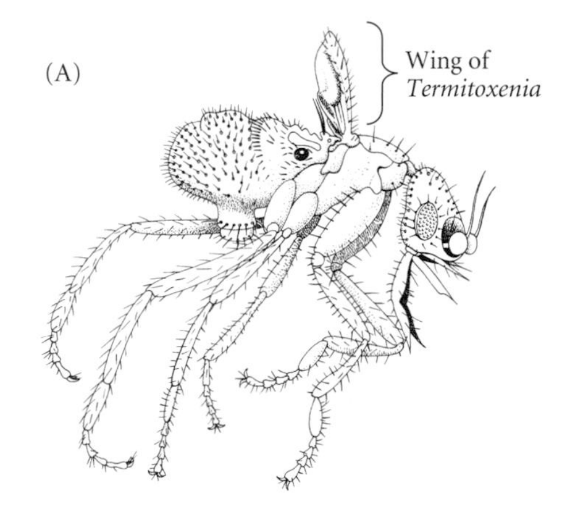

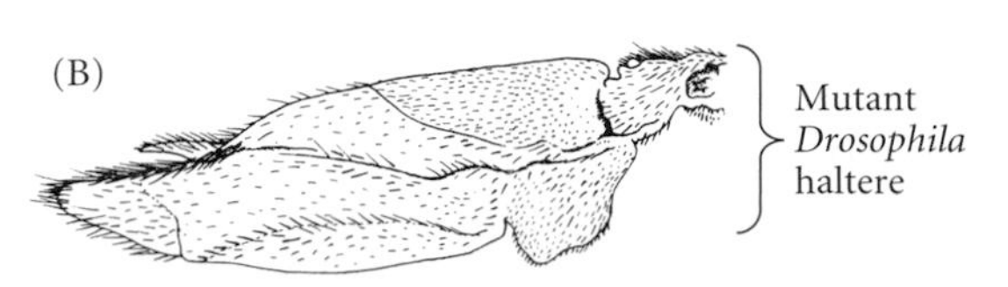

Figure 21.4 An example offered by Goldschmidt as a possible case of saltational evolution. (A) The size and venation of the wing is greatly reduced in *Termitoxenia*, a fly that inhabits termite nests. (B) The mutation *tetraptera* in *Drosophila* changes the haltere into a wing that resembles the wing of *Termitoxenia*. Goldschmidt suggested that a similar mutation caused saltational evolution of the wing of *Termitoxenia*. (After Goldschmidt 1940.)

stages, but were so rapid and so geographically localized that the fossil record presents the appearance of a discontinuous change. The saltation hypothesis, in contrast, holds that intermediates never existed—that mutant individuals differed drastically from their parents.

In judging gradualism and saltationism, we must distinguish between the evolution of *taxa* and of their *characters*. Higher taxa often differ in many characters: for example, a great many features distinguish modern birds from Cretaceous dinosaurs. Gradualists hold that many characters of higher taxa evolved independently and sequentially (MOSAIC EVOLUTION). Both comparisons of living species and the fossil record provide abundant evidence of mosaic evolution (see Chapters 3 and 4). There is some disagreement, however, on whether each of the distinguishing *characters* of a higher taxon—the reduction and the fusion of birds' tail vertebrae, for example—might have evolved discontinuously.

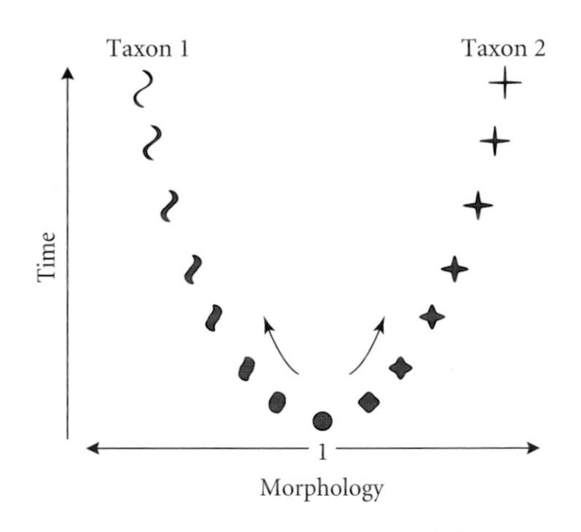

Figure 21.5 Two very different taxa may have evolved gradually from a common ancestor, even though no form precisely intermediate between them ever existed.

Certainly many mutations arise that have discontinuous, large, even drastic effects on the phenotype. Many of these, however, have such important pleiotropic effects that they greatly reduce viability. A mutation of the *Ultrabithorax* gene in *Drosophila*, for example, converts the halteres into wings. It may be tempting to think that this mutation reverses evolution, and that a mutation in this gene caused the evolutionary transformation of the second pair of wings into halteres in the ancestor of the Diptera—but the *Ultrabithorax* mutation is lethal in homozygous condition. Moreover, we understand that the normal form of the *Ultrabithorax* gene regulates the many genes that determine the distinctive development of the third thoracic segment, including the form of the halteres (see Chapter 20). Thus mutations that reduce the function of this master gene interfere with a complex developmental pathway, and development is routed into a "default" pathway that produces the features of the second thoracic segment (including wings). The whole system can be shut down in a single step by turning a master switch, but that does not mean that the system came into existence by a single step.

Neo-Darwinians have always accepted, however, that characters can evolve by minor jumps—that is, by mutations that have fairly large (but not huge) effects. For example, variation both within and among species in characters such as bristle number in *Drosophila* is often caused by a mixture of quantitative trait loci with both small and quite large effects (Orr and Coyne 1992; see Chapters 13 and 16). Alleles with large effects contribute importantly to mimetic polymorphisms in butterflies such as the African swallowtail *Papilio dardanus* (see Figure 9.2A), in which each of several very different forms is a mimic of a different unpalatable species (or model). Were phenotypes to arise that deviated only slightly from one mimetic pattern toward another, they would lack protective resemblance to either unpalatable model and presumably would suffer a disadvantage. Thus it is likely that the evolution of one mimetic pattern from another was initiated by a mutation of large enough effect to provide substantial resemblance to a different model species, followed by selection of alleles with smaller effects that "fine-tuned" the phenotype (Figure 21.6). Genetic analysis of the color patterns of *P. dardanus* supports this hypothesis (Ford 1971).

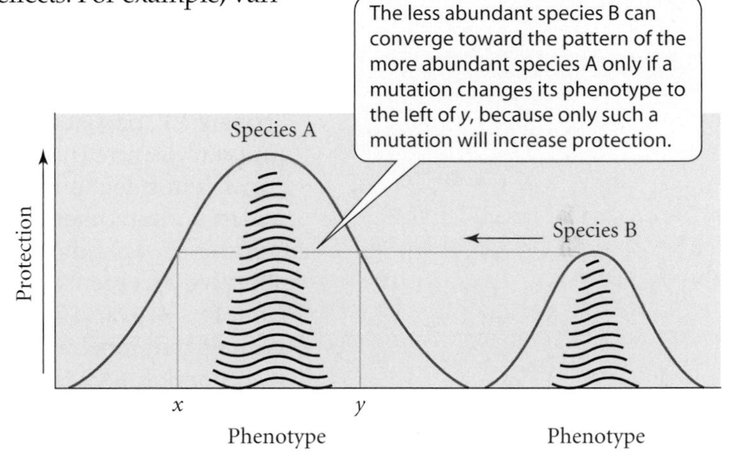

Figure 21.6 A model of the evolution of color pattern in Müllerian mimics such as *Heliconius* butterflies. The red curves represent the degree of protection against predators for two species, A and B, that differ in phenotype and abundance. The wavy lines indicate the distribution of phenotypes within each species. Selection favors convergence of the less abundant species B toward the pattern of the more abundant species A because predators more often learn to avoid the more abundant species. A mutation of small effect that slightly alters the phenotype of species B will be selectively disadvantageous. However, a mutation of large effect that causes members of species B to acquire a phenotype with a modest resemblance to species A (e.g., just left of phenotype *y*) will be selectively advantageous. Subsequently, allele substitutions with small effects that bring B closer to the peak for A will be advantageous. (After Charlesworth 1990.)

Perhaps the most dramatic single-gene differences yet described are those with strong heterochronic effects. The axolotl (*Ambystoma mexicanum*) differs from the closely related tiger salamander (*A. tigrinum*) by being paedomorphic: it does not metamorphose, but retains larval characteristics throughout its adult life (see Figure 3.13). Crosses between *A. tigrinum* and a laboratory stock of axolotl showed that a single-gene difference causes paedomorphosis. But this allele has not engendered a whole new complex morphology: it merely truncates a complex pathway of development that evolved by many small steps. Interestingly, paedomorphosis in wild axolotls is not caused by a single gene, but has a more complex genetic basis (Voss and Shaffer 2000).

Phylogenetic Conservatism and Change

Given the great diversity of phenotypes among even closely related species, and given the rapidity with which evolution can occur, biologists are challenged to understand the existence of "living fossils"—organisms such as the ginkgo (see Figure 5.18B), the tadpole shrimp (Figure 21.7A), and the coelacanth (Figure 21.7B) that have changed so little over many millions of years that they closely resemble their Mesozoic or even Paleozoic relatives. Such evolutionary "conservatism" often extends to ecological characteristics as well: for instance, the larvae of all species of the butterfly tribe Heliconiini feed on passion flower plants (Passifloraceae), and there is no reason to think that any members of this clade have evolved a different feeding habit since the Oligocene. The synapomorphies of large clades also represent conservatism: almost all mammals, no matter how long or short their necks, have seven neck vertebrae, and almost no tetrapods have had more than five digits per limb. (The earliest amphibians had more, but soon settled on five.)

We may postulate that such conservatism has been caused by consistent stabilizing selection or by inadequate variation for selection to act on.

Stabilizing selection

Many characters must remain relatively unchanged throughout clades simply because the optimum character value has remained unchanged. An important reason for this lack of change is **niche conservatism**: long-continued dependence of related species on much the same resources and environmental conditions, as the similar host plant associations of all species of heliconiine butterflies illustrate (Holt 1996; Travis and Futuyma 1993). Similarly, closely related species of plants have climatically similar geographic ranges in Asia and North America (Ricklefs and Latham 1992), and sister species of birds, mammals, and butterflies occupy climatically similar areas in Mexico (Peterson et al. 1999).

By occupying one niche (e.g., host plant, climatic zone) rather than another, a species subjects itself to some selective pressures and screens off others; it may even be said to "construct" or determine its own niche, and therefore many aspects of its potential evolutionary future (Lewontin 2000; Odling-Smee et al. 2003). Niche conservatism implies consistent selection. Niches remain conservative for two major, sometimes interacting, reasons: First, other species, often by acting as competitors, may prevent a species from shifting or expanding its niche. We have seen how alleviation of competition may release adaptive divergence and radiation, for example, on islands and after mass extinctions (see Chapters 3, 4, and 7). Second, and more generally, if there is gene exchange among individuals that inhabit the ancestral niche (e.g., microhabitat) and those that inhabit a novel niche, and if there is a fitness trade-off between character states that improve fitness in the two environments, then selection will generally favor the ancestral character state (i.e., stabilizing selection will prevail) simply because most of the population occupies the an-

(A)

(B)

Figure 21.7 Two "living fossils." (A) The tadpole shrimp, *Triops cancriformis*, found in temporary pools in arid regions of Eurasia and northern Africa, has undergone no evident morphological evolution since the Triassic. (B) Coelacanths are lobe-finned fishes that originated in the Devonian and were thought to have become extinct in the Cretaceous, until this living species, *Latimeria chalumnae*, was discovered in 1938. (A © OSF/photolibrary.com; B © The Natural History Museum, London.)

cestral environment (Holt 1996). In organisms that lack habitat selection behavior, selection for ancestral character states reduces fitness in the novel environment, so that a population in that environment may be a "sink" population, incapable of persistence. Moreover, in animals that are capable of habitat selection, the differential in fitness between individuals that occupy ancestral versus novel environments favors choice of the ancestral environment. Both factors will result in the habitat tracking described above.

As the degree of adaptation to any one environment increases, the differential in fitness between that environment and a new environment also increases, so that adaptation to an alternative environment may become steadily less likely. In extreme cases, a lineage becomes very highly specialized. Whether or not specialization is then a "dead end," precluding shifts to different environments or ways of life, is an old question that we will treat below.

Limitations on variation

Most characters display heritable phenotypic variation, as we have seen, so the origin of variation seems unlikely to limit evolution (Barton and Partridge 2000). However, not all features are equally variable, and genetic correlations can make some character combinations very unlikely to arise, at least for moderate periods of time. There is some evidence that the direction of divergence between species can be affected by the availability of genetic variation (see Chapter 13).

Perhaps most intriguing are features that display little phenotypic variation despite underlying genetic variation. For example, *Drosophila melanogaster* and related species almost always have four bristles on the scutellum, but a mutation that perturbs development reveals plenty of hidden polygenic variation for bristle number (see Figure 13.21). Development appears ordinarily to be "buffered" against phenotypic expression of this genetic variation. This buffering, or CANALIZATION, in some cases can evolve by natural selection for a consistent phenotype, although such evolution may occur only under restricted conditions (Wagner et al. 1997). In other cases, canalization may be an incidental effect of a mechanism that evolved for other reasons. For example, the phylogenetically universal heat-shock protein Hsp90 is a "chaperone" protein that binds with many other proteins (such as transcription factors and kinases) and keeps them ready for activation. When the function of Hsp90 is reduced by a mutation of the encoding gene or by environmental stress, previously undetected genetic variation is revealed in a great number of phenotypic traits, in both *Drosophila* and *Arabidopsis*. Thus the normal Hsp90 prevents expression of much genetic variation, and indeed, may enable such variation to accumulate. It has been suggested that some features might ordinarily evolve slowly, but be capable of rapid change if a population experienced changes at the *Hsp90* locus (Rutherford and Lindquist 1998). Computer models suggest that many genes could have similar effects if they are part of a complex net of gene interactions of the kind that commonly underlie morphological development (Bergman and Siegal 2003). Whatever the cause may be, it is clear that some developmental pathways produce highly buffered, almost invariant characteristics.

When variation does occur, it may be directional: reduction or loss tends to be more common than elaboration or gain. In tetrapod vertebrates, for example, variants with fewer than five digits arise more often, may be more viable, and have been fixed during evolution much more frequently than variants with more than five. Moreover, differences among clades in developmental pathways dictate the kinds of variation that can arise: salamanders are more likely to lose preaxial toes, and frogs postaxial toes, corresponding to the order in which the digits develop in the two groups (see Figure 20.19).

Development can therefore impose constraints on the rate or direction of evolution of a character. The consequences of such constraints are made clear when an adaptive function is performed not by the structure we might expect, but by another structure that has been modified instead. The giant panda, for example, has six apparent fingers, evidently useful for manipulating the bamboo on which it feeds. The outermost "finger" (or "thumb"), however, is not a true digit, but a sesamoid bone that develops from cartilage (Figure 21.8). Likewise, developmental constraints are striking (and hardly surprising)

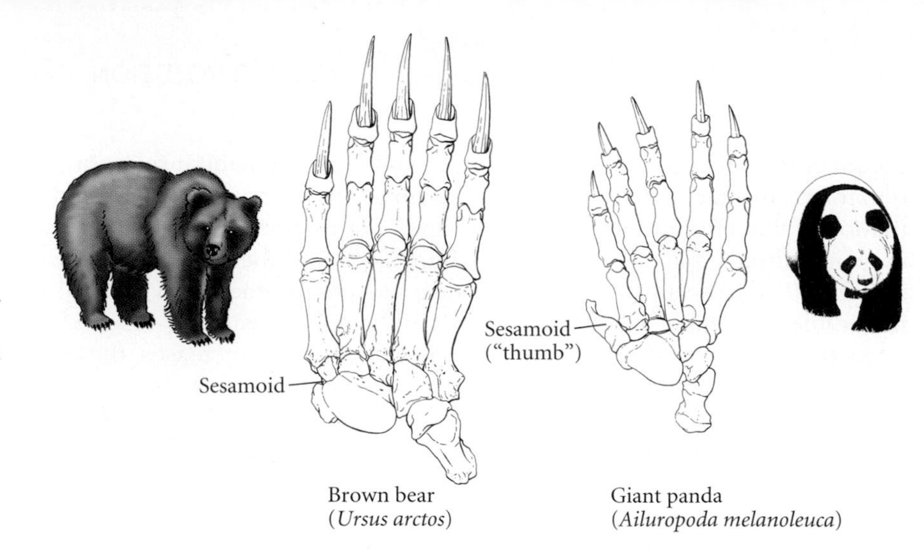

Figure 21.8 The right hand (in dorsal view) of two members of the bear family, a brown bear (left) and the giant panda (right). A small sesamoid bone in bears has been modified into a false finger ("thumb") in the panda, which uses it to help manipulate the bamboo on which it feeds. Developmental constraints probably prevented the evolution of a sixth true digit. The panda's "thumb" is an example of what has been called "tinkering" by natural selection: evolving adaptations from whatever variable characters a lineage happens to already have. (After Davis 1964.)

when a character and the underlying developmental genetic pathway have been lost in evolution. As DOLLO'S LAW states, a complex character, if lost, is seldom, if ever, regained in its original form. Instead, other features are often modified to serve its function. Birds have not had teeth since the end of the Cretaceous, but mergansers have evolved tooth-like serrations of the bill margin that enable these ducks to catch fish (Figure 21.9).

The Evolution of Novelty

How do major changes in characters evolve, and how do new features originate? These questions have two distinct meanings. First, we can ask what the genetic and developmental bases of such changes are—the subject of Chapter 20. Second, we can ask what role natural selection plays in their evolution. For instance, we may well ask whether each step, from the slightest initial alteration of a feature to the full complexity of form displayed by later descendants, could have been guided by selection. What functional advantage can there be, skeptics ask, in an incompletely developed eye? And we can ask how complex characters could have evolved if their proper function depends on the mutually adjusted form of each of their many components.

Accounting for incipient and novel features

Several pathways of evolutionary change account for the macroevolution of phenotypic characters (see Mayr 1960; Nitecki 1990; Müller and Wagner 1991; Galis 1996):

1. *A feature originates as a new structure, or a new modification of an existing structure.* For example, sesamoid bones often develop in connective tissue in response to embryonic movement. Such bones are the origin of novel skeletal elements, such as the extra "finger" of the giant panda (see Figure 21.8) and the patella (kneecap) in the leg of mammals, which is lacking in reptiles (Müller and Wagner 1991). One can often envision the advantages of new features on their first appearance. For example, a slight thickening of bone in the nasal region of the earliest titanotheres could have been advantageous in the butting contests that the males of many hoofed mammals engage in; once this thickening originated, sexual selec-

(A)

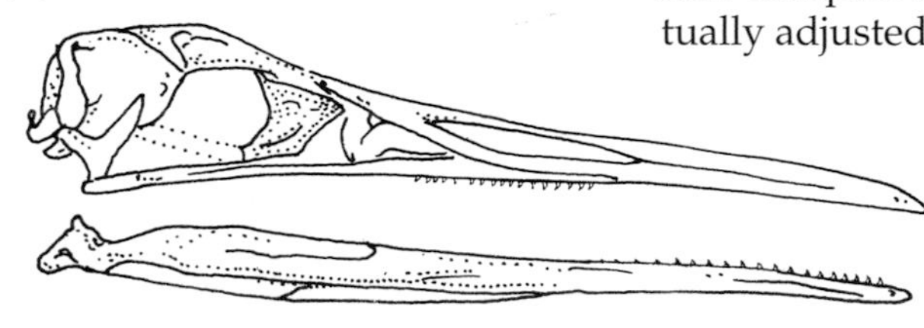

(B)

(C)

Figure 21.9 Complex structures, if lost, are generally not regained, but their function may be. (A) *Hesperornis*, a marine bird of the late Cretaceous, had teeth that enabled it to grip fish. (B) The bill of a typical living fish-eating bird, the anhinga (darter), as drawn by John James Audubon. No living bird has teeth. (C) The bill of the merganser, a fish-eating duck, has substitutes for teeth: serrations that enable it to grip fish. (A courtesy of Larry Martin in Feduccia 1999; C art by Nancy Haver.)

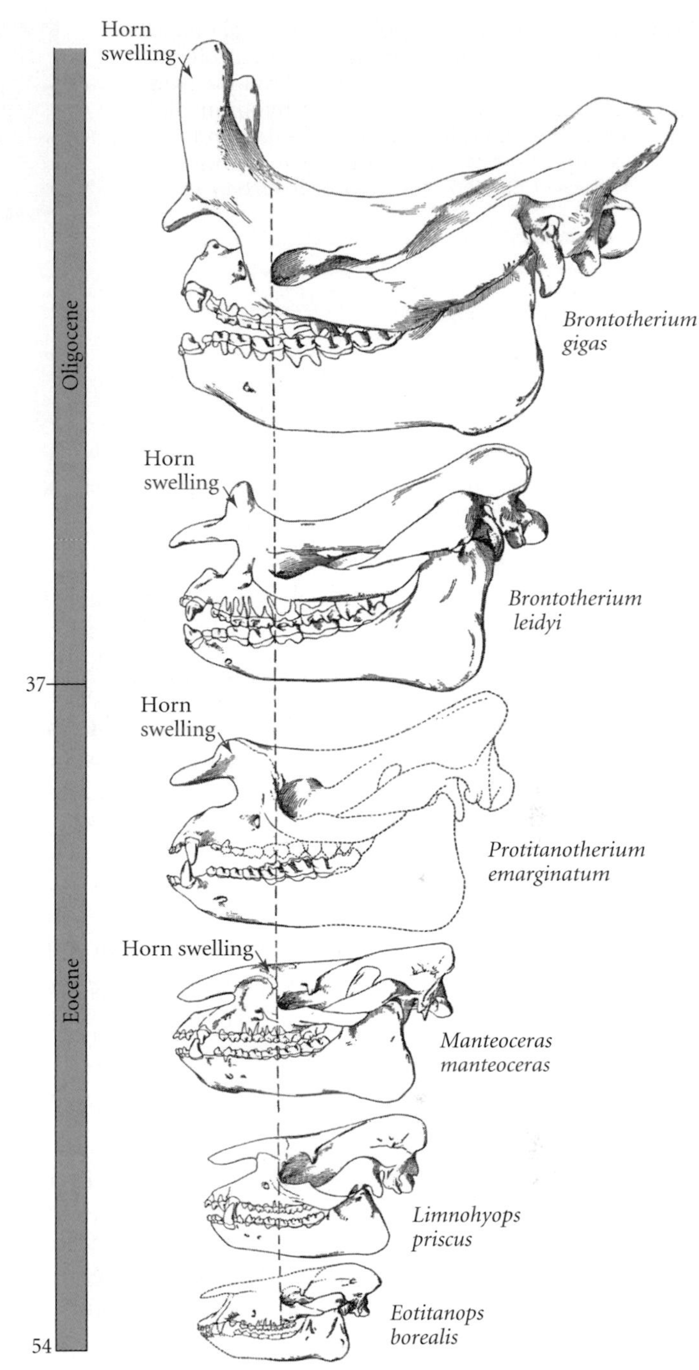

Figure 21.10 Gradual evolution of a character that may have been selectively advantageous from its inception. The paired nasal horn of a lineage of titanotheres (Brontotheriidae), an extinct group of hoofed mammals related to horses, was probably used in fighting, and any thickening of the bone probably reduced the harmful effects of impacts. The largest titanotheres, such as *Brontotherium gigas* (top), were the size of rhinoceroses. (From Osborn 1929.)

tion could have favored a larger hornlike structure (Figure 21.10).

2. *A feature is a developmental by-product* of other adaptive features. It may not be adaptive initially, but at some later point, it may be recruited or modified to serve an adaptive function. For instance, by excreting nitrogenous wastes as crystalline uric acid, insects lose less water than if they excreted ammonia or urea. Excreting uric acid is surely an adaptation, but the white color of uric acid is not. However, pierine butterflies such as the cabbage white butterfly (*Pieris rapae*) sequester uric acid in their wing scales, imparting to the wings a white color that plays a role in thermoregulation and probably in other functions.
3. *An ancestral function of a feature may become accentuated*, leading to its modification. This is especially likely if any other function of the ancestral feature is lost. For example, lizards primitively use the tongue to capture prey. The lizard lineage that includes the monitors captures prey with the jaws, which has freed the tongue to be used (and modified) mostly for chemosensory function (Schwenk 1993). In most snakes, a system of muscles and bones can move the tooth-bearing maxillary bone forward and back to pull prey into the mouth. These features are accentuated in vipers such as rattlesnakes, in which the same apparatus rotates a shortened maxilla that bears a single large, hollow fang, which is used to deadly effect on prey and sometimes on predators (see Figure 11.3)
4. *Decoupling the multiple functions* of an ancestral feature frees it from functional constraints and may lead to its elaboration. For example, the locomotory muscles of many "reptiles" insert on the ribs, so that these animals cannot breathe effectively while running. In birds and mammals, the muscle insertions have shifted to processes on the vertebrae, so that breathing and running (or flying) are decoupled, and many lineages have evolved features associated with rapid locomotion (Galis 1996). David Wake (1982) has proposed that the loss of lungs in the largest family of salamanders (Plethodontidae) has relieved a functional constraint on the evolution of the tongue. In other salamanders, the bones that support the tongue are also used for moving air in and out of the lungs. In plethodontids, this hyobranchial skeleton, no longer used for ventilating the lungs, has been modified into a set of long elements that can be greatly extended from a folded configuration. This modification enables plethodontids to catch prey by projecting the tongue. Members of one clade, the bolitoglossines, can shoot their extraordinarily long tongues at a greater speed than any other movement known in vertebrates (Figure 21.11).
5. *Duplication with divergence* gives rise to diversity at the morphological level, as it does at the level of genes and proteins (see Chapter 19). The diversity of teeth in a mammal, for example, provides enormous functional possibilities: molars can slice or grind, while canines stab. Some teeth—such as elephants' tusks—are used more for social interactions than for eating.

Figure 21.11 A lungless bolitoglossine salamander (*Hydromantes supramontis*) captures prey with its extraordinarily long tongue. The rapid tongue extension is accomplished with a modified hyobranchial apparatus, which in other families of salamanders plays an important role in ventilating the lungs. (From Deban et al. 1997, courtesy of S. Deban.)

6. A *change in the function* of a feature *alters the selective regime*, leading to its modification. This principle, already recognized by Darwin, is one of the most important in macroevolution (Mayr 1960), and every group of organisms presents numerous examples. A bee's sting is a modified ovipositor, or egg-laying device. The wings of auks and several other aquatic birds are used in the same way in both air and water; in penguins, the wings have become entirely modified for underwater flight (see Figure 11.17). The ability of an electric eel (*Electrophorus electricus*) to kill prey and defend itself by electric shock is an elaboration of the capacity of other members of the knifefish family to generate much weaker electric fields for orientation and communication in murky waters.

Complex characteristics

A common argument against Darwinian evolution is based on what is sometimes termed "irreducible complexity": the proposition that a complex organismal feature cannot function effectively except by the coordinated action of all its components, so that an organism's fitness would be reduced by elimination or alteration of any one of the interacting components. From this proposition, opponents of evolution argue that the feature must have required all of its components from the beginning, for the entire functional complex could not have arisen in a single mutational step.

The first person to recognize this potential problem was Darwin himself, in *The Origin of Species:* "That the eye, with all its inimitable contrivances for adjusting the focus to different distances, for admitting different amounts of light, and for the correction of spherical and chromatic aberration, could have been formed by natural selection seems, I freely confess, absurd in the highest possible degree." But he then proceeded to supply examples of animals' eyes as evidence that "if numerous gradations from a perfect and complex eye to one very imperfect and simple, each grade being useful to its possessor, can be shown to exist; if further, the eye does vary ever so slightly, and the variations be inherited, which is certainly the case; and if any variation or modification in the organ be ever useful to an animal under changing conditions of life, then the difficulty of believing that a perfect and complex eye could be formed by natural selection, though insuperable by our imagination, can hardly be considered real."

Darwin's claim has been fully supported by later research (e.g., Osorio 1994; Nilsson and Pelger 1994). The eyes of various animals range from small groups of merely light-sensitive cells (in some flatworms, annelid worms, and others), to cuplike or "pinhole camera" eyes (in cnidarians, molluscs, cephalochordates, and others), to the "closed" eyes, capable of registering precise images, that have evolved independently in cnidarians, snails, bivalves, polychaete worms, arthropods, and vertebrates (Figure 21.12). The evolution of eyes is apparently not so improbable! Each of the many grades of photorecep-

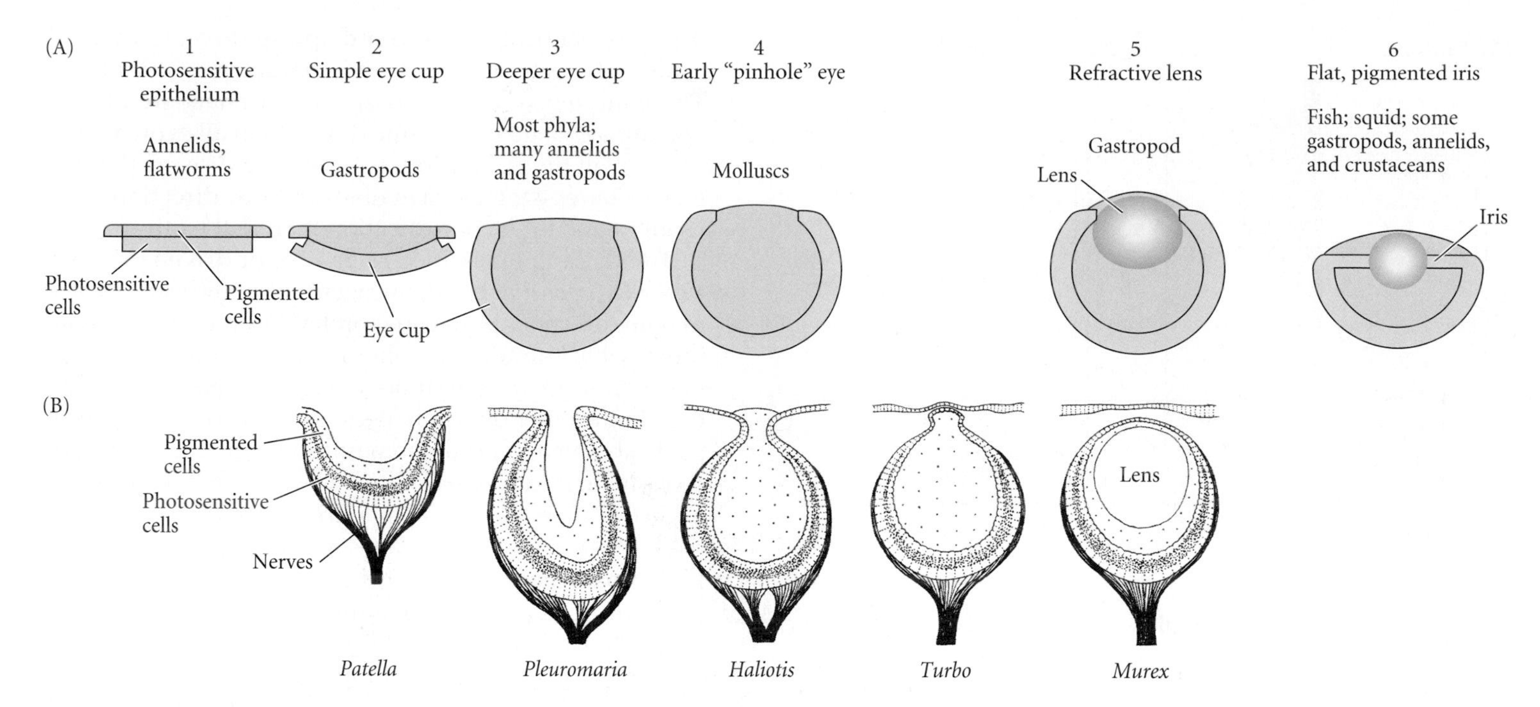

Figure 21.12 Intermediate stages in the evolution of complex eyes. (A) Schematic diagrams of stages of eye development in various animals, from a simple photosensitive epithelium, through the deepening of the eye cup (providing progressively more information on the direction of the light source). There is gradual evolution toward a "pinhole camera" eye, eventually including a refractive lens and a pigmented iris for more perfect focusing. (B) Most of these stages can be found among various gastropod species, as shown in these drawings. (A after Osorio 1994; B after Salvini-Plawen and Mayr 1977.)

tors, from the simplest to the most complex, serves an adaptive function. Simple epidermal photoreceptors and cups are most common in slowly moving or burrowing animals; highly elaborated structures are typical of more mobile animals. The mystery of how a simple eye could be adaptive is no great mystery after all.

The antievolutionary argument, then, ignores evidence that intermediate stages in the evolution of complex systems exist and have adaptive value. It also fails to recognize that a component of a functional complex that was initially merely superior can *become* indispensable because other characters evolve to *become functionally integrated* with it. Although the eyes of many animals do not have a lens, these animals do quite well without the visual acuity that a lens can provide. But a lens is indispensable for eagles, since their way of hunting prey has been acquired and made possible only by such acuity. Eagles and mammals have *acquired* dependence on the elements of a complex eye. Such dependence, indeed, is often lost: many burrowing and cave-dwelling vertebrates have degenerate eyes.

Trends and Progress

For many decades after the publication of *The Origin of Species*, many of those who accepted the historical reality of evolution viewed it as a cosmic history of progress. As humanity had been the highest earthly link in the pre-evolutionary Great Chain of Being, just below the angels (see Chapter 1), so humans were seen as the supreme achievement of the evolutionary process (and Western Europeans as the pinnacle of human evolution). Darwin distinguished himself from his contemporaries by denying the necessity of progress or improvement in evolution (Fisher 1986), but almost everyone else viewed progress as an intrinsic, even defining, property of evolution. In this section, we shall examine the nature and possible causes of trends in evolution and ask whether the concept of evolutionary progress is meaningful.

A **trend** may be described objectively as a directional shift over time. "Progress," in contrast, implies betterment, which requires a value judgment of what "better" might mean. We will come to terms with evolutionary "progress" after discussing trends.

Trends: Kinds and causes

We will consider a **trend** to be a persistent, directional change in the average value of a feature, or perhaps its maximum (or minimum) value, in a clade over the course of time.

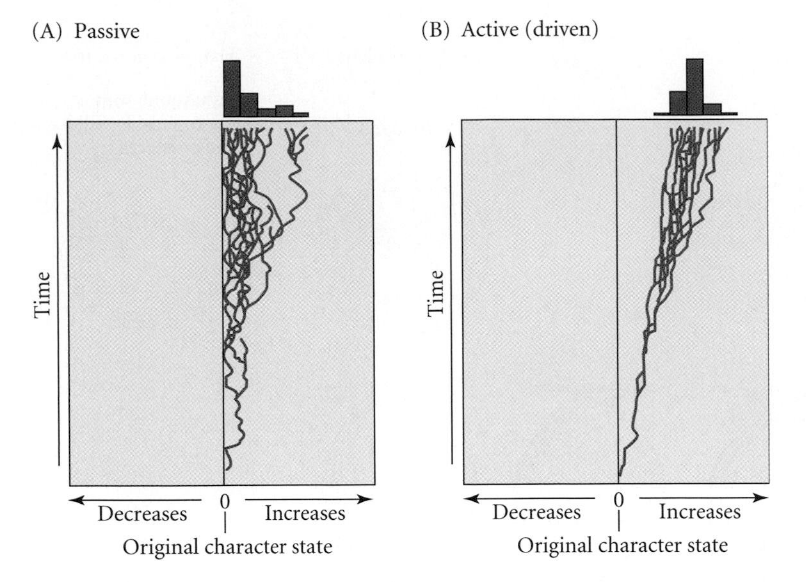

Figure 21.13 Computer simulations of the diversification of a clade. (A) A passive trend. A character shift in either direction is equally likely, but the character value cannot go beyond the boundary at left. The mean increases, but many lineages retain the original character value. (B) A driven trend. The entire distribution of character values is shifted by a bias in the direction of change, caused by a factor such as natural selection. (After McShea 1994.)

A phylogenetically **local trend** applies to an individual clade, whereas a **global trend** characterizes all of life. There are few, if any, UNIFORM evolutionary trends, involving a change in the same direction in all evolving lineages at all times. NET trends have been discerned, however, whereby a feature evolves in both directions, but more frequently in one direction than in the other.

Trends can be classified as passive or driven (McShea 1994). In a **passive trend**, lineages within the clade evolve in both directions with equal probability, but if there is an impassable boundary on one side (e.g., a minimal possible body size), the variation among lineages can expand only in the other direction. Because the variance expands, so do the mean and the maximum (Figure 21.13A). Although the mean increases, some lineages remain near the ancestral character value. In a **driven trend**, on the other hand, changes within lineages in one direction are more likely than changes in the other (i.e., there is a "bias" in direction), so both the maximum and the minimum character values change along with the mean (Figure 21.13B).

Either driven or passive trends could have any of several causes. Neutral evolution by *mutation and genetic drift* results in increasing variance among lineages (see Chapter 12) and could produce a passive trend if variation were bounded as in Figure 21.13A. *Individual selection* could be responsible for all the changes within lineages and could result in either a passive or a driven trend, depending on whether or not ancestral character states remained advantageous for some lineages. The mean character state among species in a clade can also change due to a correlation with speciation or extinction rates (see Figure 11.16). The character might cause a rate difference (SPECIES SELECTION in the broad sense; see Chapter 11) or might simply be correlated with another character that causes a rate difference. This process has been called **species hitchhiking** (Levinton 2001) by analogy with the hitchhiking of linked genes within populations.

Examples of trends

Increases in the body size of mammals may represent a passive trend. Paleontologists noticed long ago that the maximum body size of species in many animal groups has tended to increase over time, a trend dubbed COPE'S RULE. A plot of the body sizes of 1534 species of North American late Cretaceous and Cenozoic mammals (Figure 21.14A) against their dates in the fossil record shows such a passive trend (Alroy 1998). Mammals were small before the K/T mass extinction, and the

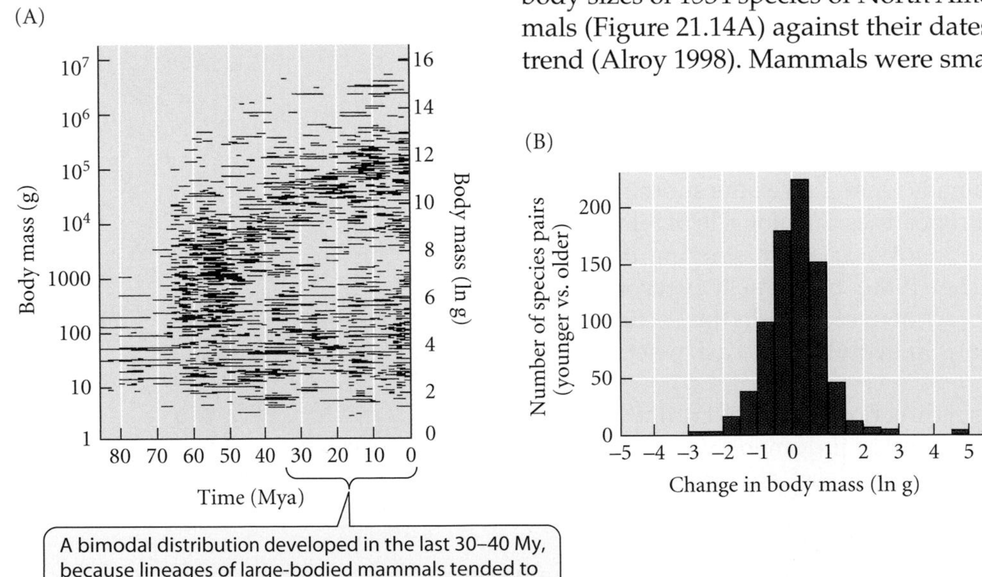

Figure 21.14 A passive trend: Cope's rule in late Cretaceous American mammals. (A) Each of 1,534 species is plotted as a line showing its temporal extension and its body mass (estimated from tooth size). Although small mammals persist throughout the Cenozoic, there is an increasing number of large species over the course of time. (B) Change in body mass (negative or positive) is plotted for 779 pairs of older and younger species in the same genera (likely ancestor-descendant pairs). There are significantly more positive than negative changes, implicating natural selection as the cause of the trend toward increased size. (After Alroy 1998.)

lower size limit has remained nearly the same ever since. However, mean and maximal sizes have increased, especially after the K/T extinction, when the explosive diversification of mammals began. Changes in body size between 779 matched pairs of older and younger species in the same genera (likely ancestor-descendant pairs) occurred in both directions, but were significantly biased toward increases (Figure 21.14B). This nonrandomness strongly suggests that the trend was caused by natural selection rather than genetic drift.

Whereas data for all mammals taken together suggest a passive trend, body size in the horse family (Equidae) conforms to a driven trend (MacFadden 1986; McShea 1994). Not only the maximum and the mean, but also the minimum size, increased during the Cenozoic (Figure 21.15). Ancestor-descendant pairs showed an increase in size much more often than a decrease.

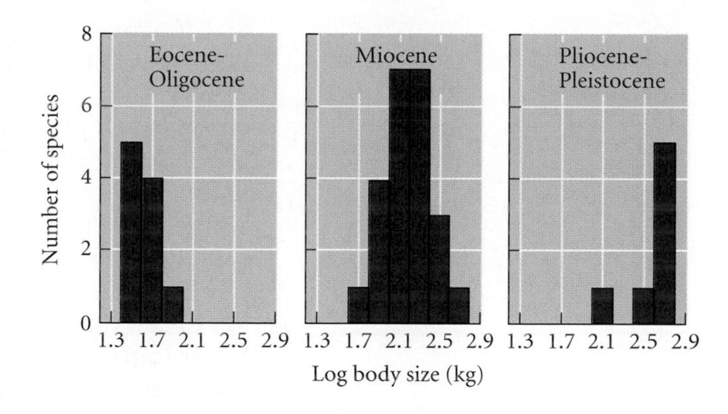

Figure 21.15 A driven trend: Cope's rule in the horse family, Equidae. The entire distribution of body sizes in the family shifted toward larger sizes during the Cenozoic. (After McShea 1994.)

These trends can be attributed to individual selection. A good example of a trend due to species selection is the increase in the ratio of nonplanktotrophic to planktotrophic species in several clades of Cenozoic gastropods (Figure 21.16). Species that lack a planktotrophic dispersal stage are more susceptible to extinction than planktotrophic species (species that feed as planktonic larvae). However, they more than compensate by their higher rate of speciation, probably because their lower rate of dispersal reduces the rate of gene flow among populations (Hansen 1980; Jablonski and Lutz 1983).

Trends due to lineage sorting by species hitchhiking are probably very common because if any one character causes one clade to become richer in species than other clades due to its effect on the rate of speciation or extinction, then all the other features of that clade will also tend toward greater frequencies. For example, coiled, sucking adult mouthparts may have become more prevalent among insects because they are a feature of the extremely diverse Lepidoptera (moths and butterflies). Lepidoptera larvae are herbivorous, and the herbivorous habit has consistently been associated with a high diversification rate in insects (Mitter et al. 1988; see Chapter 7). So the increased frequency of sucking mouthparts may be a result of hitchhiking with herbivory.

The boundaries that enforce passive trends may be due to either functional or developmental genetic constraints. For example, the smallest birds and mammals may have reached the lower functional limit of body size because a smaller animal might be unable to maintain a high body temperature due to its greater surface/volume ratio. In addition, developmental pathways may evolve that act as RATCHETS—mechanisms that make reversal unlikely. The thoracic segments of insects, for example, have acquired individual developmental identities and are unlikely to regain the homogeneous condition postulated for ancestral arthropods. The most extreme epigenetic ratchets probably are represented by evolutionary losses of complex features, which are unlikely to evolve again in the same form (Dollo's law). Some such evolutionary changes may be *irreversible* (Bull and Charnov 1985). Among lizards and snakes, for example, egg-laying ancestors have given rise to live-bearing (viviparous) descendants many times, but there is not a single unequivocal example of the reverse change (Lee and Shine 1998).

Figure 21.16 A trend caused by species selection. The bars show the stratigraphic distributions of fossil species of volutid snails. Although nonplanktotrophic species had shorter durations, they arose by speciation at a higher rate, so the ratio of nonplanktotrophic to planktotrophic species increased over time. (After Hansen 1980.)

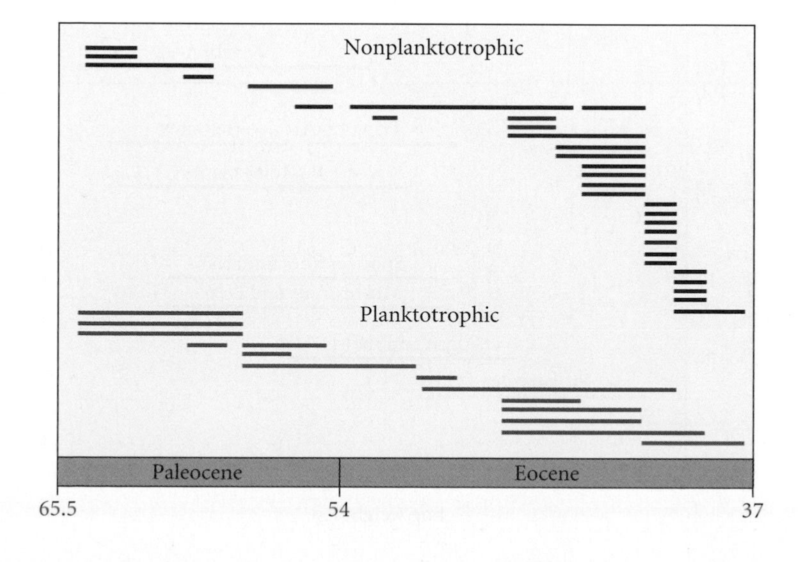

Are there major trends in the history of life?

Do any trends or directions characterize the entire evolutionary history of life? Although many have been postulated, it is probably safe to say that no uniform driven trends can be discerned, because exceptions can be cited for every proposed trend. Still, one might ask if there is any feature that, on the whole, has evolved with enough consistency of direction that one would be able to tell, from snapshots of life at different times in the past, which was taken earlier and which later. Let us consider some possibilities (see McShea 1998; Knoll and Bambach 2000).

COMPLEXITY. It is difficult to define, measure, and compare complexity among very different organisms, although anatomical complexity may be considered to be proportional to the number of different kinds of parts of which an organism is composed and to the irregularity of their arrangement (McShea 1991).

Among the few features that we can imagine comparing among all forms of life are DNA content, the number of different coding sequences (genes) in the DNA, and the number of cell types. The genomes of Bacteria and Archaea have a lower DNA content and fewer functional genes than those of most eukaryotes. Genome size has a much smaller range among bacterial species, about 0.6 to 8.6 Mb (megabases), than among eukaryotes, and the bacterial taxa with the smallest genomes are parasites or endosymbionts that have lost many genes that perform functions supplied by their host (Moran 2002). There is no evidence, so far, of a driven trend toward greater gene number in prokaryotes. Among eukaryotes, DNA content varies 200,000-fold and is not correlated with any indicators of phenotypic complexity. Although organisms such as yeast and the nematode *C. elegans* have much smaller genomes than vertebrates, the human genome, with 2910 Mb, has only about 18 percent as much DNA as some salamanders and only about 2 percent as much as the African lungfish—and it is dwarfed by the protist *Amoeba dubia*, with 670,000 Mb.

So far, too few eukaryotic genomes have been sequenced to look for phylogenetic trends in the number of distinct functional genes, but it is already clear that this number is not correlated with total DNA content (Saccone and Pesole 2002). The yeast *Saccharomyces cerevisiae*, the nematode *C. elegans*, and the fruit fly *Drosophila melanogaster* have rather low gene numbers (estimated at about 5,885, 19,100, and 13,600, respectively), but vertebrates appear not to vary greatly: the estimate for the human genome (30,000) is not greatly different from that for the puffer fish (31,000) or the mouse (22,444). Plants may have more genes than animals, if the estimates for rice (32,000 to 50,000) are a valid indication. On the whole, there is little evidence that the number of distinct genes in eukaryotic genomes has increased much since the diversification of vertebrates, or perhaps of animals generally, began. Similarly, although the number of recognizably different cell types is greater in animals and plants than in protists and fungi, and may be greater in vertebrates than in most invertebrate phyla (Figure 21.17), we do not know whether this feature has increased within any phyla, much less in the animal kingdom as a whole, since the phyla first appeared in the Precambrian.

At the level of morphology, there has likewise been a passive trend toward greater complexity, if we compare Cambrian or post-Cambrian organisms with early Precambrian life, which included only prokaryotes. There has been an increase in the maximal level of HIERARCHICAL ORGANIZATION during the history of life, whereby entities have emerged that consist of functionally integrated associations of lower-level individuals (Maynard Smith and Szathmáry 1995; McShea 2001): the eukaryotic cell evolved from an association of prokaryotic cells, multicellular organisms with differentiated cell types evolved from aggregations of unicellular ancestors, and aggregates of a few kinds of multicellular organisms (e.g., social insects, humans) form highly integrated colonies. These changes represent only a few evolutionary events, in which the great majority of lineages did not participate. In contrast, the anatomical complexity of Cambrian animals was arguably as great as that of living forms, and many characteristics have evolved toward simplification or loss in innumerable clades (see Chapter 3). In one of the few tests for net trends within individual higher taxa, the complexity of the vertebral column of mammals was found to have increased and decreased about equally often (McShea 1993).

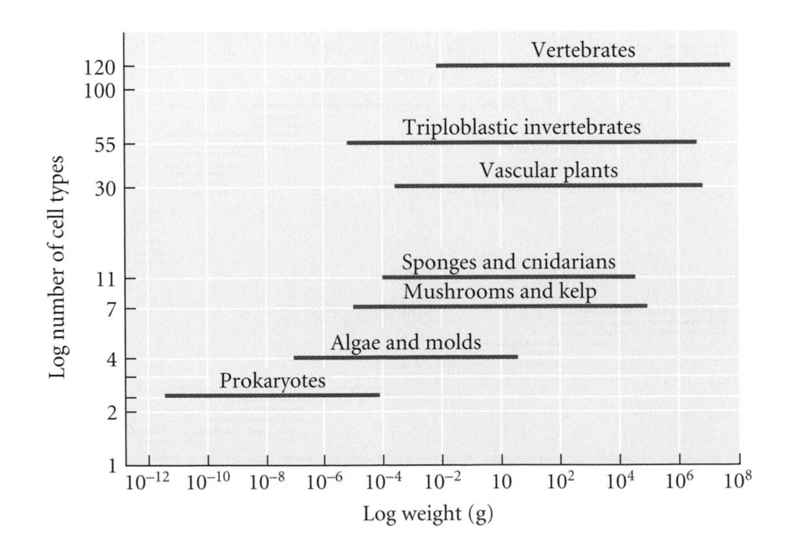

Figure 21.17 The number of recognizably different types of cells, plotted against the range of body weights, in various groups of organisms. Both variables are plotted on a logarithmic scale. Although some groups, such as vertebrates, have more kinds of cells, there is little evidence that the number has increased within these groups over evolutionary time. (After Bonner 1988.)

If anatomy has not become steadily more complex, has behavior? Although there now exist species that surely can learn more, or engage in more complex social interactions, than those of the Cambrian, few phylogenetic analyses that might shed light on overall changes in behavioral complexity have been carried out. Social behavior has reverted to solitary behavior (which is presumably simpler) in various groups of birds, spiders, and insects, including the eusocial insects that have sterile worker castes (Wcislo and Danforth 1997). Eusocial behavior has evolved several times in the Hymenoptera, but has been lost even more often; in the bee genus *Lasioglossum*, for example, it evolved once, but has been lost at least six times (Danforth et al. 2003). The advantage of eusociality, or probably the capacity for any other complex behavior, must depend on the environment, and there is no guarantee that it will always increase.

EVOLVABILITY. Some authors have suggested that **evolvability**—the capacity of lineages to adapt to changing conditions—might itself evolve. For example, genetic correlations among characters may constrain some features from evolving rapidly to a new optimum (see Chapter 13). We can envision that functionally related features might evolve to be more highly correlated (INTEGRATION) so that they are functionally matched even if they vary. They would then be less capable of independent evolution (i.e., less "evolvable"). On the other hand, selection might favor PARCELLATION of a highly integrated network of characters into different modules, each of a few functionally related characters, and each able to evolve independently of other modules (Figure 21.18). Wagner and Altenberg (1996) suggest that metazoan animals have tended to evolve toward greater modularity—hence greater evolvability—but this idea has not yet been tested.

Evolvability might, however, decrease over evolutionary time. For example, lineages that become highly ecologically specialized might lose the capacity to adapt to different resources or environmental conditions (i.e., specialization may be irreversible: Futuyma and Moreno 1988). We noted above, for example, that larvae of heliconiine butterflies seem to have retained a specialized diet of Passifloraceae (passion flowers) since the Oligocene. Nevertheless, recent phylogenetic analyses of herbivorous insects and other groups suggest that, at least on a relatively short evolutionary time scale, transitions between ecological specialization and generalization have been almost equally frequent in both directions (e.g., Nosil 2002). Still, specialization might become irreversible on a much longer time scale because it is often associated with the loss of structures or biochemical capacities. Ancient lineages of parasites invariably show losses of various sensory, locomotory, and biochemical characteristics, and the likelihood that dolphins or whales could ever return to terrestrial habitats seems remote.

Another possible reason for the loss of evolvability is that the genetic and developmental bases of different characters might become more integrated over time (the opposite of parcellation; see Figure 21.18), leading to greater genetic correlations or to longer or more intricate developmental pathways. It has been proposed that early steps in a developmental pathway may become more phylogenetically conservative over evolutionary time because they carry a heavy "burden": later steps depend on them and could easily go awry if the earlier steps were altered (Riedl 1978; Wimsatt 1986). This hypothesis predicts that although the phenotypic diversity among species in a clade tends to increase over time (as we will see below), it should do so at a decelerating rate.

At this time, much more research is needed to determine whether evolvability generally changes over the course of evolutionary time.

Figure 21.18 Two ways in which interactions among suites of genes and characters can evolve by changes in pleiotropic effects. In parcellation, the pleiotropic effects of genes on different phenotypic characters are reduced. Such changes increase the possibility of independent character evolution. In integration, the pleiotropic effects of genes are increased, molding initially independent characters into modular complexes. Such changes may act as constraints because independent evolution becomes less likely. (After G. P. Wagner 1996.)

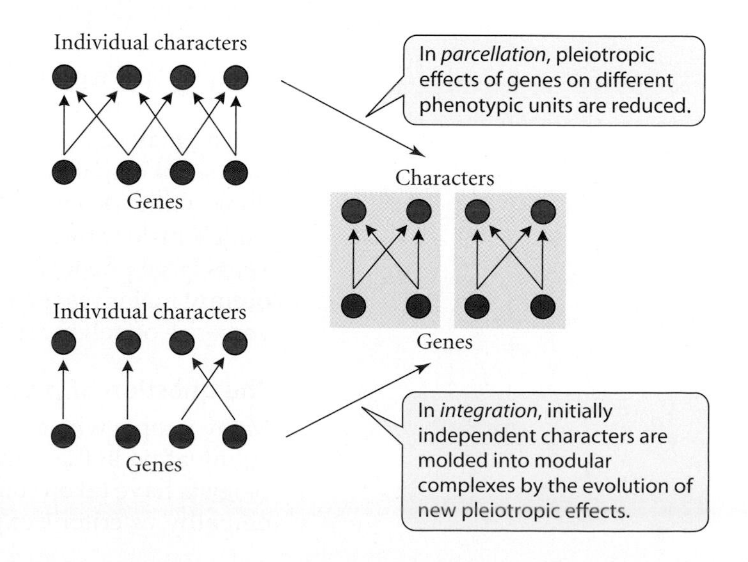
Individual characters
Genes
In *parcellation*, pleiotropic effects of genes on different phenotypic units are reduced.
Characters
Genes
Individual characters
Genes
In *integration*, initially independent characters are molded into modular complexes by the evolution of new pleiotropic effects.

EFFICIENCY AND ADAPTEDNESS. Innumerable examples exist of improvements in the design of features that serve a specific function. The mammal-like reptiles, for example, show trends in feeding and locomotory structures associated with higher metabolism and activity levels, culminating in the typical body plan of mammals (see Chapter 4). Swiftly moving falcons, soaring condors, and hovering hummingbirds all fly more efficiently than did early flying dinosaurs such as *Archaeopteryx*—although they fly in very different ways, befitting their different feeding habits and ways of life. There may well be a global trend toward greater efficiency (Ghiselin 1995). However, efficiency and effectiveness are difficult to measure, and they must always be defined relative to the task set by the context—by the organism's environment and way of life, which differ from species to species.

If efficiency of design has increased, does that mean that organisms are more highly adapted than in the past? Darwin thought this likely; he imagined that if long-extinct species were revived and had to compete with today's species, they would lose the competition badly. We might find evidence of the evolution of competitive superiority if the fossil record provided many examples of competitive *displacement* of early by later taxa—but as we have seen (in Chapter 7), this pattern is much less common than *replacement* by later taxa well after the earlier ones became extinct. And even though natural selection within populations increases mean fitness (specifically, *relative* fitness), fitness values are always context-dependent. We cannot meaningfully compare the level of adaptedness of a shark and a falcon, or even of a bird-hunting falcon and a bat-hunting falcon, since they are as adapted to different tasks as are flat-head and Phillips-head screwdrivers. And it may be equally difficult to compare the adaptedness of a species with that of its long-extinct ancestor, since they may have experienced quite different selective regimes.

Likewise, although we might suppose that species longevity would be a measure of increase in adaptedness, this need not be so. The consequence of natural selection is the adaptation of a population to the currently prevailing environment, not to future environments, so selection does not imbue a species with insurance against environmental change. We have seen that within many clades, the age of a genus or family does not influence its probability of extinction, implying that a lineage does not become more extinction-resistant over time (see Figure 7.9). It is true that the rate of "background" extinction of taxa has declined through the Phanerozoic (see Figure 7.6). This decline might indicate increasing adaptedness, but it might also be explained by the greater numbers of species in later taxa, or by the demise of major extinction-prone clades in the Paleozoic. Neither of these hypotheses requires us to postulate that extinction resistance evolves within lineages.

DIVERSITY AND DISPARITY. Although many higher taxa have dwindled to extinction, and although several mass extinctions have temporarily reduced diversity, the total number of taxa (species, genera, families) on Earth has increased fairly steadily since the beginning of the Mesozoic era (251 Mya). The likely causes of this increase are discussed in Chapter 7. Note that the trend toward greater diversity describes a collective property of life and thus is qualitatively different from trends in the properties of individual organisms or species, such as structural complexity.

In many clades, and in life as a whole, not only the number of species and higher taxa, but also the amount of phenotypic variation among species, has increased (Foote 1997). Increases in the phenotypic diversity (often referred to as the **disparity**) of a taxon or clade are often strongly correlated with increases in the number of taxa within that clade, which indeed may be used as an indicator of disparity. Morphological disparity, in turn, reflects ecological disparity—the variety of adaptive zones that have been occupied over evolutionary time (Knoll and Bambach 2000). Disparity often increases more rapidly early in a clade's history than later. It is not known if developmental integration increases and prevents large phenotypic changes from evolving later, or if increasing numbers of species occupy major resources, so that there is less room for ecological and morphological divergence of newly arisen species (Foote 1997).

The question of progress

Many people who accept evolution still conceive it as a purposive, progressive process, culminating in the emergence of consciousness and intellect. Even some evolutionary biologists have taken some of the qualities we most prize in ourselves, such as intellect or empathy, as criteria of progress, and, quite ignoring the innumerable lineages that have not evolved at all in these directions, have seen in evolution a history of progress toward the emergence of humankind (see Ruse 1996).

The word **progress** usually implies movement toward a goal, as well as improvement or betterment. The processes of evolution, such as mutation and natural selection, cannot imbue evolution with a goal. We cannot suppose that humans, or any creatures with our mental faculties, were destined to evolve. Every step in the long evolutionary history of humans—or any other species—was contingent on genetic and environmental events that could well have been otherwise (Gould 1989). For this reason, reflective evolutionary biologists have often concluded that humanity's existence has never been predetermined in any sense, and, moreover, that it is very unlikely that any other "humanoid" species exists elsewhere in the near reaches of the universe—especially any with which we have the faintest hope of communicating (Simpson 1964; Mayr 1988b).

Progress in the sense of betterment implies a value judgment, and here we must guard against the parochial, anthropocentric view that human features are better than those of other species. Were a rattlesnake or a knifefish capable of conscious reflection, it doubtless would measure evolutionary progress by the elegance of an animal's venom delivery system or its ability to communicate by electrical signals. The great majority of animal lineages—to say nothing of plants and fungi—show no evolutionary trend toward greater "intelligence" (however it might be defined and measured), which must be seen as a special adaptation appropriate to some ways of life, but not others. It is difficult, if not impossible, to specify a universal criterion by which to measure "improvement" that is not laden with our human-centered values.

Most evolutionary biologists have therefore concluded that we cannot objectively find progress in evolutionary history (Ruse 1996). As we have seen, it is hard to document even objective trends, especially driven trends, in any feature such as complexity or adaptedness. The most characteristic feature of the history of evolution, rather, is the unceasing proliferation of new forms of life, of new ways of living, of seemingly boundless, exquisite diversity. The majesty of this history inspired Darwin to end *The Origin of Species* by reflecting on the "grandeur in this view of life," that "whilst this planet has gone cycling on according to the fixed law of gravity, from so simple a beginning endless forms most beautiful and most wonderful have been, and are being, evolved."

Summary

1. The average rate of evolution of most characters is very low because long periods of little change (stasis) are averaged with short periods of rapid evolution, or the character mean fluctuates without long-term directional change. The highest rates of character evolution in the fossil record are comparable to rates observed in contemporary populations and can readily be explained by known processes such as mutation, genetic drift, natural selection, and speciation. The duration of speciation is short enough to account for the observed rates of increase in species diversity.
2. The fossil record provides examples of both gradual change and the pattern called punctuated equilibrium: a rapid shift from one static phenotype to another. The hypothesis that such shifts require the occurrence of peripatric speciation is not widely accepted because responses to selection do not depend on speciation.
3. Stasis can be explained by genetic constraints, stabilizing selection (owing largely to habitat tracking), or the erasure of divergence by episodic massive gene flow among populations with ancestral and derived character states.
4. Higher taxa arise not in single steps, by macromutational jumps (saltation), but by multiple changes in genetically independent characters (mosaic evolution). Most such characters evolve gradually, through intermediate stages, but some characters evolve discontinuously due to mutations with moderately large effects.
5. Characters may be phylogenetically conservative because of limits on the origin of variation (genetic and developmental constraints) or because of niche conservatism (resulting in stabilizing selection).
6. New features often are advantageous even at their inception. They often evolve by modification of pre-existing characters to serve accentuated or new functions, or sometimes as by-

products of the development of other structures. Evolutionary novelties often result when two or more functions of a structure are decoupled, or when structures are duplicated and diverge in structure and function.

7. Complex structures such as eyes evolve by rather small, individually advantageous steps. They may acquire functional integration with other structures so that they become indispensable.
8. Long-term trends may result from individual selection or from species hitchhiking, the phylogenetic association of a character with other characters that affect speciation or extinction rates. Driven trends, whereby the entire frequency distribution of a character among species in a clade shifts in a consistent direction over time, are less common than passive trends, in which variation among species (and therefore the mean of the clade) expands from an ancestral state that is located near a boundary (e.g., the clade may begin near a minimal body size).
9. Probably no feature exhibits a trend common to all living things. Features such as genome size and structural complexity display passive trends, in that the maximum has increased since very early in evolutionary history, but such changes have been relatively uncommon and inconsistent among lineages. There is no clear evidence of trends in evolvability or indices of adaptedness such as the longevity of species or higher taxa in geological time. The most conspicuous directional change in evolutionary history is an increase (with setbacks due to major extinctions) in the species diversity and the phenotypic and ecological disparity of organisms taken as a whole.
10. If "progress" implies a goal, then there can be no progress in evolution. If it implies betterment or improvement, we still cannot identify objective criteria by which the history of evolution can be shown to be one of "improvement." Characters improve in their capacity to serve certain functions, but these functions are specific to the ecological context of each species.

Terms and Concepts

disparity
driven trend
evolvability
global trend
habitat tracking
local trend
macroevolution
microevolution
niche conservatism
passive trend
progress
punctuated equilibrium
saltation
species hitchhiking
stasis
trend

Suggestions for Further Reading

Tempo and mode in evolution, by G. G. Simpson (Columbia University Press, New York, 1944), and *Evolution above the species level*, by B. Rensch (Columbia University Press, New York, 1959) are classic works of the evolutionary synthesis, in which the authors reconcile macroevolutionary phenomena with neo-Darwinian theory. Stephen Jay Gould's magnum opus, *The structure of evolutionary theory* (Belknap Press of Harvard University Press, Cambridge, MA, 2002) is a huge (and controversial) treatment of punctuated equilibrium and other macroevolutionary themes on which Gould influenced modern thought.

J. S. Levinton, in *Genetics, paleontology, and macroevolution* (second edition, Cambridge University Press, Cambridge, 2001), discusses many of the topics of this chapter, taking a strong stand on controversial issues. D. W. McShea provides a good introduction to contemporary research on evolutionary trends in "Possible largest-scale trends in organismal evolution: Eight "live hypotheses" (1998, *Annual Review of Ecology and Systematics* 29: 293–318).

Problems and Discussion Topics

1. Snapdragons (*Antirrhinum*) and their relatives in the traditionally recognized family Scrophulariaceae have bilaterally symmetrical flowers, derived from the radially symmetrical condition of their ancestors. A mutation in the *cycloidea* gene makes snapdragon flowers radially symmetrical. The *cycloidea* gene has been mapped and sequenced, and was found

to be asymmetrically expressed in the developing snapdragon flower, where it probably regulates other genes that contribute to flower development (Luo et al. 1996). Is it possible to determine whether this gene was the primary basis of the original evolution of bilateral symmetry in the Scrophulariaceae? How might one determine whether bilateral symmetry evolved continuously through intermediate stages or by a single discrete ("saltational") change at this locus?

2. Despite the close relationship between humans and chimpanzees, and despite the genetic variation in most of their morphological characteristics, there are no records of humans giving birth to babies with chimpanzee morphology. Would you expect this to occur, if humans have evolved gradually from apelike ancestors? If not, why not?
3. Find and evaluate the evolutionary literature that describes how wings could have evolved gradually, through advantageous intermediate steps.
4. How might one test the three hypotheses proposed to explain stasis that are mentioned in this chapter?
5. How can we tell whether a trend was caused by species selection or individual selection? How might we tell whether a passive trend was due to natural selection or genetic drift?
6. Paleontologists and some biologists commonly infer function, and even behavior, from anatomical details. Skeletal features, for example, are often used to infer that an extinct mammal (such as an early hominin) was highly, somewhat, or not at all arboreal. This inference assumes a nearly perfect fit of form to function (i.e., optimal form). Can this assumption be justified?
7. Would you expect "living fossils," such as the tadpole shrimp, to differ from other species in amount of genetic variation, genetic correlations among characters, canalization, or any other feature that might affect "evolvability"?
8. Are you surprised that the number of distinct genes in the human genome seems not to differ greatly from that in some other animals? Why or why not?
9. List some possible criteria for complexity and for progress. Now compare lobsters, earthworms, honeybees, sheep, rattlesnakes, and humans. Can these organisms be objectively arrayed along one or more scales according to these criteria? (You may need to do background research on their characteristics.)
10. Stephen Jay Gould (1989) and others have argued that the evolution of a self-conscious, intelligent species (i.e., humans) was historically contingent: it would not have occurred had any of a great many historical events been different. The philosopher Daniel Dennett (1995) and others have disagreed, arguing that convergent evolution is so common that if humans had not evolved, some other lineage would probably have given rise to a species with similar mental abilities. What do you think, and why? If Gould's position is right, what are its philosophical implications, if any?

22

Evolutionary Science, Creationism, and Society

We cannot pretend to know more than a small fraction of the evolutionary history of life, or to understand fully all the mechanisms of evolution, and there are many differences of opinion among evolutionary biologists about details of both history and mechanisms. However, the historical reality of evolution—the descent, with modification, of all organisms from common ancestors—has not been in question among scientists for well over a century. It is as much a scientific fact as the atomic constitution of matter or the revolution of the Earth around the Sun. Nevertheless, many people do not believe in evolution. More than 40 percent of Americans believe that the human species was created directly by God, rather than evolving from a common ancestor shared with other primates (and from a much more remote common ancestor shared with all other species). People who hold this belief are often referred to as **creationists**. In contrast, most people in Europe (even in countries such as Italy with an officially established religion) do not question the reality of evolution, and are often astonished that antiscientific attitudes on evolution flourish in the technologically and scientifically most prominent country in the world.

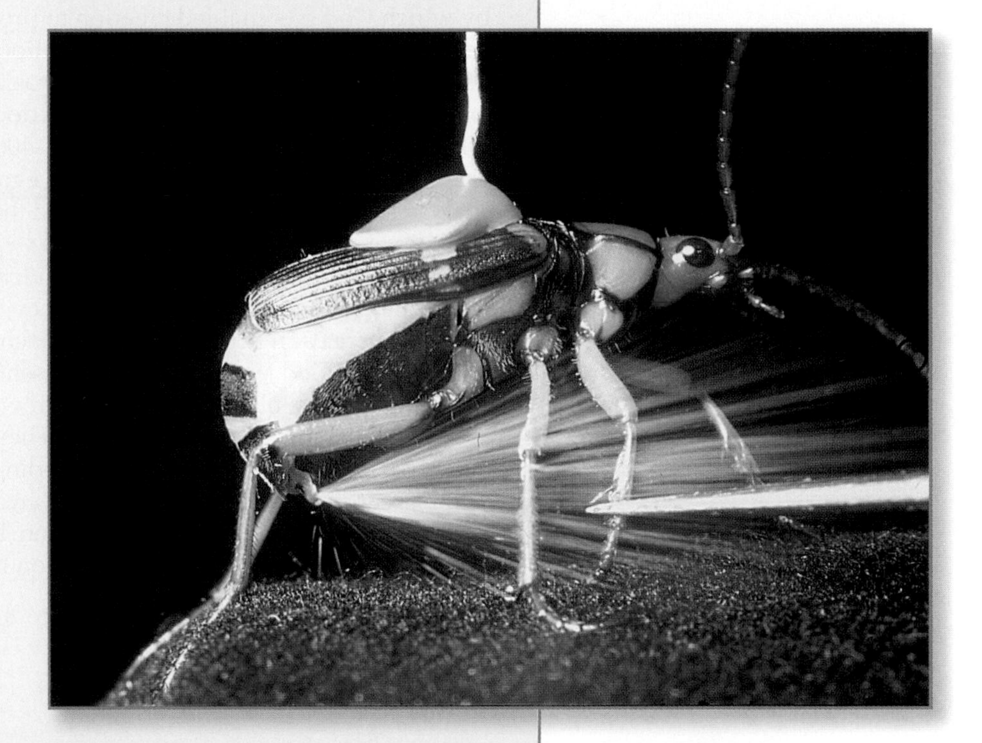

Too complex? This bombardier beetle (*Stenaptinus insignis*) defends itself by discharging a toxic chemical that explodes from a boiling, enzyme-catalyzed reaction in its special abdominal chamber, as this tethered laboratory specimen demonstrates. Creationists have argued that this defense system is too complex to have evolved. See page 534 for a refutation of this position. (Photo courtesy of Thomas Eisner.)

In this chapter, we will look at the beliefs of creationists and their arguments against evolution, and we will review the evidence for evolution presented throughout this book. We will also look at the many ways in which human society depends upon an understanding of evolutionary science.

Creationists and Other Skeptics

Most disbelievers in evolution reject the idea because they think it conflicts with their religious beliefs. Many, perhaps most, are Christian, Muslim, or (rarely) Jewish fundamentalists, who adhere to a literal or almost literal reading of sacred texts. For Christian and Jewish fundamentalists, evolution conflicts with their interpretation of the Bible, especially the first chapters of Genesis, which portray God's creation of the heavens, the Earth, plants, animals, and humans within six days. However, many Western religions understand these biblical descriptions to contain symbolic truths, not literal or scientific ones. Many deeply religious people believe in evolution, viewing it as the natural mechanism by which God has enabled creation to proceed. Some religious leaders have made clear their acceptance of the reality of evolution. For example, Pope John Paul II affirmed the validity of evolution in 1996 and emphasized that there is no conflict between evolution and the Catholic Church's theological doctrines. (The text of his letter was reprinted in *The Quarterly Review of Biology* 72: 381–406 [1977].) The Pope's position was close to the argument generally known as **theistic evolution**, which holds that God established natural laws (such as natural selection) and then let the universe run on its own, without further supernatural intervention.

The beliefs of creationists vary considerably (Figure 22.1). The most extreme interpret every statement in the Bible literally. They include "young Earth" creationists who believe in **special creation** (the doctrine that each species, living and extinct, was created independently by God, essentially in its present form) and in a young universe and Earth (less than 10,000 years old), a deluge that drowned the Earth, and an ark in which Noah preserved a pair of every living species. They must therefore deny not only evolution, but also most of geology and physics (including radioactive dating and astronomical evidence of the great age of the universe). Some creationists share many or most of the biblical literalists' beliefs, including special creation, but grant the antiquity of the Earth and the rest of the universe. (The "day-age" position holds that each of the days of creation cited in

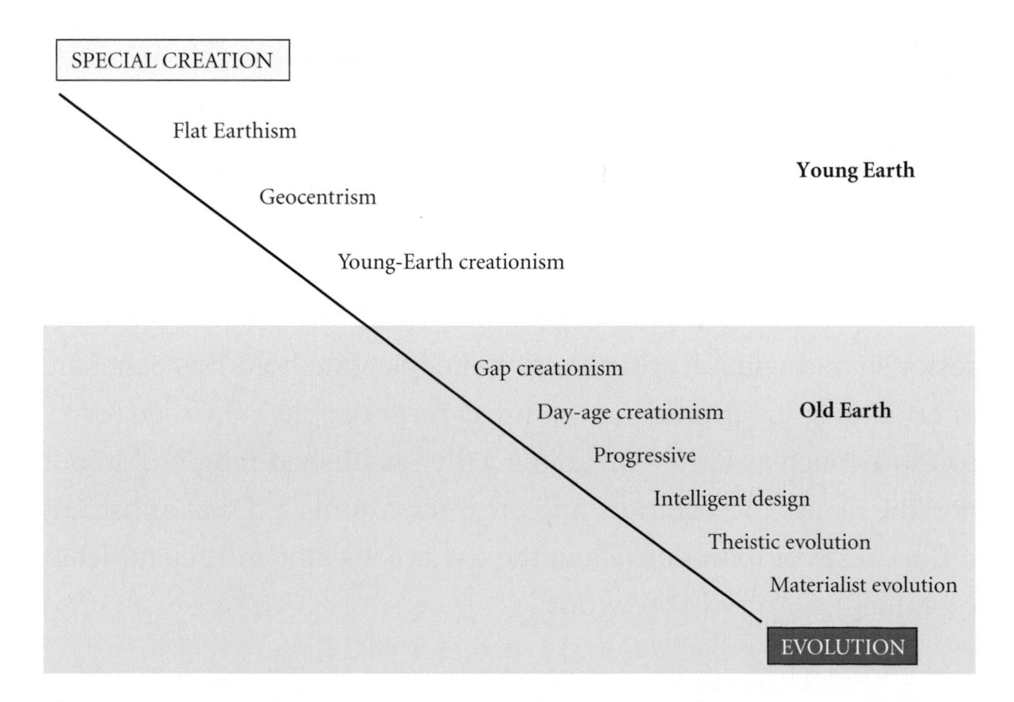

Figure 22.1 The special creation-evolution continuum. Young-Earth proponents affirm that the Earth is young (thousands of years old) not old (billions of years). Adherents of "special creation" tend to consider the Bible as literally true. (After Scott 1997.)

Genesis was millions of years long.) Other creationists allow that mutation and natural selection can occur, and even that very similar species can arise from a common ancestor. However, they deny that higher taxa (genera, families, etc.) have evolved from common ancestors, and most of them vigorously assert that the human species was specially created by God, in His image—which is their major reason for caring about evolution at all.

Some deniers of evolution, including a very few scientists, do not explicitly invoke special creation by God. Some of them even seem to accept certain aspects of evolution, such as development of different species from common ancestors. However, they argue that many biological phenomena are too complicated to have arisen by natural processes, and can only be explained by **intelligent design (ID)**. Proponents claim that intelligent design is a scientific, not a religious, concept. However, the designer that they envision is a supernatural rather than a material being (i.e., a being equivalent to God).

Most of the efforts of activist creationists are devoted to suppressing the teaching of evolution in schools, or at least insisting on "equal time" for their views. In the United States, however, the Constitutional prohibition of state sponsorship of religion has been interpreted by the courts to mean that the biblical version, or any other explicitly religious version, of the origin of life's diversity cannot be promulgated in public schools. The activists have therefore adopted several forms of camouflage. One is **"scientific creationism"** or "**creation science**," which consists of attacks on, and supposedly scientific disproofs of, evolution. These arguments have not succeeded, chiefly because they do not have, and cannot have, any scientific content. The more recent camouflage is "intelligent design theory," which, as we have seen, teaches that the complexity of living things can be explained only by intelligent design, without specifically mentioning God.

Bills requiring public schools to give "creation science" equal time have been introduced in state legislatures, but the U.S. Supreme Court in 1987 found one such law unconstitutional because it "endorses religion by advocating the religious belief that a supernatural being created humankind," and because it was written "to restructure the science curriculum to conform with a particular religious viewpoint." As a result, local and state legislators today use different, sometimes innocuous-looking wording to get creationism into school science curricula. For example, a bill introduced in the Missouri legislature in 2003 required that "if scientific theory concerning biological origin is taught, biological evolution and biological intelligent design shall be taught and given equal treatment." (The implication is that ID is a scientific theory—which it isn't, as we will see). A bill introduced in the Kansas senate in 2003 would have mandated schools to "encourage the presentation of scientific evidence supporting the origins of life and its diversity, objectively and without religious, naturalistic or philosophical bias or assumption." (This language sounds marvelously enlightened—except that antievolutionists commonly claim that science has an inherently "naturalistic" bias because, by definition, it explains phenomena by natural processes and laws, and even claim that evolution is a religious dogma.)

Science, Belief, and Education

Biologists and others who argue against including any form of creationism in science curricula are not against free speech, and they are not trying to extinguish religious belief. They simply hold that although it might be acceptable to teach about creation stories in classes on, say, history or contemporary society, such beliefs are not valid *scientific* hypotheses and have no place in *science* classes. Unfortunately, most people have very little understanding of what science is and how it works, even if they have had science courses—and this understanding is critically important to the evolution-versus-creationism debate.

Science is not a collection of facts, contrary to popular belief, but rather a *process* of acquiring understanding of natural phenomena. This process consists largely of posing hypotheses and testing them with observational or experimental evidence. Despite loose talk about "proving" hypotheses, most scientists agree with philosophers of science that scientists cannot absolutely prove hypotheses; that is, they cannot attain absolute, guaranteed proof of the kind that can be obtained in mathematics. Rather, the hypothesis that currently best explains the data is *provisionally* accepted, with the understanding that it may be altered, expanded, or rejected if subsequent evidence warrants doing so, or if a better hypothesis, not yet imagined, is devised. Sometimes, indeed, a radically new "paradigm" replaces an old one; for example, plate tectonics revolutionized geology in the 1950s, replacing the conviction that continents are fixed in position. More often, old hypotheses are incrementally modified and expanded over time. For instance, Mendel's laws of assortment and independent segregation, which initiated modern genetics, were modified when phenomena such as linkage and meiotic drive were discovered, but his underlying principle of inheritance based on "particles" (genes) holds true today.

This process reflects one of the most important and valuable features of science: even if individual scientists may be committed to a hypothesis, scientists as a group are not irrevocably committed to any belief, and do not maintain it in the face of convincing contrary evidence. They must, and do, change their minds if the evidence so warrants. Indeed, much of science consists of seeking chinks in the armor of established ideas, and few successes will burnish a scientist's reputation more than showing that an important orthodox hypothesis is inadequate or flawed. Thus science, as a social process, is tentative; it questions belief and authority; it continuously tests its views against evidence. Scientific claims, in fact, are the outcome of a process of natural selection, for ideas (and scientists) compete with one another, so that the body of ideas in a scientific field grows in explanatory content and power (Hull 1998). Science differs in this way from creationism, which does not use evidence to test its claims, does not allow evidence to shake its a priori commitment to certain beliefs, and does not grow in its capacity to explain the natural world.

How could it? Suppose an advocate of "intelligent design" says that multicellular organisms are so complex, compared with single-celled organisms, that they must have originated by the intervention of an intelligent designer. Unless the ID advocate proposes that extraterrestrial creatures are responsible, this designer must be a supernatural rather than a material being. So what was this designer, how did it equip organisms with new features, how long did it take, and why did it do this? Natural science can at least imagine ways to address such questions (for example, we can look for phylogenetic intermediates, analyze sequence differences in genes that encode the relevant character differences, look for fossils, do experiments on the selective advantage of multicellularity). But the ID hypothesis generates no research ideas.

Scientific research requires that we have some way of testing hypotheses based on experimental or observational data. *The most important feature of scientific hypotheses is that they are testable*, at least in principle. Sometimes we can test a hypothesis by direct observation, but more often we do not see processes or causes directly (for example, electrons, atoms, hydrogen bonds, molecules, and genes are not directly visible, and we cannot watch the occurrence of mutation during DNA replication). Rather, we infer such processes by comparing the outcome of observations or experiments with predictions made from competing hypotheses. In order to make such inferences, we must assume that the processes obey **natural laws**: statements that certain patterns of events will always occur if certain conditions hold. That is, science depends on the consistency, or predictability (at least in a statistical sense), of natural phenomena, as exemplified by the laws of physics and chemistry. Because supernatural events or agents are supposed to suspend or violate natural laws, science cannot infer anything about them, and indeed, cannot judge the validity of any hypotheses that involve them.

It is important to understand that, just as religion does not provide scientific, mechanistic explanations for natural phenomena, science cannot provide answers to any questions that are not about natural phenomena: it cannot tell us what is beautiful or ugly, good or bad, moral or immoral. It cannot tell us what the meaning of life is, and it cannot tell us whether or not supernatural beings exist (see Gould 1999; Pigliucci 2002).

Scientists can test and falsify some specific creationist claims, such as the occurrence of a worldwide flood or the claim that the Earth and all organisms are less than 10,000 years old, but scientists cannot test the hypothesis that God exists, or that He created anything, because we do not know what consistent patterns these hypotheses might predict. (Try to think of any observation at all that would definitively rule out these supernatural possibilities.) Science must therefore adopt the position that natural causes are responsible for whatever we wish to explain about the natural world. This is not necessarily a commitment to METAPHYSICAL NATURALISM (the assumption that everything *truly* has natural rather than supernatural causes), but it is a commitment to METHODOLOGICAL NATURALISM (the *working principle* that we can entertain only natural causes when we seek scientific explanations). The fundamental claim of creationism—that biological diversity is the result of supernatural powers—is not testable. This is equally true of "intelligent design" theory; it cannot be evaluated by the methods of science.

We have been using the words "hypothesis," "theory," and "fact," and it is imperative that we understand what they mean. A **hypothesis** is a proposition, a supposition. Before 1944, the idea that the genetic material is DNA was a plausible hypothesis that had little evidence to support it. Since 1944, this hypothesis became stronger and stronger as evidence grew. Now we consider it a fact. A **fact** is, simply, a hypothesis that has become so well supported by evidence that we feel safe in acting as if it were true. To use a courtroom analogy, it has been "proven" beyond reasonable doubt. Not beyond any conceivable doubt, but reasonable doubt.

A **theory**, as the word is used in science, doesn't mean an unsupported speculation or hypothesis (the popular use of the word). A theory is, instead, a big idea that encompasses other ideas and hypotheses and weaves them into a coherent fabric. It is a mature, interconnected body of statements, based on reasoning and evidence, that explains a wide variety of observations. It is, in one of the definitions offered by the *Oxford English Dictionary*, "a scheme or system of ideas and statements held as an explanation or account of a group or ideas or phenomena; ... a statement of what are known to be the general laws, principles, or causes of something known or observed." Thus atomic theory, quantum theory, and plate tectonic theory are not mere speculations or opinions, but strongly supported ideas that explain a great variety of phenomena. There are few theories in biology, and among them evolution is surely the most important.

So is evolution a fact or a theory? In light of these definitions, evolution is a scientific fact. That is, the descent of all species, with modification, from common ancestors is a hypothesis that in the last 150 years or so has been supported by so much evidence, and has so successfully resisted all challenges, that is has become a fact. But this history of evolutionary change is explained by evolutionary theory, the body of statements (about mutation, selection, genetic drift, developmental constraints, and so forth) that together account for the various changes that organisms have undergone.

Because creationist explanations for the diversity and characteristics of living things cannot be evaluated by the methods of science, they should not be given "equal time" in a science classroom. Neither should other hypotheses that are either unscientific or demonstrably wrong. Chemistry teachers do not and should not teach about alchemy (the old idea that one element, such as lead, can be magically transformed into another, such as gold); earth science classes should not even mention the hypothesis that Earth is flat; history and psychology teachers should not consider astrological explanations for historical events or personality traits—even though there are people who believe in all of these pseudoscientific ideas. The ideal of democracy doesn't extend to ideas—some are simply wrong, and as a purely practical matter it is imperative that we recognize them as such. In everyday life, we assume and depend on natural, not supernatural, explanations. Unlike the Puritans of Salem, Massachusetts, who in 1692 condemned people for witchcraft, we no longer seriously entertain the notion that someone can be victimized by a witch's spell or possessed by devils, and we would be outraged if a criminal successfully avoided conviction because he claimed that "the Devil made me do it." Even those who most devoutly believe that God sustains them in the palm of His hand will panic if the airplane's wing flaps malfunction. We depend on scientific explanations, and we know that science has proven its ability—because it works.

The Evidence for Evolution

The evidence for evolution has been presented throughout the preceding chapters of this book. The examples provided represent only a very small percentage of the studies that might be cited for each particular line of evidence. In this section, we will simply review the sources of evidence for evolution and refer back to earlier chapters for detailed examples.

The fossil record

The fossil record is extremely incomplete, for reasons that geologists understand well (see Chapter 4). Consequently, the transitional stages that we postulate in the origin of many higher taxa have not (yet) been found. But there is absolutely no truth to the claim, made by many creationists, that the fossil record does not provide any intermediate forms. There are many examples of such forms, both at low and high taxonomic levels; Chapter 4 provides several examples in the evolution of the classes of tetrapod vertebrates. Critically important intermediates are still being found: just in the last few years, several Chinese fossils, including feathered dinosaurs, have greatly expanded the record of the origin of birds. The fossil record, moreover, documents two important aspects of character evolution: mosaic evolution (e.g., the more or less independent evolution of different features in the evolution of mammals) and gradual change of individual features (e.g., cranial capacity and other features of hominins).

Many discoveries in the fossil record fit predictions made based on phylogenetic or other evidence. The earliest fossil ants, for instance, have the wasplike features that had been predicted by entomologists, and the discovery of feathered dinosaurs was to be expected, given the consensus that birds are modified dinosaurs. Likewise, phylogenetic analyses of living organisms imply a sequence of branching events, as well as a sequence of origin of the diagnostic characters of those branches. The fossil record often matches the predicted sequences (as we saw in Chapters 4 and 5): for example, prokaryotes precede eukaryotes in the fossil record, wingless insects (the phylogenetically basal bristletails) precede winged insects, fishes precede tetrapods, amphibians precede amniotes, algae precede vascular plants, ferns and "gymnosperms" precede flowering plants.

Phylogenetic and comparative studies

Although many uncertainties about phylogenetic relationships persist (e.g., the branching order of the major groups of birds), phylogenies that are well supported by one class of characters usually match the relationships implied by other evidence quite well (see Chapter 2). For example, molecular phylogenies support many of the relationships that have long been postulated from morphological data. These two data sets are entirely independent (the molecular phylogenies are often based on sequences that have no biological function), so their correspondence justifies confidence that the relationships are real: that the lineages form monophyletic groups and have indeed descended from common ancestors.

The largest monophyletic group encompasses all organisms. Although Darwin allowed that life might have originated from a few original ancestors, we can be confident today that all known living things stem from a single ancestor because of the many features that are universally shared. These features include most of the codons in the genetic code, the machinery of nucleic acid replication, the mechanisms of transcription and translation, proteins composed only of "left-handed" (L-isomer) amino acids, and many aspects of fundamental biochemistry. Many genes are shared among all organisms, including the three major "empires" (Bacteria, Archaea, and Eukaryota; see Chapters 4 and 19), and these genes have been successfully used to infer the deepest branches in the tree of life.

Systematists have shown that the differences among related species often form gradual series, ranging from slight differences to great differences with stepwise intermediates (e.g., Figure 3.21). Such intermediates often make it difficult to establish clear-cut families or other higher taxa, so that classification often becomes a somewhat arbitrary choice between "splitting" species among many taxa and "lumping" them into few. Systematic studies have also demonstrated the common origin, or homology, of characteristics that may differ greatly among taxa—the most familiar examples being the radically different forms of limbs among tetrapod vertebrates. Homology of structures is often more evident in early developmental stages than in adult organisms, and contemporary developmental biology demonstrates that Hox genes and other developmental mechanisms are shared among animal phyla that diverged from common ancestors a billion or more years ago (see Chapter 20).

Genes and genomes

The revolution in molecular biology and genomics is yielding data about evolution on a larger scale than ever before. These data increasingly show the extraordinary commonality of all living things. Because of this commonality, the structure and function of genes and genomes can be understood through comparisons among species and evolutionary models. (Indeed, it is only because of this common ancestry that there has ever been any reason to think that human biochemistry, physiology, or brain function, much less genome function, could be understood by studying yeast, flies, rats, or cats!)

Molecular studies show that the genomes of most organisms have similar elements, such as a great abundance of noncoding pseudogenes and satellite DNA and a plethora of "selfish" transposable elements that generally provide no advantage to the organism. These features are readily understandable under evolutionary theory, but would hardly be expected of an intelligent, omnipotent designer. Molecular evolutionary analyses have shown in great detail how new genes arise by processes such as unequal crossing over, and how duplicate genes diverge in function, increasing the genetic repertoire (see Chapter 19). Some DNA polymorphisms are shared between species, so that, for example, some major histocompatibility sequences of humans are more similar and more closely related to chimpanzee sequences than to other human sequences (see Figure 12.23B). What more striking evidence of common ancestry could there be?

Among the many other ways in which molecular studies affirm the reality of evolution, consider just one more: molecular clocks. They are far from perfectly accurate, but sequence differences between species nevertheless are roughly correlated with the time since common ancestry, judged from other evidence such as biogeography or the fossil record (see Chapters 2 and 4). The sequence differences do not simply encode the phenotypic differences that the organisms manifest, and the phenotypic differences themselves are much less correlated with time since common ancestry. No theory but evolution makes sense of these patterns of DNA differences among species.

Biogeography

We noted in Chapter 6 that the geographic distributions of organisms provided Darwin with abundant evidence of evolution, and they have continued to do so. For example, the distributions of many taxa correspond to geological events such as the movement of land masses and the formation and dissolution of connections between them. We saw that the phylogeny of Hawaiian species matches the sequence by which the islands came into existence. We saw that diverse ecological niches in a region are typically filled not by the same taxa that occupy similar niches in other parts of the world, but by different monophyletic groups that have undergone independent adaptive radiation (such as the anoles of the Greater Antilles). We saw, as did Darwin, that an isolated region such as an island is not populated by all the kinds of organisms that could thrive there, as we might suppose a thoughtful designer could arrange. Instead, whole groups are commonly missing, and human-introduced species often come to dominate.

Failures of the argument from design

Since God cannot be known directly, theologians such as Thomas Aquinas have long attempted to infer His characteristics from His works. Theologians have argued, for instance, that order in the universe, such as the predictable movement of celestial bodies, implied that God must be orderly and rational, and that He created according to a plan. From the observation that organisms have characteristics that serve their survival, it could similarly be inferred that God is a rational, intelligent designer who, furthermore, is beneficent: He not only conferred on living things the boon of existence, but equipped them for all their needs. Such a beneficent God would not create an imperfect world; so, as the philosopher Leibniz said, this must be "the best of all possible worlds." (Leibniz's position was actually more complicated, but his phrase was mercilessly ridiculed by Voltaire in his marvelous satire *Candide*.) The adaptive design of organisms, in fact, has long been cited as evidence of an intelligent designer. This was the thrust of William Paley's (1831) famous example: as the design evident in a watch implies a watchmaker, so the design evident in organisms implies a designer of life. This "argument from design" has been renewed in the "intelligent design" version of creationism, and it is apparently the most frequently cited reason people give for believing in God (Pigliucci 2002).

Of course, Darwin made this particular theological argument passé by providing a natural mechanism of design: natural selection. Moreover, Darwin and subsequent evolutionary biologists have described innumerable examples of biological phenomena that are hard to reconcile with beneficent intelligent design. Just as Voltaire showed (in *Candide*) that cruelties and disasters make a mockery of the idea that this is "the best of all possible worlds," biology has shown that organisms have imperfections and anomalies that can be explained only by the contingencies of history, and characteristics that make sense only if natural selection has produced them. If "good design" were evidence of a kindly, omnipotent designer, would "inferior design" be evidence of an unkind, incompetent, or handicapped designer?

Only evolutionary history can explain vestigial organs—the rudiments of once-functional features, such as the tiny, useless pelvis and femur of whales, the reduced wings under the fused wing covers of some flightless beetles, and the nonfunctional stamens or pistils of plants that have evolved separate-sexed flowers from an ancestral hermaphroditic condition. Likewise, only history can explain why the genome is full of "fossil" genes: pseudogenes that have lost their function. Only the contingencies of history can explain the arbitrary nature of some adaptations. For instance, whereas ectothermic ("cold-blooded") tetrapods have two aortic arches, "warm-blooded" endotherms have only one. This difference is probably adaptive, but can anything except historical chance explain why birds have retained the right arch and mammals the left?

Because characteristics evolve from pre-existing features, often undergoing changes in function, many features are poorly engineered, as anyone who has suffered lower back pain or wisdom teeth can testify. Once the pentadactyl limb became developmentally canalized, tetrapods could not evolve more than five digits even if they would be useful: the extra "finger" of the giant panda's hand is not a true digit at all, and lacks the flexibility of true fingers because it is not jointed (see Figure 21.8). Similarly, animals would certainly be better off if they could synthesize their own food, and corals do so by harboring endosymbiotic algae—but no animal is capable of photosynthesis.

If a designer were to equip species with a way to survive environmental change, it might make sense to devise a Lamarckian mechanism, whereby genetic changes would occur in response to need. Instead, adaptation is based on a combination of a random process (mutation) that cannot be trusted to produce the needed variation (and often does not) and a process that is the very epitome of waste and seeming cruelty (natural selection, which requires that great numbers of organisms fail to survive or reproduce). It would be hard to imagine a crueler instance of natural selection than sickle-cell anemia, whereby part of the human population is protected against malaria at the expense of hundreds of thousands of other people, who are condemned to die because they are homozygous for a gene that happens to be worse for the malarial parasite than for heterozygous carriers (see Chapter 12). Indeed, Darwin's theory of the cause of evolution was widely rejected just because people found it so distasteful, even horrifying, to contemplate. And, of course, this process often does not preserve species in the face of change: more than 99 percent of all species that have ever lived are extinct. Were they the products of an incompetent designer? Or one that couldn't foresee that species would have to adapt to changing circumstances?

Many species become extinct because of competition, predation, and parasitism. Some of these interactions are so appalling that Darwin was led to write, "What a book a devils' chaplain might write on the clumsy, wasteful, blundering, low, and horribly cruel works of Nature!" Darwin knew of maggots that work their way up the nasal passages into the brains of sheep, and wasp larvae that, having consumed the internal organs of a living caterpillar, burst out like the monsters in the movie *Alien*. The life histories of parasites, whether parasitic wasp or human immunodeficiency virus, ill fit our concept of an intelligent, kindly designer, but are easily explained by natural selection (see Chapter 18).

No one has yet demonstrated a characteristic of any species that serves only to benefit a different species, or only to enhance the so-called balance of nature—for, as Darwin saw, "such could not have been produced through natural selection." Because natural selection consists only of differential reproductive success, it results in "selfish" genes and genotypes, some of which have results that are inexplicable by intelligent design (see Chapter 14). We have seen that genomes are brimming with sequences such as transposable elements that increase their own numbers without benefiting the organism. We have seen maternally transmitted cytoplasmic genes that cause male sterility in many plants, and nuclear genes that have evolved to override them and restore male fertility. Such conflicts among genes in a genome are widespread. Are they predicted by intelligent design theory? Likewise, no theory of design can predict or explain features that we ascribe to sexual selection, such as males that remove the sperm of other males from the female's reproductive tract, or chemicals that enhance a male's reproductive success but shorten his mate's life span. Nor can we rationalize why a beneficent designer would shape the many other selfish behaviors that natural selection explains, such as cannibalism, siblicide, and infanticide.

Evolution and its mechanisms, observed

Anyone can observe erosion, and geologists can measure the movement of continental plates, which travel at up to 10 centimeters per year. No geologist doubts that these mechanisms, even if they accomplish only slight changes on the scale of human generations, have shaped the Grand Canyon and have separated South America from Africa over the course of millions of years. Likewise, biologists do not expect to see anything like the origin of mammals played out on a human time scale, but they have documented the mechanisms that will yield such grand changes, given enough time.

Evolution requires genetic variation, which originates by mutation. From decades of genetic study of initially homozygous laboratory populations, we know that mutations arise that have effects, ranging from very slight to drastic, on all kinds of phenotypic characters (see Chapter 8). These mutations can provide new variation in quantitative characters, seemingly without limit. This variation has been used for millennia to develop strains of domesticated plants and animals that differ in morphology more than whole families of natural organisms do. In experimental studies of laboratory populations of microorganisms, we have seen new advantageous mutations arise and enable rapid adaptation to temperature changes, toxins, or other environmental stresses. Laboratory studies have documented the occurrence of the same kinds of mutations, at the molecular level, that are found in natural populations and distinguish species. These mutations include base pair substitutions, gene duplications, chromosome rearrangements, and transposition of transposable elements. No geneticist or molecular biologist doubts that the differences among species in their genes and genomes originated by natural mutational processes that, by and large, are well understood.

We know also that most natural populations carry a great deal of genetic variation that can yield rapid responses to artificial or natural selection (see Chapters 9, 11, and 12). We have seen allele frequency differences among recently established populations that can be confidently attributed to genetic drift (see Chapter 10). Evolutionary biologists have documented literally hundreds of examples of natural selection acting on genetic and phenotypic variation (see Chapters 12 and 13). They have described hundreds of cases in which populations have responded to directional selection and have adapted to new environmental factors, ranging from the evolution of resistance to insecticides, herbicides, and antibiotics to the evolution of different diets (see Chapter 13).

Speciation generally takes a very long time, but some processes of speciation can also be observed. Substantial reproductive isolation has evolved in laboratory populations, and species of plants that apparently originated by polyploidy and by hybridization have been "re-created" de novo by crossing their suspected parent forms and selecting for the species' diagnostic characters (see Chapter 16).

In summary, the major causes of evolution are known, and they have been extensively documented. The two major aspects of long-term evolution, anagenesis (changes of characters within lineages) and cladogenesis (origin of two or more lineages from common ancestors), are abundantly supported by evidence from every possible source, ranging from molecular biology to paleontology. Over the past century, we have certainly learned of evolutionary processes that were formerly unknown; we now know, for example, that some species may arise from hybridization, and that some DNA sequences are mobile and can cause mutations in other genes. But no scientific observations have ever cast serious doubt on the reality of the basic mechanisms of evolution, such as natural selection, nor on the reality of the basic historical patterns, such as transformation of characters and the origin of all known forms of life from common ancestors. Contrast this mountain of evidence with the evidence for supernatural creation or intelligent design: *there is no such evidence whatever*.

Refuting Creationist Arguments

Creationists attribute the existence of diverse organisms and their characteristics to miracles: direct supernatural intervention. As we have seen, it is impossible to predict miracles or to do experiments on supernatural processes, so creationists do not do original research in support of their theory.* Thus "creation science," rather than providing positive evidence of creation, consists entirely of attempts to demonstrate the falsehood or inadequacy of evolutionary science and to show that biological phenomena must, by default, be the products of intelligent design. Here are some of the most commonly encountered creationist arguments, together with capsule counterarguments.

1. ***Evolution is outside the realm of science because it cannot be observed.***

   Evolutionary changes have indeed been observed, as we saw earlier in this chapter. In any case, most of science depends not on direct observation, but on testing hypotheses against the predictions they make about what we should observe. Observation of the postulated processes or entities is not required in science.

2. ***Evolution cannot be proved.***

   Nothing in science is ever absolutely proved. "Facts" are hypotheses in which we can have very high confidence because of massive evidence in their favor and the absence of contradictory evidence. Abundant evidence from every area of biology and paleontology supports evolution, and there exists no contradictory evidence.

3. ***Evolution is not a scientific hypothesis because it is not testable: no possible observations could refute it.***

   Many conceivable observations could refute or cast serious doubt on evolution, such as finding incontrovertibly mammalian fossils in incontrovertibly Precambrian rocks. In contrast, any puzzling quirk of nature could be attributed to the inscrutable will and infinite power of a supernatural intelligence, so creationism is untestable.

---

*About the only quasi-exception to this statement was their claim to have found commingled human and dinosaur footprints in fossilized sediments in a riverbed in Texas, supposedly showing that these organisms were contemporaneous. Even if this claim had proved to be true, it would not have falsified evolution; but in any case, most creationists now acknowledge that the "human" prints are a mixture of fraudulent carvings and natural depressions.

4. ***The orderliness of the universe, including the order manifested in organisms' adaptations, is evidence of intelligent design.***

   Order in nature, such as the structure of crystals, arises from natural causes and is not evidence of intelligent design. The order displayed by the correspondence between organisms' structures and their functions is the consequence of natural selection acting on genetic variation, as has been observed in many experimental and natural populations (see Chapters 13 and 14). Darwin's realization that the combination of a random process (the origin of genetic variation) and a nonrandom process (natural selection) can account for adaptations provided a natural explanation for the apparent design and purpose in the living world and made a supernatural account unnecessary and obsolete.

5. ***Evolution of greater complexity violates the second law of thermodynamics, which holds that entropy (disorder) increases.***

   This is a common misrepresentation of one of the most important laws of physics. The second law applies only to closed systems, such as the universe as a whole. Order and complexity can increase in local, open systems due to an influx of energy. This is evident in the development of individual organisms, in which biochemical reactions are powered by energy derived ultimately from the Sun.

6. ***It is almost infinitely improbable that even the simplest life could arise from nonliving matter. The probability of random assembly of a functional nucleotide sequence only 100 bases long is $1/4^{100}$, an exceedingly small number. And scientists have never synthesized life from nonliving matter.***

   It is true that a fully self-replicating system of nucleic acids and replicase enzymes has not yet arisen from simple organic constituents in the laboratory, but the history of scientific progress shows that it would be foolish and arrogant to assert that what science has not accomplished in a few decades cannot be accomplished. (And even if, given our human limitations, we should never succeed in this endeavor, why should that require us to invoke the supernatural?) Critical steps in the probable origin of life, such as abiotic synthesis of purines, pyrimidines, and amino acids and self-replication of short RNAs, have been demonstrated in the laboratory (see Chapter 5). And there is no reason to think that the first self-replicating or polypeptide-encoding nucleic acids had to have had any particular sequence. If there are many possible sequences with such properties, the probability of their formation rises steeply. Moreover, the origin of life is an entirely different problem from the modification and diversification of life once it has arisen. We do not need to know anything about the origin of life in order to understand and document the evolution of different life forms from their common ancestor.

7. ***Mutations are harmful and do not give rise to complex new adaptive characteristics.***

   Most mutations are indeed harmful and are purged from populations by natural selection. Some, however, are beneficial, as shown in many experiments (see Chapters 8 and 13). Complex adaptations usually are based not on single mutations, but on combinations of mutations that jointly or successively increase in frequency due to natural selection.

8. ***Natural selection merely eliminates unfit mutants, rather than creating new characters.***

   "New" characters, in most cases, are modifications of pre-existing characters, which are altered in size, shape, developmental timing, or organization (see Chapters 3, 4, and 21). This is true at the molecular level as well (see Chapter 19). Natural selection "creates" such modifications by increasing the frequencies of alleles at several or many loci so that combinations of alleles, initially improbable because of their rarity, become probable. Observations and experiments on both laboratory and natural populations have demonstrated the efficacy of natural selection.

9. ***Chance could not produce complex structures.***

   This is true, but natural selection is a deterministic, not a random, process. The random processes of evolution—mutation and genetic drift—do not result in the evolution of complexity, as far as we know. Indeed, when natural selection is relaxed, complex structures, such as the eyes of cave-dwelling animals, slowly degenerate, due in part to fixation of neutral mutations by genetic drift.

10. ***Complex adaptations such as wings, eyes, and biochemical pathways could not have evolved gradually because the first stages would not have been adaptive. The full complexity of such an adaptation is necessary, and this could not arise in a single step by evolution.***

    This was one of the first objections that greeted *The Origin of Species*, and it has recently been christened "irreducible complexity" by advocates of intelligent design. Our answer has two parts. First, many such features, such as hemoglobins and eyes, do show various stages of increasing complexity among different organisms (see Chapters 3, 4, 5, and 21). "Half an eye"—an eye capable of discriminating light from dark, but incapable of forming a focused image—is indeed better than no eye at all. Second, many structures have been modified for a new function after being elaborated to serve a different function (see Chapters 3 and 21). The "finished version" of an adaptation that we see today may indeed require precise coordination of many components in order to perform its current function, but the earlier stages, performing different or less demanding functions, and performing them less efficiently, are likely to have been an improvement on the ancestral feature. The evolution of the mammalian skull and jaw (see Chapter 4) provides a good example.

    Bombardier beetles such as the one pictured at the opening of this chapter have been a favorite example used by intelligent design proponents; they insist these beetles' unique "boiling chemical explosion" defense system could not possibly have come into being through evolution. In fact, however, many beetle species possess various components of the bombardiers' adaptive defense system, and related beetles show intermediate stages of the system's complex chemical and anatomical elements. An interesting site on the World Wide Web is www.talkorigins.org/faqs/bombardier.html. Compiled by Mark Isaak, it is a meticulously referenced refutation of the intelligent design position on this organism.

11. ***If an altered structure, such as the long neck of the giraffe, is advantageous, why don't all species have that structure?***

    This naive question ignores the fact that different species and populations have different ecological niches and environments, for which different features are adaptive. This principle holds for all features, including "intelligence."

12. ***If gradual evolution had occurred, there would be no phenotypic gaps among species, and classification would be impossible.***

    Many disparate organisms are connected by intermediate species, and in such cases, classification into higher taxa is indeed rather arbitrary (see Chapter 3). In other cases, gaps exist because of the extinction of intermediate forms (see Chapter 4). Moreover, although much of evolution is gradual, some advantageous mutations with large, discrete effects on the phenotype have probably played a role. Whether or not evolution has been entirely gradual is an empirical question, not a theoretical necessity.

13. ***The fossil record does not contain any transitional forms representing the origin of major new forms of life.***

    This very common claim is flatly false, for there are many such intermediates (see Chapter 4). Creationists sometimes use rhetorical subterfuge in presenting this argument, such as defining *Archaeopteryx* as a bird because of its feathers and then claiming that there are no known intermediates between reptiles and birds.

14. The fossil record does not objectively represent a time series because strata are ordered by their fossil contents, and then are assigned different times on the assumption that evolution has occurred.

Even before *The Origin of Species* was published, geologists who did not believe in evolution recognized the temporal order of fossils that are characteristic of different periods, and named most of the geological periods. Since then, radioactive dating and other methods have established the absolute dates of geological strata.

15. Vestigial structures are not vestigial, but functional.

According to creationist thought, an intelligent Creator must have had a purpose, or design, in each element of His creation. Thus all features of organisms must be functional. For this reason, creationists view adaptations as support for their position. However, nonfunctional, imperfect, and even maladaptive structures are expected if evolution is true, especially if a change in an organism's environment or way of life has rendered them superfluous or harmful. As noted earlier, organisms display many such features at both the morphological and molecular levels.

16. The classic examples of evolution are false.

Some creationists have charged that some of best-known studies of evolution, cited in almost every textbook, are flawed and that evolutionary biologists have dishonestly perpetuated these supposed falsehoods. For example, it has been charged that Ernst Haeckel, in the nineteenth century, falsely exaggerated the similarities among vertebrate embryos and that biologists have no basis for using these similarities as evidence for evolution. Another example is the classic study of industrial melanism in the peppered moth by H. B. D. Kettlewell, who is accused of having obtained spurious evidence for natural selection by predatory birds by pinning moths to unnatural resting sites (tree trunks).

Suppose these and the other cited studies were indeed flawed. First, it does not follow that textbook authors and other contemporary biologists have dishonestly perpetuated falsehood; they simply might not have checked and analyzed the original studies, but instead borrowed from other books and "secondary" sources. They might be charged with laziness or sloppiness, but no textbook author can check every study in depth. In any case, there is no reason to suspect intellectual dishonesty. Second, whether or not Haeckel or Kettlewell is guilty as charged is completely irrelevant to the validity of the basic claims involved. Evolutionary biologists (not creationists) first revealed that Haeckel did "improve" his drawings of embryos. But the various vertebrate embryos really do share profoundly important similarities (such as the notochord and pharyngeal pouches, often misnamed "gill slits") and really are more similar, overall, than the animals are later in development. Moreover, extraordinary similarities in both the superficial morphology of embryos and the underlying developmental mechanisms are found not just among vertebrate embryos, but in many other groups of animals and in plants. Likewise, both natural selection and rapid evolutionary changes have been demonstrated in so many species that these principles would stand firmly even if the peppered moth story were completely false. (Kettlewell's evidence that birds differentially attack dark and light peppered moths was based on a variety of experiments, and other investigators have added to this evidence since then. They have shown that other selective factors also affect the evolution and loss of melanism in this species.) Whatever flaws there may be in a few old studies, they do not invalidate decades of subsequent research by hundreds of scientists.

Of course, the creationists who cite these examples of supposed flaws and frauds realize that the strength of evolutionary biology does not rest on these studies. After all, most creationists accept natural selection and "microevolution," such as changes in moth coloration. Rather, these attacks enable their readers to doubt the truthfulness of evolutionary scientists and to justify their disbelief in evolution. But remember that even if individual scientists are stupid (which a few are) or dishonest (which almost none are), the social process of science uncovers errors and justifies our confidence in its major claims.

17. ***Disagreements among evolutionary biologists show that Darwin was wrong. Even prominent evolutionists have abandoned the theory of natural selection, and the entire study of evolution is in disarray.***

Disagreements among scientists exist in every field of inquiry, and are in fact the fuel of scientific progress. They stimulate research and are a sign of vitality. Creationists misunderstand or misinterpret evolutionary biologists who have argued *(a)* that the fossil record displays abrupt shifts rather than gradual change (punctuated equilibrium); *(b)* that many characteristics of species may not be adaptations; *(c)* that evolution may involve mutations with large effects as well as those with small effects; and *(d)* that natural selection does not explain certain major events and trends in the history of life. In fact, none of the evolutionary biologists who hold these positions deny the central proposition that adaptive characteristics evolve by the action of natural selection on random mutations. All these debates arise from differing opinions on the relative frequency and importance of factors known to influence evolution: large-effect versus small-effect mutations, genetic drift versus natural selection, individual selection versus species selection, adaptation versus constraint, and so forth (see Chapters 11, 20, and 21). These arguments about the relative importance of different processes do not at all undermine the strength of the evidence for the historical fact of evolution—i.e., descent, with modification from common ancestors. On this point, there is no disagreement among evolutionary biologists.

18. ***There are no fossil intermediates between apes and humans; australopithecines were merely apes. And there exists an unbridgeable gap between humans and all other animals in cognitive abilities.***

This is a claim about one specific detail in evolutionary history, but it is the issue about which creationists care most. This claim is simply false. The array of fossil hominids shows numerous stages in the evolution of posture, hands and feet, teeth, facial structure, brain size, and other features. Both functional and nonfunctional DNA sequences of humans and African apes are extremely similar and clearly demonstrate that they share a recent common ancestor. The mental abilities of humans are indeed developed to a far greater degree than those of other species, but many of our mental faculties seem to be present in more rudimentary form in other primates and mammals.

19. ***As a matter of fairness, alternative theories, such as supernatural creation and intelligent design, should be taught, so that students can make their own decisions.***

This train of thought, if followed to its logical conclusion, would have teachers presenting hundreds of different creation myths, in fairness to the peoples who hold them, and it would compel teachers to entertain supernatural explanations of everything in earth science and astronomy, in chemistry and physics—for anything explained by these sciences, too, could have a supernatural cause instead. It would imply teaching students that to do a proper job of investigating an airplane crash, federal agencies should consider the possibility of mechanical failure, a bomb carried by a terrorist on board, a missile impact—or supernatural intervention (Alters and Alters 2001).

Science teachers should be expected to teach the content of contemporary science—which means the hypotheses that have been strongly supported and the ideas that are subjects of ongoing research. That is, they should teach what scientists do. Several scientists have searched the scientific literature for research reports on intelligent design and creation science, and have found no such reports at all. Nor is there any evidence that creation scientists have carried out scientific research that a biased community of scientists has refused to publish. As noted earlier in this chapter, there is no way of testing hypotheses about the supernatural, so there cannot be any scientific research on the subject. And that means that the subject should not be taught in a science course.

On arguing for evolution

How, and whether or not, to convince people to accept evolution is not a subject for this book, which is devoted to the content of evolutionary science. Brian and Sandra Alters have addressed this topic in *Defending Evolution: A Guide to the Creation/Evolution Controversy* (2001). As these authors point out, for many people, religious beliefs take precedence over scientific evidence, especially among people who believe that their fate for all eternity depends on adhering to their belief system. The points made in this chapter will do little to persuade such people. Presenting the nature of science and the evidence for evolution will, however, have an impact on people who genuinely question what the truth is; but the way in which it is presented is important, as Alters and Alters describe. In this context, anyone who is considering holding a public debate on evolution with a creationist should first become thoroughly familiar with creationist arguments and strategies, and especially with the intelligent design movement, which aggressively markets itself as science when it is not. As Barbara Forrest and Paul Gross emphasize in *Creationism's Trojan Horse: The Wedge of Intelligent Design* (2004), this movement aims not merely to establish ID along with evolutionary theory in the classroom, but further, to replace naturalistic scientific methodology with a religiously framed version of science.

Why Should We Teach Evolution?

If evolution is so controversial, why invite trouble? Why not just drop it from the science curriculum? After all, it's an academic subject that doesn't really affect people's lives, right?

Wrong. Evolution is a foundation of all of biology. Understanding evolution is as relevant to everyday life as understanding physics or chemistry, and research in evolutionary biology affects our lives both directly and indirectly.

Successful application of science to human affairs depends completely on sound basic science. The perils of ignoring or distorting fundamental science became tragically evident in the Soviet Union between the 1930s and the late 1960s. Before that time, Russia and the early Soviet Union had led the world in genetics, including evolutionary genetics. Under Stalin, however, Mendelism and Darwinism came to be viewed as unwelcome or dangerous Western ideologies. Trofim Lysenko, an agronomist with no scientific training, gained Stalin's favor by declaring genetics a capitalist, bourgeois threat to the state, and led a campaign in which geneticists were persecuted and imprisoned and genetics research and training were extinguished. Rejecting the Mendelian-Darwinian view of adaptation, Lysenko preached a Lamarckian doctrine that exposing organisms to an environmental factor, such as low temperature, would induce inherited adaptive changes in their offspring. Lysenko completely altered the practice of Soviet agriculture to fit this notion. The result was a disaster for Soviet food production and the Soviet people. Moreover, the Soviet Union was left out of major advances in food production that were developed by Western research (Soyfer 1994). There is no more dramatic lesson on the importance of free scientific inquiry and on the danger that ideological control of science can pose to the welfare of society.

Educating students about evolution should mean teaching them not only the basic principles and facts of evolutionary processes and evolutionary history, but also the concepts and ways of testing hypotheses that are used in evolutionary science. These approaches have many applications. For example, evolutionary biologists specialize in studying variation, both within and among species, and the conceptual approaches and methods they employ are broadly useful. Simply being aware of the importance of genetic and nongenetic variation is a useful lesson; for example, everyone, whether patient or doctor, should be aware that people may vary in their reaction to a drug (or to a disease, for that matter). Understanding the difference between genetic and nongenetic variation is profoundly important for interpreting claims that are made about differences among ethnic or "racial" groups (see Chapter 9). Evolutionary biology shows, moreover, that a trait may have high heritability and nevertheless be readily altered by environmental change (e.g., medical or educational intervention).

Let us very briefly consider some of the practical applications of evolutionary biology.

Health and medicine

Genetic diseases are caused by variant genes or chromosomes. Population genetic studies (e.g., of variant hemoglobins and the *G6PD* locus; see Chapter 12) can enable us to explain their prevalence and provide the basis for GENETIC COUNSELING—advising people about the likelihood of bearing genetically impaired children. Determining whether or not an individual carries a particular mutation requires, first, that the gene and the mutated site be mapped. One of the major methods for doing this is linkage disequilibrium mapping (using the consistency of association between a deleterious allele and a linked marker; see Chapter 9), which is based on population genetics. Furthermore, evolutionary biologists have developed methods for determining where natural selection has acted on a DNA sequence (see Chapters 12 and 19), a first step in understanding how the gene and its protein product malfunction. Comparisons among species can also be useful: in several genes, mutations that cause severe genetic disease in humans occur at those amino acid positions that have been most conserved among animal phyla, and cause amino acid changes that are more functionally radical than the variation observed among species (Miller and Kumar 2001). This finding suggests that genetic comparisons among species may be useful in determining the causes of genetic diseases that are not already well understood.

Evolutionary biology is crucial for understanding and combating infectious diseases. It enables us to study disease organisms' mode of reproduction, population structure, and genetic variation—and, most importantly, the rapid evolution of both antibiotic resistance and virulence in

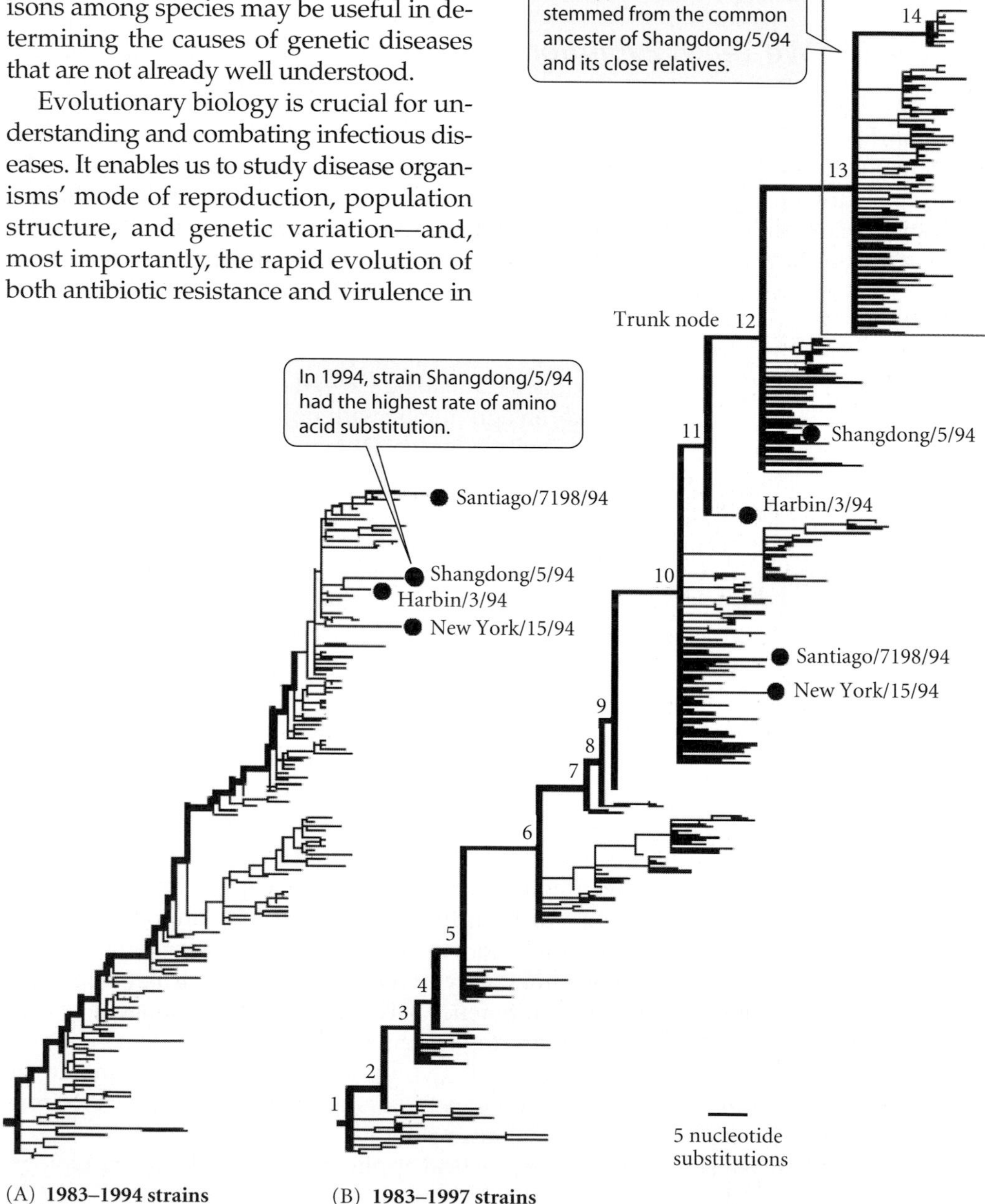

Figure 22.2 Prediction of epidemic strains of influenza A virus by phylogenetic analysis. (A) A phylogeny of influenza virus strains from 1983 through the 1993–1994 influenza season. The bold line, or "trunk," traces the single lineage that was successful in leaving descendants through this period. The labeled dots mark lineages that underwent exceptionally high rates of amino acid substitution at positively selected sites in the hemagglutinin gene. Strain Shangdong/5/94 had the highest rate of substitution, and the researchers predicted that future epidemics would stem from this or a closely related lineage. (B) A phylogeny of virus strains from 1983 through the 1996–1997 influenza season, during which the genotypes stemming from nodes 13 and 14 were collected. These genotypes are descended from trunk node 12, which is the common ancestor of Shangdong/5/94 and its close relatives, as the researchers had predicted. (From Bush et al. 1999 ©AAAS.)

many pathogenic microorganisms (see Chapters 12 and 18). Evolutionary studies may also prove useful in vaccine development, as new strains of pathogens continually arise that are resistant to existing vaccines. Robin Bush and colleagues (1999) tested the hypothesis that new strains of influenza A virus spread because their fitness is increased by mutations at codons in the hemagglutinin gene (the main target of human antibodies) that have a history of positive selection. New vaccines are required to provide protection against these strains. Bush et al. determined phylogenetic relationships among strains that had been preserved in 1993–1994 and identified those strains (e.g., Shangdong/5/94) that had undergone the most amino acid substitutions at positively selected sites in the hemagglutinin gene (Figure 22.2A). They predicted that Shangdong/5/94 or its close relatives would be ancestral to strains that later spread widely. Their prediction was correct (Figure 22.2B). This method may provide a way of estimating which currently rare strains are likely to give rise to epidemic strains, so that vaccines can be developed before those strains spread.

In some cases, we can learn where and when an infectious disease arose. For example, there are two major types of human immunodeficiency virus, HIV-2, which is largely restricted to western Africa, and HIV-1, which has become a worldwide epidemic. Both types include a number of distinct groups of strains; the M group of HIV-1, which itself includes several subgroups, represents the vast majority of infections. At least 18 kinds of related viruses (lentiviruses) are known to infect at least 20 species of African primates, each of which carries a different virus. Phylogenetic analyses have shown that HIV-2 evolved from the simian immunodeficiency virus SIVsm, which infects the sooty mangabey monkey, and that HIV-1 evolved several times from SIVcpz, the virus carried by two subspecies of the chimpanzee (Figure 22.3). Moreover, the rate of sequence evolution of HIV genes is fairly accurately known, enabling researchers to confidently date the last common ancestor of the M-group strains of HIV-1 to the 1930s. This conclusion conflicts with one hypothesis for the origin of human HIV-1—that it was caused by errors in poliovirus vaccination trials in 1957–1960—but it supports the hypothesis that HIV-1 was transmitted several times from chimpanzees to humans in Africa, probably by infected blood of chimpanzees that were butchered for food (Hahn et al. 2000; Korber et al. 2000). Cross-species infection has been documented for other primate lentiviruses, so there is concern that humans could acquire still others in the future.

Evolutionary studies also shed light on many normal physiological functions, such as those of the major histocompatibility complex of genes (see Chapter 12). DARWINIAN MEDICINE develops and tests hypotheses about human function and malfunction; for example, there is evidence that fever is an adaptive response that enables the body to combat infection (Nesse and Williams 1994). (On evolution and medicine, see also Trevathan et al. 1999; Stephens and Price 1996.)

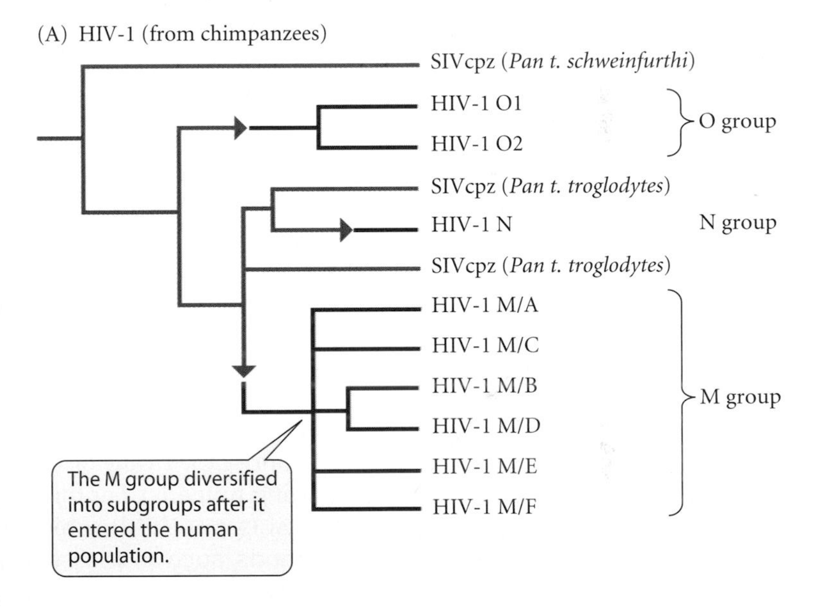

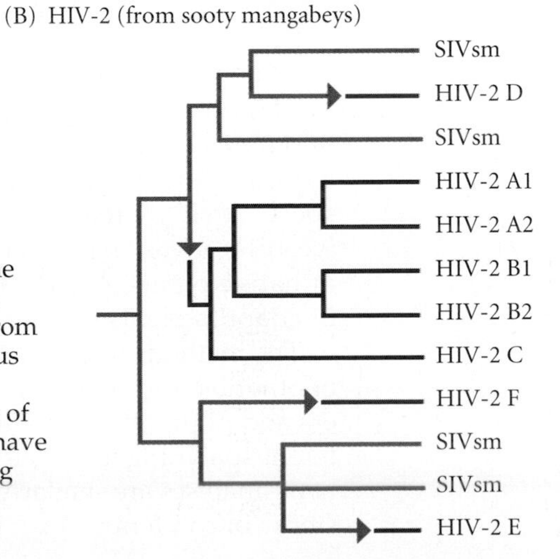

Figure 22.3 Schematic phylogenetic relationships of groups of the two major types of the human immunodeficiency virus, HIV-1 and HIV-2, showing multiple transmission of both viruses to humans from chimpanzees and from sooty mangabeys, respectively. The lentivirus SIVcpz is found in two chimpanzee subspecies (*Pan troglodytes troglodytes* and *Pan troglodytes schweinfurthi*). Three different groups of HIV-1 (O, N, and M) are derived from SIVcpz ancestors and must have been independently transmitted to humans (shown by arrows along branches). Similarly, HIV-2 has evolved from SIVsm, and has been transmitted to humans, at least four times. (After Hahn et al. 2000.)

Agriculture and natural resources

The development of improved varieties of domesticated plants and animals is evolution by artificial selection. Evolutionary genetics and plant and animal breeding have had an intimate, mutually beneficial relationship for more than a century. Both theoretical evolutionary methods and experimental studies of *Drosophila* and other model organisms have contributed both to traditional breeding and to modern QTL analysis, which is used to locate and characterize genes that contribute to traits of interest (see Chapter 13).

Agronomists have learned by bitter experience what evolutionary biologists have long known: that genetic diversity is essential for a population's long-term success. For example, more than 85 percent of the seed corn planted in the United States in the late 1960s was uniform for a genetic factor that suppresses the development of male flowers (which was useful for producing genetically uniform hybrid varieties). This factor turned out to make the corn susceptible to a mutant race of a fungus, leading to the southern corn leaf blight that caused losses of $1 billion in 1970. This example shows why it is important to build "germ plasm" banks of different crop strains, containing genes for various characteristics such as disease resistance. Useful genes can also be found in wild species that are related to crop plants. Genetic and evolutionary studies of the tomato, for example, led to study of the many related species in South America. At least 40 genes for resistance to various diseases have been found in these species, of which about 20 have been crossed into commercial tomato stocks. Studies of wild species' genes for other traits, such as salinity and drought tolerance, are under way. This approach depends on phylogenetics, evolutionary genetics, and evolutionary ecology.

Genes from wild species can also be transferred to crops by molecular methods of genetic engineering. Evolutionary biology contributes to this revolution in agronomics by contributing to gene mapping methods, by identifying likely sources of genes for useful characteristics, and by evaluating possible risks posed by transgenic organisms. Genetic engineering is controversial for several reasons; for example, there is concern that transgenes may spread from crop plants to wild species, which then could become more vigorous weeds. Phylogenetic studies help to indicate which wild species might hybridize with crop plants, and population genetic methods can estimate the fitness effects of transgenes and the chances of gene flow into natural plant populations (Ellstrand 2003).

Insects, weeds, and other organisms cause billions of dollars' worth of crop losses. Much of this loss is due to the evolution of resistance to chemical insecticides and herbicides by crop pests (see Chapter 12). This resistance not only increases the costs of agriculture, but also results in a steady increase in the amount of toxic chemicals sprayed on the landscape (some of which find their way into the higher levels of the food chain, including humans). For this reason, many crop pests are now managed by INTEGRATED PEST MANAGEMENT (IPM), which uses a combination of chemical pesticides and nonchemical methods, such as release of natural enemies of the pest (BIOLOGICAL CONTROL). In some places, regulations on the use of pesticides follow recommendations made by evolutionary biologists about how to manage pest populations in order to keep them susceptible to pesticides (see Gould 1998; Gould et al. 2002). The development of biological controls also benefits from evolutionary analysis. When a new insect species suddenly appears on a crop, phylogenetic systematics is the first approach to identifying the pest and determining where in the world it has come from. That is where entomologists will search for natural enemies, scrutinizing in particular those that are related to known enemies of species that are related to the new pest. Likewise, herbivorous insects used to control weeds or invasive plants must be screened to be sure that they will not also attack crops or native plants. A good approach is to see whether they have the potential to feed on or adapt to plants that are related to the target plant species (Futuyma 2000).

The methods of evolutionary biology contribute to the management and development of other biological resources as well. For example, fisheries biologists use genetic markers and population genetic methods to distinguish stocks of fish that migrate from different spawning locations (an issue with great political and economic implications). Genetic analyses are similarly important in forestry, for example, in developing commercial stocks of conifers.

Environment and conservation

In a few instances, evolutionary biology has contributed to environmental remediation and restoration of degraded land. For example, populations of grasses that have evolved tolerance to nickel and other heavy metals (see Chapter 13) can be used for revegetating land made barren by mining activities, and some such plants may even be used to remove contaminants from soil and water.

There is little doubt that a major extinction event has been initiated by the huge and accelerating impact of human activities on every aspect of the environment. The most important means of conservation are obvious and require mostly political, legal, and economic expertise: save natural habitats in preserves, establish and enforce limits on the exploitation of fish populations and other biological resources, reduce pollution and global warming. But biologists, including evolutionary biologists, also make indispensable contributions to conservation efforts. They use phylogenetic information to determine where potential nature reserves should be located to protect the greatest variety of biologically different species; they use evolutionary biogeography to identify regions with many endemic species (e.g., Madagascar); they use genetic methods and theory to prevent inbreeding depression in rare species and to distinguish genetically unique populations; and they use genetic markers to identify illegal traffic in endangered species (see Baker et al. 2000).

Understanding nature and humanity

"All art," said Oscar Wilde, "is perfectly useless." He meant that as high praise: art is a human creation that needs no utilitarian justification, a creation that is justified simply by being an expression—indeed, one of the defining characteristics—of humanity.

Much of what is most meaningful to us is "perfectly useless": music, sunsets, walking on a clean beach, baseball, soccer, movies, gardening, spiritual inspiration—and understanding. Whether the subject be mathematics, the natural world, philosophy, or human nature, attempting to understand is rewarding in itself, quite aside from whatever practical consequences it may yield.

To know about the extraordinary diversity of organisms, about the complexities of the cell, of development, or of our brains, and about how these marvels came to be is deeply rewarding to anyone with a sense of curiosity and wonder. To have achieved such knowledge is, like other advances in science and technology, among humanity's great accomplishments. Likewise, to have some understanding, however imperfect, of what we humans are and how we came into existence is richly rewarding. It is fascinating and ennobling to learn of our 3.5-billion-year-old pedigree, of when and how and possibly why our ancestors evolved the characteristics that led to our present condition, of how and when modern humans emerged from Africa and colonized the rest of the Earth, of how genetically unified all humans are with one another, and yet how genetically diverse we are. It is both challenging and important to try to understand "human nature"—to understand how our behavior is shaped by our genes and therefore by our evolutionary past, and how it is shaped by culture, social forces, and our unique individual history of learning and experience.

No understanding of human nature and history can be complete without a scientific perspective, including some consideration of evolution. Conversely, just as the Bible has been used to justify crusades, inquisitions, and witch hunts, scientific ideas have been used to justify social inequity and even persecution. From its earliest days, evolutionary science, under the banner of "social Darwinism," was misappropriated to justify racism and imperialist domination, to exclude women from political and economic power, and to ascribe poverty, illiteracy, and crime not to the social conditions that exclude much of society from education and economic self-sufficiency, but to genetic inferiority (Hofstadter 1955). Evolution was used to justify the American eugenics movement, which advocated encouraging "superior" people to reproduce and discouraging or preventing "inferior" people from doing so; to justify discriminatory quotas in United States immigration policy; and, worst of all, to justify the racism that found its most horrifying expression in Nazi Germany. All these abuses were based on misunderstanding or twisting of the data and theory of evolution and genetics (and, to their credit, some evolutionary biologists and geneticists said so). A proper understanding of evolutionary biology, as of any science, is necessary to prevent it from being misused.

Evolutionary biology is, therefore, an important foundation for the "human sciences": medicine, psychology, sociology. Although psychologists and anthropologists may differ among themselves on the role of evolution in determining "human nature," most of them will agree that some knowledge of evolutionary principles is essential. And although evolutionary biologists and social scientists do not set social policy, they can speak out against abuses of their science. They can point out misunderstandings of evolutionary theory, such as the "naturalistic fallacy" that what is natural is good and can therefore guide human conduct. Social Darwinism was based on this fallacy; so is the belief that homosexuality is morally wrong because it does not lead to reproduction, and the belief that women should be subservient to men because of their "natural," evolved sex roles.

Evolution lacks either moral or immoral content, and evolutionary biology provides no philosophical basis for aesthetics or ethics. But evolutionary biology, like other knowledge, can serve the cause of human freedom and dignity by helping to relieve disease and hunger and by helping us to understand and appreciate both the unity and diversity of humankind. And it can enhance our appreciation of life in all its diversity: in Darwin's phrase, "endless forms most beautiful and most wonderful."

Summary

1. Evolution is a fact—a hypothesis that is so thoroughly supported that it is extremely unlikely to be false. The theory of evolution is not a speculation, but rather a complex set of well-supported hypotheses that explain how evolution happens.
2. Although many people do not think there is any necessary incompatibility between evolution and religion, many others reject evolution, and instead accept divine creation, because they think evolution conflicts with their religious beliefs. The positions taken by creationists on issues such as the age of the Earth and of life vary.
3. Science is tentative, it accepts hypotheses provisionally and changes in the face of convincing new evidence, and it is concerned only with testable hypotheses; it depends on empirical studies that are subject to peer scrutiny and that can be verified and repeated by others. Supernatural hypotheses, in contrast, cannot be tested. Creationism has none of the features of science, so it has no claim to be taught in science classes.
4. The evidence for evolution comes from all realms of biology and geology, including comparative studies of morphology, development, life histories, and other features, molecular biology, genomics, paleontology, and biogeography. Evolutionary principles can explain features of organisms that would not be expected of a beneficent intelligent designer, such as imperfect adaptation, useless or vestigial features, extinction, selfish DNA, sexually selected characteristics, conflicts between genes within the genome, and infanticide. Furthermore, all the proposed mechanisms of evolution have been thoroughly documented, and evolution has been observed.
5. The arguments used by creationists are all logically refutable in scientific terms, and are contradicted by data.
6. It is important to understand evolution because it has broad implications for how we think about nature and humanity, and because it has many practical ramifications. Evolutionary science contributes to many aspects of medicine and public health, agriculture and the management of natural resources, pest management, and conservation.

Terms and Concepts

creationists
fact
hypothesis
intelligent design (ID)
natural laws
"scientific creationism" ("creation science")
special creation
theistic evolution
theory

Suggestions for Further Reading

Several excellent Web sites provide information about evolution and can serve as valuable teaching aids. "Understanding Evolution" (http://evolution.berkeley.edu) is an outstanding site, developed by the University of California Museum of Paleontology to provide content and resources for teachers at all grade levels. "Evolution" (www.pbs.org/wgbh/evolution) is an authoritative and enjoyable tour through many evolutionary topics, including religious issues and the practical relevance of evolution, based on a superb series of WGBH/NOVA programs on the Public Broadcasting System. It is available on CD from WGBH, 125 Western Avenue, Boston, MA, 02134. The National Center for Science Education (P.O. Box 9477, Berkeley, CA 94709-0477; www.ncseweb.org) actively supports the teaching of evolution and combats creationism. It is a very useful source of information, and issues a newsletter, *NCSE Reports*.

Eugenie Scott, director of the National Center for Science Education, probably knows more about the evolution-creationism controversy than anyone else. Her *Evolution vs. creationism: An introduction* (Greenwood Press, Westport, CT, 2004) is an ideal introduction to the subject. *Defending evolution in the classroom: A guide to the creation/evolution controversy*, by B. J. Alters and S. M. Alters (Jones and Bartlett, Sudbury, MA 2001) is a perceptive analysis of how to present and teach evolution.

The National Academy of Sciences has published two documents addressed to teachers and policy makers. *Teaching about evolution and the nature of science* (National Academy Press, Washington, DC, 1998), prepared by scientific leaders, provides "dialogues" for use in teaching, describes what students should know about evolution, suggests activities and materials for teaching, and summarizes court decisions and statements from various organizations about teaching evolution. *Science and creationism: A view from the National Academy of Sciences* (second edition, National Academy Press, Washington, DC, 1999) summarizes evidence for evolution, explains how science differs from other human endeavors, and analyzes the claims of creationists. (Both documents are available from the Press at 2101 Constitution Avenue, NW, Box 285, Washington, DC 20055, or online at www.nap.edu.)

M. Pigliucci's *Denying evolution: Creationism, scientism, and the nature of science* (Sinauer Associates, Sunderland, MA, 2002) is an excellent analysis of creationism and science from the perspective of both biology and philosophy of science. *Tower of Babel: The evidence against the new creationism*, by R. T. Pennock (MIT Press, Cambridge, MA, 1999) is a detailed critique of contemporary creationism, largely from the perspective of philosophy. See also *Intelligent design and its critics: Philosophical, theological, and scientific perspectives*, edited by R. T. Pennock (MIT Press, Cambridge, MA, 2001), which includes essays by both opponents and adherents of "intelligent design."

In *Finding Darwin's God: A scientist's search for common ground between God and evolution* (HarperCollins, New York, 1999) K. Miller, a cell biologist at a leading university, argues that one can fully accept evolution and reconcile it with religion.

"Applied evolution," by J. J. Bull and H. A. Wichman (2001, *Annual Review of Ecology and Systematics* 32: 183–217) is a concise and highly informative review of many of the social applications of evolutionary biology, ranging from medicine to computer technology. Another source of information on this topic is "Evolution, Science, and Society," at http://www.amnat.org or in *The American Naturalist* [2001, 158 (supplement): S1–S47.]

Glossary

Most of the terms in this glossary appear at several or many places in the text of this book. Most terms that are used broadly in biology or are used in this book only near their definition in the text are not included here.

A

adaptation A process of genetic change in a population whereby, as a result of natural selection, the average state of a character becomes improved with reference to a specific function, or whereby a population is thought to have become better suited to some feature of its environment. Also, *an* adaptation: a feature that has become prevalent in a population because of a selective advantage conveyed by that feature in the improvement in some function. A complex concept; see Chapter 11.

adaptive peak That allele frequency, or combination of allele frequencies at two or more loci, at which the mean fitness of a population has a (local) maximum. Also, the mean phenotype (for one or more characters) that maximizes mean fitness. An **adaptive valley** is a set of allele frequencies at which mean fitness has a minimum.

adaptive radiation Evolutionary divergence of members of a single phylogenetic lineage into a variety of different adaptive forms; usually the taxa differ in the use of resources or habitats, and have diverged over a relatively short interval of geological time. The term **evolutionary radiation** describes a pattern of rapid diversification without assuming that the differences are adaptive.

adaptive zone A set of similar **ecological niches** occupied by a group of (usually) related species, often constituting a higher taxon.

additive effect The magnitude of the effect of an allele on a character, measured as half the phenotypic difference between homozygotes for that allele compared with homozygotes for a different allele.

additive genetic variance That component of the **genetic variance** in a character that is attributable to additive effects of alleles.

allele One of several forms of the same gene, presumably differing by mutation of the DNA sequence. Alleles are usually recognized by their phenotypic effects; DNA sequence variants, which may differ at several or many sites, are usually called **haplotypes**.

allele frequency The proportion of gene copies in a population that are a given allele; i.e., the probability of finding this allele when a gene is taken randomly from the population; also called **gene frequency**.

allometric growth Growth of a feature during ontogeny at a rate different from that of another feature with which it is compared.

allopatric Of a population or species, occupying a geographic region different from that of another population or species. *Cf.* **parapatric**, **sympatric**.

allopolyploid A **polyploid** in which the several chromosome sets are derived from more than one species.

allozyme One of several forms of an enzyme encoded by different alleles at a locus.

alternative splicing Splicing of different sets of exons from mRNA to form mature transcripts that are translated into different proteins (thus allowing the same gene to encode different proteins).

altruism Conferral of a benefit on other individuals at an apparent cost to the donor.

anagenesis Evolution of a feature within a lineage over an arbitrary period of time.

aneuploid Of a cell or organism, possessing too many or too few of one or more chromosomes.

antagonistic selection A source of natural selection that opposes another source of selection on a trait.

apomixis Parthenogenetic reproduction in which an individual develops from one or more mitotically produced cells that have not experienced recombination or syngamy.

apomorphic Having a **derived** character or state, with reference to another character or state. *See* **synapomorphy**.

aposematic Coloration or other features that advertise noxious properties; warning coloration.

artificial selection Selection by humans of a deliberately chosen trait or combination of traits in a (usually captive) population; differing from natural selection in that the criterion for survival and reproduction is the trait chosen, rather than fitness as determined by the entire genotype.

asexual Pertaining to reproduction that does not entail meiosis and syngamy.

assortative mating Nonrandom mating on the basis of phenotype; usually refers to positive assortative mating, the propensity to mate with others of like phenotype.

autopolyploid A **polyploid** in which the several chromosome sets are derived from the same species.

autosome A chromosome other than a sex chromosome.

B

background extinction A long-prevailing rate at which taxa become extinct, in contrast to the highly elevated rates that characterize **mass extinction**.

background selection Elimination of deleterious mutations in a region of the genome; may explain low levels of neutral sequence variation.

balancing selection A form of natural selection that maintains polymorphism at a locus within a population.

benthic Inhabiting the bottom, or substrate, of a body of water. *Cf.* **planktonic**.

biogeography The study of the geographic distribution of organisms.

biological species A population or group of populations within which genes are actually or potentially exchanged by interbreeding, and which are reproductively isolated from other such groups.

bottleneck A severe, temporary reduction in population size.

C

canalization The evolution of internal factors during development that reduce the effect of perturbing environmental and genetic influences, thereby constraining variation and consistently producing a particular (usually wild-type) phenotype.

candidate gene A gene thought to be involved in the evolution of a particular trait based on its mutant phenotype or the function of the protein it encodes.

carrying capacity The population density that can be sustained by limiting resources.

category In taxonomy, one of the ranks of classification (e.g., genus, family). *Cf.* **taxon**.

character A feature, or trait. *Cf.* **character state.**

character displacement Usually refers to a pattern of geographic variation in which a character differs more greatly between sympatric than between allopatric populations of two species; sometimes used for the evolutionary process of accentuation of differences between sympatric populations of two species as a result of the reproductive or ecological interactions between them.

character state One of the variant conditions of a character (e.g., yellow versus brown as state of the character "color of snail shell").

chronospecies A segment of an evolving lineage preserved in the fossil record that differs enough from earlier or later members of the lineage to be given a different binomial (name). Not equivalent to biological species.

***cis*-regulatory element** A noncoding DNA sequence in or near a gene required for proper spatiotemporal expression of that gene, often containing binding sites for transcription factors. Often used interchangeably with **enhancer**.

clade The set of species descended from a particular ancestral species.

cladistic Pertaining to branching patterns; a cladistic classification classifies organisms on the basis of the historical sequences by which they have diverged from common ancestors.

cladogenesis Branching of lineages during phylogeny.

cladogram A branching diagram depicting relationships among taxa; i.e., an estimated history of the relative sequence in which they have evolved from common ancestors. Used by some authors to mean a branching diagram that displays the hierarchical distribution of derived character states among taxa.

cline A gradual change in an allele frequency or in the mean of a character over a geographic transect.

clone A lineage of individuals reproduced asexually, by mitotic division.

coadapted gene pool A population or set of populations in which prevalent genotypes are composed of alleles at two or more loci that confer high fitness in combination with each other, but not with alleles that are prevalent in other such populations.

coalescence Derivation of the gene copies in one or more populations from a single ancestral copy, viewed retrospectively (from the present back into the past).

coevolution Strictly, the joint evolution of two (or more) ecologically interacting species, each of which evolves in response to selection imposed by the other. Sometimes used loosely to refer to evolution of one species caused by its interaction with another, or simply to a history of joint divergence of ecologically associated species.

commensalism An ecological relationship between species in which one is benefited but the other is little affected.

common garden A place in which (usually conspecific) organisms, perhaps from different geographic populations, are reared together, enabling the investigator to ascribe variation among them to genetic rather than environmental differences. Originally applied to plants, but now more generally used to describe any experiment of this design.

comparative method A procedure for inferring the adaptive function of a character by correlating its states in various taxa with one or more variables, such as ecological factors hypothesized to affect its evolution.

compartment A contiguous group of cells, descended from the same progenitor cell, that form a spatially discrete part of a developing organ or structure and often act as a discrete developmental unit. Cells from one compartment typically do not intermix with cells from other compartments.

competition An interaction between individuals of the same species or different species whereby resources used by one are made unavailable to others.

competitive exclusion Extinction of a population due to competition with another species.

concerted evolution Maintenance of a homogeneous nucleotide sequence among the members of a gene family, which evolves over time.

conspecific Belonging to the same species.

convergent evolution (**convergence**) Evolution of similar features independently in different evolutionary lineages, usually from different antecedent features or by different developmental pathways.

co-option The evolution of a function for a gene, tissue, or structure other than the one it was originally adapted for. At the gene level, used interchangeably with **recruitment** and, occasionally, **exaptation**.

cost A reduction in fitness caused by a correlated effect of a feature that provides an increment in fitness (i.e., a benefit).

correlation A statistical relationship that quantifies the degree to which two variables are associated. For *phenotypic correlation*, *genetic correlation*, *environmental correlation* as applied to the relationship between two traits, see Chapter 13.

creationism The doctrine that each species (or perhaps higher taxon) was created separately, essentially in its present form, by a supernatural Creator.

C-value paradox The lack of correlation between the DNA content of eukaryotic genomes and a given organism's phenotypic complexity (i.e., the genome of a less complex eukaryotic organism, such as a plant, may contain far more DNA than that of a more complex organism, such as a human being). The paradox is explained by the amount of noncoding repetitive DNA sequences in a genome.

D

deme A local population; usually, a small, panmictic population.

demographic Pertaining to processes that change the size of a population (i.e., birth, death, dispersal).

density-dependent Affected by population density.

derived character (state) A character (or character state) that has evolved from an antecedent (ancestral) character or state.

deterministic Causing a fixed outcome, given initial conditions. *Cf.* **stochastic**.

differential gene expression Differences in the time, location, and/or quantitative level at which a gene expresses the protein it encodes. Differential gene expression involves differences between species, developmental stages, or physiological states in the specific cells, tissues, structures, or body segments that express a given gene; it is believed to be a significant agent of morphological change over evolutionary time.

diploid Of a cell or organism, possessing two chromosome complements. *See also* **haploid**, **polyploid**.

direct development A life history in which the intermediate larval stage is omitted and development proceeds directly from an embryonic form to an adult-like form. *Cf.* **indirect development**.

directional selection Selection for a value of a character that is higher or lower than its current mean value.

dispersal In population biology, movement of individual organisms to different localities; in biogeography, extension of the geographic range of a species by movement of individuals.

disruptive selection Selection in favor of two or more modal phenotypes and against those intermediate between them; also called **diversifying selection**.

divergence The evolution of increasing difference between lineages in one or more characters.

diversification An evolutionary increase in the number of species in a clade, usually accompanied by divergence in phenotypic characters.

diversifying selection *See* **disruptive selection.**

domain A relatively small protein segment or module (100 amino acids or less) that can fold into a specific three-dimensional structure independently of other domains.

dominance Of an allele, the extent to which it produces when heterozygous the same phenotype as when homozygous. Of a species, the extent to which it is numerically (or otherwise) predominant in a community.

duplication The production of another copy of a locus (or other sequence) that is inherited as an addition to the genome.

E

ecological niche The range of combinations of all relevant environmental variables under which a species or population can persist; often more loosely used to describe the "role" of a species, or the resources it utilizes.

ecological release The expansion of a population's niche (e.g., range of habitats or resources used) where competition with other species is alleviated.

ecotype A genetically determined phenotype of a species that is found as a local variant associated with certain ecological conditions.

effective population size The effective size of a real population is equal to the number of individuals in an ideal population (i.e., a population in which all individuals reproduce equally) that produces the rate of genetic drift seen in the real population.

enhancer A DNA sequence that, when acted on by **transcription factors** controls transcription of an associated gene. *Cf.* ***cis*-regulatory element, promoter**.

endemic Of a species, restricted to a specified region or locality.

environment Usually, the complex of external physical, chemical, and biotic factors that may affect a population, an organism, or the expression of an organism's genes; more generally, anything external to the object of interest (e.g., a gene, an organism, a population) that may influence its function or activity. Thus, other genes within an organism may be part of a gene's environment, or other individuals in a population may be part of an organism's environment.

environmental variance Variation among individuals in a phenotypic trait that is caused by variation in the environment rather than by genetic differences.

epistasis An effect of the interaction between two or more gene loci on the phenotype or fitness whereby their joint effect differs from the sum of the loci taken separately.

equilibrium An unchanging condition, as of population size or genetic composition. Also, the value (e.g., of population size, allele frequency) at which this condition occurs. An equilibrium need not be stable. *See* **stability**, **unstable equilibrium**.

ESS *See* **evolutionarily stable strategy**.

essentialism The philosophical view that all members of a class of objects (such as a species) share certain invariant, unchanging properties that distinguish them from other classes.

evolution In a broad sense, the origin of entities possessing different states of one or more characteristics and changes in the proportions of those entities over time. *Organic evolution*, or *biological evolution*, is a change over time in the proportions of individual organisms differing genetically in one or more traits. Such changes transpire by the origin and subsequent alteration of the frequencies of genotypes from generation to generation within populations, by alteration of the proportions of genetically differentiated populations within a species, or by changes in the numbers of species with different characteristics, thereby altering the frequency of one or more traits within a higher taxon.

evolutionarily stable strategy (ESS) A phenotype such that, if almost all individuals in a population have that phenotype, no alternative phenotype can invade the population or replace it.

evolutionary radiation *See* **adaptive radiation**.

evolutionary reversal The evolution of a character from a derived state back toward a condition that resembles an earlier state.

evolutionary synthesis The reconciliation of Darwin's theory with the findings of modern genetics, which gave rise to a theory that emphasized the coaction of random mutation, selection, genetic drift, and gene flow; also called the **modern synthesis**.

exaptation The evolution of a function of a gene, tissue, or structure other than the one it was originally adapted for; can also refer to the adaptive use of a previously nonadaptive trait.

exon That part of a gene that is translated into a polypeptide (protein). *Cf.* **intron**.

exon shuffling The formation of new genes by assembly of exons from two or more preexisting genes. The classical model of exon shuffling generates new combinations of exons mediated via recombination of intervening introns; however, exon shuffling can also come about by retrotransposition of exons into preexisting genes.

F

fecundity The quantity of gametes (usually eggs) produced by an individual.

fitness The success of an entity in reproducing; hence, the average contribution of an allele or genotype to the next generation or to succeeding generations. *See also* **relative fitness**.

fixation Attainment of a frequency of 1 (i.e., 100 percent) by an allele in a population, which thereby becomes **monomorphic** for the allele.

founder effect The principle that the founders of a new population carry only a fraction of the total genetic variation in the source population.

frameshift mutation An insertion or deletion of base pairs in a translated DNA sequence that alters the reading frame, resulting in multiple downstream changes in the gene product.

frequency In this book, usually used to mean *proportion* (e.g., the frequency of an allele is the proportion of gene copies having that allelic state).

frequency-dependent selection A mode of natural selection in which the fitness of each genotype varies as a function of its frequency in the population.

function The way in which a character contributes to the fitness of an organism.

G

gene The functional unit of heredity. A complex concept.

gene complex A group of two or more genes that are members of the same family and in most cases are located in close proximity to one another in the genome, often in tandem separated by various amounts of intergenic, noncoding DNA.

gene conversion A process involving the unidirectional transfer of DNA information from one gene to another. In a typical conversion event, a gene or part of a gene acquires the same sequence as the other allele at that locus (intralocus or intra-allelic conversion), or the same sequences as a different, usually paralogous, locus (interlocus conversion). One consequence of gene conversion may be the homogenization of sequences among members of a gene family.

gene duplication When new genes arise as copies of preexisting gene sequences. The result can be a **gene family**.

gene family Two or more loci with similar nucleotide sequences that have been derived from a common ancestral sequence.

gene flow The incorporation of genes into the gene pool of one population from one or more other populations.

gene frequency *See* **allele frequency**.

gene pool The totality of the genes of a given sexual population.

gene tree A diagram representing the history by which gene copies have been derived from ancestral gene copies in previous generations.

genetic conflict Antagonistic fitness relationships between alleles at different loci in a genome.

genetic correlation Correlated differences among genotypes in two or more phenotypic characters, due to pleiotropy or linkage disequilibrium.

genetic distance Any of several measures of the degree of genetic difference between populations, based on differences in allele frequencies.

genetic drift Random changes in the frequencies of two or more alleles or genotypes within a population.

genetic load Any reduction of the mean fitness of a population resulting from the existence of genotypes with a fitness lower than that of the most fit genotype.

genetic variance Variation in a trait within a population, as measured by the variance that is due to genetic differences among individuals.

genic selection A form of selection in which the single gene is the unit of selection, such that the outcome is determined by fitness values assigned to different alleles. *See* **individual selection**, **kin selection**, **natural selection**.

genome The entire complement of DNA sequences in a cell or organism. A distinction may be made between the nuclear genome and organelle genomes, such as those of mitochondria and plastids.

genotype The set of genes possessed by an individual organism; often, its genetic composition at a specific locus or set of loci singled out for discussion.

genotype × environment interaction Phenotypic variation arising from the difference in the effect of the environment on the expression of different genotypes.

geographic variation Differences among spatially distributed populations of a species.

grade A group of species that have evolved the same state in one or more characters and typically constitute a **paraphyletic** group relative to other species that have evolved further in the same direction.

gradualism The proposition that large differences in phenotypic characters have evolved through many slightly different intermediate states.

group selection The differential rate of origination or extinction of whole populations (or species, if the term is used broadly) on the basis of differences among them in one or more characteristics. *See also* **interdemic selection**, **species selection**.

H

habitat selection The capacity of an organism (usually an animal) to choose a habitat in which to perform its activities. Habitat selection is not a form of natural selection.

haploid Of a cell or organism, possessing a single chromosome complement, hence a single gene copy at each locus.

haplotype A DNA sequence that differs from homologous sequences at one or more base pair sites.

Hardy-Weinberg Pertaining to the genotype frequencies expected at a locus under ideal equilibrium conditions in a randomly mating population.

heritability The proportion of the **variance** in a trait among individuals that is attributable to differences in genotype. Heritability in the narrow sense is the ratio of **additive genetic variance** to phenotypic variance.

heterochrony An evolutionary change in phenotype caused by an alteration of timing of developmental events.

heterokaryotype A genome or individual that is heterozygous for a chromosomal rearrangement such as an inversion. *Cf.* **homokaryotype**.

heterozygosity In a population, the proportion of loci at which a randomly chosen individual is heterozygous, on average.

heterozygote An individual organism that possesses different alleles at a locus.

heterozygous advantage The manifestation of higher fitness by heterozygotes than by homozygotes at a specific locus.

hitchhiking Change in the frequency of an allele due to linkage with a selected allele at another locus.

homeobox genes A large family of eukaryotic genes that contain a DNA sequence known as the **homeobox**. The homeobox sequence encodes a protein **homeodomain** about 60 amino acids in length that binds DNA. Most homeobox genes are transcriptional regulators. *Cf.* **domain**; **Hox genes**.

homeostasis Maintenance of an equilibrium state by some self-regulating capacity of an individual.

homeotic mutation A mutation that causes a transformation of one structure into another of the organism's structures.

homokaryotype A genome or individual that is homozygous for a chromosomal rearrangement such as an inversion. *Cf.* **heterokaryotype**.

homology Possession by two or more species of a character state derived, with or without modification, from their common ancestor. **Homologous chromosomes** are those members of a chromosome complement that bear the same genes.

homonymous Pertaining to biological structures that occur repeatedly within one segment of the organism, such as teeth or bristles.

homoplasy Possession by two or more species of a similar or identical character state that has not been derived by both species from their common ancestor; embraces **convergence**, **parallel evolution**, and **evolutionary reversal**.

homozygote An individual organism that has the same allele at each of its copies of a genetic locus.

horizontal transmission Movement of genes or symbionts (such as parasites) between individual organisms other than by transmission from parents to their offspring (which is **vertical transmission**). Horizontal transmission of genes is also called **lateral gene transfer**.

Hox genes A subfamily of **homeobox genes**, conserved in all metazoan animals, that controls anterior-posterior segment identity by regulating the transcription of many genes during development.

hybrid An individual formed by mating between unlike forms, usually genetically differentiated populations or species.

hybrid zone A region in which genetically distinct populations come into contact and produce at least some offspring of mixed ancestry.

hypermorphosis An evolutionary increase in the duration of ontogenetic development, resulting in features that are exaggerated compared to those of the ancestor.

I

identical by descent Of two or more gene copies, being derived from a single gene copy in a specified common ancestor of the organisms that carry the copies.

inbreeding Mating between relatives that occurs more frequently than if mates were chosen at random from a population.

inbreeding depression Reduction, in inbred individuals, of the mean value of a character (usually one correlated with fitness).

inclusive fitness The fitness of a gene or genotype as measured by its effect on the survival or reproduction of both the organism bearing it and the genes, identical by descent, borne by the organism's relatives.

indirect development A life history consisting of a larval stage between embryo and adult stages. *Cf.* **direct development**.

individual selection A form of natural selection consisting of nonrandom differences among different genotypes (or phenotypes) within a population in their contribution to subsequent generations. *See also* **genic selection, natural selection**.

inter-, intra- Prefixes meaning, respectively, "between" and "within." For example, "interspecific" differences are differences between species and "intraspecific" differences are differences among individuals within a species.

interaction Strictly, the dependence of an outcome on a combination of causal factors, such that the outcome is not predictable from the average effects of the factors taken separately. More loosely, an interplay between entities that affects one or more of them (as in interactions between species). *See also* **genotype × environment interaction.**

interdemic selection **Group selection** of populations within a species.

intrinsic rate of natural increase The potential per capita rate of increase of a population with a stable age distribution whose growth is not depressed by the negative effects of density.

introgression Movement of genes from one species or population into another by hybridization and backcrossing; carries the implication that some genes in a genome undergo such movement, but others do not.

intron A part of a gene that is not translated into a polypeptide. *Cf.* **exon**.

inversion A 180° reversal of the orientation of a part of a chromosome, relative to some standard chromosome.

isolating barrier, isolating mechanism A genetically determined difference between populations that restricts or prevents gene flow between them. The term does not include spatial segregation by extrinsic geographic or topographic barriers.

iteroparous Pertaining to a life history in which individuals reproduce more than once. *Cf.* **semelparous**.

K

karyotype The chromosome complement of an individual.

key adaptation An adaptation that provides the basis for using a new, substantially different habitat or resource.

kin selection A form of selection whereby alleles differ in their rate of propagation by influencing the impact of their bearers on the reproductive success of individuals (kin) who carry the same alleles by common descent.

L

Lamarckism The theory that evolution is caused by inheritance of character changes acquired during the life of an individual due to its behavior or to environmental influences.

lateral gene transfer *See* **horizontal transmission**.

lethal allele An allele (usually recessive) that causes virtually complete mortality, usually early in development.

lineage A series of ancestral and descendant populations through time; usually refers to a single evolving species, but may include several species descended from a common ancestor.

lineage sorting The process by which each of several descendant species, carrying several gene lineages inherited from a common ancestral species, acquires a single gene lineage; hence, the derivation of a monophyletic gene tree, in each species, from the paraphyletic gene tree inherited from their common ancestor.

linkage Occurrence of two loci on the same chromosome: the loci are functionally linked only if they are so close together that they do not segregate independently in meiosis.

linkage disequilibrium The association of two alleles at two or more loci more frequently (or less frequently) than predicted by their individual frequencies.

linkage equilibrium The association of two alleles at two or more loci at the frequency predicted by their individual frequencies.

locus (plural: **loci**) A site on a chromosome occupied by a specific gene; more loosely, the gene itself, in all its allelic states.

logistic equation An equation describing the idealized growth of a population subject to a density-dependent limiting factor. As density increases, the rate of growth gradually declines until population growth stops.

M

macroevolution A vague term, usually meaning the evolution of substantial phenotypic changes, usually great enough to place the changed lineage and its descendants in a distinct genus or higher taxon. *Cf.* **microevolution**.

mass extinction A highly elevated rate of extinction of species, extending over an interval that is relatively short on a geological time scale (although still very long on a human time scale).

maternal effect A nongenetic effect of a mother on the phenotype of her offspring, stemming from factors such as cytoplasmic inheritance, transmission of symbionts from mother to offspring, or nutritional conditions.

maximum parsimony *See* **parsimony**.

mean Usually the arithmetic mean or average; the sum of n values, divided by n. The mean value of x, symbolized as $\bar{x}$, equals $(x_1 + x_2 + \ldots + x_n)/n$.

mean fitness The arithmetic average fitness of all individuals in a population, usually relative to some standard.

meiotic drive Used broadly to denote a preponderance (> 50 percent) of one allele among the gametes produced by a heterozygote; results in genic selection.

metapopulation A set of local populations, among which there may be gene flow and patterns of extinction and recolonization.

microevolution A vague term, usually referring to slight, short-term evolutionary changes within species. *Cf.* **macroevolution**.

microsatellite A short, highly repeated, untranslated DNA sequence.

migration Used in theoretical population genetics as a synonym for gene flow among populations; in other contexts, refers to directed large-scale movements of organisms that do not necessarily result in gene flow.

mimicry Similarity of certain characters in two or more species due to convergent evolution when there is an advantage conferred by the resemblance. Common types include *Batesian* mimicry, in which a palatable *mimic* experiences lower predation because of its resemblance to an unpalatable *model*; and *Müllerian* mimicry, in which two or more unpalatable species enjoy reduced predation due to their similarity.

modern synthesis *See* **evolutionary synthesis**.

modularity The ability of individual parts of an organism, such as segments or organs, to develop or evolve independently from one another; the ability of developmental regulatory genes and pathways to be regulated independently in different tissues and developmental stages.

molecular clock The concept of a steady rate of change in DNA sequences over time, providing a basis for dating the time of divergence of lineages if the rate of change can be estimated.

monomorphic Having one form; refers to a population in which virtually all individuals have the same genotype at a locus. *Cf.* **polymorphism**.

monophyletic Refers to a taxon, phylogenetic tree, or gene tree whose members are all derived from a common ancestral taxon. In cladistic taxonomy, the term describes a taxon consisting of all the known species descended from a single ancestral species. *Cf.* **paraphyletic**, **polyphyletic**.

mosaic evolution Evolution of different characters within a lineage or clade at different rates, hence more or less independently of one another.

mutation An error in the replication of a nucleotide sequence, or any other alteration of the genome that is not manifested as reciprocal recombination.

mutational variance The increment in the genetic variance of a phenotypic character caused by new mutations in each generation.

mutualism A symbiotic relation in which each of two species benefits by their interaction.

N

natural selection The differential survival and/or reproduction of classes of entities that differ in one or more characteristics. To constitute natural selection, the difference in survival and/or reproduction cannot be due to chance, and it must have the potential consequence of altering the proportions of the different entities. Thus natural selection is also definable as a deterministic difference in the contribution of different classes of entities to subsequent generations. Usually the differences are inherited. The entities may be alleles, genotypes or subsets of genotypes, populations, or, in the broadest sense, species. A complex concept; see Chapter 11. *See also* **genic selection**, **individual selection**, **kin selection**, **group selection**.

neo-Darwinism The modern belief that natural selection, acting on randomly generated genetic variation, is a major, but not the sole, cause of evolution.

neofunctionalization Divergence of duplicate genes whereby one acquires a new function. *Cf.* **subfunctionalization**.

neoteny Heterochronic evolution whereby development of some or all somatic features is retarded relative to sexual maturation, resulting in sexually mature individuals with juvenile features. *See also* **paedomorphosis**, **progenesis**.

neutral alleles Alleles that do not differ measurably in their effect on fitness.

nonsynonymous substitution A base pair substitution in DNA that results in an amino acid substitution in the protein product; also called **replacement substitution**. *Cf.* **synonymous substitution**.

norm of reaction The set of phenotypic expressions of a genotype under different environmental conditions. *See also* **phenotypic plasticity**.

normal distribution A bell-shaped frequency distribution of a variable; the expected distribution if many factors with independent, small effects determine the value of a variable; the basis for many statistical formulations.

nucleotide substitution The complete replacement of one nucleotide base pair by another within a lineage over evolutionary time.

O

ontogeny The development of an individual organism, from fertilized zygote until death.

organism Usually used in this book to refer to an individual member of a species.

orthologous Refers to corresponding (homologous) members of a gene family in two or more species. *Cf.* **paralogous**.

outcrossing Mating with another genetic individual. *Cf.* **selfing**.

outgroup A taxon that diverged from a group of other taxa before they diverged from one another.

overdominance The expression by two alleles in heterozygous condition of a phenotypic value for some character that lies outside the range of the two corresponding homozygotes.

P

paedomorphosis Possession in the adult stage of features typical of the juvenile stage of the organism's ancestor.

panmixia Random mating among members of a population.

parallel evolution (parallelism) The evolution of similar or identical features independently in related lineages, thought usually to be based on similar modifications of the same developmental pathways.

paralogous Refers to the homologous relationship between two different members of a gene family, within a species or in a comparison of different species. *Cf.* **orthologous**.

parapatric Of two species or populations, having contiguous but non-overlapping geographic distributions.

paraphyletic Refers to a taxon, phylogenetic tree, or gene tree whose members are all derived from a single ancestor, but which does not include all the descendants of that ancestor. *Cf.* **monophyletic**.

parental investment Parental activities or processes that enhance the survival of existing offspring but whose **costs** reduce the parent's subsequent reproductive success.

parsimony Economy in the use of means to an end (*Webster's New Collegiate Dictionary*); the principle of accounting for observations by that hypothesis requiring the fewest or simplest assumptions that lack evidence; in systematics, the principle of invoking the minimal number of evolutionary changes to infer phylogenetic relationships.

parthenogenesis Virgin birth; development from an egg to which there has been no paternal contribution of genes.

PCR (polymerase chain reaction) A laboratory technique by which the number of copies of a DNA sequence is increased by replication in vitro.

peak shift Change in allele frequencies within a population from one to another local maximum of mean fitness by passage through states of lower mean fitness.

peripatric Of a population, peripheral to most of the other populations of a species.

peripatric speciation Speciation by evolution of reproductive isolation in peripatric populations as a consequence of a combination of genetic drift and natural selection.

phenetic Pertaining to phenotypic similarity, as in a phenetic classification.

phenotype The morphological, physiological, biochemical, behavioral, and other properties of an organism manifested throughout its life; or any subset of such properties, especially those affected by a particular allele or other portion of the **genotype**.

phenotypic plasticity The capacity of an organism to develop any of several phenotypic states, depending on the environment; usually this capacity is assumed to be adaptive.

phylogenetic species concept Any of several related concepts of species as sets of populations that are diagnosably different from other populations.

phylogeny The history of descent of a group of taxa such as species from their common ancestors, including the order of branching and sometimes the absolute times of divergence; also applied to the genealogy of genes derived from a common ancestral gene.

planktonic Living in open water. *Cf.* **benthic**.

pleiotropy A phenotypic effect of a gene on more than one character.

ploidy The number of chromosome complements in an organism.

polygenic character A character whose variation is based wholly or in part on allelic variation at more than a few loci.

polymorphism The existence within a population of two or more genotypes, the rarest of which exceeds some arbitrarily low frequency (say, 1 percent); more rarely, the existence of phenotypic variation within a population, whether or not genetically based. *Cf.* **monomorphic**.

polyphenism The capacity of a species or genotype to develop two or more forms, with the specific form depending on specific environmental conditions or cues, such as temperature or day length. A polyphenism is distinct from a **polymorphism** in that the former is the property of a single genotype, whereas the latter refers to multiple forms encoded by two or more different genotypes.

polyphyletic Refers to a taxon, phylogenetic tree, or gene tree composed of members derived by evolution from ancestors in more than one ancestral taxon; hence, composed of members that do not share a unique common ancestor. *Cf.* **monophyletic**.

polyploid Of a cell or organism, possessing more than two chromosome complements.

population A group of conspecific organisms that occupy a more or less well defined geographic region and exhibit reproductive continuity from generation to generation; ecological and reproductive interactions are more frequent among these individuals than with members of other populations of the same species.

positive selection Selection for an allele that increases fitness. *Cf.* **purifying selection**.

postzygotic Occurring after union of the nuclei of uniting gametes; usually refers to inviability or sterility that confers reproductive isolation.

preadaptation Possession of the necessary properties to permit a shift to a new niche, habitat, or function. A structure is preadapted for a new function if it can assume that function without evolutionary modification.

prezygotic Occurring before union of the nuclei of uniting gametes; usually refers to events in the reproductive process that cause reproductive isolation.

primordium A group of embryonic or larval cells destined to give rise to a particular adult structure.

processed pseudogene A **pseudogene** that has arisen via the retrotransposition of mRNA into cDNA.

progenesis A decrease during evolution of the duration of ontogenetic development, resulting in retention of juvenile features in the sexually mature adult. *See also* **neoteny**, **paedomorphosis**.

promoter Usually refers to the DNA sequences immediately 5′ to (upstream of) a gene that are bound by the RNA polymerase and its cofactors and/or are required in order to transcribe the gene. Sometimes used interchangeably with **enhancer**.

provinciality The degree to which the taxonomic composition of a biota is differentiated among major geographic regions.

pseudogene A nonfunctional member of a gene family that has been derived from a functional gene. *Cf.* **processed pseudogene**.

punctuated equilibria A pattern of rapid evolutionary change in the phenotype of a lineage separated by long periods of little change; also, a hypothesis intended to explain such a pattern, whereby phenotypic change transpires rapidly in small populations, in concert with the evolution of reproductive isolation.

purifying selection Elimination of deleterious alleles from a population. *Cf.* **positive selection**.

Q

quantitative trait A phenotypic character that varies continuously rather than as discretely different character states.

QTL Quantitative trait locus (or loci); a chromosome region containing at least one gene that contributes to variation in a quantitative trait. **QTL mapping** is a procedure for determining the map positions of QTL on chromosomes.

R

race A poorly defined term for a set of populations occupying a particular region that differ in one or more characteristics from populations elsewhere; equivalent to **subspecies.** In some writings, a distinctive phenotype, whether or not allopatric from others.

radiation *See* **adaptive radiation**.

random genetic drift *See* **genetic drift**.

recruitment (1) In evolutionary genetics, the evolution of a new function for a gene other than the function for which that gene was originally adapted. (2) In population biology, refers to the addition of new adult (breeding) individuals to a population via reproduction (i.e., individuals born into the population that reach reproductive age).

recurrent mutation Repeated origin of mutations of a particular kind within a species.

refugia Locations in which species have persisted while becoming extinct elsewhere.

regression In geology, withdrawal of sea from land, accompanying lowering of sea level; in statistics, a function that best predicts a dependent from an independent variable.

reinforcement Evolution of enhanced reproductive isolation between populations due to natural selection for greater isolation.

relative fitness The fitness of a genotype relative to (as a proportion of) the fitness of a reference genotype, which is often set at 1.0.

relict A species that has been "left behind"; for example, the last survivor of an otherwise extinct group. Sometimes, a species or population left in a locality after extinction throughout most of the region.

replacement substitution *See* **nonsynonymous substitution**.

reporter construct A DNA segment in which a putative *cis*-regulatory sequence is spliced upstream of a gene whose expression can be easily assayed, such as β-galactosidase or green fluorescent protein.

reproductive effort The proportion of energy or materials that an organism allocates to reproduction rather than to growth and maintenance.

response to selection The change in the mean value of a character over one or more generations due to selection.

restriction enzyme An enzyme that cuts double-stranded DNA at specific short nucleotide sequences. Genetic variation within a population results in variation in DNA sequence lengths after treatment with a restriction enzyme, or **restriction fragment length polymorphism** (**RFLP**).

reticulate evolution Union of different lineages of a clade by hybridization.

RFLP *See* **restriction enzyme**.

S

saltation A jump; a discontinuous mutational change in one or more phenotypic traits, usually of considerable magnitude.

scala naturae The "scale of nature," or Great Chain of Being: the pre-evolutionary concept that all living things were created in an orderly series of forms, from lower to higher.

selection Nonrandom differential survival or reproduction of classes of phenotypically different entities. *See* **natural selection**, **artificial selection**.

selection coefficient The difference between the mean relative fitness of individuals of a given genotype and that of a reference genotype.

selective advantage The increment in fitness (survival and/or reproduction) provided by an allele or a character state.

selective sweep Reduction or elimination of DNA sequence variation in the vicinity of a mutation that has been fixed by natural selection relatively recently.

selfing Self-fertilization; union of female and male gametes produced by the same genetic individual. *Cf.* **outcrossing**.

"selfish DNA" A DNA sequence that has the capacity for its own replication, or replication via other self-replicating elements, but has no immediate function (or is deleterious) for the organism in which it resides.

semelparous Pertaining to a life history in which individuals (especially females) reproduce only once. *Cf.* **iteroparous**.

semispecies One of several groups of populations that are partially but not entirely isolated from one another by biological factors (**isolating mechanisms**).

serial homology A relationship among repeated, often differentiated, structures of a single organism, defined by their similarity of developmental origin; for example, the several legs and other appendages of an arthropod.

sex-linked Of a gene, being carried by one of the sex chromosomes; it may be expressed phenotypically in both sexes.

sexual reproduction Production of offspring whose genetic constitution is a mixture of those of two potentially genetically different gametes.

sexual selection Differential reproduction as a result of variation in the ability to obtain mates.

sibling species Species that are difficult or impossible to distinguish by morphological characters, but may be discerned by differences in ecology, behavior, chromosomes, or other such characters.

silent substitution *See* **synonymous substitution**.

sister taxa Two species or higher taxa that are derived from an immediate common ancestor, and are therefore each other's closest relatives.

speciation Evolution of reproductive isolation within an ancestral species, resulting in two or more descendant species.

species In the sense of biological species, the members of a group of populations that interbreed or potentially interbreed with one another under natural conditions; a complex concept (see Chapter 15). Also, a fundamental taxonomic category to which individual specimens are assigned, which often but not always corresponds to the biological species. *See also* **biological species**, **phylogenetic species concept**.

species selection A form of **group selection** in which species with different characteristics increase (by speciation) or decrease (by extinction) in number at different rates because of a difference in their characteristics.

stability Often used to mean constancy; more often in this book, the propensity to return to a condition (a stable equilibrium) after displacement from that condition.

stabilizing selection Selection against phenotypes that deviate in either direction from an optimal value of a character.

standard deviation The square root of the **variance**.

stasis Absence of evolutionary change in one or more characters for some period of evolutionary time.

stochastic Random. *Cf.* **deterministic**.

strata Layers of sedimentary rock that were deposited at different times.

subfunctionalization Divergence of duplicate genes whereby each retains only a subset of the several functions of the ancestral gene. *Cf.* **neofunctionalization**.

subspecies A named geographic race; a set of populations of a species that share one or more distinctive features and occupy a different geographic area from other subspecies.

substitution The complete replacement of one allele by another within a population or species over evolutionary time. *Cf.* **fixation**.

superspecies A group of semispecies.

symbiosis An intimate, usually physical, association between two or more species.

sympatric Of two species or populations, occupying the same geographic locality so that the opportunity to interbreed is presented.

synapomorphy A derived character state that is shared by two or more taxa and is postulated to have evolved in their common ancestor.

synonymous substitution Fixation of a base pair change that does not alter the amino acid in the protein product of a gene; also called **silent substitution**. *Cf.* **nonsynonymous substitution**.

T

target gene In developmental genetics, a gene regulated by a transcription factor of interest. This regulation may be direct or indirect.

taxon (plural: **taxa**) The named taxonomic unit (e.g., *Homo sapiens*, Hominidae, or Mammalia) to which individuals, or sets of species, are assigned. **Higher taxa** are those above the species level. *Cf.* **category**.

teleology The belief that natural events and objects have purposes and can be explained by their purposes.

territory An area or volume of habitat defended by an organism or a group of organisms against other individuals, usually of the same species; **territorial behavior** is the behavior by which the territory is defended.

trade-off The existence of both a fitness benefit and a fitness cost of a mutation or character state, relative to another.

transcription factor A protein that interacts with a regulatory DNA sequence and affects the transcription of the associated gene.

transition A mutation that changes a nucleotide to another nucleotide in the same class (purine or pyrimidine). *Cf.* **transversion**.

translocation The transfer of a segment of a chromosome to another, nonhomologous, chromosome; the chromosome formed by the addition of such a segment.

transposable element A DNA sequence, copies of which become inserted into various sites in the genome.

transversion A mutation that changes a nucleotide to another nucleotide in the opposite class (purine or pyrimidine). *Cf.* **transition**.

U

unstable equilibrium An **equilibrium** to which a system does not return if disturbed.

V

variance (σ^2, s^2, V) The average squared deviation of an observation from the arithmetic mean; hence, a measure of variation. $s^2 = [\Sigma(x_i - \bar{x})^2]/(n - 1)$, where $\bar{x}$ is the mean and n the number of observations.

vertical transmission *See* **horizontal transmission**.

vestigial Occurring in a rudimentary condition as a result of evolutionary reduction from a more elaborated, functional character state in an ancestor.

viability Capacity for survival; often refers to the fraction of individuals surviving to a given age, and is contrasted with inviability due to deleterious genes.

vicariance Separation of a continuously distributed ancestral population or species into separate populations due to the development of a geographic or ecological barrier.

virulence Usually, the damage inflicted on a host by a pathogen or parasite; sometimes, the capacity of a pathogen or parasite to infect and develop in a host.

W

wild-type The allele, genotype, or phenotype that is most prevalent (if there is one) in wild populations; with reference to the wild-type allele, other alleles are often termed mutations.

Z

zygote A single-celled individual formed by the union of gametes. Occasionally used more loosely to refer to an offspring produced by sexual reproduction.

Literature Cited

A

Abbott, R. J. 1992. Plant invasions, interspecific hybridization and the evolution of new plant taxa. *Trends Ecol. Evol.* 7: 401–405. [15, 16]

Abrams, P. A. 2000. The evolution of predator-prey interactions: Theory and evidence. *Annu. Rev. Ecol. Syst.* 31: 79–108. [18]

Abrams, P. A., and H. Matsuda. 1997. Fitness minimization and dynamic instability as a consequence of predator-prey coevolution. *Evol. Ecol.* 11: 1–20. [18]

Adey, N. B., T. O. Tollefsbol, A. B. Sparks, M. H. Edgell, and C. A. Hutchison III. 1994. Molecular resurrection of an extinct ancestral promoter for mouse L1. *Proc. Natl. Acad. Sci. USA* 91: 1569–1573. [3]

Adoutte, A., G. Balavoine, N. Lartillot, O. Lespinet, B. Prud'homme, and R. de Rosa. 2000. The new animal phylogeny: reliability and implications. *Proc. Nat. Acad. Sci. USA* 97: 4453–4456. [5]

Agrawal, A., Q. M. Eastman, and D. G. Schatz. 1998. Implications of transposition mediated by *V(D)J*-recombination proteins RAG1 and RAG2 for origins of antigen-specific immunity. *Nature* 392: 744–751. [19]

Alberch, P., and E. A. Gale. 1985. A developmental analysis of an evolutionary trend: Digital reduction in amphibians. *Evolution* 39: 8–23. [20, 21]

Alberch, P., S. J. Gould, G. F. Oster, and D. B. Wake. 1979. Size and shape in ontogeny and phylogeny. *Paleobiology* 5: 296–317. [3]

Allison, A. C. 1955. Aspects of polymorphism in man. *Cold Spring Harbor Symp. Quant. Biol.* 20: 239–255. [9]

Alroy, J. 1998. Cope's rule and the dynamics of body mass evolution in North American fossil mammals. *Science* 280: 731–734. [7, 21]

Alroy, J. 2001. A multispecies overkill simulation of the end-Pleistocene megafaunal mass extinction. *Science* 292: 1893–1896. [5]

Alroy, J., P. L. Koch, and J. C. Zachos. 2000. Global climate change and North American mammal evolution. In D. H. Erwin and S. L. Wing (eds.), *Deep Time: Paleobiology's Perspective*, pp. 259–288. *Paleobiology* 25 (4), supplement. [7]

Alroy, J., C. R. Marshall, R. K. Bambach, and 22 others. 2001. Effects of sampling standardization on estimates of Phanerozoic marine diversification. *Proc. Nat. Acad. Sci. USA* 98: 6261–6266. [7]

Alters, B. J., and S. M. Alters. 2001. *Defending Evolution: A Guide to the Creation/Evolution Controversy*. Jones and Bartlett, Sudbury, MA. [22]

Alvarez, L. W., W. Alvarez, F. Asaro, and H. V. Michel. 1980. Extraterrestrial cause for the Cretaceous-Tertiary extinction. *Science* 208: 1095–1108. [7]

Ambrose, B. A., D. R. Lerner, P. Ciceri, C. M. Padilla, M. F. Yanofsky, and R. J. Schmidt. 2000. Molecular and genetic analyses of the *silky1* gene reveal conservation in floral organ specification between eudicots and monocots. *Molec. Cell* 5: 569–579. [20]

Anderson, E. 1949. *Introgressive Hybridization*. Wiley, New York. [15, 16]

Andersson, M. B. 1982. Female choice selects for extreme tail length in a widowbird. *Nature* 299: 818–820. [11, 12]

Andersson, M. B. 1994. *Sexual Selection*. Princeton University Press, Princeton, NJ. [14, 17, 18]

Andersson, M. B., and Y. Iwasa. 1996. Sexual selection. *Trends Ecol. Evol.* 11: 53–58. [14]

Andolfatto, P., and M. Przeworski. 2001. Regions of lower crossing over harbor more rare variants in African populations of *Drosophila melanogaster*. *Genetics* 158: 657–665. [12]

Andrews, S. M., and T. S. Westoll. 1970. The postcranial skeleton of *Eusthenopteron foordi* Whiteaves. *Trans. R. Soc. Edinburgh* 68: 207–329. [4]

Antonovics, J. 1968. Evolution in closely adjacent plant populations. V. Evolution of self-fertility. *Heredity* 23: 219–238. [16]

Antonovics, J., A. D. Bradshaw, and R. G. Turner. 1971. Heavy metal tolerance in plants. *Adv. Ecol. Res.* 7: 1–85. [13]

Aquadro, C. F., D. J. Begun, and E. C. Kindahl. 1994. Selection, recombination, and DNA polymorphism in *Drosophila*. In B. Golding (ed.), *Non-neutral Evolution: Theories and Molecular Data*, pp. 46–56. Chapman & Hall, New York. [12]

Armbruster, P., W. E. Bradshaw, and C. M. Holzapfel. 1998. Effects of postglacial range expansion on allozyme and quantitative genetic variation of the pitcher-plant mosquito, *Wyeomyia smithii*. *Evolution* 52: 1697–1704. [9]

Armbruster, W. S., V. S. Di Stilio, J. D. Tuxill, T. C. Flores, and J. L. Velasquez Runk. 2000. Covariance and decoupling of floral and vegetative traits in nine neotropical plants: A reevaluation of Berg's correlation-pleiades concept. *Am. J. Bot.* 86: 39–55. [13]

Armbruster, W. S., C. Pélabon, T. F. Hansen, and C. P. H. Mulder. 2004. Floral integration, modularity, and accuracy. In M. Pigliucci and K. Preston (eds.), *Phenotypic Integration: Studying the Ecology and Evolution of Complex Phenotypes*, pp. 23–49. Oxford University Press, Oxford. [13]

Arnold, M. J. 1997. *Natural Hybridization and Evolution*. Oxford University Press, Oxford. [15]

Arnold, S. J. 1981. Behavioral variation in natural populations. I. Phenotypic, genetic, and environmental correlations between chemoreceptive responses to prey in the garter snake, *Thamnophis elegans*. *Evolution* 35: 489–509. [13]

Arnqvist, G., and L. Rowe. 2002. Correlated evolution of male and female morphologies in water striders. *Evolution* 56: 936–947. [14]

Atlan, A., H. Merçot, C. Landre, and C. Montchamp-Moreau. 1997. The sex-ratio trait in *Drosophila simulans*: Geographical distribution of distortion and resistance. *Evolution* 51: 1886–1895. [14]

Atwood, R. C., L. K. Schneider, and F. J. Ryan. 1951. Selective mechanisms in bacteria. *Cold Spring Harbor Symp. Quant. Biol.* 16: 345–355. [11]

Ausich, W. I., and N. G. Lane. 1999. *Life of the past*, fourth edition. Prentice-Hall, Upper Saddle River, NJ. [5]

Averoff, M., and N. H. Patel. 1997. Crustacean appendage evolution associated with changes in Hox gene expression. *Nature* 388: 682–686. [3]

Avise, J. C. 1998. *The Genetic Gods: Evolution and Belief in Human Affairs*. Harvard University Press, Cambridge, MA. [8]

Avise, J. C. 2000. *Phylogeography*. Harvard University Press, Cambridge, MA. [6]

Avise, J. C. 1994. *Molecular Markers, Natural History, and Evolution*. Chapman & Hall, New York. [15, 17]

Avise, J. C. 2004. *Molecular Markers, Natural History, and Evolution*, second edition. Sinauer Associates, Sunderland, MA. [9]

Avise, J. C., and R. M. Ball, Jr. 1990. Principles of genealogical concordance in species concepts and biological taxonomy. *Oxford Surv. Evol. Biol.* 7: 45–67. [15]

Avise, J. C., B. C. Bowen, T. Lamb, A. B. Meylan, and E. Bermingham. 1992. Mitochondrial DNA evolution at a turtle's pace: Evidence for low genetic variability and reduced microevolutionary rate in Testudines. *Mol. Biol. Evol.* 9: 433–446. [10]

Axelrod, R., and W. D. Hamilton. 1981. The evolution of cooperation. *Science* 211: 1390–1396. [14]

Ayala, F. J., M. L. Tracey, D. Hedgecock, and R. C. Richmond. 1974. Genetic differentiation during the speciation process in *Drosophila*. *Evolution* 28: 576–592. [15]

B

Backwell, P. R. Y., J. H. Christy, S. R. Telford, M. D. Jennions, and N. I. Passmore. 2000. Dishonest signaling in a fiddler crab. *Proc. R. Soc. Lond. B* 267: 719–724. [14]

Bailey, W. J., D. H. A. Fitch, D. A. Tagle, J. Czelusniak, J. L. Slightom, and M. Goodman. 1991. Molecular evolution of the ψη-globin gene locus: Gibbon phylogeny and the hominoid slowdown. *Mol. Biol. Evol.* 8: 155–184. [2]

Baker, C. S., et al. 2000. Predicted decline of protected whales based on molecular genetic monitoring of Japanese and Korean markets. *Proc. R. Soc. Lond. B* 267: 1191–1199. [22]

Baldauf, S. L., A. J. Roger, I. Wenk-Siefert, and W. F. Doolittle. 2000. A kingdom-level phylogeny of eukaryotes based on combined protein data. *Science* 290: 972–977. [5]

Baldauf, S. L., D. Bhattacharya, J. Cockrill, P. Hugenholtz, J. Pawlowski, and A. G. B. Simpson. 2004. The tree of life: An overview. In J. Cracraft and M. J. Donoghue (eds.), *Assembling the Tree of Life*, pp. 44–75. Oxford University Press, New York. [2]

Bambach, R. K. 1985. Classes and adaptive variety: The ecology of diversification in marine faunas through the Phanerozoic. In J. W. Valentine (ed.), *Phanerozoic Diversity Patterns: Profiles in Macroevolution*, pp. 191–253. Princeton University Press, Princeton, NJ. [7]

Bambach, R. K., A. H. Knoll, and J. J. Sepkoski, Jr. 2002. Anatomical and ecological constraints on Phanerozoic animal diversity in the marine realm. *Proc. Nat. Acad. Sci. USA* 99: 6854–6859. [7]

Barber, I., S. A. Arnott, V. A. Braithwaite, J. Andrew, and F. Huntingford. 2001. Indirect fitness consequences of mate choice in sticklebacks: offspring of brighter males grow slowly but resist parasitic infections. *Proc. R. Soc. Lond. B* 268: 71–76. [14]

Barraclough, T. G., and A. P. Vogler. 2000. Detecting the geographical pattern of speciation from species–level phylogenies. *Am. Nat.* 155: 419–434. [16]

Barrett, S. C. H., and D. Charlesworth. 1991. Effects of a change in the level of inbreeding on the genetic load. *Nature* 352: 522–524. [17]

Bartel, D. P. 2004. MicroRNAs: Genomics, biogenesis, mechanism, and function. *Cell* 116: 281–297. [19]

Barton, N. H., and B. Charlesworth. 1984. Genetic revolutions, founder effects, and speciation. *Annu. Rev. Ecol. Syst.* 15: 133–164. [12]

Barton, N. H., and B. Charlesworth. 1998. Why sex and recombination? *Science* 281: 1986–1990. [17]

Barton, N. H., and K. S. Gale. 1993. Genetic analysis of hybrid zones. In R. G. Harrison (ed.), *Hybrid Zones and the Evolutionary Process*, pp. 13–45. Oxford University Press, New York. [15, 16]

Barton, N. H., and G. M. Hewitt. 1985. Analysis of hybrid zones. *Annu. Rev. Ecol. Syst.* 16: 113–148. [16]

Barton, N., and L. Partridge. 2000. Limits to natural selection. *BioEssays* 22: 1075–1084. [21]

Barton, N. H., and M. Turelli. 1989. Evolutionary quantitative genetics: How little do we know? *Annu. Rev. Genet.* 23: 337–370. [13]

Basolo, A. L. 1994. The dynamics of Fisherian sex-ratio evolution: Theoretical and experimental investigations. *Am. Nat.* 144: 473–490. [12]

Basolo, A. L. 1995. Phylogenetic evidence for the role of a preexisting bias in sexual selection. *Proc. R. Soc. Lond.* B 259: 307–311. [14]

Basolo, A. L. 1998. Evolutionary change in a receiver bias: a comparison of female preference functions. *Proc. R. Soc. Lond. B* 265: 2223–2228. [14]

Bateman, A. J. 1947. Contamination of seed crops. II. Wind pollination. *Heredity* 1: 235–246. [9]

Baum, D. A., and K. L. Shaw. 1995. Genealogical perspectives on the species problem. In P. C. Hoch and A. G. Stephenson (eds.), *Experimental and Molecular Approaches to Plant Biosystematics*, pp. 289–303. Monographs in Systematic Botany. Missouri Botanical Garden, St. Louis, MO. [15]

Begun, D. R. 2004. The earliest hominins: Is less more? *Science* 303: 1478–1480. [4]

Behrensmeyer, A. K., J. D. Damuth, W. A. DiMichele, R. Potts, H.-D. Sues, and S. L. Wing (eds.). 1992. *Terrestrial Ecosystems Through Time: Evolutionary Paleoecology of Terrestrial Plants and Animals*. University of Chicago Press, Chicago. [5]

Bell, G. 1982. *The Masterpiece of Nature: The Evolution and Genetics of Sexuality*. University of California Press, Berkeley. [17]

Bell, G., and V. Koufopanou. 1986. The cost of reproduction. *Oxford Surv. Evol. Biol.* 3: 83–131. [17]

Bell, M. A., J. V. Baumgartner, and E. C. Olson. 1985. Patterns of temporal change in single morphological characters of a Miocene stickleback fish. *Paleobiology* 11: 258–271. [4]

Bennett, A. F., R. E. Lenski, and J. E. Mittler. 1992. Evolutionary adaptation to temperature. I. Fitness responses of *Escherichia coli* to changes in its thermal environment. *Evolution* 46: 16–30. [8]

Bennetzen, J. L. 2000. Transposable element contributions to plant gene and genome evolution. *Plant Mol. Biol.* 42: 251–269. [8]

Benton, M. J. (ed.). 1988. *The Phylogeny and Classification of the Tetrapods*. Clarendon Press, Oxford. [5]

Benton, M. J. 1990. The causes of the diversification of life. In P. D. Taylor and G. P. Larwood (eds.), *Major Evolutionary Radiations*, pp. 409–430. Clarendon Press, Oxford. [7]

Benton, M. J. 1996. On the nonprevalence of competitive replacement in the evolution of tetrapods. In D. Jablonski, D. H. Erwin, and J. Lipps (eds.), *Evolutionary Paleobiology*, pp. 185–210. University of Chicago Press, Chicago. [7]

Benton, M. J., and R. J. Twitchett. 2003. How to kill (almost) all life: The end-Permian extinction event. *Trends Ecol. Evol.* 18: 358–365. [7]

Berenbaum, M. R. 1983. Coumarins and caterpillars: A case for coevolution. *Evolution* 39: 163–179. [18]

Berenbaum, M. R., and A. R. Zangerl. 1988. Stalemates in the coevolutionary arms race: Synthesis, synergisms, and sundry other sins. In K. C. Spencer (ed.), *Chemical Mediation of Coevolution*, pp. 113–132. Academic Press, San Diego, CA. [18]

Bergman, A. and M. L. Siegal. 2003. Evolutionary capacitance as a general feature of complex gene networks. *Nature* 424: 549–552. [21]

Bermingham, E., and J. C. Avise. 1986. Molecular zoogeography of freshwater fishes in the southeastern United States. *Genetics* 113: 939–965. [16]

Bernays, E. A., and M. Graham. 1988. On the evolution of host specificity by phytophagous arthropods. *Ecology* 69: 886–892. [18]

Bernstein, C., and H. Bernstein. 1991. *Aging, Sex, and DNA Repair*. Academic Press, San Diego, CA. [17]

Berthold, P., A. J. Heibig, G. Mohr, and U. Querner. 1992. Rapid microevolution of migratory behavior in a wild bird species. *Nature* 360: 668–670. [13]

Berven, K. A., D. E. Gill, and S. J. Smith-Gill. 1979. Countergradient selection in the green frog, *Rana clamitans*. *Evolution* 33: 609–623. [9]

Bharathan, G., and R. Sinha, R. 2001. The regulation of compound leaf development. *Plant Physiol.* 127: 1–5. [20]

Birch, L. C., T. Dobzhansky, P. D. Elliott, and R. C. Lewontin. 1963. Relative fitness of geographic races of *Drosophila serrata*. *Evolution* 17: 72–83. [17]

Birkhead, T. R. 2000. *Promiscuity: An Evolutionary History of Sperm Competition*. Harvard University Press, Cambridge, MA. [14]

Birkhead, T. R., and A. P. Møller. 1992. *Sperm Competition in Birds: Evolutionary Causes and Consequences*. Academic Press, London. [14]

Bishop, J. A. 1981. A neo-Darwinian approach to resistance: Examples from mammals. In J. A. Bishop and L. M. Cook (eds.), *Genetic Consequences of Man Made Change*, pp. 37–51. Academic Press, London. [12]

Boag, P. T. 1983. The heritability of external morphology in Darwin's ground finches (*Geospiza*) on Isla Daphne Major, Galápagos. *Evolution* 37: 877–894. [9]

Bodmer, W., and M. Ashburner. 1984. Conservation and change in the DNA sequences coding for alcohol dehydrogenase in sibling species of *Drosophila*. *Nature* 309: 425–540. [8]

Bonaparte, J. F. 1978. El Mesozoico de America del Sur y sus tetrapodos. *Opera Lilloana* 26: 1–596. [5]

Bonnell, M. L., and R. K. Selander. 1974. Elephant seals: Genetic variation and near extinction. *Science* 184: 908–909. [10]

Bonner, J. T. 1988. *The Evolution of Complexity*. Princeton University Press, Princeton, NJ. [21]

Borries, C., K. Launhardt, C. Epplen, J. T. Epplen, and P. Winkler. 1999. DNA analyses support the hypothesis that infanticide is adaptive in langur monkeys. *Proc. R. Soc. Lond. B* 266: 901–904. [14]

Bouchard, T. J. Jr., D. T. Lykken, M. McGue, N. L. Segal, and A. Tellegen. 1990. Sources of human psychological differences: The Minnesota study of twins reared apart. *Science* 250: 223–228. [9]

Boucot, A. J. 1975. *Evolution and Extinction Rate Controls*. Elsevier, Amsterdam. [7]

Boureau, E. 1964. *Traité de paléobotanique*, Vol. III. Masson, Paris. [5]

Bourke, A. F. G., and N. R. Franks. 1995. *Social Evolution in Ants*. Princeton University Press, Princeton, NJ. [14]

Bowler, P. J. 1989. *Evolution: The History of an Idea*. University of California Press, Berkeley. [1]

Boyce, M. S., and C. J. Perrins. 1987. Optimizing great tit clutch size in a fluctuating environment. *Ecology* 68: 142–153. [14, 17]

Bradbury, J. W., and S. L. Vehrencamp. 1998. *Principles of Animal Communication*. Sinauer Associates, Sunderland, MA. [14]

Bradshaw, A. D. 1991. Genostasis and the limits to evolution. The Croonian Lecture, 1991. *Phil. Trans. R. Soc. Lond.* B 333: 289–305. [8, 13]

Bradshaw, H. D., S. M. Wilbert, K. G. Otto, and D. W. Schemske. 1998. Quantitative trait loci affecting differences in floral morphology between two species of monkeyflower. *Genetics* 149: 367–382. [15]

Briggs, D. E. G., and P. R. Crowther (eds.). 1990. *Palaeobiology: A Synthesis*. Blackwell Scientific, Oxford. [4]

Brisson, D. 2003. The directed mutation controversy in evolutionary context. *Crit. Rev. Microbiol.* 29: 25–35. [8]

Britten, R. J. 2002. Divergence between samples of chimpanzee and human DNA sequences is 5%, counting indels *Proc. Natl. Acad. Sci. USA* 99: 13633–13635. [20]

Broadhead, T. W., and J. A. Waters. 1980. *Echinoderms: Notes for a Short Course*. Studies in Geology 3. University of Tennessee Dept. of Geological Sciences, Knoxville. [5]

Brodie, E. D. III. 1992. Correlational selection for color pattern and antipredator behavior in the garter snake *Thamnophis ordinoides*. *Evolution* 46: 1284–1298. [13]

Brodie, E. D. III. 1993. Homogeneity of the genetic variance-covariance matrix for antipredator traits in two natural populations of the garter snake *Thamnophis ordinoides*. *Evolution* 47: 844–854. [13, 16]

Brodie, E. D. III, and E. D. Brodie, Jr. 1999. Predator-prey arms races. *BioScience* 49: 557–568. [18]

Brodie, E. D., Jr., B. J. Ridenhour, and E. D. Brodie III. 2002. The evolutionary response of predators to dangerous prey: Hotspots and coldspots in the geographic mosaic of coevolution between garter snakes and newts. *Evolution* 56: 2067–2082. [18]

Brooks, D. R. 1990. Parsimony analysis in historical biogeography and coevolution: methodological and theoretical update. *Syst. Zool.* 39: 14–30. [6]

Brooks, D. R., and D. A. McLennan. 2002. *The Nature of Diversity: An Evolutionary Voyage of Discovery*. University of Chicago Press, Chicago. [3]

Brooks, L. D. 1988. The evolution of recombination rates. In B. R. Levin and R. E. Michod (eds.), *The Evolution of Sex*, pp. 87–105. Sinauer Associates, Sunderland, MA. [18]

Brooks, R. 2000. Negative genetic correlation between male sexual attractiveness and survival. *Nature* 406: 67–70. [14]

Brown, J. H., and A. C. Gibson. 1983. *Biogeography*. Mosby, St. Louis. [6]

Brown, J. H., and M. V. Lomolino. 1998. *Biogeography*, second edition. Sinauer Associates, Sunderland, MA [5, 6]

Brown, W. L. Jr., and E. O. Wilson. 1956. Character displacement. *Syst. Zool.* 5: 49–64. [16, 18]

Brown, W. M., J. M. George, and A. C. Wilson. 1979. Rapid evolution of animal mitochondrial DNA. *Proc. Natl. Acad. Sci. USA* 76: 1967–1971. [10]

Brunet, M., and 37 others. 2002. A new hominid from the Upper Miocene of Chad, Central Africa. *Nature* 418: 145–151. [4]

Brusca, G. J., and R. C. Brusca. 1978. *A Naturalist's Seashore Guide.* Mad River Press, Eureka, CA. [7]

Brusca, R. C., and G. J. Brusca. 1990. *Invertebrates*. Sinauer Associates, Sunderland, MA. [3]

Bull, J. J. 1983. *Evolution of Sex Determining Mechanisms*. Benjamin Cummings, Menlo Park, CA. [18]

Bull, J. J. 1994. Perspective: Virulence. *Evolution* 48: 1423–1437. [18]

Bull, J. J., and E. L. Charnov. 1985. On irreversible evolution. *Evolution* 39: 1149–1155. [21]

Bull, J. J., and W. R. Rice. 1991. Distinguishing mechanisms for the evolution of cooperation. *J. Theor. Biol.* 149: 63–74. [18]

Bull, J. J., and H. A. Wichman. 2001. Applied evolution. *Annu. Rev. Ecol. Syst.* 32:183–217. [1]

Bull, J. J., I. J. Molineaux, and W. R. Rice. 1991. Selection of benevolence in a host–parasite system. *Evolution* 45:875–882. [18]

Burbrink, F. T., R. Lawson, and J. B. Slowinski. 2000. Mitochondrial DNA phylogeography of the polytypic North American rat snake (*Elaphe obsoleta*): a critique of the subspecies concept. *Evolution* 54:2107–2118. [9]

Bürger, R., and A. Gimelfarb. 1999. Genetic variation maintenance in multilocus models of additive quantitative traits under stabilizing selection. *Genetics* 152:807–820. [13]

Bürger, R., G. P. Wagner, and F. Stettinger. 1989. How much heritable variation can be maintained in finite populations by mutation-selection balance? *Evolution* 43: 1748–1766. [13]

Buri, P. 1956. Gene frequency drift in small population of mutant *Drosophila*. *Evolution* 10: 367–402. [10]

Burns, K. J., S. J. Hackett, and N. K. Klein. 2002. Phylogenetic relationships and morphological diversity in Darwin's finches and their relatives. *Evolution* 56: 1240–1252. [3]

Bush, G. L. 1969. Sympatric host race formation and speciation in frugivorous flies of the genus *Rhagoletis* (Diptera, Tephritidae). *Evolution* 23: 237–251. [16]

Bush, R. M., C. A. Bender, K. Subbarao, N. J. Cox, and W. M. Fitch. 1999. Predicting the evolution of human influenza A. *Science* 286: 1921–1925. [22]

Buss, L. W. 1987. *The Evolution of Individuality*. Princeton University Press, Princeton, NJ. [14]

Butlin, R. 1989. Reinforcement of premating isolation. In D. Otte and J. A. Endler (eds.), *Speciation and Its Consequences*, pp. 158–179. Sinauer Associates, Sunderland, MA. [16]

Byrne, M., M. Timmermans, C. Kidner, and R. Martienssen. 2001. Development of leaf shape *Curr. Opin. Plant Biol.* 4: 38–43. [20]

C

Cain, M. L., V. Andreasen, and D. J. Howard. 1999. Reinforcing selection is effective under a relatively broad set of conditions in a mosaic hybrid zone. *Evolution* 53: 1343–1353. [16]

Cann, R. L., M. Stoneking, and A. C. Wilson. 1987. Mitochondrial DNA and human evolution. *Nature* 325: 31–36. [6]

Carlquist, S., B. G. Baldwin, and G. E. Carr (eds.). 2003. *Tarweeds and Silverswords: Evolution of the Madiinae (Asteraceae)*. Missouri Botanical Garden Press, St. Louis, MO. [3]

Carroll, R. L. 1988. *Vertebrate Paleontology and Evolution*. W. H. Freeman, New York. [4, 5, 8]

Carroll, S. B. 2003. Genetics and the making of *Homo sapiens. Nature* 422: 849–857. [20]

Carroll, S. B., and C. Boyd. 1992. Host race radiation in the soapberry bug: Natural history with the history. *Evolution* 46: 1052–1069. [13]

Carroll, S. B., S. D. Weatherbee, and J. A. Langeland. 1995. Homeotic genes and the regulation and evolution of insect wing number. *Nature* 375: 58–61. [20]

Carroll, S. P., H. Dingle, and S. P. Klassen. 1997. Genetic differentiation of fitness-associated traits among rapidly evolving populations of the soapberry bug. *Evolution* 51: 1182–1188. [13]

Carroll, S. B., J. K. Grenier, and S. D. Weatherbee. 2001. *From DNA to Diversity: Molecular Genetics and the Evolution of Animal Design.* Blackwell Science, Malden, MA. [20]

Carson, H. L. 1967. Selection for parthenogenesis in *Drosophila mercatorum*. *Genetics* 55: 157–171. [17]

Carson, H. L. 1975. The genetics of speciation at the diploid level. *Am. Nat.* 109: 73–92. [16]

Carson, H. L., and B. A. Clague. 1995. Geology and biogeography of the Hawaiian Islands. In W. L. Wagner and V. A. Funk (eds.), *Hawaiian Biogeography*, pp. 14–29. Smithsonian Institution Press, Washington, DC. [6]

Carson, H. L., and K. Y. Kaneshiro. 1976. *Drosophila* of Hawaii: Systematics and ecological genetics. *Annu. Rev. Ecol. Syst.* 7: 311–345. [3]

Cavalli-Sforza, L. L., and W. F. Bodmer. 1971. *The Genetics of Human Populations*. W. H. Freeman, San Francisco. [9, 12, 13]

Cavalli-Sforza, L. L., P. Menozzi, and A. Piazza. 1994. *The History and Geography of Human Genes*. Princeton University Press, Princeton, NJ. [6, 9]

Chaline, J., and B. Laurin. 1986. Phyletic gradualism in a European Plio–Pleistocene *Mimomys* lineage (Arvicolidae, Rodentia). *Paleobiology* 12: 203–216. [4]

Chapman, T., L. F. Liddle, J. Kalb, M. Wolfner, and L. Partridge. 1995. Cost of mating in *Drosophila melanogaster* females is mediated by male accessory gland products. *Nature* 373: 241–244. [14]

Chapman, T., G. Arnqvist, J. Bangham, and L. Rowe. 2003. Sexual conflict. *Trends Ecol. Evol.* 18: 41–47. [14]

Charlesworth, B. 1978. The population genetics of anisogamy. *J. Theor. Biol.* 73: 347–357. [18]

Charlesworth, B. 1982. Hopeful monsters cannot fly. *Paleobiology* 8: 469–474. [21]

Charlesworth, B. 1989. The evolution of sex and recombination. *Trends Ecol. Evol.* 4: 264–267. [17]

Charlesworth, B. 1990. The evolutionary genetics of adaptation. In M. Nitecki (ed.), *Evolutionary Innovations*, pp. 47–70. University of Chicago Press, Chicago. [21]

Charlesworth, B. 1994a. The effect of background selection against deleterious mutations on weakly selected, linked variants. *Genet. Res.* 63: 213–227. [12]

Charlesworth, B. 1994b. *Evolution in Age-Structured Populations*. Cambridge University Press, Cambridge. [17]

Charlesworth, B. 1997. Is founder-flush speciation defensible? *Am. Nat.* 149: 600–603. [17]

Charlesworth, B., and D. Charlesworth. 1978. A model for the evolution of dioecy and gynodioecy. *Am. Nat.* 112: 975–997. [17]

Charlesworth, B., and C. H. Langley. 1989. The population genetics of *Drosophila* transposable elements. *Annu. Rev. Genet.* 23: 251–287. [19, 20]

Charlesworth, B., and S. Rouhani. 1988. The probability of peak shifts in a founder population. II. An additive polygenic trait. *Evolution* 42: 1129–1145. [16]

Charlesworth, B., R. Lande, and M. Slatkin. 1982. A neo-Darwinian commentary on macroevolution. *Evolution* 36: 474–498. [16]

Charlesworth, B., M. T. Morgan, and D. Charlesworth. 1993. The effect of deleterious mutations on neutral molecular variation. *Genetics* 134: 1289–1303. [12]

Charnov, E. L. 1982. *The Theory of Sex Allocation*. Princeton University Press, Princeton, NJ. [12, 17]

Cheetham, A. H. 1987. Tempo of evolution in a Neogene bryozoan: Are trends in single morphological characters misleading? *Paleobiology* 13: 286–296. [4, 21]

Chen, F.-C., and W.-H. Li. 2001. Genomic divergences between humans and other hominoids and the effective population size of the common ancestor of humans and chimpanzees. *Am. J. Hum. Genet.* 68: 444–456. [20]

Cheng, C.-H., and L. Chen. 1999. Evolution of an antifreeze glycoprotein. *Nature* 401: 443–444. [19]

Chesser, R. T., and R. M. Zink. 1994. Modes of speciation in birds: a test of Lynch's method. *Evolution* 48: 490–497. [17]

Chiappe, L. M., and G. J. Dyke. 2002. The Mesozoic radiation of birds. *Annu. Rev. Ecol. Syst.* 33: 91–124. [4]

Christiansen, F. B. 1984. The definition and measurement of fitness. In B. Shorrocks (ed.), *Evolutionary Ecology*, pp. 65–79. Blackwell Scientific, Oxford. [12, 14]

Christiansen, F. B. 1990. Natural selection: Measures and modes. In K. Wöhrmann and S. K. Jain (eds.), *Population Biology: Ecological and Evolutionary Viewpoints*, pp. 27–81. Springer-Verlag, Berlin. [12]

Ciochon, R. L., and J. G. Fleagle (eds.). 1993. *The Human Evolution Sourcebook*. Prentice-Hall, Englewood Cliffs, NJ. [4]

Civetta, A., and A. G. Clark. 2000. Correlated effects of sperm competition and postmating female mortality. *Proc. Natl. Acad. Sci. USA* 97: 13162–13165. [14]

Clack, J. A. 2002a. *Gaining Ground: The Origin and Evolution of Tetrapods*. Indiana University Press, Bloomington. [4]

Clack, J. A. 2002b. An early tetrapod from "Romer's Gap." *Nature* 418: 72–76. [4]

Clark, A. G. 1993. Evolutionary inferences from molecular characterization of self-incompatibility alleles. In N. Takahata and A. G. Clark (eds.), *Mechanisms of Molecular Evolution*, pp. 79–108. Sinauer Associates, Sunderland, MA. [17]

Clark, A. G., and D. J. Begun. 1998. Female genotypes affect sperm displacement in *Drosophila*. *Genetics* 149: 1487–1493. [14]

Clark, A. G., M. Aguadé, T. Prout, L. G. Harshman, and C. H. Langley. 1994. Variation in sperm displacement and its association with accessory gland protein loci in *Drosophila melanogaster*. *Genetics* 139: 189–201. [14]

Clark, A. G., S. Glanowski, R. Nielsen, P. D. Thomas, A. Kejariwal, M. A. Todd, D. M. Tanenbaum, D. Civello, F. Lu, B. Murphy, S. Ferriera, G. Wang, X. Zheng, T. J. White, J. J. Sninsky, M. D. Adams, and M. Cargill. 2003. Inferring non-neutral evolution from human-chimp-mouse orthologous gene trios. *Science* 302: 1960–1963. [19]

Clarke, P. H. 1974. The evolution of enzymes for the utilisation of novel substrates. In M. J. Carlile and J. J. Skehel (eds.), *Evolution in the Microbial World*, pp. 183–217. Cambridge University Press, Cambridge. [8]

Clarkson, E. N. K. 1993. *Invertebrate Palaeontology and Evolution*. Chapman & Hall, London. [5]

Clausen, J., D. D. Keck, and W. M. Hiesey. 1940. Experimental studies on the nature of species. I. Effect of varied environments on western North American plants. Carnegie Institution of Washington Publication no. 520: 1–452. [9]

Clausen, J., D. D. Keck, and W. M. Hiesey. 1947. Heredity of geographically and ecologically isolated races. *Am. Nat.* 81: 114–133. [9]

Clegg, J. B., and D. J. Weatherall. 1999. Thalassaemia and malaria: new insights into an old problem. *Proc. Assoc. Am. Physicians* 111: 278–282. [12]

Clutton-Brock, T. H. 1991. *The Evolution of Parental Care*. Princeton University Press, Princeton, NJ. [14]

Clutton-Brock, T., and C. Godfray. 1991. Parental investment. In J. R. Krebs and N. B. Davies (eds.), *Behavioural Ecology: An Evolutionary Approach*, third edition, pp. 234–262. Blackwell Scientific, Oxford. [14]

Coates, M. I., and J. A. Clack. 1990. Polydactyly in the earliest known tetrapod limbs. *Nature* 347: 66–69. [4]

Cockburn, A. 1998. *Evolution* of helping behavior in cooperatively breeding birds. *Annu. Rev. Ecol. Syst.* 29: 141–177. [14]

Cohan, F. M. 1984. Can uniform selection retard random genetic divergence between isolated conspecific populations? *Evolution* 38: 495–504. [8]

Colbert, E. H. 1980. *Evolution of the Vertebrates*, third edition. Wiley, New York. [4]

Colegrave, N. 2002. Sex releases the speed limit on evolution. *Nature* 420: 664–666. [17]

Coley, P. D., J. P. Bryant, and F. S. Chapin III. 1985. Resource availability and plant antiherbivore defense. *Science* 230: 895–899. [18]

Conant, R. 1958. *A Field Guide to Reptiles and Amphibians*. Houghton Mifflin, Boston, MA. [9]

Connell, J. H. 1970. A predator-prey system in the marine intertidal region. I. *Balanus glandula* and several predatory species of *Thais*. *Ecol. Monogr.* 40: 49–78. [18]

Conner, J. K. 2002. Genetic mechanisms of floral trait correlation in a natural population. *Nature* 420: 407–410. [13]

Cook, C. D. K. 1968. Phenotypic plasticity with particular reference to three amphibious plant species. Pp. 97–111in *Modern Methods in Plant Taxonomy*, V. Heywood (ed.) Academic Press, London. [13]

Cook, L. M. 2003. The rise and fall of the *carbonaria* form of the peppered moth. *Q. Rev. Biol.* 78: 399–417. [12]

Coope, G. R. 1979. Late Cenozoic fossil Coleoptera: Evolution, biogeography, and ecology. *Annu. Rev. Ecol. Syst.* 10: 249–267. [5, 16, 18]

Coope, G. R. 1995. Insect faunas in ice age environments: why so little extinction? In J. H. Lawton and R. M. May (eds.), *Extinction Rates*, pp. 55–74. Oxford University Press, Oxford. [21]

Cooper, S. B. J., K. M. Ibrahim, and G. M. Hewitt. 1995. Postglacial expansion and genome subdivision in the European grasshopper *Chorthippus parallelus*. *Mol. Ecol.* 4: 49–60. [6]

Cooper, V. S., and R. E. Lenski. 2000. The population genetics of ecological specialization in evolving *Escherichia coli* populations. *Nature* 407: 736–739. [8]

Coyne, J. A. 1974. The evolutionary origin of hybrid inviability. *Evolution* 28: 505–506. [16]

Coyne, J. A. 1984. Genetic basis of male sterility in hybrids between two closely related species of *Drosophila*. *Proc. Natl. Acad. Sci. USA* 81: 4444–4447. [15]

Coyne, J. A. 1992. Genetics and speciation. *Nature* 355: 511–515. [16]

Coyne, J. A., and H. A. Orr. 1989a. Patterns of speciation in *Drosophila*. *Evolution* 43: 362–381. [15]

Coyne, J. A., and H. A. Orr. 1989b. Two rules of speciation. In D. Otte and J. A. Endler (eds.), *Speciation and Its Consequences*, pp. 180–207. Sinauer Associates, Sunderland, MA. [15, 15]

Coyne, J. A., and H. A. Orr. 1997. "Patterns of speciation in *Drosophila*" revisited. *Evolution* 51: 295–303. [15, 16]

Coyne, J. A., and H. A. Orr. 2004. *Speciation*. Sinauer Associates, Sunderland, MA. [16]

Coyne, J. A., and T. D. Price. 2000. Little evidence for sympatric speciation in island birds. *Evolution* 54: 2166–2171. [16]

Coyne, J. A., I. A. Boussy, T. Prout, S. H. Bryant, J. S. Jones, and J. A. Moore. 1982. Long-distance migration of *Drosophila*. *Am. Nat.* 119: 589–595. [9]

Cracraft, J. 1989. Speciation and its ontology: The empirical consequences of alternative species concepts for understanding patterns and processes of differentiation. In D. Otte and J. A. Endler (eds.), *Speciation and Its Consequences*, pp. 29–59. Sinauer Associates, Sunderland, MA. [15]

Cracraft, J. 1991. Patterns of diversification within continental biotas: Hierarchical congruence among the areas of endemism of Australian vertebrates. *Austral. Syst. Bot.* 4: 211–227. [6]

Cracraft, J. 2001. Avian evolution, Gondwana biogeography and the Cretaceous-Tertiary mass extinction event. *Proc. R. Soc. Lond. B* 268: 459–469. [6]

Creel, S., And P. M. Waser. 1991. Failures of reproductive suppression In dwarf mongooses: Accident or adaptation? *Behav. Ecol.* 2: 7–15. [14]

Crow, J. F. 1993. Mutation, mean fitness, and genetic load. *Oxford Surv. Evol. Biol.* 9: 3–42. [8, 12]

Crow, J. F., and M. Kimura. 1965. Evolution in sexual and asexual populations. *Am. Nat.* 99: 439–450. [17]

Crow, J. F., and M. Kimura. 1970. *An Introduction to Population Genetics Theory*. Harper & Row, New York. [10]

Crozier, R. H., and P. Pamilo. 1996. *Evolution of Social Insect Colonies*. Oxford University Press, Oxford. [14]

Cruzan, M. B., and M. L. Arnold. 1993. Ecological and genetic associations in an *Iris* hybrid zone. *Evolution* 47: 1432–1445. [15]

CSIRO (Commonwealth Scientific and Industrial Research Organisation). 1991. *The Insects of Australia,* second edition. Cornell University Press, Ithaca, NY. [4]

Culver, D. C. 1982. *Cave Life: Evolution and Ecology*. Harvard University Press, Cambridge, MA. [17]

Cunningham, C., W.-H. Zhu, and D. M. Hillis. 1998. Best-fit maximum-likelihood models for phylogenetic inference: empirical tests with known phylogenies. *Evolution* 52: 978–987. [2]

Curtsinger, J. W., P. M. Service, and T. Prout. 1994. Antagonistic pleiotropy, reversal of dominance, and genetic polymorphism. *Am. Nat.* 144: 210–228. [12]

D

Dacosta, C. P., and C. M. Jones. 1971. Cucumber beetle resistance and mite susceptibility controlled by the bitter gene in *Cucumis sativus*. *Science* 172: 1145–1146. [18]

Daly, M., and M. Wilson. 1983. *Sex, Evolution, and Behavior*. Willard Grant Press, Boston, MA. [14]

Danforth, B. N., L. Conway, and S. Ji. 2003. Phylogeny of eusocial *Lasioglossum* reveals multiple losses of eusociality within a primitively eusocial clade of bees (Hymenoptera: Halictidae). *Syst. Biol.* 52: 23–36. [21]

Darlington, C. D. 1939. *The Evolution of Genetic Systems*. Cambridge University Press, Cambridge. [16]

Darwin, C. 1854. *A Monograph of the Sub-class Cirripedia, with Figures of All the Species.* The Ray Society, London. [3, 4]

Darwin, C. 1859. *The Origin of Species by Means of Natural Selection, or the Preservation of Favored Races in the Struggle for Life.* Modern Library, New York. [1, 2, 22]

Davies, N. B. 1991. Mating systems. In J. R. Krebs and N. B. Davies (eds.), *Behavioural Ecology: An Ecological Approach*, third edition, pp. 263–294. Blackwell Scientific, Oxford. [14]

Davies, N. B. 1992. *Dunnock Behaviour and Social Evolution*. Oxford University Press, Oxford. [14]

Davies, N. B., and M. de L. Brooke. 1998. Cuckoos versus hosts: Experimental evidence for coevolution. In S. I. Rothstein and S. K. Robinson (eds.), *Parasitic Birds and Their Hosts: Studies in Coevolution*, pp. 59–79. Oxford University Press, New York. [18]

Davies, N. B., and T. R. Halliday. 1978. Deep croaks and fighting assessment in toads *Bufo bufo*. *Nature* 275: 683–685. [14]

Dawkins, R. 1986. *The Blind Watchmaker*. W. W. Norton, New York. [11]

Dawkins, R. 1989. *The Selfish Gene.* Revised edition. Oxford University Press, Oxford. [11]

Dawkins, R., and J. R. Krebs. 1979. Arms races between and within species. *Proc. R. Soc. London B* 205: 489–511. [18]

Dean, A. M., D. E. Dykhuizen, and D. L. Hartl. 1986. Fitness as a function of β-galactosidase activity in *Escherichia coli*. *Genet. Res.* 48: 1–8. [11]

Deban, S. M., D. B. Wake, and G. Roth. 1997. Salamander with a ballistic tongue. *Nature* 389: 27–28. [21]

Deininger, P. L., and M. A. Batzer. 2002. Mammalian retroelements. *Genome Res* 12: 1455–65. [19]

Delson, E., I. Tattersall, J. A. Van Couvering, and A. S. Brooks (eds.) 2000. *Encyclopedia of Human Evolution and Prehistory*, second edition. Garland Press, New York. [2]

de Muizon, C. 2001. Walking with whales. *Nature* 413: 259–161. [4]

Dennett, D. C. 1995. *Darwin's Dangerous Idea: Evolution and the Meanings of Life.* Simon & Schuster, New York. [11, 21]

Depew, D. J. and B. H. Weber. 1994. *Darwinism Evolving: Systems Dynamics and the Genealogy of Natural Selection*. MIT Press, Cambridge, MA. [20]

de Queiroz, K., and M. J. Donoghue. 1990. Phylogenetic systematics or Nelson's version of cladistics? *Cladistics* 6: 61–75. [15]

Dessain, S., C. T. Gross, M. A. Kuziora, and W. McGinnis. 1992. Antp-type homeodomains have distinct DNA-binding specificities that correlate with their different regulatory functions in embryos. *EMBO J.* 11: 991–1002. [20]

DeRisi, J. L., V. R. Iyer, and P. O. Brown. 1997. Exploring the metabolic and genetic control of gene expression on a genomic scale. *Science* 278: 680–686. [19]

Desmond, A., and J. Moore. 1991. *Darwin*. Warner Books, New York. [1]

Diamond, J. M. 1975. Assembly of species communities. In M. L. Cody and J. M. Diamond (eds.), *Ecology and Evolution of Communities*, pp. 342–444. Harvard University Press, Cambridge, MA. [6]

Dieckmann, U., and M. Doebeli. 1999. On the origin of species by sympatric speciation. *Nature* 400: 354–357. [16]

Diehl, S. R., and G. L. Bush. 1989. The role of habitat preference in adaptation and speciation. In D. Otte and J. A. Endler (eds.), *Speciation and Its Consequences*, pp. 345–365. Sinauer Associates, Sunderland, MA. [16]

Dilda, C. L., and T. F. C. Mackay. 2002. The genetic architecture of *Drosophila* sensory bristle number. *Genetics* 162: 1655–1674. [13]

Dobzhansky, Th. 1934. Studies on hybrid sterility. I. Spermatogenesis in pure and hybrid *Drosophila pseudoobscura*. *Z. Zellforsch. Mikrosk. Anat.* 21: 169–221. [15]

Dobzhansky, Th. 1936. Studies on hybrid sterility. II. Localization of sterility factors in *Drosophila pseudoobscura* hybrids. *Genetics* 21: 113–135. [15, 16]

Dobzhansky, Th. 1937. *Genetics and the Origin of Species*. Columbia University Press, New York. [1, 16]

Dobzhansky, Th. 1948. Genetics of natural populations. XVIII. Experiments on chromosomes of *Drosophila pseudoobscura* from different geographic regions. *Genetics* 33: 588–602. [11]

Dobzhansky, Th. 1951. *Genetics and the Origin of Species*, third edition. Columbia University Press, New York. [15]

Dobzhansky, Th. 1955. A review of some fundamental concepts and problems of population genetics. *Cold Spring Harbor Symp. Quant. Biol.* 20: 1–15. [15]

Dobzhansky, Th. 1956. What is an adaptive trait? *Am. Nat.* 90: 337–347. [13]

Dobzhansky, Th. 1970. *Genetics of the Evolutionary Process*. Columbia University Press, New York. [8, 9, 11]

Dobzhansky, Th., and B. Spassky. 1969. Artificial and natural selection for two behavioral traits in *Drosophila pseudoobscura*. *Proc. Natl. Acad. Sci. USA* 62: 75–80. [9]

Dobzhansky, Th., and S. Wright. 1943. Genetics of natural populations. X. Dispersion rates in *Drosophila pseudoobscura*. *Genetics* 28: 304–340. [9]

Doebley, J., A. Stec, and C. Gustus. 1995. *teosinte branched1* and the origin of maize: Evidence for epistasis and the evolution of dominance. *Genetics* 141: 333–346. [20]

Donoghue, M. J. 1992. Homology. In E. F. Keller and E. A. Lloyd (eds.), *Keywords in Evolutionary Biology*, pp. 170–179. Harvard University Press, Cambridge, MA. [20]

Donoghue, M. J., J. A. Doyle, J. Gauthier, and A. G. Kluge. 1989. The importance of fossils in phylogeny reconstruction. *Annu. Rev. Ecol. Syst.* 20: 431–460. [4]

Doolittle, W. F. 1999. Phylogenetic classification and the universal tree. *Science* 284: 2124–9. [19]

Drake, J. W., B. Charlesworth, D. Charlesworth, and J. F. Crow. 1998. Rates of spontaneous mutation. *Genetics* 148: 1667–1686. [8]

Dujon, B., and 66 others. 2004. Genome evolution in yeasts. *Nature* 430: 35–44. [19]

Dybdahl, M. F., and C. M. Lively. 1998. Host-parasite coevolution: evidence for rare advantage and time-lagged selection in a natural population. *Evolution* 52: 1057–1066. [18]

Dykhuizen, D. E. 1990. Experimental studies of natural selection in bacteria. *Annu. Rev. Ecol. Syst.* 21: 393–398. [8]

E

Easteal, S., and C. Collett. 1994. Consistent variation in amino-acid substitution rate, despite uniformity of mutation rate: Protein evolution in mammals is not neutral. *Mol. Biol. Evol.* 11: 643–647. [10]

Eberhard, W. G. 1996. *Female Control: Sexual Selection by Cryptic Female Choice*. Princeton University Press, Princeton, NJ. [15]

Ebert, D. 1994. Virulence and local adaptation of a horizontally transmitted parasite. *Science* 265: 1084–1086. [18]

Ehrlich, P. R., and P. H. Raven. 1964. Butterflies and plants: A study in coevolution. *Evolution* 18: 586–608. [7, 18]

Eicher, D. L. 1976. *Geologic Time*. Prentice-Hall, Englewood Cliffs, NJ. [4]

Eldredge, N. 1989. *Macroevolutionary Patterns and Evolutionary Dynamics: Species, Niches and Adaptive Peaks*. McGraw-Hill, New York. [21]

Eldredge, N., and S. J. Gould. 1972. Punctuated equilibria: an alternative to phyletic gradualism. In T. J. M. Schopf (ed.), *Models in Paleobiology*, pp. 82–115. Freeman, Cooper and Co., San Francisco. [4, 16, 21]

Elena, S. F., and R. E. Lenski. 1997. Test of synergistic interactions among deleterious mutations in bacteria. *Nature* 390: 395–398. [17]

Elena, S. F., and R. E. Lenski. 2003. *Evolution* experiments with microorganisms: The dynamics and genetic bases of adaptation. *Nature Rev. Genetics* 4: 457–469. [8]

Ellstrand, N., and J. Antonovics. 1984. Experimental studies of the evolutionary significance of sexual reproduction. I. A test of the frequency-dependent selection hypothesis. *Evolution* 38: 103–115. [12]

Emlen, S. T. 1991. Evolution of cooperative breeding in birds and mammals. In J. R. Krebs and N. B. Davies (eds.), *Behavioural Ecology*, third edition, pp. 301–337. Blackwell Scientific, Oxford. [14]

Emlen, S. T. 1997. Predicting family dynamics in social vertebrates. In J. R. Krebs and N. B. Davies (eds.), *Behavioural Ecology: An Ecological Approach*, fourth edition, pp. 228–253. Blackwell Scientific, Oxford. [14]

Emmons, L. H. 1990. *Neotropical Rain Forest Mammals: A Field Guide*. University of Chicago Press, Chicago. [6]

Enard, W., M. Przeworski, S. E. Fisher, C. S. Lai, V. Wiebe, T. Kitano, A. P. Monaco, and S. Pääbo. 2002. Molecular evolution of *FOXP2*, a gene involved in speech and language. *Nature* 418: 869–872. [8, 19]

Endler, J. A. 1973. Gene flow and population differentiation. *Science* 179: 243–250. [12]

Endler, J. A. 1977. *Geographic Variation, Speciation, and Clines*. Princeton University Press, Princeton, NJ. [15, 16]

Endler, J. A. 1980. Natural selection on color patterns in *Poecilia reticulata*. *Evolution* 34: 76–91. [11]

Endler, J. A. 1983. Testing causal hypotheses in the study of geographic variation. In J. Felsenstein (ed.), *Numerical Taxonomy*, pp. 424–443. Springer-Verlag, Berlin. [6]

Endler, J. A. 1986. *Natural Selection in the Wild*. Princeton University Press, Princeton, NJ. [11, 12, 13]

Erwin, D. H. 1991. Metazoan phylogeny and the Cambrian radiation. *Trends Ecol. Evol.* 6: 131–134. [5]

Erwin, D. H. 1993. *The Great Paleozoic Crisis: Life and Death in the Early Permian*. Columbia University Press, New York. [5, 7]

Erwin, D. H., and R. L. Anstey (eds.). 1995. *New Approaches to Speciation in the Fossil Record*. Columbia University Press, New York. [21]

Erwin, D. H., J. W. Valentine, and J. J. Sepkoski, Jr. 1987. A comparative study of diversification events: The early Paleozoic versus the Mesozoic. *Evolution* 41: 1177–1186. [7]

Ewald, P. W. 1994. *Evolution of Infectious Disease*. Oxford University Press, Oxford. [18]

Eyre-Walker, A., and P. D. Keightley. 1999. High genomic deleterious mutation rates in hominids. *Nature* 397: 344–347. [8]

F

Falconer, D. S., and T. F. C. Mackay. 1996. *Introduction to Quantitative Genetics*, fourth edition. Longman, Harlow, U.K. [13]

Farrell, B. D. 1998. "Inordinate fondness" explained: Why are there so many beetles? *Science* 281: 555–559. [18]

Farrell, B., D. Dussourd, and C. Mitter. 1991. Escalation of plant defenses: Do latex and resin canals spur plant diversification? *Am. Nat.* 138: 881–900. [7]

Feder, J. L. 1998. The apple maggot fly *Rhagoletis pomonella*: Flies in the face of wisdom about speciation? In D. J. Howard and S. H. Berlocher (eds.), *Endless Forms: Species and Speciation*, pp. 130–144. Oxford University Press, New York. [16]

Feder, J. L., C. A. Chilcote, and G. L. Bush. 1990. Geographic pattern of genetic differentiation between host-associated populations of

Rhagoletis pomonella (Diptera: Tephritidae) in the eastern United States and Canada. *Evolution* 44: 570–594. [16]

Feder, J. L., S. H. Berlocher, J. B. Roethele, H. Dambroski, J. J. Smith, W. L. Perry, V. Gavrilovic, K. E. Filchak, J. Rull, and M. Aluja. 2003. Allopatric genetic origins for sympatric host-plant shifts and race formation in *Rhagoletis*. *Proc. Natl. Acad. Sci. USA* 100: 10314–10319. [16]

Fedigan, L. M. 1986. The changing role of women in models of human evolution. *Annu. Rev. Anthropol.* 15: 25–66. [4]

Feduccia, A. 1999. *The Origin and Evolution of Birds*, second edition. Yale University Press, New Haven, CT. [4, 21]

Felsenstein, J. 1976. The theoretical population genetics of variable selection and migration. *Annu. Rev. Genet.* 10: 253–280. [12]

Felsenstein, J. 1981. Skepticism towards Santa Rosalia, or why are there so few kinds of animals? *Evolution* 35: 124–138. [16]

Felsenstein, J. 1985. Phylogenies and the comparative method. *Am. Nat.* 125: 1–15. [11]

Felsenstein, J. 2004. *Inferring Phylogenies*. Sinauer Associates, Sunderland, MA. [2]

Fenner, F., and F. N. Ratcliffe. 1965. *Myxomatosis*. Cambridge University Press, Cambridge. [18]

Fenster, C. B., L. F. Galloway, and L. Chao. 1997. Epistasis and its consequences for the evolution of natural populations. *Trends Ecol. Evol.* 12: 282–286. [13]

Fernald, R. D. 2000. Evolution of eyes. *Curr. Opin. Neurobiol.* 10: 444–450. [20]

ffrench-Constant, R. H., R. T. Roush, D. Mortlock, and G. P. Dively. 1990. Isolation of dieldrin resistance from field populations of *Drosophila melanogaster* (Diptera: Drosophilidae). *J. Econ. Entomol.* 83: 1733–1737. [8]

Field, J., B. Rosenthal, and J. Samuelson J. 2000. Early lateral transfer of genes encoding malic enzyme, acetyl-CoA synthetase and alcohol dehydrogenases from anaerobic prokaryotes to *Entamoeba histolytica*. *Molec. Microbiol.* 38: 446–455. [19]

Fischer, C. S., M. Hout, M. S. Jankowski, S. R. Lucas, A. Swidler, and K. Voss. 1996. *Inequality by Design: Cracking the Bell Curve Myth*. Princeton University Press, Princeton, NJ. [9]

Fisher, D. C. 1986. Progress in organismal design. In D. M. Raup and D. Jablonski (eds.), *Patterns and Processes in the History of Life*, pp. 99–117. Springer-Verlag, Berlin. [21]

Fisher, R. A. 1930. *The genetical theory of natural selection*. Clarendon Press, Oxford. [12, 14]

Fisher, R. A. 1941. Average excess and average effect of a gene substitution. *Ann. Eugenics* 11: 53–63. [17]

Flessa, K. W., and D. Jablonski. 1985. Declining Phanerozoic background extinction rates: Effect of taxonomic structure? *Nature* 313: 216–218. [7]

Foote, M. 1997. The evolution of morphological diversity. *Annu. Rev. Ecol. Syst.* 28: 129–152. [21]

Foote, M. 2000a. Origination and extinction components of diversity: general problems. In D. H. Erwin and S. L. Wing (eds.), *Deep Time: Paleobiology's Perspective*, pp. 74–102. *Paleobiology* 26 (4), supplement. [7]

Foote, M. 2000b. Origination and extinction components of taxonomic diversity: Paleozoic and post-Paleozoic dynamics. *Paleobiology* 26: 578–605. [7]

Fong, D. W., T. C. Kane, and J. C. Culver. 1995. Vestigialization and loss of nonfunctional characters. *Annu. Rev. Ecol. Syst.* 26: 249–268. [13]

Force, A., M. Lynch, F. B. Pickett, A. Amores A., Yan, and J.Postlethwait. 1999. Preservation of duplicate genes by complementary, degenerative mutations. *Genetics* 151: 1531–1545. [19]

Ford, E. B. 1971. *Ecological Genetics*. Chapman & Hall, London. [9, 21]

Forrest, B., and P. R. Gross. 2002. *Creationism's Trojan Horse: The Wedge of Intelligent Design*. Oxford University Press, New York. [22]

Fowler, K., and L. Partridge. 1989. A cost of mating in female fruitflies. *Nature* 338: 760–761. [17]

Fowler, N. L., and D. A. Levin. 1984. Ecological constraints on the establishment of a novel polyploid in competition with its diploid progenitor. *Am. Nat.* 124: 703–711. [17]

Frank, S. A. 1990. Sex allocation theory for birds and mammals. *Annu. Rev. Ecol. Syst.* 21: 13–55. [17]

Frank, S. A. 1992. Models of plant-pathogen coevolution. *Trends Genet.* 8: 213–219. [18]

Frank, S. A. 1996. Models of parasite virulence. *Q. Rev. Biol.* 71: 37–78. [18]

Frank, S. A. 1997. Models of symbiosis. *Am. Nat.* 150: S80–S99. [14]

Frankham, R., J. D. Ballou, and D. A. Briscoe. 2002. *Introduction to Conservation Genetics*. Cambridge University Press, Cambridge. [9]

Fraser, H. B., A. E. Hirsch, L. M. Steinmetz, C. Scharfe, and M. W. Feldman. 2002. Evolutionary rate in the protein interaction network. *Science* 296: 750–752. [19]

Fraser, S. (ed.). 1995. *The Bell Curve Wars: Race, Intelligence, and the Future of America*. Basic Books, New York. [9]

Fry, J. D. 2003. Multilocus models of sympatric speciation: Bush vs. Rice vs. Felsenstein. *Evolution* 57: 1735–1746. [16]

Fryer, G., and T. D. Iles. 1972. *The Cichlid Fishes of the Great Lakes of Africa*. T. F. H. Publications, Neptune City, NJ. [3, 26]

Funk, D. J. 1998. Isolating a role for natural selection in speciation: host adaptation and sexual isolation in *Neochlamisus bebbianae* leaf beetles. *Evolution* 52: 1744–1759. [16]

Fürsich, F. T., and D. Jablonski. 1984. Lake Triassic naticid drillholes: Carnivorous gastropods gain a major adaptation but fail to radiate. *Science* 224: 78–80. [7]

Futuyma, D. J. 1987. On the role of species in anagenesis. *Am. Nat.* 130: 465–473. [15, 21]

Futuyma, D. J. 1991. A new species of *Ophraella* Wilcox (Coleoptera: Chrysomelidae) from the southeastern United States. *J. New York Entomol. Soc.* 99: 643–653. [16]

Futuyma, D. J. 1995. *Science on Trial: The Case for Evolution*. Sinauer Associates, Sunderland, MA. [3, 4]

Futuyma, D. J. 2000. Some current approaches to the evolution of plant-herbivore interactions. *Plant Species Biol.* 15: 1–9. [22]

Futuyma, D. J. 2004. The fruit of the tree of life: Insights into evolution and ecology. In J. Cracraft and M. J. Donoghue (eds.), *Assembling the Tree of Life*, pp. 25–39. Oxford University Press, New York. [3]

Futuyma, D. J., and G. Moreno. 1988. The evolution of ecological specialization. *Annu. Rev. Ecol. Syst.* 19: 207–233. [21]

Futuyma, D. J., and M. Slatkin (eds.). 1983. *Coevolution*. Sinauer Associates, Sunderland, MA. [18]

Futuyma, D. J., M. C. Keese, and D. J. Funk. 1995. Genetic constraints on macroevolution: The evolution of host affiliation in the leaf beetle genus *Ophraella*. *Evolution* 49: 797–809. [13]

G

Galindo, B. E., V. D. Vacquier, and W. J. Swanson. 2003. Positive selection in the egg receptor for abalone sperm lysin. *Proc. Natl. Acad. Sci. USA* 100: 4639–4643. [15]

Galis, F. 1996. The application of functional morphology to evolutionary studies. *Trends Ecol. Evol.* 11: 124–129. [21]

Gao, F., E. Bailes, D. L. Robertson et al. 1999. Origin of HIV–1 in the chimpanzee *Pan troglodytes troglodytes*. *Nature* 397: 436–441. [1]

Gao, L. Z., and H. Innan. 2004. Very low gene duplication rate in the yeast genome. *Science* 306: 1367–1370. [19]

Gatesy, J., M. Milinkovitch, V. Waddell, and M. Stanhope. 1999. Stability of cladistic relationships between Cetacea and higher-level artiodactyls taxa. *Syst. Biol.* 48: 6–20. [4]

Gavrilets, S. 2004. *Fitness Landscapes and the Origin of Species*. Princeton University Press, Princeton, NJ. [16]

Gavrilets, S., and A. Hastings. 1996. Founder effect speciation: A theoretical reassessment. *Am. Nat.* 147: 466–491. [16]

Gehring, W. J., and K. Ikeo. 1999. *Pax6* mastering eye morphogenesis and evolution. *Trends Genet.* 15: 371–377. [8]

Gentles, A. J., and S. Karlin. 1999. Why are human G-protein-coupled receptors predominantly intronless? *Trends Genet.* 15: 47–49. [19]

Ghiselin, M. T. 1969. *The Triumph of the Darwinian Method*. University of California Press, Berkeley. [11]

Ghiselin, M. T. 1995. Perspective: Darwin, progress, and economic principles. *Evolution* 49: 1029–1037. [21]

Gilad, Y., O. Man, S. Pääbo, and D. Lancet. 2003. Human-specific loss of olfactory receptor genes. *Proc. Natl. Acad. Sci. USA* 100: 3324–3327. [19]

Gill, F. B. 1995. *Ornithology*, second edition. W. H. Freeman, New York. [11]

Gilles, A., and L. F. Randolph. 1951. Reduction in quadrivalent frequency in autotetraploid maize during a period of 10 years. *Am. J. Bot.* 38: 12–16. [16]

Gingerich, P. D. 1993. Quantification and comparison of evolutionary rates. *Am. J. Sci.* 293A: 453–478. [4]

Gingerich, P. D. 2001. Rates of evolution on the time scale of the evolutionary process. *Genetica* 112–113: 127–144. [4]

Gingerich, P. D. 2003. Land-to-sea transition of early whales: Evolution of Eocene Archaeoceti (Cetacea) in relation to skeletal proportions and locomotion of living semiaquatic mammals. *Paleobiology* 29: 429–454. [4]

Gingerich, P. D. , M. ul Haq, I. S. Zalmout, I. H. Khan, and M. S, Malkani. 2001. Origin of whales from early artiodactyls: Hands and feet of Eocene Protocetidae from Pakistan. *Science* 293: 2239–2242. [4]

Giot., L. et al. 2003. A protein interaction map of *Drosophila melanogaster*. *Science* 302: 1727–1736. [19]

Gislason, D., M. M. Ferguson, S. Skúlason, and S. S. Snorasson. 1999. Rapid and coupled phenotypic differentiation in Icelandic Arctic charr (*Salvelinus alpinus*). *Can. J. Fish. Aquat. Sci.* 56: 2229–2234. [16]

Godfray, H. C. J. 1999. Parent–offspring conflict. In L. Keller (ed.), *Levels of Selection in Evolution*, pp. 100–120. Princeton University Press, Princeton, NJ. [14]

Goldblatt, P. 1979. Polyploidy in angiosperms: Monocotyledons. In W. H. Lewis (ed.), *Polyploidy: Biological Relevance*, pp. 219–239. Plenum, New York. [16]

Goldblatt, P. (ed.). 1993. *Biological Relationships Between Africa and South America*. Yale University Press, New Haven, CT. [6]

Goldschmidt, R. B. 1940. *The Material Basis of Evolution*. Yale University Press, New Haven, CT. [15, 21]

Goldstein, D. B., and K. E. Holsinger. 1992. Maintenance of polygenic variation in spatially structured populations: Roles for local mating and genetic redundancy. *Evolution* 46: 412–429. [13]

Goldstein, D. B., A. Ruiz-Linares, L. L. Cavalli–Sforza, and M. W. Feldman. 1995. Genetic absolute dating based on microsatellites and the origin of modern humans. *Proc. Natl. Acad. Sci. USA* 92: 6723–6727. [10]

Goliber, T., S. Kessler, J.-J. Chen, G. Bharathan, and N. Sinha. 1999. Genetic, molecular, and morphological analysis of compound leaf development. *Curr. Topics Dev. Biol.* 43: 260–290. [20]

Gompel, N., and S. B. Carroll. 2003. Genetic mechanisms and constraints governing the evolution of correlated traits in drosophilid flies. *Nature* 424: 931–935. [20]

Goodman, M., B. F. Koop, J. Czelusniak, D. H. A. Fitch, D. A. Tagle, and J. L. Slightom. 1989. Molecular phylogeny of the family of apes and humans. *Genome* 31: 316-335. [2]

Goodwin, D. 1986. *Crows of the World*. British Museum (Natural History), London. [15]

Gould, F. 1998. Sustainability of transgenic insecticidal cultivars: Integrating pest genetics and ecology. *Annu. Rev. Entomol.* 443: 701–726. [22]

Gould, F., N. Blair, M. Reid, T. L. Rennie, J. Lopez, and S. Micinski. 2002. *Bacillus thuringiensis*-toxin resistance management: Stable isotope assessment of alternate host use by *Helicoverpa zea*. *Proc. Natl. Acad. Sci. USA* 99: 16581–16586. [22]

Gould, J. and C. G. Gould. 1989. *Sexual Selection*. Scientific American Library, New York. [14]

Gould, S. J. 1974. The origin and function of "bizarre" structures: Antler size and skull size in the "Irish elk," *Megaloceros giganteus*. *Evolution* 28: 191–220. [3]

Gould, S. J. 1977. *Ontogeny and Phylogeny*. Harvard University Press, Cambridge, MA. [3, 20]

Gould, S. J. 1981. *The Mismeasure of Man*. Norton, New York. [9]

Gould, S. J. 1982. The meaning of punctuated equilibrium and its role in validating a hierarchical approach to macroevolution. In R. Milkman (ed.), *Perspectives on Evolution*, pp. 83–104. Sinauer Associates, Sunderland, MA. [11]

Gould, S. J. 1985. The paradox of the first tier: An agenda for paleobiology. *Paleobiology* 11: 2–12. [7]

Gould, S. J. 1989. *Wonderful Life: The Burgess Shale and the Nature of History*. W. W. Norton, New York. [5, 21]

Gould, S. J. 1999. *Rocks of Ages*. Ballantine, New York. [22]

Gould, S. J. 2002. *The Structure of Evolutionary Theory*. Belknap Press of Harvard University Press, Cambridge, MA. [16, 21]

Gould, S. J., and N. Eldredge. 1993. Punctuated equilibrium comes of age. *Nature* 366: 223–227. [4, 16]

Gould, S. J., and E. S. Vrba. 1982. Exaptation: A missing term in the science of form. *Paleobiology* 8: 4–15. [11, 20]

Grafen, A. 1990. Biological signals as handicaps. *J. Theor. Biol.* 144: 517–546. [14]

Grafen, A. 1991. Modelling in behavioural ecology. In J. R. Krebs and N. B. Davies (eds.), *Behavioural Ecology*, third edition, pp. 5–31. Blackwell Scientific, Oxford. [14]

Grant, B. S. 2002. Sour grapes of wrath. *Science* 297: 940–941. [12]

Grant, B. S., and L. L. Wiseman. 2002. Recent history of melanism in the American peppered moth. *J. Hered.* 93: 86–90. [12]

Grant, B. R., and P. R. Grant. 1989. *Evolutionary Dynamics of a Natural Population: The Large Cactus Finch of the Galápagos*. University of Chicago Press, Chicago. [13]

Grant, P. R. 1986. *Ecology and Evolution of Darwin's Finches*. Princeton University Press, Princeton, NJ. [9, 13, 11, 18]

Grant, V. 1966. The selective origin of incompatibility barriers in the plant genus *Gilia*. *Am. Nat.* 100: 99–118. [16]

Grant, V. 1981. *Plant Speciation*. Columbia University Press, New York. [15, 16]

Graur, D., and W.-H. Li. 2000. *Fundamentals of Molecular Evolution*, second editon. Sinauer Associates, Sunderland, MA. [10, 19]

Green, P. M., J. A. Naylor, and F. Giannelli. 1995. The hemophilias. *Adv. Genet.* 32: 99–139. [8]

Gregory, T. 2001. Coincidence, coevolution, or causation? DNA content, cell size, and the C-value enigma. *Biol. Rev.* 76: 65–101. [19]

Gregory, W. K. 1951. *Evolution Emerging*. Macmillan, New York. [21]

Grimaldi, D. A. 1987. Phylogenetics and taxonomy of *Zygothrica* (Diptera: Drosophilidae). *Bull. Am. Mus. Nat. Hist.* 186: 103–268. [3]

Gross, M. 1984. Sunfish, salmon, and the evolution of alternative reproductive strategies and tactics in fishes. In G. W. Potts and R. J. Wootton (eds.), *Fish Reproduction: Strategies and Tactics*, pp. 55–75. Academic Press, London. [17]

Gu, X., Y. Wang, and J. Gu. 2002. Age distribution of human gene families shows significant roles of both large- and small-scale duplications in vertebrate evolution. *Nature Genet.* 31: 205–209. [19]

Gupta, A. P., and R. C. Lewontin. 1982. A study of reaction norms in natural populations of *Drosophila pseudoobscura*. *Evolution* 36: 934–948. [9]

Guss, K. A., C. E. Nelson, A. Hudson, M. E. Kraus, and S. B. Carroll. 2001. Control of a genetic regulatory network by a selector gene. *Science* 292: 1164–1167. [20]

H

Haddrath, O., and A. J. Baker. 2001. Complete mitochondrial DNA genome sequences of extinct birds: ratite phylogenetics and the vicariance biogeography hypothesis. *Proc. R. Soc. Lond. B* 268: 939–945. [6]

Hafner, M. S., J. W. Demastes, T. A. Spradling, and D. L. Reed. 2003. Cophylogeny between pocket gophers and chewing lice. In R. D. M. Page (ed.), *Tangled Trees: Phylogeny, Cospeciation, and Coevolution*, pp. 195–220. University of Chicago Press, Chicago. [18]

Hahn, B. H., G. M. Shaw, K. M. De Cock, and P. M. Sharp. 2000. AIDS as a zoonosis: Scientific and public health implications. *Science* 287: 607–614. [22]

Haig, D. 1993. Genetic conflicts in human pregnancy. *Q. Rev. Biol.* 68: 495–532. [14]

Haig, D. 1997. Parental antagonism, relatedness asymmetries, and genomic imprinting. *Proc. R. Soc. Lond. B* 264: 1657–1662. [14]

Haldane, J. B. S. 1932. *The Causes of Evolution*. Longmans, Green, New York. [12]

Hall, B. G. 1982. Evolution on a petri dish: The evolved β-galactosidase system as a model for studying acquisitive evolution in the laboratory. *Evol. Biol.* 15: 85–150. [8]

Hall, B. G. 2001. *Phylogenetic Trees Made Easy. A How-To Manual for Molecular Biologists*. Sinauer Associates, Sunderland, MA. [2]

Hall, E. R., and K. R. Kelson. 1959. *The Mammals of North America*. Ronald Press, New York. [20]

Hamilton, W. D. 1964. The genetical evolution of social behavior, I and II. *J. Theor. Biol.* 7: 1–52. [11, 14]

Hamilton, W. D. 1967. Extraordinary sex ratios. *Science* 156: 477–488. [17]

Hamilton, W. D. 1971. Geometry for the selfish herd. *J. Theor. Biol.* 31: 295–311. [14]

Hamilton, W. D., and M. Zuk. 1982. Heritable true fitness and bright birds: A role for parasites? *Science* 218: 384–387. [14]

Hammer, M. F. 1995. A recent common ancestry for human Y chromosomes. *Nature* 378: 376–378. [10]

Hanken, J. 1984. Miniaturization and its effects on cranial morphology in plethodontid salamanders, genus *Thorius* (Amphibia: Plethodontidae). I. Osteological variation. *Biol. J. Linn. Soc.* 23: 55–75. [3]

Hansen, T. A. 1980. Influence of larval dispersal and geographic distribution on species longevity in neogastropods. *Paleobiology* 6: 193–207. [21]

Hansen, T. F., and D. Houle. 2004. Evolvability, stabilizing selection, and the problem of stasis. In M. Pigliucci and K. Preston (eds.), *Phenotypic Integration: Studying the Ecology and Evolution of Complex Phenotypes*, pp 131–150. Oxford University Press, Oxford. [13]

Harland, S. C. 1936. The genetical conception of the species. *Biol. Rev.* 11: 83–112. [15]

Harris, H. 1966. Enzyme polymorphisms in man. *Proc. R. Soc. Lond.* B 164: 298–310. [9]

Harris, H., and D. A. Hopkinson. 1972. Average heterozygosity in man. *J. Hum. Genet.* 36: 9–20. [9]

Harrison, R. G. 1979. Speciation in North American field crickets: Evidence from electrophoretic comparisons. *Evolution* 33: 1009–1023. [15]

Harrison, R. G. 1990. Hybrid zones: Windows on evolutionary process. *Oxford Surv. Evol. Biol.* 7: 69–128. Oxford University Press, New York. [15]

Harrison, R. G. (ed.). 1993. *Hybrid Zones and the Evolutionary Process*. Oxford University Press, New York. [15]

Harrison, R. G. 1998. Linking evolutionary pattern and process: the relevance of species concepts for the study of speciation. In D. J. Howard and S. H. Berlocher (eds.), *Endless Forms: Species and Speciation*, pp. 19–31. Oxford University Press, New York. [15]

Hartl, D. L., and A. G. Clark. 1989. *Principles of Population Genetics*, second edition. Sinauer Associates, Sunderland, MA. [9, 10, 12, 13]

Hartl, D. L., and A. G. Clark. 1997. *Principles of Population Genetics*, third edition. Sinauer Associates, Sunderland, MA. [9, 12]

Hartl, D. L., and E. W. Jones. 2001. *Genetics: Analysis of Genes and Genomes*. Jones and Bartlett, Sudbury, MA. [8, 9]

Hartwell, L. H., L. Hood, M. L. Goldberg, A. E. Reynolds, L. M. Silver, and R. C. Veres. 2000. *Genetics: From Genes to Genomes*. McGraw-Hill Higher Education, Boston. [8]

Harvey, P. H., and M. D. Pagel. 1991. *The Comparative Method in Evolutionary Biology*. Oxford University Press, Oxford. [11]

Hausfater, G., and S. Hrdy (eds.). 1984. *Infanticide: Comparative and Evolutionary Perspectives.* Aldine Publishing Co., New York. [14]

Haverschmidt, F. 1968. *Birds of Surinam*. Oliver & Boyd, London. [6]

Hayman, P., J. Marchant, and T. Prater. 1986. *Shorebirds: An Identification Guide to the Waders of the World*. Houghton Mifflin, Boston, MA. [3]

Hedrick, P. W. 1986. Genetic polymorphisms in heterogeneous environments: A decade later. *Annu. Rev. Ecol. Syst.* 17: 535–566. [12]

Hendry, A. P., and M. T. Kinnison. 1999. Perspective: The pace of modern life: Measuring rates of contemporary microevolution. *Evolution* 53: 1637–1653. [4]

Hennig, W. 1966. *Phylogenetic Systematics*. University of Illinois Press, Urbana. [2]

Herre, E. A., N. Knowlton, U. G. Mueller, and S. A. Rehner. 1999. The evolution of mutualism: Exploring the paths between conflict and cooperation. *Trends Ecol. Evol.* 14: 49–53. [18]

Herrera, C. M., and O. Pellmyr (eds.). 2002. *Plant–Animal Interactions: An Evolutionary Approach*. Blackwell Science, Malden, MA. [18]

Herrera, C. M., X. Cerdá, M. B. García, J. Guitán, M. Medrano, P. J. Rey and A. M. Sánche-Lafuente. 2002. Floral integration, phenotypic covariance structure and pollinator variation in bumblebee-pollinated *Helleborus foetidus*. *J. Evol. Biol.* 15: 108–121. [13]

Herrnstein, R. J., and C. Murray. 1994. *The Bell Curve: Intelligence and Class Structure in American Life*. The Free Press, New York. [9]

Hershkovitz, P. 1977. *Living New World Monkeys (Platyrrhini)*. University of Chicago Press, Chicago. [6]

Hewitt, G. M. 2000. The genetic legacy of the Quaternary ice ages. *Nature* 405: 907–913. [6]

Hill, G. E. 1991. Plumage coloration is a sexually selected indicator of male quality. *Nature* 350: 337–339. [14]

Hill, W. G., and A. Caballero. 1992. Artificial selection experiments. *Annu. Rev. Ecol. Syst.* 23: 287–310. [13]

Hillis, D. M. 1988. Systematics of the *Rana pipiens* complex: Puzzle and paradigm. *Annu. Rev. Ecol. Syst.* 19: 39–63. [16]

Hillis, D. M., J. J. Bull, M. E. White, M. R. Badgett, and I. J. Molineaux. 1992. Experimental phylogenetics: Generation of a known phylogeny. *Science* 255: 589–592. [2]

Hobbs, H. H., M. S. Brown, J. L. Goldstein, and D. W. Russell. 1986. Deletion of exon encoding cysteine-rich repeat of low density lipoprotein receptor alters its binding specificity in a subject with familial hypercholesterolemia. *J. Biol. Chem.* 261: 13114–13120. [8]

Höbel, G., and H. C. Gerhardt. 2003. Reproductive character displacement in the acoustic communication of green tree frogs (*Hyla cinerea*). *Evolution* 57: 894–904. [16]

Hofstadter, R. 1955. *Social Darwinism in American Thought.* Beacon Press, Boston, MA. [1, 22]

Holland, B. and W. R. Rice. 1998. Chase-away selection: Antagonistic seduction versus resistance. *Evolution* 52: 1–7. [14, 16]

Holland, B. and W. R. Rice. 1999. Experimental removal of sexual selection reverses intersexual antagonistic coevolution and removes a reproductive load. *Proc. Natl. Acad. Sci. USA* 96: 5083–5088. [14]

Hölldobler, B., and E. O. Wilson. 1983. The evolution of communal nest-weaving in ants. *Am. Sci.* 71: 490–499. [11]

Holsinger, K. E. 1991. Inbreeding depression and the evolution of plant mating systems. *Trends Ecol. Evol.* 6: 307–308. [17]

Holt, R. D. 1996. Demographic constraints in evolution: Towards unifying the evolutionary theories of senescence and niche conservatism. *Evol. Ecol.* 10: 1–11. [21]

Holt, R. D., and M. S. Gaines. 1992. Analysis of adaptation in heterogeneous landscapes: Implications for the evolution of fundamental niches. *Evol. Ecol.* 6: 433–447. [21]

Holt, S. B. 1955. Genetics of dermal ridges: Frequency distribution of total finger ridge count. *Ann. Hum. Genet.* 20: 270–281. [9]

Hooper, J. 2002. *Of Moths and Men: Intrigue, Tragedy and the Peppered Moth*. Fourth Estate, London. [12]

Horai, S., K. Hayasaka, R. Kondo, K. Tsugane, and N. Takahata. 1995. Recent African origin of humans revealed by complete sequences of hominoid mitochondrial DNAs. *Proc. Natl. Acad. Sci. USA* 92: 532–536. [10]

Hori, M. 1993. Frequency-dependent natural selection in the handedness of scale-eating cichlid fish. *Science* 260: 216–219. [12]

Hougen-Eitzman, D., and M. D. Rausher. 1994. Interactions between herbivorous insects and plant-insect coevolution. *Am. Nat.* 143: 677–697. [18]

Houle, D. 1989. The maintenance of polygenic variation in finite populations. *Evolution* 43: 1767–1780. [13]

Houle, D. 1992. Comparing evolvability and variability of quantitative traits. *Genetics* 130: 195–204. [13]

Houle, D., and A. S. Kondrashov. 2002. Coevolution of costly mate choice and condition–dependent display of good genes. *Proc. R. Soc. Lond. B* 269: 97–104. [14]

Houle, D., B. Morikawa, and M. Lynch. 1996. Comparing mutational variabilities. *Genetics* 143: 1467–1483. [13]

Howard, D. J. 1993. Reinforcement: Origin, dynamics, and fate of an evolutionary hypothesis. In R. G. Harrison (ed.), *Hybrid Zones and the Evolutionary Process*, pp. 46–69. Oxford University Press, New York. [16]

Howard, D. J. 1999. Conspecific sperm and pollen precedence and speciation. *Annu. Rev. Ecol. Syst.* 30: 109–132. [15]

Howard, D. J., and S. H. Berlocher (eds.). 1998. *Endless Forms: Species and Speciation.* Oxford University Press, New York. [15]

Howell, F. C. 1978. Hominidae. In V. J. Maglio and H. B. S. Cooke (eds.), *Evolution of African Mammals*, pp. 154–248. Harvard University Press, Cambridge, MA. [4]

Hubbard, L., P. McSteen, J. Doebley, and S. Hake. 2002. Expression patterns and mutant phenotype of *teosinte branched1* correlate with growth suppression in maize and teosinte. *Genetics* 162: 1927–1935. [20]

Hudson, R. R. 1990. Gene genealogies and the coalescent process. In D. J. Futuyma and J. Antonovics (eds.), *Oxford Surv. Evol. Biol.*, pp. 1–44. Oxford University Press, New York. [10]

Huelsenbeck, J. P., F. Ronquist, R. Nielsen, and J. P. Bollback. 2001. Bayesian inference of phylogeny and its impact on evolutionary biology. *Science* 294: 2310–2314. [2]

Hull, D. L. 1988. *Science as a Process*. University of Chicago Press, Chicago. [22]

Humphries, C. J., and L. R. Parenti. 1986. *Cladistic biogeography.* Clarendon Press, Oxford. [6]

Hurst, G. D. D., and J. H. Werren. 2001. The role of selfish genetic elements in eukaryotic evolution. *Nature Rev. Genet.* 2: 597–606. [11]

Hurst, L. D., A. Atlan, and B. D. Bengtsson. 1996. Genetic conflicts. *Q. Rev. Biol.* 7: 317–364. [14]

Husband, B. C. 2000. Constraints on polyploid evolution: A test of the minority cytotype exclusion principle. *Proc. R. Soc. Lond. B* 267: 217–223. [16]

Hutchinson, J. 1969. *Evolution and Phylogeny of Flowering Plants*. Academic Press, New York. [3]

I

Ingman, M., H. Kaessmann, S. Pääbo, and U. Gyllensten. 2000. Mitochondrial genome variation and the origin of modern humans. *Nature* 408: 708–713. [6, 10]

International Human Genome Sequencing Consortium. 2001. Initial sequencing and analysis of the human genome. *Nature* 409: 860–921. [3, 8, 19]

Ioerger, T. R., A. G. Clark, and T.-H. Kao. 1990. Polymorphism at the self-incompatibility locus in Solanaceae predates speciation. *Proc. Natl. Acad. Sci. USA* 87: 9732–9735. [12]

J

Jablonski, D. 1995. Extinctions in the fossil record. In J. H. Lawton and R. M. May (eds.), *Extinction Rates*, pp. 25–44. Oxford University Press, Oxford. [7]

Jablonski, D. 2002. Survival without recovery after mass extinctions. *Proc. Nat. Acad. Sci. USA* 99: 8139–8144. [7]

Jablonski, D., and D. J. Bottjer. 1990. Onshore-offshore trends in marine invertebrate evolution. In R. M. Ross and W. D. Allmon (eds.), *Causes of Evolution: A Paleontological Perspective*, pp 21–75. University of Chicago Press, Chicago. [7]

Jablonski, D., and R. A. Lutz. 1983. Larval ecology of marine benthic invertebrates: Paleobiological implications. *Biol. Rev.* 58: 21–89. [21]

Jablonski, D., and K. Roy. 2003. Geographical range and speciation in fossil and living molluscs. *Proc. R. Soc. Lond. B* 270: 401–406. [7]

Jablonski, D., S. J. Gould, and D. M. Raup. 1986. The nature of the fossil record: A biological perspective. In D. M. Raup and D. Jablonski (eds.), *Patterns and Processes in the History of Life*, pp. 7–22. Springer-Verlag, Berlin. [4]

Jablonski, D., D. H. Erwin, and J. H. Lipps (eds.). 1996. *Evolutionary Paleobiology: Essays in Honor of James W. Valentine*. University of Chicago Press, Chicago. [4]

Jackson, J. B. C. 1974. Biogeographic consequences of eurytopy and stenotopy among marine bivalves and their biogeographic significance. *Am. Nat.* 104: 541–560. [7]

Jacob, F. 1977. Evolution and tinkering. *Science* 196: 1161–1166. [20]

Jain, S. K., and D. R. Marshall. 1967. Population studies on predominantly self-pollinating species. X. Variation in natural populations of *Avena fatua* and *A. barbata*. *Am. Nat.* 101: 19–33. [9]

Janis, C. M. 1993. Tertiary mammal evolution in the context of changing climates, vegetation, and tectonic events. *Annu. Rev. Ecol. Syst.* 24: 467–500. [7]

Jansson, R., and M. Dynesius. 2002. The fate of clades in a world of recurrent climatic change: Milankovitch oscillations and evolution. *Annu. Rev. Ecol. Syst.* 33: 741–777. [21]

Jarne, P., and D. Charlesworth. 1993. The evolution of the selfing rate in functionally hermaphroditic plants and animals. *Annu. Rev. Ecol. Syst.* 24: 441–466. [17]

Jarvik, E. 1955. The oldest tetrapods and their forerunners. *Sci. Monthly* 80: 141–154. [4]

Jarvik, E. 1980. *Basic Structure and Evolution of Vertebrates*. Academic Press, London. [4]

Jensen, A. R. 1973. *Educability and Group Differences*. Harper & Row, New York. [9]

Johnstone, R. A. 1995. Sexual selection, honest advertisement and the handicap principle: Reviewing the evidence. *Biol. Rev.* 70: 1–65. [14]

Johnstone, R. A., and K. Norris. 1993. Badges of status and the cost of aggression. *Behav. Ecol. Sociobiol.* 32: 127–134. [14]

Jokela, J., and C. M. Lively. 1995. Parasites, sex, and early reproduction in a mixed population of freshwater snails. *Evolution* 49: 1268–1271. [17]

Jones, I. L., and F. M. Hunter. 1999. Experimental evidence for mutual inter- and intrasexual selection favouring a crested auklet ornament. *Animal Behaviour* 57: 521–528. [14]

Jones, R., and D. C. Culver. 1989. Evidence for selection on sensory structures in a cave population of *Gammarus minus* Say (Amphipoda). *Evolution* 43: 688–693. [13]

Jones, S., R. Martin, and D. Pilbeam (eds.). 1992. *The Cambridge Encyclopedia of Human Evolution*. Cambridge University Press, Cambridge. [4]

Jones, T. M., R. J. Quinnell, and A. Balmford. 1998. Fisherian flies: Benefits of female choice in a lekking sandfly. *Proc. R. Soc. Lond. B* 265: 1651–1657. [14]

Jordan, D. S., and B. W. Evermann. 1973. *The Shore Fishes of Hawaii.* Charles E. Tuttle, Rutland, VT. [3]

Jorde, L. B., M. Bamshad. and A. R. Rogers. 1998. Using mitochondrial and nuclear DNA markers to reconstruct human evolution. *BioEssays* 20: 126–136. [6]

Joron, M., and J. L. B. Mallet. 1998. Diversity in mimicry: Paradox or paradigm? *Trends Ecol. Evol.* 13: 461–466. [8, 18]

Joy, D. A., X.-R. Feng, J.-B. Mu, T. Furuya, K. Chotinavich, A. U. Krettli, M. Ho, A. Wang, N. J. White, E. Suh, P. Beerli, and X.-Z. Su. 2003. Early origin and recent expansion of *Plasmodium falciparum*. *Science* 300: 318–321. [12]

K

Kareiva, P. M., J. G. Kingsolver, and R. B. Huey (eds.). 1993. *Biotic Interactions and Global Change*. Sinauer Associates, Sunderland, MA. [5, 7]

Karn, M. N., and L. S. Penrose. 1951. Birth weight and gestation time in relation to maternal age, parity, and infant survival. *Ann. Eugenics* 16: 147–164. [13]

Kasahara, M. 1997. New insights into the genome organization and origin of the major histocompatibility complex: role of chromosomal (genome) duplication and emergence of the adaptive immune system. *Hereditas* 127: 59–66. [19]

Katakura, H., and T. Hosogai. 1994. Performance of hybrid ladybird beetles (*Epilachna*) on the host plants of parental species. *Entomol. Exp. Appl.* 71: 81–85. [15]

Kaufman, T. C., M. A. Seeger, and G. Olsen. 1990. Molecular and genetic organization of the Antennapedia gene complex of *Drosophila melanogaster*. *Adv. Genet.* 27: 309–362. [20]

Kawecki, T. J. 2000. The evolution of genetic canalization under fluctuating selection. *Evolution* 54: 1–12. [13]

Kazazian, H. H., Jr. 2004. Mobile elements: Drivers of genome evolution. *Science* 303: 1626–1632. [8]

Kearsey, M. J., and B. W. Barnes. 1970. Variation for metrical characters in *Drosophila* populations. II. Natural selection. *Heredity* 25: 11–21. [13]

Keller, L. (ed.) 1993. *Queen Number and Sociality in Insects*. Oxford University Press, Oxford. [14]

Keller, L. (ed.) 1999. *Levels of Selection in Evolution*. Princeton University Press, Princeton, NJ. [14]

Keller, L., and H. K. Reeve. 1994. Partitioning of reproduction in animal societies. *Trends Ecol. Evol.* 9: 98–102. [14]

Keller, L., and H. K. Reeve. 1999. Dynamics of conflicts within insect societies. In L. Keller (ed.), *Levels of Selection in Evolution*, pp. 153–175. Princeton University Press, Princeton, NJ. [14]

Kemp, T. S. 1982. *Mammal-like Reptiles and the Origin of Mammals*. Academic Press, London. [4]

Kenrick, P., and P. R. Crane. 1997a. The origin and early evolution of plants on land. *Nature* 389: 33–39. [5]

Kenrick, P., and P. R. Crane. 1997b. *The Origin and Early Diversification of Land Plants*. Smithsonian Institution Press, Washington. [5]

Kidston, R., and W. H. Lang. 1921. On Old Red Sandstone plants showing structure from the Rhynie chert bed, Aberdeenshire, Part IV. Restorations of the vascular cryptogams, and discussion of their bearing on the general morphology of Pteridophyta and the origin of the organization of land plants. *Trans. R. Soc. Edinburgh* 32: 477–487. [5]

Kimura, M. 1955. Solution of a process of random genetic drift with a continuous model. *Proc. Natl. Acad. Sci. USA* 41: 144–150. [10]

Kimura, M. 1968. Evolutionary rate at the molecular level. *Nature* 217: 624–626. [10]

Kimura, M. 1983. *The Neutral Theory of Molecular Evolution*. Cambridge University Press, Cambridge. [10]

King, J. L., and T. H. Jukes. 1969. Non-Darwinian evolution. *Science* 164: 788–798. [10]

King, M. 1993. *Species Evolution: The Role of Chromosome Change*. Cambridge University Press, Cambridge. [15]

King, M.-C., and A. C. Wilson. 1975. Evolution at two levels in humans and chimpanzees. *Science* 188: 107–116. [19]

Kingsolver, J. G., H. E. Hoekstra, J. M. Hoekstra, D. Berrigan, S. N. Vignieri, C. E. Hill, A. Hoang, P, Gilbert, and P. Beerli. 2001. The strength of phenotypic selection in natural populations. *Am. Nat.* 157: 245–261. [13]

Kirkpatrick, M. 1982. Sexual selection and the evolution of female choice. *Evolution* 3: 1–12. [14]

Kirkpatrick, M., and N. H. Barton. 1997. The strength of indirect selection on female mating preferences. *Proc. Nat. Acad. Sci. USA* 94: 1282–1286. [14]

Kirkpatrick, M., and M. R. Servedio. 1999. The reinforcement of mating preferences on an island. *Genetics* 151: 865–884. [16]

Kitcher, P. 1985. *Vaulting Ambition*. MIT Press, Cambridge, MA. [14]

Klein, J., and N. Takahata. 2002. *Where Do We Come From? The Molecular Evidence for Human Descent*. Springer-Verlag, Berlin. [6]

Klein, R. G. 2003. Whither the Neanderthals? *Science* 2999: 1525–1527. [6]

Klicka, J., and R. M. Zink. 1997.The importance of recent ice ages in speciation: A failed paradigm. *Science* 277: 1666–1669 [16]

Kliman, R. M., P. Andolfatto, J. A. Coyne, F. Depaulis, M. Kreitman, A. J. Berry, M. McCarter, J. Wakeley, and J. Hey. 2000. The population genetics of the origin and divergence of the *Drosophila simulans* complex species. *Genetics* 156: 1913–1931. [16]

Knoll, A. H. 2003. *Life on a Young Planet*. Princeton University Press, Princeton, NJ. [5]

Knoll, A. H., and R. K. Bambach. 2000. Directionality in the history of life: Diffusion from the left wall or repeated scaling of the right? In D. H. Erwin and S. L. Wing (eds.), *Deep Time: Paleobiology's Perspective*, pp. 1–14. The Paleontological Society, Allen Press, Lawrence, Kan. [21]

Knoll, A. H., and S. B. Carroll. 1999. Early animal evolution: emerging views from comparative biology and geology. *Science* 284: 2129–2137. [5]

Knowlton, N., L. A. Weigt, L. A. Solórzano, D. K. Mills, and E. Bermingham. 1993. Divergence in proteins, mitochondrial DNA, and reproductive compatibility across the Isthmus of Panama. *Science* 260: 1629–1632. [16]

Koehn, R. K., and J. J. Hilbish. 1987. The adaptive importance of genetic variation. *Am. Sci.* 75: 134–141. [12]

Kokko, H., R. Brooks, J. M. McNamara, and A. Houston. 2002. The sexual selection continuum. *Proc. R. Soc. Lond. B* 269: 1331–1340. [14]

Kondrashov, A. S. 1988. Deleterious mutations and the evolution of sexual reproduction. *Nature* 336: 435–441. [17]

Kondrashov, A. S. 1993. Classification of hypotheses on the advantage of amphimixis. *J. Hered.* 84: 372–387. [17]

Kondrashov, A. S., and F. A. Kondrashov. 1999. Interactions among quantitative traits in the course of sympatric speciation. *Nature* 400: 351–354. [16]

Kopp, A., I. Duncan, and S. B. Carroll. 2000. Genetic control and evolution of sexually dimorphic characters in *Drosophila. Nature* 408: 553–559. [20]

Kopp, A., R. M. Graze, S. Xu, S. B. Carroll, and S. V. Nuzhdin. 2003. Quantitative trait loci responsible for variation in sexually dimorphic traits in *Drosophila melanogaster. Genetics* 163: 771–787. [20]

Korber, B., M. Muldoon, J. Theiler, F. Gao, R. Gupta, A. Lapedes, B. H. Hahn, S. Wolinksy, and T. Bhattacharya. 2000. Timing the ancestor of the HIV-1 pandemic strains. *Science* 288: 1789–1796. [1, 22]

Kornegay, J. R., J. W. Schilling, and A. C. Wilson. 1994. Molecular adaptation of a leaf-eating bird: Stomach lysozyme of the hoatzin. *Mol. Biol. Evol.* 11: 921–928. [19]

Kornfield, I., and P. F. Smith. 2000. African cichlid fishes: Model systems for evolutionary biology. *Annu. Rev. Ecol. Syst.* 31: 163–196. [3]

Kotiaho, J. S., L. W. Simmons, and J. L. Tomkins. 2001. Towards a resolution of the lek paradox. *Nature* 410: 684–686. [14]

Koufopanou, V., and G. Bell. 1991. Developmental mutants of *Volvox*: Does mutation recreate patterns of phylogenetic diversity? *Evolution* 45: 1806–1822. [8]

Kreitman, M. 1983. Nucleotide polymorphism at the alcohol dehydrogenase locus of *Drosophila melanogaster*. *Nature* 304: 412–417. [9]

Kreitman, M., and R. R. Hudson. 1991. Inferring the evolutionary histories of the *Adh* and *Adh-dup* loci in *Drosophila melanogaster* from patterns of polymorphism and divergence. *Genetics* 127: 565–582. [12]

Kurtèn, B. 1963. Return of a lost structure in the evolution of the felid dentition. *Soc. Scient. Fenn. Comm. Biol.* 26: 1–12. [20]

L

Labandeira, C. C. 2002. The history of associations between plants and animals. In C. M. Herrera and O. Pellmyr (eds.), *Plant-Animal Interactions: An Evolutionary Approach*, pp. 26–74. Blackwell Science, London. [18]

Labandeira, C. C., and J. J. Sepkoski Jr. 1993. Insect diversity in the fossil record. *Science* 261: 310–315. [7]

Lack, D. 1954. *The Natural Regulation of Animal Numbers*. Oxford University Press, Oxford. [17]

Lai, C., R. F. Lyman, A. D. Long, C. H. Langley, and T. F. C. Mackay. 1994. Naturally occurring variation in bristle number and DNA polymorphisms at the *scabrous* locus of *Drosophila melanogaster*. *Science* 266: 1697–1702. [13, 20]

Lande, R. 1976a. The maintenance of genetic variability by mutation in a polygenic character with linked loci. *Genet. Res.* 26: 221–235. [13]

Lande, R. 1976b. Natural selection and random genetic drift in phenotypic evolution. *Evolution* 30: 314–334. [13]

Lande, R. 1979. Effective deme sizes during long-term evolution estimated from rates of chromosome rearrangement. *Evolution* 33: 234–251. [8]

Lande, R. 1980. Sexual dimorphism, sexual selection and adaptation in polygenic characters. *Evolution* 34: 292–305. [13]

Lande, R. 1981. Models of speciation by sexual selection on polygenic traits. *Proc. Natl. Acad. Sci. USA* 78: 3721–3725. [14, 16]

Lande, R. 1982. Rapid origin of sexual isolation and character divergence in a cline. *Evolution* 36: 213–223. [16]

Lande, R., and S. J. Arnold. 1983. The measurement of selection on correlated characters. *Evolution* 37: 1210–1226. [11, 13]

Lande, R., and D. W. Schemske. 1985. The evolution of self-fertilization and inbreeding depression. I. Genetic models. *Evolution* 39: 24–40. [17]

Langley, C. H., and W. M. Fitch. 1974. An examination of the constancy of the rate of molecular evolution. *J. Mol. Evol.* 3: 161–177. [2]

Latham, R. E., and R. E. Ricklefs. 1993. Continental comparisons of temperate-zone tree species diversity. In R. E. Ricklefs and D. Schluter (eds.), *Species Diversity in Ecological Communities*, pp. 294–314. University of Chicago Press, Chicago. [6]

Lauder, G. V. 1981. Form and function: Structural analysis in evolutionary morphology. *Paleobiology* 7: 430–442. [3]

Law, R., A. D. Bradshaw, and P. D. Putwain. 1979. The cost of reproduction in annual meadow grass. *Am. Nat.* 113: 3–16. [17]

Lawton, J. H., and R. M. May (eds.). 1995. *Extinction Rates*. Oxford University Press, Oxford. [7]

Lederberg, J., and E. M. Lederberg. 1952. Replica plating and indirect selection of bacterial mutants. *J. Bacteriol.* 63: 399–406. [8]

Lee, D. W., and J. H. Richards. 1991. Heteroblastic development in vines. In F. E. Putz and H. A. Mooney (eds.), *The Biology of Vines*, pp. 205–243. Cambridge University Press, Cambridge. [11]

Lee, M. S. Y., and R. Shine. 1998. Reptilian viviparity and Dollo's law. *Evolution* 52: 1441–1450. [21]

Lee, M. S. Y., T. W. Reeder, J. B. Slowinski, and R. Lawson. 2004. Resolving reptile relationships: Molecular and morphological markers. In J. Cracraft and M. J. Donoghue (eds.), *Assembling the Tree of Life*, pp. 451–467. Oxford University Press, New York. [5]

Leigh, E. G. Jr. 1973. The evolution of mutation rates. *Genetics* Suppl. 73: 1–18. [17]

Lenormand, T. 2002. Gene flow and the limits to natural selection. *Trends Ecol. Evol.* 17: 183–189. [12]

Lessios, H. A. 1998. The first stage of speciation as seen in organisms separated by the Isthmus of Panama. In D. J. Howard and S. H. Berlocher (eds.), *Endless Forms: Species and Speciation*, pp. 186–201. Oxford University Press, Oxford. [6]

Levin, B. R., and R. E. Michod. (eds.). 1988. *The Evolution of Sex: An Examination of Current Ideas*. Sinauer Associates, Sunderland, MA. [17]

Levin, D. A. 1979. The nature of plant species. *Science* 204: 381–384. [15]

Levin, D. A. 1983. Polyploidy and novelty in flowering plants. *Am. Nat.* 122: 1–25. [16]

Levin, D. A. 1984. Immigration in plants: An exercise in the subjunctive. In R. Dirzo and J. Sarukhán (eds.), *Perspectives on Plant Population Ecology*, pp. 242–260. Sinauer Associates, Sunderland, MA. [9]

Levinton, J. S. 2001. *Genetics, Paleontology, and Macroevolution*, second edition. Cambridge University Press, Cambridge. [4, 21]

Lewin, B. 1985. *Genes II*. Wiley, New York. [8]

Lewin, R. *Human Evolution: An Illustrated Introduction*. Blackwell Publishing, Oxford. [4]

Lewontin, R. C. 1974. *The Genetic Basis of Evolutionary Change*. Columbia University Press, New York. [9]

Lewontin, R. C. 2000. *The Triple Helix: Gene, Organism, and Environment*. Harvard University Press, Cambridge, MA. [21]

Lewontin, R. C., and J. L. Hubby. 1966. A molecular approach to the study of genic heterozygosity in natural populations. II. Amount of variation and degree of heterozygosity in natural populations of *Drosophila pseudoobscura*. *Genetics* 54: 595–609. [9, 10]

Lewontin, R. C., S. Rose, and L. Kamin. 1984. *Not in Our Genes: Biology, Ideology, and Human Nature*. Pantheon, New York. [9]

Li, W.-H. 1997. *Molecular Evolution*. Sinauer Associates, Sunderland, MA. [2, 8, 10]

Li, W.-H., and D. Graur. 1991. *Fundamentals of Molecular Evolution*. Sinauer Associates, Sunderland, MA. [2, 8]

Lieberman, B. S. 2003. Paleobiogeography: The relevance of fossils to biogeography. *Annu. Rev. Ecol. Evol. Syst.* 34: 51–69. [6]

Liou, L. W., and T. D. Price. 1994. Speciation by reinforcement of premating isolation. *Evolution* 48: 1451–1459. [16]

Lipps, J. H., and P. W. Signor III. 1992. *Origin and Early Evolution of the Metazoa*. Plenum, New York. [5]

Lively, C. M., and M. F. Dybdahl. 2000. Parasite adaptation to locally common host genotypes. *Nature* 405: 679–681. [17]

Lloyd, D. G. 1992. Evolutionarily stable strategies of reproduction in plants: Who benefits and how? In R. Wyatt (ed.), *Ecology and Evolution of Plant Reproduction*, pp. 137–168. Chapman & Hall, New York. [17]

Locascio, A., M. Manzanares, M. J. Blanco, and M. A. Nieto. 2002. Modularity and reshuffling of *snail* and *slug* expression during vertebrate evolution. *Proc. Natl. Acad. Sci. USA* 99: 16841–16846. [20]

Long, A. D., S. L. Mullaney, L. A. Reid, J. D. Fry, C. H. Langley, and T. F. C. Mackay. 1995. High resolution mapping of genetic factors affecting abdominal bristle number in *Drosophila melanogaster*. *Genetics* 139: 1273–1291. [13]

Long, A. D., R. F. Lyman, C. H. Langley, and T. F. C. Mackay. 1998. Two sites in the *Delta* gene region contribute to naturally occurring variation in bristle number in *Drosophila melanogaster. Genetics* 149: 999–1017. [20]

Long, A. D., R. F. Lyman, A. H. Morgan, C. H. Langley, and T. F. C. Mackay. 2000. Both naturally occurring inserts of transposable elements and intermediate frequency polymorphisms at the *achaete-scute* complex are associated with variation in bristle number in *Drosophila melanogaster. Genetics* 154: 1255–1269. [20]

Long, M., S. J. de Souza, C. Rosenberg, and W. Gilbert. 1996. Exon shuffling and the origin of the mitochondrial targeting function in plant cytochrome *c1* precursor. *Proc. Natl. Acad. Sci. USA* 93: 7727–7731. [19]

Long, M., E. Betran, K. Thornton, and W. Wang. 2003a. The origin of new genes: Glimpses of the young and old. *Nature Rev. Genet.* 4: 865–875. [19]

Long, M., M. Deutsch, W. Wang, E. Betran, F. G. Brunet, and J. Zhang. 2003b. Origin of new genes: Evidence from experimental and computational analyses. *Genetica* 118: 171–182. [19]

López, M. A., and C. López-Fanjul. 1993. Spontaneous mutation for a quantitative trait in *Drosophila melanogaster*. I. Response to artificial selection. *Genet. Res.* 61: 107–116. [13]

Losos, J. B. 1990. A phylogenetic analysis of character displacement in Caribbean *Anolis* lizards. *Evolution* 44: 558–569. [6]

Losos, J. B. 1992. The evolution of convergent structure in Caribbean *Anolis* communities. *Syst. Biol.* 41: 403–420. [6, 18]

Losos, J. B., T. R. Jackman, A. Larson, K. de Queiroz, and L. Rodríguez-Schettino. 1998. Contingency and determinism in replicated adaptive radiations of island lizards. *Science* 279: 2115–2118. [6, 18]

Lovejoy, C. O. 1981. The origins of man. *Science* 211: 341–350. [4]

Lucas, S. G. 1994. *Dinosaurs: The Textbook*. Wm. C. Brown, Dubuque, IA. [5]

Ludwig, M. Z. and M. Kreitman. 1995. Evolutionary dynamics of the enhancer region of *even-skipped* in *Drosophila. Mol. Biol. Evol.* 12: 1002–1011. [20]

Ludwig, M. Z., N. H. Patel, and M. Kreitman. 1998. Functional analysis of *eve stripe 2* enhancer evolution in *Drosophila*: Rules governing conservation and change. *Development* 125: 949–958. [20]

Ludwig, M. Z., C. Bergman, N. H. Patel, and M. Kreitman. 2000. Evidence for stabilizing selection in a eukaryotic enhancer element. *Nature* 403: 564–567. [20]

Luo, D., R. Carpenter, C. Vincent, L. Copsey, and E. Coen. 1996. Origin of floral asymmetry in *Antirrhinum*. *Nature* 383: 794-799. [21]

Luo, Z.-X., A. W. Crompton, and A.-L. Sun. 2001. A new mammaliaform from the early Jurassic and evolution of mammalian characteristics. *Science* 292: 1535–1540. [4]

Lupia, R., S. Lidgard, and P. R. Crane. 1999. Comparing palynological abundance and diversity: implications for biotic replacement during the Cretaceous angiosperm radiation. *Paleobiology* 25: 305–340. [7]

Lyman, R. F., F. Lawrence, S. V. Nuzhdin and T. F. C. Mackay. 1996. Effects of single *P*-element insertions on bristle number and viability in *Drosophila melanogaster*. Genetics 143: 277–292. [8]

Lynch, M. 1988. The rate of polygenic mutation. *Genet. Res.* 51: 137–148. [8, 13]

Lynch, M. 1990. The rate of morphological evolution in mammals from the standpoint of the neutral expectation. *Am. Nat.* 136: 727–741. [13]

Lynch, M., and J. S. Conery. 2000. The evolutionary fate and consequences of duplicate genes. *Science* 290: 1151–1155. [19]

Lynch, M., and J. S. Conery. 2003. The origins of genome complexity. *Science* 302: 1401–1404. [19]

Lynch, M., and R. Lande. 1993. Evolution and extinction in response to environmental change. In P. M. Kareiva, J. G. Kingsolver, and R. B. Huey (eds.), *Biotic Interactions and Global Change*, pp. 234–250. Sinauer Associates, Sunderland, MA. [7, 17]

Lynch, M., and J. B. Walsh. 1998. *Genetics and Analysis of Quantitative Traits*. Sinauer Associates, Sunderland, MA. [9, 13]

Lynch, M., R. Bürger, D. Butcher, and W. Gabriel. 1993. The mutational meltdown in an asexual population. *J. Hered.* 84: 339–344. [17]

Lynch, M., J. Blanchard, D. Houle, T. Kibota, S. Schultz, L. Vassilieva, and J. Willis. 1999. Perspective: Spontaneous deleterious mutation. *Evolution* 53: 645–663. [8]

Lyon, B. E., J. M. Eadie, and L. D. Hamilton. 1994. Parental choice selects for ornamental plumage in American coot chicks. *Nature* 371: 240–243. [14]

M

Ma, H. and C. dePamphilis. 2000. The ABCs of floral evolution. *Cell* 101: 5–8. [20]

MacArthur, R. H., and E. O. Wilson. 1967. *The Theory of Island Biogeography*. Princeton University Press, Princeton, NJ. [6]

MacFadden, B. J. 1986. Fossil horses from "Eohippus" (*Hyracotherium*) to *Equus*: Scaling, Cope's law, and the evolution of body size. *Paleobiology* 12: 355–369. [4, 21]

Mackay, T. F. C. 2001. The genetic architecture of quantitative traits. *Annu. Rev. Genet.* 35: 303–339. [13]

Mackay, T. F. C., R. F. Lyman, aand M. S. Jackson. 1992. Effects of P-element insertion on quantitative traits in *Drosophila melanogaster*. *Genetics* 130: 315–332. [8, 13]

Mackay, T. F. C., J. D. Fry, R. F. Lyman, and S. V. Nuzhdin. 1994. Polygenic mutation in *Drosophila melanogaster*: Estimates from response to selection in inbred strains. *Genetics* 136: 937–951. [13]

MacLeod, N. 1996. K/T redux. *Paleobiology* 22: 311–317. [7]

Macnair, M. R. 1981. Tolerance of higher plants to toxic materials. In J. A. Bishop and L. M. Cook (eds.), *Genetic Consequences of Man Made Change*, pp. 177–207. Academic Press, New York. [13]

Macnair, M. R., and P. Christie. 1983. Reproductive isolation as a pleiotropic effect of copper tolerance in *Mimulus guttatus*. *Heredity* 50: 295–302. [16]

Maddison, W. 1995. Phylogenetic histories within and among species. In P. C. Hoch and A. G. Stevenson (eds.), *Experimental and Molecular Approaches to Plant Biosystematics*, pp. 273–287. Monographs in Systematic Botany, 53. Missouri Botanical Garden, St. Louis. [2]

Maddison, W. P., and D. R. Maddison. 1992. *MacClade*, Version 3.0. Sinauer Associates, Sunderland, MA. [2]

Madsen, O., M. Scally, C. J. Douady, D. J. Kao, R. W. DeBry, R. Adkins, H. M. Amrine, M. J. Stanhope, W. W. de Jong, and M. S. Springer. 2001. Parallel adaptive radiations in two major clades of placental mammals. *Nature* 409: 610–614. [5]

Madsen, T., B. Stille, and R. Shine. 1995. Inbreeding depression in an isolated population of adders *Vipera berus*. *Biol. Conserv.* 75: 113–118. [9]

Madsen, T., R. Shine, M. Olsson, and H. Wittzell. 1999. Restoration of an inbred adder population. *Nature* 402: 34–35. [9]

Majerus, M. E. N. 1998. *Melanism: Evolution in Action*. Oxford University Press, Oxford. [12]

Makova, K. D., and W.-H. Li. 2002. Strong male-driven evolution of DNA sequences in humans and apes. *Nature* 416: 624–626. [8]

Mallet, J., and N. Barton. 1989. Strong natural selection in a warning-color hybrid zone. *Evolution* 43: 421–431. [12]

Mallet, J., and M. Joron. 1999. Evolution of diversity in warning color and mimicry: polymorphisms, shifting balance, and speciation. *Annu. Rev. Ecol. Syst.* 30: 201–233. [3, 8, 18]

Malmgren, B. A., W. A. Berggren, and G. P. Lohmann. 1983. Evidence for punctuated gradualism in the Late Neogene *Globorotalia tumida* lineage of planktonic Foraminifera. *Paleobiology* 9: 377–389. [4]

Margulis, L. 1993. *Symbiosis in Cell Evolution*. Second edition. W. H. Freeman, San Francisco. [5]

Martin, P. S., and R. G. Klein (eds.). 1984. *Quaternary Extinctions: A Prehistoric Revolution*. University of Arizona, Tucson. [5]

Martínez Wells, M., and C. S. Henry. 1992a. Behavioural responses of green lacewings (Neuroptera: Chrysopidae: *Chrysoperla*) to synthetic mating songs. *Anim. Behav.* 44: 641–652. [15]

Martínez Wells, M., and C. S. Henry. 1992b. The role of courtship songs in reproductive isolation among populations of green lacewings of the genus *Chrysoperla* (Neuroptera: Chrysopidae). *Evolution* 46: 31–43. [15]

Mather, K. 1949. *Biometrical Genetics: The Study of Continuous Variation*. Methuen, London. [8]

Matton, D. P., N. Nass, A. G. Clark, and E. Newbigin. 1994. Self-incompatibility: How plants avoid illegitimate offspring. *Proc. Natl. Acad. Sci. USA* 91: 1992–1997. [17]

May, R. M., and R. M. Anderson. 1983. Parasite-host coevolution. In D. J. Futuyma and M. Slatkin (eds.), *Coevolution*, pp. 186–206. Sinauer Associates, Sunderland, MA. [18]

Maynard Smith, J. 1978. *The Evolution of Sex*. Cambridge University Press, Cambridge. [17]

Maynard Smith, J. 1980. Selection for recombination in a polygenic model. *Genet. Res.* 35: 269–277. [17]

Maynard Smith, J. 1982. *Evolution and the Theory of Games*. Cambridge University Press, Cambridge. [14]

Maynard Smith, J. 1988. The evolution of recombination. In B. R. Levin and R. E. Michod (eds.), *The Evolution of Sex*, pp. 106–125. Sinauer Associates, Sunderland, MA. [17]

Maynard Smith, J., and E. Szathmáry. 1995. *The Major Transitions in Evolution*. W. H. Freeman, San Francisco. [5, 14, 21]

Maynard Smith, J., R. Burian, S. Kauffman, P. Alberch, et al. 1985. Developmental constraints and evolution. *Q. Rev. Biol.* 60: 265–287. [20]

Mayr, E. 1942. *Systematics and the Origin of Species*. Columbia University Press, New York. [1, 15, 21]

Mayr, E. 1954. Change of genetic environment and evolution. In J. Huxley, A. C. Hardy, and E. B. Ford (eds.), *Evolution as a Process*, pp. 157–180. Allen and Unwin, London. [4, 16]

Mayr, E. 1960. The emergence of evolutionary novelties. In S. Tax (ed.), *The Evolution of Life*, pp. 157–180. University of Chicago Press, Chicago. [21]

Mayr, E. 1963. *Animal Species and Evolution*. Harvard University Press, Cambridge, MA. [9, 15, 16]

Mayr, E. 1982a. *The Growth of Biological Thought: Diversity, Evolution, and Inheritance*. Harvard University Press, Cambridge, MA. [1]

Mayr, E. 1982b. Processes of speciation in animals. In C. Barigozzi (ed.), *Mechanisms of Speciation*, pp. 1–19. Alan R. Liss, New York. [16]

Mayr, E. 1988a. Cause and effect in biology. In E. Mayr (ed.), *Toward a New Philosophy of Biology*, pp. 24–37. Harvard University Press, Cambridge, MA. [11]

Mayr, E. 1988b. The probability of extraterrestrial intelligent life. In E. Mayr (ed.), *Toward a New Philosophy of Biology*, pp. 67–74. Harvard University Press, Cambridge, MA. [21]

Mayr, E. 2004. *What Makes Biology Unique? Considerations on the Autonomy of a Scientific Discipline*. Cambridge University Press, Cambridge.

Mayr, E., and W. B. Provine (eds.). 1980. *The Evolutionary Synthesis: Perspectives on the Unification of Biology*. Harvard University Press, Cambridge, MA. [1]

McCarty, D. R., and J. Chory. 2000. Conservation and innovation in plant signaling pathways. *Cell* 103: 201–209. [20]

McCauley, D. E. 1993. Evolution in metapopulations with frequent local extinction and recolonization. In D. J. Futuyma and J. Antonovics (eds.), *Oxford Surv. Evol. Biol.*, pp. 109–134. Oxford University Press, Oxford. [9]

McClearn, G. E., B. Johansson, S. Berg, N. L. Pedersen, F. Ahern, S. A. Petrill, and R. Plomin. 1997. Substantial genetic influence on cognitive abilities in twins 80 or more years old. *Science* 276: 1560–1563. [9]

McClintock, W. J., and G. W. Uetz. 1996. Female choice and pre-existing bias: visual cues during courtship in two *Schizocosa* wolf spiders (Araneae: Lycosidae). *Animal Behaviour* 52: 167–181. [14]

McCommas, S. A., and E. H. Bryant. 1990. Loss of electrophoretic variation in serially bottlenecked populations. *Heredity* 64: 315–321. [10]

McCune, A. R., and N. J. Lovejoy. 1998. The relative rate of sympatric and allopatric speciation in fishes: tests using DNA sequence divergence between sister species and among clades. In D. J. Howard and S. H. Berlocher (eds.), *Endless Forms: Species and Speciation*, pp. 172–185. Oxford University Press, New York. [15, 16]

McDonald, J. H., and M. Kreitman. 1991. Adaptive protein evolution at the *Adh* locus in *Drosophila*. *Nature* 351: 652–654. [10]

McGinnis, W., and R. Krumlauf. 1992. Homeobox genes and axial patterning. *Cell* 68: 283–302. [20]

McGrath, C. L., and L. A. Katz. 2004. Genome diversity in microbial eukaryotes. *Trends Ecol. Evol.* 19: 32–38. [5]

McKaye, K. R., T. Kocher, P. Reinthal, R. Harrison, and I. Kornfield. 1984. Genetic evidence for allopatric and sympatric differentiation among color morphs of a Lake Malawi cichlid fish. *Evolution* 38: 215–219. [16]

McKenzie, J. A., and G. M. Clarke. 1988. Diazinon resistance, fluctuating asymmetry and fitness in the Australian sheep blowfly, *Lucilia cuprina*. *Genetics* 120: 213–220. [13]

McKinney, M. L., and K. J. McNamara. 1991. *Heterochrony: The Evolution of Ontogeny*. Plenum, New York. [3]

McNamara, K. J. 1997. *Shapes of Time*. Johns Hopkins University Press, Baltimore. [3]

McNeilly, T., and J. Antonovics. 1968. Evolution in closely adjacent plant populations. IV. Barriers to gene flow. *Heredity* 23: 205–218. [16]

McShea, D. W. 1991. Complexity and evolution: What everybody knows. *Biol. Phil.* 6: 303–324. [21]

McShea, D. W. 1993. Evolutionary change in the morphological complexity of the mammalian vertebral column. *Evolution* 47: 730–740. [21]

McShea, D. W. 1994. Mechanisms of large-scale evolutionary trends. *Evolution* 48: 1747–1763. [21]

McShea, D. W. 1998. Possible largest–scale trends in organismal evolution: eight "live hypotheses." *Annu. Rev. Ecol. Syst.* 29: 293–318. [21]

McShea, D. W. 2001. The hierarchical structure of organisms: a scale and documentation of a trend in the maximum. *Paleobiology* 27: 405–423. [21]

Meagher, T. R., and D. Futuyma. 2001. Evolution, science, and society. *Am. Nat.* 158: 1–46. [1]

Medawar, P. B. 1952. *An Unsolved Problem of Biology*. H. K. Lewis, London. [17]

Meffert, L. M., and E. H. Bryant. 1991. Mating propensity and courtship behavior in serially bottlenecked lines of the housefly. *Evolution* 45: 293–306. [16]

Mehdiabadi, N. J., H. K. Reeve, and U. G. Mueller. 2003. Queens versus workers: sex–ratio conflict in eusocial Hymenoptera. *Trends Ecol. Evol.* 18: 88–93. [14]

Messier, W., and C.-B. Stewart. 1997. Episodic adaptive evolution of primate lysozymes. *Nature* 385: 151–154. [19]

Metcalf, R. L., and W. H. Luckmann (eds.). 1994. *Introduction to Insect Pest Management*, third edition. Wiley, New York. [12]

Meyer, A., and R. Zardoya. 2003. Recent advances in the (molecular) phylogeny of vertebrates. *Annu. Rev. Ecol. Evol. Syst.* 34: 311–338. [2]

Michod, R. E. 1982. The theory of kin selection. *Annu. Rev. Ecol. Syst.* 13: 23–55. [14]

Michod, R. E. 1997. *Evolution* of the individual. *Am. Nat.* 150: S5–S21. [14]

Milá, B., D. J. Girman, M. Kimura, and T. B. Smith. 2000. Genetic evidence for the effect of a postglacial population expansion on the phylogeography of a North American songbird. *Proc. R. Soc. Lond. B* 267: 1033–1040. [9]

Millais, J. G. 1897. *British Deer and Their Horns*. Henry Sotheran and Co., London. [3]

Miller, M. P. and S. Kumar. 2001. Understanding human disease mutations through the use of interspecific genetic variation. *Hum. Molec. Genet.* 21: 2319–2328. [22]

Miller, S. L. 1953. Production of amino acids under possible primitive earth conditions. *Science* 117: 528–529. [5]

Mindell, D. F., and C. E. Thacker. 1996. Rates of molecular evolution: phylogenetic issues and applications. *Annu. Rev. Ecol. Syst.* 27: 279–303. [2]

Mitchell-Olds, T. 1995. The molecular basis of quantitative genetic variation in natural populations. *Trends Ecol. Evol.* 10: 324–328. [12]

Mitter, C., B. Farrell, and B. Wiegmann. 1988. The phylogenetic study of adaptive zones: Has phytophagy promoted insect diversification? *Am. Nat.* 132: 107–128. [7, 21]

Mitra, S., H. Landel, and S. Pruett-Jones. 1996. Species richness covaries with mating systems in birds. *Auk* 113: 544–551. [16]

Mitton, J. B., and M. C. Grant. 1984. Associations among protein heterozygosity, growth rate, and developmental homeostasis. *Annu. Rev. Ecol. Syst.* 15: 479–499. [12]

Mochizuki, A., Y. Takeda, and Y. Iwasa. 1996. The evolution of genomic imprinting. *Genetics* 144: 1283–1295. [14]

Mock, D. W. 1984. Infanticide, siblicide,and avian nestlingt mortality. In G. Hausfater and S. B. Hrdy (eds.), *Infanticide: Comparative and Evolutionary Perspectives*, pp. 3–30. Aldine, New York. [14]

Moczek, A. P., J. Hunt, D. J. Emlen, and L. W. Simmons. 2002. Threshold evolution in exotic populations of a polyphenic beetle. *Evol. Ecol. Res.* 4: 587–601. [20]

Møller, A. P., and R. V. Alatalo. 1999. Good-genes effects in sexual selection. *Proc. R. Soc. Lond. B* 266: 85–91. [14]

Møller, A. P., and J. J. Cuervo. 1998. Speciation and feather ornamentation in birds. *Evolution* 52: 859–869. [16]

Montgomery, S. L. 1982. Biogeography of the moth genus *Eupithecia* in Oceania and the evolution of ambush predation in Hawaiian caterpillars (Lepidoptera: Geometridae). *Entomologia Generalis* 8: 27–34. [7]

Moore, F. B. G., D. E. Rozen, and R. E. Lenski. 2000. Pervasive compensatory adaptation in *Escherichia coli. Proc. R. Soc. Lond. B* 267: 515–522. [8]

Moore, R. C. (ed.) 1966. *Treatise on Invertebrate Paleontology, Part U, Echinodermata*. Geological Society of America and University of Kansas Press 3(2), Boulder, CO. [5]

Moore, W. S., and J. T. Price. 1993. Nature of selection in the northern flicker hybrid zone and its implications for speciation theory. In R. G. Harrison (ed.), *Hybrid Zones and the Evolutionary Process*, pp. 196–225. Oxford University Press, New York. [9]

Moran, N. 2002. Micobial minimalism: Genome reduction in bacterial pathogens. *Cell* 108: 583–586. [21]

Moran, N. 2003. Tracing the evolution of gene loss in obligate bacterial symbionts. *Curr. Opin. Microbiol.* 6: 512–518. [19]

Moran, N. A., and P. Baumann. 1994. Phylogenetics of cytoplasmically inherited microorganisms of arthropods. *Trends Ecol. Evol.* 9: 15–20. [18]

Moran, P., and I. Kornfield. 1993. Retention of an ancestral polymorphism in the mbuna species flock (Teleostei: Cichlidae) of Lake Malawi. *Mol. Biol. Evol.* 10: 1015–1029. [2, 15]

Morton, W. F., J. F. Crow, and H. J. Muller. 1956. An estimate of the mutational damage in man from data on consanguineous marriages. *Proc. Natl. Acad. Sci. USA* 42: 855–863. [9]

Mousseau, T. A., and D. A. Roff. 1987. Natural selection and the heritability of fitness components. *Heredity* 59: 181–197. [9, 13]

Moyle, P. B., and J. J. Cech Jr. 1983. *Fishes: An Introduction to Ichthyology*. Prentice-Hall, Englewood Cliffs, NJ. [6]

Moy-Thomas, J. A., and R. S. Miles. 1971. *Palaeozoic fishes*. Second edition. Chapman & Hall, London. [4]

Mukai, T., S. I. Chigusa, L. E. Mettler, and J. F. Crow. 1972. Mutation rate and dominance of genes affecting viability in *Drosophila melanogaster*. *Genetics* 72: 335–355. [8]

Müller, G. B. 1990. Developmental mechanisms at the origin of morphological novelty: A side-effect hypothesis. In M. H. Nitecki (ed.), *Evolutionary Innovations*, pp. 99–130. University of Chicago Press, Chicago. [3, 21]

Müller, G. B., and G. P. Wagner. 1991. Novelty in evolution: Restructuring the concept. *Annu. Rev. Ecol. Syst.* 23: 229–256. [21]

Müller, G. B., and G. P. Wagner. 1996. Homology, *Hox* genes, and developmental integration. *Am. Zool.* 36: 4–13. [3]

Muller, H. J. 1940. Bearing of the *Drosophila* work on systematics. In J. S. Huxley (ed.), *The New Systematics*, pp. 185–268. Clarendon Press, Oxford. [15, 16]

Muller, H. J. 1964. The relation of recombination to mutational advance. *Mutat. Res.* 1: 2–9. [17]

Müntzing, A. 1930. Über Chromosomenvermehrung in Galeopsis-Kreuzungen und ihre phylogenetische Bedeutung. *Hereditas* 14: 153–172. [16]

Myers, A. A., and P. S. Giller (eds.). 1988. *Analytical Biogeography*. Chapman & Hall, London. [6]

N

Nason, J. D., N. C. Ellstrand, and M. L. Arnold. 1992. Patterns of hybridization and introgression in populations of oaks, manzanitas, and irises. *Am. J. Bot.* 79: 101–111. [15]

Nei, M. 1987. *Molecular Evolutionary Genetics*. Columbia University Press, New York. [9]

Nei, M. 1995. Genetic support for the out of Africa theory of human evolution. *Proc. Natl. Acad. Sci. USA* 92: 6720–6722. [6]

Nei, M., and A. L. Hughes. 1991. Polymorphism and evolution of the major histocompatibility complex loci in mammals. In R. K. Selander, A. G. Clark, and T. S. Whittam (eds.), *Evolution at the Molecular Level*, pp. 222–247. Sinauer Associates, Sunderland, MA. [12]

Nei, M., and A. K. Roychoudhury. 1982. Genetic relationship and evolution of human races. *Evol. Biol.* 14: 1–59. [9]

Nei, M., T. Maruyama, and R. Chakraborty. 1975. The bottleneck effect and genetic variability in populations. *Evolution* 29: 1–10. [10]

Neigel, J. E., and J. C. Avise. 1986. Phylogenetic relationship of mitochondrial DNA under various demographic models of speciation. In E. Nevo and S. Karlin (eds.), *Evolutionary Processes and Theory*, pp. 515–534. Academic Press, London. [15]

Nesse, R., and G. C. Williams. 1994. *Why We Get Sick: The New Science of Darwinian Medicine*. Times Books, New York. [22]

Nestmann, E. R., and R. F. Hill. 1973. Population genetics in continuously growing mutator cultures of *Escherichia coli*. *Genetics* 73: 41–44. [11]

Nevo, E. 1991. Evolutionary theory and processes of active speciation and adaptive radiation in subterranean mole rats, *Spalax ehrenbergi* superspecies, in Israel. *Evol. Biol.* 25: 1–125. [15]

Niklas, K. J., B. H. Tiffney, and A. H. Knoll. 1983. Patterns in vascular land plant diversification. *Nature* 303: 614–616. [7]

Nilsson, D. E., and S. Pelger. 1994. A pessimistic estimate of the time required for an eye to evolve. *Proc. R. Soc. Lond. B* 256: 59–65. [21]

Nilsson, L. A. 1983. Processes of isolation and introgressive interplay between *Platanthera bifolia* (L.) Rich. and *P. chlorantha* (Custer) Reichb. (Orchidaceae). *Bot. J. Linn. Soc.* 87: 325–350. [15]

Nilsson, L. A., L. Jonsson, L. Ralison, and E. Randrianjohany. 1985. Monophily and pollination mechanisms in *Angraecum arachnites* Schltr. (Orchidaceae) in a guild of long-tongued hawkmoths (Sphingidae). *Biol. J. Linn. Soc.* 26: 1–19. [18]

Nisbett, R. 1995. Race, IQ, and scientism. In S. Fraser (ed.), *The Bell Curve Wars*, pp. 36–57. BasicBooks, New York. [9]

Nitecki, M. H. (ed.). 1988. *Evolutionary Progress*. University of Chicago Press, Chicago. [11]

Nitecki, M. H. (ed.). 1990. *Evolutionary Innovations*. University of Chicago Press, Chicago. [21]

Noble, G. K. 1931. *The Biology of the Amphibia*. McGraw-Hill, New York. [3]

Noor, M. A. F. 1995. Speciation driven by natural selection in *Drosophila*. *Nature* 375: 674–675. [16]

Noor, M. A. F. 1999. Reinforcement and other consequences of sympatry. *Heredity* 83: 503–508. [16]

Norell, M., Q. Ji, K. Gao, C. Yuan, Y. Zhao, and L. Wang. 2002. "Modern" feathers on a non-avian dinosaur. *Nature* 416: 36–37. [4]

Normark, B. B., O. P. Judson, and N. A. Moran. 2003. Genomic signatures of ancient asexual lineages. *Biol. J. Linn. Soc.* 79: 69–84. [11, 17]

Nosil, P. 2002. Transition rates between specialization and generalization. *Evolution* 56: 1701–1706. [21]

Nuzhdin, S. V., and T. F. C. Mackay. 1994. Direct determination of retrotransposon transposition rates in *Drosophila melanogaster*. *Genet. Res.* 63: 139–144. [8]

Nuzhdin, S. V., C. L. Dilda, and T. F. C. Mackay. 1999. The genetic architecture of selection response: inferences from fine-scale mapping of bristle number quantitative trait loci in *Drosophila melanogaster*. *Genetics* 153: 1317–1331. [13]

O

Oakeshott, J. G., J. B. Gibson, P. R. Anderson, W. R. Knib, D. G. Anderson, and G. K. Chambers. 1982. Alcohol dehydrogenase and glycerol-3-phosphate dehydrogenase clines in *Drosophila melanogaster* on different continents. *Evolution* 36: 86–96. [9]

Ochman, H., J. G. Lawrence, and E. A. Groisman. 2000. Lateral gene transfer and the nature of bacterial innovation. *Nature* 405: 299–304. [2, 19]

Odling-Smee, F. J., K. N. Laland, and M. W. Feldman. 2003. *Niche Construction: The Neglected Process in Evolution*. Princeton University Press, Princeton, NJ. [21]

Ohno, S. 1970. *Evolution by Gene Duplication*. Springer-Verlag, Berlin. [19]

Oliver, G., A. Mailhos, R. Wehr, N. G. Copeland, N. A. Jenkins, and P. Gruss. 1995. *Six3*, a murine homolog of the *sine oculis* gene, demarcates the most anterior border of the developing neural plate and is expressed during eye development. *Development* 121: 4045–4055. [20]

Olivieri, I., Y. Michalakis, and P.-H. Gouyon. 1995. Metapopulation genetics and the evolution of dispersal. *Am. Nat.* 146: 202–228. [17]

Olmstead, R. G., and J. A. Sweere. 1994. Combining data in phylogenetic systematics: an empirical approach using three molecular data sets in the Solanaceae. *Syst. Biol.* 43: 467–481. [2]

Olson, E. C., and R. L. Miller. 1958. *Morphological Integration*. University of Chicago Press, Chicago. [13]

Orgel, L. E. 1994. The origin of life on earth. *Sci. Am.* October: 77–91. [5]

Orr, H. A. 1995. The population genetics of speciation the evolution of hybrid incompatibilities. *Genetics* 139: 1805–1813. [16]

Orr, H. A. 1998. The population genetics of adaptation: the distribution of factors fixed during adaptive evolution. *Evolution* 52: 935–949. [13]

Orr, H. A., and J. Coyne. 1992. The genetics of adaptation: A reassessment. *Am. Nat.* 140: 725–742. [21]

Orr, H. A., and S. Irving. 2001. Complex epistasis and the genetic basis of hybrid sterility in the *Drosophila pseudoobscura* Bogotá-USA hybridization. *Genetics* 158: 1089–1100. [15]

Osborn, H. F. 1929. *The Titanotheres of Ancient Wyoming, Dakota, and Nebraska*. United States Geological Survey Monograph 55 vol. 1. U. S. Government Printing Office, Washington, D.C. [21]

Osorio, D. 1994. Eye evolution: Darwin's shudder stilled. *Trends Ecol. Evol.* 9: 241–242. [21]

Ostrom, J. H. 1976. On a new specimen of the Lower Cretaceous theropod dinosaur *Deinonychus antirrhopus*. *Breviora* 439: 1–21. [4]

Ota, T., and M. Nei. 1994. Divergent evolution by the birth-and-death process in the immunoglobulin V_H gene family. *Molec. Biol. Evol.* 11:469–482. [19]

Otto, S. P., and S. L. Nuismer. 2004. Species interactions and the evolution of sex. *Science* 304: 1018–1020. [17]

Ovchinnikov, I. V., A. Götherström, G. P. Romanova, V. M. Kharitonov, K. Lidén, and W. Goodwin. 2000. Molecular analysis of Neanderthal DNA from the northern Caucasus. *Nature* 404: 490–493. [6]

Ownbey, M. 1950. Natural hybridization and amphiploidy in the genus *Tragopogon*. *Am. J. Bot.* 37: 489–499. [16]

P

Page, R. D. M. 1994. Maps between trees and cladistic analyses of historical associations among genes, organisms, and areas. *Syst. Biol.* 43: 58–77. [6]

Page, R. D. M. 2002. Introduction. In R. D. M. Page (ed.), *Tangled Trees: Phylogeny, Cospeciation, and Coevolution*, pp. 1–21. University of Chicago Press, Chicago. [6, 18]

Palmer, A. R. 1982. Predation and parallel evolution: Recurrent parietal plate reduction in balanomorph barnacles. *Paleobiology* 8: 31–44. [4]

Palmer, J. D., and J. M. Logsdon, Jr. 1991. The recent origin of introns. *Curr. Opin. Genet. Dev.* 1: 470–477. [19]

Palumbi, S. R. 1998. Species formation and the evolution of gamete recognition loci. In D. J. Howard and S. H. Berlocher (eds.), *Endless Forms: Species and Speciation*, pp. 271–278. Oxford University Press, New York. [14, 15]

Palumbi, S. R. 2001. *The Evolution Explosion: How Humans Cause Rapid Evolutionary Change*. W. W. Norton, New York [1]

Pamilo, P., and M. Nei. 1988. Relationships between gene trees and species trees. *Mol. Biol. Evol.* 5: 568–583. [15]

Panganiban, G., S. M. Irvine, C. Lowe, H. Roehl, L. S. Corley, B. Sherbon, J. K. Grenier, J. F. Fallon, J. Kimble, M. Walker, G. A. Wray, B. J. Swalla, M. Q. Martindale, and S. B. Carroll. 1997. The origin and evolution of animal appendages. *Proc. Natl. Acad. Sci. USA* 94: 5162–5166. [20]

Panhuis, T. M., R. Butlin, M. Zuk, and T. Tregenza. 2001. Sexual selection and speciation. *Trends Ecol. Evol.* 16: 364–371. [16]

Paradis, J., and G. C. Williams. 1989. *Evolution and Ethics: T. H. Huxley's Evolution & Ethics with New Essays on its Victorian and Sociobiological Context*. Princeton University Press, Princeton, NJ. [1]

Parker, G. A. 1970. Sperm competition and its evolutionary consequences in the insects. *Biol. Rev.* 45: 525–567. [14]

Parker, G. A. 1979. Sexual selection and sexual conflict. In M. S. Blum and N. A. Blum (eds.), *Sexual Selection and Reproduction*, pp. 123–166. Academic Press, New York. [14]

Parmesan, C., and G. Yohe. 2003. A globally coherent fingerprint of climate change impacts across natural systems. *Nature* 421: 37–42. [7]

Parmesan, C., N. Ryrholm, C. Stefanescu, J. K. Hill, C. D. Thomas, H. Descimon, B. Huntley, L. Kaila, J. Kullberg, T. Tammaru, J. Tennent, J. A. Thomas, and M. Warren. 1999. Poleward shift of butterfly species' ranges associated with regional warming. *Nature* 399: 579–583. [7]

Partridge, L., and L. D. Hurst. 1998. Sex and conflict. *Science* 281: 2003–2008. [14]

Partridge, L., N. Prowse, and P. Pignatelli. 1999. Another set of responses and correlated responses to selection on age at reproduction in *Drosophila melanogaster*. *Proc. R. Soc. Lond. B* 266: 255–261. [17]

Paterson, H. E. H. 1985. The recognition concept of species. In E. S. Vrba (ed.), *Species and Speciation*, pp. 21–29. Transvaal Museum Monograph No. 4, Pretoria, South Africa. [15]

Patterson, C., D. M. Williams, and C. J. Humphries. 1993. Congruence between molecular and morphological phylogenies. *Annu. Rev. Ecol. Syst.* 24: 153–188. [2]

Patton, J. L., and S. Y. Yang. 1977. Genetic variation in *Thomomys bottate* pocket gophers: Macrogeographic patterns. *Evolution* 31: 697–720. [10]

Pellmyr, O., and C. J. Huth. 1994. Evolutionary stability of mutualism between yuccas and yucca moths. *Nature* 372: 257–260. [18]

Pellmyr, O., and J. Leebens-Mack. 1999. Forty million years of mutualism: evidence for Eocene origin of the yucca-yucca moth association. *Proc. Natl. Acad. Sci. USA* 96: 9178–9183. [18]

Pennock, R. T. 1999. *Tower of Babel: The Evidence against the New Creationism*. M.I.T. Press, Cambridge, MA. [22]

Pennock, R. T. 2001. *Intelligent Design and Its Critics: Philosophical, Theological, and Scientific Perspectives*. M.I.T. Press, Cambridge, MA. [22]

Peterson, A. T., J. Soberón, and V. Sánchez-Cordeiro. 1999. Conservatism of ecological niches in evolutionary time. *Science* 285: 1265–1267. [21]

Pfennig, D. W. 1997. Kinship and cannibalism. *BioScience* 47: 667–675. [14]

Pielou, E. C. 1991. *After the Ice Age: The Return of Life to Glaciated North America*. University of Chicago Press, Chicago. [5]

Pigliucci, M. 2002. *Denying Evolution: Creationism, Scientism, and the Nature of Science*. Sinauer, Sunderland, MA. [1, 22]

Pigliucci, M., and K. Preston (eds.). 2004. *Phenotypic Integration: Studying the Ecology and Evolution of Complex Phenotypes*. Oxford University Press, Oxford. [13]

Plomin, R., J. C. DeFries, G. E. McClearn, and M. Rutter. 1997. *Behavioral Genetics*, third edition. W. H. Freeman, New York. [9]

Pomiankowski, A. 1988. The evolution of female mate preferences for male genetic quality. *Oxford Surv. Evol. Biol.* 5: 136–184. [14]

Pomiankowski, A., and Y. Iwasa. 1998. Runaway ornament diversity caused by Fisherian sexual selection. *Proc. Natl. Acad. Sci. USA* 95: 5106–5111. [14, 16]

Poole, A. M., D. C. Jeffares, and D. Penny. 1998. The path from the RNA world. *J. Molec. Evol.* 46: 1–17. [19]

Porter, K. R. 1972. *Herpetology*. W. B. Saunders, Philadelphia, PA. [11]

Potts, R. 1988. *Early Hominid Activity at Olduvai*. Aldine de Gruyter, New York. [4]

Potts, W. K., C. J. Manning, and E. K. Wakeland. 1991. Mating patterns in seminatural populations of mice influenced by MHC genotype. *Nature* 352: 619–621. [17]

Presgraves, D. C., L. Balagopalan, S. M. Abmayr, and H. A. Orr. 2003. Adaptive evolution drives divergence of a hybrid inviability gene between two species of *Drosophila*. *Nature* 423: 715–719. [15]

Price, T., M. Turelli, and M. Slatkin. 1993. Peak shifts produced by correlated responses to selection. *Evolution* 47: 280–290. [16]

Prince, V. E. 2002. The Hox paradox: More complex(es) than imagined. *Dev. Biol.* 249: 1–15. [19]

Prince, V. E., and F. B. Pickett. 2002. Splitting pairs: The diverging fates of duplicated genes. *Nature Rev. Genet.* 3: 827–837. [19]

Promislow, D. E. L., and P. H. Harvey. 1991. Mortality rates and the evolution of mammalian life histories. *Acta Oecologica* 220: 417–437. [17]

Prum, R. O. 2003. Dinosaurs take to the air. *Nature* 421: 323–324. [4]

Purugganan, M. D. 1997. The MADS-box floral homeotic gene lineages predate the origin of seed plants: Phylogenetic and molecular clock estimates. *J. Mol. Evol.* 45: 392–396. [20]

Pyke, G. 1982. Local geographic distribution of bumblebees near Crested Butte, Colorado: competition and community structure. *Ecology* 63: 555–573. [18]

Q

Quattrocchio, F., J. Wing, K. Van der Woude, E. Souer, N. van Netten, J. Mol, and R. Koes. 1999. Molecular analysis of the *ANTHOCYANIN2* gene of petunia and its role in the evolution of flower color. *Plant Cell* 11: 1433–1444. [8]

Queller, D. C., F. Zacchi, R. Cervo, S. Turillazzi, M. T. Henshaw, L. A. Santorelli, and J. E. Strassmann. 2000. Unrelated helpers in a social insect. *Nature* 405: 784–787. [14]

R

Radinsky, L. B. 1984. Ontogeny and phylogeny in horse skull evolution. *Evolution* 38: 1–15. [20]

Raff, R. A. 1996. *The Shape of Life: Genes, Development, and the Evolution of Animal Form*. University of Chicago Press, Chicago. [3, 20]

Ralls, K., P. H. Harvey, and A. M. Lyles. 1986. Inbreeding in natural populations of birds and mammals. In M. E. Soulé (ed.), *Conservation Biology: The Science of Scarcity and Diversity*, pp. 35–56. Sinauer Associates, Sunderland, MA. [17]

Ramsey, J., and D. W. Schemske. 1998. Pathways, mechanisms, and rates of polyploid formation in flowering plants. *Annu. Rev. Ecol. Syst.* 29: 467–501. [8, 16]

Rasmussen, S., L. Chen, D. Deamer, D. . Krakauer, N. H. Packard, P. F. Stadler, and M. A. Bedau. 2004. Transitions from nonliving to living matter. *Science* 303: 963–965. [5]

Raup, D. M. 1972. Taxonomic diversity during the Phanerozoic. *Science* 177: 1065–1071. [7]

Raup, D. M., and J. J. Sepkoski Jr. 1982. Mass extinctions in the marine fossil record. *Science* 215: 1501–1503. [7]

Raven, P. H. 1976. Systematics and plant population biology. *Syst. Bot.* 1: 284–316. [15]

Raxworthy, C. J., M. R. J. Forstner, and R. A. Nussbaum. 2002. Chamaeleon radiation by oceanic dispersal. *Nature* 415: 784–787. [6]

Razin, S. 1997. The minimal cellular genome of mycoplasma. *Indian J. Biochem. Biophys.* 34: 124–130. [19]

Reeve, H. K., and L. Keller. 1999. Levels of selection: burying the units-of-selection debate and unearthing the crucial new issues. In L. Keller (ed.), *Levels of Selection in Evolution*, pp. 3–14. Princeton University Press, Princeton, NJ. [14]

Reeve, H. K., and P. W. Sherman. 1993. Adaptation and the goals of evolutionary research. *Q. Rev. Biol.* 68: 1–32. [11]

Relethford, J. H. 2001. *Genetics and the Search for Modern Human Origins*. Wiley-Liss, New York. [6]

Rendel, J. M., B. L. Sheldon, and D. E. Finlay. 1966. Selection for canalization of the scute phenotype. II. *Am. Nat.* 100: 13–31. [13]

Rensch, B. 1959. *Evolution Above the Species Level*. Columbia University Press, New York. [1, 3, 21]

Reyer, H.-U. 1990. Pied kingfishers: Ecological causes and reproductive consequences of cooperative breeding. In P. B. Stacey and W. D. Koenig (eds.), *Cooperative Breeding in Birds*, pp. 527–587. Cambridge University Press, Cambridge. [14]

Reznick, D. 1985. Cost of reproduction: An evaluation of the empirical evidence. *Oikos* 44: 257–267. [17]

Reznick, D., and J. Travis. 2002. Adaptation. In C.W. Fox, D. A. Roff, and D. J. Fairbairn (eds.), *Evolutionary Ecology: Concepts and Case Studies*, pp. 44–57. Oxford University Press, New York. [17]

Reznick, D., H. Bryga, and J. A. Endler. 1990. Experimentally induced life-history evolution in a natural population. *Nature* 346: 357–359. [17]

Rice, W. R. 1992. Sexually antagonistic genes: experimental evidence. *Science* 256: 1436–1439. [14]

Rice, W. R. 1996. Sexually antagonistic male adaptation triggered by experimental arrest of female evolution. *Nature* 381: 232–234. [14]

Rice, W. R., and B. Holland. 1997. The enemies within: Intergenomic conflict, interlocus contest evolution (ICE), and the intraspecific Red Queen. *Behav. Ecol. Sociobiol.* 41: 1–10. [14]

Rice, W. R., and E. E. Hostert. 1993. Laboratory experiments on speciation: What have we learned in forty years? *Evolution* 47: 1637–1653. [16]

Rice, W. R., and G. W. Salt. 1990. The evolution of reproductive isolation as a correlated character under sympatric conditions: Experimental evidence. *Evolution* 44: 1140–1152. [16]

Richardson, M. K., J. Hanken, L. Selwood, G. M. Wright, R. J. Richards, C. Pieau, and A. Raynaud. 1998. Haeckel, embryos, and evolution. *Science* 280: 983–984. [3]

Ricklefs, R. E., and R. E. Latham. 1992. Intercontinental correlation of geographical ranges suggests stasis in ecological traits of relict genera of temperate perennial herbs. *Am. Nat.* 139: 1305–1321. [21]

Ricklefs, R. E., and D. Schluter (eds.). 1993. *Species Diversity in Ecological Communities*. University of Chicago Press, Chicago. [6, 7]

Rico, P., P. Bouteillon, M. J. H. van Oppen, M. E. Knight, G. M. Hewitt, and G. F. Turner. 2003. No evidence for parallel sympatric speciation in cichlid species of the genus *Pseudotropheus* from northwestern Lake Malawi. *J. Evol. Biol.* 16: 37–46. [16]

Ridley, M. 1983. *The Explanation of Organic Diversity: The Comparative Method and Adaptations for Mating*. Oxford University Press, Oxford. [11]

Riedl, R. 1978. *Order in Living Organisms: A Systems Analysis of Evolution*. Wiley, New York. [4, 20, 21]

Rieseberg, L. H. 1997. Hybrid origins of plant species. *Annu. Rev. Ecol. Syst.* 28: 359–389. [16]

Rieseberg, L. H. 2001. Chromosomal arrangements and speciation. *Trends Ecol. Evol.* 16: 351–358. [15]

Rieseberg, L. H., and J. F. Wendel. 1993. Introgression and its consequences in plants. In R. G. Harrison (ed.), *Hybrid Zones and the Evolutionary Process*, pp. 70–109. Oxford University Press, New York. [16]

Rieseberg, L. H., J. Whitten, and K. Gardner. 1999. Hybrid zones and the genetic architecture of a barrier to gene flow between two sunflower species. *Genetics* 152: 713–727. [15]

Rieseberg, L. H., D. M. Raymond, Z. Lai, K. Livingstone, J. L. Durphy, A. E. Schwarzbach, L. A. Donovan, and C. Lexer. 2003. Major ecological transitions in wild sunflowers facilitated by hybridization. *Science* 301: 1211–1216. [16]

Ritchie, M. G. 2000. The inheritance of female preference functions in a mate recognition system. *Proc. R. Soc. Lond. B* 267: 327–332. [16]

Ritchie, M. G., and S. D. F. Phillips. 1998. The genetics of sexual isolation. In D. J. Howard and S. H. Berlocher (eds.), *Endless Forms: Species and Speciation*, pp. 291–308. Oxford University Press, New York. [15]

Roberts, G. C., and C. W. Smith. 2002. Alternative splicing: Combinatorial output from the genome. *Curr. Opin. Chem. Biol.* 6: 375–383. [19]

Roberts, R. G., T. F. Flannery, L. K. Ayliffe, and 8 others. 2001. New ages for the last Australian megafauna: Continent-wide extinction about 46,000 years ago. *Science* 292: 1888–1892. [5]

Rodríguez, D. J. 1996. A model for the establishment of polyploidy in plants. *Am. Nat.* 147: 33–46. [16]

Roelofs, W. L., T. J. Glover, X.-H. Tang, I. Sreng, P. S. Robbins, E. E. Eckenrode, C. Löfstedt, B. S. Hansson, and B. O. Bengtsson. 1987. Sex pheromone production and perception in European corn borer moths is determined by both autosomal and sex-linked genes. *Proc. Natl. Acad. Sci. USA* 84: 7585–7589. [15]

Roff, D. A. 2002. *Life History Evolution*. Sinauer Associates, Sunderland, MA. [17]

Rogers, A. R., and H. Harpending. 1992. Population growth makes waves in the distribution of pairwise differences. *Mol. Biol. Evol.* 9: 552–569. [10]

Romer, A. S. 1966. *Vertebrate Paleontology*. University of Chicago Press, Chicago. [3, 5]

Romer, A. S., and T. S. Parsons. 1986. *The Vertebrate Body*. Saunders College Publishing, Philadelphia, PA. [3, 5]

Ronquist, F. 1997. Dispersal-vicariance analysis: A new approach to the quantification of historical biogeography. *Syst. Biol.* 46: 195–203. [6]

Root, T., J. T. Price, K. R. Hall, S. Schnider, C. Rosenzweig, and J. A. Pounds. 2003. Fingerprints of global warming on wild animals and plants. *Nature* 421: 57–60. [7]

Rose, M. R. 1991. *The Evolutionary Biology of Aging*. Oxford University Press, Oxford. [17]

Rosenberg, N. A., J. K. Pritchard, J. L. Webert, H. M. Cann, K. K. Kidd, L. A. Zhivotovsky, and M. W. Feldman. 2002. Genetic structure of human populations. *Science* 298: 2381–2385. [9]

Rosenzweig, M. L., and R. D. McCord. 1991. Incumbent replacement: Evidence for long-term evolutionary progress. *Paleobiology* 17: 202–213. [7]

Roth, V. L. 1988. The biological basis of homology. In C. J. Humphries (ed.), *Ontogeny and Systematics*, pp. 1–26. British Museum (Natural History). [20]

Roth, V. L. 1991 Homology and hierarchies: Problems solved and unresolved. *J. Evol. Biol.* 4: 167–194. [20]

Rothstein, S. I., and S. K. Robinson (eds.). 1998. *Parasitic Birds and Their Hosts: Studies in Coevolution*. Oxford University Press, New York. [18]

Roughgarden, J. 1971. Density-dependent natural selection. *Ecology* 52: 453–468. [17]

Roughgarden, J. 2004. *Evolution's Rainbow: Diversity, Gender, and Sexuality in Nature and People*. University of California Press, Berkeley. [17]

Roush, R. T., and J. A. McKenzie. 1987. Ecological genetics of insecticide and acaricide resistance. *Annu. Rev. Entomol.* 32: 361–380. [12]

Roush, R. T., and B. E. Tabashnik (eds.). 1990. *Pesticide Resistance in Arthropods*. Chapman and Hall, New York. [12]

Rowe, L. E., and D. Houle. 1996. The lek paradox and the capture of genetic variance by condition dependent traits. *Proc. R. Soc. Lond. B* 263: 1415–1421. [14]

Rundle, H. D. 2003. Divergent environments and population bottlenecks fail to generate premating isolation in *Drosophila pseudoobscura*. *Evolution* 57: 2557–2565. [16]

Rundle, H. D., A. O. Mooers, and M. C. Whitlock. 1998. Single founder-flush events and the evolution of reproductive isolation. *Evolution* 52: 1850–1855. [16]

Rundle, H. D., L. Nagel, J. W. Boughman, and D. Schluter. 2000. Natural selection and parallel speciation in sympatric sticklebacks. *Science* 287: 306–308. [17]

Ruse, M. 1979. *The Darwinian Revolution*. University of Chicago Press, Chicago. [11]

Ruse, M. 1996. *Monad to Man: The Concept of Progress in Evolutionary Biology*. Harvard University Press, Cambridge, MA. [11, 21]

Rutherford, S. L., and S. Lindquist. 1998. Hsp90 as a capacitor for morphological evolution. *Nature* 396: 336–342. [21]

Rutledge, R., S. Regan, O. Nicolas, P. Fobert et al. 1998. Characterization of an *AGAMOUS* homologue from the conifer black spruce (*Picea mariana*) that produces floral homeotic conversions when expressed in *Arabidopsis*. *Plant J.* 15: 625–634. [20]

Ruvolo, M. 1997. Molecular phylogeny of the hominoids: Inferences from multiple independent DNA sequence data sets. *Mol. Biol. Evol.* 14: 248–265. [2]

Ryan, M. J. 1985. *The Túngara Frog: A Study in Sexual Selection and Communication*. University of Chicago Press, Chicago. [14]

Ryan, M. J. 1998. Sexual selection, receiver biases, and the evolution of sex differences. *Science* 281: 1999–2003. [14]

Ryan, M. L., and A. S. Rand. 1993. Species recognition and sexual selection as a unitary problem in animal communication. *Evolution* 47: 647–657. [16]

S

Saccone, C., and G. Pesole. 2003. *Handbook of Comparative Genomics: Principles and Methodology*. Wiley-Liss, Hoboken, NJ. [21]

Salvini-Plawen, L. V., and E. Mayr. 1977. On the evolution of photoreceptors and eyes. *Evol. Biol.* 10: 207–263. [21]

Salzberg, S. L., O. White, J. Peterson, and J. A. Eisen. 2001. Microbial genes in the human genome: Lateral transfer or gene loss? *Science* 292: 1903–1906. [19]

Sanderson, N. 1989. Can gene flow prevent reinforcement? *Evolution* 43: 1223–1235. [16]

Sang, T., and Y. Zhong. 2000. Testing hybridization hypotheses based on incongruent gene trees. *Syst. Biol.* 49: 422–434. [2]

Sanmartín, I., H. Enghoff, and F. Ronquist. 2001. Patterns of animal dispersal, vicariance and diversification in the Holarctic. *Biol. J. Linn. Soc.* 73: 345–390. [6]

Santos, F. R., A. Pandya, C. Tyler-Smith, S. D. J. Pena, M. Schanfield, W. R. Leonard, L. Osipova, M. H. Crawford, and R. J. Mitchell. 1999. The central Siberian origin for Native American Y chromosomes. *Am. J. Hum. Genet.* 64: 619–628. [6]

Sasa, M., P. T. Chippendale, and N. A. Johnson. 1998. Patterns of postzygotic isolation in frogs. *Evolution* 52: 1811–1820. [16]

Scharloo, W. 1991. Canalization: Genetic and developmental aspects. *Annu. Rev. Ecol. Syst.* 22: 65–93. [13]

Schemske, D. W., and H. D. Bradshaw, Jr. 1999. Pollinator preference and the evolution of floral traits in monkeyflkowers (*Mimulus*). *Proc. Natl. Acad. Sci. USA* 96: 11910–11915. [15]

Schindewolf, O. H. 1950. *Grundfrage der Paläontologie*. Schweitzerbart, Jena, Germany. [21]

Schlichting, C. D., and M. Pigliucci. 1998. *Phenotypic Evolution: A Reaction Norm Perspective*. Sinauer Associates, Sunderland, MA. [13]

Schliewen, U. K., D. Tautz, and S. Pääbo. 1994. Sympatric speciation suggested by monophyly of crater lake cichlids. *Nature* 368: 629–632. [16]

Schluter, D. 1996. Adaptive radiation along genetic lines of least resistance. *Evolution* 50: 1766–1774. [13]

Schluter, D. 2000. *The Ecology of Adaptive Radiation*. Oxford University Press, Oxford. [3, 13, 16, 18]

Schluter, D. 2001. Ecology and the origin of species. *Trends Ecol. Evol.* 16: 372–380. [18]

Schluter, D., and J. D. McPhail. 1992. Ecological character displacement and speciation in sticklebacks. *Am. Nat.* 140: 85–108. [18]

Schluter, D., and L. M. Nagel. 1995. Parallel speciation by natural selection. *Am. Nat.* 146: 292–301. [16]

Schluter, D., and R. E. Ricklefs. 1993. Convergence and the regional component of species diversity. In R. E. Ricklefs and D. Schluter (eds.), *Species Diversity in Ecological Communities*, pp. 230–240. University of Chicago Press, Chicago. [18]

Schoener, T. W. 1984. Size differences among sympatric, bird-eating hawks: A worldwide survey. In D. R. Strong, Jr., D. Simberloff, L. G. Abele, and A. B. Thistle (eds.), *Ecological Communities: Conceptual Issues and the Evidence*, pp. 254–281. Princeton University Press, Princeton, NJ. [18]

Schwenk, K. 1993. The evolution of chemoreception in squamate reptiles: A phylogenetic approach. *Brain Behav. Evol.* 41: 124–137. [21]

Scott, E. C. 1997. Antievolution and creationism in the United States. *Annu. Rev. Anthropol.* 26: 263–289. [22]

Scott, E. C. 2004. *Evolution versus Creationism: An Introduction*. Greenwood Press, Westport, CT. [22]

Searle, J. B. 1993. Chromosomal hybrid zones in eutherian mammals. In R. G. Harrison (ed.), *Hybrid Zones and the Evolutionary Process*, pp. 309–353. Oxford University Press, New York. [17]

Seehausen, O., P. J. Mayhew, and J. J. van Alphen. 1999. *Evolution* of colour patterns in East African cichlid fish. *J. Evol. Biol.* 12: 514–534. [16]

Seger, J. 1992. Evolution of exploiter-victim relationships. In M. J. Crawley (ed.), *Natural Enemies: The Population Biology of Predators, Parasites, and Disease*, pp. 3–25. Blackwell Scientific, Oxford. [18]

Selander, R. K. 1966. Sexual dimorphism and differential niche utilization in birds. *Condor* 68: 113–151. [18]

Selander, R. K. 1970. Behavior and genetic variation in natural populations. *Am. Zool.* 10: 53–66. [10]

Sepkoski, J. J. Jr. 1984. A kinetic model of Phanerozoic taxonomic diversity. III. Post-Paleozoic families and mass extinctions. *Paleobiology* 10: 246–267. [7]

Sepkoski, J. J. Jr. 1993. Ten years in the library: New data confirm paleontological patterns. *Paleobiology* 19: 43–51. [7]

Sepkoski, J. J. Jr. 1996a. Competition in macroevolution: The double wedge revisited. In D. Jablonski, D. H. Erwin, and J. Lipps (eds.), *Evolutionary Paleobiology*, pp. 211–255. University of Chicago Press, Chicago. [7]

Sepkoski, J. J. Jr. 1996b. Large-scale history of biodiversity. In V. H. Heywood (ed.), *Global Biodiversity Assessment*, pp. 202–212. United Nations Environmental Programme. Cambridge University Press, Cambridge. [7]

Sereno, P. C. 1999. The evolution of dinosaurs. *Science* 284: 2137–2147. [4, 5]

Sgrò, C. M., and L. Partridge. 1999. A delayed wave of death from reproduction in *Drosophila*. *Science* 286: 2521–2524. [17]

Shabalina, S. A., L. Y. Yampolsky, and A. S. Kondrashov. 1997. Rapid decline of fitness in panmictic populations of *Drosophila melanogaster* maintained under relaxed natural selection. *Proc. Natl. Acad. Sci. USA* 94: 13034–13039. [8]

Shabalina, S. A., A. Y. Ogurtsov, V. A. Kondrashov, and A. S. Kondrashov. 2001. Selective constraint in intergenic regions of human and mouse genomes. *Trends Genet* 17: 373–376. [19]

Shaffer, H. B. and S. R. Voss. 1996. Phylogenetic and mechanistic analysis of a developmentally integrated character complex: Alternate life history modes in ambystomatid salamanders. *Am. Zool.* 36: 24–35. [20]

Shaver, A. C., P. G. Dombrowski, J. Y. Sweeney, T. Treis, R. M. Zappala, and P. D. Sniegowski. 2002. Fitness evolution and the rise of mutator alleles in experimental *Escherichia coli* populations. *Genetics* 162: 557–566. [17]

Shaw, K. L. 1995. Biogeographic patterns of two independent Hawaiian cricket radiations (*Laupala* and *Prognathogryllus*). In W. L. Wagner and V. A. Funk (eds.), *Hawaiian Biogeography*, pp. 39–56. Smithsonian Institution Press, Washington, DC. [6]

Sheldon, P. R. 1987. Parallel gradualistic evolution of Ordovician trilobites. *Nature* 330: 561–563. [4, 21]

Sheldon, P. R. 1990. Shaking up evolutionary patterns. *Nature* 345: 772. [16]

Sherman, P. W., H. K. Reeve, and D. W. Pfennig. 1997. Recognition systems. In J. R. Krebs and N. B. Davies (eds.), *Behavioural Ecology: An Evolutionary Approach*, fourth edition, pp. 69–96. Blackwell Scientific, Oxford. [14]

Shine, R., and E. L. Charnov. 1992. Patterns of survival, growth, and maturation in snakes and lizards. *Am. Nat.* 139: 1257–1269. [17]

Short, L. L. 1965. Hybridization in the flickers (*Colaptes*) of North America. *Bull. Am. Mus. Nat. Hist.* 129: 307–428. [9]

Shoshani, J., C. P. Groves, E. L. Simons, and G. F. Gunnell. 1996. Primate phylogeny: Morphological and molecular results. *Mol. Phyl. Evol.* 5: 102–154. [2]

Shu, D.-G., S. Conway Morris, J. Han, Z.-F. Zhang, K. Yasul, P. Janvier, L. Chen, X.-L. Zhang, J.-N. Liu, Y. Li, and H.-Q. Liu. 2003. Head and backbone of the Early Cambrian vertebrate *Haikouichthys*. *Nature* 421: 526–529. [5]

Sibley, C. G. 1950. Species formation in the red-eyed towhees of Mexico. *Univ. Calif. Publ. Zool.* 50: 109–193. [15]

Sibley, C. G. 1954. Hybridization in the red-eyed towhees of Mexico. *Evolution* 8: 252–290. [15]

Sidor, C. A., and J. A. Hopson. 1998. Ghost lineages and "mammalness:" Assessing the temporal pattern of character acquisition in the Synapsida. *Paleobiology* 24: 254–273. [4]

Signor, P. W. III. 1985. Real and apparent trends in species richness through time. In J. W. Valentine (ed.), *Phanerozoic Diversity Patterns: Profiles in Macroevolution*, pp. 129–150. Princeton University Press, Princeton, NJ. [7]

Signor, P. W. III. 1990. The geological history of diversity. *Annu. Rev. Ecol. Syst.* 21: 509–539. [7]

Simons, E. L. 1979. The early relatives of man. In G. Isaac and R. E. F Leakey (eds.), *Human Ancestors*, pp. 22–42. W. H Freeman, San Francisco. [5]

Simpson, G. G. 1944. *Tempo and Mode in Evolution*. Columbia University Press, New York. [1, 21]

Simpson, G. G. 1953. *The Major Features of Evolution*. Columbia University Press, New York. [1, 21]

Simpson, G. G. 1964. *This View of Life: The World of an Evolutionist*. Harcourt, Brace and World, New York. [21]

Sinervo, B. 1990. The evolution of maternal investment in lizards: An experimental and comparative analysis of egg size and its effect on offspring performance. *Evolution* 44: 279–294. [17]

Skutch, A. F. 1973. *The Life of the Hummingbird*. Crown Publishers, New York. [15]

Slatkin, M. 1980. Ecological character displacement. *Ecology* 61: 163–177. [18]

Slatkin, M. 1985. Gene flow in natural populations. *Annu. Rev. Ecol. Syst.* 16: 393–430. [10]

Slatkin, M., and W. P. Maddison. 1989. A cladistic measure of gene flow inferred from the phylogenies of alleles. *Genetics* 123: 603–613. [10]

Smith, A. B., and K. J. Peterson. 2002. Dating the time of origin of major clades: Molecular clocks and the fossil record. *Annu. Rev. Earth Planet. Sci.* 30: 65–88. [2]

Smith, N. G., and A. Eyre-Walker. 2002. Adaptive protein evolution in *Drosophila*. *Nature* 415: 1022–1024. [19]

Smith, T. B. 1993. Disruptive selection and the genetic basis of bill size polymorphism in the African finch *Pyrenestes*. *Nature* 363: 618–620. [12]

Smocovitis, V. B. 1996. *Unifying Biology: The Evolutionary Synthesis and Evolutionary Biology*. Princeton University Press, Princeton, NJ. [1]

Sniegowski, P. D., and R. E. Lenski. 1995. Mutation and adaptation: The directed mutation controversy in evolutionary perspective. *Annu. Rev. Ecol. Syst.* 26: 553–578. [8]

Sniegowski, P. D., P. J. Gerrish, T. Johnson, and A. Shaver. 2000. The evolution of mutation rates: Separating causes from consequences. *Bioessays* 22: 1057–1066. [17]

Sparks, J. S. 2003. Molecular phylogeny and biogeography of Malagasy and South Asian cichlids (Teleostei: Perciformes: Cichlidae). *Mol. Phyl. Evol.* 30: 599–614. [6]

Springer, M. S., W. J. Murphy, E. Eizirik, and S. J. O'Brien. 2003. Placental mammal diversification and the Cretaceous-Tertiary boundary. *Proc. Natl. Acad. Sci. USA* 100: 1056–1061. [5]

Sober, E. 1984. *The Nature of Selection: Evolutionary Theory in Philosophical Focus*. MIT Press, Cambridge, MA. [11]

Soltis, P. S., and D. E. Soltis. 1991. Multiple origins of the allotetraploid *Tragopogon mirus* (Compositae): rRNA evidence. *Syst. Bot.* 16: 407–413. [16]

Somers, C. M., B. E. McCarry, F. Malek and J. S. Quinn. 2004. Reduction of particulate air pollution lowers the risk of heritable mutations in mice. *Science* 304: 1008–1010. [8]

Soulé, M. 1966. Trends in the insular radiation of a lizard. *Am. Nat.* 100: 47–64. [16]

Soyfer, V. 1994. *Lysenko and the Tragedy of Soviet Science*. Rutgers University Press, New Brunswick, NJ. [22]

Spassky, B., N. Spassky, H. Levene, and T. Dobzhansky. 1958. Release of genetic variability through recombination. I. *Drosophila pseudoobscura*. *Genetics* 43: 844–867. [8]

Spiegelman, S. 1970. Extracellular evolution of replicating molecules. In F. O. Schmitt (ed.), *The Neuro Sciences: A Second Study Program*, pp. 927–945. Rockefeller University Press, New York. [5]

Stace, C. A. 1989. Hybridization and the plant species. In K. M. Urbanska (ed.), *Differentiation Patterns in Higher Plants*, pp. 115–127. Academic Press, New York. [8]

Stanley, S. M. 1973. An explanation for Cope's rule. *Evolution* 27: 1–26. [21]

Stanley, S. M. 1979. *Macroevolution: Pattern and Process*. W. H. Freeman, San Francisco. [7, 11, 16, 21]

Stanley, S. M. 1990. The general correlation between rate of speciation and rate of extinction: Fortuitous causal linkages. In R. M. Ross and W. D. Allmon (eds.), *Causes of Evolution: A Paleontological Perspective*, pp. 103–127. University of Chicago Press, Chicago. [7]

Stanley, S. M. 1993. *Earth and Life Through Time*, second edition. W. H. Freeman, New York. [4, 5, 8]

Stanley, S. M. 2005. *Earth System History*. W. H. Freeman, New York. [4]

Stanley, S. M., and X. Yang. 1987. Approximate evolutionary stasis for bivalve morphology over millions of years: A multivariate, multilineage study. *Paleobiology* 13: 113–139. [21]

Stark, J., J. Bonacum, J. Remsen, and R. DeSalle. 1999. The evolution and development of dipteran wing veins: A systematic approach. *Annu. Rev. Entomol.* 44: 97–129. [20]

Stearns, S. C. 1992. *The Evolution of Life Histories*. Oxford University Press, Oxford. [17]

Stebbins, G. L. 1950. *Variation and Evolution in Plants*. Columbia University Press, New York. [1, 15, 16]

Stebbins, G. L. 1974. *Flowering Plants: Evolution Above the Species Level*. Belknap Press of Harvard University Press, Cambridge, MA. [13]

Stebbins, G. L., and P. Basile. 1985. Phyletic phenocopies: A useful technique for probing the genetic developmental basis of evolutionary change. *Evolution* 40: 422–425. [20]

Stebbins, R. C. 1954. *Amphibians and Reptiles of Western North America*. McGraw-Hill, New York. [15]

Stedman, H. H., B. W. Kozyak, A. Nelson, D. M. Thesier, L. T. Su, D. W. Low, C. R. Bridges, J. B. Shrager, N. Minugh-Purvis, and M. A.

Mitchell. 2004. Myosin gene mutation correlates with anatomical changes in the human lineage. *Nature* 428: 415–418. [19]

Stephens, J. C., and 27 others. 2001. Haplotype variation and linkage disequilibrium in 313 human genes. *Science* 293: 489–493. [19]

Steppan, S. J., P. C. Phillips, and D. Houle. 2002. Comparative quantitative genetics: evolution of the G matrix. *Trends Ecol. Evol.* 17: 320–327. [13]

Stern, C. 1973. *Principles of Human Genetics*. W. H. Freeman, San Francisco. [9]

Stern, D. L. 1998 A role of Ultrabithorax in morphological differences between *Drosophila* species. *Nature* 396: 463–466. [20]

Stevens, A., and J. Price. 1996. *Evolutionary Psychiatry: A New Beginning*. Routledge, London. [22]

Stewart, C.-B., J. W. Schilling, and A. C. Wilson. 1987. Adaptive evolution in the stomach lysozymes of foregut fermenters. *Nature* 330: 401–404. [19]

Stewart, W. N. 1983. *Paleobotany and the Evolution of Plants*. Cambridge University Press, Cambridge. [5]

Stock, C. 1925. Cenozoic gravigrade edentates of western North America with special reference to the Pleistocene Megalonychinae and Mylodontidae of Rancho La Brea. Carnegie Institution of Washington Publication no. 331: 1–206. [5]

Storfer, A., and A. Sih. 1998. Gene flow and ineffective antipredator behavior in a stream-breeding salamander. *Evolution* 52: 558–565. [12]

Storfer, A., J. Cross, V. Rush, and J. Caruso. 1999. Adaptive coloration and gene flow as a constraint to local adaptation in the streamside salamander, *Ambystoma barbouri*. *Evolution* 53: 889–898. [12]

Strait, D. S., F. E. Grine,and M. A. Moniz. 1997. A reappraisal of early hominid phylogeny. *J. Hum. Evol.* 32: 17–82. [4]

Strickberger, M. W. 1968. *Genetics*. Macmillan, New York. [8]

Strobeck, C. 1983. Expected linkage disequilibrium for a neutral locus linked to a chromosomal arrangement. *Genetics* 103: 545–555. [12]

Stutt, A. D., and M. T. Siva-Jothy. 2001. Traumatic insemination and sexual conflict in the bed bug *Cimex lectularius*. *Proc. Natl. Acad. Sci. USA* 98: 5683–5687. [14]

Sundström, L., M. Chapuisat, and L. Keller. 1996. Conditional manipulation of sex ratio by ant workers: A test of kin selection theory. *Science* 274: 993–995. [14]

Swanson, W. J., and V. D. Vacquier. 2002. The rapid evolution of reproductive proteins. *Nature Rev. Genet.* 3: 137–144. [19]

Szathmáry, E. 1993. Coding coenzyme handles: A hypothesis for the origin of the genetic code. *Proc. Natl. Acad. Sci. USA* 90: 9916–9920. [5]

Szymura, J. M. 1993. Analysis of hybrid zones with *Bombina*. In R. G. Harrison (ed.), *Hybrid Zones and the Evolutionary Process*, pp. 261–289. Oxford University Press, New York. [15]

T

Taberlet, P., L. Fumagalli, A.-G. Wust-Saucy, and J.-F. Cosson. 1998. Comparative phylogeography and postglacial colonization routes in Europe. *Mol. Ecol.* 7: 453–464. [6]

Tagle, D. A., J. L. Slightom, R. T. Jones, and M. Goodman. 1991. Concerted evolution led to high expression of a prosimian primate δ-globin gene locus. *J. Biol. Chem.* 266: 7469–7480. [19]

Tallamy, D. W., M. B. Darlington, J. D. Pesek, and B. E. Powell. 2003. Copulatory courtship signals male genetic quality in cucumber beetles. *Proc. R. Soc. Lond. B* 270: 77–82. [14]

Taper, M. L., and T. J. Case. 1992. Coevolution among competitors. *Oxford Surv. Evol. Biol.* 8: 63–109. [18]

Tauber, M. J., C. A. Tauber, and S. Masaki. 1986. *Seasonal Adaptations of Insects*. Oxford University Press, New York. [13]

Taylor, G. E. J., L. E. Pitelka, and M. J. Clegg (eds.). 1991. *Ecological Genetics and Air Pollution*. Springer-Verlag, Berlin. [13]

Taylor, P. D., and G. P. Larwood (eds.). 1990. *Major Evolutionary Radiations*. Clarendon Press, Oxford. [5, 7]

Templeton, A. R. 1980. The theory of speciation via the founder principle. *Genetics* 94: 1011–1038. [16]

Templeton, A. R. 1989. The meaning of species and speciation: A genetic perspective. In D. Otte and J. A. Endler (eds.), *Speciation and Its Consequences*, pp. 3–27. Sinauer Associates, Sunderland, MA. [15]

Templeton, A. R. 2002. Out of Africa again and again. *Nature* 416: 45–51. [6]

Thewissen, J. G. M., and S. Bajpai. 2001. Whale origins as a poster child for macroevolution. *BioScience* 51: 1017–1029. [4]

Thewissen, J. G. M., and E. M. Williams, 2002. The early radiations of Cetacea (Mammalia): Evolutionary pattern and developmental correlations. *Annu. Rev. Ecol. Syst.* 33: 73–90. [4]

Thomas, C. D., A. Cameron, R. E. Green, and 16 others. 2004. Extinction risk from climate change. *Nature* 427: 145–148. [5, 7]

Thompson, J. N. 1994. *The Coevolutionary Process*. University of Chicago Press, Chicago. [18]

Thompson, J. N. 1999. Specific hypotheses on the geographic mosaic of coevolution. *Am. Nat.* 153: S1–S14. [18]

Thompson, J. N. 2004. *The Geographic Mosaic of Coevolution*. University of Chicago Press, Chicago. [18]

Thomson, J. D., and J. Brunet. 1990. Hypotheses for the evolution of dioecy in seed plants. *Trends Ecol. Evol.* 5: 11–16. [17]

Thornhill, N. W. (ed.). 1993. *The Natural History of Inbreeding and Outbreeding*. University of Chicago Press, Chicago. [17]

Thornhill, R., and J. Alcock. 1983. *The Evolution of Insect Mating Systems*. Harvard University Press, Cambridge, MA. [14]

Thrall, P. H., and J. J. Burdon. 2003. Evolution of virulence in a plant host-pathogen metapopulation. *Science* 299: 1735–1737. [18]

Tilley, S. G., P. A. Verrell, and S. J. Arnold. 1990. Correspondence between sexual isolation and allozyme differentiation: A test in the salamander *Desmognathus ochrophaeus*. *Proc. Natl. Acad. Sci. USA* 87: 2715–2719. [16]

Tishkoff, S. A., R. Varkonyi, N. Cahinhinan, and 14 others. 2001. Haplotype diversity and linkage disequilibrium at human *G6PD*: Recent origin of alleles that confer malarial resistance. *Science* 293: 455–462. [12]

Tizard, B. 1973. IQ and race. *Nature* 247: 316. [9]

Torrents, D., M. Suyama, E. Zdobnov, and P. Bork. 2003. A genome-wide survey of human pseudogenes. *Genome Res.* 13: 2559–2567. [19]

Travis, J. 1989. The role of optimizing selection in natural populations. *Annu. Rev. Ecol. Syst.* 20: 279–296. [13]

Travis, J., and D. J. Futuyma. 1993. Global change: Lessons from and for evolutionary biology. In P. M. Kareiva, J. G. Kingsolver, and R. B. Huey (eds.), *Biotic Interactions and Global Change*, pp. 251–263. Sinauer Associates, Sunderland, MA. [21]

Trevathan, W. R., E. O. Smith, and J. J. McKenna (eds). 1999. *Evolutionary Medicine*. Oxford Universitty Press, New York. [22]

Trivers, R. L. 1971. The evolution of reciprocal altruism. *Q. Rev. Biol.* 46: 35–57. [14]

Trivers, R. L. 1972. Parental investment and sexual selection. In B. Campbell (ed.), *Sexual Selection and the Descent of Man*, pp. 136–179. Heinemann, London. [14, 18]

Trivers, R. L. 1974. Parent- offspring conflict. *Am. Zool.* 11: 249–264. [14]

Trivers, R. L., and H. Hare. 1976. Haplodiploidy and the evolution of the social insects. *Science* 191: 249–263. [14]

True, J. R., and S. B. Carroll. 2002. Gene co-option in physiological and morphological evolution. *Annu. Rev. Cell. Dev. Biol.* 18: 53–80. [20]

True, J. R., B. S. Weir, and C. C. Laurie. 1996. A genome-wide survey of hybrid incompatibility factors by the introgression of marked segments of *Drosophila mauritiana* chromosomes into *Drosophila simulans*. *Genetics* 142: 819–837. [16]

Tsutsui, N. D., A. V. Suarez, D. A. Holway, and T. J. Case. 2000. Reduced genetic variation and the success of an invasive species. *Proc. Natl. Acad. Sci. USA* 97: 5948–5953. [10]

Turelli, M. 1984. Heritable genetic variation via mutation-selection balance: Lerch's zeta meets the abdominal bristle. *Theor. Pop. Biol.* 25: 138–193. [13]

Turelli, M. 1988. Phenotypic evolution, constant covariances, and the maintenance of additive variance. *Evolution* 42: 1342–1347. [13]

Turelli, M., N. H. Barton, and J. A. Coyne. 2001. Theory and speciation. *Trends Ecol. Evol.* 16: 330–343. [16]

Turner, J. R. G. 1977. Butterfly mimicry: The genetical evolution of an adaptation. *Evol. Biol.* 10: 163–206. [18]

U

Underhill, P. A., G. Passarino, A. A. Lin, P. Shen, M. Mirazón Lahr, R. A. Foley, P. J. Oefner, and L. L. Cavalli-Sforza. 2001. The phylogeography of Y chromosome binary haplotypes and the origins of human populations. *Ann. Hum. Genet.* 65: 43–62. [6]

Ungerer, M. C., S. J. E. Baird, J. Pan, and L. H. Rieseberg. 1998. Rapid hybrid speciation in wild sunflowers. *Proc. Natl. Acad. Sci. USA* 95: 11757–11762. [16]

Unwin, D. M. 1998. Feathers, filaments, and theropod dinosaurs. *Nature* 391: 119–120. [4]

Uyenoyama, M. K., K. E. Holsinger, and D. M. Waller. 1993. Ecological and genetic factors directing the evolution of self-fertilization. *Oxford Surv. Evol. Biol.* 9: 327–381. [17]

V

Vacquier, V. D. 1998. Evolution of gamete recognition proteins. *Science* 281: 1995–1998. [14]

Valentine, J. W. (ed.). 1985. *Phanerozoic Diversity Patterns: Profiles in Macroevolution*. Princeton University Press, Princeton, NJ. [7]

Valentine, J. W., T. C. Foin, and D. Peart. 1978. A provincial model of Phanerozoic marine diversity. *Paleobiology* 4: 55–66. [7]

van Ham, R. C., J. Kamerbeek, C. Palacios, C. Rausell, F. Abascal, U. Bastolla, J. M. Fernandez, L. Jimenez, M. Postigo, F. J. Silva, J. Tamames, E. Viguera, A. Latorre, A. Valencia, F. Moran, and A. Moya. 2003. Reductive genome evolution in *Buchnera aphidicola*. *Proc. Natl. Acad. Sci. USA* 100: 581–586. [19]

van Noordwijk, A. J., and G. deJong. 1986. Acquisition and allocation of resources: Their influence on variation in life history tactics. *Am. Nat.* 128: 137–142. [17]

Van Tyne, J., and A. J. Berger. 1959. *Fundamentals of Ornithology*. Wiley, New York. [6]

Van Valen, L. 1973. A new evolutionary law. *Evol. Theory* 1: 1–30. [7]

Vaughan, T. A. 1986. *Mammalogy*. Third edition. Saunders College Publishing, Philadelphia, PA. [3]

Vehrencamp, S. L. 1983. Optimal degree of skew in cooperative societies. *Am. Zool.* 23: 327–335. [14]

Vehrencamp, S. L., J. W. Bradbury, and R. M. Gibson. 1989. The energetic cost of display in male sage grouse. *Anim. Behav.* 38: 885–896. [17]

Vences, M., J. Freyhof, R. Sonnenberg, J. Kosuch, and M. Veith. 2001. Reconciling fossils and molecules: Cenozoic divergence of cichlid fishes and the biogeography of Madagascar. *J. Biogeog.* 28: 1095–1099. [6]

Venter, J. C. et al. 2001. The sequence of the human genome. *Science* 291: 1304–1351. [3, 8]

Vermeij, G. J. 1987. *Evolution and Escalation: An Ecological History of Life*. Princeton University Press, Princeton, NJ. [5, 7, 18]

Vigilant, L., M. Stoneking, H. Harpending, K. Hawkes, and A. C. Wilson. 1991. African populations and the evolution of human mitochondrial DNA. *Science* 253: 1503–1507. [6, 10]

Voelker, R. A., H. E. Schaffer, and T. Mukai. 1980. Spontaneous allozyme mutations in *Drosophila melanogaster*: Rate of occurrence and nature of the mutants. *Genetics* 94: 961–968. [10]

Volkman, S. K., A. E. Barry, E. J. Lyons, K. M. Nielsen, S. M. Thomas, M. Choi, S. S. Thakore, K. P. Day, D. F. Wirth, and D. L. Hartl. 2001. Recent origin of *Plasmodium falciparum* from a single ancestor. *Science* 293: 482–484. [12]

von Baer, K. E. 1828. *Entwicklungsgeschichte der Thiere: Beobachtung und Reflexion*. Bornträger, Konigsberg. [3]

Voss, S. R., and H. B. Shaffer. 2000. Evolutionary genetics of metamorphic failure using wild-caught vs. laboratory axolotls (*Ambystoma mexicanum*). *Molec. Ecol.* 9: 1401–1407. [20, 21]

Voss, S. R., K. L. Prudic, J. C. Oliver, and H. B. Shaffer. 2003. Candidate gene analysis of metamorphic timing in ambystomatid salamanders. *Molec. Ecol.* 12: 1217–1223. [20]

W

Waddington, C. H. 1942. Canalization of development and the inheritance of acquired characters. *Nature* 150: 563–565. [20]

Waddington, C. H. 1953. Genetic assimilation of an acquired character. *Evolution* 7: 118–126. [13]

Wade, M. J. 1977. An experimental study of group selection. *Evolution* 31: 134–153. [11]

Wade, M. J. 1979. The primary characteristics of *Tribolium* populations group selected for increased and decreased population size. *Evolution* 33: 749–764. [11]

Wade, M. J., and S. J. Arnold. 1980. The intensity of sexual selection in relation to male sexual behavior, female choice, and sperm precedence. *Anim. Behav.* 28: 446–461. [14]

Wagner, A., G. P. Wagner, and P. Similion. 1994. Epistasis can facilitate the evolution of reproductive isolation by peak shifts: A two-locus two-allele model. *Genetics* 138: 533–545. [16]

Wagner, G. P. 1988. The influence of variation and of developmental constraints on the rate of multivariate phenotypic evolution. *J. Evol. Biol.* 1: 45–66. [13]

Wagner, G. P. 1989a. The biological homology concept. *Annu. Rev. Ecol. Syst.* 20: 51–69. [3, 20]

Wagner, G. P. 1989b. The origin of morphological characters and the biological basis of homology. *Evolution* 43: 1157–1171. [20]

Wagner, G. P. 1996. Homologues, natural kinds and the evolution of modularity. *Am. Zool.* 36: 36–43. [3, 21]

Wagner, G. P., and L. Altenberg. 1996. Perspective: Complex adaptations and the evolution of evolvability. *Evolution* 50: 967–976. [13, 21]

Wagner, G. P., G. Booth, and H. Bagheri-Chaichian. 1997. A population genetic theory of canalization. *Evolution* 51: 329–347. [13, 21]

Wagner, P. J. 1996. Contrasting the underlying patterns of active trends in morphologic evolution. *Evolution* 50: 990–1007. [20]

Wake, D. B. 1982. Functional and developmental constraints and opportunities in the evolution of feeding systems in urodeles. In D. Mossakowski and G. Roth (eds.), *Environmental Adaptation and Evolution*, pp. 51–66. G. Fischer, Stuttgart. [21]

Walker, T. D., and J. W. Valentine. 1984. Equilibrium models of evolutionary species diversity and the number of empty niches. *Am. Nat.* 124: 887–899. [7]

Walther, F. R. 1984. *Communication and Expression in Hoofed Mammals.* Indiana University Press, Bloomington. [14]

Walther, G.-R., E. Post, P. Convey, A. Menzel, C. Parmesan, T. J. Beebee, J.-M. Fromentin, O. Hoegh-Guldberg, and F. Bairlein. 2002. Ecological responses to recent climate change. *Nature* 416: 389–395. [7]

Warner, R. R. 1984. Mating behavior and hermaphroditism in coral reef fishes. *Am. Sci.* 72: 128–136. [17]

Warren, R. W., L. Nagy, J. Selegue, J. Gates, and S. B. Carroll. 1994. Evolution of homeotic gene function in flies and butterflies. *Nature* 372: 458–461. [20]

Waser, N. M., and M. V. Price. 1989. Optimal outcrossing in *Ipomopsis aggregata*: Seed set and offspring fitness. *Evolution* 43: 1097–1109. [17]

Watanabe, H., and 44 others. (2004). DNA sequence and comparative analysis of chimpanzee chromosome 22. *Nature* 429: 382–388. [19]

Wcislo, W. T., and B. N. Danforth. 1997. Secondarily solitary: The evolutionary loss of social behavior. *Trends Ecol. Evol.* 12: 468–474. [21]

Weatherbee, S. D., G. Halder, J. Kim, A. Hudson, and S. B. Carroll. 1998 Ultrabithorax regulates genes at several levels of the wing-patterning hierarchy to shape the development of the *Drosophila* haltere. *Genes Devel.* 12: 1474–1482. [20]

Weber, K. E., and L. T. Diggins. 1990. Increased selection response in larger populations. II. Selection for ethanol vapor resistance in *Drosophila melanogaster* at two population sizes. *Genetics* 125: 585–597. [13]

Weis, A. E., W. G. Abrahamson, and M. C. Andersen. 1992. Variable selection on *Eurosta's* gall size. I. The extent and nature of variation in phenotypic selection. *Evolution* 46: 1674–1697. [13]

Weishampel, D. B., P. Dodson, and H. Osmólska. 2004. *The Dinosauria.* Second edition. University of California Press, Berkeley. [5]

Welch, A. M., R. D. Semlitsch, and H. C. Gerhardt. 1998. Call duration as an indicator of genetic quality in male gray tree frogs. *Science* 280: 1928–1930.

Wellman, C. H., P. L. Osterloff, and U. Mohiuddin. 2003. Fragments of the earliest land plants. *Nature* 425: 282–290. [5]

Wells, J. 2000. *Icons of Evolution: Science or Myth?* Regnery, Washington, D.C. [12]

Wen, J. 1999. *Evolution* of eastern Asian and eastern North American disjunct distributions of flowering plants. *Annu. Rev. Ecol. Syst.* 30: 421–455. [6]

Werner, P. A., and W. J. Platt. 1976. Ecological relationships of co-occurring goldenrods (*Solidago*: Compositae). *Am. Nat.* 110: 959–971. [17]

Werner, T. K., and T. W. Sherry. 1987. Behavioral feeding specialization in *Pinaroloxias inornata*, the "Darwin's Finch" of Cocos Island, Costa Rica. *Proc. Natl. Acad. Sci. USA* 84: 5506–5510. [16, 18]

Werren, J. H. 1980. Sex ratio adaptations to local mate competition in a parasitic wasp. *Science* 208: 1157–1159. [17]

Werren, J. H. 1998. *Wolbachia* and speciation. In D. J. Howard and S. H. Berlocher (eds.), *Endless Forms: Species and Speciation*, pp. 245–260. Oxford University Press, New York. [16]

West, S. A., C. M. Lively, and A. F. Read. 1999. A pluralist approach to sex and recombination. *J. Evol. Biol.* 12: 1003–1012. [18]

West-Eberhard, M. J. 1983. Sexual selection, social competition, and speciation. *Q. Rev. Biol.* 58: 155–183. [14]

West-Eberhard, M. J. 2003. *Developmental Plasticity and Evolution.* Oxford University Press, New York. [13]

Westoby, M., M. R. Leishman, and J. M. Lord. 1997. On misinterpreting the "phylogenetic correlation." *J. Ecol.* 83: 531–534. [11]

White, M. J. D. 1978. *Modes of Speciation.* W. H. Freeman, San Francisco. [12, 17]

Whittingham, L. A., P. D. Taylor, and R. J. Robertson. 1992. Confidence of paternity and male parental care. *Am. Nat.* 139: 1115–1125. [14]

Wichman, H. A., L. A. Scott, C. D. Yarber, and J. J. Bull. 2000. Experimental evolution recapitulates natural evolution. *Phil. Trans. R. Soc. Lond. B* 355: 1677–1684. [8]

Wiens, J. J. 2001. Widespread loss of sexually selected traits: How the peacock lost its spots. *Trends Ecol. Evol.* 16: 517–523. [14]

Wiley, E. O. 1978. The evolutionary species concept reconsidered. *Syst. Zool.* 27: 17–26. [15]

Wilkins, A. S. 2002. *The Evolution of Developmental Pathways.* Sinauer Associates, Sunderland, MA. [20]

Willig, M. R., D. M. Kaufman, and R. D. Stevens. 2003. Latitudinal gradients of biodiversity: pattern, process, scale, and synthesis. *Annu. Rev. Ecol. Evol. Syst.* 34: 273–309. [6]

Wilkinson, G. S. 1988. Reciprocal altruism in bats and other mammals. *Ethol. Sociobiol.* 9: 85–100. [14]

Williams, E. E. 1972. The origin of faunas: Evolution of lizard congeners in a complex island fauna—a trial analysis. *Evol. Biol.* 6: 47–89. [6]

Williams, G. C. 1957. Pleiotropy, natural selection, and the evolution of senescence. *Evolution* 11: 398–411. [17]

Williams, G. C. 1966. *Adaptation and Natural Selection.* Princeton University Press, Princeton, NJ. [11, 14, 17]

Williams, G. C. 1975. *Sex and Evolution.* Princeton University Press, Princeton, NJ. [17]

Williams, G. C. 1992a. *Gaia*, nature worship and biocentric fallacies. *Q. Rev. Biol.* 67: 479–486. [11]

Williams, G. C. 1992b. *Natural Selection: Domains, Levels, and Challenges.* Oxford University Press, New York. [11, 20]

Williston, S. W. 1925. *The Osteology of the Reptiles.* Harvard University Press, Cambridge, MA. [5]

Wilson, A. C., S. S. Carlson, and T. J. White. 1977. Biochemical evolution. *Annu. Rev. Biochem.* 46: 573–639. [2, 22]

Wilson, A. M., and K. Thompson. 1989. A comparative study of reproductive allocation in 40 British grasses. *Functional Ecology* 3: 297–302. [17]

Wilson, D. S. 1983. The group selection controversy: History and current status. *Annu. Rev. Ecol. Syst.* 14: 159–187. [11]

Wilson, D. S., and R. K. Colwell. 1981. The evolution of sex ratio in structured demes. *Evolution* 35: 882–897. [17]

Wilson, E. O. 1971. *The Insect Societies.* Harvard University Press, Cambridge, MA. [14]

Wilson, E. O. 1975. *Sociobiology: The New Synthesis.* Harvard University Press, Cambridge, MA. [11]

Wilson, E. O. 1992. *The Diversity of Life.* Harvard University Press, Cambridge, MA. [5, 7]

Wilson, E. O., F. M. Carpenter, and W. L. Brown Jr. 1967. The first Mesozoic ants. *Science* 157: 1038–1040. [4]

Wimsatt, W. C. 1986. Developmental constraints, generative entrenchment, and the innate-acquired distinction. In W. Bechtel (ed.),

Integrating Scientific Disciplines, pp. 185–208. Martinus-Nijhoff, Dordrecht, The Netherlands. [21]

Winter, K. U., A. Becker, T. Munster, J. T. Kim, H. Saedler, and G. Theissen. 1999. MADS-box genes reveal that gnetophytes are more closely related to conifers than to flowering plants. *Proc. Natl. Acad. Sci. USA* 96: 7342–7347. [20]

Wistow, G. 1993. Lens crystallins: Gene recruitment and evolutionary dynamism. *Trends Biochem. Sci.* 18: 301–306. [20]

Wistow, G., and J. Piatigorsky, J. 1988. Lens crystallins: The evolution and expression of proteins for a highly specialized tissue. *Annu. Rev. Biochem.* 57: 479–504. [20]

Wittkopp, P. J., B. K. Haerum, and A. G. Clark. 2004. Evolutionary changes in *cis* and *trans* gene regulation. *Nature* 430: 85–88. [20]

Woese, C. R. 2000. Interpreting the universal phylogenetic tree. *Proc. Nat. Acad. Sci. USA* 97: 8392–8396. [5]

Wood, B. A. 2002. Hominid revelations from Chad. *Nature* 418: 133–135. [4]

Wood, B. A., and M. C. Collard. 1999. The human genus. *Science* 284: 65–71. [4]

Wood, R. J. 1981. Insecticide resistance: Genes and mechanisms. In J. A. Bishop and L. M. Cook (eds.), *Genetic consequences of man made change*, pp. 53–96. Academic Press, London. [12]

Woodruff, R. C., H. Huai and J. N. Thonpson, Jr. 1996. Clusters of identical new mutation in the evolutionary landscape. *Genetica* 98: 149–160. [8]

Wray, G. A., J. S. Levinton, and L. H. Shapiro. 1996. Molecular evidence for deep pre-Cambrian divergences among metazoan phyla. *Science* 274: 568–573. [5]

Wright, S. 1935. The analysis of variance and the correlations between relatives with respect to deviations from an optimum. *J. Genet.* 30: 243–256. [10]

Wu, C.-I., and H. Hollocher. 1998. Subtle is nature: Differentiation and speciation. In D. J. Howard and S. H. Berlocher (eds.), *Endless Forms: Species and Speciation*, pp. 339–351. Oxford University Press, New York. [16]

Wyatt, R. 1988. Phylogenetic aspects of the evolution of self-pollination. In L. D. Gottlieb and S. K. Jain (eds.), *Plant Evolutionary Biology*, pp. 109–131. Chapman & Hall, London. [17]

X

Xu, X., Z. Zhou, X. Wang, X. Kuang, F. Zhang, and X. Du. 2003. Four-winged dinosaurs from China. *Nature* 421: 335–340. [4]

Xu, Z. P., I. Woo, H. Her, D. R. Beier, and R. L. Maas. 1997. Mouse *Eya* homologues of the *Drosophila eyes absent* gene require *Pax6* for expression in lens and nasal placode. *Development* 124: 219–231. [20]

Y

Yang, Z., and J. P. Bielawski. 2000. Statistical tests of adaptive molecular evolution. *Trends Ecol. Evol.* 15: 496–502. [19]

Yoder, A. B., M. M. Burns, S. Zehr, T. Delefosse, G. Veron, S. M. Goodman and J. J. Flynn. 2003. Single origin of Malagasy Carnivora from an African ancestor. *Nature* 421: 734–737. [6]

Yoo, B. H. 1980. Long-term selection for a quantitative character in large replicate populations of *Drosophila melanogaster*. I. Response to selection. II. Lethals and visible mutants with large effects. *Genet. Res.* 35: I. 1–17, II. 19–31. [13]

Z

Zahavi, A. 1975. Mate selection: A selection for a handicap. *J. Theor. Biol.* 53: 205–214. [14]

Zeh, J. A. 2004. Sexy sons: A dead end for cytoplasmic genes. *Proc. R. Soc. Lond. B* 271: S306–S309. [14]

Zeng, Z.-B., J. Liu, L. F. Stam, C.-H. Kao, J. M. Mercer, and C. C. Laurie. 2000. Genetic architecture of a morphological shape difference between two *Drosophila* species. *Genetics* 154: 299–310. [16]

Zhang, J., and D. M. Webb. 2003. Evolutionary deterioration of the vomeronasal pheromone transduction pathway in catarrhine primates. *Proc. Natl. Acad. Sci. USA* 100: 8337–8341. [19]

Zhang, J., H. F. Rosenberg, and M. Nei. 1998. Positive Darwinian selection after gene duplication in primate ribonuclease genes. *Proc. Natl. Acad. Sci. USA* 95: 3708–13. [19]

Zhang, J., D. M. Webb and O. Podlaha. 2002. Accelerated protein evolution and origins of human-specific features: FOXP2 as an example. *Genetics* 162: 1825–1835. [8, 19]

Zhang, J., A. M. Dean, F. Brunet, and M. Long. 2004. Evolving protein functional diversity in new genes of *Drosophila. Proc. Natl. Acad. Sci. USA* 101: 16246–16250. [19]

Zhang, Z., P. M. Harrison, Y. Liu, and M. Gerstein. 2003. Millions of years of evolution preserved: A comprehensive catalog of the processed pseudogenes in the human genome. *Genome Res.* 13: 2541–2558. [19]

Zielenski, J., and L.-C. Tsui. 1995. Cystic fibrosis: Genotypic and phenotypic variations. *Annu. Rev. Genet.* 29: 777–807. [8]

Zouros, E. 1987. On the relation between heterozygosity and heterosis: An evaluation of the evidence from marine mollusks. In M. C. Rattazi, J. G. Scandalios, and G. S. Whitt (eds.), *Isozymes: Current Topics in Biological and Medical Research*, pp. 255–270. Alan R. Liss, New York. [12]

Zouros, E., K. Lofdahl, and P. A. Martin. 1988. Male hybrid sterility in *Drosophila*: Interactions between autosomes and sex chromosomes in crosses of *D. mojavensis* and *D. arizonensis*. *Evolution* 42: 1321–1331. [12]

Zuckerkandl, E., and L. Pauling. 1965. Evolutionary divergence and convergence of proteins. In V. Bryson and H. J. Vogel (eds.), *Evolving Genes and Proteins*, pp. 97–166. Academic Press, New York. [2]

Index

Numbers in *italic* indicate information in an illustration or illustration caption.

A

B

C

D

E

H

I

J

K

L

M

Q

R

S

T

U

V

About the Book

Editor: Andrew D. Sinauer

Project Editor and Art Development: Carol J. Wigg

Review Coordinator and Academic Liaison: Susan McGlew

Copy Editor: Norma Roche

Photo Research: David McIntyre

Production Manager: Christopher Small

Book Design and Layout: Jefferson Johnson

Cover Design: Jefferson Johnson

Illustration Program: Elizabeth Morales

Index: Acorn Indexing

Book and Cover Manufacture: Courier Companies, Inc.